T0189085

Lecture Notes in Computer Science **14220**

The series Lecture Notes in Computer Science (LNCS), including its subseries Lecture Notes in Artificial Intelligence (LNAI) and Lecture Notes in Bioinformatics (LNBI), has established itself as a medium for the publication of new developments in computer science and information technology research, teaching, and education.

LNCS enjoys close cooperation with the computer science R & D community, the series counts many renowned academics among its volume editors and paper authors, and collaborates with prestigious societies. Its mission is to serve this international community by providing an invaluable service, mainly focused on the publication of conference and workshop proceedings and postproceedings. LNCS commenced publication in 1973.

Hayit Greenspan · Anant Madabhushi ·
Parvin Mousavi · Septimiu Salcudean ·
James Duncan · Tanveer Syeda-Mahmood ·
Russell Taylor
Editors

Medical Image Computing and Computer Assisted Intervention – MICCAI 2023

26th International Conference Vancouver, BC, Canada, October 8–12, 2023 Proceedings, Part I

Editors
Hayit Greenspan
Icahn School of Medicine, Mount Sinai,
NYC, NY, USA

Tel Aviv University
Tel Aviv, Israel

Parvin Mousavi
Queen's University
Kingston, ON, Canada

James Duncan
Yale University
New Haven, CT, USA

Russell Taylor
Johns Hopkins University
Baltimore, MD, USA

Anant Madabhushi
Emory University
Atlanta, GA, USA

Septimiu Salcudean
The University of British Columbia
Vancouver, BC, Canada

Tanveer Syeda-Mahmood
IBM Research
San Jose, CA, USA

ISSN 0302-9743 ISSN 1611-3349 (electronic)
Lecture Notes in Computer Science
ISBN 978-3-031-43906-3 ISBN 978-3-031-43907-0 (eBook)
https://doi.org/10.1007/978-3-031-43907-0

This Springer imprint is published by the registered company Springer Nature Switzerland AG
The registered company address is: Gewerbestrasse 11, 6330 Cham, Switzerland

Paper in this product is recyclable.

Preface

We are pleased to present the proceedings for the 26th International Conference on Medical Image Computing and Computer-Assisted Intervention (MICCAI). After several difficult years of virtual conferences, this edition was held in a mainly in-person format with a hybrid component at the Vancouver Convention Centre, in Vancouver, BC, Canada October 8–12, 2023. The conference featured 33 physical workshops, 15 online workshops, 15 tutorials, and 29 challenges held on October 8 and October 12. Co-located with the conference was also the 3rd Conference on Clinical Translation on Medical Image Computing and Computer-Assisted Intervention (CLINICCAI) on October 10.

MICCAI 2023 received the largest number of submissions so far, with an approximately 30% increase compared to 2022. We received 2365 full submissions of which 2250 were subjected to full review. To keep the acceptance ratios around 32% as in previous years, there was a corresponding increase in accepted papers leading to 730 papers accepted, with 68 orals and the remaining presented in poster form. These papers comprise ten volumes of Lecture Notes in Computer Science (LNCS) proceedings as follows:

- Part I, LNCS Volume 14220: Machine Learning with Limited Supervision and Machine Learning – Transfer Learning
- Part II, LNCS Volume 14221: Machine Learning – Learning Strategies and Machine Learning – Explainability, Bias, and Uncertainty I
- Part III, LNCS Volume 14222: Machine Learning – Explainability, Bias, and Uncertainty II and Image Segmentation I
- Part IV, LNCS Volume 14223: Image Segmentation II
- Part V, LNCS Volume 14224: Computer-Aided Diagnosis I
- Part VI, LNCS Volume 14225: Computer-Aided Diagnosis II and Computational Pathology
- Part VII, LNCS Volume 14226: Clinical Applications – Abdomen, Clinical Applications – Breast, Clinical Applications – Cardiac, Clinical Applications – Dermatology, Clinical Applications – Fetal Imaging, Clinical Applications – Lung, Clinical Applications – Musculoskeletal, Clinical Applications – Oncology, Clinical Applications – Ophthalmology, and Clinical Applications – Vascular
- Part VIII, LNCS Volume 14227: Clinical Applications – Neuroimaging and Microscopy
- Part IX, LNCS Volume 14228: Image-Guided Intervention, Surgical Planning, and Data Science
- Part X, LNCS Volume 14229: Image Reconstruction and Image Registration

The papers for the proceedings were selected after a rigorous double-blind peer-review process. The MICCAI 2023 Program Committee consisted of 133 area chairs and over 1600 reviewers, with representation from several countries across all major continents. It also maintained a gender balance with 31% of scientists who self-identified as women. With an increase in the number of area chairs and reviewers, the reviewer load on the experts was reduced this year, keeping to 16–18 papers per area chair and about 4–6 papers per reviewer. Based on the double-blinded reviews, area chairs' recommendations, and program chairs' global adjustments, 308 papers (14%) were provisionally accepted, 1196 papers (53%) were provisionally rejected, and 746 papers (33%) proceeded to the rebuttal stage. As in previous years, Microsoft's Conference Management Toolkit (CMT) was used for paper management and organizing the overall review process. Similarly, the Toronto paper matching system (TPMS) was employed to ensure knowledgeable experts were assigned to review appropriate papers. Area chairs and reviewers were selected following public calls to the community, and were vetted by the program chairs.

Among the new features this year was the emphasis on clinical translation, moving Medical Image Computing (MIC) and Computer-Assisted Interventions (CAI) research from theory to practice by featuring two clinical translational sessions reflecting the real-world impact of the field in the clinical workflows and clinical evaluations. For the first time, clinicians were appointed as Clinical Chairs to select papers for the clinical translational sessions. The philosophy behind the dedicated clinical translational sessions was to maintain the high scientific and technical standard of MICCAI papers in terms of methodology development, while at the same time showcasing the strong focus on clinical applications. This was an opportunity to expose the MICCAI community to the clinical challenges and for ideation of novel solutions to address these unmet needs. Consequently, during paper submission, in addition to MIC and CAI a new category of "Clinical Applications" was introduced for authors to self-declare.

MICCAI 2023 for the first time in its history also featured dual parallel tracks that allowed the conference to keep the same proportion of oral presentations as in previous years, despite the 30% increase in submitted and accepted papers.

We also introduced two new sessions this year focusing on young and emerging scientists through their Ph.D. thesis presentations, and another with experienced researchers commenting on the state of the field through a fireside chat format.

The organization of the final program by grouping the papers into topics and sessions was aided by the latest advancements in generative AI models. Specifically, Open AI's GPT-4 large language model was used to group the papers into initial topics which were then manually curated and organized. This resulted in fresh titles for sessions that are more reflective of the technical advancements of our field.

Although not reflected in the proceedings, the conference also benefited from keynote talks from experts in their respective fields including Turing Award winner Yann LeCun and leading experts Jocelyne Troccaz and Mihaela van der Schaar.

We extend our sincere gratitude to everyone who contributed to the success of MICCAI 2023 and the quality of its proceedings. In particular, we would like to express our profound thanks to the MICCAI Submission System Manager Kitty Wong whose meticulous support throughout the paper submission, review, program planning, and proceeding preparation process was invaluable. We are especially appreciative of the effort and dedication of our Satellite Events Chair, Bennett Landman, who tirelessly coordinated the organization of over 90 satellite events consisting of workshops, challenges and tutorials. Our workshop chairs Hongzhi Wang, Alistair Young, tutorial chairs Islem

Rekik, Guoyan Zheng, and challenge chairs, Lena Maier-Hein, Jayashree Kalpathy-Kramer, Alexander Seitel, worked hard to assemble a strong program for the satellite events. Special mention this year also goes to our first-time Clinical Chairs, Drs. Curtis Langlotz, Charles Kahn, and Masaru Ishii who helped us select papers for the clinical sessions and organized the clinical sessions.

We acknowledge the contributions of our Keynote Chairs, William Wells and Alejandro Frangi, who secured our keynote speakers. Our publication chairs, Kevin Zhou and Ron Summers, helped in our efforts to get the MICCAI papers indexed in PubMed. It was a challenging year for fundraising for the conference due to the recovery of the economy after the COVID pandemic. Despite this situation, our industrial sponsorship chairs, Mohammad Yaqub, Le Lu and Yanwu Xu, along with Dekon's Mehmet Eldegez, worked tirelessly to secure sponsors in innovative ways, for which we are grateful.

An active body of the MICCAI Student Board led by Camila Gonzalez and our 2023 student representatives Nathaniel Braman and Vaishnavi Subramanian helped put together student-run networking and social events including a novel Ph.D. thesis 3-minute madness event to spotlight new graduates for their careers. Similarly, Women in MICCAI chairs Xiaoxiao Li and Jayanthi Sivaswamy and RISE chairs, Islem Rekik, Pingkun Yan, and Andrea Lara further strengthened the quality of our technical program through their organized events. Local arrangements logistics including the recruiting of University of British Columbia students and invitation letters to attendees, was ably looked after by our local arrangement chairs Purang Abolmaesumi and Mehdi Moradi. They also helped coordinate the visits to the local sites in Vancouver both during the selection of the site and organization of our local activities during the conference. Our Young Investigator chairs Marius Linguraru, Archana Venkataraman, Antonio Porras Perez put forward the startup village and helped secure funding from NIH for early career scientist participation in the conference. Our communications chair, Ehsan Adeli, and Diana Cunningham were active in making the conference visible on social media platforms and circulating the newsletters. Niharika D'Souza was our cross-committee liaison providing note-taking support for all our meetings. We are grateful to all these organization committee members for their active contributions that made the conference successful.

We would like to thank the MICCAI society chair, Caroline Essert, and the MICCAI board for their approvals, support and feedback, which provided clarity on various aspects of running the conference. Behind the scenes, we acknowledge the contributions of the MICCAI secretariat personnel, Janette Wallace, and Johanne Langford, who kept a close eye on logistics and budgets, and Diana Cunningham and Anna Van Vliet for including our conference announcements in a timely manner in the MICCAI society newsletters. This year, when the existing virtual platform provider indicated that they would discontinue their service, a new virtual platform provider Conference Catalysts was chosen after due diligence by John Baxter. John also handled the setup and coordination with CMT and consultation with program chairs on features, for which we are very grateful. The physical organization of the conference at the site, budget financials, fund-raising, and the smooth running of events would not have been possible without our Professional Conference Organization team from Dekon Congress & Tourism led by Mehmet Eldegez. The model of having a PCO run the conference, which we used at

MICCAI, significantly reduces the work of general chairs for which we are particularly grateful.

Finally, we are especially grateful to all members of the Program Committee for their diligent work in the reviewer assignments and final paper selection, as well as the reviewers for their support during the entire process. Lastly, and most importantly, we thank all authors, co-authors, students/postdocs, and supervisors for submitting and presenting their high-quality work, which played a pivotal role in making MICCAI 2023 a resounding success.

With a successful MICCAI 2023, we now look forward to seeing you next year in Marrakesh, Morocco when MICCAI 2024 goes to the African continent for the first time.

October 2023

Tanveer Syeda-Mahmood
James Duncan
Russ Taylor
General Chairs

Hayit Greenspan
Anant Madabhushi
Parvin Mousavi
Septimiu Salcudean
Program Chairs

Organization

General Chairs

Tanveer Syeda-Mahmood	IBM Research, USA
James Duncan	Yale University, USA
Russ Taylor	Johns Hopkins University, USA

Program Committee Chairs

Hayit Greenspan	Tel-Aviv University, Israel and Icahn School of Medicine at Mount Sinai, USA
Anant Madabhushi	Emory University, USA
Parvin Mousavi	Queen's University, Canada
Septimiu Salcudean	University of British Columbia, Canada

Satellite Events Chair

Bennett Landman	Vanderbilt University, USA

Workshop Chairs

Hongzhi Wang	IBM Research, USA
Alistair Young	King's College, London, UK

Challenges Chairs

Jayashree Kalpathy-Kramer	Harvard University, USA
Alexander Seitel	German Cancer Research Center, Germany
Lena Maier-Hein	German Cancer Research Center, Germany

Tutorial Chairs

Islem Rekik — Imperial College London, UK
Guoyan Zheng — Shanghai Jiao Tong University, China

Clinical Chairs

Curtis Langlotz — Stanford University, USA
Charles Kahn — University of Pennsylvania, USA
Masaru Ishii — Johns Hopkins University, USA

Local Arrangements Chairs

Purang Abolmaesumi — University of British Columbia, Canada
Mehdi Moradi — McMaster University, Canada

Keynote Chairs

William Wells — Harvard University, USA
Alejandro Frangi — University of Manchester, UK

Industrial Sponsorship Chairs

Mohammad Yaqub — MBZ University of Artificial Intelligence, Abu Dhabi
Le Lu — DAMO Academy, Alibaba Group, USA
Yanwu Xu — Baidu, China

Communication Chair

Ehsan Adeli — Stanford University, USA

Publication Chairs

Ron Summers	National Institutes of Health, USA
Kevin Zhou	University of Science and Technology of China, China

Young Investigator Chairs

Marius Linguraru	Children's National Institute, USA
Archana Venkataraman	Boston University, USA
Antonio Porras	University of Colorado Anschutz Medical Campus, USA

Student Activities Chairs

Nathaniel Braman	Picture Health, USA
Vaishnavi Subramanian	EPFL, France

Women in MICCAI Chairs

Jayanthi Sivaswamy	IIIT, Hyderabad, India
Xiaoxiao Li	University of British Columbia, Canada

RISE Committee Chairs

Islem Rekik	Imperial College London, UK
Pingkun Yan	Rensselaer Polytechnic Institute, USA
Andrea Lara	Universidad Galileo, Guatemala

Submission Platform Manager

Kitty Wong	The MICCAI Society, Canada

Virtual Platform Manager

John Baxter	INSERM, Université de Rennes 1, France

Cross-Committee Liaison

Niharika D'Souza	IBM Research, USA

Program Committee

Sahar Ahmad	University of North Carolina at Chapel Hill, USA
Shadi Albarqouni	University of Bonn and Helmholtz Munich, Germany
Angelica Aviles-Rivero	University of Cambridge, UK
Shekoofeh Azizi	Google, Google Brain, USA
Ulas Bagci	Northwestern University, USA
Wenjia Bai	Imperial College London, UK
Sophia Bano	University College London, UK
Kayhan Batmanghelich	University of Pittsburgh and Boston University, USA
Ismail Ben Ayed	ETS Montreal, Canada
Katharina Breininger	Friedrich-Alexander-Universität Erlangen-Nürnberg, Germany
Weidong Cai	University of Sydney, Australia
Geng Chen	Northwestern Polytechnical University, China
Hao Chen	Hong Kong University of Science and Technology, China
Jun Cheng	Institute for Infocomm Research, A*STAR, Singapore
Li Cheng	University of Alberta, Canada
Albert C. S. Chung	University of Exeter, UK
Toby Collins	Ircad, France
Adrian Dalca	Massachusetts Institute of Technology and Harvard Medical School, USA
Jose Dolz	ETS Montreal, Canada
Qi Dou	Chinese University of Hong Kong, China
Nicha Dvornek	Yale University, USA
Shireen Elhabian	University of Utah, USA
Sandy Engelhardt	Heidelberg University Hospital, Germany
Ruogu Fang	University of Florida, USA

Jianming Liang	Arizona State University, USA
Jianfei Liu	National Institutes of Health Clinical Center, USA
Mingxia Liu	University of North Carolina at Chapel Hill, USA
Xiaofeng Liu	Harvard Medical School and MGH, USA
Herve Lombaert	École de technologie supérieure, Canada
Ismini Lourentzou	Virginia Tech, USA
Le Lu	Damo Academy USA, Alibaba Group, USA
Dwarikanath Mahapatra	Inception Institute of Artificial Intelligence, United Arab Emirates
Saad Nadeem	Memorial Sloan Kettering Cancer Center, USA
Dong Nie	Alibaba (US), USA
Yoshito Otake	Nara Institute of Science and Technology, Japan
Sang Hyun Park	Daegu Gyeongbuk Institute of Science and Technology, South Korea
Magdalini Paschali	Stanford University, USA
Tingying Peng	Helmholtz Munich, Germany
Caroline Petitjean	LITIS Université de Rouen Normandie, France
Esther Puyol Anton	King's College London, UK
Chen Qin	Imperial College London, UK
Daniel Racoceanu	Sorbonne Université, France
Hedyeh Rafii-Tari	Auris Health, USA
Hongliang Ren	Chinese University of Hong Kong, China and National University of Singapore, Singapore
Tammy Riklin Raviv	Ben-Gurion University, Israel
Hassan Rivaz	Concordia University, Canada
Mirabela Rusu	Stanford University, USA
Thomas Schultz	University of Bonn, Germany
Feng Shi	Shanghai United Imaging Intelligence, China
Yang Song	University of New South Wales, Australia
Aristeidis Sotiras	Washington University in St. Louis, USA
Rachel Sparks	King's College London, UK
Yao Sui	Peking University, China
Kenji Suzuki	Tokyo Institute of Technology, Japan
Qian Tao	Delft University of Technology, Netherlands
Mathias Unberath	Johns Hopkins University, USA
Martin Urschler	Medical University Graz, Austria
Maria Vakalopoulou	CentraleSupelec, University Paris Saclay, France
Erdem Varol	New York University, USA
Francisco Vasconcelos	University College London, UK
Harini Veeraraghavan	Memorial Sloan Kettering Cancer Center, USA
Satish Viswanath	Case Western Reserve University, USA
Christian Wachinger	Technical University of Munich, Germany

Hua Wang	Colorado School of Mines, USA
Qian Wang	ShanghaiTech University, China
Shanshan Wang	Paul C. Lauterbur Research Center, SIAT, China
Yalin Wang	Arizona State University, USA
Bryan Williams	Lancaster University, UK
Matthias Wilms	University of Calgary, Canada
Jelmer Wolterink	University of Twente, Netherlands
Ken C. L. Wong	IBM Research Almaden, USA
Jonghye Woo	Massachusetts General Hospital and Harvard Medical School, USA
Shandong Wu	University of Pittsburgh, USA
Yutong Xie	University of Adelaide, Australia
Fuyong Xing	University of Colorado, Denver, USA
Daguang Xu	NVIDIA, USA
Yan Xu	Beihang University, China
Yanwu Xu	Baidu, China
Pingkun Yan	Rensselaer Polytechnic Institute, USA
Guang Yang	Imperial College London, UK
Jianhua Yao	Tencent, China
Chuyang Ye	Beijing Institute of Technology, China
Lequan Yu	University of Hong Kong, China
Ghada Zamzmi	National Institutes of Health, USA
Liang Zhan	University of Pittsburgh, USA
Fan Zhang	Harvard Medical School, USA
Ling Zhang	Alibaba Group, China
Miaomiao Zhang	University of Virginia, USA
Shu Zhang	Northwestern Polytechnical University, China
Rongchang Zhao	Central South University, China
Yitian Zhao	Chinese Academy of Sciences, China
Tao Zhou	Nanjing University of Science and Technology, USA
Yuyin Zhou	UC Santa Cruz, USA
Dajiang Zhu	University of Texas at Arlington, USA
Lei Zhu	ROAS Thrust HKUST (GZ), and ECE HKUST, China
Xiahai Zhuang	Fudan University, China
Veronika Zimmer	Technical University of Munich, Germany

Reviewers

Alaa Eldin Abdelaal
John Abel
Kumar Abhishek
Shahira Abousamra
Mazdak Abulnaga
Burak Acar
Abdoljalil Addeh
Ehsan Adeli
Sukesh Adiga Vasudeva
Seyed-Ahmad Ahmadi
Euijoon Ahn
Faranak Akbarifar
Alireza Akhondi-asl
Saad Ullah Akram
Daniel Alexander
Hanan Alghamdi
Hassan Alhajj
Omar Al-Kadi
Max Allan
Andre Altmann
Pablo Alvarez
Charlems Alvarez-Jimenez
Jennifer Alvén
Lidia Al-Zogbi
Kimberly Amador
Tamaz Amiranashvili
Amine Amyar
Wangpeng An
Vincent Andrearczyk
Manon Ansart
Sameer Antani
Jacob Antunes
Michel Antunes
Guilherme Aresta
Mohammad Ali Armin
Kasra Arnavaz
Corey Arnold
Janan Arslan
Marius Arvinte
Muhammad Asad
John Ashburner
Md Ashikuzzaman
Shahab Aslani
Mehdi Astaraki
Angélica Atehortúa
Benjamin Aubert
Marc Aubreville
Paolo Avesani
Sana Ayromlou
Reza Azad
Mohammad Farid Azampour
Qinle Ba
Meritxell Bach Cuadra
Hyeon-Min Bae
Matheus Baffa
Cagla Bahadir
Fan Bai
Jun Bai
Long Bai
Pradeep Bajracharya
Shafa Balaram
Yaël Balbastre
Yutong Ban
Abhirup Banerjee
Soumyanil Banerjee
Sreya Banerjee
Shunxing Bao
Omri Bar
Adrian Barbu
Joao Barreto
Adrian Basarab
Berke Basaran
Michael Baumgartner
Siming Bayer
Roza Bayrak
Aicha BenTaieb
Guy Ben-Yosef
Sutanu Bera
Cosmin Bercea
Jorge Bernal
Jose Bernal
Gabriel Bernardino
Riddhish Bhalodia
Jignesh Bhatt
Indrani Bhattacharya
Binod Bhattarai
Lei Bi
Qi Bi
Cheng Bian
Gui-Bin Bian
Carlo Biffi
Alexander Bigalke
Benjamin Billot
Manuel Birlo
Ryoma Bise
Daniel Blezek
Stefano Blumberg
Sebastian Bodenstedt
Federico Bolelli
Bhushan Borotikar
Ilaria Boscolo Galazzo
Alexandre Bousse
Nicolas Boutry
Joseph Boyd
Behzad Bozorgtabar
Nadia Brancati
Clara Brémond Martin
Stéphanie Bricq
Christopher Bridge
Coleman Broaddus
Rupert Brooks
Tom Brosch
Mikael Brudfors
Ninon Burgos
Nikolay Burlutskiy
Michal Byra
Ryan Cabeen
Mariano Cabezas
Hongmin Cai
Tongan Cai
Zongyou Cai
Liane Canas
Bing Cao
Guogang Cao
Weiguo Cao
Xu Cao
Yankun Cao
Zhenjie Cao

Jaime Cardoso
M. Jorge Cardoso
Owen Carmichael
Jacob Carse
Adrià Casamitjana
Alessandro Casella
Angela Castillo
Kate Cevora
Krishna Chaitanya
Satrajit Chakrabarty
Yi Hao Chan
Shekhar Chandra
Ming-Ching Chang
Peng Chang
Qi Chang
Yuchou Chang
Hanqing Chao
Simon Chatelin
Soumick Chatterjee
Sudhanya Chatterjee
Muhammad Faizyab Ali Chaudhary
Antong Chen
Bingzhi Chen
Chen Chen
Cheng Chen
Chengkuan Chen
Eric Chen
Fang Chen
Haomin Chen
Jianan Chen
Jianxu Chen
Jiazhou Chen
Jie Chen
Jintai Chen
Jun Chen
Junxiang Chen
Junyu Chen
Li Chen
Liyun Chen
Nenglun Chen
Pingjun Chen
Pingyi Chen
Qi Chen
Qiang Chen
Runnan Chen
Shengcong Chen
Sihao Chen
Tingting Chen
Wenting Chen
Xi Chen
Xiang Chen
Xiaoran Chen
Xin Chen
Xiongchao Chen
Yanxi Chen
Yixiong Chen
Yixuan Chen
Yuanyuan Chen
Yuqian Chen
Zhaolin Chen
Zhen Chen
Zhenghao Chen
Zhennong Chen
Zhihao Chen
Zhineng Chen
Zhixiang Chen
Chang-Chieh Cheng
Jiale Cheng
Jianhong Cheng
Jun Cheng
Xuelian Cheng
Yupeng Cheng
Mark Chiew
Philip Chikontwe
Eleni Chiou
Jungchan Cho
Jang-Hwan Choi
Min-Kook Choi
Wookjin Choi
Jaegul Choo
Yu-Cheng Chou
Daan Christiaens
Argyrios Christodoulidis
Stergios Christodoulidis
Kai-Cheng Chuang
Hyungjin Chung
Matthew Clarkson
Michaël Clément
Dana Cobzas
Jaume Coll-Font
Olivier Colliot
Runmin Cong
Yulai Cong
Laura Connolly
William Consagra
Pierre-Henri Conze
Tim Cootes
Teresa Correia
Baris Coskunuzer
Alex Crimi
Can Cui
Hejie Cui
Hui Cui
Lei Cui
Wenhui Cui
Tolga Cukur
Tobias Czempiel
Javid Dadashkarimi
Haixing Dai
Tingting Dan
Kang Dang
Salman Ul Hassan Dar
Eleonora D'Arnese
Dhritiman Das
Neda Davoudi
Tareen Dawood
Sandro De Zanet
Farah Deeba
Charles Delahunt
Herve Delingette
Ugur Demir
Liang-Jian Deng
Ruining Deng
Wenlong Deng
Felix Denzinger
Adrien Depeursinge
Mohammad Mahdi Derakhshani
Hrishikesh Deshpande
Adrien Desjardins
Christian Desrosiers
Blake Dewey
Neel Dey
Rohan Dhamdhere

Maxime Di Folco
Songhui Diao
Alina Dima
Hao Ding
Li Ding
Ying Ding
Zhipeng Ding
Nicola Dinsdale
Konstantin Dmitriev
Ines Domingues
Bo Dong
Liang Dong
Nanqing Dong
Siyuan Dong
Reuben Dorent
Gianfranco Doretto
Sven Dorkenwald
Haoran Dou
Mitchell Doughty
Jason Dowling
Niharika D'Souza
Guodong Du
Jie Du
Shiyi Du
Hongyi Duanmu
Benoit Dufumier
James Duncan
Joshua Durso-Finley
Dmitry V. Dylov
Oleh Dzyubachyk
Mahdi (Elias) Ebnali
Philip Edwards
Jan Egger
Gudmundur Einarsson
Mostafa El Habib Daho
Ahmed Elazab
Idris El-Feghi
David Ellis
Mohammed Elmogy
Amr Elsawy
Okyaz Eminaga
Ertunc Erdil
Lauren Erdman
Marius Erdt
Maria Escobar
Hooman Esfandiari
Nazila Esmaeili
Ivan Ezhov
Alessio Fagioli
Deng-Ping Fan
Lei Fan
Xin Fan
Yubo Fan
Huihui Fang
Jiansheng Fang
Xi Fang
Zhenghan Fang
Mohammad Farazi
Azade Farshad
Mohsen Farzi
Hamid Fehri
Lina Felsner
Chaolu Feng
Chun-Mei Feng
Jianjiang Feng
Mengling Feng
Ruibin Feng
Zishun Feng
Alvaro Fernandez-Quilez
Ricardo Ferrari
Lucas Fidon
Lukas Fischer
Madalina Fiterau
Antonio Foncubierta-Rodríguez
Fahimeh Fooladgar
Germain Forestier
Nils Daniel Forkert
Jean-Rassaire Fouefack
Kevin François-Bouaou
Wolfgang Freysinger
Bianca Freytag
Guanghui Fu
Kexue Fu
Lan Fu
Yunguan Fu
Pedro Furtado
Ryo Furukawa
Jin Kyu Gahm
Mélanie Gaillochet
Francesca Galassi
Jiangzhang Gan
Yu Gan
Yulu Gan
Alireza Ganjdanesh
Chang Gao
Cong Gao
Linlin Gao
Zeyu Gao
Zhongpai Gao
Sara Garbarino
Alain Garcia
Beatriz Garcia Santa Cruz
Rongjun Ge
Shiv Gehlot
Manuela Geiss
Salah Ghamizi
Negin Ghamsarian
Ramtin Gharleghi
Ghazal Ghazaei
Florin Ghesu
Sayan Ghosal
Syed Zulqarnain Gilani
Mahdi Gilany
Yannik Glaser
Ben Glocker
Bharti Goel
Jacob Goldberger
Polina Golland
Alberto Gomez
Catalina Gomez
Estibaliz Gómez-de-Mariscal
Haifan Gong
Kuang Gong
Xun Gong
Ricardo Gonzales
Camila Gonzalez
German Gonzalez
Vanessa Gonzalez Duque
Sharath Gopal
Karthik Gopinath
Pietro Gori
Michael Götz
Shuiping Gou

Maged Goubran
Sobhan Goudarzi
Mark Graham
Alejandro Granados
Mara Graziani
Thomas Grenier
Radu Grosu
Michal Grzeszczyk
Feng Gu
Pengfei Gu
Qiangqiang Gu
Ran Gu
Shi Gu
Wenhao Gu
Xianfeng Gu
Yiwen Gu
Zaiwang Gu
Hao Guan
Jayavardhana Gubbi
Houssem-Eddine Gueziri
Dazhou Guo
Hengtao Guo
Jixiang Guo
Jun Guo
Pengfei Guo
Wenzhangzhi Guo
Xiaoqing Guo
Xueqi Guo
Yi Guo
Vikash Gupta
Praveen Gurunath Bharathi
Prashnna Gyawali
Sung Min Ha
Mohamad Habes
Ilker Hacihaliloglu
Stathis Hadjidemetriou
Fatemeh Haghighi
Justin Haldar
Noura Hamze
Liang Han
Luyi Han
Seungjae Han
Tianyu Han
Zhongyi Han
Jonny Hancox
Lasse Hansen
Degan Hao
Huaying Hao
Jinkui Hao
Nazim Haouchine
Michael Hardisty
Stefan Harrer
Jeffry Hartanto
Charles Hatt
Huiguang He
Kelei He
Qi He
Shenghua He
Xinwei He
Stefan Heldmann
Nicholas Heller
Edward Henderson
Alessa Hering
Monica Hernandez
Kilian Hett
Amogh Hiremath
David Ho
Malte Hoffmann
Matthew Holden
Qingqi Hong
Yoonmi Hong
Mohammad Reza Hosseinzadeh Taher
William Hsu
Chuanfei Hu
Dan Hu
Kai Hu
Rongyao Hu
Shishuai Hu
Xiaoling Hu
Xinrong Hu
Yan Hu
Yang Hu
Chaoqin Huang
Junzhou Huang
Ling Huang
Luojie Huang
Qinwen Huang
Sharon Xiaolei Huang
Weijian Huang
Xiaoyang Huang
Yi-Jie Huang
Yongsong Huang
Yongxiang Huang
Yuhao Huang
Zhe Huang
Zhi-An Huang
Ziyi Huang
Arnaud Huaulmé
Henkjan Huisman
Alex Hung
Jiayu Huo
Andreas Husch
Mohammad Arafat Hussain
Sarfaraz Hussein
Jana Hutter
Khoi Huynh
Ilknur Icke
Kay Igwe
Abdullah Al Zubaer Imran
Muhammad Imran
Samra Irshad
Nahid Ul Islam
Koichi Ito
Hayato Itoh
Yuji Iwahori
Krithika Iyer
Mohammad Jafari
Srikrishna Jaganathan
Hassan Jahanandish
Andras Jakab
Amir Jamaludin
Amoon Jamzad
Ananya Jana
Se-In Jang
Pierre Jannin
Vincent Jaouen
Uditha Jarayathne
Ronnachai Jaroensri
Guillaume Jaume
Syed Ashar Javed
Rachid Jennane
Debesh Jha
Ge-Peng Ji

Luping Ji
Zexuan Ji
Zhanghexuan Ji
Haozhe Jia
Hongchao Jiang
Jue Jiang
Meirui Jiang
Tingting Jiang
Xiajun Jiang
Zekun Jiang
Zhifan Jiang
Ziyu Jiang
Jianbo Jiao
Zhicheng Jiao
Chen Jin
Dakai Jin
Qiangguo Jin
Qiuye Jin
Weina Jin
Baoyu Jing
Bin Jing
Yaqub Jonmohamadi
Lie Ju
Yohan Jun
Dinkar Juyal
Manjunath K N
Ali Kafaei Zad Tehrani
John Kalafut
Niveditha Kalavakonda
Megha Kalia
Anil Kamat
Qingbo Kang
Po-Yu Kao
Anuradha Kar
Neerav Karani
Turkay Kart
Satyananda Kashyap
Alexander Katzmann
Lisa Kausch
Maxime Kayser
Salome Kazeminia
Wenchi Ke
Youngwook Kee
Matthias Keicher
Erwan Kerrien
Afifa Khaled
Nadieh Khalili
Farzad Khalvati
Bidur Khanal
Bishesh Khanal
Pulkit Khandelwal
Maksim Kholiavchenko
Ron Kikinis
Benjamin Killeen
Daeseung Kim
Heejong Kim
Jaeil Kim
Jinhee Kim
Jinman Kim
Junsik Kim
Minkyung Kim
Namkug Kim
Sangwook Kim
Tae Soo Kim
Younghoon Kim
Young-Min Kim
Andrew King
Miranda Kirby
Gabriel Kiss
Andreas Kist
Yoshiro Kitamura
Stefan Klein
Tobias Klinder
Kazuma Kobayashi
Lisa Koch
Satoshi Kondo
Fanwei Kong
Tomasz Konopczynski
Ender Konukoglu
Aishik Konwer
Thijs Kooi
Ivica Kopriva
Avinash Kori
Kivanc Kose
Suraj Kothawade
Anna Kreshuk
AnithaPriya Krishnan
Florian Kromp
Frithjof Kruggel
Thomas Kuestner
Levin Kuhlmann
Abhay Kumar
Kuldeep Kumar
Sayantan Kumar
Manuela Kunz
Holger Kunze
Tahsin Kurc
Anvar Kurmukov
Yoshihiro Kuroda
Yusuke Kurose
Hyuksool Kwon
Aymen Laadhari
Jorma Laaksonen
Dmitrii Lachinov
Alain Lalande
Rodney LaLonde
Bennett Landman
Daniel Lang
Carole Lartizien
Shlomi Laufer
Max-Heinrich Laves
William Le
Loic Le Folgoc
Christian Ledig
Eung-Joo Lee
Ho Hin Lee
Hyekyoung Lee
John Lee
Kisuk Lee
Kyungsu Lee
Soochahn Lee
Woonghee Lee
Étienne Léger
Wen Hui Lei
Yiming Lei
George Leifman
Rogers Jeffrey Leo John
Juan Leon
Bo Li
Caizi Li
Chao Li
Chen Li
Cheng Li
Chenxin Li
Chnegyin Li

Dawei Li
Fuhai Li
Gang Li
Guang Li
Hao Li
Haofeng Li
Haojia Li
Heng Li
Hongming Li
Hongwei Li
Huiqi Li
Jian Li
Jieyu Li
Kang Li
Lin Li
Mengzhang Li
Ming Li
Qing Li
Quanzheng Li
Shaohua Li
Shulong Li
Tengfei Li
Weijian Li
Wen Li
Xiaomeng Li
Xingyu Li
Xinhui Li
Xuelu Li
Xueshen Li
Yamin Li
Yang Li
Yi Li
Yuemeng Li
Yunxiang Li
Zeju Li
Zhaoshuo Li
Zhe Li
Zhen Li
Zhenqiang Li
Zhiyuan Li
Zhjin Li
Zi Li
Hao Liang
Libin Liang
Peixian Liang
Yuan Liang
Yudong Liang
Haofu Liao
Hongen Liao
Wei Liao
Zehui Liao
Gilbert Lim
Hongxiang Lin
Li Lin
Manxi Lin
Mingquan Lin
Tiancheng Lin
Yi Lin
Zudi Lin
Claudia Lindner
Simone Lionetti
Chi Liu
Chuanbin Liu
Daochang Liu
Dongnan Liu
Feihong Liu
Fenglin Liu
Han Liu
Huiye Liu
Jiang Liu
Jie Liu
Jinduo Liu
Jing Liu
Jingya Liu
Jundong Liu
Lihao Liu
Mengting Liu
Mingyuan Liu
Peirong Liu
Peng Liu
Qin Liu
Quan Liu
Rui Liu
Shengfeng Liu
Shuangjun Liu
Sidong Liu
Siyuan Liu
Weide Liu
Xiao Liu
Xiaoyu Liu
Xingtong Liu
Xinwen Liu
Xinyang Liu
Xinyu Liu
Yan Liu
Yi Liu
Yihao Liu
Yikang Liu
Yilin Liu
Yilong Liu
Yiqiao Liu
Yong Liu
Yuhang Liu
Zelong Liu
Zhe Liu
Zhiyuan Liu
Zuozhu Liu
Lisette Lockhart
Andrea Loddo
Nicolas Loménie
Yonghao Long
Daniel Lopes
Ange Lou
Brian Lovell
Nicolas Loy Rodas
Charles Lu
Chun-Shien Lu
Donghuan Lu
Guangming Lu
Huanxiang Lu
Jingpei Lu
Yao Lu
Oeslle Lucena
Jie Luo
Luyang Luo
Ma Luo
Mingyuan Luo
Wenhan Luo
Xiangde Luo
Xinzhe Luo
Jinxin Lv
Tianxu Lv
Fei Lyu
Ilwoo Lyu
Mengye Lyu

Qing Lyu
Yanjun Lyu
Yuanyuan Lyu
Benteng Ma
Chunwei Ma
Hehuan Ma
Jun Ma
Junbo Ma
Wenao Ma
Yuhui Ma
Pedro Macias Gordaliza
Anant Madabhushi
Derek Magee
S. Sara Mahdavi
Andreas Maier
Klaus H. Maier-Hein
Sokratis Makrogiannis
Danial Maleki
Michail Mamalakis
Zhehua Mao
Jan Margeta
Brett Marinelli
Zdravko Marinov
Viktoria Markova
Carsten Marr
Yassine Marrakchi
Anne Martel
Martin Maška
Tejas Sudharshan Mathai
Petr Matula
Dimitrios Mavroeidis
Evangelos Mazomenos
Amarachi Mbakwe
Adam McCarthy
Stephen McKenna
Raghav Mehta
Xueyan Mei
Felix Meissen
Felix Meister
Afaque Memon
Mingyuan Meng
Qingjie Meng
Xiangzhu Meng
Yanda Meng
Zhu Meng
Martin Menten
Odyssée Merveille
Mikhail Milchenko
Leo Milecki
Fausto Milletari
Hyun-Seok Min
Zhe Min
Song Ming
Duy Minh Ho Nguyen
Deepak Mishra
Suraj Mishra
Virendra Mishra
Tadashi Miyamoto
Sara Moccia
Marc Modat
Omid Mohareri
Tony C. W. Mok
Javier Montoya
Rodrigo Moreno
Stefano Moriconi
Lia Morra
Ana Mota
Lei Mou
Dana Moukheiber
Lama Moukheiber
Daniel Moyer
Pritam Mukherjee
Anirban Mukhopadhyay
Henning Müller
Ana Murillo
Gowtham Krishnan Murugesan
Ahmed Naglah
Karthik Nandakumar
Venkatesh Narasimhamurthy
Raja Narayan
Dominik Narnhofer
Vishwesh Nath
Rodrigo Nava
Abdullah Nazib
Ahmed Nebli
Peter Neher
Amin Nejatbakhsh
Trong-Thuan Nguyen
Truong Nguyen
Dong Ni
Haomiao Ni
Xiuyan Ni
Hannes Nickisch
Weizhi Nie
Aditya Nigam
Lipeng Ning
Xia Ning
Kazuya Nishimura
Chuang Niu
Sijie Niu
Vincent Noblet
Narges Norouzi
Alexey Novikov
Jorge Novo
Gilberto Ochoa-Ruiz
Masahiro Oda
Benjamin Odry
Hugo Oliveira
Sara Oliveira
Arnau Oliver
Jimena Olveres
John Onofrey
Marcos Ortega
Mauricio Alberto Ortega-Ruíz
Yusuf Osmanlioglu
Chubin Ou
Cheng Ouyang
Jiahong Ouyang
Xi Ouyang
Cristina Oyarzun Laura
Utku Ozbulak
Ece Ozkan
Ege Özsoy
Batu Ozturkler
Harshith Padigela
Johannes Paetzold
José Blas Pagador Carrasco
Daniel Pak
Sourabh Palande
Chengwei Pan
Jiazhen Pan

Jin Pan
Yongsheng Pan
Egor Panfilov
Jiaxuan Pang
Joao Papa
Constantin Pape
Bartlomiej Papiez
Nripesh Parajuli
Hyunjin Park
Akash Parvatikar
Tiziano Passerini
Diego Patiño Cortés
Mayank Patwari
Angshuman Paul
Rasmus Paulsen
Yuchen Pei
Yuru Pei
Tao Peng
Wei Peng
Yige Peng
Yunsong Peng
Matteo Pennisi
Antonio Pepe
Oscar Perdomo
Sérgio Pereira
Jose-Antonio Pérez-Carrasco
Mehran Pesteie
Terry Peters
Eike Petersen
Jens Petersen
Micha Pfeiffer
Dzung Pham
Hieu Pham
Ashish Phophalia
Tomasz Pieciak
Antonio Pinheiro
Pramod Pisharady
Theodoros Pissas
Szymon Płotka
Kilian Pohl
Sebastian Pölsterl
Alison Pouch
Tim Prangemeier
Prateek Prasanna
Raphael Prevost
Juan Prieto
Federica Proietto Salanitri
Sergi Pujades
Elodie Puybareau
Talha Qaiser
Buyue Qian
Mengyun Qiao
Yuchuan Qiao
Zhi Qiao
Chenchen Qin
Fangbo Qin
Wenjian Qin
Yulei Qin
Jie Qiu
Jielin Qiu
Peijie Qiu
Shi Qiu
Wu Qiu
Liangqiong Qu
Linhao Qu
Quan Quan
Tran Minh Quan
Sandro Queirós
Prashanth R
Febrian Rachmadi
Daniel Racoceanu
Mehdi Rahim
Jagath Rajapakse
Kashif Rajpoot
Keerthi Ram
Dhanesh Ramachandram
João Ramalhinho
Xuming Ran
Aneesh Rangnekar
Hatem Rashwan
Keerthi Sravan Ravi
Daniele Ravì
Sadhana Ravikumar
Harish Raviprakash
Surreerat Reaungamornrat
Samuel Remedios
Mengwei Ren
Sucheng Ren
Elton Rexhepaj
Mauricio Reyes
Constantino Reyes-Aldasoro
Abel Reyes-Angulo
Hadrien Reynaud
Razieh Rezaei
Anne-Marie Rickmann
Laurent Risser
Dominik Rivoir
Emma Robinson
Robert Robinson
Jessica Rodgers
Ranga Rodrigo
Rafael Rodrigues
Robert Rohling
Margherita Rosnati
Łukasz Roszkowiak
Holger Roth
José Rouco
Dan Ruan
Jiacheng Ruan
Daniel Rueckert
Danny Ruijters
Kanghyun Ryu
Ario Sadafi
Numan Saeed
Monjoy Saha
Pramit Saha
Farhang Sahba
Pranjal Sahu
Simone Saitta
Md Sirajus Salekin
Abbas Samani
Pedro Sanchez
Luis Sanchez Giraldo
Yudi Sang
Gerard Sanroma-Guell
Rodrigo Santa Cruz
Alice Santilli
Rachana Sathish
Olivier Saut
Mattia Savardi
Nico Scherf
Alexander Schlaefer
Jerome Schmid

Adam Schmidt
Julia Schnabel
Lawrence Schobs
Julian Schön
Peter Schueffler
Andreas Schuh
Christina Schwarz-Gsaxner
Michaël Sdika
Suman Sedai
Lalithkumar Seenivasan
Matthias Seibold
Sourya Sengupta
Lama Seoud
Ana Sequeira
Sharmishtaa Seshamani
Ahmed Shaffie
Jay Shah
Keyur Shah
Ahmed Shahin
Mohammad Abuzar Shaikh
S. Shailja
Hongming Shan
Wei Shao
Mostafa Sharifzadeh
Anuja Sharma
Gregory Sharp
Hailan Shen
Li Shen
Linlin Shen
Mali Shen
Mingren Shen
Yiqing Shen
Zhengyang Shen
Jun Shi
Xiaoshuang Shi
Yiyu Shi
Yonggang Shi
Hoo-Chang Shin
Jitae Shin
Keewon Shin
Boris Shirokikh
Suzanne Shontz
Yucheng Shu
Hanna Siebert
Alberto Signoroni
Wilson Silva
Julio Silva-Rodríguez
Margarida Silveira
Walter Simson
Praveer Singh
Vivek Singh
Nitin Singhal
Elena Sizikova
Gregory Slabaugh
Dane Smith
Kevin Smith
Tiffany So
Rajath Soans
Roger Soberanis-Mukul
Hessam Sokooti
Jingwei Song
Weinan Song
Xinhang Song
Xinrui Song
Mazen Soufi
Georgia Sovatzidi
Bella Specktor Fadida
William Speier
Ziga Spiclin
Dominik Spinczyk
Jon Sporring
Pradeeba Sridar
Chetan L. Srinidhi
Abhishek Srivastava
Lawrence Staib
Marc Stamminger
Justin Strait
Hai Su
Ruisheng Su
Zhe Su
Vaishnavi Subramanian
Gérard Subsol
Carole Sudre
Dong Sui
Heung-Il Suk
Shipra Suman
He Sun
Hongfu Sun
Jian Sun
Li Sun
Liyan Sun
Shanlin Sun
Kyung Sung
Yannick Suter
Swapna T. R.
Amir Tahmasebi
Pablo Tahoces
Sirine Taleb
Bingyao Tan
Chaowei Tan
Wenjun Tan
Hao Tang
Siyi Tang
Xiaoying Tang
Yucheng Tang
Zihao Tang
Michael Tanzer
Austin Tapp
Elias Tappeiner
Mickael Tardy
Giacomo Tarroni
Athena Taymourtash
Kaveri Thakoor
Elina Thibeau-Sutre
Paul Thienphrapa
Sarina Thomas
Stephen Thompson
Karl Thurnhofer-Hemsi
Cristiana Tiago
Lin Tian
Lixia Tian
Yapeng Tian
Yu Tian
Yun Tian
Aleksei Tiulpin
Hamid Tizhoosh
Minh Nguyen Nhat To
Matthew Toews
Maryam Toloubidokhti
Minh Tran
Quoc-Huy Trinh
Jocelyne Troccaz
Roger Trullo

Chialing Tsai
Apostolia Tsirikoglou
Puxun Tu
Samyakh Tukra
Sudhakar Tummala
Georgios Tziritas
Vladimír Ulman
Tamas Ungi
Régis Vaillant
Jeya Maria Jose Valanarasu
Vanya Valindria
Juan Miguel Valverde
Fons van der Sommen
Maureen van Eijnatten
Tom van Sonsbeek
Gijs van Tulder
Yogatheesan Varatharajah
Madhurima Vardhan
Thomas Varsavsky
Hooman Vaseli
Serge Vasylechko
S. Swaroop Vedula
Sanketh Vedula
Gonzalo Vegas Sanchez-Ferrero
Matthew Velazquez
Archana Venkataraman
Sulaiman Vesal
Mitko Veta
Barbara Villarini
Athanasios Vlontzos
Wolf-Dieter Vogl
Ingmar Voigt
Sandrine Voros
Vibashan VS
Trinh Thi Le Vuong
An Wang
Bo Wang
Ce Wang
Changmiao Wang
Ching-Wei Wang
Dadong Wang
Dong Wang
Fakai Wang
Guotai Wang
Haifeng Wang
Haoran Wang
Hong Wang
Hongxiao Wang
Hongyu Wang
Jiacheng Wang
Jing Wang
Jue Wang
Kang Wang
Ke Wang
Lei Wang
Li Wang
Liansheng Wang
Lin Wang
Ling Wang
Linwei Wang
Manning Wang
Mingliang Wang
Puyang Wang
Qiuli Wang
Renzhen Wang
Ruixuan Wang
Shaoyu Wang
Sheng Wang
Shujun Wang
Shuo Wang
Shuqiang Wang
Tao Wang
Tianchen Wang
Tianyu Wang
Wenzhe Wang
Xi Wang
Xiangdong Wang
Xiaoqing Wang
Xiaosong Wang
Yan Wang
Yangang Wang
Yaping Wang
Yi Wang
Yirui Wang
Yixin Wang
Zeyi Wang
Zhao Wang
Zichen Wang
Ziqin Wang
Ziyi Wang
Zuhui Wang
Dong Wei
Donglai Wei
Hao Wei
Jia Wei
Leihao Wei
Ruofeng Wei
Shuwen Wei
Martin Weigert
Wolfgang Wein
Michael Wels
Cédric Wemmert
Thomas Wendler
Markus Wenzel
Rhydian Windsor
Adam Wittek
Marek Wodzinski
Ivo Wolf
Julia Wolleb
Ka-Chun Wong
Jonghye Woo
Chongruo Wu
Chunpeng Wu
Fuping Wu
Huaqian Wu
Ji Wu
Jiangjie Wu
Jiong Wu
Junde Wu
Linshan Wu
Qing Wu
Weiwen Wu
Wenjun Wu
Xiyin Wu
Yawen Wu
Ye Wu
Yicheng Wu
Yongfei Wu
Zhengwang Wu
Pengcheng Xi
Chao Xia
Siyu Xia
Wenjun Xia
Lei Xiang

Tiange Xiang
Deqiang Xiao
Li Xiao
Xiaojiao Xiao
Yiming Xiao
Zeyu Xiao
Hongtao Xie
Huidong Xie
Jianyang Xie
Long Xie
Weidi Xie
Fangxu Xing
Shuwei Xing
Xiaodan Xing
Xiaohan Xing
Haoyi Xiong
Yujian Xiong
Di Xu
Feng Xu
Haozheng Xu
Hongming Xu
Jiangchang Xu
Jiaqi Xu
Junshen Xu
Kele Xu
Lijian Xu
Min Xu
Moucheng Xu
Rui Xu
Xiaowei Xu
Xuanang Xu
Yanwu Xu
Yanyu Xu
Yongchao Xu
Yunqiu Xu
Zhe Xu
Zhoubing Xu
Ziyue Xu
Kai Xuan
Cheng Xue
Jie Xue
Tengfei Xue
Wufeng Xue
Yuan Xue
Zhong Xue
Ts Faridah Yahya
Chaochao Yan
Jiangpeng Yan
Ming Yan
Qingsen Yan
Xiangyi Yan
Yuguang Yan
Zengqiang Yan
Baoyao Yang
Carl Yang
Changchun Yang
Chen Yang
Feng Yang
Fengting Yang
Ge Yang
Guanyu Yang
Heran Yang
Huijuan Yang
Jiancheng Yang
Jiewen Yang
Peng Yang
Qi Yang
Qiushi Yang
Wei Yang
Xin Yang
Xuan Yang
Yan Yang
Yanwu Yang
Yifan Yang
Yingyu Yang
Zhicheng Yang
Zhijian Yang
Jiangchao Yao
Jiawen Yao
Lanhong Yao
Linlin Yao
Qingsong Yao
Tianyuan Yao
Xiaohui Yao
Zhao Yao
Dong Hye Ye
Menglong Ye
Yousef Yeganeh
Jirong Yi
Xin Yi
Chong Yin
Pengshuai Yin
Yi Yin
Zhaozheng Yin
Chunwei Ying
Youngjin Yoo
Jihun Yoon
Chenyu You
Hanchao Yu
Heng Yu
Jinhua Yu
Jinze Yu
Ke Yu
Qi Yu
Qian Yu
Thomas Yu
Weimin Yu
Yang Yu
Chenxi Yuan
Kun Yuan
Wu Yuan
Yixuan Yuan
Paul Yushkevich
Fatemeh Zabihollahy
Samira Zare
Ramy Zeineldin
Dong Zeng
Qi Zeng
Tianyi Zeng
Wei Zeng
Kilian Zepf
Kun Zhan
Bokai Zhang
Daoqiang Zhang
Dong Zhang
Fa Zhang
Hang Zhang
Hanxiao Zhang
Hao Zhang
Haopeng Zhang
Haoyue Zhang
Hongrun Zhang
Jiadong Zhang
Jiajin Zhang
Jianpeng Zhang

Jiawei Zhang
Jingqing Zhang
Jingyang Zhang
Jinwei Zhang
Jiong Zhang
Jiping Zhang
Ke Zhang
Lefei Zhang
Lei Zhang
Li Zhang
Lichi Zhang
Lu Zhang
Minghui Zhang
Molin Zhang
Ning Zhang
Rongzhao Zhang
Ruipeng Zhang
Ruisi Zhang
Shichuan Zhang
Shihao Zhang
Shuai Zhang
Tuo Zhang
Wei Zhang
Weihang Zhang
Wen Zhang
Wenhua Zhang
Wenqiang Zhang
Xiaodan Zhang
Xiaoran Zhang
Xin Zhang
Xukun Zhang
Xuzhe Zhang
Ya Zhang
Yanbo Zhang
Yanfu Zhang
Yao Zhang
Yi Zhang
Yifan Zhang
Yixiao Zhang
Yongqin Zhang
You Zhang
Youshan Zhang
Yu Zhang
Yubo Zhang
Yue Zhang
Yuhan Zhang
Yulun Zhang
Yundong Zhang
Yunlong Zhang
Yuyao Zhang
Zheng Zhang
Zhenxi Zhang
Ziqi Zhang
Can Zhao
Chongyue Zhao
Fenqiang Zhao
Gangming Zhao
He Zhao
Jianfeng Zhao
Jun Zhao
Li Zhao
Liang Zhao
Lin Zhao
Mengliu Zhao
Mingbo Zhao
Qingyu Zhao
Shang Zhao
Shijie Zhao
Tengda Zhao
Tianyi Zhao
Wei Zhao
Yidong Zhao
Yiyuan Zhao
Yu Zhao
Zhihe Zhao
Ziyuan Zhao
Haiyong Zheng
Hao Zheng
Jiannan Zheng
Kang Zheng
Meng Zheng
Sisi Zheng
Tianshu Zheng
Yalin Zheng
Yefeng Zheng
Yinqiang Zheng
Yushan Zheng
Aoxiao Zhong
Jia-Xing Zhong
Tao Zhong
Zichun Zhong
Hong-Yu Zhou
Houliang Zhou
Huiyu Zhou
Kang Zhou
Qin Zhou
Ran Zhou
S. Kevin Zhou
Tianfei Zhou
Wei Zhou
Xiao-Hu Zhou
Xiao-Yun Zhou
Yi Zhou
Youjia Zhou
Yukun Zhou
Zongwei Zhou
Chenglu Zhu
Dongxiao Zhu
Heqin Zhu
Jiayi Zhu
Meilu Zhu
Wei Zhu
Wenhui Zhu
Xiaofeng Zhu
Xin Zhu
Yonghua Zhu
Yongpei Zhu
Yuemin Zhu
Yan Zhuang
David Zimmerer
Yongshuo Zong
Ke Zou
Yukai Zou
Lianrui Zuo
Gerald Zwettler

Outstanding Area Chairs

Mingxia Liu	University of North Carolina at Chapel Hill, USA
Matthias Wilms	University of Calgary, Canada
Veronika Zimmer	Technical University Munich, Germany

Outstanding Reviewers

Kimberly Amador	University of Calgary, Canada
Angela Castillo	Universidad de los Andes, Colombia
Chen Chen	Imperial College London, UK
Laura Connolly	Queen's University, Canada
Pierre-Henri Conze	IMT Atlantique, France
Niharika D'Souza	IBM Research, USA
Michael Götz	University Hospital Ulm, Germany
Meirui Jiang	Chinese University of Hong Kong, China
Manuela Kunz	National Research Council Canada, Canada
Zdravko Marinov	Karlsruhe Institute of Technology, Germany
Sérgio Pereira	Lunit, South Korea
Lalithkumar Seenivasan	National University of Singapore, Singapore

Honorable Mentions (Reviewers)

Kumar Abhishek	Simon Fraser University, Canada
Guilherme Aresta	Medical University of Vienna, Austria
Shahab Aslani	University College London, UK
Marc Aubreville	Technische Hochschule Ingolstadt, Germany
Yaël Balbastre	Massachusetts General Hospital, USA
Omri Bar	Theator, Israel
Aicha Ben Taieb	Simon Fraser University, Canada
Cosmin Bercea	Technical University Munich and Helmholtz AI and Helmholtz Center Munich, Germany
Benjamin Billot	Massachusetts Institute of Technology, USA
Michal Byra	RIKEN Center for Brain Science, Japan
Mariano Cabezas	University of Sydney, Australia
Alessandro Casella	Italian Institute of Technology and Politecnico di Milano, Italy
Junyu Chen	Johns Hopkins University, USA
Argyrios Christodoulidis	Pfizer, Greece
Olivier Colliot	CNRS, France

Lei Cui	Northwest University, China
Neel Dey	Massachusetts Institute of Technology, USA
Alessio Fagioli	Sapienza University, Italy
Yannik Glaser	University of Hawaii at Manoa, USA
Haifan Gong	Chinese University of Hong Kong, Shenzhen, China
Ricardo Gonzales	University of Oxford, UK
Sobhan Goudarzi	Sunnybrook Research Institute, Canada
Michal Grzeszczyk	Sano Centre for Computational Medicine, Poland
Fatemeh Haghighi	Arizona State University, USA
Edward Henderson	University of Manchester, UK
Qingqi Hong	Xiamen University, China
Mohammad R. H. Taher	Arizona State University, USA
Henkjan Huisman	Radboud University Medical Center, the Netherlands
Ronnachai Jaroensri	Google, USA
Qiangguo Jin	Northwestern Polytechnical University, China
Neerav Karani	Massachusetts Institute of Technology, USA
Benjamin Killeen	Johns Hopkins University, USA
Daniel Lang	Helmholtz Center Munich, Germany
Max-Heinrich Laves	Philips Research and ImFusion GmbH, Germany
Gilbert Lim	SingHealth, Singapore
Mingquan Lin	Weill Cornell Medicine, USA
Charles Lu	Massachusetts Institute of Technology, USA
Yuhui Ma	Chinese Academy of Sciences, China
Tejas Sudharshan Mathai	National Institutes of Health, USA
Felix Meissen	Technische Universität München, Germany
Mingyuan Meng	University of Sydney, Australia
Leo Milecki	CentraleSupelec, France
Marc Modat	King's College London, UK
Tiziano Passerini	Siemens Healthineers, USA
Tomasz Pieciak	Universidad de Valladolid, Spain
Daniel Rueckert	Imperial College London, UK
Julio Silva-Rodríguez	ETS Montreal, Canada
Bingyao Tan	Nanyang Technological University, Singapore
Elias Tappeiner	UMIT - Private University for Health Sciences, Medical Informatics and Technology, Austria
Jocelyne Troccaz	TIMC Lab, Grenoble Alpes University-CNRS, France
Chialing Tsai	Queens College, City University New York, USA
Juan Miguel Valverde	University of Eastern Finland, Finland
Sulaiman Vesal	Stanford University, USA

Contents – Part I

Machine Learning with Limited Supervision

PET-Diffusion: Unsupervised PET Enhancement Based on the Latent Diffusion Model

Caiwen Jiang[1], Yongsheng Pan[1], Mianxin Liu[3], Lei Ma[1], Xiao Zhang[1], Jiameng Liu[1], Xiaosong Xiong[1], and Dinggang Shen[1,2,4](✉)

[1] School of Biomedical Engineering, ShanghaiTech University, Shanghai, China
{jiangcw,panysh,dgshen}@shanghaitech.edu.cn
[2] Shanghai United Imaging Intelligence Co., Ltd., Shanghai 200230, China
[3] Shanghai Artificial Intelligence Laboratory, Shanghai 200232, China
[4] Shanghai Clinical Research and Trial Center, Shanghai 201210, China

Abstract. Positron emission tomography (PET) is an advanced nuclear imaging technique with an irreplaceable role in neurology and oncology studies, but its accessibility is often limited by the radiation hazards inherent in imaging. To address this dilemma, PET enhancement methods have been developed by improving the quality of low-dose PET (LPET) images to standard-dose PET (SPET) images. However, previous PET enhancement methods rely heavily on the paired LPET and SPET data which are rare in clinic. Thus, in this paper, we propose an unsupervised PET enhancement (uPETe) framework based on the latent diffusion model, which can be trained only on SPET data. Specifically, our SPET-only uPETe consists of an encoder to compress the input SPET/LPET images into latent representations, a latent diffusion model to learn/estimate the distribution of SPET latent representations, and a decoder to recover the latent representations into SPET images. Moreover, from the theory of actual PET imaging, we improve the latent diffusion model of uPETe by 1) adopting **PET image compression** for reducing the computational cost of diffusion model, 2) using **Poisson diffusion** to replace Gaussian diffusion for making the perturbed samples closer to the actual noisy PET, and 3) designing **CT-guided cross-attention** for incorporating additional CT images into the inverse process to aid the recovery of structural details in PET. With extensive experimental validation, our uPETe can achieve superior performance over state-of-the-art methods, and shows stronger generalizability to the dose changes of PET imaging. The code of our implementation is available at https://github.com/jiang-cw/PET-diffusion.

Keywords: Positron emission tomography (PET) · Enhancement · Latent diffusion model · Poisson diffusion · CT-guided cross-attention

1 Introduction

Positron emission tomography (PET) is a sensitive nuclear imaging technique, and plays an essential role in early disease diagnosis, such as cancers and

H. Greenspan et al. (Eds.): MICCAI 2023, LNCS 14220, pp. 3–12, 2023.
https://doi.org/10.1007/978-3-031-43907-0_1

Alzheimer's disease [8]. However, acquiring high-quality PET images requires injecting a sufficient dose (standard dose) of radionuclides into the human body, which poses unacceptable radiation hazards for pregnant women and infants even following the As Low As Reasonably Achievable (ALARA) principle [19]. To reduce the radiation hazards, besides upgrading imaging hardware, designing advanced PET enhancement algorithms for improving the quality of low-dose PET (LPET) images to standard-dose PET (SPET) images is a promising alternative.

In recent years, many enhancement algorithms have been proposed to improve PET image quality. Among the earliest are filtering-based methods such as non-local mean (NLM) filter [1], block-matching 3D filter [4], bilateral filter [7], and guided filter [22], which are quite robust but tend to over-smooth images and suppress the high-frequency details. Subsequently, with the development of deep learning, the end-to-end PET enhancement networks [9,14,21] were proposed and achieved significant performance improvement. But these supervised methods relied heavily on the paired LPET and SPET data that are rare in actual clinic due to radiation exposure and involuntary motions (e.g., respiratory and muscle relaxation). Consequently, unsupervised PET enhancement methods such as deep image prior [3], Noise2Noise [12,20], and their variants [17] were developed to overcome this limitation. However, these methods still require LPET to train models, which contradicts with the fact that only SPET scans are conducted in clinic.

Fortunately, the recent glowing diffusion model [6] provides us with the idea for proposing a clinically-applicable PET enhancement approach, whose training only relies on SPET data. Generally, the diffusion model consists of two reversible processes, where the forward diffusion adds noise to a clean image until it becomes pure noise, while the reverse process removes noise from pure noise until the clean image is recovered. By combining the mechanics of diffusion model with the observation that the main differences between LPET and SPET are manifested as levels of noises in the image [11], we can view LPET and SPET as results at different stages in an integrated diffusion process. Therefore, when a diffusion model (trained only on SPET) can recover noisy samples to SPET, this model can also recover LPET to SPET. However, extending the diffusion model developed for 2D photographic images to PET enhancement still faces two problems: a) three-dimensionsal (3D) PET images will dramatically increase the computational cost of diffusion model; b) PET is the detail-sensitive images and may be introduced/lost some details during the procedure of adding/removing noise, which will affect the downstream diagnosis.

Taking all into consideration, we propose the SPET-only unsupervised PET enhancement (uPETe) framework based on the latent diffusion model. Specifically, uPETe has an encoder-<diffusion model>-decoder structure that first uses the encoder to compress input the LPET/SPET images into latent representations, then uses the latent diffusion model to learn/estimate the distribution of SPET latent representations, and finally uses the decoder to recover SPET images from the estimated SPET latent representations. The keys of our uPETe

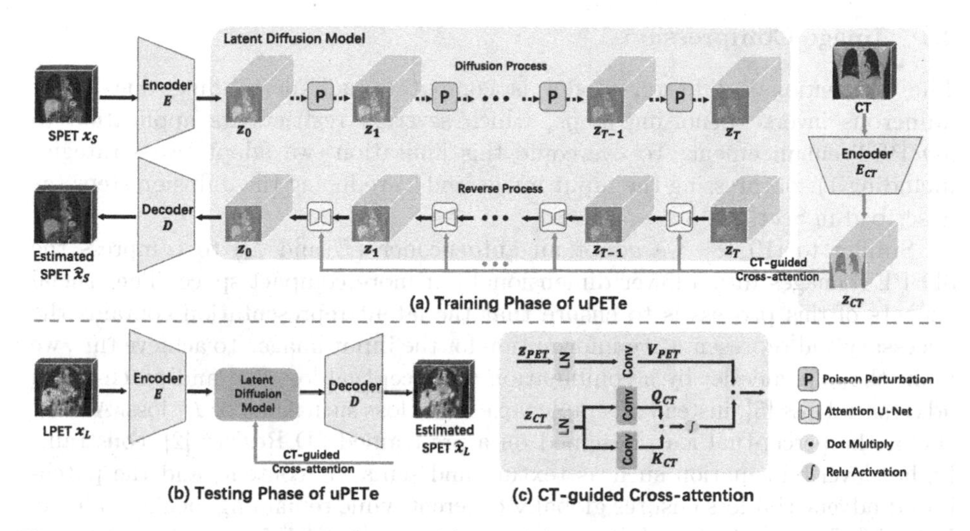

Fig. 1. Overview of proposed uPETe. (a) and (b) provide the framework of uPETe as well as depict its implementation during both the training and testing phases, and (c) illustrates the details of CT-guided cross-attention.

include 1) compressing the 3D PET images into a lower dimensional space for reducing the computational cost of diffusion model, 2) adopting the Poisson noise, which is the dominant noise in PET imaging [20], to replace the Gaussian noise in the diffusion process for avoiding the introduction of details that are not existing in PET images, and 3) designing CT-guided cross-attention to incorporate additional CT images into the inverse process for helping the recovery of structural details in PET.

Our work had three main features/contributions: i) proposing a clinically-applicable unsupervised PET enhancement framework, ii) designing three targeted strategies for improving the diffusion model, including PET image compression, Poisson diffusion, and CT-guided cross-attention, and iii) achieving better performance than state-of-the-art methods on the collected PET datasets.

2 Method

The framework of uPETe is illustrated in Fig. 1. When given an input PET image x (i.e., SPET for training and LPET for testing), x is first compressed into the latent representation z_0 by the encoder E. Subsequently, z_0 is fed into a latent diffusion model followed by the decoder D to output the expected SPET image $\hat{x}$. In addition, a specialized encoder E_{CT} is used to compress the CT image corresponding to the input PET image into the latent representation z_{CT}, which is fed into each denoising network for CT-guided cross-attention. In the following, we introduce the details of image compression, latent diffusion model, and implementation.

2.1 Image Compression

The conventional diffusion model is computationally-demanding due to its numerous inverse denoising steps, which severely restricts its application to 3D PET enhancement. To overcome this limitation, we adopt two strategies including 1) compressing the input image and 2) reducing the diffusion steps (as described in Sect. 2.3).

Similar to [10,18], we adopt an autoencoder (E and D) to compress the 3D PET images into a lower dimensional but more compact space. The crucial aspects of this process is to ensure that the latent representation contains the necessary and representative information for the input image. To achieve this, we train the autoencoder by a combination of perceptual loss [24] and patch-based adversarial loss [5], instead of simple voxel-level loss such as L_2 or L_1 loss. Among them, the perceptual loss, designed on a pre-trained 3D ResNet [2], constrains higher-level information such as texture and semantic content, and the patch-based adversarial loss ensures globally coherent while remaining locally realistic. Let $x \in \mathbb{R}^{H,W,Z}$ denote the input image and $z_0 \in \mathbb{R}^{h,w,z,c}$ denote the latent representation. The compression process can be formulated as $\hat{x} = D(z_0) = D(E(x))$. In this way, we compress the input image by a factor of $f = H/h = W/w = Z/z$. The results of SPET estimation under different compression rates f are provided in the supplement.

2.2 Latent Diffusion Model

After compressing the input PET image, its latent representation is fed into the latent diffusion model, which is the key to achieving the SPET-only unsupervised PET enhancement. As described above, the LPET can be viewed as noisy SPET (even in the compressed space), so the diffusion process from SPET to pure noise actually covers the situations of LPET. That is, the diffusion model trained with SPET is capable of estimating SPET from the noisy sample (diffused from LPET). But the diffusion model is developed from photographic images, which have significant difference with the detail-sensitive PET images. To improve its applicability for PET images, we design several targeted strategies for the diffusion process and inverse process, namely *Poisson diffusion* and *CT-guided cross-attention*, respectively.

Poisson Diffusion. In conventional diffusion models, the forward process typically employs Gaussian noise to gradually perturb input samples. However, in PET images, the dominant source of noise is Poisson noise, rather than Gaussian noise. Considering this, in our uPETe we choose to adopt Poisson diffusion to perturb the input samples, which facilitates the diffusion model for achieving better performance on the PET enhancement task.

Let z_t be the perturbation sample in Poisson diffusion, where $t = 0, 1, ..., T$. Then the Poisson diffusion can be formulate as follows:

$$z_t = perturb(z_{t-1}, \lambda_t), \quad \lambda_1 < \lambda_2 < ... < \lambda_T. \tag{1}$$

At each diffusion step, we apply the *perturb* function to the previous perturbed sample z_{t-1} by imposing a Poisson noise with an expectation of λ_t, which is linearly interpolated from $[0, 1]$ and incremented with t. In our implementation, we apply the same Poisson noise imposition operation as in [20], i.e., applying Poisson deviates on the projected sinograms, to generate a sequence of perturbed samples with increasing Poisson noise intensity as the step number t increases.

CT-Guided Cross-Attention. The attenuation correction of PET typically relies on the corresponding anatomical image (CT or MR), resulting in a PET scan usually accompanied by a CT or MR scan. To fully utilize the extra-modality images (i.e., CT in our work) as well as improve the applicability of diffusion models, we design a CT-guided cross-attention to incorporate the CT images into the reverse process for assisting the recovery of structural details.

As shown in Fig. 1, to achieve a particular SPET estimation, the corresponding CT image is first compressed into the latent representation z_{CT} by encoder E_{CT}. Then z_{CT} is fed into a denoising attention U-Net [16] at each step for calculation of cross-attention, where the query Q and key K are calculated from z_{CT} while the value V is still calculated from the output of the previous layer because our final goal is SPET estimation. Denoting the output of previous layer as z_{PET}, the CT-guided cross-attention can be formulated as follows:

$$\begin{gathered} Output = softmax(\frac{Q_{CT}K_{CT}^T}{\sqrt{d}} + B) \cdot V_{PET}, \\ Q_{CT} = Conv_Q(z_{CT}), \quad K_{CT} = Conv_K(z_{CT}), \quad V_{PET} = Conv_V(z_{PET}), \end{gathered} \tag{2}$$

where d is the number of channels, B is the position bias, and $Conv(\cdot)$ denotes the $1 \times 1 \times 1$ convolution with stride of 1.

2.3 Implementation Details

Typically, the trained diffusion model generates target images from random noise, requiring a large number of steps T to make the final perturbed sample (z_T) close to pure noise. However, in our task, the target SPET image is generated from a given LPET image during testing, and making z_T as close to pure noise as possible is not necessary since the remaining PET-related information can also benefit the image recovery. Therefore, we can considerably reduce the number of diffusion steps T to accelerate the model training, and T is set to 400 in our implementation. We evaluate the quantitative results using two metrics, including Peak Signal to Noise Ratio (PSNR) and Structural Similarity Index Measure (SSIM).

Table 1. Quantitative results of ablation analysis, in terms of PSNR and SSIM.

Method	PSNR [dB]↑	SSIM ↑
LDM	23.732 ± 1.264	0.986 ± 0.010
LDM-P	24.125 ± 1.072	0.987 ± 0.009
LDM-CT	25.348 ± 0.822	0.990 ± 0.006
LDM-P-CT	**25.817 ± 0.675**	**0.992 ± 0.004**

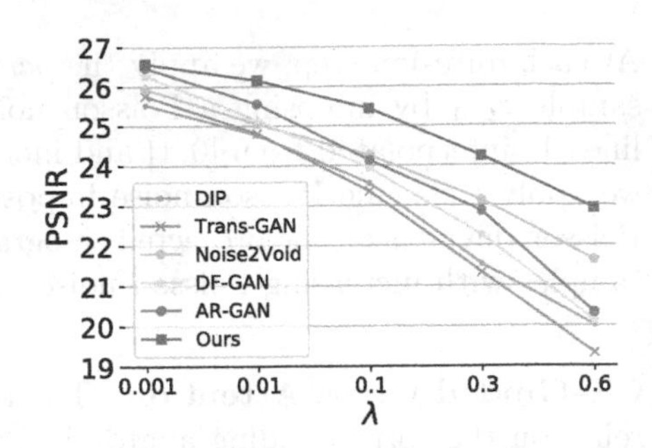

Fig. 2. Generalizability to dose changes.

3 Experiments

3.1 Dataset

Our dataset consists of 100 SPET images for training and 30 paired LPET and SPET images for testing. Among them, 50 chest-abdomen SPET images are collected from (total-body) uEXPLORER PET/CT scanner [25], and 20 paired chest-abdomen images are collected by list mode of the scanner with 256 MBq of [^{18}F]-FDG injection. Specifically, the SPET images are reconstructed by using the 1200 s data between 60–80 min after tracer injection, while the corresponding LPET images are simultaneously reconstructed by 120 s data uniformly sampled from 1200 s data.

As a basic data preprocessing, all images are resampled to voxel spacing of $2 \times 2 \times 2$ mm^3 and resolution of $256 \times 256 \times 160$, while their intensity range is normalized to $[0, 1]$ by min-max normalization. For increasing the training samples and reducing the dependence on GPU memory, we extract the overlapped patches of size $96 \times 96 \times 96$ from every whole PET image.

3.2 Ablation Analysis

To verify the effectiveness of our proposed strategies, i.e. *Poisson diffusion process* and *CT-guided cross-attention*, we design another four variant latent diffusion models (LDMs) with the same compression model, including: 1) LDM: standard LDM; 2) LDM-P: LDM with Poisson diffusion process; 3) LDM-CT: LDM with CT-guided cross-attention; 4) LDM-P-CT: LDM with Poisson diffusion process and CT-guided cross-attention. All methods use the same experimental settings, and their quantitative results are given in Table 1.

From Table 1, we can have the following observations. (1) LDM-P achieves better performance than LDM. This proves that the Poisson diffusion is more appropriate than the Gaussian diffusion for PET enhancement. (2) LDM-CT with the corresponding CT image for assisting denoising achieves better results than LDM. This can be reasonable as the CT image can provide anatomical information, thus benefiting the recovery of structural details (e.g., organ boundaries) in SPET images. (3) LDM-P-CT achieves better results than all other variants

Table 2. Quantitative comparison of our uPETe with several state-of-the-art PET enhancement methods, in terms of PSNR and SSIM, where * denotes unsupervised method and † denotes fully-supervised method.

Method	PSNR [dB]↑	SSIM ↑
DIP* [3]	22.538 ± 2.136	0.981 ± 0.015
Noisier2Noise * [23]	22.932 ± 1.983	0.983 ± 0.014
LA-GAN † [21]	23.351 ± 1.725	0.984 ± 0.012
MR-GDD * [17]	23.628 ± 1.655	0.985 ± 0.011
Trans-GAN† [13]	23.852 ± 1.522	0.985 ± 0.009
Noise2Void * [20]	24.263 ± 1.351	0.987 ± 0.009
DF-GAN† [9]	24.821 ± 0.975	0.989 ± 0.007
AR-GAN† [14]	25.217 ± 0.853	0.990 ± 0.006
uPETe*	**25.817 ± 0.675**	**0.992 ± 0.004**

on both PSNR and SSIM, which shows both of our proposed strategies contribute to the final performance. These three comparisons conjointly verify the effective design of our proposed uPETe, where the *Poisson diffusion process* and *CT-guided cross-attention* both benefit the PET enhancement.

3.3 Comparison with State-of-the-Art Methods

We further compare our uPETe with several state-of-the-art PET enhancement methods, which can be divided into two classes: 1) fully-supervised methods, including LA-GAN [21], Transformer-GAN (Trans-GAN) [13], Dual-frequency GAN (DF-GAN) [9], and AR-GAN [14]; 2) unsupervised methods, including deep image prior (DIP) [3], Noisier2Noise [23], magnetic resonance guided deep decoder (MR-GDD) [17], and Noise2Void [20]. The quantitative and qualitative results are provided in Table 2 and Fig. 3, respectively.

Quantitative Comparison: Table 2 shows that our uPETe outperforms all competing methods. Compared to the fully-supervised method AR-GAN which achieves sub-optimal performance, our uPETe does not require paired LPET and SPET, yet still achieves improvement. Additionally, uPETe also achieves noticeable performance improvement to Noise2Void (which is a supervised method). Specifically, the average improvement in PSNR and SSIM on SPET estimation are 1.554 dB and 0.005, respectively. This suggests that our uPETe can generate promising results without relying on paired data, demonstrating its potential for clinical applications.

Qualitative Comparison: In Fig. 3, we provide a visual comparison of SPET estimation for two typical cases. First, compared to unsupervised methods such as DIP and Noise2Void, the SPET images estimated by our uPETe have less noise but clearer boundaries. Second, our uPETe performs better on the structural details compared to the fully-supervised methods, i.e., missing unclear tissue (Trans-GAN) or introducing non-existing artifacts in PET image (DF-GAN).

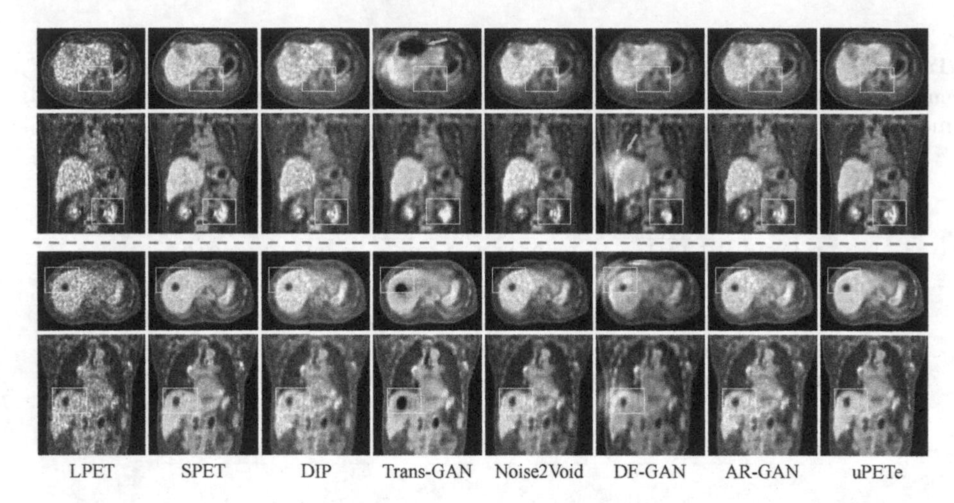

Fig. 3. Visual comparison of estimated SPET images on two typical cases. In each case, the first and second rows show the axial and coronal views, respectively, and from left to right are the input (LPET), ground truth (SPET), results by five other methods (3rd–7th columns), and the result by our uPETe (last column). Red boxes and arrows show areas for detailed comparison. (Color figure online)

Overall, these pieces of evidence demonstrate the superiority of our uPETe over state-of-the-art methods.

3.4 Generalization Evaluation

We further evaluate the generalizability of our uPETe to tracer dose changes by simulating Poisson noise on SPET to produce different doses for LPET, which is a common way to generate noisy PET data [20]. Notably, we do not need to retrain the models since they have been trained in Sect. 3.3. The quantitative results of our uPETe and five state-of-the-art methods are provided in Fig. 2.

As shown in Fig. 2, our uPETe outperforms the other five methods at all doses and exhibits a lower PSNR descent slope as dose decreases (i.e., λ increases), demonstrating its superior generalizability to dose changes. This is because uPETe is based on diffusion model, which simplifies the complex distribution prediction task into a series of simple denoising tasks and thus has strong generalizability. Moreover, we also find that the unsupervised methods (i.e., uPETe, Noise2Void, and DIP) have stronger generalizability than fully-supervised methods (i.e., AR-GAN, DF-GAN, and Trans-GAN) as they have a smoother descent slope. The main reason is that the unsupervised learning has the ability to extract patterns and features from the data based on the inherent structure and distribution of the data itself [15].

4 Conclusion and Limitations

In this paper, we have developed a clinically-applicable unsupervised PET enhancement framework based on the latent diffusion model, which uses only the clinically-available SPET data for training. Meanwhile, we adopt three strategies to improve the applicability of diffusion models developed from photographic images to PET enhancement, including 1) compressing the size of the input image, 2) using Poisson diffusion, instead of Gaussian diffusion, and 3) designing CT-guided cross-attention to enable additional anatomical images (e.g., CT) to aid the recovery of structural details in PET. Validated by extensive experiments, our uPETe achieved better performance than both state-of-the-art unsupervised and fully-supervised PET enhancement methods, and showed stronger generalizability to the tracer dose changes.

Despite the advance of uPETe, our current work still suffers from a few limitations such as (1) lacking theoretical support for our Poisson diffusion, which is just an engineering attempt, and 2) only validating the generalizability of uPETe on a simulated dataset. In our future work, we will complete the design of Poisson diffusion from theoretical perspective, and collect more real PET datasets (e.g., head datasets) to comprehensively validate the generalizability of our uPETe.

Acknowledgment. This work was supported in part by National Natural Science Foundation of China (No. 62131015), Science and Technology Commission of Shanghai Municipality (STCSM) (No. 21010502600), The Key R&D Program of Guangdong Province, China (No. 2021B0101420006), and the China Postdoctoral Science Foundation (Nos. BX2021333, 2021M703340).

References

1. Buades, A., Coll, B., Morel, J.: A non-local algorithm for image denoising. In: 2005 IEEE Computer Society Conference on Computer Vision and Pattern Recognition, vol. 2, pp. 60–65 (2005)
2. Chen, S., Ma, K., Zheng, Y.: Med3D: transfer learning for 3D medical image analysis. arXiv preprint arXiv:1904.00625 (2019)
3. Cui, J., et al.: PET image denoising using unsupervised deep learning. Eur. J. Nucl. Med. Mol. Imaging **46**(13), 2780–2789 (2019)
4. Dabov, K., Foi, A., Katkovnik, V., Egiazarian, K.: Image denoising with block-matching and 3D filtering. Image Process. Algorithms Syst. Neural Netw. Mach. Learn. **6064**, 354–365 (2006)
5. Dosovitskiy, A., Brox, T.: Generating images with perceptual similarity metrics based on deep networks. In: Advances in Neural Information Processing Systems, vol. 29 (2016)
6. Ho, J., Jain, A., Abbeel, P.: Denoising diffusion probabilistic models. In: Advances in Neural Information Processing Systems, vol. 33, pp. 6840–6851 (2020)
7. Hofheinz, F., et al.: Suitability of bilateral filtering for edge-preserving noise reduction in PET. EJNMMI Res. **1**(1), 1–9 (2011)

8. Jiang, C., Pan, Y., Cui, Z., Nie, D., Shen, D.: Semi-supervised standard-dose PET image generation via region-adaptive normalization and structural consistency constraint. IEEE Trans. Med. Imaging (2023)
9. Jiang, C., Pan, Y., Cui, Z., Shen, D.: Reconstruction of standard-dose PET from low-dose PET via dual-frequency supervision and global aggregation module. In: 2022 IEEE 19th International Symposium on Biomedical Imaging (ISBI), pp. 1–5 (2022)
10. Khader, F., et al.: Medical diffusion-denoising diffusion probabilistic models for 3D medical image generation. arXiv preprint arXiv:2211.03364 (2022)
11. Lu, W., et al.: An investigation of quantitative accuracy for deep learning based denoising in oncological PET. Phys. Med. Biol. **64**(16), 165019 (2019)
12. Lu, Z., Li, Z., Wang, J., Shen, D.: Two-stage self-supervised cycle-consistency network for reconstruction of thin-slice MR images. arXiv preprint arXiv:2106.15395 (2021)
13. Luo, Y., et al.: 3D transformer-GAN for high-quality PET reconstruction. In: de Bruijne, M., et al. (eds.) MICCAI 2021. LNCS, vol. 12906, pp. 276–285. Springer, Cham (2021). https://doi.org/10.1007/978-3-030-87231-1_27
14. Luo, Y., et al.: Adaptive rectification based adversarial network with spectrum constraint for high-quality PET image synthesis. Med. Image Anal. **77**, 102335 (2022)
15. Noroozi, M., Favaro, P.: Unsupervised learning of visual representations by solving jigsaw puzzles. In: Leibe, B., Matas, J., Sebe, N., Welling, M. (eds.) ECCV 2016. LNCS, vol. 9910, pp. 69–84. Springer, Cham (2016). https://doi.org/10.1007/978-3-319-46466-4_5
16. Oktay, O., et al.: Attention U-Net: learning where to look for the pancreas. arXiv preprint arXiv:1804.03999 (2018)
17. Onishi, Y., et al.: Anatomical-guided attention enhances unsupervised PET image denoising performance. Med. Image Anal. **74**, 102226 (2021)
18. Rombach, R., Blattmann, A., Lorenz, D., Esser, P., Ommer, B.: High-resolution image synthesis with latent diffusion models. In: Proceedings of the IEEE/CVF Conference on Computer Vision and Pattern Recognition, pp. 10684–10695 (2022)
19. Slovis, T.L.: The ALARA concept in pediatric CT: myth or reality? Radiology **223**(1), 5–6 (2002)
20. Song, T., Yang, F., Dutta, J.: Noise2Void: unsupervised denoising of PET images. Phys. Med. Biol. **66**(21), 214002 (2021)
21. Wang, Y., et al.: 3D auto-context-based locality adaptive multi-modality GANs for PET synthesis. IEEE Trans. Med. Imaging **38**(6), 1328–1339 (2019)
22. Yan, J., Lim, J., Townsend, D.: MRI-guided brain PET image filtering and partial volume correction. Phys. Med. Biol. **60**(3), 961 (2015)
23. Yie, S., Kang, S., Hwang, D., Lee, J.: Self-supervised PET denoising. Nucl. Med. Mol. Imaging **54**(6), 299–304 (2020)
24. Zhang, R., Isola, P., Efros, A.A., Shechtman, E., Wang, O.: The unreasonable effectiveness of deep features as a perceptual metric. In: Proceedings of the IEEE Conference on Computer Vision and Pattern Recognition, pp. 586–595 (2018)
25. Zhang, X., et al.: Total-body dynamic reconstruction and parametric imaging on the uEXPLORER. J. Nucl. Med. **61**(2), 285–291 (2020)

MedIM: Boost Medical Image Representation via Radiology Report-Guided Masking

Yutong Xie[1], Lin Gu[2,3], Tatsuya Harada[2,3], Jianpeng Zhang[4], Yong Xia[4], and Qi Wu[1(✉)]

[1] Australian Institute for Machine Learning, The University of Adelaide, Adelaide, Australia
qi.wu01@adelaide.edu.au
[2] RIKEN AIP, Tokyo, Japan
[3] RCAST, The University of Tokyo, Tokyo, Japan
[4] School of Computer Science and Engineering, Northwestern Polytechnical University, Xi'an, China

Abstract. Masked image modelling (MIM)-based pre-training shows promise in improving image representations with limited annotated data by randomly masking image patches and reconstructing them. However, random masking may not be suitable for medical images due to their unique pathology characteristics. This paper proposes Masked medical Image Modelling (MedIM), a novel approach, to our knowledge, the first research that masks and reconstructs discriminative areas guided by radiological reports, encouraging the network to explore the stronger semantic representations from medical images. We introduce two mutual comprehensive masking strategies, knowledge word-driven masking (KWM) and sentence-driven masking (SDM). KWM uses Medical Subject Headings (MeSH) words unique to radiology reports to identify discriminative cues mapped to MeSH words and guide the mask generation. SDM considers that reports usually have multiple sentences, each of which describes different findings, and therefore integrates sentence-level information to identify discriminative regions for mask generation. MedIM integrates both strategies by simultaneously restoring the images masked by KWM and SDM for a more robust and representative medical visual representation. Our extensive experiments on various downstream tasks covering multi-label/class image classification, medical image segmentation, and medical image-text analysis, demonstrate that MedIM with report-guided masking achieves competitive performance. Our method substantially outperforms ImageNet pre-training, MIM-based pre-training, and medical image-report pre-training counterparts. Codes are available at https://github.com/YtongXie/MedIM.

Supplementary Information The online version contains supplementary material available at https://doi.org/10.1007/978-3-031-43907-0_2.

H. Greenspan et al. (Eds.): MICCAI 2023, LNCS 14220, pp. 13–23, 2023.
https://doi.org/10.1007/978-3-031-43907-0_2

1 Introduction

Accurate medical representation is crucial for clinical decision-making. Deep learning has shown promising results in medical image analysis, but the accuracy of these models heavily relies on the quality and quantity of data and annotations [21]. Masked image modelling (MIM)-based pre-training approach [3,8,23] such as masked autoencoders (MAE) [8] has shown prospects in improving the image representation under limited annotated data. MIM masks a set of image patches before inputting them into a network and then reconstructs these masked patches by aggregating information from the surrounding context. This ability to aggregate contextual information is essential for vision tasks and understanding medical image analysis [24]. Recently, MIM has witnessed much success in medical domain [4–6,11,20,24] such as chest X-ray and CT image analysis.

While the random masking strategy is commonly used in current MIM-based works, randomly selecting a percentage of patches to mask. We argue that such a strategy may not be the most suitable approach for medical images due to the domain particularity. Medical images commonly present relatively fixed anatomical structures, while subtle variations between individuals, such as sporadic lesions that alter the texture and morphology of surrounding tissues or organs, may exist. These pathology characteristics may be minute and challenging to perceive visually but are indispensable for early screening and clinical diagnosis. Representation learning should capture these desired target representations to improve downstream diagnosis models' reliability, interpretability, and generalizability. Random masking is less likely to deliberately focus on these important parts. We put forward a straightforward principle, *i.e., masking and reconstructing meaningful characteristics*, encouraging the network to explore stronger representations from medical images.

We advocate utilising radiological reports to locate relevant characteristics and guide mask generation. These reports are routinely produced in clinical practice by expert medical professionals such as radiologists, and can provide a valuable source of semantic knowledge at little to no additional cost [9,17]. When medical professionals read a medical image, they will focus on areas of the image that are relevant to the patient's or clinical conditions. These areas are then recorded in a report, along with relevant information such as whether they are normal or abnormal, the location and density of abnormal areas, and any other materials about the patient's condition. By incorporating reports into the medical image representation learning, the models can simulate the professionals' gaze and learn to focus on the pathology characteristics of images.

In this paper, we propose a new approach called MedIM (**M**asked m**ed**ical **I**mage **M**odelling). MedIM aligns semantic correspondences between medical images and radiology reports and reconstructs regions masked by the guidance of learned correspondences. Especially we introduce two masking strategies: knowledge word-driven masking (KWM) and sentence-driven masking (SDM). KWM uses Medical Subject Headings (MeSH) words [14] as the domain knowledge. MeSH words provide a standardized language for medical concepts and conditions. In radiology reports, MeSH words describe imaging modalities, anatomic locations, and pathologic findings, such as "Heart", "Pulmonary", "Vascular", and "Pneumothorax" in Fig. 1, and are important semantic components. This inspired KWM to identify regions mapped to MeSH words and generate an attention map, where the highly activated tokens indicate more discriminative cues. We utilize this attention map to selectively mask then restore the high-activated regions, stimulating the network to focus more on regions related to MeSH words during the modelling process. SDM considers multiple sentences in reports, each potentially providing independent information about different aspects of the image. It generates an attention map by identifying regions mapped to one selected sentence, enabling the network to focus on specific aspects of the image mentioned in that sentence during modelling. KWM and SDM identify different sources of discriminative cues and are therefore complementary. MedIM leverages the superiority of both strategies by simultaneously restoring images masked by KWM and SDM in each iteration. This integration creates a more challenging and comprehensive modelling task, which encourages the network to learn more robust and representative medical visual representations. Our MedIM approach is pre-trained on a large chest X-ray dataset of image-report pairs. The learned image representations are transferred to several medical image analysis downstream tasks: multi-label/class image classification and pneumothorax segmentation. Besides, our MedIM pre-trained model can be freely applied to image-text analysis downstream tasks such as image-to-text/text-to-image retrieval.

Our contributions mainly include three-fold: (1) we present a novel masking approach MedIM, which is the first work to explore the potential of radiology reports in mask generation for medical images, offering a new perspective to enhance the accuracy and interpretability of medical image representation; (2) we propose two mutual comprehensive masking strategies, KWM and SDM, that effectively identify word-level and sentence-level of discriminative cues to guide the mask generation; (3) we conduct extensive experiments on medical image and image-text downstream tasks, and the performance beats strong competitors like ImageNet pre-training, MIM-based pre-training and advanced medical image-report pre-training counterparts.

2 Approach

As shown in Fig. 1, our MedIM framework has dual encoders that map images and reports to a latent representation, a report-guided mask generation module, and a decoder that reconstructs the images from the masked representation.

2.1 Image and Text Encoders

Image Encoder. We use the vision Transformer (ViT) [7] as the image encoder$\mathcal{F}(\cdot)$. For an input medical image $\boldsymbol{x}$, it is first reshaped into a sequence of flattened patches that are then embedded and fed into stacked Transformer layers to obtain the encoded representations of visual tokens $\boldsymbol{E}_{img} = \mathcal{F}(\boldsymbol{x}) \in \mathbb{R}^{N_{img} \times C}$, where C is the encoding dimension and N_{img} denotes the number of patches.

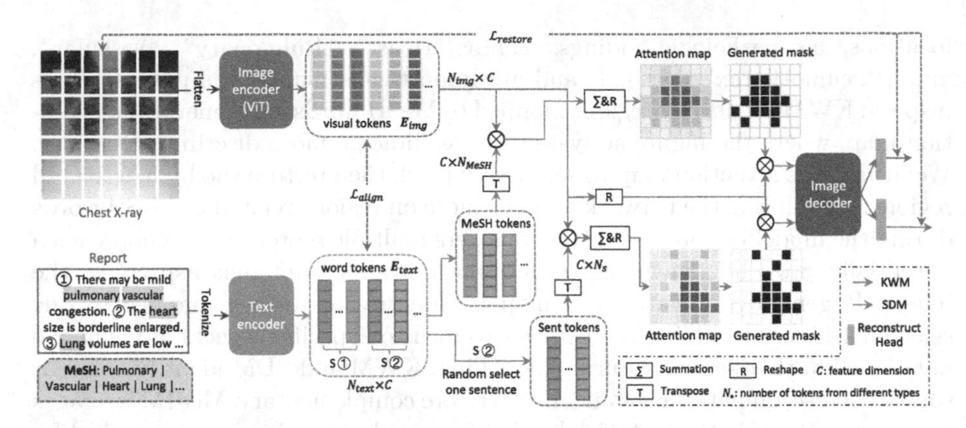

Fig. 1. Illustration of our MedIM framework. It includes dual encoders to obtain latent representations. Two report-guided masking strategies, KWM and SDM, are then introduced to generate the masked representations. The decoder is built to reconstruct the original images from the masked representation. Noted that the back regions in the generated mask will be masked.

Text Encoder. We use the BioClinicalBERT [2] model, pre-trained on the MIMIC III dataset [13], as our text encoder $\mathcal{T}(\cdot)$. We employ WordPiece [19] for tokenizing free-text medical reports. This technique is particularly useful for handling the large and diverse vocabularies that are common in the medical language. For an input medical report $\boldsymbol{r}$ with N_{text} words, the tokenizer segments each word to sub-words and generates word piece embeddings as the input to the text encoder. The text encoder extracts features for word pieces, which are aggregated to generate the word representations $\boldsymbol{E}_{text} = \mathcal{T}(\boldsymbol{r}) \in \mathbb{R}^{N_{text}\times C}$.

2.2 Report-Guided Mask Generation

We introduce two radiology report-guided masking strategies, *i.e.*, KWM and SDM, identifying different cues to guide the mask generation.

Knowledge Word-Driven Masking (KWM). MeSH words shown in Fig. 1 are important for accurately describing medical images, as they provide a standardized vocabulary to describe the anatomical structures and pathologies observed in the images. Hence the KWM is proposed to focus on the MeSH word tokens during mask generation. Given a report $\boldsymbol{r}$ and its text representations $\boldsymbol{E}_{text}$, we first match MeSH words in the report based on the MeSH Table [14] and extract the representations of MeSH word tokens, formally as

$$\boldsymbol{E}_{\mathrm{MeSH}} = \left\{\boldsymbol{E}_{text}^{j}, \boldsymbol{r}^{j} \in \mathrm{MeSH}, \mathrm{j} \in \{1, ..., \mathrm{N_{text}}\}\right\} \in \mathbb{R}^{N_{\mathrm{MeSH}}\times C}, \tag{1}$$

where N_{MeSH} represents the number of MeSH words in the report $\boldsymbol{r}$. Then, we compute an attention map $\mathbb{C}_{\mathrm{MeSH}}$ to identify image regions mapped to MeSH

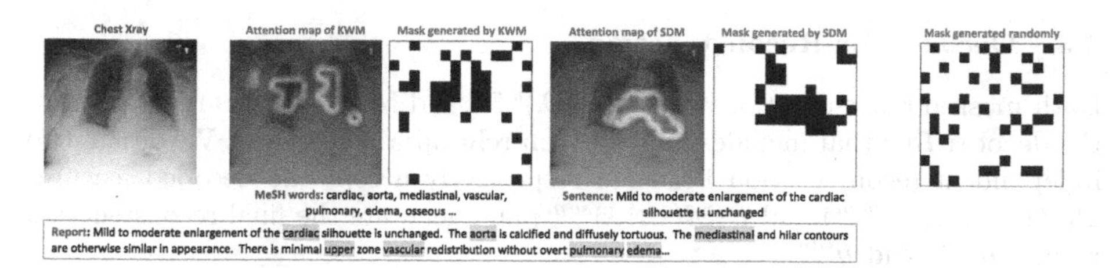

Fig. 2. A image-report pair and the corresponding attention map and mask generated by KWM and SDM. The black regions in the generated mask will be masked.

words as follows

$$\mathbb{C}_{\text{MeSH}} = R(\sum \text{softmax}(\boldsymbol{E}_{img} \cdot \boldsymbol{E}_{\text{MeSH}}^T)) \in \mathbb{R}^{H \times W}, \tag{2}$$

where $H = W = \sqrt{N_{img}}$, T and R represent the transpose and reshape functions, and the softmax function normalizes the elements along the image dimension to find the focused region matched to each MeSH word. The summation operation $\sum$ performs on the text dimension to aggregate the attentions related to all MeSH words.

Subsequently, the high-activated masking is presented to remove the discovered attention regions. Here, we define a corresponding binary mask $\boldsymbol{m} \in \{0,1\}^{H \times W}$ formulated as $\boldsymbol{m}^{(i,j)} = \mathbb{I}(\mathbb{C}_{\text{MeSH}}^{(i,j)} \leqslant \mathbb{C}_{\text{MeSH}}^{[\gamma * N_{img}]})$. Here $\mathbb{C}_{\text{MeSH}}^{[\gamma * N_{img}]}$ refers to the $(\gamma * N_{img})$-th largest activation in $\mathbb{C}_{\text{MeSH}}$, andγ is the masking ratio that determines how many activations would be suppressed. With this binary mask, we can compute the masked representations produced by KWM as

$$\mathbb{M}(\mathbb{C}_{\text{MeSH}}; \lambda)^{kwm} = \{\boldsymbol{z}^{(i,j)} | \boldsymbol{m}^{(i,j)} \cdot R(\boldsymbol{F}_{img})^{(i,j)} + (1 - \boldsymbol{m}^{(i,j)}) \cdot \texttt{[MASK]}\}_{i=1}^{H} {}_{j=1}^{W}, \tag{3}$$

where [MASK] is a masked placeholder.

Sentence-Driven Masking (SDM). Medical reports often contain multiple sentences that describe different findings related to the image, which inspires SDM to introduce sentence-level information during mask generation. For the report $\boldsymbol{r}$, we randomly select a sentence $\boldsymbol{s}$ and extract its representations as

$$\boldsymbol{E}_s = \left\{ \boldsymbol{E}_{text}^j, \boldsymbol{r}^j \in s, j \in \{1, ..., N_{text}\} \right\} \in \mathbb{R}^{N_s \times C} \tag{4}$$

where N_s represents the length of $\boldsymbol{s}$. Then, an attention map $\mathbb{C}_s$ can be computed to identify regions mapped to this sentence as

$$\mathbb{C}_s = R(\sum \text{softmax}(\boldsymbol{E}_{img} \cdot \boldsymbol{E}_s^T)) \in \mathbb{R}^{H \times W}, \tag{5}$$

After that, the high-activated masking is performed based on $\mathbb{C}_s$ to compute the masked representations $\mathbb{M}(\mathbb{C}_s; \lambda)^{sdm}$.

We also select an image-report pair and visualize the corresponding attention map and generated mask procured by KWM and SDM in Fig. 2 to show the superiority of our masking strategies.

2.3 Decoder for Reconstruction

Both masked representations $\mathbb{M}(\mathbb{C}_{\text{MeSH}};\lambda)^{kwm}$ and $\mathbb{M}(\mathbb{C}_s;\lambda)^{sdm}$ are mapped to the decoder $\mathcal{D}(\cdot)$ that includes four conv-bn-relu-upsample blocks. We design two independent reconstruction heads to respectively accept the decoded features $\mathcal{D}(\mathbb{M}(\mathbb{C}_{\text{MeSH}};\lambda)^{kwm})$ and $\mathcal{D}(\mathbb{M}(\mathbb{C}_s;\lambda)^{sdm})$ and generate the final reconstruction results $\boldsymbol{y}^{kwm}$ and $\boldsymbol{y}^{sdm}$.

2.4 Objective Function

MedIM creates a more challenging reconstruction objective by removing then restoring the most discriminative regions guided by radiological reports. We optimize this reconstruction learning process with the mean square error (MSE) loss function, expressed as

$$\mathcal{L}_{restore} = \left\|\boldsymbol{y}^{kwm}, \boldsymbol{x}\right\|^2 + \left\|\boldsymbol{y}^{sdm}, \boldsymbol{x}\right\|^2 \tag{6}$$

MedIM also combines the cross-modal alignment constraint, which aligns medical images' visual and semantic aspects with their corresponding radiological reports, benefiting in better identifying the reported-guided discriminative regions during mask generation. We follow the work [17] and compute the objective alignment function $\mathcal{L}_{align}$ by exploiting the fine-grained correspondences between images and reports. The final objective of our MedIM is the combination of reconstruction and alignment objectives as $\mathcal{L}_{\text{MedIM}} = \alpha\mathcal{L}_{\text{restore}} + \mathcal{L}_{align}$, where α is a weight factor to balance both objectives.

2.5 Downstream Transfer Learning

After pre-training, we can transfer the weight parameters of the MedIM to various downstream tasks. For the classification task, we use the commonly used Linear probing, *i.e.*, freezing the pre-trained image encoder and solely training a randomly initialized linear classification head. For the segmentation task, the encoder and decoder are first initialized with the MedIM pre-trained weights, and a downstream-specific head is added to the network. The network is then fine-tuned end-to-end. For the retrieval task, we take an image or report as an input query and retrieve target reports or images by computing the similarity between the query and all candidates using the learned image and text encoders.

3 Experiments and Results

3.1 Experimental Details

Pre-training Setup. We use the MIMIC-CXR-JPG dataset [12] to pre-train our MedIM framework. Following [17], we only include frontal-view chest images from the dataset and extract the impression and finding sections from radiological reports. As a result, over 210,000 radiograph-report pairs are available. We manually split 80% of pairs for pre-training and 20% of pairs used for downstream to validate in-domain transfer learning. We set the input size to 224×224 adopt the AdamW optimizer [16] with a cosine decaying learning rate [15], a momentum of 0.9, and a weight decay of 0.05. We set the initial learning rate to 0.00002, batch size to 144, and maximum epochs to 50. Through the ablation study, we empirically set the mask ratio to 50% and loss weight α to 10.

Downstream Setup. We validate the transferability of learned MedIM representations on four X-ray-based downstream tasks: (1) multi-label classification on CheXpert [10] dataset using its official split, which contains five individual binary labels: atelectasis, cardiomegaly, consolidation, edema, and pleural effusion; (2) multi-class classification on COVIDx [18] dataset with over 30k chest X-ray images, which aims to classify each radiograph into COVID-19, non-COVID pneumonia or normal, and is split into training, validation, and test set with 80%/10%/10% ratio; (3) pneumothorax segmentation on SIIM-ACR Pneumothorax Segmentation dataset [1] with over 12k chest radiographs, which is split into training, validation, and test set with 70%/15%/15% ratio; and (4) image-text/report-text retrieval on the MIMIC-CXR validation dataset. We use the Dice coefficient score (Dice) to measure the segmentation performance, use the mean area under the receiver operator curve (mAUC) to measure the multi-label classification performance, and use the accuracy to measure the multi-class classification performance. We use the recall of the corresponding image/report that appears in the top-k ranked images/reports (denoted by R@k) to measure the retrieval performance [9]. Each downstream experiment is conducted three times and the average performance is reported. More details are in the Appendix.

3.2 Comparisons with Different Pre-training Methods

We compare the downstream performance of our MedIM pre-training with five pre-training methods in Table 1 and Table 2. Our MedIM achieves state-of-the-art results on all downstream datasets, outperforming ImageNet pre-training [7], MIM-based pre-training MAE [8] and three medical image-report pre-training approaches, GLoRIA [9], MRM [22] and MGCA [17], under different labelling ratios. The superior performance corroborates the effectiveness of our report-guided masking pre-training strategy over other pre-training strategies in learning discriminative information. Besides, our MedIM achieves 88.91% when using only 1% downstream labelled data on CheXpert, better than other competitors with 100% labelled data. These convincing results have demonstrated the enormous potential of MedIM for annotation-limited medical image tasks.

3.3 Discussions

Ablation Study. Ablation studies are performed over each component of MedIM, including knowledge word-driven masking (KWM) and Sentence-driven masking (SDM), as listed in Table 3. We sequentially add each component to

Table 1. Classification and segmentation results of different pre-training methods on three downstream test sets under different ratios of available labelled data. All methods were evaluated with the ViT-B/16 backbone. * denotes our implementation of on same pre-training dataset and backbone due to the lack of available pre-trained weights.

Methods	CheXpert			COVIDx			SIIM	
	1%	10%	100%	1%	10%	100%	10%	100%
Random Init	68.11	71.17	71.91	67.01	79.68	82.71	19.13	60.97
ImageNet [7]	73.52	80.38	81.84	71.56	84.28	89.74	55.06	76.02
MAE* [8]	82.36	85.22	86.69	73.31	87.67	91.79	57.68	77.16
GLoRIA* [9]	86.50	87.53	88.24	75.79	88.68	92.11	57.67	77.23
MRM [22]	88.50	88.50	88.70	76.11	88.92	92.21	61.21	79.45
MGCA* [17]	88.11	88.29	88.88	76.29	89.04	92.47	60.64	79.31
MedIM	**88.91**	**89.25**	**89.65**	**77.22**	**90.34**	**93.57**	**63.50**	**81.32**

the vanilla baseline, $\mathcal{L}_{align}$ only, thus the downstream performance is gradually improved in Table 3. First, by reconstructing the masked representations produced by KWM, the total performance of three tasks is increased by 3.28 points. This indicates that using MeSH words as knowledge to guide the mask generation can improve the model representations and generalization. Equipped with KWM and SDM, our MedIM can surpass the baseline model by a total of 5.12 points on three tasks, suggesting the superiority of adding the SDM strategy and integrating these two masking strategies.

Masking Strategies. To demonstrate the effectiveness of the High-activated masking strategy, we compare it with three counterparts, No masking, Random masking, and Low-activated masking. Here No masking means that the recon-

Table 2. Image-to-text (I2T) and text-to-image (T2I) retrieval results on the MIMIC-CXR test set.

Methods	T2I			I2T		
	R@1	R@5	R@10	R@1	R@5	R@10
MGCA [17]	5.74	22.91	31.90	6.22	23.61	32.51
MedIM	**7.67**	**23.96**	**33.55**	**8.70**	**24.63**	**34.27**

Table 3. Ablation study of different components in MedIM.

Different components			Tasks		
$\mathcal{L}_{align}$	KWM	SDM	COVIDx	CheXpert	SIIM
✓	×	×	89.04	88.29	60.64
✓	✓	×	89.85	88.86	62.54
✓	✓	✓	**90.34**	**89.25**	**63.50**

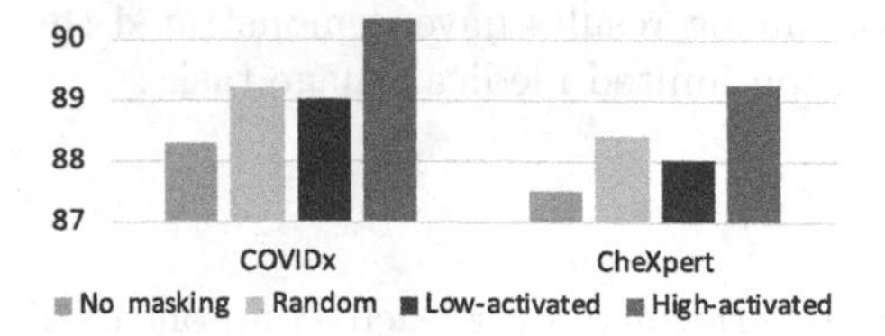

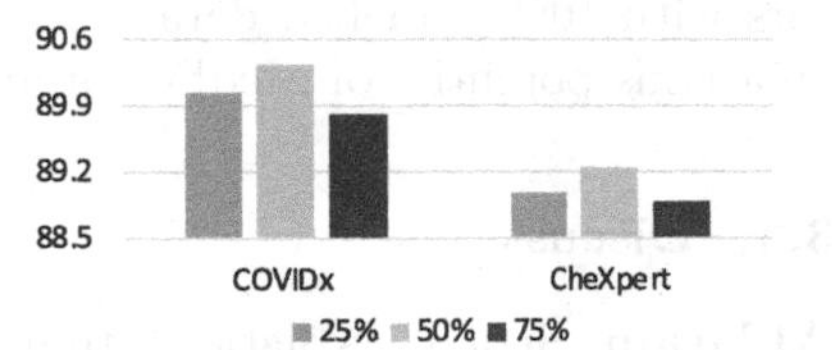

Fig. 3. Left: Results when using different masking strategies. **Right**: Results when using different masking ratios.

struction is performed based on the complete image encoder representations instead of the masked one. Low-activated masking refers to masking the tokens exhibiting a low response in both KWM and SDM strategies. The comparison on the left side of Fig. 3 reveals that all masking strategies are more effective in improving the accuracy than No masking. Benefiting from mining more discriminative information, our High-activated masking performs better than the Random and Low-activated masking. Besides, we also compare different masking ratios, varying from 25% to 75%, on the right side of Fig. 3.

4 Conclusion

We propose a new masking approach called MedIM that uses radiological reports to guide the mask generation of medical images during the pre-training process. We introduce two masking strategies KWM and SDM, which effectively identify different sources of discriminative cues to generate masked inputs. MedIM is pre-trained on a large dataset of image-report pairs to restore the masked regions, and the learned image representations are transferred to three medical image analysis tasks and image-text/report-text retrieval tasks. The results demonstrate that MedIM outperforms strong pre-training competitors and the random masking method. In the future, we will extend our MedIM to handle other modalities, *e.g.*, 3D medical image analysis.

Acknowledgments. Dr. Lin Gu was supported by JST Moonshot R&D Grant Number JPMJMS2011, Japan. Prof. Yong Xia was supported in part by the Key Research and Development Program of Shaanxi Province, China, under Grant 2022GY-084, in part by the National Natural Science Foundation of China under Grants 62171377, and in part by the National Key R&D Program of China under Grant 2022YFC2009903/2022YFC2009900.

References

1. Siim-acr pneumothorax segmentation. Society for Imaging Informatics in Medicine (2019)
2. Alsentzer, E., et al.: Publicly available clinical BERT embeddings. arXiv preprint arXiv:1904.03323 (2019)
3. Bao, H., Dong, L., Piao, S., Wei, F.: Beit: BERT pre-training of image transformers. In: International Conference on Learning Representations (ICLR) (2022)
4. Cai, Z., Lin, L., He, H., Tang, X.: Uni4Eye: unified 2D and 3D self-supervised pre-training via masked image modeling transformer for ophthalmic image classification. In: Wang, L., Dou, Q., Fletcher, P.T., Speidel, S., Li, S. (eds.) MICCAI 2022. LNCS, vol. 13438, pp. 88–98. Springer, Cham (2022). https://doi.org/10.1007/978-3-031-16452-1_9

5. Chen, Z., Agarwal, D., Aggarwal, K., Safta, W., Balan, M.M., Brown, K.: Masked image modeling advances 3D medical image analysis. In: Proceedings of the IEEE/CVF Winter Conference on Applications of Computer Vision, pp. 1970–1980 (2023)
6. Chen, Z., et al.: Multi-modal masked autoencoders for medical vision-and-language pre-training. In: Wang, L., Dou, Q., Fletcher, P.T., Speidel, S., Li, S. (eds.) MICCAI 2022. LNCS, vol. 13435, pp. 679–689. Springer, Cham (2022). https://doi.org/10.1007/978-3-031-16443-9_65
7. Dosovitskiy, A., et al.: An image is worth 16x16 words: transformers for image recognition at scale. In: International Conference on Learning Representations (ICLR) (2021)
8. He, K., Chen, X., Xie, S., Li, Y., Dollár, P., Girshick, R.: Masked autoencoders are scalable vision learners. In: Proceedings of the IEEE/CVF Conference on Computer Vision and Pattern Recognition, pp. 16000–16009 (2022)
9. Huang, S.C., Shen, L., Lungren, M.P., Yeung, S.: Gloria: a multimodal global-local representation learning framework for label-efficient medical image recognition. In: Proceedings of the IEEE/CVF International Conference on Computer Vision, pp. 3942–3951 (2021)
10. Irvin, J., et al.: CheXpert: a large chest radiograph dataset with uncertainty labels and expert comparison. In: Proceedings of the AAAI Conference on Artificial Intelligence, vol. 33, pp. 590–597 (2019)
11. Jiang, J., Tyagi, N., Tringale, K., Crane, C., Veeraraghavan, H.: Self-supervised 3D anatomy segmentation using self-distilled masked image transformer (smit). In: Wang, L., Dou, Q., Fletcher, P.T., Speidel, S., Li, S. (eds.) MICCAI 2022. LNCS, vol. 13434, pp. 556–566. Springer, Cham (2022). https://doi.org/10.1007/978-3-031-16440-8_53
12. Johnson, A.E., et al.: Mimic-CXR, a de-identified publicly available database of chest radiographs with free-text reports. Sci. Data **6**(1), 1–8 (2019)
13. Johnson, A.E., et al.: Mimic-III, a freely accessible critical care database. Sci. Data **3**(1), 1–9 (2016)
14. Lipscomb, C.E.: Medical subject headings (mesh). Bull. Med. Libr. Assoc. **88**(3), 265 (2000)
15. Loshchilov, I., Hutter, F.: SGDR: stochastic gradient descent with warm restarts. In: ICLR (2017)
16. Loshchilov, I., Hutter, F.: Fixing weight decay regularization in Adam (2018)
17. Wang, F., Zhou, Y., Wang, S., Vardhanabhuti, V., Yu, L.: Multi-granularity cross-modal alignment for generalized medical visual representation learning. In: Advances in Neural Information Processing Systems (2022)
18. Wang, L., Lin, Z.Q., Wong, A.: COVID-net: a tailored deep convolutional neural network design for detection of COVID-19 cases from chest x-ray images. Sci. Rep. **10**(1), 1–12 (2020)
19. Wu, Y., et al.: Google's neural machine translation system: bridging the gap between human and machine translation. arXiv preprint arXiv:1609.08144 (2016)
20. Xiao, J., Bai, Y., Yuille, A., Zhou, Z.: Delving into masked autoencoders for multi-label thorax disease classification. In: Proceedings of the IEEE/CVF Winter Conference on Applications of Computer Vision, pp. 3588–3600 (2023)
21. Xie, Y., Zhang, J., Xia, Y., Wu, Q.: UniMISS: universal medical self-supervised learning via breaking dimensionality barrier. In: Avidan, S., Brostow, G., Cissé, M., Farinella, G.M., Hassner, T. (eds.) ECCV 2022. LNCS, vol. 13681, pp. 558–575. Springer, Cham (2022). https://doi.org/10.1007/978-3-031-19803-8_33

22. Zhou, H.Y., Lian, C., Wang, L., Yu, Y.: Advancing radiograph representation learning with masked record modeling. In: International Conference on Learning Representations (ICLR) (2023)
23. Zhou, J., et al.: Image BERT pre-training with online tokenizer. In: International Conference on Learning Representations (ICLR) (2022)
24. Zhou, L., Liu, H., Bae, J., He, J., Samaras, D., Prasanna, P.: Self pre-training with masked autoencoders for medical image analysis. arXiv preprint arXiv:2203.05573 (2022)

UOD: Universal One-Shot Detection of Anatomical Landmarks

Heqin Zhu[1,2,3], Quan Quan[3], Qingsong Yao[3], Zaiyi Liu[4,5], and S. Kevin Zhou[1,2(✉)]

[1] School of Biomedical Engineering, Division of Life Sciences and Medicine, University of Science and Technology of China, Hefei 230026, Anhui, People's Republic of China
skevinzhou@ustc.edu.cn
[2] Suzhou Institute for Advanced Research, University of Science and Technology of China, Suzhou 215123, Jiangsu, People's Republic of China
[3] Key Lab of Intelligent Information Processing of Chinese Academy of Sciences (CAS), Institute of Computing Technology, CAS, Beijing 100190, China
[4] Department of Radiology, Guangdong Provincial People's Hospital, Guangdong Academy of Medical Sciences, Guangzhou, China
[5] Guangdong Provincial Key Laboratory of Artificial Intelligence in Medical Image Analysis and Application, Guangdong Provincial People's Hospital, Guangdong Academy of Medical Sciences, Guangzhou, China

Abstract. One-shot medical landmark detection gains much attention and achieves great success for its label-efficient training process. However, existing one-shot learning methods are highly specialized in a single domain and suffer domain preference heavily in the situation of multi-domain unlabeled data. Moreover, one-shot learning is not robust that it faces performance drop when annotating a sub-optimal image. To tackle these issues, we resort to developing a domain-adaptive one-shot landmark detection framework for handling multi-domain medical images, named **Universal One-shot Detection (UOD)**. UOD consists of two stages and two corresponding universal models which are designed as combinations of domain-specific modules and domain-shared modules. In the first stage, a domain-adaptive convolution model is self-supervised learned to generate pseudo landmark labels. In the second stage, we design a domain-adaptive transformer to eliminate domain preference and build the global context for multi-domain data. Even though only one annotated sample from each domain is available for training, the domain-shared modules help UOD aggregate all one-shot samples to detect more robust and accurate landmarks. We investigated both qualitatively and quantitatively the proposed UOD on three widely-used public X-ray datasets in different anatomical domains (i.e., head, hand, chest) and obtained state-of-the-art performances in each domain. The code is at https://github.com/heqin-zhu/UOD_universal_oneshot_detection.

Keywords: One-shot learning · Domain-adaptive model · Anatomical landmark detection · Transformer network

H. Greenspan et al. (Eds.): MICCAI 2023, LNCS 14220, pp. 24–34, 2023.
https://doi.org/10.1007/978-3-031-43907-0_3

1 Introduction

Robust and accurate detecting of anatomical landmarks is an essential task in medical image applications [24,25], which plays vital parts in varieties of clinical treatments, for instance, vertebrae localization [20], orthognathic and orthodontic surgeries [9], and craniofacial anomalies assessment [4]. Moreover, anatomical landmarks exert their effectiveness in other medical image tasks such as segmentation [3], registration [5], and biometry estimation [1].

In the past years, lots of fully supervised methods [4,8,11,11,20,21,26,27] have been proposed to detect landmarks accurately and automatically. To relieve the burden of experts and reduce the amount of annotated labels, various one-shot and few-shot methods have been come up with. Zhao et al. [23] demonstrate a model which learns transformations from the images and uses the labeled example to synthesize additional labeled examples, where each transformation is composed of a spatial deformation field and an intensity change. Yao et al. [22] develop a cascaded self-supervised learning framework for one-shot medical landmark detection. They first train a matching network to calculate the cosine similarity between features from an image and a template patch, then fine-tune the pseudo landmark labels from coarse to fine. Browatzki et al. [2] propose a semi-supervised method that consists of two stages. They first employ an adversarial auto-encoder to learn implicit face knowledge from unlabeled images and then fine-tune the decoder to detect landmarks with few-shot labels.

However, one-shot methods are not robust enough because they are dependent on the choice of labeled template and the accuracy of detected landmarks may decrease a lot when choosing a sub-optimal image to annotate. To address this issue, Quan et al. [12] propose a novel Sample Choosing Policy (SCP) to select the most worthy image to annotate. Despite the improved performance, SCP brings an extra computation burden. Another challenge is the scalability of model building when facing multiple domains (such as different anatomical regions). While conventional wisdom is to independently train a model for each domain, Zhu et al. [26] propose a universal model YOLO for detecting landmarks across different anatomies and achieving better performances than a collection of single models. YOLO is regularly supervised using the CNN as backbone and it is unknown if the YOLO model works for a one-shot scenario and with a modern transformer architecture.

Motivated by above challenges, to detect robust multi-domain label-efficient landmarks, we design domain-adaptive models and propose a universal one-shot landmark detection framework called **Universal One-shot Detection (UOD)**, illustrated in Fig. 1. A universal model is comprised of domain-specific modules and domain-shared modules, learning the specified features of each domain and common features of all domains to eliminate domain preference and extract representative features for multi-domain data. Moreover, one-shot learning is not robust enough because of the sample selection while multi-domain one-shot learning reaps benefit from different one-shot samples from various domains, in which cross-domain features are excavated by domain-shared modules. Our proposed UOD framework consists of two stages: 1) Contrastive learning for

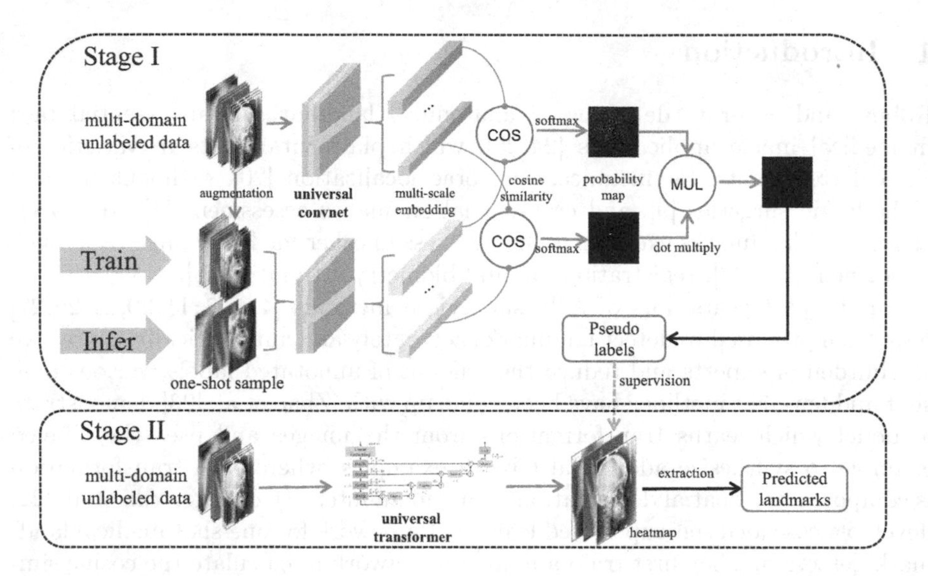

Fig. 1. Overview of UOD framework. In stage I, two universal models are learned via contrastive learning for matching similar patches from original image and augmented one-shot sample image and generating pseudo labels. In stage II, DATR is designed to better capture global context information among all domains for detecting more accurate landmarks.

training a universal model with multi-domain data to generate pseudo landmark labels. 2) Supervised learning for training domain-adaptive transformer (DATR) to avoid domain preference and detect robust and accurate landmarks.

In summary, our contributions can be categorized into three parts: **1)** We design the first universal framework for multi-domain one-shot landmark detection, which improves detecting accuracy and relieves domain preference on multi-domain data from various anatomical regions. **2)** We design a domain-adaptive transformer block (DATB), which is effective for multi-domain learning and can be used in any other transformer network. **3)** We carry out comprehensive experiments to demonstrate the effectiveness of UOD for obtaining SOTA performance on three publicly used X-ray datasets of head, hand, and chest.

2 Method

As Fig. 1 shows, UOD consists of two stages: 1) Contrastive learning and 2) Supervised learning. In stage I, to learn the local appearance of each domain, a universal model is trained via self-supervised learning, which contains domain-specific VGG [15] and UNet [13] decoder with each standard convolution replaced by a domain adaptor [7]. In stage II, to grasp the global constraint and eliminate domain preference, we designed a domain-adaptive transformer (DATR).

2.1 Stage I: Contrastive Learning

As Fig. 1 shows, following Yao et al. [22], we employ contrastive learning to train siamese network for matching similar patches of original image and augmented image. Given a multi-domain input image $X^d \in R^{H^d \times W^d \times C^d}$ belongs to domain d from multi-domain data, we randomly select a target point P and crop a half-size patch X_p^d which contains P. After applying data augmentation on X_p^d, the target point is mapped to P_p. Then we feed X^d and X_p^d into the siamese network respectively and obtain the multi-scale feature embeddings. We compute cosine similarity of two feature embeddings from each scale and apply softmax to the cosine similarity map to generate a probability matrix. Finally, we calculate the cross entropy loss of the probability matrix and ground truth map which is produced with the one-hot encoding of P_p^d to optimize the siamese network for learning the latent similarities of patches. At inferring stage, we replace augmented patch X_p^d with the augmented one-shot sample patch X_s^d. We use the annotated one-shot landmarks as target points to formulate the ground truth maps. After obtaining probability matrices, we apply $\arg\max$ to extract the strongest response points as the pseudo landmarks, which will be used in UOD Stage II.

2.2 Stage II: Supervised Learning

In stage II, we design a universal transformer to capture global relationship of multi-domain data and train it with the pseudo landmarks generated in stage I. The universal transformer has a domain-adaptive transformer encoder and domain-adaptive convolution decoder. The decoder is based on a U-Net [13] decoder with each standard convolution replaced by a domain adaptor [7]. The encoder is based on Swin Transformer [10] with shifted window and limited self-attention within non-overlapping local windows for computation efficiency. Different from Swin Transformer [10], we design a domain-adaptive transformer block (DATB) and use it to replace the original transformer block.

Domain-Adaptive Transformer Encoder. As Fig. 2(a) shows, the transformer encoder is built up with DATB, making full use of the capability of transformer for modeling global relationship and extracting multi-domain representative features. As in Fig. 2(b), a basic transformer block [17] consists of a multi-head self-attention module (MSA), followed by a two-layer MLP with GELU activation. Furthermore, layer normalization (LN) is adopted before each MSA and MLP and a residual connection is adopted after each MSA and MLP. Given a feature map $x^d \in R^{h \times w \times c}$ from domain d with height h, width w, and c channels, the output feature maps of MSA and MLP, denoted by $\hat{y}^d$ and y^d, respectively, are formulated as:

$$\begin{aligned} \hat{y}^d &= \mathrm{MSA}(\mathrm{LN}(x^d)) + x^d \\ y^d &= \mathrm{MLP}(\mathrm{LN}(\hat{y}^d)) + \hat{y}^d \end{aligned} \quad (1)$$

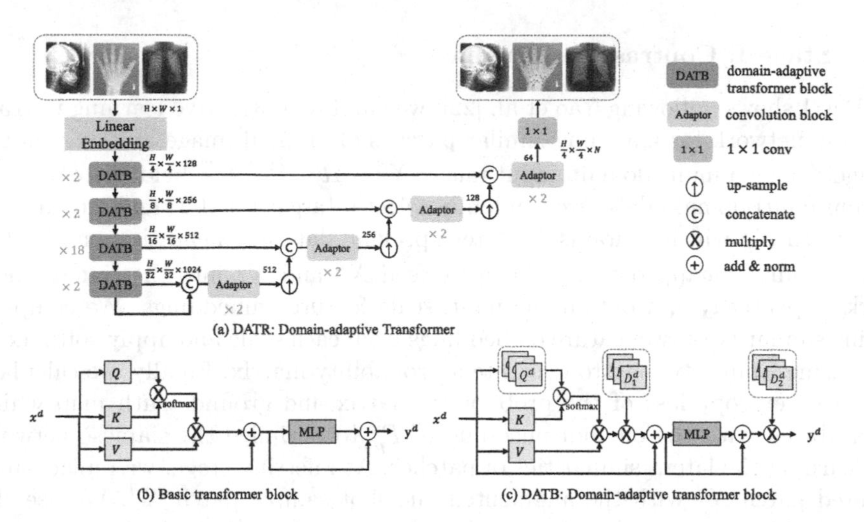

Fig. 2. (a) The architecture of DATR in stage II, which is composed of domain-adaptive transformer encoder and convolution adaptors [7]. (b) Basic transformer block. (c) Domain-adaptive transformer block. Each domain-adaptive transformer is a basic transformer block with query matrix duplicated and domain-adaptive diagonal for each domain. The batch-normalization, activation, and patch merging are omitted.

where MSA $=$ softmax$(QK^T)V$.

As illustrated in Fig. 2(b)(c), DATB is based on Eq. (1). Similar to U2Net [7] and GU2Net [26], we adopt domain-specific and domain-shared parameters in DATB. Since the attention probability is dependent on query and key matrix which are symmetrical, we duplicate the query matrix for each domain to learn domain-specific query features and keep key and value matrix domain-shared to learn common knowledge and reduce parameters. Inspired by LayerScale [16], we further adopt learnable diagonal matrix [16] after each MSA and MLP module to facilitate the learning of domain-specific features, which costs few parameters ($O(N)$ for $N \times N$ diagonal). Different from LayerScale [16], proposed domain-adaptive diagonal D_1^d and D_2^d are applied for each domain with D_2^d applied after residual connection for generating more representative and direct domain-specific features. The above process can be formulated as:

$$\begin{aligned} \hat{y}^d &= D_1^d \times \mathrm{MSA}_{Q^d}(\mathrm{LN}(x^d)) + x^d \\ y^d &= D_2^d \times (\mathrm{MLP}(\mathrm{LN}(\hat{y}^d)) + \hat{y}^d) \end{aligned} \tag{2}$$

where $\mathrm{MSA}_{Q^d} = \mathrm{softmax}(Q^d K^T)V$.

Overall Pipeline. Given that a random input $X^d \in R^{H^d \times W^d \times C^d}$ belongs to domain d from mixed datasets on various anatomical regions, which contains N^d landmarks with corresponding coordinates being $\{(i_1^d, j_1^d), (i_2^d, j_2^d), \ldots, (i_{N_d}^d, j_{N_d}^d)\}$, we set the n-th $\in \{1, 2, \ldots, N^d\}$ initial heatmap

$\tilde{Y}_n^d \in R^{H^d \times W^d \times C^d}$ with Gaussian function to be $\tilde{Y}_n^d = \frac{1}{\sqrt{2\pi}\sigma} e^{-\frac{(i-i_n^d)^2+(j-j_n^d)^2}{2\sigma^2}}$ if $\sqrt{(i-i_n^d)^2+(j-j_n^d)^2} \leq \sigma$ and 0 otherwise. We further add an exponential weight to the Gaussian distribution to distinguish close heatmap pixels and obtain the ground truth heatmap $Y_n^d(i,j) = \alpha^{\tilde{Y}_n^d(i,j)}$.

As illustrated in Fig. 2, firstly, the input image from a random batch is partitioned into non-overlapping patches and linearly embedded. Next, these patches are fed into cascaded transformer blocks at each stage, which are merged except in the last stage. Finally, a domain-adaptive convolution decoder makes dense prediction to generate heatmaps, which is further used to extract landmarks via threshold processing and connected components filtering.

3 Experiment

Datasets. For performance evaluation, we adopt three public X-ray datasets from different domains on various anatomical regions of head, hand, and chest. (i) Head dataset is a widely-used dataset for IEEE ISBI 2015 challenge [18,19] which contains 400 X-ray cephalometric images with 150 images for training and 250 images for testing. Each image is of size 2400 × 1935 with a resolution of 0.1 mm × 0.1 mm, which contains 19 landmarks manually labeled by two medical experts and we use the average labels same as Payer et al. [11]. (ii) Hand dataset is collected by [6] which contains 909 X-ray images and 37 landmarks annotated by [11]. We follow [26] to split this dataset into a training set of 609 images and a test set of 300 images. Following [11] we assume the distance between two endpoints of wrist is 50 mm and calculate the physical distance as $\text{distance}_{\text{physical}} = \text{distance}_{\text{pixel}} \times \frac{50}{\|p-q\|_2}$ where p, q are the two endpoints of the wrist respectively. (iii) Chest dataset [26] is a popular chest radiography database collected by Japanese Society of Radiological Technology (JSRT) [14] which contains 247 images. Each image is of size 2048 × 2048 with a resolution of 0.175 mm × 0.175 mm. We split it into a training set of 197 images and a test set of 50 images and select 6 landmarks from landmark labels at the boundary of the lung as target landmarks.

Implementation Details. UOD is implemented in Pytorch and trained on a TITAN RTX GPU with CUDA version being 11. All encoders are initialized with corresponding pre-trained weights. We set batch size to 8, σ to 3, and α to 10. We adopt binary cross-entropy (BCE) as loss function for both stages. In stage I, we resize each image to the same shape of 384 × 384 and train universal convolution model by Adam optimizer for 1000 epochs with a learning rate of 0.00001. In stage II, we resize each image to the same shape of 576 × 576 and optimize the universal transformer by Adam optimizer for 300 epochs with a learning rate of 0.0001. When calculating metrics, all predicted landmarks are resized back to the original size. For evaluation, we choose model with minimum validation loss as the inference model and adopt two metrics: mean radial error (MRE) $\text{MRE} = \frac{1}{N}\sum_i^N \sqrt{(x_i-\tilde{x}_i)^2+(y_i-\tilde{y}_i)^2}$ and successful detection rates (SDR) within different thresholds t: $\text{SDR}(t) = \frac{1}{N}\sum_i^N \delta(\sqrt{(x_i-\tilde{x}_i)^2+(y_i-\tilde{y}_i)^2} \leq t)$.

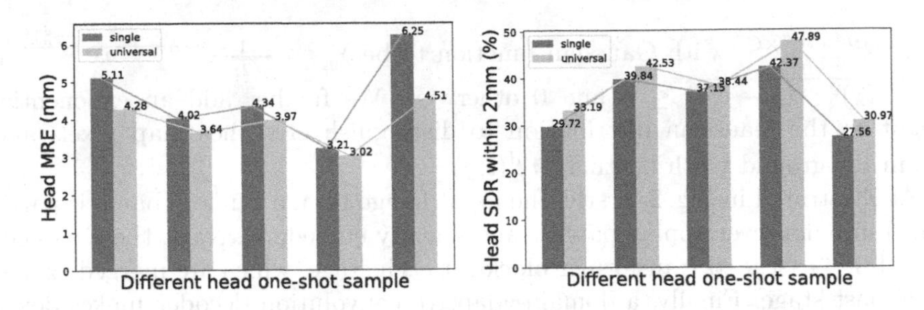

Fig. 3. Comparison of single model and universal model on head dataset.

Table 1. Quantitative comparison of UOD with SOTA methods on head, hand, and chest datasets. * denotes the method is trained on every single dataset respectively while †denotes the method is trained on mixed data.

Method	Label	Head [19]					Hand [6]				Chest [14]			
		MRE↓	SDR↑ (%)				MRE↓	SDR↑ (%)			MRE↓	SDR↑ (%)		
		(mm)	2 mm	2.5 mm	3 mm	4 mm	(mm)	2 mm	4 mm	10 mm	(mm)	2 mm	4 mm	10 mm
YOLO [26]†	all	1.32	81.14	87.85	92.12	96.80	0.85	94.93	99.14	99.67	4.65	31.00	69.00	93.67
YOLO [26]†	25	1.96	62.05	77.68	88.21	97.11	2.88	72.71	92.32	97.65	7.03	19.33	51.67	89.33
YOLO [26]†	10	2.69	47.58	66.47	78.42	90.89	9.70	48.66	76.69	90.52	16.07	11.67	33.67	76.33
YOLO [26]†	5	5.40	26.16	41.32	54.42	73.74	24.35	20.59	48.91	72.94	34.81	4.33	19.00	56.67
CC2D [22]*	1	2.76	42.36	51.82	64.02	78.96	2.65	51.19	82.56	95.62	10.25	11.37	35.73	68.14
Ours†	1	**2.43**	**51.14**	**62.37**	**74.40**	**86.49**	**2.52**	**53.37**	**84.27**	**97.59**	**8.49**	**14.00**	**39.33**	**76.33**

3.1 Experimental Results

The Effectiveness of Universal Model: To demonstrate the effectiveness of universal model for multi-domain one-shot learning, we adopt head and hand datasets for evaluation. In stage I, the convolution models are trained in two ways: 1) single: trained on every single dataset respectively, and 2) universal: trained on mixed datasets together. With a fixed one-shot sample for the hand dataset, we change the one-shot sample for the head dataset and report the MRE and SDR of the head dataset. As Fig. 3 shows, universal model performs much better than single model on various one-shot samples and metrics. It is proved that universal model learns domain-shared knowledge and promotes domain-specific learning. Furthermore, the MRE and SDR metrics of universal model have a smaller gap among various one-shot samples, which demonstrates the robustness of universal model learned on multi-domain data.

Comparisons with State-of-the-Art Methods: As Table 1 shows, we compare UOD with two open-source landmark detection methods, i.e., YOLO [26] and CC2D [22]. YOLO is a multi-domain supervised method while CC2D is a single-domain one-shot method. UOD achieves SOTA results on all datasets under all metrics, outperforming the other one-shot method by a big margin. On the head dataset, benefiting from multi-domain learning, UOD achieves an MRE of 2.43 mm and an SDR of 86.49% within 4 mm, which is comparative with supervised method YOLO trained with at least 10 annotated labels, and much

Table 2. Ablation study of different components of our DATR. Base is the basic transformer block; MSA_{Q^d} denotes the domain-adaptive self-attention and D^d denotes the domain-adaptive diagonal matrix. In each column, the best results are in **bold**.

Transformer	Head [19]					Hand [6]				Chest [14]			
	MRE↓	SDR↑ (%)				MRE↓	SDR↑ (%)			MRE↓	SDR↑ (%)		
	(mm)	2 mm	2.5 mm	3 mm	4 mm	(mm)	2 mm	4 mm	10 mm	(mm)	2 mm	4 mm	10 mm
(a) Base	24.95	2.02	3.17	4.51	5.85	9.83	5.33	16.79	58.64	58.11	0.37	1.96	3.85
(b) $+D^d$	22.75	2.13	3.24	4.61	6.96	7.52	6.13	20.66	68.43	52.98	0.59	2.17	4.68
(c) $+MSA_{Q^d}$	2.51	49.29	60.89	72.17	84.36	2.72	48.56	80.44	94.38	9.09	12.00	19.33	74.00
(d) $+MSA_{Q^d}+D^d$	**2.43**	**51.14**	**62.37**	**74.40**	**86.49**	**2.52**	**53.37**	**84.27**	**97.59**	**8.49**	**14.00**	**39.33**	**76.33**

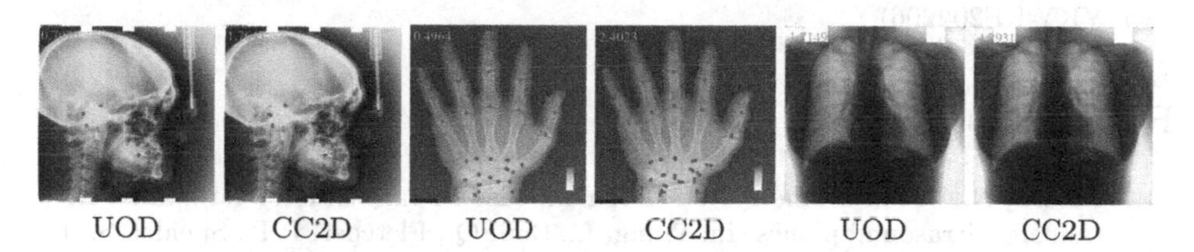

Fig. 4. Qualitative comparison of UOD and CC2D [22] on head, hand, and chest datasets. The red points • indicate predicted landmarks while the green points • indicate ground truth landmarks. The MRE value is displayed in the top left corner of the image. (Color figure online)

better than CC2D. On the hand dataset, there are some performance improvements in all metrics compared to CC2D, outperforming the supervised method YOLO trained with 25 annotated images. On the chest dataset, UOD shows the superiority of DATR which eliminates domain preference and balances the performance of all domains. In contrast, the performance of YOLO on chest dataset suffers a tremendous drop when the available labels are reduced to 25, 10, and 5. Figure 4 visualizes the predicted landmarks by UOD and CC2D.

Ablation Study: We compare various components of the proposed domain-adaptive transformer. The experiments are carried out in UOD Stage II. As presented in Table 2, the domain-adaptive transformer has two key components: domain-adaptive self-attention MSA_{Q^d} and domain-adaptive diagonal matrix D^d. The performances of (b) and (c) are much superior to those of (a) which demonstrates the effectiveness of D^d and MSA_{Q^d}. Further, (d) combines the two components and achieves much better performances, which illustrates that domain-adaptive transformer improves the accuracy of detecting via cross-domain knowledge and global context information. We take (d) as the final transformer block.

4 Conclusion

To improve the robustness and reduce domain preference of multi-domain one-shot learning, we design a universal framework in that we first train a universal model via contrastive learning to generate pseudo landmarks and further use these labels to learn a universal transformer for accurate and robust detection of landmarks. UOD is the first universal framework of one-shot landmark detection on multi-domain data, which outperforms other one-shot methods on three public datasets from different anatomical regions. We believe UOD will significantly reduce the labeling burden and pave the path of developing more universal framework for multi-domain one-shot learning.

Acknowledgment. Supported by Natural Science Foundation of China under Grant 62271465 and Open Fund Project of Guangdong Academy of Medical Sciences, China (No. YKY-KF202206).

References

1. Avisdris, N., et al.: BiometryNet: landmark-based fetal biometry estimation from standard ultrasound planes. In: Wang, L., Dou, Q., Fletcher, P.T., Speidel, S., Li, S. (eds.) MICCAI 2022. LNCS, vol. 13434, pp. 279–289. Springer, Cham (2022). https://doi.org/10.1007/978-3-031-16440-8_27
2. Browatzki, B., Wallraven, C.: 3fabrec: fast few-shot face alignment by reconstruction. In: Proceedings of the IEEE/CVF Conference on Computer Vision and Pattern Recognition, pp. 6110–6120 (2020)
3. Chen, Z., Qiu, T., Tian, Y., Feng, H., Zhang, Y., Wang, H.: Automated brain structures segmentation from PET/CT images based on landmark-constrained dual-modality atlas registration. Phys. Med. Biol. **66**(9), 095003 (2021)
4. Elkhill, C., LeBeau, S., French, B., Porras, A.R.: Graph convolutional network with probabilistic spatial regression: Application to craniofacial landmark detection from 3D photogrammetry. In: Wang, L., Dou, Q., Fletcher, P.T., Speidel, S., Li, S. (eds.) MICCAI 2022. LNCS, vol. 13433, pp. 574–583. Springer, Cham (2022). https://doi.org/10.1007/978-3-031-16437-8_55
5. Espinel, Y., Calvet, L., Botros, K., Buc, E., Tilmant, C., Bartoli, A.: Using multiple images and contours for deformable 3D-2D registration of a preoperative ct in laparoscopic liver surgery. In: de Bruijne, M., et al. (eds.) MICCAI 2021. LNCS, vol. 12904, pp. 657–666. Springer, Cham (2021). https://doi.org/10.1007/978-3-030-87202-1_63
6. Gertych, A., Zhang, A., Sayre, J., Pospiech-Kurkowska, S., Huang, H.: Bone age assessment of children using a digital hand atlas. Comput. Med. Imaging Graph. **31**(4–5), 322–331 (2007)
7. Huang, C., Han, H., Yao, Q., Zhu, S., Zhou, S.K.: 3D U^2-Net: a 3D universal U-net for multi-domain medical image segmentation. In: Shen, D., et al. (eds.) MICCAI 2019. LNCS, vol. 11765, pp. 291–299. Springer, Cham (2019). https://doi.org/10.1007/978-3-030-32245-8_33
8. Jiang, Y., Li, Y., Wang, X., Tao, Y., Lin, J., Lin, H.: CephalFormer: incorporating global structure constraint into visual features for general cephalometric landmark detection. In: Wang, L., Dou, Q., Fletcher, P.T., Speidel, S., Li, S. (eds.) MICCAI 2022. LNCS, vol. 13433, pp. 227–237. Springer, Cham (2022). https://doi.org/10.1007/978-3-031-16437-8_22
9. Lang, Y., et al.: DentalPointNet: landmark localization on high-resolution 3d digital dental models. In: Wang, L., Dou, Q., Fletcher, P.T., Speidel, S., Li, S. (eds.)

MICCAI 2022. LNCS, vol. 13432, pp. 444–452. Springer, Cham (2022). https://doi.org/10.1007/978-3-031-16434-7_43
10. Liu, Z., et al.: Swin transformer: hierarchical vision transformer using shifted windows. In: Proceedings of the IEEE/CVF International Conference on Computer Vision, pp. 10012–10022 (2021)
11. Payer, C., Štern, D., Bischof, H., Urschler, M.: Integrating spatial configuration into heatmap regression based CNNs for landmark localization. Med. Image Anal. **54**, 207–219 (2019)
12. Quan, Q., Yao, Q., Li, J., Zhou, S.K.: Which images to label for few-shot medical landmark detection? In: Proceedings of the IEEE/CVF Conference on Computer Vision and Pattern Recognition, pp. 20606–20616 (2022)
13. Ronneberger, O., Fischer, P., Brox, T.: U-Net: convolutional networks for biomedical image segmentation. In: Navab, N., Hornegger, J., Wells, W.M., Frangi, A.F. (eds.) MICCAI 2015. LNCS, vol. 9351, pp. 234–241. Springer, Cham (2015). https://doi.org/10.1007/978-3-319-24574-4_28
14. Shiraishi, J., et al.: Development of a digital image database for chest radiographs with and without a lung nodule: receiver operating characteristic analysis of radiologists' detection of pulmonary nodules. Am. J. Roentgenol. **174**(1), 71–74 (2000)
15. Simonyan, K., Zisserman, A.: Very deep convolutional networks for large-scale image recognition. In: International Conference on Learning Representations (ICLR), pp. 1–14 (2015)
16. Touvron, H., Cord, M., Sablayrolles, A., Synnaeve, G., Jégou, H.: Going deeper with image transformers. In: Proceedings of the IEEE/CVF International Conference on Computer Vision, pp. 32–42 (2021)
17. Vaswani, A., et al.: Attention is all you need. In: Advances in Neural Information Processing Systems, vol. 30 (2017)
18. Wang, C.W., et al.: Evaluation and comparison of anatomical landmark detection methods for cephalometric x-ray images: a grand challenge. IEEE Trans. Med. Imaging **34**(9), 1890–1900 (2015)
19. Wang, C.W., et al.: A benchmark for comparison of dental radiography analysis algorithms. Med. Image Anal. **31**, 63–76 (2016)
20. Wang, Z., et al.: Accurate scoliosis vertebral landmark localization on x-ray images via shape-constrained multi-stage cascaded CNNs. Fundam. Res. (2022)
21. Yao, Q., He, Z., Han, H., Zhou, S.K.: Miss the point: targeted adversarial attack on multiple landmark detection. In: Martel, A.L., et al. (eds.) MICCAI 2020. LNCS, vol. 12264, pp. 692–702. Springer, Cham (2020). https://doi.org/10.1007/978-3-030-59719-1_67
22. Yao, Q., Quan, Q., Xiao, L., Kevin Zhou, S.: One-shot medical landmark detection. In: de Bruijne, M., et al. (eds.) MICCAI 2021. LNCS, vol. 12902, pp. 177–188. Springer, Cham (2021). https://doi.org/10.1007/978-3-030-87196-3_17
23. Zhao, A., Balakrishnan, G., Durand, F., Guttag, J.V., Dalca, A.V.: Data augmentation using learned transformations for one-shot medical image segmentation. In: Proceedings of the IEEE/CVF Conference on Computer Vision and Pattern Recognition, pp. 8543–8553 (2019)
24. Zhou, S.K., et al.: A review of deep learning in medical imaging: imaging traits, technology trends, case studies with progress highlights, and future promises. Proc. IEEE (2021)
25. Zhou, S.K., Rueckert, D., Fichtinger, G.: Handbook of Medical Image Computing and Computer Assisted Intervention. Academic Press, Cambridge (2019)

26. Zhu, H., Yao, Q., Xiao, L., Zhou, S.K.: You only learn once: universal anatomical landmark detection. In: de Bruijne, M., et al. (eds.) MICCAI 2021. LNCS, vol. 12905, pp. 85–95. Springer, Cham (2021). https://doi.org/10.1007/978-3-030-87240-3_9
27. Zhu, H., Yao, Q., Xiao, L., Zhou, S.K.: Learning to localize cross-anatomy landmarks in x-ray images with a universal model. BME Front. **2022** (2022)

S^2ME: Spatial-Spectral Mutual Teaching and Ensemble Learning for Scribble-Supervised Polyp Segmentation

An Wang[1], Mengya Xu[2], Yang Zhang[1,3], Mobarakol Islam[4], and Hongliang Ren[1,2](✉)

[1] Department of Electronic Engineering, Shun Hing Institute of Advanced Engineering (SHIAE), The Chinese University of Hong Kong, Hong Kong, Hong Kong SAR, China
wa09@link.cuhk.edu.hk, yzhangcst@hbut.edu.cn, hlren@ee.cuhk.edu.hk
[2] Department of Biomedical Engineering, National University of Singapore, Singapore, Singapore
mengya@u.nus.edu
[3] School of Mechanical Engineering, Hubei University of Technology, Wuhan, China
[4] Department of Medical Physics and Biomedical Engineering, Wellcome/EPSRC Centre for Interventional and Surgical Sciences (WEISS), University College London, London, UK
mobarakol.islam@ucl.ac.uk

Abstract. Fully-supervised polyp segmentation has accomplished significant triumphs over the years in advancing the early diagnosis of colorectal cancer. However, label-efficient solutions from weak supervision like scribbles are rarely explored yet primarily meaningful and demanding in medical practice due to the expensiveness and scarcity of densely-annotated polyp data. Besides, various deployment issues, including data shifts and corruption, put forward further requests for model generalization and robustness. To address these concerns, we design a framework of **S**patial-**S**pectral Dual-branch **M**utual Teaching and Entropy-guided Pseudo Label **E**nsemble Learning (S^2ME). Concretely, for the first time in weakly-supervised medical image segmentation, we promote the dual-branch co-teaching framework by leveraging the intrinsic complementarity of features extracted from the spatial and spectral domains and encouraging cross-space consistency through collaborative optimization. Furthermore, to produce reliable mixed pseudo labels, which enhance the effectiveness of ensemble learning, we introduce a novel adaptive pixel-wise fusion technique based on the entropy guidance from the spatial and spectral branches. Our strategy efficiently mitigates the deleterious effects of uncertainty and noise present in pseudo labels and surpasses previous alternatives in terms of efficacy. Ultimately, we formulate a holistic optimization objective to learn from the hybrid supervision of

Supplementary Information The online version contains supplementary material available at https://doi.org/10.1007/978-3-031-43907-0_4.

H. Greenspan et al. (Eds.): MICCAI 2023, LNCS 14220, pp. 35–45, 2023.
https://doi.org/10.1007/978-3-031-43907-0_4

scribbles and pseudo labels. Extensive experiments and evaluation on four public datasets demonstrate the superiority of our method regarding in-distribution accuracy, out-of-distribution generalization, and robustness, highlighting its promising clinical significance. Our code is available at https://github.com/lofrienger/S2ME.

Keywords: Polyp Image Segmentation · Weakly-supervised Learning · Spatial-Spectral Dual Branches · Mutual Teaching · Ensemble Learning

1 Introduction

Colorectal cancer is a leading cause of cancer-related deaths worldwide [1]. Early detection and efficient diagnosis of polyps, which are precursors to colorectal cancer, is crucial for effective treatment. Recently, deep learning has emerged as a powerful tool in medical image analysis, prompting extensive research into its potential for polyp segmentation. The effectiveness of deep learning models in medical applications is usually based on large, well-annotated datasets, which in turn necessitates a time-consuming and expertise-driven annotation process. This has prompted the emergence of approaches for annotation-efficient weakly-supervised learning in the medical domain with limited annotations like points [8], bounding boxes [12], and scribbles [15]. Compared with other sparse labeling methods, scribbles allow the annotator to annotate arbitrary shapes, making them more flexible than points or boxes [13]. Besides, scribbles provide a more robust supervision signal, which can be prone to noise and outliers [5]. Hence, this work investigates the feasibility of conducting polyp segmentation using scribble annotation as supervision. The effectiveness of medical applications during in-site deployment depends on their ability to generalize to unseen data and remain robust against data corruption. Improving these factors is crucial to enhance the accuracy and reliability of medical diagnoses in real-world scenarios [22,27,28]. Therefore, we comprehensively evaluate our approach on multiple datasets from various medical sites to showcase its viability and effectiveness across different contexts.

Dual-branch learning has been widely adopted in annotation-efficient learning to encourage mutual consistency through co-teaching. While existing approaches are typically designed for learning in the spatial domain [21,25,29, 30], a novel spatial-spectral dual-branch structure is introduced to efficiently leverage domain-specific complementary knowledge with synergistic mutual teaching. Furthermore, the outputs from the spatial-spectral branches are aggregated to produce mixed pseudo labels as supplementary supervision. Different from previous methods, which generally adopt the handcrafted fusion strategies [15], we design to aggregate the outputs from spatial-spectral dual branches with an entropy-guided adaptive mixing ratio for each pixel. Consequently, our incorporated tactic of pseudo-label fusion aptly assesses the pixel-level ambiguity emerging from both spatial and frequency domains based on their entropy maps, thereby allocating substantially assured categorical labels to individual pixels and facilitating effective pseudo label ensemble learning.

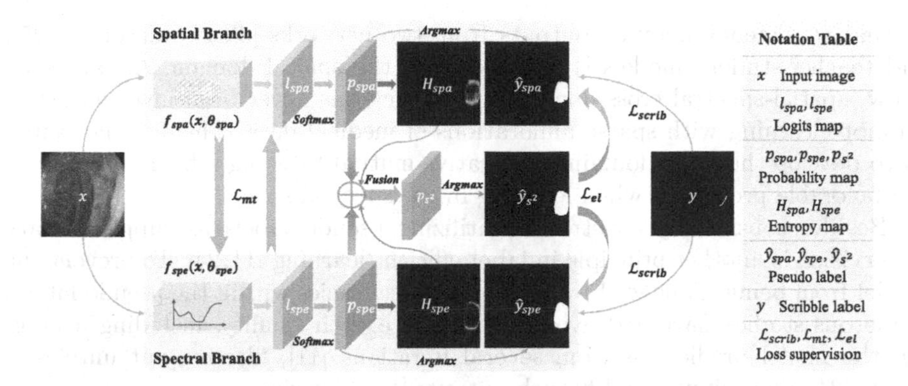

Fig. 1. Overview of our **S**patial-**S**pectral Dual-branch **M**utual Teaching and Pixel-level Entropy-guided Pseudo Label **E**nsemble Learning (S^{2}ME) for scribble-supervised polyp segmentation. Spatial-spectral cross-domain consistency is encouraged through mutual teaching. High-quality mixed pseudo labels are generated with pixel-level guidance from the dual-space entropy maps, ensuring more reliable supervision for ensemble learning.

Contributions. Overall, the contributions of this work are threefold: First, we devise a spatial-spectral dual-branch structure to leverage cross-space knowledge and foster collaborative mutual teaching. To our best knowledge, this is the first attempt to explore the complementary relations of the spatial-spectral dual branch in boosting weakly-supervised medical image analysis. Second, we introduce the pixel-level entropy-guided fusion strategy to generate mixed pseudo labels with reduced noise and increased confidence, thus enhancing ensemble learning. Lastly, our proposed hybrid loss optimization, comprising scribbles-supervised loss, mutual training loss with domain-specific pseudo labels, and ensemble learning loss with fused-domain pseudo labels, facilitates obtaining a generalizable and robust model for polyp image segmentation. An extensive assessment of our approach through the examination of four publicly accessible datasets establishes its superiority and clinical significance.

2 Methodology

2.1 Preliminaries

Spectral-domain learning [26] has gained increasing popularity in medical image analysis [23] for its ability to identify subtle frequency patterns that may not be well detected by the pure spatial-domain network like UNet [20]. For instance, a recent dual-encoder network, YNet [6], incorporates a spectral encoder with Fast Fourier Convolution (FFC) [4] to disentangle global patterns across varying frequency components and derives hybrid feature representation. In addition, spectrum learning also exhibits advantageous robustness and generalization against adversarial attacks, data corruption, and distribution shifts [19]. In label-efficient learning, some preliminary works have been proposed to encourage mutual consistency between outputs from two networks [3], two decoders [25], and teacher-student models [14], yet only in the spatial domain. As far as we know, spatial-spectral cross-domain consistency has never been investigated to promote learning with sparse annotations of medical data. This has motivated us to develop the cross-domain cooperative mutual teaching scheme to leverage the favorable properties when learning in the spectral space.

Besides consistency constraints, utilizing pseudo labels as supplementary supervision is another principle in label-efficient learning [11,24]. To prevent the model from being influenced by noise and inaccuracies within the pseudo labels, numerous studies have endeavored to enhance their quality, including averaging the model predictions from several iterations [11], filtering out unreliable pixels [24], and mixing dual-branch outputs [15] following

$$p_{mix} = \alpha \times p_1 + (1 - \alpha) \times p_2, \alpha = random(0, 1), \tag{1}$$

where α is the random mixing ratio. p_1, p_2, and p_{mix} denote the probability maps from the two spatial decoders and their mixture. These approaches only operate in the spatial domain, regardless of single or dual branches, while we consider both spatial and spectral domains and propose to adaptively merge dual-branch outputs with respective pixel-wise entropy guidance.

2.2 S^{2}ME: Spatial-Spectral Mutual Teaching and Ensemble Learning

Spatial-Spectral Cross-domain Mutual Teaching. In contrast to prior weakly-supervised learning methods that have merely emphasized spatial considerations, our approach designs a dual-branch structure consisting of a spatial branch $f_{spa}(x, \theta_{spa})$ and a spectral branch $f_{spe}(x, \theta_{spe})$, with x and θ being the input image and randomly initialized model parameters. As illustrated in Fig. 1, the spatial and spectral branches take the same training image as the input and extract domain-specific patterns. The raw model outputs, *i.e.*, the logits l_{spa} and l_{spe}, will be converted to probability maps p_{spa} and p_{spe} with *Softmax* normalization, and further to respective pseudo labels $\hat{y}_{spa}$ and $\hat{y}_{spe}$ by $\hat{y} = \arg\max p$. The spatial and spectral pseudo labels supervise the other branch collaboratively during mutual teaching and can be expressed as

$$\hat{y}_{spa} \rightarrow f_{spe} \text{ and } \hat{y}_{spe} \rightarrow f_{spa}, \tag{2}$$

where "$\rightarrow$" denotes supervision[1]. Through cross-domain engagement, these two branches complement each other, with each providing valuable domain-specific insights and feedback to the other. Consequently, such a scheme can lead to better feature extraction, more meaningful data representation, and domain-specific knowledge transmission, thus boosting model generalization and robustness.

Entropy-Guided Pseudo Label Ensemble Learning. In addition to mutual teaching, we consider aggregating the pseudo labels from the spatial and spectral branches in ensemble learning, aiming to take advantage of the distinctive

[1] For convenience, we omit the input x and model parameters θ.

yet complementary properties of the cross-domain features. As we know, a pixel characterized by a higher entropy value indicates elevated uncertainty in terms of its corresponding prediction. We can observe from the entropy maps $\mathcal{H}_{spa}$ and $\mathcal{H}_{spe}$ in Fig. 1 that the pixels of the polyp boundary exhibit greater difficulties in accurate segmentation, presenting with higher entropy values (the white contours). Considering such property, we propose a novel adaptive strategy to automatically adjust the mixing ratio for each pixel based on the entropy of its categorical probability distribution. Hence, the mixed pseudo labels are more reliable and beneficial for ensemble learning. Concretely, with the spatial and spectral probability maps p_{spa} and p_{spe}, the corresponding entropy maps $\mathcal{H}_{spa}$ and $\mathcal{H}_{spe}$ can be computed with

$$\mathcal{H} = -\sum_{c=0}^{C-1} p(c) \times \log p(c), \tag{3}$$

where C is the number of classes that equals 2 in our task. Unlike previous image-level fixed-ratio mixing or random mixing as Eq. (1), we can update the mixing ratio between the two probability maps p_{spa} and p_{spe} with the weighted entropy guidance at each pixel location by

$$p_{s^2} = \frac{\mathcal{H}_{spe}}{\mathcal{H}_{spa} + \mathcal{H}_{spe}} \otimes p_{spa} + \frac{\mathcal{H}_{spa}}{\mathcal{H}_{spa} + \mathcal{H}_{spe}} \otimes p_{spe}, \tag{4}$$

where "$\otimes$" denotes pixel-wise multiplication. p_{s^2} is the merged probability map and can be further converted to the pseudo label by $\hat{y}_{s^2} = \arg\max p_{s^2}$ to supervise the spatial and spectral branch in the context of ensemble learning following

$$\hat{y}_{s^2} \rightarrow f_{spa} \text{ and } \hat{y}_{s^2} \rightarrow f_{spe}. \tag{5}$$

By absorbing strengths from the spatial and spectral branches, ensemble learning from the mixed pseudo labels facilitates model optimization with reduced overfitting, increased stability, and improved generalization and robustness.

Hybrid Loss Supervision from Scribbles and Pseudo Labels. Besides the scribble annotations for partial pixels, the aforementioned three types of pseudo labels $\hat{y}_{spa}$, $\hat{y}_{spe}$, and $\hat{y}_{s^2}$ can offer complementary supervision for every pixel, with different learning regimes. Overall, our hybrid loss supervision is based on Cross Entropy loss ℓ_{CE} and Dice loss ℓ_{Dice}. Specifically, we employ the partial Cross Entropy loss [13] ℓ_{pCE}, which only calculates the loss on the labeled pixels, for learning from scribbles following

$$\mathcal{L}_{scrib} = \ell_{pCE}(l_{spa}, y) + \ell_{pCE}(l_{spe}, y), \tag{6}$$

where y denotes the scribble annotations. Furthermore, the mutual teaching loss with supervision from domain-specific pseudo labels is

$$\mathcal{L}_{mt} = \underbrace{\{\ell_{CE}(l_{spa}, \hat{y}_{spe}) + \ell_{Dice}(p_{spa}, \hat{y}_{spe})\}}_{\hat{y}_{spe} \rightarrow f_{spa}} + \underbrace{\{\ell_{CE}(l_{spe}, \hat{y}_{spa}) + \ell_{Dice}(p_{spe}, \hat{y}_{spa})\}}_{\hat{y}_{spa} \rightarrow f_{spe}}. \tag{7}$$

Likewise, the ensemble learning loss with supervision from the enhanced mixed pseudo labels can be formulated as

$$\mathcal{L}_{el} = \underbrace{\{\ell_{CE}(l_{spa}, \hat{y}_{s^2}) + \ell_{Dice}(p_{spa}, \hat{y}_{s^2})\}}_{\hat{y}_{s^2} \rightarrow f_{spa}} + \underbrace{\{\ell_{CE}(l_{spe}, \hat{y}_{s^2}) + \ell_{Dice}(p_{spe}, \hat{y}_{s^2})\}}_{\hat{y}_{s^2} \rightarrow f_{spe}}. \quad (8)$$

Holistically, our hybrid loss supervision can be stated as

$$\mathcal{L}_{hybrid} = \mathcal{L}_{scrib} + \lambda_{mt} \times \mathcal{L}_{mt} + \lambda_{el} \times \mathcal{L}_{el}, \quad (9)$$

where λ_{mt} and λ_{el} serve as weighting coefficients that regulate the relative significance of various modes of supervision. The hybrid loss considers all possible supervision signals in the spatial-spectral dual-branch network and exceeds partial combinations of its constituent elements, as evidenced in the ablation study.

3 Experiments

3.1 Experimental Setup

Datasets. We employ the SUN-SEG [10] dataset with scribble annotations for training and assessing the in-distribution performance. This dataset is based on the SUN database [16], which contains 100 different polyp video cases. To reduce data redundancy and memory consumption, we choose the first of every five consecutive frames in each case. We then randomly split the data into 70, 10, and 20 cases for training, validation, and testing, leaving 6677, 1240, and 1993 frames in the respective split. For out-of-distribution evaluation, we utilize three public datasets, namely Kvasir-SEG [9], CVC-ClinicDB [2], and PolypGen [1] with 1000, 612, and 1537 polyp frames, respectively. These datasets are collected from diversified patients in multiple medical centers with various data acquisition systems. Varying data shifts and corruption like motion blur and specular reflections[2] pose significant challenges to model generalization and robustness.

Implementation Details. We implement our method with PyTorch [18] and run the experiments on a single NVIDIA RTX3090 GPU. The SGD optimizer is utilized for training 30k iterations with a momentum of 0.9, a weight decay of 0.0001, and a batch size of 16. The execution time for each experiment is approximately 4 h. The initial learning rate is 0.03 and updated with the poly-scheduling policy [15]. The loss weighting coefficients λ_{mt} and λ_{el} are empirically set the same and exponentially ramped up [3] from 0 to 5 in 25k iterations. All the images are randomly cropped at the border with maximally 7 pixels and resized to 224×224 in width and height. Besides, random horizontal and vertical flipping are applied with a probability of 0.5, respectively.

We utilize UNet [20] and YNet [6] as the respective segmentation model in the spatial and spectral branches. The performance of the scribble-supervised model with partial Cross Entropy [13] loss (Scrib-pCE) and the fully-supervised

[2] Some exemplary polyp frames are presented in the supplementary materials.

model with Cross Entropy loss (Fully-CE) are treated as the lower and upper bound, respectively. Five classical and relevant methods, including EntMin [7], GCRF [17], USTM [14], CPS [3], and DMPLS [15] are employed as the comparative baselines and implemented with UNet [20] as the segmentation backbone referring to the WSL4MIS[3] repository. For a fair comparison, the output from the spatial branch is taken as the final prediction and utilized in evaluation without post-processing. In addition, statistical evaluations are conducted with multiple seeds, and the mean and standard deviations of the results are reported.

3.2 Results and Analysis

Table 1. Quantitative comparison of the in-distribution segmentation performance. The shaded grey and blue rows are the lower and upper bound. The best results of the scribble-supervised methods are in bold.

Method	SUN-SEG [10]			
	DSC ↑	IoU ↑	Prec ↑	HD ↓
Scrib-pCE [13]	0.633±0.010	0.511±0.012	0.636±0.021	5.587±0.149
EntMin [7]	0.642±0.012	0.519±0.013	0.666±0.016	5.277±0.063
GCRF [17]	0.656±0.019	0.541±0.022	0.690±0.017	4.983±0.089
USTM [14]	0.654±0.008	0.533±0.009	0.663±0.011	5.207±0.138
CPS [3]	0.658±0.004	0.539±0.005	0.676±0.005	5.092±0.063
DMPLS [15]	0.656±0.006	0.539±0.005	0.659±0.011	5.208±0.061
S²ME (Ours)	**0.674±0.003**	**0.565±0.001**	**0.719±0.003**	**4.583±0.014**
Fully-CE	0.713±0.021	0.617±0.023	0.746±0.027	4.405±0.119

The performance of weakly-supervised methods is assessed with four metrics, *i.e.*, Dice Similarity Coefficient (DSC), Intersection over Union (IoU), Precision (Prec), and a distance-based measure of Hausdorff Distance (HD). As shown in Table 1 and Fig. 2, our S²ME achieves superior in-distribution performance quantitatively and qualitatively compared with other baselines on the SUN-SEG [10] dataset. Regarding generalization and robustness, as indicated in Table 2, our method outperforms other weakly-supervised methods by a significant margin on three unseen datasets, and even exceeds the fully-supervised upper bound on two of them[4]. These results suggest the efficacy and reliability of the proposed solution S²ME in fulfilling polyp segmentation tasks with only scribble annotations. Notably, the encouraging performance on unseen datasets exhibits promising clinical implications in deploying our method to real-world scenarios.

[3] https://github.com/HiLab-git/WSL4MIS.

[4] Complete results of all four metrics are present in the supplementary materials.

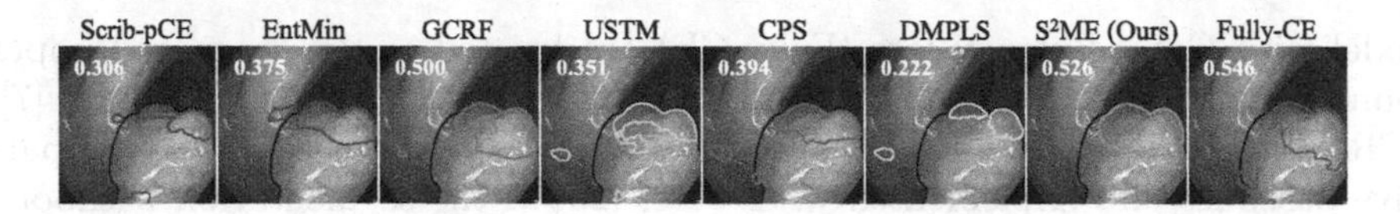

Fig. 2. Qualitative performance comparison of one camouflaged polyp image with DSC values on the left top. The contour of the ground-truth mask is displayed in black, in comparison with that of each method shown in different colors.

Table 2. Generalization comparison on three unseen datasets. The underlined results surpass the upper bound.

Method	Kvasir-SEG [9]		CVC-ClinicDB [2]		PolypGen [1]	
	DSC ↑	HD ↓	DSC ↑	HD ↓	DSC ↑	HD ↓
Scrib-pCE [13]	0.679±0.010	6.565±0.173	0.573±0.016	6.497±0.156	0.524±0.012	6.084±0.189
EntMin [7]	0.684±0.004	6.383±0.110	0.578±0.016	6.308±0.254	0.542±0.003	5.887±0.063
GCRF [17]	0.702±0.004	6.024±0.014	0.558±0.008	6.192±0.290	0.530±0.006	5.714±0.133
USTM [14]	0.693±0.005	6.398±0.138	0.587±0.019	5.950±0.107	0.538±0.007	5.874±0.068
CPS [3]	0.703±0.011	6.323±0.062	0.591±0.017	6.161±0.074	0.546±0.013	5.844±0.065
DMPLS [15]	0.707±0.006	6.297±0.077	0.593±0.013	6.194±0.028	0.547±0.007	5.897±0.045
S²ME (Ours)	**0.750±0.003**	**5.449±0.150**	**0.632±0.010**	**5.633±0.008**	**0.571±0.002**	**5.247±0.107**
Fully-CE	0.758±0.013	5.414±0.097	0.631±0.026	6.017±0.349	0.569±0.016	5.252±0.128

3.3 Ablation Studies

Network Structures. We first conduct the ablation analysis on the network components. As shown in Table 3, the spatial-spectral configuration of our S^{2}ME yields superior performance compared to single-domain counterparts with ME, confirming the significance of utilizing cross-domain features.

Table 3. Ablation comparison of dual-branch network architectures. Results are from outputs of Model-1 on the SUN-SEG [10] dataset.

Model-1	Model-2	Method	DSC ↑	IoU ↑	Prec ↑	HD ↓
UNet [20]	UNet [20]	ME (Ours)	0.666 ± 0.002	0.557 ± 0.002	0.715 ± 0.008	4.684 ± 0.034
YNet [6]	YNet [6]		0.648 ± 0.004	0.538 ± 0.005	0.695 ± 0.004	4.743 ± 0.006
UNet [20]	YNet [6]	S²ME (Ours)	**0.674 ± 0.003**	**0.565 ± 0.001**	**0.719 ± 0.003**	**4.583 ± 0.014**

Pseudo Label Fusion Strategies. To ensure the reliability of the mixed pseudo labels for ensemble learning, we present the pixel-level adaptive fusion strategy according to entropy maps of dual predictions to balance the strengths and weaknesses of spatial and spectral branches. As demonstrated in Table 4, our method achieves improved performance compared to two image-level fusion strategies, *i.e.*, random [15] and equal mixing.

Hybrid Loss Supervision. We decompose the proposed hybrid loss $\mathcal{L}_{hybrid}$ in Eq. (9) to demonstrate the effectiveness of holistic supervision from scribbles,

Table 4. Ablation on the pseudo label fusion strategies on the SUN-SEG [10] dataset.

Fusion		Metrics	
Strategy	Level	DSC ↑	HD ↓
Random [15]	Image	0.665 ± 0.008	4.750 ± 0.169
Equal (0.5)	Image	0.667 ± 0.001	4.602 ± 0.013
Entropy (Ours)	Pixel	**0.674 ± 0.003**	**4.583 ± 0.014**

Table 5. Ablation study on the loss components on the SUN-SEG [10] dataset.

Loss			Metrics	
$\mathcal{L}_{scrib}$	$\mathcal{L}_{mt}$	$\mathcal{L}_{el}$	DSC ↑	HD ↓
✓	✗	✗	0.627 ± 0.004	5.580 ± 0.112
✓	✓	✗	0.668 ± 0.007	4.782 ± 0.020
✓	✗	✓	0.662 ± 0.004	4.797 ± 0.146
✓	✓	✓	**0.674 ± 0.003**	**4.583 ± 0.014**

mutual teaching, and ensemble learning. As shown in Table 5, our proposed hybrid loss, involving $\mathcal{L}_{scrib}$, $\mathcal{L}_{mt}$, and $\mathcal{L}_{el}$, achieves the optimal results.

4 Conclusion

To our best knowledge, we propose the first spatial-spectral dual-branch network structure for weakly-supervised medical image segmentation that efficiently leverages cross-domain patterns with collaborative mutual teaching and ensemble learning. Our pixel-level entropy-guided fusion strategy advances the reliability of the aggregated pseudo labels, which provides valuable supplementary supervision signals. Moreover, we optimize the segmentation model with the hybrid mode of loss supervision from scribbles and pseudo labels in a holistic manner and witness improved outcomes. With extensive in-domain and out-of-domain evaluation on four public datasets, our method shows superior accuracy, generalization, and robustness, indicating its clinical significance in alleviating data-related issues such as data shift and corruption which are commonly encountered in the medical field. Future efforts can be paid to apply our approach to other annotation-efficient learning contexts like semi-supervised learning, other sparse annotations like points, and more medical applications.

Acknowledgements. This work was supported by Hong Kong Research Grants Council (RGC) Collaborative Research Fund (CRF C4063-18G), the Shun Hing Institute of Advanced Engineering (SHIAE project BME-p1-21) at the Chinese University of Hong Kong (CUHK), General Research Fund (GRF 14203323), Shenzhen-Hong Kong-Macau Technology Research Programme (Type C) STIC Grant SGDX20210823103535014 (202108233000303), and (GRS) #3110167.

References

1. Ali, S., et al.: a multi-centre polyp detection and segmentation dataset for generalisability assessment. Sci. Data **10**(1), 75 (2023)
2. Bernal, J., Sánchez, F.J., Fernández-Esparrach, G., Gil, D., Rodríguez, C., Vilariño, F.: WM-DOVA maps for accurate polyp highlighting in colonoscopy: Validation vs. saliency maps from physicians. Comput. Med. Imaging Graph. **43**, 99–111 (2015)

3. Chen, X., Yuan, Y., Zeng, G., Wang, J.: Semi-supervised semantic segmentation with cross pseudo supervision. In: Proceedings of the IEEE/CVF Conference on Computer Vision and Pattern Recognition, pp. 2613–2622 (2021)
4. Chi, L., Jiang, B., Mu, Y.: Fast Fourier convolution. Adv. Neural. Inf. Process. Syst. **33**, 4479–4488 (2020)
5. Cinbis, R.G., Verbeek, J., Schmid, C.: Weakly supervised object localization with multi-fold multiple instance learning. IEEE Trans. Pattern Anal. Mach. Intell. **39**(1), 189–203 (2016)
6. Farshad, A., Yeganeh, Y., Gehlbach, P., Navab, N.: Y-net: a spatiospectral dual-encoder network for medical image segmentation. In: Wang, L., Dou, Q., Fletcher, P.T., Speidel, S., Li, S. (eds.) Medical Image Computing and Computer Assisted Intervention - MICCAI 2022. Lecture Notes in Computer Science, vol. 13432, pp. 582–592. Springer, Cham (2022)
7. Grandvalet, Y., Bengio, Y.: Semi-supervised learning by entropy minimization. In: Advances in Neural Information Processing Systems, vol. 17 (2004)
8. He, X., Fang, L., Tan, M., Chen, X.: Intra-and inter-slice contrastive learning for point supervised oct fluid segmentation. IEEE Trans. Image Process. **31**, 1870–1881 (2022)
9. Jha, D., et al.: Kvasir-SEG: a segmented polyp dataset. In: Ro, Y.M., et al. (eds.) MMM 2020. LNCS, vol. 11962, pp. 451–462. Springer, Cham (2020). https://doi.org/10.1007/978-3-030-37734-2_37
10. Ji, G.P., et al.: Video polyp segmentation: a deep learning perspective. Mach. Intell. Res. 1–19 (2022)
11. Lee, H., Jeong, W.-K.: Scribble2Label: scribble-supervised cell segmentation via self-generating pseudo-labels with consistency. In: Martel, A.L., et al. (eds.) MICCAI 2020. LNCS, vol. 12261, pp. 14–23. Springer, Cham (2020). https://doi.org/10.1007/978-3-030-59710-8_2
12. Li, Y., Xue, Y., Li, L., Zhang, X., Qian, X.: Domain adaptive box-supervised instance segmentation network for mitosis detection. IEEE Trans. Med. Imaging **41**(9), 2469–2485 (2022)
13. Lin, D., Dai, J., Jia, J., He, K., Sun, J.: Scribblesup: scribble-supervised convolutional networks for semantic segmentation. In: Proceedings of the IEEE Conference on Computer Vision and Pattern Recognition, pp. 3159–3167 (2016)
14. Liu, X., et al.: Weakly supervised segmentation of covid19 infection with scribble annotation on CT images. Pattern Recogn. **122**, 108341 (2022)
15. Luo, X., et al.: Scribble-supervised medical image segmentation via dual-branch network and dynamically mixed pseudo labels supervision. In: Wang, L., Dou, Q., Fletcher, P.T., Speidel, S., Li, S. (eds.) MICCAI 2022. LNCS, vol. 13431, pp. 528–538. Springer, Cham (2022). https://doi.org/10.1007/978-3-031-16431-6_50
16. Misawa, M., et al.: Development of a computer-aided detection system for colonoscopy and a publicly accessible large colonoscopy video database (with video). Gastrointest. Endosc. **93**(4), 960–967 (2021)
17. Obukhov, A., Georgoulis, S., Dai, D., Van Gool, L.: Gated CRF loss for weakly supervised semantic image segmentation. arXiv preprint arXiv:1906.04651 (2019)
18. Paszke, A., et al.: Automatic differentiation in pyTorch. In: NIPS-W (2017)
19. Rao, Y., Zhao, W., Zhu, Z., Lu, J., Zhou, J.: Global filter networks for image classification. Adv. Neural. Inf. Process. Syst. **34**, 980–993 (2021)
20. Ronneberger, O., Fischer, P., Brox, T.: U-net: convolutional networks for biomedical image segmentation (2015)

21. Valvano, G., Leo, A., Tsaftaris, S.A.: Learning to segment from scribbles using multi-scale adversarial attention gates. IEEE Trans. Med. Imaging **40**(8), 1990–2001 (2021)
22. Wang, A., Islam, M., Xu, M., Ren, H.: Rethinking surgical instrument segmentation: a background image can be all you need. In: Wang, L., Dou, Q., Fletcher, P.T., Speidel, S., Li, S. (eds.) MICCAI 2022. LNCS, vol. 13437, pp. 355–364. Springer, Cham (2022). https://doi.org/10.1007/978-3-031-16449-1_34
23. Wang, K.N., et al.: Ffcnet: Fourier transform-based frequency learning and complex convolutional network for colon disease classification. In: Wang, L., Dou, Q., Fletcher, P.T., Speidel, S., Li, S. (eds.) MICCAI 2022. LNCS, vol. 13433, pp. 78–87. Springer, Cham (2022). https://doi.org/10.1007/978-3-031-16437-8_8
24. Wang, Y., et al.: Freematch: self-adaptive thresholding for semi-supervised learning. arXiv preprint arXiv:2205.07246 (2022)
25. Wu, Y., Xu, M., Ge, Z., Cai, J., Zhang, L.: Semi-supervised left atrium segmentation with mutual consistency training. In: de Bruijne, M., et al. (eds.) MICCAI 2021. LNCS, vol. 12902, pp. 297–306. Springer, Cham (2021). https://doi.org/10.1007/978-3-030-87196-3_28
26. Xu, K., Qin, M., Sun, F., Wang, Y., Chen, Y.K., Ren, F.: Learning in the frequency domain. In: Proceedings of the IEEE/CVF Conference on Computer Vision and Pattern Recognition, pp. 1740–1749 (2020)
27. Xu, M., Islam, M., Lim, C.M., Ren, H.: Class-incremental domain adaptation with smoothing and calibration for surgical report generation. In: de Bruijne, M., et al. (eds.) MICCAI 2021. LNCS, vol. 12904, pp. 269–278. Springer, Cham (2021). https://doi.org/10.1007/978-3-030-87202-1_26
28. Xu, M., Islam, M., Lim, C.M., Ren, H.: Learning domain adaptation with model calibration for surgical report generation in robotic surgery. In: 2021 IEEE International Conference on Robotics and Automation (ICRA), pp. 12350–12356. IEEE (2021)
29. Zhang, K., Zhuang, X.: Cyclemix: a holistic strategy for medical image segmentation from scribble supervision. In: Proceedings of the IEEE/CVF Conference on Computer Vision and Pattern Recognition, pp. 11656–11665 (2022)
30. Zhang, K., Zhuang, X.: ShapePU: a new PU learning framework regularized by global consistency for scribble supervised cardiac segmentation. In: Wang, L., Dou, Q., Fletcher, P.T., Speidel, S., Li, S. (eds.) MICCAI 2022. LNCS, vol. 13438, pp. 162–172. Springer, Cham (2022). https://doi.org/10.1007/978-3-031-16452-1_16

Modularity-Constrained Dynamic Representation Learning for Interpretable Brain Disorder Analysis with Functional MRI

Qianqian Wang[1], Mengqi Wu[1], Yuqi Fang[1], Wei Wang[2], Lishan Qiao[3(✉)], and Mingxia Liu[1(✉)]

[1] Department of Radiology and BRIC, University of North Carolina at Chapel Hill, Chapel Hill, NC 27599, USA
mingxia_liu@med.unc.edu

[2] Department of Radiology, Beijing Youan Hospital, Capital Medical University, Beijing 100069, China

[3] School of Mathematics Science, Liaocheng University, Shandong 252000, China
qiaolishan@lcu.edu.cn

Abstract. Resting-state functional MRI (rs-fMRI) is increasingly used to detect altered functional connectivity patterns caused by brain disorders, thereby facilitating objective quantification of brain pathology. Existing studies typically extract fMRI features using various machine/deep learning methods, but the generated imaging biomarkers are often challenging to interpret. Besides, the brain operates as a modular system with many cognitive/topological modules, where each module contains subsets of densely inter-connected regions-of-interest (ROIs) that are sparsely connected to ROIs in other modules. However, current methods cannot effectively characterize brain modularity. This paper proposes a modularity-constrained dynamic representation learning (MDRL) framework for interpretable brain disorder analysis with rs-fMRI. The MDRL consists of 3 parts: (1) dynamic graph construction, (2) modularity-constrained spatiotemporal graph neural network (MSGNN) for dynamic feature learning, and (3) prediction and biomarker detection. In particular, the MSGNN is designed to learn spatiotemporal dynamic representations of fMRI, constrained by 3 functional modules (*i.e.*, central executive network, salience network, and default mode network). To enhance discriminative ability of learned features, we encourage the MSGNN to reconstruct network topology of input graphs. Experimental results on two public and one private datasets with a total of 1,155 subjects validate that our MDRL outperforms several state-of-the-art methods in fMRI-based brain disorder analysis. The detected fMRI biomarkers have good explainability and can be potentially used to improve clinical diagnosis.

Supplementary Information The online version contains supplementary material available at https://doi.org/10.1007/978-3-031-43907-0_5.

H. Greenspan et al. (Eds.): MICCAI 2023, LNCS 14220, pp. 46–56, 2023.
https://doi.org/10.1007/978-3-031-43907-0_5

Keywords: Functional MRI · Modularity · Biomarker · Brain disorder

1 Introduction

Resting-state functional magnetic resonance imaging (rs-fMRI) has been increasingly used to help us understand pathological mechanisms of neurological disorders by revealing abnormal or dysfunctional brain connectivity patterns [1–4]. Brain regions-of-interest (ROIs) or functional connectivity (FC) involved in these patterns can be used as potential biomarkers to facilitate objective quantification of brain pathology [5]. Previous studies have designed various machine and deep learning models to extract fMRI features and explore disease-related imaging biomarkers [6,7]. However, due to the complexity of brain organization and black-box property of many learning-based models, the generated biomarkers are usually difficult to interpret, thus limiting their utility in clinical practice [8,9].

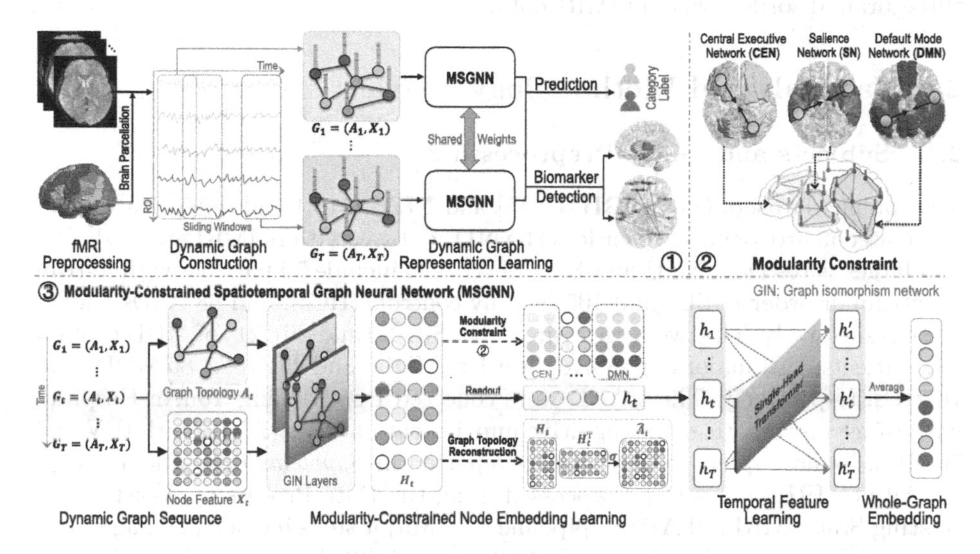

Fig. 1. Illustration of our modularity-constrained dynamic representation learning (MDRL) framework, with 3 components: (1) dynamic graph construction via sliding windows, (2) modularity-constrained spatiotemporal graph neural network (MSGNN) for dynamic representation learning, and (3) prediction and biomarker detection. The MSGNN is designed to learn spatiotemporal features via GIN and transformer layers, constrained by 3 neurocognitive modules (*i.e.*, central executive network, salience network, and default mode network) and graph topology reconstruction.

On the other hand, the human brain operates as a modular system, where each module contains a set of ROIs that are densely connected within the module but sparsely connected to ROIs in other modules [10,11]. In particular, the central executive network (CEN), salience network (SN), and default mode network (DMN) are three prominent resting-state neurocognitive modules in the brain, supporting efficient cognition [12]. Unfortunately, existing learning-based fMRI studies usually ignore such inherent modular brain structures [13,14].

To this end, we propose a modularity-constrained dynamic representation learning (**MDRL**) framework for interpretable brain disorder analysis with rs-fMRI. As shown in Fig. 1, the MDRL consists of (1) dynamic graph construction, (2) modularity-constrained spatiotemporal graph neural network (MSGNN) for dynamic graph representation learning, and (3) prediction and biomarker detection. The MSGNN is designed to learn spatiotemporal features via graph isomorphism network and transformer layers, constrained by 3 neurocognitive modules (*i.e.*, central executive network, salience network, and default mode network). To enhance discriminative ability of learned fMRI embeddings, we also encourage the MSGNN to reconstruct topology of input graphs. To our knowledge, this is among the first attempts to incorporate modularity prior to graph learning models for fMRI-based brain disorder analysis. Experimental results on two public and one private datasets validate the effectiveness of the MDRL in detecting three brain disorders with rs-fMRI data.

2 Materials and Methodology

2.1 Subjects and Image Preprocessing

Two public datasets (*i.e.*, **ABIDE** [15] and **MDD** [16]) and one private HIV-associated neurocognitive disorder (**HAND**) dataset with rs-fMRI are used. The two largest sites (*i.e.*, NYU and UM) of ABIDE include 79 patients with autism spectrum disorder (ASD) and 105 healthy controls (HCs), and 68 ASDs and 77 HCs, respectively. The two largest sites (*i.e.*, Site 20 and Site 21) of MDD contain 282 patients with major depressive disorder (MDD) and 251 HCs, 86 MDDs and 70 HCs, respectively. The HAND were collected from Beijing YouAn Hospital, with 67 asymptomatic neurocognitive impairment patients (ANIs) with HIV and 70 HCs. Demographics of subjects are reported in *Supplementary Materials*.

All rs-fMRI data were preprocessed using the Data Processing Assistant for Resting-State fMRI (DPARSF) pipeline [17]. Major steps include (1) magnetization equilibrium by trimming the first 10 volumes, (2) slice timing correction and head motion correction, (3) regression of nuisance covariates (*e.g.*, white matter signals, ventricle, and head motion parameters), (4) spatial normalization to the MNI space, (5) bandpass filtering (0.01–0.10 Hz). The average rs-fMRI time series of 116 ROIs defined by the AAL atlas are extracted for each subject.

2.2 Proposed Method

As shown in Fig. 1, the proposed MDRL consists of (1) dynamic graph construction via sliding windows, (2) MSGNN for dynamic graph representation learning, and (3) prediction and biomarker detection, with details introduced below.

Dynamic Graph Construction. Considering that brain functional connectivity (FC) patterns change dynamically over time [18], we propose to first construct dynamic networks/graphs using a sliding window strategy for each subject. Denote original fMRI time series as $S \in R^{N \times M}$, where N is the number of ROIs and M is the number of time points of blood-oxygen-level-dependent (BOLD) signals in rs-fMRI. We first divide the original time series into T segments along the temporal dimension via overlapped sliding windows, with the window size of Γ and the step size of τ. Then, we construct an FC network by calculating Pearson correlation (PC) coefficients between time series of pairwise ROIs for each of T segments, denoted as $X_t \in R^{N \times N} (t = 1, \cdots, T)$. The original feature for the j-th node is represented by the j-th row in X_t for segment t. Considering all connections in an FC network may include some noisy or redundant information, we retain the top 30% strongest edges in each FC network to generate an adjacent matrix $A_t \in \{0, 1\}^{N \times N}$ for segment t. Thus, the obtained dynamic graph sequence of each subject can be described as $G_t = \{A_t, X_t\}$ $(t = 1, \cdots, T)$.

Modularity-Constrained Spatiotemporal GNN. With the dynamic graph sequence $\{G_t\}_{t=1}^{T}$ as input, we design a modularity-constrained spatiotemporal graph neural network (MSGNN) to learn interpretable and discriminative graph embeddings, with two unique constraints: 1) a *modularity constraint*, and 2) a *graph topology reconstruction constraint*. In MSGNN, we first stack two graph isomorphism network (GIN) layers [19] for node-level feature learning. The node-level embedding H_t at the segment t learned by GIN layers is formulated as:

$$H_t = \psi(\varepsilon^{(1)} I + A_t)[\psi(\varepsilon^{(0)} I + A_t) X_t W^{(0)}] W^{(1)} \tag{1}$$

where ψ is nonlinear activation, $\varepsilon^{(i)}$ is a parameter at the i-th GIN layer, I is an identity matrix, and $W^{(i)}$ is the weight for the fully connected layers in GIN.

1) Modularity Constraint. It has been demonstrated that the central executive network (CEN), salience network (SN) and default mode network (DMN) are three crucial neurocognitive modules in the brain and these three modules have been consistently observed across different individuals and experimental paradigms, where CEN performs high-level cognitive tasks (*e.g.*, decision-making and rule-based problem-solving), SN mainly detects external stimuli and coordinates brain neural resources, and DMN is responsible for self-related cognitive functions [10–12]. The ROIs/nodes within a module are densely inter-connected, resulting in a high degree of clustering between nodes from the same module. Based on such prior knowledge and clinical experience, we reasonably assume that the *learned embeddings of nodes within the same neurocognitive module tend to be similar*. We develop a novel modularity constraint to encourage similarity between paired node-level embeddings in the same module. Mathematically, the proposed *modularity constraint* is formulated as:

$$L_M = -\sum_{t=1}^{T} \sum_{k=1}^{K} \sum_{i,j=1}^{N_k} \frac{h_i^{t,k} \cdot h_j^{t,k}}{\|h_i^{t,k}\| \cdot \|h_j^{t,k}\|} \tag{2}$$

where $h_i^{t,k}$ and $h_j^{t,k}$ are embeddings of two nodes in the k-th module (with N_k ROIs) at segment t, and K is the number of modules ($K = 3$ in this work). With Eq. (2), we encourage the MSGNN to focus on modular brain structures during representation learning, thus improving discriminative ability of fMRI features.

2) Graph Topology Reconstruction Constraint. To further enhance discriminative ability of learned embeddings, we propose to preserve graph topology by reconstructing adjacent matrices. For the segment t, its adjacent matrix A_t can be reconstructed through $\hat{A}_t = \sigma(H_t \cdot H_t^\top)$, where σ is a nonlinear mapping function. The *graph topology reconstruction constraint* is then formulated as:

$$L_R = \sum_{t=1}^{T} \Psi(A_t, \hat{A}_t) \tag{3}$$

where Ψ is a cross-entropy loss function. We then apply an SERO operation [20] to generate graph-level embeddings based on node-level embeddings, formulated as $h_t = H_t \Phi(P^{(2)} \sigma(P^{(1)} H_t \phi_{mean}))$, where Φ is a sigmoid function, $P^{(1)}$ and $P^{(2)}$ are learnable weight matrices, and ϕ_{mean} is average operation.

3) Temporal Feature Learning. To further capture temporal information, a single-head transformer is used to fuse features derived from T segments, with a self-attention mechanism to model temporal dynamics across segments. We then sum the learned features $\{h_i^{'}\}_{i=1}^{T}$ to obtain the whole-graph embedding.

Prediction and Biomarker Detection. The whole-graph embedding is fed into a fully connected layer with Softmax for prediction, with final loss defined as:

$$L = L_C + \lambda_1 L_R + \lambda_2 L_M \tag{4}$$

where L_C is a cross-entropy loss for prediction, and λ_1 and λ_2 are two hyper-parameters. To facilitate interpretation of our learned graph embeddings, we calculate PC coefficients between paired node embeddings for each segment and average them across segments to obtain an FC network for each subject. The upper triangle of each FC network is flattened into a vector and Lasso [21] (with default parameter) is used to select discriminative features. Finally, we map these features to the original feature space to detect disease-related ROIs and FCs.

Implementation. The MDRL is implemented in PyTorch and trained using an Adam optimizer (with learning rate of 0.001, training epochs of 30, batch size of 8 and $\tau = 20$). We set window size $\Gamma = 40$ for NYU and $\Gamma = 70$ for the rest, and results of MDRL with different Γ values are shown in *Supplementary Materials.* In the modularity constraint, we randomly select $m = 50\%$ of all $\frac{N_k(N_k-1)}{2}$ paired ROIs in the k-th module (with N_k ROIs) to constrain the MDRL.

3 Experiment

Competing Methods. We compare the MDRL with 2 shallow methods: 1) linear **SVM** with node-level statistics (*i.e.*, degree centrality, clustering coefficient, betweenness centrality, and eigenvector centrality) of FC networks as fMRI features (with each FC network constructed using PC), 2) **XGBoost** with the same features as SVM; and 4 state-of-the-art (SOTA) deep models with default architectures: 3) **GCN** [22], 4) **GAT** [23], 5) **BrainGNN** [9], and 6) **STGCN** [18].

Table 1. Results of seven methods on ABIDE.

Method	ASD vs. HC classification on NYU					ASD vs. HC classification on UM				
	AUC (%)	ACC (%)	SEN (%)	SPE (%)	BAC (%)	AUC (%)	ACC (%)	SEN (%)	SPE (%)	BAC (%)
SVM	56.6(2.9)*	54.8(3.1)	51.5(4.6)	57.9(4.7)	54.7(3.1)	53.6(4.3)*	53.3(3.6)	50.3(4.5)	56.6(4.7)	53.5(3.8)
XGBoost	61.9(0.6)*	63.0(1.6)	48.0(2.7)	**75.9(3.7)**	61.9(0.6)	58.8(0.8)*	58.6(1.9)	47.6(3.6)	70.0(3.9)	58.8(0.8)
GCN	67.5(3.3)*	63.6(3.1)	51.0(5.1)	73.5(4.7)	62.3(2.8)	66.7(2.6)	60.0(3.0)	54.6(4.4)	66.5(5.2)	60.6(2.6)
GAT	64.9(2.6)*	60.1(2.6)	53.0(4.9)	66.1(3.2)	59.5(2.8)	66.5(3.5)*	60.4(3.1)	**56.1(3.1)**	65.4(6.9)	60.7(2.8)
BrainGNN	66.9(2.9)*	63.2(3.2)	**57.1(4.8)**	68.5(3.0)	62.8(3.2)	65.9(2.5)	62.7(2.6)	55.5(3.3)	68.1(5.7)	61.8(2.1)
STGCN	66.6(0.8)*	61.5(1.5)	53.6(2.3)	68.4(1.7)	61.0(1.5)	64.0(0.1)*	63.9(0.1)	55.9(1.1)	72.1(0.5)	64.0(0.1)
MDRL (Ours)	**72.6(1.7)**	**65.6(2.1)**	57.0(2.8)	74.1(3.1)	**65.6(1.9)**	**67.1(2.3)**	**64.5(1.4)**	55.6(3.9)	**72.7(2.5)**	**64.1(2.4)**

The term '*' denotes the results of MDRL and a competing method are statistically significantly different ($p < 0.05$).

Table 2. Results of seven methods on MDD.

Method	MDD vs. HC classification on Site 20					MDD vs. HC classification on Site 21				
	AUC (%)	ACC (%)	SEN (%)	SPE (%)	BAC (%)	AUC (%)	ACC (%)	SEN (%)	SPE (%)	BAC (%)
SVM	53.7(2.1)	53.0(2.6)	54.0(3.1)	51.5(1.8)	53.0(2.4)	52.8(3.0)*	52.7(2.7)	59.0(3.8)	45.8(4.7)	52.4(2.8)
XGBoost	55.9(1.9)*	55.9(2.2)	**64.4(4.1)**	47.4(1.3)	55.9(1.9)	52.0(2.3)*	52.5(2.7)	**66.2(4.7)**	37.8(4.6)	52.0(2.3)
GCN	55.7(2.7)	54.9(2.1)	59.6(4.4)	50.1(4.4)	54.8(2.0)	54.8(3.1)*	54.0(3.0)	60.9(6.0)	46.6(7.5)	53.8(3.3)
GAT	57.8(1.3)*	55.7(1.3)	61.4(5.7)	49.5(4.4)	55.4(1.1)	53.2(3.1)*	52.8(2.4)	62.0(4.7)	42.7(7.1)	52.3(3.4)
BrainGNN	56.3(2.4)*	52.8(2.1)	51.7(7.6)	**55.0(8.5)**	53.4(2.0)	53.9(3.3)	53.5(8.6)	58.4(12.8)	45.7(2.0)	52.1(5.4)
STGCN	54.2(4.5)*	54.6(4.1)	56.6(5.2)	52.2(2.9)	54.4(4.0)	54.9(0.3)*	53.4(0.1)	61.4(1.8)	44.2(3.7)	52.7(0.9)
MDRL (Ours)	**60.9(2.6)**	**57.4(1.9)**	62.2(5.0)	51.6(3.3)	**56.9(1.4)**	**56.6(9.4)**	**55.2(7.8)**	58.1(10.2)	**51.3(7.1)**	**54.6(7.8)**

The term '*' denotes the results of MDRL and a competing method are statistically significantly different ($p < 0.05$).

Experimental Setting. Three classification tasks are performed: 1) ASD vs. HC on ABIDE, 2) MDD vs. HC on MDD, and 3) ANI vs. HC on HAND. A 5-fold cross-validation (CV) strategy is employed. Within each fold, we also perform an inner 5-fold CV to select optimal parameters. Five evaluation metrics are used: area under ROC curve (AUC), accuracy (ACC), sensitivity (SEN), specificity (SPE), and balanced accuracy (BAC). Paired sample t-test is performed to evaluate whether the MDRL is significantly different from a competing method.

Classification Results. Results achieved by different methods in three classification tasks on three datasets are reported in Tables 1–2 and Fig. 2. It can be seen that our MDRL generally outperforms two shallow methods (*i.e.*, SVM and XGBoost) that rely on handcrafted node features without modeling whole-graph topological information. Compared with 4 SOTA deep learning methods, our MDRL achieves superior performance in terms of most metrics in three tasks. For instance, for ASD vs. HC classification on NYU of ABIDE (see Table 1), the

AUC value of MDRL is improved by 5.7% compared with BrainGNN (a SOTA method designed for brain network analysis). This implies the MDRL can learn discriminative graph representations to boost fMRI-based learning performance.

Ablation Study. We compare the proposed MDRL with its three degenerated variants: 1) **MDRLw/oM** without the modularity constraint, 2) **MDRLw/oR** without the graph topology reconstruction constraint, and 3) **MDRLw/oMR** without the two constraints. The results are reported in Fig. 3, from which one can see that MDRL is superior to its three variants, verifying the effectiveness of the two constraints defined in Eqs. (2)–(3). Besides, MDRLw/oM is generally inferior to MDRLw/oR in three tasks, implying that the modularity constraint may contribute more to MDRL than the graph reconstruction constraint.

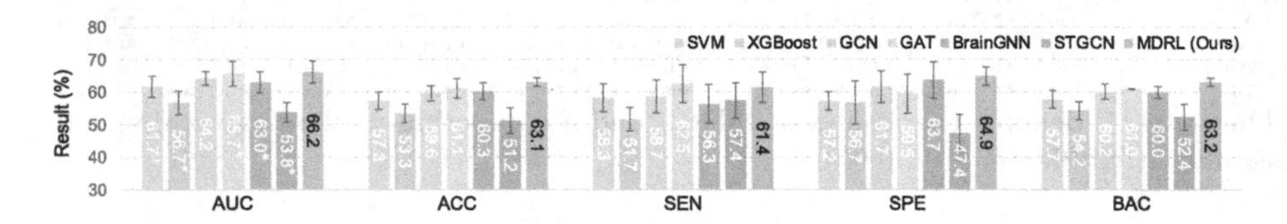

Fig. 2. Results of seven methods in ANI vs. HC classification on HAND.

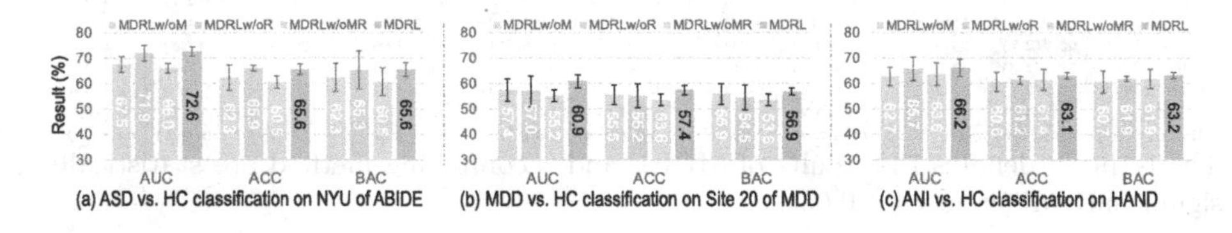

Fig. 3. Performance of the MDRL and its variants in three tasks on three datasets.

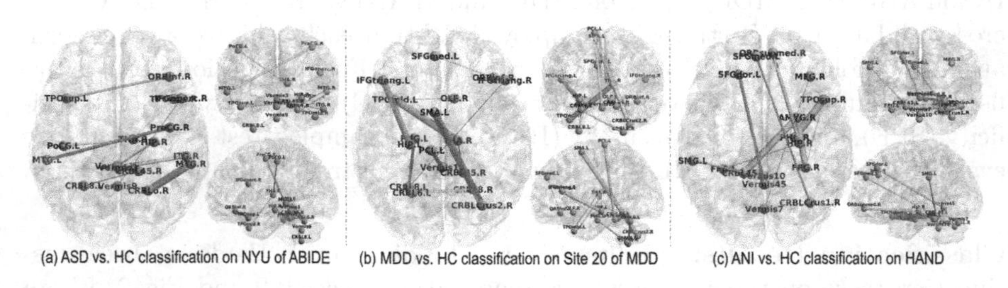

Fig. 4. Visualization of the top 10 most discriminative functional connectivities identified by the MDRL in identifying 3 diseases on 3 datasets (with AAL for ROI partition).

Discriminative ROIs and Functional Connectivities. The top 10 discriminative FCs detected by the MDRL in three tasks are shown in Fig. 4. The thickness of each line represents discriminative power that is proportional to the corresponding Lasso coefficient. For ASD identification (see Fig. 4 (a)), the FCs involved in *thalamus* and *middle temporal gyrus* are frequently identified, which complies with previous findings [24,25]. For MDD detection (see Fig. 4 (b)), we find that several ROIs (*e.g.*, *hippocampus*, *supplementary motor area* and *thalamus*) are highly associated with MDD identification, which coincides with previous studies [26–28]. For ANI identification (see Fig. 4 (c)), the detected ROIs such as *amygdala*, *right temporal pole: superior temporal gyrus* and *parahippocampal gyrus*, are also reported in previous research [29–31]. This further demonstrates the effectiveness of the MDRL in disease-associated biomarker detection.

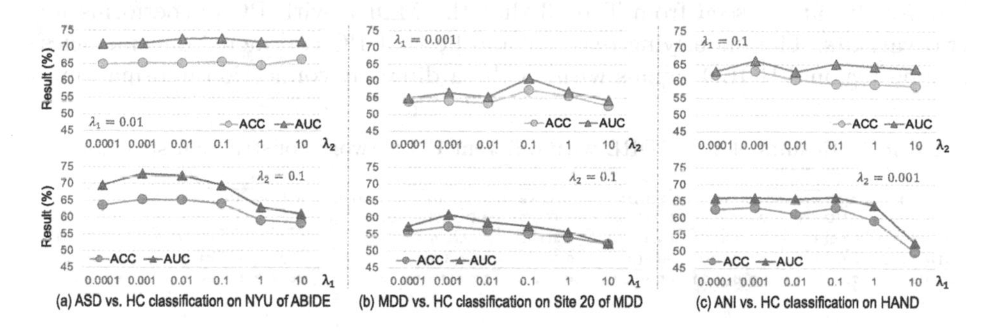

Fig. 5. Results of our MDRL with different hyperparameters (*i.e.*, λ_1 and λ_2).

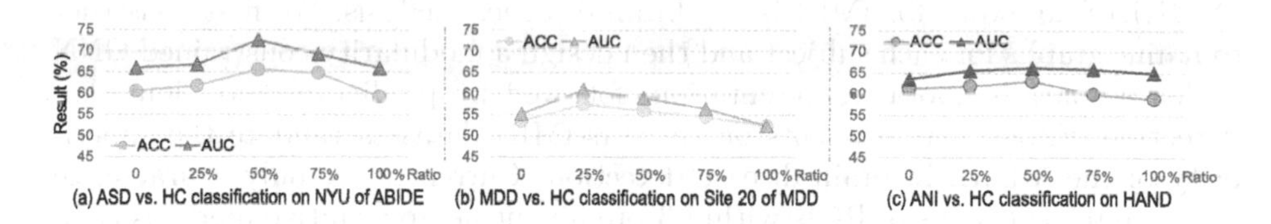

Fig. 6. Results of the proposed MDRL with different modularity ratios.

4 Discussion

Parameter Analysis. To investigate the influence of hyperparameters, we vary the values of two parameters (*i.e.*, λ_1 and λ_2) in Eq. (4) and report the results of MDRL in Fig. 5. It can be seen from Fig. 5 that, with λ_1 fixed, the performance of MDRL exhibits small fluctuations with the increase of parameter values of λ_2, implying that MDRL is not very sensitive to λ_2 in three tasks. With λ_2 fixed, the MDRL with a large λ_1 (*e.g.*, $\lambda_1 = 10$) achieves worse performance. The possible reason could be that using a strong graph reconstruct constraint will make the model difficult to converge, thus degrading its learning performance.

Influence of Modularity Ratio. In the main experiments, we randomly select $m = 50\%$ of all $\frac{N_k(N_k-1)}{2}$ paired ROIs in the k-th module (with N_k ROIs) to constrain the MDRL. We now vary the modularity ratio m within $[0\%, 25\%, \cdots, 100\%]$ and record the results of MDRL in three tasks in Fig. 6. It can be seen from Fig. 6 that, when $m < 75\%$, the ACC and AUC results generally increase as m increases. But when using a large modularity ratio (*e.g.*, $m = 100\%$), the MDRL cannot achieve satisfactory results. This may be due to the over-smoothing problem caused by using a too-strong modularity constraint.

Influence of Network Construction. We use PC to construct the original FC networks in MDRL. We also use sparse representation (SR) and low-rank representation (LR) for network construction in MDRL and report results in Table 3. It can be seen from Table 3 that the MDRL with PC outperforms its two variants. The underlying reason could be that PC can model dependencies among regional BOLD signals without discarding any connection information.

Table 3. Results of the MDRL with different FC network construction strategies.

Method	ASD vs. HC on NYU of ABIDE			MDD vs. HC on Site 20 of MDD			ANI vs. HC on HAND		
	AUC (%)	ACC (%)	BAC (%)	AUC (%)	ACC (%)	BAC (%)	AUC (%)	ACC (%)	BAC (%)
MDRL_LR	62.2(3.5)	59.2(4.6)	59.4(4.6)	54.4(5.0)	53.5(4.9)	53.2(4.8)	58.2(3.3)	60.6(2.0)	60.7(1.8)
MDRL_SR	60.9(1.9)	62.7(1.9)	60.8(2.7)	55.5(7.4)	52.7(3.8)	53.2(4.6)	64.6(4.0)	61.0(2.2)	61.2(2.0)
MDRL	**72.6(1.7)**	**65.6(2.1)**	**65.6(2.0)**	**60.9(2.6)**	**57.4(1.9)**	**56.9(1.4)**	**66.2(3.4)**	**63.1(1.3)**	**63.2(1.3)**

5 Conclusion and Future Work

In this work, we propose a modularity-constrained dynamic graph representation (MDRL) framework for fMRI-based brain disorder analysis. We first construct dynamic graphs for each subject and then design a modularity-constrained GNN to learn spatiotemporal representation, followed by prediction and biomarker detection. Experimental results on three rs-fMRI datasets validate the superiority of the MDRL in brain disease detection. Currently, we only characterize pairwise relationships of ROIs within 3 prominent neurocognitive modules (*i.e.*, CEN, SN, and DMN) as prior knowledge to design the modularity constraint in MDRL. Fine-grained modular structure and disease-specific modularity constraint will be considered. Besides, we will employ advanced harmonization methods [32] to reduce inter-site variance, fully utilizing multi-site fMRI data for model training.

Acknowledgment. M. Wu, Y. Fang, and M. Liu were supported by an NIH grant RF1AG073297.

References

1. Pagani, M., et al.: mTOR-related synaptic pathology causes autism spectrum disorder-associated functional hyperconnectivity. Nat. Commun. **12**(1), 6084 (2021)

2. Sezer, I., Pizzagalli, D.A., Sacchet, M.D.: Resting-state fMRI functional connectivity and mindfulness in clinical and non-clinical contexts: a review and synthesis. Neurosci. Biobehav. Rev. (2022) 104583
3. Liu, J., et al.: Astrocyte dysfunction drives abnormal resting-state functional connectivity in depression. Sci. Adv. **8**(46), eabo2098 (2022)
4. Sahoo, D., Satterthwaite, T.D., Davatzikos, C.: Hierarchical extraction of functional connectivity components in human brain using resting-state fMRI. IEEE Trans. Med. Imaging **40**(3), 940–950 (2020)
5. Traut, N., et al.: Insights from an autism imaging biomarker challenge: promises and threats to biomarker discovery. Neuroimage **255**, 119171 (2022)
6. Azevedo, T., et al.: A deep graph neural network architecture for modelling spatio-temporal dynamics in resting-state functional MRI data. Med. Image Anal. **79**, 102471 (2022)
7. Bessadok, A., Mahjoub, M.A., Rekik, I.: Graph neural networks in network neuroscience. IEEE Trans. Pattern Anal. Mach. Intell. (2022)
8. Zhang, Z., Xie, Y., Xing, F., McGough, M., Yang, L.: MDNet: a semantically and visually interpretable medical image diagnosis network. In: CVPR, pp. 6428–6436 (2017)
9. Li, X., et al.: BrainGNN: interpretable brain graph neural network for fMRI analysis. Med. Image Anal. **74**, 102233 (2021)
10. Sporns, O., Betzel, R.F.: Modular brain networks. Annu. Rev. Psychol. **67**, 613–640 (2016)
11. Bertolero, M.A., Yeo, B.T., D'Esposito, M.: The modular and integrative functional architecture of the human brain. Proc. Natl. Acad. Sci. **112**(49), E6798–E6807 (2015)
12. Goulden, N., et al.: The salience network is responsible for switching between the default mode network and the central executive network: replication from DCM. Neuroimage **99**, 180–190 (2014)
13. Geirhos, R., et al.: Shortcut learning in deep neural networks. Nat. Mach. Intell. **2**(11), 665–673 (2020)
14. Knyazev, B., Taylor, G.W., Amer, M.: Understanding attention and generalization in graph neural networks. In: Advances in Neural Information Processing Systems, vol. 32 (2019)
15. Di Martino, A., et al.: The autism brain imaging data exchange: Towards a large-scale evaluation of the intrinsic brain architecture in autism. Mol. Psychiatry **19**(6), 659–667 (2014)
16. Yan, C.G., et al.: Reduced default mode network functional connectivity in patients with recurrent major depressive disorder. Proc. Natl. Acad. Sci. **116**(18), 9078–9083 (2019)
17. Yan, C., Zang, Y.: DPARSF: a MATLAB toolbox for "pipeline" data analysis of resting-state fMRI. Front. Syst. Neurosci. **4**, 13 (2010)
18. Gadgil, S., Zhao, Q., Pfefferbaum, A., Sullivan, E.V., Adeli, E., Pohl, K.M.: Spatio-temporal graph convolution for resting-state fMRI analysis. In: Martel, A.L., et al. (eds.) MICCAI 2020. LNCS, vol. 12267, pp. 528–538. Springer, Cham (2020). https://doi.org/10.1007/978-3-030-59728-3_52
19. Kim, B.H., Ye, J.C.: Understanding graph isomorphism network for rs-fMRI functional connectivity analysis. Front. Neurosci. 630 (2020)
20. Hu, J., Shen, L., Sun, G.: Squeeze-and-excitation networks. In: CVPR, pp. 7132–7141 (2018)
21. Tibshirani, R.: Regression shrinkage and selection via the Lasso. J. Roy. Stat. Soc.: Ser. B (Methodol.) **58**(1), 267–288 (1996)

22. Kipf, T.N., Welling, M.: Semi-supervised classification with graph convolutional networks. arXiv preprint arXiv:1609.02907 (2016)
23. Veličković, P., Cucurull, G., Casanova, A., Romero, A., Lio, P., Bengio, Y.: Graph attention networks. arXiv preprint arXiv:1710.10903 (2017)
24. Ayub, R., et al.: Thalamocortical connectivity is associated with autism symptoms in high-functioning adults with autism and typically developing adults. Transl. Psychiatry **11**(1), 93 (2021)
25. Xu, J., et al.: Specific functional connectivity patterns of middle temporal gyrus subregions in children and adults with autism spectrum disorder. Autism Res. **13**(3), 410–422 (2020)
26. MacQueen, G., Frodl, T.: The hippocampus in major depression: evidence for the convergence of the bench and bedside in psychiatric research? Mol. Psychiatry **16**(3), 252–264 (2011)
27. Sarkheil, P., Odysseos, P., Bee, I., Zvyagintsev, M., Neuner, I., Mathiak, K.: Functional connectivity of supplementary motor area during finger-tapping in major depression. Compr. Psychiatry **99**, 152166 (2020)
28. Batail, J.M., Coloigner, J., Soulas, M., Robert, G., Barillot, C., Drapier, D.: Structural abnormalities associated with poor outcome of a major depressive episode: the role of thalamus. Psychiatry Res. Neuroimaging **305**, 111158 (2020)
29. Clark, U.S., et al.: Effects of HIV and early life stress on amygdala morphometry and neurocognitive function. J. Int. Neuropsychol. Soc. **18**(4), 657–668 (2012)
30. Zhan, Y., et al.: The resting state central auditory network: a potential marker of HIV-related central nervous system alterations. Ear Hear. **43**(4), 1222 (2022)
31. Sarma, M.K., et al.: Regional brain gray and white matter changes in perinatally HIV-infected adolescents. NeuroImage Clin. **4**, 29–34 (2014)
32. Guan, H., Liu, M.: Domain adaptation for medical image analysis: a survey. IEEE Trans. Biomed. Eng. **69**(3), 1173–1185 (2021)

Anatomy-Driven Pathology Detection on Chest X-rays

Philip Müller[1(✉)], Felix Meissen[1], Johannes Brandt[1], Georgios Kaissis[1,2], and Daniel Rueckert[1,3]

[1] Institute for AI in Medicine, Technical University of Munich, Munich, Germany
philip.j.mueller@tum.de
[2] Helmholtz Zentrum Munich, Munich, Germany
[3] Department of Computing, Imperial College London, London, UK

Abstract. Pathology detection and delineation enables the automatic interpretation of medical scans such as chest X-rays while providing a high level of explainability to support radiologists in making informed decisions. However, annotating pathology bounding boxes is a time-consuming task such that large public datasets for this purpose are scarce. Current approaches thus use weakly supervised object detection to learn the (rough) localization of pathologies from image-level annotations, which is however limited in performance due to the lack of bounding box supervision. We therefore propose anatomy-driven pathology detection (ADPD), which uses easy-to-annotate bounding boxes of anatomical regions as proxies for pathologies. We study two training approaches: supervised training using anatomy-level pathology labels and multiple instance learning (MIL) with image-level pathology labels. Our results show that our anatomy-level training approach outperforms weakly supervised methods and fully supervised detection with limited training samples, and our MIL approach is competitive with both baseline approaches, therefore demonstrating the potential of our approach.

Keywords: Pathology detection · Anatomical regions · Chest X-rays

1 Introduction

Chest radiographs (chest X-rays) represent the most widely utilized type of medical imaging examination globally and hold immense significance in the detection of prevalent thoracic diseases, including pneumonia and lung cancer, making them a crucial tool in clinical care [10,15]. Pathology detection and localization – for brevity we will use the term *pathology detection* throughout this work – enables the automatic interpretation of medical scans such as chest X-rays by predicting bounding boxes for detected pathologies. Unlike classification, which only predicts the presence of pathologies, it provides a high level of explainability supporting radiologists in making informed decisions.

Supplementary Information The online version contains supplementary material available at https://doi.org/10.1007/978-3-031-43907-0_6.

H. Greenspan et al. (Eds.): MICCAI 2023, LNCS 14220, pp. 57–66, 2023.
https://doi.org/10.1007/978-3-031-43907-0_6

However, while image classification labels can be automatically extracted from electronic health records or radiology reports [7,20], this is typically not possible for bounding boxes, thus limiting the availability of large datasets for pathology detection. Additionally, manually annotating pathology bounding boxes is a time-consuming task, further exacerbating the issue. The resulting scarcity of large, publicly available datasets with pathology bounding boxes limits the use of supervised methods for pathology detection, such that current approaches typically follow weakly supervised object detection approaches, where only classification labels are required for training. However, as these methods are not guided by any form of bounding boxes, their performance is limited.

We, therefore, propose a novel approach towards pathology detection that uses anatomical region bounding boxes, solely defined on anatomical structures, as proxies for pathology bounding boxes. These region boxes are easier to annotate – the physiological shape of a healthy subject's thorax can be learned relatively easily by medical students – and generalize better than those of pathologies, such that huge labeled datasets are available [21]. In summary:

- We propose anatomy-driven pathology detection (ADPD), a pathology detection approach for chest X-rays, trained with pathology classification labels together with anatomical region bounding boxes as proxies for pathologies.
- We study two training approaches: using localized (anatomy-level) pathology labels for our model *Loc-ADPD* and using image-level labels with multiple instance learning (MIL) for our model *MIL-ADPD*.
- We train our models on the Chest ImaGenome [21] dataset and evaluate on NIH ChestX-ray 8 [20], where we found that our Loc-ADPD model outperforms both, weakly supervised methods and fully supervised detection with a small training set, while our MIL-ADPD model is competitive with supervised detection and slightly outperforms weakly supervised approaches.

2 Related Work

Weakly Supervised Pathology Detection. Due to the scarcity of bounding box annotations, pathology detection on chest X-rays is often tackled using weakly supervised object detection with Class Activation Mapping (CAM) [25], which only requires image-level classification labels. After training a classification model with global average pooling (GAP), an activation heatmap is computed by classifying each individual patch (extracted before pooling) with the trained classifier, before thresholding this heatmap for predicting bounding boxes. Inspired by this approach, several methods have been developed for chest X-rays [6,14,20,23]. While CheXNet [14] follows the original approach, the method provided with the NIH ChestX-ray 8 dataset [20] and the STL method [6] use Logsumexp (LSE) pooling [13], while the MultiMap model [23] uses max-min pooling as first proposed for the WELDON [3] method. Unlike our method, none of these methods utilize anatomical regions as proxies for predicting pathology bounding boxes, therefore leading to inferior performance.

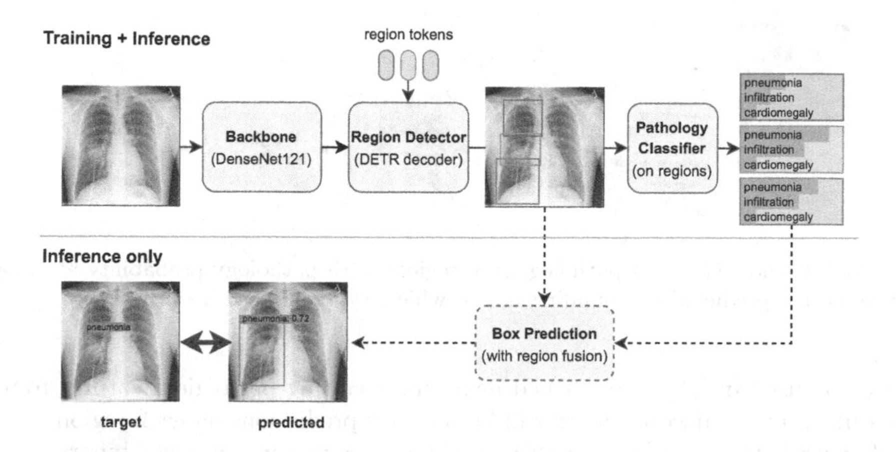

Fig. 1. Overview of our method. Anatomical regions are first detected using a CNN backbone and a shallow detector. For each region, observed pathologies are predicted using a shared classifier. Bounding boxes for each pathology are then predicted by considering regions with positive predictions and fusing overlapping boxes.

Localized Pathology Classification. Anatomy-level pathology labels have been utilized before to train localized pathology classifiers [1,21] or to improve weakly supervised pathology detection [24]. Along with the Chest ImaGenome dataset [21] several localized pathology classification models have been proposed which use a Faster R-CNN [16] to extract anatomical region features before predicting observed pathologies for each region using either a linear model or a GCN model based on pathology co-occurrences. This approach has been further extended to use GCNs on anatomical region relationships [1]. While utilizing the same form of supervision as our method, these methods do not tackle pathology detection.

In AGXNet [24], anatomy-level pathology classification labels are used to train a weakly-supervised pathology detection model. Unlike our and the other described methods, it does however not use anatomical region bounding boxes.

3 Method

3.1 Model

Figure 1 provides an overview of our method. Given a chest X-ray, we apply a DenseNet121 [5] backbone and extract patch-wise features by using the feature map after the last convolutional layer (before GAP). We then apply a lightweight object detection model consisting of a single DETR [2] decoder layer to detect anatomical regions. Following [2], we use learned query tokens attending to patch features in the decoder layer, where each token corresponds to one predicted bounding box. As no anatomical region can occur more than once in each chest X-ray, each query token is assigned to exactly one pre-defined anatomical region, such that the number of tokens equals the number of anatomical regions. This one-to-one assignment of tokens and regions allows us to remove the Hungarian

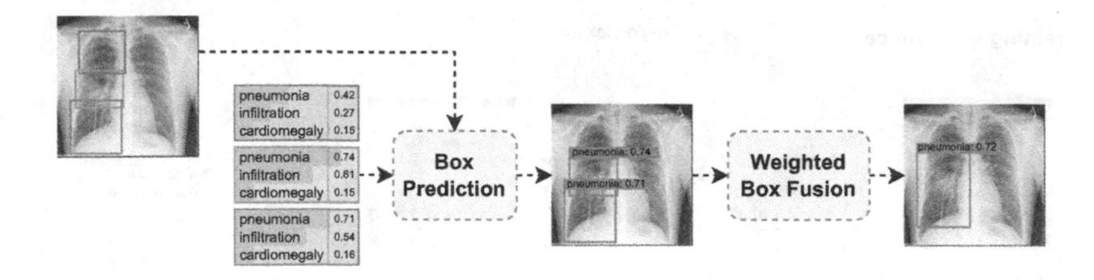

Fig. 2. Inference. For each pathology, the regions with pathology probability above a threshold are predicted as bounding boxes, which are then fused if overlapping.

matching used in [2]. As described next, the resulting per-region features from the output of the decoder layer will be used for predictions on each region.

For predicting whether the associated region is present, we use a binary classifier with a single linear layer, for bounding box prediction we use a three-layer MLP followed by sigmoid. We consider the prediction of observed pathologies as a multi-label binary classification task and use a single linear layer (followed by sigmoid) to predict the probabilities of all pathologies. Each of these predictors is applied independently to each region with their weights shared across regions.

We experimented with more complex pathology predictors like an MLP or a transformer layer but did not observe any benefits. We also did not observe improvements when using several decoder layers and observed degrading performance when using ROI pooling to compute region features.

3.2 Inference

During inference, the trained model predicts anatomical region bounding boxes and per-region pathology probabilities, which are then used to predict pathology bounding boxes in two steps, as shown in Fig. 2. In step (i), pathology probabilities are first thresholded and for each positive pathology (with probability larger than the threshold) the bounding box of the corresponding anatomical region is predicted as its pathology box, using the pathology probability as box score. This means, if a region contains several predicted pathologies, then all of its predicted pathologies share the same bounding box during step (i). In step (ii), weighted box fusion (WBF) [19] merges bounding boxes of the same pathology with IoU-overlaps above 0.03 and computes weighted averages (using box scores as weights) of their box coordinates. As many anatomical regions are at least partially overlapping, and we use a small IoU-overlap threshold, this allows the model to either pull the predicted boxes to relevant subparts of an anatomical region or to predict that pathologies stretch over several regions.

3.3 Training

The anatomical region detector is trained using the DETR loss [2] with fixed one-to-one matching (i.e. without Hungarian matching). For training the pathology classifier, we experiment with two different levels of supervision (Fig. 3).

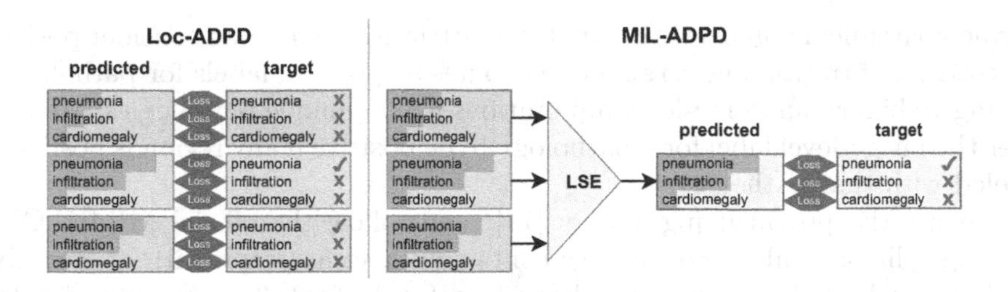

Fig. 3. Training. **Loc-ADPD:** Pathology predictions of regions are directly trained using anatomy-level supervision. **MIL-ADPD:** Region predictions are first aggregated using LSE pooling and then trained using image-level supervision.

For our *Loc-ADPD* model, we utilize anatomy-level pathology classification labels. Here, the target set of observed pathologies is available for each anatomical region individually such that the pathology observation prediction can directly be trained for each anatomical region. We apply the ASL [17] loss function independently on each region-pathology pair and average the results over all regions and pathologies. The decoder feature dimension is set to 512.

For our *MIL-ADPD* model, we experiment with a weaker form of supervision, where pathology classification labels are only available on the per-image level. We utilize multiple instance learning (MIL), where an image is considered a bag of individual instances (i.e. the anatomical regions), and only a single label (per pathology) is provided for the whole bag, which is positive if any of its instances is positive. To train using MIL, we first aggregate the predicted pathology probabilities of each region over all detected regions in the image using LSE pooling [13], acting as a smooth approximation of max pooling. The resulting per-image probability for each pathology is then trained using the ASL [17] loss. In this model, the decoder feature dimension is set to 256.

In both models, the ASL loss is weighted by a factor of 0.01 before adding it to the DETR loss. We train using AdamW [12] with a learning rate of 3e−5 (Loc-ADPD) or 1e−4 (MIL-ADPD) and weight decay 1e−5 (Loc-ADPD) or 1e−4 (MIL-ADPD) in batches of 128 samples with early stopping (with 20 000 steps patience) for roughly 7 h on a single Nvidia RTX A6000.

3.4 Dataset

Training Dataset. We train on the Chest ImaGenome dataset [4,21,22][1], consisting of roughly 240 000 frontal chest X-ray images with corresponding scene graphs automatically constructed from free-text radiology reports. It is derived from the MIMIC-CXR dataset [9,10], which is based on imaging studies from 65 079 patients performed at Beth Israel Deaconess Medical Center in Boston, US. Amongst other information, each scene graph contains bounding boxes for 29

[1] https://physionet.org/content/chest-imagenome/1.0.0 (PhysioNet Credentialed Health Data License 1.5.0).

unique anatomical regions with annotated attributes, where we consider positive `anatomical finding` and `disease` attributes as positive labels for pathologies, leading to binary anatomy-level annotations for 55 unique pathologies. We consider the image-level label for a pathology to be positive if any region is positively labeled with that pathology.

We use the provided jpg-images [11][2] and follow the official MIMIC-CXR training split but only keep samples containing a scene graph with at least five valid region bounding boxes, resulting in a total of 234 307 training samples. During training, we use random resized cropping with size 224×224, apply contrast and brightness jittering, random affine augmentations, and Gaussian blurring.

Evaluation Dataset and Class Mapping. We evaluate our method on the subset of 882 chest X-ray images with pathology bounding boxes, annotated by radiologists, from the NIH ChestXray-8 (CXR8) dataset [20][3] from the National Institutes of Health Clinical Center in the US. We use 50% for validation and keep the other 50% as a held-out test set. Note that for evaluation only pathology bounding boxes are required (to compute the metrics), while during training only anatomical region bounding boxes (without considering pathologies) are required. All images are center-cropped and resized to 224×224.

The dataset contains bounding boxes for 8 unique pathologies. While partly overlapping with the training classes, a one-to-one correspondence is not possible for all classes. For some evaluation classes, we therefore use a many-to-one mapping where the class probability is computed as the mean over several training classes. We refer to the supp. material for a detailed study on class mappings.

4 Experiments and Results

4.1 Experimental Setup and Baselines

We compare our method against several weakly supervised object detection methods (CheXNet [14], STL [6], GradCAM [18], CXR [20], WELDON [3], MultiMap Model [23], LSE Model [13]), trained on the CXR8 training set using only image-level pathology labels. Note that some of these methods focus on (image-level) classification and do not report quantitative localization results. Nevertheless, we compare their localization approaches quantitatively with our method. We also use AGXNet [24] for comparison, a weakly supervised method trained using anatomy-level pathology labels but without any bounding box supervision. It was trained on MIMIC-CXR (sharing the images with our method) with labels from RadGraph [8] and finetuned on the CXR8 training set with image-level labels. Additionally, we also compare with a Faster-RCNN [16] trained on a small subset of roughly 500 samples from the CXR8 training set that have been

[2] https://physionet.org/content/mimic-cxr-jpg/2.0.0/ (PhysioNet Credentialed Health Data License 1.5.0).

[3] https://www.kaggle.com/datasets/nih-chest-xrays/data (CC0: Public Domain).

annotated with pathology bounding boxes by two medical experts, including one board-certified radiologist.

Table 1. Results on the NIH ChestX-ray 8 dataset [20]. Our models Loc-ADPD and MIL-ADPD, trained using anatomy (An) bounding boxes, both outperform all weakly supervised methods trained with image-level pathology (Pa) and anatomy-level pathology (An-Pa) labels by a large margin. MIL-ADPD is competitive with the supervised baseline trained with pathology (Pa) bounding boxes, while Loc-ADPD outperforms it by a large margin.

Method	Supervision		IoU@10-70	IoU@10		IoU@30		IoU@50	
	Box	Class	mAP	AP	loc-acc	AP	loc-acc	AP	loc-acc
MIL-ADPD (ours)	An	Pa	7.84	14.01	0.68	8.85	0.65	7.03	0.65
w/o WBF			5.42	11.05	0.67	7.97	0.65	3.44	0.64
Loc-ADPD (ours)	An	An-Pa	**10.89**	**19.99**	**0.85**	**12.43**	**0.84**	**8.72**	**0.83**
w/o WBF			8.88	17.02	0.84	9.65	0.83	7.36	0.83
w/ MIL			10.29	19.16	0.84	10.95	0.83	8.00	0.82
CheXNet [14]	–	Pa	5.80	12.87	0.58	8.23	0.55	3.12	0.52
STL [6]	–	Pa	5.61	12.76	0.57	7.94	0.54	2.45	0.50
GradCAM [18]	–	Pa	4.43	12.53	0.58	6.67	0.54	0.13	0.51
CXR [20]	–	Pa	5.61	13.91	0.59	8.01	0.55	1.24	0.51
WELDON [3]	–	Pa	4.76	14.57	0.61	6.18	0.56	0.34	0.51
MultiMap [23]	–	Pa	4.91	12.36	0.61	7.13	0.57	1.35	0.53
LSE Model [13]	–	Pa	3.77	14.49	0.61	2.62	0.56	0.42	0.54
AGXNet [24]	–	An-Pa	5.30	11.39	0.59	6.58	0.56	4.14	0.54
Faster R-CNN	Pa	–	7.36	9.11	0.79	7.62	0.79	7.26	0.78

For all models, we only consider the predicted boxes with the highest box score per pathology, as the CXR8 dataset never contains more than one box per pathology. We report the standard object detection metrics *average precision (AP)* at different IoU-thresholds and the *mean AP (mAP)* over thresholds $(0.1, 0.2, \ldots, 0.7)$, commonly used thresholds on this dataset [20]. Additionally, we report the localization accuracy (loc-acc) [20], a common localization metric on this dataset, where we use a box score threshold of 0.7 for our method.

4.2 Pathology Detection Results

Comparison with Baselines. Table 1 shows the results of our MIL-ADPD and Loc-ADPD models and all baselines on the CXR8 test set. Compared to the best weakly supervised method with image-level supervision (CheXNet) our methods improve by large margins (MIL-ADPD by Δ+35.2%, Loc-ADPD by Δ+87.8% in mAP). Improvements are especially high when considering larger IoU-thresholds and huge improvements are also achieved in loc-acc at all thresholds. Both models also outperform AGXNet (which uses anatomy-level supervision) by large

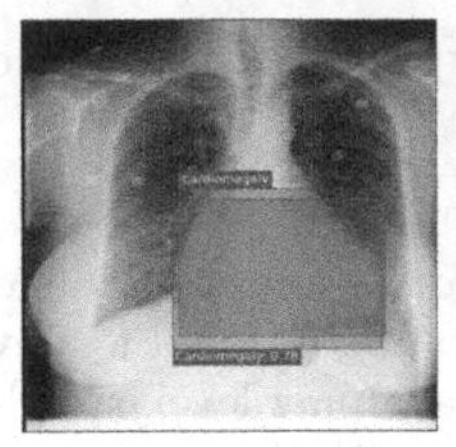
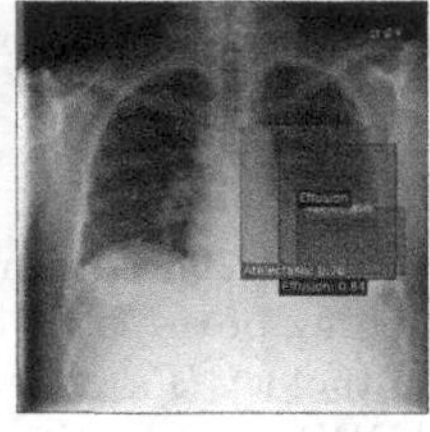
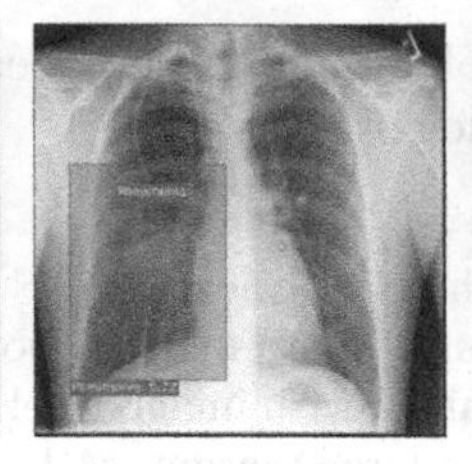
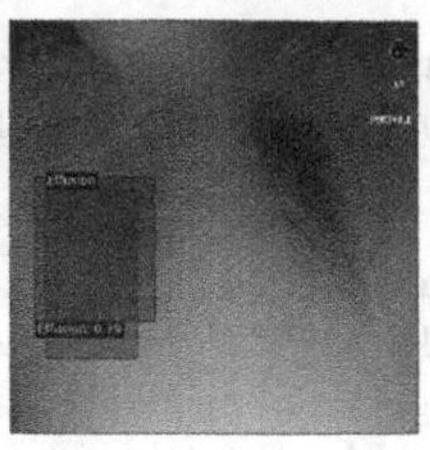

Fig. 4. Qualitative results of Loc-ADPD, with predicted (solid) and target (dashed) boxes. Cardiomegaly (red) is detected almost perfectly, as it is always exactly localized at one anatomical region. Other pathologies like atelectasis (blue), effusion (green), or pneumonia (cyan) are detected but often with non-perfect overlapping boxes. Detection also works well for predicting several overlapping pathologies (second from left). (Color figure online)

margins (MIL-ADPD by $\Delta + 47.9\%$ and Loc-ADPD by $\Delta + 105.5\%$ mAP), while improvements on larger thresholds are smaller here. Even when compared to Faster R-CNN trained on a small set of fully supervised samples, MIL-ADPD is competitive ($\Delta + 6.5\%$), while Loc-ADPD improves by $\Delta + 48.0\%$. However, on larger thresholds (IoU@50) the supervised baseline slightly outperforms MIL-ADPD, while Loc-ADPD is still superior. This shows that using anatomical regions as proxies is an effective approach to tackle pathology detection. While using image-level annotations (MIL-ADPD) already gives promising results, the full potential is only achieved using anatomy-level supervision (Loc-ADPD). Unlike Loc-ADPD and MIL-ADPD, all baselines were either trained or fine-tuned on the CXR8 dataset, showing that our method generalizes well to unseen datasets and that our class mapping is effective.

For detailed results per pathology we refer to the supp. material. We found that the improvements of MIL-ADPD are mainly due to improved performance on Cardiomegaly and Mass detection, while Loc-ADPD consistently outperforms all baselines on all classes except Nodule, often by a large margin.

Ablation Study. In Table 1 we also show the results of different ablation studies. Without WBF, results degrade for both of our models, highlighting the importance of merging region boxes. Combining the training strategies of Loc-ADPD and MIL-ADPD does not lead to an improved performance. Different class mappings between training and evaluation set are studied in the supp. material.

Qualitative Results. As shown in Fig. 4 Loc-ADPD detects cardiomegaly almost perfectly, as it is always exactly localized at one anatomical region. Other pathologies are detected but often with too large or too small boxes as they only cover parts of anatomical regions or stretch over several of them, which cannot be completely corrected using WBF. Detection also works well for predicting several overlapping pathologies. For qualitative comparisons between Loc-ADPD and MIL-ADPD, we refer to the supp. material.

5 Discussion and Conclusion

Limitations. While our proposed ADPD method outperforms all competing models, it is still subject to limitations. First, due to the dependence on region proxies, for pathologies covering only a small part of a region, our models predict the whole region, as highlighted by their incapability to detect nodules. We however note that in clinical practice, chest X-rays are not used for the final diagnosis of such pathologies and even rough localization can be beneficial. Additionally, while not requiring pathology bounding boxes, our models still require supervision in the form of anatomical region bounding boxes, and Loc-ADPD requires anatomy-level labels. However, anatomical bounding boxes are easier to annotate and predict than pathology bounding boxes, and the used anatomy-level labels were extracted automatically from radiology reports [21]. While our work is currently limited to chest X-rays, we see huge potential for modalities where abnormalities can be assigned to meaningful regions.

Conclusion. We proposed a novel approach tackling pathology detection on chest X-rays using anatomical region bounding boxes. We studied two training approaches, using anatomy-level pathology labels and using image-level labels with MIL. Our experiments demonstrate that using anatomical regions as proxies improves results compared weakly supervised methods and supervised training on little data, thus providing a promising direction for future research.

References

1. Agu, N.N., et al.: AnaXNet: anatomy aware multi-label finding classification in chest X-ray. In: de Bruijne, M., et al. (eds.) MICCAI 2021. LNCS, vol. 12905, pp. 804–813. Springer, Cham (2021). https://doi.org/10.1007/978-3-030-87240-3_77
2. Carion, N., Massa, F., Synnaeve, G., Usunier, N., Kirillov, A., Zagoruyko, S.: End-to-end object detection with transformers. In: Vedaldi, A., Bischof, H., Brox, T., Frahm, J.-M. (eds.) ECCV 2020. LNCS, vol. 12346, pp. 213–229. Springer, Cham (2020). https://doi.org/10.1007/978-3-030-58452-8_13
3. Durand, T., Thome, N., Cord, M.: Weldon: weakly supervised learning of deep convolutional neural networks. In: CVPR, pp. 4743–4752 (2016). https://doi.org/10.1109/CVPR.2016.513
4. Goldberger, A.L., et al.: PhysioBank, PhysioToolkit, and PhysioNet: components of a new research resource for complex physiologic signals. Circulation **101**(23), e215–e220 (2000)
5. Huang, G., Liu, Z., Van Der Maaten, L., Weinberger, K.Q.: Densely connected convolutional networks. In: CVPR, pp. 2261–2269 (2017). https://doi.org/10.1109/CVPR.2017.243
6. Hwang, S., Kim, H.-E.: Self-transfer learning for weakly supervised lesion localization. In: Ourselin, S., Joskowicz, L., Sabuncu, M.R., Unal, G., Wells, W. (eds.) MICCAI 2016. LNCS, vol. 9901, pp. 239–246. Springer, Cham (2016). https://doi.org/10.1007/978-3-319-46723-8_28
7. Irvin, J., et al.: CheXpert: a large chest radiograph dataset with uncertainty labels and expert comparison. In: AAAI, pp. 590–597 (2019). https://doi.org/10.1609/aaai.v33i01.3301590

8. Jain, S., et al.: Radgraph: extracting clinical entities and relations from radiology reports. In: NeurIPS (2021)
9. Johnson, A.E.W., et al.: Mimic-cxr, a de-identified publicly available database of chest radiographs with free-text reports. Sci. Data **6**(1), 1–8 (2019)
10. Johnson, A.E.W., et al.: Mimic-cxr database (version 2.0.0). PhysioNet (2019)
11. Johnson, A.E.W., et al.: Mimic-cxr-jpg, a large publicly available database of labeled chest radiographs. arXiv preprint arXiv:1901.07042 (2019)
12. Loshchilov, I., Hutter, F.: Decoupled weight decay regularization. In: ICLR (2019)
13. Pinheiro, P.O., Collobert, R.: From image-level to pixel-level labeling with convolutional networks. In: CVPR, pp. 1713–1721 (2015). https://doi.org/10.1109/CVPR.2015.7298780
14. Rajpurkar, P., et al.: CheXnet: radiologist-level pneumonia detection on chest x-rays with deep learning. arXiv preprint arXiv:1711.05225 (2017). https://doi.org/10.48550/arXiv.1711.05225
15. Raoof, S., Feigin, D., Sung, A., Raoof, S., Irugulpati, L., Rosenow, E.C., III.: Interpretation of plain chest roentgenogram. Chest **141**(2), 545–558 (2012)
16. Ren, S., He, K., Girshick, R., Sun, J.: Faster R-CNN: towards real-time object detection with region proposal networks. In: NIPS, vol. 28 (2015)
17. Ridnik, T., et al.: Asymmetric loss for multi-label classification. In: ICCV, pp. 82–91 (2021). https://doi.org/10.1109/ICCV48922.2021.00015
18. Selvaraju, R.R., Cogswell, M., Das, A., Vedantam, R., Parikh, D., Batra, D.: Grad-CAM: visual explanations from deep networks via gradient-based localization. In: CVPR, pp. 618–626 (2017). https://doi.org/10.1109/ICCV.2017.74
19. Solovyev, R., Wang, W., Gabruseva, T.: Weighted boxes fusion: ensembling boxes from different object detection models. Image Vis. Comput. **107**, 104117 (2021). https://doi.org/10.1016/j.imavis.2021.104117
20. Wang, X., Peng, Y., Lu, L., Lu, Z., Bagheri, M., Summers, R.M.: Chestx-ray8: hospital-scale chest x-ray database and benchmarks on weakly-supervised classification and localization of common thorax diseases. In: CVPR, pp. 2097–2106 (2017). https://doi.org/10.1109/CVPR.2017.369
21. Wu, J., et al.: Chest imagenome dataset for clinical reasoning. In: NIPS (2021)
22. Wu, J.T., et al.: Chest imagenome dataset (version 1.0.0). PhysioNet (2021)
23. Yan, C., Yao, J., Li, R., Xu, Z., Huang, J.: Weakly supervised deep learning for thoracic disease classification and localization on chest x-rays. In: ACM BCB, pp. 103–110 (2018)
24. Yu, K., Ghosh, S., Liu, Z., Deible, C., Batmanghelich, K.: Anatomy-guided weakly-supervised abnormality localization in chest x-rays. In: Wang, L., et al. (eds.) MICCAI 2022. LNCS, vol. 13435, pp. 658–668. Springer, Cham (2022). https://doi.org/10.1007/978-3-031-16443-9_63
25. Zhou, B., Khosla, A., Lapedriza, A., Oliva, A., Torralba, A.: Learning deep features for discriminative localization. In: CVPR, pp. 2921–2929 (2016). https://doi.org/10.1109/CVPR.2016.319

VesselVAE: Recursive Variational Autoencoders for 3D Blood Vessel Synthesis

Paula Feldman[1,3](✉), Miguel Fainstein[3], Viviana Siless[3], Claudio Delrieux[1,2], and Emmanuel Iarussi[1,3]

[1] Consejo Nacional de Investigaciones Científicas y Técnicas, Buenos Aires, Argentina
paulafeldman@conicet.gov.ar
[2] Universidad Nacional del Sur, Bahía Blanca, Argentina
[3] Universidad Torcuato Di Tella, Buenos Aires, Argentina

Abstract. We present a data-driven generative framework for synthesizing blood vessel 3D geometry. This is a challenging task due to the complexity of vascular systems, which are highly variating in shape, size, and structure. Existing model-based methods provide some degree of control and variation in the structures produced, but fail to capture the diversity of actual anatomical data. We developed VesselVAE, a recursive variational Neural Network that fully exploits the hierarchical organization of the vessel and learns a low-dimensional manifold encoding branch connectivity along with geometry features describing the target surface. After training, the VesselVAE latent space can be sampled to generate new vessel geometries. To the best of our knowledge, this work is the first to utilize this technique for synthesizing blood vessels. We achieve similarities of synthetic and real data for radius (.97), length (.95), and tortuosity (.96). By leveraging the power of deep neural networks, we generate 3D models of blood vessels that are both accurate and diverse, which is crucial for medical and surgical training, hemodynamic simulations, and many other purposes.

Keywords: Vascular 3D model · Generative modeling · Neural Networks

1 Introduction

Accurate 3D models of blood vessels are increasingly required for several purposes in Medicine and Science [25]. These meshes are typically generated using either image segmentation or synthetic methods. Despite significant advances in vessel segmentation [26], reconstructing thin features accurately from medical images remains challenging [2]. Manual editing of vessel geometry is a tedious and error prone task that requires expert medical knowledge, which explains the

Supplementary Information The online version contains supplementary material available at https://doi.org/10.1007/978-3-031-43907-0_7.

H. Greenspan et al. (Eds.): MICCAI 2023, LNCS 14220, pp. 67–76, 2023.
https://doi.org/10.1007/978-3-031-43907-0_7

scarcity of curated datasets. As a result, several methods have been developed to adequately synthesize blood vessel geometry [29].

Within the existing literature on generating vascular 3D models, we identified two primary types of algorithms: fractal-based, and space-filling algorithms. Fractal-based algorithms use a set of fixed rules that include different branching parameters, such as the ratio of asymmetry in arterial bifurcations and the relationship between the diameter of the vessel and the flow [7,33]. On the other hand, space-filling algorithms allow the blood vessels to grow into a specific perfusion volume while aligning with hemodynamic laws and constraints on the formation of blood vessels [9,17,21,22,25]. Although these *model-based* methods provide some degree of control and variation in the structures produced, they often fail to capture the diversity of real anatomical data.

In recent years, deep neural networks led to the development of powerful generative models [30], such as Generative Adversarial Networks [8,12] and Diffusion Models [11], which produced groundbreaking performance in many applications, ranging from image and video synthesis to molecular design. These advances have inspired the creation of novel network architectures to model 3D shapes using voxel representations [28], point clouds [31], signed distance functions [19], and polygonal meshes [18]. In particular, and close to our aim, Wolterink et al. [27] propose a GAN model capable of generating coronary artery anatomies. However, this model is limited to generating single-channel blood vessels and thus does not support the generation of more complex, tree-like vessel topologies.

In this work we propose a novel *data-driven* framework named VesselVAE for synthesizing blood vessel geometry. Our generative framework is based on a Recursive variational Neural Network (RvNN), that has been applied in various contexts, including natural language [23,24], shape semantics modeling [14,15], and document layout generation [20]. In contrast to previous data-driven methods, our recursive network fully exploits the hierarchical organization of the vessel and learns a low-dimensional manifold encoding branch connectivity along with geometry features describing the target surface. Once trained, the VesselVAE latent space is sampled to generate new vessel geometries. To the best of our knowledge, this work is the first to synthesize multi-branch blood vessel trees by learning from real data. Experiments show that synth and real blood vessel geometries are highly similar measured with the cosine similarity: radius (.97), length (.95), and tortuosity (.96).

2 Methods

Input. The network input is a binary tree representation of the blood vessel 3D geometry. Formally, each tree is defined as a tuple $(T, \mathcal{E})$, where T is the set of nodes, and $\mathcal{E}$ is the set of directed edges connecting a pair of nodes (n, m), with $n, m \in T$. In order to encode a 3D model into this representation, vessel segments V are parameterized by a central axis consisting of ordered points in Euclidean space: $V = v_1, v_2, \ldots, v_N$ and a radius r, assuming a piece-wise tubular vessel for simplicity. We then construct the binary tree as a set of nodes $T = n_1, n_2, \ldots, n_N$,

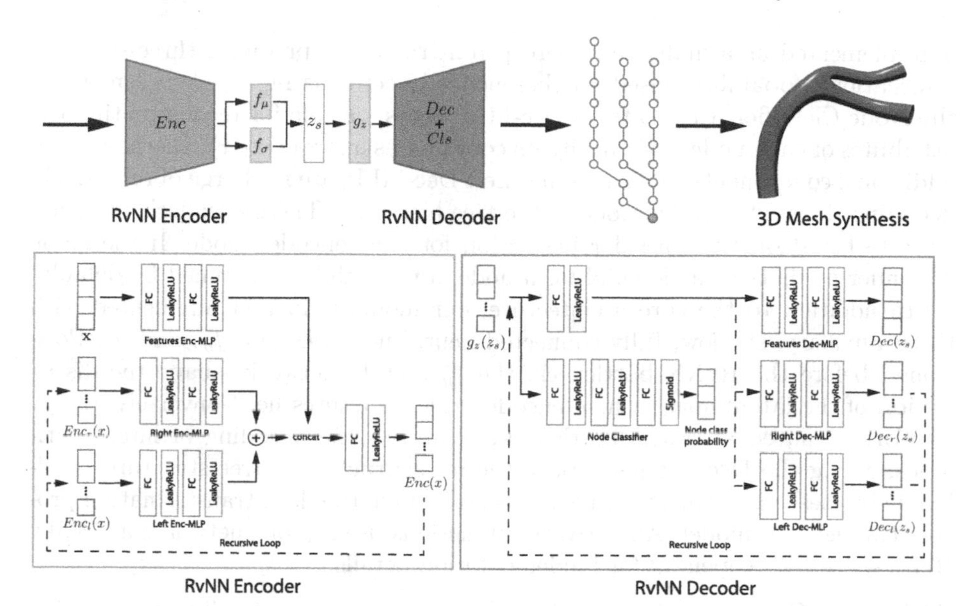

Fig. 1. Top: Overview of the Recursive variational Neural Network for synthesizing blood vessel structures. The architecture follows an Encoder-Decoder framework which can handle the hierarchical tree representation of the vessels. VesselVAE learns to generate the topology and attributes for each node in the tree, which is then used to synthesize 3D meshes. Bottom: Layers of the Encoder and Decoder networks comprising branches of fully-connected layers followed by leaky ReLU activations. Note that the right/left Enc-MLPs within the Encoder are triggered respectively when the incoming node in the tree is identified as a right or left child. Similarly, the Decoder only uses right/left Dec-MLPs when the Node Classifier predicts bifurcations.

where each node n_i represents a vessel segment v and contains an attribute vector $\mathbf{x}_i = [x_i, y_i, z_i, r_i] \in \mathbb{R}^4$ with the coordinates of the corresponding point and its radius r_i. See Sect. 3 for details.

Network Architecture. The proposed generative model is a Recursive variational Neural Network (RvNN) consisting of two main components: the Encoder (*Enc*) and the Decoder (*Dec*) networks. The Encoder transforms a tree structure into a hierarchical encoding on the learned manifold. The Decoder network is capable of sampling from this encoded space to decode tree structures, as depicted in Fig. 1. The encoding and decoding processes are achieved through a depth-first traversal of the tree, where each node is combined with its parent node recursively. The model outputs a hierarchy of vessel branches, where each internal node in the hierarchy is represented by a vector that encodes its own attributes and the information of all subsequent nodes in the tree.

Within the RvNN Decoder network there are two essential components: the Node Classifier (*Cls*) and the Features Decoder Multi-Layer Perceptron (Features Dec-MLP). The Node Classifier discerns the type of node to be decoded, whether it is a leaf node or an internal node with one or two bifurcations. This is implemented as a multi-layer perceptron trained to predict a three-category bifurcation probability based on the encoded vector as input. Complementing the Node Classifier, the Features Dec-MLP is responsible for reconstructing the attributes of each node, specifically its coordinates and radius. Furthermore, two additional components, the Right and Left Dec-MLP, are in charge of recursively decoding the next encoded node in the tree hierarchy. These decoder's branches execute based on the classifier prediction for that encoded node. If the Node Classifier predicts a single child for a node, a right child is assumed by default.

In addition to the core architecture, our model is further augmented with three auxiliary, shallow, fully-connected neural networks: f_μ, f_σ, and g_z. Positioned before the RvNN bottleneck, the f_μ and f_σ networks shape the distribution of the latent space where encoded tree structures lie. Conversely, the g_z network, situated after the bottleneck, facilitates the decoding of latent variables, aiding the Decoder network in the reconstruction of tree structures. Collectively, these supplementary networks streamline the data transformation process through the model. All activation functions used in our networks are leaky ReLUs. See the Appendix for implementation details.

Objective. Our generative model is trained to learn a probability distribution over the latent space that can be used to generate new blood vessel segments. After encoding, the decoder takes samples from a multivariate Gaussian distribution: $z_s(x) \sim N(\mu, \sigma)$ with $\mu = f_\mu(Enc(x))$ and $\sigma = f_\sigma(Enc(x))$, where Enc is the recursive encoder and f_μ, f_σ are two fully-connected neural networks. In order to recover the feature vectors $\mathbf{x}$ for each node along with the tree topology, we simultaneously train the regression network (Features Dec-MLP in Fig. 1) on a reconstruction objective L_{recon}, and the Node Classifier using L_{topo}. Additionally, in line with the general framework proposed by β-VAE [10], we incorporated a Kullback-Leibler (KL) divergence term encouraging the distribution $p(z_s(x))$ over all training samples x to move closer to the prior of the standard normal distribution $p(z)$. We therefore minimize the following equation:

$$L = L_{recon} + \alpha L_{topo} + \gamma L_{KL}, \tag{1}$$

where the reconstruction loss is defined as $L_{recon} = \|Dec(z_s(x)) - x\|_2$, the Kullback-Leibler divergence loss is $L_{\mathrm{KL}} = D_{\mathrm{KL}}(p(z_s(x)) \| p(z))$, and the topology objective is a three-class cross entropy loss $L_{topo} = \Sigma_{c=1}^{3} x_c \log(Cls(Dec(x))_c)$. Notice that x_c is a binary indicator (0 or 1) for the true class of the sample x. Specifically, $x_c = 1$ if the sample belongs to class c and 0 otherwise. $Cls(Dec(x))_c$ is the predicted probability of the sample x belonging to class c (zero, one, or two bifurcations), as output by the classifier. Here, $Dec(x)$ denotes the encoded-decoded node representation of the input sample x.

3D Mesh Synthesis. Several algorithms have been proposed in the literature to generate a surface 3D mesh from a tree-structured centerline [29]. For simplicity and efficiency, we chose the approach described in [6], which produces good quality meshes from centerlines with a low sample rate. The implemented method iterates through the points in the curve generating a coarse quadrilateral

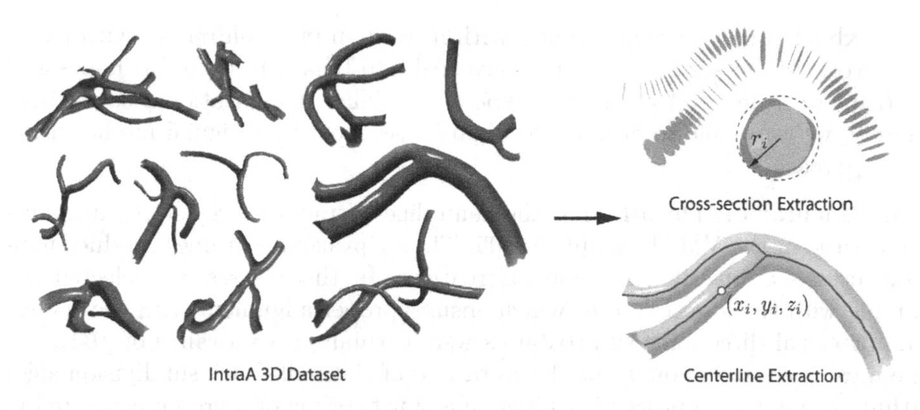

Fig. 2. Dataset and pre-processing overview: The raw meshes from the IntraA 3D collection undergo pre-processing using the VMTK toolkit. This step is crucial for extracting centerlines and cross-sections from the meshes, which are then used to construct their binary tree representations.

mesh along the segments and joints. The centerline sampling step is crucial for a successful reconstruction outcome. Thus, our re-sampling is not equispaced but rather changes with curvature and radius along the centerline, increasing the frequency of sampling near high-curvature regions. This results in a better quality and more accurate mesh. Finally, Catmull-Clark subdivision algorithm [5] is used to increase mesh resolution and smooth out the surface.

3 Experimental Setup

Materials. We trained our networks using a subset of the open-access IntrA dataset[1] published by Yang et al. in 2020 [32]. This subset consisted of 1694 healthy vessel segments reconstructed from 2D MRA images of patients. We converted 3D meshes into a binary tree representation and used the *network extraction* script from the VMTK toolkit[2] to extract the centerline coordinates of each vessel model. The centerline points were determined based on the ratio between the sphere step and the local maximum radius, which was computed using the advancement ratio specified by the user. The radius of the blood vessel conduit at each centerline sample was determined using the computed cross-sections assuming a maximal circular shape (See Fig. 2). To improve computational efficiency during recursive tree traversal, we implemented an algorithm that balances each tree by identifying a new root. We additionally trimmed trees to a depth of ten in our experiments. This decision reflects a balance between the computational demands of depth-first tree traversal in each training step and the complexity of the training meshes. We excluded from our study trees

[1] https://github.com/intra3d2019/IntrA.
[2] http://www.vmtk.org/vmtkscripts/vmtknetworkextraction.

that exhibited greater depth, nodes with more than two children, or with loops. However, non-binary trees can be converted into binary trees and it is possible to train with deeper trees at the expense of higher computational costs. Ultimately, we were able to obtain 700 binary trees from the original meshes using this approach.

Implementation Details. For the centerline extraction, we set the advancement ratio in the VMTK script to 1.05. The script can sometimes produce multiple cross-sections at centerline bifurcations. In those cases, we selected the sample with the lowest radius, which ensures proper alignment with the centerline principal direction. All attributes were normalized to a range of $[0, 1]$. For the mesh reconstruction we used 4 iterations of Catmull-Clark subdivision algorithm. The data pre-processing pipeline and network code were implemented in Python and PyTorch Framework.

Training. In all stages, we set the batch size to 10 and used the ADAM optimizer with $\beta_1 = 0.9$, $\beta_2 = 0.999$, and a learning rate of 1×10^{-4}. We set $\alpha = .3$ and $\gamma = .001$ for Eq. 1 in our experiments. To enhance computation speed, we implemented dynamic batching [16], which groups together operations involving input trees of dissimilar shapes and different nodes within a single input graph. It takes approximately 12 h to train our models on a workstation equipped with an NVIDIA A100 GPU, 80 GB VRAM, and 256 GB RAM. However, the memory footprint during training is very small ($\leq$1 GB) due to the use of a lightweight tree representation. This means that the amount of memory required to store and manipulate our training data structures is minimal. During training, we ensure that the reconstructed tree aligns with the original structure, rather than relying solely on the classifier's predictions. We train the classifier using a cross-entropy loss that compares its predictions to the actual values from the original tree. Since the number of nodes in each class is unbalanced, we scale the weight given to each class in the cross-entropy loss using the inverse of each class count. During preliminary experiments, we observed that accurately classifying nodes closer to the tree root is critical. This is because a miss-classification of top nodes has a cascading effect on all subsequent nodes in the tree (i.e. skip reconstructing a branch). To account for this, we introduce a weighting scheme that for each node, assigns a weight to the cross-entropy loss based on the number of total child nodes. The weight is normalized by the total number of nodes in the tree.

Metrics. We defined a set of metrics to evaluate our trained network's performance. By using these metrics, we can determine how well the generated 3D models of blood vessels match the original dataset distribution, as well as the diversity of the generated output. The chosen metrics have been widely used in the field of blood vessel 3D modeling, and have shown to provide reliable and accurate quantification of blood vessels main characteristics [3,13]. We analyzed tortuosity per branch, the vessel centerline total length, and the average radius of the tree. Tortuosity distance metric [4] is a widely used metric in the field of blood vessel analysis, mainly because of its clinical importance. It measures the amount of twistiness in each branch of the vessel. Vessel's total length and

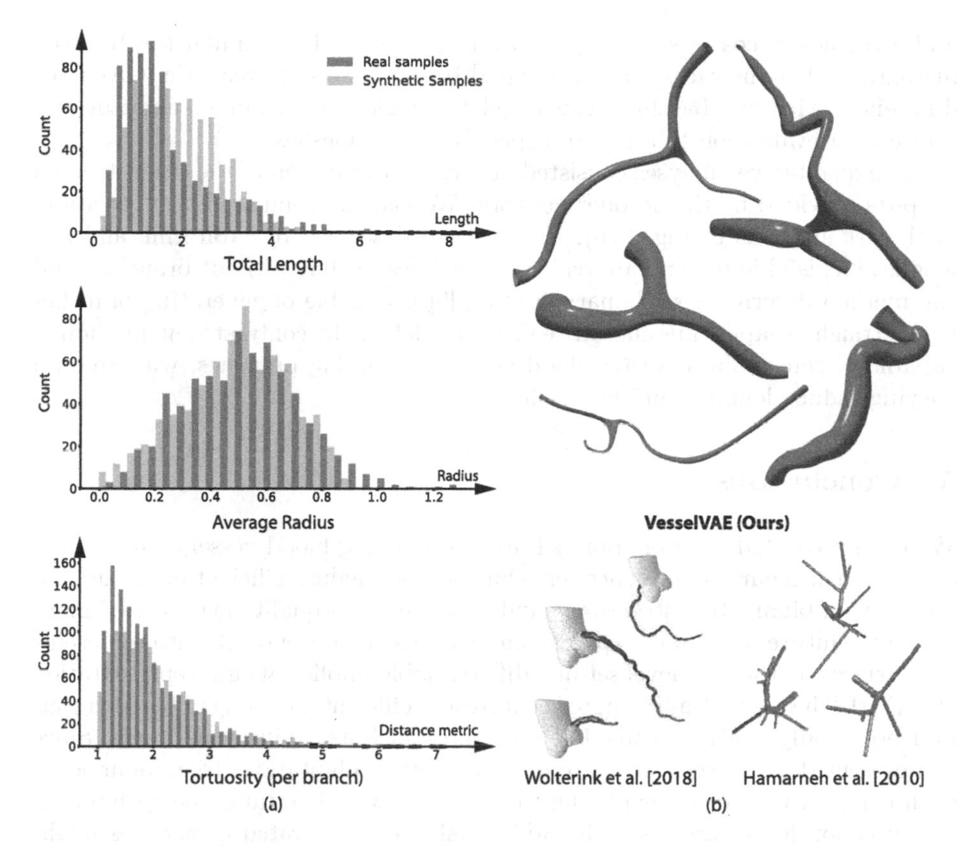

Fig. 3. (a) shows the histograms of total length, average radius and tortuosity per branch for both, real and synthetic samples. (b) shows a visual comparison among our method and two baselines [9,27].

average radius were used in previous work to distinguish healthy vasculature from cancerous malformations. Finally, in order to measure the distance across distributions for each metric, we compute the cosine similarity.

4 Results

We conducted both quantitative and qualitative analyses to evaluate the model's performance. For the quantitative analyses, we implemented a set of metrics commonly used for characterizing blood vessels. We computed histograms of the radius, total length, and tortuosity for the real blood vessel set and the generated set (700 samples) in Fig. 3 (a). The distributions are aligned and consistent. We measured the closeness of histograms with the cosine similarity by projecting the distribution into a vector of n-dimensional space (n is the number of bins in the histogram). Since our points are positive, the results range from 0 to 1. We obtain a radius cosine similarity of .97, a total length cosine similarity of .95, and a tortuosity cosine similarity of .96. Results show high similarities between histograms demonstrating that generated blood vessels are realistic. Given the differences with the baselines generated topologies, for a fair comparison, we limited our evaluation to a visual inspection of the meshes.

The qualitative analyses consisted of a visual evaluation of the reconstructed outputs provided by the decoder network. We visually compared them to state-of-the-art methods in Fig. 3 (b). The method described by Wolterink and colleagues [27] is able to generate realistic blood vessels but without branches, and the method described by Hamarneh et al. [9] is capable of generating branches with straight shapes, missing on realistic modeling. In contrast, our method is capable of generating realistic blood vessels containing branches, with smooth varying radius, lengths, and tortuosity.

5 Conclusions

We have presented a novel approach for synthesizing blood vessel models using a variational recursive autoencoder. Our method enables efficient encoding and decoding of binary tree structures, and produces high-quality synthesized models. In the future, we aim to explore combinations of our approach with representing surfaces by the zero level set in a differentiable implicit neural representation (INR) [1]. This could lead to more accurate and efficient modeling of blood vessels and potentially other non-tree-like structures such as capillary networks. Since the presented framework would require significant adaptations to accommodate such complex topologies, exploring this problem would certainly be an interesting direction for future research. Additionally, the generated geometries might show self-intersections. In the future, we would like to incorporate restrictions into the generative model to avoid such artifacts. Overall, we believe that our proposed approach holds great promise for advancing 3D blood vessel geometry synthesis and contributing to the development of new clinical tools for healthcare professionals.

Acknowledgements. This project was supported by grants from Salesforce, USA (Einstein AI 2020), National Scientific and Technical Research Council (CONICET), Argentina (PIP 2021-2023 GI - 11220200102981CO), and Universidad Torcuato Di Tella, Argentina.

References

1. Alblas, D., Brune, C., Yeung, K.K., Wolterink, J.M.: Going off-grid: continuous implicit neural representations for 3D vascular modeling. In: Camara, O., et al. (eds.) STACOM 2022. LNCS, vol. 13593, pp. 79–90. Springer, Cham (2022). https://doi.org/10.1007/978-3-031-23443-9_8
2. Alblas, D., Brune, C., Wolterink, J.M.: Deep learning-based carotid artery vessel wall segmentation in black-blood MRI using anatomical priors. arXiv preprint arXiv:2112.01137 (2021)

3. Bullitt, E., et al.: Vascular attributes and malignant brain tumors. In: Ellis, R.E., Peters, T.M. (eds.) MICCAI 2003. LNCS, vol. 2878, pp. 671–679. Springer, Heidelberg (2003). https://doi.org/10.1007/978-3-540-39899-8_82
4. Bullitt, E., Gerig, G., Pizer, S.M., Lin, W., Aylward, S.R.: Measuring tortuosity of the intracerebral vasculature from MRA images. IEEE Trans. Med. Imaging **22**(9), 1163–1171 (2003)
5. Catmull, E., Clark, J.: Recursively generated b-spline surfaces on arbitrary topological meshes. Comput. Aided Des. **10**(6), 350–355 (1978)
6. Felkel, P., Wegenkittl, R., Buhler, K.: Surface models of tube trees. In: Proceedings Computer Graphics International, pp. 70–77. IEEE (2004)
7. Galarreta-Valverde, M.A., Macedo, M.M., Mekkaoui, C., Jackowski, M.P.: Three-dimensional synthetic blood vessel generation using stochastic l-systems. In: Medical Imaging 2013: Image Processing, vol. 8669, pp. 414–419. SPIE (2013)
8. Goodfellow, I., et al.: Generative adversarial networks. Commun. ACM **63**(11), 139–144 (2020)
9. Hamarneh, G., Jassi, P.: Vascusynth: simulating vascular trees for generating volumetric image data with ground-truth segmentation and tree analysis. Comput. Med. Imaging Graph. **34**(8), 605–616 (2010)
10. Higgins, I., et al.: beta-VAE: learning basic visual concepts with a constrained variational framework. In: International Conference on Learning Representations (2017)
11. Ho, J., Jain, A., Abbeel, P.: Denoising diffusion probabilistic models. Adv. Neural. Inf. Process. Syst. **33**, 6840–6851 (2020)
12. Kazeminia, S., et al.: GANs for medical image analysis. Artif. Intell. Med. **109**, 101938 (2020)
13. Lang, S., et al.: Three-dimensional quantification of capillary networks in healthy and cancerous tissues of two mice. Microvasc. Res. **84**(3), 314–322 (2012)
14. Li, J., Xu, K., Chaudhuri, S., Yumer, E., Zhang, H., Guibas, L.: Grass: generative recursive autoencoders for shape structures. ACM Trans. Graph. (TOG) **36**(4), 1–14 (2017)
15. Li, M., et al.: Grains: generative recursive autoencoders for indoor scenes. ACM Trans. Graph. (TOG) **38**(2), 1–16 (2019)
16. Looks, M., Herreshoff, M., Hutchins, D., Norvig, P.: Deep learning with dynamic computation graphs. arXiv preprint arXiv:1702.02181 (2017)
17. Merrem, A., Bartzsch, S., Laissue, J., Oelfke, U.: Computational modelling of the cerebral cortical microvasculature: effect of x-ray microbeams versus broad beam irradiation. Phys. Med. Biol. **62**(10), 3902 (2017)
18. Nash, C., Ganin, Y., Eslami, S.A., Battaglia, P.: Polygen: an autoregressive generative model of 3D meshes. In: International Conference on Machine Learning, pp. 7220–7229. PMLR (2020)
19. Park, J.J., Florence, P., Straub, J., Newcombe, R., Lovegrove, S.: DeepSDF: learning continuous signed distance functions for shape representation. In: Proceedings of the IEEE/CVF Conference on Computer Vision and Pattern Recognition, pp. 165–174 (2019)
20. Patil, A.G., Ben-Eliezer, O., Perel, O., Averbuch-Elor, H.: Read: recursive autoencoders for document layout generation. In: Proceedings of the IEEE/CVF Conference on Computer Vision and Pattern Recognition Workshops, pp. 544–545 (2020)
21. Rauch, N., Harders, M.: Interactive synthesis of 3D geometries of blood vessels. In: Theisel, H., Wimmer, M. (eds.) Eurographics 2021 - Short Papers. The Eurographics Association (2021)

22. Schneider, M., Reichold, J., Weber, B., Székely, G., Hirsch, S.: Tissue metabolism driven arterial tree generation. Med. Image Anal. **16**(7), 1397–1414 (2012)
23. Socher, R.: Recursive deep learning for natural language processing and computer vision. Stanford University (2014)
24. Socher, R., Lin, C.C., Manning, C., Ng, A.Y.: Parsing natural scenes and natural language with recursive neural networks. In: Proceedings of the 28th International Conference on Machine Learning (ICML-11), pp. 129–136 (2011)
25. Talou, G.D.M., Safaei, S., Hunter, P.J., Blanco, P.J.: Adaptive constrained constructive optimisation for complex vascularisation processes. Sci. Rep. **11**(1), 1–22 (2021)
26. Tetteh, G., et al.: Deepvesselnet: vessel segmentation, centerline prediction, and bifurcation detection in 3-D angiographic volumes. Front. Neurosci. 1285 (2020)
27. Wolterink, J.M., Leiner, T., Isgum, I.: Blood vessel geometry synthesis using generative adversarial networks. arXiv preprint arXiv:1804.04381 (2018)
28. Wu, J., Zhang, C., Xue, T., Freeman, B., Tenenbaum, J.: Learning a probabilistic latent space of object shapes via 3D generative-adversarial modeling. In: Advances in Neural Information Processing Systems, vol. 29 (2016)
29. Wu, J., Hu, Q., Ma, X.: Comparative study of surface modeling methods for vascular structures. Comput. Med. Imaging Graph. **37**(1), 4–14 (2013)
30. Xu, M., et al.: Generative AI-empowered simulation for autonomous driving in vehicular mixed reality metaverses. arXiv preprint arXiv:2302.08418 (2023)
31. Yang, G., Huang, X., Hao, Z., Liu, M.Y., Belongie, S., Hariharan, B.: Pointflow: 3D point cloud generation with continuous normalizing flows. In: Proceedings of the IEEE/CVF International Conference on Computer Vision, pp. 4541–4550 (2019)
32. Yang, X., Xia, D., Kin, T., Igarashi, T.: Intra: 3D intracranial aneurysm dataset for deep learning. In: Proceedings of the IEEE/CVF Conference on Computer Vision and Pattern Recognition, pp. 2656–2666 (2020)
33. Zamir, M.: Arterial branching within the confines of fractal l-system formalism. J. Gen. Physiol. **118**(3), 267–276 (2001)

Dense Transformer based Enhanced Coding Network for Unsupervised Metal Artifact Reduction

Wangduo Xie(✉) and Matthew B. Blaschko

Center for Processing Speech and Images, Department of ESAT, KU Leuven, Leuven, Belgium
{wangduo.xie,matthew.blaschko}@esat.kuleuven.be

Abstract. CT images corrupted by metal artifacts have serious negative effects on clinical diagnosis. Considering the difficulty of collecting paired data with ground truth in clinical settings, unsupervised methods for metal artifact reduction are of high interest. However, it is difficult for previous unsupervised methods to retain structural information from CT images while handling the non-local characteristics of metal artifacts. To address these challenges, we proposed a novel ***D**ense **T**ransformer based **E**nhanced **C**oding **Net**work* (**DTEC-Net**) for unsupervised metal artifact reduction. Specifically, we introduce a Hierarchical Disentangling Encoder, supported by the high-order dense process, and transformer to obtain densely encoded sequences with long-range correspondence. Then, we present a second-order disentanglement method to improve the dense sequence's decoding process. Extensive experiments and model discussions illustrate DTEC-Net's effectiveness, which outperforms the previous state-of-the-art methods on a benchmark dataset, and greatly reduces metal artifacts while restoring richer texture details.

Keywords: Metal artifact reduction · CT image restoration · Unsupervised learning · Enhanced coding

1 Introduction

CT technology can recover the internal details of the human body in a non-invasive way and has been widely used in clinical practice. However, if there is metal in the tissue, metal artifacts (MA) will appear in the reconstructed CT image, which will corrupt the image and affect the medical diagnosis [1,6].

In light of the clinical need for MA reduction, various traditional methods [5,6,12,19] have been proposed to solve the problem by using interpolation and iterative optimization. As machine learning research increasingly impacts medical imaging, deep learning based methods have been proposed for MA reduction. Specifically, these methods can be roughly divided into supervised and unsupervised categories according to the degree of supervision. In the supervised category, the methods [9,14,16,18] based on the dual domain (sinogram and image

---

Supplementary Information The online version contains supplementary material available at https://doi.org/10.1007/978-3-031-43907-0_8.

H. Greenspan et al. (Eds.): MICCAI 2023, LNCS 14220, pp. 77–86, 2023.
https://doi.org/10.1007/978-3-031-43907-0_8

domains) can achieve good performance for MA reduction. However, supervised learning methods are hindered by the lack of large-scale real-world data pairs consisting of "images with MA" and "images without MA" representing the same region. The lack of such data can lead algorithms trained on synthetic data to over-fit simulated data pairs, resulting in difficulties in generalizing to clinical settings [11]. Furthermore, although sinogram data can bring additional information, it is difficult to collect in realistic settings [8,15]. Therefore, *unsupervised methods based only on the image domain* are strongly needed in practice.

For unsupervised methods in the image domain, Liao *et al.* [8] used Generative Adversarial Networks (GANs) [2] to disentangle the MA from the underlying clean structure of the artifact-affected image in latent space by using unpaired data with and without MA. Although the method can separate the artifact component in the latent space, the features from the latent space can't represent rich low-level information of the original input. Further, it's also hard for the encoder to represent long-range correspondence across different regions. Accordingly, the restored image loses texture details and can't retain structure from the CT image. In the same unsupervised setting, Lyu *et al.* [11] directly separate the MA component and clean structure in image space using a CycleGAN-based method [25]. Although implementation in the image space makes it possible to construct dual constraints, directly operating in the image space affects the algorithm's performance upper limit, because it is difficult to encode in the image space as much low-level information as the feature space.

Considering the importance of low-level features in the latent space for generating the artifact-free component, we propose a novel *Dense Transformer based Enhanced Coding Network*(DTEC-Net) for unsupervised metal artifact reduction, which can obtain low-level features with hierarchical information and map them to a clean image space through adversarial training. DTEC-Net contains our developed Hierarchical Disentangling Encoder (HDE), which utilizes long-range correspondences obtained by a lightweight transformer and a high-order dense process to produce the enhanced coded sequence. To ease the burden of decoding the sequence, we also propose a second-order disentanglement method to finish the sequence decomposition. Extensive empirical results show that our method can not only reduce the MA greatly and generate high-quality images, but also surpasses the competing unsupervised approaches.

2 Methodology

We design a Hierarchical Disentangling Encoder(HDE) that can capture low-level sequences and enable high-performance restoration. Moreover, to reduce the burden of the decoder group brought by the complicated sequences, we propose a second-order disentanglement mechanism. The intuition is shown in Fig. 1.

2.1 Hierarchical Disentangling Encoder (HDE)

As shown in Fig. 2(a), the generator of DTEC-Net consists of three encoders and four decoders. We design the HDE to play the role of Encoder1 for enhanced

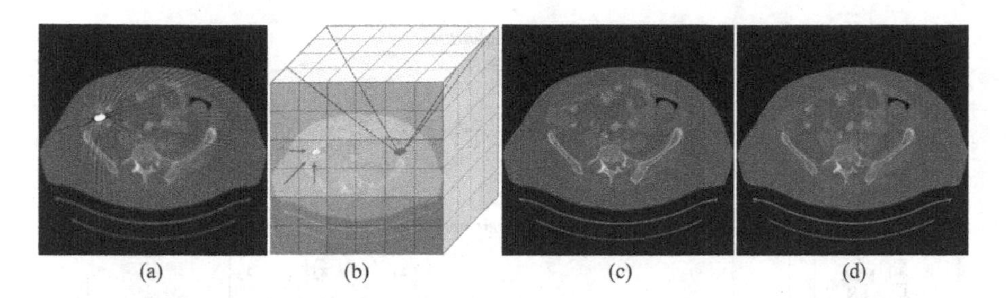

Fig. 1. (a) CT image with metal artifacts. (b) Blue/Orange arrows: Reuse of low-level features/Long-range correspondence. (c) Output of our DTEC-Net. (d) Ground truth. (Color figure online)

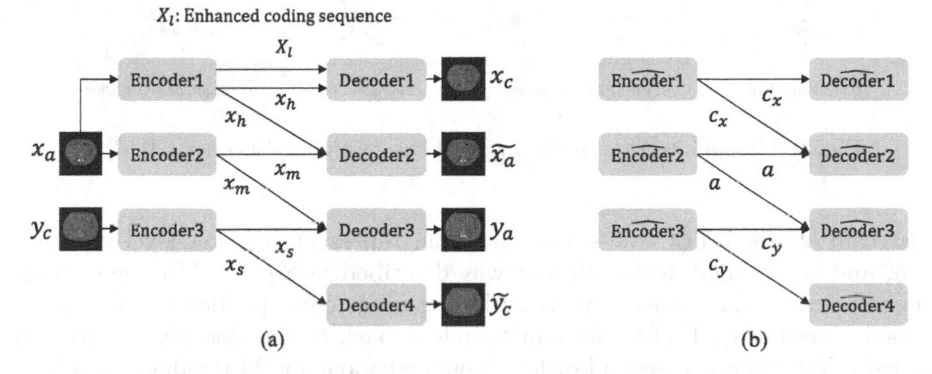

Fig. 2. (a) Generator of DTEC-Net. x_a: input with artifacts. y_c: unpaired input without artifacts. X_l and x_h are defined in the Sect. 2.1. x_m: the MA parts in latent space. x_s: the overall "structural part" in latent space. x_c (or y_a): the output after removing (or adding) the artifacts. $\widetilde{x_a}$ (or $\widetilde{y_c}$): the output of the identity map. (b) Generator of ADN [8]. The data relationship is shown in [8]. In addition to the difference in disentanglement, DTEC-Net and ADN also have different inner structures.

coding. Specifically, for the HDE's input image $x_a \in R^{1\times H\times W}$ with MA, HDE first uses a convolution for the preliminary feature extraction and produces a high-dimensional tensor x_{l_0} with c channels. Then, x_{l_0} will be encoded by three *Dense Transformers for Disentanglement* (DTDs) in a first-order reuse manner [4,23]. Specifically, the output x_{l_i} of the ith DTD can be characterized as:

$$x_{l_i} = \begin{cases} f_{\text{DTD}_i}(f_{\text{s-hde}}(\text{cat}(x_{l_{i-1}}, x_{l_{i-2}}..., , x_{l_0}))), i = 2, ..., N. \\ f_{\text{DTD}_i}(x_{l_{i-1}}), i = 1. \end{cases} \tag{1}$$

In Eq. (1), $f_{\text{s-hde}}$ represents the channel compression of the concatenation of multiple DTDs' outputs, and N represents the total number of DTDs in the HDE. As shown in Fig. 3, HDE can obtain the hierarchical information sequence $X_l \triangleq \{x_{l_0}, ...x_{l_N}\}$ and high-level semantic features $x_h \triangleq x_{l_N}$.

As shown in Fig. 2(b), Encoder1 of ADN cannot characterize upstream low-level information, and results in limited performance. By using HDE, the

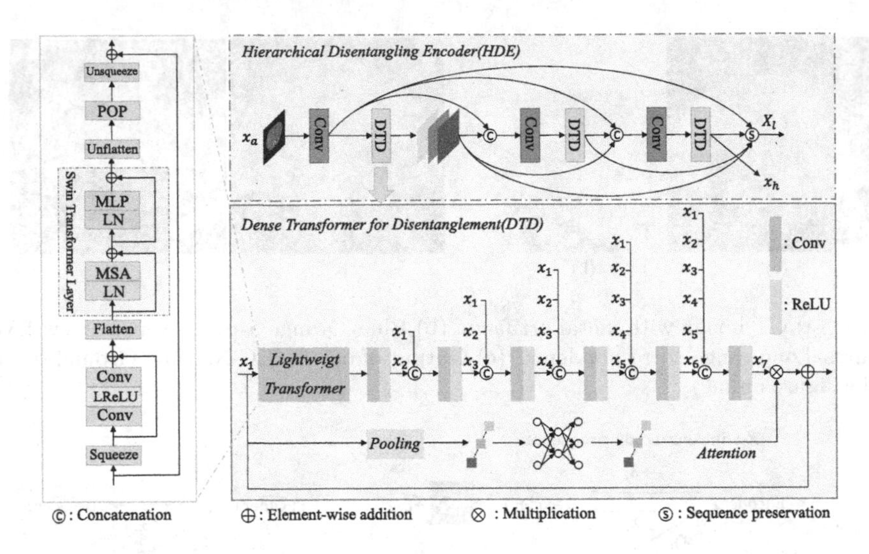

Fig. 3. The architecture of Hierarchical Disentangling Encoder (HDE).

upstream of the DTEC-Net's Encoder1 can represent rich low-level information, and be encoded in the efficient way described in Eq. (1). After generating the enhanced coding sequences X_l with long-range correspondence and densely reused information, DTEC-Net can decode it back to the clean image domain by using the proposed second-order disentanglement for MA reduction, which reduces the decoder group's burden to a large extent.

2.2 Dense Transformer for Disentanglement (DTD)

In addition to the first-order feature multiplexing given in Eq. (1), HDE also uses the DTD to enable second-order feature reuse. The relationship between HDE and DTD is shown in Fig. 3. Inspired by [7,20], DTD first uses a lightweight transformer based on the Swin transformer [7,10] to represent content-based information with long-range correspondence inside of every partition window. It then performs in-depth extraction and second-order reuse.

Specifically, the input $x_1 \in R^{C\times H\times W}$ of the DTD will be processed sequentially by the lightweight transformer and groups of convolutions in the form of second-order dense connections. The output x_{j+1} of the jth convolution with ReLU, which is connected in a second-order dense pattern, can be expressed as:

$$x_{j+1} = \begin{cases} f_{c_j}(\text{cat}(x_1, x_2, ..., x_j)), j = 2, 3, ..., J. \\ f_{c_j}(f_{\text{transformer-light}}(x_j)), j = 1. \end{cases} \tag{2}$$

In Eq. (2), f_{c_j} indicates the jth convolution with ReLU after the lightweight transformer, and the J indicates the total number of convolutions after the lightweight transformer and is empirically set to six. The dense connection method can effectively reuse low-level features [22,23] so that the latent space including these type of features will help the decoder to restore clean images without metal artifacts. Because the low-level information on different channels has different importance to the final restoration task, we use the channel attention mechanism [3] to filter the output of the final convolution layer:

$$x_{\text{out}} = x_{J+1} \odot f_{\text{MLP}}(f_{\text{pooling}}(x_1)) + x_1, \tag{3}$$

where $\odot$ represents the Hadamard product, f_{MLP} indicates a multi-layer perceptron with only one hidden layer, and f_{pooling} represents global pooling.

Because the transformer usually requires a large amount of data for training and CT image datasets are usually smaller than those for natural images, we do lightweight processing for the Swin transformer. Specifically, for an input tensor $x \in R^{C \times H \times W}$ of the lightweight transformer, the number of channels will be reduced from C to C_{in} to lighten the burden of the attention matrix. Then, a residual block is employed to extract information with low redundancy.

After completing lightweight handling, the tensor will first be partitioned into multiple local windows and flattened to $x_{\text{in}} \in R^{(\frac{HW}{P^2}) \times P^2 \times C_{in}}$ according the pre-operation [7,10] of the Swin transformer. $P \times P$ represents the window size for partitioning as shown in Fig. 1(b). Then, the attention matrix belonging to the ith window can be calculated by pairwise multiplication between converted vectors in $S_i \triangleq \{x_{\text{in}}(i, j, :)|j = 0, 1, ..., P^2 - 1\}$. Specifically, by using a linear map from $R^{C_{\text{in}}}$ to R^{C_a} for every vector in S_i, the query key and value: Q, K, $V \in R^{P^2 \times C_a}$ can be derived. Afterwards, the attention matrix for each window can be obtained by the following formula [10]:

$$\text{Attention}(Q, K, V) = \text{SoftMax}(QK^T/\sqrt{C_a})V. \tag{4}$$

In actual operation, we use window-based multi-head attention (MSA) [7,10,13] to replace the single-head attention because of the performance improvement [7]. The output of the Swin transformer layer will be unflattened and operated by post processing (POP) which consists of a classic convolution and layer norm (LN) with flatten and unflatten operations. After POP, the lightweight tensor with fewer channels will be unsqueezed to expand the channels, and finally added to the original input x in the form of residuals.

2.3 Second-Order Disentanglement for MA Reduction (SOD-MAR)

As mentioned in Sect. 2.1, X_l represents the hierarchical sequence and facilitates the generator's representation. However, X_l needs to be decoded by a high-capacity decoder to match the encoder. Considering that Decoder2 does not directly participate in the restoration branch and already loaded up the complicated artifact part x_m in traditional first-order disentanglement learning [8], to reduce the burden of the decoder group, we propose and analyze SOD-MAR.

Specifically, Decoder2 of DTEC-Net doesn't decode sequence X_l, it only decodes the combination of second-order disentangled information $x_h \in X_l$ and the latent feature x_m representing the artifact parts shown in Fig. 2(a). In order

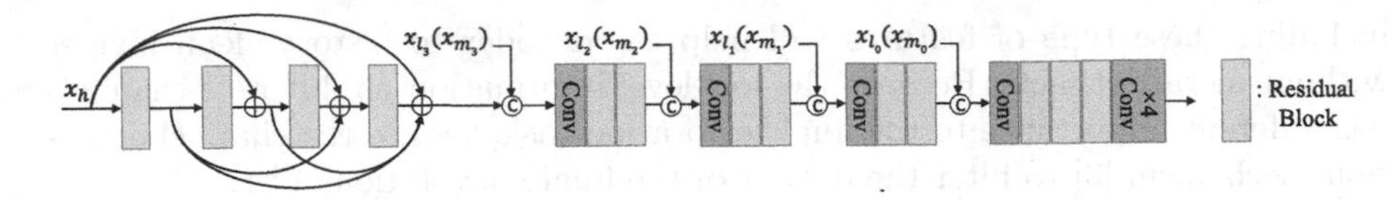

Fig. 4. The architecture of Decoder1 (Decoder2). (x_{m_*}) represents the Decoder2 case.

to complete the process, Decoder2 uses the structure shown in Fig. 4 to finish the decoding step, which is also used by Decoder1 to decode the sequence X_l.

Moreover, we don't only map the x_h into Decoder1 and Decoder2 while dropping the $X_l \backslash \{x_h\}$ to implement the burden reduction, because the low-level information in $X_l \backslash \{x_h\}$ is vital for restoring artifact-free images. Furthermore, x_h will be disturbed by noise from the approaching target x_a of Decoder2 while information $X_l \setminus \{x_h\}$ upstream from the HDE can counteract the noise disturbance to a certain extent. The reason behind the counteraction is that the update to upstream parameters is not as large as that of the downstream parameters.

2.4 Loss Function

Following [8], we use discriminators D_0, D_1 to constrain the output x_c and y_a:

$$L_{adv} = \mathbb{E}[\log(1 - D_0(x_c)) + \log(D_0(y_c))] + \mathbb{E}[\log(D_1(x_a)) + \log(1 - D_1(y_a))]. \tag{5}$$

The above x_a, y_c represent the input as shown in Fig. 2(a). Following [8], we use the reconstruction loss L_{rec} to constrain the identity map, and also use the artifact consistency loss L_{art} and self-reduction loss L_{self} to control the optimization process. The coefficients for each of these losses are set as in [8].

3 Empirical Results

Synthesized DeepLesion Dataset. Following [8], we randomly select 4186 images from DeepLesion [17] and 100 metal templates [21] to build a dataset. The simulation is consistent with [8,21]. For training, we randomly select 3986 images from DeepLesion combined with 90 metal templates for simulation. The 3986 images will be divided to two disjoint image sets with and without MA after simulation. Then a random combination can form the physically unpaired data with and without MA in the training process. Besides, another 200 images combined with the remaining 10 metal templates are used for the testing process.

Real Clinic Dataset. We randomly combine 6165 artifacts-affected images and 20729 artifacts-free images from SpineWeb[1] [8] for training, and 105 artifacts-affected images from SpineWeb for testing.

[1] spineweb.digitalimaginggroup.ca.

Implementation Details. We use peak signal-to-noise ratio (PSNR) and structural similarity index (SSIM) to measure performance. We use mean squared error (MSE) only for measuring ablation experiments. For Synthesized DeepLesion dataset (and Real Clinic dataset), we set the batch size to 2 (and 2) and trained the network for 77 (and 60) epochs using the Adam optimizer. Our DTEC-Net was implemented in Pytorch using an Nvidia Tesla P100.

Table 1. Ablation study on Synthesized DeepLesion under different settings. ↑: Higher value is better; ↓: Lower value is better. The best values are in **bold**.

Model	PSNR↑	SSIM↑	MSE↓
HDE with one DTD	34.46	0.937	27.12
HDE with two DTD	34.71	0.938	24.96
HDE with three DTD (only Transformer)	34.31	0.936	27.40
HDE with three DTD (without SOD-MAR)	34.91	0.940	24.36
HDE with three DTD (with SOD-MAR, Ours)	**35.11**	**0.941**	**22.89**

3.1 Ablation Study

To verify the effectiveness of the proposed methods, ablation experiments were carried out on Synthesized DeepLesion. The results are shown in Table 1.

The Impact of DTD in HDE. In this experiment, we change the encoding ability of HDE by changing the number of DTDs. We first use only one DTD to build the HDE, then the PSNR is 0.65 dB lower than our DTEC-Net using three DTDs. Additionally, the average MSE in this case is much higher than DTEC-Net. When the number of DTDs increases to two, the performance improves by 0.25 dB and is already better than the SOTA method [11]. As we further increase the number of DTDs to three, the PSNR and SSIM increase 0.4 dB and 0.003, respectively. The number of DTDs is finally set to three in a trade-off between computation and performance. To match different encoders and decoders and facilitate training, we also adjust the accept headers of Decoder1 to adapt to the sequence length determined by the different numbers of DTDs.

Only Transformer in DTD. Although the transformer can obtain better long-range correspondence than convolutions, it lacks the multiplexing of low-level information. For every DTD in DTEC-Net, we delete the second-order feature reuse pattern and only keep the lightweight transformer, the degraded version's results are 0.8 dB lower than our DTEC-Net. At the same time, great instability appears in generative adversarial training. So, only using the transformer cannot achieve good results in reducing metal artifacts.

Removing SOD-MAR. Although SOD-MAR mainly helps by easing the burden of decoding as discussed in Sect. 2.3, it also has a performance gain compared to first-order disentanglement. We delete the SOD-MAR in DTEC-Net and let x_h be the unique feature decoded by Decoder1. The Performance is 0.2 dB lower than our DTEC-Net, while MSE increases by 1.47.

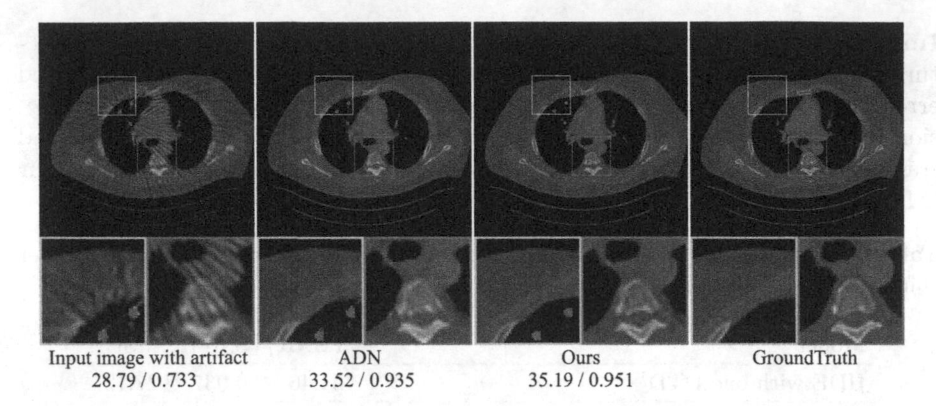

Fig. 5. Visual comparison(Metal implants are colored in red. The bottom values represent PSNR/SSIM). Our method has sharper edges and richer textures than ADN. (Color figure online)

Table 2. Quantitative comparison of different methods on Synthesized DeepLesion. The best results are in **bold**.

Method Classification	Method	PSNR↑	SSIM↑
Conventional	LI [5]	32.00 [8]	0.910 [8]
Supervised	CNNMAR [21]	32.50 [8]	0.914 [8]
Unsupervised	CycleGAN [25]	30.80 [8]	0.729 [8]
Unsupervised	RCN [24]	32.98 [11]	0.918 [11]
Unsupervised	ADN [8]	33.60 [8]	0.924 [8]
Unsupervised	U-DuDoNet [11]	34.54 [11]	0.934 [11]
Unsupervised	DTEC-Net(Ours)	**35.11**	**0.941**

3.2 Comparison to State-of-the-Art (SOTA)

For a fair comparison, we mainly compare with SOTA methods under unsupervised settings: ADN [8], U-DuDoNet [11], RCN [24], and CycleGAN [25]. We also compare with the traditional method LI [5] and classical supervised method CNNMAR [21]. The quantitative results of ADN, CycleGAN, CNNMAR and LI are taken from [8], the results of U-DuDoNet and RCN are taken from [11]. Because ADN has open-source code, we run their code for qualitative results.

Quantitative Results. As shown in Table 2. For the Synthesized DeepLesion Dataset, our method has the highest PSNR and SSIM value and outperforms the baseline ADN by 1.51 dB in PSNR and 0.017 in SSIM. At the same time, it also exceeds the SOTA method U-DuDoNet by 0.57 dB. For the Real Clinic Dataset, the numerical results can't be calculated because the ground truth does not exist. We will present the qualitative results in the appendix. Furthermore, as our work is single-domain based, it has the potential to be easily applied in clinical practice.

Qualitative Results. A visual comparison is shown in Fig. 5. Our method not only reduces artifacts to a large extent, but also has sharper edges and richer textures than the compared method. More results are shown in the appendix.

4 Conclusion

In this paper, we proposed a Dense Transformer based Enhanced Coding Network (DTEC-Net) for unsupervised metal-artifact reduction. In DTEC-Net, we developed a Hierarchical Disentangling Encoder (HDE) to represent long-range correspondence and produce an enhanced coding sequence. By using this sequence, the DTEC-Net can better recover low-level characteristics. In addition, to decrease the burden of decoding, we specifically design a Second-order Disentanglement for MA Reduction (SOD-MAR) to finish the sequence decomposition. The extensive quantitative and qualitative experiments demonstrate our DTEC-Net's effectiveness and show it outperforms other SOTA methods.

Acknowledgements. This research work was undertaken in the context of Horizon 2020 MSCA ETN project "xCTing" (Project ID: 956172).

References

1. Barrett, J.F., Keat, N.: Artifacts in CT: recognition and avoidance. Radiographics **24**(6), 1679–1691 (2004)
2. Goodfellow, I., et al.: Generative adversarial networks. Commun. ACM **63**(11), 139–144 (2020)
3. Hu, J., Shen, L., Sun, G.: Squeeze-and-excitation networks. In: Proceedings of the IEEE Conference on Computer Vision and Pattern Recognition, pp. 7132–7141 (2018)
4. Huang, G., Liu, Z., Van Der Maaten, L., Weinberger, K.Q.: Densely connected convolutional networks. In: Proceedings of the IEEE Conference on Computer Vision and Pattern Recognition, pp. 4700–4708 (2017)
5. Kalender, W.A., Hebel, R., Ebersberger, J.: Reduction of CT artifacts caused by metallic implants. Radiology **164**(2), 576–577 (1987)
6. Lemmens, C., Faul, D., Nuyts, J.: Suppression of metal artifacts in CT using a reconstruction procedure that combines map and projection completion. IEEE Trans. Med. Imaging **28**(2), 250–260 (2008)
7. Liang, J., Cao, J., Sun, G., Zhang, K., Van Gool, L., Timofte, R.: SwinIR: image restoration using swin transformer. In: Proceedings of the IEEE/CVF International Conference on Computer Vision, pp. 1833–1844 (2021)
8. Liao, H., Lin, W.A., Zhou, S.K., Luo, J.: ADN: artifact disentanglement network for unsupervised metal artifact reduction. IEEE Trans. Med. Imaging **39**(3), 634–643 (2019)
9. Lin, W.A., et al.: Dudonet: dual domain network for CT metal artifact reduction. In: Proceedings of the IEEE/CVF Conference on Computer Vision and Pattern Recognition, pp. 10512–10521 (2019)
10. Liu, Z., et al.: Swin transformer: hierarchical vision transformer using shifted windows. In: Proceedings of the IEEE/CVF International Conference on Computer Vision, pp. 10012–10022 (2021)

11. Lyu, Y., Fu, J., Peng, C., Zhou, S.K.: U-DuDoNet: unpaired dual-domain network for CT metal artifact reduction. In: de Bruijne, M., et al. (eds.) MICCAI 2021. LNCS, vol. 12906, pp. 296–306. Springer, Cham (2021). https://doi.org/10.1007/978-3-030-87231-1_29
12. Meyer, E., Raupach, R., Lell, M., Schmidt, B., Kachelrieß, M.: Normalized metal artifact reduction (NMAR) in computed tomography. Med. Phys. **37**(10), 5482–5493 (2010)
13. Vaswani, A., et al.: Attention is all you need. In: Advances in Neural Information Processing Systems, vol. 30 (2017)
14. Wang, H., Li, Y., Zhang, H., Meng, D., Zheng, Y.: Indudonet+: a deep unfolding dual domain network for metal artifact reduction in CT images. Med. Image Anal. **85**, 102729 (2023)
15. Wang, H., Xie, Q., Li, Y., Huang, Y., Meng, D., Zheng, Y.: Orientation-shared convolution representation for CT metal artifact learning. In: Wang, L., Dou, Q., Fletcher, P.T., Speidel, S., Li, S. (eds.) MICCAI 2022. LNCS, vol. 13436, pp. 665–675. Springer, Cham (2022). https://doi.org/10.1007/978-3-031-16446-0_63
16. Wang, T., et al.: Dual-domain adaptive-scaling non-local network for CT metal artifact reduction. In: de Bruijne, M., et al. (eds.) MICCAI 2021. LNCS, vol. 12906, pp. 243–253. Springer, Cham (2021). https://doi.org/10.1007/978-3-030-87231-1_24
17. Yan, K., Wang, X., Lu, L., Summers, R.M.: Deeplesion: automated mining of large-scale lesion annotations and universal lesion detection with deep learning. J. Med. Imaging **5**(3), 036501–036501 (2018)
18. Yu, L., Zhang, Z., Li, X., Xing, L.: Deep sinogram completion with image prior for metal artifact reduction in CT images. IEEE Trans. Med. Imaging **40**(1), 228–238 (2020)
19. Zhang, H., Wang, L., Li, L., Cai, A., Hu, G., Yan, B.: Iterative metal artifact reduction for x-ray computed tomography using unmatched projector/backprojector pairs. Med. Phys. **43**(6Part1), 3019–3033 (2016)
20. Zhang, J., Zhang, Y., Gu, J., Zhang, Y., Kong, L., Yuan, X.: Accurate image restoration with attention retractable transformer. arXiv preprint arXiv:2210.01427 (2022)
21. Zhang, Y., Yu, H.: Convolutional neural network based metal artifact reduction in x-ray computed tomography. IEEE Trans. Med. Imaging **37**(6), 1370–1381 (2018)
22. Zhang, Y., Li, K., Li, K., Wang, L., Zhong, B., Fu, Y.: Image super-resolution using very deep residual channel attention networks. In: Ferrari, V., Hebert, M., Sminchisescu, C., Weiss, Y. (eds.) ECCV 2018. LNCS, vol. 11211, pp. 294–310. Springer, Cham (2018). https://doi.org/10.1007/978-3-030-01234-2_18
23. Zhang, Y., Tian, Y., Kong, Y., Zhong, B., Fu, Y.: Residual dense network for image super-resolution. In: Proceedings of the IEEE Conference on Computer Vision and Pattern Recognition, pp. 2472–2481 (2018)
24. Zhao, B., Li, J., Ren, Q., Zhong, Y.: Unsupervised reused convolutional network for metal artifact reduction. In: Yang, H., Pasupa, K., Leung, A.C.-S., Kwok, J.T., Chan, J.H., King, I. (eds.) ICONIP 2020. CCIS, vol. 1332, pp. 589–596. Springer, Cham (2020). https://doi.org/10.1007/978-3-030-63820-7_67
25. Zhu, J.Y., Park, T., Isola, P., Efros, A.A.: Unpaired image-to-image translation using cycle-consistent adversarial networks. In: Proceedings of the IEEE International Conference on Computer Vision, pp. 2223–2232 (2017)

Multi-scale Cross-restoration Framework for Electrocardiogram Anomaly Detection

Aofan Jiang[1,2], Chaoqin Huang[1,2,4], Qing Cao[3], Shuang Wu[3], Zi Zeng[3], Kang Chen[3], Ya Zhang[1,2], and Yanfeng Wang[1,2](✉)

[1] Shanghai Jiao Tong University, Shanghai, China
{stillunnamed,huangchaoqin,ya_zhang,wangyanfeng}@sjtu.edu.cn
[2] Shanghai AI Laboratory, Shanghai, China
[3] Ruijin Hospital, Shanghai Jiao Tong University School of Medicine, Shanghai, China
{cq30553,ck11208}@rjh.com.cn, {shuang-renata,zengzidoct}@sjtu.edu.cn
[4] National University of Singapore, Singapore, Singapore

Abstract. Electrocardiogram (ECG) is a widely used diagnostic tool for detecting heart conditions. Rare cardiac diseases may be underdiagnosed using traditional ECG analysis, considering that no training dataset can exhaust all possible cardiac disorders. This paper proposes using anomaly detection to identify any unhealthy status, with normal ECGs solely for training. However, detecting anomalies in ECG can be challenging due to significant inter-individual differences and anomalies present in both global rhythm and local morphology. To address this challenge, this paper introduces a novel multi-scale cross-restoration framework for ECG anomaly detection and localization that considers both local and global ECG characteristics. The proposed framework employs a two-branch autoencoder to facilitate multi-scale feature learning through a masking and restoration process, with one branch focusing on global features from the entire ECG and the other on local features from heartbeat-level details, mimicking the diagnostic process of cardiologists. Anomalies are identified by their high restoration errors. To evaluate the performance on a large number of individuals, this paper introduces a new challenging benchmark with signal point-level ground truths annotated by experienced cardiologists. The proposed method demonstrates state-of-the-art performance on this benchmark and two other well-known ECG datasets. The benchmark dataset and source code are available at: https://github.com/MediaBrain-SJTU/ECGAD

Keywords: Anomaly Detection · Electrocardiogram

A. Jiang and C. Huang—Equal contribution.

Supplementary Information The online version contains supplementary material available at https://doi.org/10.1007/978-3-031-43907-0_9.

H. Greenspan et al. (Eds.): MICCAI 2023, LNCS 14220, pp. 87–97, 2023.
https://doi.org/10.1007/978-3-031-43907-0_9

1 Introduction

The electrocardiogram (ECG) is a monitoring tool widely used to evaluate the heart status of patients and provide information on cardiac electrophysiology. Developing automated analysis systems capable of detecting and identifying abnormal signals is crucial in light of the importance of ECGs in medical diagnosis and the need to ease the workload of clinicians. However, training a classifier on labeled ECGs that focus on specific diseases may not recognize new abnormal statuses that were not encountered during training, given the diversity and rarity of cardiac diseases [8,16,23]. On the other hand, anomaly detection, which is trained only on normal healthy data, can identify any potential abnormal status and avoid the failure to detect rare cardiac diseases [10,17,21].

The current anomaly detection techniques, including one-class discriminative approaches [2,14], reconstruction-based approaches [15,30], and self-supervised learning-based approaches [3,26], all operate under the assumption that models trained solely on normal data will struggle to process anomalous data and thus the substantial drop in performance presents an indication of anomalies. While anomaly detection has been widely used in the medical field to analyze medical images [12,24] and time-series data [18,29], detecting anomalies in ECG data is particularly challenging due to the substantial inter-individual differences and the presence of anomalies in both global rhythm and local morphology. So far, few studies have investigated anomaly detection in ECG [11,29]. TSL [29] uses expert knowledge-guided amplitude- and frequency-based data transformations to simulate anomalies for different individuals. BeatGAN [11] employs a generative adversarial network to separately reconstruct normalized heartbeats instead of the entire raw ECG signal. While BeatGAN alleviates individual differences, it neglects the important global rhythm information of the ECG.

This paper proposes a novel multi-scale cross-restoration framework for ECG anomaly detection and localization. To our best knowledge, this is the first work to integrate both local and global characteristics for ECG anomaly detection. To take into account multi-scale data, the framework adopts a two-branch autoencoder architecture, with one branch focusing on global features from the entire ECG and the other on local features from heartbeat-level details. A multi-scale cross-attention module is introduced, which learns to combine the two feature types for making the final prediction. This module imitates the diagnostic process followed by experienced cardiologists who carefully examine both the entire ECG and individual heartbeats to detect abnormalities in both the overall rhythm and the specific local morphology of the signal [7]. Each of the branches employs a masking and restoration strategy, *i.e.,* the model learns how to perform temporal-dependent signal inpainting from the adjacent unmasked regions within a specific individual. Such context-aware restoration has the advantage of making the restoration less susceptible to individual differences. During testing, anomalies are identified as samples or regions with high restoration errors.

To comprehensively evaluate the performance of the proposed method on a large number of individuals, we adopt the public PTB-XL database [22] with only patient-level diagnoses, and ask experienced cardiologists to provide signal

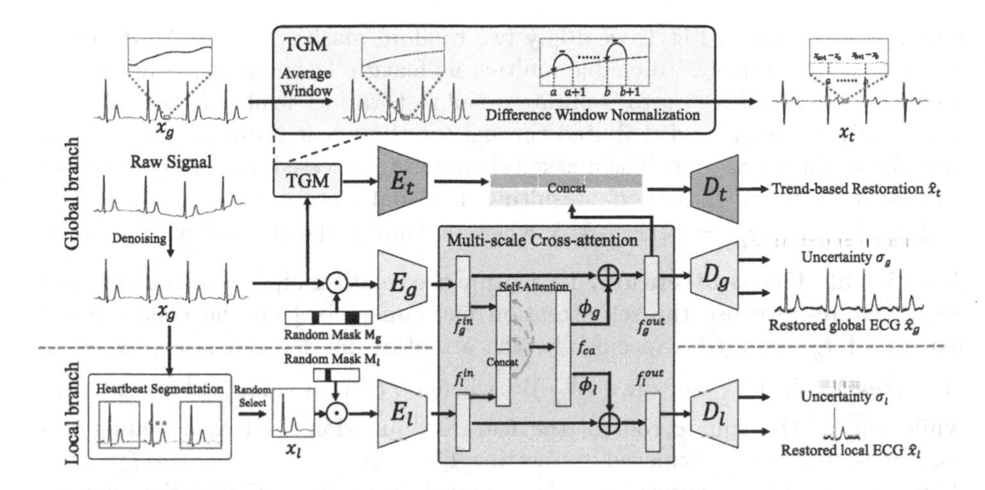

Fig. 1. The multi-scale cross-restoration framework for ECG anomaly detection.

point-level localization annotations. The resulting dataset is then introduced as a large-scale challenging benchmark for ECG anomaly detection and localization. The proposed method is evaluated on this challenging benchmark as well as on two traditional ECG anomaly detection benchmarks [6,13]. The experimental results have shown that the proposed method outperforms several state-of-the-art methods for both anomaly detection and localization, highlighting its potential for real-world clinical diagnosis.

2 Method

In this paper, we focus on unsupervised anomaly detection and localization on ECGs, training based on only normal ECG data. Formally, given a set of N normal ECGs denoted as $\{x_i, i = 1, ..., N\}$, where $x_i \in \mathbb{R}^D$ represents the vectorized representation of the i-th ECG consisting of D signal points, the objective is to train a computational model capable of identifying whether a new ECG is normal or anomalous, and localize the regions of anomalies in abnormal ECGs.

2.1 Multi-scale Cross-restoration

In Fig. 1, we present an overview of our two-branch framework for ECG anomaly detection. One branch is responsible for learning global ECG features, while the other focuses on local heartbeat details. Our framework comprises four main components: (i) masking and encoding, (ii) multi-scale cross-attention module, (iii) uncertainty-aware restoration, and (iv) trend generation module. We provide detailed explanations of each of these components in the following sections.

Masking and Encoding. Given a pair consisting of a global ECG signal $x_g \in \mathbb{R}^D$ and a randomly selected local heartbeat $x_l \in \mathbb{R}^d$ segmented from x_g for training, as shown in Fig. 1, we apply two random masks, M_g and M_l, to mask x_g and x_l, respectively. To enable multi-scale feature learning, M_l is applied to a consecutive small region to facilitate detail restoration, while M_g is applied to several distinct regions distributed throughout the whole sequence to facilitate global rhythm restoration. The masked signals are processed separately by global and local encoders, E_g and E_l, resulting in global feature $f_g^{in} = E_g(x_g \odot M_g)$ and local feature $f_l^{in} = E_l(x_l \odot M_l)$, where $\odot$ denotes the element-wise product.

Multi-scale Cross-attention. To capture the relationship between global and local features, we use the self-attention mechanism [20] on the concatenated feature of f_g^{in} and f_l^{in}. Specifically, the attention mechanism is expressed as $Attention(Q, K, V) = \text{softmax}(\frac{QK^T}{\sqrt{d_k}})V$, where Q, K, V are identical input terms, while $\sqrt{d_k}$ is the square root of the feature dimension used as a scaling factor. Self-attention is achieved by setting $Q = K = V = concat(f_g^{in}, f_l^{in})$. The cross-attention feature, f_{ca}, is obtained from the self-attention mechanism, which dynamically weighs the importance of each element in the combined feature. To obtain the final outputs of the global and local features, f_g^{out} and f_l^{out}, containing cross-scale information, we consider residual connections: $f_g^{out} = f_g^{in} + \phi_g(f_{ca})$, $f_l^{out} = f_l^{in} + \phi_l(f_{ca})$, where $\phi_g(\cdot)$ and $\phi_l(\cdot)$ are MLP architectures with two fully connected layers.

Uncertainty-Aware Restoration. Targeting signal restorations, features of f_g^{out} and f_l^{out} are decoded by two decoders, D_g and D_l, to obtain restored signals $\hat{x}_g$ and $\hat{x}_l$, respectively, along with corresponding restoration uncertainty maps σ_g and σ_l measuring the difficulty of restoration for various signal points, where $\hat{x}_g, \sigma_g = D_g(f_g^{out})$, $\hat{x}_l, \sigma_l = D_l(f_l^{out})$. An uncertainty-aware restoration loss is used to incorporate restoration uncertainty into the loss functions,

$$\mathcal{L}_{global} = \sum_{k=1}^{D} \{ \frac{(x_g^k - \hat{x}_g^k)^2}{\sigma_g^k} + \log \sigma_g^k \}, \quad \mathcal{L}_{local} = \sum_{k=1}^{d} \{ \frac{(x_l^k - \hat{x}_l^k)^2}{\sigma_l^k} + \log \sigma_l^k \}, \tag{1}$$

where for each function, the first term is normalized by the corresponding uncertainty, and the second term prevents predicting a large uncertainty for all restoration pixels following [12]. The superscript k represents the position of the k-th element of the signal. It is worth noting that, unlike [12], the uncertainty-aware loss is used for restoration, but not for reconstruction.

Trend Generation Module. The trend generation module (TGM) illustrated in Fig. 1 generates a smooth time-series trend $x_t \in \mathbb{R}^D$ by removing signal details, which is represented as the smooth difference between adjacent time-series signal points. An autoencoder (E_t and D_t) encodes the trend information into $E_t(x_t)$, which are concatenated with the global feature f_g^{out} to restore the global ECG $\hat{x}_t = D_t(\text{concat}(E_t(x_t), f_g^{out}))$. The restoration loss is defined as the Euclidean distance between x_g and $\hat{x}_t$, $\mathcal{L}_{trend} = \sum_{k=1}^{D}(x_g^k - \hat{x}_t^k)^2$. This process guides global feature learning using time-series trend information, emphasizing rhythm characteristics while de-emphasizing morphological details.

Loss Function. The final loss function for optimizing our model during the training process can be written as

$$\mathcal{L} = \mathcal{L}_{global} + \alpha\mathcal{L}_{local} + \beta\mathcal{L}_{trend}, \quad (2)$$

where α and β are trade-off parameters weighting the loss function. For simplicity, we adopt $\alpha = \beta = 1.0$ as the default.

2.2 Anomaly Score Measurement

For each test sample x, local ECGs from the segmented heartbeat set $\{x_{l,m}, m = 1, ..., M\}$ are paired with the global ECG x_g one at a time as inputs. The anomaly score $\mathcal{A}(x)$ is calculated to estimate the abnormality,

$$\mathcal{A}(x) = \sum_{k=1}^{D} \frac{(x_g^k - \hat{x}_g^k)^2}{\sigma_g^k} + \sum_{m=1}^{M}\sum_{k=1}^{d} \frac{(x_{l,m}^k - \hat{x}_{l,m}^k)^2}{\sigma_{l,m}^k} + \sum_{k=1}^{D}(x_g^k - \hat{x}_t^k)^2, \quad (3)$$

where the three terms correspond to global restoration, local restoration, and trend restoration, respectively. For localization, an anomaly score map is generated in the same way as Eq. (3), but without summing over the signal points. The anomalies are indicated by relatively large anomaly scores, and vice versa.

3 Experiments

Datasets. Three publicly available ECG datasets are used to evaluate the proposed method, including PTB-XL [22], MIT-BIH [13], and Keogh ECG [6].

- **PTB-XL** database includes clinical 12-lead ECGs that are 10 s in length for each patient, with only patient-level annotations. To build a new challenging anomaly detection and localization **benchmark**, 8167 normal ECGs are used for training, while 912 normal and 1248 abnormal ECGs are used for testing. We provide signal point-level annotations of 400 ECGs, including 22 different abnormal types, that were annotated by two experienced cardiologists. To our best knowledge, we are the first to explore ECG anomaly detection and localization across various patients on such a complex and large-scale database.
- **MIT-BIH** arrhythmia dataset divides the ECGs from 44 patients into independent heartbeats based on the annotated R-peak position, following [11]. 62436 normal heartbeats are used for training, while 17343 normal and 9764 abnormal beats are used for testing, with heartbeat-level annotations.
- **Keogh ECG** dataset includes 7 ECGs from independent patients, evaluating anomaly localization with signal point-level annotations. For each ECG, there is an anomaly subsequence that corresponds to a pre-ventricular contraction, while the remaining sequence is used as normal data to train the model. The ECGs are partitioned into fixed-length sequences of 320 by a sliding window with a stride of 40 during training and 160 during testing.

Table 1. Anomaly detection and anomaly localization results on PTB-XL database. Results are shown in the patient-level AUC for anomaly detection and the signal point-level AUC for anomaly localization, respectively. The best-performing method is in **bold**, and the second-best is underlined.

Method	Year	detection	localization
DAGMM [30]	2018	0.782	0.688
MADGAN [9]	2019	0.775	0.708
USAD [1]	2020	0.785	0.683
TranAD [18]	2022	0.788	0.685
AnoTran [25]	2022	0.762	0.641
TSL [29]	2022	0.757	0.509
BeatGAN [11]	2022	0.799	0.715
Ours	2023	**0.860**	**0.747**

Table 2. Anomaly detection results on MIT-BIH dataset, comparing with state-of-the-arts. Results are shown in terms of the AUC and F1 score for heartbeat-level classification. The best-performing method is in **bold**, and the second-best is underlined.

Method	Year	F1	AUC
DAGMM [30]	2018	0.677	0.700
MSCRED [27]	2019	0.778	0.627
USAD [1]	2020	0.384	0.352
TranAD [18]	2022	0.621	0.742
AnoTran [25]	2022	0.650	0.770
TSL [29]	2022	0.750	0.894
BeatGAN [11]	2022	0.816	0.945
Ours	2023	**0.883**	**0.969**

Table 3. Anomaly localization results on Keogh ECG [6] dataset, comparing with several state-of-the-arts. Results are shown in the signal point-level AUC. The best-performing method is in **bold**, and the second-best is underlined.

Methods	Year	A	B	C	D	E	F	G	Avg
DAGMM [30]	2018	0.672	0.612	0.805	0.713	0.457	0.662	0.676	0.657
MSCRED [27]	2019	0.667	0.633	0.798	0.714	0.461	0.746	0.659	0.668
MADGAN [9]	2019	0.688	**0.702**	**0.833**	0.664	0.463	0.692	0.678	0.674
USAD [1]	2020	0.667	0.616	0.795	0.715	0.462	0.649	0.680	0.655
GDN [4]	2021	0.695	0.611	0.790	0.674	0.458	0.648	0.671	0.650
CAE-M [28]	2021	0.657	0.618	0.802	0.715	0.457	0.708	0.671	0.661
TranAD [18]	2022	0.647	0.623	0.820	0.720	0.446	**0.780**	0.680	0.674
AnoTran [25]	2022	0.739	0.502	0.792	0.799	0.498	0.748	0.711	0.684
BeatGAN [11]	2022	0.803	0.623	0.783	0.747	0.506	0.757	**0.852**	0.724
Ours	2023	**0.832**	0.641	0.819	**0.815**	**0.543**	0.760	0.833	**0.749**

Evaluation Protocols. The performance of anomaly detection and localization is quantified using the area under the Receiver Operating Characteristic curve (AUC), with a higher AUC value indicating a better method. To ensure comparability across different annotation levels, we used patient-level, heartbeat-level, and signal point-level AUC for each respective setting. For heartbeat-level classification, the F1 score is also reported following [11].

Implementation Details. The ECG is pre-processed by a Butterworth filter and Notch filter [19] to remove high-frequency noise and eliminate ECG baseline wander. The R-peaks are detected with an adaptive threshold following [5], which

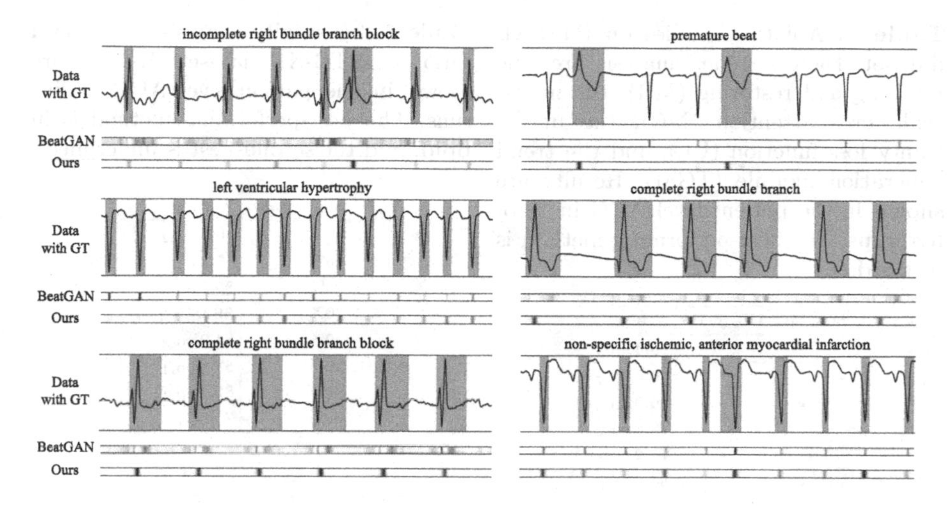

Fig. 2. Anomaly localization visualization on PTB-XL with different abnormal types. Ground truths are highlighted in red boxes on the ECG data, and anomaly localization results for each case, compared with the state-of-the-art method, are attached below. (Color figure online)

does not require any learnable parameters. The positions of the detected R-peaks are then used to segment the ECG sequence into a set of heartbeats.

We use a convolutional-based autoencoder, following the architecture proposed in [11]. The model is trained using the AdamW optimizer with an initial learning rate of 1e-4 and a weight decay coefficient of 1e-5 for 50 epochs on a single NVIDIA GTX 3090 GPU, with a single cycle of cosine learning rate used for decay scheduling. The batch size is set to 32. During testing, the model requires 2365M GPU memory and achieves an inference speed of 4.2 fps.

3.1 Comparisons with State-of-the-Arts

We compare our method with several time-series anomaly detection methods, including heartbeat-level detection method BeatGAN [11], patient-level detection method TSL [29], and several signal point-level anomaly localization methods [1,4,9,18,25,27,28,30]. For a fair comparison, we re-trained all the methods under the same experimental setup. For those methods originally designed for signal point-level tasks only [1,9,18,25,30], we use the mean value of anomaly localization results as their heartbeat-level or patient-level anomaly scores.

Anomaly Detection. The anomaly detection performance on PTB-XL is summarized in Table 1. The proposed method achieves 86.0% AUC in patient-level anomaly detection and outperforms all baselines by a large margin (10.3%). Table 2 displays the comparison results on MIT-BIH, where the proposed method achieves a heartbeat-level AUC of 96.9%, showing an improvement of 2.4% over the state-of-the-art BeatGAN (94.5%). Furthermore, the F1-score of the proposed method is 88.3%, which is 6.7% higher than BeatGAN (81.6%).

Table 4. Ablation studies on PTB-XL dataset. Factors under analysis are: the masking and restoring (MR), the multi-scale cross-attention (MC), the uncertainty loss function (UL), and the trend generation module (TGM). Results are shown in the patient-level AUC in % of five runs. The best-performing method is in **bold**.

MR	MC	UL	TGM	AUC
				$70.4_{\pm 0.3}$
✓				$80.4_{\pm 0.7}$
	✓			$80.3_{\pm 0.3}$
		✓		$72.8_{\pm 2.0}$
			✓	$71.2_{\pm 0.5}$
✓	✓			$84.8_{\pm 0.8}$
✓	✓	✓		$85.2_{\pm 0.4}$
✓	✓	✓	✓	**86.0**$_{\pm 0.1}$

Table 5. Sensitivity analysis w.r.t. mask ratio on PTB-XL dataset. Results are shown in the patient-level AUC of five runs. The best-performing method is in **bold**, and the second-best is underlined.

Mask Ratio	AUC
0%	$80.2_{\pm 0.2}$
10%	$85.2_{\pm 0.2}$
20%	$\underline{85.5}_{\pm 0.3}$
30%	**86.0**$_{\pm 0.1}$
40%	$84.9_{\pm 0.3}$
50%	$83.8_{\pm 0.1}$
60%	$82.9_{\pm 0.1}$
70%	$75.8_{\pm 1.0}$

Anomaly Localization. Table 1 presents the results of anomaly localization on our proposed benchmark for multiple individuals. The proposed method achieves a signal point-level AUC of 74.7%, outperforming all baselines (3.2% higher than BeatGAN). It is worth noting that TSL, which is not designed for localization, shows poor performance in this task. Table 3 shows the signal point-level anomaly localization results for each independent individual on Keogh ECG. Overall, the proposed method achieves the best or second-best performance compared to other methods on six subsets and the highest mean AUC among all subsets (74.9%, 2.5% higher than BeatGAN), indicating its effectiveness. The proposed method shows a lower standard deviation (±10.5) across the seven subsets compared to TranAD (±11.3) and BeatGAN (±11.0), which indicates good generalizability of the proposed method across different subsets.

Anomaly Localization Visualization. We present visualization results of anomaly localization on several samples from our proposed benchmark in Fig. 2, with ground truths annotated by experienced cardiologists. Regions with higher anomaly scores are indicated by darker colors. Our proposed method outperforms BeatGAN in accurately localizing various types of ECG anomalies, including both periodic and episodic anomalies, such as incomplete right bundle branch block and premature beats. Our method though provides narrower localization results than ground truths, as it is highly sensitive to abrupt unusual changes in signal values, but still represents the important areas for anomaly identification, a fact confirmed by experienced cardiologists.

3.2 Ablation Study and Sensitivity Analysis

Ablation studies were conducted on PTB-XL to confirm the effectiveness of individual components of the proposed method. Table 4 shows that each module contributes positively to the overall performance of the framework. When none of the modules were employed, the method becomes a ECG reconstruction approach with a naive L2 loss and lacks cross-attention in multi-scale data. When individually adding the MR, MC, UL, and TGM modules to the baseline model without any of them, the AUC values improve from 70.4% to 80.4%, 80.3%, 72.8%, and 71.2%, respectively, demonstrating the effectiveness of each module. Moreover, as the modules are added in sequence, the performance improves step by step from 70.4% to 86.0% in AUC, highlighting the combined impact of all modules on the proposed framework.

We conduct a sensitivity analysis on the mask ratio, as shown in Table 5. Restoration with a 0% masking ratio can be regarded as reconstruction, which takes an entire sample as input and its target is to output the input sample. Results indicate that the model's performance first improves and then declines as the mask ratio increases from 0% to 70%. This trend is due to the fact that a low mask ratio can limit the model's feature learning ability during restoration, while a high ratio can make it increasingly difficult to restore the masked regions. Therefore, there is a trade-off between maximizing the model's potential and ensuring a reasonable restoration difficulty. The optimal mask ratio is 30%, which achieves the highest anomaly detection result (86.0% in AUC).

4 Conclusion

This paper proposes a novel framework for ECG anomaly detection, where features of the entire ECG and local heartbeats are combined with a masking-restoration process to detect anomalies, simulating the diagnostic process of cardiologists. A challenging benchmark, with signal point-level annotations provided by experienced cardiologists, is proposed, facilitating future research in ECG anomaly localization. The proposed method outperforms state-of-the-art methods, highlighting its potential in real-world clinical diagnosis.

Acknowledgement. This work is supported by the National Key R&D Program of China (No. 2022ZD0160702), STCSM (No. 22511106101, No. 18DZ2270700, No. 21DZ1100100), 111 plan (No. BP0719010), the Youth Science Fund of National Natural Science Foundation of China (No. 7210040772) and National Facility for Translational Medicine (Shanghai) (No. TMSK-2021-501), and State Key Laboratory of UHD Video and Audio Production and Presentation.

References

1. Audibert, J., Michiardi, P., Guyard, F., Marti, S., Zuluaga, M.A.: Usad: unsupervised anomaly detection on multivariate time series. In: Proceedings of the 26th ACM SIGKDD International Conference on Knowledge Discovery & Data Mining, pp. 3395–3404 (2020)
2. Chalapathy, R., Menon, A.K., Chawla, S.: Robust, deep and inductive anomaly detection. In: Ceci, M., Hollmén, J., Todorovski, L., Vens, C., Džeroski, S. (eds.) ECML PKDD 2017. LNCS (LNAI), vol. 10534, pp. 36–51. Springer, Cham (2017). https://doi.org/10.1007/978-3-319-71249-9_3

3. Chen, L., Bentley, P., Mori, K., Misawa, K., Fujiwara, M., Rueckert, D.: Self-supervised learning for medical image analysis using image context restoration. Med. Image Anal. **58**, 101539 (2019)
4. Deng, A., Hooi, B.: Graph neural network-based anomaly detection in multivariate time series. In: Proceedings of the AAAI Conference on Artificial Intelligence, vol. 35, pp. 4027–4035 (2021)
5. van Gent, P., Farah, H., van Nes, N., van Arem, B.: Analysing noisy driver physiology real-time using off-the-shelf sensors: Heart rate analysis software from the taking the fast lane project. J. Open Res. Softw. **7**(1) (2019)
6. Keogh, E., Lin, J., Fu, A.: Hot sax: finding the most unusual time series subsequence: Algorithms and applications. In: Proceedings of the IEEE International Conference on Data Mining, pp. 440–449. Citeseer (2004)
7. Khan, M.G.: Step-by-step method for accurate electrocardiogram interpretation. In: Rapid ECG Interpretation, pp. 25–80. Humana Press, Totowa, NJ (2008)
8. Kiranyaz, S., Ince, T., Gabbouj, M.: Real-time patient-specific ECG classification by 1-d convolutional neural networks. IEEE Trans. Biomed. Eng. **63**(3), 664–675 (2015)
9. Li, D., Chen, D., Jin, B., Shi, L., Goh, J., Ng, S.-K.: MAD-GAN: multivariate anomaly detection for time series data with generative adversarial networks. In: Tetko, I.V., Kůrková, V., Karpov, P., Theis, F. (eds.) ICANN 2019. LNCS, vol. 11730, pp. 703–716. Springer, Cham (2019). https://doi.org/10.1007/978-3-030-30490-4_56
10. Li, H., Boulanger, P.: A survey of heart anomaly detection using ambulatory electrocardiogram (ECG). Sensors **20**(5), 1461 (2020)
11. Liu, S., et al.: Time series anomaly detection with adversarial reconstruction networks. IEEE Trans. Knowl. Data Eng. (2022)
12. Mao, Y., Xue, F.-F., Wang, R., Zhang, J., Zheng, W.-S., Liu, H.: Abnormality detection in chest X-Ray images using uncertainty prediction autoencoders. In: Martel, A.L., et al. (eds.) MICCAI 2020. LNCS, vol. 12266, pp. 529–538. Springer, Cham (2020). https://doi.org/10.1007/978-3-030-59725-2_51
13. Moody, G.B., Mark, R.G.: The impact of the mit-bih arrhythmia database. IEEE Eng. Med. Biol. Mag. **20**(3), 45–50 (2001)
14. Ruff, L., et al.: Deep one-class classification. In: International Conference on Machine Learning, pp. 4393–4402. PMLR (2018)
15. Schlegl, T., Seeböck, P., Waldstein, S.M., Langs, G., Schmidt-Erfurth, U.: f-anogan: fast unsupervised anomaly detection with generative adversarial networks. Med. Image Anal. **54**, 30–44 (2019)
16. Shaker, A.M., Tantawi, M., Shedeed, H.A., Tolba, M.F.: Generalization of convolutional neural networks for ECG classification using generative adversarial networks. IEEE Access **8**, 35592–35605 (2020)
17. Shen, L., Yu, Z., Ma, Q., Kwok, J.T.: Time series anomaly detection with multiresolution ensemble decoding. In: Proceedings of the AAAI Conference on Artificial Intelligence, vol. 35, pp. 9567–9575 (2021)
18. Tuli, S., Casale, G., Jennings, N.R.: Tranad: deep transformer networks for anomaly detection in multivariate time series data. In: International Conference on Very Large Databases 15(6), pp. 1201–1214 (2022)
19. Van Gent, P., Farah, H., Van Nes, N., Van Arem, B.: Heartpy: a novel heart rate algorithm for the analysis of noisy signals. Transport. Res. F: Traffic Psychol. Behav. **66**, 368–378 (2019)
20. Vaswani, A., et al.: Attention is all you need. In: Advances in Neural Information Processing Systems 30 (2017)

21. Venkatesan, C., Karthigaikumar, P., Paul, A., Satheeskumaran, S., Kumar, R.: ECG signal preprocessing and SVM classifier-based abnormality detection in remote healthcare applications. IEEE Access **6**, 9767–9773 (2018)
22. Wagner, P., et al.: Ptb-xl, a large publicly available electrocardiography dataset. Sci. Data **7**(1), 1–15 (2020)
23. Wang, J., et al.: Automated ECG classification using a non-local convolutional block attention module. Comput. Methods Programs Biomed. **203**, 106006 (2021)
24. Wolleb, J., Bieder, F., Sandkühler, R., Cattin, P.C.: Diffusion models for medical anomaly detection. In: Medical Image Computing and Computer Assisted Intervention-MICCAI 2022, pp. 35–45. Springer, Cham (2022). https://doi.org/10.1007/978-3-031-16452-1_4
25. Xu, J., Wu, H., Wang, J., Long, M.: Anomaly transformer: time series anomaly detection with association discrepancy. In: International Conference on Learning Representations (2022)
26. Ye, F., Huang, C., Cao, J., Li, M., Zhang, Y., Lu, C.: Attribute restoration framework for anomaly detection. IEEE Trans. Multimedia **24**, 116–127 (2022)
27. Zhang, C., et al.: A deep neural network for unsupervised anomaly detection and diagnosis in multivariate time series data. In: Proceedings of the AAAI Conference on Artificial Intelligence, vol. 33, pp. 1409–1416 (2019)
28. Zhang, Y., Chen, Y., Wang, J., Pan, Z.: Unsupervised deep anomaly detection for multi-sensor time-series signals. IEEE Trans. Knowl. Data Eng. **35**(2), 2118–2132 (2021)
29. Zheng, Y., Liu, Z., Mo, R., Chen, Z., Zheng, W.s., Wang, R.: Task-oriented self-supervised learning for anomaly detection in electroencephalography. In: International Conference on Medical Image Computing and Computer-Assisted Intervention, pp. 193–203. Springer, Cham (2022). https://doi.org/10.1007/978-3-031-16452-1_19
30. Zong, B., et al.: Deep autoencoding gaussian mixture model for unsupervised anomaly detection. In: International Conference on Learning Representations (2018)

Correlation-Aware Mutual Learning for Semi-supervised Medical Image Segmentation

Shengbo Gao[1], Ziji Zhang[2], Jiechao Ma[1], Zihao Li[1], and Shu Zhang[1](✉)

[1] Deepwise AI Lab, Beijing, China
zhangshu@deepwise.com
[2] School of Artificial Intelligence, Beijing University of Posts and Telecommunications, Beijing, China

Abstract. Semi-supervised learning has become increasingly popular in medical image segmentation due to its ability to leverage large amounts of unlabeled data to extract additional information. However, most existing semi-supervised segmentation methods only focus on extracting information from unlabeled data, disregarding the potential of labeled data to further improve the performance of the model. In this paper, we propose a novel Correlation Aware Mutual Learning (CAML) framework that leverages labeled data to guide the extraction of information from unlabeled data. Our approach is based on a mutual learning strategy that incorporates two modules: the Cross-sample Mutual Attention Module (CMA) and the Omni-Correlation Consistency Module (OCC). The CMA module establishes dense cross-sample correlations among a group of samples, enabling the transfer of label prior knowledge to unlabeled data. The OCC module constructs omni-correlations between the unlabeled and labeled datasets and regularizes dual models by constraining the omni-correlation matrix of each sub-model to be consistent. Experiments on the Atrial Segmentation Challenge dataset demonstrate that our proposed approach outperforms state-of-the-art methods, highlighting the effectiveness of our framework in medical image segmentation tasks. The codes, pre-trained weights, and data are publicly available.

Keywords: Semi-supervised learning · Medical Image Segmentation. · Mutual learning · Cross-sample correlation

1 Introduction

Despite the remarkable advancements achieved through the use of deep learning for automatic medical image segmentation, the scarcity of precisely annotated training data remains a significant obstacle to the widespread adoption of such

https://github.com/Herschel555/CAML
S. Gao and Z. Zhang—Both authors contributed equally to this work.
Z. Zhang—Work done as an intern in Deepwise AI Lab.

H. Greenspan et al. (Eds.): MICCAI 2023, LNCS 14220, pp. 98–108, 2023.
https://doi.org/10.1007/978-3-031-43907-0_10

techniques in clinical settings. As a solution, the concept of semi-supervised segmentation has been proposed to enable models to be trained using less annotated but abundant unlabeled data.

Recently, methods that adopt the co-teaching [3,11,19] or mutual learning [25] paradigm have emerged as a promising approach for semi-supervised learning. Those methods adopt two simultaneously updated models, each trained to predict the prediction results of its counterpart, which can be seen as a combination of the notions of consistency regularization [1,4,9,14,15] and entropy minimization [2,7,20,22,24]. In the domain of semi-supervised medical image segmentation, MC-Net [19] has shown significant improvements in segmentation performance.

With the rapid advancement of semi-supervised learning, the importance of unlabeled data has garnered increased attention across various disciplines in recent years. However, the role of labeled data has been largely overlooked, with the majority of semi-supervised learning techniques treating labeled data supervision as merely an initial step of the training pipeline or as a means to ensure training convergence [6,15,26]. Recently, methods that can leverage labeled data to directly guide information extraction from unlabeled data have attracted the attention of the community [16]. In the domain of semi-supervised medical image segmentation, there exist shared characteristics between labeled and unlabeled data that possess greater intuitiveness and instructiveness for the algorithm. Typically, partially labeled clinical datasets exhibit similar foreground features, including comparable texture, shape, and appearance among different samples. As such, it can be hypothesized that constructing a bridge across the entire training dataset to connect labeled and unlabeled data can effectively transfer prior knowledge from labeled data to unlabeled data and facilitate the extraction of information from unlabeled data, ultimately overcoming the performance bottleneck of semi-supervised learning methods.

Based on the aforementioned conception, we propose a novel *Correlation Aware Mutual Learning (CAML)* framework to explicitly model the relationship between labeled and unlabeled data to effectively utilize the labeled data. Our proposed method incorporates two essential components, namely the *Cross-sample Mutual Attention module (CMA)* and the *Omni-Correlation Consistency module (OCC)*, to enable the effective transfer of labeled data information to unlabeled data. The *CMA* module establishes mutual attention among a group of samples, leading to a mutually reinforced representation of co-salient features between labeled and unlabeled data. Unlike conventional methods, where supervised signals from labeled and unlabeled samples are separately back-propagated, the proposed *CMA* module creates a new information propagation path among each pixel in a group of samples, which synchronously enhances the feature representation ability of each intra-group sample.

In addition to the *CMA* module, we introduce the *OCC* module to regularize the segmentation model by explicitly modeling the omni-correlation between unlabeled features and a group of labeled features. This is achieved by constructing a memory bank to store the labeled features as a reference set of features or basis vectors. In each iteration, a portion of features from the memory bank is utilized to calculate the omni-correlation with unlabeled features, reflecting

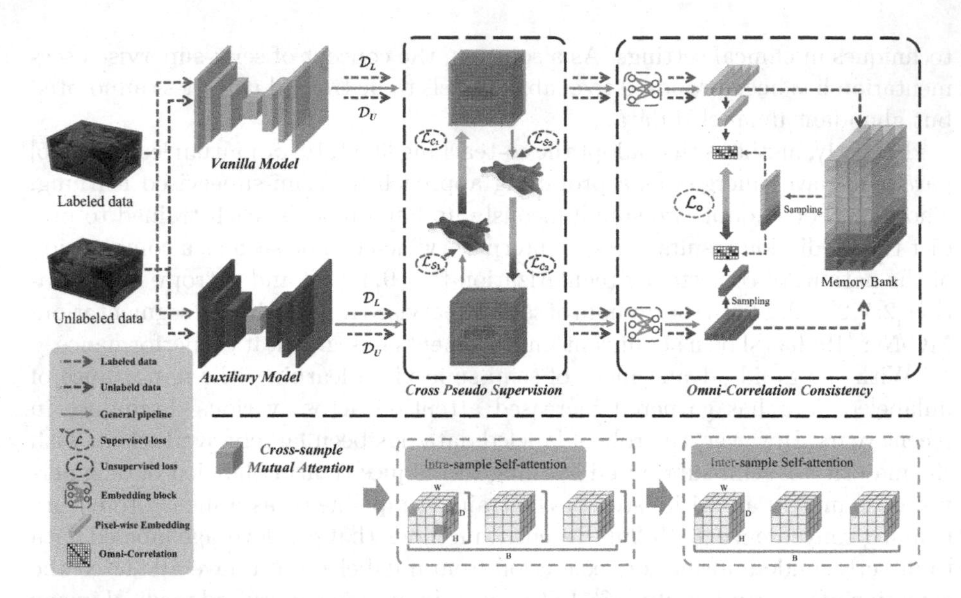

Fig. 1. Overview of our proposed *CAML*. *CAML* adopts a co-teaching scheme with cross-pseudo supervision. The *CMA* module incorporated into the auxiliary network and the *OCC* module are introduced for advanced cross-sample relationship modeling.

the similarity relationship of an unlabeled pixel with respect to a set of basis vectors of the labeled data. Finally, we constrain the omni-correlation matrix of each sub-model to be consistent to regularize the entire framework. With the proposed omni-correlation consistency, the labeled data features serve as anchor groups to guide the representation learning of the unlabeled data feature and explicitly encourage the model to learn a more unified feature distribution among unlabeled data.

In summary, our contributions are threefold: (1) We propose a novel *Correlation Aware Mutual Learning (CAML)* framework that focuses on the efficient utilization of labeled data to address the challenge of semi-supervised medical image segmentation. (2) We introduce the *Cross-sample Mutual Attention module (CMA)* and the *Omni-Correlation Consistency module (OCC)* to establish cross-sample relationships directly. (3) Experimental results on a benchmark dataset demonstrate significant improvements over previous SOTAs, especially when only a small number of labeled images are available.

2 Method

2.1 Overview

Figure 1 gives an overview of *CAML*. We adopt a co-teaching paradigm like MC-Net [19] to enforce two parallel networks to predict the prediction results of its counterpart. To achieve efficient cross-sample relationship modeling and enable information propagation among labeled and unlabeled data in a mini-batch, we incorporate a ***Cross-sample Mutual Attention*** module to the auxiliary segmentation network f_a, whereas the vanilla segmentation network f_v remains the original V-Net structure. In addition, we employ an ***Omni-Correlation Consistency*** regularization to further regularize the representation learning of the unlabeled data. Details about those two modules will be elaborated on in the following sections. The total loss of *CAML* can be formulated as:

$$L = L_s + \lambda_c l_c + \lambda_o l_o \tag{1}$$

where l_o represents the proposed omni-correlation consistency loss, while L_s and l_c are the supervised loss and the cross-supervised loss implemented in the *Cross Pseudo Supervision(CPS)* module. λ_c and λ_o are the weights to control l_c and l_o separately. During the training procedure, a batch of mixed labeled and unlabeled samples are fed into the network. The supervised loss is only applied to labeled data, while all samples are utilized to construct cross-supervised learning. Please refer to [3] for a detailed description of the *CPS* module and loss design of L_s and l_c.

2.2 Cross-Sample Mutual Attention Module

To enable information propagation through any positions of any samples in a mini-batch, one can simply treat each pixel's feature vector as a token and perform self-attentions for all tokens in a mini-batch. However, this will make the computation cost prohibitively large as the computation complexity of self-attention is $O(n^2)$ with respect to the number of tokens. We on the other hand adopt two sequentially mounted self-attention modules along different dimensions to enable computation efficient mutual attention among all pixels.

As illustrated in Fig. 1, the proposed *CMA* module consists of two sequential transformer encoder layers, termed as E_1 and E_2, each including a multi-head attention and a *MLP* block with a layer normalization after each block. For an input feature map $a_{in} \in \mathbb{R}^{b \times c \times k}$, where $k = h' \times w' \times d'$, b represents batch size and c is the dimension of a_{in}, E_1 performs intra-sample self-attention on the spatial dimension of each sample. This is used to model the information propagation paths between every pixel position within each sample. Then, to further enable information propagation among different samples, we perform an inter-sample self-attention along the batch dimension. In other words, along the b dimension, the pixels located in the same spatial position from samples are fed into a self-attention module to construct cross-sample relationships.

In *CAML*, we employ the proposed *CMA* module in the auxiliary segmentation network f_a, whereas the vanilla segmentation network f_v remains the original V-Net structure. The reasons can be summarized into two folds. From deployment perspective, the insertion of the CMA module requires a batch size of large than 1 to model the attention among samples within a mini-batch, which is not applicable for model inference(batchsize=1). From the perspective of model design, we model the vanilla and the auxiliary branch with different architectures to increase the architecture heterogeneous for better performance in a mutual learning framework.

2.3 Omni-Correlation Consistency Regularization

In this chapter, we introduce *Omni-Correlation Consistency (OCC)* to formulate additional model regularization. The core of the *OCC* module is omni-correlation, which is a kind of similarity matrix that is calculated between the feature of an unlabeled pixel and a group of prototype features sampled from labeled instances features. It reflects the similar relationship of an unlabeled pixel with respect to a set of labeled reference pixels. During the training procedure, we explicitly constrain the omni-correlation calculated using heterogeneous unlabeled features from those two separate branches to remain the same. In practice, we use an Omni-correlation matrix to formulate the similarity distribution between unlabeled features and the prototype features.

Let g_v and g_a denote two projection heads attached to the backbones of f_v and f_a separately, and $z_v \in \mathbb{R}^{m \times c'}$ and $z_a \in \mathbb{R}^{m \times c'}$ represent two sets of embeddings sampled from their projected features extracted from unlabeled samples, where m is the number of sampled features and c' is the dimension of the projected features. It should be noted that z_v and z_a are sampled from the embeddings corresponding to the same set of positions on unlabeled samples. Suppose $z_p \in \mathbb{R}^{n \times c'}$ represents a set of prototype embeddings sampled from labeled instances, where n represents the number of sampled prototype features, the omni-correlation matrix calculation between z_v and z_p can be formulated as:

$$sim_{vp_i} = \frac{exp(cos(z_v, z_{p_i}) * t)}{\sum_{j=1}^{n} exp(cos(z_v, z_{p_j}) * t)}, i \in \{1, ..., n\} \tag{2}$$

where cos means the cosine similarity and t is the temperature hyperparameter. $sim_{vp} \in \mathbb{R}^{m \times n}$ is the calculated omni-correlation matrix. Similarly, the similarity distribution sim_{ap} between z_a and z_p can be calculated by replacing z_v with z_a.

To constrain the consistency of omni-correlation between dual branches, the omni-correlation consistency regularization can be conducted with the cross-entropy loss l_{ce} as follows:

$$l_o = \frac{1}{m} \sum l_{ce} \left(sim_{vp}, sim_{ap}\right) \tag{3}$$

Memory Bank Construction. We utilize a memory bank T to iteratively update prototype embeddings for *OCC* computation. Specifically, T initializes N slots for each labeled training sample and updates prototype embeddings with filtered labeled features projected by g_v and g_a. To ensure the reliability of the features stored in T, we select embeddings on the positions where both f_v and f_a have the correct predictions and update T with the mean fusion of the projected features projected by g_v and g_a. For each training sample, following [5], T updates slots corresponding to the labeled samples in the current mini-batch in a query-like manner.

Embeddings Sampling. For computation efficiency, omni-correlation is not calculated on all labeled and unlabeled pixels. Specifically, we have developed a confidence-based mechanism to sample the pixel features from the unlabeled data. Practically, to sample z_v and z_a from unlabeled features, we first select the pixels where f_v and f_a have the same prediction. For each class, we sort the confidence scores of these pixels, and then select features of the top i pixels as the sampled unlabeled features. Thus, $m = i \times C$, where C represents the number of classes. With regards to the prototype embeddings, we randomly sample j embeddings from each class among all the embeddings contained in T and $n = j \times C$ to increase its diversity.

3 Experiments and Results

Dataset. Our method is evaluated on the Left Atrium (LA) dataset [21] from the 2018 Atrial Segmentation Challenge. The dataset comprises 100 gadolinium-enhanced MR imaging scans (GE-MRIs) and their ground truth masks, with an isotropic resolution of $0.625^3\,\text{mm}^3$. Following [23], we use 80 scans for training and 20 scans for testing. All scans are centered at the heart region and cropped accordingly, and then normalized to zero mean and unit variance.

Implementation Details. We implement our *CAML* using PyTorch 1.8.1 and CUDA 10.2 on an NVIDIA TITAN RTX GPU. For training data augmentation, we randomly crop sub-volumes of size $112 \times 112 \times 80$ following [23]. To ensure a fair comparison with existing methods, we use the V-Net [13] as the backbone for all our models. During training, we use a batch size of 4, with half of the images annotated and the other half unannotated. We train the entire framework using the SGD optimizer, with a learning rate of 0.01, momentum of 0.9, and weight decay of 1e−4 for 15000 iterations. To balance the loss terms in the training process, we use a time-dependent Gaussian warming up function for λ_U and λ_C, where $\lambda(t) = \beta * e^{-5(1-t/t_{max})^2}$, and set β to 1 and 0.1 for λ_U and λ_C, respectively. For the *OCC* module, we set c' to 64, j to 256, and i to 12800. During inference, prediction results from the vanilla V-Net are used with a general sliding window strategy without any post-processing.

Quantitative Evaluation and Comparison. Our CAML is evaluated on four metrics: Dice, Jaccard, 95% Hausdorff Distance (95HD), and Average Surface Distance (ASD). It is worth noting that the previous researchers reported results (**Reported Metrics** in Table 1) on LA can be confusing, with some studies reporting results from the final training iteration, while others report the best performance obtained during training. However, the latter approach can lead to overfitting of the test dataset and unreliable model selection. To ensure a fair comparison, we perform all experiments three times with a fixed set of randomly selected seeds on the same machine, and report the mean and standard deviation of the results from the final iteration.

The results on LA are presented in Table 1. The results of the full-supervised V-Net model trained on different ratios serve as the lower and upper bounds

Table 1. Comparison with state-of-the-art methods on the LA database. **Metrics** reported the *mean±standard* results with three random seeds, **Reported Metrics** are the results reported in the original paper.

Method	Scans used		Metrics				Reported Metrics			
	Labeled	Unlabeled	Dice(%)	Jaccard(%)	95HD(voxel)	ASD(voxel)	Dice(%)	Jaccard(%)	95HD(voxel)	ASD(voxel)
V-Net	4	0	$43.32_{\pm 8.62}$	$31.43_{\pm 6.90}$	$40.19_{\pm 1.11}$	$12.13_{\pm 0.57}$	52.55	39.60	47.05	9.87
V-Net	8	0	$79.87_{\pm 1.23}$	$67.60_{\pm 1.88}$	$26.65_{\pm 6.36}$	$7.94_{\pm 2.22}$	78.57	66.96	21.20	6.07
V-Net	16	0	$85.94_{\pm 0.48}$	$75.99_{\pm 0.57}$	$16.70_{\pm 1.82}$	$4.80_{\pm 0.62}$	86.96	77.31	11.85	3.22
V-Net	80	0	$90.98_{\pm 0.67}$	$83.61_{\pm 1.06}$	$8.58_{\pm 2.34}$	$2.10_{\pm 0.59}$	91.62	84.60	5.40	1.64
UA-MT [23] (MICCAI'19)	4(5%)	76(95%)	$78.07_{\pm 0.90}$	$65.03_{\pm 0.96}$	$29.17_{\pm 3.82}$	$8.63_{\pm 0.98}$	-	-	-	-
SASSNet [8] (MICCAI'20)			$79.61_{\pm 0.54}$	$67.00_{\pm 0.59}$	$25.54_{\pm 4.60}$	$7.20_{\pm 1.21}$	-	-	-	-
DTC [10] (AAAI'21)			$80.14_{\pm 1.22}$	$67.88_{\pm 1.82}$	$24.08_{\pm 2.63}$	$7.18_{\pm 0.62}$	-	-	-	-
MC-Net [19] (MedIA'21)			$80.92_{\pm 3.88}$	$68.90_{\pm 5.09}$	$17.25_{\pm 6.08}$	$2.76_{\pm 0.49}$	-	-	-	-
URPC [12] (MedIA'22)			$80.75_{\pm 0.21}$	$68.54_{\pm 0.34}$	$19.81_{\pm 0.67}$	$4.98_{\pm 0.25}$	-	-	-	-
SS-Net [18] (MICCAI'22)			$83.33_{\pm 1.66}$	$71.79_{\pm 2.36}$	$15.70_{\pm 0.80}$	$4.33_{\pm 0.36}$	86.33	76.15	9.97	2.31
MC-Net+ [17] (MedIA'22)			$83.23_{\pm 1.41}$	$71.70_{\pm 1.99}$	$14.92_{\pm 2.56}$	$3.43_{\pm 0.64}$	-	-	-	-
ours			$87.34_{\pm 0.05}$	$77.65_{\pm 0.08}$	$9.76_{\pm 0.92}$	$2.49_{\pm 0.22}$	-	-	-	-
UA-MT [23] (MICCAI'19)	8(10%)	72(90%)	$85.81_{\pm 0.17}$	$75.41_{\pm 0.22}$	$18.25_{\pm 1.04}$	$5.04_{\pm 0.24}$	84.25	73.48	3.36	13.84
SASSNet [8] (MICCAI'20)			$85.71_{\pm 0.87}$	$75.35_{\pm 1.28}$	$14.74_{\pm 3.14}$	$4.00_{\pm 0.86}$	86.81	76.92	3.94	12.54
DTC [10] (AAAI'21)			$84.55_{\pm 1.72}$	$73.91_{\pm 2.36}$	$13.80_{\pm 0.16}$	$3.69_{\pm 0.25}$	-	-	-	-
MC-Net [19] (MedIA'21)			$86.87_{\pm 1.74}$	$78.49_{\pm 1.06}$	$11.17_{\pm 1.40}$	$2.18_{\pm 0.14}$	87.71	78.31	9.36	2.18
URPC [12] (MedIA'22)			$83.37_{\pm 0.21}$	$71.99_{\pm 0.31}$	$17.91_{\pm 0.73}$	$4.41_{\pm 0.17}$	-	-	-	-
SS-Net [18] (MICCAI'22)			$86.56_{\pm 0.69}$	$76.61_{\pm 1.03}$	$12.76_{\pm 0.58}$	$3.02_{\pm 0.19}$	88.55	79.63	7.49	1.90
MC-Net+ [17] (MedIA'22)			$87.68_{\pm 0.56}$	$78.27_{\pm 0.83}$	$10.35_{\pm 0.77}$	$1.85_{\pm 0.01}$	88.96	80.25	7.93	1.86
ours			$89.62_{\pm 0.20}$	$81.28_{\pm 0.32}$	$8.76_{\pm 1.39}$	$2.02_{\pm 0.17}$	-	-	-	-
UA-MT [23] (MICCAI'19)	16(20%)	64(80%)	$88.18_{\pm 0.69}$	$79.09_{\pm 1.05}$	$9.66_{\pm 2.99}$	$2.62_{\pm 0.59}$	88.88	80.21	2.26	7.32
SASSNet [8] (MICCAI'20)			$88.11_{\pm 0.34}$	$79.08_{\pm 0.48}$	$12.31_{\pm 4.14}$	$3.27_{\pm 0.96}$	89.27	80.82	3.13	8.83
DTC [10] (AAAI'21)			$87.79_{\pm 0.50}$	$78.52_{\pm 0.73}$	$10.29_{\pm 1.52}$	$2.50_{\pm 0.65}$	89.42	80.98	2.10	7.32
MC-Net [19] (MedIA'21)			$90.43_{\pm 0.52}$	$82.69_{\pm 0.75}$	$6.52_{\pm 0.66}$	$1.66_{\pm 0.14}$	90.34	82.48	6.00	1.77
URPC [12] (MedIA'22)			$87.68_{\pm 0.36}$	$78.36_{\pm 0.53}$	$14.39_{\pm 0.54}$	$3.52_{\pm 0.17}$	-	-	-	-
SS-Net [18] (MICCAI'22)			$88.19_{\pm 0.42}$	$79.21_{\pm 0.63}$	$8.12_{\pm 0.34}$	$2.20_{\pm 0.12}$	-	-	-	-
MC-Net+ [17] (MedIA'22)			$90.60_{\pm 0.39}$	$82.93_{\pm 0.64}$	$6.27_{\pm 0.25}$	$1.58_{\pm 0.07}$	91.07	83.67	5.84	1.67
ours			$90.78_{\pm 0.11}$	$83.19_{\pm 0.18}$	$6.11_{\pm 0.39}$	$1.68_{\pm 0.15}$	-	-	-	-

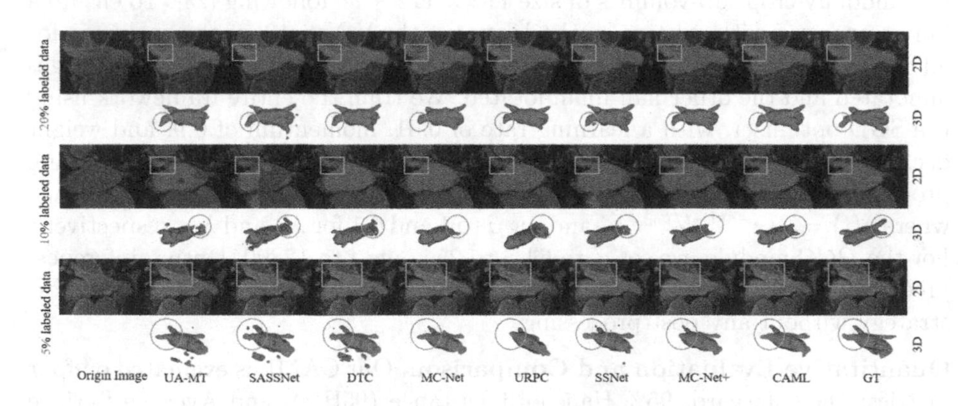

Fig. 2. Visualization of the segmentations results from different methods.

of each ratio setting. We report the reproduced results of state-of-the-art semi-supervised methods and corresponding reported results if available. By comparing the reproduced and reported results, we observe that although the performance of current methods generally shows an increasing trend with the development of algorithms, the performance of individual experiments can be unstable. and the reported results may not fully reflect the true performance.

It is evident from Table 1 that *CAML* outperforms other methods by a significant margin across all settings without incurring any additional inference or post-processing costs. With only 5% labeled data, *CAML* achieves 87.34% Dice score with an absolute improvement of 4.01% over the state-of-the-art. *CAML* also achieves 89.62% Dice score with only 10% labeled data. When the amount of labeled data is increased to 20%, the model obtains comparable results with the results of V-Net trained in 100% labeled data), achieving a Dice score of 90.78% compared to the upper-bound model's score of 90.98%. As presented in Table 1, through the effective transfer of knowledge between labeled and unlabeled data, *CAML* achieves impressive improvements.

Table 2. Ablation study of our proposed *CAML* on the LA database.

Scans used		Components			Metrics			
Labeled	Unlabeled	Baseline	OCC	CMA	Dice(%)	Jaccard(%)	95HD(voxel)	ASD(voxel)
4(5%)	76(95%)	√			$80.92_{\pm 3.88}$	$68.90_{\pm 5.09}$	$17.25_{\pm 6.08}$	$2.76_{\pm 0.49}$
		√	√		$83.12_{\pm 2.12}$	$71.73_{\pm 3.04}$	$16.94_{\pm 7.25}$	$4.51_{\pm 2.16}$
		√		√	$86.35_{\pm 0.26}$	$76.16_{\pm 0.40}$	$12.36_{\pm 0.20}$	$2.94_{\pm 0.21}$
		√	√	√	$87.34_{\pm 0.05}$	$77.65_{\pm 0.08}$	$9.76_{\pm 0.92}$	$2.49_{\pm 0.22}$
8(10%)	72(90%)	√			$86.87_{\pm 1.74}$	$78.49_{\pm 1.06}$	$11.17_{\pm 1.40}$	$2.18_{\pm 0.14}$
		√	√		$88.50_{\pm 3.25}$	$79.53_{\pm 0.51}$	$9.89_{\pm 0.83}$	$2.35_{\pm 0.21}$
		√		√	$88.84_{\pm 0.55}$	$80.05_{\pm 0.85}$	$8.50_{\pm 0.66}$	$1.97_{\pm 0.02}$
		√	√	√	$89.62_{\pm 0.20}$	$81.28_{\pm 0.32}$	$8.76_{\pm 1.39}$	$2.02_{\pm 0.17}$
16(20%)	64(80%)	√			$90.43_{\pm 0.52}$	$82.69_{\pm 0.75}$	$6.52_{\pm 0.66}$	$1.66_{\pm 0.14}$
		√	√		$90.27_{\pm 0.22}$	$82.42_{\pm 0.39}$	$6.96_{\pm 1.03}$	$1.91_{\pm 0.24}$
		√		√	$90.25_{\pm 0.28}$	$82.34_{\pm 0.43}$	$6.95_{\pm 0.09}$	$1.79_{\pm 0.18}$
		√	√	√	$90.78_{\pm 0.11}$	$83.19_{\pm 0.18}$	$6.11_{\pm 0.30}$	$1.68_{\pm 0.15}$

Table 1 also demonstrated that as the labeled data ratio declines, the model maintains a low standard deviation of results, which is significantly lower than other state-of-the-art methods. This finding suggests that *CAML* is highly stable and robust. Furthermore, the margin between our method and the state-of-the-art semi-supervised methods increases with the decline of the labeled data ratio, indicating that our method rather effectively transfers knowledge from labeled data to unlabeled data, thus enabling the model to extract more universal features from unlabeled data. Figure 2 shows the qualitative comparison results. The figure presents 2D and 3D visualizations of all the compared methods and the corresponding ground truth. As respectively indicated by the orange rectangle and circle in the 2D and 3D visualizations Our CAML achieves the best segmentation results compared to all other methods.

Ablation Study. In this section, we analyze the effectiveness of the proposed *CMA* module and *OCC* module. We implement the MC-Net as our baseline, which uses different up-sampling operations to introduce architecture heterogeneity. Table 2 presents the results of our ablation study. The results demonstrate that under 5% ratio, both *CMA* and *OCC* significantly improve the performance of the baseline. By combining these two modules, *CAML* achieves an absolute improvement of 6.42% in the Dice coefficient. Similar improvements can be observed for a data ratio of 10%. Under a labeled data ratio of 20%, the baseline performance is improved to 90.43% in the Dice coefficient, which is approximately comparable to the upper bound of a fully-supervised model. In this setting, adding the *CMA* and *OCC* separately may not achieve a significant improvement. Nonetheless, by combining these two modules in our proposed *CAML* framework, we still achieve the best performance in this setting, which further approaches the performance of a fully-supervised model.

4 Conclusion

In this paper, we proposed a novel framework named *CAML* for semi-supervised medical image segmentation. Our key idea is that cross-sample correlation should be taken into consideration for semi-supervised learning. To this end, two novel modules: *Cross-sample Mutual Attention(CMA)* and *Omni-Correlation Consistency(OCC)* are proposed to encourage efficient and direct transfer of the prior knowledge from labeled data to unlabeled data. Extensive experimental results on the LA dataset demonstrate that we outperform previous state-of-the-art results by a large margin without extra computational consumption in inference.

Acknowledgements. This work is funded by the Scientific and Technological Innovation 2030 New Generation Artificial Intelligence Project of the National Key Research and Development Program of China (No. 2021ZD0113302), Beijing Municipal Science and Technology Planning Project (No. Z201100005620008, Z211100003521009).

References

1. Berthelot, D., Carlini, N., Goodfellow, I., Papernot, N., Oliver, A., Raffel, C.A.: Mixmatch: a holistic approach to semi-supervised learning. In: Advances in Neural Information Processing Systems 32 (2019)
2. Cascante-Bonilla, P., Tan, F., Qi, Y., Ordonez, V.: Curriculum labeling: revisiting pseudo-labeling for semi-supervised learning. In: Proceedings of the AAAI Conference on Artificial Intelligence, vol. 35, pp. 6912–6920 (2021)
3. Chen, X., Yuan, Y., Zeng, G., Wang, J.: Semi-supervised semantic segmentation with cross pseudo supervision. In: Proceedings of the IEEE/CVF Conference on Computer Vision and Pattern Recognition, pp. 2613–2622 (2021)
4. French, G., Laine, S., Aila, T., Mackiewicz, M., Finlayson, G.: Semi-supervised semantic segmentation needs strong, varied perturbations. arXiv preprint arXiv:1906.01916 (2019)
5. He, K., Fan, H., Wu, Y., Xie, S., Girshick, R.: Momentum contrast for unsupervised visual representation learning. In: Proceedings of the IEEE/CVF Conference on Computer Vision and Pattern Recognition, pp. 9729–9738 (2020)

6. Kwon, D., Kwak, S.: Semi-supervised semantic segmentation with error localization network. In: Proceedings of the IEEE/CVF Conference on Computer Vision and Pattern Recognition, pp. 9957–9967 (2022)
7. Lee, D.H., et al.: Pseudo-label: the simple and efficient semi-supervised learning method for deep neural networks. In: Workshop on Challenges in Representation Learning, ICML, vol. 3, p. 896 (2013)
8. Li, S., Zhang, C., He, X.: Shape-aware semi-supervised 3D semantic segmentation for medical images. In: Martel, A.L., Abolmaesumi, P., Stoyanov, D., Mateus, D., Zuluaga, M.A., Zhou, S.K., Racoceanu, D., Joskowicz, L. (eds.) MICCAI 2020. LNCS, vol. 12261, pp. 552–561. Springer, Cham (2020). https://doi.org/10.1007/978-3-030-59710-8_54
9. Liu, Y., Tian, Y., Chen, Y., Liu, F., Belagiannis, V., Carneiro, G.: Perturbed and strict mean teachers for semi-supervised semantic segmentation. In: Proceedings of the IEEE/CVF Conference on Computer Vision and Pattern Recognition, pp. 4258–4267 (2022)
10. Luo, X., Chen, J., Song, T., Wang, G.: Semi-supervised medical image segmentation through dual-task consistency. In: AAAI Conference on Artificial Intelligence (2021)
11. Luo, X., Hu, M., Song, T., Wang, G., Zhang, S.: Semi-supervised medical image segmentation via cross teaching between CNN and transformer. In: International Conference on Medical Imaging with Deep Learning, pp. 820–833. PMLR (2022)
12. Luo, X., et al.: Semi-supervised medical image segmentation via uncertainty rectified pyramid consistency. Med. Image Anal. **80**, 102517 (2022)
13. Milletari, F., Navab, N., Ahmadi, S.A.: V-net: fully convolutional neural networks for volumetric medical image segmentation. In: 2016 Fourth International Conference on 3D Vision (3DV), pp. 565–571. IEEE (2016)
14. Mittal, S., Tatarchenko, M., Brox, T.: Semi-supervised semantic segmentation with high-and low-level consistency. IEEE Trans. Pattern Anal. Mach. Intell. **43**(4), 1369–1379 (2019)
15. Ouali, Y., Hudelot, C., Tami, M.: Semi-supervised semantic segmentation with cross-consistency training. In: Proceedings of the IEEE/CVF Conference on Computer Vision and Pattern Recognition, pp. 12674–12684 (2020)
16. Wu, L., Fang, L., He, X., He, M., Ma, J., Zhong, Z.: Querying labeled for unlabeled: Cross-image semantic consistency guided semi-supervised semantic segmentation. IEEE Trans. Pattern Anal. Mach. Intell. (2023)
17. Wu, Y., et al.: Mutual consistency learning for semi-supervised medical image segmentation. Med. Image Anal. **81**, 102530 (2022)
18. Wu, Y., Wu, Z., Wu, Q., Ge, Z., Cai, J.: Exploring smoothness and class-separation for semi-supervised medical image segmentation. In: MICCAI 2022, Part V, pp. 34–43. Springer, Cham (2022). https://doi.org/10.1007/978-3-031-16443-9_4
19. Wu, Y., Xu, M., Ge, Z., Cai, J., Zhang, L.: Semi-supervised left atrium segmentation with mutual consistency training. In: de Bruijne, M., Cattin, P.C., Cotin, S., Padoy, N., Speidel, S., Zheng, Y., Essert, C. (eds.) MICCAI 2021. LNCS, vol. 12902, pp. 297–306. Springer, Cham (2021). https://doi.org/10.1007/978-3-030-87196-3_28
20. Xie, Q., Luong, M.T., Hovy, E., Le, Q.V.: Self-training with noisy student improves imagenet classification. In: Proceedings of the IEEE/CVF Conference on Computer Vision and Pattern Recognition, pp. 10687–10698 (2020)
21. Xiong, Z., et al.: A global benchmark of algorithms for segmenting late gadolinium-enhanced cardiac magnetic resonance imaging. Medical Image Analysis (2020)

22. Yang, L., Zhuo, W., Qi, L., Shi, Y., Gao, Y.: St++: make self-training work better for semi-supervised semantic segmentation. In: Proceedings of the IEEE/CVF Conference on Computer Vision and Pattern Recognition, pp. 4268–4277 (2022)
23. Yu, L., Wang, S., Li, X., Fu, C.W., Heng, P.A.: Uncertainty-aware self-ensembling model for semi-supervised 3d left atrium segmentation. In: MICCAI (2019)
24. Yuan, J., Liu, Y., Shen, C., Wang, Z., Li, H.: A simple baseline for semi-supervised semantic segmentation with strong data augmentation. In: Proceedings of the IEEE/CVF International Conference on Computer Vision, pp. 8229–8238 (2021)
25. Zhang, P., Zhang, B., Zhang, T., Chen, D., Wen, F.: Robust mutual learning for semi-supervised semantic segmentation. arXiv preprint arXiv:2106.00609 (2021)
26. Zou, Y., et al.: Pseudoseg: designing pseudo labels for semantic segmentation. arXiv preprint arXiv:2010.09713 (2020)

TPRO: Text-Prompting-Based Weakly Supervised Histopathology Tissue Segmentation

Shaoteng Zhang[1,2,4], Jianpeng Zhang[2], Yutong Xie[3(✉)], and Yong Xia[1,2,4(✉)]

[1] Ningbo Institute of Northwestern Polytechnical University, Ningbo 315048, China
yxia@nwpu.edu.cn

[2] National Engineering Laboratory for Integrated Aero-Space-Ground-Ocean Big Data Application Technology, School of Computer Science and Engineering, Northwestern Polytechnical University, Xi'an 710072, China

[3] Australian Institute for Machine Learning, The University of Adelaide, Adelaide, SA, Australia
yutong.xie678@gmail.com

[4] Research and Development Institute of Northwestern Polytechnical University in Shenzhen, Shenzhen 518057, China

Abstract. Most existing weakly-supervised segmentation methods rely on class activation maps (CAM) to generate pseudo-labels for training segmentation models. However, CAM has been criticized for highlighting only the most discriminative parts of the object, leading to poor quality of pseudo-labels. Although some recent methods have attempted to extend CAM to cover more areas, the fundamental problem still needs to be solved. We believe this problem is due to the huge gap between image-level labels and pixel-level predictions and that additional information must be introduced to address this issue. Thus, we propose a text-prompting-based weakly supervised segmentation method (TPRO), which uses text to introduce additional information. TPRO employs a vision and label encoder to generate a similarity map for each image, which serves as our localization map. Pathological knowledge is gathered from the internet and embedded as knowledge features, which are used to guide the image features through a knowledge attention module. Additionally, we employ a deep supervision strategy to utilize the network's shallow information fully. Our approach outperforms other weakly supervised segmentation methods on benchmark datasets LUAD-HistoSeg and BCSS-WSSS datasets, setting a new state of the art. Code is available at: https://github.com/zhangst431/TPRO.

Keywords: Histopathology Tissue Segmentation · Weakly-Supervised Semantic Segmentation · Vision-Language

---

Supplementary Information The online version contains supplementary material available at https://doi.org/10.1007/978-3-031-43907-0_11.

H. Greenspan et al. (Eds.): MICCAI 2023, LNCS 14220, pp. 109–118, 2023.
https://doi.org/10.1007/978-3-031-43907-0_11

1 Introduction

Automated segmentation of histopathological images is crucial, as it can quantify the tumor micro-environment, provide a basis for cancer grading and prognosis, and improve the diagnostic efficiency of clinical doctors [6,13,19]. However, pixel-level annotation of images is time-consuming and labor-intensive, especially for histopathology images that require specialized knowledge. Therefore, there is an urgent need to pursue weakly supervised solutions for pixel-wise segmentation. Nonetheless, weakly supervised histopathological image segmentation presents a challenge due to the low contrast between different tissues, intra-class variations, and inter-class similarities [4,11]. Additionally, the tissue structures in histopathology images can be randomly arranged and dispersed, which makes it difficult to identify complete tissues or regions of interest [7].

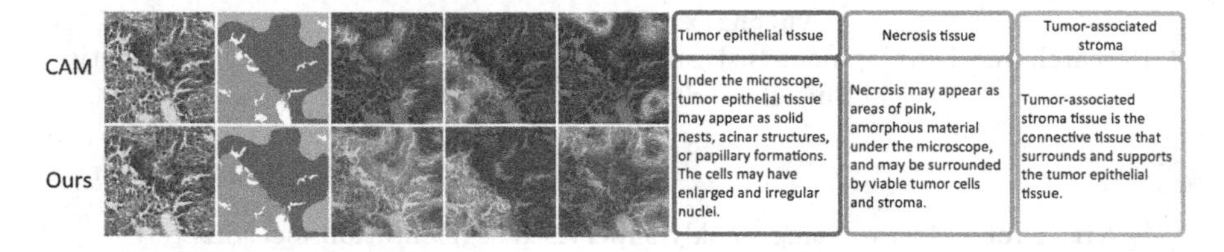

Fig. 1. Comparison of activation maps extracted from CAM and our method, from left to right: origin image, ground truth, three activation maps of tumor epithelial (red), necrosis (green), and tumor-associated stroma (orange) respectively. On the right side, there are some examples of the related language knowledge descriptions used in our method. It shows that CAM only highlights a small portion of the target, while our method, which incorporates external language knowledge, can encompass a wider and more precise target tissue. (Color figure online)

Recent studies on weakly supervised segmentation primarily follow class activation mapping (CAM) [20], which localizes the attention regions and then generates the pseudo labels to train the segmentation network. However, the CAM generated based on the image-level labels can only highlight the most discriminative region, but fail to locate the complete object, leading to defective pseudo labels, as shown in Fig. 1. Accordingly, many attempts have been made to enhance the quality of CAM and thus boost the performance of weakly supervised segmentation. Han *et al.* [7] proposed an erasure-based method that continuously expands the scope of attention areas to obtain rich content of pseudo labels. Li *et al.* [11] utilized the confidence method to remove any noise that may exist in the pseudo labels and only included the confident pixel labels for the segmentation training. Zhang *et al.* [18] leveraged the Transformer to model the long-distance dependencies on the whole histopathological images to improve the CAM's ability to find more complete regions. Lee *et al.* [10] utilized the ability of an advanced saliency detection model to assist CAM in locating more precise targets. However, these improved variants still face difficulties in capturing the complete tissues. The primary limitation is that the symptoms and manifestations of histopathological subtypes cannot be comprehensively described by an abstract semantic category. As a result, the image-level label supervision may not be sufficient to pinpoint the complete target area.

To remedy the limitations of image-level supervision, we advocate for the integration of language knowledge into weakly supervised learning to provide reliable guidance for the accurate localization of target structures. To this end, we propose a text-prompting-based weakly supervised segmentation method (TPRO) for accurate histopathology tissue segmentation. The text information originates from the task's semantic labels and external descriptions of subtype manifestations. For each semantic label, a pre-trained medical language model is utilized to extract the corresponding text features that are matched to each feature point in the image spatial space. A higher similarity represents a higher possibility of this location belonging to the corresponding semantic category. Additionally, the text representations of subtype manifestations, including tissue morphology, color, and relationships to other tissues, are extracted by the language model as external knowledge. The discriminative information can be explored from the text knowledge to help identify and locate complete tissues accurately by jointly modeling long-range dependencies between image and text. We conduct experiments on two weakly supervised histological segmentation benchmarks, LUAD-HistoSeg and BCSS-WSSS, and demonstrate the superior quality of pseudo labels produced by our TPRO model compared to other CAM-based methods.

Our contributions are summarized as follows: (1) To the best of our knowledge, this is the first work that leverages language knowledge to improve the quality of pseudo labels for weakly-supervised histopathology image segmentation. (2) The proposed text prompting models the correlation between image representations and text knowledge, effectively improving the quality of pseudo labels. (3) The effectiveness of our approach has been effectively validated by two benchmarks, setting a new state of the art.

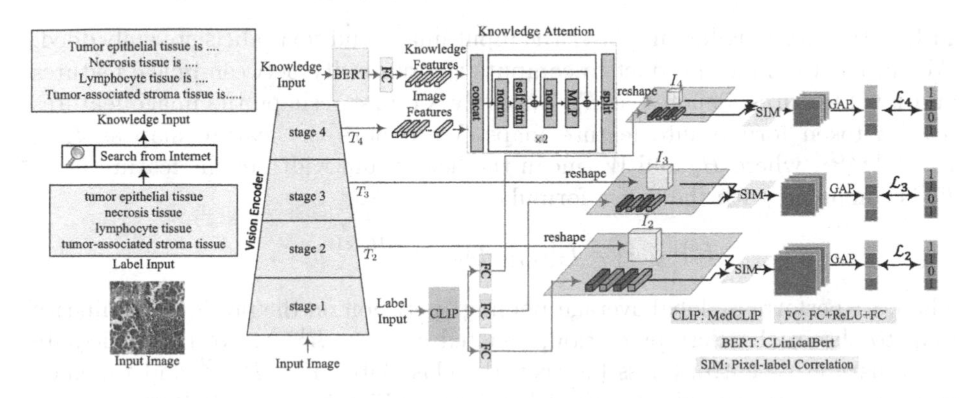

Fig. 2. The framework of the proposed TPRO.

2 Method

Figure 2 displays the proposed TPRO framework, a classification network designed to train a suitable model and extract segmentation pseudo-labels. The framework comprises a knowledge attention module and three encoders: one vision encoder and two text encoders (label encoder and knowledge encoder).

2.1 Classification with Deep Text Guidance

Vision Encoder. The vision encoder is composed of four stages that encode the input image into image features. The image features are denoted as $T_s \in R^{M_s \times C_s}$, where $2 \leq s \leq 4$ indicates the stage number.

Label Encoder. The label encoder encodes the text labels in the dataset into N label features, denoted as $L \in R^{N \times C_l}$, where N represents the number of classes in the dataset and C_l represents the dimension of label features. Since the label features will be used to calculate the similarity with image features, it is important to choose a language model that has been pre-trained on image-text pairs. Here we use MedCLIP[1] as our label encoder, which is a model fine-tuned on the ROCO dataset [12] based on CLIP [14].

Knowledge Encoder. The knowledge encoder is responsible for embedding the descriptions of subtype manifestations into knowledge features, denoted as $K \in R^{N \times C_k}$. The knowledge features guide the image features to focus on regions relevant to the target tissue. To encode the subtype manifestations description into more general semantic features, we employ ClinicalBert [2] as our knowledge encoder. ClinicalBert is a language model that has been fine-tuned on the MIMIC-III [8] dataset based on BioBert [9].

Adaptive Layer. We freeze the label and knowledge encoders for training efficiency but add an adaptive layer after the text encoders to better tailor the text features to our dataset. The adaptive layer is a simple FC-ReLU-FC block that allows for fine-tuning of the features extracted from the text encoders.

Label-Pixel Correlation. After the input image and text labels are embedded. We employ the inner product to compute the similarity between image features and label features, denoted as F_s. Specially, we first reshape the image features from a token format into feature maps. We denote the feature map as $I_s \in R^{H_s \times W_s \times C_s}$, where H_s and W_s mean the height and width of the feature map. F_s is computed with the below formula

$$F_s[i, j, k] = I_s[i, j] \cdot L[k] \in R^{H_s \times W_s \times N}. \quad (1)$$

Then, we perform a global average-pooling operation on the produced similarity map to obtain the class prediction, denoted as $P_s \in R^{1 \times N}$. We then calculate the binary cross-entropy loss between the class label $Y \in R^{1 \times N}$ and the class prediction P_s to supervise the model training, which is formulated as:

[1] https://github.com/Kaushalya/medclip.

$$\mathcal{L}_s = -\frac{1}{N}\sum_{n=1}^{N} Y[n]log\,\sigma(P_s[n]) + (1 - Y[n])log[1 - \sigma(P_s[n])] \quad (2)$$

Deep Supervision. To leverage the shallow features in the network, we employ a deep supervision strategy by calculating the similarity between the image features from different stages and the label features from different adaptive layers. Class predictions are derived from these similarity maps. The loss of the entire network is computed as:

$$\mathcal{L} = \lambda_2\mathcal{L}_2 + \lambda_3\mathcal{L}_3 + \lambda_4\mathcal{L}_4. \quad (3)$$

2.2 Knowledge Attention Module

To enhance the model's understanding of the color, morphology, and relationships between different tissues, we gather text representations of different subtype manifestations from the Internet and encode them into external knowledge via the knowledge encoder. The knowledge attention module uses this external knowledge to guide the image features toward relevant regions of the target tissues.

The knowledge attention module, shown in Fig. 2, consists of two multi-head self-attention modules. The image features $T_4 \in R^{M_4 \times C_4}$ and knowledge features after adaptive layer $K \in R^{N \times C_4}$ are concatenated in the token dimension to obtain $T_{fuse} \in R^{(M_4+N) \times C_4}$. This concatenated feature is then fed into the knowledge attention module for self-attention calculation. The output tokens are split, and the part corresponding to the image features is taken out. Noted that the knowledge attention module is added only after the last stage of the vision encoder to save computational resources.

2.3 Pseudo Label Generation

In the classification process, we calculate the similarity between image features and label features to obtain a similarity map F, and then directly use the result of global average pooling on the similarity map as a class prediction. That is, the value at position (i, j, k) of F represents the probability that pixel (i, j) is classified into the k_{th} class. Therefore we directly use F as our localization map. We first perform min-max normalization on it, the formula is as follows

$$F_{fg}^c = \frac{F^c - \min(F^c)}{\max(F^c) - \min(F^c)}, \quad (4)$$

where $1 \leq c \leq N$ means c_{th} class in the dataset. Then we calculate the background localization map by the following formula:

$$F_{bg}(i,j) = \{1 - \max_{c \in [0,C)} F_{fg}^c(i,j)\}^\alpha, \quad (5)$$

where $\alpha \geq 1$ denotes a hyper-parameter that adjusts the background confidence scores. Referring to [1] and combined with our own experiments, we set α to

10. Then we stitch together the localization map of foreground and background, denoted as $\hat{F}$. In order to make full use of the shallow information of the network, we perform weighted fusion on the localization maps from different stages by the following formula:

$$F_{all} = \gamma_2 \hat{F}_2 + \gamma_3 \hat{F}_4 + \gamma_4 \hat{F}_4. \quad (6)$$

Finally, we perform argmax operation on F_{all} to obtain the final pseudo-label.

3 Experiments

3.1 Dataset

LUAD-HistoSeg[2] [7] is a weakly-supervised histological semantic segmentation dataset for lung adenocarcinoma. There are four tissue classes in this dataset: tumor epithelial (TE), tumor-associated stroma (TAS), necrosis (NEC), and lymphocyte (LYM). The dataset comprises 17,258 patches of size 224×224. According to the official split, the dataset is divided into a training set (16,678 patch-level annotations), a validation set (300 pixel-level annotations), and a test set (307 pixel-level annotations). **BCSS-WSSS**[3] is a weakly supervised tissue semantic segmentation dataset extracted from the fully supervised segmentation dataset BCSS [3], which contains 151 representative H&E-stained breast cancer pathology slides. The dataset was randomly cut into 31826 patches of size 224 × 224 and divided into a training set (23422 patch-level annotations), a validation set (3418 pixel-level annotations), and a test set (4986 pixel-level annotations) according to the official split. There are four foreground classes in this dataset, including Tumor (TUM), Stroma (STR), Lymphocytic infiltrate (LYM), and Necrosis (NEC).

3.2 Implementation Details

For the classification part, we adopt MixTransformer [17] pretrained on ImageNet, MedCLIP, and ClinicalBert [2] as our vision encoder, label encoder, and

Table 1. Comparison of the pseudo labels generated by our proposed method and those generated by previous methods.

Dataset	LUAD-HistoSeg					BCSS-WSSS				
Method	TE	NEC	LYM	TAS	mIoU	TUM	STR	LYM	NEC	mIoU
CAM [20]	69.66	72.62	72.58	66.88	70.44	66.83	58.71	49.41	51.12	56.52
Grad-CAM [15]	70.07	66.01	70.18	64.76	67.76	65.96	56.71	43.36	30.04	49.02
TransWS (CAM) [18]	65.92	60.16	73.34	69.11	67.13	64.85	58.17	44.96	50.60	54.64
MLPS [7]	71.72	76.27	73.53	67.67	72.30	70.76	61.07	50.87	52.94	58.91
TPRO (Ours)	**74.82**	**77.55**	**76.40**	**70.98**	**74.94**	**77.18**	**63.77**	**54.95**	**61.43**	**64.33**

[2] https://drive.google.com/drive/folders/1E3Yei3Or3xJXukHIybZAgochxfn6FJpr.
[3] https://drive.google.com/drive/folders/1iS2Z0DsbACqGp7m6VDJbAcgzeXNEFr77.

knowledge encoder, respectively. The hyperparameters during training and evaluation can be found in the supplementary materials. We conduct all of our experiments on 2 NVIDIA GeForce RTX 2080 Ti GPUs.

3.3 Compare with State-of-the-Arts

Comparison on Pseudo-Labels. Table 1 compares the quality of our pseudo-labels with those generated by previous methods. CAM [20] and Grad-CAM [15] were evaluated using the same ResNet38 [16] classifier, and the results showed that CAM [20] outperformed Grad-CAM [15], with mIoU values of 70.44% and 56.52% on the LUAD-HistoSeg and BCSS-WSSS datasets, respectively. TransWS [18] consists of a classification and a segmentation branch, and Table 1 displays the pseudo-label scores generated by the classification branch. Despite using CAM [20] for pseudo-label extraction, TransWS [18] yielded inferior results compared to CAM [20]. This could be due to the design of TransWS [18] for single-label image segmentation, with the segmentation branch simplified to binary segmentation to reduce the difficulty, while our dataset consists of multi-label images. Among the compared methods, MLPS [7] was the only one to surpass CAM [20] in terms of the quality of the generated pseudo-labels, with its proposed progressive dropout attention effectively expanding the coverage of target regions beyond what CAM [20] can achieve. Our proposed method outperformed all previous methods on both LUAD-HistoSeg and BCSS-WSSS datasets, with improvements of 2.64% and 5.42% over the second-best method, respectively (Table 2).

Table 2. Comparison of the final segmentation results between our method and the methods in previous years.

Dataset	LUAD-HistoSeg					BCSS-WSSS				
Method	TE	NEC	LYM	TAS	mIoU	TUM	STR	LYM	NEC	mIoU
HistoSegNet [5]	45.59	36.30	58.28	50.82	47.75	33.14	46.46	29.05	1.91	27.64
TransWS (seg) [18]	57.04	49.98	59.46	58.59	56.27	44.71	36.49	41.72	38.08	40.25
OEEM [11]	73.81	70.49	71.89	69.48	71.42	74.86	64.68	48.91	61.03	62.37
MLPS [7]	73.90	77.48	73.61	69.53	73.63	74.54	64.45	52.54	58.67	62.55
TPRO (Ours)	**75.80**	**80.56**	**78.14**	**72.69**	**76.80**	**77.95**	**65.10**	**54.55**	**64.96**	**65.64**

Comparison on Segmentation Results. To further evaluate our proposed method, we trained a segmentation model using the extracted pseudo-labels and compared its performance with previous methods. Due to its heavy reliance on dataset-specific post-processing steps, HistoSegNet [5] failed to produce the desired results on our datasets. As we have previously analyzed since the datasets we used are all multi-label images, it was challenging for the segmentation branch of TransWS [18] to perform well, and it failed to provide an overall benefit to the model. Experimental results also indicate that the IoU scores of its segmentation

Table 3. Comparison the effectiveness of label text(LT), knowledge text(KT), and deep supervision(DS).

LT	DS	KT	TE	NEC	LYM	TAS	mIoU
			68.11	75.24	64.95	66.57	68.72
✓			72.39	72.44	71.37	68.67	71.22
✓	✓		72.41	72.11	74.21	70.07	72.20
✓	✓	✓	**74.82**	**77.55**	**76.40**	**70.98**	**74.94**

Table 4. Comparison of pseudo labels extracted from the single stage and our fused version.

	TE	NEC	LYM	TAS	mIoU
stage2	67.16	65.28	67.38	55.09	63.73
stage3	72.13	70.83	73.47	69.46	71.47
stage4	72.69	77.57	76.06	69.81	74.03
fusion	**74.82**	**77.55**	**76.40**	**70.98**	**74.94**

branch were even lower than the pseudo-labels of the classification branch. By training the segmentation model of OEEM [11] using the pseudo-labels extracted by CAM [20] in Table 1, we can observe a significant improvement in the final segmentation results. The final segmentation results of MLPS [7] showed some improvement compared to its pseudo-labels, indicating the effectiveness of the Multi-layer Pseudo Supervision and Classification Gate Mechanism strategy proposed by MLPS [7]. Our segmentation performance surpassed all previous methods. Specifically, our mIoU scores exceeded the second-best method by 3.17% and 3.09% on LUAD-HistoSeg and BCSS-WSSS datasets, respectively. Additionally, it is worth noting that we did not use any strategies specifically designed for the segmentation stage.

3.4 Ablation Study

The results of our ablation experiments are presented in Table 3. We set the baseline as the framework shown in Fig. 2 with all text information and deep supervision strategy removed. It is evident that the addition of textual information increases our pseudo-label mIoU by 2.50%. Furthermore, including the deep supervision strategy and knowledge attention module improves our pseudo-label by 0.98% and 2.74%, respectively. These findings demonstrate the significant contribution of each proposed module to the overall improvement of the results.

In order to demonstrate the effectiveness of fusing pseudo-labels from the last three stages, we have presented in Table 4 the IoU scores for each stage's pseudo-labels as well as the fused pseudo-labels. It can be observed that after fusing the pseudo-labels, not only have the IoU scores for each class substantially increased, but the mIoU score has also increased by 0.91% compared to the fourth stage.

4 Conclusion

In this paper, we propose the TPRO to address the limitation of weakly supervised semantic segmentation on histopathology images by incorporating text supervision and external knowledge. We argue that image-level labels alone cannot provide sufficient information and that text supervision and knowledge attention can provide additional guidance to the model. The proposed method achieves the best results on two public datasets, LUAD-HistoSeg and BCSS-WSSS, demonstrating the superiority of our method.

Acknowledgment. This work was supported in part by the Natural Science Foundation of Ningbo City, China, under Grant 2021J052, in part by the Ningbo Clinical Research Center for Medical Imaging under Grant 2021L003 (Open Project 2022LYK-FZD06), in part by the National Natural Science Foundation of China under Grant 62171377, in part by the Key Technologies Research and Development Program under Grant 2022YFC2009903/2022YFC2009900, in part by the Key Research and Development Program of Shaanxi Province, China, under Grant 2022GY-084, and in part by the Science and Technology Innovation Committee of Shenzhen Municipality, China, under Grants JCYJ20220530161616036.

References

1. Ahn, J., Kwak, S.: Learning pixel-level semantic affinity with image-level supervision for weakly supervised semantic segmentation. In: Proceedings of the IEEE Conference on Computer Vision and Pattern Recognition, pp. 4981–4990 (2018)
2. Alsentzer, E., et al.: Publicly available clinical bert embeddings. arXiv preprint arXiv:1904.03323 (2019)
3. Amgad, M., et al.: Structured crowdsourcing enables convolutional segmentation of histology images. Bioinformatics **35**(18), 3461–3467 (2019)
4. Chan, L., Hosseini, M.S., Plataniotis, K.N.: A comprehensive analysis of weakly-supervised semantic segmentation in different image domains. Int. J. Comput. Vision **129**, 361–384 (2021)
5. Chan, L., Hosseini, M.S., Rowsell, C., Plataniotis, K.N., Damaskinos, S.: Histosegnet: semantic segmentation of histological tissue type in whole slide images. In: Proceedings of the IEEE/CVF International Conference on Computer Vision, pp. 10662–10671 (2019)
6. Chen, H., Qi, X., Yu, L., Heng, P.A.: Dcan: deep contour-aware networks for accurate gland segmentation. In: Proceedings of the IEEE Conference on Computer Vision and Pattern Recognition, pp. 2487–2496 (2016)
7. Han, C., et al.: Multi-layer pseudo-supervision for histopathology tissue semantic segmentation using patch-level classification labels. Med. Image Anal. **80**, 102487 (2022)
8. Johnson, A.E., et al.: Mimic-iii, a freely accessible critical care database. Sci. Data **3**(1), 1–9 (2016)
9. Lee, J., et al.: Biobert: a pre-trained biomedical language representation model for biomedical text mining. Bioinformatics **36**(4), 1234–1240 (2020)
10. Lee, S., Lee, M., Lee, J., Shim, H.: Railroad is not a train: saliency as pseudo-pixel supervision for weakly supervised semantic segmentation. In: Proceedings of the IEEE/CVF Conference on Computer Vision and Pattern Recognition, pp. 5495–5505 (2021)
11. Li, Y., Yu, Y., Zou, Y., Xiang, T., Li, X.: Online easy example mining for weakly-supervised gland segmentation from histology images. In: Medical Image Computing and Computer Assisted Intervention-MICCAI 2022: 25th International Conference, Singapore, September 18–22, 2022, Proceedings, Part IV. pp. 578–587. Springer, Cham (2022). https://doi.org/10.1007/978-3-031-16440-8_55

12. Pelka, O., Koitka, S., Rückert, J., Nensa, F., Friedrich, C.M.: Radiology Objects in COntext (ROCO): a multimodal image dataset. In: Stoyanov, D., et al. (eds.) LABELS/CVII/STENT -2018. LNCS, vol. 11043, pp. 180–189. Springer, Cham (2018). https://doi.org/10.1007/978-3-030-01364-6_20
13. Qaiser, T., et al.: Fast and accurate tumor segmentation of histology images using persistent homology and deep convolutional features. Med. Image Anal. **55**, 1–14 (2019)
14. Radford, A., et al.: Learning transferable visual models from natural language supervision. In: International Conference on Machine Learning, pp. 8748–8763. PMLR (2021)
15. Selvaraju, R.R., Cogswell, M., Das, A., Vedantam, R., Parikh, D., Batra, D.: Grad-cam: visual explanations from deep networks via gradient-based localization. In: Proceedings of the IEEE International Conference on Computer Vision, pp. 618–626 (2017)
16. Wu, Z., Shen, C., Van Den Hengel, A.: Wider or deeper: revisiting the resnet model for visual recognition. Pattern Recogn. **90**, 119–133 (2019)
17. Xie, E., Wang, W., Yu, Z., Anandkumar, A., Alvarez, J.M., Luo, P.: Segformer: simple and efficient design for semantic segmentation with transformers. Adv. Neural. Inf. Process. Syst. **34**, 12077–12090 (2021)
18. Zhang, S., Zhang, J., Xia, Y.: Transws: Transformer-based weakly supervised histology image segmentation. In: Machine Learning in Medical Imaging: 13th International Workshop, MLMI 2022, Held in Conjunction with MICCAI 2022, Singapore, September 18, 2022, Proceedings. pp. 367–376. Springer, Cham (2022). https://doi.org/10.1007/978-3-031-21014-3_38
19. Zhao, B., et al.: Triple u-net: hematoxylin-aware nuclei segmentation with progressive dense feature aggregation. Med. Image Anal. **65**, 101786 (2020)
20. Zhou, B., Khosla, A., Lapedriza, A., Oliva, A., Torralba, A.: Learning deep features for discriminative localization. In: Proceedings of the IEEE Conference on Computer Vision and Pattern Recognition, pp. 2921–2929 (2016)

Additional Positive Enables Better Representation Learning for Medical Images

Dewen Zeng[1(✉)], Yawen Wu[2], Xinrong Hu[1], Xiaowei Xu[3], Jingtong Hu[2], and Yiyu Shi[1(✉)]

[1] University of Notre Dame, Notre Dame, IN, USA
{dzeng2,yshi4}@nd.edu
[2] University of Pittsburgh, Pittsburgh, PA, USA
[3] Guangdong Provincial People's Hospital, Guangzhou, China

Abstract. This paper presents a new way to identify additional positive pairs for BYOL, a state-of-the-art (SOTA) self-supervised learning framework, to improve its representation learning ability. Unlike conventional BYOL which relies on only one positive pair generated by two augmented views of the same image, we argue that information from different images with the same label can bring more diversity and variations to the target features, thus benefiting representation learning. To identify such pairs without any label, we investigate TracIn, an instance-based and computationally efficient influence function, for BYOL training. Specifically, TracIn is a gradient-based method that reveals the impact of a training sample on a test sample in supervised learning. We extend it to the self-supervised learning setting and propose an efficient batch-wise per-sample gradient computation method to estimate the pairwise TracIn for representing the similarity of samples in the mini-batch during training. For each image, we select the most similar sample from other images as the additional positive and pull their features together with BYOL loss. Experimental results on two public medical datasets (i.e., ISIC 2019 and ChestX-ray) demonstrate that the proposed method can improve the classification performance compared to other competitive baselines in both semi-supervised and transfer learning settings.

Keywords: self-supervised learning · representation learning · medical image classification

1 Introduction

Self-supervised learning (SSL) has been extremely successful in learning good image representations without human annotations for medical image applications like classification [1,23,29] and segmentation [2,4,16]. Usually, an encoder is pre-trained on a large-scale unlabeled dataset. Then, the pre-trained encoder is

Supplementary Information The online version contains supplementary material available at https://doi.org/10.1007/978-3-031-43907-0_12.

H. Greenspan et al. (Eds.): MICCAI 2023, LNCS 14220, pp. 119–129, 2023.
https://doi.org/10.1007/978-3-031-43907-0_12

used for efficient training on downstream tasks with limited annotation [19,24]. Recently, contrastive learning has become the state-of-the-art (SOTA) SSL method due to its powerful learning ability. A recent contrastive learning method learns by pulling the representations of different augmented views of the same image (a.k.a positive pair) together and pushing the representation of different images (a.k.a negative pair) apart [6]. The main disadvantage of this method is its heavy reliance on negative pairs, making it necessary to use a large batch size [6] or memory banks [15] to ensure effective training. To overcome this challenge, BYOL [12] proposes two siamese neural networks - the online and target networks. The online network is trained to predict the target network representation of the same image under a different augmented view, requiring only one positive pair per sample. This approach makes BYOL more resilient to batch size and the choice of data augmentations.

As the positive pair in BYOL is generated from the same image, the diversity of features within the positive pair could be quite limited. For example, one skin disease may manifest differently in different patients or locations, but such information is often overlooked in the current BYOL framework. In this paper, we argue that such feature diversity can be increased by adding additional positive pairs from other samples with the same label (a.k.a. True Positives). Identifying such pairs without human annotation is challenging because of the unrelated information in medical images, such as the background normal skin areas in dermoscopic images. One straightforward way to detect positive pairs is using feature similarity: two images are considered positive if their representations are close to each other in the feature space. However, samples with different labels might also be close in the feature space because the learned encoder is not perfect. Considering them as positive might further pull them together after learning, leading to degraded performance.

To solve this problem, we propose BYOL-TracIn, which improves vanilla BYOL using the TracIn influence function. Instead of quantifying the similarity of two samples based on feature similarity, we propose using TracIn to estimate their similarity by calculating the impact of training one sample on the other. TracIn [22] is a gradient-based influence function that measures the loss reduction of one sample by the training process of another sample. Directly applying TracIn in BYOL is non-trivial as it requires the gradient of each sample and careful selection of model checkpoints and data augmentations to accurately estimate sample impacts without labels. To avoid per-sample gradient computation, we introduce an efficient method that computes the pairwise TracIn in a mini-batch with only one forward pass. For each image in the mini-batch, the sample from other images with the highest TracIn values is selected as the additional positive pair. Their representation distance is then minimized using BYOL loss. To enhance positive selection accuracy, we propose to use a pre-trained model for pairwise TracIn computation as it focuses more on task-related features compared to an on-the-fly model. Light augmentations are used on the samples for TracIn computation to ensure stable positive identification. To the best of our knowledge, we are the first to incorporate additional positive pairs from different images in BYOL. Our extensive empirical results show that our proposed method outperforms other competing approaches in both semi-supervised and transfer learning settings for medical image classification tasks.

2 Related Work

Self-supervised Learning. Most SSL methods can be categorized as either generative [10,28] or discriminative [11,21], in which pseudo labels are automatically generated from the inputs. Recently, contrastive learning [6,15,27] as a new discriminative SSL method has dominated this field because of its excellent performance. SimCLR [6] and MoCo [15] are two typical contrastive learning methods that try to attract positive pairs and repulse negative pairs. However, these methods rely on a large number of negative samples to work well. BYOL [12] improves contrastive learning by directly predicting the representation output from another view and achieves SOTA performance. As such, only positive pairs are needed for training. SimSiam [7] further proves that stop-gradient plays an essential role in the learning stability of siamese neural networks. Since the positive pairs in BYOL come from the same image, the feature diversity from different images of the same label is ignored. Our method introduces a novel way to accurately identify such positive pairs and attract them in the feature space.

Influence Function. The influence function (IF) was first introduced to machine learning models in [20] to study the following question: which training points are most responsible for a given prediction? Intuitively, if we remove an important sample from the training set, we will get a large increase in the test loss. IF can be considered as an interpretability score that measures the importance of all training samples on the test sample. Aside from IF, other types of scores and variants have also been proposed in this field [3,5,14]. Since IF is extremely computationally expensive, TracIn [22] was proposed as an efficient alternative to estimate training sample influence using first-order approximation. Our method extends the normal TracIn to the SSL setting (i.e., BYOL) with a sophisticated positive pair selection schema and an efficient batch-wise per-sample gradient computation method, demonstrating that aside from model interpretation, TracIn can also be used to guide SSL pre-training.

3 Method

3.1 Framework Overview

Our BYOL-TracIn framework is built upon classical BYOL method [12]. Figure 1 shows the overview of our framework. Here, we use x_1 as the anchor sample for an explanation, and the same logic can be applied to all samples in the mini-batch. Unlike classical BYOL where only one positive pair (x_1 and x'_1) generated from the same image is utilized, we use the influence function, TracIn, to find another sample (x'_3) from the batch that has the largest impact on the anchor sample. During training, the representations distance of x_1 and x'_3 will also be minimized. We think this additional positive pair can increase the variance and diversity of

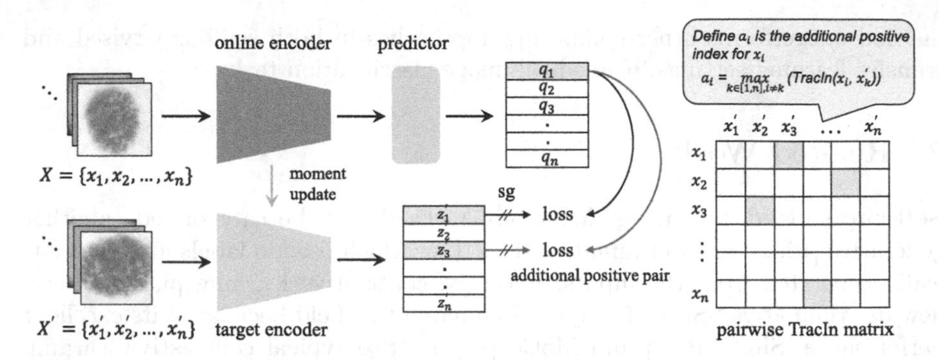

Fig. 1. Overview of the proposed BYOL-TracIn framework. X and X' represent two augmentations of the mini-batch inputs. BYOL-TracIn minimizes the similarity loss of two views of the same image (e.g., q_1 and z_1') as well as the similarity loss of the additional positive (e.g., z_3') identified by our TracIn algorithm. sg means stop-gradient.

the features of the same label, leading to better clustering in the feature space and improved learning performance. The pairwise TracIn matrix is computed using first-order gradient approximation which will be discussed in the next section. For simplicity, this paper only selects the top-1 additional sample, but our method can be easily extended to include top-k ($k > 1$) additional samples.

3.2 Additional Positive Selection Using TracIn

Idealized TracIn and Its First-order Approximation. Suppose we have a training dataset $D = \{x_1, x_2, ..., x_n\}$ with n samples. $f_w(\cdot)$ is a model with parameter $w \in \mathbb{R}$, and $\ell(w, x_i)$ is the loss function when model parameter is w and training example is x_i. The training process in iteration t can be viewed as minimizing the training loss $\ell(w_t, x_t)$ and updating parameter w_t to w_{t+1} using gradient descent (suppose only $x_t \in D$ is used for training in each iteration). Then the idealized TracIn of one sample x_i on another sample x_k can be defined as the total loss reduction by training x_i in the whole training process.

$$\text{TracInIdeal}(x_i, x_k) = \sum_{t:x_t=x_i}^{T} (\ell(w_t, x_k) - \ell(w_{t+1}, x_k)). \tag{1}$$

where T is the total number of iterations. If stochastic gradient descent is utilized as the optimization method, we can approximately express the loss reduction after iteration t as $\ell(w_{t+1}, x_k) - \ell(w_t, x_k) = \nabla\ell(w_t, x_k)\cdot(w_{t+1} - w_t) + O(||\Delta w_t||^2)$. The parameter change in iteration t is $\Delta w_t = w_{t+1} - w_t = -\eta_t \nabla\ell(w_t, x_t)$, in which η_t is the learning rate in iteration t, and x_t is the training example. Since η_t is usually small during training, we can ignore the high order term $O(||\Delta w_t||^2)$, and the first-order TracIn can be formulated as:

$$\text{TracIn}(x_i, x_k) = \sum_{t:x_t=x_i}^{T} \eta_t \nabla\ell(w_t, x_k) \cdot \nabla\ell(w_t, x_i). \tag{2}$$

The above equation reveals that we can estimate the influence of x_i on x_k by summing up their gradient dot products across all training iterations. In practical BYOL training, the optimization is usually done on mini-batches, and it is impossible to save the gradients of a sample for all iterations. However, we can use the TracIn of **the current iteration** to represent the similarity of two samples in the mini-batch because we care about the pairwise relative influences instead of the exact total values across training. Intuitively, if the TracIn of two samples is large in the current iteration, this means that the training of one sample can benefit the other sample a lot because they share some common features. Therefore, they are similar to each other.

Efficient Batch-wise TracIn Computation. Equation 2 requires the gradient of each sample in the mini-batch for pairwise TracIn computation. However, it is prohibitively expensive to compute the gradient of samples one by one. Moreover, calculating the dot product of gradients on the entire model is computationally and memory-intensive, especially for large deep-learning models where there could be millions or trillions of parameters. Therefore, we work with the gradients of the last linear layer in the online predictor.

As current deep learning frameworks (e.g., Pytorch and TensorFlow) do not support per-sample gradient when the batch size is larger than 1, we use the following method to efficiently compute the per-sample gradient of the last layer. Suppose the weight matrix of the last linear layer is $W \in \mathbb{R}^{m \times n}$, where m and n are the numbers of input and output units. $f(q) = 2 - 2 \cdot \langle q, z \rangle / (\|q\|_2 \cdot \|z\|_2)$ is the standard BYOL loss function, where q is the online predictor output (a.k.a., logits) and z is the target encoder output that can be viewed as a constant during training. We have $q = Wa$, where a is the input to the last linear layer. According to the chain rule, the gradient of the last linear layer can be computed as $\nabla_W f(q) = \nabla_q f(q) a^T$, in which the gradient of the logits can be computed by:

$$\nabla_q f(q) = 2 \cdot \left(\frac{\langle q, z \rangle \cdot q}{\|q\|_2^3 \cdot \|z\|_2} - \frac{z}{\|q\|_2 \cdot \|z\|_2}\right). \tag{3}$$

Therefore, the TracIn of sample x_i and x_k at iteration t can be computed as:

$$\begin{aligned} \text{TracIn}(x_i, x_k) &\approx \eta_t \nabla_W f(q_i) \cdot \nabla_W f(q_k) \\ &= \eta_t (\nabla_q f(q_i) \cdot \nabla_q f(q_k))(a_i \cdot a_k). \end{aligned} \tag{4}$$

Equation 3 and 4 tell us that the per-sample gradient of the last linear layer can be computed by using the inputs of this layer and the gradient of the output logits for each sample, which can be achieved with only one forward pass on the mini-batch. This technique greatly speeds up the TracIn computation and makes it possible to be used in BYOL.

Using Pre-trained Model to Increase True Positives. During the pre-training stage of BYOL, especially in the early stages, the model can be unstable and may focus on unrelated features in the background instead of the target features. This can result in the selection of wrong positive pairs while using TracIn. For example, the model may identify all images with skin diseases on the face as positive pairs, even if they are from different diagnostics, as it focuses on the face feature instead of the diseases. To address this issue, we suggest using a pre-trained model to select additional positives with TracIn to guide BYOL training. This is because a pre-trained model is more stable and well-trained to focus on the target features, thus increasing the selected true positive ratio.

4 Experiments and Results

4.1 Experimental Setups

Datasets. We evaluate the performance of the proposed BYOL-TracIn on four publicly available medical image datasets. **(1) ISIC 2019 dataset** is a dermatology dataset that contains 25,331 dermoscopic images among nine different diagnostic categories [8,9,25]. **(2) ISIC 2016 dataset** was hosted in ISBI 2016 [13]. It contains 900 dermoscopic lesion images with two classes benign and malignant. **(3) ChestX-ray dataset** is a chest X-ray database that comprises 108,948 frontal view X-ray images of 32,717 unique patients with 14 disease labels [26]. Each image may have multiple labels. **(4) Shenzhen dataset** is a small chest X-ray dataset with 662 frontal chest X-rays, of which 326 are normal cases and 336 are cases with manifestations of Tuberculosis [18].

Training Details. We use Resnet18 as the backbone. The online projector and predictor follow the classical BYOL [12], and the embedding dimension is set to 256. On both ISIC 2019 and ChestX-ray datasets, we resize all the images to 140×140 and then crop them to 128×128. Data augmentation used in pre-training includes horizontal flipping, vertical flipping, rotation, color jitter, and cropping. For TracIn computation, we use one view with no augmentation and the other view with horizontal flipping and center cropping because this setting has the best empirical results in our experiments. We pre-train the model for 300 epochs using SGD optimizer with momentum 0.9 and weight decay $1 \times e^{-5}$. The learning rate is set to 0.1 for the first 10 epochs and then decays following a concise learning rate schedule. The batch size is set to 256. The moving average decay of the momentum encoder is set to 0.99 at the beginning and then gradually updates to 1 following a concise schedule. All experiments are performed on one NVIDIA GeForce GTX 1080 GPU.

Baselines. We compare the performance of our method with a random initialization approach without pre-training and the following SOTA baselines that involve pre-training. (1) BYOL [12]: the vanilla BYOL with one positive pair from the same image. (2) FNC [17]: a false negative identification method designed to improve contrastive-based SSL framework. We adapt it to BYOL to select additional positives because false negatives are also equal to true positives for a particular anchor sample. (3) FT [30]: a feature transformation method used in contrastive learning that creates harder positives and negatives to improve the learning ability. We apply it in BYOL to create harder virtual positives. (4) FS: using feature similarity from the current mini-batch to select the top-1 additional positive. (5) FS-pretrained: different from the FS that uses the current

Table 1. Comparison of all methods on ISIC 2019 and ChestX-ray datasets in the semi-supervised setting. We also report the fine-tuning results on 100% datasets. BYOL-Sup is the upper bound of our method. BMA represents the balanced multiclass accuracy.

Method	ISIC 2019			ChestX-ray		
	10%	50%	100%	10%	50%	100%
	BMA ↑			AUC ↑		
Random	0.327(.004)	0.558(.005)	0.650(.004)	0.694(.005)	0.736(.001)	0.749(.001)
BYOL [12]	0.399(.001)	0.580(.006)	0.692(.005)	0.699(.004)	0.738(.003)	0.750(.001)
FNC [17]	0.401(.004)	0.584(.004)	0.694(.005)	0.706(.001)	0.739(.001)	0.752(.002)
FT [30]	0.405(.005)	0.588(.008)	0.695(.005)	0.708(.001)	0.743(.001)	0.751(.002)
FS	0.403(.006)	0.591(.003)	0.694(.004)	0.705(.003)	0.738(.001)	0.752(.002)
FS-pretrained	0.406(.002)	0.596(.004)	0.697(.005)	0.709(.001)	0.744(.002)	0.752(.002)
BYOL-TracIn	0.403(.003)	0.594(.004)	0.694(.004)	0.705(.001)	0.742(.003)	0.753(.002)
BYOL-TracIn-pretrained	**0.408(.007)**	**0.602(.003)**	**0.700(.006)**	**0.712(.001)**	**0.746(.002)**	**0.754(.002)**
BYOL-Sup	0.438(.006)	0.608(.007)	0.705(.005)	0.714(.001)	0.748(.001)	0.756(.003)

model to compute the feature similarity on the fly, we use a pre-trained model to test whether a well-trained encoder is more helpful in identifying the additional positives. (6) BYOL-Sup: the supervised BYOL in which we randomly select one additional positive from the mini-batch using the label information. This baseline is induced as the upper bound of our method because the additional positive is already correct. We evaluate two variants of our method, BYOL-TracIn and BYOL-TracIn-pretrained. The former uses the current training model to compute the TracIn for each iteration while the latter uses a pre-trained model. For a fair comparison, all methods use the same pre-training and finetuning setting unless otherwise specified. For FS-pretrained and BYOL-TracIn-pretrained, the pre-trained model uses the same setting as BYOL. Note that this pre-trained model is only used for positive selection and not involves in training.

4.2 Semi-supervised Learning

In this section, we evaluate the performance of our method by finetuning with the pre-trained encoder on the same dataset as pre-training with limited annotations. We sample 10% or 50% of the labeled data from ISIC 2019 and ChestX-ray training sets and finetune the model for 100 epochs on the sampled datasets. Data augmentation is the same as pre-training. Table 1 shows the comparisons of all methods. For ISIC 2019, we report the balanced multiclass accuracy (BMA, suggested by the ISIC challenge). For ChestX-ray, we report the average AUC across all diagnoses. We conduct each finetuning experiment 5 times with different random seeds and report the mean and std.

From Table 1, we have the following observations: (1) Compared to Random, all the other methods have better accuracy, which means that pre-training can indeed help downstream tasks. (2) Compared to vanilla BYOL, other pre-training methods show performance improvement on both datasets. This shows that additional positives can increase feature diversity and benefit BYOL learning. (3) Our BYOL-TracIn-pretrained consistently outperforms all other unsu-

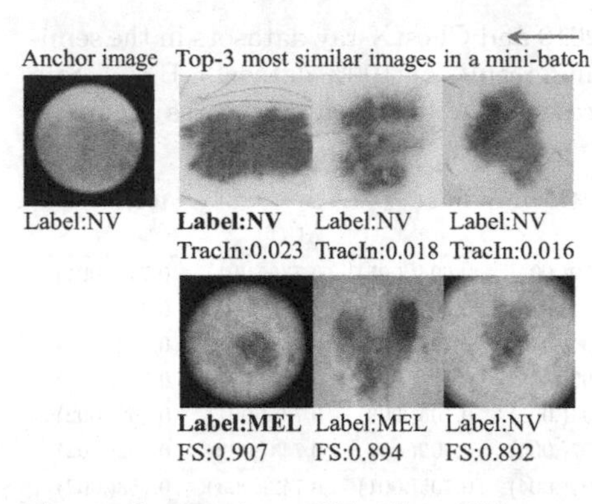

Fig. 2. Comparison of TracIn and Feature Similarity (FS) in selecting the additional positive during training on ISIC 2019.

Table 2. Transfer learning comparison of the proposed method with the baselines on ISIC 2016 and Shenzhen datasets.

Method	ISIC 2016 Precision ↑	Shenzhen AUC ↑
Random	0.400(.005)	0.835(.010)
BYOL [12]	0.541(.008)	0.858(.003)
FNC [17]	0.542(.007)	0.862(.006)
FT [30]	0.559(.011)	0.876(.005)
FS	0.551(.003)	0.877(.004)
FS-pretrained	0.556(.004)	0.877(.006)
BYOL-TracIn	0.555(.012)	0.880(.007)
BYOL-TracIn-pretrained	**0.565(.010)**	**0.883(.001)**
BYOL-Sup	0.592(.008)	0.893(.006)

pervised baselines. Although BYOL-TracIn can improve BYOL, it could be worse than other baselines like FT and FS-pretrained (e.g., 10% on ISIC 2019). This is because some additional positives identified by the on-the-fly model may be false positives, and attracting representations of such samples will degrade the learned features. However, with a pre-trained model in BYOL-TracIn-pretrained, the identification accuracy can be increased, leading to more true positives and better representations. (4) TracIn-pretrained performs better than FS-pretrained in all settings, and the improvement in BMA could be up to 0.006. This suggests that TracIn can be a more reliable metric for assessing the similarity between images when there is no human label information available. (5) Supervised BYOL can greatly increase the BYOL performance on both datasets. Yet our BYOL-TracIn-pretrained only has a marginal accuracy drop from supervised BYOL with a sufficient number of training samples (e.g., 100% on ISIC 2019).

To further demonstrate the superiority of TracIn over Feature Similarity (FS) in selecting additional positive pairs for BYOL, we use an image from ISIC 2019 as an example and visualize the top-3 most similar images selected by both metrics using a BYOL pre-trained model in Fig. 2. We can observe that TracIn accurately identifies the most similar images with the same label as the anchor image, whereas two of the images selected by FS have different labels. This discrepancy may be attributed to the fact that the FS of these two images is dominated by unrelated features (e.g., background tissue), which makes it unreliable. More visualization examples can be found in the supplementary.

4.3 Transfer Learning

To evaluate the transfer learning performance of the learned features, we use the encoder learned from the pre-training to initialize the model on the downstream datasets (ISIC 2019 transfers to ISIC 2016, and ChestX-ray transfers to Shenzhen). We finetune the model for 50 epochs and report the precision and AUC on ISIC 2016 and Shenzhen datasets, respectively. Table 2 shows the comparison results of all methods. We can see that BYOL-TracIn-pretrained always outperforms other unsupervised pre-training baselines, indicating that the additional positives can help BYOL learn better transferrable features.

5 Conclusion

In this paper, we propose a simple yet effective method, named BYOL-TracIn, to boost the representation learning performance of the vanilla BYOL framework. BYOL-TracIn can effectively identify additional positives from different samples in the mini-batch without using label information, thus introducing more variances to learned features. Experimental results on multiple public medical image datasets show that our method can significantly improve classification performance in both semi-supervised and transfer learning settings. Although this paper only discusses the situation of one additional pair for each image, our method can be easily extended to multiple additional pairs. However, more pairs will introduce more computation costs and increase the false positive rate which may degrade the performance. Another limitation of this paper is that BYOL-TracIn requires a pre-trained model to start with, which means more computation resources are needed to demonstrate its effectiveness.

References

1. Azizi, S., et al.: Big self-supervised models advance medical image classification. In: Proceedings of the IEEE/CVF International Conference on Computer Vision, pp. 3478–3488 (2021)
2. Bai, W., et al.: Self-supervised learning for cardiac MR image segmentation by anatomical position prediction. In: Shen, D., Liu, T., Peters, T.M., Staib, L.H., Essert, C., Zhou, S., Yap, P.-T., Khan, A. (eds.) MICCAI 2019. LNCS, vol. 11765, pp. 541–549. Springer, Cham (2019). https://doi.org/10.1007/978-3-030-32245-8_60
3. Barshan, E., Brunet, M.E., Dziugaite, G.K.: Relatif: identifying explanatory training samples via relative influence. In: International Conference on Artificial Intelligence and Statistics, pp. 1899–1909. PMLR (2020)
4. Chaitanya, K., Erdil, E., Karani, N., Konukoglu, E.: Contrastive learning of global and local features for medical image segmentation with limited annotations. Adv. Neural. Inf. Process. Syst. **33**, 12546–12558 (2020)
5. Chen, H., et al.: Multi-stage influence function. Adv. Neural. Inf. Process. Syst. **33**, 12732–12742 (2020)
6. Chen, T., Kornblith, S., Norouzi, M., Hinton, G.: A simple framework for contrastive learning of visual representations. In: International Conference on Machine Learning, pp. 1597–1607. PMLR (2020)
7. Chen, X., He, K.: Exploring simple siamese representation learning. In: Proceedings of the IEEE/CVF Conference on Computer Vision and Pattern Recognition, pp. 15750–15758 (2021)
8. Codella, N., et al.: Skin lesion analysis toward melanoma detection 2018: A challenge hosted by the international skin imaging collaboration (isic). arXiv preprint arXiv:1902.03368 (2019)

9. Combalia, M., et al.: Bcn20000: Dermoscopic lesions in the wild. arXiv preprint arXiv:1908.02288 (2019)
10. Doersch, C., Gupta, A., Efros, A.A.: Unsupervised visual representation learning by context prediction. In: Proceedings of the IEEE International Conference on Computer Vision, pp. 1422–1430 (2015)
11. Gidaris, S., Singh, P., Komodakis, N.: Unsupervised representation learning by predicting image rotations. arXiv preprint arXiv:1803.07728 (2018)
12. Grill, J.B., et al.: Bootstrap your own latent-a new approach to self-supervised learning. Adv. Neural. Inf. Process. Syst. **33**, 21271–21284 (2020)
13. Gutman, D., et al.: Skin lesion analysis toward melanoma detection: a challenge at the international symposium on biomedical imaging (isbi) 2016, hosted by the international skin imaging collaboration (isic). arXiv preprint arXiv:1605.01397 (2016)
14. Hara, S., Nitanda, A., Maehara, T.: Data cleansing for models trained with sgd. In: Advances in Neural Information Processing Systems 32 (2019)
15. He, K., Fan, H., Wu, Y., Xie, S., Girshick, R.: Momentum contrast for unsupervised visual representation learning. In: Proceedings of the IEEE/CVF Conference on Computer Vision and Pattern Recognition, pp. 9729–9738 (2020)
16. Hu, X., Zeng, D., Xu, X., Shi, Y.: Semi-supervised contrastive learning for label-efficient medical image segmentation. In: de Bruijne, M., Cattin, P.C., Cotin, S., Padoy, N., Speidel, S., Zheng, Y., Essert, C. (eds.) MICCAI 2021. LNCS, vol. 12902, pp. 481–490. Springer, Cham (2021). https://doi.org/10.1007/978-3-030-87196-3_45
17. Huynh, T., Kornblith, S., Walter, M.R., Maire, M., Khademi, M.: Boosting contrastive self-supervised learning with false negative cancellation. In: Proceedings of the IEEE/CVF Winter Conference on Applications of Computer Vision, pp. 2785–2795 (2022)
18. Jaeger, S., Candemir, S., Antani, S., Wáng, Y.X.J., Lu, P.X., Thoma, G.: Two public chest x-ray datasets for computer-aided screening of pulmonary diseases. Quant. Imaging Med. Surg. **4**(6), 475 (2014)
19. Jaiswal, A., Babu, A.R., Zadeh, M.Z., Banerjee, D., Makedon, F.: A survey on contrastive self-supervised learning. Technologies **9**(1), 2 (2020)
20. Koh, P.W., Liang, P.: Understanding black-box predictions via influence functions. In: International Conference on Machine Learning, pp. 1885–1894. PMLR (2017)
21. Noroozi, M., Favaro, P.: Unsupervised learning of visual representations by solving jigsaw puzzles. In: Leibe, B., Matas, J., Sebe, N., Welling, M. (eds.) ECCV 2016. LNCS, vol. 9910, pp. 69–84. Springer, Cham (2016). https://doi.org/10.1007/978-3-319-46466-4_5
22. Pruthi, G., Liu, F., Kale, S., Sundararajan, M.: Estimating training data influence by tracing gradient descent. Adv. Neural. Inf. Process. Syst. **33**, 19920–19930 (2020)
23. Sowrirajan, H., Yang, J., Ng, A.Y., Rajpurkar, P.: Moco pretraining improves representation and transferability of chest x-ray models. In: Medical Imaging with Deep Learning, pp. 728–744. PMLR (2021)
24. Tajbakhsh, N., Jeyaseelan, L., Li, Q., Chiang, J.N., Wu, Z., Ding, X.: Embracing imperfect datasets: a review of deep learning solutions for medical image segmentation. Med. Image Anal. **63**, 101693 (2020)
25. Tschandl, P., Rosendahl, C., Kittler, H.: The ham10000 dataset, a large collection of multi-source dermatoscopic images of common pigmented skin lesions. Sci. Data **5**(1), 1–9 (2018)

26. Wang, X., Peng, Y., Lu, L., Lu, Z., Bagheri, M., Summers, R.M.: Chestx-ray8: hospital-scale chest x-ray database and benchmarks on weakly-supervised classification and localization of common thorax diseases. In: Proceedings of the IEEE Conference on Computer Vision and Pattern Recognition, pp. 2097–2106 (2017)
27. Zbontar, J., Jing, L., Misra, I., LeCun, Y., Deny, S.: Barlow twins: self-supervised learning via redundancy reduction. In: International Conference on Machine Learning, pp. 12310–12320. PMLR (2021)
28. Zhang, R., Isola, P., Efros, A.A.: Colorful image colorization. In: Leibe, B., Matas, J., Sebe, N., Welling, M. (eds.) ECCV 2016. LNCS, vol. 9907, pp. 649–666. Springer, Cham (2016). https://doi.org/10.1007/978-3-319-46487-9_40
29. Zhang, Y., Jiang, H., Miura, Y., Manning, C.D., Langlotz, C.P.: Contrastive learning of medical visual representations from paired images and text. In: Machine Learning for Healthcare Conference, pp. 2–25. PMLR (2022)
30. Zhu, R., Zhao, B., Liu, J., Sun, Z., Chen, C.W.: Improving contrastive learning by visualizing feature transformation. In: Proceedings of the IEEE/CVF International Conference on Computer Vision, pp. 10306–10315 (2021)

Multi-modal Semi-supervised Evidential Recycle Framework for Alzheimer's Disease Classification

Yingjie Feng[1], Wei Chen[2], Xianfeng Gu[3], Xiaoyin Xu[4], and Min Zhang[1](✉)

[1] Collaborative Innovation Center of Artificial Intelligence, College of Computer Science and Technology, Zhejiang University, Hangzhou, China
min_zhang@zju.edu.cn
[2] Zhejiang University Affiliated Sir Run Run Shaw Hospital, Hangzhou, China
[3] Department of Computer Science, Stony Brook University, Stony Brook, NY, USA
[4] Department of Radiology, Brigham and Women's Hospital, Harvard Medical School, Boston, MA, USA

Abstract. Alzheimer's disease (AD) is an irreversible neurodegenerative disease, so early identification of Alzheimer's disease and its early stage disorder, mild cognitive impairment (MCI), is of great significance. However, currently available labeled datasets are still small, so the development of semi-supervised classification algorithms will be beneficial for clinical applications. We propose a novel uncertainty-aware semi-supervised learning framework based on the improved evidential regression. Our framework uses the aleatoric uncertainty (AU) from the data itself and the epistemic uncertainty (EU) from the model to optimize the evidential classifier and feature extractor step by step to achieve the best performance close to supervised learning with small labeled data counts. We conducted various experiments on the ADNI-2 dataset, demonstrating the effectiveness and advancement of our method.

Keywords: Semi-supervised learning · Deep evidential regression · EfficientNet-V2 · Alzheimer's disease · Multi-modality

1 Introduction

Alzheimer's disease (AD) is an irreversible neurodegenerative disease that leaves patients with impairments in memory, language and cognition [7]. Previous work of [6,22] show that the combination of image data and other related data is beneficial to the improvement of model performance, but how to efficiently combine statistical non-imaging data and medical image data is still an open question. Second, although it is not too difficult to obtain and collect patient data, subjective bias in the AD diagnosis process and the time-consuming and complicated

M. Zhang was partially supported by NSFC62202426. X. Gu was partially supported by NIH 3R01LM012434-05S1, 1R21EB029733-01A1, NSF FAIN-2115095, NSF CMMI-1762287.

H. Greenspan et al. (Eds.): MICCAI 2023, LNCS 14220, pp. 130–140, 2023.
https://doi.org/10.1007/978-3-031-43907-0_13

process of labeling the diagnostic results lead to the scarcity of labeled data [11]. Therefore, research and development of models that require only a small amount of labeled data to achieve higher accuracy has attracted great attention [14].

Semi-supervised learning (SSL) methods are commonly used in medical image analysis to address the lack of manually annotated data [24]. Hang et al. [10] proposed a contrastive self-ensembling framework by introducing the weight formula and reliability-awareness for semi-supervised medical image classification. In [3], Aviles et al., based on the diffusion model and hypergraph learning, proposed a multi-modal hypergraph diffusion network to implement semi-supervised learning for AD classification. In [4], researchers introduced CSEAL, a semi-supervised learning framework that combines consistency-based SSL with uncertainty-based active learning, for multi-label chest X-ray classification tasks. Other studies using evidential learning [5,17] have demonstrated the great potential of this theory in fitting low-dimensional manifolds in high-dimensional spaces for classification with uncertainty estimates. This feature makes the model based on evidential learning promising in the SSL field of medical images.

The proposal of evidential deep learning (EDL) [21] allows the model to better estimate the uncertainty in multi-classification tasks. On the binary classification task, controlling evidential regression to obtain a continuous probability value before 0 and 1 can often achieve more accurate results than using the Dirichlet distribution to obtain a discrete distribution of EDL [19].

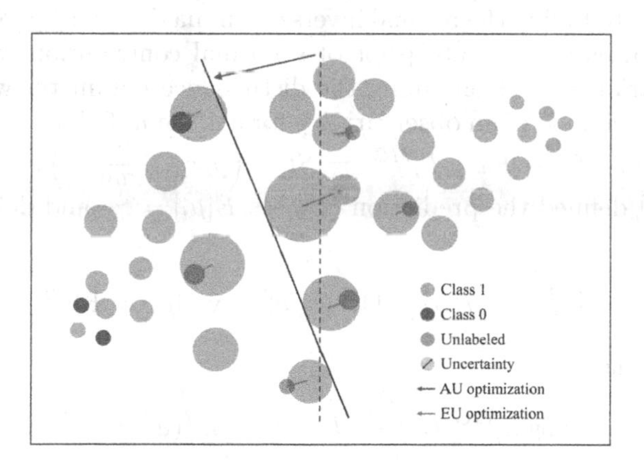

Fig. 1. An iteration of optimization of our model. The AU optimization process does not rely on any label and obtains the best classification of the current model by reducing AU. Then, on the basis of this classification, EU optimization relies on ground truth and pseudo labels to optimize the model to get better prediction results.

The residual between the prediction results of the imperfect model and the true distribution of the data can be decomposed into aleatoric uncertainty (AU) and epistemic uncertainty (EU). Theoretically, the former comes from the noise of the data, which usually does not depend on the sample size. Therefore, by iteratively reducing this part of uncertainty, the best classification results can be obtained under a given amount of data. The latter is proportional to the sample size. As sample size increases, the reduction of this part of uncertainty can make the model closer to the observed distribution or fit with more complex conditions, thereby improving the performance of the model itself. Based on this understanding, we exploit the ability of evidential regression of handling uncertainty to decompose the two parts of uncertainty, AU and EU, and proposed a method by adjusting the two parts of uncertainty to achieve semi-supervised classification which shows in Fig. 1.

Our main contributions include: 1) Adjusting the loss function of evidential regression so it can obtain more accurate results and better separate AU and EU; 2) Building a multi-layer and multi-step network to implement evidential regression and a semi-supervised learning method of step-by-step training is proposed; 3) A new SOTA of semi-supervised learning is achieved on the ADNI dataset, and performance close to supervised learning can be achieved with only a small amount of labeled data.

2 Methods

2.1 Original Deep Evidential Regression (DER)

DER [2] adopts the simplest setting: $y_i \sim N(0, \sigma_i^2)$. In a Bayesian framework, this corresponds to taking the normal inverse Gamma distribution $\text{NIG}(\mu, \sigma^2|m)$, $m = (\gamma, \nu, \alpha, \beta)$, as a conjugate prior of a normal contribution with unknown mean μ and variance σ^2. Combining the disturbance parameter with Bayesian inference, the likelihood of an observation y for a given m follows a t-distribution with $2\alpha_i$ degrees of freedom: $L_i^{NIG} = St_{2\alpha_i}\left(y_i|\gamma_i, \frac{\beta_i(1+\gamma_i)}{\gamma_i\alpha_i}\right)$. For known m, Animi et al. [2] defined the prediction of y_i as $E[\mu_i] = \gamma_i$, and defined AU and EU as u_a and u_e:

$$u_a^2 = E[\sigma_i^2] = \beta_i/(\alpha_i - 1), \qquad u_e^2 = \text{var}[\mu_i] = E[\sigma_i^2]/\nu_i.$$

And it follows that:

$$L_i'(w) = -\log L_i^{NIG}(w) + \lambda L_i^R(w), \quad L_i^R(w) = |y_i - \gamma_i| \cdot \Phi$$

where $m = NN(w)$ is specified by a neural network (NN), λ is a hyperparameter, and $\Phi = 2\gamma_i + \alpha_i$ represents the total evidence gained from training.

2.2 Evidential Regression Beyond DER

Although DER has achieved some success in both theoretical and practical applications [2,15], as pointed out by Meinert et al. [16], this theory has some major flaws. First, although the regularization part of loss function L^R is added, the constrain on parameter β_i is not enough. Second, although the two parts of AU and EU are defined separately, the correlation between them is too high.

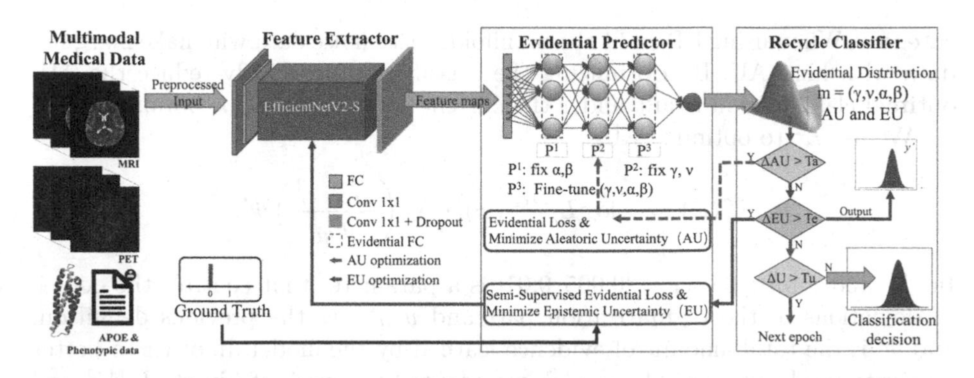

Fig. 2. Architecture of our network. The recycle classifier part judges the decrease of the uncertainty of the predicted value, and controls the framework to train the other two parts in a loop.

In practice, disentangling and effectively using uncertainty information for training remains challenging.

After practice and theoretical proof, Meinert et al. [16] states that the width of the t-distribution projected by the NIG distribution, that is, w_{St}, can better reflect the noise in data.

And, correspondingly, we use the residual $1/\sqrt{\nu_i}$ part of u_a and u_e in the original definition to represent EU:

$$u_A = w_{St} = \sqrt{\frac{\beta_i(1+\nu_i)}{\alpha_i \nu_i}}, \qquad u_E = \frac{u_e}{u_a} = \frac{1}{\sqrt{\nu_i}}$$

where ν_i, α_i, and β_i are part of the parameters of the evidence distribution $m = (\gamma, \nu, \alpha, \beta)$, and we verify the performance of this new uncertainty estimation method through experiments.

2.3 Model and Workflow

With efficient estimation of AU and EU, our model has the basis for implementation. As shown in Fig. 2, our model is divided into three parts: multimodal feature extractor, evidential predictor, and recycle classifier. The multimodal feature extractor form a high-dimensional feature space and the evidential predictor generates the evidential distribution in the feature space. After calculating the classification result and the uncertainty (AU and EU) based on the evidential distribution, the recycle classifier controls the training process and reaches the best performances through a step-by-step recurrent training workflow.

AU for Training Classifier. Based on the manifold assumption, the real data is gathered on the low-dimensional manifold of the high-dimensional space, and the noise of the data is located on the edge of the manifold for the corresponding category. When using ER to fit the manifold, these noise data will make marginal data with high AU. By optimizing the classifier to iteratively reduce the AU, optimal classification result under the current conditions can be obtained.

We use L^a to optimize AU:

$$L_i^a(w) = -\log L_i^{NIG'}(w) + \lambda_a \left(\frac{\gamma_i - \gamma_i^2}{w_{St}^2} \right) \Phi'.$$

In the above formula $\lambda_a = [0.005, 0.01]$ is a parameter that controls the degree of deviations of the regularization part and w_{St} uses the previous definition, and Φ' is the total amount of evidence learned by the model. In order to better motivate the learning of the model, we adopted the work of Liu et al. [15] and used the form of $\Phi' = \gamma_i + 2\alpha_i$. We used the expanded form of L^{NIG} with minor adjustments according to the optimization objective, specifically, $-\log L_i^{NIG'} = \frac{1}{2}\log(\frac{\pi}{\nu}) - \alpha\log(2\beta + 2\beta\nu) + (\alpha + 0.5)\log(\nu(\gamma_i - \gamma_i^2) + 2\beta + 2\beta\nu) + \log\left(\frac{\Gamma(\alpha)}{\Gamma(\alpha+0.5)}\right)$.

EU for Training Extractor. If only the AU part is optimized, there will always be this gap between the model prediction and the real data.

EU is mainly used to optimize the feature extractor since EU mainly reflects the bias of the model in the prediction. For data D_l, given groundtruth labels, we use $L_i^l(w) = -\log L_i^{NIG}(w) + \lambda_l \left(\frac{y_i - \gamma_i}{w_{St}}\right)^2 \Phi'$. In order to enhance the certainty of the data with ground truth, we set a smaller $\lambda_l = [0.005, 0.015]$. For dataset D_u without real labels, we use the prediction results y' obtained in the last iterative training to replace the real labels to get $L_i^u(w) = -\log L_i^{NIG}(w) + \lambda_u \left(\frac{y_i' - \gamma_i}{w_{St}}\right)^2 \Phi'$. In order for the model to utilize the results of the previous round of learning, we set a larger $\lambda_u = [0.015, 0.025]$, which can make the model more conservative about making predictions in the next iteration. This reduces our models being affected by misleading evidence and obtains better performance by retaining higher uncertainty to allow the model to have more room to optimize.

In order to effectively combine labeled and unlabeled data we adjust the weights of different data:

$$L_i^e(w) = \mu_l L_i^l + \mu_u L_i^u = -\log L_i^{NIG}(w) + \mu_u \lambda_u \left(\frac{y_i' - \gamma_i}{w_{St}} \right)^2 \Phi' + \mu_l \lambda_l \left(\frac{y_i - \gamma_i}{w_{St}} \right)^2 \Phi'$$

where $\mu_l + \mu_u = 1$, $\mu_l, \mu_u \in [0, 1]$, are two weight factors.

Model. In terms of the feature extractor, we use the latest EfficientNetV2, which, in Feng et al. [8], has achieved good results in combination with EDL. In order to avoid overfitting, we used the minimum model in this network and added Dropout to the output end. At the same time, in order to fill the differences between multi-modality data and model input, we have added the fully connected (FC) layer and convolutional layer (Conv) to adaptive adjust input channels. We employed three evidential FC layers proposed by Amini et al. [2] to form our evidential predictor. At the same time, in order to achieve the optimization of AU and EU separately, we froze some parameters in the first two layers and limited the range of the last layer of parameter adjustment.

Workflow of Recycle Training. First, we do not fix any weight and we use a small part of the data for warm-up training. Second, we freeze the weight update of the extractor and P^2 and P^3 in the evidential predictor and use L^a to optimize the classifier. We calculate and record the AU score after each update. When the difference between the update $|\Delta AU|$ is smaller than the threshold value $T_a = [0.0005, 0.001]$ we set, the cycle of AU optimization is over. Then, we fix the weight of P^1 and P^3 in the evidential predictor and use L^e to optimize the extractor. Similarly, when the change $|\Delta EU|$ brought by the update is less than the threshold $T_e = [0.0025, 0.005]$, end the cycle and output y'. Finally, we fix all network parameters except P^3 to fine-tune until $|\Delta U| = |\Delta AU| + |\Delta EU|$ brought by the update loss function $L = L^e + L^a$ is less than threshold $T_u = [0.002, 0.005]$. All thresholds are adjusted according to the proportion of labels and unlabeled data during training.

Table 1. Classification results of all comparison methods on ADNI-2 dataset (%). The top section of the table shows the results of supervised learning (SL), while the bottom section shows the performance of the current SSL SOTA methods.

Method	AD vs. NC			EMCI vs. LMCI			LMCI vs. NC		
	ACC	SPE	SEN	ACC	SPE	SEN	ACC	SPE	SEN
Baseline [12]	80.53	80.10	80.32	74.10	73.18	75.85	72.05	70.80	71.26
SL SOTA [18,20,23]	96.84	98.23	95.76	92.40	93.70	89.50	92.49	91.08	93.48
Upper bound (UB)	94.45	93.80	94.07	89.95	90.29	90.81	88.56	88.81	86.32
Π model [13]	90.45	85.82	90.05	81.58	80.15	84.50	80.65	83.48	78.75
DS^3L [9]	90.86	89.03	89.72	81.07	83.25	82.81	80.79	81.55	81.16
RFS-LDA [1]	92.11	89.50	88.40	80.90	81.05	83.63	81.90	84.72	80.05
Hypergraph diffusion [3]	92.11	92.80	91.33	85.22	86.40	84.02	82.01	84.01	81.80
Ours	**93.90**	**92.95**	**93.01**	**89.45**	**88.50**	**89.47**	**87.27**	**86.94**	**85.83**

3 Experiments and Results

Data Description. In this paper, we assess the effectiveness of our multi-modal semi-supervised evidential recycle framework on the ADNI-2 dataset[1], which comprises multi-center data consisting of various modalities, including imaging and multiple phenotype data. Specifically, the dataset consists of four

[1] *Data used in preparation of this article were obtained from the Alzheimer's Disease Neuroimaging Initiative (ADNI) database (adni.loni.usc.edu).

categories: normal control (NC), early mild cognitive impairment (EMCI), late mild cognitive impairment (LMCI), and Alzheimer's disease (AD). To ensure the effectiveness of our training and balance the number of categories, we used a sample of 515 patients, utilizing their MRI, PET, demographics, and APOE as inputs. On MRI images, we used 3T T1-weighted and FLAIR MR images, and the preprocessing process used CAT12 and SPM tools. All MRI data were processed using standard pipeline, including anterior commissure (AC)-posterior commissure (PC) correction, intensity correction, and skull stripping. Affine registration is performed to linearly align each MRI to the Colin27 template and resample to $224 \times 224 \times 91$ for subsequent processing. For PET images, we used the official pre-processed AV-45 PET image and resampled them in the same way as the MRIs. We chose to include APOE in our analysis, as it is a well-established genetic risk factor for developing AD.

Evaluation. We evaluated our model from three aspects. First, for the sake of comparison, we followed the technical conventions of most similar studies and selected three comparison tasks: AD vs NC, LMCI vs NC, and EMCI vs LMCI. Second, we compared and demonstrated the results of our model under different numbers of ground truth labels to verify that its performance improves as the label data volume increases. Third, we conducted different ablation experiments, which shows in Fig. 3, to prove the validity and rationality of each part of the proposed model framework. Among them, CNNs represents the performance when using only the EfficientNetV2-S model and its original supervised learning classifier without using unlabeled data for training, which is the baseline model. AU and EU represent the training process using only the corresponding parts. DER uses our proposed complete training process but does not use our improved u_a and u_e estimations, instead continuing to use the estimation method u_A and u_E which proposed in the original DER paper [2]. To compare performance fairly, we ran all techniques under the same conditions. The results were evaluated on accuracy (ACC), specificity (SPE), and sensitivity (SEN).

Implementation Details. The upper bound in performance is the result obtained when the model is trained with all the input data are labeled. In the current supervised learning algorithms, the performance of each algorithm on

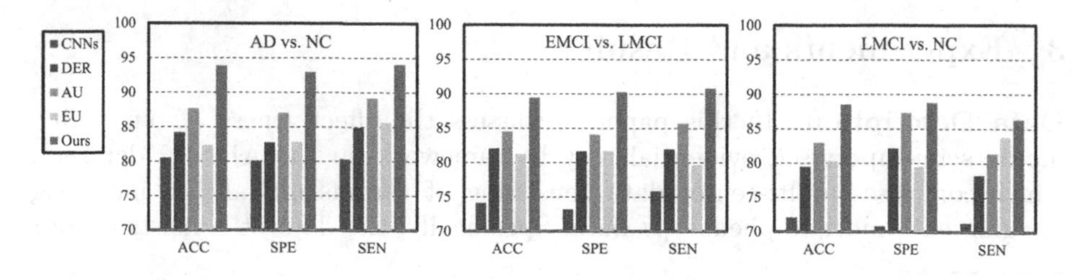

Fig. 3. Ablation experiment results.

Table 2. Classification results(%) with different percentages of labeled and unlabelled data in the training process.

Method	Labeled	AD vs. NC			EMCI vs. LMCI			LMCI vs. NC		
		ACC	SPE	SEN	ACC	SPE	SEN	ACC	SPE	SEN
Upper Bound	100%	94.45	93.80	94.07	89.95	90.29	90.81	88.56	88.81	86.32
Baseline	5%	75.11	73.92	75.79	73.17	69.47	75.14	68.95	67.81	68.53
DS^3L		78.92	76.75	78.14	74.68	71.65	72.39	74.27	70.58	73.71
Ours		82.45	81.63	84.97	73.05	70.84	74.22	73.86	74.15	72.28
Baseline	10%	78.50	78.03	77.48	73.67	70.29	74.11	69.13	68.57	68.38
DS^3L		84.73	79.67	82.69	80.93	80.13	79.54	74.57	75.39	72.88
Ours		90.18	89.73	87.42	81.67	83.45	82.29	80.01	78.25	80.89
Baseline	20%	80.53	80.10	80.32	74.10	73.18	75.85	72.05	70.80	71.26
DS^3L		90.86	89.03	89.72	81.07	83.25	82.81	80.79	81.55	81.16
Ours		**93.90**	**92.95**	**93.01**	**89.45**	**88.50**	**89.47**	**87.27**	**86.94**	**85.83**

each task is not consistent, so we selected three papers in supervised learning, each representing the SOTA performance of the three tasks [18,20,23] for comparison. Our implementation employs PyTorch v1.4.0 and utilizes the Adam optimizer with a learning rate of 1×10^{-4} and a weight decay of 1×10^{-5}. We utilize a linear decay scheduler of 0.1 based on the loss functions above. The optimizer is set with β values of [0.9, 0.999] and ϵ value of 1×10^{-8}. In terms of data, since the SSL method needs to learn from unlabeled data, 100% of the data is put into training, and some of the data have ground truth labels. In the test, only the result index of the unlabeled data is calculated, so the training set and the test set are not divided. But in order to determine the threshold of each uncertainty, we randomly selected 10% of the data as the validation set, and calculated the uncertainty independently outside the training process.

Results. We compared our model with the semi-supervised learning methods currently achieving the best performance on the ADNI-2 dataset, as well as other top models in the semi-supervised learning field. As shown in Table 1, our model achieved SOTA performance in all three tasks of the semi-supervised learning category. At the same time, compared with other semi-supervised learning algorithms, our results are unprecedentedly close to the best supervised learning methods, indicating the performance of our model under less labeled data and the feasibility of applying this algorithm in clinical settings.

Our ablation experiment results are shown in Fig. 3. Firstly, compared with the baseline, our semi-supervised learning algorithm effectively learns classification information from unlabeled data. Secondly, compared with DER, our uncertainty estimation surpasses the original DER method. The AU and EU items demonstrate the importance of optimizing both the AU and EU components in our framework.

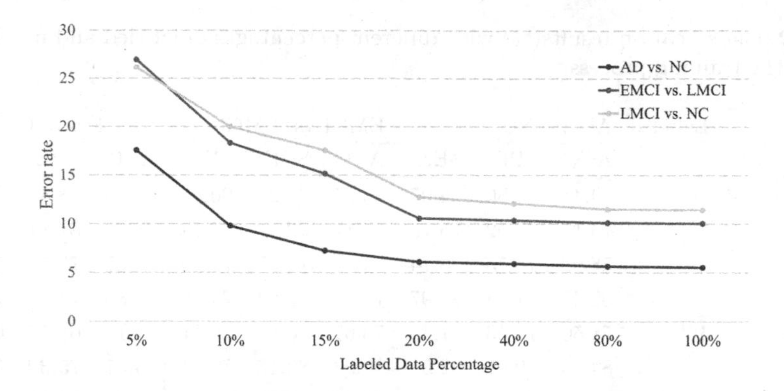

Fig. 4. Error rate of different percentages of label counts, the well-known conduction effect in the field of semi-supervised learning can be observed.

From Table 2, we can observe that we have outperformed the currently representative advanced semi-supervised learning algorithm DS^3L [9] in each labeled data count. At the same time, the superiority of our model compared to the baseline method also proves the learning efficiency of our framework. The performance of our model at 20% labeled data count is already very close to the upper bound, which is the result obtained using 100% labeled data. This indicates the strong learning ability of our model in the case of a small labeled data amount.

In addition, we have plotted the error rate of our framework under different labeled data counts in Fig. 4. It is apparent that the performance of our model improves as the labeled data amount increases from 5% to 10%, 15%, and 20%. Combining with Table 2, we can observe the well-known transductive effect in the field of semi-supervised learning, which means that beyond a certain data amount, increasing the size of the dataset can only bring marginal performance improvement. This is evident when comparing the model performance under 20%, 40%, 80%, and 100% labeled data counts.

4 Conclusions

We proposed an evidential regression-based semi-supervised learning framework, using the characteristics of AU and EU to train classifiers and extractors, respectively. Our model achieves SOTA performance on the ADNI-2 dataset. And due to the characteristics of semi-supervised learning, our model has unique advantages in adding private data, fine-tuning downstream tasks, and avoiding overfitting, which makes our model have great potential in clinical applications.

References

1. Adeli, E., et al.: Semi-supervised discriminative classification robust to sample-outliers and feature-noises. IEEE Trans. Pattern Anal. Mach. Intell. **41**(2), 515–522 (2018)
2. Amini, A., Schwarting, W., Soleimany, A., Rus, D.: Deep evidential regression. Adv. Neural. Inf. Process. Syst. **33**, 14927–14937 (2020)
3. Aviles-Rivero, A.I., Runkel, C., Papadakis, N., Kourtzi, Z., Schönlieb, C.B.: Multi-modal hypergraph diffusion network with dual prior for Alzheimer classification. In: MICCAI 2022, Part III. LNCS, pp. 717–727. Springer, Cham (2022). https://doi.org/10.1007/978-3-031-16437-8_69
4. Balaram, S., Nguyen, C.M., Kassim, A., Krishnaswamy, P.: Consistency-based semi-supervised evidential active learning for diagnostic radiograph classification. In: MICCAI 2022, Part I. LNCS, vol. 13431, pp. 675–685. Springer, Cham (2022). https://doi.org/10.1007/978-3-031-16431-6_64
5. Bengs, V., Hüllermeier, E., Waegeman, W.: Pitfalls of epistemic uncertainty quantification through loss minimisation. In: Advances in Neural Information Processing Systems (2022)
6. Cobbinah, B.M., et al.: Reducing variations in multi-center Alzheimer's disease classification with convolutional adversarial autoencoder. Med. Image Anal. **82**, 102585 (2022)
7. De Strooper, B., Karran, E.: The cellular phase of Alzheimer's disease. Cell **164**(4), 603–615 (2016)
8. Feng, Y., Wang, J., An, D., Gu, X., Xu, X., Zhang, M.: End-to-end evidential-efficient net for radiomics analysis of brain MRI to predict oncogene expression and overall survival. In: Wang, L., Dou, Q., Fletcher, P.T., Speidel, S., Li, S. (eds.) MICCAI 2022, Part III, vol. 13433, pp. 282–291. Springer, Cham (2022). https://doi.org/10.1007/978-3-031-16437-8_27
9. Guo, L.Z., Zhang, Z.Y., Jiang, Y., Li, Y.F., Zhou, Z.H.: Safe deep semi-supervised learning for unseen-class unlabeled data. In: International Conference on Machine Learning, pp. 3897–3906. PMLR (2020)
10. Hang, W., Huang, Y., Liang, S., Lei, B., Choi, K.S., Qin, J.: Reliability-aware contrastive self-ensembling for semi-supervised medical image classification. In: Medical Image Computing and Computer Assisted Intervention-MICCAI 2022: 25th International Conference, Singapore, September 18–22, 2022, Proceedings, Part I. pp. 754–763. Springer, Cham (2022). https://doi.org/10.1007/978-3-031-16431-6_71
11. Hett, K., Ta, V.T., Oguz, I., Manjón, J.V., Coupé, P., Initiative, A.D.N., et al.: Multi-scale graph-based grading for Alzheimer's disease prediction. Med. Image Anal. **67**, 101850 (2021)
12. Huang, G., Liu, Z., Van Der Maaten, L., Weinberger, K.Q.: Densely connected convolutional networks. In: Proceedings of the IEEE Conference on Computer Cision and Pattern Recognition, pp. 4700–4708 (2017)
13. Laine, S., Aila, T.: Temporal ensembling for semi-supervised learning. In: International Conference on Learning Representations (2016)
14. Li, Z., Togo, R., Ogawa, T., Haseyama, M.: Chronic gastritis classification using gastric x-ray images with a semi-supervised learning method based on tri-training. Med. Biol. Eng. Comput. **58**, 1239–1250 (2020)
15. Liu, Z., Amini, A., Zhu, S., Karaman, S., Han, S., Rus, D.L.: Efficient and robust lidar-based end-to-end navigation. In: 2021 IEEE International Conference on Robotics and Automation (ICRA), pp. 13247–13254. IEEE (2021)

16. Meinert, N., Gawlikowski, J., Lavin, A.: The unreasonable effectiveness of deep evidential regression. arXiv e-prints pp. arXiv-2205 (2022)
17. Neupane, K.P., Zheng, E., Yu, Q.: MetaEDL: Meta evidential learning for uncertainty-aware cold-start recommendations. In: 2021 IEEE International Conference on Data Mining (ICDM), pp. 1258–1263. IEEE (2021)
18. Ning, Z., Xiao, Q., Feng, Q., Chen, W., Zhang, Y.: Relation-induced multi-modal shared representation learning for Alzheimer's disease diagnosis. IEEE Trans. Med. Imaging **40**(6), 1632–1645 (2021)
19. Oh, D., Shin, B.: Improving evidential deep learning via multi-task learning. In: Proceedings of the AAAI Conference on Artificial Intelligence, vol. 36, pp. 7895–7903 (2022)
20. Pei, Z., Wan, Z., Zhang, Y., Wang, M., Leng, C., Yang, Y.H.: Multi-scale attention-based pseudo-3D convolution neural network for Alzheimer's disease diagnosis using structural MRI. Pattern Recogn. **131**, 108825 (2022)
21. Sensoy, M., Kaplan, L., Kandemir, M.: Evidential deep learning to quantify classification uncertainty. In: Advances in Neural Information Processing Systems 31 (2018)
22. Song, X., et al.: Graph convolution network with similarity awareness and adaptive calibration for disease-induced deterioration prediction. Med. Image Anal. **69**, 101947 (2021)
23. Song, X., et al.: Multi-center and multi-channel pooling GCN for early AD diagnosis based on dual-modality fused brain network. IEEE Trans. Med. Imaging (2022)
24. Yang, X., Song, Z., King, I., Xu, Z.: A survey on deep semi-supervised learning. IEEE Trans. Knowl. Data Eng. (2022)

3D Arterial Segmentation via Single 2D Projections and Depth Supervision in Contrast-Enhanced CT Images

Alina F. Dima[1,2(✉)], Veronika A. Zimmer[1,2], Martin J. Menten[1,4], Hongwei Bran Li[1,3], Markus Graf[2], Tristan Lemke[2], Philipp Raffler[2], Robert Graf[1,2], Jan S. Kirschke[2], Rickmer Braren[2], and Daniel Rueckert[1,2,4]

[1] School of Computation, Information and Technology, Technical University of Munich, Munich, Germany
alina.dima@tum.de
[2] School of Medicine, Klinikum Rechts der Isar, Technical University of Munich, Munich, Germany
[3] Department of Quantitative Biomedicine, University of Zurich, Zurich, Switzerland
[4] Department of Computing, Imperial College London, London, UK

Abstract. Automated segmentation of the blood vessels in 3D volumes is an essential step for the quantitative diagnosis and treatment of many vascular diseases. 3D vessel segmentation is being actively investigated in existing works, mostly in deep learning approaches. However, training 3D deep networks requires large amounts of manual 3D annotations from experts, which are laborious to obtain. This is especially the case for 3D vessel segmentation, as vessels are sparse yet spread out over many slices and disconnected when visualized in 2D slices. In this work, we propose a novel method to segment the 3D peripancreatic arteries **solely from one annotated 2D projection per training image** with depth supervision. We perform extensive experiments on the segmentation of peripancreatic arteries on 3D contrast-enhanced CT images and demonstrate how well we capture the rich depth information from 2D projections. We demonstrate that by annotating a single, randomly chosen projection for each training sample, we obtain comparable performance to annotating multiple 2D projections, thereby reducing the annotation effort. Furthermore, by mapping the 2D labels to the 3D space using depth information and incorporating this into training, we almost close the performance gap between 3D supervision and 2D supervision. Our code is available at: https://github.com/alinafdima/3Dseg-mip-depth.

Keywords: vessel segmentation · 3D segmentation · weakly supervised segmentation · curvilinear structures · 2D projections

---

Supplementary Information The online version contains supplementary material available at https://doi.org/10.1007/978-3-031-43907-0_14.

H. Greenspan et al. (Eds.): MICCAI 2023, LNCS 14220, pp. 141–151, 2023.
https://doi.org/10.1007/978-3-031-43907-0_14

1 Introduction

Automated segmentation of blood vessels in 3D medical images is a crucial step for the diagnosis and treatment of many diseases, where the segmentation can aid in visualization, help with surgery planning, be used to compute biomarkers, and further downstream tasks. Automatic vessel segmentation has been extensively studied, both using classical computer vision algorithms [16] such as vesselness filters [8], or more recently with deep learning [3,5,6,11,19,21], where state-of-the-art performance has been achieved for various vessel structures. Supervised deep learning typically requires large, well-curated training sets, which are often laborious to obtain. This is especially the case for 3D vessel segmentation.

Manually delineating 3D vessels typically involves visualizing and annotating a 3D volume through a sequence of 2D cross-sectional slices, which is not a good medium for visualizing 3D vessels. This is because often only the cross-section of a vessel is visible in a 2D slice. In order to segment a vessel, the annotator has to track the cross-section of that vessel through several adjacent slices, which is especially tedious for curved or branching vessel trees. Projecting 3D vessels to a 2D plane allows for the entire vessel tree to be visible within a single 2D image, providing a more robust representation and potentially alleviating the burden of manual annotation. Kozinski et al. [13] propose to annotate up to three maximum intensity projections (MIP) for the task of centerline segmentation [13], obtaining results comparable to full 3D supervision. Compared to centerline segmentation, where the vessel diameter is disregarded, training a 3D vessel segmentation model from 2D annotations poses additional segmentation-specific challenges, as 2D projections only capture the outline of the vessels, providing no information about their interior. Furthermore, the axes of projection are crucial for the model's success, given the sparsity of information in 2D annotations.

To achieve 3D vessel segmentation with only 2D supervision from projections, we first investigate which viewpoints to annotate in order to maximize segmentation performance. We show that it is feasible to segment the full extent of vessels in 3D images with high accuracy by annotating only a single randomly-selected 2D projection per training image. This approach substantially reduces the annotation effort, even compared to works training only on 2D projections. Secondly, by mapping the 2D annotations to the 3D space using the depth of the MIPs, we obtain a partially segmented 3D volume that can be used as an additional supervision signal. We demonstrate the utility of our method on the challenging task of peripancreatic arterial segmentation on contrast-enhanced arterial-phase computed tomography (CT) images, which feature large variance in vessel diameter. Our contribution to 3D vessel segmentation is three-fold:

- Our work shows that highly accurate automatic segmentation of 3D vessels can be learned by annotating single MIPs.
- Based on extensive experimental results, we determine that the best annotation strategy is to label randomly selected viewpoints, while also substantially reducing the annotation cost.

- By incorporating additional depth information obtained from 2D annotations at no extra cost to the annotator, we almost close the gap between 3D supervision and 2D supervision.

2 Related Work

Learning from Weak Annotations. Weak annotations have been used in deep learning segmentation to reduce the annotation effort through cheaper, less accurate, or sparser labeling [20]. Bai et al. [1] learn to perform aortic image segmentation by sparsely annotating only a subset of the input slices. Multiple instance learning approaches bin pixels together by only providing labels at the bin level. Jia et al. [12] use this approach to segment cancer on histopathology images successfully. Annotating 2D projections for 3D data is another approach to using weak segmentation labels, which has garnered popularity recently in the medical domain. Bayat et al. [2] propose to learn the spine posture from 2D radiographs, while Zhou et al. [22] use multi-planar MIPs for multi-organ segmentation of the abdomen. Kozinski et al.[13] propose to segment vessel centerlines using as few as 2-3 annotated MIPs. Chen et al. [4] train a vessel segmentation model from unsupervised 2D labels transferred from a publicly available dataset, however, there is still a gap to be closed between unsupervised and supervised model performance. Our work uses weak annotations in the form of annotations of 2D MIPs for the task of peripancreatic vessel segmentation, where we attempt to reduce the annotation cost to a minimum by only annotating a single projection per training input without sacrificing performance.

Incorporating Depth Information. Depth is one of the properties of the 3D world. Loss of depth information occurs whenever 3D data is projected onto a lower dimensional space. In natural images, depth loss is inherent through image acquisition, therefore attempts to recover or model depth have been employed for 3D natural data. For instance, Fu et al. [9] use neural implicit fields to semantically segment images by transferring labels from 3D primitives to 2D images. Lawin et al. [14] propose to segment 3D point clouds by projecting them onto 2D and training a 2D segmentation network. At inference time, the predicted 2D segmentation labels are remapped back to the original 3D space using the depth information. In the medical domain, depth information has been used in volume rendering techniques [7] to aid with visualization, but it has so far not been employed when working with 2D projections of 3D volumes to recover information loss. We propose to do the conceptually opposite approach from Lawin et al. [14], by projecting 3D volumes onto 2D to facilitate and reduce annotation. We use depth information to map the 2D annotations to the original 3D space at annotation time and generate partial 3D segmentation volumes, which we incorporate in training as an additional loss term.

3 Methodology

Overview. The maximum intensity projection (MIP) of a 3D volume $I \in \mathbb{R}^{N_x \times N_y \times N_z}$ is defined as the highest intensity along a given axis:

$$mip(x, y) = \max_z \; I(x, y, z) \in \mathbb{R}^{N_x \times N_y}. \tag{1}$$

For simplicity, we only describe MIPs along the z-axis, but they can be performed on any image axis.

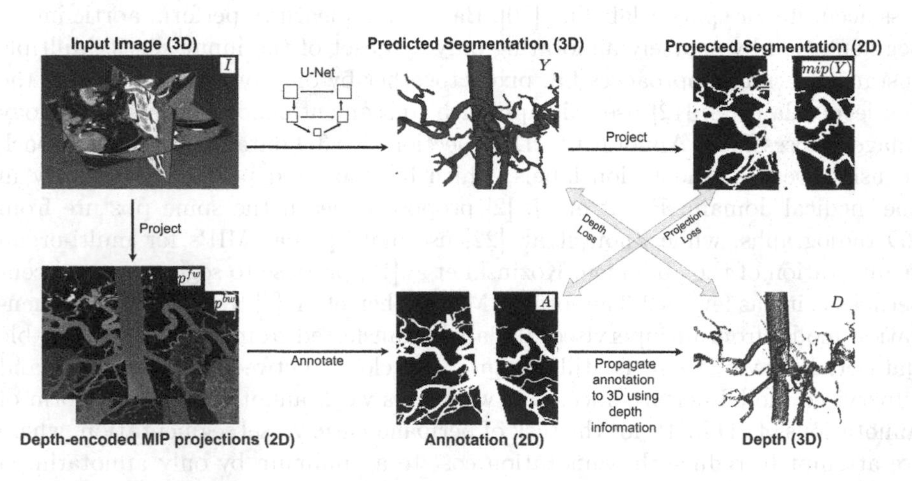

Fig. 1. Method overview. We train a 3D network to segment vessels from 2D annotations. Given an input image I, depth-encoded MIPs p^{fw}, p^{bw} are generated by projecting the input image to 2D. 2D binary labels A are generated by annotating one 2D projection per image. The 2D annotation is mapped to the 3D space using the depth information, resulting in a partially labeled 3D volume D. During training, both 2D annotations and 3D depth maps are used as supervision signals in a combined loss, which uses both predicted 3D segmentation Y and its 2D projection $mip(Y)$.

Exploiting the fact that arteries are hyperintense in arterial phase CTs, we propose to annotate MIPs of the input volume for binary segmentation. The hyperintensities of the arteries ensures their visibility in the MIP, while additional processing removes most occluding nearby tissue (Sect. 4).

Given a binary 2D annotation of a MIP $A \in \{0, 1\}^{N_x \times N_y}$, we map the foreground pixels in A to the original 3D image space. This is achieved by using the first and last z coordinates where the maximum intensity is observed along any projection ray. Owing to the fact that the vessels in the abdominal cavity are relatively sparse in 2D projections and most of the occluding tissue is removed in postprocessing, this step results in a fairly complete surface of the vessel tree. Furthermore, we can partially fill this surface volume, resulting in a 3D depth map D, which is a partial segmentation of the vessel tree. We use the 2D annotations as well as the depth map to train a 3D segmentation network in a weakly supervised manner.

An overview of our method is presented in Fig. 1. In the following, we describe these components and how they are combined to train a 3D segmentation network in more detail.

Depth Information. We can view MIP as capturing the intensity of the brightest pixel along each ray $r_{xy} \in \mathbb{R}^{N_z}$, where $r_{xy}(z) = I(x, y, z)$. Along each projection ray, we denote the first and last z coordinates which have the same intensity as the MIP to be the forward depth $z^{fw} = \arg\max_z I(x, y, z)$ and backward depth $z^{bw} = \arg\min_z I(x, y, z)$. This information can be utilized for the following: (1) enhancing the MIP visualization, or (2) providing a way to map pixels from the 2D MIP back to the 3D space (depth map). The reason why the maximum intensity is achieved multiple times along a ray is because our images are clipped, which removes a lot of the intensity fluctuations.

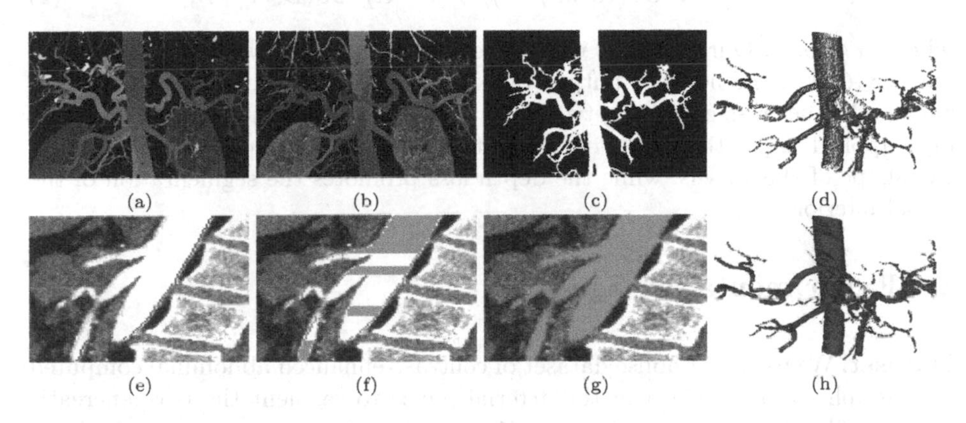

Fig. 2. Example depth-enhanced MIP using (a) forward depth z^{fw} and (b) backward depth z^{bw} visualized in color; (c) binary 2D annotation; a slice view from a 3D volume illustrating: (e) the forward – in green – and backward depth – in blue – , (f) the depth map, (g) 3D ground truth; volume rendering of (h) the depth map and (d) the depth map with only forward and backward depth pixels. The input images are contrast-enhanced.(Color figure online)

Depth-Enhanced MIP. We encode depth information into the MIPs by combining the MIP with the forward and backward depth respectively, in order to achieve better depth perception during annotation: $p^{fw} = \sqrt{mip} \cdot z^{fw}$ defines the forward projection, while $p^{bw} = \sqrt{mip} \cdot z^{bw}$ defines the backward projection. Figure 2 showcases (a) forward and (b) backward depth encoded MIPs.

Depth Map Generation. Foreground pixels from the 2D annotations are mapped to the 3D space by combining a 2D annotation with the forward and backward depth, resulting in a 3D partial vessel segmentation:

---

1. Create an empty 3D volume $D \in \mathbb{R}^{N_x \times N_y \times N_z}$.
2. For each foreground pixel in the annotation A at location (x, y), we label (x, y, z^{fw}) and (x, y, z^{bw}) as foreground pixels in D.
3. If the fluctuation in intensity between z^{fw} and z^{bw} along the ray r_{xy} is below a certain threshold in the source image I, the intermediate pixels are also labeled as foreground in D.

Training Loss. We train a 3D segmentation network to predict 3D binary vessel segmentation given a 3D input volume using 2D annotations. Our training set $\mathcal{D}_{tr}(I, A, D)$ consists of 3D volumes I paired with 2D annotations A and their corresponding 3D depth maps D. Given the 3D network output $Y = \theta(I)$, we minimize the following loss during training:

$$\mathcal{L}(Y) = \alpha \cdot \mathcal{CE}(A, mip(Y)) + (1 - \alpha) \cdot \mathcal{CE}(D, Y) \cdot D, \quad (2)$$

where $\alpha \in [0, 1]$. Our final loss is a convex combination between: **(a)** the cross-entropy($\mathcal{CE}$) of the network output projected to 2D and the 2D annotation, as well as **(b)** the cross-entropy between the network output and the depth map, but only applied to positive pixels in the depth map. Notably, the 2D loss constrains the shape of the vessels, while the depth loss promotes the segmentation of the vessel interior.

4 Experimental Design

Dataset. We use an in-house dataset of contrast-enhanced abdominal computed tomography images (CTs) in the arterial phase to segment the peripancreatic arteries [6]. The cohort consists of 141 patients with pancreatic ductal adenocarcinoma, of an equal ratio of male to female patients. Given a 3D arterial CT of the abdominal area, we automatically extract the vertebrae [15,18] and semi-automatically extract the ribs, which have similar intensities as arteries in arterial CTs and would otherwise occlude the vessels. In order to remove as much of the cluttering surrounding tissue and increase the visibility of the vessels in the projections, the input is windowed so that the vessels appear hyperintense. Details of the exact preprocessing steps can be found in Table 2 of the supplementary material. The dataset contains binary 3D annotations of the peripancreatic arteries carried out by two radiologists, each having annotated half of the dataset. The 2D annotations we use in our experiments are projections of these 3D annotations. For more information about the dataset, see [6].

Image Augmentation and Transformation. As the annotations lie on a 2D plane, 3D spatial augmentation cannot be used due to the information sparsity in the ground truth. Instead, we apply an invertible transformation $\mathcal{T}$ to the input volume and apply the inverse transformation $\mathcal{T}^{-1}$ to the network output before applying the loss, such that the ground truth need not be altered. A detailed description of the augmentations and transformations used can be found in Table 1 in the supplementary material.

Training and Evaluation. We use a 3D U-Net [17] with four layers as our backbone, together with Xavier initialization [10]. A diagram of the network architecture can be found in Fig. 2 in the supplementary material. The loss weight α is tuned at 0.5, as this empirically yields the best performance. Our experiments are averaged over 5-fold cross-validation with 80 train samples, 20 validation samples, and a fixed test set of 41 samples. The network initialization is different for each fold but kept consistent across different experiments run on the same fold. This way, both data variance and initialization variance are accounted for through cross-validation. To measure the performance of our models, we use the Dice score, precision, recall, and mean surface distance (MSD). We also compute the skeleton recall as the percentage of the ground truth skeleton pixels which are present in the prediction.

Table 1. Viewpoint ablation. We compare models trained on single random viewpoints (VPs) with (+D) or without (−D) depth against fixed viewpoint baselines without depth and full 3D supervision. We distinguish between model selection based on 2D annotations vs. 3D annotations on the validation set. The best-performing models for each model selection (2D *vs.* 3D) are highlighted in bold.

Experiment	Model Selection	Dice ↑	Precision ↑	Recall ↑	Skeleton Recall ↑	MSD ↓
3D	3D	**92.18 ± 0.35**	**93.86 ± 0.81**	90.64 ± 0.64	76.04 ± 4.51	1.15 ± 0.11
fixed 3VP	3D	92.02 ± 0.52	93.05 ± 0.61	91.13 ± 0.79	**78.61 ± 1.52**	1.13 ± 0.11
fixed 2VP	3D	91.29 ± 0.78	91.46 ± 2.13	**91.37 ± 1.45**	78.51 ± 2.78	**1.13 ± 0.09**
fixed 3VP	2D	90.78 ± 1.30	90.66 ± 1.30	91.18 ± 3.08	81.77 ± 2.13	1.16 ± 0.13
fixed 2VP	2D	90.22 ± 1.19	88.16 ± 2.86	92.74 ± 1.63	**82.18 ± 2.47**	1.14 ± 0.09
fixed 1VP	2D	60.76 ± 24.14	50.47 ± 23.21	92.52 ± 3.09	81.19 ± 2.39	2.96 ± 3.15
random 1VP−D	2D	91.29 ± 0.81	**91.42 ± 0.92**	91.45 ± 1.00	80.16 ± 2.35	**1.13 ± 0.04**
random 1VP+D	2D	**91.69 ± 0.48**	90.77 ± 1.76	**92.79 ± 0.95**	81.27 ± 2.02	1.15 ± 0.11

5 Results

The Effectiveness of 2D Projections and Depth Supervision. We compare training using single random viewpoints with and without depth information against baselines that use more supervision. Models trained on full 3D ground truth represent the upper bound baseline, which is very expensive to annotate. We implement [13] as a baseline on our dataset, training on up to 3 fixed orthogonal projections. We distinguish between models selected according to the 2D performance on the validation set (2D) which is a fair baseline, and models selected according to the 3D performance on the validation set (3D), which is an unfair baseline as it requires 3D annotations on the validation set. With the exception of the single fixed viewpoint baselines where the models have the tendency to diverge towards over- or segmentation, we perform binary hole-filling on the output of all of our other models, as producing hollow objects is a common under-segmentation issue.

In Table 1 we compare our method against the 3D baseline, as well as baselines trained on multiple viewpoints. We see that by using **depth information** paired with training using a single random viewpoint per sample performs almost at the level of models trained on 3D labels, at a very small fraction of the annotation cost. The depth information also reduces model variance compared to the same setup without depth information. Even without depth information, training the model on single **randomly** chosen viewpoints offers a robust training signal that the Dice score is on par with training on 2 fixed viewpoints under ideal model selection at only half the annotation cost. Randomly selecting viewpoints for training acts as powerful data augmentation, which is why we are able to obtain performance comparable to using more fixed viewpoints. Under ideal 3D-based model selection, three views would come even closer to full 3D performance; however, with realistic 2D-based model selection, fixed viewpoints are more prone to diverge. This occurs because sometimes 2D-based model selection favors divergent models which only segment hollow objects, which cannot be fixed in postprocessing. Single fixed viewpoints contain so little information on their own that models trained on such input fail to learn how to segment the vessels and generally converge to over-segmenting in the blind spots in the projections. We conclude that using random viewpoints is not only helpful in reducing annotation cost but also decreases model variance.

In terms of other metrics, randomly chosen projection viewpoints with and without depth improve both recall and skeleton recall even compared to fully 3D annotations, while generally reducing precision. We theorize that this is because the dataset itself contains noisy annotations and fully supervised models better overfit to the type of data annotation, whereas our models converge to following the contrast and segmenting more vessels, which are sometimes wrongfully labeled as background in the ground truth. MSD are not very telling in our dataset due to the noisy annotations and the nature of vessels, as an under- or over-segmented vessel branch can quickly translate into a large surface distance.

The Effect of Dataset Size. We vary the size of the training set from $|\mathcal{D}_{tr}| = 80$ to as little as $|\mathcal{D}_{tr}| = 10$ samples, while keeping the size of the validation and test sets constant, and train models on single random viewpoints.

In Table 2, we compare single random projections trained with and without depth information at varying dataset sizes to ilustrate the usefulness of the depth information with different amounts of training data. Our depth loss offers consistent improvement across multiple dataset sizes and reduces the overall performance variance. The performance boost is noticeable across the board, the only exception being precision. The smaller the dataset size is, the greater the performance boost from the depth. We perform a Wilcoxon rank-sum statistical test comparing the individual sample predictions of the models trained at various dataset sizes with single random orthogonal viewpoints with or without depth information, obtaining a statistically significant (p-value of < 0.0001). We conclude that the depth information complements the segmentation effectively.

Table 2. Dataset size ablation. We vary the training dataset size $|\mathcal{D}_{tr}|$ and compare models trained on single random viewpoints, with or without depth. Best performing models in each setting are highlighted.

$\|\mathcal{D}_{tr}\|$	Depth	Dice ↑	Precision ↑	Recall ↑	Skeleton Recall ↑	MSD ↓
10	−D	86.03 ± 2.94	88.23 ± 2.58	84.81 ± 6.42	78.25 ± 2.20	1.92 ± 0.55
10	+D	**89.06 ± 1.20**	**88.55 ± 1.73**	**89.91 ± 1.29**	**78.95 ± 3.62**	**1.80 ± 0.28**
20	−D	88.22 ± 3.89	**90.26 ± 1.64**	86.74 ± 6.56	80.78 ± 1.66	1.44 ± 0.20
20	+D	**90.51 ± 0.38**	89.84 ± 0.90	**91.50 ± 1.23**	**80.00 ± 1.95**	**1.33 ± 0.16**
40	−D	88.07 ± 2.34	**89.09 ± 2.01**	87.62 ± 4.43	78.38 ± 2.39	1.38 ± 0.10
40	+D	**90.21 ± 0.89**	89.08 ± 2.89	**91.82 ± 2.11**	**79.16 ± 2.36**	**1.24 ± 0.14**
80	−D	91.29 ± 0.81	**91.42 ± 0.92**	91.45 ± 1.00	80.16 ± 2.35	**1.13 ± 0.04**
80	+D	**91.69 ± 0.48**	90.77 ± 1.76	**92.79 ± 0.95**	**81.27 ± 2.02**	1.15 ± 0.11

6 Conclusion

In this work, we present an approach for 3D segmentation of peripancreatic arteries using very sparse 2D annotations. Using a labeled dataset consisting of single, randomly selected, orthogonal 2D annotations for each training sample and additional depth information obtained at no extra cost, we obtain accuracy almost on par with fully supervised models trained on 3D data at a mere fraction of the annotation cost. Limitations of our work are that the depth information relies on the assumption that the vessels exhibit minimal intensity fluctuations within local neighborhoods, which might not hold on other datasets, where more sophisticated ray-tracing methods would be more effective in locating the front and back of projected objects. Furthermore, careful preprocessing is performed to eliminate occluders, which would limit its transferability to datasets with many occluding objects of similar intensities. Further investigation is needed to quantify how manual 2D annotations compare to our 3D-derived annotations, where we expect occluders to affect the annotation process.

References

1. Bai, W., et al.: Recurrent neural networks for aortic image sequence segmentation with sparse annotations. In: Frangi, A.F., Schnabel, J.A., Davatzikos, C., Alberola-López, C., Fichtinger, G. (eds.) MICCAI 2018. LNCS, vol. 11073, pp. 586–594. Springer, Cham (2018). https://doi.org/10.1007/978-3-030-00937-3_67
2. Bayat, A.: Inferring the 3D standing spine posture from 2D radiographs. In: Martel, A.L., et al. (eds.) MICCAI 2020. LNCS, vol. 12266, pp. 775–784. Springer, Cham (2020). https://doi.org/10.1007/978-3-030-59725-2_75
3. Chen, C., Chuah, J.H., Ali, R., Wang, Y.: Retinal vessel segmentation using deep learning: a review. IEEE Access **9**, 111985–112004 (2021)

4. Chen, H., Wang, X., Wang, L.: 3D vessel segmentation with limited guidance of 2D structure-agnostic vessel annotations. arXiv preprint arXiv:2302.03299 (2023)
5. Ciecholewski, M., Kassjański, M.: Computational methods for liver vessel segmentation in medical imaging: A review. Sensors **21**(6), 2027 (2021)
6. Dima, A., et al.: Segmentation of peripancreatic arteries in multispectral computed tomography imaging. In: Lian, C., Cao, X., Rekik, I., Xu, X., Yan, P. (eds.) MLMI 2021. LNCS, vol. 12966, pp. 596–605. Springer, Cham (2021). https://doi.org/10.1007/978-3-030-87589-3_61
7. Drebin, R.A., Carpenter, L., Hanrahan, P.: Volume rendering. ACM Siggraph Comput. Graphics **22**(4), 65–74 (1988)
8. Frangi, A.F., Niessen, W.J., Vincken, K.L., Viergever, M.A.: Multiscale vessel enhancement filtering. In: Wells, W.M., Colchester, A., Delp, S. (eds.) MICCAI 1998. LNCS, vol. 1496, pp. 130–137. Springer, Heidelberg (1998). https://doi.org/10.1007/BFb0056195
9. Fu, X., et al.: Panoptic NeRF: 3D-to-2D label transfer for panoptic urban scene segmentation. In: International Conference on 3D Vision, 3DV 2022, Prague, Czech Republic, 12–16 September 2022, pp. 1–11. IEEE (2022)
10. He, K., Zhang, X., Ren, S., Sun, J.: Delving deep into rectifiers: surpassing human-level performance on imagenet classification. In: Proceedings of the IEEE International Conference on Computer Vision, pp. 1026–1034 (2015)
11. Isensee, F., Jaeger, P.F., Kohl, S.A., Petersen, J., Maier-Hein, K.H.: nnU-Net: a self-configuring method for deep learning-based biomedical image segmentation. Nat. Methods **18**(2), 203–211 (2021)
12. Jia, Z., Huang, X., Eric, I., Chang, C., Xu, Y.: Constrained deep weak supervision for histopathology image segmentation. IEEE Trans. Med. Imaging **36**(11), 2376–2388 (2017)
13. Koziński, M., Mosinska, A., Salzmann, M., Fua, P.: Tracing in 2D to reduce the annotation effort for 3D deep delineation of linear structures. Med. Image Anal. **60**, 101590 (2020)
14. Lawin, F.J., Danelljan, M., Tosteberg, P., Bhat, G., Khan, F.S., Felsberg, M.: Deep projective 3D semantic segmentation. In: Felsberg, M., Heyden, A., Krüger, N. (eds.) CAIP 2017. LNCS, vol. 10424, pp. 95–107. Springer, Cham (2017). https://doi.org/10.1007/978-3-319-64689-3_8
15. Löffler, M.T., et al.: A vertebral segmentation dataset with fracture grading. Radiol. Artifi. Intell. **2**(4), e190138 (2020)
16. Luboz, V., et al.: A segmentation and reconstruction technique for 3D vascular structures. In: Duncan, J.S., Gerig, G. (eds.) MICCAI 2005. LNCS, vol. 3749, pp. 43–50. Springer, Heidelberg (2005). https://doi.org/10.1007/11566465_6
17. Ronneberger, O., Fischer, P., Brox, T.: U-Net: convolutional networks for biomedical image segmentation. In: Navab, N., Hornegger, J., Wells, W.M., Frangi, A.F. (eds.) MICCAI 2015. LNCS, vol. 9351, pp. 234–241. Springer, Cham (2015). https://doi.org/10.1007/978-3-319-24574-4_28
18. Sekuboyina, A., et al.: VerSe: a vertebrae labelling and segmentation benchmark for multi-detector CT images. Med. Image Anal. **73**, 102166 (2021)
19. Shi, F., et al.: Intracranial vessel wall segmentation using convolutional neural networks. IEEE Trans. Biomed. Eng. **66**(10), 2840–2847 (2019)
20. Tajbakhsh, N., Jeyaseelan, L., Li, Q., Chiang, J.N., Wu, Z., Ding, X.: Embracing imperfect datasets: A review of deep learning solutions for medical image segmentation. Med. Image Anal. **63**, 101693 (2020)

21. Tetteh, G., et al.: Deepvesselnet: vessel segmentation, centerline prediction, and bifurcation detection in 3-d angiographic volumes. Front. Neurosci., 1285 (2020)
22. Zhou, Y., et al.: Semi-supervised 3D abdominal multi-organ segmentation via deep multi-planar co-training. In: 2019 IEEE Winter Conference on Applications of Computer Vision (WACV), pp. 121–140. IEEE (2019)

Automatic Retrieval of Corresponding US Views in Longitudinal Examinations

Hamideh Kerdegari[1(✉)], Nhat Phung Tran Huy[1,3], Van Hao Nguyen[2], Thi Phuong Thao Truong[2], Ngoc Minh Thu Le[2], Thanh Phuong Le[2], Thi Mai Thao Le[2], Luigi Pisani[4], Linda Denehy[5], Reza Razavi[1], Louise Thwaites[3], Sophie Yacoub[3], Andrew P. King[1], and Alberto Gomez[1]

[1] School of Biomedical Engineering and Imaging Sciences, King's College London, London, UK
hamideh.kerdegari@kcl.ac.uk
[2] Hospital for Tropical Diseases, Ho Chi Minh City, Vietnam
[3] Oxford University Clinical Research Unit, Ho Chi Minh City, Vietnam
[4] Mahidol Oxford Tropical Medicine Research Unit, Bangkok, Thailand
[5] Melbourne School of Health Sciences, The University of Melbourne, Melbourne, Australia

Abstract. Skeletal muscle atrophy is a common occurrence in critically ill patients in the intensive care unit (ICU) who spend long periods in bed. Muscle mass must be recovered through physiotherapy before patient discharge and ultrasound imaging is frequently used to assess the recovery process by measuring the muscle size over time. However, these manual measurements are subject to large variability, particularly since the scans are typically acquired on different days and potentially by different operators. In this paper, we propose a self-supervised contrastive learning approach to automatically retrieve similar ultrasound muscle views at different scan times. Three different models were compared using data from 67 patients acquired in the ICU. Results indicate that our contrastive model outperformed a supervised baseline model in the task of view retrieval with an AUC of 73.52% and when combined with an automatic segmentation model achieved 5.7% ± 0.24% error in cross-sectional area. Furthermore, a user study survey confirmed the efficacy of our model for muscle view retrieval.

Keywords: Muscle atrophy · Ultrasound view retrieval · Self-supervised contrastive learning · Classification

H. Kerdegari—This work was supported by the Wellcome Trust UK (110179/Z/15/Z, 203905/Z/16/Z, WT203148/Z/16/Z). H. Kerdegari, N. Phung, R. Razavi, A. P King and A. Gomez acknowledge financial support from the Department of Health via the National Institute for Health Research (NIHR) comprehensive Biomedical Research Centre award to Guy's and St Thomas' NHS Foundation Trust in partnership with King's College London and King's College Hospital NHS Foundation Trust.
Vital Consortium: Membership of the VITAL Consortium is provided in the Acknowledgments.

H. Greenspan et al. (Eds.): MICCAI 2023, LNCS 14220, pp. 152–161, 2023.
https://doi.org/10.1007/978-3-031-43907-0_15

1 Introduction

Muscle wasting, also known as muscle atrophy (see Fig. 1), is a common complication in critically ill patients, especially in those who have been hospitalized in the intensive care unit (ICU) for a long period [17]. Factors contributing to muscle wasting in ICU patients include immobilization, malnutrition, inflammation, and the use of certain medications [13]. Muscle wasting can result in weakness, impaired mobility, and increased morbidity and mortality. Assessing the degree of muscle wasting in ICU patients is essential for monitoring their progress and tailoring their rehabilitation program to recover muscular mass through physiotherapy before patient discharge. Traditional methods of assessing muscle wasting, such as physical examination, bioelectrical impedance analysis, and dual-energy X-ray absorptiometry, may be limited in ICUs due to the critical illness of patients [15]. Instead, ultrasound (US) imaging has emerged as a reliable, non-invasive, portable tool for assessing muscle wasting in the ICU [11].

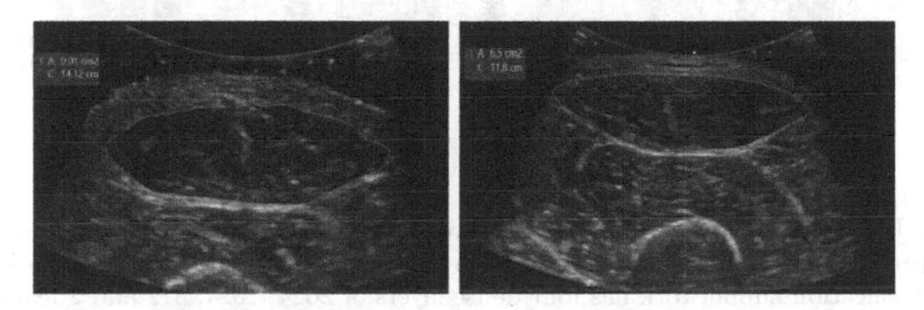

Fig. 1. Example of the cross-section of the rectus femoris (RF) on one ICU patient showing muscle mass reduction from admission (9cm^2, left) to discharge (6cm^2, right).

The accuracy and reliability of US imaging in assessing muscle wasting in ICU patients have been demonstrated by Parry et al. [12]. US imaging can provide accurate measurements of muscle size, thickness, and architecture, allowing clinicians to track changes over time. However, these measurements are typically performed manually, which is time-consuming, subject to large variability and depends on the expertise of the operator. Furthermore, operators might be different from day to day and/or start scanning from different positions in each scan which will cause further variability.

In recent years, self-supervised learning (SSL) has gained popularity for automated diagnosis in the field of medical imaging due to its ability to learn from unlabeled data [1,6,8,16]. Previous studies on SSL for medical imaging have focused on designing pretext tasks [2,9,10,18]. A class of SSL, contrastive learning (CL), aims to learn feature representations via a contrastive loss function to distinguish between negative and positive image samples. A relatively small number of works have applied CL to US imaging, for example to synchronize different cross-sectional views [7] and to perform view classification [4] in echocardiography (cardiac US).

In this paper, we focus on the underinvestigated application of view matching for longitudinal RF muscle US examinations to assess muscle wasting. Our method uses a CL approach (see Fig. 2) to learn a discriminative representation from muscle US data which facilitates the retrieval of similar muscle views from different scans.

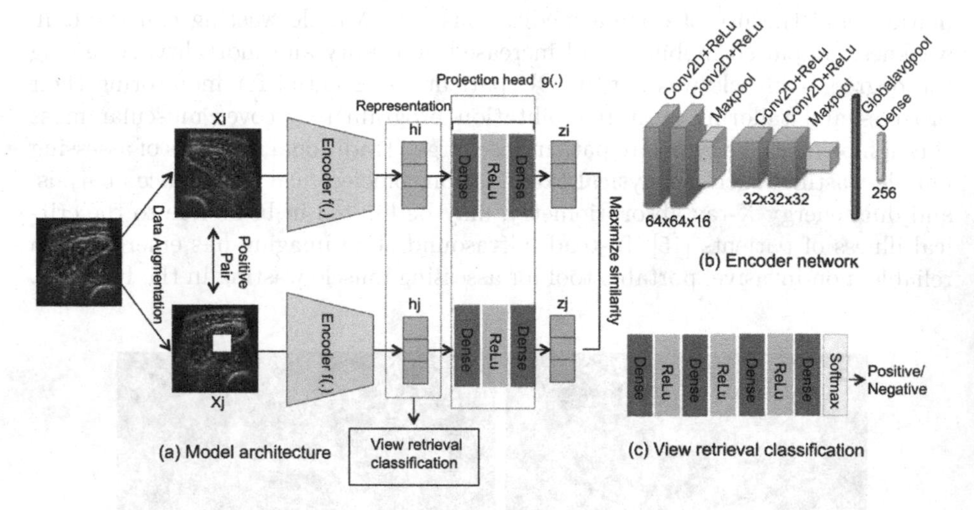

Fig. 2. Proposed architecture for US view retrieval. (a): Overview, a shared encoder and a projection head (two dense layers, each 512 nodes). (b): Encoder subnetwork. (c): The classification subnetwork has four dense layers of 2024, 1024, 512 and 2 features.

The novel contributions of this paper are: 1) the first investigation of the problem of muscle US view matching for longitudinal image analysis, and 2) our approach is able to automatically retrieve similar muscle views between different scans, as shown by quantitative validation and qualitatively through a clinical survey.

2 Method

2.1 Problem Formulation

Muscle wasting assessment requires matching of corresponding cross-sectional US views of the RF over subsequent (days to weeks apart) examinations. The first acquisition is carried out following a protocol to place the transducer half way through the thigh and perpendicular to the skin, but small variations in translation and angulation away from this standard view are common. This scan produces the reference view at time T_1 (RT_1). The problem is as follows: given RT_1, the task is to retrieve the corresponding view (VT_2) at a later time (T_2) from a sequence of US images captured by the operator using the transducer at approximately the same location and angle as for T_1. The main challenges of this problem include: (1) the transducer pose and angle might be different, (2) machine settings might be slightly different, and (3) parts of the anatomy (specifically the RF) might change in shape and size over time. As a result, our aim is to develop a model that can select the most similar view acquired during T_2 to the reference view RT_1 acquired at T_1.

2.2 Contrastive Learning Framework for Muscle View Matching

Inspired by the SimCLR algorithm [5], our model learns representations by maximizing the similarity between two different augmented views of the same muscle US image via a contrastive loss in the latent space. We randomly sample a minibatch of N images from the video sequences over three times T_1, T_2 and T_3, and define the contrastive learning on positive pairs (Xi, Xj) of augmented images derived from the minibatch, resulting in $2N$ samples. Rather than explicitly sampling negative examples, given a positive pair, we consider the other $2(N-1)$ augmented image pairs within a minibatch as negative.

The contrastive loss function for a positive pair (Xi, Xj) is defined as:

$$L_C^i = -log\frac{exp(sim(z_i, z_j)/\tau)}{\sum_{k=1}^{2n} 1_{[k \neq i]} exp(sim(z_i, z_k)/\tau)}, \tag{1}$$

where $1 \in (0, 1)$, τ is a temperature parameter and $sim(\cdot)$ denotes the pairwise cosine similarity. z is a representation vector, calculated by $z = g(f(X))$, where f($\cdot$) indicates a shared encoder and g($\cdot$) is a projection head. L_C^i is computed across all positive pairs in a mini-batch. Then $f(\cdot)$ and $g(\cdot)$ are trained to maximize similarity using this contrastive loss.

2.3 The Model Architecture

The model architecture is shown in Fig. 2a. First, we train the contrastive model to identify the similarity between two images, which are a pair of image augmentations created by horizontal flipping and random cropping (size 10×10) applied on a US image (i.e., they represent different versions of the same image). Each image of this pair (Xi, Xj) is fed into an encoder to extract representation vectors (hi, hj) from them. The encoder architecture (Fig. 2b) has four conv layers (kernel 3×3) with ReLU and two max-poolings. A projection head (a multilayer perceptron with two dense layers of 512 nodes) follows mapping these representations to the space where the contrastive loss is applied.

Second, we use the trained encoder $f(\cdot)$ for the training of our main task (i.e. the downstream task), which is the classification of positive and negative matches (corresponding and non-corresponding views) of our test set. For that, we feed a reference image X_{ref}, and a candidate frame X_j to the encoder to obtain the representations hi, hj and feed these in turn to a classification network (shown in Fig. 2c) that contains four dense layers with ReLU activation and a softmax layer.

3 Materials

The muscle US exams were performed using GE Venue Go and GE Vivid IQ machines, both with linear probes (4.2-13.0 MHz), by five different doctors. During examination, patients were in supine position with the legs in a neutral rotation with relaxed muscle and passive extension. Measurements were taken at the point three fifths of the way between the anterior superior iliac spine and the patella upper pole. The transducer was placed perpendicular to the skin and to the longitudinal axis of the thigh to get the cross-sectional area of the RF. An excess of US gel was used and pressure on the skin was kept minimal to maximise image quality. US measurements were taken at ICU admission (T_1), 2-7 d after admission (T_2) and at ICU discharge (T_3). For this study, 67 Central Nervous System (CNS) and Tetanus patients were recruited and their data were acquired between June 2020 and Feb 2022. Each patient had an average of six muscle ultrasound examinations, three scans for each leg, totalling 402 examinations. The video resolution was 1080 × 1920 with a frame rate of 30fps. This study was performed in line with the principles of the Declaration of Helsinki. Approval was granted by the Ethics Committee of the Hospital for Tropical Diseases, Ho Chi Minh City and Oxford Tropical Research Ethics Committee.

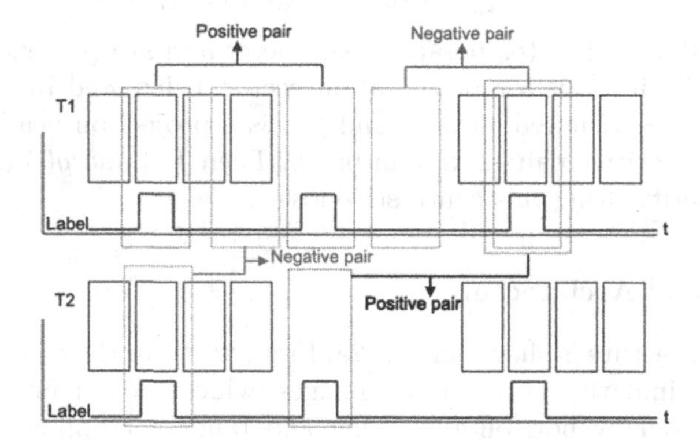

Fig. 3. An example of positive and negative pair labeling for US videos acquired at T_1 and T_2. Positive pairs are either the three views acquired consecutively at the T_i, or a view labeled at T_1 and the corresponding view on the same leg at T_2 or T_3.

The contrastive learning network was trained without any annotations. However, for the view matching classification task, our test data were annotated automatically as positive and negative pairs based upon manual frame selection by a team of five doctors comprising three radiologists and two ultrasound specialists with expertise in muscle ultrasound. Specifically, each frame in an examination was manually labelled as containing a similar view to the reference RT_1 or not. Based upon these labelings, as shown in Fig. 3, the positive pairs are combinations of similar views within each examination ($T_1/T_2/T_3$) and between examinations. The rest are considered negative pairs.

4 Experiments and Results

4.1 Implementation Details

Our model was implemented using Tensorflow 2.7. During training, input videos underwent experimentation with clip sizes of 256 × 256, 128 × 128, and 64 × 64. Eventually, they were resized to 64 × 64 clips, which yielded the best performance. All the hyperparameters were chosen using the validation set. For the CL training, the standard Adam optimizer was used with learning rate =0.00001, kernel size = 3 × 3, batch size = 128, batch normalization, dropout with p = 0.2 and L2 regularization of the model parameters with a weight = 0.00001. The CL model was trained on 80% of the muscle US data for 500 epochs. For the view retrieval model, the standard Adam optimizer with learning rate = 0.0001, batch size = 42 and dropout of p = 0.2 was used. The classifier was trained on the remaining 20% of the data (of which 80% were used for training, 10% for validation and 10% for testing) and the network converged after 60 epochs. For the supervised baseline model, the standard Adam optimizer was used with learning rate =0.00001, kernel size = 3 × 3, batch size = 40, and batch normalization. Here, we used the same data splitting as our view retrieval classifier. The code we used to train and evaluate our models is available at https://github.com/hamidehkerdegari/Muscle-view-retrieval.

4.2 Results

Quantitative Results. We carried out two quantitative experiments. First, we evaluated the performance of the view classifier. Second, we evaluated the quality of the resulting cross-sectional areas segmented using a U-Net [14].

The classifier performance was carried out by measuring, for the view retrieval task, the following metrics: Area Under the Curve (AUC), precision, recall, and F1-score. Because there is no existing state of the art for this task, we created two baseline models to compare our proposed model to: first, a naive image-space comparison using normalized cross-correlation (NCC) [3], and second, a supervised classifier. The supervised classifier has the same architecture as our CL model, but with the outputs of the two networks being concatenated after the representation h followed by a dense layer with two nodes and a softmax activation function to produce the probabilities of being a positive or negative pair. Table 1 shows the classification results on our dataset.

Table 1. AUC, precision, recall and F1 score results on the muscle video dataset.

Model	AUC	Precision	Recall	F1
Normalized cross-correlation	68.35 %	58.65 %	63.12 %	60.8 %
Supervised baseline model	69.87 %	65.81 %	60.57 %	63.08 %
Proposed model	**73.52 %**	**67.2 %**	**68.31 %**	**67.74 %**

As shown in Table 1, our proposed method achieved superior performance in terms of AUC, precision, recall, and F1-score compared to all other models. The NCC method demonstrated the lowest performance, as it lacked the capability to accurately capture dynamic changes and deformations in US images which can result in significant structural differences. A representative example of a model-retrieved view for one case is presented in Fig. 4. It shows positive, negative, and middle (i.e., images with a probability value between the highest and lowest values predicted by our model) pairs of images generated by our model from a patient's left leg. As reference, on the left we show the user pick (RT_2).

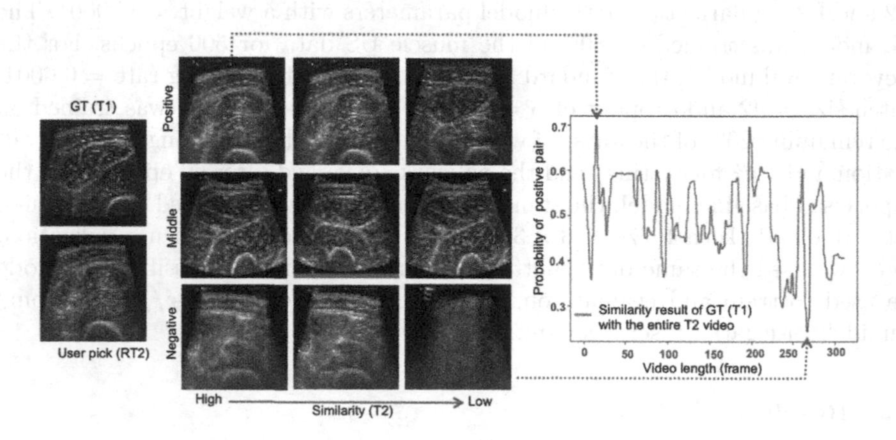

Fig. 4. Results showing three sample positive, medium and negative predicted pairs by our model when ground truth (GT) from T_1 is compared with the T_2 video.

To assess the quality of the resulting cross-sections, we calculated the mean relative absolute area difference (d) between the ground truth (a_{GT}) frame and that of the model predicted frame (a_{pred}) for each examination as follows:

$$d = \frac{|a_{GT} - a_{pred}|}{a_{GT}} \tag{2}$$

We applied a trained U-Net model (already trained with 1000 different US muscle images and manual segmentations). Results showed an overall cross-sectional mean relative absolute area error of 5.7% ± 0.24% on the test set (Full details provided in Fig. 5, right). To put this number into context, Fig. 5, left visualizes two cases where the relative error is 2.1% and 5.2%.

Qualitative Results. We conducted a user study survey to qualitatively assess our model's performance. The survey was conducted blindly and independently by four clinicians and consisted of thirty questions. In each, clinicians were shown two different series of three views of the RF: (1) RT_1, GT match from T_2 and model prediction from T_2, and (2) RT_1, a random frame from T_2 and model prediction from T_2. They were asked to indicate which (second or third) was the best match with the first image.

The first question aimed to determine if the model's performance was on par with clinicians, while the second aimed to determine if the model's selection of images was superior to a randomly picked frame. As shown in Fig. 6, left, clinicians chose the model prediction more often than the GT; however, this difference was not significant (paired Student's t-test, $p = 0.44$, significance$= 0.05$). Therefore, our model can retrieve the view as well as clinicians, and significantly better (Fig. 6, right) than randomly chosen frames (paired Student's t-test, $p = 0.02$, significance$= 0.05$).

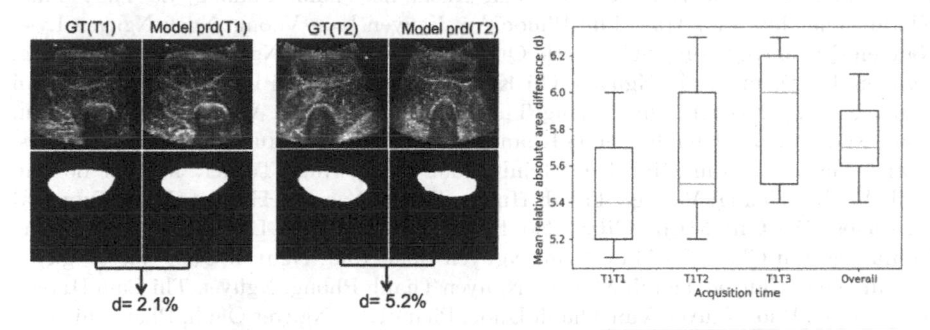

Fig. 5. Left: cross-sectional area error for T_1 and T_2 examinations (acquisition times). Right: mean relative absolute area difference (d) for T_1T_1, T_1T_2, T_1T_3 (reference frame from T1 and corresponding predicted frames from T1, T2 and T3 respectively) and overall acquisition time.

5 Discussion and Conclusion

This paper has presented a self-supervised CL approach for automatic muscle US view retrieval in ICU patients. We trained a classifier to find positive and

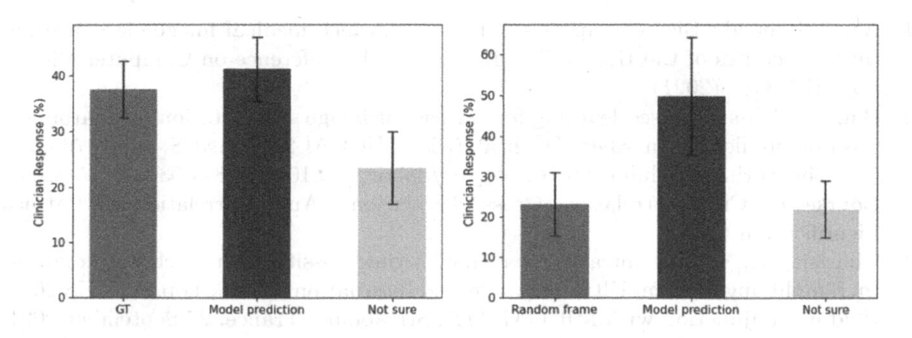

Fig. 6. User study survey results. Left: when T_1GT, T_2GT and model prediction (from T_2) are shown to the users. Right: when T_1GT, T_2-random frame and model prediction (from T_2) are shown to the users.

negative matches. We also computed the cross-sectional area error between the ground truth frame and the model prediction in each acquisition time to evaluate model performance. The performance of our model was evaluated on our muscle US video dataset and showed AUC of 73.52% and 5.7% ± 0.24% error in cross-sectional view. Results showed that our model outperformed the supervised baseline approach. This is the first work proposed to identify corresponding ultrasound views over time, addressing an unmet clinical need.

Acknowledgments. The VITAL Consortium: **OUCRU**: Dang Phuong Thao, Dang Trung Kien, Doan Bui Xuan Thy, Dong Huu Khanh Trinh, Du Hong Duc, Ronald Geskus, Ho Bich Hai, Ho Quang Chanh, Ho Van Hien, Huynh Trung Trieu, Evelyne Kestelyn, Lam Minh Yen, Le Dinh Van Khoa, Le Thanh Phuong, Le Thuy Thuy Khanh, Luu Hoai Bao Tran, Luu Phuoc An, Nguyen Lam Vuong, Ngan Nguyen Lyle, Nguyen Quang Huy, Nguyen Than Ha Quyen, Nguyen Thanh Ngoc, Nguyen Thi Giang, Nguyen Thi Diem Trinh, Nguyen Thi Kim Anh, Nguyen Thi Le Thanh, Nguyen Thi Phuong Dung, Nguyen Thi Phuong Thao, Ninh Thi Thanh Van, Pham Tieu Kieu, Phan Nguyen Quoc Khanh, Phung Khanh Lam, Phung Tran Huy Nhat, Guy Thwaites, Louise Thwaites, Tran Minh Duc, Trinh Manh Hung, Hugo Turner, Jennifer Ilo Van Nuil, Vo Tan Hoang, Vu Ngo Thanh Huyen, Sophie Yacoub. **Hospital for Tropical Diseases, Ho Chi Minh City**: Cao Thi Tam, Ha Thi Hai Duong, Ho Dang Trung Nghia, Le Buu Chau, Le Mau Toan, Nguyen Hoan Phu, Nguyen Quoc Viet, Nguyen Thanh Dung, Nguyen Thanh Nguyen, Nguyen Thanh Phong, Nguyen Thi Cam Huong, Nguyen Van Hao, Nguyen Van Thanh Duoc, Pham Kieu Nguyet Oanh, Phan Thi Hong Van, Phan Vinh Tho, Truong Thi Phuong Thao. **University of Oxford**: Natasha Ali, James Anibal, David Clifton, Mike English, Ping Lu, Jacob McKnight, Chris Paton, Tingting Zhu **Imperial College London**: Pantelis Georgiou, Bernard Hernandez Perez, Kerri Hill-Cawthorne, Alison Holmes, Stefan Karolcik, Damien Ming, Nicolas Moser, Jesus Rodriguez Manzano. **King's College London**: Liane Canas, Alberto Gomez, Hamideh Kerdegari, Andrew King, Marc Modat, Reza Razavi. **University of Ulm**: Walter Karlen. **Melbourne University**: Linda Denehy, Thomas Rollinson. **Mahidol Oxford Tropical Medicine Research Unit (MORU)**: Luigi Pisani, Marcus Schultz

References

1. Azizi, S., et al.: Big self-supervised models advance medical image classification. In: Proceedings of the IEEE/CVF International Conference on Computer Vision, pp. 3478–3488 (2021)
2. Bai, W.: Self-supervised learning for cardiac mr image segmentation by anatomical position prediction. In: Shen, D., et al. (eds.) MICCAI 2019. LNCS, vol. 11765, pp. 541–549. Springer, Cham (2019). https://doi.org/10.1007/978-3-030-32245-8_60
3. Bourke, P.: Cross correlation. Cross Correlation", Auto Correlation-2D Pattern Identification (1996)
4. Chartsias, A., et al.: Contrastive learning for view classification of echocardiograms. In: Simplifying Medical Ultrasound: Second International Workshop, ASMUS 2021, Held in Conjunction with MICCAI 2021, Strasbourg, France, 27 September 2021, Proceedings 2, pp. 149–158. Springer (2021). https://doi.org/10.1007/978-3-031-16440-8_33

5. Chen, T., Kornblith, S., Norouzi, M., Hinton, G.: A simple framework for contrastive learning of visual representations. In: International Conference on Machine Learning, pp. 1597–1607. PMLR (2020)
6. Chen, Y., et al.: USCL: pretraining deep ultrasound image diagnosis model through video contrastive representation learning. In: de Bruijne, M., et al. (eds.) MICCAI 2021. LNCS, vol. 12908, pp. 627–637. Springer, Cham (2021). https://doi.org/10.1007/978-3-030-87237-3_60
7. Dezaki, F.T.: Echo-syncnet: self-supervised cardiac view synchronization in echocardiography. IEEE Trans. Med. Imaging **40**(8), 2092–2104 (2021)
8. Hosseinzadeh Taher, M.R., Haghighi, F., Feng, R., Gotway, M.B., Liang, J.: A systematic benchmarking analysis of transfer learning for medical image analysis. In: Albarqouni, S., et al. (eds.) DART/FAIR -2021. LNCS, vol. 12968, pp. 3–13. Springer, Cham (2021). https://doi.org/10.1007/978-3-030-87722-4_1
9. Hu, S.Y., et al.: Self-supervised pretraining with dicom metadata in ultrasound imaging. In: Machine Learning for Healthcare Conference, pp. 732–749 (2020)
10. Jiao, J., Droste, R., Drukker, L., Papageorghiou, A.T., Noble, J.A.: Self-supervised representation learning for ultrasound video. In: 2020 IEEE 17th International Symposium on Biomedical Imaging (ISBI), pp. 1847–1850. IEEE (2020)
11. Mourtzakis, M., Wischmeyer, P.: Bedside ultrasound measurement of skeletal muscle. Current Opinion Clinical Nutrition Metabolic Care **17**(5), 389–395 (2014)
12. Parry, S.M., et al.: Ultrasonography in the intensive care setting can be used to detect changes in the quality and quantity of muscle and is related to muscle strength and function. J. Crit. Care **30**(5), 1151-e9 (2015)
13. Puthucheary, Z.A., et al.: Acute skeletal muscle wasting in critical illness. JAMA **310**(15), 1591–1600 (2013)
14. Ronneberger, O., Fischer, P., Brox, T.: U-Net: convolutional networks for biomedical image segmentation. In: Navab, N., Hornegger, J., Wells, W.M., Frangi, A.F. (eds.) MICCAI 2015. LNCS, vol. 9351, pp. 234–241. Springer, Cham (2015). https://doi.org/10.1007/978-3-319-24574-4_28
15. Schefold, J.C., Wollersheim, T., Grunow, J.J., Luedi, M.M., Z'Graggen, W.J., Weber-Carstens, S.: Muscular weakness and muscle wasting in the critically ill. J. Cachexia. Sarcopenia Muscle **11**(6), 1399–1412 (2020)
16. Sowrirajan, H., Yang, J., Ng, A.Y., Rajpurkar, P.: Moco pretraining improves representation and transferability of chest x-ray models. In: Medical Imaging with Deep Learning, pp. 728–744. PMLR (2021)
17. Trung, T.N., et al.: Functional outcome and muscle wasting in adults with tetanus. Trans. R. Soc. Trop. Med. Hyg. **113**(11), 706–713 (2019)
18. Zhuang, X., Li, Y., Hu, Y., Ma, K., Yang, Y., Zheng, Y.: Self-supervised feature learning for 3d medical images by playing a rubik's cube. In: Shen, D., et al. (eds.) MICCAI 2019. LNCS, vol. 11767, pp. 420–428. Springer, Cham (2019). https://doi.org/10.1007/978-3-030-32251-9_46

Many Tasks Make Light Work: Learning to Localise Medical Anomalies from Multiple Synthetic Tasks

Matthew Baugh[1(✉)], Jeremy Tan[2], Johanna P. Müller[3], Mischa Dombrowski[3], James Batten[1], and Bernhard Kainz[1,3]

[1] Imperial College London, London, United Kingdom
matthew.baugh17@imperial.ac.uk
[2] ETH Zurich, Zurich, Switzerland
[3] Friedrich–Alexander University Erlangen–Nürnberg, Erlangen, Germany

Abstract. There is a growing interest in single-class modelling and out-of-distribution detection as fully supervised machine learning models cannot reliably identify classes not included in their training. The long tail of infinitely many out-of-distribution classes in real-world scenarios, *e.g.*, for screening, triage, and quality control, means that it is often necessary to train single-class models that represent an expected feature distribution, *e.g.*, from only strictly healthy volunteer data. Conventional supervised machine learning would require the collection of datasets that contain enough samples of all possible diseases in every imaging modality, which is not realistic. Self-supervised learning methods with synthetic anomalies are currently amongst the most promising approaches, alongside generative auto-encoders that analyse the residual reconstruction error. However, all methods suffer from a lack of structured validation, which makes calibration for deployment difficult and dataset-dependant. Our method alleviates this by making use of multiple visually-distinct synthetic anomaly learning tasks for both training and validation. This enables more robust training and generalisation. With our approach we can readily outperform state-of-the-art methods, which we demonstrate on exemplars in brain MRI and chest X-rays. Code is available at https://github.com/matt-baugh/many-tasks-make-light-work.

1 Introduction

In recent years, the workload of radiologists has grown drastically, quadrupling from 2006 to 2020 in Western Europe [4]. This huge increase in pressure has led to long patient-waiting times and fatigued radiologists who make more mistakes [3]. The most common of these errors is underreading and missing anomalies (42%); followed by missing additional anomalies when concluding their search after an initial finding (22%) [10]. Interestingly, despite the challenging work environment, only 9% of errors reviewed in [10] were due to mistakes in the clinicians'

Supplementary Information The online version contains supplementary material available at https://doi.org/10.1007/978-3-031-43907-0_16.

H. Greenspan et al. (Eds.): MICCAI 2023, LNCS 14220, pp. 162–172, 2023.
https://doi.org/10.1007/978-3-031-43907-0_16

reasoning. Therefore, there is a need for automated second-reader capabilities, which brings any kind of anomalies to the attention of radiologists. For such a tool to be useful, its ability to detect rare or unusual cases is particularly important. Traditional supervised models would not be appropriate, as acquiring sufficient training data to identify such a broad range of pathologies is not feasible. Unsupervised or self-supervised methods to model an expected feature distribution, *e.g.*, of healthy tissue, is therefore a more natural path, as they are geared towards identifying any deviation from the normal distribution of samples, rather than a particular type of pathology.

There has been rising interest in using end-to-end self-supervised methods for anomaly detection. Their success is most evident at the MICCAI Medical Out-of-Distribution Analysis (MOOD) Challenge [31], where all winning methods have followed this paradigm so far (2020-2022). These methods use the variation within normal samples to generate diverse anomalies through sample mixing [7, 23–25]. However all these methods lack a key component: structured validation. This creates uncertainty around the choice of hyperparameters for training. For example, selecting the right training duration is crucial to avoid overfitting to proxy tasks. Yet, in practice, training time is often chosen arbitrarily, reducing reproducibility and potentially sacrificing generalisation to real anomalies.

Contribution: We propose a cross-validation framework, using separate self-supervision tasks to minimise overfitting on the synthetic anomalies that are used for training. To make this work effectively we introduce a number of non-trivial and seamlessly-integrated synthetic tasks, each with a distinct feature set so that during validation they can be used to approximate generalisation to unseen, real-world anomalies. To the best of our knowledge, this is the first work to train models to directly identify anomalies on tasks that are deformation-based, tasks that use Poisson blending with patches extracted from external datasets, and tasks that perform efficient Poisson image blending in 3D volumes, which is in itself a new contribution of our work. We also introduce a synthetic anomaly labelling function which takes into account the natural noise and variation in medical images. Together our method achieves an average precision score of 76.2 for localising glioma and 78.4 for identifying pathological chest X-rays, thus setting the state-of-the-art in self-supervised anomaly detection.

Related Work: The most prevalent methods for self-supervised anomaly detection are based on generative auto-encoders that analyse the residual error from reconstructing a test sample. This is built on the assumption that a reconstruction model will only be able to correctly reproduce data that is similar to the instances it has been trained on, *e.g.* only healthy samples. Theoretically, at test time, the residual reconstruction error should be low for healthy tissues but high for anomalous features. This is an active area of research with several recent improvements upon the initial idea [22], *e.g.*, [21] applied a diffusion model to a VQ-VAE [27] to resample the unlikely latent codes and [30] gradually transition from a U-Net architecture to an autoencoder over the training process in order to improve the reconstruction of finer details. Several other methods aim to ensure that the model will not reproduce anomalous regions by training it to restore samples altered by augmentations such as masking out regions [32], interpolating heavily augmented textures [29] or adding coarse noise [9]. [5] sought to identify more meaningful errors in image reconstructions by comparing the reconstructions of models trained on only healthy data against those trained on all available data.

However, the general assumption that reconstruction error is a good basis for an anomaly scoring function has recently been challenged. Auto-encoders are unable to identify anomalies with extreme textures [16], are reliant on empirical post-processing to reduce false-positives in healthy regions [2] and can be outperformed by trivial approaches like thresholding of FLAIR MRI [15].

Self-supervised methods take a more direct approach, training a model to directly predict an anomaly score using synthetic anomalies. Foreign patch interpolation (FPI) [24] was the first to do this at a pixel-level, by linearly interpolating patches extracted from other samples and predicting the interpolation factor as the anomaly score. Similar to CutPaste [11], [7] fully replaces 3D patches with data extracted from elsewhere in the same sample, but then trains the model to segment the patches. Poisson image interpolation (PII) [25] seamlessly integrates sample patches into training images, preventing the models from learning to identify the anomalies by their discontinuous boundaries. Natural synthetic anomalies (NSA) [23] relaxes patch extraction to random locations in other samples and introduces an anomaly labelling function based on the changes introduced by the anomaly.

Some approaches combine self-supervised and reconstruction-based methods by training a discriminator to compute more exact segmentations from reconstruction model errors [6,29]. Other approaches have also explored contrasting self-supervised learning for anomaly detection [12,26].

2 Method

The core idea of our method is to use synthetic tasks for both training and validation. This allows us to monitor performance and prevent overfitting, all without the need for real anomalous data. Each self-supervised task involves introducing a synthetic anomaly into otherwise normal data whilst also producing the corresponding label. Since the relevant pathologies are unknown a priori, we avoid simulating any specific pathological features. Instead, we use a wide range of subtle and well-integrated anomalies to help the model detect many different kinds of deviations, ideally including real unforeseen anomalies. In our experiments, we use five tasks, but more could be used as long as each one is sufficiently unique. Distinct tasks are vital because we want to use these validation tasks to estimate the model's generalisation to unseen classes of anomalies. If the training and validation tasks are too similar, the performance on the validation set may be an overly optimistic estimate of how the model would perform on unseen real-world anomalies.

When performing cross-validation over all synthetic tasks and data partitions independently, the number of possible train/validation splits increases significantly, requiring us to train $F \cdot (T_N \mathbf{C} T)$ independent models, where T_N is the total number of tasks, T is the number of tasks used to train each model and F is the number of data folds, which is computationally expensive. Instead, as in our case $T_N = F = 5$, we opt to associate each task with a single fold of the training data (Fig. 1). We then apply $5\mathbf{C}T$-fold cross-validation over each combination. In each iteration, the corresponding data folds are collected and used for training or validation, depending on which partition forms the majority.

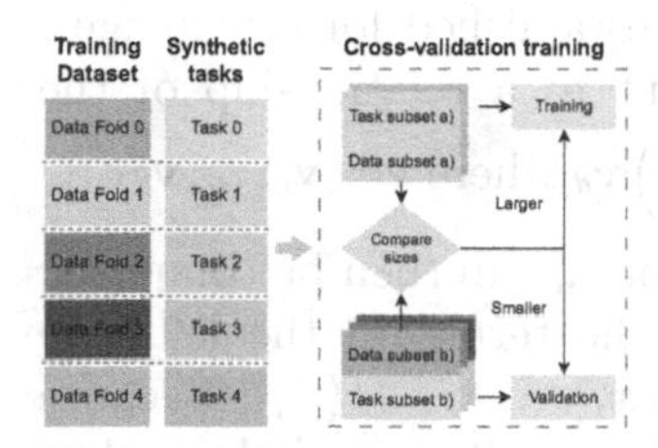

Fig. 1. Our pipeline performs syn-cross-validation over the synthetic task and data fold pairs.

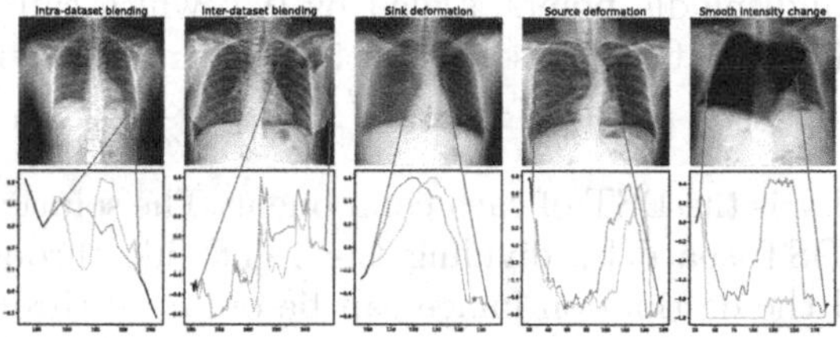

Fig. 2. Examples of changes introduced by synthetic anomalies, showing the before (grey) and after (green) of a 1D slice across the affected area. $\star$ - deformation centre for sink/source. (Color figure online)

Synthetic Tasks: Figure 2 shows examples of our self-supervised tasks viewed in both one and two dimensions. Although each task produces visually distinct anomalies, they fall into three overall categories, based on blending, deformation, or intensity variation. Also, all tasks share a common recipe: the target anomaly mask M_h is always a randomly sized and rotated ellipse or rectangle (ellipsoids/cuboids in 3D); all anomalies are positioned such that at least 50% of the mask intersects with the foreground of the image; and after one augmentation is applied, the process is randomly repeated (based on a fair coin toss, $p = 0.5$), for up to a maximum of 4 anomalies per image.

The Intra-dataset Blending Task. Poisson image blending is the current state-of-the-art for synthetic anomaly tasks [23,25], but it does not scale naturally to more than two dimensions or non-convex interpolation regions [17]. Therefore, we extend [20] and propose a D-dimensional variant of Poisson image editing following earlier ideas by [17].

Poisson image editing [20] uses the image gradient to seamlessly blend a patch into an image. It does this by combining the target gradient with Dirichlet boundary conditions to define a minimisation problem $\min_{f_{in}} \iint_h |\nabla f_{in} - \mathbf{v}|^2$ with $f_{in}|_{\partial h} = f_{out}|_{\partial h}$, and f_{in} representing the intensity values within the patch h. The goal is to find intensity values of f_{in} that will match the surrounding values, f_{out}, of the destination image x_i, along the border of the patch and follow the image gradient, $\mathbf{v} = \nabla \cdot = \left(\frac{\partial \cdot}{\partial x}, \frac{\partial \cdot}{\partial y}\right)$, of the source image x_j. Its solution is the Poisson equation $\Delta f_{in} = \mathrm{div}\mathbf{v}$ over h with $f_{in}|_{\partial h} = f_{out}|_{\partial h}$. Note that the divergence of $\mathbf{v}$ is equal to the Laplacian of the source image Δx_j. Also, by defining h as the axis-aligned bounding box of M_h, we can ensure the boundaries coincide with coordinate lines. This enables us to use the Fourier transform method to solve this partial differential equation [17], which yields a direct relationship between Fourier coefficients of Δf_{in} and $\mathbf{v}$ after padding to a symmetric image. To simplify for our use case, an image with shape $N_0 \times \cdots \times N_{D-1}$, we replace the Fourier transformation with a discrete sine transform (DST) $\hat{f}_{\mathbf{u}} = \sum_{d=0}^{D-1} \sum_{n=0}^{N_d-1} n \sin\left[\frac{\pi(n+1)(\mathbf{u}_d+1)}{N_d+1}\right]$. This follows as a DST is equivalent to a discrete Fourier transform of a real sequence that is odd around the zero-th and middle points, scaled by 0.5, which can be established for our images. With this, the Poisson equation becomes congruent to a relationship of the coefficients, $\left(\sum_{d=0}^{D-1} \left(\frac{\pi(u_d+1)}{N_d+1}\right)^2\right) \hat{f}_{\mathbf{u}} \cong \sum_{d=0}^{D-1} \left(\frac{\pi(\mathbf{u}_d+1)}{N_d+1}\right) \hat{\mathbf{v}}_d$ where $\mathbf{v}{=}(\mathbf{v}_0, ..., \mathbf{v}_{D-1})$ and $\hat{\mathbf{v}}$ is the DST of each component. The solution for $\hat{f}_{\mathbf{u}}$ can then be computed in DST space by dividing the right side through the terms on the left side and the destination image can be obtained through $x_i = DST^{-1}(\hat{f}_{\mathbf{u}})$. Because this approach uses a frequency transform-based solution, it may slightly alter areas outside of M_h (where image gradients are explicitly edited) in order to ensure the changes are seamlessly integrated. We refer to this blending process as $\tilde{x} = PoissonBlend(x_i, x_j, M_h)$ in the following. The intra-dataset blending task therefore results from $\tilde{x}_{intra} = PoissonBlend(x, x', M_h)$ with $x, x' \in \mathcal{D}$ with samples from a common dataset $\mathcal{D}$ and is therefore similar to the self-supervision task used in [23] for 2D images.

The inter-dataset blending task follows the same process as intra-dataset blending but uses patches extracted from an external dataset D', allowing for a greater variety of structures. Therefore, samples from this task can be defined as $\tilde{x}_{inter} = PoissonBlend(x, x', M_h)$ with $x \in \mathcal{D}, x' \in \mathcal{D}'$.

The sink/source tasks shift all points in relation to a randomly selected deformation centre c. For a given point p, we resample intensities from a new location $\tilde{p}$. To create a smooth displacement centred on c, we consider the distance $\|p-c\|_2$ in relation to the radius of the mask (along this direction), d. The extent of this displacement is controlled by the exponential factor $f > 1$. For example, the sink task (Eqn. 1) with a factor of $f = 2$ would take the intensity at $0.75d$ and place it at $0.5d$, effectively pulling these intensities closer to the centre. Note that unlike the sink equation in [24] this formulation cannot sample outside of the boundaries of P_{M_h} meaning it seamlessly blends into the surrounding area. The source task (Eqn. 2) performs the reverse, appearing to push the pixels away from the centre by sampling intensities towards it.

$$\tilde{x}_p = x_{\tilde{p}},\ \tilde{p} = c{+}d\frac{p-c}{\|p-c\|_2}\left(1-\left(1-\frac{\|p-c\|_2}{d}\right)^f\right), c \in P_{M_h},\ \forall\, p \in P_{M_h} \quad (1)$$

$$\tilde{x}_p = x_{\tilde{p}},\ \tilde{p} = c + d\frac{p-c}{\|p-c\|_2}\left(\frac{\|p-c\|_2}{d}\right)^f,\ c \in P_{M_h},\ \forall\, p \in P_{M_h} \quad (2)$$

The smooth intensity change task aims to either add or subtract an intensity over the entire anomaly mask. To avoid sharp discontinuities at the boundaries, this intensity change is gradually dampened for pixels within a certain margin of the boundary. This smoothing starts at a random distance from the boundary, d_s, and the change is modulated by d_p/d_s.

Anomaly Labelling: In order to train and validate with multiple tasks simultaneously we use the same anomaly labelling function across all of our tasks. The scaled logistic function, used in NSA [23], helps to translate raw intensity changes into more semantic labels. But, it also rounds imperceptible differences up to a minimum score of about 0.1. This sudden and arbitrary jump creates noisy labels and can lead to unstable training. We correct this semantic discontinuity by computing labels as $y = 1 - \frac{p_X(x)}{p_X(0)}$ with $X \sim \mathcal{N}(0, \sigma^2)$, instead of $y = \frac{1}{1+e^{-k(x-x_0)}}$ [23]. This flipped Gaussian shape is C1 continuous and smoothly approaches zero, providing consistent labels even for smaller changes.

3 Experiments and Results

Data: We evaluate our method on T2-weighted brain MR and chest X-ray datasets to provide direct comparisons to state-of-the-art methods over a wide range of real anomalies. For brain MRI we train on the Human Connectome Project (HCP) dataset [28] which consists of 1113 MRI scans of healthy, young adults acquired as part of a scientific study. To evaluate, we use the Brain Tumor Segmentation Challenge 2017 (BraTS) dataset [1], containing 285 cases with either high or low grade glioma, and the ischemic stroke lesion segmentation challenge 2015 (ISLES) dataset [13], containing 28 cases with ischemic stroke lesions. The data from both test sets was acquired as part of clinical routine. The HCP dataset was resampled to have 1mm isotropic spacing to match the test datasets. We apply z-score normalisation to each sample and then align the bounding box of each brain before padding it to a size of $160 \times 224 \times 160$. Lastly, samples are downsampled by a factor of two.

For chest X-rays we use the VinDr-CXR dataset [18] including 22 different local labels. To be able to compare with the benchmarks reported in [6] we use the same healthy subset of 4000 images for training along with their test set (DDAD_{ts}) of 1000 healthy and 1000 unhealthy samples, with some minor changes outlined as follows. First note that [6] derives VinDr-CXR labels using the majority vote of the 3 annotators. Unfortunately, this means there are 52 training samples, where 1/3 of radiologists identified an anomaly, but the majority label is counted as healthy. The same applies to 10 samples within the healthy testing subset. To avoid this ambiguity, we replace these samples with leftover training data that all radiologists have labelled as healthy. We also evaluate using the true test set (VinDr_{ts}), where two senior radiologists have reviewed and consolidated all labels. For preprocessing, we clip pixel intensities according to the window centre and width attributes in each DICOM file, and apply histogram equalisation, before scaling intensities to the range $[-1, 1]$. Finally, images are resized to 256×256.

Table 1. Upper left part: Metrics on Brain MRI, evaluated on BraTS and ISLES, presented as AP/AUROC. $\cdot^{\text{CRADL setup}}$ indicates that the metrics are evaluated over the same region and at the same resolution as CRADL [12]. **Upper right part:** Metrics on VinDr-CXR, presented as AP/AUROC on the VinDr and DDAD test splits. *Random* is the baseline performance of a random classifier. **Lower part:** a sensitivity analysis of the average AP of each individual fold (mean±s.d.) alongside that of the model ensemble, varying how many tasks we use for training versus validation. Best results are highlighted in bold.

		Brain MRI				Chest X-Ray (CXR)				
		Slice-wise		Pixel-wise		Sample-wise		Pixel-wise		
		BraTS17	ISLES	BraTS	ISLES	DDAD$_{ts}$	VinDr$_{ts}$	DDAD$_{ts}$	VinDr$_{ts}$	
Methods MRI						59.8/55.8				MemAE [8] (Methods CXR)
	VAE	80.7/83.3	51.9/71.7	29.8/92.5	7.7/87.5	74.8/76.3				f-AnoGAN [22]
	ceVAE[32]	85.6/86.5	54.1/72.7	48.3/94.8	14.5/87.9	72.8/73.8				AE-U [14]
	CRADL[12]	81.9/82.6	54.9/69.3	38.0/94.2	18.6/89.8	49.9/48.2				FPI [24]
	Ours$^{\text{CRADL setup}}$	87.6/89.4	61.3/80.2	76.2/98.7	46.5/97.1	65.8/65.9				PII [25]
	Random$^{\text{CRADL setup}}$	49.0/50.0	36.6/50.0	2.4/50.0	1.1/50.0	65.8/64.4				NSA [23]
Random		40.3/50.0	29.4/50.0	1.7/50.0	0.8/50.0	50.0/50.0	31.6/50.0	4.5/50.0	2.7/50.0	
Ours		**87.6/92.2**	**62.0/84.6**	**76.2/99.1**	**45.9/97.9**	**78.4/76.6**	**71.2/81.1**	**21.1/75.6**	**21.4/81.2**	
Train/val. split *abl.* 1/4	all	83.4±4.4	59.3±2.2	46.9±14.9	23.7±7.7	74.7±4.9	66.3±4.4	15.3±3.7	15.2±4.5	
	ens.	**87.6**	**62.0**	**76.2**	**45.9**	78.4	71.2	21.1	21.4	
2/3	all	82.5±3.3	55.9±8.5	42.8±12.8	21.2±9.3	78.6±1.4	71.0±1.4	19.2±1.7	19.5±1.8	
	ens.	85.7	58.4	72.2	41.0	**80.7**	**73.8**	24.0	**24.7**	
3/2	all	81.1±4.3	52.5±4.7	37.9±11.1	15.4±3.3	78.7±1.8	71.1±1.4	20.3±1.4	20.4±1.7	
	ens.	84.0	55.0	63.7	26.6	80.4	73.3	**24.3**	**24.7**	
4/1	all	81.5±2.7	53.1±2.3	36.1±9.0	16.5±5.0	79.2±1.3	71.8±1.3	20.4±0.9	21.1±0.9	
	ens.	83.1	54.7	52.5	23.7	80.5	73.6	23.5	24.5	

Comparison to State-of-the-Art Methods: Validating on synthetic tasks is one of our main motivations; as such, we use a 1/4 (train/val.) task split to compare with benchmark methods. For brain MRI, we evaluate results at the slice and voxel level, computing average precision (AP) and area under the receiver operating characteristic curve (AUROC), as implemented in scikit learn [19]. Note that the distribution shift between training and test data (research vs. clinical scans) adds further difficulty to this task. In spite of this, we substantially improve upon the current state-of-the-art (Table 1 upper left). In particular, we achieve a pixel-wise AP of 76.2 and 45.9 for BraTS and ISLES datasets respectively. To make our comparison as faithful as possible, we also re-evaluate after post-processing our predictions to match the region and resolution used by CRADL, where we see similar improvement. Qualitative examples are shown in Fig. 3. Note that all baseline methods use a validation set consisting of real anomalous samples from BraTS and ISLES to select which anomaly scoring function to use. We, however, only use synthetic validation data. This further verifies that our method of using synthetic data to estimate generalisation works well.

For both VinDr-CXR test sets we evaluate at a sample and pixel level, although previous publications have only reported their results at a sample level.

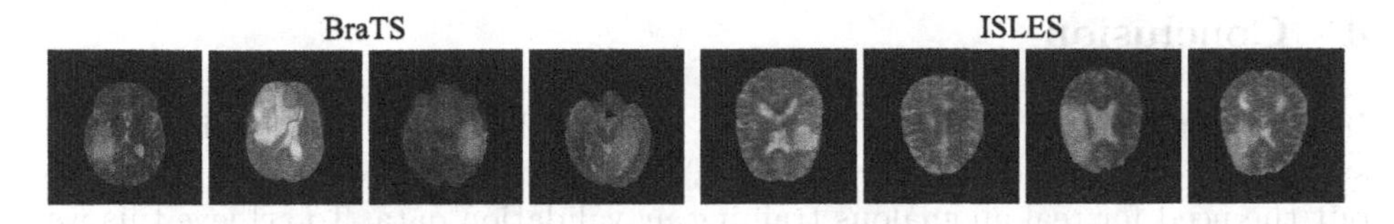

Fig. 3. Examples of predictions on randomly selected BraTS and ISLES samples after training on HCP. The red contour outlines the ground truth segmentation. (Color figure online)

We again show performance above the current state-of-the-art (Table 1 upper right). Our results are also substantially higher than previously proposed self-supervised methods, improving on the current state-of-the-art NSA [23] by 12.6 to achieve 78.4 image-level AP. This shows that our use of synthetic validation data succeeds where their fixed training schedule fails.

Ablation and Sensitivity Analysis on Cross-Validation Structure: We also investigate how performance changes as we vary the number of tasks used for training and validation (Table 1 lower). For VinDr-CXR, in an individual fold, the average performance increases as training becomes more diverse (i.e. more tasks); however, the performance of the ensemble plateaus. Having more training tasks can help the model to be sensitive to a wider range of anomalous features. But as the number of training tasks increases, so does the overlap between different models in the ensemble, diminishing the benefit of pooling predictions. This could also explain why the standard deviation (across folds) decreases as the number of training tasks increases, since the models are becoming more similar. Our best configuration is close to being competitive with the state-of-the-art *semi*-supervised method DDAD-ASR [6]. Even though their method uses twice as much training data, as well as some real anomalous data, our purely synthetic method begins to close the gap (AP of [6] 84.3 vs. ours 80.7 on $DDAD_{ts}$). For the brain datasets, all metrics generally decrease as the number of training tasks increases. This could be due to the distribution shift between training and test data. Although more training tasks may increase sensitivity to diverse irregularities, this can actually become a liability if there are differences between (healthy) training and test data (e.g. acquisition parameters). More sensitive models may then lead to more "false" positives.

Discussion: We demonstrate the effectiveness of our method in multiple settings and across different modalities. A unique aspect of the brain data is the domain shift. The HCP training data was acquired at a much higher isotropic resolution than the BraTS and ISLES test data, which are both anisotropic. Here we achieve the best performance using more tasks for validation, which successfully reduces overfitting and hypersensitivity. Incorporating greater data augmentations, such as simulating anisotropic spacing, could further improve results by training the model to ignore these transformations. We also achieve strong results for the X-ray data, although precise localisation remains a challenging task. The gap between current performance and clinicially useful localisation should therefore be high priority for future research.

4 Conclusion

In this work we use multiple synthetic tasks to both train and validate self-supervised anomaly detection models. This enables more robust training without the need for real anomalous training *or* validation data. To achieve this we propose multiple diverse tasks, exposing models to a wide range of anomalous features. These include patch blending, image deformations and intensity modulations. As part of this, we extend Poisson image editing to images of arbitrary dimensions, enabling the current state-of-the-art tasks to be applied beyond just 2D images. In order to use all of these tasks in a common framework we also design a unified labelling function, with improved continuity for small intensity changes. We evaluate our method on both brain MRI and chest X-rays and achieve state-of-the-art performance and above. We also report pixel-wise results, even for the challenging case of chest X-rays. We hope this encourages others to do the same, as accurate localisation is essential for anomaly detection to have a future in clinical workflows.

Acknowledgements. We thank EPSRC for DTP funding and HPC resources provided by the Erlangen National High Performance Computing Center (NHR @ FAU) of the Friedrich-Alexander-Universität Erlangen-Nürnberg (FAU) under the NHR project b143dc. NHR funding is provided by federal and Bavarian state authorities. NHR@FAU hardware is partially funded by the German Research Foundation (DFG) - 440719683. Support was also received by the ERC - project MIA-NORMAL 101083647 and DFG KA 5801/2-1, INST 90/1351-1.

References

1. Bakas, S., et al.: Identifying the best machine learning algorithms for brain tumor segmentation, progression assessment, and overall survival prediction in the brats challenge. arXiv:1811.02629 (2018)
2. Baur, C., Denner, S., Wiestler, B., Navab, N., Albarqouni, S.: Autoencoders for unsupervised anomaly segmentation in brain mr images: a comparative study. Med. Image Anal. **69**, 101952 (2021)
3. Brady, A.P.: Error and discrepancy in radiology: inevitable or avoidable? Insights Imaging **8**(1), 171–182 (2017)
4. Bruls, R., Kwee, R.: Workload for radiologists during on-call hours: dramatic increase in the past 15 years. Insights Imaging **11**, 1–7 (2020)
5. Cai, Y., Chen, H., Yang, X., Zhou, Y., Cheng, K.T.: Dual-distribution discrepancy for anomaly detection in chest x-rays. In: MICCAI 2022, Part III, pp. 584–593. Springer (2022). https://doi.org/10.1007/978-3-031-16437-8_56
6. Cai, Y., Chen, H., Yang, X., Zhou, Y., Cheng, K.T.: Dual-distribution discrepancy with self-supervised refinement for anomaly detection in medical images. arXiv:2210.04227 (2022)
7. Cho, J., Kang, I., Park, J.: Self-supervised 3d out-of-distribution detection via pseudoanomaly generation. In: Biomedical Image Registration, Domain Generalisation and Out-of-Distribution Analysis, pp. 95–103 (2022)

8. Gong, D., et al.: Memorizing normality to detect anomaly: memory-augmented deep autoencoder for unsupervised anomaly detection. In: CVPR 2019, pp. 1705–1714 (2019)
9. Kascenas, A., et al.: The role of noise in denoising models for anomaly detection in medical images. arXiv:2301.08330 (2023)
10. Kim, Y.W., Mansfield, L.T.: Fool me twice: delayed diagnoses in radiology with emphasis on perpetuated errors. Am. J. Roentgenol. **202**(3), 465–470 (2014)
11. Li, C.L., Sohn, K., Yoon, J., Pfister, T.: Cutpaste: self-supervised learning for anomaly detection and localization. In: CVPR 2021, pp. 9664–9674 (2021)
12. Lüth, C.T., et al.: Cradl: contrastive representations for unsupervised anomaly detection and localization. arXiv:2301.02126 (2023)
13. Maier, O., Menze, B.H., von der Gablentz, J., Häni, L., Heinrich, M.P., et al.: ISLES 2015 - a public evaluation benchmark for ischemic stroke lesion segmentation from multispectral MRI. Med. Image Anal. **35**, 250–269 (2017). https://doi.org/10.1016/j.media.2016.07.009
14. Mao, Y., Xue, F.F., Wang, R., Zhang, J., Zheng, W.S., Liu, H.: Abnormality detection in chest x-ray images using uncertainty prediction autoencoders. In: MICCAI 2020, pp. 529–538 (2020)
15. Meissen, F., Kaissis, G., Rueckert, D.: Challenging current semi-supervised anomaly segmentation methods for brain mri. In: BrainLes 2021 at MICCAI 2021, 27 Sept 2021, Part I, pp. 63–74. Springer (2022). https://doi.org/10.1007/978-3-031-08999-2_5
16. Meissen, F., Wiestler, B., Kaissis, G., Rueckert, D.: On the pitfalls of using the residual error as anomaly score. In: Proceedings of The 5th International Conference on Medical Imaging with Deep Learning. Proceedings of Machine Learning Research, vol. 172, pp. 914–928. PMLR (06–08 Jul 2022)
17. Morel, J.M., Petro, A.B., Sbert, C.: Fourier implementation of poisson image editing. Pattern Recogn. Lett. **33**(3), 342–348 (2012)
18. Nguyen, H.Q., et al.: Vindr-cxr: an open dataset of chest x-rays with radiologist's annotations. Scientific Data **9**(1), 429 (2022)
19. Pedregosa, F., et al.: Scikit-learn: machine learning in python. J. Mach. Learn. Res. **12**, 2825–2830 (2011)
20. Pérez, P., Gangnet, M., Blake, A.: Poisson image editing. In: ACM SIGGRAPH 2003 Papers, pp. 313–318 (2003)
21. Pinaya, W.H., et al.: Fast unsupervised brain anomaly detection and segmentation with diffusion models. In: MICCAI 2022, Part VIII, pp. 705–714. Springer (2022). https://doi.org/10.1007/978-3-031-16452-1_67
22. Schlegl, T., Seeböck, P., Waldstein, S.M., Langs, G., Schmidt-Erfurth, U.: f-AnoGAN: fast unsupervised anomaly detection with generative adversarial networks. Med. Image Anal. **54**, 30–44 (2019). https://doi.org/10.1016/j.media.2019.01.010
23. Schlüter, H.M., Tan, J., Hou, B., Kainz, B.: Natural synthetic anomalies for self-supervised anomaly detection and localization. In: Computer Vision - ECCV 2022, pp. 474–489. Springer Nature Switzerland, Cham (2022). https://doi.org/10.1007/978-3-031-19821-2_27
24. Tan, J., Hou, B., Batten, J., Qiu, H., Kainz, B.: Detecting outliers with foreign patch interpolation. Mach. Learn. Biomed. Imaging **1**, 1–27 (2022)
25. Tan, J., Hou, B., Day, T., Simpson, J., Rueckert, D., Kainz, B.: Detecting outliers with poisson image interpolation. In: de Bruijne, M., et al. (eds.) MICCAI 2021. LNCS, vol. 12905, pp. 581–591. Springer, Cham (2021). https://doi.org/10.1007/978-3-030-87240-3_56

26. Tian, Y., et al.: Constrained contrastive distribution learning for unsupervised anomaly detection and localisation in medical images. In: de Bruijne, M., et al. (eds.) MICCAI 2021. LNCS, vol. 12905, pp. 128–140. Springer, Cham (2021). https://doi.org/10.1007/978-3-030-87240-3_13
27. Van Den Oord, A., Vinyals, O., et al.: Neural discrete representation learning. In: Advances in Neural Information Processing Systems 30 (2017)
28. Van Essen, D., Ugurbil, K., Auerbach, E., et al.: The human connectome project: a data acquisition perspective. NeuroImage **62**(4), 2222–2231 (2012). https://doi.org/10.1016/j.neuroimage.2012.02.018
29. Zavrtanik, V., Kristan, M., Skočaj, D.: Draem-a discriminatively trained reconstruction embedding for surface anomaly detection. In: CVPR 2021, pp. 8330–8339 (2021)
30. Zhang, W., et al.: A multi-task network with weight decay skip connection training for anomaly detection in retinal fundus images. In: MICCAI 2022, Part II, pp. 656–666. Springer (2022)
31. Zimmerer, D., et al.: Mood 2020: a public benchmark for out-of-distribution detection and localization on medical images. IEEE Trans. Med. Imaging **41**(10), 2728–2738 (2022)
32. Zimmerer, D., Kohl, S.A., Petersen, J., Isensee, F., Maier-Hein, K.H.: Context-encoding variational autoencoder for unsupervised anomaly detection. arXiv:1812.05941 (2018)

AME-CAM: Attentive Multiple-Exit CAM for Weakly Supervised Segmentation on MRI Brain Tumor

Yu-Jen Chen[1](✉), Xinrong Hu[2], Yiyu Shi[2], and Tsung-Yi Ho[3]

[1] National Tsing Hua University, Hsinchu, Taiwan
yujenchen@gapp.nthu.edu.tw
[2] University of Notre Dame, Notre Dame, IN, USA
{xhu7,yshi4}@nd.edu
[3] The Chinese University of Hong Kong, Hong Kong, China
tyho@cse.cuhk.edu.hk

Abstract. Magnetic resonance imaging (MRI) is commonly used for brain tumor segmentation, which is critical for patient evaluation and treatment planning. To reduce the labor and expertise required for labeling, weakly-supervised semantic segmentation (WSSS) methods with class activation mapping (CAM) have been proposed. However, existing CAM methods suffer from low resolution due to strided convolution and pooling layers, resulting in inaccurate predictions. In this study, we propose a novel CAM method, Attentive Multiple-Exit CAM (AME-CAM), that extracts activation maps from multiple resolutions to hierarchically aggregate and improve prediction accuracy. We evaluate our method on the BraTS 2021 dataset and show that it outperforms state-of-the-art methods.

Keywords: Tumor segmentation · Weakly-supervised semantic segmentation

1 Introduction

Deep learning techniques have greatly improved medical image segmentation by automatically extracting specific tissue or substance location information, which facilitates accurate disease diagnosis and assessment. However, most deep learning approaches for segmentation require fully or partially labeled training datasets, which can be time-consuming and expensive to annotate. To address this issue, recent research has focused on developing segmentation frameworks that require little or no segmentation labels.

To meet this need, many researchers have devoted their efforts to Weakly-Supervised Semantic Segmentation (WSSS) [21], which utilizes weak supervision, such as image-level classification labels. Recent WSSS methods can be broadly categorized into two types [4]: Class-Activation-Mapping-based (CAM-based) [9,13,16,19,20,22], and Multiple-Instance-Learning-based (MIL-based) [15] methods.

Supplementary Information The online version contains supplementary material available at https://doi.org/10.1007/978-3-031-43907-0_17.

H. Greenspan et al. (Eds.): MICCAI 2023, LNCS 14220, pp. 173–182, 2023.
https://doi.org/10.1007/978-3-031-43907-0_17

The literature has not adequately addressed the issue of low-resolution Class-Activation Maps (CAMs), especially for medical images. Some existing methods, such as dilated residual networks [24] and U-Net segmentation architecture [3,7,17], have attempted to tackle this issue, but still require many upsampling operations, which the results become blurry. Meanwhile, LayerCAM [9] has proposed a hierarchical solution that extracts activation maps from multiple convolution layers using Grad-CAM [16] and aggregates them with equal weights. Although this approach successfully enhances the resolution of the segmentation mask, it lacks flexibility and may not be optimal.

In this paper, we propose an Attentive Multiple-Exit CAM (AME-CAM) for brain tumor segmentation in magnetic resonance imaging (MRI). Different from recent CAM methods, AME-CAM uses a classification model with multiple-exit training strategy applied to optimize the internal outputs. Activation maps from the outputs of internal classifiers, which have different resolutions, are then aggregated using an attention model. The model learns the pixel-wise weighted sum of the activation maps by a novel contrastive learning method.

Our proposed method has the following contributions:

- To tackle the issues in existing CAMs, we propose to use multiple-exit classification networks to accurately capture all the internal activation maps of different resolutions.
- We propose an attentive feature aggregation to learn the pixel-wise weighted sum of the internal activation maps.
- We demonstrate the superiority of AME-CAM over state-of-the-art CAM methods in extracting segmentation results from classification networks on the 2021 Brain Tumor Segmentation Challenge (BraTS 2021) [1,2,14].
- For reproducibility, we have released our code at https://github.com/windstormer/AME-CAM

Overall, our proposed method can help overcome the challenges of expensive and time-consuming segmentation labeling in medical imaging, and has the potential to improve the accuracy of disease diagnosis and assessment.

2 Attentive Multiple-Exit CAM (AME-CAM)

The proposed AME-CAM method consists of two training phases: activation extraction and activation aggregation, as shown in Fig. 1. In the activation extraction phase, we use a binary classification network, e.g., ResNet-18, to obtain the class probability $y = f(I)$ of the input image I. To enable multiple-exit training, we add one internal classifier after each residual block, which generates the activation map M_i of different resolutions. We use a cross-entropy loss to train the multiple-exit classifier, which is defined as

$$loss = \sum_{i=1}^{4} CE(GAP(M_i), L) \tag{1}$$

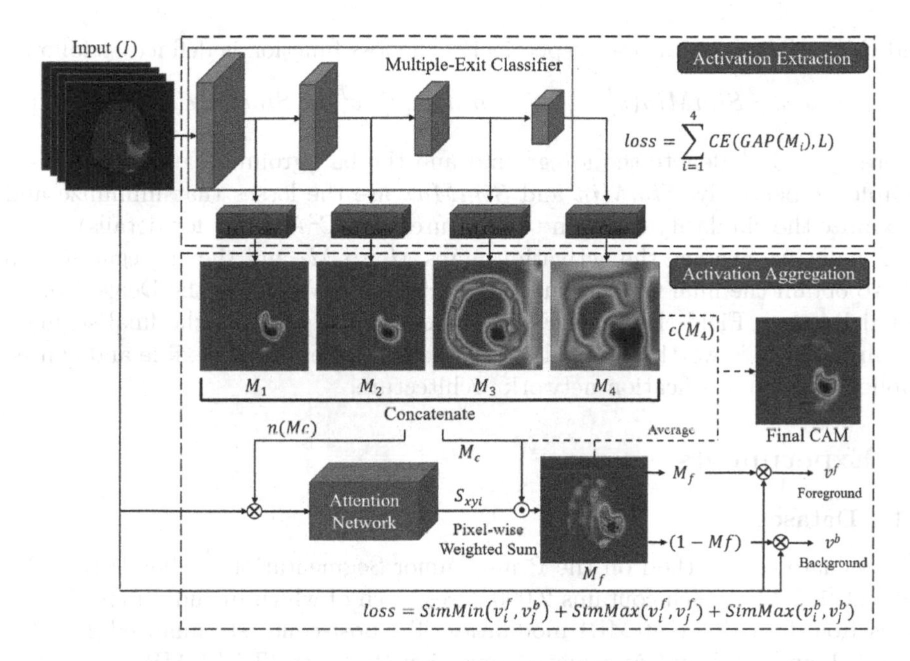

Fig. 1. An overview of the proposed AME-CAM method, which contains multiple-exit network based activation extraction phase and attention based activation aggregation phase. The operator $\odot$ and $\otimes$ denote the pixel-wise weighted sum and the pixel-wise multiplication, respectively.

where $GAP(\cdot)$ is the global-average-pooling operation, $CE(\cdot)$ is the cross-entropy loss, and L is the image-wise ground-truth label.

In the activation aggregation phase, we create an efficient hierarchical aggregation method to generate the aggregated activation map M_f by calculating the pixel-wise weighted sum of the activation maps M_i. We use an attention network $A(\cdot)$ to estimate the importance of each pixel from each activation map. The attention network takes in the input image I masked by the activation map and outputs the pixel-wised importance score S_{xyi} of each activation map. We formulate the operation as follows:

$$S_{xyi} = A([I \otimes n(M_i)]_{i=1}^{4}) \quad (2)$$

where $[\cdot]$ is the concatenate operation, $n(\cdot)$ is the min-max normalization to map the range to [0,1], and $\otimes$ is the pixel-wise multiplication, which is known as image masking. The aggregated activation map M_f is then obtained by the pixel-wise weighted sum of M_i, which is $M_f = \sum_{i=1}^{4}(S_{xyi} \otimes M_i)$.

We train the attention network with unsupervised contrastive learning, which forces the network to disentangle the foreground and the background of the aggregated activation map M_f. We mask the input image by the aggregated activation map M_f and its opposite $(1 - M_f)$ to obtain the foreground feature and the background feature, respectively. The loss function is defined as follows:

$$loss = SimMin(v_i^f, v_j^b) + SimMax(v_i^f, v_j^f) + SimMax(v_i^b, v_j^b) \quad (3)$$

where v_i^f and v_i^b denote the foreground and the background feature of the i-th sample, respectively. *SimMin* and *SimMax* are the losses that minimize and maximize the similarity between two features (see C^{2}AM [22] for details).

Finally, we average the activation maps M_1 to M_4 and the aggregated map M_f to obtain the final CAM results for each image. We apply the Dense Conditional Random Field (DenseCRF) [12] algorithm to generate the final segmentation mask. It is worth noting that the proposed method is flexible and can be applied to any classification network architecture.

3 Experiments

3.1 Dataset

We evaluate our method on the Brain Tumor Segmentation challenge (BraTS) dataset [1,2,14], which contains 2,000 cases, each of which includes four 3D volumes from four different MRI modalities: T1, post-contrast enhanced T1 (T1-CE), T2, and T2 Fluid Attenuated Inversion Recovery (T2-FLAIR), as well as a corresponding segmentation ground-truth mask. The official data split divides these cases by the ratio of 8:1:1 for training, validation, and testing (5,802 positive and 1,073 negative images). In order to evaluate the performance, we use the validation set as our test set and report statistics on it. We preprocess the data by slicing each volume along the z-axis to form a total of 193,905 2D images, following the approach of Kang et al. [10] and Dey and Hong [6]. We use the ground-truth segmentation masks only in the final evaluation, not in the training process.

3.2 Implementation Details and Evaluation Protocol

We implement our method in PyTorch using ResNet-18 as the backbone classifier. We pretrain the classifier using SupCon [11] and then fine-tune it in our experiments. We use the entire training set for both pretraining and fine-tuning. We set the initial learning rate to 1e-4 for both phases, and use the cosine annealing scheduler to decrease it until the minimum learning rate is 5e-6. We set the weight decay in both phases to 1e-5 for model regularization. We use Adam optimizer in the multiple-exit phase and SGD optimizer in the aggregation phase. We train all classifiers until they converge with a test accuracy of over 0.9 for all image modalities. Note that only class labels are available in the training set.

We use the Dice score and Intersection over Union (IoU) to evaluate the quality of the semantic segmentation, following the approach of Xu et al. [23], Tang et al. [18], and Qian et al. [15]. In addition, we report the 95% Hausdorff Distance (HD95) to evaluate the boundary of the prediction mask.

Interested readers can refer to the supplementary material for results on other network architectures.

Table 1. Comparison with weakly supervised methods (WSSS), unsupervised method (UL), and fully supervised methods (FSL) on BraTS dataset with T1, T1-CE, T2, and T2-FLAIR MRI images. Results are reported in the form of mean ± std. We mark the highest score among WSSS methods with bold text.

BraTS T1

Type	Method	Dice ↑	IoU ↑	HD95 ↓
WSSS	Grad-CAM (2016)	0.107 ± 0.090	0.059 ± 0.055	121.816 ± 22.963
	ScoreCAM (2020)	0.296 ± 0.128	0.181 ± 0.089	60.302 ± 14.110
	LFI-CAM (2021)	0.568 ± 0.167	0.414 ± 0.152	23.939 ± 25.609
	LayerCAM (2021)	0.571 ± 0.170	0.419 ± 0.161	23.335 ± 27.369
	Swin-MIL (2022)	0.477 ± 0.170	0.330 ± 0.147	46.468 ± 30.408
	AME-CAM (ours)	**0.631 ± 0.119**	**0.471 ± 0.119**	**21.813 ± 18.219**
UL	C&F (2020)	0.200 ± 0.082	0.113 ± 0.051	79.187 ± 14.304
FSL	C&F (2020)	0.572 ± 0.196	0.426 ± 0.187	29.027 ± 20.881
	Opt. U-net (2021)	0.836 ± 0.062	0.723 ± 0.090	11.730 ± 10.345

BraTS T1-CE

Type	Method	Dice ↑	IoU ↑	HD95 ↓
WSSS	Grad-CAM (2016)	0.127 ± 0.088	0.071 ± 0.054	129.890 ± 27.854
	ScoreCAM (2020)	0.397 ± 0.189	0.267 ± 0.163	46.834 ± 22.093
	LFI-CAM (2021)	0.121 ± 0.120	0.069 ± 0.076	136.246 ± 38.619
	LayerCAM (2021)	0.510 ± 0.209	0.367 ± 0.180	29.850 ± 45.877
	Swin-MIL (2022)	0.460 ± 0.169	0.314 ± 0.140	46.996 ± 22.821
	AME-CAM (ours)	**0.695 ± 0.095**	**0.540 ± 0.108**	**18.129 ± 12.335**
UL	C&F (2020)	0.179 ± 0.080	0.101 ± 0.050	77.982 ± 14.042
FSL	C&F (2020)	0.246 ± 0.104	0.144 ± 0.070	130.616 ± 9.879
	Opt. U-net (2021)	0.845 ± 0.058	0.736 ± 0.085	11.593 ± 11.120

BraTS T2

Type	Method	Dice ↑	IoU ↑	HD95 ↓
WSSS	Grad-CAM (2016)	0.049 ± 0.058	0.026 ± 0.034	141.025 ± 23.107
	ScoreCAM (2020)	0.530 ± 0.184	0.382 ± 0.174	28.611 ± 11.596
	LFI-CAM (2021)	0.673 ± 0.173	0.531 ± 0.186	18.165 ± 10.475
	LayerCAM (2021)	0.624 ± 0.178	0.476 ± 0.173	23.978 ± 44.323
	Swin-MIL (2022)	0.437 ± 0.149	0.290 ± 0.117	38.006 ± 30.000
	AME-CAM (ours)	**0.721 ± 0.086**	**0.571 ± 0.101**	**14.940 ± 8.736**
UL	C&F (2020)	0.230 ± 0.089	0.133 ± 0.058	76.256 ± 13.192
FSL	C&F (2020)	0.611 ± 0.221	0.474 ± 0.217	109.817 ± 27.735
	Opt. U-net (2021)	0.884 ± 0.064	0.798 ± 0.098	8.349 ± 9.125

BraTS T2-FLAIR

Type	Method	Dice ↑	IoU ↑	HD95 ↓
WSSS	Grad-CAM (2016)	0.150 ± 0.077	0.083 ± 0.050	110.031 ± 23.307
	ScoreCAM (2020)	0.432 ± 0.209	0.299 ± 0.178	39.385 ± 17.182
	LFI-CAM (2021)	0.161 ± 0.192	0.102 ± 0.140	125.749 ± 45.582
	LayerCAM (2021)	0.652 ± 0.206	0.515 ± 0.210	22.055 ± 33.959
	Swin-MIL (2022)	0.272 ± 0.115	0.163 ± 0.079	41.870 ± 19.231
	AME-CAM (ours)	**0.862 ± 0.088**	**0.767 ± 0.122**	**8.664 ± 6.440**
UL	C&F (2020)	0.306 ± 0.190	0.199 ± 0.167	75.651 ± 14.214
FSL	C&F (2020)	0.578 ± 0.137	0.419 ± 0.130	138.138 ± 14.283
	Opt. U-net (2021)	0.914 ± 0.058	0.847 ± 0.093	8.093 ± 11.879

4 Results

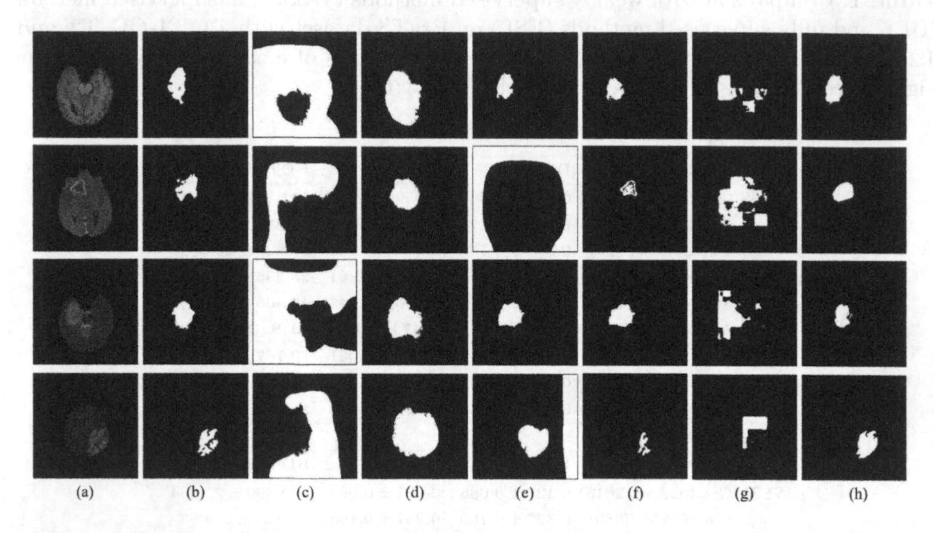

Fig. 2. Qualitative results of all methods. (a) Input Image. (b) Ground Truth. (c) Grad-CAM [16] (d) ScoreCAM [19]. (e) LFI-CAM [13]. (f) LayerCAM [9]. (g) Swin-MIL [15]. (h) AME-CAM (ours). The image modalities of rows 1-4 are T1, T1-CE, T2, T2-FLAIR, respectively from the BraTS dataset.

4.1 Quantitative and Qualitative Comparison with State-of-the-Art

In this section, we compare the segmentation performance of the proposed AME-CAM with five state-of-the-art weakly-supervised segmentation methods, namely Grad-CAM [16], ScoreCAM [19], LFI-CAM [13], LayerCAM [9], and Swin-MIL [15]. We also compare with an unsupervised approach C&F [5], the supervised version of C&F, and the supervised Optimized U-net [8] to show the comparison with non-CAM-based methods. We acknowledge that the results from fully supervised and unsupervised methods are not directly comparable to the weakly supervised CAM methods. Nonetheless, these methods serve as interesting references for the potential performance ceiling and floor of all the CAM methods.

Quantitatively, Grad-CAM and ScoreCAM result in low dice scores, demonstrating that they have difficulty extracting the activation of medical images. LFI-CAM and LayerCAM improve the dice score in all modalities, except LFI-CAM in T1-CE and T2-FLAIR. Finally, the proposed AME-CAM achieves optimal performance in all modalities of the BraTS dataset.

Compared to the unsupervised baseline (UL), C&F is unable to separate the tumor and the surrounding tissue due to low contrast, resulting in low dice scores in all experiments. With pixel-wise labels, the dice of supervised C&F

improves significantly. Without any pixel-wise label, the proposed AME-CAM outperforms supervised C&F in all modalities.

The fully supervised (FSL) Optimized U-net achieves the highest dice score and IoU score in all experiments. However, even under different levels of supervision, there is still a performance gap between the weakly supervised CAM methods and the fully supervised state-of-the-art. This indicates that there is still potential room for WSSS methods to improve in the future.

Qualitatively, Fig. 2 shows the visualization of the CAM and segmentation results from all six CAM-based approaches under four different modalities from the BraTS dataset. Grad-CAM (Fig. 2(c)) results in large false activation region, where the segmentation mask is totally meaningless. ScoreCAM eliminates false activation corresponding to air. LFI-CAM focus on the exact tumor area only in the T1 and T2 MRI (row 1 and 3). Swin-MIL can hardly capture the tumor region of the MRI image, where the activation is noisy. Among all, only LayerCAM and the proposed AME-CAM successfully focus on the exact tumor area, but AME-CAM reduces the under-estimation of the tumor area. This is attributed to the benefit provided by aggregating activation maps from different resolutions.

4.2 Ablation Study

Table 2. Ablation study for aggregation phase using T1 MRI images from the BraTS dataset. Avg. ME denotes that we directly average four activation maps generated by the multiple-exit phase. The dice score, IoU, and the HD95 are reported in the form of mean ± std.

Method	Dice ↑	IoU ↑	HD95 ↓
Avg. ME	0.617 ± 0.121	0.457 ± 0.121	23.603 ± 20.572
Avg. ME+C^{2}AM [22]	0.484 ± 0.256	0.354 ± 0.207	69.242 ± 121.163
AME-CAM (ours)	**0.631 ± 0.119**	**0.471 ± 0.119**	**21.813 ± 18.219**

Effect of Different Aggregation Approaches: In Table 2, we conducted an ablation study to investigate the impact of using different aggregation approaches after extracting activations from the multiple-exit network. We aim to demonstrate the superiority of the proposed attention-based aggregation approach for segmenting tumor regions in T1 MRI of the BraTS dataset. Note that we only report the results for T1 MRI in the BraTS dataset. Please refer to the supplementary material for the full set of experiments.

As a baseline, we first conducted the average of four activation maps generated by the multiple-level activation extraction (Avg. ME). We then applied C^{2}AM [22], a state-of-the-art CAM-based refinement approach, to refine the result of the baseline, which we call "Avg. ME+C^{2}AM". However, we observed that C^{2}AM tended to segment the brain region instead of the tumor region due to the larger contrast between the brain tissue and the air than that between the tumor region and its surrounding tissue. Any incorrect activation of C^2AM also led to inferior results, resulting in a degradation of the average dice score from 0.617 to 0.484. In contrast, the proposed attention-based approach provided a significant weighting solution that led to optimal performance in all cases.

Table 3. Ablation study for using single-exit from M_1, M_2, M_3 or M_4 of Fig. 1 and the multiple-exit using results from M_2 and M_3 and using all exits (AME-CAM). The experiments are done on the T1-CE MRI of BraTS dataset. The dice score, IoU, and the HD95 are reported in the form of mean ± std.

Selected Exit		Dice ↑	IoU ↑	HD95 ↓
Single-exit	M_1	0.144 ± 0.184	0.090 ± 0.130	74.249 ± 62.669
	M_2	0.500 ± 0.231	0.363 ± 0.196	43.762 ± 85.703
	M_3	0.520 ± 0.163	0.367 ± 0.141	43.749 ± 54.907
	M_4	0.154 ± 0.101	0.087 ± 0.065	120.779 ± 44.548
Multiple-exit	$M_2 + M_3$	0.566 ± 0.207	0.421 ± 0.186	27.972 ± 56.591
	AME-CAM (ours)	**0.695 ± 0.095**	**0.540 ± 0.108**	**18.129 ± 12.335**

Effect of Single-Exit and Multiple-Exit: Table 3 summarizes the performance of using single-exit from M_1, M_2, M_3, or M_4 of Fig. 1 and the multiple-exit using results from M_2 and M_3, and using all exits (AME-CAM) on T1-CE MRI in the BraTS dataset.

The comparisons show that the activation map obtained from the shallow layer M_1 and the deepest layer M_4 result in low dice scores, around 0.15. This is because the network is not deep enough to learn the tumor region in the shallow layer, and the resolution of the activation map obtained from the deepest layer is too low to contain sufficient information to make a clear boundary for the tumor. Results of the internal classifiers from the middle of the network (M_2 and M_3) achieve the highest dice score and IoU, both of which are around 0.5.

To evaluate whether using results from all internal classifiers leads to the highest performance, we further apply the proposed method to the two internal classifiers with the highest dice scores, i.e., M_2 and M_3, called $M_2 + M_3$. Compared with using all internal classifiers (M_1 to M_4), $M_2 + M_3$ results in 18.6% and 22.1% lower dice and IoU, respectively. In conclusion, our AME-CAM still achieves the optimal performance among all the experiments of single-exit and multiple-exit.

Other ablation studies are presented in the supplementary material due to space limitations.

5 Conclusion

In this work, we propose a brain tumor segmentation method for MRI images using only class labels, based on an Attentive Multiple-Exit Class Activation

Mapping (AME-CAM). Our approach extracts activation maps from different exits of the network to capture information from multiple resolutions. We then use an attention model to hierarchically aggregate these activation maps, learning pixel-wise weighted sums.

Experimental results on the four modalities of the 2021 BraTS dataset demonstrate the superiority of our approach compared with other CAM-based weakly-supervised segmentation methods. Specifically, AME-CAM achieves the highest dice score for all patients in all datasets and modalities. These results indicate the effectiveness of our proposed approach in accurately segmenting brain tumors from MRI images using only class labels.

References

1. Bakas, S., et al.: Advancing the cancer genome atlas glioma mri collections with expert segmentation labels and radiomic features. Scientific Data **4**(1), 1–13 (2017)
2. Bakas, S., et al.: Identifying the best machine learning algorithms for brain tumor segmentation, progression assessment, and overall survival prediction in the brats challenge. arXiv preprint arXiv:1811.02629 (2018)
3. Belharbi, S., Sarraf, A., Pedersoli, M., Ben Ayed, I., McCaffrey, L., Granger, E.: F-cam: Ffull resolution class activation maps via guided parametric upscaling. In: Proceedings of the IEEE/CVF Winter Conference on Applications of Computer Vision, pp. 3490–3499 (2022)
4. Chan, L., Hosseini, M.S., Plataniotis, K.N.: A comprehensive analysis of weakly-supervised semantic segmentation in different image domains. Int. J. Comput. Vision **129**, 361–384 (2021)
5. Chen, J., Frey, E.C.: Medical image segmentation via unsupervised convolutional neural network. arXiv preprint arXiv:2001.10155 (2020)
6. Dey, R., Hong, Y.: ASC-Net: adversarial-based selective network for unsupervised anomaly segmentation. In: de Bruijne, M., et al. (eds.) MICCAI 2021. LNCS, vol. 12905, pp. 236–247. Springer, Cham (2021). https://doi.org/10.1007/978-3-030-87240-3_23
7. Englebert, A., Cornu, O., De Vleeschouwer, C.: Poly-cam: high resolution class activation map for convolutional neural networks. arXiv preprint arXiv:2204.13359 (2022)
8. Futrega, M., Milesi, A., Marcinkiewicz, M., Ribalta, P.: Optimized u-net for brain tumor segmentation. arXiv preprint arXiv:2110.03352 (2021)
9. Jiang, P.T., Zhang, C.B., Hou, Q., Cheng, M.M., Wei, Y.: Layercam: exploring hierarchical class activation maps for localization. IEEE Trans. Image Process. **30**, 5875–5888 (2021)
10. Kang, H., Park, H.m., Ahn, Y., Van Messem, A., De Neve, W.: Towards a quantitative analysis of class activation mapping for deep learning-based computer-aided diagnosis. In: Medical Imaging 2021: Image Perception, Observer Performance, and Technology Assessment, vol. 11599, p. 115990M. International Society for Optics and Photonics (2021)
11. Khosla, P., et al.: Supervised contrastive learning. arXiv preprint arXiv:2004.11362 (2020)
12. Krähenbühl, P., Koltun, V.: Efficient inference in fully connected crfs with gaussian edge potentials. In: Advances in Neural Information Processing Systems 24 (2011)

13. Lee, K.H., Park, C., Oh, J., Kwak, N.: Lfi-cam: learning feature importance for better visual explanation. In: Proceedings of the IEEE/CVF International Conference on Computer Vision, pp. 1355–1363 (2021)
14. Menze, B.H., et al.: The multimodal brain tumor image segmentation benchmark (brats). IEEE Trans. Med. Imaging **34**(10), 1993–2024 (2014)
15. Qian, Z., et al.: Transformer based multiple instance learning for weakly supervised histopathology image segmentation. In: Medical Image Computing and Computer Assisted Intervention-MICCAI 2022: 25th International Conference, Singapore, September 18–22, 2022, Proceedings, Part II, pp. 160–170. Springer (2022). https://doi.org/10.1007/978-3-031-16434-7_16
16. Selvaraju, R.R., Cogswell, M., Das, A., Vedantam, R., Parikh, D., Batra, D.: Grad-cam: Visual explanations from deep networks via gradient-based localization. In: Proceedings of the IEEE International Conference on Computer Vision, pp. 618–626 (2017)
17. Tagaris, T., Sdraka, M., Stafylopatis, A.: High-resolution class activation mapping. In: 2019 IEEE International Conference On Image Processing (ICIP), pp. 4514–4518. IEEE (2019)
18. Tang, W., et al.: M-SEAM-NAM: multi-instance self-supervised equivalent attention mechanism with neighborhood affinity module for double weakly supervised segmentation of COVID-19. In: de Bruijne, M., et al. (eds.) MICCAI 2021. LNCS, vol. 12907, pp. 262–272. Springer, Cham (2021). https://doi.org/10.1007/978-3-030-87234-2_25
19. Wang, H., et al.: Score-cam: score-weighted visual explanations for convolutional neural networks. In: Proceedings of the IEEE/CVF Conference on Computer Vision and Pattern Recognition Workshops, pp. 24–25 (2020)
20. Wang, Y., Zhang, J., Kan, M., Shan, S., Chen, X.: Self-supervised equivariant attention mechanism for weakly supervised semantic segmentation. In: Proceedings of the IEEE/CVF Conference on Computer Vision and Pattern Recognition, pp. 12275–12284 (2020)
21. Wolleb, J., Bieder, F., Sandkühler, R., Cattin, P.C.: Diffusion models for medical anomaly detection. In: Medical Image Computing and Computer Assisted Intervention-MICCAI 2022: 25th International Conference, Singapore, 18–22 September 2022, Proceedings, Part VIII, pp. 35–45. Springer (2022). https://doi.org/10.1007/978-3-031-16452-1_4
22. Xie, J., Xiang, J., Chen, J., Hou, X., Zhao, X., Shen, L.: C2am: contrastive learning of class-agnostic activation map for weakly supervised object localization and semantic segmentation. In: Proceedings of the IEEE/CVF Conference on Computer Vision and Pattern Recognition, pp. 989–998 (2022)
23. Xu, X., et al.: Whole heart and great vessel segmentation in congenital heart disease using deep neural networks and graph matching. In: Shen, D., et al. (eds.) MICCAI 2019. LNCS, vol. 11765, pp. 477–485. Springer, Cham (2019). https://doi.org/10.1007/978-3-030-32245-8_53
24. Yu, F., Koltun, V., Funkhouser, T.: Dilated residual networks. In: Proceedings of the IEEE Conference on Computer Vision and Pattern Recognition, pp. 472–480 (2017)

Cross-Adversarial Local Distribution Regularization for Semi-supervised Medical Image Segmentation

Thanh Nguyen-Duc[1](✉), Trung Le[1], Roland Bammer[1], He Zhao[2], Jianfei Cai[1], and Dinh Phung[1]

[1] Monash University, Melbourne, Australia
{thanh.nguyen4,trunglm,roland.bammer,jianfei.cai,dinh.phung}@monash.edu
[2] CSIRO's Data61, Melbourne, Australia
he.zhao@ieee.org

Abstract. Medical semi-supervised segmentation is a technique where a model is trained to segment objects of interest in medical images with limited annotated data. Existing semi-supervised segmentation methods are usually based on the smoothness assumption. This assumption implies that the model output distributions of two similar data samples are encouraged to be invariant. In other words, the smoothness assumption states that similar samples (e.g., adding small perturbations to an image) should have similar outputs. In this paper, we introduce a novel cross-adversarial local distribution (Cross-ALD) regularization to further enhance the smoothness assumption for semi-supervised medical image segmentation task. We conducted comprehensive experiments that the Cross-ALD archives state-of-the-art performance against many recent methods on the public LA and ACDC datasets.

Keywords: Semi-supervised segmentation · Adversarial local distribution · Adversarial examples · Cross-adversarial local distribution

1 Introduction

Medical image segmentation is a critical task in computer-aided diagnosis and treatment planning. It involves the delineation of anatomical structures or pathological regions in medical images, such as magnetic resonance imaging (MRI) or computed tomography (CT) scans. Accurate and efficient segmentation is essential for various medical applications, including tumor detection, surgical planning, and monitoring disease progression. However, manual medical imaging annotation is time-consuming and expensive because it requires the domain

Supplementary Information The online version contains supplementary material available at https://doi.org/10.1007/978-3-031-43907-0_18.

H. Greenspan et al. (Eds.): MICCAI 2023, LNCS 14220, pp. 183–194, 2023.
https://doi.org/10.1007/978-3-031-43907-0_18

knowledge from medical experts. Therefore, there is a growing interest in developing semi-supervised learning that leverages both labeled and unlabeled data to improve the performance of image segmentation models [16,27].

Existing semi-supervised segmentation methods exploit smoothness assumption, e.g., the data samples that are closer to each other are more likely to to have the same label. In other words, the smoothness assumption encourages the model to generate invariant outputs under small perturbations. We have seen such perturbations being be added to natural input images at data-level [4,9,14,19,21], feature-level [6,17,23,25], and model-level [8,11,12,24,28]. Among them, virtual adversarial training (VAT) [14] is a well-known one which promotes the smoothness of the local output distribution using adversarial examples. The adversarial examples are near decision boundaries generated by adding adversarial perturbations to natural inputs. However, VAT can only create one adversarial sample in a run, which is often insufficient to completely explore the space of possible perturbations (see Sect. 2.1). In addition, the adversarial examples of VAT can also lie together and lose diversity that significantly reduces the quality of adversarial examples [15,20]. Mixup regularization [29] is a data augmentation method used in deep learning to improve model generalization. The idea behind mixup is to create new training examples by linearly interpolating between pairs of existing examples and their corresponding labels, which has been adopted in [2,3,19] to semi-supervised learning. The work [5] suggests that Mixup improves the smoothness of the neural function by bounding the Lipschitz constant of the gradient function of the neural networks. However, we show that mixing between more informative samples (e.g., adversarial examples near decision boundaries) can lead to a better performance enhancement compared to mixing natural samples (see Sect. 3.3).

In this paper, we propose a novel cross-adversarial local distribution regularization for semi-supervised medical image segmentation for smoothness assumption enhancement[1]. Our contributions are summarized as follows: **1)** To overcome the VAT's drawback, we formulate an adversarial local distribution (ALD) with Dice loss function that covers all possible adversarial examples within a ball constraint. **2)** To enhance smoothness assumption, we propose a novel cross-adversarial local distribution regularization (Cross-ALD) to encourage the smoothness assumption, which is a random mixing between two ALDs. **3)** We also propose a sufficiently approximation for the Cross-ALD by a multiple particle-based search using semantic feature Stein Variational Gradient Decent (SVGDF), an enhancement of the vanilla SVGD [10]. **4)** We conduct comprehensive experiments on ADCD [1] and LA [26] datasets, showing that our Cross-ALD regularization achieves state-of-the-art performance against existing solutions [8,11,12,14,21,22,28].

[1] The Cross-ALD implementation in https://github.com/PotatoThanh/Cross-adversarial-local-distribution-regularization.

2 Method

In this section, we begin by reviewing the minimax optimization problem of virtual adversarial training (VAT) [14]. Given an input, we then formulate a novel adversarial local distribution (ALD) with Dice loss, which benefits the medical semi-supervised image segmentation problem specifically. Next, a cross-adversarial local distribution (Cross-ALD) is constructed by randomly combining two ALDs. We approximate the ALD by a particle-based method named semantic feature Stein Variational Gradient Descent (SVGDF). Considering the resolution of medical images are usually high, we enhance the vanilla SVGD [10] from data-level to feature-level, which is named SVGDF. We finally provide our regularization loss for semi-supervised medical image segmentation.

2.1 The Minimax Optimization of VAT

Let $\mathbb{D}_l$ and $\mathbb{D}_{ul}$ be the labeled and unlabeled dataset, respectively, with $P_{\mathbb{D}_l}$ and $P_{\mathbb{D}_{ul}}$ being the corresponding data distribution. Denote $\boldsymbol{x} \in \mathbb{R}^d$ as our d-dimensional input in a space $\boldsymbol{X}$. The labeled image $\boldsymbol{x}_l$ and segmentation ground-truth $\boldsymbol{y}$ are sampled from the labeled dataset $\mathbb{D}_l$ ($\boldsymbol{x}_l, \boldsymbol{y} \sim P_{\mathbb{D}_l}$), and the unlabeled image sampled from $\mathbb{D}_{ul}$ is $\boldsymbol{x} \sim P_{\mathbb{D}_{ul}}$.

Given an input $\boldsymbol{x} \sim P_{\mathbb{D}_{ul}}$ (i.e., the unlabeled data distribution), let us denote the ball constraint around the image $\boldsymbol{x}$ as $C_\epsilon(\boldsymbol{x}) = \{\boldsymbol{x}' \in \boldsymbol{X} : ||\boldsymbol{x}' - \boldsymbol{x}||_p \leq \epsilon\}$, where ϵ is a ball constraint radius with respect to a norm $||\cdot||_p$, and $\boldsymbol{x}'$ is an adversarial example[2]. Given that f_θ is our model parameterized by θ, VAT [14] trains the model with the loss of ℓ_{vat} that a minimax optimization problem:

$$\ell_{vat} := \min_\theta \mathbb{E}_{\boldsymbol{x}\sim P_{\mathbb{D}_{ul}}} \left[\max_{\boldsymbol{x}' \in C_\epsilon(\boldsymbol{x})} D_{\mathrm{KL}}(f_\theta(\boldsymbol{x}'), f_\theta(\boldsymbol{x})) \right], \tag{1}$$

where D_{KL} is the Kullback-Leibler divergence. The inner *maximization problem* is to find an adversarial example near decision boundaries, while the *minimization problem* enforces the local smoothness of the model. However, VAT is insufficient to explore the set of of all adversarial examples within the constraint C_ϵ because it only find one adversarial example $\boldsymbol{x}'$ given a natural input $\boldsymbol{x}$. Moreover, the works [15,20] show that even solving the *maximization problem* with random initialization, its solutions can also lie together and lose diversity, which significantly reduces the quality of adversarial examples.

2.2 Adversarial Local Distribution

In order to overcome the drawback of VAT, we introduce our proposed adversarial local distribution (ALD) with Dice loss function instead of D_{KL} in [14,15]. ALD forms a set of all adversarial examples $\boldsymbol{x}'$ within the ball constraint given

[2] A sample generated by adding perturbations toward the adversarial direction.

an input $\boldsymbol{x}$. Therefore, the distribution can helps to sufficiently explore all possible adversarial examples. The adversarial local distribution $P_\theta(\boldsymbol{x}'|\boldsymbol{x})$ is defined with a ball constraint C_ϵ as follow:

$$P_\theta(\boldsymbol{x}'|\boldsymbol{x}) := \frac{e^{\ell_{Dice}(\boldsymbol{x}',\boldsymbol{x};\theta)}}{\int_{C_\epsilon(\boldsymbol{x})} e^{\ell_{Dice}(\boldsymbol{x}'',\boldsymbol{x};\theta)} d\boldsymbol{x}''} = \frac{e^{\ell_{Dice}(\boldsymbol{x}',\boldsymbol{x};\theta)}}{Z(\boldsymbol{x};\theta)}, \tag{2}$$

where $P_\theta(\cdot|\boldsymbol{x})$ is the conditional local distribution, and $Z(\boldsymbol{x};\theta)$ is a normalization function. The ℓ_{Dice} is the Dice loss function as shown in Eq. 3

$$\ell_{Dice}(\boldsymbol{x}',\boldsymbol{x};\theta) = \frac{1}{C}\sum_{c=1}^{C}[1 - \frac{2||p_\theta(\hat{\boldsymbol{y}}_c|\boldsymbol{x}) \cap p_\theta(\tilde{\boldsymbol{y}}_c|\boldsymbol{x}')||}{||p_\theta(\hat{\boldsymbol{y}}_c|\boldsymbol{x}) + p_\theta(\tilde{\boldsymbol{y}}_c|\boldsymbol{x}')||}], \tag{3}$$

where C is the number of classes. $p_\theta(\hat{\boldsymbol{y}}_c|\boldsymbol{x})$ and $p_\theta(\tilde{\boldsymbol{y}}_c|\boldsymbol{x}')$ are the predictions of input image $\boldsymbol{x}$ and adversarial image $\boldsymbol{x}'$, respectively.

2.3 Cross-Adversarial Distribution Regularization

Given two random samples $\boldsymbol{x}_i, \boldsymbol{x}_j \sim P_{\mathbb{D}}$ $(i \neq j)$, we define the cross-adversarial distribution (Cross-ALD) denoted $\tilde{P}_\theta$ as shown in Eq. 4

$$\tilde{P}_\theta(\cdot|\boldsymbol{x}_i,\boldsymbol{x}_j) = \gamma P_\theta(\cdot|\boldsymbol{x}_i) + (1-\gamma)P_\theta(\cdot|\boldsymbol{x}_j) \tag{4}$$

where $\gamma \sim \text{Beta}(\alpha,\alpha)$ for $\alpha \in (0,\infty)$, inspired by [29]. The $\tilde{P}_\theta$ is the Cross-ALD distribution, a mixture between the two adversarial local distributions.
Given Eq. 4, we propose the Cross-ALD regularization at two random input images $\boldsymbol{x}_i, \boldsymbol{x}_j \sim P_{\mathbb{D}}$ $(i \neq j)$ as

$$R(\theta,\boldsymbol{x}_i,\boldsymbol{x}_j) := \mathbb{E}_{\tilde{\boldsymbol{x}}' \sim \tilde{P}_\theta(\cdot|\boldsymbol{x}_i,\boldsymbol{x}_j)}[\log \tilde{P}_\theta(\tilde{\boldsymbol{x}}'|\boldsymbol{x}_i,\boldsymbol{x}_j)] = -H(\tilde{P}_\theta(\cdot|\boldsymbol{x}_i,\boldsymbol{x}_j)), \tag{5}$$

where H indicates the entropy of a given distribution.

When minimizing $R(\theta,\boldsymbol{x}_i,\boldsymbol{x}_j)$ or equivalently $-H(\tilde{P}_\theta(\cdot|\boldsymbol{x}_i,\boldsymbol{x}_j))$ w.r.t. θ, we encourage $\tilde{P}_\theta(\cdot|\boldsymbol{x}_i,\boldsymbol{x}_j)$ to be closer to a uniform distribution. This implies that the outputs of $f(\tilde{\boldsymbol{x}}') = f(\tilde{\boldsymbol{x}}'')$ = a constant c, where $\tilde{\boldsymbol{x}}', \tilde{\boldsymbol{x}}'' \sim \tilde{P}_\theta(\cdot|\boldsymbol{x}_i,\boldsymbol{x}_j)$. In other words, we encourages the invariant model outputs under small perturbations. Therefore, minimizing the Cross-ALD regularization loss leads to an enhancement in the model smoothness. While VAT only enforces local smoothness using one adversarial example, Cross-ALD further encourages smoothness of both local and mixed adversarial distributions to improve the model generalization.

2.4 Multiple Particle-Based Search to Approximate the Cross-ALD Regularization

In Eq. 2, the normalization $Z(\boldsymbol{x};\theta)$ in denominator term is intractable to find. Therefore, we propose a multiple particle-based search method named SVGDF to sample $\boldsymbol{x}'^{(1)}, \boldsymbol{x}'^{(2)}, \ldots, \boldsymbol{x}'^{(N)} \sim P_\theta(\cdot|\boldsymbol{x})$). N is the number of samples (or *adversarial particles*). SVGDF is used to solve the optimization problem of finding a target distribution $P_\theta(\cdot|\boldsymbol{x})$). SVGDF is a particle-based Bayesian inference algorithm that seeks a set of points (or particles) to approximate the target distribution without explicit parametric assumptions using iterative gradient-based updates. Specifically, a set of adversarial particles ($\boldsymbol{x}'^{(n)}$) is initialized by adding uniform noises, then projected onto the ball C_ϵ. These adversarial particles are then iteratively updated using a closed-form solution (Eq. 6) until reaching termination conditions (, number of iterations).

$$\boldsymbol{x}'^{(n),(l+1)} = \prod_{C_\epsilon}\Big(\boldsymbol{x}'^{(n),(l)} + \tau * (\phi(\boldsymbol{x}'^{(n),(l)}))\Big)$$
$$s.t.\ \phi(\boldsymbol{x}') = \frac{1}{N}\sum_{j=1}^{N}[k(\Phi(\boldsymbol{x}'^{(j),(l)}), \Phi(\boldsymbol{x}'))\nabla_{\boldsymbol{x}'^{(j),(l)}} \log P(\boldsymbol{x}'^{(j),(l)}|\boldsymbol{x}) \qquad (6)$$
$$+ \nabla_{\boldsymbol{x}^{(j),(l)}} k(\Phi(\boldsymbol{x}'^{(j),(l)}), \Phi(\boldsymbol{x}'))],$$

where $\boldsymbol{x}'^{(n),(l)}$ is a nth adversarial particle at l^{th} iteration ($n \in \{1, 2, ..., N\}$, and $l \in \{1, 2, ..., L\}$ with the maximum number of iteration L). $\prod_{C_\epsilon}$ is projection operator to the C_ϵ constraint. τ is the step size updating. k is the radial basis function (RBF) kernel $k(\boldsymbol{x}', \boldsymbol{x}) = \exp\left\{\frac{-||\boldsymbol{x}'-\boldsymbol{x}||^2}{2\sigma^2}\right\}$. Φ is a fixed feature extractor (e.g., encoder of U-Net/V-Net). While vanilla SVGD [10] is difficult to capture semantic meaning of high-resolution data because of calculating RBF kernel (k) directly on the data-level, we use the feature extractor Φ as a semantic transformation to further enhance the SVGD algorithm performance for medical imaging. Moreover, the two terms of ϕ in Eq. 6 have different roles: (i) the first one encourages the adversarial particles to move towards the high density areas of $P_\theta(\cdot|\boldsymbol{x})$ and (ii) the second one prevents all the particles from collapsing into the local modes of $P_\theta(\cdot|\boldsymbol{x})$ to enhance diversity (e.g.,pushing the particles away from each other). Please refer to the Cross-ALD Github repository for more details.

SVGDF approximates $P_\theta(\cdot|\boldsymbol{x}_i)$ and $P_\theta(\cdot|\boldsymbol{x}_j)$ in Eq. 4, where $\boldsymbol{x}_i, \boldsymbol{x}_j \sim P_{\mathbb{D}_{ul}}$ ($i \neq j$). We form sets of adversarial particles as $\mathbb{D}_{adv}|\boldsymbol{x}_i = \{\boldsymbol{x}_i'^{(1)}, \boldsymbol{x}_i'^{(2)}, \ldots, \boldsymbol{x}_i'^{(N)}\}$ and $\mathbb{D}_{adv}|\boldsymbol{x}_j = \{\boldsymbol{x}_j'^{(1)}, \boldsymbol{x}_j'^{(2)}, \ldots, \boldsymbol{x}_j'^{(N)}\}$. The problem (5) can then be relaxed to

$$R(\theta, \boldsymbol{x}_i, \boldsymbol{x}_j) := \mathbb{E}_{\boldsymbol{x}_i'^{(n)} \sim P_{\mathbb{D}_{adv}|\boldsymbol{x}_i}, \boldsymbol{x}_j'^{(m)} \sim P_{\mathbb{D}_{adv}|\boldsymbol{x}_j}}\Big[\ell_{Dice}(\tilde{x}', \tilde{x}; \theta)\Big] \qquad (7)$$
$$s.t.: \tilde{x}' = \gamma \boldsymbol{x}_i'^{(n)} + (1-\gamma)\boldsymbol{x}_j'^{(m)};\ \tilde{x} = \gamma \boldsymbol{x}_i + (1-\gamma)\boldsymbol{x}_j,$$

where $\gamma \sim \text{Beta}(\alpha, \alpha)$ for $\alpha \in (0, \infty)$.

2.5 Cross-ALD Regularization Loss in Medical Semi-supervised Image Segmentation

In this paper, the overall loss function ℓ_{total} consists of three loss terms. The first term is the dice loss, where labeled image $\boldsymbol{x}_l$ and segmentation ground-truth $\boldsymbol{y}$

are sampled from labeled dataset $\mathbb{D}_l$. The second term is a contrastive learning loss for inter-class separation ℓ_{cs} proposed by [21]. The third term is our Cross-ALD regularization, which is an enhancement of ℓ_{vat} to significantly improve the model performance.

$$\ell_{total} := \min_{\theta} \mathbb{E}_{(\boldsymbol{x}_l,\boldsymbol{y})\sim P_{\mathbb{D}_l}} \left[l_{Dice}(\boldsymbol{x}_l,\boldsymbol{y};\theta)\right] + \lambda_{cs}\, \mathbb{E}_{\boldsymbol{x}_l\sim P_{\mathbb{D}_l},\boldsymbol{x}\sim P_{\mathbb{D}_{ul}}} \left[\ell_{cs}(\boldsymbol{x}_l,\boldsymbol{x})\right] + \lambda_{Cross-ALD}\, \mathbb{E}_{(\boldsymbol{x}_i,\boldsymbol{x}_j)\sim P_{\mathbb{D}_{ul}}} \left[R(\theta,\boldsymbol{x}_i,\boldsymbol{x}_j)\right], \tag{8}$$

where λ_{cs} and $\lambda_{Cross-ALD}$ are the corresponding weights to balance the losses. Note that our implementation is replacing ℓ_{vat} loss with the proposed Cross-AD regularization in SS-Net code repository[3] [21] to reach the state-of-the-art performance.

3 Experiments

In this section, we conduct several comprehensive experiments using the ACDC[4] dataset [1] and the LA[5] dataset [26] for 2D and 3D image segmentation tasks, respectively. For fair comparisons, all experiments are conducted using the identical setting, following [21]. We evaluate our model in challenging semi-supervised scenarios, where only 5% and 10% of the data are labeled and the remaining data in the training set is treated as unlabeled. The Cross-ALD uses the U-Net [18] and V-Net [13] architectures for the ACDC and LA dataset, respectively. We compare the diversity between the adversarial particles generated by our method against vanilla SVGD and VAT with random initialization in Sect. 3.1 . We then illustrate the Cross-AD outperforms other recent methods on ACDC and LA datasets in Sect. 3.2. We show ablation studies in Sect. 3.3. The effect of the number particles to the model performance is studied in the Cross-ALD Github repository.

3.1 Diversity of Adversarial Particle Comparison

Settings. We fixed all the decoder models (U-Net for ACDC and V-Net for LA). We run VAT with random initialization and SVGD multiple times to produce adversarial examples, which we compared to the adversarial particles generated using SVGDF. SVGDF is the proposed algorithm, which leverages feature transformation to capture the semantic meaning of inputs. Φ is the decoder of U-Net in ACDC dataset, while Φ is the decoder of V-Net in LA dataset. We set the same radius ball constraint, updating step, and etc. We randomly pick three images from the datasets to generate adversarial particles. To evaluate their diversity, we report the sum squared error (SSE) between these particles. Higher SSE indicates more diversity, and for each number of particles, we calculate the average of the mean of SSEs.

[3] https://github.com/ycwu1997/SS-Net.
[4] https://www.creatis.insa-lyon.fr/Challenge/acdc/databases.html.
[5] http://atriaseg2018.cardiacatlas.org.

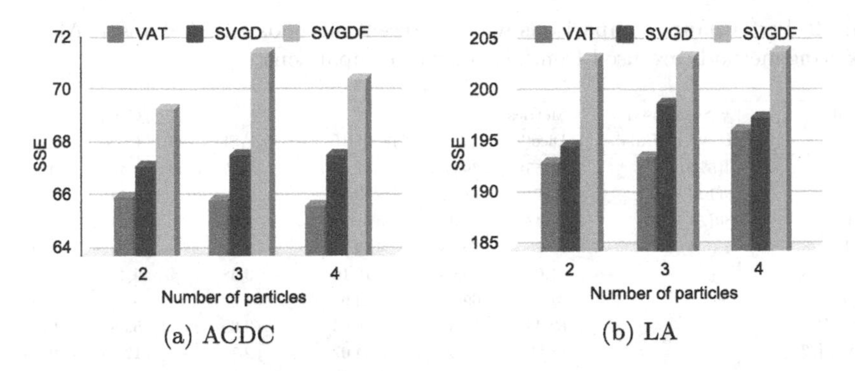

(a) ACDC (b) LA

Fig. 1. Diversity comparison of our SVGDF, SVGD and VAT with random initialization using sum of square error (SSE) of ACDC and LA datasets.

Table 1. Performance comparisons with six recent methods on ACDC dataset. All results of existing methods are used from [21] for fair comparisons.

Method	# Scans used		Metrics				Complexity	
	Labeled	Unlabeled	Dice(%)↑	Jaccard(%)↑	95HD(voxel)↓	ASD(voxel)↓	Para.(M)	MACs(G)
U-Net	3(5%)	0	47.83	37.01	31.16	12.62	1.81	2.99
U-Net	7(10%)	0	79.41	68.11	9.35	2.7	1.81	2.99
U-Net	70(All)	0	91.44	84.59	4.3	0.99	1.81	2.99
UA-MT [28]	3 (5%)	67(95%)	46.04	35.97	20.08	7.75	1.81	2.99
SASSNet [8]			57.77	46.14	20.05	6.06	1.81	3.02
DTC [11]			56.9	45.67	23.36	7.39	1.81	3.02
URPC [12]			55.87	44.64	13.6	3.74	1.83	3.02
MC-Net [22]			62.85	52.29	7.62	2.33	2.58	5.39
SS-Net [21]			65.82	55.38	6.67	2.28	1.83	2.99
Cross-ALD (**Ours**)			**80.6**	**69.08**	**5.96**	**1.9**	1.83	2.99
UA-MT [28]	7 (10%)	63(90%)	81.65	70.64	6.88	2.02	1.81	2.99
SASSNet [8]			84.5	74.34	5.42	1.86	1.81	3.02
DTC [11]			84.29	73.92	12.81	4.01	1.81	3.02
URPC [12]			83.1	72.41	4.84	1.53	1.83	3.02
MC-Net [22]			86.44	77.04	5.5	1.84	2.58	5.39
SS-Net [21]			86.78	77.67	6.07	**1.4**	1.83	2.99
Cross-ALD (**Ours**)			**87.52**	**78.62**	**4.81**	1.6	1.83	2.99

Results. Note that the advantage of SVGD over VAT is that the former generates diversified adversarial examples because of the second term in Eq. 6 while VAT only creates one example. Moreover, vanilla SVGD is difficult to capture semantic meaning of high-resolution medical imaging because it calculates kernel k on image-level. In Fig. 1, our SVGDF produces the most diverse particles compared to SVGD and VAT with random initialization.

3.2 Performance Evaluation on the ACDC and la Datasets

Settings. We use the metrics of Dice, Jaccard, 95% Hausdorff Distance (95HD), and Average Surface Distance (ASD) to evaluate the results. We compare our

Table 2. Performance comparisons with six recent methods on LA dataset. All results of existing methods are used from [21] for fair comparisons.

Method	# Scans used		Metrics				Complexity	
	Labeled	Unlabeled	Dice(%)↑	Jaccard(%)↑	95HD(voxel)↓	ASD(voxel)↓	Para.(M)	MACs(G)
V-Net	4(5%)	0	52.55	39.6	47.05	9.87	9.44	47.02
V-Net	8(10%)	0	82.74	71.72	13.35	3.26	9.44	47.02
V-Net	80(All)	0	91.47	84.36	5.48	1.51	9.44	47.02
UA-MT [28]	4 (5%)	76(95%)	82.26	70.98	13.71	3.82	9.44	47.02
SASSNet [8]			81.6	69.63	16.16	3.58	9.44	47.05
DTC [11]			81.25	69.33	14.9	3.99	9.44	47.05
URPC [12]			82.48	71.35	14.65	3.65	5.88	69.43
MC-Net [22]			83.59	72.36	14.07	2.7	12.35	95.15
SS-Net [21]			86.33	76.15	9.97	2.31	9.46	47.17
Cross-ALD (**Ours**)			**88.62**	**79.62**	**7.098**	**1.83**	9.46	47.17
UA-MT [28]	8 (10%)	72(90%)	87.79	78.39	8.68	2.12	9.44	47.02
SASSNet [8]			87.54	78.05	9.84	2.59	9.44	47.05
DTC [11]			87.51	78.17	8.23	2.36	9.44	47.05
URPC [12]			86.92	77.03	11.13	2.28	5.88	69.43
MC-Net [22]			87.62	78.25	10.03	1.82	12.35	95.15
SS-Net [21]			88.55	79.62	**7.49**	1.9	9.46	47.17
Cross-ALD (**Ours**)			**89.92**	**81.78**	7.65	**1.546**	9.46	47.17

Cross-ALD to six recent methods including UA-MT [28] (MICCAI'19), SASSNet [8] (MICCAI'20), DTC [11] (AAAI'21) , URPC [12] (MICCAI'21) , MC-Net [22] (MICCAI'21), and SS-Net [21] (MICCAI'22). The loss weights $\lambda_{Cross-ALD}$ and λ_{cs} are set as an iteration dependent warming-up function [7], and number of particles $N = 2$. All experiments are conducted using the identical settings in the Github repository[6] [21] for fair comparisons.

Results. Recall that our Cross-ALD generates diversified adversarial particles using SVGDF compared to vanilla SVGD and VAT, and further enhances smoothness of cross-adversarial local distributions. In Table 1 and 2, the Cross-ALD can significantly outperform other recent methods with only 5%/10% labeled data training based on the four metrics. Especially, our method impressively gains 14.7% and 2.3% Dice score higher than state-of-the-art SS-Net using 5% labeled data of ACDC and LA, respectively. Moreover, the visualized results of Fig. 2 shows Cross-ALD can segment the most organ details compared to other methods.

3.3 Ablation Study

Settings. We use the same network architectures and parameter settings in Sect. 3.2, and train the models with 5% labeled training data of ACDC and LA. We illustrate that crossing adversarial particles is more beneficial than random

[6] https://github.com/ycwu1997/SS-Net.

Table 3. Ablation study on ACDC and LA datasets.

Dataset	Method	# Scans used		Metrics			
		Labeled	Unlabeled	Dice(%)↑	Jaccard(%)↑	95HD(voxel)↓	ASD(voxel)↓
ACDC	U-Net	4(5%)	0	47.83	37.01	31.16	12.62
	RanMixup	4 (5%)	76(95%)	61.78	51.69	8.16	3.44
	VAT			63.87	53.18	7.61	3.38
	VAT + Mixup			66.23	56.37	7.18	2.53
	SVGD			66.53	58.09	6.41	2.4
	SVGDF			73.15	61.71	6.32	2.12
	SVGDF + ℓ_{cs}			74.89	62.61	6.52	2.01
	Cross-ALD (**Ours**)			**80.6**	**69.08**	**5.96**	**1.9**
LA	V-Net	3(5%)	0	52.55	39.6	47.05	9.87
	RanMixup	3 (5%)	67(95%)	79.82	67.44	16.52	5.19
	VAT			82.27	70.46	13.82	3.48
	VAT + Mixup			83.28	71.77	12.8	2.63
	SVGD			84.62	73.6	11.68	2.94
	SVGDF			86.3	76.17	10.01	2.11
	SVGDF + ℓ_{cs}			86.55	76.51	9.41	2.24
	Cross-ALD (**Ours**)			**87.52**	**78.62**	**4.81**	**1.6**

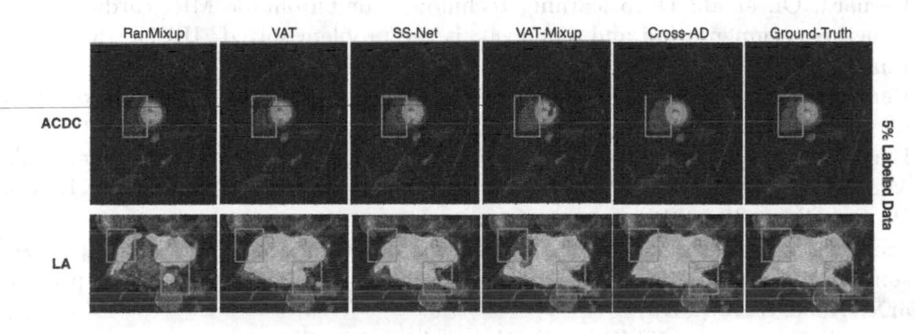

Fig. 2. Visualization results of several semi-supervised segmentation methods with 5% labeled training data and its corresponding ground-truth on ACDC and LA datasets.

mixup between natural inputs (RanMixup [29]) because these particles are near decision boundaries. Recall that our SVGDF is better than VAT and SVGD by producing more diversified adversarial particles. Applying SVGDF's particles and ℓ_{cs} (SVGDF + ℓ_{cs}) to gain the model performance in the semi-supervised segmentation task, while Cross-ALD efficiently enhances smoothness to significantly improve the generalization.

Result. Table 3 shows that mixing adversarial examples from VAT outperform those from RanMixup. While SVGDF + ℓ_{cs} is better than SVGD and VAT, the proposed Cross-ALD achieves the most outstanding performance among comparisons methods. In addition, our method produces more accurate segmentation masks compared to the ground-truth, as shown in Fig. 2.

4 Conclusion

In this paper, we have introduced a novel cross-adversarial local distribution (Cross-ALD) regularization that extends and overcomes drawbacks of VAT and Mixup techniques. In our method, SVGDF is proposed to approximate Cross-ALD, which produces more diverse adversarial particles than vanilla SVGD and VAT with random initialization. We adapt Cross-ALD to semi-supervised medical image segmentation to achieve start-of-the-art performance on the ACDC and LA datasets compared to many recent methods such as VAT [14], UA-MT [28], SASSNet [8], DTC [11], URPC [12] , MC-Net [22], and SS-Net [21].

Acknowledgements. This work was partially supported by the Australian Defence Science and Technology (DST) Group under the Next Generation Technology Fund (NGTF) scheme. Dinh Phung further gratefully acknowledges the partial support from the Australian Research Council, project ARC DP230101176.

References

1. Bernard, O., et al.: Deep learning techniques for automatic MRI cardiac multi-structures segmentation and diagnosis: is the problem solved? IEEE Trans. Med. Imaging **37**(11), 2514–2525 (2018)
2. Berthelot, D., et al.: Remixmatch: Semi-supervised learning with distribution alignment and augmentation anchoring. arXiv preprint arXiv:1911.09785 (2019)
3. Berthelot, D., Carlini, N., Goodfellow, I., Papernot, N., Oliver, A., Raffel, C.A.: Mixmatch: A holistic approach to semi-supervised learning. Adv. Neural Inform. Process. Syst. **32** (2019)
4. French, G., Laine, S., Aila, T., Mackiewicz, M., Finlayson, G.: Semi-supervised semantic segmentation needs strong, varied perturbations. arXiv preprint arXiv:1906.01916 (2019)
5. Gyawali, P., Ghimire, S., Wang, L.: Enhancing mixup-based semi-supervised learningwith explicit lipschitz regularization. In: 2020 IEEE International Conference on Data Mining (ICDM), pp. 1046–1051. IEEE (2020)
6. Lai, X., et al.: Semi-supervised semantic segmentation with directional context-aware consistency. In: Proceedings of the IEEE/CVF Conference on Computer Vision and Pattern Recognition, pp. 1205–1214 (2021)
7. Laine, S., Aila, T.: Temporal ensembling for semi-supervised learning. arXiv preprint arXiv:1610.02242 (2016)
8. Li, S., Zhang, C., He, X.: Shape-aware semi-supervised 3D semantic segmentation for medical images. In: Martel, A.L., et al. (eds.) MICCAI 2020. LNCS, vol. 12261, pp. 552–561. Springer, Cham (2020). https://doi.org/10.1007/978-3-030-59710-8_54
9. Li, X., Yu, L., Chen, H., Fu, C.W., Xing, L., Heng, P.A.: Transformation-consistent self-ensembling model for semisupervised medical image segmentation. IEEE Trans. Neural Netw. Learn. Syst. **32**(2), 523–534 (2020)
10. Liu, Q., Wang, D.: Stein variational gradient descent: A general purpose bayesian inference algorithm. In: Lee, D., Sugiyama, M., Luxburg, U., Guyon, I., Garnett, R. (eds.) Proceedings of NeurIPS. vol. 29 (2016)

11. Luo, X., Chen, J., Song, T., Wang, G.: Semi-supervised medical image segmentation through dual-task consistency. In: Proceedings of the AAAI Conference on Artificial Intelligence. vol. 35, pp. 8801–8809 (2021)
12. Luo, X., et al.: Efficient semi-supervised gross target volume of nasopharyngeal carcinoma segmentation via uncertainty rectified pyramid consistency. In: Bruijne, M., et al. (eds.) MICCAI 2021. LNCS, vol. 12902, pp. 318–329. Springer, Cham (2021). https://doi.org/10.1007/978-3-030-87196-3_30
13. Milletari, F., Navab, N., Ahmadi, S.A.: V-net: Fully convolutional neural networks for volumetric medical image segmentation. In: 2016 Fourth International Conference on 3D Vision (3DV), pp. 565–571. IEEE (2016)
14. Miyato, T., Maeda, S.i., Koyama, M., Ishii, S.: Virtual adversarial training: a regularization method for supervised and semi-supervised learning. IEEE TPAMI **41**(8), 1979–1993 (2018)
15. Nguyen-Duc, T., Le, T., Zhao, H., Cai, J., Phung, D.Q.: Particle-based adversarial local distribution regularization. In: AISTATS, pp. 5212–5224 (2022)
16. Ouali, Y., Hudelot, C., Tami, M.: An overview of deep semi-supervised learning. arXiv preprint arXiv:2006.05278 (2020)
17. Ouali, Y., Hudelot, C., Tami, M.: Semi-supervised semantic segmentation with cross-consistency training. In: Proceedings of the IEEE/CVF Conference on Computer Vision and Pattern Recognition, pp. 12674–12684 (2020)
18. Ronneberger, O., Fischer, P., Brox, T.: U-net: Convolutional networks for biomedical image segmentation. In: Medical Image Computing and Computer-Assisted Intervention–MICCAI 2015: 18th International Conference, Munich, Germany, October 5-9, 2015, Proceedings, Part III 18, pp. 234–241. Springer (2015)
19. Sohn, K., et al.: Fixmatch: simplifying semi-supervised learning with consistency and confidence. Adv. Neural. Inf. Process. Syst. **33**, 596–608 (2020)
20. Tashiro, Y., Song, Y., Ermon, S.: Diversity can be transferred: Output diversification for white-and black-box attacks. Proc. NeurIPS **33**, 4536–4548 (2020)
21. Wu, Y., Wu, Z., Wu, Q., Ge, Z., Cai, J.: Exploring smoothness and class-separation for semi-supervised medical image segmentation. In: International Conference on Medical Image Computing and Computer-Assisted Intervention, vol. 13435, pp. 34–43. Springer, Cham (2022). https://doi.org/10.1007/978-3-031-16443-9_4
22. Wu, Y., Xu, M., Ge, Z., Cai, J., Zhang, L.: Semi-supervised left atrium segmentation with mutual consistency training. In: de Bruijne, M., et al. (eds.) MICCAI 2021. LNCS, vol. 12902, pp. 297–306. Springer, Cham (2021). https://doi.org/10.1007/978-3-030-87196-3_28
23. Wu, Z., Shi, X., Lin, G., Cai, J.: Learning meta-class memory for few-shot semantic segmentation. In: Proceedings of the IEEE/CVF International Conference on Computer Vision, pp. 517–526 (2021)
24. Xia, Y., et al.: 3D semi-supervised learning with uncertainty-aware multi-view co-training. In: Proceedings of the IEEE/CVF Winter Conference on Applications of Computer Vision, pp. 3646–3655 (2020)
25. Xie, Y., Zhang, J., Liao, Z., Verjans, J., Shen, C., Xia, Y.: Intra-and inter-pair consistency for semi-supervised gland segmentation. IEEE Trans. Image Process. **31**, 894–905 (2021)
26. Xiong, Z., et al.: A global benchmark of algorithms for segmenting the left atrium from late gadolinium-enhanced cardiac magnetic resonance imaging. Med. Image Anal. **67**, 101832 (2021)
27. Yang, X., Song, Z., King, I., Xu, Z.: A survey on deep semi-supervised learning. IEEE Transactions on Knowledge and Data Engineering (2022)

28. Yu, L., Wang, S., Li, X., Fu, C.-W., Heng, P.-A.: Uncertainty-aware self-ensembling model for semi-supervised 3D left atrium segmentation. In: Shen, D., et al. (eds.) MICCAI 2019. LNCS, vol. 11765, pp. 605–613. Springer, Cham (2019). https://doi.org/10.1007/978-3-030-32245-8_67
29. Zhang, H., Cisse, M., Dauphin, Y.N., Lopez-Paz, D.: mixup: Beyond empirical risk minimization. arXiv preprint arXiv:1710.09412 (2017)

AMAE: Adaptation of Pre-trained Masked Autoencoder for Dual-Distribution Anomaly Detection in Chest X-Rays

Behzad Bozorgtabar[1,3](✉), Dwarikanath Mahapatra[2], and Jean-Philippe Thiran[1,3]

[1] École Polytechnique Fédérale de Lausanne (EPFL), Lausanne, Switzerland
{behzad.bozorgtabar,jean-philippe.thiran}@epfl.ch
[2] Inception Institute of AI (IIAI), Abu Dhabi, United Arab Emirates
dwarikanath.mahapatra@inceptioniai.org
[3] Lausanne University Hospital (CHUV), Lausanne, Switzerland

Abstract. Unsupervised anomaly detection in medical images such as chest radiographs is stepping into the spotlight as it mitigates the scarcity of the labor-intensive and costly expert annotation of anomaly data. However, nearly all existing methods are formulated as a one-class classification trained only on representations from the normal class and discard a potentially significant portion of the unlabeled data. This paper focuses on a more practical setting, dual distribution anomaly detection for chest X-rays, using the entire training data, including both normal and unlabeled images. Inspired by a modern self-supervised vision transformer model trained using partial image inputs to reconstruct missing image regions- we propose AMAE, a two-stage algorithm for adaptation of the pre-trained masked autoencoder (MAE). Starting from MAE initialization, AMAE first creates synthetic anomalies from only normal training images and trains a lightweight classifier on frozen transformer features. Subsequently, we propose an adaptation strategy to leverage unlabeled images containing anomalies. The adaptation scheme is accomplished by assigning pseudo-labels to unlabeled images and using two separate MAE based modules to model the normative and anomalous distributions of pseudo-labeled images. The effectiveness of the proposed adaptation strategy is evaluated with different anomaly ratios in an unlabeled training set. AMAE leads to consistent performance gains over competing self-supervised and dual distribution anomaly detection methods, setting the new state-of-the-art on three public chest X-ray benchmarks - RSNA, NIH-CXR, and VinDr-CXR.

Keywords: Anomaly detection · Chest X-ray · Masked autoencoder

---

Supplementary Information The online version contains supplementary material available at https://doi.org/10.1007/978-3-031-43907-0_19.

H. Greenspan et al. (Eds.): MICCAI 2023, LNCS 14220, pp. 195–205, 2023.
https://doi.org/10.1007/978-3-031-43907-0_19

1 Introduction

To reduce radiologists' reading burden and make the diagnostic process more manageable, especially when the number of experts is scanty, computer-aided diagnosis (CAD) systems, particularly deep learning-based anomaly detection [1,2,22], have witnessed the flourish due to their capability to detect rare anomalies for different imaging modalities including chest X-ray (CXR). Nonetheless, unsupervised anomaly detection methods [20,26] are strongly preferred due to the difficulties of highly class-imbalanced learning and the tedious annotation of anomaly data for developing such systems. Most current anomaly detection methods are formulated as a one-class classification (OCC) problem [18], where the goal is to model the distribution of normal images used for training and thus detect abnormal cases that deviate from normal class at test time. On this basis, image reconstruction based, e.g., autoencoder [9] or generative models [20], self-supervised learning (SSL) based, e.g., contrastive learning [26], and embedding-similarity-based methods [7] have been proposed for anomaly detection. Some recent self-supervised methods proposed synthetic anomalies using cut-and-paste data augmentation [12,19] to approximate real sub-image anomalies. Nonetheless, their performances lag due to the lack of real anomaly data. More importantly, these methods have often ignored readily available unlabeled images. More recently, similar to our method, DDAD [3] leverages readily available unlabeled images for anomaly detection, but it requires training an ensemble of several reconstruction-based networks. Self-supervised model adaptation on *unlabeled data* has been widely investigated using convolutional neural networks (CNNs) in many vision tasks via self-training [17], contrastive learning [22,26], and anatomical visual words [10]. Nonetheless, the adaptation of vision transformer (ViT) [8] architectures largely remains unexplored, particularly for anomaly detection. Recently, masked autoencoder (MAE) [11] based models demonstrated great scalability and substantially improved several self-supervised learning benchmarks [27].

In this paper, inspired by the success of the MAE approach, we propose a two-stage algorithm for "**A**daptation of pre-trained **M**asked **A**uto**E**ncoder" (AMAE) to leverage simultaneously normal and unlabeled images for anomaly detection in chest X-rays. As for **Stage 1** of our method, (i) AMAE creates synthetic anomalies from only normal training images, and the usefulness of pre-trained MAE [11] is evaluated by training a lightweight classifier using a proxy task to detect synthetic anomalies. (ii) For the **Stage 2**, AMAE customizes the recipe of MAE adaptation based on an unlabeled training set. In particular, we propose an adaptation strategy based on reconstructing the masked-out input images. The rationale behind the proposed adaptation strategy is to assign pseudo-labels to unlabeled images and train two separate modules to measure the distribution discrepancy between normal and pseudo-labeled abnormal images. (iii) We conduct extensive experiments across three chest X-ray datasets and verify the effectiveness of our adaptation strategy in apprehending anomalous features from unlabeled images. In addition, we evaluate the model with different

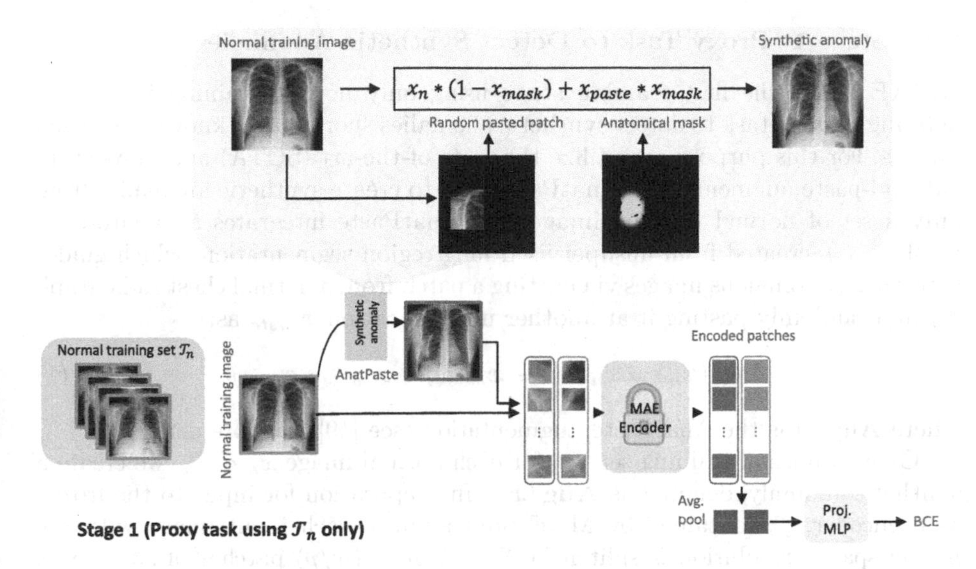

Fig. 1. Schematic overview of AMAE training (Stage 1). Top. Illustration of Anatpaste augmentation [19] generated from normal training images. **Bottom**. Starting from MAE initialization, only the MLP-based projection head (`Proj.`) is trained to classify synthetic anomalies.

anomaly ratios (ARs) in an unlabeled training set and show consistent performance improvement with increasing AR.

2 Method

Notation. We first formally define the problem setting for the proposed dual-distribution anomaly detection. Contrary to previous unsupervised anomaly detection methods, AMAE fully uses unlabeled images, yielding a training data $\mathcal{T}_{train} = \mathcal{T}_n \cup \mathcal{T}_u$ consisting of both normal $\mathcal{T}_n$ and unlabeled $\mathcal{T}_u$ training sets. We denote the normal training set as $\mathcal{T}_n = \{\boldsymbol{x}_{ni}\}_{i=1}^{N}$, with N normal images, and the unlabeled training set as $\mathcal{T}_u = \{\boldsymbol{x}_{ui}\}_{i=1}^{M}$, with M unlabeled images to be composed of both normal and abnormal images. At test time, given a test set $\mathcal{T}_{test} = \{(\boldsymbol{x}_{ti}, y_i)\}_{i=1}^{S}$ with S normal or abnormal images, where $y_i \in \{0, 1\}$ is the corresponding label to $\boldsymbol{x}_{ti}$ (0 for normal (negative) and 1 for abnormal (positive) image), the trained anomaly detection model should identify whether the test image is abnormal or not.

Architecture. Our architecture is ⧔-shaped: the ViT-small (ViT-S/16) [8] encoder f followed by a ViT head g and lightweight (3-layer) multilayer perception (MLP) based projection head h, simultaneously. Starting from the pre-trained MAE on 0.3M unlabeled chest X-rays and officially released checkpoints, we use exactly the same ViT encoder f and decoder g as MAE [27].

2.1 Stage 1- Proxy Task to Detect Synthetic Anomalies

AMAE starts the first training stage using only normal training images by defining a proxy task to detect synthetic anomalies shorn of real known abnormal images. For this purpose, we utilize the state-of-the-art (SOTA) anatomy-aware cut-and-paste augmentation, AnatPaste [19], to create synthetic anomalies from only a set of normal training images $\mathcal{T}_n$. AnatPaste integrates an anatomical mask $\boldsymbol{x}_{mask}$ created from unsupervised lung region segmentation, which guides generating anomalous images via cutting a patch from a normal chest radiograph $\boldsymbol{x}_n$ and randomly pasting it at another image location $\boldsymbol{x}_{paste}$ as:

$$\mathbf{Aug}\left(\boldsymbol{x}_n\right) = \boldsymbol{x}_n * \left(1 - \boldsymbol{x}_{mask}\right) + \boldsymbol{x}_{paste} * \boldsymbol{x}_{mask} \quad (1)$$

where $\mathbf{Aug}\left(\cdot\right)$ is the AnatPaste augmentation (see [19] for more details).

Given a normal training set $\mathcal{T}_n$, for each normal image $\boldsymbol{x}_n \sim \mathcal{T}_n$, we create a synthetic anomaly, denoted as $\mathbf{Aug}\left(\boldsymbol{x}_i\right)$. In preparation for input to the frozen ViT encoder, f_0 (obtained by MAE pre-training), each input image with the $h \times w$ spatial resolution is split into $T = (h/p) \times (w/p)$ patches of size $p \times p$. Then, for every input patch, a token is created by linear projection with an added positional embedding. The sequence of tokens is then fed to the frozen ViT encoder f_0 consisting stack of transformer blocks, yielding the embeddings of tokens $\boldsymbol{z}_i^1, \boldsymbol{z}_i^2 \in \mathbb{R}^{T \times d}$ corresponding to i^{th} normal and synthetic anomaly images. The returned embeddings $\boldsymbol{z}_i^1, \boldsymbol{z}_i^2 \in \mathbb{R}^{T \times d}$ are pooled via *average pooling* to form d-dimensional embeddings, which are fed to an MLP anomaly classifier projection head h (see Fig. 1 for schematic overview). Subsequently, we only train an anomaly classifier projection head h on top of the frozen embeddings to detect synthetic anomalies using the cross-entropy loss l_{ce} as follows:

$$h_0 = \arg\min_{h} \mathbb{E}_{\boldsymbol{x}_n \sim \mathcal{T}_n}\left[l_{ce}\left(h \circ f_0\left(\boldsymbol{x}_n\right), 0\right) + l_{ce}\left(h \circ f_0\left(\mathbf{Aug}\left(\boldsymbol{x}_n\right)\right), 1\right)\right] \quad (2)$$

We set the label for the normal image to 0 and 1 otherwise (synthetic anomaly). The above gradient-based optimization produces a trained classifier projection head h_0. Thus, the whole architecture can be trained with much fewer parameters while making only the classifier projection head specialized at recognizing anomalies without influencing the ViT encoder.

2.2 Stage 2- MAE Inter-Discrepancy Adaptation

The proposed MAE adaptation scheme is inspired by [3] to model the dual distribution of training data. Unlike [3], which treats all unlabeled images similarly, we propose assigning pseudo-labels to unlabeled images and formulating anomaly detection by measuring the distribution discrepancy between normal and pseudo-labeled abnormal images from unlabeled sets. We use a pre-trained anomaly classifier (**Stage 1**) to assign pseudo labels to unlabeled images. To begin with, for each unlabeled image $\boldsymbol{x}_u \sim \mathcal{T}_u$, we consider the anomaly detection model's confidence from **Stage 1** $\left(h_0 \circ f_0\left(\boldsymbol{x}_u\right)\right)$. Those images on which the model is highly confident (normal or abnormal) are treated as reliable images

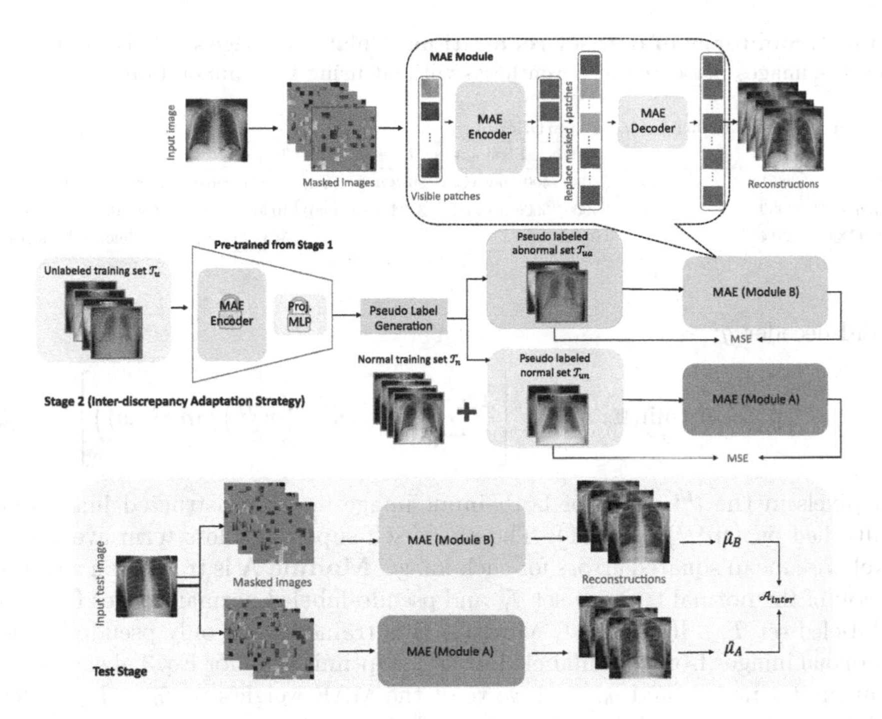

Fig. 2. Schematic overview of AMAE **training (Stage 2) and test stage. Top.** Our adaptation strategy first assigns pseudo labels to unlabeled images using a pre-trained classifier from **Stage 1** and uses two separate MAE modules to model the normative and anomalous distributions of pseudo-labeled images. **Bottom.** The discrepancy between the average of multiple reconstructions from two modules is used at test time to compute the anomaly score.

used for adaptation. For this purpose, we collect all output probabilities and opt for a threshold per class t_c corresponding to each class's top K-th percentile ($K = 50$) of all given confidence values. Those unlabeled images deemed reliable yield two subsets: a subset of pseudo-labeled normal images $\mathcal{T}_{un}$ and a subset of pseudo-labeled abnormal images $\mathcal{T}_{ua}$. We then utilize two MAE-based modules, **Module A** and **Module B** (see Fig. 2, **Top**), using the same MAE architecture and pixel-wise mean squared error (MSE) optimization in [11]. Within each module, the input patches for each image are masked out using a set of L random masks $\left\{\boldsymbol{m}^{(j)} \in \{0,1\}^T\right\}_{j=1}^{L}$ in which a different small subset of the patches (ratio of 25%) is retained each time to be fed to the ViT encoder f. The lightweight ViT decoder g receives unmasked patches' embeddings and adds learnable masked tokens to replace the masked-out patches. Subsequently, the full set of embeddings of visible patches and masked tokens with added positional embeddings to all tokens is processed by the ViT decoder g to reconstruct the missing patches of each image in pixels. This yields the reconstructed image $\hat{\boldsymbol{x}}^{(j)} = g \circ f\left(\boldsymbol{m}^{(j)}(\boldsymbol{x})\right)$, which is then compared against the input image $\boldsymbol{x}$ to optimize both ViT encoder

Table 1. Summary of dataset repartitions. Unlabeled image set $\mathcal{T}_u$ is constructed from the images presented in parentheses without using their annotations.

Dataset	Normal training set $\mathcal{T}_n$	Repartition	
		Unlabeled training set $\mathcal{T}_u$	Test set $\mathcal{T}_{test}$
RSNA	3851	4000 (4000 normal + 5012 abnormal)	1000 normal + 1000 abnormal images
VinDr-CXR	4000	4000 (5606 normal + 3394 abnormal)	1000 normal + 1000 abnormal images
NIH-CXR	3614	0	543 normal + 262 abnormal images

f and decoder g:

$$f^*, g^* = \arg\min_{f,g} \mathbb{E}_{\boldsymbol{x}\sim\mathcal{T}_{train}} \left[\frac{1}{L} \sum_{j=1}^{L} l_{mse} \left(\boldsymbol{m}^{(j)} \left(\hat{\boldsymbol{x}}^{(j)} \right), \boldsymbol{m}^{(j)} (\boldsymbol{x}) \right) \right] \tag{3}$$

All pixels in the tth patch of both input image and reconstructed images are multiplied by $\left(\boldsymbol{m}^{(j)}\right)_t \in \{0,1\}$. The above self-supervised loss term averages L pixel-wise mean squared errors for each image. **Module A** is trained on a combination of the normal training set $\mathcal{T}_n$ and pseudo-labeled normal images from the unlabeled set $\mathcal{T}_{un}$. In contrast, **Module B** is trained using only pseudo-labeled abnormal images from an unlabeled set $\mathcal{T}_{ua}$. Optimization for Eq. 3 always starts from pre-trained f_0 and g_0, and we reset the MAE weights to f_0 and g_0 before training each module. A high discrepancy between the reconstruction outputs of the two modules can indicate potential abnormal regions. Similar to the training stage, we apply L random masks to the test image $\boldsymbol{x}_t \sim \mathcal{T}_{test}$ to obtain L reconstructions (see Fig. 2, **Bottom**). Thus, the anomaly score based on the inter-discrepancy of the two MAE modules is computed as follows:

$$\mathcal{A}^p_{inter} = |\hat{\mu}^p_A - \hat{\mu}^p_B| \tag{4}$$

where p is the index of pixels, $\hat{\mu}_A$ and $\hat{\mu}_B$ are the *mean* maps of L reconstructed images from **Module A** and **Module B**, respectively. The pixel-level anomaly scores for each image are averaged, yielding the image-level anomaly score.

3 Experiments

Datasets. We evaluated our method on three public CXR datasets: 1) the RSNA Pneumonia Detection Challenge dataset[1], 2) the VinBigData Chest X-ray Abnormalities Detection Challenge dataset (VinDrCXR)[2] [15], and a subset of 3) the curated NIH dataset (NIH-CXR)[3] [21,25], by including only posteroanterior view images of both male and female patients aged over 18. We show a summary of each dataset's repartitions in Table 1. Except for NIH, where we use only the normal set $\mathcal{T}_n$ (OCC setting), for the other two datasets, we utilize both $\mathcal{T}_n$ and $\mathcal{T}_u$ and exact repartition files from [3] for model training.

[1] https://www.kaggle.com/c/rsna-pneumonia-detection-challenge.
[2] https://www.kaggle.com/c/vinbigdata-chest-xray-abnormalities-detection.
[3] https://nihcc.app.box.com/v/ChestXray-NIHCC/file/371647823217.

Table 2. Performance comparison with SOTA methods (Avg. over four replicate). The two best results for each protocol are highlighted in **bold** and underlined. Note that "IN" refers to "ImageNet-Pretrained," and "e2e" refers to end-to-end training. The experimental results of competing methods† are taken from [4].

Method	Protocol	Taxonomy	CXR Datasest & Metrics					
			RSNA		VinDr-CXR		NIH-CXR	
			AUC %	AP %	AUC %	AP %	AUC %	AP %
AE†	N	Rec.	66.9	66.1	55.9	60.3	70.0	65.4
AE-U† [14]	N	Rec.	86.7	84.7	73.8	72.8	91.0	83.2
f-AnoGAN† [20]	N	Rec.	79.8	75.6	**76.3**	**74.8**	84.1	79.2
IGD† [6]	N	Rec.	81.2	78.0	59.2	58.7	85.2	80.3
DRAEM† [28]	N	Rec.+SSL	62.3	61.6	63.0	68.3	65.5	62.4
CutPaste^{e2e†} [21]	N	SSL	55.0	58.0	54.6	55.5	58.0	-
AnatPaste [19]	N	SSL	83.1	83.7	66.0	66.2	94.0	93.5
FPI† [23]	N	SSL	47.6	55.7	48.2	49.9	70.5	-
PII† [24]	N	SSL	82.9	83.6	65.9	65.8	92.2	-
NSA† [21]	N	SSL	82.2	82.6	64.4	65.8	94.1	-
DDAD-AE [3]	N	Rec.	69.4	-	60.1	-	73.0	71.5
AMAE^{IN} - Stage 1	N	SSL	83.3	83.8	65.9	66.0	94.1	93.7
AMAE - Stage 1	N	SSL	**86.8**	**84.9**	74.2	72.9	**95.0**	**94.9**
CutPaste^{e2e†} [21]	Y	SSL	59.8	61.7	59.2	60.0	-	
AnatPaste [19]	Y	SSL	84.4	85.5	67.1	67.5	-	
FPI† [23]	Y	SSL	46.6	53.8	47.4	49.4	-	
PII† [24]	Y	SSL	84.3	85.4	66.8	67.2	-	
NSA† [21]	Y	SSL	84.2	84.3	64.4	64.8	-	
DDAD-AE [3] ($\mathcal{A}_{iner}$)	Y	Rec.	81.5	81.0	71.0	-	-	
AMAE - Stage 2 ($\mathcal{A}_{inter}$)	Y	Rec.	**91.4**	**91.7**	**86.1**	**84.5**	-	

Implementation Details. We adopt AdamW [13] optimizer and set the learning rate (lr) and batch size to 2.5e-4 and 16, where we linearly warm up the lr for the first 20 epochs and decay it following a cosine schedule thereafter till 150 epochs. We follow the exact recipe as [27] for other hyperparameters (see **Supplementary Material**). The number of generated masks L per image is set to 2 and 4 for the adaptation and test stages (see ablation in **Supplementary Material**). We use PyTorch 1.9 [16] and train each model on a single GeForce RTX 2080 Ti GPU. We use the area under the ROC curve (AUC) and average precision (AP) for the evaluation metrics.

Comparison with SOTA Methods. Table 2 compares AMAE with a comprehensive collection of SOTA methods, including self-supervised synthetic anomaly and reconstruction (Rec.) based methods using their official codes and under two experimental protocols. We use the **Y** protocol to indicate if access to

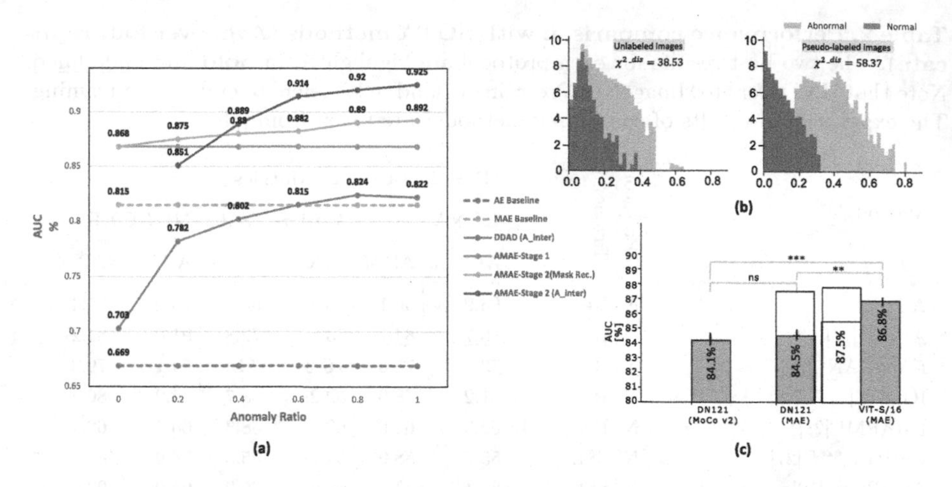

Fig. 3. Ablations on the RSNA dataset. (a) Performance comparison with a varying AR of $\mathcal{T}_u$. **(b)** The $\chi^{2-\text{distance}}$ of AS histograms *with* and *without* pseudo labeling. **(c)** Ablation for different backbones and pre-training schemes. Test AUC performances are presented as $mean \pm 1.96std$ averaged over four replicates. Samples are statistically tested for H_0: *means are similar*, using a bilateral Welch t-test. *ns* non-significant, ** *pvalue* $\in [0.05, 0.01]$, *** *pvalue* $\in [0.01, 0.001]$.

the unlabeled images is possible in which an AR of 60% of $\mathcal{T}_u$ is assumed in the experiments; otherwise, we use **N**. Under protocol **N** (OCC setting), except for VinDr-CXR, AMAE-**Stage 1** achieves SOTA results on two CXR benchmarks, demonstrating the effectiveness of pre-trained ViT using MAE and synthetic anomalies. In particular, AMAE-**Stage 1** surpasses the best-performing synthetic anomaly-based method, AnatPaste [19], with the same synthesis approach as ours but using ResNet18 as a feature extractor. Furthermore, outperforming MAE pre-training on ImageNet (e.g., improved AUC from 65.9% to 74.2% on the VinDr-CXR) indicates the importance of in-domain adaptation. Under the **Y** protocol, our adaptation strategy, AMAE- **Stage 2** ($\mathcal{A}_{inter}$), outperforms the current SSL methods by a larger margin, underlining the importance of modeling the dual distribution and leveraging unlabeled images more effectively.

Ablation Studies. To understand the effectiveness of AMAE in capturing abnormal features from unlabeled images, we conduct ablation experiments on the RSNA with the AR of $\mathcal{T}_u$ varying from 0 to 100% (Fig. 3 (**a**)). Concerning reconstruction-based methods, the baseline aggregating multiple reconstructions via MAE achieves consistent improvement compared with the AE baseline (+14.6% AUC), implying better capturing of fine-grained texture information. With an increasing AR of $\mathcal{T}_u$, our MAE adaptation strategy (AMAE - **Stage 2** ($\mathcal{A}_{inter}$)) performs favorably against the SOTA method (DDAD [3]), and AMAE without adaptation (**Stage 1**). Furthermore, we consider an additional baseline for model adaptation based on patch reconstruction in [11] on pooled normal and unlabeled images, denoted as (AMAE - **Stage 2** (Mask Rec.)). AMAE - **Stage 2** ($\mathcal{A}_{inter}$) rises with a steeper slope than AMAE - **Stage 2** (Mask Rec.), e.g., improved AUC from 89% to 92% on AR=80%, suggesting high-quality pseudo-labeled images. We also analyze the discriminative capability of our adaptation *with* and *without* pseudo labeling by levering all unlabeled images in **Module B**. We utilize the $\chi^{2-\text{distance}}$ between the histograms of anomaly scores (AS) of normal and abnormal images in the RSNA test set (Fig. 3 (**b**)), showing a more substantial discriminative capability of incorporating pseudo labeling (improved $\chi^{2-\text{distance}}$ from 38.53 to 58.37). Finally, the ViT encoder obtained by MAE pre-training (**Stage 1**) surpasses DenseNet-121 (DN121), either pre-trained by MAE [27] or an advanced contrastive learning method (MoCo v2 [5]) on the RSNA dataset (Fig. 3 (**c**)).

4 Conclusion

We present AMAE, an adaptation strategy of the pre-trained MAE for dual distribution anomaly detection in CXRs, which makes our method capable of more effectively apprehending anomalous features from unlabeled images. Experiments on the three CXR benchmarks demonstrate that AMAE is generalizable to different model architectures, achieving SOTA performance. As for the limitation, an adequate number of normal training images is still required, and we will extend our pseudo-labeling scheme in our future work for robust anomaly detection bypassing any training annotations.

References

1. Bozorgtabar, B., Mahapatra, D.: Attention-conditioned augmentations for self-supervised anomaly detection and localization. In: Proceedings of the AAAI Conference on Artificial Intelligence. vol. 37, pp. 14720–14728 (2023)
2. Bozorgtabar, B., Mahapatra, D., Vray, G., Thiran, J.-P.: SALAD: self-supervised aggregation learning for anomaly detection on X-Rays. In: Martel, A.L., et al. (eds.) MICCAI 2020. LNCS, vol. 12261, pp. 468–478. Springer, Cham (2020). https://doi.org/10.1007/978-3-030-59710-8_46
3. Cai, Y., Chen, H., Yang, X., Zhou, Y., Cheng, K.T.: Dual-distribution discrepancy for anomaly detection in chest x-rays. In: Medical Image Computing and Computer Assisted Intervention-MICCAI 2022: 25th International Conference, Singapore, September 18–22, 2022, Proceedings, Part III. pp. 584–593. Springer (2022). https://doi.org/10.1007/978-3-031-16437-8_56
4. Cai, Y., Chen, H., Yang, X., Zhou, Y., Cheng, K.T.: Dual-distribution discrepancy with self-supervised refinement for anomaly detection in medical images. Med. Image Anal. **86**, 102794 (2023)
5. Chen, X., Fan, H., Girshick, R., He, K.: Improved baselines with momentum contrastive learning. arXiv preprint arXiv:2003.04297 (2020)
6. Chen, Y., Tian, Y., Pang, G., Carneiro, G.: Deep one-class classification via interpolated gaussian descriptor. In: Proceedings of the AAAI Conference on Artificial Intelligence. vol. 36, pp. 383–392 (2022)

7. Defard, T., Setkov, A., Loesch, A., Audigier, R.: PaDiM: a patch distribution modeling framework for anomaly detection and localization. In: Del Bimbo, A., et al. (eds.) ICPR 2021. LNCS, vol. 12664, pp. 475–489. Springer, Cham (2021). https://doi.org/10.1007/978-3-030-68799-1_35
8. Dosovitskiy, A., et al.: An image is worth 16x16 words: transformers for image recognition at scale. arXiv preprint arXiv:2010.11929 (2020)
9. Gong, D., et al.: Memorizing normality to detect anomaly: Memory-augmented deep autoencoder for unsupervised anomaly detection. In: Proceedings of the IEEE/CVF International Conference on Computer Vision, pp. 1705–1714 (2019)
10. Haghighi, F., Taher, M.R.H., Zhou, Z., Gotway, M.B., Liang, J.: Transferable visual words: exploiting the semantics of anatomical patterns for self-supervised learning. IEEE Trans. Med. Imaging **40**(10), 2857–2868 (2021)
11. He, K., Chen, X., Xie, S., Li, Y., Dollár, P., Girshick, R.: Masked autoencoders are scalable vision learners. In: Proceedings of the IEEE/CVF Conference on Computer Vision and Pattern Recognition, pp. 16000–16009 (2022)
12. Li, C.L., Sohn, K., Yoon, J., Pfister, T.: Cutpaste: Self-supervised learning for anomaly detection and localization. In: Proceedings of the IEEE/CVF Conference on Computer Vision and Pattern Recognition, pp. 9664–9674 (2021)
13. Loshchilov, I., Hutter, F.: Decoupled weight decay regularization. In: International Conference on Learning Representations (2018)
14. Mao, Y., Xue, F.-F., Wang, R., Zhang, J., Zheng, W.-S., Liu, H.: Abnormality detection in chest X-ray images using uncertainty prediction autoencoders. In: Martel, A.L., et al. (eds.) MICCAI 2020. LNCS, vol. 12266, pp. 529–538. Springer, Cham (2020). https://doi.org/10.1007/978-3-030-59725-2_51
15. Nguyen, H.Q., et al.: VinDr-CXR: an open dataset of chest x-rays with radiologist's annotations. Scientific Data **9**(1), 429 (2022)
16. Paszke, A., et al.: Pytorch: an imperative style, high-performance deep learning library. Adv. Neural. Inf. Process. Syst. **32**, 8026–8037 (2019)
17. Prabhu, V., Khare, S., Kartik, D., Hoffman, J.: Sentry: Selective entropy optimization via committee consistency for unsupervised domain adaptation. In: Proceedings of the IEEE/CVF International Conference on Computer Vision, pp. 8558–8567 (2021)
18. Ruff, L., Vandermeulen, R., Goernitz, N., Deecke, L., Siddiqui, S.A., Binder, A., Müller, E., Kloft, M.: Deep one-class classification. In: International Conference on Machine Learning, pp. 4393–4402. PMLR (2018)
19. Sato, J., et al.: Anatomy-aware self-supervised learning for anomaly detection in chest radiographs. arXiv preprint arXiv:2205.04282 (2022)
20. Schlegl, T., Seeböck, P., Waldstein, S.M., Langs, G., Schmidt-Erfurth, U.: f-AnoGan: fast unsupervised anomaly detection with generative adversarial networks. Med. Image Anal. **54**, 30–44 (2019)
21. Schlüter, H.M., Tan, J., Hou, B., Kainz, B.: Natural synthetic anomalies for self-supervised anomaly detection and localization. In: Computer Vision-ECCV 2022: 17th European Conference, Tel Aviv, Israel, October 23–27, 2022, Proceedings, Part XXXI. pp. 474–489. Springer (2022). https://doi.org/10.1007/978-3-031-19821-2_27
22. Spahr, A., Bozorgtabar, B., Thiran, J.P.: Self-taught semi-supervised anomaly detection on upper limb x-rays. In: 2021 IEEE 18th International Symposium on Biomedical Imaging (ISBI), pp. 1632–1636. IEEE (2021)
23. Tan, J., Hou, B., Batten, J., Qiu, H., Kainz, B.: Detecting outliers with foreign patch interpolation. arXiv preprint arXiv:2011.04197 (2020)

24. Tan, J., Hou, B., Day, T., Simpson, J., Rueckert, D., Kainz, B.: Detecting outliers with poisson image interpolation. In: de Bruijne, M., et al. (eds.) MICCAI 2021. LNCS, vol. 12905, pp. 581–591. Springer, Cham (2021). https://doi.org/10.1007/978-3-030-87240-3_56
25. Tang, Y.X., et al.: Automated abnormality classification of chest radiographs using deep convolutional neural networks. NPJ Digital Med. **3**(1), 70 (2020)
26. Tain, Yu., et al.: Constrained contrastive distribution learning for unsupervised anomaly detection and localisation in medical images. In: de Bruijne, M., et al. (eds.) MICCAI 2021. LNCS, vol. 12905, pp. 128–140. Springer, Cham (2021). https://doi.org/10.1007/978-3-030-87240-3_13
27. Xiao, J., Bai, Y., Yuille, A., Zhou, Z.: Delving into masked autoencoders for multi-label thorax disease classification. In: Proceedings of the IEEE/CVF Winter Conference on Applications of Computer Vision, pp. 3588–3600 (2023)
28. Zavrtanik, V., Kristan, M., Skočaj, D.: Draem-a discriminatively trained reconstruction embedding for surface anomaly detection. In: Proceedings of the IEEE/CVF International Conference on Computer Vision, pp. 8330–8339 (2021)

Gall Bladder Cancer Detection from US Images with only Image Level Labels

Soumen Basu[1(✉)], Ashish Papanai[1], Mayank Gupta[1], Pankaj Gupta[1,2], and Chetan Arora[1]

[1] Indian Institute of Technology, Delhi, India
soumen.basu@cse.iitd.ac.in

[2] Postgraduate Institute of Medical Education and Research, Chandigarh, India

Abstract. Automated detection of Gallbladder Cancer (GBC) from Ultrasound (US) images is an important problem, which has drawn increased interest from researchers. However, most of these works use difficult-to-acquire information such as bounding box annotations or additional US videos. In this paper, we focus on GBC detection using only image-level labels. Such annotation is usually available based on the diagnostic report of a patient, and do not require additional annotation effort from the physicians. However, our analysis reveals that it is difficult to train a standard image classification model for GBC detection. This is due to the low inter-class variance (a malignant region usually occupies only a small portion of a US image), high intra-class variance (due to the US sensor capturing a 2D slice of a 3D object leading to large viewpoint variations), and low training data availability. We posit that even when we have only the image level label, still formulating the problem as object detection (with bounding box output) helps a deep neural network (DNN) model focus on the relevant region of interest. Since no bounding box annotations is available for training, we pose the problem as weakly supervised object detection (WSOD). Motivated by the recent success of transformer models in object detection, we train one such model, DETR, using multi-instance-learning (MIL) with self-supervised instance selection to suit the WSOD task. Our proposed method demonstrates an improvement of AP and detection sensitivity over the SOTA transformer-based and CNN-based WSOD methods. Project page is at https://gbc-iitd.github.io/wsod-gbc.

Keywords: Weakly Supervised Object Detection · Ultrasound · Gallbladder Cancer

1 Introduction

GBC is a deadly disease that is difficult to detect at an early stage [12,15]. Early diagnosis can significantly improve the survival rate [14]. Non-ionizing radiation,

---

Supplementary Information The online version contains supplementary material available at https://doi.org/10.1007/978-3-031-43907-0_20.

H. Greenspan et al. (Eds.): MICCAI 2023, LNCS 14220, pp. 206–215, 2023.
https://doi.org/10.1007/978-3-031-43907-0_20

low cost, and accessibility make US a popular non-invasive diagnostic modality for patients with suspected gall bladder (GB) afflictions. However, identifying signs of GBC from routine US imaging is challenging for radiologists [11]. In recent years, automated GBC detection from US images has drawn increased interest [3,5] due to its potential for improving diagnosis and treatment outcomes. Many of these works formulate the problem as an object detection, since training a image classification model for GBC detection seems challenging due to the reasons outlined in the abstract (also see Fig. 1).

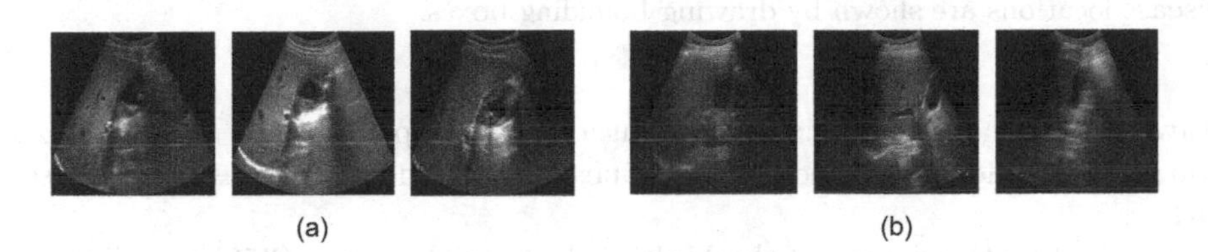

Fig. 1. (a) Low inter-class variability. The first two GBs show benign wall thickening, and the third one shows malignant thickening. However, the appearance of the GB in all three images is very similar. (b) High intra-class variability. All three images have been scanned from the same patient, but due to the sensor's scanning plane, the appearances change drastically.

Recently, GBCNet [3], a CNN-based model, achieved SOTA performance on classifying malignant GB from US images. GBCNet uses a two-stage pipeline consisting of object detection followed by classification, and requires bounding box annotations for GB as well as malignant regions for training. Such bounding box annotations surrounding the pathological regions are time-consuming and require an expert radiologist for annotation. This makes it expensive and non-viable for curating large datasets for training large DNN models. In another recent work, [5] has exploited additional unlabeled video data for learning good representations for downstream GBC classification and obtained performance similar to [3] using a ResNet50 [13] classifier. The reliance of both SOTA techniques on additional annotations or data, limits their applicability. On the other hand, the image-level malignancy label is usually available at a low cost, as it can be obtained readily from the diagnostic report of a patient without additional effort from clinicians.

Instead of training a classification pipeline, we propose to solve an object detection problem, which involves predicting a bounding box for the malignancy. The motivation is that, running a classifier on a focused attention/ proposal region in an object detection pipeline would help tackle the low inter-class and high intra-class variations. However, since we only have image-level labels available, we formulate the problem as a Weakly Supervised Object Detection (WSOD) problem. As transformers are increasingly outshining CNNs due to their ability to aggregate focused cues from a large area [6,9], we choose to use transformers in our model. However, in our initial experiments SOTA WSOD methods for transformers failed miserably. These methods primarily rely on training a classification pipeline and later generating activation heatmaps using attention and

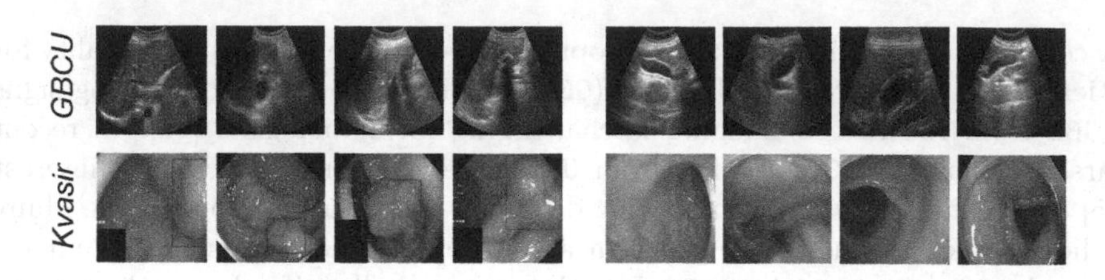

Fig. 2. Samples from the GBCU [3] and Kvasir-SEG [17] datasets. Four images from each of the disease and non-disease classes are shown on the left and right, respectively. Disease locations are shown by drawing bounding boxes.

drawing a bounding box circumscribing the heatmaps [2,10] to show localization. However, for GBC detection, this line of work is not helpful as we discussed earlier.

Inspired by the success of the Multiple Instance Learning (MIL) paradigm for weakly supervised training on medical imaging tasks [20,22], we train a detection transformer, DETR, using the MIL paradigm for weakly supervised malignant region detection. In this, one generates region proposals for images, and then considers the images as bags and region proposals as instances to solve the instance classification (object detection) under the MIL constraints [8]. At inference, we use the predicted instance labels to predict the bag labels. Our experiments validate the utility of this approach in circumventing the challenges in US images and detecting GBC accurately from US images using only image-level labels.

Contributions: The key contributions of this work are:

- We design a novel DETR variant based on MIL with self-supervised instance learning towards the weakly supervised disease detection and localization task in medical images. Although MIL and self-supervised instance learning has been used for CNNs [24], such a pipeline has not been used for transformer-based detection models.
- We formulate the GBC classification problem as a weakly supervised object detection problem to mitigate the effect of low inter-class and large intra-class variances, and solve the difficult GBC detection problem on US images without using the costly and difficult to obtain additional annotation (bounding box) or video data.
- Our method provides a strong baseline for weakly supervised GBC detection and localization in US images, which has not been tackled earlier. Further, to assess the generality of our method, we apply our method to Polyp detection from Colonoscopy images.

2 Datasets

Gallbladder Cancer Detection in Ultrasound Images: We use the public GBC US dataset [3] consisting of 1255 image samples from 218 patients. The dataset contains 990 non-malignant (171 patients) and 265 malignant (47 patients) GB images (see Fig. 2 for some sample images). The dataset contains image labels as well as bounding box annotations showing the malignant regions. Note that, we use only the image labels for training. We report results on 5-fold cross-validation. We did the cross-validation splits at the patient level, and all images of any patient appeared either in the train or validation split.

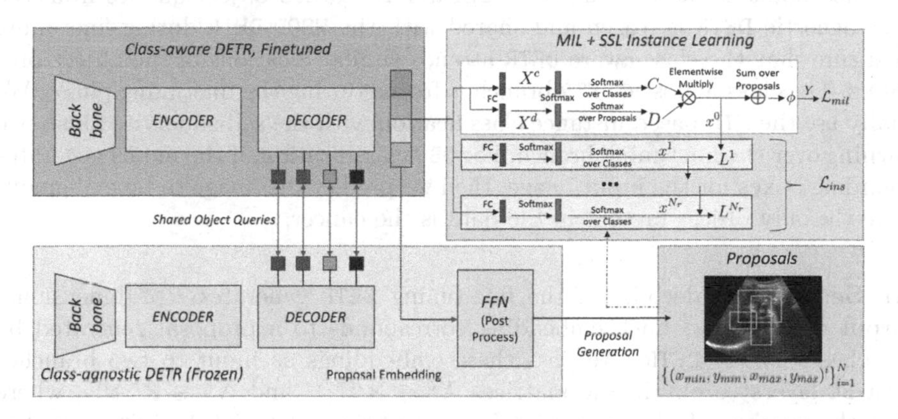

Fig. 3. Overview of the proposed Weakly Supervised DETR architecture. The location information in the object queries learned by the class-agnostic DETR ensures generation of high-quality proposals. The MIL framework uses the proposal embeddings generated at the class-aware branch.

Polyp Detection in Colonoscopy Images: We use the publicly available Kvasir-SEG [17] dataset consisting of 1000 white light colonoscopy images showing polyps (see Fig. 2). Since Kvasir-SEG does not contain any control images, we add 600 non-polyp images randomly sampled from the PolypGen [1] dataset. Since the patient information is not available with the data, we use random stratified splitting for 5-fold cross-validation.

3 Our Method

Revisiting DETR: The DETR [6] architectures utilize a ResNet [13] backbone to extract 2D convolutional features, which are then flattened and added with a positional encoding, and fed to the self-attention-based transformer encoder. The decoder uses cross-attention between learned object queries containing positional embedding, and encoder output to produce output embedding containing the class and localization information. The number of object queries, and the decoder output embeddings is set to 100 in DETR. Subsequently, a feed-forward network generates predictions for object bounding boxes with their corresponding labels and confidence scores.

Proposed Architecture: Fig. 3 gives an overview of our method. We use a COCO pre-trained class-agnostic DETR as proposal generator. The learned object queries contain the embedded positional information of the proposal. Class-agnostic indicates that all object categories are considered as a single object class, as we are only interested in the object proposals. We then finetune a regular, class-aware DETR for the WSOD task. This class-aware DETR is initialized with the checkpoint of the class-agnostic DETR. The learned object queries from the class-agnostic DETR is frozen and shared with the WSOD DETR during finetuning to ensure that the class-aware DETR attends similar locations of the object proposals. The class-agnostic DETR branch is frozen during the finetuning phase. We finally use the MIL-based instance classification with the self-supervised instance learning over the finetuning branch. For GBC classification, if the model generates bounding boxes for the input image, then we predict the image to be malignant, since the only object present in the data is the cancer.

MIL Setup: The decoder of the fine-tuning DETR generates R d-dimensional output embeddings. Each embedding corresponds to a proposal generated by the class-agnostic DETR. We pass these embeddings as input to two branches with FC layers to obtain the matrices $X^c \in \mathbb{R}^{R \times N_c}$ and $X^r \in \mathbb{R}^{R \times N_c}$, where R is the number of object queries (same as proposals) and N_c is the number of object (disease) categories. Let $\sigma(\cdot)$ denote the softmax operation. We then generate the class-wise and detection-wise softmax matrices $C \in \mathbb{R}^{R \times N_c}$ and $D \in \mathbb{R}^{R \times N_c}$, where $C_{ij} = \sigma((X^c)^T_j)i$ and $D_{ij} = \sigma(X^r_i)j$, and X_i denotes the i-th row of X. C provides classification probabilities of each proposal, and D provides the relative score of the proposals corresponding to each class. The two matrices are element-wise multiplied and summed over the proposal dimension to generate the image-level classification predictions, $\phi \in \mathbb{R}^{N_c}$:

$$\phi_j = \sum_{i=1}^{R} C_{ij} \cdot D_{ij} \tag{1}$$

Notice, $\phi_j \in (0, 1)$ since C_{ij} and D_{ij} are normalized. Finally, the negative log-likelihood loss between the predicted labels, and image labels $y \in \mathbb{R}^{N_c}$ is computed as the MIL loss:

$$\mathcal{L}_{\text{mil}} = -\sum_{i=1}^{N_c} [y_i \log \phi_i + (1 - y_i) \log (1 - \phi_i)] \tag{2}$$

The MIL classifier further suffers from overfitting to the distinctive classification features due to the mismatch of classification and detection probabilities [24]. To tackle this, we further use a self-supervised module to improve the instances.

Self-supervised Instance Learning: Inspired by [24], we design a instance learning module with N_r blocks in a self-supervised framework to refine the instance scores with instance-level supervision. Each block consists of an FC layer. A class-wise softmax is used to generate instance scores $x^n \in \mathbb{R}^{R\times(N_c+1)}$ at n-th block. $N_c + 1$ includes the background/ no-finding class. Instance supervision of each layer (n) is obtained from the scores of the previous layer ($x^{(n-1)}$). The instance supervision for the first layer is obtained from the MIL head. Suppose $\hat{y}^n \in \mathbb{R}^{R\times(N_c+1)}$ is the pseudo-labels of the instances. An instance (p_j) is labelled 1 if it overlaps with the highest-scoring instance by a chosen threshold.

Table 1. Weakly supervised disease detection performance comparison of our method and SOTA baselines in GBC and Polyps. We report Average Precision at IoU 0.25 (AP_{25}).

Method	GBC	Polyp
	AP_{25}	AP_{25}
TS-CAM [10] (ICCV 2021)	0.024 ± 0.008	0.058 ± 0.015
SCM [2] (ECCV 2022)	0.013 ± 0.001	0.082 ± 0.036
OD-WSCL [21] (ECCV 2022)	0.482 ± 0.067	0.239 ± 0.032
WS-DETR [19] (WACV 2023)	0.520 ± 0.088	0.246 ± 0.023
Point-Beyond-Class [18] (MICCAI 2022)	0.531 ± 0.070	0.283 ± 0.022
Ours	0.628 ± 0.080	0.363 ± 0.052

Table 2. Ablation study. Performance of MIL-framework variants on DETR. We compare the AP and detection sensitivity.

Design	GBC		Polyp	
	AP_{25}	Sens.	AP_{25}	Sens.
MIL + DETR	0.520 ± 0.088	0.833 ± 0.034	0.246 ± 0.023	0.882 ± 0.034
MIL + SSL + DETR (Ours)	0.628 ± 0.080	0.861 ± 0.089	0.363 ± 0.052	0.932 ± 0.022

Otherwise, the instance is labeled 0 as defined in Eq. 3:

$$m_j^n = \underset{i}{\operatorname{argmax}}\, x_{ij}^{(n-1)} \quad ; \qquad \hat{y}_{ij}^n = \begin{cases} 1, & IoU(p_j, p_{m_j^n}) \geq \tau \\ 0, & \text{otherwise} \end{cases} \tag{3}$$

The loss over the instances is given by Eq. 4:

$$\mathcal{L}_{ins} = -\frac{1}{N_r} \sum_{n=1}^{N_r} \frac{1}{R} \sum_{i=1}^{R} \sum_{j=1}^{N_c+1} w_i^n \hat{y}_{ij}^n \log x_{ij}^n \tag{4}$$

Here x_{ij}^{n} denotes the score of i-th instance for j-th class at layer n. Following [24], the loss weight $w_i^n = x_{i\,m_j^n}^{(n-1)}$ is applied to stabilize the loss. Assuming λ to be a scaling value, the overall loss function is given in Eq. 5:

$$\mathcal{L} = \mathcal{L}_{mil} + \lambda \mathcal{L}_{ins} \tag{5}$$

4 Experiments and Results

Experimental Setup: We use a machine with Intel Xeon Gold 5218@2.30GHz processor and 8 Nvidia Tesla V100 GPUs for our experiments. The model is trained using SGD with LR 0.001 (for MIL head), weight decay 10^{-6}, and momentum 0.9 for 100 epochs with batch size 32. The LR at backbone and transformer are 0.003, and 0.0003, respectively. We use a cosine annealing of the LR.

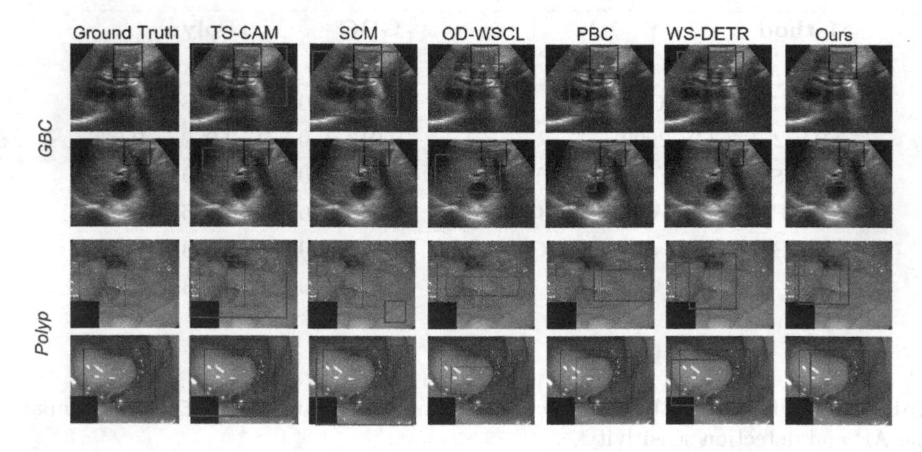

Fig. 4. Qualitative analysis of the predicted bounding boxes. Ground truths are in blue, and predictions are in green. We compare with SOTA WSOD techniques and our proposed method. Our method predicts much tighter bounding boxes that cover the clinically significant disease regions. (Color figure online)

Comparison with SOTA: Table 1 shows the bounding box localization results of the WSOD task. Our method surpasses all latest SOTA WSOD techniques by 9 points, and establishes itself as a strong WSOD baseline for GBC localization in US images. Our method also achieves 7-point higher AP score for polyp detection. We present visualizations of the predicted bounding boxes in Fig. 4 which shows that the localization by our method is more precise and clinically relevant as compared to the baselines.

Generality of the Method: We assess the generality of our method by applying it to polyp detection on colonoscopy images. The applicability of our method on two different tasks - (1) GBC detection from US and (2) Polyp detection from Colonoscopy, indicates the generality of the method across modalities.

Ablation Study: We show the detection sensitivity to the self-supervised instance learning module in Table 2 for two variants, (1) vanilla MIL head on DETR, and (2) MIL with self-supervised instance learning on DETR. Table 2 shows the Average Precision and detection sensitivity for both diseases. The results establish the benefit of using the self-supervised instance learning. Other ablations related to the hyper-parameter sensitivity is given in Supplementary Fig. S1.

Classification Performance: We compare our model with the standard CNN-based and Transformer-based classifiers, SOTA WSOD-based classifiers, and SOTA classifiers using additional data or annotations (Table 3). Our method beats the SOTA weakly supervised techniques and achieves 1.2% higher sensitivity for GBC detection. The current SOTA GBC detection models require additional bound-

Table 3. Performance comparison of our method and other SOTA methods in GBC classification. We report accuracy, specificity, and sensitivity.

Type	Method	Acc.	Spec.	Sens.
CNN Classifier	ResNet50 [13]	0.867 ± 0.031	0.926 ± 0.069	0.672 ± 0.147
	InceptionV3 [23]	0.869 ± 0.039	0.913 ± 0.032	0.708 ± 0.078
Transformer Classifier	ViT [9]	0.803 ± 0.078	0.901 ± 0.050	0.860 ± 0.068
	DEIT [25]	0.829 ± 0.030	0.900 ± 0.040	0.875 ± 0.063
	PVTv2 [26]	0.824 ± 0.033	0.887 ± 0.057	0.894 ± 0.076
	RadFormer [4]	0.921 ± 0.062	0.961 ± 0.049	0.923 ± 0.062
Additional Data/ Annotation	USCL [7]	0.889 ± 0.047	0.895 ± 0.054	0.869 ± 0.097
	US-UCL [5]	0.920 ± 0.034	0.926 ± 0.043	0.900 ± 0.046
	GBCNet [3]	0.921 ± 0.029	0.967 ± 0.023	0.919 ± 0.063
	Point-Beyond-Class [18]	0.929 ± 0.013	0.983 ± 0.042	0.731 ± 0.077
SOTA WSOD	TS-CAM [10]	0.862 ± 0.049	0.879 ± 0.049	0.751 ± 0.045
	SCM [2]	0.795 ± 0.101	0.783 ± 0.130	0.849 ± 0.072
	OD-WSCL [21]	0.815 ± 0.144	0.805 ± 0.129	0.847 ± 0.214
	WS-DETR [19]	0.839 ± 0.042	0.843 ± 0.028	0.833 ± 0.034
WSOD	Ours	0.834 ± 0.057	0.817 ± 0.061	0.861 ± 0.089

Table 4. Comparison with SOTA WSOD baselines in classifying Polyps from Colonoscopy images.

Method	Acc.	Spec.	Sens.
TS-CAM [10]	0.704 ± 0.017	0.394 ± 0.042	0.891 ± 0.054
SCM [2]	0.751 ± 0.026	0.523 ± 0.014	0.523 ± 0.016
OD-WSCL [21]	0.805 ± 0.056	0.609 ± 0.076	0.923 ± 0.034
WS-DETR [19]	0.857 ± 0.071	0.812 ± 0.088	0.882 ± 0.034
Point-Beyond-Class [18]	0.953 ± 0.007	0.993 ± 0.004	0.924 ± 0.011
Ours	0.878 ± 0.067	0.785 ± 0.102	0.932 ± 0.022

ing box annotation [3] or, US videos [5,7]. However, even without these additional annotations/ data, our method reaches 86.1% detection sensitivity. The results for polyp classification are reported in Table 4. Although our method has a slightly lower specificity, the sensitivity surpasses the baselines reported in literature [16], and the SOTA WSOD based baselines.

5 Conclusion

GBC is a difficult-to-detect disease that benefits greatly from early diagnosis. While automated GBC detection from US images has gained increasing interest from researchers, training a standard image classification model for this task is challenging due to the low inter-class variance and high intra-class variability of malignant regions. Current SOTA models for GBC detection require costly bounding box annotation of the pathological regions, or additional US video data, which limit their applicability. We proposed to formulate GBC detection as a weakly supervised object detection/ localization problem using a DETR with self-supervised instance learning in a MIL framework. Our experiments show that the approach achieves competitive performance without requiring additional annotation or data. We hope that our technique will simplify the model training at the hospitals with easily available data locally, enhancing the applicability and impact of automated GBC detection.

References

1. Ali, S., et al.: A multi-centre polyp detection and segmentation dataset for generalisability assessment. Scientific Data **10**(1), 75 (2023)
2. Bai, H., Zhang, R., Wang, J., Wan, X.: Weakly supervised object localization via transformer with implicit spatial calibration. In: ECCV. pp. 612–628. Springer (2022). https://doi.org/10.1007/978-3-031-20077-9_36
3. Basu, S., Gupta, M., Rana, P., Gupta, P., Arora, C.: Surpassing the human accuracy: Detecting gallbladder cancer from USG images with curriculum learning. In: CVPR, pp. 20886–20896 (2022)
4. Basu, S., Gupta, M., Rana, P., Gupta, P., Arora, C.: Radformer: transformers with global-local attention for interpretable and accurate gallbladder cancer detection. Med. Image Anal. **83**, 102676 (2023)
5. Basu, S., Singla, S., Gupta, M., Rana, P., Gupta, P., Arora, C.: Unsupervised contrastive learning of image representations from ultrasound videos with hard negative mining. In: MICCAI, pp. 423–433. Springer (2022). https://doi.org/10.1007/978-3-031-16440-8_41
6. Carion, N., Massa, F., Synnaeve, G., Usunier, N., Kirillov, A., Zagoruyko, S.: End-to-end object detection with transformers. In: Vedaldi, A., Bischof, H., Brox, T., Frahm, J.-M. (eds.) ECCV 2020. LNCS, vol. 12346, pp. 213–229. Springer, Cham (2020). https://doi.org/10.1007/978-3-030-58452-8_13
7. Chen, Y., et al.: USCL: pretraining deep ultrasound image diagnosis model through video contrastive representation learning. In: de Bruijne, M., et al. (eds.) MICCAI 2021. LNCS, vol. 12908, pp. 627–637. Springer, Cham (2021). https://doi.org/10.1007/978-3-030-87237-3_60

8. Dietterich, T.G., Lathrop, R.H., Lozano-Pérez, T.: Solving the multiple instance problem with axis-parallel rectangles. Artif. Intell. **89**(1–2), 31–71 (1997)
9. Dosovitskiy, A., et al.: An image is worth 16x16 words: Transformers for image recognition at scale. arXiv preprint arXiv:2010.11929 (2020)
10. Gao, W., et al.: Ts-cam: Token semantic coupled attention map for weakly supervised object localization. In: ICCV, pp. 2886–2895 (2021)
11. Gupta, P.: Imaging-based algorithmic approach to gallbladder wall thickening. World J. Gastroenterol. **26**(40), 6163 (2020)
12. Gupta, P., et al.: Locally advanced gallbladder cancer: a review of the criteria and role of imaging. Abdominal Radiol. **46**(3), 998–1007 (2021)
13. He, K., Zhang, X., Ren, S., Sun, J.: Deep residual learning for image recognition. In: CVPR, pp. 770–778 (2016)
14. Hong, E.K., et al.: Surgical outcome and prognostic factors in patients with gallbladder carcinoma. Ann. Hepato-Biliary-Pancreat. Surg. **18**(4), 129–137 (2014)
15. Howlader, N., et al.: Seer cancer statistics review, 1975–2014, national cancer institute, pp. 1–12. Bethesda, MD pp (2017)
16. Jha, D., et al.: Real-time polyp detection, localization and segmentation in colonoscopy using deep learning. IEEE Access **9**, 40496–40510 (2021)
17. Jha, D., et al.: Kvasir-SEG: a segmented polyp dataset. In: Ro, Y.M., et al. (eds.) MMM 2020. LNCS, vol. 11962, pp. 451–462. Springer, Cham (2020). https://doi.org/10.1007/978-3-030-37734-2_37
18. Ji, H., et al.: Point beyond class: A benchmark for weakly semi-supervised abnormality localization in chest x-rays. In: MICCAI. pp. 249–260. Springer (2022). https://doi.org/10.1007/978-3-031-16437-8_24
19. LaBonte, T., Song, Y., Wang, X., Vineet, V., Joshi, N.: Scaling novel object detection with weakly supervised detection transformers. In: WACV, pp. 85–96 (2023)
20. Qian, Z., et al.: Transformer based multiple instance learning for weakly supervised histopathology image segmentation. In: MICCAI, pp. 160–170. Springer Nature Switzerland Cham (2022). https://doi.org/10.1007/978-3-031-16434-7_16
21. Seo, J., Bae, W., Sutherland, D.J., Noh, J., Kim, D.: Object discovery via contrastive learning for weakly supervised object detection. In: ECCV, pp. 312–329. Springer (2022). https://doi.org/10.1007/978-3-031-19821-2_18
22. Shao, Z., Bian, H., Chen, Y., Wang, Y., Zhang, J., Ji, X., et al.: Transmil: transformer based correlated multiple instance learning for whole slide image classification. NeurIPS **34**, 2136–2147 (2021)
23. Szegedy, C., Vanhoucke, V., Ioffe, S., Shlens, J., Wojna, Z.: Rethinking the inception architecture for computer vision. In: Proceedings of the IEEE Conference on Computer Vision and Pattern Recognition, pp. 2818–2826 (2016)
24. Tang, P., Wang, X., Bai, X., Liu, W.: Multiple instance detection network with online instance classifier refinement. In: CVPR, pp. 2843–2851 (2017)
25. Touvron, H., Cord, M., Douze, M., Massa, F., Sablayrolles, A., Jégou, H.: Training data-efficient image transformers & distillation through attention. In: ICML, pp. 10347–10357. PMLR (2021)
26. Wang, W., et al.: Pvtv 2: Improved baselines with pyramid vision transformer (2021)

Dual Conditioned Diffusion Models for Out-of-Distribution Detection: Application to Fetal Ultrasound Videos

Divyanshu Mishra[1(✉)], He Zhao[1], Pramit Saha[1], Aris T. Papageorghiou[2], and J. Alison Noble[1]

[1] Institute of Biomedical Engineering, University of Oxford, Oxford, UK
divyanshu.mishra@eng.ox.ac.uk

[2] Nuffield Department of Women's and Reproductive Health, University of Oxford, Oxford, UK

Abstract. Out-of-distribution (OOD) detection is essential to improve the reliability of machine learning models by detecting samples that do not belong to the training distribution. Detecting OOD samples effectively in certain tasks can pose a challenge because of the substantial heterogeneity within the in-distribution (ID), and the high structural similarity between ID and OOD classes. For instance, when detecting heart views in fetal ultrasound videos there is a high structural similarity between the heart and other anatomies such as the abdomen, and large in-distribution variance as a heart has 5 distinct views and structural variations within each view. To detect OOD samples in this context, the resulting model should generalise to the intra-anatomy variations while rejecting similar OOD samples. In this paper, we introduce dual-conditioned diffusion models (DCDM) where we condition the model on in-distribution class information and latent features of the input image for reconstruction-based OOD detection. This constrains the generative manifold of the model to generate images structurally and semantically similar to those within the in-distribution. The proposed model outperforms reference methods with a 12% improvement in accuracy, 22% higher precision, and an 8% better F1 score.

1 Introduction

Existing out-of-distribution (OOD) detection methods work well when the in-distribution (ID) classes have low heterogeneity (low variance) but fail when in-distribution classes have high heterogeneity [23] or high spatial similarity between ID and OOD classes [9]. Fetal ultrasound (US) anatomy detection is one such application where both the challenges co-exist.

In this paper, we propose a Dual-Conditioned Diffusion Model (DCDM) to detect OOD samples when in-distribution data has high variance and test the

---

Supplementary Information The online version contains supplementary material available at https://doi.org/10.1007/978-3-031-43907-0_21.

H. Greenspan et al. (Eds.): MICCAI 2023, LNCS 14220, pp. 216–226, 2023.
https://doi.org/10.1007/978-3-031-43907-0_21

performance by detecting heart views in fetal US videos as an example application. Specifically, an Ultrasound (US) typically comprises 13 anatomies and their views. However, analysis models are usually developed for anatomy-specific tasks. Hence, to separate heart views from other 12 anatomies (head, abdomen, femur etc.) we develop an OOD detection algorithm. Our in-distribution data comprises five structurally different heart views captured across different cardiac cycles of a beating heart during obstetric US scanning. We develop a diffusion-based model for reconstruction-based OOD detection, which extends [14] with a novel dual conditioning mechanism that alleviates the influence of high inter- and intra-class variation within different classes by leveraging in-distribution class conditioning (IDCC) and latent image feature conditioning (LIFC). These conditioning mechanisms allow our model to generate images similar to the input image for in-distribution data. The primary contributions of our paper are summarized as follows: 1) We introduce a novel conditioned diffusion model for OOD detection and demonstrate that the dual conditioning mechanism is effective in tackling challenging scenarios where in-distribution data comprises multiple heterogeneous classes and there is a high spatial similarity between ID and OOD classes. 2) Two original conditions are proposed for the diffusion model, which are in-distribution class conditioning (IDCC) and latent image feature conditioning (LIFC). IDCC is proposed to handle high inter-class variance within in-distribution classes and high spatial similarity between ID and OOD classes. LIFC is introduced to counter the intra-class variance within each class. 3) We demonstrate in our experiments that DCDM can detect and separate heart views from other anatomies in fetal ultrasound videos without needing any labelled data for OOD classes. Extensive experiments and ablations demonstrate superior performance over existing OOD detection methods. Our approach is not fetal ultrasound specific and could be applied to other OOD applications.

2 Related Work

OOD detection [31] involves identifying samples that do not belong to the training distribution. Such models can be categorized into: (a) unsupervised OOD detection [23] and (b) supervised OOD detection. [5,11,34]. Unsupervised OOD detection methods can again be divided into two main categories: (i) likelihood-based approaches [12,21,30], and (ii) reconstruction-based [3,25,33]. Likelihood-based approaches suffer from several issues, including assigning higher likelihood to OOD samples [4,19], susceptibility to adversarial attacks [8], and calibration issues [28]. Current reconstruction-based approaches are sensitive to dimensions of the bottleneck layer and require rigorous tuning specific to the dataset and task [10]. Additionally, models trained using a generator-discriminator architecture and optimizing adversarial losses can be highly unstable and challenging to train [1,2]. Finally, reconstruction-based methods often rely on highly compressed latent representations, which can lead to loss of important low-level detail. This can be problematic when discriminating between classes with high spatial similarity. Recently, diffusion models have been introduced to address these limitations on tasks such as image synthesis [6], and OOD detection [10].

Denoising Diffusion Probabilistic Models (DDPMs) [14] are generative models that work by gradually adding noise to an input image through a forward diffusion process followed by gradually removing noise using a trained neural network in the backward diffusion process [32]. To guide the generative process of a diffusion model (DM), previous work [18,22,24] condition the DDPMs on task-specific conditioning. In image-to-image translation tasks like super-resolution, colourization, *etc.*, previous papers [24] condition the model by concatenating a resized or grayscale version of the input image to the noised image. This concatenation is unsuitable for reconstruction-based OOD detection as the model will generate similar images for ID and OOD samples. In the context of OOD detection using DMs, previous works [10] have trained unconditional DDPMs and, during inference, sampled using a Pseudo Linear Multi Step (PLMS) [16] sampler for varying noise levels. However, their approach generates 5500 samples to detect OOD samples for each input image which is time-consuming and impractical for settings where shorter inference times are needed. AnoDDPM [29] utilises simplex noise rather than Gaussian noise to corrupt the image (t=250 rather than t=1000) for anomaly detection. However, this approach requires data specific tuning, and is outperformed by [10].

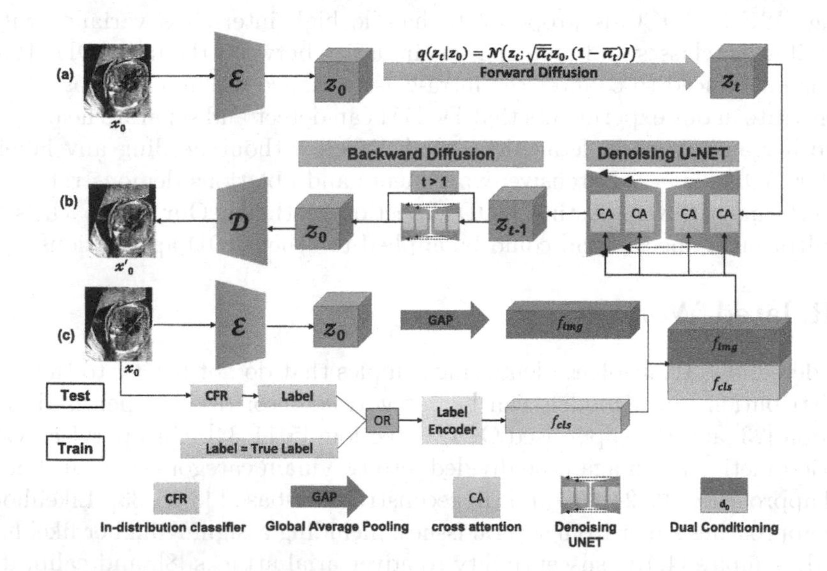

Fig. 1. DCDM architecture where (*a*) the input image x_0 is mapped to the latent vector z_0 using a pretrained encoder $\mathcal{E}$ and forward diffusion is applied, (*b*) the backward diffusion process denoises the latent vector z_t and the final denoised latent vector z_0 is mapped to pixel space by the decoder $\mathcal{D}$ (*c*) the dual-conditioning mechanism. We obtain f_{img} by passing the input image x_0 through the encoder $\mathcal{E}$. f_{cls} is obtained using the true label during training or predicted class label during testing.

3 Methods

3.1 Dual Conditioned Diffusion Models

Diffusion models are generative models that rely on two Markov processes known as forward and backward diffusion [14]. To improve efficiency during training and inference, forward and backward diffusion is applied to the latent space [22]. Autoencoder (AE $= \mathcal{E} + \mathcal{D}$) is pretrained separately on ID heart data and can successfully reconstruct the input heart images (SSIM=0.956). The latent variable z_0 is obtained by passing an input image x_0 through a pretrained encoder $\mathcal{E}$. Given the latent vector z_0 and a fixed variance schedule [14] $\{\beta_t \in (0,1)\}_{t=1}^{T}$, the forward diffusion process, defined by Eq. 1, gradually adds Gaussian noise to z_0 to give a noised latent vector z_t where $\alpha_t = 1 - \beta_t$ and $\bar{\alpha}_t = \prod_{i=1}^{t} \alpha_i$:

$$q(\boldsymbol{z}_t|\boldsymbol{z}_0) = \mathcal{N}(\boldsymbol{z}_t; \sqrt{\bar{\alpha}_t}\boldsymbol{z}_0, (1 - \bar{\alpha}_t)\mathbf{I}) \tag{1}$$

In backward diffusion, we aim to reverse the forward diffusion process and predict z_{t-1} given z_t. To predict $(z_{t-1}|z_t)$, we train a denoising U-Net [14] denoted as $\epsilon_\theta\left(z_t, t, d_0\right)$ that takes the current timestep t, noised latent vector z_t and the dual conditioning embedding vector d_0 as input and predicts the noise at timestep t as shown in Eq. 2.

$$\boldsymbol{z}_{t-1} = \mathcal{N}(\boldsymbol{z}_{t-1}; \frac{1}{\sqrt{\alpha_t}}\Big(\boldsymbol{z}_t - \frac{1-\alpha_t}{\sqrt{1-\bar{\alpha}_t}}\epsilon_\theta(\boldsymbol{z}_t, t, d_0)\Big), (1-\bar{\alpha}_t)\mathbf{I}) \tag{2}$$

The dual embedding vector d_0 is obtained by combining IDCC (f_{cls}) and LIFC (f_{img}) vectors, which we explain in Sect. 3.2. The output z_{t-1} is again input to ϵ_θ. This process is repeated until z_0 is obtained.

The final model optimisation objective is given by Eq. 3 where ϵ is the original noise added during the forward diffusion process.

$$\mathcal{L}_{DCDM} := \mathbb{E}_{\mathcal{E}(x), \epsilon \sim \mathcal{N}(0,1), t}\left[\left\|\epsilon - \epsilon_\theta\left(z_t, t, d_0\right)\right\|_2^2\right] \tag{3}$$

Once we obtain z_0 from the backward diffusion process, it is passed on to the decoder $\mathcal{D}$ and mapped back to the pixel space to give generated image x_0'.

3.2 Dual Conditioning Mechanism

Image features and in-distribution class information are utilised in our proposed dual conditioning mechanism. This guides the DCDM to generate images that are spatially and semantically similar to the input image for in-distribution samples and dissimilar for OOD samples.

Latent Image Feature Conditioning (LIFC): The image conditioning dictates the desired appearance of generated images in terms of shape and texture. In our model, we use the features extracted by a pretrained encoder for conditioning. Empirically, we use the same encoder $\mathcal{E}$ as our feature extractor to obtain latent feature vector z_0 as shown in Fig. 1. Specifically, the input image of dimension $224 \times 224 \times 3$ is passed through the encoder $\mathcal{E}$ and a feature map with the size of $7 \times 7 \times 128$ is obtained which is followed by global average pooling (GAP) resulting in a feature vector (f_{img}) with dimension 128.

In-Distribution Class Conditioning (IDCC): Given an in-distribution dataset comprising n heterogeneous classes, conditioning the model only on image-level features is insufficient. Therefore we introduce an in-distribution class conditioning (IDCC) that informs the DCDM of the class of the input image and enables it to generate samples belonging to the same class for ID. A label encoder generates a unique class conditional embedding (f_{cls}) of dimension 128 for each class label. The class label is assigned based on the ground truth label during the training phase and to the classifier's prediction during inference, as depicted in Fig. 1. In practice, we train a CNN classifier, freeze its weight and use it as our in-distribution classifier (CFR), as discussed in Sect. 3.3.

Cross Attention Guidance: To integrate the dual-conditioning guidance into the diffusion model, we use a cross-attention [27] mechanism inside the denoising U-Net rather than just concatenation [24] as it is more effective [13,17,20] and allows condition diffusion models on various input modalities [22]. Our LIFC and IDCC are first concatenated to give a feature vector with a dimension of 256. This acts as a side input to each UNet block. The features from the UNet block and the conditional features are fused by cross-attention and serve as input to the following UNet block as shown in Fig. 1. For more details,regarding cross-attention block refer to Rombach *et al.* [22].

3.3 In-Distribution Classifier

The in-distribution classifier (CFR) serves two main functions. First, it provides labels for the class conditioning during inference; second, it is utilized as a feature extractor for calculating the OOD score.

Inference Class Guidance. IDCC requires in-distribution class information to generate the class conditional embedding. However, class information is only available during training. To obtain class information during inference, we separately train a ConvNext CNN based classifier (accuracy $= 88\%$) on the in-distribution data and use its predictions as the class information. During inference, the input image x_0 is passed through the classifier, and the predicted label is used to generate the class embedding by feeding to the label encoder as shown in Fig. 1. Moreover, as the classifier is only trained on in-distribution data, it classifies an OOD sample to an in-distribution class. The classifier's prediction is utilised by the DCDM and it tries to generate an image belonging to in-distribution class for the OOD samples. This reduces the structural and semantic similarity between the input and the generated image, as demonstrated by our qualitative results (Fig. 2).

Feature-Based OOD Detection To evaluate the performance of the DCDM, the cosine similarity between features of the input image x_0 and the generated image x'_0 from the in-distribution classifier is calculated and is referred as an OOD score where f_0 and f'_0 are the features of x_0 and x'_0, respectively:

$$\text{OOD score} = sim(f_0, f'_0) = \frac{f_0 \cdot f'_0}{\|f_0\|_2 \|f'_0\|_2}, \tag{4}$$

An input image x_0 is classified as in-distribution (ID) or OOD based on Eq. 5 where τ is a pre-defined threshold and y_{pred} is the prediction of our feature-based OOD detection algorithm.

$$y_{pred} = \begin{cases} 0(ID) & if\, \text{OOD score} > \tau \\ 1\ (\text{OOD}) & otherwise \end{cases} \tag{5}$$

4 Experiments and Results

Dataset and Implementation. For our experiments, we utilized a fetal ultrasound dataset of 359 subject videos that were collected as part of the PULSE project [7]. The in-distribution dataset consisted of 5 standard heart views (3VT, 3VV, LVOT, RVOT, and 4CH), while the out-of-distribution dataset comprised of three non-heart anatomies - fetal head, abdomen, and femur. The original images were of size 1008 × 784 pixels and were resized to 224 × 224 pixels.

To train the models, we randomly sampled 5000 fetal heart images and used 500 images for evaluating image generation performance. To test the performance of our final model and compare it with other methods, we used an held-out dataset of 7471 images, comprising 4309 images of different heart views and 3162 images (about 1000 for each anatomy) of out-of-distribution classes. Further details about the dataset are given in **Supp. Fig. 2 and 3**.

All models were trained using PyTorch version 1.12 with a Tesla V100 32 GB GPU. During training, we used T=1000 for noising the input image and a linearly increasing noise schedule that varied from 0.0015 to 0.0195. To generate samples from our trained model, we used DDIM [26] sampling with T=100. All baseline models were trained and evaluated using the original implementation.

4.1 Results

We evaluated the performance of the dual-conditioned diffusion models (DCDMs) for OOD detection by comparing them with two current state-of-the-art unsupervised reconstruction-based approaches and one likelihood-based approach. The first baseline is Deep-MCDD [15], a likelihood-based OOD detection method that proposes a Gaussian discriminant-based objective to learn class conditional distributions. The second baseline is ALOCC [23] a GAN-based model that uses the confidence of the discriminator on reconstructed samples to detect OOD samples. The third baseline is the method of Graham *et al.* [10], where they use DDPM [14] to generate multiple images at varying noise levels

Table 1. Quantitative comparison of our model (DCDM) with reference methods

Method	AUC(%)	F1-Score(%)	Accuracy(%)	Precision(%)
Deep-MCDD [15]	64.58	66.23	60.41	51.82
ALOCC [23]	57.22	59.34	52.28	45.63
Graham et al. [10]	63.86	63.55	60.15	50.89
DCDM(Ours)	**77.60**	**74.29**	**77.95**	**73.34**

for each input. They then compute the MSE and LPIPS metrics for each image compared to the input, convert them to Z-scores, and finally average them to obtain the OOD score.

Quantitative Results. The performance of the DCDM, along with comparisons with the other approaches, are shown in Table 1. The GAN-based method ALOCC [23] has the lowest AUC of 57.22%, which is improved to 63.86% by the method of Graham *et al.* and further improved to 64.58% by likelihood-based Deep-MCDD. DCDM outperforms all the reference methods by 20%, 14% and 13%, respectively and has an AUC of 77.60%. High precision is essential for OOD detection as this can reduce false positives and increase trust in the model. DCDM exhibits a precision that is 22% higher than the reference methods while still having an 8% improvement in F1-Score.

Qualitative Results. Qualitative results are shown in Fig. 2. Visual comparisons show ALOCC generates images structurally similar to input images for in-distribution and OOD samples. This makes it harder for the ALOCC model to detect OOD samples. The model of Graham *et al.* generates any random heart view for a given image as a DDPM is unconditional, and our in-distribution data contains multiple heart views. For example, given a 4CH view as input, the model generates an entirely different heart view. However, unlike ALOCC, the Graham *et al.* model generates heart views for OOD samples, improving OOD detection performance. DCDM generates images with high spatial similarity to the input image and belonging to the same heart view for ID samples while structurally diverse heart views for OOD samples. In Fig. 2 (c) for OOD sample, even-though the confidence is high (0.68), the gap between ID and OOD classes is wide enough to separate the two.Additional qualitative results can be observed in **Supp. Fig. 4.**

4.2 Ablation Study

Ablation experiments were performed to study the impact of various conditioning mechanisms on the model performance both qualitatively and quantitatively. When analyzed quantitatively, as shown in Table 2, the unconditional model has the lowest AUC of 69.61%. Incorporating the IDCC guidance or LIFC

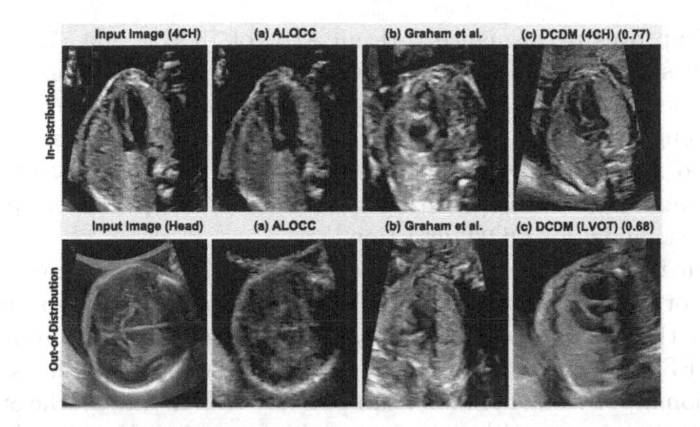

Fig. 2. Qualitative comparison of our method with (*a*) ALOCC generates similar images to the input for ID and OOD samples (*b*) Graham *et al.* generates any random heart view for a given input image (*c*) Our model generates images that are similar to the input image for ID and dissimilar for OOD samples. Classes predicted by CFR and the OOD score ($\tau = 0.73$) are mentioned in brackets.

Table 2. Ablation study of different conditioning mechanisms of DCDM.

Method	Accuracy (%)	Precision (%)	AUC (%)
Unconditional	68.16	58.44	69.61
In-Distribution Class Conditioning	74.39	66.12	75.27
Latent Image Feature Conditioning	77.02	70.02	77.40
Dual Conditioning	**77.95**	**73.34**	**77.60**

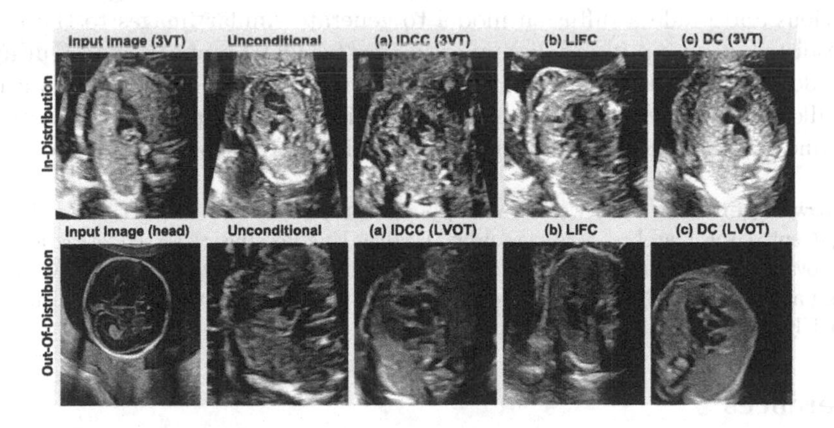

Fig. 3. Qualitative ablation study showing the effect of **(a)** IDCC, **(b)** LIFC and, **(c)** DC on generative results of DM. Brackets in IDCC, DC show labels predicted by CFR.

separately, improves performance with an AUC of 75.27% and 77.40%, respectively. The best results are achieved when both mechanisms are used (DCDM), resulting in an 11% improvement in the AUC score relative to the unconditional model. Although there is a small margin of performance improvement between the combined model (DCDM) and the LIFC model in terms of AUC, the precision improves by 3%, demonstrating the combined model is more precise and hence the best model for OOD detection.

As shown in Fig. 3, the unconditional diffusion model generates a random heart view for a given input for both in-distribution and OOD samples. The IDCC guides the model to generate a heart view according to the in-distribution classifier (CFR) prediction which leads to the generation of similar samples for in-distribution input while dissimilar samples for OOD input. On the other hand, LIFC generates an image with similar spatial information. However, heart views are still generated for OOD samples as the model was only trained on them. When dual-conditioning (DC) is used, the model generates images that are closer aligned to the input image for in-distribution input and high-fidelity heart views for OOD than those generated by a model conditioned on either IDCC or LIFC alone. **Supp. Fig. 1** presents further qualitative ablations.

5 Conclusion

We introduce novel dual-conditioned diffusion model for OOD detection in fetal ultrasound videos and demonstrate how the proposed dual-conditioning mechanisms can manipulate the generative space of a diffusion model. Specifically, we show how our dual-conditioning mechanism can tackle scenarios where the in-distribution data has high inter- (using IDCC) and intra- (using LIFC) class variations and guide a diffusion model to generate similar images to the input for in-distribution input and dissimilar images for OOD input images. Our approach does not require labelled data for OOD classes and is especially applicable to challenging scenarios where the in-distribution data comprises more than one class and there is high similarity between the in-distribution and OOD classes.

Acknowledgement. This work was supported in part by the InnoHK-funded Hong Kong Centre for Cerebro-cardiovascular Health Engineering (COCHE) Project 2.1 (Cardiovascular risks in early life and fetal echocardiography), the UK EPSRC (Engineering and Physical Research Council) Programme Grant EP/T028572/1 (VisualAI), and a UK EPSRC Doctoral Training Partnership award.

References

1. Arjovsky, M., Bottou, L.: Towards principled methods for training generative adversarial networks. arXiv preprint arXiv:1701.04862 (2017)
2. Bau, D., et al.: Seeing what a GAN cannot generate. In: Proceedings of the IEEE/CVF International Conference on Computer Vision, pp. 4502–4511 (2019)
3. Chen, X., Konukoglu, E.: Unsupervised detection of lesions in brain MRI using constrained adversarial auto-encoders. arXiv preprint arXiv:1806.04972 (2018)

4. Choi, H., Jang, E., Alemi, A.A.: Waic, but why? Generative ensembles for robust anomaly detection. arXiv preprint arXiv:1810.01392 (2018)
5. DeVries, T., Taylor, G.W.: Learning confidence for out-of-distribution detection in neural networks. arXiv preprint arXiv:1802.04865 (2018)
6. Dhariwal, P., Nichol, A.: Diffusion models beat GANs on image synthesis. Adv. Neural. Inf. Process. Syst. **34**, 8780–8794 (2021)
7. Drukker, L., et al.: Transforming obstetric ultrasound into data science using eye tracking, voice recording, transducer motion and ultrasound video. Sci. Rep. **11**(1), 14109 (2021)
8. Fort, S.: Adversarial vulnerability of powerful near out-of-distribution detection. arXiv preprint arXiv:2201.07012 (2022)
9. Fort, S., Ren, J., Lakshminarayanan, B.: Exploring the limits of out-of-distribution detection. Adv. Neural. Inf. Process. Syst. **34**, 7068–7081 (2021)
10. Graham, M.S., Pinaya, W.H., Tudosiu, P.D., Nachev, P., Ourselin, S., Cardoso, M.J.: Denoising diffusion models for out-of-distribution detection. arXiv preprint arXiv:2211.07740 (2022)
11. Guénais, T., Vamvourellis, D., Yacoby, Y., Doshi-Velez, F., Pan, W.: Bacoun: Bayesian classifers with out-of-distribution uncertainty. arXiv preprint arXiv:2007.06096 (2020)
12. Hendrycks, D., Gimpel, K.: A baseline for detecting misclassified and out-of-distribution examples in neural networks. arXiv preprint arXiv:1610.02136 (2016)
13. Hertz, A., Mokady, R., Tenenbaum, J., Aberman, K., Pritch, Y., Cohen-Or, D.: Prompt-to-prompt image editing with cross attention control. arXiv preprint arXiv:2208.01626 (2022)
14. Ho, J., Jain, A., Abbeel, P.: Denoising diffusion probabilistic models. Adv. Neural. Inf. Process. Syst. **33**, 6840–6851 (2020)
15. Lee, D., Yu, S., Yu, H.: Multi-class data description for out-of-distribution detection. In: Proceedings of the 26th ACM SIGKDD International Conference on Knowledge Discovery and Data Mining, pp. 1362–1370 (2020)
16. Liu, L., Ren, Y., Lin, Z., Zhao, Z.: Pseudo numerical methods for diffusion models on manifolds. arXiv preprint arXiv:2202.09778 (2022)
17. Margatina, K., Baziotis, C., Potamianos, A.: Attention-based conditioning methods for external knowledge integration. arXiv preprint arXiv:1906.03674 (2019)
18. Meng, C., et al.: Sdedit: guided image synthesis and editing with stochastic differential equations. In: International Conference on Learning Representations (2021)
19. Nalisnick, E., Matsukawa, A., Teh, Y.W., Gorur, D., Lakshminarayanan, B.: Do deep generative models know what they don't know? arXiv preprint arXiv:1810.09136 (2018)
20. Rebain, D., Matthews, M.J., Yi, K.M., Sharma, G., Lagun, D., Tagliasacchi, A.: Attention beats concatenation for conditioning neural fields. arXiv preprint arXiv:2209.10684 (2022)
21. Ren, J., et al.: Likelihood ratios for out-of-distribution detection. Adv. Neural Inform. Process. Syst. **32** (2019)
22. Rombach, R., Blattmann, A., Lorenz, D., Esser, P., Ommer, B.: High-resolution image synthesis with latent diffusion models. In: Proceedings of the IEEE/CVF Conference on Computer Vision and Pattern Recognition, pp. 10684–10695 (2022)
23. Sabokrou, M., Khalooei, M., Fathy, M., Adeli, E.: Adversarially learned one-class classifier for novelty detection. In: Proceedings of the IEEE Conference on Computer Vision and Pattern Recognition, pp. 3379–3388 (2018)
24. Saharia, C., et al.: Palette: Image-to-image diffusion models. In: ACM SIGGRAPH 2022 Conference Proceedings, pp. 1–10 (2022)

25. Schlegl, T., Seeböck, P., Waldstein, S.M., Schmidt-Erfurth, U., Langs, G.: Unsupervised anomaly detection with generative adversarial networks to guide marker discovery. In: Neithammer, M., et al. (eds.) IPMI 2017. LNCS, vol. 10265, pp. 146–157. Springer, Cham (2017). https://doi.org/10.1007/978-3-319-59050-9_12
26. Song, J., Meng, C., Ermon, S.: Denoising diffusion implicit models. arXiv preprint arXiv:2010.02502 (2020)
27. Vaswani, A., et al.: Attention is all you need. Adv. Neural Inform. Process. Syst. **30** (2017)
28. Wald, Y., Feder, A., Greenfeld, D., Shalit, U.: On calibration and out-of-domain generalization. Adv. Neural. Inf. Process. Syst. **34**, 2215–2227 (2021)
29. Wyatt, J., Leach, A., Schmon, S.M., Willcocks, C.G.: Anoddpm: anomaly detection with denoising diffusion probabilistic models using simplex noise. In: Proceedings of the IEEE/CVF Conference on Computer Vision and Pattern Recognition, pp. 650–656 (2022)
30. Xu, H., et al.: Unsupervised anomaly detection via variational auto-encoder for seasonal KPIs in web applications. In: Proceedings of the 2018 World Wide Web Conference, pp. 187–196 (2018)
31. Yang, J., Zhou, K., Li, Y., Liu, Z.: Generalized out-of-distribution detection: a survey. arXiv preprint arXiv:2110.11334 (2021)
32. Yang, L., et al.: Diffusion models: a comprehensive survey of methods and applications. arXiv preprint arXiv:2209.00796 (2022)
33. Zhou, Y.: Rethinking reconstruction autoencoder-based out-of-distribution detection. In: Proceedings of the IEEE/CVF Conference on Computer Vision and Pattern Recognition, pp. 7379–7387 (2022)
34. Zhou, Z., Guo, L.Z., Cheng, Z., Li, Y.F., Pu, S.: Step: out-of-distribution detection in the presence of limited in-distribution labeled data. Adv. Neural. Inf. Process. Syst. **34**, 29168–29180 (2021)

Weakly-Supervised Positional Contrastive Learning: Application to Cirrhosis Classification

Emma Sarfati[1,2(✉)], Alexandre Bône[1], Marc-Michel Rohé[1], Pietro Gori[2], and Isabelle Bloch[2,3]

[1] Guerbet Research, Villepinte, France
[2] LTCI, Télécom Paris, Institut Polytechnique de Paris, Paris, France
emma.sarfati@guerbet.com
[3] Sorbonne Université, CNRS, LIP6, Paris, France

Abstract. Large medical imaging datasets can be cheaply and quickly annotated with low-confidence, weak labels (*e.g.*, radiological scores). Access to high-confidence labels, such as histology-based diagnoses, is rare and costly. Pretraining strategies, like contrastive learning (CL) methods, can leverage unlabeled or weakly-annotated datasets. These methods typically require large batch sizes, which poses a difficulty in the case of large 3D images at full resolution, due to limited GPU memory. Nevertheless, volumetric positional information about the spatial context of each 2D slice can be very important for some medical applications. In this work, we propose an efficient weakly-supervised positional (WSP) contrastive learning strategy where we integrate both the spatial context of each 2D slice and a weak label via a generic kernel-based loss function. We illustrate our method on cirrhosis prediction using a large volume of weakly-labeled images, namely radiological low-confidence annotations, and small strongly-labeled (*i.e.*, high-confidence) datasets. The proposed model improves the classification AUC by 5% with respect to a baseline model on our internal dataset, and by 26% on the public LIHC dataset from the Cancer Genome Atlas. The code is available at: https://github.com/Guerbet-AI/wsp-contrastive.

Keywords: Weakly-supervised learning · Contrastive learning · CT · Cirrhosis prediction · Liver

1 Introduction

In the medical domain, obtaining a large amount of high-confidence labels, such as histopathological diagnoses, is arduous due to the cost and required technicality. It is however possible to obtain lower confidence assessments for a large amount of images, either by a clinical questioning, or directly by a radiological diagnosis. To take advantage of large volumes of unlabeled or weakly-labeled

---

Supplementary Information The online version contains supplementary material available at https://doi.org/10.1007/978-3-031-43907-0_22.

H. Greenspan et al. (Eds.): MICCAI 2023, LNCS 14220, pp. 227–237, 2023.
https://doi.org/10.1007/978-3-031-43907-0_22

images, pre-training encoders with self-supervised methods showed promising results in deep learning for medical imaging [1,4,21,27–29]. In particular, contrastive learning (CL) is a self-supervised method that learns a mapping of the input images to a representation space where similar (positive) samples are moved closer and different (negative) samples are pushed far apart. Weak discrete labels can be integrated into contrastive learning by, for instance, considering as positives only the samples having the same label, as in [13], or by directly weighting unsupervised contrastive and supervised cross entropy loss functions, as in [19]. In this work, we focus on the scenario where radiological meta-data (thus, low-confidence labels) are available for a large amount of images, whereas high-confidence labels, obtained by histological analysis, are scarce.

Naive extensions of contrastive learning methods, such as [5,10,11], from 2D to 3D images may be difficult due to limited GPU memory and therefore small batch size. A usual solution consists in using patch-based methods [8,23]. However, these methods pose two difficulties: they reduce the spatial context (limited by the size of the patch), and they require similar spatial resolution across images. This is rarely the case for abdominal CT/MRI acquisitions, which are typically strongly anisotropic and with variable resolutions. Alternatively, depth position of each 2D slice, within its corresponding volume, can be integrated in the analysis. For instance, in [4], the authors proposed to integrate depth in the sampling strategy for the batch creation. Likewise, in [26], the authors proposed to define as similar only 2D slices that have a *small* depth difference, using a normalized depth coordinate $d \in [0, 1]$. These works implicitly assume a certain threshold on depth to define positive and negative samples, which may be difficult to define and may be different among applications and datasets. Differently, inspired by [2,8], here we propose to use a degree of "positiveness" between samples by defining a kernel function w on depth positions. This allows us to consider volumetric depth information during pre-training *and* to use large batch sizes. Furthermore, we also propose to *simultaneously* leverage weak discrete attributes during pre-training by using a novel and efficient contrastive learning composite kernel loss function, denoting our global method Weakly-Supervised Positional (WSP).

We apply our method to the classification of histology-proven liver cirrhosis, with a large volume of (weakly) radiologically-annotated CT-scans and a small amount of histopathologically-confirmed cirrhosis diagnosis. We compare the proposed approach to existing self-supervised methods.

2 Method

Let x_t be an input 2D image, usually called *anchor*, extracted from a 3D volume, y_t a corresponding discrete weak variable and d_t a related continuous variable. In this paper, y_t refers to a weak radiological annotation and d_t corresponds to the normalized depth position of the 2D image within its corresponding 3D volume: if V_{max} corresponds to the maximal depth-coordinate of a volume V, we compute $d_t = \frac{p_t}{V_{max}}$ with $p_t \in [0, V_{max}]$ being the original depth coordinate.

Let x_j^- and x_i^+ be two semantically different (negative) and similar (positive) images with respect to x_t, respectively.

The definition of similarity is crucial in CL and is the main difference between existing methods. For instance, in unsupervised CL, methods such as SimCLR [5, 6] choose as positive samples random augmentations of the anchor $x_i^+ = t(x_t)$, where $t \sim \mathcal{T}$ is a random transformation chosen among a user-selected family $\mathcal{T}$. Negative images x_j^- are all other (transformed) images present in the batch.

Once x_j^- and x_i^+ are defined, the goal of CL is to compute a mapping function $f_\theta : \mathcal{X} \rightarrow \mathbb{S}^d$, where $\mathcal{X}$ is the set of images and $\mathbb{S}^d$ the representation space, so that similar samples are mapped closer in the representation space than dissimilar samples. Mathematically, this can be defined as looking for a f_θ that satisfies the condition:

$$s_{tj}^- - s_{ti}^+ \leq 0 \quad \forall t, j, i \tag{1}$$

where $s_{tj}^- = sim(f_\theta(x_t), f_\theta(x_j^-))$ and $s_{ti}^+ = sim(f_\theta(x_t), f_\theta(x_i^+))$, with sim a similarity function defined here as $sim(a, b) = \frac{a^T b}{\tau}$ with $\tau > 0$.

In the presence of discrete labels y, the definition of negative (x_j^-) and positive (x_i^+) samples may change. For instance, in SupCon [13], the authors define as positives all images with the same discrete label y. However, when working with continuous labels d, one cannot use the same strategy since all images are somehow positive and negative at the same time. A possible solution [26] would be to define a threshold γ on the distance between labels (*e.g.*, d_a, d_b) so that, if the distance is smaller than γ (*i.e.*, $||d_a - d_b||_2 < \gamma$), the samples (*e.g.*, x_a and x_b) are considered as positives. However, this requires a user-defined hyper-parameter γ, which could be hard to find in practice. A more efficient solution, as proposed in [8], is to define a degree of "positiveness" between samples using a normalized kernel function $w_\sigma(d, d_i) = K_\sigma(d - d_i)$, where K_σ is, for instance, a Gaussian kernel, with user defined hyper-parameter σ and $0 \leq w_\sigma \leq 1$. It is interesting to notice that, for discrete labels, one could also define a kernel as: $w_\delta(y, y_i) = \delta(y - y_i)$, δ being the Dirac function, retrieving exactly SupCon [13].

In this work, we propose to leverage both continuous d and discrete y labels, by combining (here by multiplying) the previously defined kernels, w_σ and w_δ, into a composite kernel loss function. In this way, samples will be considered as similar (positive) only if they have a *composite* degree of "positiveness" greater than zero, namely both kernels have a value greater (or different) than 0 ($w_\sigma > 0$ and $w_\delta \neq 0$). An example of resulting representation space is shown in Fig. 1. This constraint can be defined by slightly modifying the condition introduced in Eq. 1, as:

$$\underbrace{w_\delta(y_t, y_i) \cdot w_\sigma(d_t, d_i)}_{\text{composite kernel } w_{ti}}(s_{tj} - s_{ti}) \leq 0 \quad \forall t, i, j \neq i \tag{2}$$

where the indices t, i, j traverse all N images in the batch since there are no "hard" positive or negative samples, as in SimCLR or SupCon, but all images are considered as positive and negative at the same time. As commonly done in CL [3], this condition can be transformed into an optimization problem using

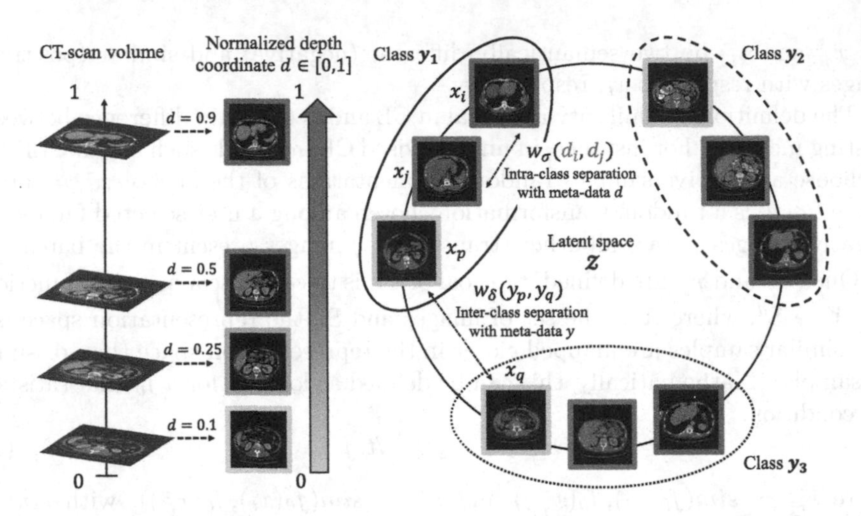

Fig. 1. Example of representation space constructed by our loss function, leveraging both continuous depth coordinate d and discrete label y (*i.e.*, radiological diagnosis y_{radio}). Samples from different radiological classes are well separated and, at the same time, samples are ordered within each class based on their depth coordinate d.

the max operator and its smooth approximation *LogSumExp*:

$$\begin{aligned}\arg\min_{f_\theta} \sum_{t,i} \max(0, w_{ti}\{s_{tj} - s_{ti}\}_{\substack{j=1\\j\neq i}}^{N}) &= \arg\min_{f_\theta} \sum_{t,i} w_{ti} \max(0, \{s_{tj} - s_{ti}\}_{\substack{j=1\\j\neq i}}^{N}) \\ &\approx \arg\min_{f_\theta} \left(- \sum_{t,i} w_{ti} \log \left(\frac{\exp(s_{ti})}{\sum_{j\neq i}^{N} \exp(s_{tj})} \right) \right)\end{aligned} \tag{3}$$

By defining $P(t) = \{i : y_i = y_t\}$ as the set of indices of images x_i in the batch with the same discrete label y_i as the anchor x_t, we can rewrite our final loss function as:

$$\mathcal{L}_{WSP} = - \sum_{t=1}^{N} \sum_{i\in P(t)} w_\sigma(d_t, d_i) \log \left(\frac{\exp(s_{ti})}{\sum_{j\neq i}^{N} \exp(s_{tj})} \right) \tag{4}$$

where $w_\sigma(d_t, d_i)$ is normalized over $i \in P(t)$. In practice, it is rather easy to find a good value of σ, as the proposed kernel method is quite robust to its variation. A robustness study is available in the supplementary material. For the experiments, we fix $\sigma = 0.1$.

3 Experiments

We compare the proposed method with different contrastive and non-contrastive methods, that either use no meta-data (SimCLR [5], BYOL [10]), or leverage only discrete labels (SupCon [13]), or continuous labels (depth-Aware [8]). The proposed method is the only one that takes simultaneously into account both discrete and continuous labels. In all experiments, we work with 2D slices rather than 3D volumes due to the anisotropy of abdominal CT-scans in the depth direction and the limited spatial context or resolution obtained with 3D patch-based or downsampling methods, respectively, which strongly impacts the cirrhosis diagnosis that is notably based on the contours irregularity. Moreover, the large batch sizes necessary in contrastive learning can not be handled in 3D due to a limited GPU memory.

3.1 Datasets

Three datasets of abdominal CT images are used in this study. One dataset is used for contrastive pretraining, and the other two for evaluation. All images have a 512×512 size, and we clip the intensity values between -100 and 400.

$\boldsymbol{\mathcal{D}_{radio}}$. First, $\mathcal{D}_{radio}$ contains 2,799 CT-scans of patients in portal venous phase with a radiological (weak) annotation, *i.e.* realized by a radiologist, indicating four different stages of cirrhosis: no cirrhosis, mild cirrhosis, moderate cirrhosis and severe cirrhosis (y_{radio}). The respective numbers are 1880, 385, 415 and 119. y_{radio} is used as the discrete label y during pre-training.

$\boldsymbol{\mathcal{D}^1_{histo}}$. It contains 106 CT-scans from different patients in portal venous phase, with an identified histopathological status (METAVIR score) obtained by a histological analysis, designated as y^1_{histo}. It corresponds to absent fibrosis (F0), mild fibrosis (F1), significant fibrosis (F2), severe fibrosis (F3) and cirrhosis (F4). This score is then binarized to indicate the absence or presence of advanced fibrosis [14]: F0/F1/F2 (N = 28) vs. F3/F4 (N = 78).

$\boldsymbol{\mathcal{D}^2_{histo}}$. This is the public LIHC dataset from the Cancer Genome Atlas [9], which presents a histological score, the Ishak score, designated as y^2_{histo}, that differs from the METAVIR score present in $\mathcal{D}^1_{histo}$. This score is also distributed through five labels: No Fibrosis, Portal Fibrosis, Fibrous Speta, Nodular Formation and Incomplete Cirrhosis and Established Cirrhosis. Similarly to the METAVIR score in $\mathcal{D}^1_{histo}$, we also binarize the Ishak score, as proposed in [16,20], which results in two cohorts of 34 healthy and 15 pathological patients.

In all datasets, we select the slices based on the liver segmentation of the patients. To gain in precision, we keep the top 70% most central slices with respect to liver segmentation maps obtained manually in $\mathcal{D}_{radio}$, and automatically for $\mathcal{D}^1_{histo}$ and $\mathcal{D}^2_{histo}$ using a U-Net architecture pretrained on $\mathcal{D}_{radio}$ [18]. For the latter pretraining dataset, it presents an average slice spacing of 3.23 mm with a standard deviation of 1.29 mm. For the x and y axis, the dimension is 0.79 mm per voxel on average, with a standard deviation of 0.10 mm.

3.2 Architecture and Optimization

Backbones. We propose to work with two different backbones in this paper: TinyNet and ResNet-18 [12]. TinyNet is a small encoder with 1.1M parameters, inspired by [24], with five convolutional layers, a representation space (for downstream tasks) of size 256 and a latent space (after a projection head of two dense layers) of size 64. In comparison, ResNet-18 has 11.2M parameters, a representation space of dimension 512 and a latent space of dimension 128. More details and an illustration of TinyNet are available in the supplementary material, as well as a full illustration of the algorithm flow.

Data Augmentation, Sampling and Optimization. CL methods [5,10,11] require strong data augmentations on input images, in order to strengthen the association between positive samples [22]. In our work, we leverage three types of augmentations: rotations, crops and flips. Data augmentations are computed on the GPU, using the Kornia library [17]. During inference, we remove the augmentation module to only keep the original input images.

For sampling, inspired by [4], we propose a strategy well-adapted for contrastive learning in 2D medical imaging. We first sample N patients, where N is the batch size, in a balanced way with respect to the radiological/histological classes; namely, we roughly have the same number of subjects per class. Then, we randomly select only one slice per subject. In this way, we maximize the slice heterogeneity within each batch. We use the same sampling strategy also for classification baselines. For $\mathcal{D}^2_{histo}$, which has fewer patients than the batch size, we use a balanced sampling strategy with respect to the radiological/histological classes with no obligation of one slice per patient in the batch. As we work with 2D slices rather than 3D volumes, we compute the average probability per patient of having the pathology. The evaluation results presented later are based on the patient-level aggregated prediction.

Finally, we run our experiments on a Tesla V100 with 16GB of RAM and a 6 CPU cores, and we used the PyTorch-Lightning library to implement our models. All models share the same data augmentation module, with a batch size of $B = 64$ and a fixed number of epochs $n_{epochs} = 200$. For all experiments, we fix a learning rate (LR) of $\alpha = 10^{-4}$ and a weight decay of $\lambda = 10^{-4}$. We add a cosine decay learning rate scheduler [15] to prevent over-fitting. For BYOL, we initialize the moving average decay at 0.996.

Evaluation Protocol. We first pretrain the backbone networks on $\mathcal{D}_{radio}$ using all previously listed contrastive and non-contrastive methods. Then, we train a regularized logistic regression on the frozen representations of the datasets $\mathcal{D}^1_{histo}$ and $\mathcal{D}^2_{histo}$. We use a stratified 5-fold cross-validation. As a baseline, we train a classification algorithm from scratch (supervised) for each dataset, $\mathcal{D}^1_{histo}$ and $\mathcal{D}^2_{histo}$, using both backbone encoders and the same 5-fold cross-validation strategy. We also train a regularized logistic regression on representations obtained with a random initialization as a second baseline (random). Finally, we report the cross-validated results for each model on the aggregated dataset $\mathcal{D}^{1+2}_{histo} = \mathcal{D}^1_{histo} + \mathcal{D}^2_{histo}$.

Table 1. Resulting 5-fold cross-validation AUCs. For each encoder, best results are in **bold**, second top results are underlined. * = We use the pretrained weights from ImageNet with ResNet-18 and run a logistic regression on the frozen representations.

Backbone	Pretraining method	Weak labels	Depth pos.	$\mathcal{D}^1_{histo}$ (N=106)	$\mathcal{D}^2_{histo}$ (N=49)	$\mathcal{D}^{1+2}_{histo}$ (N=155)
TinyNet	Supervised	✗	✗	0.79 (±0.05)	0.65 (±0.25)	0.71 (±0.04)
	None (random)	✗	✗	0.64 (±0.10)	0.75 (±0.13)	0.73 (±0.06)
	SimCLR	✗	✗	0.75 (±0.08)	0.88 (±0.16)	0.76 (±0.11)
	BYOL	✗	✗	0.75 (±0.09)	**0.95 (±0.07)**	0.77 (±0.08)
	SupCon	✓	✗	0.76 (±0.09)	0.93 (±0.07)	0.72 (±0.06)
	depth-Aware	✗	✓	0.80 (±0.13)	0.81 (±0.08)	0.77 (±0.08)
	Ours	✓	✓	**0.84 (±0.12)**	0.91 (±0.11)	**0.79 (±0.11)**
ResNet-18	Supervised	✗	✗	0.77 (±0.10)	0.56 (±0.29)	0.72 (±0.08)
	None (random)	✗	✗	0.69 (±0.19)	0.73 (±0.12)	0.68 (±0.09)
	ImageNet*	✗	✗	0.72 (±0.17)	0.76 (±0.04)	0.66 (±0.10)
	SimCLR	✗	✗	0.79 (±0.09)	0.82 (±0.14)	0.79 (±0.08)
	BYOL	✗	✗	0.78 (±0.09)	0.77 (±0.11)	0.78 (±0.08)
	SupCon	✓	✗	0.69 (±0.07)	0.69 (±0.13)	0.76 (±0.12)
	depth-Aware	✗	✓	0.83 (±0.07)	0.82 (±0.11)	0.80 (±0.07)
	Ours	✓	✓	**0.84 (±0.07)**	**0.85 (±0.10)**	**0.84 (±0.07)**

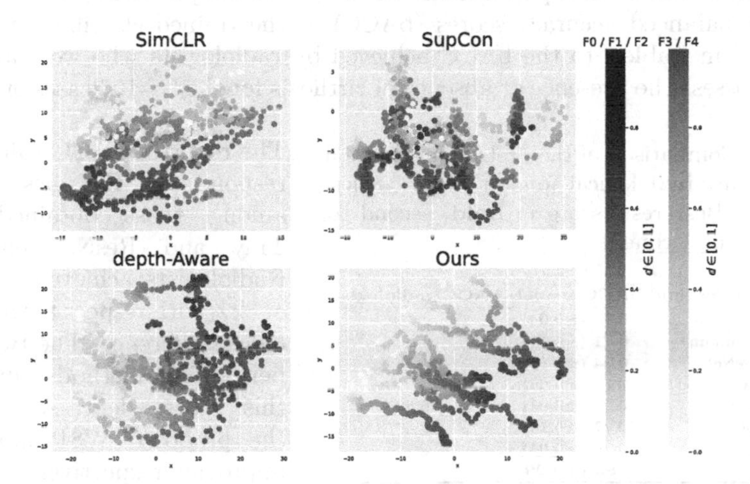

Fig. 2. Projections of the ResNet-18 representation vectors of 10 randomly selected subjects of $\mathcal{D}^1_{histo}$ onto the first two modes of a PCA. Each dot represents a 2D slice. Color gradient refers to different depth positions. Red = cirrhotic cases. Blue = healthy subjects.

4 Results and Discussion

We present in Table 1 the results of all our experiments. For each of them, we report whether the pretraining method integrates the weak label meta-data, the depth spatial encoding, or both, which is the core of our method. First, we can notice that our method outperforms all other pretraining methods in $\mathcal{D}^1_{histo}$ and $\mathcal{D}^{1+2}_{histo}$, which are the two datasets with more patients. For the latter, the proposed method surpasses the second best pretraining method, depth-Aware, by 4%. For $\mathcal{D}^1_{histo}$, it can be noticed that WSP (ours) provides the best AUC score whatever the backbone used. For the second dataset $\mathcal{D}^2_{histo}$, our method is on par with BYOL and SupCon when using a small encoder and outperforms the other methods when using a larger backbone.

To illustrate the impact of the proposed method, we report in Fig. 2 the projections of the ResNet-18 representation vectors of 10 randomly selected subjects of $\mathcal{D}^1_{histo}$ onto the first two modes of a PCA. It can be noticed that the representation space of our method is the only one where the diagnostic label (not available during pretraining) and the depth position are correctly integrated. Indeed, there is a clear separation between slices of different classes (healthy at the bottom and cirrhotic cases at the top) and at the same time it seems that the depth position has been encoded in the x-axis, from left to right. SupCon performs well on the training set of $\mathcal{D}_{radio}$ (figure available in the supplementary material), as well as $\mathcal{D}^2_{histo}$ with TinyNet, but it poorly generalizes to $\mathcal{D}^1_{histo}$ and $\mathcal{D}^{1+2}_{histo}$. The method depth-Aware manages to correctly encode the depth position but not the diagnostic class label.

To assess the clinical performance of the pretraining methods, we also compute the balanced accuracy scores (bACC) of the trained classifiers, which is compared in Table 2 to the bACC achieved by radiologists who were asked to visually assess the presence or absence of cirrhosis for the N=106 cases of $\mathcal{D}^1_{histo}$.

Table 2. Comparison of the pretraining methods with a binary radiological annotation for cirrhosis on $\mathcal{D}^1_{histo}$. Best results are in **bold**, second top results are underlined.

Pretraining method	bACC models	bACC radiologists
Supervised	0.78 (±0.04)	
None (random)	0.71 (±0.13)	
ImageNet	0.74 (±0.13)	
SimCLR	0.78 (±0.08)	
BYOL	0.77 (±0.04)	0.82
SupCon	0.77 (±0.10)	
depth-Aware	<u>0.84 (±0.04)</u>	
Ours	**0.85 (±0.09)**	

The reported bACC values correspond to the best scores among those obtained with Tiny and ResNet encoders. Radiologists achieved a bACC of 82% with respect to the histological reference. The two best-performing methods surpassed this score: depth-Aware and the proposed WSP approach, improving respectively the radiologists score by 2% and 3%, suggesting that including 3D information (depth) at the pretraining phase was beneficial.

5 Conclusion

In this work, we proposed a novel kernel-based contrastive learning method that leverages both continuous and discrete meta-data for pretraining. We tested it on a challenging clinical application, cirrhosis prediction, using three different datasets, including the LIHC public dataset. To the best of our knowledge, this is the first time that a pretraining strategy combining different kinds of meta-data has been proposed for such application. Our results were compared to other state-of-the-art CL methods well-adapted for cirrhosis prediction. The pretraining methods were also compared visually, using a 2D projection of the representation vectors onto the first two PCA modes. Results showed that our method has an organization in the representation space that is in line with the proposed theory, which may explain its higher performances in the experiments. As future work, it would be interesting to adapt our kernel method to non-contrastive methods, such as SimSIAM [7], BYOL [10] or Barlow Twins [25], that need smaller batch sizes and have shown greater performances in computer vision tasks. In terms of application, our method could be easily translated to other medical problems, such as pancreas cancer prediction using the presence of intrapancreatic fat, diabetes mellitus or obesity as discrete meta-labels.

Acknowledgments. This work was supported by Région Ile-de-France (ChoTherIA project) and ANRT (CIFRE #2021/1735).

Compliance with Ethical Standards. This research study was conducted retrospectively using human data collected from various medical centers, whose Ethics Committees granted their approval. Data was de-identified and processed according to all applicable privacy laws and the Declaration of Helsinki.

References

1. Azizi, S., et al.: Big self-supervised models advance medical image classification. In: 2021 IEEE/CVF International Conference on Computer Vision (ICCV), pp. 3458–3468 (2021)
2. Barbano, C.A., Dufumier, B., Duchesnay, E., Grangetto, M., Gori, P.: Contrastive learning for regression in multi-site brain age prediction. In: IEEE ISBI (2022)
3. Barbano, C.A., Dufumier, B., Tartaglione, E., Grangetto, M., Gori, P.: Unbiased Supervised Contrastive Learning. In: ICLR (2023)
4. Chaitanya, K., Erdil, E., Karani, N., Konukoglu, E.: Contrastive learning of global and local features for medical image segmentation with limited annotations. In: Larochelle, H., Ranzato, M., Hadsell, R., Balcan, M., Lin, H. (eds.) Advances in Neural Information Processing Systems. vol. 33, pp. 12546–12558. Curran Associates, Inc. (2020)
5. Chen, T., Kornblith, S., Norouzi, M., et al.: A simple framework for contrastive learning of visual representations. In: 37th International Conference on Machine Learning (ICML) (2020)
6. Chen, T., Kornblith, S., Swersky, K., et al.: Big self-supervised models are strong semi-supervised learners. In: NeurIPS (2020)
7. Chen, X., He, K.: Exploring simple Siamese representation learning. In: 2021 IEEE/CVF Conference on Computer Vision and Pattern Recognition (CVPR), pp. 15745–15753 (2020)
8. Dufumier, B., et al.: Contrastive learning with continuous proxy meta-data for 3D MRI classification. In: de Bruijne, M., et al. (eds.) MICCAI 2021. LNCS, vol. 12902, pp. 58–68. Springer, Cham (2021). https://doi.org/10.1007/978-3-030-87196-3_6
9. Erickson, B.J., Kirk, S., Lee, et al.: Radiology data from the cancer genome atlas colon adenocarcinoma [TCGA-COAD] collection. (2016)

10. Grill, J.B., Strub, F., Altché, F., et al.: Bootstrap your own latent - a new approach to self-supervised learning. In: Larochelle, H., Ranzato, M., Hadsell, R., Balcan, M., Lin, H. (eds.) Advances in Neural Information Processing Systems. vol. 33, pp. 21271–21284. Curran Associates, Inc. (2020)
11. He, K., Fan, H., Wu, Y., Xie, S., Girshick, R.: Momentum contrast for unsupervised visual representation learning. In: IEEE/CVF Conference on Computer Vision and Pattern Recognition (CVPR), pp. 9726–9735 (2020)
12. He, K., Zhang, X., Ren, S., et al.: Deep residual learning for image recognition. IEEE Conference on Computer Vision and Pattern Recognition (CVPR), pp. 770–778 (2016)
13. Khosla, P., Teterwak, P., Wang, C., et al.: Supervised contrastive learning. Adv. Neural. Inf. Process. Syst. **33**, 18661–18673 (2020)
14. Li, Q., Yu, B., Tian, X., Cui, X., Zhang, R., Guo, Q.: Deep residual nets model for staging liver fibrosis on plain CT images. Int. J. Comput. Assist. Radiol. Surg. **15**(8), 1399–1406 (2020). https://doi.org/10.1007/s11548-020-02206-y
15. Loshchilov, I., Hutter, F.: SGDR: Stochastic gradient descent with warm restarts. In: International Conference on Learning Representations (2017)
16. Mohamadnejad, M., et al.: Histopathological study of chronic hepatitis B: a comparative study of Ishak and METAVIR scoring systems. Int. J. Organ Transp. Med. **1** (2010)
17. Riba, E., Mishkin, D., Ponsa, D., Rublee, E., Bradski, G.: Kornia: an open source differentiable computer vision library for PyTorch. In: Winter Conference on Applications of Computer Vision (2020)
18. Ronneberger, O., Fischer, P., Brox, T.: U-Net: convolutional networks for biomedical image segmentation. In: Navab, N., Hornegger, J., Wells, W.M., Frangi, A.F. (eds.) MICCAI 2015. LNCS, vol. 9351, pp. 234–241. Springer, Cham (2015). https://doi.org/10.1007/978-3-319-24574-4_28
19. Sarfati, E., Bone, A., Rohe, M.M., Gori, P., Bloch, I.: Learning to diagnose cirrhosis from radiological and histological labels with joint self and weakly-supervised pretraining strategies. In: IEEE ISBI. Cartagena de Indias, Colombia (Apr 2023)
20. Shiha, G., Zalata, K.: Ishak versus METAVIR: Terminology, convertibility and correlation with laboratory changes in chronic hepatitis C. In: Takahashi, H. (ed.) Liver Biopsy, chap. 10. IntechOpen, Rijeka (2011)
21. Taleb, A., Kirchler, M., Monti, R., Lippert, C.: Contig: Self-supervised multimodal contrastive learning for medical imaging with genetics. In: Proceedings of the IEEE/CVF Conference on Computer Vision and Pattern Recognition (CVPR), pp. 20908–20921 (June 2022)
22. Wang, X., Qi, G.J.: Contrastive learning with stronger augmentations. CoRR abs/2104.07713 (2021)
23. Wen, J., et al.: Convolutional neural networks for classification of Alzheimer's disease: overview and reproducible evaluation. Med. Image Anal. **63**, 101694 (2020)
24. Yin, Y., Yakar, D., Dierckx, R.A.J.O., Mouridsen, K.B., Kwee, T.C., de Haas, R.J.: Liver fibrosis staging by deep learning: a visual-based explanation of diagnostic decisions of the model. Eur. Radiol. **31**(12), 9620–9627 (2021). https://doi.org/10.1007/s00330-021-08046-x
25. Zbontar, J., Jing, L., Misra, I., LeCun, Y., Deny, S.: Barlow twins: Self-supervised learning via redundancy reduction. In: International Conference on Machine Learning (2021)
26. Zeng, D., et al.: Positional contrastive learning for volumetric medical image segmentation. In: MICCAI, pp. 221–230. Springer-Verlag, Berlin, Heidelberg (2021)

27. Zhang, P., Wang, F., Zheng, Y.: Self supervised deep representation learning for fine-grained body part recognition. In: 2017 IEEE 14th International Symposium on Biomedical Imaging (ISBI 2017), pp. 578–582 (2017)
28. Zhou, Z., et al.: Models genesis: generic autodidactic models for 3D medical image analysis. In: Shen, D., et al. (eds.) MICCAI 2019. LNCS, vol. 11767, pp. 384–393. Springer, Cham (2019). https://doi.org/10.1007/978-3-030-32251-9_42
29. Zhuang, X., Li, Y., Hu, Y., Ma, K., Yang, Y., Zheng, Y.: Self-supervised feature learning for 3D medical images by playing a Rubik's cube. In: MICCAI (2019)

Inter-slice Consistency for Unpaired Low-Dose CT Denoising Using Boosted Contrastive Learning

Jie Jing[1], Tao Wang[1], Hui Yu[1], Zexin Lu[1], and Yi Zhang[2](✉)

[1] College of Computer Science, Sichuan University, Chengdu 610065, China
[2] School of Cyber Science and Engineering, Sichuan University, Chengdu 610065, China
yzhang@scu.edu.cn

Abstract. The research field of low-dose computed tomography (LDCT) denoising is primarily dominated by supervised learning-based approaches, which necessitate the accurate registration of LDCT images and the corresponding NDCT images. However, since obtaining well-paired data is not always feasible in real clinical practice, unsupervised methods have become increasingly popular for LDCT denoising. One commonly used method is CycleGAN, but the training processing of CycleGAN is memory-intensive and mode collapse may occur. To address these limitations, we propose a novel unsupervised method based on boosted contrastive learning (BCL), which requires only a single generator. Furthermore, the constraints of computational power and memory capacity often force most existing approaches to focus solely on individual slices, leading to inconsistency in the results between consecutive slices. Our proposed BCL-based model integrates inter-slice features while maintaining the computational cost at an acceptable level comparable to most slice-based methods. Two modifications are introduced to the original contrastive learning method, including weight optimization for positive-negative pairs and imposing constraints on difference invariants. Experiments demonstrate that our method outperforms existing several state-of-the-art supervised and unsupervised methods in both qualitative and quantitative metrics.

Keywords: Low-dose computed tomography · unsupervised learning · image denoising · machine learning

1 Introduction

Computed tomography (CT) is a common tool for medical diagnosis but increased usage has led to concerns about the possible risks caused by excessive radiation exposure. The well-known ALARA (as low as reasonably achievable)

This work was supported in part by the National Natural Science Foundation of China under Grant 62271335; in part by the Sichuan Science and Technology Program under Grant 2021JDJQ0024; and in part by the Sichuan University "From 0 to 1" Innovative Research Program under Grant 2022SCUH0016.

H. Greenspan et al. (Eds.): MICCAI 2023, LNCS 14220, pp. 238–247, 2023.
https://doi.org/10.1007/978-3-031-43907-0_23

[20] principle is widely adopted to reduce exposure based on the strategies such as sparse sampling and tube flux reduction. However, reducing radiation dose will degrade the imaging quality and then inevitably jeopardize the subsequent diagnoses. Various algorithms have been developed to address this issue, which can be roughly categorized into sinogram domain filtration [16], iterative reconstruction [2,9], and image post-processing [1,8,14].

Recently, deep learning (DL) has been introduced for low-dose computed tomography (LDCT) image restoration. The utilization of convolutional neural networks (CNNs) for image super-resolution, as described in [7], outperformed most conventional techniques. As a result, it was subsequently employed for LDCT in [6]. The DIP [21] method is an unsupervised image restoration technique that leverages the inherent ability of untrained networks to capture image statistics. Other methods that do not require clean images are also used in this field [11,23]. Various network architectures have been proposed, such as RED [5] and MAP-NN [18]. The choice of loss function also significantly affects model performance. Perceptual loss [12] based on the pretrained VGG [19] was proposed to mitigate over-smoothing caused by MSE. Most DL techniques for LDCT denoising are supervised models, but unsupervised learning frameworks which do not need paired data for training like GANs [3,10,26], Invertible Network [4] and CUT [25] have also been applied for LDCT [13,15,24].

This study presents a novel unsupervised framework for denoising low-dose CT (LDCT) images, which utilizes contrastive learning (CL) and doesn't require paired data. Our approach possesses three major contributions as follows: Firstly, We discard the use of CycleGAN that most unpaired frameworks employ, instead adopting contrastive learning to design the training framework. As a result, the training process becomes more stable and imposes a lesser computational burden. Secondly, our approach can adapt to almost all end-to-end image translation neural networks, demonstrating excellent flexibility. Lastly, the proposed inter-slice consistency loss makes our model generates stable output quality across slices, in contrast to most slice based methods that exhibit inter-slice instability. Our model outperforms almost all other models in this regard, making it the superior option for LDCT denoising. Further experimental data about this point will be presented in this paper.

2 Method

LDCT image denoising can be expressed as a noise reduction problem in the image domain as $\hat{x} = f(x)$, where $\hat{x}$ and x denote the denoised output and corresponding LDCT image. f represents the denoising function. Rather than directly denoising LDCT images, an encoder-decoder model is used to extract important features from the LDCT images and predict corresponding NDCT images. Most CNN-based LDCT denoising models are based on supervised learning and require both the LDCT and its perfectly paired NDCT images to learn f. However, it is infeasible in real clinical practice. Currently, some unsupervised models, including CUT and CycleGAN, relax the constraint on requiring paired data for training. Instead, these models can be trained with unpaired data.

2.1 Contrastive Learning for Unpaired Data

The task of LDCT image denoising can be viewed as an image translation process from LDCT to NDCT. CUT provides a powerful framework for training a model to complete image-to-image translation tasks. The main concept behind CUT is to use contrastive learning for enhanced feature extraction aided by an adversarial loss.

The key principle of contrastive learning is to create positive and negative pairs of samples, in order to help the model gain strong feature representation ability. The loss of contrastive learning can be formulated as:

$$l(v, v^+, v^-) = -log[\frac{exp(v \cdot v^+/\tau)}{exp(v \cdot v^+/\tau) + \sum_{n=1}^{N} exp(v \cdot v_n^-/\tau)}], \tag{1}$$

where v, v^+, v^- denote the anchors, positive and negative pairs, respectively. N is the number of negative pairs. τ is the temperature factor which is set to 0.07 in this paper. The generator G we used contains two parts, an encoder E and a decoder. A simple MLP H is used to module the features extracted from the encoder. The total loss of CUT for image translation is defined as:

$$L = L_{GAN}(G, D, X, Y) + \lambda_1 L_{PatchNCE}(G, H, X) + \lambda_2 L_{PatchNCE}(G, H, Y), \tag{2}$$

where D denotes the discriminator. X represents the input images, for which $L_{PatchNCE}(G, H, X)$ utilizes contrastive learning in the source domain (represented by noisy images). Y indicates the images in the target domain, which means NDCT images in this paper. $L_{PatchNCE}(G, H, Y)$ employs contrastive learning in this target domain. As noted in a previous study [25], this component plays a similar role as the identity loss in CycleGAN. In this work, λ_1 and λ_2 are both set to 1. Since CT images are three-dimensional data, we can identify more negative pairs between different slices. The strategy about how we design positive and negative pairs for our proposed model is illustrated in Fig. 1.

As shown in Fig. 1, we select two negative patches from the same slice as the anchor, as well as one from the previous slice and the other from the next slice. It is important to note that these patches are not adjacent, since neighbored slices are nearly identical. Similar to most contrastive learning methods, we use cosine similarity to compute the feature similarity.

2.2 Contrastive Learning for Inter-slice Consistency

Due to various constraints, most denoising methods for LDCT can only perform on the slice plane, resulting in detail loss among different slices. While 3D models can mitigate this issue to a certain degree, they require significant computational costs and are prone to model collapse during training, leading to a long training time. Additionally, most methods are unable to maintain structural consistency between slices with certain structures (e.g., bronchi and vessels) appearing continuously across several adjacent slices.

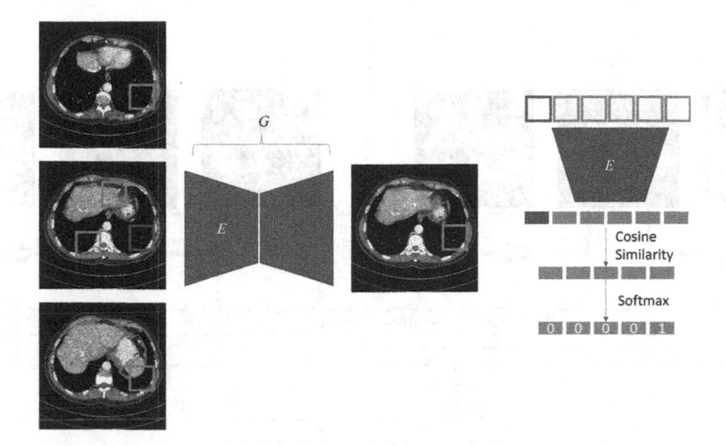

Fig. 1. The construction of training sample pairs for contrastive learning. The generator G is composed of the encoder E and the decoder. The anchor is represented by the red box, while the negative patches are indicated by blue boxes. Two of these negative patches come from the same slice but have different locations, while the other two come from different slices but have the same locations. The green box represents the positive patch, which is located in the same position as the anchor shown in the generated image. (Color figure online)

To address this issue, we design an inter-slice consistency loss based on contrastive learning. This approach helps to maintain structural consistency between slices, and then improve the overall denoising performance.

As illustrated in Fig. 2, we begin by randomly selecting the same patch from both the input (LDCT) and the generated denoised result. These patches are passed through the encoder E, allowing us to obtain the feature representation for each patch. Next, we perform a feature subtraction of each inter-slice pair. The output can be interpreted as the feature difference between slices. We assume that the feature difference between the same pair of slices should be similar, which is formulated as follows:

$$H(E(P(X_t)))-H(E(P(X_{t+1}))) = H(E(P(G(X_t))))-H(E(P(G(X_{t+1})))), \quad (3)$$

where P denotes the patch selection function. A good denoising generator can minimize the feature difference between similar slices while maximizing the feature difference between different slices. By utilizing contrastive learning, we can treat the former condition as a positive pair and the latter as a negative pair. After computing the cosine similarity of the pairs, a softmax operation is applied to assign 1 to the positive pairs and 0 to the negative pairs.

Compared to the original contrastive learning, which focuses on patch pairs, we apply this technique to measure feature differences, which stabilizes the features and improves the consistency between slices.

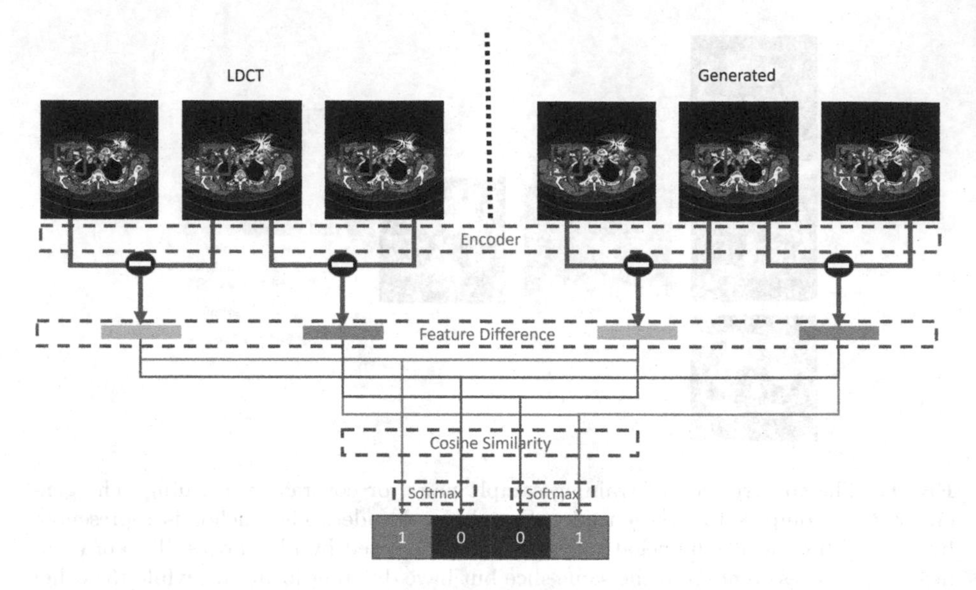

Fig. 2. Contrastive learning ultilized for stablizing inter-slice features. Patches are extracted from the same location in three consecutive slices.

2.3 Boosted Contrastive Learning

Original contrastive Learning approaches treat every positive and negative pair equally. However, in CT images, some patches may be very similar to others (e.g., patches from the same organ), while others may be completely different. Therefore, assigning the same weight to different pairs may not be appropriate. [25] demonstrated that fine-tuning the weights between pairs can significantly improve the performance of contrastive learning.

For our inter-slice consistency loss, only one positive and negative pair can be generated at a time, making it unnecessary to apply reweighting. However, we include additional negative pairs in the patchNCE loss for unpaired translation, making reweighting between pairs more critical than in the original CUT model. As a result, Eq. 1 is updated as follows:

$$l(v, v^+, v^-) = -log[\frac{exp(v \cdot v^+/\tau)}{exp(v \cdot v^+/\tau) + \sum_{n=1}^{N} w_n exp(v \cdot v^-/\tau)}], \quad (4)$$

where w stands for a weight factor for each negative patch.

According to [25], using "easy weighting" is more effective for unpaired tasks, which involves assigning higher weights to easy negative samples (i.e., samples that are easy to distinguish from the anchor). This finding contradicts most people's intuition. Nonetheless, we have demonstrated that their discovery is accurate in our specific scenario. The reweighting approach we have employed is defined as follows:

$$w_n = \frac{exp(1 - v \cdot v_n^-)/\tau)}{\sum_{\substack{j=1 \\ j \neq n}}^{N} exp((1 - v \cdot v_j^-)/\tau)}. \quad (5)$$

In summary, the less similar two patches are, the easier they can be distinguished, the more weight the pair is given for learning purposes.

3 Experiments

3.1 Dataset and Training Details

While our method only requires unpaired data for training, many of the compared methods rely on paired NDCT. We utilized the dataset provided by the Mayo Clinic called "NIH-AAPM-Mayo Clinic Low Dose CT Grand Challenge" [17], which offers paired LD-NDCT images.

The model parameters were initialized using a random Gaussian distribution with zero-mean and standard deviation of 10^{-2}. The learning rate for the optimizer was set to 10^{-4} and halved every 5 epochs for 20 epochs total. The experiments were conducted in Python on a server with an RTX 3090 GPU. Two metrics, peak signal-to-noise ratio (PSNR) and structural similarity index measure (SSIM) [22], were employed to quantitatively evaluate the image quality. The image data from five individuals were used as the training set and the data from other two individuals were used for the test set.

3.2 Comparison of Different Methods

To demonstrate the denoising performance of our model, we conducted experiments to compare our method with various types of denoising methods including unsupervised denoising methods that only use LDCT data, fully supervised methods that use perfectly registered LDCT and NDCT pairs, and semi-supervised methods, including CycleGAN and CUT, which utilize unpaired data. A representative slice processed by different methods is shown in Fig. 3. The window center is set to 40 and the window width is set to 400.

Our framework is flexible and can work with different autoencoder frameworks. In our experiments, the well-known residual encoder-decoder network (RED) was adopted as our network backbone.

The quantitative results and computational costs of unsupervised methods are presented in Table 1. It can be seen that our method produces promising denoising results, with obvious numerical improvements compared to other unsupervised and semi-supervised methods.

As shown in Table 2, our score is very close to our backbone model when trained fully supervised. Our model even got higher PSNR value.

Moreover, our framework is lightweight, which has a similar model scale to RED. It's worth noting that adding perceptual loss to our model will decrease the PSNR result, and it is consistent with the previous studies that perceptual loss may maintain more details but decrease the MSE-based metric, such as PSNR.

Furthermore, the reweighting mechanism demonstrates its effectiveness in improving our model's results. The improvement by introducing the reweighting mechanism can be easily noticed.

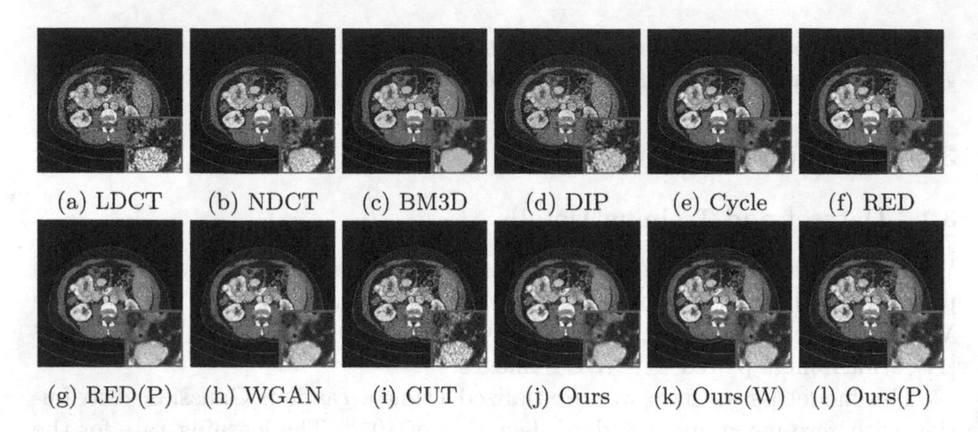

Fig. 3. Methods comparison. DIP and BM3D are fully unsupervised, RED and WGAN models will require paired dataset, CycleGAN("Cycle" in figure), CUT and ours will use unpaired dataset. "(P)" means perceptual loss is added. "(W)" means proposed re-weight mechanism is applied. (Color figure online)

Table 1. Metrics comparison for unsupervised methods. "(P)" means perceptual loss is added. "(W)" means proposed re-weight mechanism is applied.

Metrics	DIP	BM3D	CycleGAN	CUT	Ours	Ours(W)	Ours(P)	LDCT
PSNR	26.79	26.64	28.82	28.15	28.88	**29.09**	28.81	22.33
SSIM	0.86	0.82	**0.91**	0.86	0.90	**0.91**	**0.91**	0.63
MACs(G)	**75.64**	NaN	1576.09	496.87	521.36	521.36	521.36	NaN
Params(M)	2.18	NaN	7.58	3.82	**1.92**	**1.92**	**1.92**	NaN

3.3 Line Plot over Slices

Although our method may only be competitive with supervised methods, we are able to demonstrate the effectiveness of our proposed inter-slice consistency loss. The line plot in Fig. 4 shows the pixel values at point (200, 300) across different slices.

In Fig. 4, it can be observed that our method effectively preserves the inter-slice consistency of features, which is clinically important for maintaining the structural consistency of the entire volume. Although the supervised model achieves a similar overall score to our model, the results across slices of our model are closer to the ground truth (GT), especially when pixel value changes dramatically.

Table 2. Metrics comparison for supervised methods. "(P)" means perceptual loss is added. "(W)" means proposed re-weight mechanism is applied.

Metrics	RED(MSE)	RED(P)	WGAN	Ours(W)	LDCT
PSNR	29.06	28.74	27.75	**29.09**	22.33
SSIM	0.92	**0.92**	0.89	0.91	0.63
MACs(G)	**462.53**	**462.53**	626.89	521.36	NaN
Params(M)	**1.85**	**1.85**	2.52	1.92	NaN

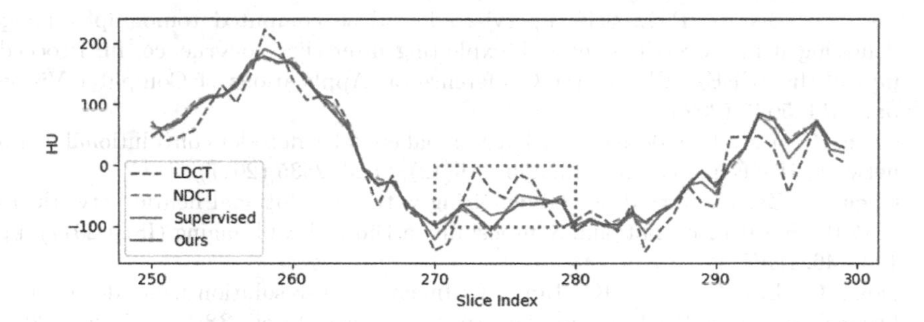

Fig. 4. Inter slice HU value line plot.

3.4 Discussion

Our method achieves competitive results and obtains the highest PSNR value in all the methods with unpaired samples. Although we cannot surpass supervised methods in terms of some metrics, our method produces promising results across consecutive slices that are more consistent and closer to the CT.

4 Conclusion

In this paper, we introduce a novel low-dose CT denoising model. The primary motivation for this work is based on the fact that most CNN-based denoising models require paired LD-NDCT images, while we usually can access unpaired CT data in clinical practice. Furthermore, many existing methods using unpaired samples require extensive computational costs, which can be prohibitive for clinical use. In addition, most existing methods focus on a single slice, which results in inconsistent results across consecutive slices. To overcome these limitations, we propose a novel unsupervised method based on contrastive learning that only requires a single generator. We also apply modifications to the original contrastive learning method to achieve SOTA denoising results using relatively a low computational cost.

Our experiments demonstrate that our method outperforms existing SOTA supervised, semi-supervised, and unsupervised methods in both qualitative and quantitative measures. Importantly, our framework does not require paired training data and is more adaptable for clinical use.

References

1. Aharon, M., Elad, M., Bruckstein, A.: K-SVD: an algorithm for designing overcomplete dictionaries for sparse representation. IEEE Trans. Signal Process. **54**(11), 4311–4322 (2006)
2. Beister, M., Kolditz, D., Kalender, W.A.: Iterative reconstruction methods in X-ray CT. Phys. Med. **28**(2), 94–108 (2012)
3. Bera, S., Biswas, P.K.: Axial consistent memory GAN with interslice consistency loss for low dose computed tomography image denoising. IEEE Trans. Radiation Plasma Med. Sci. (2023)
4. Bera, S., Biswas, P.K.: Self supervised low dose computed tomography image denoising using invertible network exploiting inter slice congruence. In: Proceedings of the IEEE/CVF Winter Conference on Applications of Computer Vision, pp. 5614–5623 (2023)
5. Chen, H., et al.: Low-dose CT with a residual encoder-decoder convolutional neural network. IEEE Trans. Med. Imaging **36**(12), 2524–2535 (2017)
6. Chen, H., Zet al.: Low-dose CT denoising with convolutional neural network. In: 2017 IEEE 14th International Symposium on Biomedical Imaging (ISBI 2017), pp. 143–146. IEEE (2017)
7. Dong, C., Loy, C.C., He, K., Tang, X.: Image super-resolution using deep convolutional networks. IEEE Trans. Pattern Anal. Mach. Intell. **38**(2), 295–307 (2015)
8. Feruglio, P.F., Vinegoni, C., Gros, J., Sbarbati, A., Weissleder, R.: Block matching 3D random noise filtering for absorption optical projection tomography. Phys. Med. Biol. **55**(18), 5401 (2010)
9. Geyer, L.L., et al.: State of the art: iterative CT reconstruction techniques. Radiology **276**(2), 339–357 (2015)
10. Goodfellow, I.J., et al.: Generative adversarial networks. arXiv preprint arXiv:1406.2661 (2014)
11. Jing, J., et al.: Training low dose CT denoising network without high quality reference data. Phys. Med. Biol. **67**(8), 084002 (2022)
12. Johnson, J., Alahi, A., Fei-Fei, L.: Perceptual losses for real-time style transfer and super-resolution. In: Leibe, B., Matas, J., Sebe, N., Welling, M. (eds.) ECCV 2016. LNCS, vol. 9906, pp. 694–711. Springer, Cham (2016). https://doi.org/10.1007/978-3-319-46475-6_43
13. Jung, C., Lee, J., You, S., Ye, J.C.: Patch-wise deep metric learning for unsupervised low-dose ct denoising. In: International Conference on Medical Image Computing and Computer-Assisted Intervention, pp. 634–643. Springer (2022). https://doi.org/10.1007/978-3-031-16446-0_60
14. Kang, D., et al.: Image denoising of low-radiation dose coronary CT angiography by an adaptive block-matching 3D algorithm. In: Medical Imaging 2013: Image Processing. vol. 8669, pp. 86692G. International Society for Optics and Photonics (2013)
15. Li, Z., Huang, J., Yu, L., Chi, Y., Jin, M.: Low-dose CT image denoising using cycle-consistent adversarial networks. In: 2019 IEEE Nuclear Science Symposium and Medical Imaging Conference (NSS/MIC), pp. 1–3 (2019). https://doi.org/10.1109/NSS/MIC42101.2019.9059965
16. Manduca, A., et al.: Projection space denoising with bilateral filtering and CT noise modeling for dose reduction in CT. Med. Phys. **36**(11), 4911–4919 (2009)
17. Moen, T.R., et al.: Low-dose CT image and projection dataset: . Med. Phys. **48**, 902–911 (2021)

18. Shan, H., et al.: Competitive performance of a modularized deep neural network compared to commercial algorithms for low-dose CT image reconstruction. Nature Mach. Intell. **1**(6), 269–276 (2019)
19. Simonyan, K., Zisserman, A.: Very deep convolutional networks for large-scale image recognition. arXiv preprint arXiv:1409.1556 (2014)
20. Smith-Bindman, R., et al.: Radiation dose associated with common computed tomography examinations and the associated lifetime attributable risk of cancer. Arch. Intern. Med. **169**(22), 2078–2086 (2009)
21. Ulyanov, D., Vedaldi, A., Lempitsky, V.S.: Deep image prior. CoRR abs/1711.10925 (2017). https://arxiv.org/abs/1711.10925
22. Wang, Z., Bovik, A., Sheikh, H., Simoncelli, E.: Image quality assessment: from error visibility to structural similarity. IEEE Trans. Image Process. **13**(4), 600–612 (2004). https://doi.org/10.1109/TIP.2003.819861
23. Wu, D., Gong, K., Kim, K., Li, X., Li, Q.: Consensus neural network for medical imaging denoising with only noisy training samples. In: Shen, D., et al. (eds.) MICCAI 2019. LNCS, vol. 11767, pp. 741–749. Springer, Cham (2019). https://doi.org/10.1007/978-3-030-32251-9_81
24. Yang, Q., et al.: Low-dose CT image denoising using a generative adversarial network with Wasserstein distance and perceptual loss. IEEE Trans. Med. Imaging **37**(6), 1348–1357 (2018). https://doi.org/10.1109/TMI.2018.2827462
25. Zhan, F., Zhang, J., Yu, Y., Wu, R., Lu, S.: Modulated contrast for versatile image synthesis. In: 2022 IEEE/CVF Conference on Computer Vision and Pattern Recognition (CVPR). pp. 18259–18269 (2022).https://doi.org/10.1109/CVPR52688.2022.01774
26. Zhu, J.Y., Park, T., Isola, P., Efros, A.A.: Unpaired image-to-image translation using cycle-consistent adversarial networks. In: Computer Vision (ICCV), 2017 IEEE International Conference on (2017)

DAS-MIL: Distilling Across Scales for MIL Classification of Histological WSIs

Gianpaolo Bontempo[1,2], Angelo Porrello[1], Federico Bolelli[1(✉)], Simone Calderara[1], and Elisa Ficarra[1]

[1] University of Modena and Reggio Emilia, Modena, Italy
{gianpaolo.bontempo,angelo.porrello,federico.bolelli, simone.calderara,elisa.ficarra}@unimore.it, gianpaolo.bontempo@phd.unipi.it
[2] University of Pisa, Pisa, Italy

Abstract. The adoption of Multi-Instance Learning (MIL) for classifying Whole-Slide Images (WSIs) has increased in recent years. Indeed, pixel-level annotation of gigapixel WSI is mostly unfeasible and time-consuming in practice. For this reason, MIL approaches have been profitably integrated with the most recent deep-learning solutions for WSI classification to support clinical practice and diagnosis. Nevertheless, the majority of such approaches overlook the multi-scale nature of the WSIs; the few existing hierarchical MIL proposals simply flatten the multi-scale representations by concatenation or summation of features vectors, neglecting the spatial structure of the WSI. Our work aims to unleash the full potential of pyramidal structured WSI; to do so, we propose a graph-based multi-scale MIL approach, termed DAS-MIL, that exploits message passing to let information flows across multiple scales. By means of a knowledge distillation schema, the alignment between the latent space representation at different resolutions is encouraged while preserving the diversity in the informative content. The effectiveness of the proposed framework is demonstrated on two well-known datasets, where we outperform SOTA on WSI classification, gaining a +1.9% AUC and +3.3% accuracy on the popular Camelyon16 benchmark. The source code is available at https://github.com/aimagelab/mil4wsi.

Keywords: Whole-slide Images · Multi-instance Learning · Knowledge Distillation

1 Introduction

Modern microscopes allow the digitalization of conventional glass slides into gigapixel Whole-Slide Images (WSIs) [18], facilitating their preservation and

Supplementary Information The online version contains supplementary material available at https://doi.org/10.1007/978-3-031-43907-0_24.

H. Greenspan et al. (Eds.): MICCAI 2023, LNCS 14220, pp. 248–258, 2023.
https://doi.org/10.1007/978-3-031-43907-0_24

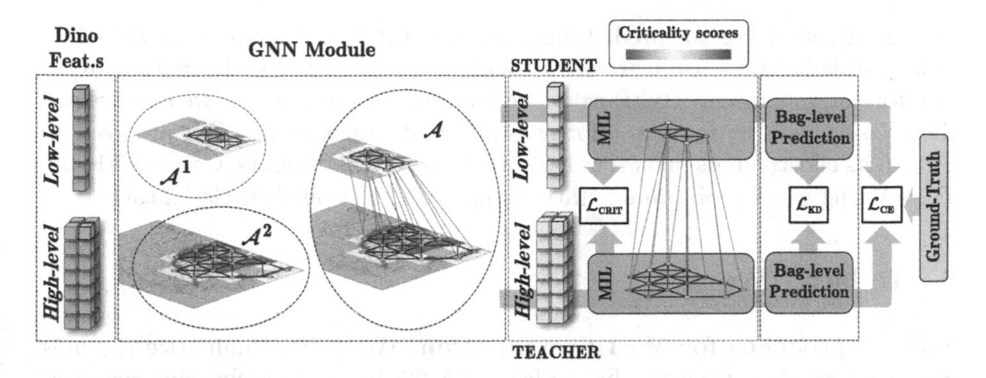

Fig. 1. Overview of our proposed framework, DAS-MIL. The features extracted at different scales are connected (8-connectivity) by means of different graphs. The nodes of both graphs are later fused into a third one, respecting the rule "part of". The contextualized features are then passed to distinct attention-based MIL modules that extract bag labels. Furthermore, a knowledge distillation mechanism encourages the agreement between the predictions delivered by different scales.

retrieval, but also introducing multiple challenges. On the one hand, annotating WSIs requires strong medical expertise, is expensive, time-consuming, and labels are usually provided at the slide or patient level. On the other hand, feeding modern neural networks with the entire gigapixel image is not a feasible approach, forcing to crop data into small patches and use them for training. This process is usually performed considering a single resolution/scale among those provided by the WSI image.

Recently, Multi-Instance Learning (MIL) emerged to cope with these limitations. MIL approaches consider the image slide as a bag composed of many patches, called instances; afterwards, to provide a classification score for the entire bag, they weigh the instances through attention mechanisms and aggregate them into a single representation. It is noted that these approaches are intrinsically flat and disregard the pyramidal information provided by the WSI [15], which have been proven to be more effective than single-resolution [4,13,15,19]. However, to the best of our knowledge, none of the existing proposals leverage the full potential of the WSI pyramidal structure. Indeed, the flat concatenation of features [19] extracted at different resolutions does not consider the substantial difference in the informative content they provide. A proficient learning approach should instead consider the heterogeneity between global structures and local cellular regions, thus allowing the information to flow effectively across the image scales.

To profit from the multi-resolution structure of WSI, we propose a pyramidal Graph Neural Network (GNN) framework combined with (self) Knowledge Distillation (KD), called DAS-MIL (Distilling Across Scales). A visual representation of the proposed approach is depicted in Fig. 1. Distinct GNNs provide contextualized features, which are fed to distinct attention-based MIL modules that compute bag-level predictions. Through knowledge distillation, we encourage agreement across the predictions delivered at different resolutions, while individual scale features are learned in isolation to preserve the diversity in terms of information content. By transferring knowledge across scales, we observe that the classifier self-improves as information flows during training. Our proposal has proven its effectiveness on two well-known histological datasets, Camelyon16 and TCGA lung cancer, obtaining state-of-the-art results on WSI classification.

2 Related Work

MIL Approaches for WSI Classification. We herein summarize the most recent approaches; we refer the reader to [11,26] for a comprehensive overview.

Single-Scale. A classical approach is represented by AB-MIL [16], which employs a side-branch network to calculate the attention scores. In [28], a similar attention mechanism is employed to support a double-tier feature distillation approach, which distills features from pseudo-bags to the original slide. Differently, DS-MIL [19] applies non-local attention aggregation by considering the distance with the most relevant patch. The authors of [20] and [25] propose variations of AB-MIL, which introduce clustering losses and transformers, respectively. In addition, SETMIL [31] makes use of spatial-encoding transformer layers to update the representation. The authors of [7] leverage DINO [5] as feature extractor, highlighting its effectiveness for medical image analysis. Beyond classical attention mechanisms, there are also algorithms based on Recurrent Neural Networks (RNN) [4], and Graphs Neural Networks (GNN) [32].

Multi-Scale. Recently, different authors focused on multi-resolution approaches. DSMIL-LC [19] merges representations from different resolutions, *i.e.*, low instance representations are concatenated with the ones obtained at a higher resolution. MS-RNNMIL [4], instead, fed an RNN with instances extracted at different scales. In [6], a self-supervised hierarchical transformer is applied at each scale. In MS-DA-MIL [13], multi-scale features are included in the same attention algorithm. [10] and [15] exploit multi-resolution through GNN architectures.

Knowledge Distillation. Distilling knowledge from a more extensive network (*teacher*) to a smaller one (*student*) has been widely investigated in recent years [21,24] and applied to different fields, ranging from model compression [3] to WSI analysis [17]. Typically, a tailored learning objective encourages the student to mimic the behaviour of its teacher. Recently, self-supervised representation learning approaches have also employed such a schema: as an example, [5,9] exploit KD to obtain an agreement between networks fed with different views of the same image. In [28], KD is used to transfer the knowledge between MIL tiers applied on different subsamples bags. Taking inspiration from [23] and [30], our model applies (self) knowledge distillation between WSI scale resolutions.

3 Method

Our approach aims to promote the information flow through the different employed resolutions. While existing works [19,20,25] take into account inter-scales interactions by mostly leveraging trivial operations (such as concatenation of related feature representations), we instead provide a novel technique that builds upon: *i)* a GNN module based on message passing, which propagates patches' representation according to the natural structure of multi-resolutions WSI; *ii)* a regulation term based on (self) knowledge distillation, which pins the most effective resolution to further guide the training of the other one(s). In the following, we are delving into the details of our architecture.

Feature Extraction. Our work exploits DINO, the self-supervised learning approach proposed in [5], to provide a relevant representation of each patch. Differently from other proposals [19,20,28], it focuses solely on aligning positive pairs during optimization (and hence avoids negative pairs), which has shown to require a lower memory footprint during training. We hence devise an initial stage with multiple self-supervised feature extractors $f(\cdot;\theta_1), \ldots, f_M(\cdot;\theta_M)$, one dedicated to each resolution: this way, we expect to promote feature diversity across scales. After training, we freeze the weights of these networks and use them as patch-level feature extractors. Although we focus only on two resolutions at time (*i.e.*, $M = 2$) the approach can be extended to more scales.

Architecture. The representations yield by DINO provide a detailed description of the local patterns in each patch; however, they retain poor knowledge of the surrounding context. To grasp a global guess about the entire slide, we allow patches to exchange local information. We achieve it through a Pyramidal Graph Neural Network (PGNN) in which each node represents an individual WSI patch seen at different scales. Each node is connected to its neighbors (8-connectivity) in the euclidean space and between scales following the relation "part of"[1]. To perform message passing, we adopt Graph ATtention layers (GAT) [27].

In general terms, such a module takes as input multi-scale patch-level representations $\mathcal{X} = [\mathcal{X}_1 \| \mathcal{X}_2]$, where $\mathcal{X}_1 \in \mathbb{R}^{N_1 \times F}$ and $\mathcal{X}_2 \in \mathbb{R}^{N_2 \times F}$ are respectively the representations of the lower and higher scale. The input undergoes two graph layers: while the former treats the two scales as independent subgraphs $\mathcal{A}_1 \in \mathbb{R}^{N_1 \times N_1}$ and $\mathcal{A}_2 \in \mathbb{R}^{N_2 \times N_2}$, the latter process them jointly by considering the entire graph $\mathcal{A}$ (see Fig. 1, left). In formal terms:

$$\begin{aligned}\mathcal{H} &= \mathrm{PGNN}(\mathcal{X}; \mathcal{A}, \mathcal{A}_1, \mathcal{A}_2, \theta_{\mathrm{PGNN}}) \\ &= \mathrm{GAT}([\mathrm{GAT}(\mathcal{X}_1; \mathcal{A}_1, \theta_1) \| \mathrm{GAT}(\mathcal{X}_2; \mathcal{A}_2, \theta_2)]; \mathcal{A}, \theta_3),\end{aligned}$$

where $\mathcal{H} \equiv [\mathcal{H}_1 \| \mathcal{H}_2]$ stands for the output of the PGNN obtained by concatenating the two scales. These new contextualized patch representations are then fed to the attention-based MIL module proposed in [19], which produces bag-level

[1] The relation "part of" connects a parent WSI patch (lying in the lower resolution) with its children, *i.e.*, the higher-scale patches it contains.

scores y_1^{BAG}, $y_2^{\text{BAG}} \in \mathbb{R}^{1\times C}$ where C equals the number of classes. Notably, such a module provides additional importance scores $z_1 \in \mathbb{R}^{N_1}$ and $z_2 \in \mathbb{R}^{N_2}$, which quantifies the importance of each original patch to the overall prediction.

Aligning Scales with (Self) Knowledge Distillation. We have hence obtained two distinct sets of predictions for the two resolutions: namely, a bag-level score (*e.g.*, a tumor is either present or not) and a patch-level one (*e.g.*, which instances contribute the most to the target class). However, as these learned metrics are inferred from different WSI zooms, a disagreement may emerge: indeed, we have observed (see Table 4) that the higher resolutions generally yield better classification performance. In this work, we exploit such a disparity to introduce two additional optimization objectives, which pin the predictions out of the higher scale as teaching signal for the lower one. Further than improving the results of the lowest scale only, we expect its benefits to propagate also to the shared message-passing module, and so to the higher resolution.

Formally, the first term seeks to align bag predictions from the two scales through (self) knowledge distillation [14,29]:

$$\mathcal{L}_{\text{KD}} = \tau^2 \ \text{KL}(\text{softmax}(\frac{y_1^{\text{BAG}}}{\tau}) \parallel \text{softmax}(\frac{y_2^{\text{BAG}}}{\tau})), \tag{1}$$

where KL stands for the Kullback-Leibler divergence and τ is a temperature that lets secondary information emerge from the teaching signal.

The second aligning term regards the instance scores. It encourages the two resolutions to assign criticality scores in a *consistent* manner: intuitively, if a low-resolution patch has been considered critical, then the average score attributed to its children patches should be likewise high. We encourage such a constraint by minimizing the Euclidean distance between the low-resolution criticality grid map z_1 and its subsampled counterpart computed by the high-resolution branch:

$$\mathcal{L}_{\text{CRIT}} = \|z_1 - \text{GraphPooling}(z_2)\|_2^2. \tag{2}$$

In the equation above, GraphPooling identifies a pooling layer applied over the higher scale: to do so, it considers the relation "part of" between scales and then averages the child nodes, hence allowing the comparison at the instance level.

Overall Objective. To sum up, the overall optimization problem is formulated as a mixture of two objectives: the one requiring higher conditional likelihood w.r.t. ground truth labels $\mathbf{y}$ and carried out through the Cross-Entropy loss $\mathcal{L}_{\text{CE}}(\cdot;\mathbf{y})$; the other one based on knowledge distillation:

$$\min_{\theta} \ (1-\lambda)\mathcal{L}_{\text{CE}}(y_2^{\text{BAG}}) + \mathcal{L}_{\text{CE}}(y_1^{\text{BAG}}) + \lambda\mathcal{L}_{\text{KD}} + \beta\mathcal{L}_{\text{CRIT}}, \tag{3}$$

where λ is a hyperparameter weighting the tradeoff between the teaching signals provided by labels and the higher resolution, while β balances the contributions of the consistency regularization introduced in Eq. (2).

4 Experiments

WSIs Pre-processing. We remove background patches through an approach similar to the one presented in the CLAM framework [20]: after an initial segmentation process based on Otsu [22] and Connected Component Analysis [2], non-overlapped patches within the foreground regions are considered.

Optimization. We use Adam as optimizer, with a learning rate of 2×10^{-4} and a cosine annealing scheduler (10^{-5} decay w/o warm restart). We set $\tau = 1.5$, $\beta = 1$, and $\lambda = 1$. The DINO feature extractor has been trained with two RTX5000 GPUs: differently, all subsequent experiments have been performed with a single RTX2080 GPU using Pytorch-Geometric [12]. To asses the performance of our approach, we adhere to the protocol of [19,28] and use the accuracy and AUC metrics. Moreover, the classifier on the higher scale has been used to make the final overall prediction. Regarding the KD loss, we apply the temperature term to both student and teacher outputs for numerical stability.

Table 1. Comparison with state-of-the-art solutions. Results marked with "†" have been calculated on our premises as the original papers lack the specific settings; all the other numbers are taken from [19,28].

Method		Camelyon16		TCGA Lung	
		Accuracy	AUC	Accuracy	AUC
Single Scale	Mean-pooling †	0.723	0.672	0.823	0.905
	Max-pooling †	0.893	0.899	0.851	0.909
	MILRNN [4]	0.806	0.806	0.862	0.911
	ABMIL [16]	0.845	0.865	0.900	0.949
	CLAM-SB [20]	0.865	0.885	0.875	0.944
	CLAM-MB [20]	0.850	0.894	0.878	0.949
	Trans-MIL † [25]	0.883	0.942	0.881	0.948
	DTFD (AFS) [28]	0.908	0.946	0.891	0.951
	DTFD (MaxMinS) [28]	0.899	0.941	0.894	0.961
	DSMIL † [19]	0.915	0.952	0.888	0.951
Multi Scale	MS-DA-MIL [13]	0.876	0.887	0.900	0.955
	MS-MILRNN [4]	0.814	0.837	0.891	0.921
	HIPT † [6]	0.890	0.951	0.890	0.950
	DSMIL-LC † [19]	0.909	0.955	0.913	0.964
	H^2-MIL † [15]	0.859	0.912	0.823	0.917
	DAS-MIL (ours)	**0.945**	**0.973**	**0.925**	**0.965**

Camelyon16. [1] We adhere to the official training/test sets. To produce the fairest comparison with the single-scale state-of-the-art solution, the 270 remaining WSIs are split into training and validation in the proportion 9:1.

TCGA Lung Dataset. It is available on the GDC Data Transfer Portal and comprises two subsets of cancer: Lung Adenocarcinoma (LUAD) and Lung Squamous Cell Carcinoma (LUSC), counting 541 and 513 WSIs, respectively. The aim is to classify LUAD *vs* LUSC; we follow the split proposed by DSMIL [19].

4.1 Comparison with the State-of-the-art

Table 1 compares our DAS-MIL approach with the state-of-the-art, including both single- and multi-scale architectures. As can be observed: *i)* the joint exploitation of multiple resolutions is generally more efficient; *ii)* our DAS-MIL yields robust and compelling results, especially on Camelyon16, where it provides 0.945 of accuracy and 0.973 AUC (*i.e.*, an improvement of +3.3% accuracy and +1.9% AUC with respect to the SOTA). Finally, we remark that most of the methods in the literature resort to different feature extractors; however, the next subsections prove the consistency of DAS-MIL benefits across various backbones.

Table 2. Impact (AUC, Camelyon16) of Eq. 3 hyperparameters.

λ	20×	10×	β	20×	10×
1.0	**0.973**	**0.974**	1.5	0.964	0.968
0.8	0.967	0.966	1.2	0.970	0.964
0.5	0.968	0.932	1.0	**0.973**	**0.974**
0.3	0.962	0.965	0.8	0.962	0.965
0.0	0.955	0.903	0.6	0.951	0.953

Table 3. Impact (Camelyon16) of KD temperature (Eq. 1), $\alpha = \beta = 1.0$.

τ	Accuracy		AUC	
	20×	10×	20×	10×
$\tau = 1$	0.883	0.962	0.906	0.957
$\tau = 1.3$	0.898	0.958	0.891	0.959
$\tau = 1.5$	**0.945**	**0.945**	**0.973**	**0.974**
$\tau = 2$	0.906	0.914	0.962	0.963
$\tau = 2.5$	0.922	0.914	0.951	0.952

4.2 Model Analysis

On the Impact of Knowledge Distillation. To assess its merits, we conducted several experiments varying the values of the corresponding balancing coefficients (see Table 2). As can be observed, lowering their values (even reaching $\lambda = 0$, *i.e.*, no distillation is performed) negatively affects the performance. Such a statement holds not only for the lower resolution (as one could expect), but also for the higher one, thus corroborating the claims we made in Sect. 3 on the bidirectional benefits of knowledge distillation in our multi-scale architecture.

We have also performed an assessment on the temperature τ, which controls the smoothing factor applied to teacher's predictions (Table 3). We found that the lowest the temperature, the better the results, suggesting that the teacher scale is naturally not over-confident about its predictions, but rather well-calibrated.

Table 4. Comparison between scales. The target column indicates the features passed to the two MIL layers: the "||" symbol indicates that they have been previously concatenated.

Input Scale	MIL Target(s)	Accuracy	AUC
10×	10×	0.818	0.816
20×	20×	0.891	0.931
5×, 20×	5×, 20×	0.891	0.938
5×, 20×	5×, [5× || 20×]	0.898	0.941
10×, 20×	10×, 20×	0.945	0.973
10×, 20×	10×, [10× || 20×]	0.922	0.953

Single-Scale *vs* Multi-Scale.
Table 4 demonstrates the contribution of hierarchical representations. For single-scale experiments, the model is fed only with patches extracted at a single reference scale. For what concerns multi-scale results, representations can be combined in different ways. Overall, the best results are obtained with 10× and 20× input resolutions; the table also highlights that 5× magnitude is less effective and presents a worst discriminative capability. We ascribe it to the specimen-level pixel size relevant for cancer diagnosis task; different datasets/tasks may benefit from different scale combinations.

Table 5. Comparison between DAS-MIL with and w/o (✗) the graph contextualization mechanism, and the most recent graph-based multi-scale approach H^2-MIL, when using different resolutions as input (5× and 20×).

Feature Extractor	Graph Mechanism	Camelyon16		TCGA Lung	
		Acc.	AUC	Acc.	AUC
SimCLR	✗	0.859	0.869	0.864	0.932
SimCLR	DAS-MIL	**0.906**	**0.928**	**0.883**	**0.9489**
SimCLR	H^2-MIL	0.836	0.857	0.826	0.916
DINO	✗	0.852	0.905	0.906	0.956
DINO	DAS-MIL	**0.891**	**0.938**	**0.925**	**0.965**
DINO	H^2-MIL	0.859	0.912	0.823	0.917

The Impact of the Feature Extractors and GNNs. Table 5 proposes an investigation of these aspects, which considers both SimCRL [8] and DINO, as well as the recently proposed graph mechanism H^2-MIL [15]. In doing so, we fix the input resolutions to 5× and 20×. We draw the following conclusions: *i)* when our DAS-MIL feature propagation layer is used, the selection of the optimal feature extractor (*i.e.*, SimCLR *vs* Dino) has less impact on performance, as the message-passing can compensate for possible lacks in the initial representation; *ii)* DAS-MIL appears a better features propagator w.r.t. H^2-MIL.

H^2-MIL exploits a global pooling layer (IHPool) that fulfils only the spatial structure of patches: as a consequence, if non-tumor patches surround a tumor patch, its contribution to the final prediction is likely to be outweighed by the IHPool module of H^2-MIL. Differently, our approach is not restricted in such a way, as it can dynamically route the information across the hierarchical structure (also based on the connections with the critical instance).

5 Conclusion

We proposed a novel way to exploit multiple resolutions in the domain of histological WSI. We conceived a novel graph-based architecture that learns spatial correlation at different WSI resolutions. Specifically, a GNN cascade architecture is used to extract context-aware and instance-level features considering the spatial relationship between scales. During the training process, this connection is further amplified by a distillation loss, asking for an agreement between the lower and higher scales. Extensive experiments show the effectiveness of the proposed distillation approach.

Acknowledgement. This project has received funding from DECIDER, the European Union's Horizon 2020 research and innovation programme under GA No. 965193, and from the Department of Engineering "Enzo Ferrari" of the University of Modena through the FARD-2022 (Fondo di Ateneo per la Ricerca 2022). We also acknowledge the CINECA award under the ISCRA initiative, for the availability of high performance computing resources and support.

References

1. Bejnordi, B.E., et al.: Diagnostic assessment of deep learning algorithms for detection of lymph node metastases in women with breast cancer. JAMA **318**(22), 2199–2210 (2017)
2. Bolelli, F., Allegretti, S., Grana, C.: One DAG to rule them all. IEEE Trans. Pattern Anal. Mach. Intell. **44**(7), 3647–3658 (2021)
3. Buciluă, C., Caruana, R., Niculescu-Mizil, A.: Model compression. In: Proceedings of the Twelfth ACM SIGKDD International Conference on Knowledge Discovery and Data Mining, pp. 535–541 (2006)
4. Campanella, G., et al.: Clinical-grade computational pathology using weakly supervised deep learning on whole slide images. Nat. Med. **25**(8), 1301–1309 (2019)
5. Caron, M., et al.: Emerging properties in self-supervised vision transformers. In: IEEE/CVF International Conference on Computer Vision (ICCV), pp. 9650–9660 (2021)
6. Chen, R.J., et al.: Scaling vision transformers to gigapixel images via hierarchical self-supervised learning. In: IEEE/CVF Conference on Computer Vision and Pattern Recognition (CVPR), pp. 16144–16155 (2022)
7. Chen, R.J., Krishnan, R.G.: Self-Supervised Vision Transformers Learn Visual Concepts in Histopathology. Learning Meaningful Representations of Life, NeurIPS (2022)

8. Chen, T., Kornblith, S., Norouzi, M., Hinton, G.: A simple framework for contrastive learning of visual representations. In: International Conference on Machine Learning, pp. 1597–1607. PMLR (2020)
9. Chen, X., He, K.: Exploring simple Siamese representation learning. In: IEEE/CVF Conference on Computer Vision and Pattern Recognition (CVPR), pp. 15750–15758 (2021)
10. Chen, Z., Zhang, J., Che, S., Huang, J., Han, X., Yuan, Y.: Diagnose like a pathologist: weakly-supervised pathologist-tree network for slide-level immunohistochemical scoring. In: Proceedings of the AAAI Conference on Artificial Intelligence (2021)
11. Dimitriou, N., Arandjelović, O., Caie, P.D.: Deep learning for whole slide image analysis: an overview. Front. Med. **6**, 264 (2019)
12. Fey, M., Lenssen, J.E.: Fast graph representation learning with pytorch geometric. In: ICLR Workshop on Representation Learning on Graphs and Manifolds (2019)
13. Hashimoto, N., et al.: Multi-scale domain-adversarial multiple-instance CNN for cancer subtype classification with unannotated histopathological images. In: IEEE/CVF Conference on Computer Vision and Pattern Recognition (CVPR), pp. 3852–3861 (2020)
14. Hinton, G., Vinyals, O., Dean, J.: Distilling the knowledge in a neural network. In: NIPS Deep Learning and Representation Learning Workshop (2015)
15. Hou, W., et al.: H2-MIL: exploring hierarchical representation with heterogeneous multiple instance learning for whole slide image analysis. In: Proceedings of the AAAI Conference on Artificial Intelligence (2022)
16. Ilse, M., Tomczak, J., Welling, M.: Attention-based deep multiple instance learning. In: International Conference on Machine Learning, vol. 80, pp. 2127–2136. PMLR (2018)
17. Ilyas, T., Mannan, Z.I., Khan, A., Azam, S., Kim, H., De Boer, F.: TSFD-Net: tissue specific feature distillation network for nuclei segmentation and classification. Neural Netw. **151**, 1–15 (2022)
18. Kumar, N., Gupta, R., Gupta, S.: Whole slide imaging (WSI) in pathology: current perspectives and future directions. J. Digit. Imaging **33**(4), 1034–1040 (2020)
19. Li, B., Li, Y., Eliceiri, K.W.: Dual-stream multiple instance learning network for whole slide image classification with self-supervised contrastive learning. In: IEEE/CVF Conference on Computer Vision and Pattern Recognition (CVPR), pp. 14318–14328 (2021)
20. Lu, M.Y., Williamson, D.F., Chen, T.Y., Chen, R.J., Barbieri, M., Mahmood, F.: Data-efficient and weakly supervised computational pathology on whole-slide images. Nat. Biomed. Eng. **5**(6), 555–570 (2021)
21. Monti, A., Porrello, A., Calderara, S., Coscia, P., Ballan, L., Cucchiara, R.: How many observations are enough? Knowledge distillation for trajectory forecasting. In: IEEE/CVF Conference on Computer Vision and Pattern Recognition (CVPR), pp. 6543–6552 (2022)
22. Otsu, N.: A threshold selection method from gray-level histograms. IEEE Trans. Syst. Man Cybern. **9**(1), 62–66 (1979)
23. Porrello, A., Bergamini, L., Calderara, S.: Robust Re-Identification by Multiple Views Knowledge Distillation. In: Computer Vision - ECCV 2020. pp. 93–110. Springer (2020)
24. Porrello, A., Bergamini, L., Calderara, S.: Robust re-identification by multiple views knowledge distillation. In: Vedaldi, A., Bischof, H., Brox, T., Frahm, J.-M. (eds.) ECCV 2020. LNCS, vol. 12355, pp. 93–110. Springer, Cham (2020). https://doi.org/10.1007/978-3-030-58607-2_6

25. Shao, Z., Bian, H., Chen, Y., Wang, Y., Zhang, J., Ji, X., et al.: Transmil: transformer based correlated multiple instance learning for whole slide image classification. Adv. Neural Inf. Process. Syst. (NeurIPS) **34**, 2136–2147 (2021)
26. Srinidhi, C.L., Ciga, O., Martel, A.L.: Deep neural network models for computational histopathology: a survey. Med. Image Anal. **67**, 101813 (2021)
27. Veličković, P., Cucurull, G., Casanova, A., Romero, A., Liò, P., Bengio, Y.: Graph attention networks. In: International Conference on Learning Representations (2018). accepted as poster
28. Zhang, H., et al.: DTFD-MIL: double-tier feature distillation multiple instance learning for histopathology whole slide image classification. In: IEEE/CVF Conference on Computer Vision and Pattern Recognition (CVPR), pp. 18802–18812 (2022)
29. Zhang, L., Bao, C., Ma, K.: Self-distillation: towards efficient and compact neural networks. IEEE Trans. Pattern Anal. Mach. Intell. **44**(8), 4388–4403 (2021)
30. Zhang, L., Song, J., Gao, A., Chen, J., Bao, C., Ma, K.: Be your own teacher: improve the performance of convolutional neural networks via self distillation. In: IEEE/CVF International Conference on Computer Vision (ICCV), pp. 3713–3722 (2019)
31. Zhao, Y. et al.: SETMIL: spatial encoding transformer-based multiple instance learning for pathological image analysis. In: Wang, L., Dou, Q., Fletcher, P.T., Speidel, S., Li, S. (eds.) Medical Image Computing and Computer Assisted Intervention – MICCAI 2022. MICCAI 2022. LNCS, vol. 13432, pp. 66–76. Springer, Cham (2022). https://doi.org/10.1007/978-3-031-16434-7_7
32. Zhao, Y., et al.: Predicting lymph node metastasis using histopathological images based on multiple instance learning with deep graph convolution. In: IEEE/CVF Conference on Computer Vision and Pattern Recognition (CVPR), pp. 4837–4846 (2020)

SLPD: Slide-Level Prototypical Distillation for WSIs

Zhimiao Yu, Tiancheng Lin, and Yi Xu(✉)

MoE Key Lab of Artificial Intelligence, AI Institute,
Shanghai Jiao Tong University, Shanghai, China
xuyi@sjtu.edu.cn

Abstract. Improving the feature representation ability is the foundation of many whole slide pathological image (WSIs) tasks. Recent works have achieved great success in pathological-specific self-supervised learning (SSL). However, most of them only focus on learning patch-level representations, thus there is still a gap between pretext and slide-level downstream tasks, *e.g.*, subtyping, grading and staging. Aiming towards slide-level representations, we propose Slide-Level Prototypical Distillation (SLPD) to explore intra- and inter-slide semantic structures for context modeling on WSIs. Specifically, we iteratively perform intra-slide clustering for the regions (4096×4096 patches) within each WSI to yield the prototypes and encourage the region representations to be closer to the assigned prototypes. By representing each slide with its prototypes, we further select similar slides by the set distance of prototypes and assign the regions by cross-slide prototypes for distillation. SLPD achieves state-of-the-art results on multiple slide-level benchmarks and demonstrates that representation learning of semantic structures of slides can make a suitable proxy task for WSI analysis. Code will be available at https://github.com/Carboxy/SLPD.

Keywords: Computational pathology · Whole slide images(WSIs) · Self-supervised learning

1 Introduction

In computational histopathology, visual representation extraction is a fundamental problem [14], serving as a cornerstone of the (downstream) task-specific learning on whole slide pathological images (WSIs). Our community has witnessed the progress of the *de facto* representation learning paradigm from the supervised ImageNet pre-training to self-supervised learning (SSL) [15,36]. Numerous

Z. Yu and T. Lin—Equal contribution.

Supplementary Information The online version contains supplementary material available at https://doi.org/10.1007/978-3-031-43907-0_25.

H. Greenspan et al. (Eds.): MICCAI 2023, LNCS 14220, pp. 259–269, 2023.
https://doi.org/10.1007/978-3-031-43907-0_25

pathological applications benefit from SSL, including classification of glioma [7], breast carcinoma [1], and non-small-cell lung carcinoma [25], mutation prediction [32], microsatellite instability prediction [31], and survival prediction from WSIs [2,16]. Among them, pioneering works [12,22,27] directly apply the SSL algorithms developed for natural images (*e.g.*, SimCLR [10], CPC [30] and MoCo [11]) to WSI analysis tasks, and the improved performance proves the effectiveness of SSL. However, WSI is quite different from natural images in that it exhibits a hierarchical structure with giga-pixel resolution. Following works turn to *designing pathological-specific tasks* to explore the inherent characteristics of WSIs for representation learning, *e.g.*, resolution-aware tasks [18,34,37] and color-aware tasks [2,38]. Since the pretext tasks encourage to mine the pathologically relevant patterns, the learned representations are expected to be more suitable for WSI analysis. Nevertheless, these works only consider learning the representations at the patch level, *i.e*, the cellular organization, but neglecting macro-scale morphological features, *e.g.*, tissue phenotypes and intra-tumoral heterogeneity. As a result, there is still a gap between the pre-trained representations and downstream tasks, as the latter is mainly at the slide level, *e.g.*, subtyping, grading and staging.

More recently, some works propose to close the gap via *directly learning slide-level representations in pre-training.* For instance, HIPT [8], a milestone work, introduces hierarchical pre-training (DINO [6]) for the patch-level (256×256) and region-level (4096×4096) in a two-stage manner, achieving superior performance on slide-level tasks. SS-CAMIL [13] uses EfficientNet-B0 for image compression in the first stage and then derives multi-task learning on the compressed WSIs, which assumes the primary site information, *e.g.*, the organ type, is always available and can be used as pseudo labels. SS-MIL [35] also proposes a two-stage pre-training framework for WSIs using contrastive learning (SimCLR [10]), where the differently subsampled bags[1] from the same WSI are positive pairs in the second stage. A similar idea can be found in Giga-SSL [20] with delicate patch- and WSI-level augmentations. The aforementioned methods share the same two-stage pre-training paradigm, *i.e.*, patch-to-region/slide. Thus broader context information is preserved to close the gap between pretext and downstream tasks. However, they are essentially *instance discrimination* where only the self-invariance of region/slide is considered, leaving the *intra- and inter-slide semantic structures* unexplored.

In this paper, we propose to encode the intra- and inter-slide semantic structures by modeling the mutual-region/slide relations, which is called SLPD: Slide-Level Prototypical Distillation for WSIs. Specifically, we perform the slide-level clustering for the 4096×4096 regions within each WSI to yield the prototypes, which characterize the medically representative patterns of the tumor (*e.g.*, morphological phenotypes). In order to learn this intra-slide semantic structure, we encourage the region representations to be closer to the assigned prototypes. By representing each slide with its prototypes, we further select semantically simi-

[1] By formulating WSI tasks as a multi-instance learning problem, the WSI is treated as a bag with corresponding patches as instances.

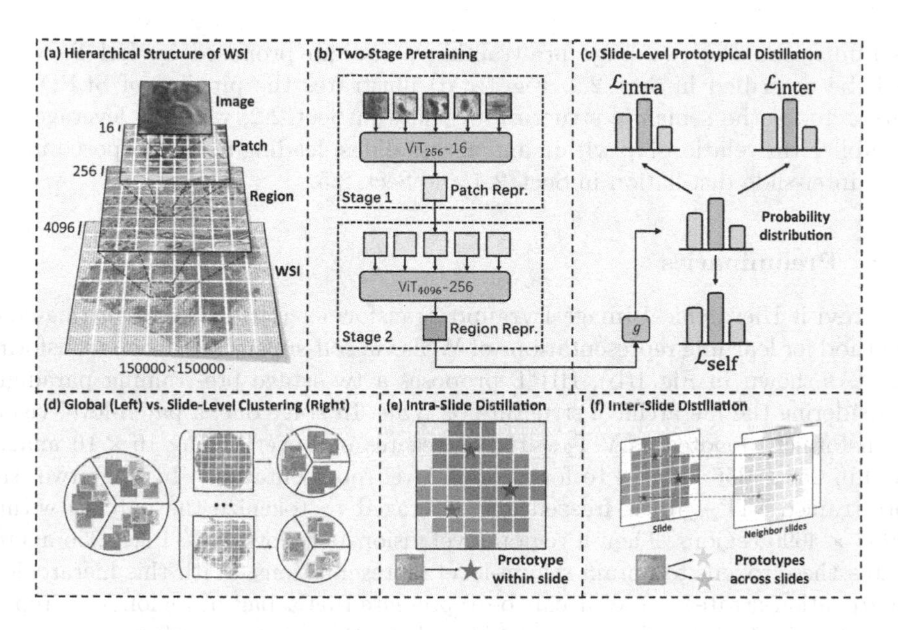

Fig. 1. (a) A WSI possesses the hierarchical structure of WSI-region-patch-image, from coarse to fine. (b) Two-stage pre-training paradigm successively performs the image-to-patch and patch-to-region aggregations. (c-e) The proposed SLPD. SLPD explores the semantic structure by slide-level clustering. Besides self-distillation, region representations are associated with the prototypes within and across slides to comprehensively understand WSIs.

lar slides by the set-to-set distance of prototypes. Then, we learn the inter-slide semantic structure by building correspondences between region representations and cross-slide prototypes. We conduct experiments on two benchmarks, NSCLC subtyping and BRCA subtyping. SLPD achieves state-of-the-art results on multiple slide-level tasks, demonstrating that representation learning of semantic structures of slides can make a suitable proxy task for WSI analysis. We also perform extensive ablation studies to verify the effectiveness of crucial model components.

2 Method

2.1 Overview

As shown in Fig. 1(a), a WSI exhibits hierarchical structure at varying resolutions under 20× magnification: 1) the 4096×4096 regions describing macro-scale organizations of cells, 2) the 256×256 patches capturing local clusters of cells, 3) and the 16×16 images characterizing the fine-grained features at the cell-level. Given N unlabeled WSIs $\{w_1, w_2, \cdots, w_N\}$, consisting of numerous regions $\{\{x_n^l\}_{l=1}^{L_n}\}_{n=1}^{N}$, where L_n denotes the number of regions of WSI w_n, we aim to learn a powerful encoder that maps each x_n^l to an embedding $z_n^l \in \mathbb{R}^D$. SLPD is built upon the two-stage pre-training paradigm proposed by HIPT, which will be described in Sect. 2.2. Fig 1(c-d) illustrates the pipeline of SLPD. We characterize the semantic structure of slides in Sect. 2.2, which is leveraged to establish the relationship within and across slides, leading to the proposed intra- and inter-slide distillation in Sect. 2.4 and Sect. 2.5.

2.2 Preliminaries

We revisit Hierarchical Image Pyramid Transformer (HIPT) [8], a cutting-edge method for learning representations of WSIs via self-supervised vision transformers. As shown in Fig. 1(b), HIPT proposes a two-stage pre-training paradigm considering the hierarchical structure of WSIs. In stage one, a patch-level vision transformer, denoted as ViT_{256}-16, aggregates non-overlapping 16×16 images within 256×256 patches to form patch-level representations. In stage two, the pre-trained ViT_{256}-16 is freezed and leveraged to tokenize the patches within 4096×4096 regions. Then a region-level vision transformer ViT_{4096}-256 aggregates these tokens to obtain region-level representations. With this hierarchical aggregation strategy, a WSI can be represented as a bag of region-level representations, which are then aggregated with another vision transformer, ViT_{WSI}-4096, to perform slide-level prediction tasks.

HIPT leverages DINO [6] to pre-train ViT_{256}-16 and ViT_{4096}-256, respectively. The learning objective of DINO is self-distillation. Taking stage two as an example, DINO distills the knowledge from teacher to student by minimizing the cross-entropy between the probability distributions of two views at region-level:

$$\mathcal{L}_{\text{self}} = \mathbb{E}_{x \sim p_d} H(g_t(\hat{z}), g_s(z)), \tag{1}$$

where $H(a, b) = -a \log b$, and p_d is the data distribution that all regions are drawn from. The teacher and the student share the same architecture consisting of an encoder (*e.g.*, ViT) and a projection head g_t/g_s. $\hat{z}$ and z are the embeddings of two views at region-level yielded by the encoder. The parameters of the student are exponentially moving averaged to the parameters of the teacher.

2.3 Slide-Level Clustering

Many histopathologic features have been established based on the morphologic phenotypes of the tumor, such as tumor invasion, anaplasia, necrosis and mitoses, which are then used for cancer diagnosis, prognosis and the estimation of response-to-treatment in patients [3,9]. To obtain meaningful representations of slides, we aim to explore and maintain such histopathologic features in the latent space. Clustering can reveal the representative patterns in the data and has achieved success in the area of unsupervised representation learning [4,5,24,26]. To characterize the histopathologic features underlying the slides, a straightforward practice is the global clustering, *i.e.*, clustering the region embeddings from all the WSIs, as shown in the left of Fig. 1(d). However, the obtained clustering centers, *i.e.*, the prototypes, are inclined to represent the visual bias related to staining or scanning procedure rather than medically relevant features [33]. Meanwhile, this clustering strategy ignores the hierarchical structure "region→WSI→whole dataset" underlying the data, where the ID of the WSI can be served as an extra learning signal. Therefore, we first consider the slide-level clustering that clusters the embeddings within each WSI, which is shown in the right of Fig. 1(d). Specifically, we conduct k-means algorithm before the start of each epoch over L_n region embeddings $\{z_n^l\}_{l=1}^{L_n}$ of w_n to obtain M prototypes $\{c_n^m \in \mathbb{R}^D\}_{m=1}^M$. Similar operations are applied across other slides, and then we acquire N groups of prototypes $\{\{c_n^m\}_{m=1}^M\}_{n=1}^N$. Each group of prototypes is expected to encode the semantic structure (*e.g.*, the combination of histopathologic features) of the WSI.

2.4 Intra-Slide Distillation

The self-distillation utilized by HIPT in stage two encourages the correspondence between two views of a region at the macro-scale because the organizations of cells share mutual information spatially. However, the self-distillation, which solely mines the spatial correspondences inside the 4096 × 4096 region, cannot comprehensively understand the histopathologic consistency at the slide-level. In order to achieve better representations, the histopathologic connections between the WSI and its regions should be modeled and learned, which is called intra-slide correspondences. With the proposed slide-level clustering in Sect. 2.3, a slide can be abstracted by a group of prototypes, which capture the semantic structure of the WSI. As shown in Fig. 1(e), we assume that the representation z and its assigned prototype c also share mutual information and encourage z to be closer to c with the intra-slide distillation:

$$\mathcal{L}_{\text{intra}} = \mathbb{E}_{x \sim p_d} H\left(g_t(c), g_s(z)\right), \tag{2}$$

We omit super-/sub-scripts of z for brevity. Through Eq. 2, we can leverage more intra-slide correspondences to guide the learning process. For further understanding, a prototype can be viewed as an augmented representation aggregating the slide-level information. Thus this distillation objective is encoding such information into the corresponding region embedding, which makes the learning process semantic structure-aware at the slide-level.

2.5 Inter-Slide Distillation

Tumors of different patients can exhibit morphological similarities in some respects [17,21], so the correspondences across slides should be characterized during learning. Previous self-supervised learning methods applied to histopathologic images only capture such correspondences with positive pairs at the patch-level [22,23], which overlooks the semantic structure of the WSI. We rethink this problem from the perspective how to measure the similarity between two slides accurately. Due to the heterogeneity of the slides, comparing them with the local crops or the averaged global features are both susceptible to being one-sided. To address this, we bridge the slides with their semantic structures and define the semantic similarity between two slides w_i and w_j through an optimal bipartite matching between two sets of prototypes:

$$D(w_i, w_j) = \max\{\frac{1}{M}\sum_{m=1}^{M} \cos(c_i^m, c_j^{\sigma(m)}) \mid \sigma \in \mathfrak{S}_M\},\ D(w_i, w_j) \in [-1, 1], \quad (3)$$

where $\cos(\cdot,\cdot)$ measures the cosine similarity between two vectors, and $\mathfrak{S}_M$ enumerates the permutations of M elements. The optimal permutation σ^* can be computed efficiently with the Hungarian algorithm [19]. With the proposed set-to-set distance, we can model the inter-slide correspondences conveniently and accurately. Specifically, for a region embedding z belonging to the slide w and assigned to the prototype c, we first search the top-K nearest neighbors of w in the dataset based on the semantic similarity, denoted as $\{\hat{w}_k\}_{k=1}^{K}$. Second, we also obtain the matched prototype pairs $\{(c, \hat{c}_k)\}_{k=1}^{K}$ determined by the optimal permutation, where $\hat{c}_k$ is the prototype of $\hat{w}_k$. Finally, we encourage z to be closer to $\hat{c}_k$ with the inter-slide distillation:

$$\mathcal{L}_{\text{inter}} = \mathbb{E}_{x \sim p_d}[\frac{1}{K}\sum_{k=1}^{K} H\left(g_t(\hat{c}_k), g_s(z)\right)]. \quad (4)$$

The inter-slide distillation can encode the sldie-level information complementary to that of intra-slide distillation into the region embeddings.

The overall learning objective of the proposed SLPD is defined as:

$$\mathcal{L}_{\text{total}} = \mathcal{L}_{\text{self}} + \alpha_1 \mathcal{L}_{\text{intra}} + \alpha_2 \mathcal{L}_{\text{inter}}, \quad (5)$$

where the loss scale is simply set to $\alpha_1 = \alpha_2 = 1$. We believe the performance can be further improved by tuning this.

3 Experimental Results

Datasets. We conduct experiments on two public WSI datasets[2]. *TCGA-NSCLC* dataset includes two subtypes in lung cancer, Lung Squamous Cell Carcinoma and Lung Adenocarcinoma, with a total of 1,054 WSIs. *TCGA-BRCA* dataset includes two subtypes in breast cancer, Invasive Ductal and Invasive Lobular Carcinoma, with a total of 1,134 WSIs.

Pre-training. We extract 62,852 and 60,153 regions at 20× magnification from TCGA-NSCLC and TCGA-BRCA for pre-training ViT_{4096}-256 in stage two. We leverage the pre-trained ViT_{256}-16 in stage one provided by HIPT to tokenize the patches within each region. Following the official code of HIPT, ViT_{4096}-256 is optimized for 100 epochs with optimizer of AdamW, base learning rate of 5e-4 and batch size of 256 on 4 GTX3090 GPUs.

[2] The data is released under a CC-BY-NC 4.0 international license.

Fine-tuning. We use the pre-trained ViT_{256}-16 and ViT_{4096}-256 to extract embeddings at the *patch-level* (256 × 256) and the *region-level* (4096 × 4096) for downstream tasks. With the pre-extracted embeddings, we fine-tune three aggregators (*i.e.*, MIL [28], DS-MIL [22] and ViT_{WSI}-4096 [8]) for 20 epochs and follow other settings in the official code of HIPT.

Evaluation Metrics. We adopt the 10-fold cross validated Accuracy (Acc.) and area under the curve (AUC) to evaluate the weakly-supervised classification performance. The data splitting scheme is kept consistent with HIPT.

Table 1. Slide-level classification. "Mean" leverages the averaged pre-extracted embeddings to evaluate KNN performance. Bold and underlined numbers highlight the best and second best performance

#	Feature Aggragtor	Feature Extraction	Pretrain Method	NSCLC		BRCA	
				Acc.	AUC	Acc.	AUC
Weakly supervised classification							
1	MIL [28]	patch-level	DINO	$0.780_{\pm 0.126}$	$0.864_{\pm 0.089}$	$0.822_{\pm 0.047}$	$0.783_{\pm 0.056}$
2		region-level	SLPD	$0.856_{\pm 0.025}$	$0.926_{\pm 0.017}$	$\mathbf{0.879}_{\pm 0.035}$	$0.863_{\pm 0.076}$
3	DS-MIL [22]	patch-level	DINO	$0.825_{\pm 0.054}$	$0.905_{\pm 0.059}$	$0.847_{\pm 0.032}$	$0.848_{\pm 0.075}$
4		region-level	DINO	$0.841_{\pm 0.036}$	$0.917_{\pm 0.035}$	$0.854_{\pm 0.032}$	$0.848_{\pm 0.075}$
5		region-level	SLPD	$\underline{0.858}_{\pm 0.040}$	$\underline{0.938}_{\pm 0.026}$	$0.854_{\pm 0.039}$	$0.876_{\pm 0.050}$
6	ViT_{WSI}-4096 [8]	region-level	DINO	$0.843_{\pm 0.044}$	$0.926_{\pm 0.032}$	$0.849_{\pm 0.037}$	$0.854_{\pm 0.069}$
7		region-level	DINO+$\mathcal{L}_{intra}$	$0.850_{\pm 0.042}$	$0.931_{\pm 0.041}$	$0.866_{\pm 0.030}$	$\underline{0.881}_{\pm 0.069}$
8		region-level	DINO+$\mathcal{L}_{inter}$	$0.850_{\pm 0.043}$	$\underline{0.938}_{\pm 0.028}$	$0.860_{\pm 0.030}$	$0.874_{\pm 0.059}$
9		region-level	SLPD	$\mathbf{0.864}_{\pm 0.042}$	$\mathbf{0.939}_{\pm 0.022}$	$\underline{0.869}_{\pm 0.039}$	$\mathbf{0.886}_{\pm 0.057}$
K-nearest neighbors (KNN) evaluation							
10	Mean	region-level	DINO	$0.770_{\pm 0.031}$	$0.840_{\pm 0.038}$	$0.837_{\pm 0.014}$	$0.724_{\pm 0.055}$
11		region-level	DINO+$\mathcal{L}_{intra}$	$0.776_{\pm 0.030}$	$0.850_{\pm 0.023}$	$0.841_{\pm 0.012}$	$0.731_{\pm 0.064}$
12		region-level	DINO+$\mathcal{L}_{inter}$	$\underline{0.782}_{\pm 0.027}$	$\underline{0.854}_{\pm 0.025}$	$\underline{0.845}_{\pm 0.014}$	$\underline{0.738}_{\pm 0.080}$
13		region-level	SLPD	$\mathbf{0.792}_{\pm 0.035}$	$\mathbf{0.863}_{\pm 0.024}$	$\mathbf{0.849}_{\pm 0.014}$	$\mathbf{0.751}_{\pm 0.079}$

3.1 Weakly-Supervised Classification

We conduct experiments on two slide-level classification tasks, NSCLC subtyping and BRCA subtyping, and report the results in Table 1. The region-level embeddings generated by SLPD outperform the patch-level embeddings across two aggregators[3] and two tasks (#1∼ 5). This illustrates that learning representations with broader image contexts is more suitable for WSI analysis. Compared with the strong baseline, *i.e.*, the two-stage pre-training method proposed by HIPT (#6), SLPD achieves performance increases of 1.3% and 3.2% AUC on NSCLC and BRCA (#9). Nontrivial performance improvements are also observed under KNN evaluation (#10 vs.#13): +2.3% and +3.1% AUC on NSCLC and BRCA. The superior performance of SLPD demonstrates that learning representations with slide-level semantic structure appropriately can significantly narrow the gap between pre-training and downstream slide-level

[3] The feature extraction of the patch-level is impracticable for the ViT-based model due to its quadratic complexity in memory usage.

tasks. Moreover, intra-slide and inter-slide distillation show consistent performance over the baseline, corroborating the effectiveness of these critical components of SLPD.

3.2 Ablation Study

Different Clustering Methods. As discussed in Sect. 2.3, we can alternatively use the global clustering to obtain prototypes and then optimize the network with a similar distillation objective as Eq. 2. For a fair comparison, the total number of prototypes of the two clustering methods is approximately the same. Table 2(#1,2) reports the comparative results, where the slide-level clustering surpasses the global clustering by 0.6% and 1.8% of AUC on NSCLC and BRCA, which verifies the effectiveness of the former. The inferior performance of the global clustering is due to the visual bias underlying the whole dataset.

Table 2. Ablation studies of SLPD. ViT_{WSI}-4096 is the aggregator with region-level embeddings.

#	Ablation	Method	NSCLC		BRCA	
			Acc.	AUC	Acc.	AUC
1	Different clustering methods	global	$0.848_{\pm 0.045}$	$0.925_{\pm 0.033}$	$0.842_{\pm 0.048}$	$0.863_{\pm 0.060}$
2		slide-level	$0.850_{\pm 0.042}$	$0.931_{\pm 0.041}$	$0.866_{\pm 0.030}$	$0.881_{\pm 0.069}$
3	Different inter-slide distillations	region	$0.828_{\pm 0.040}$	$0.915_{\pm 0.025}$	$0.843_{\pm 0.024}$	$0.849_{\pm 0.067}$
4		prototype	$0.850_{\pm 0.043}$	$0.938_{\pm 0.028}$	$0.860_{\pm 0.030}$	$0.874_{\pm 0.059}$
5	Number of prototypes	$M = 2$	$0.859_{\pm 0.036}$	$0.936_{\pm 0.021}$	$0.869_{\pm 0.039}$	$0.886_{\pm 0.057}$
6		$M = 3$	$0.864_{\pm 0.035}$	$0.938_{\pm 0.022}$	$0.861_{\pm 0.056}$	$0.878_{\pm 0.069}$
7		$M = 4$	$0.864_{\pm 0.042}$	$0.939_{\pm 0.022}$	$0.860_{\pm 0.031}$	$0.872_{\pm 0.060}$
8	Number of slide neighbors	$K = 1$	$0.864_{\pm 0.042}$	$0.939_{\pm 0.022}$	$0.869_{\pm 0.039}$	$0.886_{\pm 0.057}$
9		$K = 2$	$0.862_{\pm 0.039}$	$0.938_{\pm 0.029}$	$0.875_{\pm 0.038}$	$0.889_{\pm 0.057}$
10		$K = 3$	$0.869_{\pm 0.034}$	$0.936_{\pm 0.024}$	$0.873_{\pm 0.051}$	$0.880_{\pm 0.058}$

Different Inter-slide Distillations. The proposed inter-slide distillation is semantic structure-aware at the slide-level, since we build the correspondence between the region embedding and the matched prototype (#4 in Table 2). To verify the necessity of this distillation method, we turn to another design where the inter-slide correspondence is explored through two nearest region embeddings across slides (#3 in Table 2). As can be seen, the region-level correspondences lead to inferior performances, even worse than the baseline (#5 in Table 1), because the learning process is not guided by the slide-level information.

Number of Prototypes. As shown in Table 2(#5∼7), the performance of SLPD is relatively robust to the number of prototypes on NSCLC, but is somewhat affected by it on BRCA. One possible reason is that the heterogeneity of invasive breast carcinoma is low [29], and thus the excessive number of prototypes cannot obtain medically meaningful clustering results. Empirically, we set $M = 4$ on NSCLC and $M = 2$ on BRCA as the default configuration. We suggest the optimal number of prototypes should refer to clinical practice, by considering tissue types, cell morphology, gene expression and other factors.

Number of Slide Neighbors. As demonstrated in Table 2(#5∼7), the performance of SLPD is robust to the number of slide neighbors. Considering that more slide neighbors require more computation resources, we set $K = 1$ as the default configuration. For more results, please refer to the Supplementary.

4 Conclusion

This paper reflects on slide-level representation learning from a novel perspective by considering the intra- and inter-slide semantic structures. This leads to the proposed Slide-Level Prototypical Distillation (SLPD), a new self-supervised learning approach achieving the more comprehensive understanding of WSIs. SLPD leverages the slide-level clustering to characterize semantic structures of slides. By representing slides as prototypes, the mutual-region/slide relations are further established and learned with the proposed intra- and inter-slide distillation. Extensive experiments have been conducted on multiple WSI benchmarks and SLPD achieves state-of-the-art results. Though SLPD is distillation-based, we plan to apply our idea to other pre-training methods in the future, *e.g.*, contrastive learning [10,11].

Acknowledgement. This work was supported in part by NSFC 62171282, Shanghai Municipal Science and Technology Major Project (2021SHZDZX0102), 111 project BP0719010, STCSM 22DZ2229005, and SJTU Science and Technology Innovation Special Fund YG2022QN037.

References

1. Abbasi-Sureshjani, S., Yüce, A., Schönenberger, et al.: Molecular subtype prediction for breast cancer using H&E specialized backbone. In: MICCAI, pp. 1–9. PMLR (2021)
2. Abbet, C., Zlobec, I., Bozorgtabar, B., Thiran, J.-P.: Divide-and-rule: self-supervised learning for survival analysis in colorectal cancer. In: Martel, A.L., et al. (eds.) MICCAI 2020. LNCS, vol. 12265, pp. 480–489. Springer, Cham (2020). https://doi.org/10.1007/978-3-030-59722-1_46
3. Amin, M.B., Greene, F.L., Edge, S.B., et al.: The eighth edition AJCC cancer staging manual: continuing to build a bridge from a population-based to a more personalized approach to cancer staging. CA Cancer J. Clin. **67**(2), 93–99 (2017)
4. Caron, M., Bojanowski, P., Joulin, A., Douze, M.: Deep clustering for unsupervised learning of visual features. In: ECCV, pp. 132–149 (2018)
5. Caron, M., Misra, I., Mairal, J., Goyal, P., Bojanowski, P., Joulin, A.: Unsupervised learning of visual features by contrasting cluster assignments. Adv. Neural Inf. Process. Syst. **33**, 9912–9924 (2020)
6. Caron, M., Touvron, H., Misra, I., et al.: Emerging properties in self-supervised vision transformers. In: ICCV, pp. 9650–9660 (2021)

7. Chen, L., Bentley, P., et al.: Self-supervised learning for media using image context restoration. MedIA **58**, 101539 (2019)
8. Chen, R.J., et al.: Scaling vision transformers to gigapixel images via hierarchical self-supervised learning. In: CVPR, pp. 16144–16155 (2022)
9. Chen, R.J., Lu, M.Y., Williamson, D.F.K., et al.: Pan-cancer integrative histology-genomic analysis via multimodal deep learning. Cancer Cell **40**(8), 865–878 (2022)
10. Chen, T., Kornblith, S., Norouzi, M., Hinton, G.: A simple framework for contrastive learning of visual representations. In: ICML, pp. 1597–1607. PMLR (2020)
11. Chen, X., Fan, H., Girshick, R., He, K.: Improved baselines with momentum contrastive learning. arXiv preprint arXiv:2003.04297 (2020)
12. Dehaene, O., Camara, A., Moindrot, O., et al.: Self-supervision closes the gap between weak and strong supervision in histology. arXiv preprint arXiv:2012.03583 (2020)
13. Fashi, P.A., Hemati, S., Babaie, M., Gonzalez, R., Tizhoosh, H.: A self-supervised contrastive learning approach for whole slide image representation in digital pathology. J. Pathol. Inform. **13**, 100133 (2022)
14. Gurcan, M.N., Boucheron, L.E., Can, A., et al.: Histopathological image analysis: a review. IEEE Rev. Biomed. Eng. **2**, 147–171 (2009)
15. He, K., Fan, H., Wu, Y., Xie, S., Girshick, R.: Momentum contrast for unsupervised visual representation learning. In: CVPR, pp. 9729–9738 (2020)
16. Huang, Z., Chai, H., Wang, R., Wang, H., Yang, Y., Wu, H.: Integration of patch features through self-supervised learning and transformer for survival analysis on whole slide images. In: de Bruijne, M., et al. (eds.) MICCAI 2021. LNCS, vol. 12908, pp. 561–570. Springer, Cham (2021). https://doi.org/10.1007/978-3-030-87237-3_54
17. Jass, J.R.: HNPCC and sporadic MSI-H colorectal cancer: a review of the morphological similarities and differences. Fam. Cancer **3**, 93–100 (2004)
18. Koohbanani, N.A., Unnikrishnan, B., Khurram, S.A., Krishnaswamy, P., Rajpoot, N.: Self-path: self-supervision for classification of pathology images with limited annotations. IEEE Trans. Med. Imaging **40**(10), 2845–2856 (2021)
19. Kuhn, H.W.: The Hungarian method for the assignment problem. Nav. Res. Logist. Q. **2**(1–2), 83–97 (1955)
20. Lazard, T., Lerousseau, M., Decencière, E., Walter, T.: Self-supervised extreme compression of gigapixel images
21. Levy-Jurgenson, A., et al.: Spatial transcriptomics inferred from pathology whole-slide images links tumor heterogeneity to survival in breast and lung cancer. Sci. Rep. **10**(1), 1–11 (2020)
22. Li, B., Li, Y., Eliceiri, K.W.: Dual-stream multiple instance learning network for whole slide image classification with self-supervised contrastive learning. In: CVPR, pp. 14318–14328 (2021)
23. Li, J., Lin, T., Xu, Y.: SSLP: spatial guided self-supervised learning on pathological images. In: de Bruijne, M., et al. (eds.) MICCAI 2021. LNCS, vol. 12902, pp. 3–12. Springer, Cham (2021). https://doi.org/10.1007/978-3-030-87196-3_1
24. Li, J., Zhou, P., Xiong, C., Hoi, S.C.: Prototypical contrastive learning of unsupervised representations. arXiv preprint arXiv:2005.04966 (2020)
25. Li, L., Liang, Y., Shao, M., et al.: Self-supervised learning-based multi-scale feature fusion network for survival analysis from whole slide images. Comput. Biol. Med. **153**, 106482 (2023)
26. Li, Y., Hu, P., Liu, Z., Peng, D., Zhou, J.T., Peng, X.: Contrastive clustering. In: AAAI, vol. 35, pp. 8547–8555 (2021)

27. Lu, M.Y., Chen, R.J., Mahmood, F.: Semi-supervised breast cancer histology classification using deep multiple instance learning and contrast predictive coding (conference presentation). In: Medical Imaging 2020: Digital Pathology, vol. 11320, p. 113200J. International Society for Optics and Photonics (2020)
28. Lu, M.Y., Williamson, D.F., Chen, T.Y., et al.: Data-efficient and weakly supervised computational pathology on whole-slide images. Nat. Biomed. Eng. **5**(6), 555–570 (2021)
29. Öhlschlegel, C., Zahel, K., Kradolfer, D., Hell, M., Jochum, W.: Her2 genetic heterogeneity in breast carcinoma. J. Clin. Pathol. **64**(12), 1112–1116 (2011)
30. Oord, A.V.D., Li, Y., Vinyals, O.: Representation learning with contrastive predictive coding. arXiv preprint arXiv:1807.03748 (2018)
31. Saillard, C., Dehaene, et al.: Self supervised learning improves dMMR/MSI detection from histology slides across multiple cancers. arXiv preprint arXiv:2109.05819 (2021)
32. Saldanha, O.L., Loeffler, et al.: Self-supervised deep learning for pan-cancer mutation prediction from histopathology. bioRxiv, pp. 2022–09 (2022)
33. Sharma, Y., Shrivastava, A., Ehsan, L., et al.: Cluster-to-conquer: a framework for end-to-end multi-instance learning for whole slide image classification. In: Medical Imaging with Deep Learning, pp. 682–698. PMLR (2021)
34. Srinidhi, C.L., Kim, S.W., Chen, F.D., Martel, A.L.: Self-supervised driven consistency training for annotation efficient histopathology image analysis. Media **75**, 102256 (2022)
35. Tavolara, T.E., Gurcan, M.N., Niazi, M.K.K.: Contrastive multiple instance learning: an unsupervised framework for learning slide-level representations of whole slide histopathology images without labels. Cancers **14**(23), 5778 (2022)
36. Wu, Z., Xiong, Y., Yu, S.X., Lin, D.: Unsupervised feature learning via non-parametric instance discrimination. In: CVPR, pp. 3733–3742 (2018)
37. Xie, X., Chen, J., Li, Y., Shen, L., Ma, K., Zheng, Y.: Instance-aware self-supervised learning for nuclei segmentation. In: Martel, A.L., et al. (eds.) MICCAI 2020. LNCS, vol. 12265, pp. 341–350. Springer, Cham (2020). https://doi.org/10.1007/978-3-030-59722-1_33
38. Yang, P., et al.: CS-CO: a hybrid self-supervised visual representation learning method for H&E-stained histopathological images. Media **81**, 102539 (2022)

PET Image Denoising with Score-Based Diffusion Probabilistic Models

Chenyu Shen[1,2], Ziyuan Yang[2], and Yi Zhang[1(✉)]

[1] School of Cyber Science and Engineering, Sichuan University, Chengdu, China
yzhang@scu.edu.cn

[2] College of Computer Science, Sichuan University, Chengdu, China

Abstract. Low-count positron emission tomography (PET) imaging is an effective way to reduce the radiation risk of PET at the cost of a low signal-to-noise ratio. Our study aims to denoise low-count PET images in an unsupervised mode since the mainstream methods usually rely on paired data, which is not always feasible in clinical practice. We adopt the diffusion probabilistic model in consideration of its strong generation ability. Our model consists of two stages. In the training stage, we learn a score function network via evidence lower bound (ELBO) optimization. In the sampling stage, the trained score function and low-count image are employed to generate the corresponding high-count image under two handcrafted conditions. One is based on restoration in latent space, and the other is based on noise insertion in latent space. Thus, our model is named the bidirectional condition diffusion probabilistic model (BC-DPM). Real patient whole-body data are utilized to evaluate our model. The experiments show that our model achieves better performance in both qualitative and quantitative aspects compared to several traditional and recently proposed learning-based methods.

Keywords: PET denoising · diffusion probabilistic model · latent space conditions

1 Introduction

Positron emission tomography (PET) is an imaging modality in nuclear medicine that has been successfully applied in oncology, neurology, and cardiology. By injecting a radioactive tracer into the human body, the molecular-level activity in tissues can be observed. To mitigate the radiation risk to the human body, it is essential to reduce the dose or shorten the scan time, leading to a low signal-to-noise ratio and further negatively influencing the accuracy of diagnosis.

Recently, the denoising diffusion probabilistic model (DDPM) [6,9,11] has become a hot topic in the generative model community. The original DDPM was designed for generation tasks, and many recent works have proposed extending it for image restoration or image-to-image translation. In supervised mode, Saharia

Supplementary Information The online version contains supplementary material available at https://doi.org/10.1007/978-3-031-43907-0_26.

H. Greenspan et al. (Eds.): MICCAI 2023, LNCS 14220, pp. 270–278, 2023.
https://doi.org/10.1007/978-3-031-43907-0_26

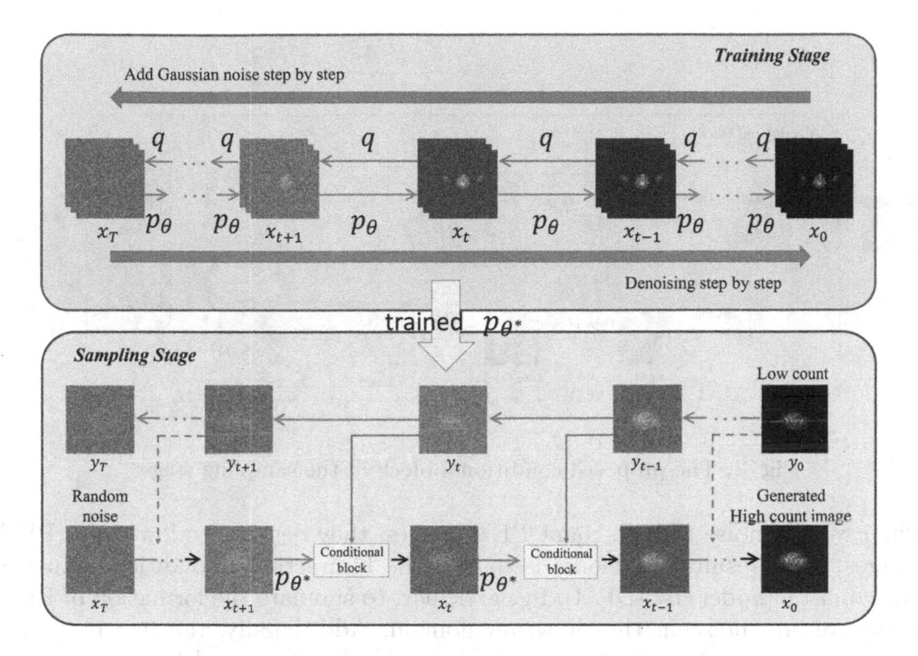

Fig. 1. Overview of our proposed BC-DPM model.

et al. [8] proposed a conditional DDPM to perform single-image super-resolution, which integrates a low-resolution image into each reverse step. In unsupervised mode, to handle the stochasticity of the generative process, Choi proposed iterative latent variable refinement (ILVR) [1] to guarantee the given condition in each transition, thus generating images with the desired semantics. DDPM has also been applied in medical imaging. To explore its generalization ability, Song *et al.* [12] proposed a fully unsupervised model for medical inverse problems, providing the measuring process and the prior distribution learned with a score-based generative model. For PET image denoising, Gong *et al.* [4] proposed two paradigms. One is directly feeding noisy PET images and anatomical priors (if available) into the score function network, which relies on paired high-quality and low-quality PET images. The other is feeding only MR images into the score function network while using noisy PET images in the inference stage under the assumption that PET image noise obeys a Gaussian distribution.

In this paper, we propose a conditional diffusion probabilistic model for low-count PET image denoising in an unsupervised manner without the Gaussian noise assumption or paired datasets. Our model is divided into two stages. In the training stage, we leverage the standard DDPM to train the score function network to learn a prior distribution of PET images. Once the network is trained, we transplant it into the sampling stage, in which we design two conditions to control the generation of high-count PET images given corresponding low-count PET images. One condition is that the denoised versions of low-count PET images are similar to high-count PET images. The other condition is that

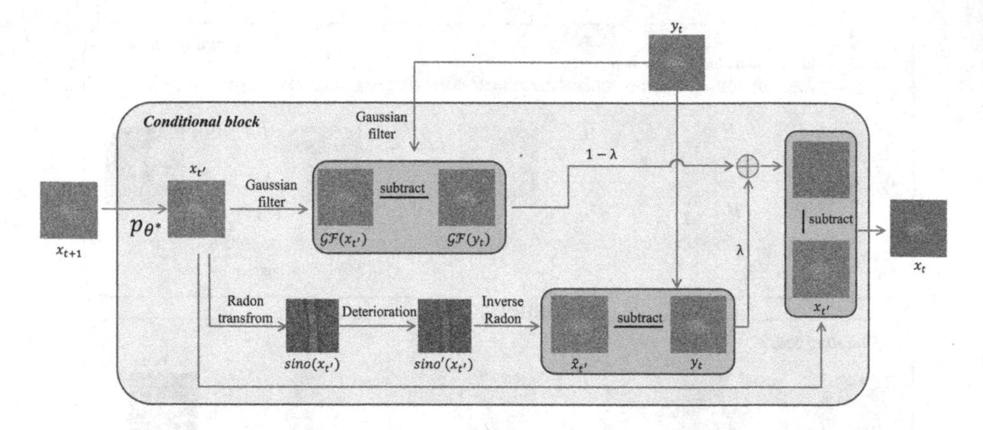

Fig. 2. The proposed conditional block in the sampling stage.

when we add noise to high-count PET images, they degrade to low-count PET images. As a result, our model is named the bidirectional condition diffusion probabilistic model (BC-DPM). In particular, to simulate the formation of PET noise, we add noise in the sinogram domain. Additionally, the two proposed conditions are implemented in latent space. Notably, Our model is 'one for all', that is, once we have trained the score network, we can utilize this model for PET images with different count levels.

2 Method

Letting $X \subset \mathcal{X}$ be a high-count PET image dataset and $Y \subset \mathcal{Y}$ be a low-count PET image dataset, x_0 and y_0 denote instances in X and Y, respectively. Our goal is to estimate a mapping $\mathcal{F}(\mathcal{Y}) = \mathcal{X}$, and the proposed BC-DPM provides an unsupervised technique to solve this problem. BC-DPM includes two stages. In the training stage, it requires only X without paired (X, Y), and in the sampling stage, it produces the denoised x_0 for a given y_0.

2.1 Training Stage

BC-DPM acts the same as the original DDPM in the training stage, it consists of a forward process and a reverse process. In the forward process, x_0 is gradually contaminated by fixed Gaussian noise, producing a sequence of latent space data $\{x_1, x_2, ..., x_T\}$, where $x_T \sim \mathcal{N}(0, \mathbf{I})$. The forward process can be described formally by a joint distribution $q(x_{1:T}|x_0)$ given x_0. Under the Markov property, it can be defined as:

$$q(x_{1:T}|x_0) := \prod_{t=1}^{T} q(x_t|x_{t-1}), \quad q(x_t|x_{t-1}) := \mathcal{N}(x_t; \sqrt{1-\beta_t}x_{t-1}, \beta_t\mathbf{I}), \tag{1}$$

where $\{\beta_1, \beta_2, ..., \beta_T\}$ is a fixed variance schedule with small positive constants and $\mathbf{I}$ represents the identity matrix. Notably, the forward process allows x_t to be sampled directly from x_0:

$$x_t = \sqrt{\bar{\alpha}_t} x_0 + \sqrt{1 - \bar{\alpha}_t}\epsilon, \tag{2}$$

where $\bar{\alpha}_t := \prod_{s=1}^{t} \alpha_s$, $\alpha_t := 1 - \beta_t$ and $\epsilon \sim \mathcal{N}(0, \mathbf{I})$.

The reverse process is defined by a Markov chain starting with $p(x_T) = \mathcal{N}(x_T; 0, \mathbf{I})$:

$$p_\theta(x_{0:T}) := p(x_T) \prod_{t=1}^{T} p_\theta(x_{t-1}|x_t), \quad p_\theta(x_{t-1}|x_t) := \mathcal{N}(x_{t-1}; \mu_\theta(x_t, t), \sigma_\theta(x_t, t)\mathbf{I}). \tag{3}$$

Given the reverse process, $p_\theta(x_0)$ can be expressed by setting up an integral over the $x_{1:T}$ variables $p_\theta(x_0) := \int p_\theta(x_{0:T}) dx_{1:T}$, and the parameter θ can be updated by optimizing the following simple loss function:

$$L_{simple}(\theta) = \mathbb{E}_{t,x_0,\epsilon}[\left\| \epsilon - \epsilon_\theta(\sqrt{\bar{\alpha}_t} x_0 + \sqrt{1 - \bar{\alpha}_t}\epsilon, t) \right\|^2]. \tag{4}$$

The $\epsilon_\theta(x_t, t)$ used in this paper heavily relies on that proposed by Dhariwal *et al.* [3]. The pseudocode for the training stage is given in Algorithm 1.

Algorithm 1: Training stage.

---

repeat
  $x_0 \sim q(x_0)$
  $t \sim Uniform(1, 2, ..., T)$
  $\epsilon \sim \mathcal{N}(0, \mathbf{I})$
  Update θ by optimizing
    $\mathbb{E}_{t,x_0,\epsilon}[\left\| \epsilon - \epsilon_\theta(\sqrt{\bar{\alpha}_t} x_0 + \sqrt{1 - \bar{\alpha}_t}\epsilon, t) \right\|^2]$
until convergence

2.2 Sampling Stage

The main difference between BC-DPM and the original DDPM lies in the sampling stage. Due to the stochasticity of the reverse process $p_\theta(x_{0:T})$, it is difficult for the original DDPM to generate images according to our expectation. To overcome this obstacle, the proposed BC-DPM models $p_\theta(x_0|c)$ given condition c instead of modeling $p_\theta(x_0)$ as

$$p_\theta(x_0|c) = \int p_\theta(x_{0:T}|c) dx_{1:T}, \quad p_\theta(x_{0:T}|c) = p(x_T) \prod_{t=1}^{T} p_\theta(x_{t-1}|x_t, c). \tag{5}$$

Condition c derives from specific prior knowledge from the high-count PET image x_0 and the low-count PET image y_0. With c, BC-DPM can control the generation of x_0 given y_0.

Then, the core problem is to design a proper condition c. A natural choice is $\mathcal{D}(y_0) \approx x_0$, that is, the restoration task itself. We must clarify that it will not cause a 'deadlock' for the following two reasons. One is that the final form of the condition $\mathcal{D}(y_0) \approx x_0$ does not involve x_0, and the other is that we choose a relatively simple denoiser in the condition, which can be viewed as a 'coarse to fine' operation. In practice, we utilize a Gaussian filter $\mathcal{GF}(\cdot)$ as the denoiser in this condition. However, the Gaussian filter usually leads to smoothed images. Based on this property, we observe that the PSNR value between $\mathcal{GF}(y_0)$ and x_0 is usually inferior to that between $\mathcal{GF}(y_0)$ and $\mathcal{GF}(x_0)$, which means that the condition $\mathcal{GF}(y_0) \approx \mathcal{GF}(x_0)$ is more accurate than $\mathcal{GF}(y_0) \approx x_0$. Thus, we choose $\mathcal{GF}(y_0) \approx \mathcal{GF}(x_0)$ in our experiments.

However, if we only utilize the above condition, the training is unstable, and distortion may be observed. To address this problem, another condition needs to be introduced. The above condition refers to denoising, so conversely, we can consider adding noise to x_0; that is, $y_0 \approx \mathcal{A}(x_0)$. According to the characteristics of PET noise, Poisson noise is used in the sinogram domain instead of the image domain. We define this condition as $\mathcal{P}^{\dagger}(Po(\mathcal{P}(x_0) + r + s)) \approx y_0$, where $\mathcal{P}$, Po, $\mathcal{P}^{\dagger}$, r and s represent the Radon transform, Poisson noise insertion, inverse Radon transform, random coincidence and scatter coincidence, respectively.

Now, we have two conditions $\mathcal{GF}(y_0) \approx \mathcal{GF}(x_0)$ and $\mathcal{P}^{\dagger}(Po(\mathcal{P}(x_0)+r+s)) \approx y_0$ from the perspectives of denoising and noise insertion, respectively. Since the conditions involve x_0, we have to convert the conditions from the original data space into latent space under certain circumstances to avoid estimating x_0. Let us denote each transition in the reverse process under global conditions as:

$$\begin{aligned} p_\theta(x_{t-1}|x_t, c_1, c_2) = p_\theta(x_{t-1}|x_t, &\mathcal{GF}(x_0) = \mathcal{GF}(y_0), \\ &\mathcal{P}^{\dagger}(Po(\mathcal{P}(x_0) + r + s)) = y_0). \end{aligned} \tag{6}$$

In Eq. (2), x_t can be represented by a linear combination of x_0 and ϵ. Then, we can express x_0 with x_t and ϵ:

$$x_0 \approx f_\theta(x_t, t) = (x_t - \sqrt{1 - \bar{\alpha}_t}\epsilon_\theta(x_t, t))/\sqrt{\bar{\alpha}_t}. \tag{7}$$

Similarly, applying the same diffusion process to y_0, we have $\{y_1, y_2, ..., y_T\}$, and y_0 can be expressed with y_t and ϵ:

$$y_0 \approx f_\theta(y_t, t) = (y_t - \sqrt{1 - \bar{\alpha}_t}\epsilon_\theta(y_t, t))/\sqrt{\bar{\alpha}_t}. \tag{8}$$

Replacing x_0 and y_0 with $f_\theta(x_t, t)$ and $f_\theta(y_t, t)$ in Eq. (6), respectively, we have:

$$\begin{aligned} p_\theta(x_{t-1}|x_t, c_1, c_2) \approx \mathbb{E}_{q(y_{t-1}|y_0)}[p_\theta(x_{t-1}|x_t, &\mathcal{GF}(x_{t-1}) = \mathcal{GF}(y_{t-1}), \\ &\mathcal{P}^{\dagger}(Po(\mathcal{P}(x_{t-1}) + r + s)) = y_{t-1})]. \end{aligned} \tag{9}$$

Assume that

$$\begin{aligned} x_{t-1} = &(1 - \lambda)(\mathcal{GF}(y_{t-1}) + (\mathcal{I} - \mathcal{GF})(x'_{t-1})) \\ &+ \lambda(y_{t-1} + x'_{t-1} - \mathcal{P}^{\dagger}(Po(\mathcal{P}(x_{t-1}) + r + s)), \end{aligned} \tag{10}$$

where x'_{t-1} is sampled from $p_\theta(x'_{t-1}|x_t)$, and $\lambda \in [0,1]$ is a balancing factor between the two conditions. Thus, we have

$$\begin{aligned}&\mathbb{E}_{q(y_{t-1}|y_0)}[p_\theta(x_{t-1}|x_t, \mathcal{GF}(x_{t-1}) = \mathcal{GF}(y_{t-1}), \mathcal{P}^\dagger(Po(\mathcal{P}(x_{t-1}) + r + s)) = y_{t-1})]\\ \approx\, &p_\theta(x_{t-1}|x_t, \mathcal{GF}(x_{t-1}) = \mathcal{GF}(y_{t-1}), \mathcal{P}^\dagger(Po(\mathcal{P}(x_{t-1}) + r + s)) = y_{t-1}).\end{aligned} \tag{11}$$

Finally, we have

$$\begin{aligned}&p_\theta(x_{t-1}|x_t, \mathcal{GF}(x_0) = \mathcal{GF}(y_0), \mathcal{P}^\dagger(Po(\mathcal{P}(x_0) + r + s)) = y_0)\\ =\, &p_\theta(x_{t-1}|x_t, \mathcal{GF}(x_{t-1}) = \mathcal{GF}(y_{t-1}), \mathcal{P}^\dagger(Po(\mathcal{P}(x_{t-1}) + r + s)) = y_{t-1}),\end{aligned} \tag{12}$$

which indicates that under the assumption of Eq. (10), the global conditions on (x_0, y_0) can be converted to local conditions on (x_{t-1}, y_{t-1}) in each transition from x_t to x_{t-1}.

Now, given a low-count PET image y_0, to estimate x_0, we can sample from white noise x_T using the following two steps iteratively. The first step is to generate an immediate x'_{t-1} from $p_\theta(x'_{t-1}|x_t)$. The second step is to generate x_{t-1} from x'_{t-1} using Eq. (10). In practice, we note that there is no need to operate the two local conditions in each transition; instead, we only need the last l transitions. Generally speaking, The larger l is, the more blurred the image will be. As l decreases, the image gets more noisy. We provide the sampling procedure of BC-DPM in Algorithm 2.

Algorithm 2: Sampling stage.

Input: low-count PET image y_0, parameter θ from the training stage, hyper-parameters λ and l
Output: high-count PET image x_0
$x_T \sim \mathcal{N}(0, \mathbf{I})$
for $t = T, ..., 1$ **do**
  if $t <= l$ **then**
    sample x'_{t-1} from $p_\theta(x'_{t-1}|x_t)$
    sample y_{t-1} from $q(y_{t-1}|y_0)$
    update x_{t-1} using Equation (10)
  else
    sample x_{t-1} from $p_\theta(x'_{t-1}|x_t)$
end for
return x_0

Figure 1 illustrates the whole model. In the training stage, q denotes fixed Gaussian noise for the forward process, and p_θ denotes a learned transition in the reverse process. Once p_θ is trained, it is moved to the sampling stage. In the sampling stage, we first use the same q to diffuse y_0 to $\{y_1, y_2, ..., y_T\}$. Then, we start with white noise x_T followed by a transition from x_{t+1} to x_t for each $t \in \{0, 1, 2, ..., T-1\}$. Each transition consists of p_θ and a conditional block. p_θ is responsible for sampling an immediate x'_t from x_{t+1}. Then, the conditional block takes x'_t and y_t as inputs and outputs x_t. Figure 2 shows the detailed structure of the conditional block. There are two parallel branches. One calculates the difference between $\mathcal{GF}(x'_t)$ and $\mathcal{GF}(y_t)$, which represents the condition of denoising, and the other computes the difference between $\hat{x}'_t$ and y_t, where $\hat{x}'_t$ is derived by adding noise to x'_t in the sinogram domain. Then, we sum the two branches weighted by λ and subtract x'_t to output the final result x_t.

3 Experiment

3.1 Experimental Setup

To evaluate the proposed method, real clinical data downloaded from TCIA were tested [2]. The computer simulation modeled the geometry of a CTI ECAT 921 PET scanner, and the system matrix P was modeled using Siddon's refined method to calculate the ray path integral [5]. To simulate low-count PET images, we first generated a noise-free sinogram by forward projecting the original data, obtaining a sinogram with a matrix size of 160 (radial bins) $\times$ 192 (azimuthal angles). Then, uniform random events were added to the noise-free sinogram as background, which accounted for 20% of total true coincidences. Independent Poisson noise with different levels was injected, raising the total number of events to 1M, 0.3M, and 0.1M, respectively. Finally, these sinograms were reconstructed by the ML-EM algorithm with 100 iterations. We used 3000 2D slices from 60 patients as the training set and 400 slices from another 10 patients as the test set.

Our method was implemented with PyTorch on a GeForce GTX 1080Ti GPU. We trained the network using the AdamW algorithm with $\beta_1 = 0.9$, $\beta_2 = 0.999$, and $weight\_decay = 0.01$. The learning rate was set to 0.0001, and the batch size was 8. In our experiments, similar to DDPM, we set the number of diffusion steps to $T = 1000$. For the variance schedule in the forward process, we employed a linear schedule from $\beta_1 = 0.0001$ to $\beta_T = 0.02$. In the sampling stage, we evenly sampled 100 steps from 1 to T and then performed generation only on these 100 steps, reducing the number of steps from 1000 to 100 by employing the trick in [7]. For the count levels of 1M, 0.3M, and 0.1M in the real clinical data study, we set l to 5, 10, and 15, respectively. In all cases, we set $\lambda = 0.2$ to balance the two conditions. As the diffusion model can generate different results due to stochasticity, we ran the model five times and used the average of the five results as the final result.

We compared our method with two conventional methods, Gaussian Filter and BM3D, and two unsupervised/unpaired methods, Noise2Void with parameter transfer (N2V-PT) [10] and unsupervised CycleWGAN [13].

3.2 Experimental Results

Figure 3 shows the results using various methods at three count levels. It can be observed that our method obtains the best performance in all cases. At the 1M

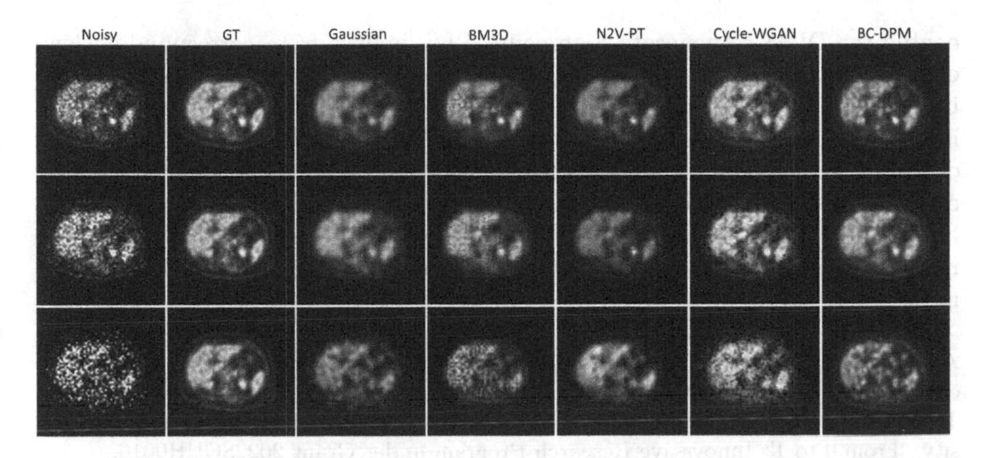

Fig. 3. Denoising results from different methods. The first row is under a count level of 1M, the second row is under a count level of 0.3M, and the third row is under a count level of 0.1M.

Table 1. PSNR and SSIM values of various methods of patient data under different count levels.

Methods	1M		0.3M		0.1M	
	PSNR	SSIM	PSNR	SSIM	PSNR	SSIM
Gaussian	32.9406	0.9713	31.5443	0.9611	30.9818	0.9531
BM3D	32.0151	0.9564	31.6693	0.9527	28.2746	0.9117
N2V-PT	34.4333	0.9767	33.4353	0.9695	31.0254	0.9545
CycleWGAN	35.1976	0.9821	33.6778	0.9653	31.6183	0.9500
BC-DPM	**35.6297**	**0.9831**	**33.9297**	**0.9700**	**32.1648**	**0.9621**

count level, the noise is small, and all methods obtain adequate results. At the 0.3M count level, the noise becomes higher. The Gaussian filter compromises the details for noise reduction. N2V-PT exhibits strange patterns due to the violation of the pixel independence assumption in PET noise. At the extremely low-count level, 0.1M, the Gaussian filter and N2V-PT cannot obtain clinically useful results, while our method can accurately recover some details due to its strong capacity for generation under these conditions. Table 1 reports the quantitative results, showing that our proposed BC-DPM outperforms other methods in terms of both PSNR and SSIM.

4 Conclusion

In conclusion, a PET denoising model based on diffusion probabilistic models is proposed in this paper. Our model is trained in an unsupervised manner and denoises low-count PET images without any anatomical prior as a reference. To enable the DPM to generate high-count PET images from corresponding low-count PET images, we design bidirectional conditions derived from relations between the low-count image and the potential high-count image. One condition is that the denoised low-count image approximates the high-count image. The other is that after adding noise, the high-count image approximates the low-count image. For implementation, we transfer the bidirectional conditions to latent space, which helps free the model from its dependence on the high-count image. Experiments on real clinical data demonstrate that our model is superior in noise suppression and detail preservation to other state-of-the-art methods.

Acknowledgement. This work was supported in part by the National Natural Science Foundation of China under Grant 62271335; in part by the Sichuan Science and Technology Program under Grant 2021JDJQ0024; and in part by the Sichuan University "From 0 to 1" Innovative Research Program under Grant 2022SCUH0016.

References

1. Choi, J., et al.: ILVR: conditioning method for denoising diffusion probabilistic models. In: Proceedings of the IEEE Conference on Computer Vision and Pattern Recognition, pp. 14347–14356 (2021)
2. Clark, K., et al.: The cancer imaging archive (TCIA): maintaining and operating a public information repository. J. Digit. Imag. **26**, 1045–1057 (2013)
3. Dhariwal, P., Nichol, A.: Diffusion models beat GANs on image synthesis. Proc. Adv. Neural Inf. Process. Syst. **34**, 8780–8794 (2021)
4. Gong, K., et al.: PET image denoising based on denoising diffusion probabilistic models. arXiv preprint arXiv:2209.06167 (2022)
5. Han, G., Liang, Z., You, J.: A fast ray-tracing technique for TCT and ECT studies. In: Proceedings of the IEEE Nuclear Science Symposium, vol. 3, pp. 1515–1518 (1999)
6. Ho, J., Jain, A., Abbeel, P.: Denoising diffusion probabilistic models. Adv. Neural Inf. Process. Syst. **33**, 6840–6851 (2020)
7. Nichol, A.Q., Dhariwal, P.: Improved denoising diffusion probabilistic models. In: Proceedings of the International Conference on Machine Learning, vol. 139, pp. 8162–8171 (2021)
8. Saharia, C., et al.: Image super-resolution via iterative refinement. IEEE Trans. Pattern Anal. Mach. Intell. (2022)
9. Sohl-Dickstein, J., et al.: Deep unsupervised learning using nonequilibrium thermodynamics. In: Proceedings of the International Conference on Machine Learning, pp. 2256–2265 (2015)
10. Song, T.A., et al.: Noise2Void: unsupervised denoising of PET images. Phys. Med. Biol. **66** (2021)
11. Song, Y., et al.: Score-based generative modeling through stochastic differential equations. In: Proceedings of the International Conference on Learning Representations (2021)
12. Song, Y., et al.: Solving inverse problems in medical imaging with score-based generative models. In: Proceedings of the International Conference on Learning Representations (2022)
13. Zhou, L., et al.: Supervised learning with cyclegan for low-dose FDG PET image denoising. Med. Image Anal. **65**, 101770 (2020)

LSOR: Longitudinally-Consistent Self-Organized Representation Learning

Jiahong Ouyang, Qingyu Zhao, Ehsan Adeli, Wei Peng, Greg Zaharchuk, and Kilian M. Pohl(✉)

Stanford University, Stanford, CA 94305, USA
kilian.pohl@stanford.edu

Abstract. Interpretability is a key issue when applying deep learning models to longitudinal brain MRIs. One way to address this issue is by visualizing the high-dimensional latent spaces generated by deep learning via self-organizing maps (SOM). SOM separates the latent space into clusters and then maps the cluster centers to a discrete (typically 2D) grid preserving the high-dimensional relationship between clusters. However, learning SOM in a high-dimensional latent space tends to be unstable, especially in a self-supervision setting. Furthermore, the learned SOM grid does not necessarily capture clinically interesting information, such as brain age. To resolve these issues, we propose the first self-supervised SOM approach that derives a high-dimensional, interpretable representation stratified by brain age solely based on longitudinal brain MRIs (i.e., without demographic or cognitive information). Called **L**ongitudinally-consistent **S**elf-**O**rganized **R**epresentation learning (LSOR), the method is stable during training as it relies on soft clustering (vs. the hard cluster assignments used by existing SOM). Furthermore, our approach generates a latent space stratified according to brain age by aligning trajectories inferred from longitudinal MRIs to the reference vector associated with the corresponding SOM cluster. When applied to longitudinal MRIs of the Alzheimer's Disease Neuroimaging Initiative (ADNI, $N = 632$), LSOR generates an interpretable latent space and achieves comparable or higher accuracy than the state-of-the-art representations with respect to the downstream tasks of classification (static vs. progressive mild cognitive impairment) and regression (determining ADAS-Cog score of all subjects). The code is available at https://github.com/ouyangjiahong/longitudinal-som-single-modality.

1 Introduction

The interpretability of deep learning models is especially a concern for applications related to human health, such as analyzing longitudinal brain MRIs. To avoid interpretation during post-hoc analysis [6,14], some methods strive for

G. Zaharchuk—Co-founder, equity Subtle Medical.

Supplementary Information The online version contains supplementary material available at https://doi.org/10.1007/978-3-031-43907-0_27.

H. Greenspan et al. (Eds.): MICCAI 2023, LNCS 14220, pp. 279–289, 2023.
https://doi.org/10.1007/978-3-031-43907-0_27

an interpretable latent representation [9]. One example is self-organizing maps (SOM) [5], which cluster the latent space so that the SOM representations (i.e., the 'representatives of the clusters) can be arranged in a discrete (typically 2D) grid while preserving high-dimensional relationships between clusters. Embedded in unsupervised deep learning models, SOMs have been used to generate interpretable representations of low-resolution natural images [3,8].

Intriguing as it sounds, we found their application to (longitudinal) 3D brain MRIs unstable during training and resulted in uninformative SOMs. These models get stuck in local minima so that only a few SOM representations are updated during backpropagation. The issue has been less severe in prior applications [3,8] as their corresponding latent space is of much lower dimension than the task at hand, which requires a high dimension latent space so that it can accurately encode the fine-grained anatomical details in brain MRIs [12,17]. To ensure all SOM representations can be updated during backpropagation, we propose a soft weighing scheme that not only updates the closest SOM representation for a given MRI but also updates all other SOM representations based on their distance to the closest SOM representation [3,8]. Moreover, our model relies on a stop-gradient operator [16], which sets the gradient of the latent representation to zero so that it only focuses on updating the SOM representations. It is especially crucial at the beginning of the training when the (randomly initialized) SOM representations are not good representatives of their clusters. Finally, the latent representations of the MRIs are updated via a commitment loss, which encourages the latent representation of an MRI sample to be close to its nearest SOM representation. In practice, these three components ensure stability during the self-supervised training of the SOM on high-dimensional latent spaces.

To generate SOMs informative to neuroscientists, we extend SOMs to the longitudinal setting such that the latent space and corresponding SOM grid encode brain aging. Inspired by [12], we encode pairs of MRIs from the same longitudinal sequence (i.e., same subject) as a trajectory and encourage the latent space to be a smooth trajectory (vector) field. We enforce smoothness by computing for each SOM cluster a reference trajectory, which represents the average aging of that cluster with respect to the training set. The reference trajectories are updated by the exponential moving average (EMA) such that, in each iteration, it aggregates the average trajectory of a cluster with respect to the corresponding training batch (i.e., batch-wise average trajectory). In doing so, the model ensures longitudinal consistency as the (subject-specific) trajectories of a cluster are maximally aligned with the reference trajectory of that cluster.

Named **L**ongitudinally-consistent **S**elf-**O**rganized **R**epresentation learning (LSOR), we evaluate our method on a longitudinal T1-weighted MRI dataset of 632 subjects from ADNI to encode the brain aging of Normal Controls (NC) and patients diagnosed with static Mild Cognitive Impairment (sMCI), progressive Mild Cognitive Impairment (pMCI), and Alzheimer's Disease (AD). LSOR clusters the latent representations of all MRIs into 32 SOM representations. The resulting 4-by-8 SOM grid is organized by both chronological age and cognitive measures that are indicators of brain age. Note, such an organization solely relies

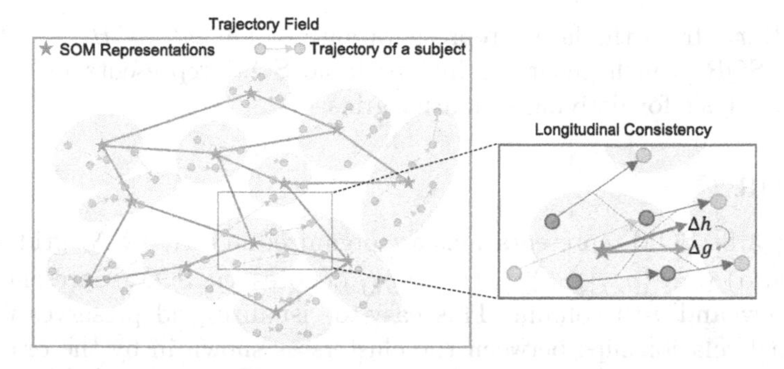

Fig. 1. Overview of the latent space derived from LSOR. All trajectories (Δz) form a trajectory field (blue box) modeling brain aging. SOM representations in $\mathcal{G}$ (orange star) are organized as a 2D grid (orange grid). As shown in the black box, reference trajectories $\Delta\mathcal{G}$ (collection of all Δg, green arrow) are iteratively updated by EMA using the aggregated trajectory Δh (purple arrow) across all trajectories of the corresponding SOM cluster within a training batch. (Color figure online)

on longitudinal MRIs, i.e., without using any tabular data such as age, cognitive measure, or diagnosis. To visualize aging effects on the grid, we compute (post-hoc) a 2D similarity grid for each MRI that stores the similarity scores between the latent representation of that MRI and all SOM representations. As the SOM grid is an encoding of brain aging, the similarity grid indicates the likelihood of placing the MRI within the "spectrum" of aging. Given all MRIs of a longitudinal scan, the change across the corresponding similarity grids over time represents the brain aging process of that individual. Furthermore, we infer brain aging on a group-level by first computing the average similarity grid for an age group and then visualizing the difference of those average similarity grids across age groups. With respect to the downstream tasks of classification (sMCI vs. pMCI) and regression (i.e., estimating the Alzheimer's Disease Assessment Scale-Cognitive Subscale (ADAS-Cog) on all subjects), our latent representations of the MRIs is associated with comparable or higher accuracy scores than representations learned by other state-of-the-art self-supervised methods.

2 Method

As shown in Fig. 1, the longitudinal 3D MRIs of a subject are encoded as a series of trajectories (blue vectors) in the latent space. Following [12,17], we consider a pair of longitudinal MRIs (that corresponds to a blue vector) as a training sample. Specifically, let $\mathcal{S}$ denote the set of image pairs of the training cohort, where the MRIs x^u and x^v of a longitudinal pair (x^u, x^v) are from the same subject and x^v was acquired Δt years after x^u. For simplicity, $\times$ refers to u or v when a function is separately applied to both time points. The MRIs are then mapped to the latent space by an encoder F, i.e., $z^\times := F(x^\times)$. On the latent space, the trajectory of the pair is denoted as $\Delta z := (z^v - z^u)/\Delta t$, which represents morphological changes. Finally, decoder H reconstructs the input MRI $x^\times$ from the latent representation $z^\times$, i.e., $\tilde{x}^\times := H(z^\times)$. Next, we describe LSOR, which generates interpretable SOM representations, and the post-hoc analysis for deriving similarity grids.

2.1 LSOR

Following [3,8], SOM representations are organized in a N_r by N_c grid (denoted as SOM grid) $\mathcal{G} = \{g_{i,j}\}_{i=1,j=1}^{N_r,N_c}$, where $g_{i,j}$ denotes the SOM representation on the i-th row and j-th column. This easy-to-visualize grid preserves the high-dimensional relationships between the clusters as shown in by the orange lines in Fig. 1. Given the latent representation $z^\times$, its closest SOM representation is denoted as $g_{\epsilon^\times}$, where $\epsilon^\times := argmin_{(i,j)} \parallel z^\times - g_{i,j} \parallel_2$ is its 2D grid index in $\mathcal{G}$ and $\parallel \cdot \parallel_2$ is the Euclidean norm. This SOM representation is also used to reconstruct the input MRI by the decoder, i.e., $\tilde{x}_g^\times = H(g_{\epsilon^\times})$. To do so, the reconstruction loss encourages both the latent representation $z^\times$ and its closet SOM representation $g_{\epsilon^\times}$ to be descriptive of the input MRI $x^\times$, i.e.,

$$L_{recon} := \mathbb{E}_{(x^u,x^v)\sim\mathcal{S}} \left(\sum_{\times\in\{x,v\}} \parallel x^\times - \tilde{x}^\times \parallel_2^2 + \parallel x^\times - \tilde{x}_g^\times \parallel_2^2 \right), \quad (1)$$

where $\mathbb{E}$ defines the expected value. The remainder describes the three novel components of our SOM representation.

Explicitly Regularizing Closeness. Though L_{recon} implicitly encourages close proximity between $z^\times$ and $g_{\epsilon^\times}$, it does not inherently optimize $g_{\epsilon^\times}$ as $z^\times$ is not differentiable with respect to $g_{\epsilon^\times}$. Therefore, we introduce an additional 'commitment' loss explicitly promoting closeness between them:

$$L_{commit} := \mathbb{E}_{(x^u,x^v)\sim\mathcal{S}} \left(\parallel z^u - g_{\epsilon^u} \parallel_2^2 + \parallel z^v - g_{\epsilon^v} \parallel_2^2 \right).$$

Soft Weighting Scheme. In addition to update $z^\times$'s closest SOM representation $g_{\epsilon^\times}$, we also update all SOM representations $g_{i,j}$ by introducing a soft weighting scheme as proposed in [10]. Specifically, we design a weight $w_{i,j}^\times$ to regularize how much $g_{i,j}$ should be updated with respect to $z^\times$ based on its proximity to the grid location $\epsilon^\times$ of $g_{\epsilon^\times}$, i.e.,

$$w_{i,j}^\times := \delta\left(e^{-\frac{\parallel \epsilon^\times - (i,j) \parallel_1^2}{2\tau}}\right), \quad (2)$$

where $\delta(w) := \frac{w}{\sum_{i,j} w_{i,j}}$ ensures that the scale of weights is constant during training and $\tau > 0$ is a scaling hyperparameter. Now, we design the following loss L_{som} so that SOM representations close to $\epsilon^\times$ on the grid are also close to $z^\times$ in the latent space (measured by the Euclidean distance $\parallel z^\times - g_{i,j} \parallel_2$):

$$L_{som} := \mathbb{E}_{(x^u,x^v)\sim\mathcal{S}} \left(\sum_{g_{i,j}\sim\mathcal{G}} (w_{i,j}^u \cdot \parallel z^u - g_{i,j} \parallel_2^2 + w_{i,j}^v \cdot \parallel z^v - g_{i,j} \parallel_2^2) \right). \quad (3)$$

To improve robustness, we make two more changes to Eq. 3. First, we account for SOM representations transitioning from random initialization to becoming meaningful cluster centers that preserve the high-dimensional relationships within the 2D SOM grid. We do so by decreasing τ in Eq. 2 with each iteration so that the weights gradually concentrate on SOM representations closer to $g_{\epsilon^\times}$ as training proceeds: $\tau(t) := N_r \cdot N_c \cdot \tau_{max} \left(\frac{\tau_{min}}{\tau_{max}}\right)^{t/T}$ with τ_{min} being the minimum and τ_{max} the maximum standard deviation in the Gaussian kernel, and t represents the current and T the maximum iteration.

The second change to Eq. 3 is to apply the stop-gradient operator $sg[\cdot]$ [16] to $z^\times$, which sets the gradients of $z^\times$ to 0 during the backward pass. The stop-gradient operator prevents the undesirable scenario where $z^\times$ is pulled towards a naive solution, i.e., different MRI samples are mapped to the same weighted average of all SOM representations. This risk of deriving the naive solution is especially high in the early stages of the training when the SOM representations are randomly initialized and may not accurately represent the clusters.

Longitudinal Consistency Regularization. We derive a SOM grid related to brain aging by generating an age-stratified latent space. Specifically, the latent space is defined by a smooth trajectory field (Fig. 1, blue box) characterizing the morphological changes associated with brain aging. The smoothness is based on the assumption that MRIs with similar appearances (close latent representations on the latent space) should have similar trajectories. It is enforced by modeling the similarity between each subject-specific trajectory Δz with a reference trajectory that represents the average trajectory of the cluster. Specifically, $\Delta g_{i,j}$ is the reference trajectory (Fig. 1, green arrow) associated with $g_{i,j}$ then the reference trajectories of all clusters $\mathcal{G}_\Delta = \{\Delta g_{i,j}\}_{i=1,j=1}^{N_r,N_c}$ represent the average aging of SOM clusters with respect to the training set. As all subject-specific trajectories are iteratively updated during the training, it is computationally infeasible to keep track of $\mathcal{G}_\Delta$ on the whole training set. We instead propose to compute the exponential moving average (EMA) (Fig. 1, black box), which iteratively aggregates the average trajectory with respect to a training batch to $\mathcal{G}_\Delta$:

$$\Delta g_{i,j} \leftarrow \begin{cases} \Delta h_{i,j} & t = 0 \\ \Delta g_{i,j} & t > 0 \text{ and } |\Omega_{i,j}| = 0 \\ \alpha \cdot \Delta g_{i,j} + (1-\alpha) \cdot \Delta h_{i,j} & t > 0 \text{ and } |\Omega_{i,j}| > 0 \end{cases}$$

$$\text{with } \Delta h_{i,j} := \frac{1}{|\Omega_{i,j}|} \sum_{k=1}^{N_{bs}} \mathbb{1}[\epsilon_k^u = (i,j)] \cdot \Delta z_k \text{ and } |\Omega_{i,j}| := \sum_{k=1}^{N_{bs}} \mathbb{1}[\epsilon_k^u = (i,j)].$$

α is the EMA keep rate, k denotes the index of the sample pair, N_{bs} symbolizes the batch size, $\mathbb{1}[\cdot]$ is the indicator function, and $|\Omega_{i,j}|$ denotes the number of sample pairs with $\epsilon^u = (i,j)$ within a batch. Then in each iteration, $\Delta h_{i,j}$ (Fig. 1, purple arrow) represents the batch-wise average of subject-specific trajectories for sample pairs with $\epsilon^u = (i,j)$. By iteratively updating $\mathcal{G}_\Delta$, $\mathcal{G}_\Delta$ then approximate the average trajectories derived from the entire training set. Lastly, inspired by [11,12], the longitudinal consistency regularization is formulated as

$$L_{dir} := \mathbb{E}_{(x^u, x^v) \sim \mathcal{S}} \left(1 - cos(\theta[\Delta z, sg[\Delta g_{\epsilon^u}]])\right),$$

where $\theta[\cdot, \cdot]$ denotes the angle between two vectors. Since Δg is optimized by EMA, the stop-gradient operator is again incorporated to only compute the gradient with respect to Δz in L_{dir}.

Objective Function. The complete objective function is the weighted combination of the prior losses with weighing parameters λ_{commit}, λ_{som}, and λ_{dir}:

$$L := L_{recon} + \lambda_{commit} \cdot L_{commit} + \lambda_{som} \cdot L_{som} + \lambda_{dir} \cdot L_{dir}$$

The objective function encourages a smooth trajectory field of aging on the latent space while maintaining interpretable SOM representations for analyzing brain age in a pure self-supervised fashion.

2.2 SOM Similarity Grid

During inference, a (2D) similarity grid ρ is computed by the closeness between the latent representation z of an MRI sample and the SOM representations:

$$\rho := softmax(- \parallel z - \mathcal{G} \parallel_2^2 / \gamma) \text{ with } \gamma := std(\parallel z - \mathcal{G} \parallel_2^2)$$

std denotes the standard deviation of the distance between z to all SOM representations. As the SOM grid is learned to be associated with brain age (e.g., represents aging from left to right), the similarity grid essentially encodes a "likelihood function" of the brain age in z. Given all MRIs of a longitudinal scan, the change across the corresponding similarity grids over time represents the brain aging process of that individual. Furthermore, brain aging on the group-level is captured by first computing the average similarity grid for an age group and then visualizing the difference of those average similarity grids across age groups.

3 Experiments

3.1 Experimental Setting

Dataset. We evaluated the proposed method on all 632 longitudinal T1-weighted MRIs (at least two visits per subject, 2389 MRIs in total) from ADNI-1 [13]. The data set consists of 185 NC (age: 75.57 $\pm$ 5.06 years), 193 subjects diagnosed with sMCI (age: 75.63 $\pm$ 6.62 years), 135 subjects diagnosed with pMCI (age: 75.91 $\pm$ 5.35 years), and 119 subjects with AD (age: 75.17 $\pm$ 7.57 years). There was no significant age difference between the NC and AD cohorts ($p = 0.55$, two-sample t-test) as well as the sMCI and pMCI cohorts ($p = 0.75$). All MRI images were preprocessed by a pipeline including denoising, bias field correction, skull stripping, affine registration to a template, re-scaling to 64 $\times$ 64 $\times$ 64 volume, and transforming image intensities to z-scores.

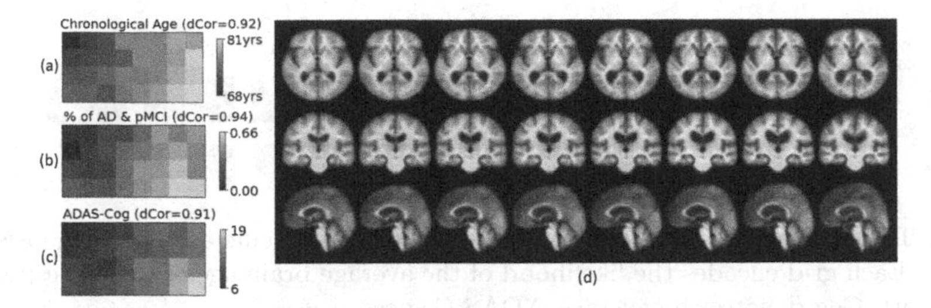

Fig. 2. The color at each SOM representation encodes the average value of (a) chronological age, (b) % of AD and pMCI, and (c) ADAS-Cog score across the training samples of that cluster; (d) Confined to the last row of the grid, the average MRI of 20 latent representations closest to the corresponding SOM representation. (Color figure online)

Implementation Details. Let C_k denote a Convolution(kernel size of $3 \times 3 \times 3$, Conv_k)-BatchNorm-LeakyReLU(slope of 0.2)-MaxPool(kernel size of 2) block with k filters, and CD_k an Convolution-BatchNorm-LeakyReLU-Upsample block. The architecture was designed as C_{16}-C_{32}-C_{64}-C_{16}-Conv_{16}-CD_{64}-CD_{32}-CD_{16}-CD_{16}-Conv_1, which results in a latent space of 1024 dimensions. The training of SOM is difficult in this high-dimensional space with random initialization in practice, thus we first pre-trained the model with only L_{recon} for 10 epochs and initialized the SOM representations by doing k-means of all training samples using this pre-trained model. Then, the network was further trained for 40 epochs with regularization weights set to $\lambda_{recon} = 1.0$, $\lambda_{commit} = 0.5$, $\lambda_{som} = 1.0$, $\lambda_{dir} = 0.2$. Adam optimizer with learning rate of 5×10^{-4} and weight decay of 10^{-5} were used. τ_{min} and τ_{max} in L_{som} were set as 0.1 and 1.0 respectively. An EMA keep rate of $\alpha = 0.99$ was used to update reference trajectories. A batch size $N_{bs} = 64$ and the SOM grid size $N_r = 4, N_c = 8$ were applied.

Evaluation. We performed five-fold cross-validation (folds split based on subjects) using 10% of the training subjects for validation. The training data was augmented by flipping brain hemispheres and random rotation and translation. To quantify the interpretability of the SOM grid, we correlated the coordinates of the SOM grid with quantitative measures related to brain age, e.g., chronological age, the percentage of subjects with severe cognitive decline, and Alzheimer's Disease Assessment Scale-Cognitive Subscale (ADAS-Cog). We illustrated the interpretability with respect to brain aging by visualizing the changes in the SOM similarity maps over time. We further visualized the trajectory vector field along with SOM representations by projecting the 1024-dimensional representations to the first two principal components of SOM representations. Lastly, we quantitatively evaluated the quality of the representations by applying them to the downstream tasks of classifying sMCI vs. pMCI and ADAS-Cog prediction. We measured the classification accuracy via Balanced accuracy (BACC) and Area Under Curve (AUC) and the prediction accuracy via R2 and root-mean-square error (RMSE). The classifier and predictor were multi-layer per-

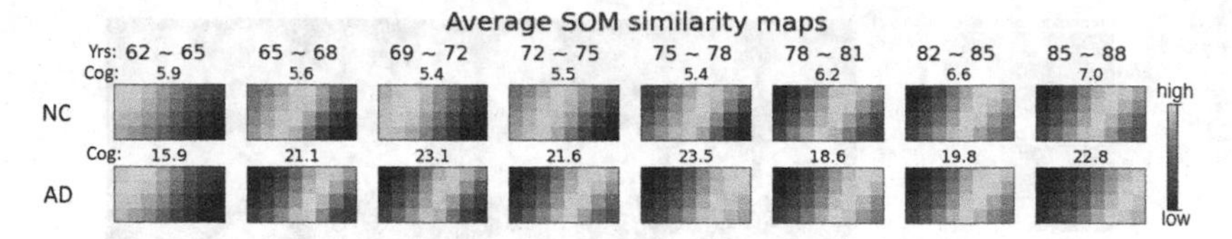

Fig. 3. The average similarity grid ρ over subjects of a specific age and diagnosis (NC vs AD). Each grid encodes the likelihood of the average brain age of the corresponding sub-cohort. Cog denotes the average ADAS-Cog score.

ceptrons containing two fully connected layers of dimensions 1024 and 64 with a LeakyReLU activation. We compared the accuracy metrics to models using the same architecture with encoders pre-trained by other representation learning methods, including unsupervised methods (AE, VAE [4]), self-supervised method (SimCLR [1]), longitudinal self-supervised method (LSSL [17]), and longitudinal neighborhood embedding (LNE [12]). All comparing methods used the same experimental setup (e.g., encoder-decoder, learning rate, batch size, epochs, etc.), and the method-specific hyperparameters followed [12].

3.2 Results

Interpretability of SOM Embeddings. Fig. 2 shows the stratification of brain age over the SOM grid $\mathcal{G}$. For each grid entry, we show the average value of chronological age (Fig. 2(a)), % of AD & pMCI (Fig. 2(b)), and ADAS-Cog score (Fig. 2(c)) over samples of that cluster. We observed a trend of older brain age (yellow) from the upper left towards the lower right, corresponding to older chronological age and worse cognitive status. The SOM grid index strongly correlated with these three factors (distance correlation of 0.92, 0.94, and 0.91 respectively). Figure 2(d) shows the average brain over 20 input images with representations that are closest to each SOM representation of the last row of the grid (see Supplement Fig. S1 for all rows). From left to right the ventricles are enlarging and the brain is atrophying, which is a hallmark for brain aging effects.

Interpretability of Similarity Grid. Visualizing the average similarity grid ρ of the NC and AD at each age range in Fig. 3, we observed that higher similarity (yellow) gradually shifts towards the right with age in both NC and AD (see Supplemental Fig. S2 for sMCI and pMCI cohorts). However, the shift is faster for AD, which aligns with AD literature reporting that AD is linked to accelerated brain aging [15]. Furthermore, the subject-level aging effects shown in Supplemental Fig. S3 reveal that the proposed visualization could capture subtle morphological changes caused by brain aging.

Interpretability of Trajectory Vector Field. Fig. 4 plots the PCA projections of the latent space in 2D, which shows a smooth trajectory field (gray arrows) and reference trajectories $\mathcal{G}_\Delta$ (blue arrows) representing brain aging.

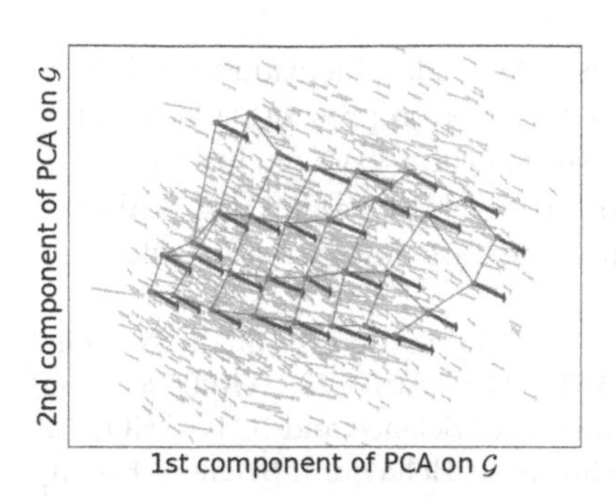

Fig. 4. 2D PCA of the LSOR's latent space. Light gray arrows represent Δz. The orange grid represents the relationships between SOM representations and associated reference trajectory $\Delta \mathcal{G}$ (blue arrow). (Color figure online)

Table 1. Supervised downstream tasks using the learned representations z (without fine-tuning the encoder). LSOR achieved comparable or higher accuracy scores than other state-of-the-art self- and un-supervised methods.

Methods	sMCI/pMCI		ADAS-Cog	
	BACC	AUC	R2	RMSE
AE	62.6	65.4	0.26	6.98
VAE [4]	61.3	64.8	0.23	7.17
SimCLR [1]	63.3	66.3	0.26	6.79
LSSL [17]	69.4	71.8	0.29	6.49
LNE [12]	**70.6**	72.1	0.30	6.46
LSOR	69.8	**72.4**	**0.32**	**6.31**

This projection also preserved the 2D grid structure (orange) of the SOM representations suggesting that aging was the most important variation in the latent space.

Downstream Tasks. To evaluate the quality of the learned representations, we froze encoders trained by each method without fine-tuning and utilized their representations for the downstream tasks (Table 1). On the task of sMCI vs. pMCI classification (Table 1 (left)), the proposed method achieved a BACC of 69.8 and an AUC of 72.4, a comparable accuracy ($p > 0.05$, DeLong's test) with LSSL [17] and LNE [12], two state-of-the-art self-supervised methods on this task. On the ADAS-Cog score regression task, the proposed method obtained the best accuracy with an R2 of 0.32 and an RMSE of 6.31. It is worth mentioning that an accurate prediction of the ADAS-Cog score is very challenging due to its large range (between 0 and 70) and its subjectiveness resulting in large variability across exams [2] so that even larger RMSEs have been reported for this task [7]. Furthermore, our representations were learned in an unsupervised manner so that further fine-tuning of the encoder would improve the prediction accuracy.

4 Conclusion

In this work, we proposed LSOR, the first SOM-based learning framework for longitudinal MRIs that is self-supervised and interpretable. By incorporating a soft SOM regularization, the training of the SOM was stable in the high-dimensional latent space of MRIs. By regularizing the latent space based on longitudinal consistency as defined by longitudinal MRIs, the latent space formed a smooth trajectory field capturing brain aging as shown by the resulting SOM grid. The interpretability of the representations was confirmed by the correlation between the SOM grid and cognitive measures, and the SOM similarity map.

When evaluated on downstream tasks sMCI vs. pMCI classification and ADAS-Cog prediction, LSOR was comparable to or better than representations learned from other state-of-the-art self- and un-supervised methods. In conclusion, LSOR is able to generate a latent space with high interpretability regarding brain age purely based on MRIs, and valuable representations for downstream tasks.

Acknowledgement. This work was partly supported by funding from the National Institute of Health (MH113406, DA057567, AA017347, AA010723, AA005965, and AA028840), the DGIST R&D program of the Ministry of Science and ICT of KOREA (22-KUJoint-02), Stanford's Department of Psychiatry & Behavioral Sciences Faculty Development & Leadership Award, and by Stanford HAI Google Cloud Credit.

References

1. Chen, T., Kornblith, S., Norouzi, M., Hinton, G.: A simple framework for contrastive learning of visual representations. In: International Conference on Machine Learning, pp. 1597–1607. PMLR (2020)
2. Connor, D.J., Sabbagh, M.N.: Administration and scoring variance on the ADAS-Cog. J. Alzheimers Dis. **15**(3), 461–464 (2008)
3. Fortuin, V., Hüser, M., Locatello, F., Strathmann, H., Rätsch, G.: SOM-VAE: interpretable discrete representation learning on time series. In: International Conference on Learning Representations (2019)
4. Kingma, D.P., Welling, M.: Auto-encoding variational bayes. arXiv preprint arXiv:1312.6114 (2013)
5. Kohonen, T.: The self-organizing map. Proc. IEEE **78**(9), 1464–1480 (1990)
6. Li, O., Liu, H., Chen, C., Rudin, C.: Deep learning for case-based reasoning through prototypes: a neural network that explains its predictions. In: Proceedings of the AAAI Conference on Artificial Intelligence, vol. 32 (2018)
7. Ma, D., Pabalan, C., Interian, Y., Raj, A.: Multi-task learning and ensemble approach to predict cognitive scores for patients with Alzheimer's disease. bioRxiv, pp. 2021–12 (2021)
8. Manduchi, L., Hüser, M., Vogt, J., Rätsch, G., Fortuin, V.: DPSOM: deep probabilistic clustering with self-organizing maps. In: Conference on Neural Information Processing Systems Workshop on Machine Learning for Health (2019)
9. Molnar, C.: Interpretable machine learning (2020)
10. Mulyadi, A.W., Jung, W., Oh, K., Yoon, J.S., Lee, K.H., Suk, H.I.: Estimating explainable Alzheimer's disease likelihood map via clinically-guided prototype learning. Neuroimage **273**, 120073 (2023)
11. Ouyang, J., Zhao, Q., Adeli, E., Zaharchuk, G., Pohl, K.M.: Self-supervised learning of neighborhood embedding for longitudinal MRI. Med. Image Anal. **82**, 102571 (2022)
12. Ouyang, J., et al.: Self-supervised longitudinal neighbourhood embedding. In: de Bruijne, M., et al. (eds.) MICCAI 2021. LNCS, vol. 12902, pp. 80–89. Springer, Cham (2021). https://doi.org/10.1007/978-3-030-87196-3_8
13. Petersen, R.C., et al.: Alzheimer's disease neuroimaging initiative (ADNI): clinical characterization. Neurology **74**(3), 201–209 (2010)
14. Rudin, C.: Stop explaining black box machine learning models for high stakes decisions and use interpretable models instead. Nat. Mach. Intell. **1**(5), 206–215 (2019)

15. Toepper, M.: Dissociating normal aging from Alzheimer's disease: a view from cognitive neuroscience. J. Alzheimers Dis. **57**(2), 331–352 (2017)
16. Van Den Oord, A., Vinyals, O., et al.: Neural discrete representation learning. Adv. Neural Inf. Process. Syst. **30** (2017)
17. Zhao, Q., Liu, Z., Adeli, E., Pohl, K.M.: Longitudinal self-supervised learning. Med. Image Anal. **71**, 102051 (2021)

Self-supervised Learning for Physiologically-Based Pharmacokinetic Modeling in Dynamic PET

Francesca De Benetti[1(✉)], Walter Simson[2], Magdalini Paschali[3], Hasan Sari[4], Axel Rominger[5], Kuangyu Shi[1,5], Nassir Navab[1], and Thomas Wendler[1]

[1] Chair for Computer-Aided Medical Procedures and Augmented Reality, Technische Universität München, Garching, Germany
francesca.de-benetti@tum.de

[2] Department of Radiology, Stanford University School of Medicine, Stanford, USA

[3] Department of Psychiatry and Behavioral Sciences, Stanford University School of Medicine, Stanford, USA

[4] Advanced Clinical Imaging Technology, Siemens Healthcare AG, Lausanne, Switzerland

[5] Department of Nuclear Medicine, Bern University Hospital, Bern, Switzerland

Abstract. Dynamic Positron Emission Tomography imaging (dPET) provides temporally resolved images of a tracer. Voxel-wise physiologically-based pharmacokinetic modeling of the Time Activity Curves (TAC) extracted from dPET can provide relevant diagnostic information for clinical workflow. Conventional fitting strategies for TACs are slow and ignore the spatial relation between neighboring voxels. We train a spatio-temporal UNet to estimate the kinetic parameters given TAC from dPET. This work introduces a self-supervised loss formulation to enforce the similarity between the measured TAC and those generated with the learned kinetic parameters. Our method provides quantitatively comparable results at organ level to the significantly slower conventional approaches while generating pixel-wise kinetic parametric images which are consistent with expected physiology. To the best of our knowledge, this is the first self-supervised network that allows voxel-wise computation of kinetic parameters consistent with a non-linear kinetic model.

Keywords: Kinetic modelling · PBPK models · Dynamic PET

1 Introduction

Positron Emission Tomography (PET) is a 3D imaging modality using radiopharmaceuticals, such as F-18-fluorodeoxyglucose (FDG), as tracers. Newly introduced long axial field-of-view PET scanners have enabled dynamic PET (dPET)

Supplementary Information The online version contains supplementary material available at https://doi.org/10.1007/978-3-031-43907-0_28.

H. Greenspan et al. (Eds.): MICCAI 2023, LNCS 14220, pp. 290–299, 2023.
https://doi.org/10.1007/978-3-031-43907-0_28

with frame duration < 1 min [18], allowing the observation of dynamic metabolic processes throughout the body. For a given voxel in space, the radioactivity concentration over time can be described by a characteristic curve known as Time Activity Curve (TAC), measured in [Bq/ml]. TACs can be described by mathematical functions, called physiologically-based pharmacokinetic (PBPK) models or kinetic models (KM) [14]. The parameters of the KM represent physiologically relevant quantities and are often called *micro-parameters*, whereas their combinations are called *macro-parameters* [14,20]. While the former can be retrieved only by methods that directly use the KM function, the latter can be computed by simplified linearized methods (such as the Logan and the Patlak-Gjedde plots). The approaches to extract KM parameters can be split in two categories: Volume of Interest (VoI) methods, in which the average TAC in a VoI is used, or voxel-based methods. Despite the former displaying less noise and, therefore, lower variance in the kinetic parameters (KPs), VoI-based methods only provide organ-wise information. On the other hand, voxel-based methods allow the generation of parametric images (KPIs), in which the KPs are visualized at a voxel level, but suffer from motion and breathing artifacts, and require more computational power or simplified linearized methods.

Parametric images are reported to be superior in lesion detection and delineation when compared to standard-of-care activity- and weight-normalized static PET volumes, known as Standard Uptake Value (SUV) volumes [6,7]. Changes in the KPs during oncological therapy are associated with pathological response to treatment, whereas this is not true for changes in SUV [15]. Despite the advantages of KPIs in diagnosis, the generation of accurate micro-parametric images is not yet possible in clinical practice.

To address the problem of the generation of micro-parametric images, we propose a custom 3D UNet [3] to estimate kinetic *micro-parameters* in an unsupervised setting drawing inspiration from physics-informed neural networks (PINN). The main contributions of this work are:

- A self-supervised formulation of the problem of kinetic *micro-parameters* estimation
- A spatio-temporal deep neural network for parametric images estimation
- A quantitative and qualitative comparison with conventional methods for PBPK modeling

The code is available at:
https://github.com/FrancescaDB/self_supervised_PBPK_modelling.

1.1 Related Work

Finding the parameters of a KM is a classical optimization problem [2,19,21] solved by fitting the KM equation to a measured TAC in a least squares sense [1, 14,17]. The non-linearity of the KM functions makes this approach prone to overfitting and local minima, and sensitive to noise [14]. Therefore, non-linear parametric imaging is still too noisy for clinical application [20].

To limit the drawbacks of the non-linear parameter fitting, the identification of KPs is commonly performed using simplified linearized versions of the KM [6,20], such as the Logan and the Patlak-Gjedde plots [5,16], which are often included in clinical software for KM such as PMOD[1].

Preliminary works towards KM parameter estimation in dPET imaging have recently begun to be explored. Moradi et al. used an auto-encoder along with a Gaussian process regression block to select the best KM to describe simulated kinetic data [13]. A similar approach was presented for the quantification of myocardial blood flow from simulated PET sinograms [11]. Huang et al. used a supervised 3D U-Net to predict a macro-parametric image using an SUV image as input, and a Patlak image derived from dPET acquisition as ground truth [9]. Cui et al. proposed a conditional deep image prior framework to predict a macro-parametric image using a DNN in an unsupervised setting [4]. Finally, a supervised DNN was used to predict Patlak KPIs from Dynamic PET sinograms [12]. Until now, methods used simulated data [11,13] or static PET [9], were supervised [9,11–13] or predicted *macro-parameters* [4,9,12].

2 Methodology and Materials

We propose to compute the kinetic *micro-parameters* in a self-supervised setting by directly including the KM function in the loss function and comparing the predicted TAC to the measured TAC. For this reason, an understanding of the KM is fundamental to describing our pipeline.

2.1 Kinetic Modelling

The concentration of the tracer $\tilde{C}(t)$ [Bq/ml] in each tissue can be described as a set of ordinary differential equations [20]. It represents the interaction of two compartments, $F(t)$ (free) and $B(t)$ (bound), and takes as input the radioactivity concentration in blood or plasma $A(t)$ [14]:

$$\begin{aligned}\frac{dF(t)}{dt} &= K_1 A(t) - (k_2 + k_3)F(t) \\ \frac{dB(t)}{dt} &= k_3 F(t) - k_4 B(t) \\ \tilde{C}(t) &= F(t) + B(t)\end{aligned} \tag{1}$$

where K_1 [ml/cm^3/min], k_2 [1/min], k_3 [1/min] and k_4 [1/min] are the *micro-parameters* [6,20]. Equation 1 describes a general two-tissue compartment (2TC) kinetic model. However, an FDG-TAC is conventionally described by an irreversible 2TC, in which k_4 is set to 0 [20]. Therefore, in the following, we will use $k_4 = 0$.

Moreover, including the blood fraction volume V_B [$\cdot$] allows to correctly model the contribution to the radioactivity in a voxel coming from vessels that are too small to be resolved by the PET scanner [14].

[1] https://www.pmod.com.

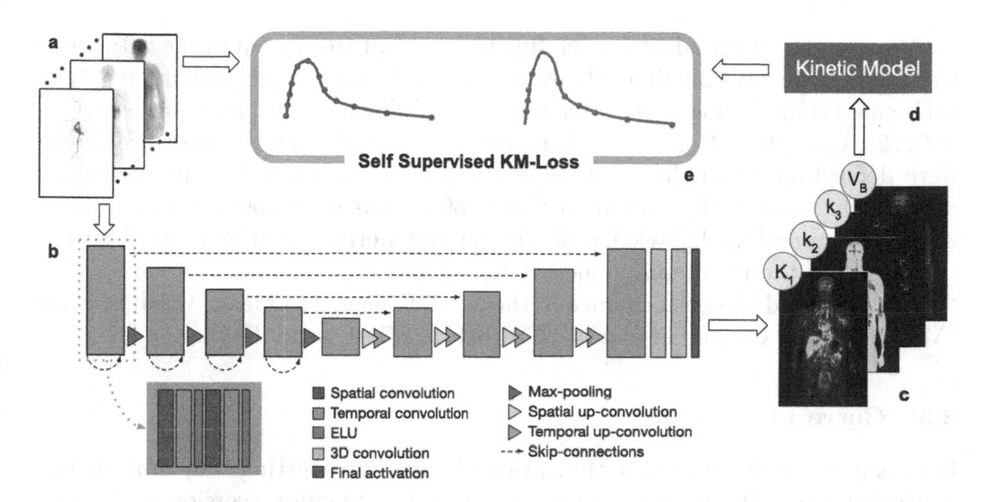

Fig. 1. Proposed Pipeline: a sequence of dPET slices (a) is processed by the proposed DNN (b) to estimate the 4 KPIs (c). During training, the predicted TACs are computed by the KM (d) and compared to the measured TACs using the mean square error (e).

Together, the TAC of each voxel in an FDG dPET acquisition can be modeled as $C(t) = (1 - V_B)\tilde{C}(t) + V_B A(t)$, and solved using the Laplace transform [20]:

$$C(t) = (1 - V_B)\left[\frac{K_1}{k_2 + k_3}\left[k_3 + k_2 e^{-(k_2+k_3)t}\right] * A(t)\right] + V_B A(t). \tag{2}$$

2.2 Proposed Pipeline

Our network takes as input a sequence of 2D axial slices and returns a 4-channel output representing the spatial distribution of the KM parameters of a 2TC for FDG metabolisation [14]. The network has a depth of four, with long [3] and short skip connections [10]. The kernel size of the max-pooling is [2, 2, 2]. After the last decoder block, two 3D convolutional layers (with kernel size [3, 3, 3] and [64, 1, 1]) estimate the KPs per voxel given the output feature of the network. Inside the network the activation function is ELU and critically batch normalization is omitted. The network was trained with an initial learning rate of 10^{-4}, which was divided by half every 25 epochs, for a maximum of 500 epochs.

Following the approach taken by Küstner et al. for motion correction of 4D spatio-temporal CINE MRI [10], we replaced a conventional 3D convolutional layer with (2+1)D spatial and temporal convolutional layers. The spatial convolutional layer is a 3D convolutional layer with kernel size [1, 3, 3] in [t, x, y]. Similarly, the temporal convolutional layer has a kernel size of [3, 1, 1].

We imposed that the KPs predicted by the network satisfy Eq. 2 by including it in the computation of the loss. At a pixel level, we computed the mean squared error between the TAC estimated using the corresponding predicted parameters ($\tilde{\text{TAC}}_i$) and the measured one (TAC_i), as seen in Fig. 1.

We introduced a final activation function to limit the output of the network to the valid parameter domain of the KM function. Using the multi-clamp function, each channel of the logits is restricted to the following parameter spaces: $K_1 \in [0.01, 2]$, $k_2 \in [0.01, 3]$, $k_3 \in [0.01, 1]$, and $V_B \in [0, 1]$. The limits of the ranges were defined based on the meaning of the parameter (as in V_B), mathematical requirements (as in the minimum values of k_2 and k_3, whose sum can not be zero) [6] or previous knowledge on the dataset derived by the work of Sari et al. [16] (as in the maximum values of K_1, k_2 and k_3).

We evaluated the performance of the network using the Mean Absolute Error (MAE) and the Cosine Similarity (CS) between TAC_i and $\tilde{\text{TAC}}_i$.

2.3 Curve Fit

For comparison, parameter optimization via non-linear fitting was implemented in Python using the `scipy.optimize.curve_fit` function (version 1.10), with step equal to 0.001. The bounds were the same as in the DNN.

2.4 Dataset

The dataset is composed of 23 oncological patients with different tumor types. dPET data was acquired on a Biograph Vision Quadra for 65 min, over 62 frames. The exposure duration of the frames were 2×10 s, 30×2 s, 4×10 s, 8×30 s, 4×60 s, 5×120 s and 9×300 s. The PET volumes were reconstructed with an isotropic voxel size of 1.6 mm. The dataset included the label maps of 7 organs (bones, lungs, heart, liver, kidneys, spleen, aorta) and one image-derived input function $A(t)$ [Bq/ml] from the descending aorta per patient. Further details on the dataset are presented elsewhere [16].

The PET frames and the label map were resampled to an isotropic voxel size of 2.5 mm. Then, the dataset was split patient-wise into training, validation, and test set, with 10, 4, and 9 patients respectively. Details on the dataset split are available in the Supplementary Material (Table 1). The training set consisted of 750 slices and the validation consisted of 300. In both cases, 75 axial slices per patient were extracted in a pre-defined patient-specific range from the lungs to the bladder (included) and were cropped to size 112×112 pixels.

3 Results

Table 1 shows the results of the 8 ablation studies we performed to find the best model. We evaluated the impact of the design of the convolutional and max-pooling kernels, as well as the choice of the final activation function. The design of the max pooling kernel (i.e., kernel size equal to [2, 2, 2] or [1, 2, 2]) had no measurable effects in terms of CS in most of the experiments, with the exception of Exp. 3.2, where max-pooling only in space resulted in a drop of 0.06. When evaluating the MAE, the use of 3D max-pooling was generally better.

Table 1. Configurations and metrics of ablation studies for architecture optimization.

Exp	Convolution	Pooling	Final activation	CS ↑	MAE ↓
1.1	3D	3D	Absolute	0.74 ± 0.05	3.55 ± 2.12
1.2	3D	space	Absolute	0.74 ± 0.05	3.64 ± 2.21
2.1	space + time	3D	Absolute	0.75 ± 0.05	3.59 ± 2.33
2.2	space + time	space	Absolute	0.75 ± 0.05	3.67 ± 2.20
3.1	space + time	3D	Clamp	0.75 ± 0.05	3.48 ± 2.04
3.2	space + time	space	Clamp	0.69 ± 0.05	3.55 ± 2.07
4.1	space + time	3D	Multi-clamp	**0.78 ± 0.05**	3.28 ± 2.03
4.2	space + time	space	Multi-clamp	0.77 ± 0.05	**3.27 ± 2.01**

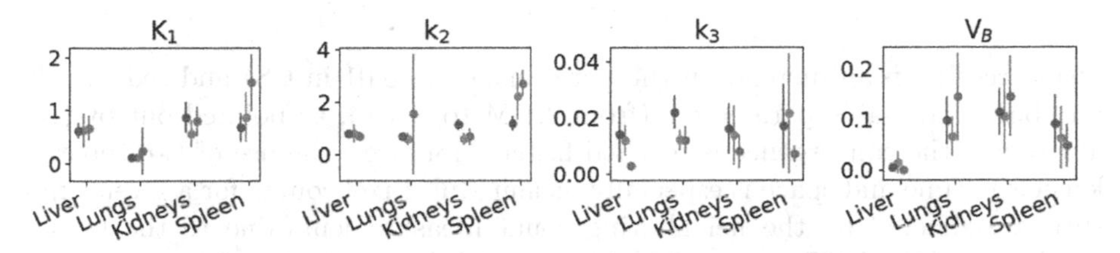

Fig. 2. Comparison between the kinetic parameters obtained with different methods: KP_{DNN} in blue, KP_{CF} in orange and, as plausibility check, KP_{CF}^{ref} [16] in green. The exact values are reported in the Supplementary Material (Table 3 and 4) and in [16]. (Color figure online)

The most important design choice is the selection of the final activation function. Indeed, the multi-clamp final activation function was proven to be the best both in terms of CS (Exp 4.1: CS = 0.78 ± 0.05) and MAE (Exp 4.2: MAE = 3.27 ± 2.01). Compared to the other final activation functions, when the multi-clamp is used the impact of the max-pooling design is negligible also in terms of MAE. For the rest of the experiments, the selected configuration is the one from Exp. 4.1 (see Table 1).

Figure 2 shows the KPs for four selected organs as computed with the proposed DNN (KP_{DNN}), as computed with curve fit using only the 9 patients of the test set (KP_{CF}) and using all 23 patients (KP_{CF}^{ref}) [16]. The voxel-wise KPs predicted by the DNN were averaged over the available organ masks.

In terms of run-time, the DNN needed ≈ 1 min to predict the KPs of the a whole-body scan (≈ 400 slices), whereas curve fit took 8.7 min for a single slice: the time reduction of the DNN is expected to be ≈ 3.500 times.

4 Discussion

Even though the choice of the final activation function has a greater impact, the selection of the kernel design is important. Using spatial and temporal convo-

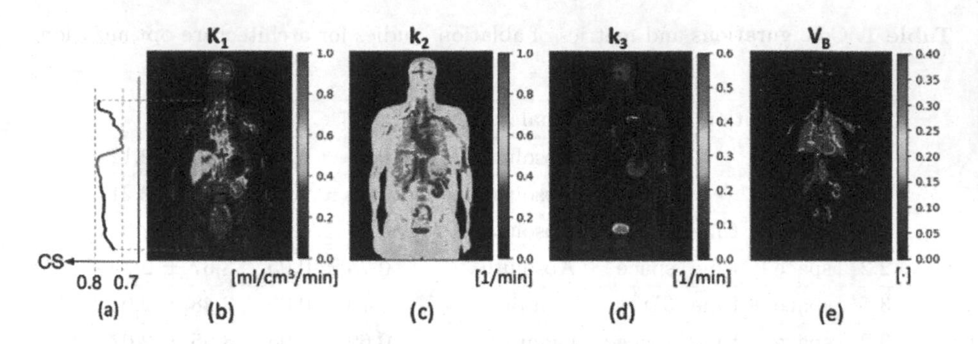

Fig. 3. (a) Cosine similarity (CS) per slice in patient 23 (blue: lungs; red: lungs and heart; green: liver). (b–e) Parametric images of a coronal slice for the same patient. (Color figure online)

lution results in an increase in the performance (+0.01 in CS) and reduces the number of trainable parameters (from 2.1 M to 8.6 K), as pointed out by [10]. Therefore, the convergence is reached faster. Moreover, the use of two separate kernels in time and space is especially meaningful. Pixel counts for a given exposure are affected by the neighboring count measurements due to the limited resolution of the PET scanner [20]. The temporally previous or following counts are independent. In general, there is good agreement between KP_{DNN}, KP_{CF} and KP_{CF}^{ref}. The DNN prediction of K_1 and k_2 in the spleen and k_3 in the lungs is outside the confidence interval of the results published by Sari et al. [16].

An analysis per slice of the metrics shows that the CS between TAC_i and $\tilde{TAC}_i$ changes substantially depending on the region: $CS_{max} = 0.87$ within the liver boundaries and $CS_{min} = 0.71$ in the region corresponding to the heart and lungs (see Fig. 3a). This can be explained by the fact that V_B is underestimated for the heart and aorta. The proposed network predicts $V_B^{heart} = 0.376 \pm 0.133$ and $V_B^{aorta} = 0.622 \pm 0.238$ while values of nearly 1 are to be expected. This is likely due to breathing and heartbeat motion artifacts, which cannot be modeled properly with a 2TC KM that assumes no motion between frames.

Figure 3b–e shows the central coronal slice of the four KPIs in an exemplary patient. As expected, K_1 is high in the heart, liver, and kidney. Similarly, the blood fraction volume V_B is higher in the heart, blood vessels, and lungs.

The KP_{DNN} are more homogeneous than KP_{CF}, as can be seen in the exemplary K_1 axial slice shown in Fig. 4. A quantitative evaluation of the smoothness of the images is reported in the Supplementary Material (Fig. 1). Moreover, the distribution in the liver is more realistic in KP_{DNN}, where the gallbladder can be seen as an ellipsoid between the right and left liver lobes. High K_1 regions are mainly within the liver, spleen, and kidney for KP_{DNN}, while they also appear in unexpected areas in the KP_{CF} (e.g., next to the spine or in the region of the stomach).

The major limitation of this work is the lack of ground truth and a canonical method to evaluate quantitatively its performance. This limitation is inherent

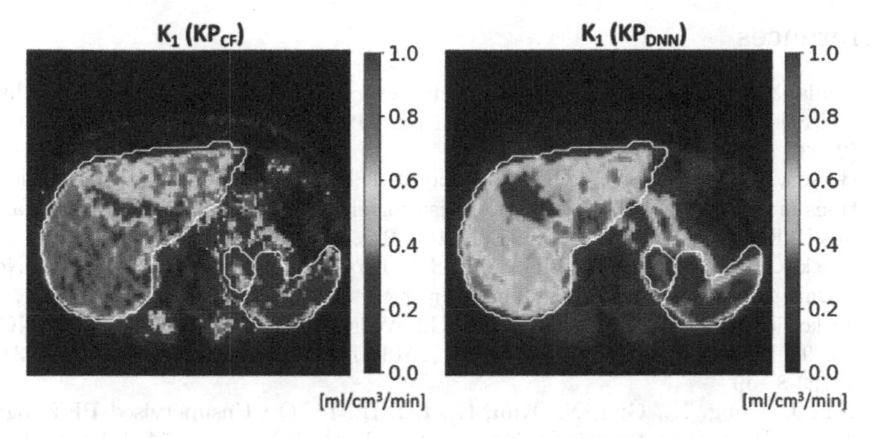

Fig. 4. Comparison of K_1 parametric images for an axial slice of patient 2, with contours of the liver (left), the spleen (center) and the left kidney (right).

to PBPK modeling and results in the need for qualitative analyses based on expected physiological processes. A possible way to leverage this would be to work on simulated data, yet the validity of such evaluations strongly depends on how realistic the underlying simulation models are. As seen in Fig. 3a, motion (gross, respiratory, or cardiac) has a major impact on the estimation quality. Registering different dPET frames has been shown to improve conventional PBPK models [8] and would possibly have a positive impact on our approach.

5 Conclusion

In this work, inspired by PINNs, we combine a self-supervised spatio-temporal DNN with a new loss formulation considering physiology to perform kinetic modeling of FDG dPET. We compare the best DNN model with the most commonly used conventional PBPK method, curve fit. While no ground truth is available, the proposed method provides similar results to curve fit but qualitatively more plausible images in physiology and with a radically shorter run-time.

Further, our approach can be applied to other KMs without significantly increasing the complexity and the need for computational power. In general, Eq. 2 should be modified to represent the desired KM [20], and the number of channels of the output of the network should be the same as the KP to be predicted.

Overall, this work offers scalability and a new research direction for analysing pharmacokinetics.

Acknowledgements. This work was partially funded by the German Research Foundation (DFG, grant NA 620/51-1).

References

1. Avula, X.J.: Mathematical modeling. In: Meyers, R.A. (ed.) Encyclopedia of Physical Science and Technology, 3rd edn., pp. 219–230. Academic Press, New York (2003)
2. Besson, F.L., et al.: 18F-FDG PET and DCE kinetic modeling and their correlations in primary NSCLC: first voxel-wise correlative analysis of human simultaneous [18F] FDG PET-MRI data. EJNMMI Res. **10**(1), 1–13 (2020)
3. Çiçek, Ö., Abdulkadir, A., Lienkamp, S.S., Brox, T., Ronneberger, O.: 3D U-Net: learning dense volumetric segmentation from sparse annotation. In: Ourselin, S., Joskowicz, L., Sabuncu, M.R., Unal, G., Wells, W. (eds.) MICCAI 2016. LNCS, vol. 9901, pp. 424–432. Springer, Cham (2016). https://doi.org/10.1007/978-3-319-46723-8_49
4. Cui, J., Gong, K., Guo, N., Kim, K., Liu, H., Li, Q.: Unsupervised PET logan parametric image estimation using conditional deep image prior. Med. Image Anal. **80**, 102519 (2022)
5. Dias, A.H., Hansen, A.K., Munk, O.L., Gormsen, L.C.: Normal values for 18F-FDG uptake in organs and tissues measured by dynamic whole body multiparametric FDG PET in 126 patients. EJNMMI Res. **12**(1), 1–14 (2022)
6. Dimitrakopoulou-Strauss, A., Pan, L., Sachpekidis, C.: Kinetic modeling and parametric imaging with dynamic PET for oncological applications: general considerations, current clinical applications, and future perspectives. Eur. J. Nucl. Med. Mol. Imaging **48**, 21–39 (2021). https://doi.org/10.1007/s00259-020-04843-6
7. Fahrni, G., Karakatsanis, N.A., Di Domenicantonio, G., Garibotto, V., Zaidi, H.: Does whole-body Patlak ^{18}F-FDG PET imaging improve lesion detectability in clinical oncology? Eur. Radiol. **29**, 4812–4821 (2019). https://doi.org/10.1007/s00330-018-5966-1
8. Guo, X., Zhou, B., Chen, X., Liu, C., Dvornek, N.C.: MCP-Net: inter-frame motion correction with Patlak regularization for whole-body dynamic PET. In: Wang, L., Dou, Q., Fletcher, P.T., Speidel, S., Li, S. (eds.) MICCAI 2022, Part IV. LNCS, vol. 13434, pp. 163–172. Springer, Cham (2022). https://doi.org/10.1007/978-3-031-16440-8_16
9. Huang, Z., et al.: Parametric image generation with the uEXPLORER total-body PET/CT system through deep learning. Eur. J. Nucl. Med. Mol. Imaging **49**(8), 2482–2492 (2022). https://doi.org/10.1007/s00259-022-05731-x
10. Küstner, T., et al.: CINENet: deep learning-based 3D cardiac CINE MRI reconstruction with multi-coil complex-valued 4D spatio-temporal convolutions. Sci. Rep. **10**(1), 13710 (2020)
11. Li, A., Tang, J.: Direct parametric image reconstruction for dynamic myocardial perfusion PET using artificial neural network representation (2022)
12. Li, Y., et al.: A deep neural network for parametric image reconstruction on a large axial field-of-view PET. Eur. J. Nucl. Med. Mol. Imaging **50**(3), 701–714 (2023). https://doi.org/10.1007/s00259-022-06003-4
13. Moradi, H., Vegh, V., O'Brien, K., Hammond, A., Reutens, D.: FDG-PET kinetic model identifiability and selection using machine learning (2022)
14. Pantel, A.R., Viswanath, V., Muzi, M., Doot, R.K., Mankoff, D.A.: Principles of tracer kinetic analysis in oncology, part I: principles and overview of methodology. J. Nucl. Med. **63**(3), 342–352 (2022)
15. Pantel, A.R., Viswanath, V., Muzi, M., Doot, R.K., Mankoff, D.A.: Principles of tracer kinetic analysis in oncology, part II: examples and future directions. J. Nucl. Med. **63**(4), 514–521 (2022)

16. Sari, H., et al.: First results on kinetic modelling and parametric imaging of dynamic ^{18}F-FDG datasets from a long axial FOV PET scanner in oncological patients. Eur. J. Nucl. Med. Mol. Imaging **49**, 1997–2009 (2022). https://doi.org/10.1007/s00259-021-05623-6
17. Snyman, J.A., Wilke, D.N., et al.: Practical Mathematical Optimization. Springer, New York (2005). https://doi.org/10.1007/b105200
18. Surti, S., Pantel, A.R., Karp, J.S.: Total body PET: why, how, what for? IEEE Trans. Radiat. Plasma Med. Sci. **4**(3), 283–292 (2020)
19. Wang, G., et al.: Total-body PET multiparametric imaging of cancer using a voxelwise strategy of compartmental modeling. J. Nucl. Med. **63**(8), 1274–1281 (2022)
20. Watabe, H.: Compartmental modeling in PET kinetics. In: Khalil, M.M. (ed.) Basic Science of PET Imaging, pp. 323–352. Springer, Cham (2017). https://doi.org/10.1007/978-3-319-40070-9_14
21. Zuo, Y., Sarkar, S., Corwin, M.T., Olson, K., Badawi, R.D., Wang, G.: Structural and practical identifiability of dual-input kinetic modeling in dynamic PET of liver inflammation. Phys. Med. Biol. **64**(17), 175023 (2019)

Geometry-Invariant Abnormality Detection

Ashay Patel(✉), Petru-Daniel Tudosiu, Walter Hugo Lopez Pinaya, Olusola Adeleke, Gary Cook, Vicky Goh, Sebastien Ourselin, and M. Jorge Cardoso

King's College London, London WC2R 2LS, UK
ashay.patel@kcl.ac.uk

Abstract. Cancer is a highly heterogeneous condition best visualised in positron emission tomography. Due to this heterogeneity, a general-purpose cancer detection model can be built using unsupervised learning anomaly detection models. While prior work in this field has showcased the efficacy of abnormality detection methods (e.g. Transformer-based), these have shown significant vulnerabilities to differences in data geometry. Changes in image resolution or observed field of view can result in inaccurate predictions, even with significant data pre-processing and augmentation. We propose a new spatial conditioning mechanism that enables models to adapt and learn from varying data geometries, and apply it to a state-of-the-art Vector-Quantized Variational Autoencoder + Transformer abnormality detection model. We showcase that this spatial conditioning mechanism statistically-significantly improves model performance on whole-body data compared to the same model without conditioning, while allowing the model to perform inference at varying data geometries.

1 Introduction

The use of machine learning for anomaly detection in medical imaging analysis has gained a great deal of traction over previous years. Most recent approaches have focused on improvements in performance rather than flexibility, thus limiting approaches to specific input types – little research has been carried out to generate models unhindered by variations in data geometries. Often, research assumes certain similarities in data acquisition parameters, from image dimensions to voxel dimensions and fields-of-view (FOV). These restrictions are then carried forward during inference [5,25]. This strong assumption can often be complex to maintain in the real-world and although image pre-processing steps can mitigate some of this complexity, test error often largely increases as new data variations arise. This can include variances in scanner quality and resolution, in addition to the FOV selected during patient scans. Usually training data, especially when acquired from differing sources, undergoes significant pre-processing such that data showcases the same FOV and has the same input

---

Supplementary Information The online version contains supplementary material available at https://doi.org/10.1007/978-3-031-43907-0_29.

H. Greenspan et al. (Eds.): MICCAI 2023, LNCS 14220, pp. 300–309, 2023.
https://doi.org/10.1007/978-3-031-43907-0_29

dimensions, e.g. by registering data to a population atlas. Whilst making the model design simpler, these pre-processing approaches can result in poor generalisation in addition to adding significant pre-processing times [11,13,26]. Given this, the task of generating an anomaly detection model that works on inputs with a varying resolution, dimension and FOV is a topic of importance and the main focus of this research.

Unsupervised methods have become an increasingly prominent field for automatic anomaly detection by eliminating the necessity of acquiring accurately labelled data [4,7] therefore relaxing the stringent data requirements of medical imaging. This approach consists of training generative models on healthy data, and defining anomalies as deviations from the defined model of normality during inference. Until recently, the variational autoencoder (VAE) and its variants held the state-of-the-art for the unsupervised approach. However, novel unsupervised anomaly detectors based on autoregressive Transformers coupled with Vector-Quantized Variational Autoencoders (VQ-VAE) have overcome issues associated with autoencoder-only methods [21,22]. In [22], the authors explore the advantage of tractably maximizing the likelihood of the normal data to model the long-range dependencies of the training data. The work in [21] takes this method a step further through multiple samplings from the Transformer to generate a non-parametric Kernel Density Estimation (KDE) anomaly map.

Even though these methods are state-of-the-art, they have stringent data requirements, such as having a consistent geometry of the input data, *e.g.*, in a whole-body imaging scenario, it is not possible to crop a region of interest and feed it to the algorithm, as this cropped region will be wrongly detected as an anomaly. This would happen even in the case that a scan's original FOV was restricted [17].

As such, we propose a geometric-invariant approach to anomaly detection, and apply it to cancer detection in whole-body PET via an unsupervised anomaly detection method with minimal spatial labelling. Through adapting the VQ-VAE Transformer approach in [21], we showcase that we can train our model on data with varying fields of view, orientations and resolutions by adding spatial conditioning in both the VQ-VAE and Transformer. Furthermore, we show that the performance of our model with spatial conditioning is at least equivalent to, and sometimes better, than a model trained on whole-body data in all testing scenarios, with the added flexibility of a "one model fits all data" approach. We greatly reduce the pre-processing requirements for generating a model (as visualised in Fig. 1), demonstrating the potential use cases of our model in more flexible environments with no compromises on performance.

2 Background

The main building blocks behind the proposed method are introduced below. Specifically, a VQ-VAE plus a Transformer are jointly used to learn the probability density function of 3D PET images as explored in prior research [21,22,24].

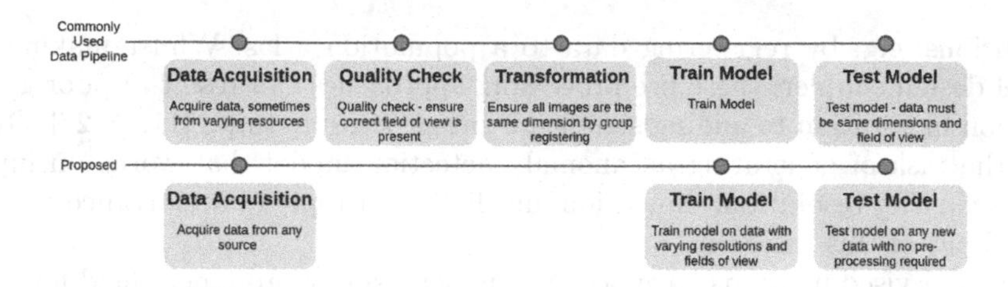

Fig. 1. Flowchart showcasing traditional data pipelines for developing machine learning models in medical imaging (top) vs. the reduced pipeline for our approach (bottom)

2.1 Vector-Quantized Variational Autoencoder

The VQ-VAE model provides a data-efficient encoding mechanism—enabling 3D inputs at their original resolution—while generating a discrete latent representation that can trivially be learned by a Transformer network [20]. The VQ-VAE is composed of an encoder that maps an image $X \in \mathbb{R}^{H \times W \times D}$ onto a compressed latent representation $Z \in \mathbb{R}^{h \times w \times d \times n_z}$ where n_z is the latent embedding vector dimension. Z is then passed through a quantization block where each feature column vector is mapped to its nearest codebook vector. Each spatial code $Z_{ijl} \in \mathbb{R}^{n_z}$ is then replaced by its nearest codebook element $e_k \in \mathbb{R}^{n_z}, k \in 1, ..., K$ where K denotes the codebook vocabulary size, thus obtaining Z_q. Given Z_q, the VQ-VAE decoder then reconstructs the observations $\hat{X} \in \mathbb{R}^{H \times W \times D}$. The architecture used for the VQ-VAE model used an encoder consisting of three downsampling layers that contain a convolution with stride 2 and kernel size 4 followed by a ReLU activation and 3 residual blocks. Each residual block consists of a kernel of size 3, followed by a ReLU activation, a convolution of kernel size 1 and another ReLU activation. Similar to the encoder, the decoder has 3 layers of 3 residual blocks, each followed by a transposed convolutional layer with stride 2 and kernel size 4. Finally, before the last transposed convolutional layer, a Dropout layer with a probability of 0.05 is added. The VQ-VAE codebook used had 256 atomic elements (vocabulary size), each of length 128. The CT VQ-VAE was identical in hyperparameters except each codebook vector has length 64. See Appendix A for implementation details.

2.2 Transformer

After training a VQ-VAE model, the next stage is to learn the probability density function of the discrete latent representations. Using the VQ-VAE, we can obtain a discrete representation of the latent space by replacing the codebook elements in Z_q with their respective indices in the codebook yielding Z_{iq}. To model the imaging data, we require the discretized latent space Z_{iq} to take the form of a 1D sequence s, which we achieve via a raster scan of the latent. The Transformer is then trained to maximize the log-likelihoods of the latent tokens sequence in an autoregressive manner. By doing this, the Transformer can learn the codebook distribution for position i within s with respect to previous codes

$p(s_i) = p(s_i|s_{<i})$. As with [21], we additionally use CT data to condition the Transformer via cross-attention using a separate VQ-VAE to encode the CT. This transforms the problem to learning the codebook distribution at position i as $p(s_i) = p(s_i|s_{<i}, c)$ where c is the entire CT latent sequence. The performer used in this work corresponds to a decoder Transformer architecture with 14 layers, each with 8 heads, and an embedding dimension of 256. Similarly the embedding dimension for the CT data and the spatial conditioning data had an embedding dimension of 256. See Appendix B for implementation details.

2.3 Anomaly Detection via Kernel Density Estimation Maps

Building on [21], given a sample for inference, a tokenized representation Z_{iq} is extracted from the VQ-VAE. Then, the representation is flattened into s where the trained Transformer model obtains the likelihoods for each token. These inferred likelihoods represent the probability of each token appearing at a certain position in the sequence - $p(s_i) = p(s_i|s_{<i}, c)$. This can then be used to single out tokens with low probability, i.e. anomalous tokens. We then resample anomalous tokens $p(s_i) < t$ where t is the resampling threshold chosen empirically using the validation set performance. Anomalous tokens are then replaced with higher likelihood (normal) tokens by resampling from the Transformer. We can then reshape the "healed" sequence back into its 3D quantized representation to feed into the VQ-VAE to generate a healed reconstruction X_r without anomalies.

In this work, abnormalities are defined as deviations between the distribution of "healed" reconstructions and the observed data, measured using a Kernel Density Estimation (KDE) approach. We generate multiple healed latent sequences by sampling multiple times for each position i with a likelihood $p(s_i) < t$. In each resampling, the Transformer outputs the likelihood for every possible token at position i. Based on these probabilities, we can create a multinomial distribution showcasing the probability of each token. We can then randomly sample multiple tokens. Each of these healed latent spaces is then decoded via the VQ-VAE multiple times with dropout. This generates multiple healed representations of the original image. A voxel-wise KDE anomaly map is generated by fitting a KDE independently at each voxel position to estimate the probability density function f across reconstructions. This is then scored at the original intensity of that voxel in the scan. Our KDE implementation used 60 samples for each anomalous token in s, followed by five decodings with dropout, yielding 300 "healed" reconstructions that are then used to calculate the KDE.

3 Method

3.1 VQ-VAE Spatial Conditioning

To date, there has been little research on generating autoencoder models capable of using images of varying sizes and resolutions (i.e. the input tensor shape to a autoencoder is assumed to be fixed). Although fully convolutional models can ingest images of varying dimensions, we have found that using training data with varying resolutions resulted in poor auto-encoder reconstructions. In this work, we take inspiration from CoordConv [19] as a mechanism to account for some level of spatial awareness, an approach which has been applied to various tasks in medical imaging scenarios with ranging levels of success [1,18].

A CoordConv layer is a concatenation of channels to the input image referencing a predefined coordinate system. After concatenation, the input is simply fed through a standard convolutional layer. For a 3D scan, we would have 3 coordinates, ijk, where the i coordinate channel is an $h \times w \times d$ rank-1 matrix with its first row filled with 0's, its second row with 1's, and so on. This would be the same for the j coordinate channel, except the columns would be filled with constant values, not the rows, and likewise for the k coordinate channel in a depth-wise fashion. These channels are then normalised between [0, 1].

The advantage of the CoordConv implementation is the constant scale of 0–1 across the channels regardless of image resolution. For example, two whole-body images with large differences in voxel-size will have CoordConv channels from 0–1 along each axis, thus conveying the notion of spatial resolution to the network. We found when training the VQ-VAE model on data with varying resolutions and dimensions that reconstructions showcased unwanted and significant artifacts, while by adding the CoordConv channels this issue was not present (See Appendix C for examples). Furthermore, when dealing with images of a ranging FOV, we adapted the [0, 1] channel values to convey the image's FOV. For example, suppose a whole body image (neck to upper leg) represented our range [0, 1] where 0 is the upper leg, and 1 is the neck. In that case, we can contract this range to represent the area displayed in the image (Fig. 2). In doing so, we convey information about the FOV to the VQ-VAE through CoordConv layers. Note that while the proposed model assumes only translation and scale changes between samples, it can be trivially extended to a full affine mapping of the coordinate system (including rotations/shearing between samples).

We used random crops during training to simulate varying FOVs of whole-body data. The random crop parameters are then used to define the coordinate system. For the implementation of the CoordConv layer, these channels are added once to the original input image and at the beginning of the VQ-VAE decoder, concatenated to the latent space, using the same value ranges but at a lower resolution given the reduced spatial dimension of the latent space.

3.2 Transformer Spatial Conditioning

Numerous approaches have used Transformers in the visual domain [7,8]. Given that Transformers work natively on 1D sequences, the spatial information in images is often lost. While various works have aimed to convey the spatial information of the original image when projected onto a 1D sequence [14,28], we require our spatial positioning to encode both where in the image ordering a token belongs, and where the token belongs in the context of the whole body. As the images have different FOVs and the image resolution, this results in

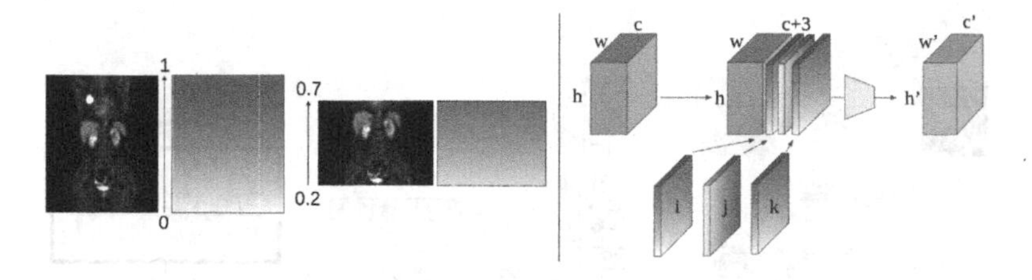

Fig. 2. CoordConv example showing whole-body image with values from 0 to 1 vs. a cropped image with values from 0.2 to 0.7 to reflect the field of view

varying token sequence lengths. As such, the Transformer must be informed of the location of a given token in relationship to the whole-body.

To do this, we use the same CoordConv principle applied to the input fed to the VQ-VAE. In order to map image coordinates to the token latent representation, we apply average pooling to each CoordConv channel separately, with kernel size and stride equal to the downsampling used in the VQ-VAE (8 used in this research). This gives us three channels i, j, k in the range of [0, 1], the same dimension as our latent space, but at lower spatial resolution to the original input. We then bin each value in each channel and combine the three values using base notation. For example, we use 20 bins (equal bins of 0.05), to which the final quantized spatial value for a given token is given as $sp_{ijk} = b_i + b_j \times B + b_k * B^2$ where sp is the quantized spatial value allocated to a given token at position ijk in the latent space, and b represents the binned value along a given channel for that token, and B is a pre-defined bin size. The choice of $B = 20$ bins was empirically chosen to closely resemble the average latent dimension of images.

During training, whole-body images and random crops are used. The spatial conditioning tokens are then generated and fed through an embedding layer of equal dimension to the CT embedding. The two embedded sequences (CT and spatial) are then added together and fed to the Transformer via cross-attention. For reference, this mechanism can be visualised in Fig. 3.

3.3 Data

For this work we leveraged whole-body PET/CT data from different sources to explore the efficacy of our approach for varying image geometries. 211 scans from NSCLC Radiogenomics [2,3,10,16] combined with 83 scans from a proprietary dataset constitute our lower resolution dataset with voxel dimensions of $3.6 \times 3.6 \times 3$ mm. From this, we split the data to give 210 training samples, 34 validation and 50 testing. Our higher resolution dataset uses AutoPET [10,15] (1014 scans) with voxel dimensions of $2.036 \times 2.036 \times 3$ mm. From this, 850 scans are used for training, 64 for validation and 100 for testing.

All baseline models work in a single space with constant dimensions, obtained by registering the AutoPET images to the space of the NSCLC dataset.

For evaluation, we use four testing sets: a lower resolution set derived from both the NSCLC and the private dataset; a higher resolution set from AutoPET;

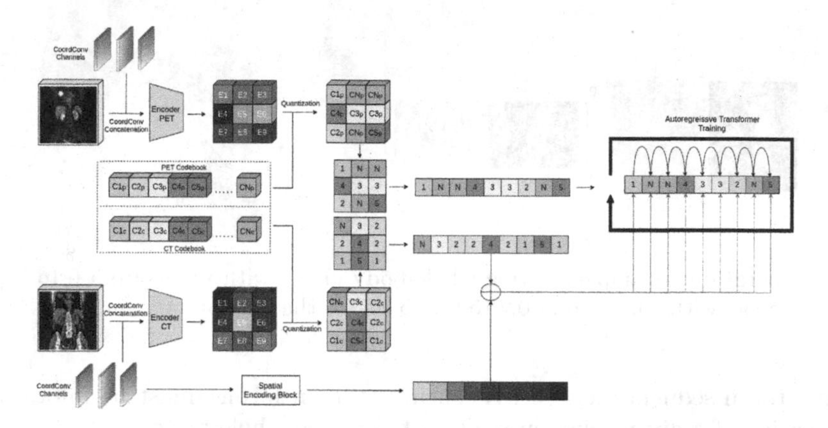

Fig. 3. Pipeline for Transformer training. PET and CT are encoded to generate a discrete latent space. CoordConv layers are used to generate the spatial conditionings that are added to the CT conditioning and fed to the Transformer via cross-attention

a testing set with random crops of the same NSCLC/private testing dataset and finally a testing set that has been rotated through 90° using the high resolution testing data. As the cropped and rotated dataset cannot be fed into the baseline models, we pad the images to the common image sizing before inference.

4 Results

The proposed model was trained on the data described in Sect. 3.3, with random crops applied while training. Model and anomaly detection hyperparameter tuning was done on our validation samples using the best DICE scores. We then test our model and baselines on 4 hold-out test sets: a low-resolution whole-body set, a low-resolution cropped set, a high-resolution rotated set and a high-resolution test set of PET images with varying cancers. The visual results shown in Fig. 4 show outputs rotated back to the original orientation. We measure our models'

Table 1. Anomaly detection results with best achievable DICE-score ($\lceil DICE \rceil$) and AUPRC on test sets. Bold values indicate best performing model with underlined values showcasing statistically significant results to the next best alternative $P < 0.05$

Model	$\lceil DICE \rceil$				AUPRC			
Whole Body	Low Res	High Res	Cropped	Rotated	Low Res	High Res	Cropped	Rotated
AE Dense [4]	0.22 ±0.15	0.25 ±0.17	0.30 ±0.19	0.25 ±0.19	0.18 ±0.12	0.26 ±0.16	0.23 ±0.14	0.23 ±0.13
AE Spatial [4]	0.32 ±0.13	0.48 ±0.21	0.34 ±0.16	0.14 ±0.08	0.26 ±0.12	0.45 ±0.20	0.33 ±0.14	0.10 ±0.07
AE SSIM [6]	0.28 ±0.16	0.30 ±0.19	0.27 ±0.17	0.18 ±0.07	0.20 ±0.15	0.26 ±0.18	0.21 ±0.12	0.15 ±0.09
VAE [4]	0.35 ±0.19	0.48 ±0.22	0.34 ±0.21	0.19 ±0.08	0.33 ±0.18	0.45 ±0.20	0.35 ±0.17	0.18 ±0.09
F-Anogan [23]	0.30 ±0.18	0.42 ±0.19	0.31 ±0.15	0.20 ±0.11	0.26 ±0.15	0.40 ±0.21	0.31 ±0.18	0.19 ±0.09
VQ-VAE + Transformer [21]	0.57 ±0.07	0.65 ±0.10	0.59 ±0.10	0.31 ±0.16	0.55 ±0.09	0.64 ±0.11	0.57 ±0.10	0.29 ±0.13
Geometry-Invariant (proposed)								
VQ-VAE CoordConv	0.57 ±0.09	0.65 ±0.08	0.63 ±0.12	0.32 ±0.17	0.55 ±0.09	0.64 ±0.09	0.61 ±0.13	0.30 ±0.15
Full CoordConv	**0.58** ±0.08	**0.68** ±0.10	**0.67** ±0.10	**0.65** ±0.12	**0.56** ±0.09	**0.66** ±0.11	**0.64** ±0.11	**0.62** ±0.12

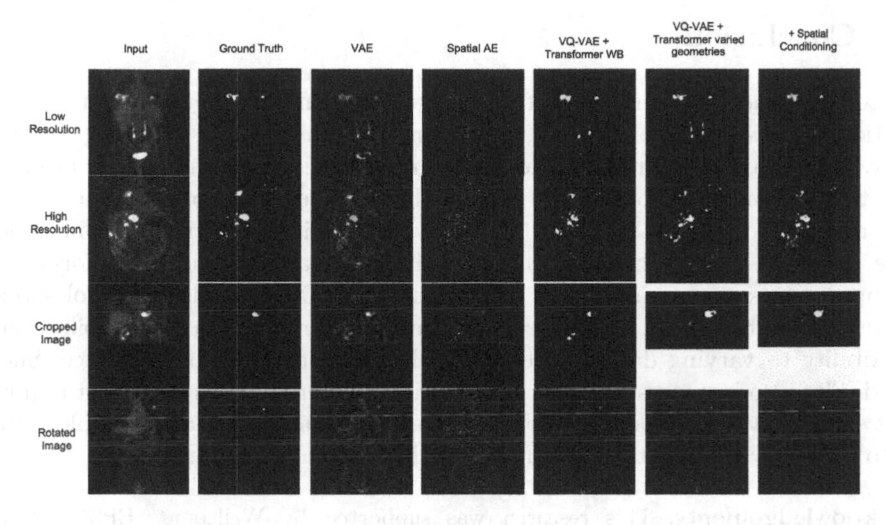

Fig. 4. Columns display (1st) the input image; (2nd) the gold standard segmentation; (3rd) residual for the VAE, (4th) AE Spatial, (5th) a KDE anomaly map for VQ-VAE Transformer trained on the whole body, (6th) trained with varied geometries, (7th) with spatial conditioning. Results are provided for a random subject in each test set.

performance using the DICE score, obtained by thresholding the residual/density score maps. In addition, we calculate the area under the precision-recall curve (AUPRC) as a suitable measure for segmentation performance under class imbalance. We additionally showcase the performance of the classic VQ-VAE + Transformer approach trained on whole-body data only (without the proposed spatial conditioning), as well as the proposed CoordConv model trained with varying image geometries but without the transformer spatial conditioning to explicitly showcase the added contribution of both spatial conditionings. The full results are presented in Table 1 with visual examples shown in Fig. 4. We can observe that the addition of spatial conditioning improves performance even against the same model without conditioning trained on whole-body data (Mann Whitney U test, $P < 0.01$ on high resolution and $P < 0.001$ on cropped data for DICE and AUPRC). For cropped data, models trained on whole-body data fail around cropping borders, as showcased in Fig. 4. This is not the case for the models trained on varying geometries. Note that the VQ-VAE + Transformer trained on varying geometries still shows adequate performance, highlighting the resilience of the Transformer network to varying sequence lengths without any form of spatial conditioning. However, by adding the transformer spatial conditioning, we see improvements across all test sets (most significantly on cropped data and the rotated data $P < 0.001$) for both evaluation metrics. For the rotated data, we see little performance degradation in the conditioned model thanks to the spatial conditioning. The same model without conditioning showed much lower performance with higher false positives likely due to the model's inability to comprehend the anatomical structures present due to the rotated orientation.

5 Conclusion

Detection and segmentation of anomalous regions, particularly for cancer patients, is essential for staging, treatment and intervention planning. Generally, the variation scanners and acquisition protocols can cause failures in models trained on data from single sources. In this study, we proposed a system for anomaly detection that is robust to variances in geometry. Not only does the proposed model showcase strong and statistically-significant performance improvements on varying image resolutions and FOV, but also on whole-body data. Through this, we demonstrate that one can improve the adaptability and flexibility to varying data geometries while also improving performance. Such flexibility also increases the pool of potential training data, as they dont require the same FOV. We hope this work serves as a foundation for further exploration into geometry-invariant deep-learning methods for medical-imaging.

Acknowledgements. This research was supported by Wellcome/ EPSRC Centre for Medical Engineering (WT203148/Z/16/Z), Wellcome Flagship Programme (WT213038/Z/18/Z), The London AI Centre for Value-based Heathcare and GE Healthcare. The models were trained on the NVIDIA Cambridge-1. Private dataset was obtained through King's College London (14/LO/0220 ethics application number).

References

1. An, C.H., Lee, J.S., Jang, J.S., Choi, H.C.: Part affinity fields and CoordConv for detecting landmarks of lumbar vertebrae and sacrum in X-ray images. Sensors **22**, 8628 (2022). https://doi.org/10.3390/s22228628
2. Bakr, S., et al.: Data for NSCLC radiogenomics collection (2017)
3. Bakr, S., et al.: A radiogenomic dataset of non-small cell lung cancer. Sci. Data **5**, 180202 (2018). https://doi.org/10.1038/sdata.2018.202
4. Baur, C., Denner, S., Wiestler, B., Albarqouni, S., Navab, N.: Autoencoders for unsupervised anomaly segmentation in brain MR images: a comparative study. Med. Image Anal. **69**, 101952 (2020)
5. Ben-David, S., Blitzer, J., Crammer, K., Pereira, F.: Analysis of representations for domain adaptation, vol. 19. MIT Press (2006). https://proceedings.neurips.cc/paper/2006/file/b1b0432ceafb0ce714426e9114852ac7-Paper.pdf
6. Bergmann, P., Fauser, M., Sattlegger, D., Steger, C.: MVTec AD - a comprehensive real-world dataset for unsupervised anomaly detection, pp. 9584–9592. IEEE (2019). https://doi.org/10.1109/CVPR.2019.00982
7. Chen, M., Radford, A., Wu, J., Heewoo, J., Dhariwal, P.: Generative pretraining from pixels (2020)
8. Child, R., Gray, S., Radford, A., Sutskever, I.: Generating long sequences with sparse transformers (2019)
9. Choromanski, K., et al.: Rethinking attention with performers (2020)
10. Clark, K., et al.: The Cancer Imaging Archive (TCIA): maintaining and operating a public information repository. J. Digit. Imaging **26**(6), 1045–1057 (2013). https://doi.org/10.1007/s10278-013-9622-7

11. Decuyper, M., Maebe, J., Van Holen, R., Vandenberghe, S.: Artificial intelligence with deep learning in nuclear medicine and radiology. EJNMMI Phys. **8**(1), 1–46 (2021). https://doi.org/10.1186/s40658-021-00426-y
12. Dhariwal, P., Jun, H., Payne, C., Kim, J.W., Radford, A., Sutskever, I.: Jukebox: a generative model for music (2020)
13. Dinsdale, N.K., Bluemke, E., Sundaresan, V., Jenkinson, M., Smith, S.M., Namburete, A.I.: Challenges for machine learning in clinical translation of big data imaging studies. Neuron **110**, 3866–3881 (2022). https://doi.org/10.1016/j.neuron.2022.09.012
14. Dosovitskiy, A., et al.: An image is worth 16 × 16 words: transformers for image recognition at scale (2020)
15. Gatidis, S., et al.: A whole-body FDG-PET/CT Dataset with manually annotated Tumor Lesions. Sci. Data **9**, 601 (2022). https://doi.org/10.1038/s41597-022-01718-3
16. Gevaert, O., et al.: Non-small cell lung cancer: identifying prognostic imaging biomarkers by leveraging public gene expression microarray data-methods and preliminary results. Radiology **264**, 387–396 (2012). https://doi.org/10.1148/radiol.12111607
17. Graham, M.S., et al.: Transformer-based out-of-distribution detection for clinically safe segmentation (2022)
18. Jurdi, R.E., Petitjean, C., Honeine, P., Abdallah, F.: CoordConv-Unet: investigating CoordConv for organ segmentation. IRBM **42**, 415–423 (2021). https://doi.org/10.1016/j.irbm.2021.03.002
19. Liu, R., et al.: An intriguing failing of convolutional neural networks and the CoordConv solution (2018)
20. van den Oord, A., Vinyals, O., Kavukcuoglu, K.: Neural discrete representation learning (2017)
21. Patel, A., et al.: Cross attention transformers for multi-modal unsupervised whole-body pet anomaly detection. In: Mukhopadhyay, A., Oksuz, I., Engelhardt, S., Zhu, D., Yuan, Y. (eds.) DGM4MICCAI 2022. LNCS, vol. 13609, pp. 14–23. Springer, Cham (2022). https://doi.org/10.1007/978-3-031-18576-2_2
22. Pinaya, W.H.L., et al.: Unsupervised brain anomaly detection and segmentation with transformers (2021)
23. Schlegl, T., Seeböck, P., Waldstein, S.M., Schmidt-Erfurth, U., Langs, G.: Unsupervised anomaly detection with generative adversarial networks to guide marker discovery. In: Niethammer, M., et al. (eds.) IPMI 2017. LNCS, vol. 10265, pp. 146–157. Springer, Cham (2017). https://doi.org/10.1007/978-3-319-59050-9_12
24. Tudosiu, P.D., et al.: Morphology-preserving autoregressive 3D generative modelling of the brain. In: Zhao, C., Svoboda, D., Wolterink, J.M., Escobar, M. (eds.) SASHIMI 2022. LNCS, vol. 13570, pp. 66–78. Springer, Cham (2022). https://doi.org/10.1007/978-3-031-16980-9_7
25. Valiant, L.G.: A theory of the learnable. Commun. ACM **27**(11), 1134–1142 (1984)
26. Varoquaux, G., Cheplygina, V.: Machine learning for medical imaging: methodological failures and recommendations for the future. NPJ Digit. Med. **5**, 48 (2022). https://doi.org/10.1038/s41746-022-00592-y
27. Vaswani, A., et al.: Attention is all you need (2017)
28. Wu, K., Peng, H., Chen, M., Fu, J., Chao, H.: Rethinking and improving relative position encoding for vision transformer (2021)

Modeling Alzheimers' Disease Progression from Multi-task and Self-supervised Learning Perspective with Brain Networks

Wei Liang[1,2], Kai Zhang[1,2], Peng Cao[1,2,3](✉), Pengfei Zhao[4], Xiaoli Liu[5], Jinzhu Yang[1,2,3](✉), and Osmar R. Zaiane[6]

[1] Computer Science and Engineering, Northeastern University, Shenyang, China
[2] Key Laboratory of Intelligent Computing in Medical Image of Ministry of Education, Northeastern University, Shenyang, China
caopeng@mail.neu.edu.cn,yangjinzhu@cse.neu.edu.cn
[3] National Frontiers Science Center for Industrial Intelligence and Systems Optimization, Shenyang 110819, China
[4] The Affiliated Brain Hospital of Nanjing Medical University, Nanjing, China
[5] DAMO Academy, Alibaba Group, Hangzhou, China
[6] Alberta Machine Intelligence Institute, University of Alberta, Edmonton, AB, Canada

Abstract. Alzheimer's disease (AD) is a common irreversible neurodegenerative disease among elderlies. Establishing relationships between brain networks and cognitive scores plays a vital role in identifying the progression of AD. However, most of the previous works focus on a single time point, without modeling the disease progression with longitudinal brain networks data. Besides, the longitudinal data is insufficient for sufficiently modeling the predictive models. To address these issues, we propose a **S**elf-supervised **M**ulti-Task learning **P**rogression model SMP-Net for modeling the relationship between longitudinal brain networks and cognitive scores. Specifically, the proposed model is trained in a self-supervised way by designing a masked graph auto-encoder and a temporal contrastive learning that simultaneously learn the structural and evolutional features from the longitudinal brain networks. Furthermore, we propose a temporal multi-task learning paradigm to model the relationship among multiple cognitive scores prediction tasks. Experiments on the Alzheimer's Disease Neuroimaging Initiative (ADNI) dataset show the effectiveness of our method and achieve consistent improvements over state-of-the-art methods in terms of Mean Absolute Error (MAE), Pearson Correlation Coefficient (PCC) and Concordance Correlation Coefficient (CCC). Our code is available at https://github.com/IntelliDAL/Graph/tree/main/SMP-Net.

Keywords: Self-supervised learning · Multi-task learning · Cognitive scores · Brain networks · Longitudinal prediction

Supplementary Information The online version contains supplementary material available at https://doi.org/10.1007/978-3-031-43907-0_30.

H. Greenspan et al. (Eds.): MICCAI 2023, LNCS 14220, pp. 310–319, 2023.
https://doi.org/10.1007/978-3-031-43907-0_30

1 Introduction

Alzheimer's disease (AD) is a progressive neurodegenerative disease, which affects the quality of life as it causes memory loss, difficulty in thinking and learning [12,19,21,23]. Establishing relationships between brain networks and cognitive scores plays a vital role in identifying the early stage of AD [15,17]. Though there has been substantial progress in AD diagnostics with brain networks [1,3,5,14], most of the current studies focus on a single time point, without exploring longitudinal modeling for disease progression with brain networks. Some learning-based methods are proposed for the longitudinal prediction of AD progression with multi-modal data but generally fail in utilizing brain networks due to the large heterogeneity of brain networks between individuals as well as developmental stages [2,6,8,20].

The cognitive scores prediction with longitudinal brain networks via deep learning models faces many challenges as follows: (*i*) The available longitudinal brain networks are scarce due to few volunteers or subject dropout [10]. Predicting cognitive scores with limited data is extremely challenging for the deep learning model training. (*ii*) Longitudinal brain networks provide rich structure information and disease progression characteristics, accounting for poor generalization for the pure supervised learning due to the insufficient supervision. (*iii*) The relationship between the brain networks and cognitive scores at multiple time points is varied, hindering the accurate prediction performance at multiple time points with a single task model.

To cope with the above challenges, we propose a self-supervised multi-task learning paradigm for AD progression modeling with longitudinal brain networks. The proposed paradigm consists of a self-supervised spatio-temporal representation learning module for exploiting the spatio-temporal characteristics of longitudinal brain networks and a temporal multi-task module for modeling the relationship among cognitive scores prediction tasks at multiple time points. In summary, our contributions are threefold:

1) To the best of our knowledge, our work is the first attempt to predict cognitive scores with longitudinal brain networks through a self-supervised multi-task paradigm.
2) We design a self-supervised spatio-temporal representation learning module (SSTR), involving masked graph auto-encoder and temporal contrastive learning are jointly pre-trained to capture the structural and evolutional features of longitudinal brain networks simultaneously. The SSTR module can lead to more robust high-level representations for longitudinal brain networks.
3) We assume that inherent correlations exist among the prediction tasks at multiple future time points. Consequently, we propose a temporal multi-task learning paradigm to assist multiple time points cognitive scores prediction, which enhances the model generalization by exploiting the commonalities and differences among different prediction tasks when limited data is available.

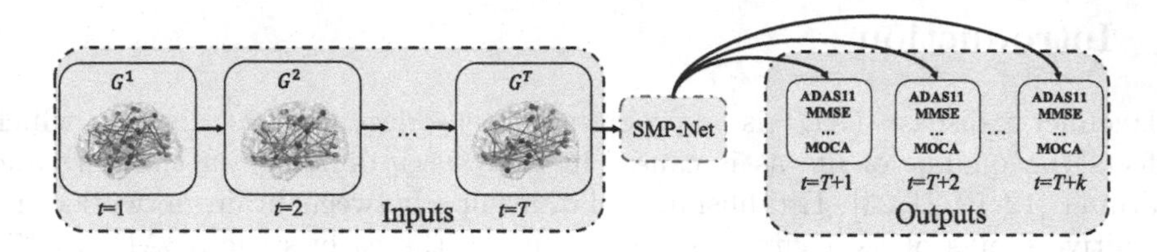

Fig. 1. Illustration of our task. The inputs of the model are T history brain networks, and the outputs are the predicted cognitive scores at the multiple future time points $(T+1, T+2, ..., T+k)$.

2 Method

2.1 Problem Formalization

The input to the proposed model is a set of N subjects, each of which has T longitudinal brain networks. Let $X_i = [G_i^1, ..., G_i^t, ..., G_i^T] \in \mathbb{R}^{T*M*M}$ represent the input longitudinal brain networks, where M denotes the number of brain regions based on a specific brain parcellation. $G_i^t = (V_i^t, A_i^t)$ is the brain network of subject i at time t; V_i^t and A_i^t are ROIs (nodes) and Pearson correlations between ROIs (edges), respectively. The model outputs are the predicted cognitive scores at k time points $Y_i = [Y_i^{T+1}, ..., Y_i^{T+k}]$ for subject i, where $Y_i^{T+k} = [Y_i^{T+k,1}, ..., Y_i^{T+k,p}]$ and $Y_i^{T+k,p}$ is the p-th cognitive score of subject i at time $T+k$. As illustrated in Fig. 1, our aim is to build a model f to predict cognitive scores at time $[T+1, ..., T+k]$ with brain networks at time $[1, ..., T]$, which is formulated as: $\{Y^{T+1}, Y^{T+2}, ..., Y^{T+k}\} = f(G^1, G^2, ..., G^T)$.

2.2 Overview

The overview of the proposed SMP-Net is shown in Fig. 2. As shown in Fig. 2, the proposed SMP-Net consists of two modules: Self-supervised spatio-temporal representation learning module (SSTR) for exploiting the spatio-temporal characteristics of longitudinal brain network data itself and a temporal multi-task learning module for modeling the relationship among cognitive scores prediction tasks at multiple time points. SSTR involves a masked graph auto-encoder and a temporal contrastive learning, both of which are jointly pre-trained to learn the structural and evolutional brain networks representation.

2.3 Self-supervised Spatio-Temporal Representation Learning

Although brain networks provide rich structure information, the pure supervised learning scheme limits the representation capacity of the models due to insufficient supervision. To solve this problem, we introduce the self-supervised spatio-temporal representation learning module, SSTR. The procedure of SSTR involves two stages as follow:

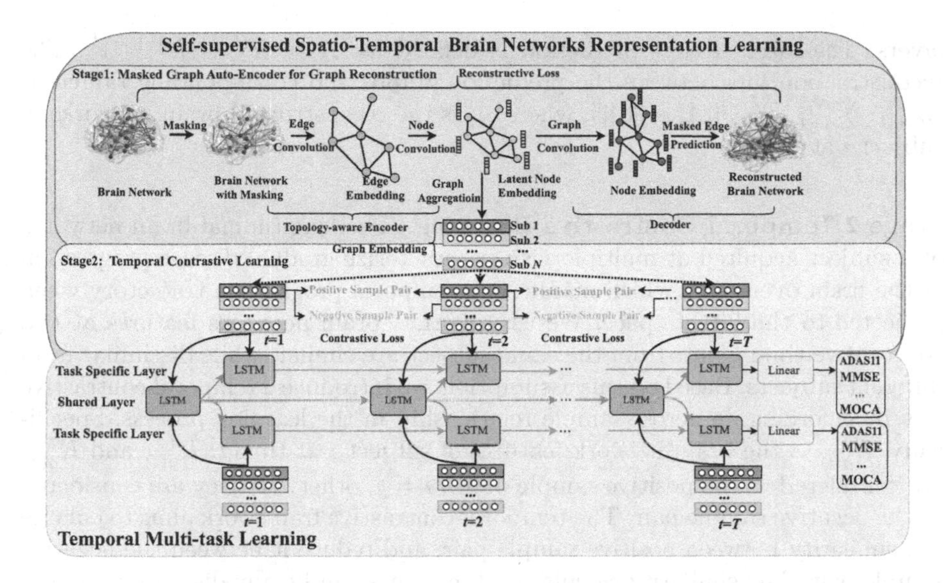

Fig. 2. Illustration of the proposed SMP-Net framework.

Stage 1 Masked Graph Auto-encoder for Graph Reconstruction. In stage 1, a masked graph auto-encoder, containing a topology-aware encoder and a decoder, is designed to exploit the crucial structural information in brain networks. To sufficiently exploit the graph structure, we randomly mask some nodes and the associated edges. The unmasked nodes and edges fed into the topology-aware encode to learn the latent representations. Let H_u indicate the feature map in the encoding stage. We define the adjacent matrix of the unmasked nodes as A_u, which is taken as the input of the topology-aware encoder, that is $H_u^{(0)} = A_u$. The topology-aware encoder consists of three parts: **1)** The edge convolution with multiple cross-shaped filters for capturing the locality in the graph according to $H_u^{(l)} = EC(H_u^{(l-1)}) = \sum_{i=0}^{M}\sum_{j=0}^{M} H_{u}{}_{(i,\cdot)}^{(l-1)} \boldsymbol{w}_r + H_{u}{}_{(\cdot,j)}^{(l-1)} \boldsymbol{w}_c$, where $\boldsymbol{w}_r \in \mathbb{R}^{1\times M}$ and $\boldsymbol{w}_c \in \mathbb{R}^{M\times 1}$ are convolution kernels. **2)** The node convolution for learning the latent node embedding. It is defined as: $H_n^{(l)} = NC(H_u^{(l-1)}) = \sum_{i=1}^{M} H_{u}{}_{(i,\cdot)}^{(l-1)} \boldsymbol{w}_n^{l-1}$, where $\boldsymbol{w}_n$ is the learned filter vector, $H_n^{(l)} \in \mathbb{R}^{M\times D_n}$ is the latent unmasked node embedding and D_n is the channels in NC. **3)** The graph aggregation for achieving the global graph embedding through: $H_g^{(l)} = GA(H_n^{(l-1)}) = \sum_{i=1}^{M} H_{n}{}_{(i,\cdot)}^{(l-1)} \boldsymbol{w}_g$, where $\boldsymbol{w}_g$ is the learned filter vector, $H_g \in \mathbb{R}^{M\times D_g}$ is the graph embedding and D_g is the dimensionality in GA.

The decoder takes the masked nodes and the latent unmasked node embeddings as inputs, and then produces predictions for the masked nodes and edges by graph convolution operations and the masked edge prediction. The graph convolution is defined as: $H^{(l+1)} = \sigma(\widetilde{A}H_n^{(l)}W^{(l)}) + b^{(l)}$, where $\widetilde{A} \in \mathbb{R}^{M\times M}$ is the binary adjacency matrix, W denotes trainable weight, $H \in \mathbb{R}^{M\times D_n'}$ is the node embedding and D_n' is the hidden layer size of graph convolutional layers. The masked edge prediction is defined as: $\hat{A} = H^{(l+1)}(H^{(l+1)})^T$. The reconstruction loss between the prediction graphs and corresponding targets is $L_{rec} = \sum_{i=1}^{N}\sum_{t=1}^{T} ||A_i^t - \hat{A}_i^t||_2^2$, where $\hat{A}_i^t$ is the reconstructed brain networks of subject i at time t.

Stage 2 Temporal Contrastive Learning. The longitudinal brain networks of a subject acquired at multiple visits characterize gradual disease progression of the brain over time, which manifests a temporal progression trajectory when projected to the latent space. We assume that brain networks features at two consecutive time points from the same subject are similar, while dissimilar from different subjects. Based on this assumption, we introduce a temporal contrastive loss by enforcing an across-sample relationship in the learning process. Specifically, $H_{g(i)}^t$ is the brain network features of subject i at time t, $H_{g(i)}^t$ and $H_{g(j)}^{t+1}$ are considered as the positive sample pair if $i = j$, otherwise they are considered as the negative sample pair. The temporal contrastive framework aims to enlarge the similarity between positive sample pair, and reduce it between the negative sample pair. The similarity calculation function s can be any distance function, and here we utilize cosine similarity. The loss for temporal contrastive learning can be represented as:

$$L_{con} = -log\sum_{i=1}^{N}\sum_{t=1}^{T-1}\frac{exp(s(H_{g(i)}^t, H_{g(i)}^{t+1})/\tau)}{\sum_{j=1,j\neq i}^{N} exp(s(H_{g(i)}^t, H_{g(j)}^{t+1})/\tau)}, \tag{1}$$

where τ is a temperature factor that controls the model's discrimination against negative sample pair and $exp(.)$ is an exponential function.

2.4 Temporal Multi-task Learning

Existing studies have demonstrated the effectiveness of multi-task learning for the extraction of a robust feature representation [9,24]. In this regard, to further exploit the correlation among the prediction tasks at multiple future time points, we design a temporal multi-task learning paradigm. Specifically, the temporal multi-task learning module consists of a shared network and multiple task-specific networks, all of which are designed with a Long Short-Term Memory (LSTM) architecture [4]. The shared network is trained for modeling the shared information h_s^t among cognitive scores prediction tasks at multiple time points. The q-th task-specific network aims to capture the task-specific information h_q^t from the shared network and the brain networks features at time t. The temporal multi-task learning module can be seen as an end-to-end architecture with the shared and task-specific parameters of W_s, W_q. By learning these parameters jointly, we arrive at a collaborative learning method to jointly improve the performance of the prediction tasks at multiple time points. The shared information h_s^t and task-specific information h_q^t are formulated as $h_s^t = LSTM(Hg^t, h_s^{t-1}, W_s)$ and $h_q^t = LSTM([Hg^t, h_s^t], h_q^{t-1}, W_q)$. The output of the temporal multi-task learning module is formulated as: $\hat{Y}^t = W_2(W_1 h_q^t + b_1) + b_2$, where W_1, W_2, b_1, b_2 are learnable parameters of LSTM. Errors between the actual observations Y^t and predictions $\hat{Y}^t$ are used to update the model parameters through the regression loss as follow:

$$L_{reg} = \sum_{i=1}^{k} \sum_{t=2}^{T} (||Y^t - \hat{Y}^t||_1 + ||Y^{T+i} - \hat{Y}^{T+i}||_1) \quad (2)$$

The overall loss function L is described as Eq. (3), where λ_{con} and λ_{rec} are the weights for contrastive loss and reconstruction loss, respectively.

$$L = L_{reg} + \lambda_{con} L_{con} + \lambda_{rec} L_{rec} \quad (3)$$

3 Experiments

3.1 Dataset and Experimental Settings

In this work, we choose 219 longitudinal resting-state fMRI scans of 73 subjects from the Alzheimer's Disease Neuroimaging Initiative (ADNI) dataset [11][1]. AAL template is used to obtain 90 ROIs for every subject [18]. We predict nine cognitive scores at time M24, M36 and M48 with brain networks times of M0, M6 and M12 to evaluate our proposed SMP-Net. The number of samples for the three tasks are 73, 35 and 31, respectively.

During the model training, the Adam optimizer is used with a momentum of 0.9 and a weight decay of 0.01. The learning rate is set to 10^{-3}. The hidden layer size of LSTM and graph convolutional layers are set to 64 and 48, respectively. The values of hyperparameter λ_c and λ_r are set to 1. The model is trained with 20 epochs in the self-supervised spatio-temporal representation learning stage and 300 epochs in the temporal multi-task learning stage with a batch size of 16. To avoid over-fitting duc to the limited subjects, in all experiments, we repeat the 5-fold cross-validation 10 times with different random seeds. We finally report the average results. Three commonly used metrics are adopted to evaluate all methods, including Mean Absolute Error (MAE), Pearson Correlation Coefficient (PCC) and Concordance Correlation Coefficient (CCC). CCC reflects both the correlation and the absolute error between the true and the predicted cognitive scores. Due to limited space, we report the results in terms of CCC in this paper. The results in terms of MAE and CC are shown in the supplementary material. To ensure a fair comparison, the hyperparameters of comparable methods are optimized to achieve their best performance.

3.2 Effectiveness Evaluation

We compare the performance of our SMP-Net with three state-of-the-art (SOTA) sequential graph learning methods: evolveGCN [13], stGCN [22] and DySAT [16] as well as a baseline method: GCN [7]. Table 1 summarizes the results of all

[1] http://adni.loni.usc.edu/.

Table 1. Experimental results in terms of CCC. The best results are bold and the superscript symbol * indicates that the proposed method significantly outperformed that method with p-value = 0.01.

	Cognitive scores	CDRSB	ADAS11	MMSE	RAVLT imm	learn	ADAS13	RAVLT forget	RAVLT perc forget	MOCA	Average
M24	GCN	0.190*	0.286*	0.139*	0.346*	0.280*	0.319*	0.320*	0.311*	0.201*	0.266 (0.045)*
	evolveGCN	0.351*	0.415*	0.141*	0.487*	0.399*	0.427*	0.283*	0.433*	0.272*	0.356 (0.078)*
	stGCN	0.310*	0.391*	0.265*	0.458*	0.323*	0.415*	0.279*	0.405*	0.373*	0.358 (0.078)*
	DySAT	0.574*	0.416*	0.438*	0.342*	**0.588**	0.353*	0.597*	0.230*	0.531*	0.452 (0.051)*
	SMP-Net	**0.617**	**0.798**	**0.658**	**0.798**	0.525	**0.809**	**0.654**	**0.819**	**0.666**	**0.705 (0.038)**
M36	GCN	0.152*	0.229*	0.122*	0.314*	0.055*	0.254*	0.099*	0.214*	0.213*	0.184 (0.076)*
	evolveGCN	0.350*	0.450*	0.114*	0.563*	0.508*	0.450*	0.341*	0.431*	0.219*	0.381 (0.124)*
	stGCN	0.246*	0.410*	0.323*	0.439*	0.333*	0.430*	0.226*	0.491*	0.341*	0.360 (0.131)*
	DySAT	**0.556**	0.355*	0.561*	0.543*	**0.577**	0.281*	**0.629**	0.140*	0.569*	0.468 (0.048)*
	SMP-Net	0.490	**0.754**	**0.593**	**0.801**	0.496	**0.788**	0.571	**0.832**	**0.663**	**0.665 (0.060)**
M48	GCN	0.144*	0.313*	0.159*	0.431*	0.141*	0.316*	0.168*	0.324*	0.097*	0.233 (0.091)*
	evolveGCN	0.471*	0.342*	0.103*	0.484*	0.397*	0.368*	0.512*	0.542*	0.108 *	0.370 (0.084)*
	stGCN	0.276*	0.300*	0.296*	0.424*	0.392*	0.321*	0.450*	0.557*	0.299*	0.368 (0.096)*
	DySAT	**0.609**	0.347*	0.416*	0.378*	0.640*	0.268*	0.675*	0.146*	0.441*	0.436 (0.085)*
	SMP-Net	0.561	**0.694**	**0.554**	**0.798**	**0.518**	**0.752**	**0.749**	**0.869**	**0.570**	**0.674 (0.083)**

methods in terms of CCC on the ADNI dataset. As reported in the supplementary material, consistent conclusions are obtained by SMP-Net in terms of MAE and CC. Based on the experimental results, we have the following observations: First, evolveGCN, stGCN and DySAT consistently outperform GCN, indicating that evolveGCN, stGCN and DySAT are able to capture the dynamism underlying a brain networks sequence through a recurrent model, which contributes to improve performance in disease prediction. Second, DySAT shows a higher average CCC than evolveGCN and stGCN. One possible reason is that DySAT utilizes joint structural and temporal self-attention, which enables it to learn more efficient dynamic graph representation compared with evolveGCN and stGCN. Finally, our proposed SMP-Net maintains a stable and competitive performance at all the time points, demonstrating that 1) SMP-Net can learn more expressive representations of brain networks structure by masked graph auto-encoder for graph reconstruction module. 2) SMP-Net sufficiently takes advantage of the temporal and subject correlation in disease progression by temporal contrastive learning. 3) The temporal multi-task paradigm of SMP-Net effectively exploits the inherent correlation among multiple prediction tasks at different time points, which facilitate to improve the model performance.

3.3 Discussion

Ablation Analysis. To valid the effect of each proposed module, we consider the following variants for evaluation: 1) SMP-Net-c: the temporal contrastive loss is removed; 2) SMP-Net-r: the reconstruction loss is removed; 3) SMP-Net-rc: both temporal contrastive loss and graph reconstruction loss are removed; 4) SMP-Net-m: the temporal multi-task paradigm is ignored. Table 2 summarizes the results of ablation studies in terms of CCC. It is apparent that SMP-Net outperforms all of the variants. Specifically, SMP-Net consistently outperforms SMPT-Net-m 11.0%, 12.7% and 23.8% at time M24, M36 and M48, respectively,

Table 2. Average CCC results of ablation studies. The best results are bold and the superscript symbol * indicates that the proposed method significantly outperformed that method with p-value = 0.01.

Methods	SMP-Net-rc	SMP-Net-r	SMP-Net-c	SMP-Net-m	**SMP-Net**
M24	0.299*	0.382*	0.472*	0.635*	**0.705**
M36	0.319*	0.410*	0.490*	0.590*	**0.665**
M48	0.313*	0.434*	0.468*	0.544*	**0.674**

indicating the effectiveness of the temporal multi-task paradigm. It also indicates that the multi-task paradigm in SMP-Net is more helpful for the prediction at farther time points. The reason is that prediction tasks at farther time points are more difficult due to the insignificant relationship between the brain networks and the cognitive scores. Temporal multi-task paradigm enforces the long-term prediction to benefit from short-term prediction, making the prediction tasks at farther time points gain more improvements. Moreover, we can observe that models with SSTR perform better than the ones without SSTR. For instance, SMP-Net-m and SMP-Net show superior performance than SMP-Net-r, SMP-Net-c and SMP-Net-rc. This demonstrates that SSTR facilitates the learning of structural and evolutional features in the condition of limited samples and insufficient supervision, thereby leading to more robust high-level representations for downstream tasks.

Evaluating Robustness. To evaluate the robustness of the SSTR module, we pre-train SMP-Net with fMRI at three time points (M0, M6, M12) and fine-tune it with different downstream tasks of predicting cognitive scores at different time points. As shown in Fig. 3, MSP-Net provides comparatively stable performance on different fine-tuning tasks, demonstrating that features learned with our pre-trained model are robust to the different fine-tuning tasks.

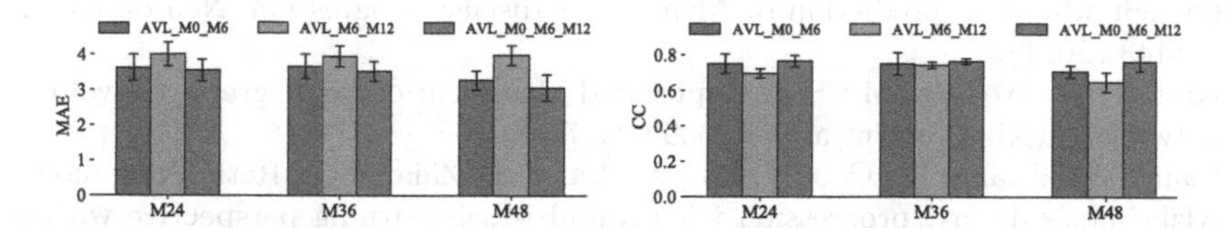

Fig. 3. Average MAE and CC of the pre-trained model fine-tuning on three different downstream tasks. AVL_M0_M6 denotes that fMRI data at M0 and M6 are available.

4 Conclusion

This paper proposes an AD progression model SMP-Net from multi-task and self-supervised learning perspective with longitudinal brain networks. In the proposed SMP-Net, self-supervised spatio-temporal representation learning is designed to learn more robust structural and evolutional features from longitudinal brain networks. The temporal multi-task paradigm is designed for boosting the ability of cognitive score prediction at multiple time points. Experimental results on the ADNI dataset with fewer samples demonstrate the advantage of self-supervised spatio-temporal representation learning and temporal multi-task learning.

Acknowledgements. This research was supported by the National Natural Science Foundation of China (No. 62076059), the Science Project of Liaoning Province (2021-MS-105) and the 111 Project (B16009).

References

1. Aviles-Rivero, A.I., Runkel, C., Papadakis, N., Kourtzi, Z., Schönlieb, C.B.: Multimodal hypergraph diffusion network with dual prior for Alzheimer classification. In: Wang, L., Dou, Q., Fletcher, P.T., Speidel, S., Li, S. (eds.) MICCAI 2022, Part III. LNCS, vol. 13433, pp. 717–727. Springer, Cham (2022). https://doi.org/10.1007/978-3-031-16437-8_69
2. Brand, L., Wang, H., Huang, H., Risacher, S., Saykin, A., Shen, L.: Joint high-order multi-task feature learning to predict the progression of Alzheimer's disease. In: Frangi, A.F., Schnabel, J.A., Davatzikos, C., Alberola-López, C., Fichtinger, G. (eds.) MICCAI 2018, Part I. LNCS, vol. 11070, pp. 555–562. Springer, Cham (2018). https://doi.org/10.1007/978-3-030-00928-1_63
3. Chen, Z., Liu, Y., Zhang, Y., Li, Q., Initiative, A.D.N., et al.: Orthogonal latent space learning with feature weighting and graph learning for multimodal Alzheimer's disease diagnosis. Med. Image Anal. **84**, 102698 (2023)
4. Graves, A.: Long short-term memory. In: Graves, A. (ed.) Supervised Sequence Labelling with Recurrent Neural Networks. SCI, vol. 385, pp. 37–45. Springer, Heidelberg (2012). https://doi.org/10.1007/978-3-642-24797-2_4
5. Huang, Y., Chung, A.C.: Disease prediction with edge-variational graph convolutional networks. Med. Image Anal. **77**, 102375 (2022)
6. Jung, W., Jun, E., Suk, H.I., Initiative, A.D.N., et al.: Deep recurrent model for individualized prediction of Alzheimer's disease progression. Neuroimage **237**, 118143 (2021)
7. Kipf, T.N., Welling, M.: Semi-supervised classification with graph convolutional networks. arXiv preprint arXiv:1609.02907 (2016)
8. Liang, W., Zhang, K., Cao, P., Liu, X., Yang, J., Zaiane, O.: Rethinking modeling Alzheimer's disease progression from a multi-task learning perspective with deep recurrent neural network. Comput. Biol. Med. **138**, 104935 (2021)
9. Liao, W., et al.: MUSCLE: multi-task self-supervised continual learning to pre-train deep models for X-ray images of multiple body parts. In: Wang, L., Dou, Q., Fletcher, P.T., Speidel, S., Li, S. (eds.) MICCAI 2022, Part VIII. LNCS, vol. 13438, pp. 151–161. Springer, Cham (2022). https://doi.org/10.1007/978-3-031-16452-1_15
10. Logothetis, N.K.: What we can do and what we cannot do with fMRI. Nature **453**(7197), 869–878 (2008)
11. Marinescu, R.V., et al.: Tadpole challenge: prediction of longitudinal evolution in Alzheimer's disease. arXiv preprint arXiv:1805.03909 (2018)

12. Nguyen, H.D., Clément, M., Mansencal, B., Coupé, P.: Interpretable differential diagnosis for Alzheimer's disease and frontotemporal dementia. In: Wang, L., Dou, Q., Fletcher, P.T., Speidel, S., Li, S. (eds.) MICCAI 2022, Part I. LNCS, vol. 13431, pp. 55–65. Springer, Cham (2022). https://doi.org/10.1007/978-3-031-16431-6_6
13. Pareja, A., et al.: EvolveGCN: evolving graph convolutional networks for dynamic graphs. In: Proceedings of the AAAI Conference on Artificial Intelligence, vol. 34, pp. 5363–5370 (2020)
14. Parisot, S., et al.: Disease prediction using graph convolutional networks: application to autism spectrum disorder and Alzheimer's disease. Med. Image Anal. **48**, 117–130 (2018)
15. Petersen, E., et al.: Feature robustness and sex differences in medical imaging: a case study in MRI-based Alzheimer's disease detection. In: Wang, L., Dou, Q., Fletcher, P.T., Speidel, S., Li, S. (eds.) MICCAI 2022, Part I. LNCS, vol. 13431, pp. 88–98. Springer, Cham (2022). https://doi.org/10.1007/978-3-031-16431-6_9
16. Sankar, A., Wu, Y., Gou, L., Zhang, W., Yang, H.: DySAT: deep neural representation learning on dynamic graphs via self-attention networks. In: Proceedings of the 13th International Conference on Web Search and Data Mining, pp. 519–527 (2020)
17. Seyfioğlu, M.S., et al.: Brain-aware replacements for supervised contrastive learning in detection of Alzheimer's disease. In: Wang, L., Dou, Q., Fletcher, P.T., Speidel, S., Li, S. (eds.) MICCAI 2022, Part I. LNCS, vol. 13431, pp. 461–470. Springer, Cham (2022). https://doi.org/10.1007/978-3-031-16431-6_44
18. Tzourio-Mazoyer, N., et al.: Automated anatomical labeling of activations in SPM using a macroscopic anatomical parcellation of the MNI MRI single-subject brain. Neuroimage **15**(1), 273–289 (2002)
19. Xiao, T., Zeng, L., Shi, X., Zhu, X., Wu, G.: Dual-graph learning convolutional networks for interpretable Alzheimer's disease diagnosis. In: Wang, L., Dou, Q., Fletcher, P.T., Speidel, S., Li, S. (eds.) MICCAI 2022, Part VIII. LNCS, vol. 13438, pp. 406–415. Springer, Cham (2022). https://doi.org/10.1007/978-3-031-16452-1_39
20. Xu, L., et al.: Multi-modal sequence learning for Alzheimer's disease progression prediction with incomplete variable-length longitudinal data. Med. Image Anal. **82**, 102643 (2022)
21. Yang, F., Meng, R., Cho, H., Wu, G., Kim, W.H.: Disentangled sequential graph autoencoder for preclinical Alzheimer's disease characterizations from ADNI study. In: de Bruijne, M., et al. (eds.) MICCAI 2021, Part II. LNCS, vol. 12902, pp. 362–372. Springer, Cham (2021). https://doi.org/10.1007/978-3-030-87196-3_34
22. Yu, B., Yin, H., Zhu, Z.: Spatio-temporal graph convolutional networks: a deep learning framework for traffic forecasting. arXiv preprint arXiv:1709.04875 (2017)
23. Zhang, S., et al.: 3D global Fourier network for Alzheimer's disease diagnosis using structural MRI. In: Wang, L., Dou, Q., Fletcher, P.T., Speidel, S., Li, S. (eds.) MICCAI 2022, Part I. LNCS, vol. 13431, pp. 34–43. Springer, Cham (2022). https://doi.org/10.1007/978-3-031-16431-6_4
24. Zhu, J., Li, Y., Ding, L., Zhou, S.K.: Aggregative self-supervised feature learning from limited medical images. In: Wang, L., Dou, Q., Fletcher, P.T., Speidel, S., Li, S. (eds.) MICCAI 2022, Part VIII. LNCS, vol. 13438, pp. 57–66. Springer, Cham (2022). https://doi.org/10.1007/978-3-031-16452-1_6

Unsupervised Discovery of 3D Hierarchical Structure with Generative Diffusion Features

Nurislam Tursynbek(✉) and Marc Niethammer

University of North Carolina, Chapel Hill, NC, USA
nurislam@cs.unc.edu

Abstract. Inspired by findings that generative diffusion models learn semantically meaningful representations, we use them to discover the intrinsic hierarchical structure in biomedical 3D images using *unsupervised* segmentation. We show that features of diffusion models from different stages of a U-Net-based ladder-like architecture capture different hierarchy levels in 3D biomedical images. We design three losses to train a predictive unsupervised segmentation network that encourages the decomposition of 3D volumes into meaningful nested subvolumes that represent a hierarchy. First, we pretrain 3D diffusion models and use the consistency of their features across subvolumes. Second, we use the visual consistency between subvolumes. Third, we use the invariance to photometric augmentations as a regularizer. Our models perform better than prior unsupervised structure discovery approaches on challenging biologically-inspired synthetic datasets and on a real-world brain tumor MRI dataset. Code is available at github.com/uncbiag/diffusion-3D-discovery.

1 Introduction

Deep neural networks (DNNs) have been successfully applied to various supervised 3D biomedical image analysis tasks, such as classification [11], segmentation [7], and registration [35]. Acquiring volumetric annotations manually to supervise deep learning models is costly and labor intensive. For example, the supervised training of 3D DNNs for segmentation requires the manual labeling of every voxel of the structures of interest for the entire training set. Additionally, the diversity of existing biomedical 3D volumetric image types (e.g. MRI, CT, electron tomography) and different tasks associated with them precludes image annotations for all existing problems in practice. Furthermore, experts may focus on annotating objects they are already aware of, thereby restricting the possibility of new structural discoveries in large datasets using deep learning. We hypothesize that the nested hierarchical structure intrinsic to many 3D biomedical images [13] might be useful for unsupervised segmentation. As a step in this direction, our goal in this work is to develop a computational approach for unsupervised structure discovery.

Recently, unsupervised part discovery in 2D natural images has gained significant attention [6,8,15]. These methods are based on the finding that intermediate

H. Greenspan et al. (Eds.): MICCAI 2023, LNCS 14220, pp. 320–330, 2023.
https://doi.org/10.1007/978-3-031-43907-0_31

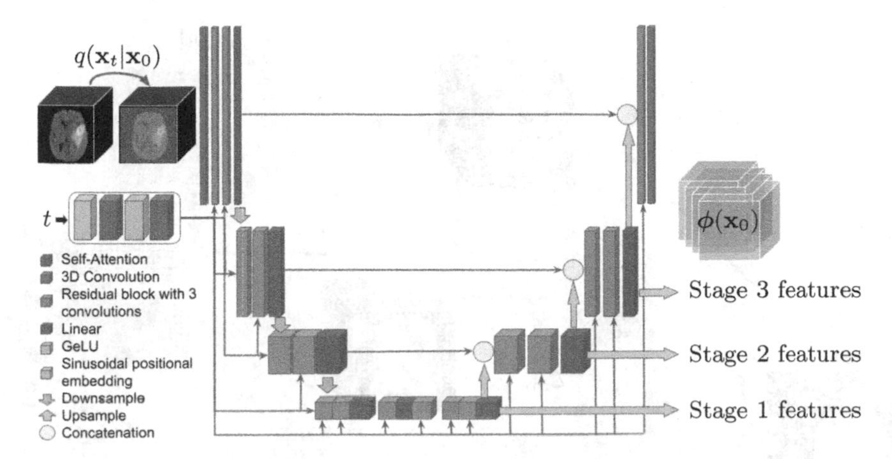

Fig. 1. Feature extractor. Given a clean 3D image $\mathbf{x}_0$, we add Gaussian noise corresponding to diffusion timestep t to the image following the distribution $q(x_t|x_0)$ using Eq. (2). The noisy image $\mathbf{x}_t$ is passed to our pretrained 3D diffusion model. We upsample intermediate activations to the original image size and use them as feature extractor ϕ for each voxel. Features from different stages of a U-Net-based ladder-like architecture for a diffusion model capture different hierarchy levels.

activations of deep ImageNet-pre-trained classification models capture semantically meaningful conceptual regions [8]. These regions are robust to pose and viewpoint variations and help high-level image understanding by providing local object representations, leading to more explainable recognition [15]. However, a naive application of part discovery methods to 3D volumetric segmentation is not feasible, due to the lack of good feature extractors for 3D biomedical images [5] and ImageNet-pretrained networks operate only on 2D images.

We hypothesize that deep generative models are good feature extractors for unsupervised structure discovery for the following reasons. First, these models do not require expert labels as they are trained in a self-supervised way. Second, the ability to generate high-quality images suggests that these models capture semantically meaningful information. Third, generative representation learning has been successfully applied to global and dense prediction tasks in 2D images [9] and has shown improvements in label efficiency and generalization [19].

Besides creating stunning image generation results, diffusion-based generative models [12] are applied to other downstream tasks. Several works use pretrained diffusion models for 2D label-efficient semantic segmentation of natural images [1,4]. In 2D medical imaging, diffusion models are used for self-supervised vessel segmentation [18], anomaly detection [27,29,31,34], denoising [14], and improving supervised segmentation models [32,33]. In 3D medical imaging, diffusion models are used for CT and MR image synthesis [10,33]. Inspired by the success of unsupervised part discovery methods in 2D images and the effective abilities of diffusion models for many downstream tasks we hypothesize that feature representations of generative diffusion models discover intrinsic hierarchical structures in 3D biomedical images. Our work explores this hypothesis.

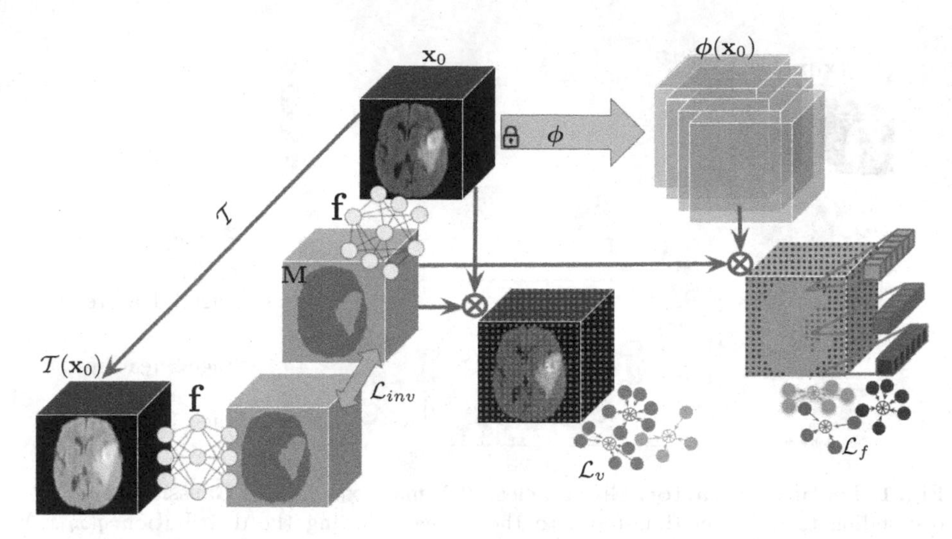

Fig. 2. Predictive unsupervised structure discovery. Our unsupervised segmentation network $\mathbf{f} : \mathbf{x}_0 \rightarrow \mathbf{M}$ is trained with three losses. Feature consistency loss $\mathcal{L}_f$ encourages features $\phi(\mathbf{x}_0)$, extracted using diffusion models (see Fig. 1), of voxels belonging to the same parts to be similar to each other. Visual consistency loss $\mathcal{L}_v$ encourages models to learn parts that align with image boundaries. Photometric invariance loss $\mathcal{L}_{inv}$ encourages invariance in models to photometric transformation $\mathcal{T}$.

Our Contributions Are:

1) We pretrain 3D diffusion models, use them as feature extractors (Fig. 1), and design losses (Fig. 2) for unsupervised 3D structure discovery.
2) We show that features from different stages of ladder-like U-Net-based diffusion models capture different hierarchy levels in 3D biomedical volumes.
3) Our approach outperforms previous 3D unsupervised discovery methods on challenging synthetic datasets and on a real-world brain tumor segmentation (BraTS'19) dataset.

2 Background on Diffusion Models

Diffusion models [12] consist of two parts: a forward pass and a reverse pass. The forward pass is a T-step process of adding a small Gaussian noise, gradually destroying image information and transforming a clean image $\mathbf{x}_0$ into pure Gaussian noise $\mathbf{x}_T$. Each step $t \in [\![1, T]\!]$ is:

$$q(\mathbf{x}_t|\mathbf{x}_{t-1}) := \mathcal{N}(\mathbf{x}_t; \sqrt{1-\beta_t}\mathbf{x}_{t-1}, \sqrt{\beta_t}\mathbf{I}), \tag{1}$$

where $\{\beta_t\}_{t=1}^T$ is a variance schedule. With $\overline{\alpha}_t = \prod_{i=1}^t (1-\beta_i)$, the noisy image $\mathbf{x}_t$ at a timestep t, following $q(\mathbf{x}_t|\mathbf{x}_0) = \prod_{i=1}^t q(\mathbf{x}_i|\mathbf{x}_{i-1})$, can be written as:

$$\mathbf{x}_t = \sqrt{\overline{\alpha}_t}\mathbf{x}_0 + \sqrt{1-\overline{\alpha}_t}\epsilon, \qquad \epsilon \sim \mathcal{N}(\mathbf{0}, \mathbf{I}). \tag{2}$$

The reverse pass is a corresponding T-step denoising process using a neural network (usually, U-Net [28]) with parameters θ. For small noises, the reverse pass is also Gaussian:

$$p_\theta(\mathbf{x}_{t-1}|\mathbf{x}_t) := \mathcal{N}(\mathbf{x}_{t-1}; \boldsymbol{\mu}_\theta(\mathbf{x}_t, t), \boldsymbol{\Sigma}_\theta(\mathbf{x}_t, t)). \tag{3}$$

Practically, instead of $\boldsymbol{\mu}_\theta(x_t, t)$ and $\boldsymbol{\Sigma}_\theta(\mathbf{x}_t, t)$, models are designed to predict either the noise ϵ_t at timestep t, or a less noisier version of image $\mathbf{x}_{t-1}$ directly.

3 Method

We formulate the 3D structure discovery task in biomedical images as an unsupervised segmentation into K parts. Given a one-channel 3D image $\mathbf{x}_0 \in \mathbb{R}^{1\times H\times W\times D}$, our segmentation model $\mathbf{f}$ predicts a mask $\mathbf{M} \in [0,1]^{K\times H\times W\times D}$. For all voxels $u \in [\![0, H-1]\!] \times [\![0, W-1]\!] \times [\![0, D-1]\!]$, we have $\sum_{k=1}^{K} \mathbf{M}_{ku} = 1$. We use three losses for unsupervised training (see Fig. 2):

$$\mathcal{L} = \lambda_v \mathcal{L}_v + \lambda_f \mathcal{L}_f + \lambda_{inv} \mathcal{L}_{inv}. \tag{4}$$

For an arbitrary representation $\boldsymbol{h}(\mathbf{x}_0)$ of an image $\mathbf{x}_0$ with voxels u, the consistency of this representation $C(\boldsymbol{h}(\mathbf{x}_0))$ across K predicted parts in the form of segmentation $\mathbf{M}$ is defined as:

$$C(\boldsymbol{h}(\mathbf{x}_0)) = \frac{1}{N}\sum_{k=1}^{K}\sum_{u} \mathbf{M}_{ku} \|\mathbf{z}_k - [\boldsymbol{h}(\mathbf{x}_0)]_u\|_2^2, \quad \text{where} \quad \mathbf{z}_k = \frac{\sum_u \mathbf{M}_{ku}[\boldsymbol{h}(\mathbf{x}_0)]_u}{\sum_u \mathbf{M}_{ku}}, \tag{5}$$

where N is the number of voxels. This is a form of volume-normalized K-means loss with z_k describing the mean feature value of partition k.

Feature Consistency. We pretrain generative 3D diffusion models and use them as feature extractors [4]. Noise is added to a clean image $\mathbf{x}_0$ based on Eq. (2) and the noisy image $\mathbf{x}_t \in \mathbb{R}^{1\times H\times W\times D}$ is passed to the 3D diffusion model. Intermediate activations (either from different stages of ladder-like U-Nets or their concatenation, see Fig. 1) upsampled to the original image size serve as a p-dimensional feature extractor $\boldsymbol{\phi}(\mathbf{x}_0) \in \mathbb{R}^{p\times H\times W\times D}$. The feature consistency loss encourages voxels corresponding to the same parts to have similar features:

$$\mathcal{L}_f = C(\boldsymbol{\phi}(\mathbf{x}_0)). \tag{6}$$

Visual Consistency. The extracted features are upsampled from low spatial resolutions and therefore do not accurately align with image boundaries. To alleviate this problem, we use a voxel visual consistency loss:

$$\mathcal{L}_v = C(\boldsymbol{I}(\mathbf{x}_0)) = C(\mathbf{x}_0) \tag{7}$$

where $\boldsymbol{I}(\mathbf{x}_0)$ is the identity feature extractor, i.e. $\boldsymbol{I}(\mathbf{x}_0) = \mathbf{x}_0$.

Photometric Invariance. As biomedical images often show acquisition differences (e.g., based on MR or CT scanner), they can be heterogeneous in their voxel intensities [25]. Therefore, robustness of models to voxel-level photometric perturbations might be helpful for *unsupervised* discovery. We use the Dice loss [22] to encourage invariance to such a photometric transformation $\mathcal{T}$:

$$\mathcal{L}_{inv} = 1 - \frac{2\sum_u [\mathbf{f}(\mathbf{x}_0)]_u [\mathbf{f}(\mathcal{T}(\mathbf{x}_0))]_u}{\sum_u [\mathbf{f}(\mathbf{x}_0)]_u^2 + \sum_u [\mathbf{f}(\mathcal{T}(\mathbf{x}_0))]_u^2}. \tag{8}$$

We assume our images are min-max normalized ($\mathbf{x}_0 \in [0, 1]$). We then use gamma-correction of the form $\mathcal{T}(\mathbf{x}_0) = \mathbf{x}_0^\gamma$ as a photometric transformation. We draw γ from the uniform distribution: $\gamma \sim U[\gamma_{min}, \gamma_{max}]$.

Table 1. Dice scores on the biologically inspired synthetic datasets. Our method outperforms all previous work on unsupervised 3D segmentation for all levels of hierarchy.

	Regular				Irregular		
	Level 1	*Level* 2	*Level* 3		*Level* 1	*Level* 2	*Level* 3
Çiçek et al. [7]	0.968	0.829	0.668	Semi-supervised	0.970	0.825	0.601
Zhao et al. [36]	0.989	0.655	0.357	Semi-supervised	0.978	0.641	0.333
Nalepa et al. [24]	0.530	0.276	0.112	Unsupervised	0.527	0.280	0.144
Ji et al. [16]	0.589	0.291	0.150	Unsupervised	0.527	0.280	0.144
Moriya et al. [23]	0.628	0.311	0.141	Unsupervised	0.525	0.232	0.094
Hsu et al. [13]	0.952	0.541	0.216	Unsupervised	0.953	0.488	0.199
Ours	**0.986**	**0.577**	**0.397**	Unsupervised	**0.967**	**0.565**	**0.382**
k-means	0.808	0.326	0.149	Non-DL	0.771	0.299	0.118
BM4D+k-means	0.949	0.529	0.335	Non-DL	0.950	0.533	0.324

4 Experiments

4.1 Datasets

To compare with state-of-the-art unsupervised 3D segmentation methods we follow [13] and evaluate our method on challenging biologically inspired 3D synthetic datasets and a real-world brain tumor segmentation (BraTS'19) dataset.

The synthetic dataset of [13], consists of 120 volumes (80–20–20 split) of size 50 × 50 × 50. Inspired by cryo-electron tomography images, it contains a three-level structure, representing a biological cell, vesicles and mitochondria, as well as protein aggregates. The intensities and locations of the objects are randomized without destroying the hierarchy. The regular variant of the dataset contains cubical and spherical objects, while the irregular variant contains more complex shapes. Pink noise of magnitude $m = 0.25$ which is commonly seen in biological data [30] is applied to the volume. Figure 3 shows sample slices of both variants.

The BraTS'19 dataset [2,3,21] is an established benchmark for 3D tumor segmentation of brain MRIs. Volumes are co-registered to the same template, interpolated to $(1\,\text{mm})^3$ resolution and brain-extracted. Following [13], images are cropped to volumes of size $200 \times 200 \times 155$. As in [13], FLAIR images and corresponding whole tumor (WT) annotations are used for unsupervised segmentation evaluation with the same split of 259 high grade glioma training examples into 180 train, 39 validation, and 40 test samples. The official BraTS'19 validation and test sets are not used as their segmentation masks are not available.

4.2 Implementation Details

All diffusion models use the same architecture shown in Fig. 1. We pretrain them for $50k$ epochs with batch size 4, using an $L1$ loss between the denoised and the original images. We use the Adam optimizer, a cosine noise schedule, learning rate 10^{-4} and $T = 250$ steps. The first layer has 64 channels and this number is doubled for the proceeding downsampling layers. Due to memory constraints for BraTS'19, we trained diffusion models at $128 \times 128 \times 128$ resolution. However, the extracted features are upsampled to the original $200 \times 200 \times 155$ resolution.

Our segmentation networks (**f** in Fig. 2) use a 3D U-Net architecture [7,28]. We trained them for 100 epochs using the Adam optimizer, a learning rate of $3{*}10^{-4}$ and the losses in Eq. (4). We selected the epoch that gave the best average probability of the segmentation mask for all inputs [26] as our final model. Noisy images at timestep $t = 25$ are used as input to the diffusion models. Due to the

Table 2. Dice scores when using features from different stages of ladder-like U-Net-based diffusion models (see Fig. 1) for unsupervised segmentation of the different hierarchy levels of the synthetic dataset. Features at lower resolutions (Stage 1) are more suitable for discovering larger objects (Level 1). Intermediate features (Stage 2) are more suitable for intermediate discoveries (Level 2). Features at higher resolutions (Stage 3) are more suitable for more detailed discoveries (Level 3).

	Predictions		
	Level 1	Level 2	Level 3
Stage 1 features	**0.986**	0.366	0.273
Stage 2 features	0.923	**0.577**	0.327
Stage 3 features	0.878	0.489	**0.397**

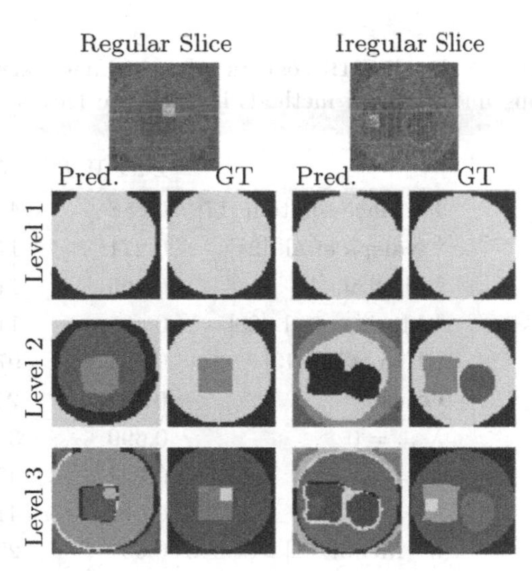

Fig. 3. Examples of unsupervised 3D structure discovery with our method on the biologically-inspired synthetic datasets. GT indicates ground truth.

memory limits, for BraTS'19, we used Stage 2 features, as they have the least number of channels. We set $\lambda_f = \lambda_v = \lambda_{inv} = 1$ and $\gamma \sim U[0.9, 1.1]$ for all cases. For all experiments we used Pytorch and 4 NVIDIA A6000 GPUs (48 Gb).

4.3 Results

We compare our method with state-of-the-art unsupervised 3D structure discovery approaches including clustering using 3D feature learning [23], a 3D convolutional autoencoder [24], and self-supervised hyperbolic representations [13].

For the synthetic datasets, we used $K = 2$ (background and cell) for Level 1, $K = 4$ (background, cell, vesicle, mitochondria) for Level 2, and $K = 8$ (background, cell, vesicle, mitochondria, and 4 small protein aggregates) for Level 3 predictions. The evaluation metric is the average Dice score on the annotated test labels. As the label order may differ we use the Hungarian algorithm to match the predicted masks with the ground truth segmentations. Table 1 shows the results for the regular and irregular variants of the cryo-ET-inspired synthetic dataset. Our models outperform all previous unsupervised work at all hierarchy levels. For some levels, our models even outperform semi-supervised methods (Çiçek et al. [7] used 2% of annotated data, Zhao et al. [36] used one annotated volume). We found that simple unsupervised denoising (BM4D [20]) followed by k-means clustering provides a good baseline, although vanilla k-means clustering on voxel intensities does not perform well due to noise. Results in Fig. 3 demonstrate that our proposed unsupervised method indeed discovers

Table 3. BraTS'19 results with ablation studies. Our method outperforms previous unsupervised methods in both the Dice score and the 95% Hausdorff distance.

	Dice WT ↑	*HD95 WT* ↓	
1st place solution [17]	0.888	4.618	Supervised
[a] Nalepa et al. [24]	0.211	170.434	Unsupervised
[a] Ji et al. [16]	0.425	114.400	Unsupervised
[a] Moriya et al. [23]	0.495	110.803	Unsupervised
[a] Hsu et al. [13]	0.684	97.641	Unsupervised
Ours	**0.719**	**27.838**	Unsupervised
$\lambda_{inv} = 0$	0.696	38.645	Unsupervised
$\lambda_f = 0$	0.677	42.318	Unsupervised
$\lambda_v = 0$	0.671	41.801	Unsupervised
w/ Med3D [5] features	0.657	29.906	Unsupervised
k-means	0.439	63.811	Non-DL
Features k-means	0.471	45.917	Unsupervised
Med3D [5] k–means	0.231	55.846	Unsupervised

[a] Dice and HD95 numbers for these models are taken from [13].

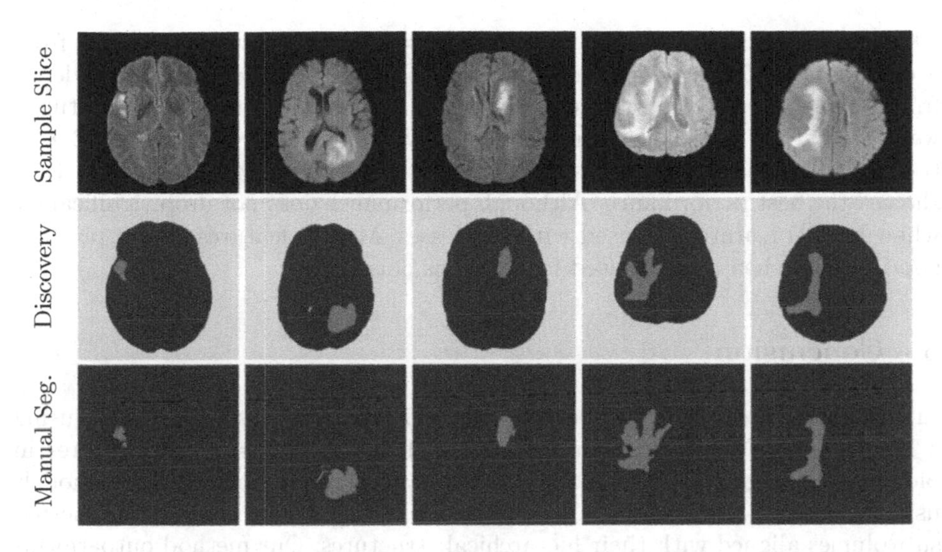

Fig. 4. Examples of discovered structures on BraTS'19. Our method discovers meaningful regions and detects tumors of different sizes in an unsupervised manner.

the hierarchical structure of different levels. We also show in Table 2 that features from early decoder stages of the U-Net-based diffusion models better discover larger objects in the hierarchy, features at intermediate stages better capture intermediate objects, and features at later stages better find smaller objects.

For the Brain Tumor Segmentation (BraTS'19) dataset, we use the whole tumor (WT) segmentation mask for evaluation, which is detectable based on the FLAIR images alone. We train segmentation models with $K = 3$ parts (background, brain, tumor). The evaluation metric, as in the BraTS'19 challenge [21], is Dice score and the 95th percentile of the symmetric Hausdorff distance, which quantifies the surface distance of the predicted segmentation from the manual tumor segmentation in millimeters. Table 3 shows that our model outperforms all prior unsupervised methods for both evaluation metrics. As an approximate upper bound we show for reference the reported results of the $1st$ place solution [17] on BraTS'19 which is based on supervised training on the full train set and evaluated on the BraTS'19 test set. The qualitative results in Fig. 4 show that our model can detect tumors of different sizes. Our predictions look smoother and do not capture fine details of tumor segmentations.

We perform ablation studies on the BraTS'19 dataset (Table 3: below the line). Measuring the impact of each loss, we see that the smallest performance drop is due to a deactivated invariance loss ($\lambda_{inv} = 0$) while deactivating the visual consistency ($\lambda_v = 0$) and feature consistency ($\lambda_f = 0$) losses results in larger, but similar performance drops. However, to achieve best performance all three components are necessary. We also perform k-means clustering on intensities and features. We observe that using our deep network model dramatically improves performance, although our losses are similar to k–means clustering.

This might be due to the fact that predictive modeling involves learning from a distribution of images and a model may therefore extract useful knowledge from a collection of images. To evaluate the significance of the diffusion features, we replaced our diffusion feature extractor with a 3D ResNet from Med3D [5] trained on 23 medical datasets. We use the "layer1_2_conv2" features as they showed the best performance. Although performance does not drop significantly when Med3D features are used with our losses, Med3D features do not produce good results when directly used for k-means clustering.

5 Conclusion

In this work, we showed that features from 3D generative diffusion models using a ladder-like U-Net-based architecture can discover *intrinsic* 3D structures in biomedical images. We trained predictive *unsupervised* segmentation models using losses that encourage the decomposition of biomedical volumes into nested subvolumes aligned with their hierarchical structures. Our method outperforms existing unsupervised segmentation approaches and discovers meaningful hierarchical concepts on challenging biologically-inspired synthetic datasets and on the BraTS brain tumor dataset. While we tested our approach for unsupervised image segmentation it is conceivable that it could also be useful in semi-supervised settings and that could be applied to data types other than images.

Acknowledgements. This work was supported by NIH grants 1R01AR072013, 1R01HL149877, and R41MH118845. The work expresses the views of the authors, not of NIH.

References

1. Asiedu, E.B., Kornblith, S., Chen, T., Parmar, N., Minderer, M., Norouzi, M.: Decoder denoising pretraining for semantic segmentation. arXiv:2205.11423 (2022)
2. Bakas, S., et al.: Advancing the cancer genome atlas glioma MRI collections with expert segmentation labels and radiomic features. Sci. Data **4**(1), 1–13 (2017)
3. Bakas, S., et al.: Identifying the best machine learning algorithms for brain tumor segmentation, progression assessment, and overall survival prediction in the BraTS challenge. arXiv:1811.02629 (2018)
4. Baranchuk, D., Rubachev, I., Voynov, A., Khrulkov, V., Babenko, A.: Label-efficient semantic segmentation with diffusion models. In: ICLR (2021)
5. Chen, S., Ma, K., Zheng, Y.: Med3D: transfer learning for 3D medical image analysis. arXiv:1904.00625 (2019)
6. Choudhury, S., Laina, I., Rupprecht, C., Vedaldi, A.: Unsupervised part discovery from contrastive reconstruction. In: NeurIPS, vol. 34, pp. 28104–28118 (2021)
7. Çiçek, Ö., Abdulkadir, A., Lienkamp, S.S., Brox, T., Ronneberger, O.: 3D U-Net: learning dense volumetric segmentation from sparse annotation. In: Ourselin, S., Joskowicz, L., Sabuncu, M.R., Unal, G., Wells, W. (eds.) MICCAI 2016. LNCS, vol. 9901, pp. 424–432. Springer, Cham (2016). https://doi.org/10.1007/978-3-319-46723-8_49

8. Collins, E., Achanta, R., Süsstrunk, S.: Deep feature factorization for concept discovery. In: Ferrari, V., Hebert, M., Sminchisescu, C., Weiss, Y. (eds.) Computer Vision – ECCV 2018. LNCS, vol. 11218, pp. 352–368. Springer, Cham (2018). https://doi.org/10.1007/978-3-030-01264-9_21
9. Donahue, J., Simonyan, K.: Large scale adversarial representation learning. In: NeurIPS, vol. 32 (2019)
10. Dorjsembe, Z., Odonchimed, S., Xiao, F.: Three-dimensional medical image synthesis with denoising diffusion probabilistic models. In: MIDL (2022)
11. Gao, X.W., Hui, R., Tian, Z.: Classification of CT brain images based on deep learning networks. Comput. Methods Programs Biomed. **138**, 49–56 (2017)
12. Ho, J., Jain, A., Abbeel, P.: Denoising diffusion probabilistic models. In: NeurIPS, vol. 33, pp. 6840–6851 (2020)
13. Hsu, J., Gu, J., Wu, G., Chiu, W., Yeung, S.: Capturing implicit hierarchical structure in 3D biomedical images with self-supervised hyperbolic representations. In: NeurIPS, vol. 34, pp. 5112–5123 (2021)
14. Hu, D., Tao, Y.K., Oguz, I.: Unsupervised denoising of retinal OCT with diffusion probabilistic model. In: Medical Imaging 2022: Image Processing, vol. 12032, pp. 25–34 (2022)
15. Hung, W.C., Jampani, V., Liu, S., Molchanov, P., Yang, M.H., Kautz, J.: SCOPS: self-supervised co-part segmentation. In: CVPR, pp. 869–878 (2019)
16. Ji, X., Henriques, J.F., Vedaldi, A.: Invariant information clustering for unsupervised image classification and segmentation. In: ICCV, pp. 9865–9874 (2019)
17. Jiang, Z., Ding, C., Liu, M., Tao, D.: Two-stage cascaded U-Net: 1st place solution to BraTS challenge 2019 segmentation task. In: Crimi, A., Bakas, S. (eds.) BrainLes 2019. LNCS, vol. 11992, pp. 231–241. Springer, Cham (2020). https://doi.org/10.1007/978-3-030-46640-4_22
18. Kim, B., Oh, Y., Ye, J.C.: Diffusion adversarial representation learning for self-supervised vessel segmentation. arXiv:2209.14566 (2022)
19. Li, D., Yang, J., Kreis, K., Torralba, A., Fidler, S.: Semantic segmentation with generative models: semi-supervised learning and strong out-of-domain generalization. In: CVPR, pp. 8300–8311 (2021)
20. Maggioni, M., Katkovnik, V., Egiazarian, K., Foi, A.: Nonlocal transform-domain filter for volumetric data denoising and reconstruction. IEEE TIP **22**(1), 119–133 (2012)
21. Menze, B.H., et al.: The multimodal brain tumor image segmentation benchmark (BraTS). IEEE TMI **34**(10), 1993–2024 (2014)
22. Milletari, F., Navab, N., Ahmadi, S.A.: V-Net: fully convolutional neural networks for volumetric medical image segmentation. In: International Conference on 3D Vision (3DV), pp. 565–571 (2016)
23. Moriya, T., et al.: Unsupervised segmentation of 3D medical images based on clustering and deep representation learning. In: Medical Imaging 2018: Biomedical Applications in Molecular, Structural, and Functional Imaging, vol. 10578, pp. 483–489 (2018)
24. Nalepa, J., Myller, M., Imai, Y., Honda, K., Takeda, T., Antoniak, M.: Unsupervised segmentation of hyperspectral images using 3-D convolutional autoencoders. IEEE Geosci. Remote Sens. Lett. **17**(11), 1948–1952 (2020)
25. Nalepa, J., Marcinkiewicz, M., Kawulok, M.: Data augmentation for brain-tumor segmentation: a review. Front. Comput. Neurosci. **13**, 83 (2019)
26. Park, J., Yang, H., Roh, H.J., Jung, W., Jang, G.J.: Encoder-weighted W-Net for unsupervised segmentation of cervix region in colposcopy images. Cancers **14**(14), 3400 (2022)

27. Pinaya, W.H., et al.: Fast unsupervised brain anomaly detection and segmentation with diffusion models. In: Wang, L., Dou, Q., Fletcher, P.T., Speidel, S., Li, S. (eds.) MICCAI 2022. LNCS, vol. 13438, pp. 705–714. Springer, Cham (2022). https://doi.org/10.1007/978-3-031-16452-1_67
28. Ronneberger, O., Fischer, P., Brox, T.: U-Net: convolutional networks for biomedical image segmentation. In: Navab, N., Hornegger, J., Wells, W.M., Frangi, A.F. (eds.) MICCAI 2015. LNCS, vol. 9351, pp. 234–241. Springer, Cham (2015). https://doi.org/10.1007/978-3-319-24574-4_28
29. Sanchez, P., Kascenas, A., Liu, X., O'Neil, A.Q., Tsaftaris, S.A.: What is healthy? Generative counterfactual diffusion for lesion localization. In: Mukhopadhyay, A., Oksuz, I., Engelhardt, S., Zhu, D., Yuan, Y. (eds.) DGM4MICCAI 2022. LNCS, vol. 13609, pp. 34–44. Springer, Cham (2022). https://doi.org/10.1007/978-3-031-18576-2_4
30. Sejdić, E., Lipsitz, L.A.: Necessity of noise in physiology and medicine. Comput. Methods Programs Biomed. **111**(2), 459–470 (2013)
31. Wolleb, J., Bieder, F., Sandkühler, R., Cattin, P.C.: Diffusion models for medical anomaly detection. In: Wang, L., Dou, Q., Fletcher, P.T., Speidel, S., Li, S. (eds.) MICCAI 2022. LNCS, vol. 13438, pp. 35–45. Springer, Cham (2022). https://doi.org/10.1007/978-3-031-16452-1_4
32. Wolleb, J., Sandkühler, R., Bieder, F., Valmaggia, P., Cattin, P.C.: Diffusion models for implicit image segmentation ensembles. In: MIDL, pp. 1336–1348 (2022)
33. Wu, J., Fang, H., Zhang, Y., Yang, Y., Xu, Y.: MedSegDiff: medical image segmentation with diffusion probabilistic model. arXiv:2211.00611 (2022)
34. Wyatt, J., Leach, A., Schmon, S.M., Willcocks, C.G.: AnoDDPM: anomaly detection with denoising diffusion probabilistic models using simplex noise. In: CVPR, pp. 650–656 (2022)
35. Yang, X., Kwitt, R., Styner, M., Niethammer, M.: Quicksilver: fast predictive image registration-a deep learning approach. Neuroimage **158**, 378–396 (2017)
36. Zhao, A., Balakrishnan, G., Durand, F., Guttag, J.V., Dalca, A.V.: Data augmentation using learned transformations for one-shot medical image segmentation. In: CVPR, pp. 8543–8553 (2019)

Domain Adaptation for Medical Image Segmentation Using Transformation-Invariant Self-training

Negin Ghamsarian[1(✉)], Javier Gamazo Tejero[1], Pablo Márquez-Neila[1], Sebastian Wolf[2], Martin Zinkernagel[2], Klaus Schoeffmann[3], and Raphael Sznitman[1]

[1] Center for AI in Medicine, Faculty of Medicine, University of Bern, Bern, Switzerland
negin.ghamsarian@unibe.ch
[2] Department of Ophthalmology, Inselspital, Bern, Switzerland
[3] Department of Information Technology, Klagenfurt University, Klagenfurt, Austria

Abstract. Models capable of leveraging unlabelled data are crucial in overcoming large distribution gaps between the acquired datasets across different imaging devices and configurations. In this regard, self-training techniques based on pseudo-labeling have been shown to be highly effective for semi-supervised domain adaptation. However, the unreliability of pseudo labels can hinder the capability of self-training techniques to induce abstract representation from the unlabeled target dataset, especially in the case of large distribution gaps. Since the neural network performance should be invariant to image transformations, we look to this fact to identify uncertain pseudo labels. Indeed, we argue that transformation invariant detections can provide more reasonable approximations of ground truth. Accordingly, we propose a semi-supervised learning strategy for domain adaptation termed transformation-invariant self-training (TI-ST). The proposed method assesses pixel-wise pseudo-labels' reliability and filters out unreliable detections during self-training. We perform comprehensive evaluations for domain adaptation using three different modalities of medical images, two different network architectures, and several alternative state-of-the-art domain adaptation methods. Experimental results confirm the superiority of our proposed method in mitigating the lack of target domain annotation and boosting segmentation performance in the target domain.

Keywords: Semi-Supervised Learning · Domain Adaptation · Semantic Segmentation · Self Training · Cataract Surgery · MRI · OCT

This work was funded by Haag-Streit Switzerland.

Supplementary Information The online version contains supplementary material available at https://doi.org/10.1007/978-3-031-43907-0_32.

H. Greenspan et al. (Eds.): MICCAI 2023, LNCS 14220, pp. 331–341, 2023.
https://doi.org/10.1007/978-3-031-43907-0_32

1 Introduction

Semantic segmentation is a prerequisite for a broad range of medical imaging applications, including disease diagnosis and treatment [13], surgical workflow analysis [6,9], operation room planning, and surgical outcome prediction [7]. While supervised deep learning approaches have yielded satisfactory performance in semantic segmentation [8,10], their performance is heavily limited by the labeled training dataset distribution. Indeed, a network trained on a dataset acquired with a specific device or configuration can dramatically underperform when evaluated on a different device or conditions. Overcoming this entails new annotations per device, a demand that is hard to meet, especially for semantic segmentation, and even more so in the medical domain, where expert knowledge is essential.

Driven by the need to overcome this challenge, numerous semi-supervised learning paradigms have looked to alleviate annotation requirements in the target domain. Semi-supervised learning refers to methods that encourage learning abstract representations from an unlabeled dataset and extending the decision boundaries towards a more-generalized or target dataset distribution. These techniques can be categorized into (i) consistency regularization [4,15–17,19,22], (ii) contrastive learning [2,11], (iii) adversarial learning [22], and (iv) self-training [24–26]. Consistency regularization techniques aim to inject knowledge via penalizing inconsistencies for identical images that have undergone different distortions, such as transformations or dropouts, or fed into networks with different initializations [4]. Specifically, the Π model [15] penalizes differences between the predictions of two transformed versions of each input image to reinforce consistent and augmentation-invariant predictions. Temporal ensembling [15] is designed to alleviate the negative effect of noisy predictions by integrating predictions of consecutive training iterations. Cross-pseudo supervision regularizes the networks by enforcing similar predictions from differently initialized networks.

More recent deep self-training approaches based on pseudo labels have emerged as promising techniques for unsupervised domain adaptation. These techniques assume that a trained network can approximate the ground-truth labels for unlabeled images. Since no metric guarantees pseudo-label reliability, several methods have been developed to alleviate pseudo-label error back-propagation. To progressively improve pseudo-labeling performance, reciprocal learning [25] adopts a teacher-student framework where the student network performance on the source domain drives the teacher network weights updates. ST++ [24] proposes to evaluate the reliability of image-based pseudo labels based on the consistency of predictions in different network checkpoints. Subsequently, half of the more reliable images are utilized to re-train the network in the first step, and the trained network is used for pseudo-labeling the whole dataset for a second re-training step. Despite the effectiveness of state-of-the-art pseudo-labeling strategies, we argue that one important aspect has been underexplored: how can a trained network self-assess the reliability of its pixel-level predictions?

To this end, we propose a novel self-training framework with a self-assessment strategy for pseudo-label reliability. The proposed framework uses transformation-invariant highly-confident predictions in the target dataset for

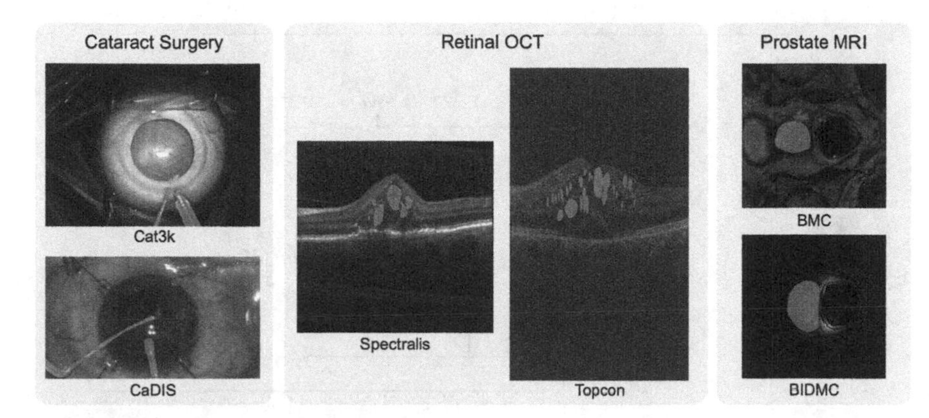

Fig. 1. Example images from the three adopted datasets: (1) cross-device-and-center instrument segmentation in cataract surgery videos (Cat101 vs. CaDIS), cross-device fluid segmentation in OCT (Spectralis vs. Topcon), and cross-institution prostate segmentation in MRI (BMC vs. BIDMC).

self-training. This objective is achieved by considering an ensemble of high-confidence predictions from transformed versions of identical inputs. To validate the effectiveness of our proposed framework on a variety of tasks, we evaluate our approach on three different semantic segmentation imaging modalities, including video (cataract surgery), optical coherence tomography (retina), and MRI (prostate), as shown in Fig. 1. We perform comprehensive experiments to validate the performance of the proposed framework, namely "Transformation-Invariant Self-Training"[1] (TI-ST). The experimental results indicate that TI-ST significantly improves segmentation performance for unlabeled target datasets compared to numerous state-of-the-art alternatives.

2 Methodology

Consider a labeled source dataset, $\mathcal{S}$, with training images $\mathcal{X}_\mathcal{S}$ and corresponding segmentation labels $\mathcal{Y}_\mathcal{S}$, while we denote a target dataset $\mathcal{T}$, containing only target images $\mathcal{X}_\mathcal{T}$. We aim to train a network using $\mathcal{X}_\mathcal{S}$, $\mathcal{Y}_\mathcal{S}$, and $\mathcal{X}_\mathcal{T}$ for semantic segmentation in the target dataset.

We propose to train the model using a self-supervised approach on the images $\mathcal{X}_\mathcal{T}$ by assigning pseudo labels during training. Typical pseudo labels are computed from independent predictions of unlabeled images. Instead, our proposed framework adopts a self-assessment strategy to determine the reliability of predictions in an unsupervised fashion. Specifically, we propose to target highly-reliable predictions generated by a network aiming for transformation-invariant confidence. Compared to self-ensembling strategies that penalize the distant predictions corresponding to the transformed versions of identical inputs, our goal is to filter out transformation-variant predictions. Indeed, our method reinforces

[1] The PyTorch implementation of TI-ST is publicly available at https://github.com/Negin-Ghamsarian/Transformation-Invariant-Self-Training-MICCAI23.

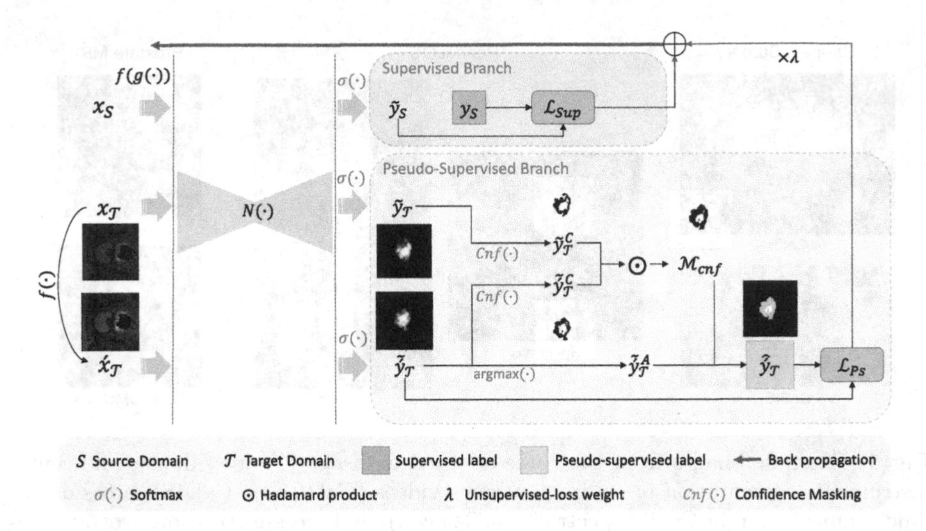

Fig. 2. Overview of the proposed semi-supervised domain adaptation framework based on transformation-invariant self-training (TI-ST). Ignored pseudo-labels during unsupervised loss computation are shown in turquoise.

the ensemble of high-confidence predictions from two versions of the same target sample. Our proposed TI-ST framework simultaneously trains on the source and target domains, so as to progressively bridge the intra-domain distribution gap. Figure 2 depicts our TI-ST framework, which we detail in the following sections.

2.1 Model

At training time, images from the source dataset are augmented using spatial $g(\cdot)$ and non-spatial $f(\cdot)$ transformations and passed through a segmentation network, $N(\cdot)$, by which the network is trained using a standard supervision loss. At the same time, images from the target dataset are also passed to the network. Specifically, we feed two versions of each target image to the network: (1) the original target image $x_{\mathcal{T}}$, and (2) its non-spatially transformed version, $\acute{x_{\mathcal{T}}} = f(x_{\mathcal{T}})$. Once fed through the network, the corresponding predictions can be defined as $\tilde{y_{\mathcal{T}}} = \sigma(N(x_{\mathcal{T}}))$ and $\tilde{\acute{y_{\mathcal{T}}}} = \sigma(N(\acute{x_{\mathcal{T}}}))$, where $\sigma(\cdot)$ is the Softmax operation. We then define a confidence-mask ensemble as

$$\mathcal{M}_{cnf} = Cnf(\tilde{y_{\mathcal{T}}}) \odot Cnf(\tilde{\acute{y_{\mathcal{T}}}}), \tag{1}$$

where $\odot$ refers to Hadamard product used for element-wise multiplication, and Cnf is the high confidence masking function,

$$Cnf_{\in (W \times H)}(y) = \begin{cases} 1, & \text{if } \max_C(y) > \tau \\ 0, & \text{else.} \end{cases} \tag{2}$$

where $\tau \in (0.5, 1)$ is the confidence threshold, and H, W, and C are the height, width, and number of classes in the output, respectively. Specifically, $\mathcal{M}_{cnf}$

encodes regions of confident predictions that are invariant to transformations. We can then compute the pseudo-ground-truth mask for each input from the target dataset as

$$\hat{y_T} = \begin{cases} \text{argmax}_C(\tilde{y_T}), & \text{if } \mathcal{M}_{cnf} = 1 \\ \text{ignore}, & \text{else.} \end{cases} \tag{3}$$

2.2 Training

To train our model, we simultaneously consider both the source and target samples by minimizing the following loss,

$$\mathcal{L}_{overall} = \mathcal{L}_{Sup}(\tilde{y_S}, y_S) + \lambda\Big(\mathcal{L}_{Ps}(\tilde{y_T}, \hat{y_T})\Big), \tag{4}$$

where $\mathcal{L}_{Sup}$ and $\mathcal{L}_{Ps}$ indicate the supervised and pseudo-supervised loss functions used, respectively. We set λ as a time-dependent weighing function that gradually increases the share of pseudo-supervised loss. Intuitively, our pseudo-supervised loss enforces predictions on transformation-invariant highly-confident regions for unlabeled images.

Discussion: The quantity and distribution of supervised data are determining factors in neural networks' performance. With highly distributed large-scale supervisory data, neural networks converge to an optimal state efficiently. However, when only limited supervisory data with heterogeneous distribution from the inference dataset are available, using more sophisticated methods to leverage a priori knowledge is essential. Our proposed use of invariance of network predictions with respect to data augmentation is a strong form of knowledge that can be learned through dataset-dependent augmentations. The trained network is then expected to provide consistent predictions under diverse transformations. Hence, the transformation variance of the network predictions can indicate the network's prediction doubt and low confidence correspondingly. We take advantage of this characteristic to assess the reliability of predictions and filter out unreliable pseudo-labels.

3 Experimental Setup

Datasets: We validate our approach on three cross-device/site datasets for three different modalities:

- **Cataract:** instrument segmentation in cataract surgery videos [12,21]. We set the "Cat101" [21] as the source dataset and the "CaDIS" as the target domain dataset [12].
- **OCT:** IRF Fluid segmentation in retinal OCTs [1]. We use the high-quality "Spectralis" dataset as the source and the lower-quality "Topcon" dataset as the target domain.

- **MRI:** multi-site prostate segmentation [18]. We sample volumes from "BMC" and "BIDMC" as the source and target domain, respectively.

We follow a four-fold validation strategy for all three cases and report the average results over all folds. The average number of labeled training images (from the source domain), unlabeled training images (from the target domain), and test images per fold are equal to $(207, 3189, 58)$ for Cataract, $(391, 569, 115)$ for OCT, and $(273, 195, 65)$ for MRI dataset.

Baseline Methods: We compare the performance of our proposed transformation-invariant self-training (SI-ST) method against seven state-of-the-art semi-supervised learning methods: Π models [15], temporal ensembling [15], mean teacher [19], cross pseudo supervision (CSP) [4], reciprocal learning (RL) [25], self-training (ST) [24], and mutual correction framework (MCF) [23].

Networks and Training Settings: We evaluate our TI-ST framework using two different architectures: (1) DeepLabV3+ [3] with ResNet50 backbone [14] and (2) scSE [20] with VGG16 backbone. Both backbones are initialized with the ImageNet [5] pre-trained parameters. We use a batch size of four for the Cataract and MRI datasets and a batch size of two for the OCT dataset. For all training strategies, we set the number of epochs to 100. The initial learning rate is set to 0.001 and decayed by a factor of $\gamma = 0.8$ every two epochs. The input size of the networks is 512×512 for cataract and OCT and 384×384 for the MRI dataset. As spatial transformations $g(\cdot)$, we apply cropping and random rotation (up to 30 degrees). The non-spatial transformations, $f(\cdot)$, include color jittering (brightness $= 0.7$, contrast $= 0.7$, saturation $= 0.7$), Gaussian blurring, and random sharpening. The confidence threshold τ for the self-training framework and the proposed TI-ST framework is set to 0.85 except in the ablation studies (See the next section). In Eq. (4), the weighting function λ ramps up from the first epoch along a Gaussian curve equal to $\exp[-5(1 - \text{current-epoch}/\text{total-epochs})]$. The self-supervised loss is set to the cross-entropy loss, and the supervised loss is set to the *cross entropy log dice* loss, which is a weighted sum of cross-entropy and the logarithm of soft dice coefficient. For the TI-ST framework, we only use non-spatial transformations for the self-training branch for simplicity.

4 Results

Table 1 compares the performance of our transformation-invariant self-training (TI-ST) approach with alternative methods across three tasks and using two network architectures. According to the quantitative results, TI-ST, RL, ST, and CPS are the best-performing methods. Nevertheless, our proposed TI-ST achieves the highest average relative improvement in dice score compared to naive supervised learning (16.18% average improvement). Considering our main competitor (RL), we note that our proposed TI-ST method is a one-stage framework using one network. In contrast, RL is a two-stage framework (requiring a

Table 1. Quantitative comparisons in Dice score (%) among the proposed (TI-ST) and alternative methods for DeepLabV3+ [3] (DLV3+) and scSENet [20] and the three datasets. Relative Dice computed over the Supervised baseline. The best results are shown in green.

Modality	Cataract Surgery		OCT		MRI		Avg. Rel.
Network	DLV3+	scSENet	DLV3+	scSENet	DLV3+	scSENet	
Supervised	15.42	37.67	22.87	24.08	52.39	65.93	N/A
Π Model [15]	27.55	35.56	1.12	0.00	10.00	6.87	-22.88
TE [15]	33.10	42.32	42.13	39.86	63.41	67.25	11.62
Mean Teacher [19]	11.06	39.54	19.11	4.70	64.82	66.87	-2.04
RL [25]	34.40	45.13	48.73	47.70	60.79	70.20	14.77
CPS [4]	36.24	39.40	47.31	14.71	76.00	68.80	10.68
ST [24]	34.34	41.10	36.84	33.01	68.63	71.97	11.26
MCF [23]	26.97	40.19	40.12	36.52	54.17	50.23	7.46
TI-ST	37.69 (+22.27)	45.31 (+7.46)	50.93 (+28.06)	40.87 (+16.79)	66.56 (+14.17)	74.07 (+8.14)	16.18

pre-training stage) and uses a teacher-student network. Hence, TI-ST is also more efficient than RL in terms of time and computation. Furthermore, the proposed strategy demonstrates the most consistent results when evaluated on different tasks, regardless of the utilized neural network architecture.

Figure 3-(a–b) demonstrates the effect of the pseudo-labeling threshold on TI-ST performance compared with regular ST. We observe that filtering out unreliable pseudo-labels based on transformation variance can remarkably boost pseudo-supervision performance regardless of the threshold. Figure 3-(c) compares the performance of the supervised baseline, ST, and TI-ST with respect to the number of source-domain labeled training images. While ST performance converges when the number of labeled images increases, our TI-ST pushes deci-

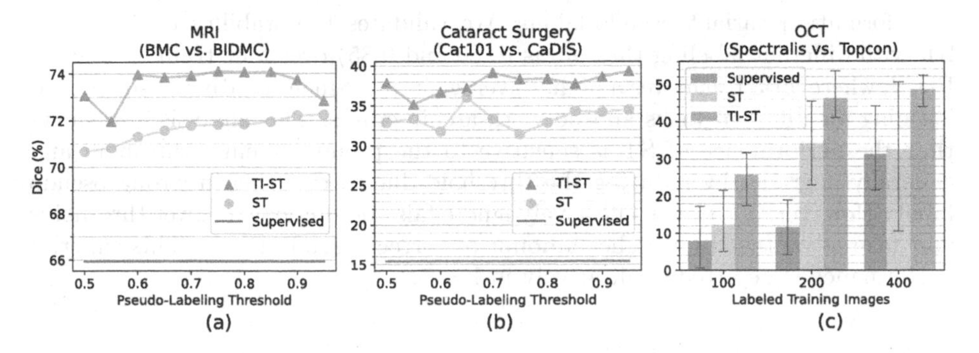

Fig. 3. Ablation studies on the pseudo-labeling threshold and size of the labeled dataset.

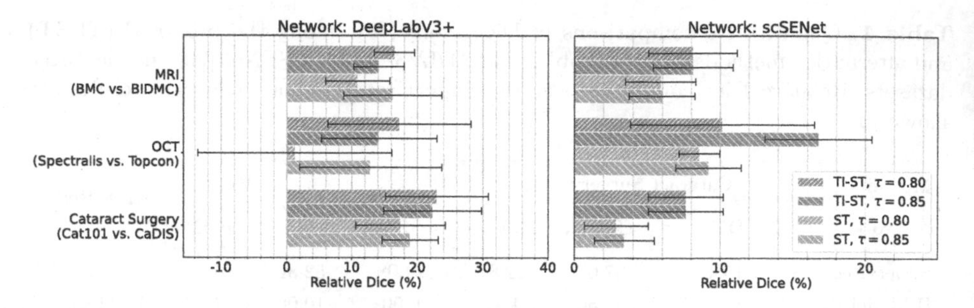

Fig. 4. Ablation study on the performance stability of TI-ST vs. ST across the different experimental segmentation tasks.

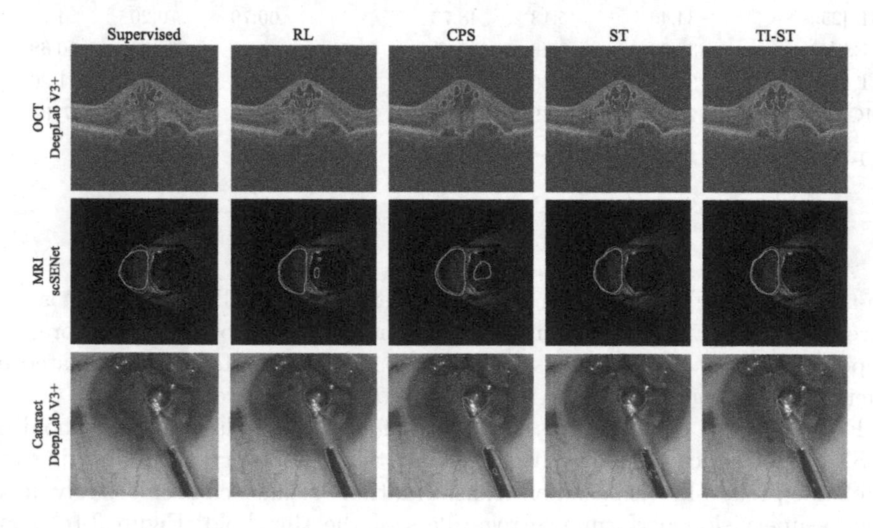

Fig. 5. Qualitative comparisons between the performance of TI-ST and four existing methods.

sion boundaries toward the target domain dataset by avoiding training with transformation variant pseudo-labels. We validates the stability of TI-ST vs. ST with different labeling thresholds (0.80 and 0.85) over four training folds in Fig. 4, where TI-ST achieves a higher average improvement relative to supervised learning for different tasks and network architectures. This analysis also shows that the performance of ST is sensitive to the pseudo-labeling threshold and generally degrades by reducing the threshold due to resulting in wrong pseudo labels. However, TI-ST can effectively ignore false predictions in lower thresholds and take advantage of a higher amount of correct pseudo labels. This superior performance is depicted qualitatively in Fig. 5.

5 Conclusion

We proposed a novel self-training framework with a self-assessment strategy for pseudo-label reliability, namely "Transformation-Invariant Self-Training" (TI-ST). This method uses transformation-invariant highly-confident predictions in the target dataset by considering an ensemble of high-confidence predictions from transformed versions of identical inputs. We experimentally show the effectiveness of our approach against numerous existing methods across three different source-to-target segmentation tasks, and when using different model architectures. Beyond this, we show that our approach is resilient to changes in the methods hyperparameter, making it well-suited for different applications.

References

1. Bogunović, H., et al.: RETOUCH: the retinal oct fluid detection and segmentation benchmark and challenge. IEEE Trans. Med. Imaging **38**(8), 1858–1874 (2019)
2. Chaitanya, K., Erdil, E., Karani, N., Konukoglu, E.: Contrastive learning of global and local features for medical image segmentation with limited annotations. In: Larochelle, H., Ranzato, M., Hadsell, R., Balcan, M., Lin, H. (eds.) Advances in Neural Information Processing Systems, vol. 33, pp. 12546–12558. Curran Associates, Inc. (2020)
3. Chen, L.C., Zhu, Y., Papandreou, G., Schroff, F., Adam, H.: Encoder-decoder with Atrous separable convolution for semantic image segmentation. In: Proceedings of the European conference on computer vision (ECCV), pp. 801–818 (2018)
4. Chen, X., Yuan, Y., Zeng, G., Wang, J.: Semi-supervised semantic segmentation with cross pseudo supervision. In: Proceedings of the IEEE/CVF Conference on Computer Vision and Pattern Recognition, pp. 2613–2622 (2021)
5. Deng, J., Dong, W., Socher, R., Li, L.J., Li, K., Fei-Fei, L.: ImageNet: a large-scale hierarchical image database. In: 2009 IEEE Conference on Computer Vision and Pattern Recognition, pp. 248–255. IEEE (2009)
6. Ghamsarian, N.: Enabling relevance-based exploration of cataract videos. In: Proceedings of the 2020 International Conference on Multimedia Retrieval, ICMR '20, pp. 378–382 (2020). https://doi.org/10.1145/3372278.3391937
7. Ghamsarian, N., Taschwer, M., Putzgruber-Adamitsch, D., Sarny, S., El-Shabrawi, Y., Schoeffmann, K.: LensID: a CNN-RNN-based framework towards lens irregularity detection in cataract surgery videos. In: de Bruijne, M., et al. (eds.) MICCAI 2021. LNCS, vol. 12908, pp. 76–86. Springer, Cham (2021). https://doi.org/10.1007/978-3-030-87237-3_8
8. Ghamsarian, N., Taschwer, M., Putzgruber-Adamitsch, D., Sarny, S., El-Shabrawi, Y., Schöffmann, K.: ReCal-Net: joint region-channel-wise calibrated network for semantic segmentation in cataract surgery videos. In: Mantoro, T., Lee, M., Ayu, M.A., Wong, K.W., Hidayanto, A.N. (eds.) ICONIP 2021. LNCS, vol. 13110, pp. 391–402. Springer, Cham (2021). https://doi.org/10.1007/978-3-030-92238-2_33
9. Ghamsarian, N., Taschwer, M., Putzgruber-Adamitsch, D., Sarny, S., Schoeffmann, K.: Relevance detection in cataract surgery videos by Spatio- temporal action localization. In: 2020 25th International Conference on Pattern Recognition (ICPR), pp. 10720–10727 (2021)

10. Ghamsarian, N., Taschwer, M., Sznitman, R., Schoeffmann, K.: DeepPyramid: enabling pyramid view and deformable pyramid reception for semantic segmentation in cataract surgery videos. In: Wang, L., Dou, Q., Fletcher, P.T., Speidel, S., Li, S. (eds.) MICCAI 2022. Lecture Notes in Computer Science, vol. 13435, pp. 276–286. Springer, Cham (2022). https://doi.org/10.1007/978-3-031-16443-9_27
11. Gomariz, A., et al.: Unsupervised domain adaptation with contrastive learning for OCT segmentation. In: Wang, L., Dou, Q., Fletcher, P.T., Speidel, S., Li, S. (eds.) MICCAI 2022. Lecture Notes in Computer Science, vol. 13438, pp. 351–361. Springer Nature Switzerland, Cham (2022). https://doi.org/10.1007/978-3-031-16452-1_34
12. Grammatikopoulou, M., et al.: CaDIS: cataract dataset for surgical RGB-image segmentation. Med. Image Anal. **71**, 102053 (2021)
13. Guo, R., et al.: Using domain knowledge for robust and generalizable deep learning-based CT-free PET attenuation and scatter correction. Nat. Commun. **13**(1), 5882 (2022)
14. He, K., Zhang, X., Ren, S., Sun, J.: Deep residual learning for image recognition. In: Proceedings of the IEEE Conference on Computer Vision and Pattern Recognition, pp. 770–778 (2016)
15. Laine, S., Aila, T.: Temporal ensembling for semi-supervised learning. CoRR abs/1610.02242 (2016). http://arxiv.org/abs/1610.02242
16. Li, C., Zhou, Y., Shi, T., Wu, Y., Yang, M., Li, Z.: Unsupervised domain adaptation for the histopathological cell segmentation through self-ensembling. In: Atzori, M., et al. (eds.) Proceedings of the MICCAI Workshop on Computational Pathology. Proceedings of Machine Learning Research, vol. 156, pp. 151–158. PMLR (2021)
17. Li, X., Yu, L., Chen, H., Fu, C.W., Xing, L., Heng, P.A.: Transformation-consistent self-ensembling model for semisupervised medical image segmentation. IEEE Trans. Neural Netw. Learn. Syst. **32**(2), 523–534 (2021)
18. Liu, Q., Dou, Q., Yu, L., Heng, P.A.: MS-Net: multi-site network for improving prostate segmentation with heterogeneous MRI data. IEEE Trans. Med. Imaging **39**, 2713–2724 (2020)
19. Perone, C.S., Ballester, P., Barros, R.C., Cohen-Adad, J.: Unsupervised domain adaptation for medical imaging segmentation with self-ensembling. Neuroimage **194**, 1–11 (2019)
20. Roy, A.G., Navab, N., Wachinger, C.: Recalibrating fully convolutional networks with spatial and channel "squeeze and excitation" blocks. IEEE Trans. Med. Imaging **38**(2), 540–549 (2019)
21. Schoeffmann, K., Taschwer, M., Sarny, S., Münzer, B., Primus, M.J., Putzgruber, D.: Cataract-101: video dataset of 101 cataract surgeries. In: Proceedings of the 9th ACM Multimedia Systems Conference, pp. 421–425 (2018)
22. Varsavsky, T., Orbes-Arteaga, M., Sudre, C.H., Graham, M.S., Nachev, P., Cardoso, M.J.: Test-time unsupervised domain adaptation. In: Martel, A.L., et al. (eds.) MICCAI 2020. LNCS, vol. 12261, pp. 428–436. Springer, Cham (2020). https://doi.org/10.1007/978-3-030-59710-8_42
23. Wang, Y., Xiao, B., Bi, X., Li, W., Gao, X.: MCF: mutual correction framework for semi-supervised medical image segmentation. In: Proceedings of the IEEE/CVF Conference on Computer Vision and Pattern Recognition, pp. 15651–15660 (2023)
24. Yang, L., Zhuo, W., Qi, L., Shi, Y., Gao, Y.: ST++: make self-training work better for semi-supervised semantic segmentation. In: Proceedings of the IEEE/CVF Conference on Computer Vision and Pattern Recognition, pp. 4268–4277 (2022)

25. Zeng, X., et al.: Reciprocal learning for semi-supervised segmentation. In: de Bruijne, M., et al. (eds.) MICCAI 2021. LNCS, vol. 12902, pp. 352–361. Springer, Cham (2021). https://doi.org/10.1007/978-3-030-87196-3_33
26. Zou, Y., Yu, Z., Vijaya Kumar, B.V.K., Wang, J.: Unsupervised domain adaptation for semantic segmentation via class-balanced self-training. In: Ferrari, V., Hebert, M., Sminchisescu, C., Weiss, Y. (eds.) ECCV 2018. LNCS, vol. 11207, pp. 297–313. Springer, Cham (2018). https://doi.org/10.1007/978-3-030-01219-9_18

Multi-IMU with Online Self-consistency for Freehand 3D Ultrasound Reconstruction

Mingyuan Luo[1,2,3], Xin Yang[1,2,3], Zhongnuo Yan[1,2,3], Junyu Li[1,2,3], Yuanji Zhang[1,2,3], Jiongquan Chen[1,2,3], Xindi Hu[4], Jikuan Qian[4], Jun Cheng[1,2,3], and Dong Ni[1,2,3](✉)

[1] National-Regional Key Technology Engineering Laboratory for Medical Ultrasound, School of Biomedical Engineering, Health Science Center, Shenzhen University, Shenzhen, China
nidong@szu.edu.cn

[2] Medical Ultrasound Image Computing (MUSIC) Lab, Shenzhen University, Shenzhen, China

[3] Marshall Laboratory of Biomedical Engineering, Shenzhen University, Shenzhen, China

[4] Shenzhen RayShape Medical Technology Inc., Shenzhen, China

Abstract. Ultrasound (US) imaging is a popular tool in clinical diagnosis, offering safety, repeatability, and real-time capabilities. Freehand 3D US is a technique that provides a deeper understanding of scanned regions without increasing complexity. However, estimating elevation displacement and accumulation error remains challenging, making it difficult to infer the relative position using images alone. The addition of external lightweight sensors has been proposed to enhance reconstruction performance without adding complexity, which has been shown to be beneficial. We propose a novel online self-consistency network (OSCNet) using multiple inertial measurement units (IMUs) to improve reconstruction performance. OSCNet utilizes a modal-level self-supervised strategy to fuse multiple IMU information and reduce differences between reconstruction results obtained from each IMU data. Additionally, a sequence-level self-consistency strategy is proposed to improve the hierarchical consistency of prediction results among the scanning sequence and its sub-sequences. Experiments on large-scale arm and carotid datasets with multiple scanning tactics demonstrate that our OSCNet outperforms previous methods, achieving state-of-the-art reconstruction performance.

Keywords: Multiple IMU · Online Learning · Freehand 3D Ultrasound

1 Introduction

Ultrasound (US) imaging has been widely used in clinical diagnosis due to its advantages of safety, repeatability, and real-time imaging. Compared with 2D

M. Luo and X. Yang—contribute equally to this work.

H. Greenspan et al. (Eds.): MICCAI 2023, LNCS 14220, pp. 342–351, 2023.
https://doi.org/10.1007/978-3-031-43907-0_33

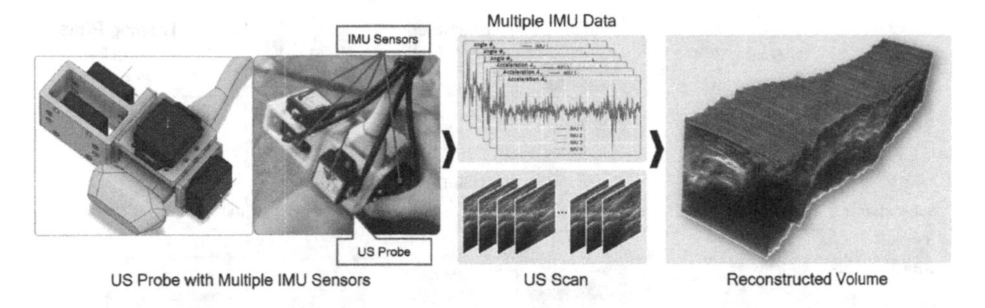

Fig. 1. Pipeline of freehand 3D US reconstruction with multiple lightweight inertial measurement unit (IMU) sensors.

US, 3D US can provide more comprehensive spatial information. Freehand 3D US can enhance the understanding of physicians about the scanned region of interest without increasing the complexity of scanning [12–14]. However, the difficulty in estimating elevation displacement and accumulation error makes it very challenging to infer the relative position only from images. In this regard, it is expected to improve the reconstruction performance with the help of external lightweight sensors, which will not significantly increase the scanning complexity.

Sensorless freehand 3D US reconstructs the volume by calculating the relative transformation of a series of US images. Previous studies were mainly based on speckle decorrelation [1,17], which estimates out-of-plane motion through the correlation of speckle patterns in two successive frames. With the development of deep learning technology, recent studies were mainly based on convolutional neural network (CNN). Prevost et al. [15] proposed an end-to-end method based on CNN to estimate the relative motion of US images. Guo et al. [4] proposed a deep contextual learning network (DCL-Net) based on 3D CNN to estimate the trajectory of US probe, and in a more recent study [3], they proposed a deep contextual-contrastive network (DC^2-Net), which introduced a contrastive learning strategy to improve the reconstruction performance by leveraging the label efficiently. Luo et al. [10,11] proposed an online learning framework (OLF) that improves reconstruction performance by online learning and shape priors.

Due to the low cost, small size, and low power consumption of micro-electro-mechanical-systems (MEMS), the sensor called inertial measurement unit (IMU) has been widely used in navigation systems. Prevost et al. [14] incorporated IMU orientation into neural network to improve reconstruction performance. Luo et al. [12] proposed a deep motion network (MoNet) to mine the valuable information of low signal-to-noise acceleration, and an online self-supervised strategy was designed to further improve reconstruction performance. However, the main disadvantage of IMU is that its measurement noise can not be completely eliminated by calibration. Existing studies have shown that combining multiple IMUs may help reduce noise and improve accuracy [2,16].

In this study, we propose a multi-IMU-based online self-consistency network (OSCNet) for freehand 3D US reconstruction. Our contribution is two-fold. First, we equip multiple IMUs (see Fig. 1) to reduce the influence of noise in individual IMU data. We propose a modal-level self-supervised strategy (MSS) to fuse the

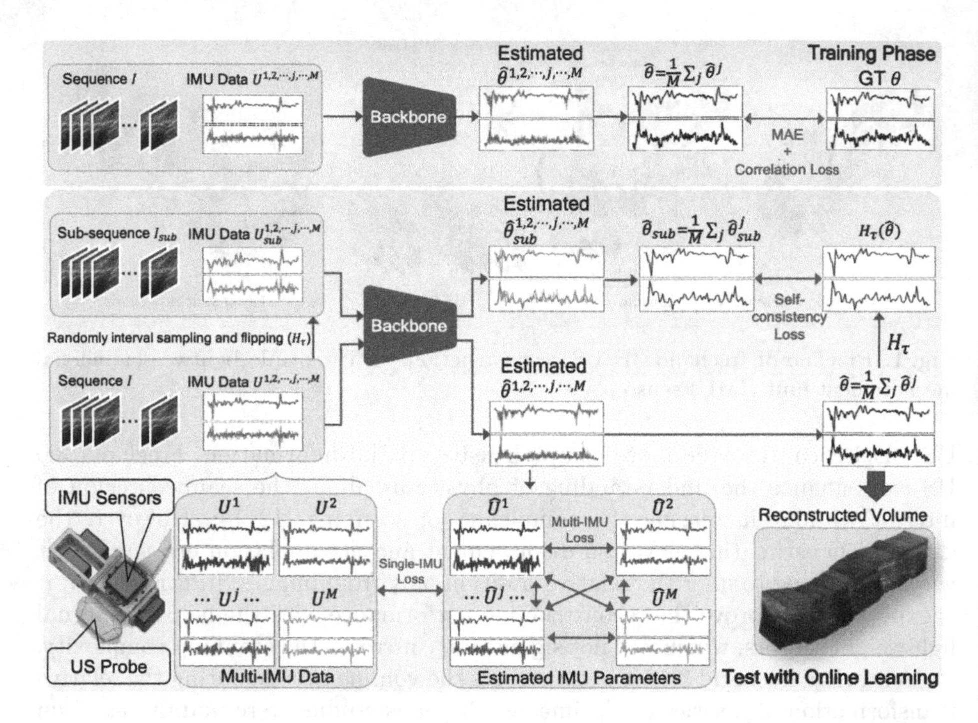

Fig. 2. Overview of our proposed multi-IMU online self-consistency network (OSCNet). IMU data diagrams ($U/\hat{U}$) show angle curves ($\Phi/\hat{\Phi}$, top) and acceleration curves ($A/\hat{A}$, bottom). Relative transformation parameter diagrams ($\theta/\hat{\theta}$) show angle curves ($\phi/\hat{\phi}$, top) and translation curves ($t/\hat{t}$, bottom).

information from multiple IMUs. MSS improves reconstruction performance by reducing the differences between reconstruction results obtained from each IMU data. Second, to reduce the estimation instability caused by scanning differences such as frame rates, we propose a sequence-level self-consistency strategy (SCS), which improves the hierarchical consistency of prediction results among the scanning sequence and its sub-sequences based on a consistent context. Experimental results show that the proposed OSCNet can effectively fuse the information of multiple IMUs and achieve state-of-the-art reconstruction performance.

2 Methodology

Figure 2 illustrates the proposed OSCNet, which consists of two essential components: backbone and online learning. We construct a backbone using the temporal and multi-branch structure from [12]. The main branch in the backbone consists of ResNet [5] for feature extraction and LSTM [6] for processing temporal information. It aids future estimation by leveraging temporal contextual information. Additionally, there is an independent motion branch in the backbone that fuses IMU information from a motion perspective with US images. For more details, please refer to [12].

In the training phase, we input an N-length scanning sequence $I = \{I_i | i = 1, 2, \cdots, N\}$ and corresponding multiple IMU data $U = \{U_i | i = 1, 2, \cdots, N-1\}$ into the backbone to estimate the 3D relative transformation parameters $\theta = \{\theta_i | i = 1, 2, \cdots, N-1\}$, where θ_i includes 3-axis translations $t_i = (t_x, t_y, t_z)_i$ and rotation angles $\phi_i = (\phi_x, \phi_y, \phi_z)_i$ between image I_i and I_{i+1}. The multiple IMU data consists of M independent IMU data $U_i = \{U_i^j | j = 1, 2, \cdots, M\}$, where U_i^j consists of 3-axis angles $\Phi_i^j = (\Phi_x, \Phi_y, \Phi_z)_i^j$ and accelerations $A_i^j = (A_x, A_y, A_z)_i^j$. The pre-processing process for Φ_i and A_i is consistent with [12]. Compared to traditional offline inference strategies, online learning can leverage valuable information from unlabeled data to improve the model's performance [10,12]. In the testing phase, we propose two online self-supervised strategies based on both the multiple IMU data (modal-level) and the scanning sequence itself (sequence-level) to improve the performance of the backbone's estimations.

2.1 Modal-Level Self-supervised Strategy

Multiple IMUs mounted in different directions provide diverse measurement constraints for the model's estimation, as shown in Fig. 1. This makes it possible to reduce the influence of noise in individual IMU data while adaptively optimizing for estimation. We construct an online modal-level self-supervised strategy (MSS) that leverages the consistency between the backbone's estimation and multiple IMU data to improve the reconstruction performance.

As shown in the top of Fig. 2, during the training phase, we repeatedly input the US images and M different IMU data into the backbone to obtain M estimated parameters. We use the average of the M estimated parameters $\hat{\theta} = \frac{1}{M} \sum_{j=1}^{M} \hat{\theta}^j$ as the final output of the backbone. We then calculate training loss between $\hat{\theta}$ and ground truth θ using mean absolute error (MAE) and Pearson correlation loss [4]:

$$L = \|\hat{\theta} - \theta\|_1 + (1 - \frac{\mathbf{Cov}(\hat{\theta}, \theta)}{\sigma(\hat{\theta})\sigma(\theta)}), \tag{1}$$

where $\mathbf{Cov}$, σ and $\|\cdot\|_1$ denote the covariance, the standard deviation, and L1 normalization, respectively.

As shown in the bottom of Fig. 2, during the testing phase, we use each IMU data U^j $(j = 1, 2, \cdots, M)$ as a weak label to constrain the corresponding estimated parameters $\hat{\theta}^j$. We calculate the estimated acceleration $\hat{A}^j$ at the center point of each image using the estimated $\hat{\theta}^j$. To reduce the influence of acceleration noise, we scale the $\hat{A}^j$ to match the mean-zeroed IMU acceleration.

$$\hat{A}_i^j = ((\hat{t}_{i-1}^j)^{-1} + \hat{t}_i^j) - \frac{1}{N-2} \sum_i ((\hat{t}_{i-1}^j)^{-1} + \hat{t}_i^j), \quad i = 2, 3, \cdots, N-1, \tag{2}$$

where $(\hat{t}_{i-1}^j)^{-1}$ represents the translations in the inversion of $\hat{\theta}_{i-1}^j$. Similar to [12], we use Pearson correlation loss to measure the difference between the estimated and IMU acceleration, while the angle is measured using MAE. Therefore, the single-IMU consistency constraint between the estimated parameters and corresponding IMU data can be expressed as:

$$L_{single-IMU} = \sum_{j=1}^{M}(1 - \frac{\mathbf{Cov}(\hat{A}^j, A^j)}{\sigma(\hat{A}^j)\sigma(A^j)}) + \|\hat{\phi}^j - \Phi^j\|_1. \quad (3)$$

In addition, the consistency among multiple IMU data itself also provides the possibility to improve the reconstruction performance. It constrains the backbone to obtain similar estimated parameters for different IMU data inputs from the same scan. Specifically, we construct multi-IMU consistency constraints as:

$$L_{multi-IMU} = \sum_{j,k=1,2,\cdots,M,j<k}^{M} (1 - \frac{\mathbf{Cov}(\hat{A}^j, \hat{A}^k)}{\sigma(\hat{A}^j)\sigma(\hat{A}^k)}) + \|\hat{\phi}^j - \hat{\phi}^k\|_1. \quad (4)$$

2.2 Sequence-Level Self-consistency Strategy

Consistent context should lead to consistent parameter estimation, which constrains the model at the sequence level, reducing the estimation instability caused by scanning differences such as frame rates. Inspired by contrastive learning [7], we construct an online sequence-level self-consistency strategy (SCS). SCS randomly generates sub-sequences with consistent context for each scan. The hierarchical consistency constraint among the generated sub-sequences and the original sequence improves the reconstruction performance of the backbone. Specifically, as shown in Fig. 2, we randomly interval sample and flip each scanning sequence I and its IMU data U to generate a sub-sequence I_{sub} (U_{sub}) with consistent context. In the testing phase, we obtain the estimated parameters $\hat{\theta}_{sub}$ of I_{sub} (U_{sub}) using the trained backbone. Then compare $\hat{\theta}_{sub}$ with the original estimated parameters $\hat{\theta}$ after the same interval sampling and flipping to construct the self-consistency constraint:

$$\begin{aligned} L_{self-consistency} &= \|\hat{\theta}_{sub} - H_\tau(\hat{\theta})\|_1 \\ &= \|\frac{1}{M}\sum_{j=1}^{M} B(H_\tau(I), H_\tau(U^j)) - H_\tau(\frac{1}{M}\sum_{j=1}^{M} B(I, U^j))\|_1, \end{aligned} \quad (5)$$

where H_τ converts the parameters, sequences, or IMU data under interval sampling and flipping operation τ. B denotes the backbone.

3 Experiments

Materials and Implementation. The equipment we used to collect data includes a portable US machine, four IMU sensors (WT901C-232, WitMotion) and an electromagnetic (EM) positioning system. The US images were acquired with a linear probe at 10 MHz, and the depth was set at 4 cm. As shown in Fig. 1,

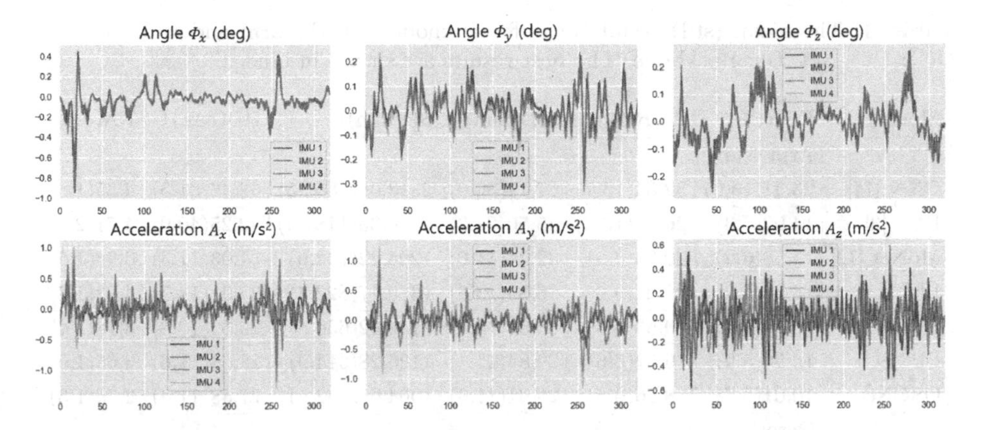

Fig. 3. Comparison of multiple IMU data. The abscissa of each subfigure indicates the image index.

we bound four IMU sensors to the probe in different orientations (three for 3-axis directions and one for redundancy) using a 3D-printed bracket, which can compensate for errors and reduce measurement singularities [9]. The resolutions of the IMU acceleration and angle are 5×10^{-4} g/LSB and 0.5°, respectively. We used the EM positioning system to trace the scan route accurately. The direction and angle resolutions of the EM positioning system are 1.4 mm and 0.5°, respectively. We calibrated the multiple IMU sensors and the EM positioning system using the Levenberg-Marquardt algorithm [8] to ensure accurate measurements and minimise system errors. As shown in Fig. 3, the calibrated IMU data exhibits a generally consistent overall trend, although differences still exist.

We constructed two datasets, including arm and carotid, from 36 volunteers. The arm dataset contains 288 scans, with scanning tactics including linear, curved, loop, and sector scan. The carotid dataset contains 216 scans, with scanning tactics including linear, loop, and sector scan. The average lengths of the arm and carotid scans are 323.96 mm and 203.25 mm, respectively. The size of scanned images is 248×260 pixels, and the image spacing is 0.15×0.15 mm^2. The collection and use of the data are approved by the local IRB.

The arm and carotid datasets were randomly divided into 200/40/48 and 150/30/36 scans based on volunteer level to construct training/validation/test set. To prevent overfitting and enhance the model's robustness, we performed random augmentations on each scan, including sub-sequence intercepting, interval sampling, and sequence inversion. We randomly augmented each training scan to 20 sequences and each test scan to 10 sequences to simulate complex real-world situations. We used the Adam optimizer to optimize the OSCNet. During the training phase, the epochs and batch size are set to 200 and 1, respectively. To avoid overfitting, we set the initial learning rate to 2×10^{-4} and used a learning rate decay strategy that halves the learning rate every 30 epochs. During the online learning phase, the iteration epoch and learning rate

Table 1. The mean (std) results of different models on the arm and carotid scans. DC^2: DC^2-Net, Bk: Backbone. The best results are shown in blue.

Models	FDR(%)↓	ADR(%)↓	MD(mm)↓	SD(mm)↓	HD(mm)↓	EA(deg)↓
	Arm scans					
CNN [14]	23.31(13.0)	32.54(13.7)	67.79(30.0)	2313.08(1852.5)	62.48(31.5)	4.35(1.8)
DC^2 [3]	14.02(7.3)	26.15(10.2)	45.50(21.3)	1560.30(1181.4)	42.25(40.9)	4.71(2.3)
MoNet [12]	11.58(6.2)	20.35(6.8)	32.21(11.0)	1205.42(742.6)	31.03(11.3)	3.98(1.5)
Bk	13.32(8.2)	23.21(9.6)	36.39(13.6)	1339.17(822.4)	34.91(13.7)	4.32(1.7)
Bk+MSS	10.78(5.6)	19.53(6.3)	30.52(10.5)	1142.42(636.8)	29.32(10.8)	3.18(1.4)
Bk+SCS	10.56(5.9)	19.57(6.6)	29.84(11.1)	1126.28(614.9)	28.64(11.6)	3.65(1.9)
OSCNet	10.01(5.7)	18.86(6.5)	28.61(11.0)	1064.06(582.5)	27.38(11.4)	2.76(1.3)
	Carotid scans					
CNN [14]	25.85(15.0)	33.95(16.8)	49.64(25.5)	1944.72(1485.1)	39.30(18.7)	3.73(2.3)
DC^2 [3]	13.54(7.1)	21.68(9.2)	26.47(9.6)	1025.06(622.8)	24.49(10.3)	4.30(3.0)
MoNet [12]	11.80(5.7)	20.42(8.8)	23.48(8.7)	894.39(381.0)	20.78(9.2)	3.67(2.1)
Bk	12.85(6.5)	21.78(10.5)	24.65(9.1)	965.12(466.6)	21.81(9.5)	3.83(2.0)
Bk+MSS	11.31(5.4)	20.04(8.8)	22.72(8.1)	850.68(321.7)	20.02(8.5)	3.16(1.8)
Bk+SCS	11.30(5.4)	20.16(8.6)	23.01(8.4)	863.48(320.9)	20.56(8.7)	3.36(1.8)
OSCNet	10.90(5.3)	19.61(8.5)	21.81(7.2)	804.27(282.8)	19.30(7.6)	2.60(1.6)

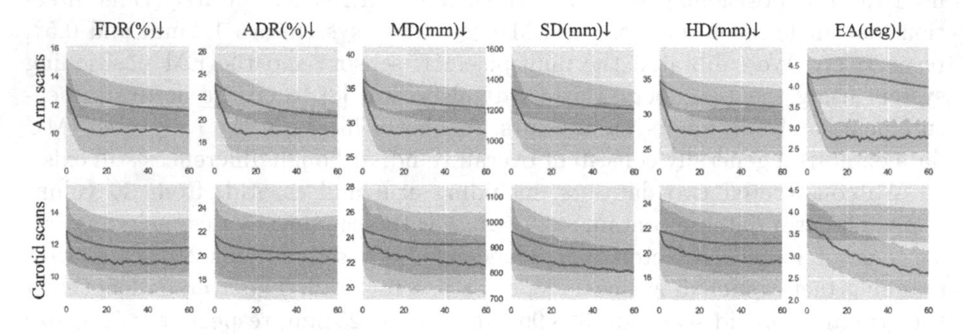

Fig. 4. Metric decline curves (with 95% confidence interval). Red: MoNet, Blue: OSCNet. The abscissa and ordinate of each subfigure represent the number of iterations and the value of metrics, respectively. (Color figure online)

are set to 60 and 2×10^{-6}, respectively. All code was implemented in PyTorch and executed on an RTX 3090 GPU.

Quantitative and Qualitative Analysis. To demonstrate the effectiveness of our OSCNet, we compared it with three state-of-the-art methods, including CNN [14], DC^2-Net [3] and MoNet [12]. All comparison methods were trained to convergence using the experimental settings given in the corresponding papers. We adopt six metrics from [12] to evaluate reconstruction performance: final drift rate (FDR), average drift rate (ADR), maximum drift (MD), sum of drift (SD), symmetric Hausdorff distance (HD), and mean error of angle (EA). In addition, ablation experiments are conducted to validate the effectiveness of MSS and SCS as proposed in our OSCNet.

Table 1 shows that our OSCNet significantly outperforms CNN, DC^2-Net, MoNet, and our Backbone in all metrics for both arm and carotid scans (t-test, $p < 0.05$), except for MoNet's ADR on the carotid scans (t-test, $p = 0.10$). Notably, sensor-based methods (MoNet and OSCNet) have exhibited improvements in all metrics compared to sensorless methods (DC^2-Net and CNN). The multi-IMU-based OSCNet outperforms the single-IMU-based MoNet, verifying the effectiveness of multiple IMU integration. Moreover, the ablation experiments further demonstrate that both multiple IMU integration (MSS) and self-consistency constraint (SCS) greatly improve the reconstruction performance.

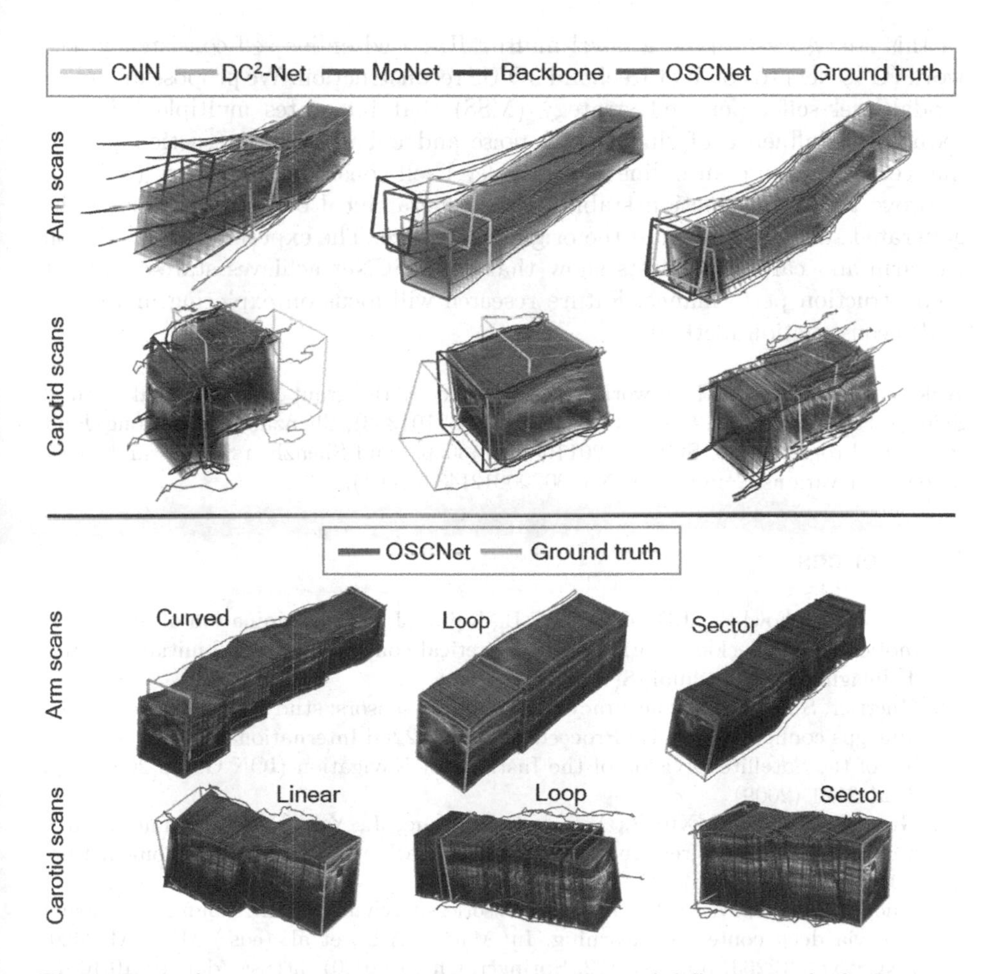

Fig. 5. Typical reconstruction cases. Top: comparison of different methods, Bottom: comparison of different scanning tactics. At the bottom, all of the estimated image positions of OSCNet are marked with red boxes to visualize the scanning tactics. (Color figure online)

In addition, Fig. 4 displays the metric decline curves during the online learning phase of MoNet and OSCNet on the arm and carotid datasets. All metric curves exhibit a decreasing trend followed by stabilization. We note that our OSCNet has achieved further improvements compared to MoNet, with 13.56% /7.32%/30.65% and 7.62%/4.00%/29.16% improvement in FDR/ADR/EA on the arm and carotid datasets, respectively. Figure 5 presents several typical reconstruction results of all methods. It can be observed that our OSCNet outperforms other methods in reconstruction results and closely approximates the ground truth across all scanning tactics.

4 Conclusion

In this study, we propose a novel multi-IMU-based online self-consistency network (OSCNet) to conduct freehand 3D US reconstruction. We propose an online modal-level self-supervised strategy (MSS) that integrates multiple IMUs to reduce the influence of single IMU noise and enhance reconstruction performance. We propose an online sequence-level self-consistency strategy (SCS) to improve the reconstruction stability using hierarchical consistency among the generated sub-sequences and the original sequence. The experimental results on the arm and carotid datasets show that our OSCNet achieves state-of-the-art reconstruction performance. Future research will focus on exploring more general reconstruction methods.

Acknowledgements. This work was supported by the grant from National Natural Science Foundation of China (Nos. 62171290, 62101343), Shenzhen-Hong Kong Joint Research Program (No. SGDX20201103095613036), and Shenzhen Science and Technology Innovations Committee (No. 20200812143441001).

References

1. Chen, J.F., Fowlkes, J.B., Carson, P.L., Rubin, J.M.: Determination of scan-plane motion using speckle decorrelation: theoretical considerations and initial test. Int. J. Imaging Syst. Technol. **8**(1), 38–44 (1997)
2. Guerrier, S.: Improving accuracy with multiple sensors: study of redundant mems-imu/gps configurations. In: Proceedings of the 22nd International Technical Meeting of the Satellite Division of the Institute of Navigation (ION GNSS 2009), pp. 3114–3121 (2009)
3. Guo, H., Chao, H., Xu, S., Wood, B.J., Wang, J., Yan, P.: Ultrasound volume reconstruction from freehand scans without tracking. IEEE Trans. Biomed. Eng. **70**(3), 970–979 (2023)
4. Guo, H., Xu, S., Wood, B., Yan, P.: Sensorless freehand 3D ultrasound reconstruction via deep contextual learning. In: Martel, A.L., et al. (eds.) MICCAI 2020. LNCS, vol. 12263, pp. 463–472. Springer, Cham (2020). https://doi.org/10.1007/978-3-030-59716-0_44
5. He, K., Zhang, X., Ren, S., Sun, J.: Deep residual learning for image recognition. In: 2016 IEEE Conference on Computer Vision and Pattern Recognition (CVPR), pp. 770–778. IEEE (2016)

6. Hochreiter, S., Schmidhuber, J.: Long short-term memory. Neural Comput. **9**(8), 1735–1780 (1997)
7. Le-Khac, P.H., Healy, G., Smeaton, A.F.: Contrastive representation learning: a framework and review. IEEE Access **8**, 193907–193934 (2020)
8. Levenberg, K.: A method for the solution of certain non-linear problems in least squares. Q. Appl. Math. **2**(2), 164–168 (1944)
9. Liang, S., Dong, X., Guo, T., Zhao, F., Zhang, Y.: Peripheral-free calibration method for redundant IMUs based on array-based consumer-grade MEMS information fusion. Micromachines **13**(8), 1214 (2022)
10. Luo, M., et al.: Self context and shape prior for sensorless freehand 3D ultrasound reconstruction. In: de Bruijne, M., et al. (eds.) MICCAI 2021. LNCS, vol. 12906, pp. 201–210. Springer, Cham (2021). https://doi.org/10.1007/978-3-030-87231-1_20
11. Luo, M., et al.: RecON: online learning for sensorless freehand 3D ultrasound reconstruction. Med. Image Anal. **87**, 102810 (2023)
12. Luo, M., Yang, X., Wang, H., Du, L., Ni, D.: Deep motion network for freehand 3D ultrasound reconstruction. In: Wang, L., Dou, Q., Fletcher, P.T., Speidel, S., Li, S. (eds.) MICCAI 2022. Lecture Notes in Computer Science, vol. 13434. Springer, Cham (2022). https://doi.org/10.1007/978-3-031-16440-8_28
13. Mohamed, F., Siang, C.V.: A survey on 3D ultrasound reconstruction techniques. In: Aceves-Fernandez, M.A. (ed.) Artificial Intelligence, chap. 4. IntechOpen, Rijeka (2019)
14. Prevost, R., et al.: 3d freehand ultrasound without external tracking using deep learning. Med. Image Anal. **48**, 187–202 (2018)
15. Prevost, R., Salehi, M., Sprung, J., Ladikos, A., Bauer, R., Wein, W.: Deep learning for sensorless 3D freehand ultrasound imaging. In: Descoteaux, M., Maier-Hein, L., Franz, A., Jannin, P., Collins, D.L., Duchesne, S. (eds.) MICCAI 2017. LNCS, vol. 10434, pp. 628–636. Springer, Cham (2017). https://doi.org/10.1007/978-3-319-66185-8_71
16. Rasoulzadeh, R., Shahri, A.M.: Implementation of a low-cost multi-IMU hardware by using a homogenous multi-sensor fusion. In: 2016 4th International Conference on Control, Instrumentation, and Automation (ICCIA), pp. 451–456 (2016)
17. Tuthill, T.A., Krücker, J., Fowlkes, J.B., Carson, P.L.: Automated three-dimensional us frame positioning computed from elevational speckle decorrelation. Radiology **209**(2), 575–582 (1998)

Deblurring Masked Autoencoder Is Better Recipe for Ultrasound Image Recognition

Qingbo Kang[1,3], Jun Gao[1,4], Kang Li[1,3(✉)], and Qicheng Lao[2,3(✉)]

[1] West China Biomedical Big Data Center, West China Hospital, Sichuan University, Chengdu, China
likang@wchscu.cn
[2] School of Artificial Intelligence, Beijing University of Posts and Telecommunications, Beijing, China
qicheng.lao@bupt.edu.cn
[3] Shanghai Artificial Intelligence Laboratory, Shanghai, China
[4] College of Computer Science, Sichuan University, Chengdu, China

Abstract. Masked autoencoder (MAE) has attracted unprecedented attention and achieves remarkable performance in many vision tasks. It reconstructs random masked image patches (known as proxy task) during pretraining and learns meaningful semantic representations that can be transferred to downstream tasks. However, MAE has not been thoroughly explored in ultrasound imaging. In this work, we investigate the potential of MAE for ultrasound image recognition. Motivated by the unique property of ultrasound imaging in high noise-to-signal ratio, we propose a novel deblurring MAE approach that incorporates deblurring into the proxy task during pretraining. The addition of deblurring facilitates the pretraining to better recover the subtle details presented in the ultrasound images, thus improving the performance of the downstream classification task. Our experimental results demonstrate the effectiveness of our deblurring MAE, achieving state-of-the-art performance in ultrasound image classification. Overall, our work highlights the potential of MAE for ultrasound image recognition and presents a novel approach that incorporates deblurring to further improve its effectiveness.

Keywords: Image Deblurring · Masked Autoencoders · Self-Supervised Learning · Ultrasound Recognition

1 Introduction

Recently, as representative of generative self-supervised learning (SSL) methods, masked autoencoder (MAE) [8] has achieved great success in many vision

Code will be available at: https://github.com/MembrAI/DeblurringMIM.

Supplementary Information The online version contains supplementary material available at https://doi.org/10.1007/978-3-031-43907-0_34.

H. Greenspan et al. (Eds.): MICCAI 2023, LNCS 14220, pp. 352–362, 2023.
https://doi.org/10.1007/978-3-031-43907-0_34

tasks [10,11,24]. In general, MAE belongs to the masked image modeling (MIM) paradigm [29], where some parts of the image are randomly masked, and the purpose of pretraining (i.e., proxy or pretext task) is to recover the missing pixels. After the pretraining, the learned image representation can be transferred to downstream tasks for improved performance. With the advent of MAE, many MAE variants have been proposed [7,22,25]. Tian *et al.* [22] investigate other image degradation methods during MAE pretraining and find that the optimal practice is enriching masking with spatial misalignment for nature images. Wu *et al.* [25] design a denoising MAE by introducing Gaussian noising into MAE pretraining, showing that their denoising MAE is robust to additive noises.

On the other hand, although numerous work has been proposed for applying MAE to medical imaging across different modalities including pathological images [1,14,19], X-rays [26,31], electrocardiogram [30], immunofluorescence images [15], MRI and CT [4,27,31]. However, the majority of them have not fully exploited the characteristics of medical images and instead, focus on vanilla applications [4,26,30,31]. This is especially problematic given the domain gap between medical and natural images, as well as the unique imaging properties associated with each medical imaging modality [16,18,20]. Furthermore, as an important and widely used medical imaging modality, ultrasound has not been extensively explored in the context of MAE-based approaches.

Based on the aforementioned analysis, in this paper, we propose a deblurring masked auto-encoder framework, which is specifically designed for ultrasound image recognition. The primary motivation for the deblurring comes from the unique imaging properties of ultrasound, e.g., high noise-to-signal ratio. Compared with nature images, the subtle details within ultrasound are particularly important for downstream analysis (e.g., microcalcifications is an important sign for malignant nodules, which is represented as tiny bright spots in ultrasound [17,21]). Moreover, the motivation also stems from the findings of our preliminary experiments, which suggest that denoising may not be appropriate for inherently noisy ultrasound images. Therefore, we introduce the opposite direction with a deblurring approach for ultrasound images. Specifically, we first apply blurring operations to the ultrasound images prior to the random masking during pretraining, enabling the model to learn how to de-blur and reconstruct the original image. It should be emphasized that denoising and deblurring are two opposite directions, i.e., denoising first adds noise to the clean image and learns to remove the noise, while deblurring blurs the noisy ultrasound image and learns to sharpen the image. The deblurring facilitates the pretraining in recovering the subtle details within the image, which is crucial for ultrasound image recognition. It should be emphasized that while blurring operation has been shown ineffective for natural images [22], ultrasound images are fundamentally different and may benefit from blurring operation.

Furthermore, to the best of our knowledge, this paper is the first attempt to apply the MAE approach to ultrasound image recognition. Our work also addresses some fundamental concerns that are of great interest to the medical imaging community with the example of ultrasound, such as the importance of in-domain data pretraining for MAE in ultrasound, as well as the finding that SSL pretraining is consistently better than the supervised pretraining as with nature images. To conclude, our contributions can be summarized as follows:

1. We propose a deblurring MAE framework that is specifically designed for ultrasound images by incorporating a deblurring task into MAE pretraining. This is motivated by the fact that ultrasound images have a high noise-to-signal ratio, and in contrast to denoising for natural images, we demonstrate that deblurring is a better recipe for ultrasound images.
2. We explore the effectiveness of various image blurring methods in our deblurring MAE and find that a simple Gaussian blurring performs the best, showing superior transferability compared with the vanilla MAE.
3. We conduct experiments on more than 10k ultrasound images for pretraining and 4,494 images for downstream thyroid nodule classification. The results demonstrate the effectiveness of the proposed deblurring MAE, achieving state-of-the-art classification performance for ultrasound images.

Note that, as a representative MIM approach, the MAE is adopted to validate our proposed deblurring pretraining in this work, our method can also be seamlessly integrated with other MIM-based approaches such as ConvMAE [7].

2 Method

2.1 Preliminary: MAE

The MAE pipeline consists of two primary stages: self-supervised pretraining and transferring for downstream tasks. During the self-supervised pretraining, the model is trained to reconstruct masked input image patches using an asymmetric encoder-decoder architecture. The encoder is typically a ViT [6], which compresses the input image into a latent representation, while the decoder is a lightweight Transformer that reconstructs the original image from the latent representation. The loss used during pretraining is the mean squared error (MSE) between the reconstructed and original images. In the transfer stage, the weights of the pre-trained ViT encoder are transferred and used as a feature extractor, to which task-specific heads are appended for learning various downstream tasks. Typically, there are two common practices in the transfer stage: 1) end-to-end fine-tuning which tunes the entire model, and 2) linear probing, which only tunes the task-specific head.

2.2 Our Proposed Deblurring MAE

Similar to MAE, our proposed deblurring MAE also contains pretraining and transfer learning for downstream tasks. We employ the same asymmetric encoder-decoder architecture as the original MAE.

Deblurring MAE Pretraining. For the pretraining, besides the original masked image modeling task in the MAE, we introduce one additional task, i.e., deblurring, into the pretraining thus making the pretraining as deblurring pretraining. As shown in Fig. 1, the deblurring is achieved by simply inserting an image blurring operation prior to random masking. The pipeline of our deblurring MAE pretraining is illustrated in Eq. 1:

$$x \xrightarrow{Blurring} x_b \xrightarrow{Masking} x_b^m \xrightarrow{\text{ViT Encoder}} h \xrightarrow{\text{Decoder}} \hat{x}. \tag{1}$$

Specifically, the original ultrasound image x is first blurred by a chosen image blurring operation *Blurring* to obtain x_b. After that, several patches in the blurred image x_b are randomly masked by the *Masking* operation with a predefined ratio to obtain x_b^m. Next, the masked blurred image x_b^m is passed as input to the ViT Encoder, which generates a latent representation h. Finally, the Decoder receives the representation h and outputs reconstructed image $\hat{x}$.

The image blurring operation *Blurring* is a commonly used technique for reducing the sharpness or details of an image, resulting in a smoother, less-detailed appearance. There exist many different methods for image blurring, with most of them involving the averaging of neighboring pixels in some way. In Fig. 1, we provide examples of two representative blurring methods: Gaussian blur and speckle reducing anisotropic diffusion (SRAD) [28].

Gaussian blur involves convolving an input image with a Gaussian kernel $G(\sigma)$, which is a two-dimensional Gauss function that represents a normal distribution with standard deviation of σ. Mathematically, Gaussian blur can be defined as follows:

$$x_b = Gaussian(x, \sigma) = x * G(\sigma) = x * \frac{1}{2\pi\sigma^2} e^{-(u^2+v^2)/2\sigma^2}, \tag{2}$$

where $*$ denotes the convolution operation, and (u, v) represents the coordinates in the kernel. The degree of blurring (i.e., blurriness) in the resulting image is determined by the standard deviation σ.

The SRAD is a nonlinear anisotropic diffusion technique for removing speckled noises, which has been extensively used in medical ultrasound images, due to its edge-sensitivity for speckled images and powerful preservation of useful information. The SRAD operation is implemented by repeating an anisotropic diffusion equation for N iterations. It can be formally given as:

$$x_b = SRAD(x, N, t) = x(i, j, 0) + \Delta t * \sum_{k=0}^{N-1} \mathrm{div}(\mathrm{c}(i, j, k) \nabla x(i, j, k)), \tag{3}$$

where x is the original image, N stands for the number of iterations, t means time. $x(i, j, k)$ and $\mathrm{c}(i, j, k)$ represent the image and diffusivity coefficient at iteration k, respectively. ∇x is the gradient of x and div is the divergence operator. The larger N or t leads to a blurrier resulting image.

The pixel-wise MSE between the reconstructed image $\hat{x}$ and the original image x is utilized as the loss function during pretraining: $\mathcal{L}_{MSE} = ||\hat{x} - x||_2$.

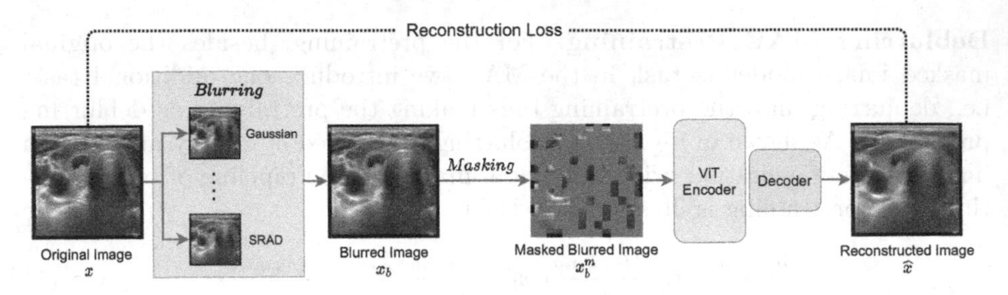

Fig. 1. Illustration of our proposed deblurring MAE pretraining.

It should be noted that a key difference from MAE is that we compute the loss across all patches, including the masked ones. This operation is necessary due to the fact that our blurring operation covers the entire image. Through the use of the proposed deblurring MAE pretraining, we aim to leverage both masked image modeling and deblurring in order to learn a robust and effective latent representation that could be successfully applied to a range of downstream tasks.

Deblurring MAE Transfer. After the deblurring MAE pretraining, only the pre-trained encoder is transferred to the downstream thyroid nodule classification task. One multi-layer perceptron (MLP) head is appended after the pre-trained encoder. The transfer learning pipeline is shown in Eq. 4:

$$x \xrightarrow{Blurring} x_b \xrightarrow{\text{ViT Encoder}} h \xrightarrow{\text{MLP}} \hat{y}, \tag{4}$$

It should be noted here that, in order to prevent data distribution shift between pretraining and transfer stages, the original image x also needs to be blurred before fed into the pre-trained encoder during transfer learning. The cross-entropy loss between ground-truth classification label y and predicted label $\hat{y}$ is used as the loss function: $\mathcal{L}_{CE} = -[y\log(\hat{y}) + (1-y)\log(1-\hat{y})]$.

3 Experiments and Results

3.1 Experimental Settings

Dataset. All thyroid ultrasound images used in our study for both pretraining and downstream classification were acquired at West China Hospital with ethical approval. We use a total of 10,675 images for pretraining and 4,493 images for the downstream classification. To avoid any potential data leakage, the images used in pretraining were not included in the test set for thyroid nodule classification. The downstream classification dataset contains 2,576 benign and 1,917 malignant cases. We randomly split the dataset into train/validation/test subsets with a 3:1:1 ratio. The classification ground-truth labels were obtained either from the fine-needle aspiration for malignant nodules or clinical diagnosis by senior radiologists for benign nodules.

Table 1. Performance comparison of different methods.

	Method	Architecture	Pretraining	ACC (%)	F1 (%)	AUROC (%)
Supervised	ResNet [9]	ResNet-101	-	86.06 ± 0.87	83.18 ± 1.15	91.96 ± 1.47
	ConvNeXt [13]	ConvNeXt-L	ImageNet	87.76 ± 0.66	85.47 ± 0.91	93.22 ± 1.10
	Swin Transformer [12]	Swin-L	ImageNet	87.43 ± 0.68	84.92 ± 0.82	92.83 ± 1.02
	Wang *et al.*[23]	-	-	87.44 ± 0.75	85.16 ± 0.87	93.11 ± 1.09
	Zhou *et al.*[32]	-	-	88.15 ± 0.67	86.09 ± 0.74	94.17 ± 1.21
	ViT [6]	ViT-B	-	80.60 ± 1.62	76.98 ± 2.05	83.89 ± 2.89
			ImageNet	86.38 ± 0.74	84.17 ± 0.98	92.69 ± 0.48
SSL	SimCLR [2]	ResNet-50	ImageNet	86.21 ± 0.96	83.81 ± 1.24	92.16 ± 1.08
	MoCo v3 [3]	ViT-B	ImageNet	86.96 ± 0.85	84.48 ± 1.12	92.77 ± 0.67
			Ultrasound	87.08 ± 0.78	84.55 ± 1.04	92.95 ± 0.59
	MAE [8]		ImageNet	87.25 ± 0.51	85.23 ± 0.57	93.71 ± 0.60
			Ultrasound	89.45 ± 0.53	87.54 ± 0.62	95.54 ± 0.46
	Denoising MAE		Ultrasound	80.38 ± 1.37	77.99 ± 1.75	84.38 ± 2.13
	Ours [SRAD]		Ultrasound	90.07 ± 0.47	88.13 ± 0.51	95.87 ± 0.45
	Ours [Gaussian]		Ultrasound	$\mathbf{90.19 \pm 0.47}$	$\mathbf{88.48 \pm 0.50}$	$\mathbf{96.08 \pm 0.41}$

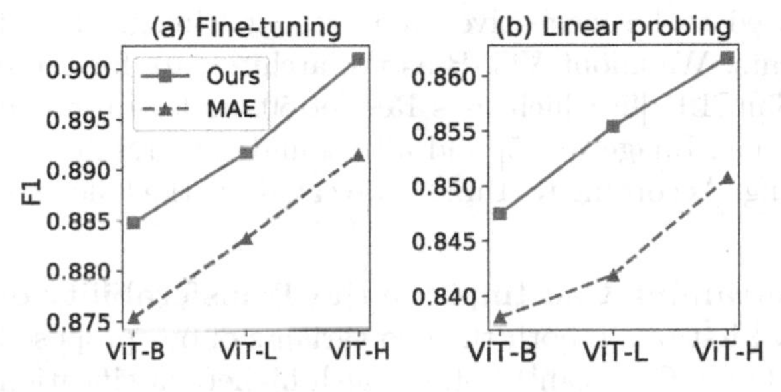

Fig. 2. Our deblurring MAE vs. vanilla MAE. Pretrained with the same ultrasound data.

Table 2. Ablation study.

Image Blurring		Blurriness	
Method	F1 (%)	σ	F1 (%)
Gaussian	**88.48**	0.8	88.03
SRAD	88.13	1.1	**88.48**
Mean	88.11	1.4	87.56
Median	87.63	1.7	85.89
Motion	78.33	2.0	84.74
Defocus	85.37	2.3	84.62
Baseline	87.54		

Implementation Details. We use a mask ratio of 75% during the pretraining. We set the batch size to 256 for both pretraining and end-to-end fine-tuning, and 1024 for linear probing. The epochs of pretraining is 12,000 due to our relatively small data. The full detailed experimental settings are presented in the appendix. We implement our approach based on PyTorch. The image size for both pretraining and transfer learning is 224 × 224. For classification, we choose the model that performs the best on the validation set as the final model to evaluate on the test set. Three widely used metrics accuracy (ACC), F1-score (F1), and the area under the receiver operating characteristic (AUROC) are utilized for classification performance evaluation.

3.2 Results and Comparisons

Our Deblurring MAE vs. Vanilla MAE. First of all, in order to evaluate the effectiveness of the proposed deblurring MAE for ultrasound images, we compare the transfer learning performance between our deblurring MAE and the vanilla MAE. Table 1 and Fig. 2 give the classification performance comparison of these two approaches. For our deblurring MAE, we use Gaussian blurring with σ equal to 1.1 as the blurring operation. In Fig. 2, we report the experimental results of three models: ViT-Base (ViT-B), ViT-Large (ViT-L) and ViT-Huge (ViT-H), and two transfer learning paradigms: end-to-end fine-tuning and linear probing. As shown in the figure, both the fine-tuning and linear probing performance of our proposed deblurring MAE is consistently better than that of the vanilla MAE, which indicates the effectiveness of deblurring for enhancing the transferability of learned representations during ultrasound pretraining.

Comparison with State-of-the-Art Approaches. Secondly, we also compare our approach with more approaches and the results are listed in Table 1. We implement two variants of our deblurring MAE which differ in blurring operation: the SRAD with N equals to 40 and t equals to 0.1, and the Gaussian blur with σ equals to 1.1. We compare with methods based on supervised learning or self-supervised learning. In addition, we still add the denoising MAE for comparison, although it has proved to be ineffective for ultrasound images based on our preliminary experiments. We adopt ViT-B as the architecture for these SSL-based methods except SimCLR [2] which uses ResNet-50, and we use two types of data for pretraining, i.e., ImageNet [5] and ultrasound. The results are based on end-to-end fine-tuning. According to Table 1, we can draw the following conclusions:

The Deblurring MAE Pretraining Can Improve the Transferability of Learned Representations. First of all, both the two variants of our proposed approach (Ours [SRAD] and Ours [Gaussian]) obtain much higher classification metrics compared with the MAE pretrained using ultrasound, which indicates the learned representation of our deblurring MAE is more effective than the vanilla MAE when transferred to downstream classification. In addition, Table 1 also shows that the performance of our proposed deblurring MAE with Gaussian blurring achieves state-of-the-art performance in terms of all metrics, surpassing all competing SSL or supervised-based approaches, which further demonstrates the superior performance of our proposed deblurring MAE. It is noteworthy that, as the opposite approach to our deblurring MAE, the denoising MAE obtains worse performance compared with the vanilla MAE, suggesting that adding noise to ultrasound images during MAE pretraining is unfavorable.

Ultrasound Pretraining is Better than ImageNet Pretraining, Better than Supervised Pretraining. Table 1 shows that the performance of MAE with ultrasound pretraining is better than the ImageNet pretraining, which underlines the importance of in-domain self-supervised pretraining in MAE. In contrast to MAE, our experiments show that the MoCo v3 [3] achieves only marginal improvement with ultrasound pretraining. Furthermore, the MAE ImageNet pretraining also performs much better than the ImageNet supervised pretraining. These two conclusions are consistent with other works [8, 26].

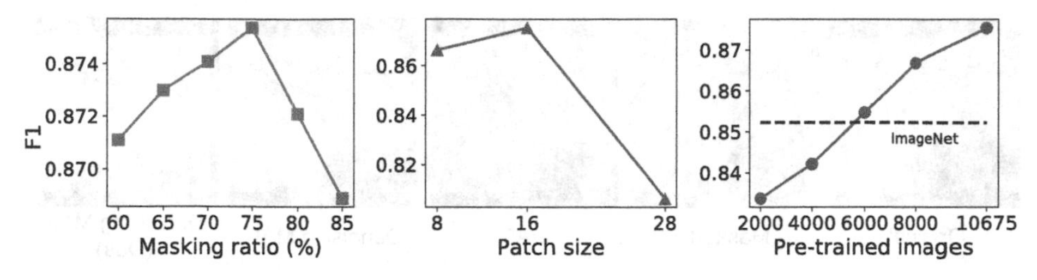

Fig. 3. Hyper-parameter choices for MAE pretraining.

Ablation Study. We design two sets of ablation studies, i.e., different image blurring methods, and the degree of blurring (blurriness) used in our deblurring MAE. We adopt the ViT-B as the architecture and end-to-end fine-tuning in transfer learning for the ablation experiments. Table 2 presents the performance results of the ablation study, where the 'Baseline' represents the vanilla MAE.

Firstly, besides the Gaussian and SRAD, we also try several other blurring methods that are commonly used in the fields including mean, median, motion and defocus blur. We set the kernel size to 5 in mean, median and motion blur, and the radius of defocusing is set to 5 for defocus blur. The performance results are presented on the left side of Table 2. From this table, we can observe that the Gaussian blurring achieves the best F1. And these six blurring methods are not all beneficial for pretraining, where some of them (motion, defocus) perform even worse than the baseline. Secondly, to investigate the effect of blurriness on the pretraining, we conduct ablation experiments on blurriness based on Gaussian blurring. The right side of Table 2 reports the performance results and we can see that the σ with 1.1 obtains the highest F1. In addition, as the σ, i.e., the blurriness continues to increase, the performance drops rapidly, which indicates that only a limited range of blurriness has a positive effect on the pretraining.

Hyper-parameter Choices for MAE Pretraining. We conduct experiments to explore hyper-parameter choices for MAE pretraining based on ViT-B, and the results are presented in Fig. 3. Our findings indicate that a masking ratio of 75% and a patch size of 16 achieve the best transfer performance, which is consistent with MAE for natural images [8]. Additionally, we observed that transfer performance improves with an increase in pre-trained images, surpassing ImageNet transfer only when a substantial amount of pre-trained images is used.

Visualization. The comparisons of reconstructed image examples among MAE, denoising MAE, and our proposed deblurring MAE are illustrated in Fig. 4. Although there is no strong evidence that reveals the relationship between reconstruction quality in pretraining and downstream task performance in MAE-based approaches, we can still obtain some insights from the reconstruction quality. As shown in Fig. 4, we can clearly observe that the reconstructed images of the

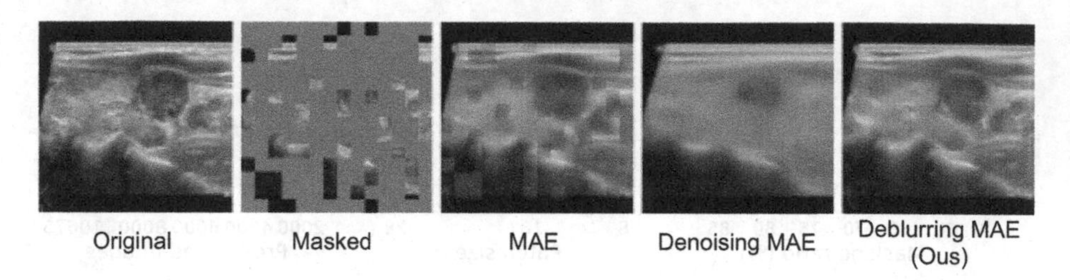

Fig. 4. Comparisons of reconstructed images.

denoising MAE are the smoothest and lost most details among all the three approaches, followed by the vanilla MAE, and our deblurring MAE achieves the best reconstruction quality with much finer details. The comparisons indicate that our deblurring MAE can capture critical details that are beneficial for downstream classification. More comparisons can be found in the appendix.

4 Conclusion and Future Work

In this paper, we propose a novel deblurring MAE by incorporating deblurring into the proxy task during MAE pretraining for ultrasound image recognition. The deblurring task is implemented by inserting image blurring operation prior to the random masking during pretraining. The integration of deblurring enables the pretraining pay more attention to recovering the intricate details presented in ultrasound images, which are critical for downstream image classification. We explore the effect of several different image blurring methods and find that Gaussian blurring achieves the best performance and only a limited range of blurriness has a beneficial effect for pretraining. Based on the optimal blurring method and blurriness, our deblurring MAE achieves state-of-the-art performance in the downstream classification of ultrasound images, indicating the effectiveness of incorporating deblurring into MAE pretraining for ultrasound image recognition. However, this work has some limitations. For example, only one downstream task: nodule classification is evaluated in this study. We plan to extend our approach to include more tasks such as segmentation in the future.

Acknowledgment. This work was supported by Natural Science Foundation of Sichuan Province under Grant NO. 2022NSFSC1855.

References

1. An, J., Bai, Y., Chen, H., Gao, Z., Litjens, G.: Masked autoencoders pre-training in multiple instance learning for whole slide image classification. In: Medical Imaging with Deep Learning (2022)
2. Chen, T., Kornblith, S., Norouzi, M., Hinton, G.: A simple framework for contrastive learning of visual representations. In: International Conference on Machine Learning, pp. 1597–1607. PMLR (2020)

3. Chen, X., Xie, S., He, K.: An empirical study of training self-supervised vision transformers. In: Proceedings of the IEEE/CVF International Conference on Computer Vision, pp. 9640–9649 (2021)
4. Chen, Z., Agarwal, D., Aggarwal, K., Safta, W., Balan, M.M., Brown, K.: Masked image modeling advances 3D medical image analysis. In: Proceedings of the IEEE/CVF Winter Conference on Applications of Computer Vision, pp. 1970–1980 (2023)
5. Deng, J., Dong, W., Socher, R., Li, L.J., Li, K., Fei-Fei, L.: ImageNet: a large-scale hierarchical image database. In: 2009 IEEE Conference on Computer Vision and Pattern Recognition, pp. 248–255. IEEE (2009)
6. Dosovitskiy, A., et al.: An image is worth 16 × 16 words: transformers for image recognition at scale. arXiv preprint arXiv:2010.11929 (2020)
7. Gao, P., Ma, T., Li, H., Dai, J., Qiao, Y.: ConvMAE: masked convolution meets masked autoencoders. arXiv preprint arXiv:2205.03892 (2022)
8. He, K., Chen, X., Xie, S., Li, Y., Dollár, P., Girshick, R.: Masked autoencoders are scalable vision learners. In: Proceedings of the IEEE/CVF Conference on Computer Vision and Pattern Recognition, pp. 16000–16009 (2022)
9. He, K., Zhang, X., Ren, S., Sun, J.: Deep residual learning for image recognition. In: Proceedings of the IEEE Conference on Computer Vision and Pattern Recognition, pp. 770–778 (2016)
10. Ke, L., Danelljan, M., Li, X., Tai, Y.W., Tang, C.K., Yu, F.: Mask transfiner for high-quality instance segmentation. In: Proceedings of the IEEE/CVF Conference on Computer Vision and Pattern Recognition, pp. 4412–4421 (2022)
11. Li, Y., Mao, H., Girshick, R., He, K.: Exploring plain vision transformer backbones for object detection. In: Avidan, S., Brostow, G., Cissé, M., Farinella, G.M., Hassner, T. (eds.) ECCV 2022. Lecture Notes in Computer Science, vol. 13669, pp. 280–296. Springer, Cham (2022). https://doi.org/10.1007/978-3-031-20077-9_17
12. Liu, Z., et al.: Swin transformer: hierarchical vision transformer using shifted windows. In: Proceedings of the IEEE/CVF International Conference on Computer Vision, pp. 10012–10022 (2021)
13. Liu, Z., Mao, H., Wu, C.Y., Feichtenhofer, C., Darrell, T., Xie, S.: A convnet for the 2020s. In: Proceedings of the IEEE/CVF Conference on Computer Vision and Pattern Recognition, pp. 11976–11986 (2022)
14. Luo, Y., Chen, Z., Gao, X.: Self-distillation augmented masked autoencoders for histopathological image classification. arXiv preprint arXiv:2203.16983 (2022)
15. Ly, S.T., Lin, B., Vo, H.Q., Maric, D., Roysam, B., Nguyen, H.V.: Student collaboration improves self-supervised learning: dual-loss adaptive masked autoencoder for brain cell image analysis. arXiv preprint arXiv:2205.05194 (2022)
16. Niu, S., Liu, M., Liu, Y., Wang, J., Song, H.: Distant domain transfer learning for medical imaging. IEEE J. Biomed. Health Inform. **25**(10), 3784–3793 (2021)
17. Park, M., et al.: Sonography of thyroid nodules with peripheral calcifications. J. Clin. Ultrasound **37**(6), 324–328 (2009)
18. Qin, Z., Yi, H., Lao, Q., Li, K.: Medical image understanding with pretrained vision language models: a comprehensive study. arXiv preprint arXiv:2209.15517 (2022)
19. Quan, H., et al.: Global contrast masked autoencoders are powerful pathological representation learners. arXiv preprint arXiv:2205.09048 (2022)
20. Raghu, M., Zhang, C., Kleinberg, J., Bengio, S.: Transfusion: understanding transfer learning for medical imaging. In: Advances in Neural Information Processing Systems, vol. 32 (2019)

21. Taki, S., et al.: Thyroid calcifications: sonographic patterns and incidence of cancer. Clin. Imaging **28**(5), 368–371 (2004)
22. Tian, Y., et al.: Beyond masking: demystifying token-based pre-training for vision transformers. arXiv preprint arXiv:2203.14313 (2022)
23. Wang, P., Patel, V.M., Hacihaliloglu, I.: Simultaneous segmentation and classification of bone surfaces from ultrasound using a multi-feature guided CNN. In: Frangi, A.F., Schnabel, J.A., Davatzikos, C., Alberola-López, C., Fichtinger, G. (eds.) MICCAI 2018. LNCS, vol. 11073, pp. 134–142. Springer, Cham (2018). https://doi.org/10.1007/978-3-030-00937-3_16
24. Wang, X., Zhao, K., Zhang, R., Ding, S., Wang, Y., Shen, W.: ContrastMask: Contrastive learning to segment every thing. In: Proceedings of the IEEE/CVF Conference on Computer Vision and Pattern Recognition, pp. 11604–11613 (2022)
25. Wu, Q., Ye, H., Gu, Y., Zhang, H., Wang, L., He, D.: Denoising masked autoencoders are certifiable robust vision learners. arXiv preprint arXiv:2210.06983 (2022)
26. Xiao, J., Bai, Y., Yuille, A., Zhou, Z.: Delving into masked autoencoders for multi-label thorax disease classification. In: Proceedings of the IEEE/CVF Winter Conference on Applications of Computer Vision, pp. 3588–3600 (2023)
27. Xu, Z., et al.: Swin MAE: masked autoencoders for small datasets. arXiv preprint arXiv:2212.13805 (2022)
28. Yu, Y., Acton, S.T.: Speckle reducing anisotropic diffusion. IEEE Trans. Image Process. **11**(11), 1260–1270 (2002)
29. Zhang, C., Zhang, C., Song, J., Yi, J.S.K., Zhang, K., Kweon, I.S.: A survey on masked autoencoder for self-supervised learning in vision and beyond. arXiv preprint arXiv:2208.00173 (2022)
30. Zhang, H., et al.: MaeFE: masked autoencoders family of electrocardiogram for self-supervised pretraining and transfer learning. IEEE Trans. Instrum. Meas. **72**, 1–15 (2022)
31. Zhou, L., Liu, H., Bae, J., He, J., Samaras, D., Prasanna, P.: Self pre-training with masked autoencoders for medical image analysis. arXiv preprint arXiv:2203.05573 (2022)
32. Zhou, Y., et al.: Multi-task learning for segmentation and classification of tumors in 3D automated breast ultrasound images. Med. Image Anal. **70**, 101918 (2021)

You've Got Two Teachers: Co-evolutionary Image and Report Distillation for Semi-supervised Anatomical Abnormality Detection in Chest X-Ray

Jinghan Sun[1,2], Dong Wei[2], Zhe Xu[2,3], Donghuan Lu[2], Hong Liu[1,2], Liansheng Wang[1(✉)], and Yefeng Zheng[2]

[1] National Institute for Data Science in Health and Medicine, Xiamen University, Xiamen, China
{jhsun,liuhong}@stu.xmu.edu.cn, lswang@xmu.edu.cn
[2] Tencent Healthcare (Shenzhen) Co., LTD, Tencent Jarvis Lab, Shenzhen, China
{donwei,caleblu,yefengzheng}@tencent.com
[3] The Chinese University of Hong Kong, Hong Kong, China
jackxz@link.cuhk.edu.hk

Abstract. Chest X-ray (CXR) anatomical abnormality detection aims at localizing and characterising cardiopulmonary radiological findings in the radiographs, which can expedite clinical workflow and reduce observational oversights. Most existing methods attempted this task in either fully supervised settings which demanded costly mass per-abnormality annotations, or weakly supervised settings which still lagged badly behind fully supervised methods in performance. In this work, we propose a co-evolutionary image and report distillation (CEIRD) framework, which approaches semi-supervised abnormality detection in CXR by grounding the visual detection results with text-classified abnormalities from paired radiology reports, and vice versa. Concretely, based on the classical teacher-student pseudo label distillation (TSD) paradigm, we additionally introduce an auxiliary report classification model, whose prediction is used for report-guided pseudo detection label refinement (RPDLR) in the primary vision detection task. Inversely, we also use the prediction of the vision detection model for abnormality-guided pseudo classification label refinement (APCLR) in the auxiliary report classification task, and propose a co-evolution strategy where the vision and report models mutually promote each other with RPDLR and APCLR performed alternatively. To this end, we effectively incorporate the weak supervision by reports into the semi-supervised TSD pipeline. Besides

J. Sun and D. Wei—Contributed equally; J. Sun contributed to this work during an internship at Tencent.

Supplementary Information The online version contains supplementary material available at https://doi.org/10.1007/978-3-031-43907-0_35.

H. Greenspan et al. (Eds.): MICCAI 2023, LNCS 14220, pp. 363–373, 2023.
https://doi.org/10.1007/978-3-031-43907-0_35

the cross-modal pseudo label refinement, we further propose an intra-image-modal self-adaptive non-maximum suppression, where the pseudo detection labels generated by the teacher vision model are dynamically rectified by high-confidence predictions by the student. Experimental results on the public MIMIC-CXR benchmark demonstrate CEIRD's superior performance to several up-to-date weakly and semi-supervised methods.

Keywords: Anatomical abnormality detection · Semi-supervised learning · Co-evolution · Visual and textual grounding · Chest X-ray

1 Introduction

Chest X-ray (CXR) is the most commonly performed diagnostic radiograph in medicine, which helps spot abnormalities or diseases of the airways, blood vessels, bones, heart, and lungs. Given the complexity and workload of clinical CXR reading, there is a growing interest in developing automated methods for anatomical abnormality detection in CXR [19]—especially using deep neural networks (DNNs) [14,17,20], which are expected to expedite clinical workflow and reduce observational oversights. Here, the detection task involves both localization (*e.g.*, with bounding boxes) and characterization (*e.g.*, cardiomegaly) of the abnormalities. However, training accurate DNN-based detection models usually requires large-scale datasets with high-quality per-abnormality annotations, which is costly in time, effort, and expense.

To completely relieve the burden of annotation, a few works [2,22,25] resorted to the radiology reports as a form of weak supervision for localization of pneumonia and pneumothorax in CXR. The text report describes important findings in each CXR and is available for most archive radiographs, thus is a valuable source of image-level supervision signal unique to medical image data. However, studies have shown that there are still apparent gaps in performance between image-level weakly supervised and bounding-box-level fully supervised detection [1,11]. Alternatively, seeking for a trade-off between annotation effort and model performance, semi-supervised learning aims to achieve reasonable performance with an acceptable quantity of manual annotations. Semi-supervised object detection methods have achieved noteworthy advances in the natural image domain [4,21,23]. Most of these methods were built on the teacher-student distillation (TSD) paradigm [10], where a teacher model is firstly trained on the labeled data, and then a student model is trained on both the labeled data with real annotations and the unlabeled data with pseudo labels generated (predicted) by the teacher. However, compared with objects in natural images, the abnormalities in CXR can be subtle and less well-defined with ambiguous boundaries, thus likely to introduce great noise to the pseudo labels and eventually lead to suboptimal performance of semi-supervised learning with TSD.

In this paper, we present a co-evolutionary image and report distillation (CEIRD) framework for semi-supervised anatomical abnormality detection in CXR, incorporating the weak supervision by radiology reports. Above all, on the basis of TSD [10], CEIRD introduces an auxiliary, also semi-supervised, multi-label report classification natural language processing task, whose prediction is used for noise reduction in the pseudo labels of the primary vision detection task, *i.e.*, report-guided pseudo detection label refinement (RPDLR). Then, noting that the performance of the auxiliary language task is crucial to RPDLR, we inversely use the abnormalities detected in the vision task to filter the pseudo labels in the language task, for abnormality-guided pseudo classification label refinement (APCLR). In addition, we implement an iterative co-evolution strategy where RPDLR and APCLR are performed alternatively in a loop, where either model is trained while fixing the other and using the other's prediction for pseudo label refinement. To the best of our knowledge, this is the first work that approaches semi-supervised abnormality detection in CXR by grounding report-classified abnormalities with the visual detection results in the paired radiograph [5,24], and vice versa. Besides the cross-modal pseudo label refinement, we additionally propose self-adaptive non-maximum suppression (SA-NMS) for intra-(image-)modal refinement, too, where the predictions by both the teacher and student vision models go through NMS together to produce new pseudo detection labels for training. In this way, the pseudo labels generated by the teacher are dynamically rectified by high-confidence predictions of the student who is getting better as training goes. Experimental results on the MIMIC-CXR [12,13] public benchmark show that our CEIRD outperforms various up-to-date weakly and semi-supervised methods, and that its building elements are effective.

To summarize, our contributions include: (1) the complementary RPDLR and APCLR for noise reduction in both vision and language pseudo labels for improved semi-supervised training via mutual grounding, (2) the co-evolution strategy for joint optimization of the primary and auxiliary tasks, and (3) the SA-NMS for dynamic intra-image-modal pseudo label refinement, all contributing to the superior performance of the proposed CEIRD framework.

2 Method

Problem Setting. In semi-supervised anatomical abnormality localization, a data set comprising both unlabeled samples $D_u = \{(x_i^u, r_i^u)\}_{i=1}^{N_u}$ and labeled ones $D_l = \{(x_i^l, r_i^l, A_i)\}_{i=1}^{N_l}$ is provided for training, where x and r are a CXR and accompanying report, respectively, $A_i = \{(y^l, B^l)\}$ is the annotation for a labeled sample including both bounding boxes $\{B^l\}$ and corresponding categories $\{y^l\}$, and $N^l \ll N^u$ for practical use scenario. It is worth noting that $\{y^l\}$ can also be considered as classification labels for the report. The objective is to obtain a detection model that can accurately localize and classify the abnormalities in any testing CXR (without report in practice), by making good use of both the labeled and unlabeled CXRs plus the accompanying reports in the training set.

Method Overview. Figure 1 provides an overview of our framework. Suppose a pretrained teacher vision model F_t^I (*e.g.*, on labeled data) for abnormality

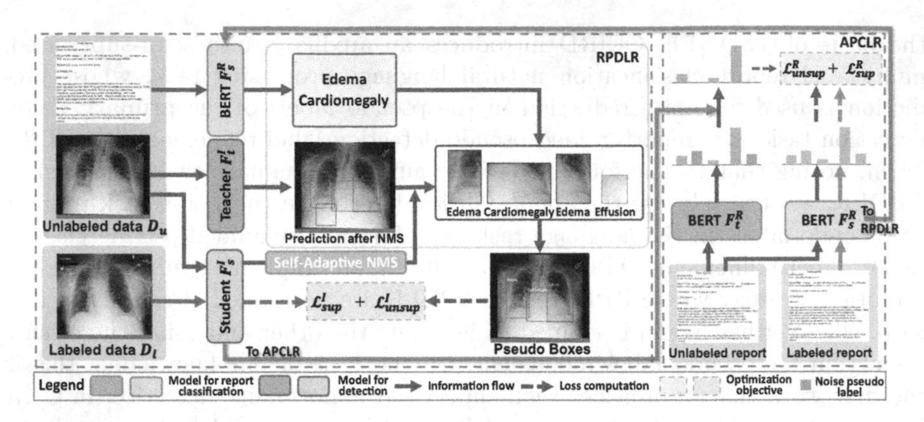

Fig. 1. Overview of the proposed framework. RPDLR: report-guided pseudo detection label refinement; APCLR: abnormality-guided pseudo classification label refinement.

detection in CXR is given, together with a pretrained language model F_s^R for multi-label abnormality classification of reports. On the one hand, we generate for an unlabeled image x_i^u pseudo detection labels with F_t^I and filter the pseudo labels by self-adaptive non-maximum suppression (NMS). Meanwhile, we feed the corresponding report r_i^u into F_s^R and use the prediction for report-guided pseudo detection label refinement (RPDLR). To this end, we obtain refined pseudo labels to supervise the student vision model F_s^I toward better anatomical abnormality localization. On the other hand, we also pass the detection predictions by F_s^I to a teacher language model F_t^R for abnormality-guided pseudo classification label refinement (APCLR), to better supervise the student language model F_s^R on unlabeled data for report-based abnormality classification. In turn, the better language model F_s^R helps train better vision models via RPDLR, thus both types of models co-evolve during training. Note that the real labels are used to train both student models along with the pseudo ones. After training, we only need the student vision model F_s^I for abnormality localization in testing CXRs.

Preliminary Pseudo Label Distillation for Semi-supervised Learning. Both of our baseline semi-supervised vision and language models follow the teacher-student knowledge distillation (TSD) procedure [10]. For report classification, we first train a teacher model F_t^R on labeled reports, and then train a student model F_s^R to predict real labels on labeled reports and pseudo labels produced by F_t^R on unlabeled ones with the loss function:

$$\mathcal{L}^R = \mathcal{L}_{\text{sup}}^R + \mathcal{L}_{\text{unsup}}^R = \sum\nolimits_{D_l} \mathcal{L}_{\text{cls}}^R \left(\hat{y}^l, y^l\right) + \sum\nolimits_{D_u} \mathcal{L}_{\text{cls}}^R \left(\hat{y}, y^{pr}\right), \tag{1}$$

where $\mathcal{L}_{\text{cls}}^R$ is the cross-entropy loss, $\{\hat{y}\} = F_s^R(r)$ is the prediction by the student model, $\{y^{pr}\} = F_t^R(r^u)$ is the set of pseudo labels generated by the teacher model. In each batch, labeled and unlabeled instances are sampled according to a controlled ratio. The resulting report classification model F_s^R will be utilized later to help with the primary task of abnormality detection in CXR. Similarly, a student vision model F_s^I for abnormality detection in CXR is trained in semi-supervised setting by distilling from a teacher vision model F_t^I trained on labeled CXRs, with the loss function:

$$\begin{aligned}\mathcal{L}^I = \mathcal{L}_{\text{sup}}^I + \mathcal{L}_{\text{unsup}}^I = \sum\nolimits_{D_l} \left[\mathcal{L}_{\text{cls}}^I \left(\hat{y}^l, y^l\right) + \mathcal{L}_{\text{reg}}^I (\hat{B}^l, B^l)\right] \\ + \sum\nolimits_{D_u} \left[\mathcal{L}_{\text{cls}}^I \left(\hat{y}, y^{pv}\right) + \mathcal{L}_{\text{reg}}^I (\hat{B}, B^{pv})\right],\end{aligned} \tag{2}$$

where $\{(\hat{y}, \hat{B})\} = F_s^I(x)$ are the predictions by the student model, $\{(y^{pv}, B^{pv})\} = F_t^I(x^u)$ are the pseudo class and bounding box labels generated by the teacher model, $\mathcal{L}_{\text{cls}}^I$ is the focal loss [16] for abnormality classification, and $\mathcal{L}_{\text{reg}}^I$ is the smooth L1 loss for bounding box regression.

Self-adaptive Non-maximum Suppression. During the TSD, the teacher vision model F_t^I is kept fixed. While its knowledge suffices for guiding the student vision model F_s^I in the early stage of TSD, it may somehow impede the learning of F_s^I when F_s^I gradually improves by also learning from the large amount of unlabeled data. Therefore, to gradually improve quality and robustness of the pseudo detection labels as F_s^I learns, we propose to perform self-adaptive non-maximum suppression (SA-NMS) to combine the pseudo labels $\{(y^{pv}, B^{pv})\}$ output by F_t^I and the predictions $\{(\hat{y}, \hat{B})\}$ by F_s^I in each mini batch. Specifically, we perform NMS on the combined set of the pseudo labels and predictions: $\{(y^{cv}, B^{cv})\} = \text{NMS}(\{(y^{pv}, B^{pv})\} \bigcup \{(\hat{y}, \hat{B})\})$, and replace $\{(y^{pv}, B^{pv})\}$ in Eq. (2) with $\{(y^{cv}, B^{cv})\}$ for supervision by unlabeled CXRs. In this way, highly confident predictions by the maturing student can rectify imprecise ones by the teacher, leading to better supervision signals stemming from unlabeled data.

Report Guided Pseudo Label Refinement. In routine clinics, almost every radiograph in archive is accompanied by a report describing findings, abnormalities (if any), and diagnosis. Compared with the captions of natural images, the report texts constitute a unique (to medical image analysis) and rich source of extra information in addition to the image modality. To this end, we propose report-guided pseudo detection label refinement (RPDLR) to make use of this cross-modal information for semi-supervised anatomical abnormality detection in CXR. Specifically, we use the student language model F_s^R (trained with Eq. (1)) to refine the pseudo detection labels. Given a pair of unlabeled image x^u and report r^u, we obtain the set of abnormalities $\{(y^{cv}, B^{cv})\}$ detected in x^u after SA-NMS, and the set of abnormalities $\{\hat{y}\}$ classified in r^u by F_s^R. Then, we only keep the pseudo detection labels whose categories are in the report-classified abnormalities:

$$\{(y^v, B^v)\} = \left\{\left(y_j^{cv}, B_j^{cv}\right) \middle| \, y_j^{cv} \in \{\hat{y}\}\right\}. \tag{3}$$

Eventually, we train the student vision model F_s^I using $\{(y^v, B^v)\}$ in Eq. (2).

Co-evolutionary Semi-supervised Learning with Cycle Pseudo Label Refinement. As the auxiliary student language model F_s^R plays an important role in RPDLR, it is reasonable to optimize its performance which in turn would

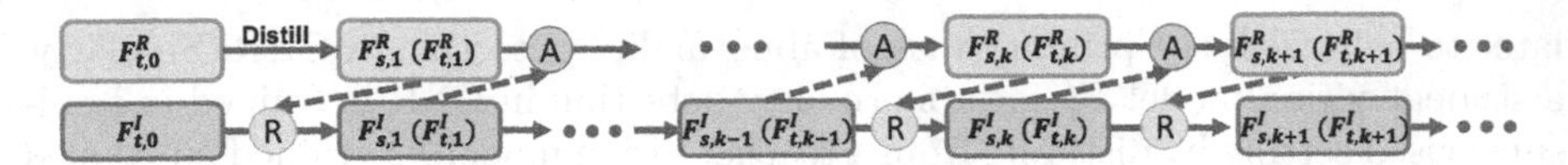

Fig. 2. Illustration of the co-evolution strategy. "R" and "A" represent report- and abnormality-guided pseudo label refinements (RPDLR and APCLR), respectively.

benefit the primary task of abnormality detection. Therefore, we further propose an inverse, abnormality-guided pseudo classification labels refinement (APCLR) to help with semi-supervised training of the report classification model. Similarly in concept to the RPDLP, given a pair of unlabeled image x^u and report r^u, we obtain the set of abnormalities $\{(\hat{y}, \hat{B})\}$ detected in x^u by the student vision model F^I_s, and the set of classification pseudo labels $\{y^{pr}\}$ generated for r^u by the teacher language model F^R_t. We retain only the pseudo labels $\{y^{pr}_j | y^{pr}_j \in \{\hat{y}\}\}$, by excluding the report-classified abnormalities not detected in the paired CXR.

Ideally, one should use an optimal report classification model for refinement of the abnormality detection pseudo labels, and vice versa. However, the two models are mutually dependent on each other in a circle. To solve this dilemma, we implement an alternative co-evolution strategy to refine the abnormality detection and report classification pseudo labels iteratively, in *generations*. As shown in Fig. 2, the kth generation student vision model $F^I_{s,k}$ is distilled from the teacher $F^I_{t,k-1}$, whose pseudo labels are refined by the prediction of the frozen student language model $F^R_{s,k}$ via RPDLR. Subsequently, $F^I_{s,k}$ is frozen and used to 1) help train the $(k+1)^{\text{th}}$ student language model $F^R_{s,k+1}$ via APCLR, and 2) serve as the teacher vision model in next generation: $F^I_{s,k} \rightarrow F^I_{t,k}$.[1] Note that in each generation the students are reborn with random initialization [8]. Thus the co-evolution continues to optimize the vision and report models cyclically with cross-modal mutual promotion. After training, we only need the Kth generation student vision model $F^I_{s,K}$ for abnormality detection in upcoming test CXRs.

3 Experiments

Dataset and Evaluation Metrics. We conduct experiments on the chest radiography dataset MIMIC-CXR [12,13], with the detection annotations provided by MS-CXR [3]. MIMIC-CXR is a large publicly available dataset of CXR and free-text radiology reports. MS-CXR provides bounding box annotations for part of the CXRs in MIMIC-CXR (1,026 samples). MIMIC-CXR includes 14 categories of anatomical abnormalities for multi-label classification of the reports, while there are only eight categories in the bounding box annotations of MS-CXR. For consistency, we exclude samples in MIMIC-CXR that have abnormality labels outside the eight categories of MS-CXR, leaving 112,425 samples. Thus, in our semi-supervised setting, 1,026 samples are labeled and the rest are

[1] The initial teachers $F^I_{t,0}$ and $F^R_{t,0}$ are obtained by training on the labeled data only.

Table 1. Abnormality detection results on the test data, using mAP (%) with the IoU thresholds of 0.25, 0.5, and 0.75. TSD: teacher-student distillation.

mAP	CAM [26]	AGXNet [25]	Sup.	TSD [10]	STAC [21]	LabelMatch [4]	Soft Teacher [23]	Ours	Semi-oracle
@0.25	20.47	29.96	37.91	38.29	39.26	39.92	40.17	**41.93**	42.61
@0.5	11.20	15.62	32.84	33.95	35.01	36.40	36.59	**37.20**	37.39
@0.75	3.05	7.44	19.21	19.51	23.90	24.06	24.78	**25.12**	25.66

not.[2] We split the labeled samples for training, validation, and testing according to the ratio of 7:1:2, and use the remaining samples as our unlabeled training data. We focus on the frontal views in this work. The mean average precision (mAP) [7] with the intersection over union (IoU) threshold of 0.25, 0.5, and 0.75 is employed to evaluate the performance of abnormality detection in CXR.

Implementation. The PyTorch [18] framework (1.4.0) is used for experiments. For report classification, we employ the BERT-base uncased model [6] with eight linear heads. Stochastic gradient descent with the momentum of 0.9 and learning rate of 10^{-4} is used for optimization. The batch size is set to 16 reports. For abnormality detection, we employ RetinaNet [16] with FPN [15]+ResNet-101 [9] as backbone. We resize all images to 512×512 pixels and use a batch size of 16. Data augmentation including random cropping and flipping is performed. Our implementation and hyper-parameters follow the official settings [16]. Unless otherwise stated, we evolve the vision and language models for two generations, and train both models for 2000 iterations in each generation (including initial training of the teacher models). The ratio of labeled to unlabeled samples in each mini batch during the semi-supervised training is empirically set to 1:1 and 2:1 for the language and vision models, respectively. The source code is available at: https://github.com/jinghanSunn/CEIRD.

Comparison with State-of-the-Art (SOTA) Methods. We compare our proposed co-evolution image and report distillation (CEIRD) framework with several up-to-date detection methods, including weakly supervised: CAM [26] (locating objects based on class activation maps), AGXNet [25] (aiding CAM-based localization with report representations), fully supervised on labeled training data only (Sup.), baseline semi-supervised via teacher-student pseudo-label distillation (TSD; see Eq. (2)) [10], and three SOTA semi-supervised (STAC [21], LabelMatch [4], and Soft Teacher [23]) object detection methods.

The results are shown in Table 1, from which we have the following observations. First, both fully supervised (by the labeled data only) and semi-supervised methods outperform the weakly-supervised by large margins, proving the efficacy of using limited annotations. Second, all semi-supervised methods outperform the fully supervised (by the labeled data only) by various margins, demonstrating

[2] In this work, we deliberately construct the semi-supervised setting by ignoring the report labels provided in MIMIC-CXR for methodology development. When it comes to a practical new application with no such label available, *e.g.*, semi-supervised lesion detection in color fundus photography, our method can be readily applied.

Table 2. Ablation study results on the validation data.

	Baseline	(a)	(b)	(c)
RPDLR	-	✓	✓	✓
CoE+APCLR	-	-	✓	✓
SA-NMS	-	-	-	✓
mAP@0.25	37.31	39.10	39.72	**41.86**
mAP@0.5	32.76	35.87	35.90	**36.18**
mAP@0.75	17.96	22.02	23.55	**24.49**

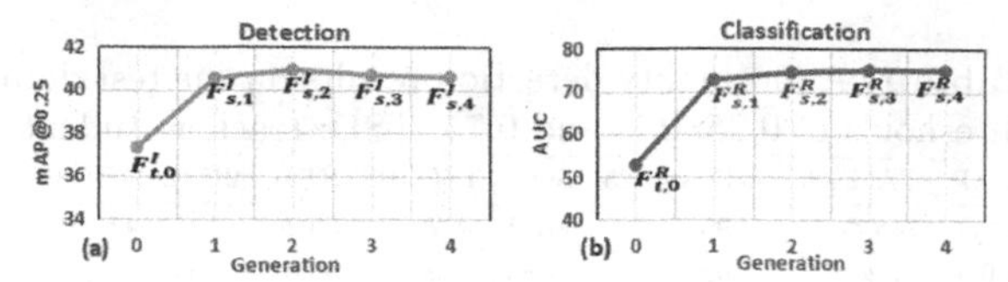

Fig. 3. Performance of (a) vision and (b) report models as a function of generations. AUC: area under the receiver operating characteristic curve.

apparent benefit of making use of the unlabeled data, too. Third, our CEIRD achieves the best performance of all the semi-supervised methods for the mAPs evaluated at three different IoU thresholds, outperforming the second best (Soft Teacher) by up to 1.76%. These results clearly demonstrate the advantage of our method which innovatively integrates the semi-supervision by unlabeled images and the weak supervision by texts. In addition, we evaluate a semi-oracle of our method, where the ground truth report labels provided in MIMIC-CXR are used for RPDLR, instead of the auxiliary model's prediction. As we can see, our method is marginally short of the semi-oracle, *e.g.*, 37.20% versus 37.39% for mAP@0.5, suggesting that our co-evolution strategy can effectively mine the relevant information from the reports. We provide in the supplementary material visualizations of example detection results by Soft Teacher and our method, where ours are visually superior with fewer misses (left), fewer false positives (middle), and better localization (right). Lastly, we also provide performance evaluation of the auxiliary report classification task in the supplement.

Ablation Study. We conduct ablation studies on the validation data to investigate efficacy of the novel building elements of our CEIRD framework, including: report-guided pseudo detection label refinement (RPDLR), co-evolution strategy (CoE) with abnormality-guided pseudo classification label refinement (APCLR), and self-adaptive non-maximum suppression (SA-NMS). We use the preliminary teacher-to-student pseudo label distillation as baseline (Eq. (2)). As shown in Table 2, RPDLR (column (a)) substantially boosts performance upon the baseline by 1.79–4.06% in mAPs thanks to the refined pseudo detection labels. By adopting CoE+APCLR (column (b)), we achieve further performance improvements up to 1.53% as the auxiliary report classification model gets better together. Last but not the least, the introduction of SA-NMS (column (c)) also brings improvements up to 2.14%. These results validate the novel design of our framework. In addition, we conduct experiments to empirically determine the optimal number of generations for the co-evolution. The results are shown in Fig. 3, where the 0th generation means fully supervised models trained on the labeled data only (*i.e.*, the initial teacher models $F^I_{t,0}$ and $F^R_{t,0}$). As we can see, the vision and report models improve in the first two and three generations, respectively, and then remain stable in the following ones, confirming that both models promote each other with the co-evolution strategy. Since our ultimate objective is abnormality detection in CXR, we select two generations for comparison with other methods.

4 Conclusion

In this work, we proposed a new co-evolutionary image and report distillation (CEIRD) framework for semi-supervised anatomical abnormality detection in chest X-ray. On the basis of a preliminary teacher-student pseudo label distillation, we first presented self-adaptive NMS to mingle highly confident predictions by both the teacher and student for improved pseudo labels. We then proposed report-guided pseudo detection label refinement (RPDLR) that used abnormalities classified from the accompanying radiology reports by an auxiliary language model to eliminate unmatched pseudo labels. Meanwhile, we further proposed an inverse, abnormality-guided pseudo classification label refinement (APCLR) making use of the abnormalities detected in X-ray images for better language model training. In addition, we implemented a co-evolution strategy that looped the RPDLR and APCLR to iteratively optimize the main vision detection model and auxiliary report classification model in an alternative manner. Experimental results showed that our CEIRD framework achieved superior performance to up-to-date semi-/weakly-supervised methods.

Acknowledgement. This work was supported by the National Key R&D Program of China under Grant 2020AAA0109500/2020AAA0109501 and the National Key Research and Development Program of China (2019YFE0113900).

References

1. Bearman, A., Russakovsky, O., Ferrari, V., Fei-Fei, L.: What's the point: semantic segmentation with point supervision. In: Leibe, B., Matas, J., Sebe, N., Welling, M. (eds.) ECCV 2016. LNCS, vol. 9911, pp. 549–565. Springer, Cham (2016). https://doi.org/10.1007/978-3-319-46478-7_34
2. Bhalodia, R., et al.: Improving pneumonia localization via cross-attention on medical images and reports. In: de Bruijne, M., et al. (eds.) MICCAI 2021. LNCS, vol. 12902, pp. 571–581. Springer, Cham (2021). https://doi.org/10.1007/978-3-030-87196-3_53
3. Boecking, B., et al.: Making the most of text semantics to improve biomedical vision-language processing. arxiv Preprint: arxiv:2204.09817 (2022)
4. Chen, B., et al.: Label matching semi-supervised object detection. In: Proceedings of the IEEE/CVF Conference on Computer Vision and Pattern Recognition, pp. 14381–14390 (2022)
5. Datta, S., Sikka, K., Roy, A., Ahuja, K., Parikh, D., Divakaran, A.: Align2Ground: weakly supervised phrase grounding guided by image-caption alignment. In: Proceedings of the IEEE/CVF International Conference on Computer Vision, pp. 2601–2610 (2019)
6. Devlin, J., Chang, M.W., Lee, K., Toutanova, K.: BERT: pre-training of deep bidirectional transformers for language understanding. arxiv Preprint: arxiv:1810.04805 (2018)

7. Everingham, M., Van Gool, L., Williams, C.K., Winn, J., Zisserman, A.: The PASCAL visual object classes (VOC) challenge. Int. J. Comput. Vision **88**, 303–308 (2009)
8. Furlanello, T., Lipton, Z., Tschannen, M., Itti, L., Anandkumar, A.: Born again neural networks. In: International Conference on Machine Learning, pp. 1607–1616. PMLR (2018)
9. He, K., Zhang, X., Ren, S., Sun, J.: Deep residual learning for image recognition. In: Proceedings of the IEEE Conference on Computer Vision and Pattern Recognition, pp. 770–778 (2016)
10. Hinton, G., Vinyals, O., Dean, J.: Distilling the knowledge in a neural network. arxiv Preprint: arxiv:1503.02531 (2015)
11. Ji, H., et al.: A benchmark for weakly semi-supervised abnormality localization in chest X-rays. In: Wang, L., Dou, Q., Fletcher, P.T., Speidel, S., Li, S. (eds.) MICCAI 2022. Lecture Notes in Computer Science, vol. 13433, pp. 249–260. Springer, Cham (2022). https://doi.org/10.1007/978-3-031-16437-8_24
12. Johnson, A., et al.: MIMIC-CXR-JPG-chest radiographs with structured labels. PhysioNet (2019)
13. Johnson, A.E., et al.: MIMIC-CXR-JPG, a large publicly available database of labeled chest radiographs. arxiv Preprint: arxiv:1901.07042 (2019)
14. Lakhani, P., Sundaram, B.: Deep learning at chest radiography: automated classification of pulmonary tuberculosis by using convolutional neural networks. Radiology **284**(2), 574–582 (2017)
15. Lin, T.Y., Dollár, P., Girshick, R., He, K., Hariharan, B., Belongie, S.: Feature pyramid networks for object detection. In: Proceedings of the IEEE Conference on Computer Vision and Pattern Recognition, pp. 2117–2125 (2017)
16. Lin, T.Y., Goyal, P., Girshick, R., He, K., Dollár, P.: Focal loss for dense object detection. In: Proceedings of the IEEE International Conference on Computer Vision, pp. 2980–2988 (2017)
17. Oğul, B.B., Koşucu, P., Özçam, A., Kanik, S.D.: Lung nodule detection in X-ray images: a new feature set. In: Lacković, I., Vasic, D. (eds.) 6th European Conference of the International Federation for Medical and Biological Engineering. IP, vol. 45, pp. 150–155. Springer, Cham (2015). https://doi.org/10.1007/978-3-319-11128-5_38
18. Paszke, A., et al.: PyTorch: an imperative style, high-performance deep learning library. In: Advances in Neural Information Processing Systems, vol. 32 (2019)
19. Qin, C., Yao, D., Shi, Y., Song, Z.: Computer-aided detection in chest radiography based on artificial intelligence: a survey. Biomed. Eng. Online **17**(1), 1–23 (2018)
20. Rajpurkar, P., et al.: CheXNet: radiologist-level pneumonia detection on chest X-rays with deep learning. arxiv preprint: arxiv:1711.05225 (2017)
21. Sohn, K., Zhang, Z., Li, C.L., Zhang, H., Lee, C.Y., Pfister, T.: A simple semi-supervised learning framework for object detection. arxiv Preprint: arxiv:2005.04757 (2020)
22. Tam, L.K., Wang, X., Turkbey, E., Lu, K., Wen, Y., Xu, D.: Weakly supervised one-stage vision and language disease detection using large scale pneumonia and pneumothorax studies. In: Martel, A.L., et al. (eds.) MICCAI 2020. LNCS, vol. 12264, pp. 45–55. Springer, Cham (2020). https://doi.org/10.1007/978-3-030-59719-1_5
23. Xu, M., et al.: End-to-end semi-supervised object detection with soft teacher. In: Proceedings of the IEEE/CVF International Conference on Computer Vision, pp. 3060–3069 (2021)

24. Yang, Z., Gong, B., Wang, L., Huang, W., Yu, D., Luo, J.: A fast and accurate one-stage approach to visual grounding. In: Proceedings of the IEEE/CVF International Conference on Computer Vision, pp. 4683–4693 (2019)
25. Yu, K., Ghosh, S., Liu, Z., Deible, C., Batmanghelich, K.: Anatomy-guided weakly-supervised abnormality localization in chest X-rays. In: Wang, L., Dou, Q., Fletcher, P.T., Speidel, S., Li, S. (eds.) MICCAI 2022. Lecture Notes in Computer Science, vol. 13435, pp. 658–668. Springer, Cham (2022). https://doi.org/10.1007/978-3-031-16443-9_63
26. Zhou, B., Khosla, A., Lapedriza, A., Oliva, A., Torralba, A.: Learning deep features for discriminative localization. In: Proceedings of the IEEE Conference on Computer Vision and Pattern Recognition, pp. 2921–2929 (2016)

Masked Vision and Language Pre-training with Unimodal and Multimodal Contrastive Losses for Medical Visual Question Answering

Pengfei Li[1], Gang Liu[1(✉)], Jinlong He[1], Zixu Zhao[1], and Shenjun Zhong[2(✉)]

[1] College of Computer Science and Technology, Harbin Engineering University, Harbin, China
liugang@hrbeu.edu.com

[2] Monash Biomedical Imaging, Monash University, Clayton, Australia
shenjun.zhong@monash.edu

Abstract. Medical visual question answering (VQA) is a challenging task that requires answering clinical questions of a given medical image, by taking consider of both visual and language information. However, due to the small scale of training data for medical VQA, pre-training fine-tuning paradigms have been a commonly used solution to improve model generalization performance. In this paper, we present a novel self-supervised approach that learns unimodal and multimodal feature representations of input images and text using medical image caption datasets, by leveraging both unimodal and multimodal contrastive losses, along with masked language modeling and image text matching as pre-training objectives. The pre-trained model is then transferred to downstream medical VQA tasks. The proposed approach achieves state-of-the-art (SOTA) performance on three publicly available medical VQA datasets with significant accuracy improvements of 2.2%, 14.7%, and 1.7% respectively. Besides, we conduct a comprehensive analysis to validate the effectiveness of different components of the approach and study different pre-training settings. Our codes and models are available at https://github.com/pengfeiliHEU/MUMC.

Keywords: Medical Visual Question Answering · Masked Vision Language Pre-training · Unimodal and Multimodal Contrastive Losses

1 Introduction

Medical VQA is a specialized domain of VQA that aims to generate answers to natural language questions about medical images. It is very challenging to train deep learning based medical VQA models from scratch, since the medical VQA datasets available for research are relatively small in scale. Many existing works are proposed to leverage pre-trained visual encoders with external datasets to solve downstream medical VQA tasks, such as utilizing denoising autoencoders [1] and meta-models [2]. These methods mainly transfer feature encoders that are separately pre-trained on unimodal (image or text) tasks.

H. Greenspan et al. (Eds.): MICCAI 2023, LNCS 14220, pp. 374–383, 2023.
https://doi.org/10.1007/978-3-031-43907-0_36

Unlike unimodal pretraining approaches, both image and text feature presentations can be enhanced by learning through the visual and language interactions, given relatively richer resources of medical image caption datasets [3–5]. Liu et al. followed the work of MOCO [19] that trained teacher model for visual encoder via contrastive loss of different image views (by data augmentations) to improve the generalization of medical VQA [6]. Eslami et al. utilized CLIP [7] for visual model initialization, and learned cross-modality representations from medical image-text pairs by maximizing the cosine similarity between the extracted features of medical images and their corresponding captions [8]. Cong et al. devised an innovative framework, which featured a semantic focusing module to emphasize image regions that were pertinent to the caption and a progressive cross-modality comprehension module that iteratively enhanced the comprehension of the correlation between the image and caption [9]. Chen et al. proposed a medical vision language pre-training approach that used both masked image modelling and masked language modelling to jointly learn representations of medical images and their corresponding descriptions [10]. However, to the best of our knowledge, there have been no existing methods that explore learning both unimodal and multimodal features at the pre-training stage for downstream medical VQA tasks.

In this paper, we proposed a new self-supervised vision language pre-training (VLP) approach that applied **M**asked image and text modeling with **U**nimodal and **M**ultimodal **C**ontrastive losses (MUMC) in the pre-training phase for solving downstream medical VQA tasks. The model was pretrained on image caption datasets for aligning visual and text information, and transferred to downstream VQA datasets. The unimodal and multimodal contrastive losses in our work are applied to (1) align image and text features; (2) learn unimodal image encoders via momentum contrasts of different views of the same image (i.e. different views are generated by different image masks); (3) learn unimodal text encoder via momentum contrasts. We also introduced a new masked image strategy by randomly masking the patches of the image with a probability of 25%, which serves as a data augmentation technique to further enhance the performance of the model. Our approach outperformed existing methods and sets new benchmarks on three medical VQA datasets [11–13], with significant enhancements of 2.2%, 14.7%, and 1.7% respectively. Besides, we conducted an analysis to verify the effectiveness of different components and find the optimal masking probability. We also conducted a qualitative analysis on the attention maps using Grad-CAM [14] to validate whether the corresponding part of the image is attended when answering a question.

2 Methods

In this section, we provide the detailed description of the proposed approach, which includes the network architectures, self-supervised pre-training objectives, and the way to fine-tune on downstream medical VQA tasks.

2.1 Model Architecture

In the pre-training phase, the network architecture comprises an image encoder, a text encoder, and a multimodal encoder, which are all based on the transformer architecture

[15]. As shown in Fig. 1(a), the image encoder leverages a 12-layer Vision Transformer (ViT) [16] to extract visual features from the input images, while the text encoder employs a 6-layer transformer which is initialized by the first 6 layers of pre-trained BERT [17]. The last 6 layers of BERT are utilized as the multimodal encoder and incorporated cross-attention at each layer, which fuses the visual and linguistic features to facilitate learning of multimodal interactions. The model is trained on medical image-caption pairs. An image is partitioned into patches of size 16 × 16, and 25% of the patches are randomly masked. The remaining unmasked image patches are converted into a sequence of embeddings by an image encoder. The text, i.e. the image caption is tokenized into a sequence of tokens using a WordPiece [18] tokenizer and fed into the BERT-based text encoder. In addition, the special tokens, [CLS] are appended to the beginning of both the image and text sequence.

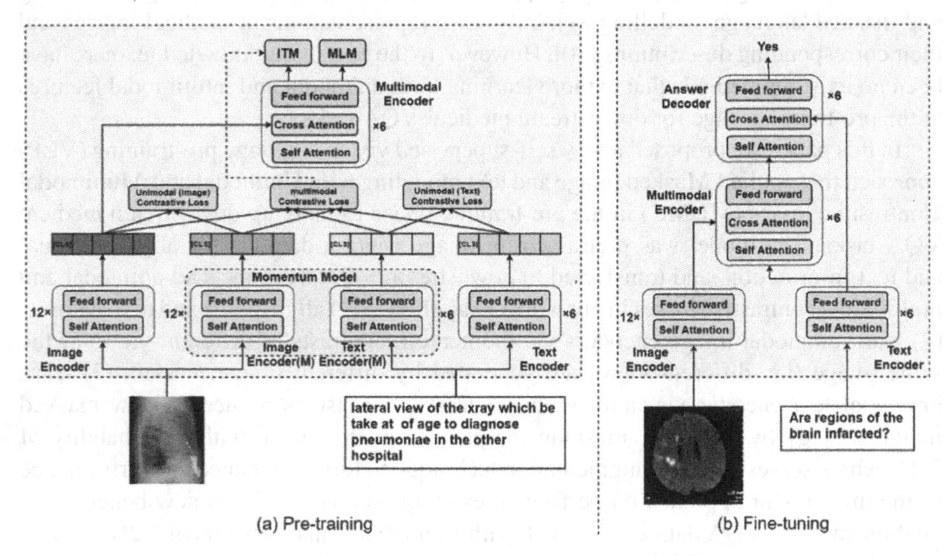

Fig. 1. Overview of the network architecture in both pre-training and fine-tuning phases.

To transfer the models trained on image caption datasets to the downstream medical VQA tasks, we utilize the weights from the pre-training stage to initialize the image encoder, text encoder and multimodal encoder, as shown in Fig. 1(b). To generate answers, we add an answering decoder with a 6-layer transformer-based decoder to the model, which receives the multimodal embeddings and output text tokens. A [CLS] token serves as the initial input token for the decoder, and a [SEP] token is appended to signify the end of the generated sequence. The downstream VQA model is fine-tuned via the masked language model (MLM) loss [17], using ground-truth answers as targets.

2.2 Unimodal and Multimodal Contrastive Losses

The proposed self-supervised objective attempts to capture the semantic discrepancy between positive and negative samples across both unimodal and multimodal domains at the same time. The unimodal contrastive loss (UCL) aims to differentiate between examples of one modality, such as images or text, in a latent space to make similar examples close. And the multimodal contrastive loss (MCL) learns the alignments between both modalities by maximizing the similarity between images and their corresponding text captions, while separating from the negative examples. In the implementation, we maintain two momentum models for image and text encoders respectively to generate different perspectives or representations of the same input sample, which serve as positive samples for contrastive learning.

In detail, we denote the image and caption embeddings from the unimodal image encoder and text encoder as v_{cls} and t_{cls}, which are further processed through the transformations g_v and g_t, to normalize and map the image and text embeddings to be lower-dimensional representations. The embeddings are inserted into a lookup table, and only the most recent 65,535 pairs of image-text embedding are stored for contrastive learning. We utilize the momentum update technique originally proposed in MoCo [19], which is updated every k iterations where k is a hyperparameter. We denote the ground-truth one-hot similarity by $y_{i2i}(V), y_{t2t}(T), y_{i2t}(V)$, and $y_{t2i}(T)$, where the probability of negative pairs is 0 and the probability of the positive pair is 1. The unimodal contrastive losses and multimodal contrastive losses can be defined as the cross-entropy H given as follows:

$$\begin{aligned} L_{ucl} = \frac{1}{2}\mathbb{E}_{(V,T)D}\Bigg[H\left(y_{i2i}(V), \frac{\exp(s(V, V_i)/\tau)}{\sum_{n=1}^{N}\exp(s(V, V_i)/\tau)}\right) \\ + H\left(y_{t2t}(T), \frac{exp(s(T, T_i)/\tau)}{\sum_{n=1}^{N} exp(s(T, T_i)/\tau)}\right)\Bigg] \end{aligned} \tag{1}$$

$$\begin{aligned} L_{mcl} = \frac{1}{2}\mathbb{E}_{(V,T)D}\Bigg[H\left(y_{i2t}(V), \frac{exp(s(V, T_i)/\tau)}{\sum_{n=1}^{N} exp(s(V, T_i)/\tau)}\right) \\ + H\left(y_{t2i}(T), \frac{exp(s(T, V_i)/\tau)}{\sum_{n=1}^{N} exp(s(T, V_i)/\tau)}\right)\Bigg] \end{aligned} \tag{2}$$

where s denotes cosine similarity function, $s(\mathrm{V}, V_i) = g_v(v_{cls})^T g_v(v_{cls})_i$, $s(\mathrm{T}, T_i) = g_t(t_{cls})^T g_t(t_{cls})_i$, $s(\mathrm{V}, T_i) = g_v(v_{cls})^T g_t(t_{cls})_i$, $s(\mathrm{T}, V_i) = g_t(t_{cls})^T g_v(v_{cls})_i$ and τ is a learnable temperature parameter.

2.3 Image Text Matching

We adopt the image text matching (ITM) strategy similar to prior works [20, 21] as one of the training objectives, by creating a binary classification task with negative text labels randomly sampled from the same minibatch. The joint representation of the image and text are encoded by the multimodal encoder, and utilized as input to the binary classification head. The ITM task is optimized using the cross-entropy loss:

$$\mathcal{L}_{itm} = \mathbb{E}_{(V,T)D} H(y_{itm}, p_{itm}(V, T)) \tag{3}$$

the function $H(,)$ represents a cross-entropy computation, where y_{itm} denotes the ground-truth label and $p_{itm}(V, T)$ is a function for predicting the class.

2.4 Masked Language Modeling

Masked Language Modeling (MLM) is another pre-trained objective in our approach, that predicts masked tokens in text based on both the visual and unmasked contextual information. For each caption text, 15% of tokens are randomly masked and replaced with the special token, [MASK]. Predictions of the masked tokens are conditioned on both unmasked text and image features. We minimize the cross-entropy loss for MLM:

$$\mathcal{L}_{mlm} = \mathbb{E}_{\left(V,\hat{T}\right)D} H(y_{mlm}, p_{mlm}(V, \hat{T})) \tag{4}$$

where $H(,)$ is a cross-entropy calculation, $\widehat{T}$ denotes the masked text token, y_{mlm} represents the ground-truth of the masked text token and $p_{mlm}(V, \widehat{T})$ is the predicted probability of a masked token.

2.5 Masked Image Strategy

Besides the training objectives, we introduce a masked image strategy as a data augmentation technique. In our experiment, input images are partitioned into patches which are randomly masked with a probability of 25%, and only the unmasked patches are passed through the network. Unlike the previous methods [10, 22], we do not utilize reconstruction loss [23], but use this only as a data augmentation method. This enables us to process more samples at each step, resulting in a more efficient pre-training of vision-language models with a similar memory footprint.

3 Experiments

3.1 Datasets

Our model is pre-trained on three datasets: ROCO [3], MedICaT [4], and the ImageCLEF2022 Image Caption Dataset [5]. ROCO comprises over 80,000 image-caption pairs. MedICaT includes over 217,000 medical images and their corresponding captions. ImageCLEF2022 is another well-known dataset that has nearly 90,000 pairs of medical images and captions.

For the downstream medical VQA task, we fine-tune and validate the model on three public medical VQA datasets: VQA-RAD [11], PathVQA [12] and SLAKE [13]. VQA-RAD has 315 radiology images with 3064 question-answer pairs, with 451 pairs used for testing. SLAKE has 14,028 pairs of samples which are further divided into 70% training, 15% validation, and 15% testing subsets. PathVQA is the largest dataset, containing 32,799 pairs of data that are split into training (50%), validation (30%), and test (20%) sets.

There are two types of questions: closed-ended questions that have limited answer choices (e.g. "yes" or "no") and open-ended questions that VQA models are required to generate answers in free text, which are more challenging.

3.2 Implementation Details

Our method was implemented in Python 3.8 and PyTorch 1.10. The experiments were conducted on a server with an Intel Xeon(R) Platinum 8255C and 2 NVIDIA Tesla V100 GPUs with 32 GB memory each. We pre-trained our model on three medical image caption datasets for 40 epochs with a batch size of 64. AdamW [24] optimizer was used with a weight decay of 0.002 and an initial learning rate of $1e^{-4}$, which decayed to $2e^{-5}$ by following the cosine schedule. We utilized randomly cropped images of 256×256 resolution as input, and also applied RandAugment to augment more training samples [25].

For downstream medical VQA tasks, we fine-tuned our model for 30 epochs with a batch size of 8. We used the AdamW optimizer with a reduced learning rate of $2e^{-5}$, which decayed to $1e^{-8}$. Besides, we increased image inputs from a resolution of 256×256 to 384×384 and interpolated the positional encoding following [16].

3.3 Comparison with the State-of-the-Arts

We performed a comparative evaluation of our model against the existing approaches [10, 26] on three benchmark datasets, VQA-RAD, PathVQA and SLAKE. Consistent with previous research [1, 2, 6, 8–10, 26, 27], we adopt accuracy as the performance metric. We treated VQA as a generative task by calculating similarities between the generated answers and candidate list answers, selecting the highest score as the final answer. As shown in Table 1, our approach outperformed all other methods on all the three datasets in terms of overall performance, and yielded the best accuracy for open-ended or closed-ended answers. On the VQA-RAD dataset [11], our method achieved an absolute margin of 2.2% overall over the current state-of-the-art method, M3AE, with improvements of 4.3% and 0.7% on open-ended and closed-ended answers respectively.

Table 1. Comparisons with the state-of-the-art methods on the VQA-RAD, PathVQA and SLAKE test set.

Methods	VQA-RAD			PathVQA			SLAKE
	Open	Closed	**Overall**	Open	Closed	**Overall**	**Overall**
MEVF [1]	43.9	75.1	62.6	8.1	81.4	44.8	78.6
MMQ [2]	52.0	72.4	64.3	11.8	82.1	47.1	-
VQAMix [27]	56.6	79.6	70.4	13.4	83.5	48.6	-
AMAM [26]	63.8	80.3	73.3	18.2	84.4	50.4	-
CPRD [6]	61.1	80.4	72.7	-	-	-	82.1
PubMedCLIP [8]	60.1	80.0	72.1	-	-	-	80.1
MTL [9]	69.8	79.8	75.8	-	-	-	82.5
M3AE [10]	67.2	83.5	77.0	-	-	-	83.2
MUMC (Ours)	**71.5**	**84.2**	**79.2**	**39.0**	**90.4**	**65.1**	**84.9**

On the largest dataset, PathVQA [12], our method significantly outperformed the previous state-of-the-art model, AMAM [26], by a substantial margin with improvements of 20.8%, 6.0% and 14.7% on the closed-ended, open-ended, and overall answers, respectively. Moreover, on the SLAKE dataset [13], the proposed approach exhibited superior performance compared to the existing state-of-the-art model, M3AE, by a margin of 1.7% in terms of overall answer accuracy.

3.4 Ablation Study

To further verify the effectiveness of the proposed methods in learning multimodal representations, we conducted an ablation study across all three medical VQA datasets. Table 2 shows the overall performance of the medical VQA tasks using various pre-training approaches. Compared to the baseline pre-training tasks (i.e., MLM + ITM), integrating either UCL or MCL significantly improved the performance of the pre-trained model across all medical VQA datasets. Notably, the simultaneous use of UCL and MCL achieved a performance increase of 1.1%, 1.0%, and 0.9% on VQA-RAD, PathVQA, and SLAKE dataset, respectively.

Table 2. Ablation Study on Different Pre-training Objective Settings.

Training tasks	VQA-RAD (Overall)	PathVQA (Overall)	SLAKE (Overall)
ITM + MLM	74.5	61.5	82.0
ITM + MLM + UCL	77.3	63.5	83.2
ITM + MLM + MCL	78.1	64.1	84.0
MUMC(ITM + MLM + UCL + MCL)	**79.2**	**65.1**	**84.9**

Furthermore, to assess the performance of the proposed masked image strategy and identify the optimal masking probability, experiments were conducted by varying the masking probabilities of input images at levels of 0%, 25%, 50% and 75%. As presented in Table 3, the results are consistent among all the three datasets. With 25% masking probability, the model yielded the best results, compared to no masking applied. The performance decreased if 50% and 75% masking probabilities were used.

3.5 Visualization

We utilized Grad-CAM [14] to visualize the cross-attention maps between the questions and images, and analyzed the relevance of the attended image regions for generating the answers. In Fig. 2, it showed some attention maps that overlayed on the original images. For answering open-ended questions, the model accurately attended to the relevant infarct regions, as shown in Fig. 2a and Fig. 2b. In Fig. 2a, to answer the question, "Where is/are the infarct located?", the model highlighted the areas that well covered the infarction. Interestingly, the model attended to infarct areas on both hemispheres (Fig. 2b) and

Table 3. Study of different masked image probabilities.

Masking probability	VQA-RAD (Overall)	PathVQA (Overall)	SLAKE (Overall)
75%	76.9	63.4	82.6
50%	78.6	64.3	83.7
25%	**79.2**	**65.1**	**84.9**
0%	77.8	64.0	83.2

generated the answer, "Bilateral". Besides the position-related questions, in Fig. 2c, it showed the attention map to answer the closed form question, "Is there any region in the brain that is lesioned?". The model successfully attended to the lesion area and provided the correct answer of "Yes". Moreover, the model demonstrated its ability to attend to the regions of ribs to answer the counting-related question in Fig. 2d, where the question was "Are there more than 12 ribs?", and the model accurately outputted the answer "Yes".

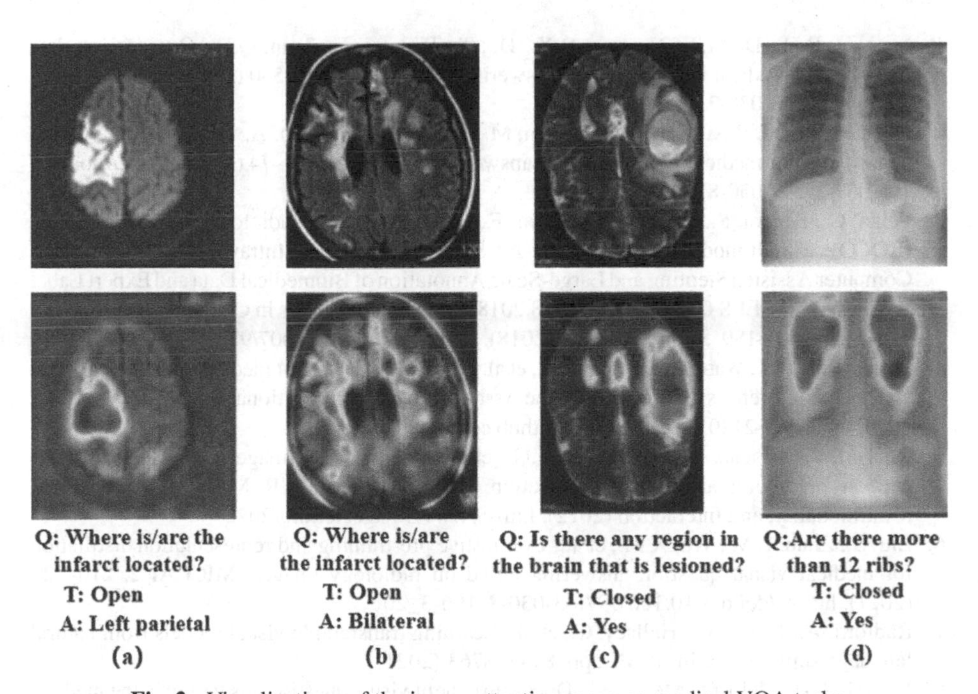
Q: Where is/are the infarct located?
T: Open
A: Left parietal
(a)
Q: Where is/are the infarct located?
T: Open
A: Bilateral
(b)
Q: Is there any region in the brain that is lesioned?
T: Closed
A: Yes
(c)
Q:Are there more than 12 ribs?
T: Closed
A: Yes
(d)

Fig. 2. Visualizations of the image attention maps on medical VQA tasks.

4 Conclusion

In this paper, we propose a new method to tackle the challenge of medical VQA tasks, which is pre-trained on the medical image caption datasets and then transferred to the downstream medical VQA tasks. The proposed self-supervised pre-training approach with unimodal and multimodal contrastive losses leads to significant performance improvement on three public VQA datasets. Also, using masked images as a data augmentation technique is proven to be effective for learning representations on medical visual and language tasks. As a result, our proposed method not only outperformed the state-of-the-art methods by a significant margin, but also demonstrated the potential for model interpretability.

Acknowledgement. This work is supported by Natural Science Foundation of Heilongjiang Province under grant number LH2021F015.

References

1. Nguyen, B.D., Do, T.T., Nguyen, B.X., Do, T., Tjiputra, E., Tran, Q.D.: Overcoming data limitation in medical visual question answering. MICCAI **4**, 522–530 (2019). https://doi.org/10.1007/978-3-030-32251-9_57
2. Do, T., Nguyen, B.X., Tjiputra, E., Tran, M., Tran, Q.D., Nguyen, A.: Multiple meta-model quantifying for medical visual question answering. MICCAI **5**, 64–74 (2021). https://doi.org/10.1007/978-3-030-87240-3_7
3. Pelka, O., Koitka, S., Rückert, J., Nensa, F., Friedrich, C.M.: Radiology objects in context (ROCO): a multimodal image dataset. In: Stoyanov, D., et al. Intravascular Imaging and Computer Assisted Stenting and Large-Scale Annotation of Biomedical Data and Expert Label Synthesis. LABELS CVII STENT 2018 2018 2018. Lecture Notes in Computer Science, vol. 11043, pp. 180–189. Springer, Cham (2018). https://doi.org/10.1007/978-3-030-01364-6_20
4. Subramanian, S., Wang, L.L., Bogin, B., et al.: Medicat: a dataset of medical images, captions, and textual references. In: Findings of the Association for Computational Linguistics, EMNLP 2020, pp. 2112–2120 (2020). https://github.com/allenai/medicat
5. Ruckert, J., Abacha, A.B., Herrera, A.G., et al.: Overview of ImageCLEF medical 2022-caption prediction and concept detection. In: Experimental IR Meets Multilinguality, Multimodality, and Interaction (2022). https://www.imageclef.org/2022
6. Liu, B., Zhan, L.M., Wu, X.M., et al.: Contrastive pre-training and representation distillation for medical visual question answering based on radiology images. MICCAI **2**, 210–220 (2021). https://doi.org/10.1007/978-3-030-87196-3_20
7. Radford, A., Kim, J.W., Hallacy, C., et al.: Learning transferable visual models from natural language supervision. In: ICML, pp. 8748–8763 (2021)
8. Eslami, S., de Melo, G., Meinel, C.: Does clip benefit visual question answering in the medical domain as much as it does in the general domain? arXiv preprint arXiv: 2112.13906 (2021)
9. Cong, F., Xu, S., Li, G., et al.: Caption-aware medical VQA via semantic focusing and progressive cross-modality comprehension. ACM Multimedia, pp. 3569–3577 (2022)
10. Chen, Z., Du, Y., Hu, J., et al.: Multi-modal masked autoencoders for medical vision-and-language pre-training. MICCAI **5**, 679–689 (2022). https://doi.org/10.1007/978-3-031-16443-9_65

11. Lau, J.J., Gayen, S., Abacha, A.B., Demner-Fushman, D.: A dataset of clinically generated visual questions and answers about radiology images. Sci. Data **5**, 1–10 (2018). https://osf.io/bd96f
12. He, X., et al.: Towards visual question answering on pathology images. In: Proceedings of the 59th Annual Meeting of the Association for Computational Linguistics and the 11th International Joint Conference on Natural Language Processing, ACL/IJCNLP, pp. 708–718 (2021). https://github.com/UCSD-AI4H/PathVQA
13. Liu, B., Zhan, L.M., Xu, L., et al.: Slake: a semantically-labeled knowledge-enhanced dataset for medical visual question answering. In: ISBI, IEEE, pp. 1650–1654 (2021). https://www.med-vqa.com/slake/
14. Selvaraju, R.R., Cogswell, M., Das, A., et al.: Grad-cam: visual explanations from deep networks via gradient-based localization. In: ICCV, pp. 618–626 (2017)
15. Vaswani, A., Shazeer, N., Parmar, N., et al.: Attention is all you need. In: NeurIPS **30** (2017)
16. Dosovitskiy, A., Beyer, L., Kolesnikov, A., et al.: An image is worth 16×16 words: Transformers for image recognition at scale. In: ICLR (2021)
17. Devlin, J., Chang, M.W., Lee, K., Toutanova, K.: BERT: pre-training of deep bidirectional transformers for language understanding. In: NAACL, pp. 4171–4186 (2019)
18. Sennrich, R., Haddow, B., Birch, A.: Neural machine translation of rare words with subword units. arXiv preprint arXiv: 1508.07909 (2015)
19. He, K., Fan, H., Wu, Y., et al.: Girshick: momentum contrast for unsupervised visual representation learning. In: Proceedings of the IEEE/CVF Conference on Computer Vision and Pattern Recognition, pp. 9726–9735 (2020)
20. Li, J., Selvaraju, R.R., Gotmare, A., et al.: Align before fuse: Vision and language representation learning with momentum distillation. Adv. Neural. Inf. Process. Syst. **34**, 9694–9705 (2021)
21. Li, J., Li, D, Xiong, C, et al.: Blip: bootstrapping language-image pre-training for unified vision-language understanding and generation. In: International Conference on Machine Learning. PMLR, pp. 12888–12900 (2022)
22. He, K., Chen, X., Xie, S., et al.: Masked autoencoders are scalable vision learners. In: Proceedings of the IEEE/CVF Conference on Computer Vision and Pattern Recognition, pp. 15979–15988 (2022)
23. Li, Y., Fan, H., Hu, R., et al.: Scaling Language-Image Pre-training via Masking. arXiv preprint arXiv: 2212.00794 (2022)
24. Cubuk, E.D., Zoph, B., Shlens, J., et al.: Randaugment: practical automated data augmentation with a reduced search space. In: CVPR Workshops, pp. 3008–3017 (2020)
25. Loshchilov, I., Hutter, F.: Decoupled weight decay regularization. arXiv preprint arXiv: 1711.05101 (2017)
26. Pan, H., He, S., Zhang, K., et al.: AMAM: an attention-based multimodal alignment model for medical visual question answering. Knowl.-Based Syst. **255**, 109763 (2022)
27. Gong, H., Chen, G., Mao, M., Li, Z., Li, G.: VQAMix: conditional triplet mixup for medical visual question answering. IEEE Trans. Med. Imaging **41**(11), 3332–3343 (2022)

CL-ADDA: Contrastive Learning with Amplitude-Driven Data Augmentation for fMRI-Based Individualized Predictions

Jiangcong Liu[1,2], Le Xu[1], Yun Guan[1], Hao Ma[1], and Lixia Tian[1](✉)

[1] School of Computer and Information Technology, Beijing Jiaotong University, Beijing 100044, China
lxtian@bjtu.edu.cn

[2] Beijing Key Laboratory of Traffic Data Analysis and Mining, Beijing Jiaotong University, Beijing 100044, China

Abstract. Effective representations of human brain function are essential for fMRI-based predictions of individual traits and classifications of neuropsychiatric disorders. Contrastive learning techniques can be favorable choices for representations of human brain function, if it were not for their requirement of large batch sizes. In this study, we proposed a novel method, namely, contrastive learning with amplitude-driven data augmentation (CL-ADDA), for effective representations of human brain function and ultimately fMRI-based individualized predictions. *SimSiam*, which sets no requirement on large batches, was used in this study to obtain discriminative representations among subjects to facilitate later predictions of individuals' traits. The fMRI data in this study was augmented based on recent neuroscience findings that fMRI frames with high- and low-amplitude are of quite different functional significance. Accordingly, we generated a positive pair by concatenating the fMRI frames with high-amplitude into one augmented sample and the frames with low-amplitude into another sample. The two augmented samples were used as inputs for CL-ADDA, and individualized predictions were made in an end-to-end way. The performance of the proposed CL-ADDA was evaluated with individualized age and IQ predictions based on a public dataset (Cam-CAN). The experimental results demonstrate that the proposed CL-ADDA can substantially improve the prediction performance as compared to the existing methods.

Keywords: *SimSiam* · Individualized Prediction · fMRI · Functional Connectivity

1 Introduction

In combination with machine learning techniques, functional magnetic resonance imaging (fMRI) has recently been widely used in predictions of individual traits (e.g., age and intelligence quotient (IQ)) [8, 12] and classifications of neuropsychiatric disorders (e.g.,

Supplementary Information The online version contains supplementary material available at https://doi.org/10.1007/978-3-031-43907-0_37.

H. Greenspan et al. (Eds.): MICCAI 2023, LNCS 14220, pp. 384–393, 2023.
https://doi.org/10.1007/978-3-031-43907-0_37

Alzheimer's disease). Representations of human brain function are essential for such predictions, as effective representations can provide information discriminative among individuals and facilitate the final predictions.

Contrastive learning techniques can be favorable choices for representations of human brain function [1, 3, 4, 10, 11], if the available samples size is large enough for large-batch training (this is not the fact for most fMRI datasets). Contrastive learning can learn effective representations through minimizing/maximizing the distance between similar/dissimilar samples in the representation space [4, 11]. *SimSiam* [5] is a contrastive learning framework that maximizes the similarity between two augmentations of one image. With the involvement of a Siamese structure, *SimSiam* does not rely on large-batch training. Accordingly, *SimSiam* can be a favorable choice for fMRI-based representations of human brain function.

Generation of similar/dissimilar samples is critical for contrastive learning [4]. Among the few studies on fMRI-based individualized predictions in which contrastive learning is involved [7, 9, 13, 21], fMRI data has often been augmented using classic methods in the region of computer vision, such as random erasing, random cropping and adversarial generation. In the pioneering study [21], two highly similar augmented samples were generated for each subject by excerpting two non-overlapping fMRI temporal segments. The highly similar pairs containing redundancy information can lead to poor performance of contrastive learning [2, 24]. Recent neuroscience findings provide an intuitive idea regarding data augmentation for RS-fMRI data. Specifically, fMRI frames with high- and low-amplitude were reported to be of quite different functional significance [23]. Accordingly, a "discrepant-enough" positive pair can be generated by acquiring one sample based on fMRI frames with high-amplitude and the other based on frames with low-amplitude.

In this study, we proposed a framework named contrastive learning with amplitude-driven data augmentation (CL-ADDA) for effective representations of human brain function and ultimately fMRI-based individualized predictions. Two augmented samples of CL-ADDA were generated through excerpting fMRI frames with relatively high amplitude and those with relatively low amplitude. With two augmented samples of the same subject used as inputs, a *SimSiam*-based contrastive learning framework was used to learn effective representations of human brain function. For the consideration that label information can guide *SimSiam* to learn more prediction-relevant representation [15, 25], individualized predictions were performed in an end-to-end way through concatenated fully connected layers.

Our major contributions are as follows:

- *SimSiam* was utilized to learn representations of human brain function.
- A neuroscience-oriented amplitude-driven data augmentation method was introduced to generate positive pairs.
- Predictions were made in an end-to-end way to improve the generalizability of the predictive models.
- CL-ADDA outperformed a variety of state-of-the-art methods for fMRI-based individualized predictions.

2 Method

2.1 Overall Workflow of CL-ADDA

Figure 1 shows the workflow of the proposed CL-ADDA. fMRI data of one subject is first excerpted into two segments, one composed of frames with high-amplitude and the other composed of frames with low-amplitude. Two functional connectivity (FC) maps (FC_{high} and FC_{low}) can then be obtained based on the augmented samples, and the two FC maps are used as inputs for the contrastive learning module. The *SimSiam* structure is used to perform contrastive learning in this study, and the classic convolution in *SimSiam* is replaced by row and column convolutions to adapt to FC maps. *SimSiam* learns representations of brain function (r_{high} and r_{low}) based on the two FC maps, using encoders (F) with shared parameters. Individualized predictions can be made through a predictor (Φ) (three fully connected layers in this study) based on the learned representations. The whole model is trained though simultaneously minimizing the distance between the two representations (r_{high} and r_{low}) and the difference between the predicted and actual label (individual traits in this study).

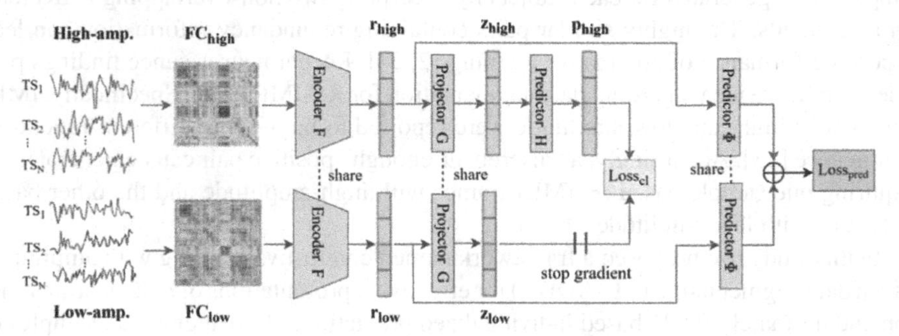

Fig. 1. Workflow of CL-ADDA. The high- and low-amplitude time series are generated from the same subject. TS - time series; FC - functional connectivity; amp. - amplitude.

2.2 Amplitude-Driven Data Augmentation

Figure 2 provides an illustration of the amplitude-driven data augmentation method. In this study, we generated two augmented samples for each subject based on the amplitude of fMRI frames following [23]. Specifically, we first defined N ROIs and extracted the mean time series of each ROI. We then z-scored each time series and obtained frame-wise co-fluctuation of each ROI pair as follows:

$$e_{jk} = \left[e_{jk}^{1}, e_{jk}^{2}, \ldots, e_{jk}^{T}\right],\ e_{jk}^{t} = x_{j}^{t} \cdot x_{k}^{t} \tag{1}$$

where e_{jk}^{t} is the co-fluctuation of ROIs-j and k at time t; x_j^t $\left(x_k^t\right)$ is the z-scored fMRI signal amplitude of ROI-j (-k) at time t; e_{jk} is obtained the co-fluctuation time series, and T is the length of fMRI time series. Accordingly, a $C_N^2 \times T$ co-fluctuation matrix

$E = [e_{12}; e_{13}; ...e_{1N}; e_{23}; e_{24}; ...e_{2N}; ...e_{(N-1)N}]$ can be obtained. We computed the root sum square (RSS) of the co-fluctuation matrix E at each time point and finally generated one high-amplitude sample by excerpting the top 50% of frames with high co-fluctuation RSS, and one low-amplitude sample by concatenating the remaining frames (with low co-fluctuation RSS).

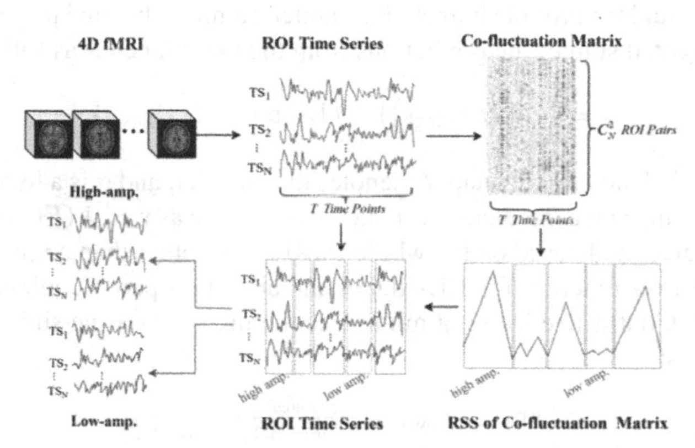

Fig. 2. Illustration of the amplitude-driven data augmentation method. TS - time series; amp. - amplitude; RSS - root sum square.

2.3 Contrastive Learning on Functional Connectivity Maps

We constructed the contrastive learning model based mainly on *SimSiam.* Setting no requirement on large batches, *SimSiam* can be a favorable choice for fMRI-based representation learning [5]. As shown in Fig. 1, the *SimSiam* structure in this study consists of two branches of the same encoders (F) with shared weights for encoding the two input FC maps in parallel, followed by the same nonlinear projectors (G) with share weights for further processing the brain function representations from the encoders, and one predictor (H) on one branch for transforming the output of the branch to match it to the output of the other branch. A stop-gradient operation is applied on the branch without predictor to avoid model collapsing [5].

For the consideration that spatial locality does not exist among adjacent elements on FC maps [14], row and column convolutions were used to construct the backbone of the encoder (F) and extract effective information from the FC maps. Specifically, each encoder (F) in this study consists of one row convolution layer (with $C_r@1 \times N$ row convolution filter, C_r is the channel number, N is the ROI number) and one column convolution layer (with $C_c@N \times 1$ column convolution filters, C_c is the channel number, outputs $C_c@1 \times 1$ representation). Information throughout the brain is expected to be integrated with the use of row and column convolutions. The row and column convolution layers in CL-ADDA are each followed by a batch normalization layer and a *Leaky_ReLU* layer.

2.4 Individualized Prediction and Loss Function

As shown in Fig. 1, individualized prediction in this study was implemented using two parallel predictors (Φ) with shared weights. Each predictor ((Φ)) consists of three fully connected layers, which transform the C_c outputs from the corresponding encoder (F) (learned representations of brain function, r_{high} and $r_{low} \in R^{C_c}$ in Fig. 1) into the predicted individual trait for the branch. For model training, the final prediction is made based on a weighted sum of the predictions from the two branches as follows:

$$\hat{y} = \alpha \cdot \Phi\big(F\big(\xi_{\text{high}}\big)\big) + (1 - \alpha) \cdot \Phi(F(\xi_{\text{low}})) \tag{2}$$

where $\xi \in \mathrm{R}^{\mathrm{N} \times \mathrm{N}}$ denotes a FC map, F denotes the encoder, and α is a hyper-parameter. In the model testing stage, prediction can directly be made as $\hat{y} = \Phi(F(\xi))$, where ξ can be a FC map calculated based on the whole fMRI scan (rather than augmented data).

The whole loss function for CL-ADDA includes two parts: contrastive loss and prediction loss. Contrastive learning minimizes the negative cosine similarity between the two branches:

$$\begin{aligned} \mathrm{D}(\mathrm{p}_{high}, \mathrm{z}_{\text{low}}) &= -\frac{p_{high}}{\|p_{high}\|_2} \cdot \frac{z_{low}}{\|z_{low}\|_2} \\ \mathrm{D}(\mathrm{p}_{low}, \mathrm{z}_{\text{high}}) &= -\frac{p_{low}}{\|p_{low}\|_2} \cdot \frac{z_{high}}{\|z_{high}\|_2} \end{aligned} \tag{3}$$

where $p_{low} = H(G(F(\xi_{low})))$, $p_{high} = H(G(F(\xi_{high})))$, $z_{low} = G(F(\xi_{low}))$, $z_{high} = G(F(\xi_{\text{high}}))$, G denotes the projector in Fig. 1, and H denotes the predictor in the contrastive learning (Fig. 1); $\|.\|_2$ is L2-norm.

Following [25], we defined contrastive loss as:

$$Loss_{cl} = \frac{1}{2}D(p_{low}, stopgrad(z_{high})) + \frac{1}{2}D(p_{high}, stopgrad(z_{low})) \tag{4}$$

L1 loss was used as the prediction loss:

$$Loss_{pred} = \|y - \hat{y}\|_1 \tag{5}$$

where y and $\hat{y}$ are the actual and predicted labels, respectively. The total loss was defined as follows:

$$Loss = \lambda \cdot Loss_{pred} + (1 - \lambda) \cdot Loss_{cl} \tag{6}$$

where λ is hyper-parameter.

3 Experiments and Results

3.1 Dataset

The resting-state fMRI data included in the dataset collected and released by the Cambridge Centre for Ageing and Neuroscience (Cam-CAN) [19] was used in this study. The public dataset contains multi-modal data from a large cohort of adult lifespan population-based samples. After removing the subjects with excessive head motions (translation/rotation more than 2.0 mm/2.0° in/around any of the x, y, or z directions) throughout the scan, 600 subjects remained (18–87, 53.900 ± 18.549 years), and IQ scores of 568 of them were available (11–44, 32.032 ± 6.762). We preprocessed the data to remove the spatial and temporal artifacts and register the images to standard space (MNI) using FSL. We defined 200 ROIs based on independent component analysis and extracted the ROI time series through regressing the spatial maps of the independent components released by the Human Connectome Project (HCP) [20] against the preprocessed fMRI data using the *dual_regress* command included in FSL. Later analyses were all based on the extracted ROI time series.

3.2 The Performance of CL-ADDA

Age and IQ predictions were taken as test cases to evaluate the performance of the proposed method, based on the Cam-CAN dataset. Amplitude-driven data augmentation was performed on the 200 ROI time series of each subject, and two FC maps were obtained based on two augmented samples for each subject. The two augmented FC maps were used as inputs for *SimSiam* for representation learning and later individualized predictions. We performed 1000 epochs of model training, with the batch size set to 128. For the consideration that high-amplitude FC maps (FC_{high}) may carry more detailed information about individuals' brain function, we empirically weight the predictions based on FC_{high} more, by setting the α in Eq. (2) to 0.8 The hyper-parameter λ in Eq. (6) was set to 0.5. Adam optimizer with a learning rate of 0.001 was used.

Ten-fold cross-validation was used to evaluate the performance of CL-ADDA, and Pearson's correlation coefficient (r) and mean absolute error (MAE) between the predicted and actual labels were used to quantitatively measure this performance. The results show that CL-ADDA performed well on both age and IQ predictions, as indicated by an r-value (MAE) of 0.886 (6.992 years) for age prediction, and 0.620 (4.531) for IQ prediction.

3.3 Comparison Experiments

We compared the performance of our proposed method with six deep learning methods for fMRI-based individualized predictions, namely, spatial-temporal graph convolutional network (ST-GCN) [22], RNN based on gated recurrent units (GRU) [6], pooling regularized graph neural network (PR-GNN) [16], brain graph neural network (BrainGNN) [17], simple fully convolutional network (SFCN) [18], and BrainNetCNN [14]. Each of the six methods has been reported to perform well on fMRI-based individualized predictions, or even provide state-of-the-art results. Each method was implemented based on its online code, with the hyper-parameters set according to its original paper.

Table 1 is a list of prediction accuracies based on the seven methods (including CL-ADDA). According to Table 1, CL-ADDA outperformed the methods for comparison by large margins. For instance, compared with the second best (ST-GCN), CL-ADDA demonstrated an r-value increase of 0.085 for IQ prediction.

Table 1. Age and IQ prediction accuracies based on different deep learning methods.

Age Prediction							
Method	ST-GCN	GRU	PR-GNN	BrainGNN	SFCN	BrainNetCNN	CL-ADDA (ours)
r	0.848	0.789	0.808	0.811	0.738	0.711	**0.886**
MAE (years)	7.702	9.733	9.347	9.008	9.815	11.566	**6.992**
IQ Prediction							
Method	ST-GCN	GRU	PR-GNN	BrainGNN	SFCN	BrainNetCNN	CL-ADDA (ours)
r	0.535	0.458	0.488	0.515	0.402	0.412	**0.620**
MAE	4.602	5.572	5.263	4.614	5.53	6.935	**4.531**

3.4 Ablation Experiments

To evaluate the effectiveness of the proposed amplitude-driven data augmentation strategy, we performed age and IQ predictions with the data augmented using classic methods, and the strategy of excerpting non-overlapping segments as proposed in [21]. For classic augmentation, two 175 × 175 augmented FC maps were generated by applying random cropping and Gaussian blurring on the 200 × 200 FC matrix calculated based on the whole fMRI scan [5]. For the non-overlapping segment excerpting strategy, we generated two 200 × 200 augmented FC matrices based on the two non-overlapping fMRI data segments excerpted from the same scan [21]. According to Table 2, the proposed amplitude-driven data augmentation is obviously superior to the other two data augmentation strategies.

We further evaluated the effectiveness of contrastive learning, as well as the end-to-end individualized prediction strategy. Specifically, (1) we performed individualized age and IQ predictions based on a network composed of one encoder (F) and one predictor (Φ) to imitate a network with contrastive learning removed. (2) We pre-trained CL-ADDA and then predicted age and IQ using the representations based on the pre-trained CL-ADDA to imitate abandoning the end-to-end individualized prediction strategy. According to Table 3, both contrastive learning and the end-to-end individualized prediction strategy were critical for CL-ADDA. The results indicate that supervised contrastive learning can be a favorable choice for neuroimage-based individualized predictions and neuropsychiatric disease classifications.

Table 2. Age and IQ prediction accuracies based on different data augmentation strategies.

Age Prediction			
Method	Classic Augmentation	Non-Overlapping Segments	Amplitude-Driven (ours)
r	0.349	0.831	**0.886**
MAE (years)	14.849	9.167	**6.992**
IQ Prediction			
Method	Classic Augmentation	Non-overlapping Segments	Amplitude-Driven (ours)
r	0.288	0.589	**0.620**
MAE	5.333	5.034	**4.531**

Table 3. Age and IQ prediction accuracies based on CL-ADDA with different modules being removed.

Age Prediction			
Method	with Individualized Prediction Removed	with Contrastive Learning Removed	CL-ADDA (ours)
r	0.782	0.873	**0.886**
MAE (years)	9.467	7.377	**6.992**
IQ Prediction			
Method	with Individualized Prediction Removed	with Contrastive Learning Removed	CL-ADDA
r	0.573	0.589	**0.620**
MAE	4.779	5.042	**4.531**

4 Conclusion

In this study, we proposed CL-ADDA for effective representation learning and ultimately precise fMRI-based individualized predictions. Originating from a recent neuroscientific finding, the proposed amplitude-driven data augmentation method provides the contrastive learning module discrepant-enough positive pairs for effective representation learning. *SimSiam*-based contrastive learning enables effective representation learning on fMRI dataset including limited samples. We evaluated the performance of CL-ADDA with age and IQ predictions based on a public dataset, and the experiments demonstrate that CL-ADDA achieved state-of-the-art predictions for both age and IQ.

Acknowledgement. We thank investigators from Cambridge Centre for Ageing and Neuroscience for sharing the public dataset.

References

1. Bachman, P., Hjelm, R.D., Buchwalter, W.: Learning representations by maximizing mutual information across views. Adv. Neural Inf. Process. Syst. **32** (2019)
2. Barlow, H.B.: Possible principles underlying the transformation of sensory messages. Sens. Commun. **1**(01), 217–233 (1961)
3. Caron, M., Misra, I., Mairal, J., Goyal, P., Bojanowski, P., et al.: Unsupervised learning of visual features by contrasting cluster assignments. Adv. Neural. Inf. Process. Syst. **33**, 9912–9924 (2020)
4. Chen, T., Kornblith, S., Norouzi, M., Hinton, G.: A simple framework for contrastive learning of visual representations. In: International Conference on Machine Learning, pp. 1597–1607. PMLR (2020)
5. Chen, X., He, K.: Exploring simple siamese representation learning. In: Proceedings of the IEEE/CVF International Conference on Computer Vision. pp. 15750–15758 (2021)
6. Chung, J., Gulcehre, C., Cho, K., Bengio, Y.: Empirical evaluation of gated recurrent neural networks on sequence modeling. arXiv preprint arXiv:1412.3555 (2014)
7. Dufumier, B., Gori, P., Victor, J., Grigis, A., Wessa, M., Brambilla, P., Favre, P., Polosan, M., McDonald, C., Piguet, C.M., Phillips, M., Eyler, L., Duchesnay, E.: Contrastive learning with continuous proxy meta-data for 3D MRI classification. In: de Bruijne, M., Cattin, P.C., Cotin, S., Padoy, N., Speidel, S., Zheng, Y., Essert, C. (eds.) MICCAI 2021. LNCS, vol. 12902, pp. 58–68. Springer, Cham (2021). https://doi.org/10.1007/978-3-030-87196-3_6
8. Gadgil, S., Zhao, Q., Pfefferbaum, A., Sullivan, E.V., Adeli, E., Pohl, K.M.: Spatio-temporal graph convolution for resting-state fMRI analysis. In: Martel, A.L., Abolmaesumi, P., Stoyanov, D., Mateus, D., Zuluaga, M.A., Zhou, S.K., Racoceanu, D., Joskowicz, L. (eds.) MICCAI 2020. LNCS, vol. 12267, pp. 528–538. Springer, Cham (2020). https://doi.org/10.1007/978-3-030-59728-3_52
9. Grigis, A., Gomez, C., Tasserie, J., Ambroise, C., Frouin, V., et al.: Predicting cortical signatures of consciousness using dynamic functional connectivity graph-convolutional neural networks. BioRxiv, pp. 2020–2005 (2020)
10. Grill, J.B., Strub, F., Altché, F., Tallec, C., Richemond, P., et al.: Bootstrap your own latent-a new approach to self-supervised learning. Adv. Neural. Inf. Process. Syst. **33**, 21271–21284 (2020)
11. He, K., Fan, H., Wu, Y., Xie, S., Girshick, R.: Momentum contrast for unsupervised visual representation learning. In: Proceedings of the IEEE/CVF International Conference on Computer Vision, pp. 9729–9738 (2020)
12. He, T., Kong, R., Holmes, A.J., Nguyen, M., Sabuncu, M.R., et al.: Deep neural networks and kernel regression achieve comparable accuracies for functional connectivity prediction of behavior and demographics. Neuroimage **206**, 116276 (2020)
13. Hsieh, W.T., Lefort-Besnard, J., Yang, H.C., Kuo, L.W., Lee, C.C.: Behavior score-embedded brain encoder network for improved classification of Alzheimer disease using resting state fMRI. In: International Conference of the IEEE Engineering in Medicine & Biology Society, pp. 5486–5489. IEEE (2020)
14. Kawahara, J., Brown, C.J., Miller, S.P., Booth, B.G., Chau, V., et al.: BrainNetCNN: convolutional neural networks for brain networks; towards predicting neurodevelopment. Neuroimage **146**, 1038–1049 (2017)
15. Li, J., Zhao, G., Tao, Y., Zhai, P., Chen, H., et al.: Multi-task contrastive learning for automatic CT and X-ray diagnosis of COVID-19. Pattern Recogn. **114**, 107848 (2021)
16. Li, X., Zhou, Y., Dvornek, N.C., Zhang, M., Zhuang, J., Ventola, P., Duncan, J.S.: Pooling regularized graph neural network for fmri biomarker analysis. In: Martel, A.L., Abolmaesumi, P., Stoyanov, D., Mateus, D., Zuluaga, M.A., Zhou, S.K., Racoceanu, D., Joskowicz, L. (eds.)

MICCAI 2020. LNCS, vol. 12267, pp. 625–635. Springer, Cham (2020). https://doi.org/10.1007/978-3-030-59728-3_61
17. Li, X., Zhou, Y., Dvornek, N., Zhang, M., Gao, S., et al.: Braingnn: interpretable brain graph neural network for FMRI analysis. Med. Image Anal. **74**, 102233 (2021)
18. Peng, H., Gong, W., Beckmann, C.F., Vedaldi, A., Smith, S.M.: Accurate brain age prediction with lightweight deep neural networks. Med. Image Anal. **68**, 101871 (2021)
19. Taylor, J.R., Williams, N., Cusack, R., Auer, T., Shafto, M.A., et al.: The Cambridge Centre for Ageing and Neuroscience (Cam-CAN) data repository: Structural and functional MRI, MEG, and cognitive data from a cross-sectional adult lifespan sample. Neuroimage **144**, 262–269 (2017)
20. Van Essen, D.C., Ugurbil, K., Auerbach, E., Barch, D., Behrens, T.E., et al.: The human connectome project: a data acquisition perspective. Neuroimage **62**(4), 2222–2231 (2012)
21. Wang, X., Yao, L., Rekik, I., Zhang, Y.: Contrastive functional connectivity graph learning for population-based fMRI classification. In: Wang, L., Dou, Q., Fletcher, P.T., Speidel, S., Li, S. (eds.) Medical Image Computing and Computer Assisted Intervention – MICCAI 2022. MICCAI 2022. Lecture Notes in Computer Science, vol. 13431, pp. 221–230. Springer, Cham.https://doi.org/10.1007/978-3-031-16431-6_21
22. Yan, S., Xiong, Y., Lin, D.: Spatial temporal graph convolutional networks for skeleton-based action recognition. In: Proceedings of the AAAI Conference on Artificial Intelligence vol. 32, no. 1 (2018)
23. Zamani Esfahlani, F., Jo, Y., Faskowitz, J., Byrge, L., Kennedy, D.P., et al.: High-amplitude cofluctuations in cortical activity drive functional connectivity. Proc. Natl. Acad. Sci. **117**(45), 28393–28401 (2020)
24. Zbontar, J., Jing, L., Misra, I., LeCun, Y., Deny, S.: Barlow twins: Self-supervised learning via redundancy reduction. In: International Conference on Machine Learning, pp. 12310–12320. PMLR (2021)
25. Zhao, Z., Liu, H.: Semi-supervised feature selection via spectral analysis. In: Proceedings of the SIAM international conference on data mining. pp. 641–646. Society for Industrial and Applied Mathematics (2007)

An Auto-Encoder to Reconstruct Structure with Cryo-EM Images via Theoretically Guaranteed Isometric Latent Space, and Its Application for Automatically Computing the Conformational Pathway

Kimihiro Yamazaki[1], Yuichiro Wada[1,2], Atsushi Tokuhisa[3], Mutsuyo Wada[1], Takashi Katoh[1], Yuhei Umeda[1], Yasushi Okuno[3,4], and Akira Nakagawa[1](✉)

[1] Fujitsu Ltd., Kanagawa, Japan
anaka@fujitsu.com
[2] RIKEN Center for Advanced Intelligence Project, Tokyo, Japan
[3] RIKEN Center for Computational Science, Hyogo, Japan
[4] Graduate School of Medicine, Kyoto University, Kyoto, Japan

Abstract. Structural analysis by cryo-electron microscopy (Cryo-EM) has become well-established in the field of structural biology. Recently, cutting-edge methods have been proposed for the purpose of reconstructing either a small set of structures or a conformational pathway (continuous structural change), where a 3D density map represents the structure. However, we usually perform heavy manual labor to define the plausible pathway related to biological significance. In this study, for automatizing such manual labor, we propose a deep Auto-Encoder (AE) with a trainable prior. The AE is trained using only a set of single particle Cryo-EM images. The trained AE reconstructs the corresponding structures for the latent variables of the Cryo-EM images. The latent distribution can not only be theoretically proportional to a distribution of the structure but also consistent with the trained prior. Taking advantage of this property, we can automatically compute the pathway by only accessing the latent space as follows: i) generating a ridgeline on the latent distribution and ii) defining the conformational pathway as a sequence of the reconstructed structures along the ridgeline using the trained decoder. In our numerical experiments, we evaluate the computed pathways by comparing them with existing ones that were manually determined by other researchers, and confirm that they are sufficiently consistent.

Keywords: Cryo-EM · Conformational Pathway · Deep Learning · 3D Reconstruction · Structural Biology

K. Yamazaki, Y. Wada, A. Tokuhisa and M. Wada—Represents equal contribution

Supplementary Information The online version contains supplementary material available at https://doi.org/10.1007/978-3-031-43907-0_38.

H. Greenspan et al. (Eds.): MICCAI 2023, LNCS 14220, pp. 394–404, 2023.
https://doi.org/10.1007/978-3-031-43907-0_38

1 Introduction

Over the past few years, Cryo Electron Microscopy (Cryo-EM) has made remarkable progress in biomolecule structure analysis, becoming a major structural analysis technique along with X-ray crystallography. Based on a brief history of the early period of the technique, researchers had proposed methods that reconstruct the static structures of protein molecules from a set of single particle Cryo-EM images, e.g., EMAN2 [24], RELION [20,21,31], and cryoSPARC [17,18]. Methods that reconstruct nonstatic structures have also been proposed, e.g., cryoDRGN [28–30], e2gmm [2], 3DVA [16], and 3DFlex [15]. In these methods, a 3D density map expresses the structure and helps to find biological significance [25,27].

Recently, taking advantage of method capturing the nonstatic structures, researchers have tried to define a plausible conformational pathway (continuous change with 3D density map) from only single particle Cryo-EM images [10,26] to efficiently determine the biological significance (e.g., new drug). The pathway is usually defined via the latent space of a trained deep Auto-Encoder (AE) by the Cryo-EM images. For example, in [10], the authors first collect the latent variables, which are the outputs of the trained encoder in cryoDRGN [28], as inputs of the Cryo-EM images. Then, the variables are mapped into a 2D space by UMAP [13]. Thereafter, a sequence of the 2D points is manually constructed via qualitatively evaluating the corresponding structures to those points. The structure is reconstructed by the trained decoder of cryoDRGN. Lastly, the conformational pathway is defined as a sequence of the reconstructed structures along the sequence of the 2D points. Although the protocol of [10] is state-of-the-art, it can be tedious as it involves the manual construction with qualitative evaluation. The following are two of the most important causes: i) for a trained AE of cryoDRGN, there are no theoretical insights into the relation between the latent distribution and the distribution of the structure and ii) the latent distribution of the trained AE is not consistent with the fixed prior, i.e., standard normal Gaussian.

In this study, we propose a deep AE with a trainable prior that is expressed by a Gaussian Mixture Model (GMM). We name the AE *cryoTWIN*[1]. CryoTWIN is trained by single particle Cryo-EM images with the estimated pose orientations under an objective inspired by RaDOGAGA [8]. The trained AE reconstructs the nonstatic structures for the latent variables of the Cryo-EM images. A property of the trained AE is that its latent space is theoretically isometric with a space of the structure, where the latent distribution can be proportional to the distribution of the structure. Additionally, the trained prior can fit the latent distribution. This useful property helps us compute a ridgeline on the trained GMM as a sequence of the latent variables and automatically define the conformational pathway by a sequence of the reconstructed structure along the ridgeline.

Our main contributions are as follows: i) we propose cryoTWIN: a deep AE model with the beneficial property for computing the conformational pathway and ii) in our numerical experiments, we confirm that the pathway computed using cryoTWIN is sufficiently consistent with an existing one, that was manually determined by researchers.

[1] The name "TWIN" comes from the isometricity.

In Sect. 2, we describe representatives of methods introduced at the beginning of this section in detail. Furthermore, we explain the differences of cryoTWIN compared with RaDOGAGA and cryoDRGN, as our method is partly inspired by them. In Sect. 3, we give a detailed account of cryoTWIN with theoretical guarantees. Additionally, we introduce an algorithm to compute the conformational pathway. In Sect. 4, we present the results obtained in our numerical experiments using a ribosome dataset. Finally, in Sect. 5, we conclude this study and discuss our future work.

2 Related Work

2.1 Existing Reconstruction Methods

The representative method from the static structure category is cryoSPARC [17,18]. Given a set of single particle Cryo-EM images and the number of structures, N, this method reconstructs N structures using an efficient stochastic gradient descent technique. Cryo-SPARC has a resolution similar to that of RELION but a smaller computational complexity [20,21,31]. The representative method from the nonstatic structure category is cryoDRGN [28–30], whose statistical model is based on spatial-VAE [1] (a variant of VAE [9]). The training objective is to maximize the variational lower-bound. After the training, cryoDRGN can reconstruct continuous structures using the continuous latent variables and the trained decoder.

2.2 RaDOGAGA Revisit

RaDOGAGA [8] is an AE whose latent space is isometric [5] to the input space, inspired by the rate-distortion theory [3]. Let f_θ, g_ϕ, and P_ψ denote an encoder, a decoder, and a trainable prior distribution of latent variables, respectively, where θ, ϕ, and ψ are a set of trainable parameters. Then, the training objective is $\arg\min_{\theta,\phi,\psi} \frac{1}{n}\sum_{i=1}^{n}\{\|\boldsymbol{x}_i - \tilde{\boldsymbol{x}}_{\boldsymbol{z}_i}\|_2^2 - \beta \log Q_{\boldsymbol{z}_i}\}$, where $\|\boldsymbol{x}_i - \tilde{\boldsymbol{x}}_{\boldsymbol{z}_i}\|_2$ (resp. $-\log Q_{\boldsymbol{z}_i}$) corresponds to the distortion (resp. rate). Here, $\boldsymbol{x}_i, (i = 1, ..., n)$ is the i-th original data. Additionally, $\tilde{\boldsymbol{x}}_{\boldsymbol{z}_i} = g_\phi(\boldsymbol{z}_i + \boldsymbol{\epsilon}), \boldsymbol{z}_i = f_\theta(\boldsymbol{x}_i)$, and each element of $\boldsymbol{\epsilon}$ is uniformly sampled from $[-T/2, T/2]$. Moreover, $\beta > 0$ is a hyper-parameter, and $Q_{\boldsymbol{z}_i}$ is given by

$$Q_{\boldsymbol{z}_i} = \int_{-\infty}^{\infty} U(\boldsymbol{z} - \boldsymbol{z}_i) P_\psi(\boldsymbol{z}) \mathrm{d}\boldsymbol{z}, \tag{1}$$

where $U(\boldsymbol{z})$ is a rectangular window function: $U(\boldsymbol{z}) = 1$ if $-T/2 \le z_j \le T/2$ holds for all j (z_j is the j-th element of $\boldsymbol{z}$), and $U(\boldsymbol{z}) = 0$ otherwise. Let $\boldsymbol{\delta}_1$ and $\boldsymbol{\delta}_2$ be infinitesimal vectors with arbitrary directions. Then, the optimally trained AE, whose decoder is g_{ϕ^*}, has the following isometric property at all $\boldsymbol{z}$ for any $\boldsymbol{\delta}_1$ and $\boldsymbol{\delta}_2$:

$$\underbrace{\langle(\boldsymbol{z} + \boldsymbol{\delta}_1) - \boldsymbol{z}, (\boldsymbol{z} + \boldsymbol{\delta}_2) - \boldsymbol{z}\rangle}_{\text{abbreviated as } \mathrm{d}z_1 \cdot \mathrm{d}z_2} \propto \underbrace{\langle \hat{\boldsymbol{x}}_{\boldsymbol{z}+\boldsymbol{\delta}_1} - \hat{\boldsymbol{x}}_{\boldsymbol{z}}, \hat{\boldsymbol{x}}_{\boldsymbol{z}+\boldsymbol{\delta}_2} - \hat{\boldsymbol{x}}_{\boldsymbol{z}}\rangle}_{\text{abbreviated as } \mathrm{d}\hat{x}_1 \cdot \mathrm{d}\hat{x}_2}, \tag{2}$$

where $\hat{\boldsymbol{x}}_{\boldsymbol{z}} = g_{\phi^*}(\boldsymbol{z})$. Because of Eq. (2), $P(\boldsymbol{z}) \propto P(\hat{\boldsymbol{x}}_{\boldsymbol{z}})$ holds, where $P(\boldsymbol{z})$ (resp. $P(\hat{\boldsymbol{x}}_{\boldsymbol{z}})$) is the latent distribution (resp. the probability density function of $\hat{\boldsymbol{x}}_{\boldsymbol{z}}$). The following equation is derived by applying $\boldsymbol{\delta}_1 = \boldsymbol{z}' - \boldsymbol{z}$ and $\boldsymbol{\delta}_2 = \boldsymbol{\delta}_1$ to Eq. (2) :

$$\|\boldsymbol{z} - \boldsymbol{z}'\|_2 \approx 0 \Rightarrow \|\boldsymbol{z} - \boldsymbol{z}'\|_2 \propto \|\hat{\boldsymbol{x}}_{\boldsymbol{z}} - \hat{\boldsymbol{x}}_{\boldsymbol{z}'}\|_2.$$

2.3 Differences Between cryoTWIN and Existing Methods

To understand the difference between cryoTWIN and RaDOGAGA, let V (resp. I) denote a 3D density map (resp. the Cryo-EM image). For cryoTWIN, only I is required to make the latent space isometric to a space of V, whereas V (usually inaccessible) is required to make RaDOGAGA have the same isometricity. Differences between cryoTWIN and cryoDRGN are i) the latent distribution of cryoTWIN is theoretically proportional to a distribution of V, whereas cryoDRGN does not hold such property and ii) cryoTWIN can fit the prior to the latent distribution, whereas cryoDRGN can not.

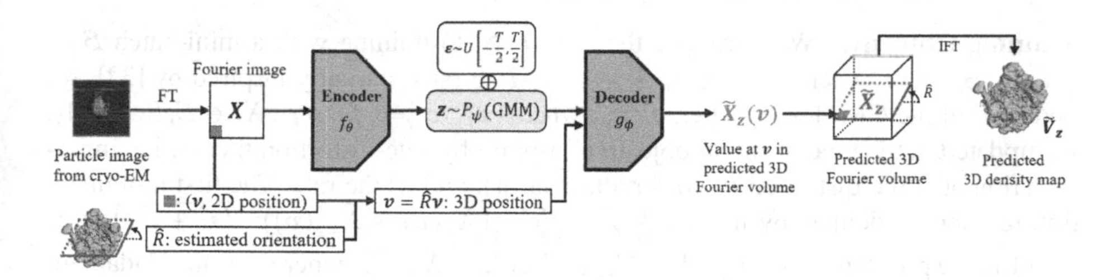

Fig. 1. Diagram of cryoTWIN during training.

3 Proposed Method

We first describe our deep AE, cryoTWIN, with the theoretical guarantees before we explain how to compute the conformational pathway using cryoTWIN. Given a set of single particle Cryo-EM images $\mathcal{I} = \{I_i\}_{i=1}^n$, the AE is trained via two steps: the first step is preprocessing and the second one is to train the AE under a rate distortion theory-based objective, such as RaDOGAGA (see Sect. 2.2). The diagram of cryoTWIN during the training is shown in Fig. 1. In the preprocessing, we estimate the orientation of the Cryo-EM image I_i in the 3D density map, as our training requires the estimated orientation $\hat{R}_i \in \mathbb{R}^{3\times3}$. Such $\hat{R}_i$ can be obtained by setting one as the number of models in RELION or cryoSPARC. For the prediction after training, given the latent variable z_i of the image I_i, the trained decoder predicts the corresponding 3D density map to I_i. Further details for the diagram, the training objective, and the prediction after training are presented in Sect. 3.1. The theoretical guarantees are presented in Sect. 3.2. In Sect. 3.3, we describe an algorithm to compute the conformational pathway.

3.1 cryoTWIN

As shown in Fig. 1, a deep AE of cryoTWIN consists of P_ψ, f_θ, and g_ϕ, which denote a trainable GMM (prior), an encoder, and the decoder, respectively. Symbols ψ, θ, and ϕ represent a set of trainable parameters. The parameter ψ is expressed as $\psi := \{(\pi_c, \boldsymbol{\mu}_c, \Sigma_c)\}_{c=1}^C$, where $\boldsymbol{\mu}_c, \Sigma_c$ respectively, indicate the mean and variance of the c-th Gaussian distribution and $\pi_c \in [0, 1], \sum_{c=1}^C \pi_c = 1$ is a weight for the c-th distribution. Additionally, C denotes the number of components.

Diagram During Training (Fig. 1): Let $\mathcal{F}$ be Fourier Transform (FT), and let $X_i = \mathcal{F}I_i \in \mathbb{R}^{D\times D}$ denote the Fourier image of the Cryo-EM image $I_i \in \mathcal{I}$. Given X_i, the encoder f_θ firstly returns $(\boldsymbol{z}_i, \boldsymbol{w}_i) = f_\theta(X_i), \boldsymbol{z}_i \in \mathbb{R}^d, \boldsymbol{w}_i \in \mathbb{R}^C$, where $\boldsymbol{z}_i$ is the latent variable of X_i. The parameter $\boldsymbol{w}_i$ is used to enforce the latent empirical distribution to follow P_ψ. Secondly, using $\boldsymbol{z}_i$, $\hat{R}_i$, and a 2D position $\boldsymbol{\nu} = (s,t,0)^\top$ in X_i, the decoder g_ϕ returns $\tilde{X}_{\boldsymbol{z}_i}(\boldsymbol{v}) = g_\phi(\boldsymbol{z}_i + \boldsymbol{\epsilon}, \boldsymbol{v}), \boldsymbol{v} = \hat{R}_i\boldsymbol{\nu}$, where $\tilde{X}_{\boldsymbol{z}_i}(\boldsymbol{v})$ means the predicted value at the 3D position $\boldsymbol{v}$ in a 3D Fourier volume and $\boldsymbol{\epsilon} \in \mathbb{R}^d$ is a random noise vector, each of whose elements is uniformly sampled from $[-T/2, T/2]$. By computing $\tilde{X}_{\boldsymbol{z}_i}(\boldsymbol{v}), \boldsymbol{v} = \hat{R}_i\boldsymbol{\nu}$ for all 2D positions $\boldsymbol{\nu}$ (i.e., for all pairs (s,t)), the predicted Fourier image $\tilde{X}_{\boldsymbol{z}_i}$ is defined via the set $\{\tilde{X}_{\boldsymbol{z}_i}(\boldsymbol{v})\}_{s,t}$.

Training Objective: We introduce the ℓ-th ($\ell \geq 1$) training with a mini-batch $\bar{\mathcal{B}} = \{X_i\}_i, |\bar{\mathcal{B}}| = m$, where $X_i = \mathcal{F}I_i, I_i \in \mathcal{B} \subset \mathcal{I}$. Firstly, partially inspired by [32], we update ψ of the GMM using $\{(\boldsymbol{z}_i, \boldsymbol{w}_i)\}_i$, where $(\boldsymbol{z}_i, \boldsymbol{w}_i) = f_\theta(X_i), X_i \in \bar{\mathcal{B}}$. Secondly, we update θ and ϕ based on an objective inspired by rate-distortion theory, i.e., minimization of i) the distortion (reconstruction error), and ii) the rate. The first minimization problem is defined by $\min_{\theta,\phi} \frac{1}{m}\sum_{i=1}^m \sum_{s,t}(X_i(\boldsymbol{v}) - \tilde{X}_{\boldsymbol{z}_i}(\boldsymbol{v}))^2\sqrt{s^2+t^2}$. Let us simplify the problem by $\min_{\theta,\phi} \frac{1}{m}\sum_{i=1}^m \|W \odot (X_i - \tilde{X}_{\boldsymbol{z}_i})\|_2^2$, where $\odot$ is the Hadamard product, and the (s,t)-th element of $W \in \mathbb{R}^{D\times D}$ is $\sqrt[4]{s^2+t^2}$. The second minimization problem is defined by $\min_\theta -\frac{1}{m}\sum_{i=1}^m \log Q_{\boldsymbol{z}_i}$, where $Q_{\boldsymbol{z}_i}$ is given by Eq. (2). Thus, we solve the combined minimization problem of Eq. (3):

$$\arg\min_{\theta,\phi} \frac{1}{m}\sum_{i=1}^m \Bigg[\underbrace{\left\|W \odot \left(X_i - \tilde{X}_{\boldsymbol{z}_i}\right)\right\|_2^2}_{\text{distortion (reconstruction error)}} + \beta\underbrace{\left(-\log Q_{\boldsymbol{z}_i}\right)}_{\text{rate}}\Bigg], \tag{3}$$

where $\beta > 0$ is a hyper-parameter whose appropriate value depends on $\mathcal{I}$. For the solver, we use the RAdam optimizer [11].

Prediction After Training: Let ψ^*, θ^*, and ϕ^* be the trained parameters in the deep AE, where $\psi^* := \{(\pi_c^*, \boldsymbol{\mu}_c^*, \Sigma_c^*\}_{c=1}^C$. Let $\mathcal{F}^{-1}$ denote the Inverse FT (IFT). Given a latent variable $\boldsymbol{z}$, the trained decoder g_{ϕ^*} predicts the corresponding 3D density map similar to cryoDRGN [28] by the following procedure. Firstly, $g_{\phi^*}(\boldsymbol{z}, \boldsymbol{v})$ is computed for all possible 3D positions $\boldsymbol{v} \in \mathbb{R}^3$. Secondly, using $\{g_{\phi^*}(\boldsymbol{z}, \boldsymbol{v})\}_{\boldsymbol{v}}$, the predicted 3D Fourier volume is defined before applying $\mathcal{F}^{-1}$ to the Fourier volume. We define the predicted 3D density map by $\hat{V}_{\boldsymbol{z}}$. Note that the reconstruction error used in Eq. (3) is not a naive one, since we theoretically need to guarantee the isometric property between the latent space and a space of $\hat{V}_{\boldsymbol{z}}$ in line with Theorem 1.

3.2 Analysis of CryoTWIN

Theorem 1. *If we assume i) $n \gg 1$ for $\mathcal{I} = \{I_i\}_{i=1}^n$ (a set of Cryo-EM images), then Eq.* (3) *is approximately equivalent to* $\arg\min_{\theta,\phi} \frac{1}{m}\sum_{i=1}^m \{\|V_{\boldsymbol{z}_i} - \tilde{V}_{\boldsymbol{z}_i}\|_2^2 - \beta' \log Q_{\boldsymbol{z}_i}\}$, *where $V_{\boldsymbol{z}}$ denotes the true 3D density map with a latent variable $\boldsymbol{z}$, and $\beta' > 0$ is a hyper-parameter. The symbol $\tilde{V}_{\boldsymbol{z}_i}$ means the predicted 3D density map during the training and it is defined by $\{g_\phi(\boldsymbol{z}_i + \boldsymbol{\epsilon}, \boldsymbol{v})\}_{\boldsymbol{v}}$ and $\mathcal{F}^{-1}$. Therefore, from Sect. 2.2, the*

Algorithm 1: Pseudocode for computing conformational pathway

Input: Two means of the GMM P_{ψ^*}: $\boldsymbol{\mu}_i^*$ as the start and $\boldsymbol{\mu}_j^*$ as the end point,
Hyper-parameters: $\omega > 1$ (Larger ω returns less continuous pathway), $K, M \in \mathbb{N}$

Output: A sequence of reconstructed structures from $\hat{V}_{\mu_i^*}$ to $\hat{V}_{\mu_j^*}$

1 Set $\boldsymbol{\mu}_i^*$ and $\boldsymbol{\mu}_j^*$ as to $\boldsymbol{z}^{i\to j}(0)$ and $\boldsymbol{z}^{i\to j}(K)$, respectively.

2 **for** $k = 1, ..., K-1$ **do**

3 Fix α_i and α_j to $\cos(\frac{\pi k}{2K})$ and $(1-\cos(\frac{\pi k}{2K}))^\omega$ respectively. Let $\boldsymbol{\alpha}_{-i,j}$ denote $(C-2)$-dimensional vector made by removing the i-th and j-th element from the C-dimensional vector $(\alpha_1, ..., \alpha_C)^\top$. Then, generate M samples for $\boldsymbol{\alpha}_{-i,j}$ under the constraint $\sum_{c\neq i,j} \alpha_c = 1 - \alpha_i - \alpha_j$ and $\alpha_c \geq 0$. Let $\boldsymbol{\alpha}_{-i,j}^{(m)}$ $(m = 1, ..., M)$ denote the m-th sample. For all m, define $\boldsymbol{z}^{(m)}$ using $\alpha_i, \alpha_j, \boldsymbol{\alpha}_{-i,j}^{(m)}$ by
$\boldsymbol{z}^{(m)} = \left(\sum_{c=1}^{C} \alpha_c \Sigma_c^{*-1}\right)^{-1} \left(\sum_{c=1}^{C} \alpha_c \Sigma_c^{*-1} \boldsymbol{\mu}_c^*\right)$. Then,

$$\boldsymbol{z}^{i\to j}(k) = \underset{\boldsymbol{z}^{(m)}}{\operatorname{argmin}} \underbrace{\frac{P_{\psi^*}(\boldsymbol{z}^{i\to j}(0))}{P_{\psi^*}(\boldsymbol{z}^{(m)})} \|\boldsymbol{z}^{(m)} - \boldsymbol{z}^{i\to j}(k-1)\|_2}_{\text{Cost function for } \boldsymbol{z}^{(m)} (m=1,\ldots,M) \text{ in the latent space}}.$$

4 Using $\{\boldsymbol{z}^{i\to j}(k)\}_{k=0}^K$ and the trained decoder g_{ϕ^*}, compute the following sequence;
$\hat{V}_{\mu_i^*}(=\hat{V}_{\boldsymbol{z}^{i\to j}(0)}) \to \hat{V}_{\boldsymbol{z}^{i\to j}(1)} \to ... \to \hat{V}_{\boldsymbol{z}^{i\to j}(k)} \to ... \to \hat{V}_{\mu_j^*}(=\hat{V}_{\boldsymbol{z}^{i\to j}(K)})$.

latent space of the trained cryoTWIN can be isometric with a space for the predicted 3D density map, i.e., $\forall(\boldsymbol{z}, \boldsymbol{z}')$; $\|\boldsymbol{z}-\boldsymbol{z}'\|_2 \approx 0 \Rightarrow \|\boldsymbol{z}-\boldsymbol{z}'\|_2 \propto \|\hat{V}_{\boldsymbol{z}} - \hat{V}_{\boldsymbol{z}'}\|_2$ & $\forall \boldsymbol{z}$; $P(\boldsymbol{z}) \propto P(\hat{V}_{\boldsymbol{z}})$ *hold, where* $P(\boldsymbol{z})$ *(resp.* $P(\hat{V}_{\boldsymbol{z}})$*) is the latent distribution (resp. the probability density function of* $\hat{V}_{\boldsymbol{z}}$*). Additionally, after the training, if we assume ii)* $\forall \boldsymbol{z}$; $P_{\psi^*}(\boldsymbol{z}) \approx \hat{P}(\boldsymbol{z})$ *holds, and iii)* $\forall \boldsymbol{z}$; $\hat{V}_{\boldsymbol{z}} \approx V_{\boldsymbol{z}}$ *holds, where* $\hat{P}(\boldsymbol{z})$ *is an empirical latent distribution, then, as* $\hat{P}(\boldsymbol{z}) \approx P(\boldsymbol{z})$ *holds due to* $n \gg 1$*, the following equation approximately holds:*

$$\forall(\boldsymbol{z}, \boldsymbol{z}');\ \|\boldsymbol{z}-\boldsymbol{z}'\|_2 \approx 0 \Rightarrow \|\boldsymbol{z}-\boldsymbol{z}'\|_2 \propto \|V_{\boldsymbol{z}} - V_{\boldsymbol{z}'}\|_2\ \&\ \forall \boldsymbol{z};\ P_{\psi^*}(\boldsymbol{z}) \propto P(V_{\boldsymbol{z}}). \quad (4)$$

A brief derivation of the theorem is given in Appendix A. The assumption ii) of the theorem is realizable if we set a large integer as the number of components C in the GMM P_ψ. This is empirically confirmed in Sect. 4. For assumption iii), as Eq. (3) implies the minimization of $\frac{1}{m}\sum_{i=1}^{m} \|V_{\boldsymbol{z}_i} - \tilde{V}_{\boldsymbol{z}_i}\|_2^2$ if $n \gg 1$, Eq. (3) tends to lead the AE to have the property of $\forall \boldsymbol{z}$; $\hat{V}_{\boldsymbol{z}} \approx V_{\boldsymbol{z}}$. Thus, if we collect a sufficient amount of the Cryo-EM images, Eq. (4) is realizable for cryoTWIN with large C.

3.3 Computation for Conformational Pathway

We compute the plausible conformational pathway that is defined as a sequence of 3D density maps. If we assume that an AE of cryoTWIN with large C is trained by the Cryo-EM images $\{I_i\}_{i=1}^n, n \gg 1$, then because of Eq. (4), the pathway can be defined by the following two steps: i) generating a ridgeline from $\boldsymbol{\mu}_i^*$ to $\boldsymbol{\mu}_j^*$ on the GMM P_{ψ^*} as a sequence of the latent variables (see the second and third lines in Algorithm 1) and ii) reconstructing the corresponding 3D density maps along the ridgeline using the trained decoder g_{ϕ^*} (see the fourth line in Algorithm 1).

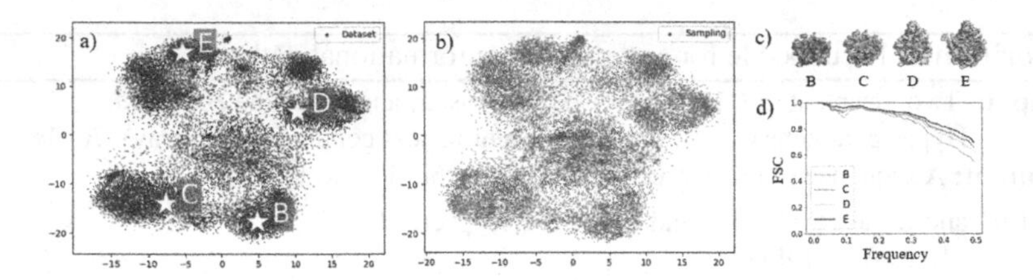

Fig. 2. a) The latent variables (blue) from the trained encoder of cryoTWIN using the ribosomal dataset are visualized by t-SNE. The white star indicates one of means with the trained GMM P_{ψ^*}, and the "B" to "E" indicate the known structural label. We annotate the labels to the corresponding structures (as shown in c)) to the stars via visual examination. b) The sampled points (green) from P_{ψ^*} of cryoTWIN as indicated by t-SNE. d) The similarity score between the two reconstructed structures by cryoTWIN and cryoDRGN with the same label as measured by Fourier Shell Correlation (FSC).

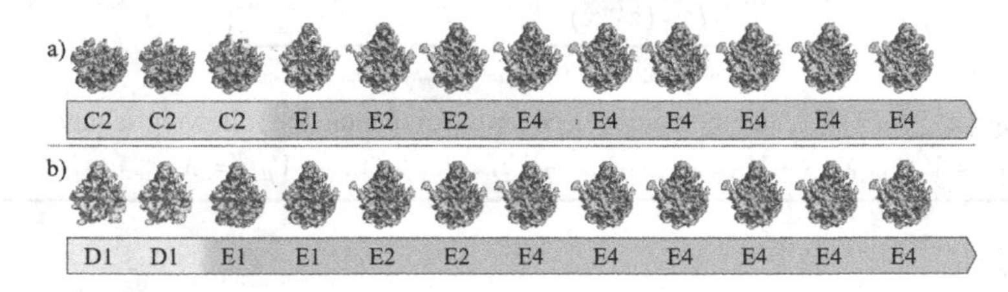

Fig. 3. Two conformational pathways via Algorithm 1 using the ribosomal dataset are shown. In a), the pathway starts from C2 and ends at E4; C2 and E4 indicate the known structural label, and annotated to the structure via visual examination.

A brief explanation of the algorithm is as follows for a sample $z^{(m)}$ of the third line, the sample always ranges a set of candidate points achieving a local maximum with P_{ψ^*}, according to the results of [19]. Additionally, the cost function used to define $z^{i \to j}(k)$ is inspired by the objective of constructing Max-Flux path [7]. Moreover, from Eq. (4), the cost is proportional to $\frac{P(V_{z^{i \to j}(0)})}{P(V_{z^{(m)}})} \| V_{z^{(m)}} - V_{z^{i \to j}(k-1)} \|_2$. Thus, an output of Algorithm 1 can be interpreted as a sequence of 3D density maps, which is generated directly in the space of the 3D density map under Max-Flux objective. We usually cannot access the space of the 3D density map.

4 Numerical Experiment

Using common single particle Cryo-EM images of ribosomes[2], we conduct two experiments: Expt1 and Expt2. The motivation of Expt1 (resp. Expt2) is to examine whether our theoretical claims work for the ribosomal dataset (resp. whether our computed pathways using the ribosomal dataset are consistent with the existing manually constructed pathways [4]). For both experiments, we employ cryoDRGN [28] as our baseline method. Let $\{I_i\}_{i=1}^n$ denote a set of the Cryo-EM images, where $n = 131899$,

[2] https://www.ebi.ac.uk/empiar/EMPIAR-10076/.

and the image size is 128×128. A set of the corresponding estimated orientations $\{\hat{R}_i\}_{i=1}^n$ is available here[3], which are the same ones used in the cryoDRGN. Throughout both experiments, we set $(100, 3, 1/8) = (C, T, \beta)$ with cryoTWIN of Sect. 3.1. The dimension of the latent space is fixed to eight. For Algorithm 1, we set $(11, 2, 1000) = (K, \omega, M)$. Our computational environment is four NVIDIA V100 GPU accelerators with two Intel Xeon Gold 6148 processors.

Expt1: ***Setting):*** Firstly, we evaluate the isometricity of cryoTWIN by the correlation coefficient between $\mathrm{d}\boldsymbol{z}_1 \cdot \mathrm{d}\boldsymbol{z}_2$ and $\mathrm{d}\hat{V}_1 \cdot \mathrm{d}\hat{V}_2$ as described by [8] (see Eq. (2)). Secondly, we evaluate the gap between $P_{\psi^*}(\boldsymbol{z})$ of cryoTWIN and $\hat{P}(\boldsymbol{z})$ using both Kullback-Leibler (KL) divergence based on KLIEP [23] and t-SNE visualization [12]. ***Results):*** Firstly, cryoTWIN achieves a correlation coefficient of 0.77, showing closeness to isometricity (by contrast, cryoDRGN has a correlation coefficient of only 0.52); see details in Fig 4 of Appendix B. Secondly, the KL divergence of cryoTWIN (resp. cryoDRGN) is only 0.25 (resp. 5.7), indicating that $P_{\psi^*}(\boldsymbol{z}) \approx \hat{P}(\boldsymbol{z})$ holds (resp. does not hold). Additionally, as shown in Fig. 2 a) and b), the visualized distributions with $P_{\psi^*}(\boldsymbol{z})$ and $\hat{P}(\boldsymbol{z})$ of cryoTWIN seem similar, whereas those of cryoDRGN are different; see Fig. 5 of Appendix B. Notably, the large KL divergence with cryoDRGN implies that the conformational pathways based on the fixed prior of cryoDRGN are implausible.

Expt2: ***Setting):*** Firstly, as the preliminary experiment of Expt2, using structural labels of [4], we examine how diversified the reconstructed structures of cryoTWIN are by comparing[4] them with those of cryoDRGN. We annotate the labels to reconstructed structures of cryoTWIN via visual examination and then quantitatively evaluate the accuracies of the annotated labels via Fourier Shell Correlation (FSC) between the two structures reconstructed by cryoTWIN and cryoDRGN with the same label. The annotation is performed by our experts using PyMOL [22]. Additionally, if the FSC is close to one in high and low frequency areas, the two structures are similar enough; see details in the third footnote of [29]. Secondly, we visually examine whether some of the conformational pathways computed by Algorithm 1 can be sufficiently consistent with the sub-paths in the four manually constructed pathways presented in [4]. For the computation, we prepare several pairs $(\boldsymbol{\mu}_i^*, \boldsymbol{\mu}_j^*)$ satisfying i) both π_i^* and π_j^* are large, and ii) $\hat{V}_{\boldsymbol{\mu}_i^*}$ and $\hat{V}_{\boldsymbol{\mu}_j^*}$ have different labels. The motivation to focus on the sub-paths comes from the following preliminary observations: Given "B" and "E" in Fig. 2 a) as start and end points, Algorithm 1 estimated only one of the four (B-C2-E1-E2-E4-E5 of Fig. 7 in [4]) as the most probable ridgeline. Thus, to reproduce the four pathways, we needed to aggregate sub-paths generated by the algorithm, whose start and end points were significant. ***Results):*** Firstly, by observing both i) the labeled structures shown in Fig. 2 c) and Fig. 3 (and Fig. 6 of Appendix B), and ii) the labeled structures presented in the original study of cryoDRGN [29] (and Fig. 5 c) of Appendix B), we confirm that cryoTWIN is as diversified as cryoDRGN in terms of the reconstructed structure. Additionally, considering the large FSC value in Fig. 2 d), the accuracies of the annotated labels are high. Secondly, considering the four pathways in [4], we confirm that

[3] https://github.com/zhonge/cryodrgn_empiar/tree/main/empiar10076.

[4] CryoSPARC was used for manually constructing the pathways in [4], but we here employ cryoDRGN, which has the higher structural diversity.

the computed two pathways in Fig. 3 are consistent with the sub-paths in the main four pathways. Note that all the four pathways were successfully reproduced by the aforementioned aggregation in our preliminary experiments.

Time and Memory Complexities: The training time of cryoTWIN (resp. cryoDRGN) is eleven hours (resp. four hours), whereas their memory complexities are comparable. After training cryoTWIN, the running time to compute the pathways including the evaluation time is around one hour, which is much shorter than the running time of the state-of-the-art protocol of a few days [10].

5 Conclusion and Future Work

We propose cryoTWIN for computing plausible pathways from Cryo-EM images, and the efficiency is demonstrated in our numerical experiments. For further research, it would be better to estimate the orientation of the image simultaneously in the training of cryoTWIN, since the preliminary estimation gives an bias to the predicted structure, as explained in [28]. Additionally, it is interesting to combine cryoTWIN and molecular dynamics simulators such as flexible fitting [14], as cryoTWIN provides envelopes for molecular structures in various intermediate states. The combined method could be a powerful tool for more practical applications, such as drug discovery.

Acknowledgements. This work was supported by the FOCUS Establishing Supercomputing Center of Excellence project subject6; MEXT as Simulation- and AI-driven next-generation medicine and drug discovery based on "Fugaku" (JPMXP1020230120). This work used the supercomputer Fugaku provided by RIKEN R-CCS through the HPCI System Research Projects (IDs: hp220078, hp230102, hp230216, ra000018); the supercomputer system at Hokkaido University through the HPCI System Research Projects (IDs: hp220078, hp230102); ABCI provided by AIST was also used.

References

1. Bepler, T., Zhong, E., Kelley, K., Brignole, E., Berger, B.: Explicitly disentangling image content from translation and rotation with spatial-VAE. In: Advances in Neural Information Processing Systems, vol. 32. Curran Associates, Inc. (2019)
2. Chen, M., Ludtke, S.J.: Deep learning-based mixed-dimensional gaussian mixture model for characterizing variability in cryo-EM. Nat. Methods **18**(8), 930–936 (2021)
3. Cover, T.M.: Elements of information theory. John Wiley & Sons (1999)
4. Davis, J.H., Tan, Y.Z., Carragher, B., Potter, C.S., Lyumkis, D., Williamson, J.R.: Modular assembly of the bacterial large ribosomal subunit. Cell **167**(6), 1610–1622 (2016)
5. Han, Q.: Isometric embedding of Riemannian manifolds in Euclidean spaces. American Mathematical Society (2006)
6. Hsieh, J.: Computed tomography: principles, design, artifacts, and recent advances. SPIE Press (2003)
7. Huo, S., Straub, J.E.: The MaxFlux algorithm for calculating variationally optimized reaction paths for conformational transitions in many body systems at finite temperature. J. Chem. Phys. **107**(13), 5000–5006 (1997)

8. Kato, K., Zhou, J., Sasaki, T., Nakagawa, A.: Rate-distortion optimization guided autoencoder for isometric embedding in euclidean latent space. In: Proceedings of the 37th International Conference on Machine Learning, ICML 2020, vol. 119, pp. 5166–5176 (2020)
9. Kingma, D.P., Welling, M.: Auto-encoding variational bayes. In: International Conference on Learning Representations, ICLR 2014 (2014)
10. Kinman, L., Powell, B., Zhong, E., Berger, B., Davis, J.: Uncovering structural ensembles from single-particle cryo-em data using cryodrgn. Nat. Protoc. **18**(2), 319–339 (2023)
11. Liu, L., et al.: On the variance of the adaptive learning rate and beyond. In: International Conference on Learning Representations, ICLR 2020 (2020)
12. van der Maaten, L., Hinton, G.: Visualizing data using t-SNE. J. Mach. Learn. Res. **9**, 2579–2605 (2008)
13. McInnes, L., Healy, J., Saul, N., Großberger, L.: UMAP: uniform manifold approximation and projection. J. Open Source Softw. **3**(29), 861 (2018)
14. Orzechowski, M., Tama, F.: Flexible fitting of high-resolution x-ray structures into cryoelectron microscopy maps using biased molecular dynamics simulations. Biophys. J . **95**(12), 5692–5705 (2008)
15. Punjani, A., Fleet, D.: 3D flexible refinement: Structure and motion of flexible proteins from cryo-EM. Microsc. Microanal. **28**(S1), 1218–1218 (2022)
16. Punjani, A., Fleet, D.J.: 3D variability analysis: Resolving continuous flexibility and discrete heterogeneity from single particle cryo-EM. J. Struct. Biol. **213**(2), 107702 (2021)
17. Punjani, A., Rubinstein, J.L., Fleet, D.J., Brubaker, M.A.: cryoSPARC: algorithms for rapid unsupervised cryo-EM structure determination. Nat. Methods **14**(3), 290–296 (2017)
18. Punjani, A., Zhang, H., Fleet, D.J.: Non-uniform refinement: adaptive regularization improves single-particle cryo-EM reconstruction. Nat. Methods **17**(12), 1214–1221 (2020)
19. Ray, S., Lindsay, B.G.: The topography of multivariate normal mixtures. Ann. Stat. **33**(5), 2042–2065 (2005)
20. Scheres, S.H.: A bayesian view on cryo-EM structure determination. J. Mol. Biol. **415**(2), 406–418 (2012)
21. Scheres, S.H.: RELION: implementation of a bayesian approach to cryo-EM structure determination. J. Struct. Biol. **180**(3), 519–530 (2012)
22. Schrödinger LLC: The PyMOL molecular graphics system, version 1.8 (2015)
23. Sugiyama, M., Nakajima, S., Kashima, H., Buenau, P., Kawanabe, M.: Direct importance estimation with model selection and its application to covariate shift adaptation. In: Advances in Neural Information Processing Systems 20 (2007)
24. Tang, G., et al.: EMAN2: an extensible image processing suite for electron microscopy. J. Struct. Biol. **157**(1), 38–46 (2007)
25. Wrapp, D., et al.: Cryo-em structure of the 2019-ncov spike in the prefusion conformation. Science **367**(6483), 1260–1263 (2020)
26. Wu, Z., Chen, E., Zhang, S., Ma, Y., Mao, Y.: Visualizing conformational space of functional biomolecular complexes by deep manifold learning. Int. J. Mol. Sci. **23**(16), 8872 (2022)
27. Yuan, J., Chen, K., Zhang, W., Chen, Z.: Structure of human chromatin-remodelling PBAF complex bound to a nucleosome. Nature **605**(7908), 166–171 (2022)
28. Zhong, E.D., Bepler, T., Berger, B., Davis, J.H.: CryoDRGN: reconstruction of heterogeneous cryo-EM structures using neural networks. Nat. Methods **18**(2), 176–185 (2021)
29. Zhong, E.D., Bepler, T., Davis, J.H., Berger, B.: Reconstructing continuous distributions of 3D protein structure from cryo-EM images. In: International Conference on Learning Representations, ICLR 2020 (2020)

30. Zhong, E.D., Lerer, A., Davis, J.H., Berger, B.: CryoDRGN2: Ab initio neural reconstruction of 3D protein structures from real cryo-EM images. In: International Conference on Computer Vision, ICCV 2021, pp. 4046–4055. IEEE (2021)
31. Zivanov, J., et al.: New tools for automated high-resolution cryo-EM structure determination in RELION-3. elife **7**, e42166 (2018)
32. Zong, B., et al.: Deep autoencoding gaussian mixture model for unsupervised anomaly detection. In: International Conference on Learning Representations, ICLR 2018 (2018)

Knowledge Boosting: Rethinking Medical Contrastive Vision-Language Pre-training

Xiaofei Chen[1], Yuting He[1], Cheng Xue[1], Rongjun Ge[2], Shuo Li[3], and Guanyu Yang[1,4,5](✉)

[1] Key Laboratory of New Generation Artificial Intelligence Technology and Its Interdisciplinary Applications (Southeast University), Ministry of Education, Dhaka, Bangladesh
yang.list@seu.edu.cn

[2] Nanjing University of Aeronautics and Astronautics, Nanjing, China

[3] Department of Biomedical Engineering, Case Western Reserve University, Cleveland, OH, USA

[4] Joint International Research Laboratory of Medical Information Processing, Southeast University, Nanjing 210096, China

[5] Centre de Recherche en Information Biomédicale Sino-Français (CRIBs), Nanjing, China

Abstract. The foundation models based on pre-training technology have significantly advanced artificial intelligence from theoretical to practical applications. These models have facilitated the feasibility of computer-aided diagnosis for widespread use. Medical contrastive vision-language pre-training, which does not require human annotations, is an effective approach for guiding representation learning using description information in diagnostic reports. However, the effectiveness of pre-training is limited by the large-scale semantic overlap and shifting problems in medical field. To address these issues, we propose the **Kno**wledge-**Bo**osting Contrastive Vision-Language Pre-training framework (KoBo), which integrates clinical knowledge into the learning of vision-language semantic consistency. The framework uses an unbiased, open-set sample-wise knowledge representation to measure negative sample noise and supplement the correspondence between vision-language mutual information and clinical knowledge. Extensive experiments validate the effect of our framework on eight tasks including classification, segmentation, retrieval, and semantic relatedness, achieving comparable or better performance with the zero-shot or few-shot settings. Our code is open on https://github.com/ChenXiaoFei-CS/KoBo.

1 Introduction

Foundation models have become a significant milestone in artificial intelligence, from theoretical research to practical applications [2], like world-impacting large language model ChatGPT [5] and art-history-defining large generative model

H. Greenspan et al. (Eds.): MICCAI 2023, LNCS 14220, pp. 405–415, 2023.
https://doi.org/10.1007/978-3-031-43907-0_39

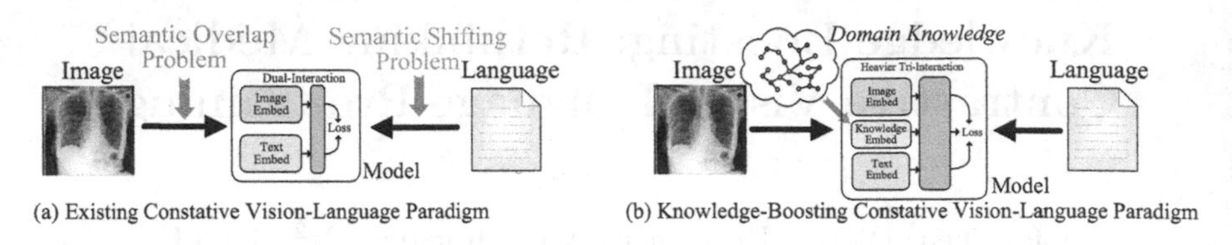

Fig. 1. Our knowledge boosting innovates the paradigm of medical vision-language contrastive learning, inspired by two problems in the existing architecture.

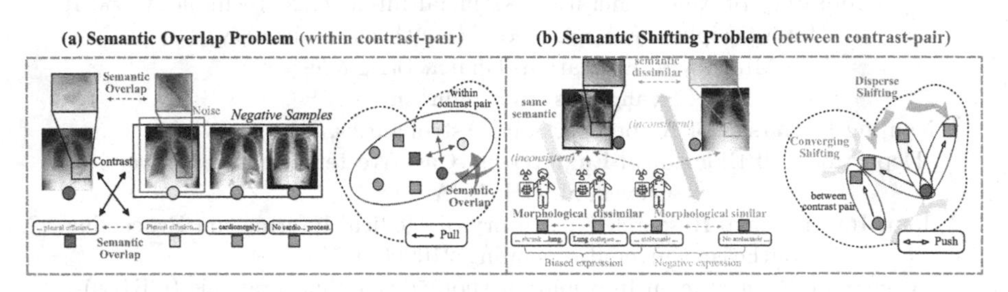

Fig. 2. Two key challenges in medical contrastive vision-language pre-training: **(a)** Semantic overlap exists between negative samples, falsely pulling apart samples with similar semantics. **(b)** Biased expression and negative expression of radiologists cause the inconsistency of semantics and text morphology between sample pairs, causing disperse and converging semantic shifting.

DALL-E [20]. In medical image analysis, foundation models are showing promising future, and pre-training technologies [3,4,8], as the cornerstone of foundation models, facilitated feasibility of computer-aided diagnosis for widespread use.

Medical contrastive vision-language pre-training [10,15,21,23,25] has shown great superiority in medical image analysis, because it utilizes easy-accessible expert interpretation from reports to precisely guide the understanding of image semantics. Therefore, contrastive vision-language pre-training will break through the bottleneck of time-consuming and expensive expert annotation [26] and difficulty in learning fine-grained clinical features with pure-image self-supervised methods [28]. It will improve data efficiency, and achieve comparable or better performance when transferred with the zero-shot or few-shot setting, demonstrating the potential of promoting the ecology of medical artificial intelligence.

However, semantic overlap and semantic shifting are two significant challenges in medical vision-language contrastive learning (Fig. 2). **(a) Semantic Overlap Problem:** There is overlapping semantics between negative samples which should be semantic-distinct, e.g. two medical images sharing the same disease are contrasted which brings noise [25]. Once directly learning, cross-modal representations of the same disease are falsely pulled apart, making the model unable to capture the disease-corresponding image feature. **(b) Semantic Shifting Problem:** Radiologists have writing preferences, e.g. biased for their own familiar concepts and observation view towards similar visual features, and inclined for negation expression towards opposite visual features. Distinct

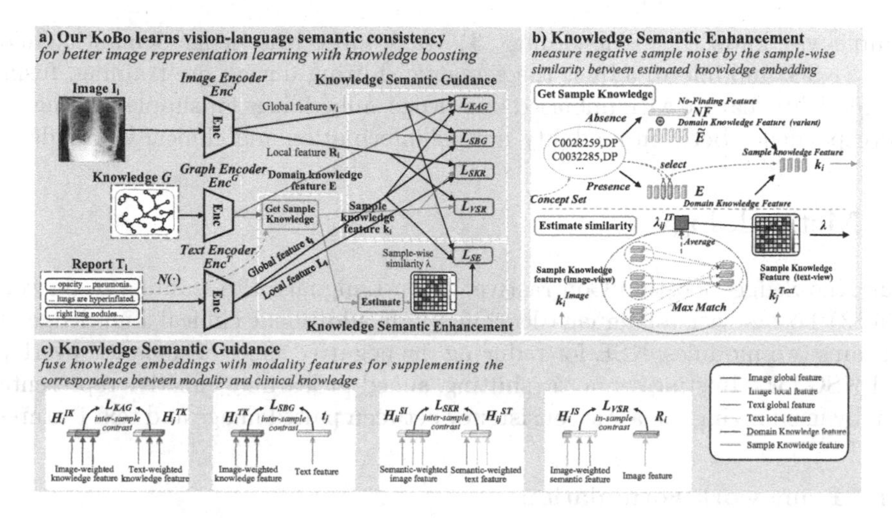

Fig. 3. Overview of our proposed architecture, where additional clinical knowledge is embedded in. Image encoder, text encoder, graph encoder, knowledge semantic enhancement module, and knowledge semantic guidance module are presented.

concepts describing the same image are morphologically dissimilar for text encoder, while the negation expression of concepts is morphologically similar [17]. Once lack of concept correlation and negation identification, representations with similar semantics are falsely pushed apart and those with opposite semantics are falsely pushed together, interfering with the learning of significant representation [7].

Rethinking the existing methods and challenges of medical contrastive vision-language pre-training [10,21,23,25,26], the lack of clinical knowledge constraints in dual-free-encoding contrastive learning structure is the key problem. Existing methods utilize sample-wise differences to learn mutual information between modalities, improving the representation quality based on the correspondence of learned mutual information and clinical knowledge. However, semantic overlap reduces the learning efficiency of mutual information with the noisy difference, and the mentioned correspondence is vulnerable to semantic shifting. Therefore, if we are able to embed an unbiased, comprehensive representation as knowledge boosting, it will reduce the negative noise and supplement the lacking correspondence. It motivates us to measure the noise with similarities between knowledge representation, and fuse the correspondence between knowledge and modality.

In this paper, we propose a novel knowledge-boosting medical contrastive vision-language pre-training framework (KoBo). Our contributions are as followed. **1)** Our KoBo pre-trains a powerful image encoder including visual information corresponding with the disease described in texts, where knowledge is embedded in our paradigm (Fig. 1) to boost the learning of vision-language consistency. **2)** We propose Knowledge Semantic Enhancement (KSE) module to reduce the negative sample noise with the similarity between open-set sample-wise knowledge embeddings. **3)** We propose Knowledge Semantic Guidance (KSG) module to adjust the semantic shifting during pre-training, fusing the modality feature with unbiased knowledge embeddings for supplementing the correspondence between modality mutual information and clinical knowledge.

2 Methodology

Our Knowledge-Boosting Contrastive Vision-Language Pre-training framework (Fig. 3) boosts vision-language learning with additional clinical knowledge. It contains two modules: KSE for reducing the negative effect of semantic overlap, and KSG for adjusting semantic shifting, aimed at learning effective representation by maximizing semantic consistency between paired image and text features.

2.1 Framework Formulation

In the framework, a powerful image encoder Enc^I and text encoder Enc^T is pre-trained, alongside a graph encoder Enc^G. Given a pair of medical image and diagnostic report $\{I_i, T_i^{Report}\}, I_i \in \mathbb{R}^{H\times W\times C}$, a sentence T_i^{Sent} is randomly selected from T_i^{Report} as a caption comprised of several tokens $\{w_1, w_2, ..., w_{N_L}\}$. Enc^I outputs global feature $z_i^{I,G}$ and local feature $z_i^{I,L}$ for N_I sub-regions, which is from the intermediate feature map. T_i^{Sent} is fed into Enc^T, obtaining global sentence feature $z_i^{T,G}$, and local token feature $z_i^{T,L}$. Distinct projectors are applied to map features into embeddings with lower semantic dim D_S, finally getting global and local image embeddings $v_i \in \mathbb{R}^{D_S}, R_i = \{r_{i1}, r_{i2}, ..., r_{iN_I}\} \in \mathbb{R}^{N_I\times D_S}$, and text embedding $t_i \in \mathbb{R}^{D_S}, L_i = \{l_{i1}, l_{i2}, ..., l_{iN_L}\} \in \mathbb{R}^{N_L\times D_S}$.

Besides using reports and images as the input for our pre-training network, we also input an external knowledge graph to the whole framework for improving the correspondence of modality features and clinical knowledge. The knowledge refers to relations between clinical pathology concepts in the radiology domain in the format of triplet $\mathcal{G} = \{(c_{h_k}, r_k, c_{t_k})\}_{k=1}^{N_G}$, such as UMLS [14]. Domain knowledge embedding for each concept $E = \{e_s\}_{s=1}^{N_E} \in \mathbb{R}^{N_E\times D_S}$ is the output of $Enc^G(\mathcal{G})$.

2.2 Knowledge Semantic Enhancement

To relieve the semantic overlap problem, where negative sample noise harms the effective learning of vision-language mutual information, we propose a semantic enhancement module to identify the noise using sample-wise similarities. The similarity is estimated upon sample knowledge k_i, calculated from domain knowledge embedding E and concept set from texts with negation marker.

Getting Sample knowledge: Firstly, we acquire a concept set that contains pathology concepts extracted from texts with Negbio $\mathcal{N}(\cdot)$ [17]. The image-view concept set which involves the overall observation is from the whole report, while the text-view set only covers the chosen sentence. Secondly, the image and text sample knowledge, as an auxiliary semantic estimation, is selected from domain knowledge embedding E according to the corresponding concept set from the report and sentence respectively, if not considering the negation problem.

Furthermore, considering the challenge that negation expression of concepts commonly exists in radiology reports, which has opposite semantics with similar morphology for text encoder (converging shifting), we randomly generate a *No Finding* embedding $\mathcal{NF}$ and a variant of domain knowledge embedding $\widetilde{E} = \{\widetilde{e}_1, \widetilde{e}_2, ..., \widetilde{e}_{N_E}\}$ of the same size as E with Xavier distribution. Upon the negation mark of concept, sample knowledge embedding $k_i = \{k_{i,s}\}_{s=1}^{N_{ES}}$ is denoted below:

$$k_{i,s} = \begin{cases} e_{i,s} & c_{i,s} \in \mathcal{N}(T_i), P(c_{i,s}) \neq Neg \\ \epsilon \cdot \mathcal{NF} + (1-\epsilon)\widetilde{e}_{i,s} & c_{i,s} \in \mathcal{N}(T_i), P(c_{i,s}) = Neg \end{cases} \tag{1}$$

where P is the negation mark of concepts, and $e_{i,s}, \widetilde{e}_{i,s}$ is the corresponding position of $c_{i,s}$ in E and $\widetilde{E}$. ϵ tunes the variance of negative sample knowledge. $k_{i,s}^{Image}$ and $k_{i,s}^{Text}$ are k_i from the image-view and text-view concept set.

Estimation of Similarities: The semantic similarity is calculated upon sample knowledge. For each image-text pair, a max-match strategy is adopted to match each two sample knowledge embedding with the most similar one for calculating cosine similarities. Sample-wise similarities are aggregated with averages.

$$\lambda_{ij}^{IT} = \frac{1}{N_{ES'}} \sum_{s=1}^{N_{ES'}} \max_{s'=1}^{N_{ES}} (k_{i,s}^{Image})^T k_{j,s'}^{Text}, \lambda_{ij}^{TI} = \frac{1}{N_{ES}} \sum_{s=1}^{N_{ES}} \max_{s'=1}^{N_{ES'}} (k_{i,s}^{Text})^T k_{j,s'}^{Image} \tag{2}$$

where N_{ES} is the number of concepts in T_i^{Sent}, while $N_{ES'}$ is that in T_i^{Report}.
Knowledge Semantic Enhancement Loss: We utilize the sample-wise semantic similarity to estimate negative sample noise, placed in the sample weight of the contrastive loss [18,26], where paired cross-modal embedding are pushed together and unpaired ones are pulled apart. The importance of estimated noisy negative samples is relatively smaller for a subtle pulling between cross-modal embeddings. The semantic enhancement loss is below:

$$\mathcal{L}_{SE} = -\frac{1}{N} \sum_{i=1}^{N} (\log \frac{\exp(v_i^T t_i/\tau_G)}{\sum_{j=1}^{N} (1-\lambda_{ij}^{IT}) \exp(v_i^T t_j/\tau_G)} + \log \frac{\exp(t_i^T v_i/\tau_G)}{\sum_{j=1}^{N} (1-\lambda_{ij}^{TI}) \exp(t_i^T v_j/\tau_G)}) \tag{3}$$

where τ_G is the global temperature, and λ^{IT}, λ^{TI} is the sample similarity measurement. specifically, $\lambda_{i,i}$ is fixed to zero to persist the positive sample weight.

2.3 Knowledge Semantic Guidance

In this section, we propose a semantic guidance module to solve the semantic shifting problem. Utilizing sample knowledge from Sect. 2.2 which contains concept correlation and negation information, the adverse effects of both disperse and converging shifting are alleviated by fusing domain-sample knowledge with global-local modality embeddings. We design four contrast schemes: knowledge anchor guidance for adjusting disperse shifting, semantic knowledge refinement for filtering converging shifting, vision semantic response for consolidating knowledge fusion, and semantic bridge guidance for narrowing the modality gap.

Knowledge Anchor Guidance: Disperse shifting will be adjusted if there are unbiased anchors in semantic space as priors to attract modality embeddings towards clinical semantics, and domain knowledge embedding does a good job. We define knowledge fused embeddings $H_i^{IK} = ATTN(v_i, E, E)$ and $H_i^{TK} = ATTN(t_i, E, E)$, and $ATTN(Q, K, V)$ means the attention function [10]:

$$\mathcal{L}_{KAG} = -\frac{1}{N}\sum_{i=1}^{N}(\log\frac{exp(H_i^{IK}\cdot H_i^{TK}/\tau_G)}{\sum_{j=1}^{N}exp(H_i^{IK}\cdot H_j^{TK}/\tau_G)} + \log\frac{exp(H_i^{TK}\cdot H_i^{IK}/\tau_G)}{\sum_{j=1}^{N}exp(H_i^{TK}\cdot H_j^{IK}/\tau_G)}) \quad (4)$$

where image-weighted and text-weighted knowledge is globally contrasted.

Semantic Knowledge Refinement: Wrong-converging pairs have distinct intrinsic responses on sample knowledge from image and text. Hence, we propose to utilize sample knowledge to refine these falsely gathered dissimilar pairs. We define $H_{ij}^{SI} = ATTN(k_i^{Text}, R_j, R_j)$ and $H_{ij}^{ST} = ATTN(k_i^{Text}, L_j, L_j)$:

$$\mathcal{L}_{SKR} = -\frac{1}{N}\sum_{i=1}^{N}\log\frac{exp(\frac{1}{N_{ES}\cdot\tau_L}\sum_{k=1}^{N_{ES}}H_{iik}^{SI}\cdot H_{iik}^{ST})}{\sum_{j=1}^{N}exp(\frac{1}{N_{ES}\cdot\tau_L}\sum_{k=1}^{N_{ES}}H_{ijk}^{SI}\cdot H_{ijk}^{ST})} \quad (5)$$

where local semantic-weighted image and text embeddings are contrasted.

Vision Semantic Response: Instead of matching single token with image sub-regions in [10], we propose to match the concept with sub-regions. As the concept is a more complete and atomic semantic unit, local response upon concept will better guide the representation learning with a fine-grained semantic match through an in-sample contrast. We define $H_i^{IS} = ATTN(R_i, k_i^{Text}, k_i^{Text})$, and the fusion of knowledge will be consolidated as below:

$$\mathcal{L}_{VSR} = -\frac{1}{N\cdot N_I}\sum_{i=1}^{N}\sum_{k=1}^{N_I}\log\frac{exp(H_{ik}^{IS}\cdot r_{ik}/\tau_L)}{\sum_{k'=1}^{N_I}exp(H_{ik}^{IS}\cdot r_{ik'}/\tau_L)} \quad (6)$$

where there is an in-sample local contrast between H_i^{IS} and vision features.

Semantic Bridge Guidance: We propose to narrow disperse shifting enlarged by the modality gap between vision and language. Specifically, the gap is bridged by the fusion of domain knowledge which is better compatible with text:

$$\mathcal{L}_{SBG} = -\frac{1}{N}\sum_{i=1}^{N}(\log\frac{exp(H_i^{IK}\cdot t_i/\tau_G)}{\sum_{j=1}^{N}exp(H_i^{IK}\cdot t_j/\tau_G)} + \log\frac{exp(t_i\cdot H_i^{IK}/\tau_G)}{\sum_{j=1}^{N}exp(t_i\cdot H_j^{IK}/\tau_G)}) \quad (7)$$

where the image-weighted domain knowledge is contrasted with text features between samples. Finally, $\mathcal{L}_{SG}$ is aggregated by these four parts as below:

$$\mathcal{L}_{SG} = \lambda_1\mathcal{L}_{KAG} + \lambda_2\mathcal{L}_{SKR} + \lambda_3\mathcal{L}_{VSR} + \lambda_4\mathcal{L}_{SBG} \quad (8)$$

3 Experiment

Experiment Protocol: Pre-training performs on MIMIC-CXR [12] following the pre-process style of [9]. The impression section of reports and frontal view of images are selected to generate 203k image-report pair. Five downstream task datasets (CheXpert [11], Covidx [24], MIMIC-CXR, UMNSRS [16] and SIIM [22]) are applied on eight tasks. Semantic relatedness is to verify the text understanding of radiology concepts, where text embedding with certain prompts predicts the relatedness. A new semantic relatedness benchmark is generated from MIMIC-CXR, adding in the extra negation discriminating. CheXpert5X200 [10](Multi-classification) is from CheXpert, and CheXpert-labeller[11] generates retrieval labels in MIMIC-CXR. More details are in appendix.

Table 1. Comparison results in eight downstream tasks. (*) defines that official pre-trained weight is used, and the remaining methods are reproduced with the same batch size, pre-processing and the evaluation. CLS, RR, SR, and SEG mean classification, retrieval, semantic relatedness and semantic segmentation, V or L means vision and language tasks. Few-shot-Frozen means the frozen encoder of the backbone and only 1% of total training data. ResNet-50 is the equal-comparing backbone except for KoBo-Vit. The best two results are highlighted in underlined red and violet.

Method	Zero-shot					Few-shot-Frozen		
	CLS(V+L) CheXpert (Auroc)	**RR(V)** CheXpert5X200 (mAP)	**RR(V+L)** MIMIC (mAP)	**SR(L)** UMNSRS (Pearson)	**SR(L)** MIMIC (Pearson)	**CLS(V)** CheXpert (Auroc)	**SEG(V)** SIIM (Dice)	**CLS(V)** Covidx (Acc)
CLIP [18](*)	0.4702	0.2544	0.7577	0.1985	-0.2879	0.5748	/	0.8975
ConVIRT [26]	0.8252	0.3808	0.8482	**0.2506**	0.1429	0.8548	0.4992	0.9475
Gloria [10]	0.8257	0.3875	0.8300	0.2294	0.1100	0.8492	0.5479	0.9250
MGCA [23]	0.8496	0.3906	0.8428	0.1889	0.1809	0.8616	0.5696	0.9375
MedCLIP [25](*)	0.7805	0.4298	0.7258	0.2032	-0.1321	0.8214	0.5619	0.9325
KoBo	**0.8590**	0.3918	**0.8467**	0.2563	**0.3712**	**0.8628**	**0.6393**	0.9550
KoBo-Vit	0.8635	**0.4123**	0.8455	0.1824	0.4229	0.8660	0.6554	**0.9525**

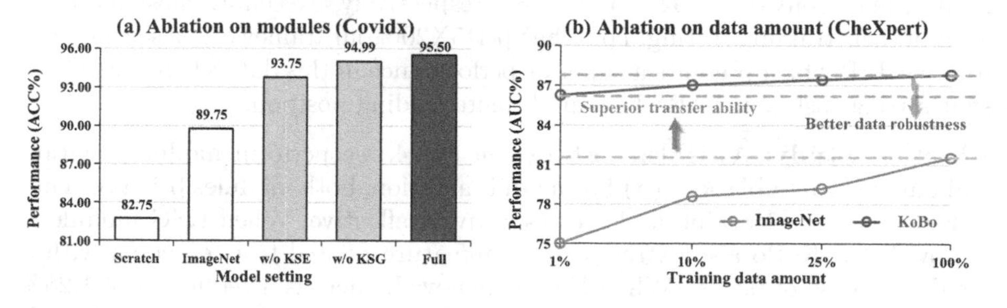

Fig. 4. (a) Module ablation study of our KoBo framework is performed on Covidx dataset compared with ImageNet and random initialization, upon few-shot frozen setting. **(b)** Data ablation study is performed on CheXpert with frozen setting when classification training data amount reduces to 25%, 10% and 1%.

For implementation, ResNet50 [6] and Vit [13] are image encoder, and Bio-ClinicalBERT [1] is the text encoder. CompGCN with LTE [27] is our graph encoder, and domain knowledge contains 10,244 concepts in UMLS which exist in MIMIC-CXR. Negbio [17] combined with UMLS disambiguation tool [14] serves as $\mathcal{N}(\cdot)$. Embeddings are projected into the dim of 256. Pre-training has the batch size of 100 and max epochs of 50 based on Pytorch on two RTX3090 GPUs. Adam optimizer with the learning rate of 5e-5 and ReduceLR scheduler are applied. τ_G is 0.07 and τ_L is 0.1. λ in KSG loss are all 0.25, while ϵ in KSE loss is 0.1.

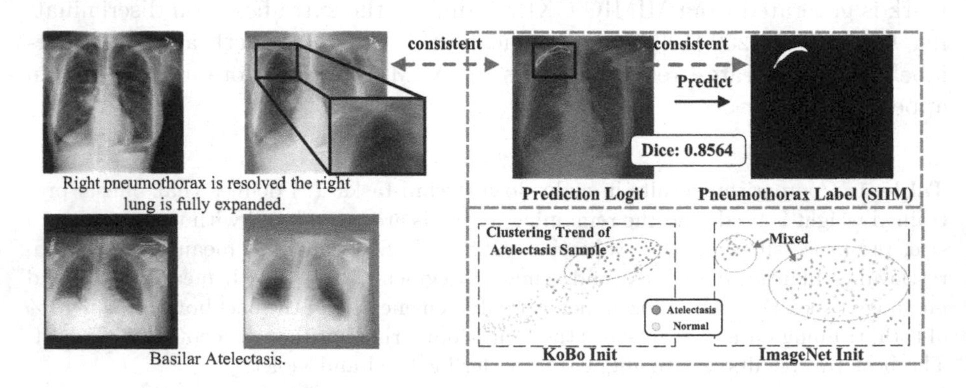

Fig. 5. Visualization of pneumothorax and atelectasis. AblationCAM [19] generates the class activate map (CAM) upon last layer of Kobo-ResNet. There is strong consistency between CAM, prediction logits and segmentation label. t-SNE [23] is applied on image embedding from CheXpert-valid, showing gathering cluster trend of disease samples.

Comparison Study: Table 1 verifies our powerful representation ability, reaching state-of-art in classification, segmentation, and semantic relatedness compared with existing vision-language pre-training tasks, while our method is also top two for retrieval. In zero-shot classification tasks, our KoBo outperforms MGCA and ConVIRT 0.94% and 3.38% respectively, exceeding most methods even in their training setting. For CheXpert5X200, our framework is second only to MedCLIP which presents a superior performance in this dataset. In three few-shot setting task, our KoBo has an absolute leading position.

Ablation Study: As is demonstrated in Fig. 4, we perform module ablation and data amount ablation. **(a)** For module ablation, both modules bring benefits in representation learning and are respectively effective. When KSG module is removed, our KoBo also extracts effective feature related to pneumonia with a subtle decrease of 0.51%. When KSE is removed, there is a reduction of 1.25% accuracy. **(b)** For data amount ablation, KoBo has better data robustness with a subtle decrease when training data reduce to 1%. KoBo also has a superior transfer ability with an absolutely better AUC with 1% data than ImageNet with all training data than ImageNet with all training data.

Qualitative Analysis: In Fig. 5, our Kobo has learned fine-grained and effective image feature with the fusion of knowledge modeling. The deepest region in the first image gathered on the top left side, showing an obvious expansion on the right lung. There is consistency with the expert annotation and our output logit. The precise location of atelectasis region in CAM of second image and clustering trend interpret for the increase in zero-shot classification.

4 Conclusion

In our paper, we propose a Knowledge-Boosting Contrastive Vision-Language Pre-traing framework (KoBo). Sample and domain knowledge are used to differentiate noisy negative samples and supplement the correspondence between modality and clinical knowledge. Our experiments on eight tasks verify the effectiveness of our framework. We hope that our work will encourage more research on knowledge-granularity alignment in medical vision-language learning.

Acknowledgements. This research was supported by the Intergovernmental Cooperation Project of the National Key Research and Development Program of China(2022YFE0116700). We thank the Big Data Computing Center of Southeast University for providing the facility support.

References

1. Alsentzer, E., et al.: Publicly available clinical bert embeddings. In: Proceedings of the 2nd Clinical Natural Language Processing Workshop, pp. 72–78 (2019)
2. Bommasani, R., et al.: On the opportunities and risks of foundation models. arXiv preprint arXiv:2108.07258 (2021)
3. Chen, Z., et al.: Multi-modal masked autoencoders for medical vision-and-language pre-training. In: Medical Image Computing and Computer Assisted Intervention-MICCAI 2022: 25th International Conference, Singapore, 18–22 September 2022, Proceedings, Part V, pp. 679–689. Springer (2022). https://doi.org/10.1007/978-3-031-16443-9_65
4. Chen, Z., Li, G., Wan, X.: Align, reason and learn: enhancing medical vision-and-language pre-training with knowledge. In: Proceedings of the 30th ACM International Conference on Multimedia, pp. 5152–5161 (2022)
5. van Dis, E.A., Bollen, J., Zuidema, W., van Rooij, R., Bockting, C.L.: Chatgpt: five priorities for research. Nature **614**(7947), 224–226 (2023)
6. He, K., Zhang, X., Ren, S., Sun, J.: Deep residual learning for image recognition. In: Proceedings of the IEEE Conference on Computer Vision and Pattern Recognition, pp. 770–778 (2016)
7. He, Y., et al.: Learning better registration to learn better few-shot medical image segmentation: Authenticity, diversity, and robustness. IEEE Trans. Neural Netw. Learn. Syst. (2022)
8. He, Y., et al.: Geometric visual similarity learning in 3d medical image self-supervised pre-training. In: Proceedings of the IEEE/CVF Conference on Computer Vision and Pattern Recognition, pp. 9538–9547 (2023)

9. Hou, B., Kaissis, G., Summers, R.M., Kainz, B.: RATCHET: medical transformer for Chest X-ray diagnosis and reporting. In: de Bruijne, M., et al. (eds.) MICCAI 2021. LNCS, vol. 12907, pp. 293–303. Springer, Cham (2021). https://doi.org/10.1007/978-3-030-87234-2_28
10. Huang, S.C., Shen, L., Lungren, M.P., Yeung, S.: Gloria: a multimodal global-local representation learning framework for label-efficient medical image recognition. In: Proceedings of the IEEE/CVF International Conference on Computer Vision, pp. 3942–3951 (2021)
11. Irvin, J., et al.: Chexpert: a large chest radiograph dataset with uncertainty labels and expert comparison. In: Proceedings of the AAAI Conference on Artificial Intelligence, vol. 33, pp. 590–597 (2019)
12. Johnson, A.E., et al.: Mimic-cxr-jpg, a large publicly available database of labeled chest radiographs. arXiv preprint arXiv:1901.07042 (2019)
13. Kolesnikov, A., et al.: An image is worth 16x16 words: Transformers for image recognition at scale (2021)
14. Mao, Y., Fung, K.W.: Use of word and graph embedding to measure semantic relatedness between unified medical language system concepts. J. Am. Med. Inform. Assoc. **27**(10), 1538–1546 (2020)
15. Müller, P., Kaissis, G., Zou, C., Rueckert, D.: Radiological reports improve pre-training for localized imaging tasks on chest x-rays. In: Medical Image Computing and Computer Assisted Intervention-MICCAI 2022: 25th International Conference, Singapore, 18–22 September 2022, Proceedings, Part V, pp. 647–657. Springer (2022). https://doi.org/10.1007/978-3-031-16443-9_62
16. Pakhomov, S.: Semantic relatedness and similarity reference standards for medical terms (2018)
17. Peng, Y., Wang, X., Lu, L., Bagheri, M., Summers, R., Lu, Z.: Negbio: a high-performance tool for negation and uncertainty detection in radiology reports. AMIA Summits Trans. Sci. Proc. **2018**, 188 (2018)
18. Radford, A., et al.: Learning transferable visual models from natural language supervision. In: International Conference on Machine Learning, pp. 8748–8763. PMLR (2021)
19. Ramaswamy, H.G., et al.: Ablation-cam: visual explanations for deep convolutional network via gradient-free localization. In: Proceedings of the IEEE/CVF Winter Conference on Applications of Computer Vision, pp. 983–991 (2020)
20. Reddy, M.D.M., Basha, M.S.M., Hari, M.M.C., Penchalaiah, M.N.: Dall-e: creating images from text. UGC Care Group I J. **8**(14), 71–75 (2021)
21. Seibold, C., Reiß, S., Sarfraz, M.S., Stiefelhagen, R., Kleesiek, J.: Breaking with fixed set pathology recognition through report-guided contrastive training. In: Medical Image Computing and Computer Assisted Intervention-MICCAI 2022: 25th International Conference, Singapore, 18–22 September 2022, Proceedings, Part V, pp. 690–700. Springer (2022). https://doi.org/10.1007/978-3-031-16443-9_66
22. Viniavskyi, O., Dobko, M., Dobosevych, O.: Weakly-supervised segmentation for disease localization in Chest X-Ray images. In: Michalowski, M., Moskovitch, R. (eds.) AIME 2020. LNCS (LNAI), vol. 12299, pp. 249–259. Springer, Cham (2020). https://doi.org/10.1007/978-3-030-59137-3_23
23. Wang, F., Zhou, Y., Wang, S., Vardhanabhuti, V., Yu, L.: Multi-granularity cross-modal alignment for generalized medical visual representation learning. In: Advances in Neural Information Processing Systems

24. Wang, L., Lin, Z.Q., Wong, A.: Covid-net: a tailored deep convolutional neural network design for detection of Covid-19 cases from chest x-ray images. Sci. Rep. **10**(1), 1–12 (2020)
25. Wang, Z., Wu, Z., Agarwal, D., Sun, J.: Medclip: contrastive learning from unpaired medical images and text. In: Proceedings of the 2022 Conference on Empirical Methods in Natural Language Processing, pp. 3876–3887 (2022)
26. Zhang, Y., Jiang, H., Miura, Y., Manning, C.D., Langlotz, C.P.: Contrastive learning of medical visual representations from paired images and text. In: Machine Learning for Healthcare Conference, pp. 2–25. PMLR (2022)
27. Zhang, Z., Wang, J., Ye, J., Wu, F.: Rethinking graph convolutional networks in knowledge graph completion. In: Proceedings of the ACM Web Conference 2022, pp. 798–807 (2022)
28. Zhou, Z.: Models genesis: generic autodidactic models for 3D medical image analysis. In: Shen, D., et al. (eds.) MICCAI 2019. LNCS, vol. 11767, pp. 384–393. Springer, Cham (2019). https://doi.org/10.1007/978-3-030-32251-9_42

A Small-Sample Method with EEG Signals Based on Abductive Learning for Motor Imagery Decoding

Tianyang Zhong[1], Xiaozheng Wei[1], Enze Shi[2], Jiaxing Gao[1], Chong Ma[1], Yaonai Wei[1], Songyao Zhang[1], Lei Guo[1], Junwei Han[1], Tianming Liu[3], and Tuo Zhang[1](✉)

[1] School of Automation, Northwestern Polytechnical University, Xi'an 710072, China
tuozhang@nwpu.edu.cn
[2] School of Computer Science, Northwestern Polytechnical University, Xi'an 710072, China
[3] School of Computing, University of Georgia, Athens 30602, USA

Abstract. Motor imagery (MI) electroencephalogram (EEG) decoding, as a core component widely used in noninvasive brain-computer interface (BCI) system, is critical to realize the interaction purpose of physical world and brain activity. However, the conventional methods are challenging to obtain desirable results for two main reasons: there is a small amount of labeled data making it difficult to fully exploit the features of EEG signals, and lack of unified expert knowledge among different individuals. To handle these dilemmas, a novel small-sample EE -G decoding method based on abductive learning (SSE-ABL) is proposed in this paper, which integrates perceiving module that can extract multiscale features of multi-channel EEG in semantic level and knowledge base module of brain science. The former module is trained via pseudo-labels of unlabeled EEG signals generated by abductive learning, and the latter is refined via the label distribution predicted by semi-supervised learning. Experimental results demonstrate that SSE-ABL has a superior performance compared with state-of-the-art methods and is also convenient for visualizing the underlying information flow of EEG decoding.

Keywords: MI Decoding · Small-Sample · Abductive Learning · Visualization

1 Introduction

Brain computer interface (BCI) technology plays an increasingly crucial role in the rehabilitation process of patients with nerve damage. However, the lack of large-scale labeled data makes the identification results more vulnerable to be affected. BCI technology with motor imagery (MI), which can decode neural activities through EEG signals to identify the movement intention of the human being, is widely applied in the field of EEG decoding, where the substantial problem of BCI decoding is to extract as much effective information as possible from the multi-channel and non-linear EEG signals for understanding the oscillating activities in the brain [1].

H. Greenspan et al. (Eds.): MICCAI 2023, LNCS 14220, pp. 416–424, 2023.
https://doi.org/10.1007/978-3-031-43907-0_40

According to the development of EEG signal processing technology, the traditional EEG signal feature extraction methods [2, 3] rely on rich human experience or expert knowledge. Furthermore, the existing deep learning models for EEG motion decoding [4, 5] rely on a large amount of labeled data. More recently, the active learning framework [6, 7] based on semi-supervised learning is used to solve a feature extraction problem with a large amount of unlabeled data, but it is sensitive to the initial model accuracy and sampling strategy. Therefore, there is still an unsolved urgent challenge that needs to be considered further, which is how to make biologically reasonable use of the information contained in a large number of unlabeled EEG data on the above basis.

The starting point of this paper is that the following biological findings can be observed in the MI decoding [8–10]: 1) the EEG rhythm energy in the contralateral motor sensory area (MSA) of the cerebral cortex is significantly reduced, while that in the ipsilateral MSA is increased, 2) different functional brain regions activities can be reflected by different sub-bands of EEG signals, and 3) the distribution density of different rhythms are clearly distinguished. And the rules are similar to the knowledge reasoning used in human decision-making. Then, the purpose of this paper is to employ these knowledge constraints to guide the model to extract MI information from unlabeled EEG data in a mutually beneficial way.

Therefore, based on abductive learning (ABL) [11], this article proposes a small-sample EEG decoding method, which can adaptively extract abstract features from complex and dynamic EEG signals, and an efficient knowledge base is designed to constrain the training process of the model. The main contributions are listed as follows:

1) A novel EEG decoding method is proposed to tackle the MIR (motion intention recognition) problem with small-sample EEG signals. This method does not rely on strict mathematical assumptions for datasets and its accuracy and robustness are well-maintained under strong interference.
2) A multi-scale feature fusion network is designed to enhance abstract features, which can capture temporal and frequency information across multi-channel EEG signals and spatial relationships among different electrodes.
3) An effective knowledge base module of motor imagery is constructed and symbolized, which can upgrade the model space under this constraint by mining the potential facts of large-scale unlabeled EEG signals.

2 Problem Definition and Method

2.1 Problem Definition

In the SSE-ABL framework for EEG decoding, the given input is defined as $\boldsymbol{Input} = \{\boldsymbol{X}_l, \boldsymbol{X}_u, \boldsymbol{KB}_\theta\}$, where tensor $\boldsymbol{X}_l$ denotes labeled EEG data, tensor $\boldsymbol{X}_u$ denotes unlabeled EEG data, the data size of $\boldsymbol{X}_u$ is much larger than that of $\boldsymbol{X}_l$, and $\boldsymbol{KB}_\theta$ denotes knowledge base on brain science in MI task. Concretely, $\boldsymbol{X}_l = \{(\boldsymbol{x}_{l1}, \boldsymbol{y}_{l1}), (\boldsymbol{x}_{l2}, \boldsymbol{y}_{l2}), \ldots, (\boldsymbol{x}_{li}, \boldsymbol{y}_{li})\}$, where variable $\boldsymbol{x}_{li}$ represents multi-channel EEG signals, and $\boldsymbol{y}_{li}$ implies the corresponding labels. And the assignment of $\boldsymbol{X}_l$ is to learn a mapping from $\boldsymbol{x}$ to $\boldsymbol{y}$. $\boldsymbol{X}_u = \{\boldsymbol{x}_{u1}, \boldsymbol{x}_{u2}, \ldots, \boldsymbol{x}_{uj} | i \ll j\}$, where $\boldsymbol{x}_{uj}$ denotes unlabeled multi-channel EEG signals, which are utilized to boost representative capability of the above mapping. $\boldsymbol{KB}_\theta$ consists of a series of first-order logic sentences with learnable parameters $\boldsymbol{\theta}$, which integrate labels EEG data $\boldsymbol{X}_l$ and unlabeled data $\boldsymbol{X}_u$ to optimize a perceptual model and train parameters $\boldsymbol{\theta}$ with the constraint of knowledge base. The final result is regarded as $\boldsymbol{Output} = \{\boldsymbol{f}, \boldsymbol{W}, \boldsymbol{\theta}'\}$, where $\boldsymbol{f}$ is a mapping from EEG signals to motion intention, $\boldsymbol{W} \in \boldsymbol{R}^{m \times n}$ indicates the proportion of the mth sub-band to the nth channel in EEG data, $\boldsymbol{\theta}' \in \boldsymbol{R}^k$ represents the contribution rate of the kth channel to the whole in MIR. The SSE-ABL algorithm yields the corresponding pseudo-labels to the unlabeled data by the classifier optimized by a small amount of labeled EEG signals, and the produced labels may be incorrect due to the small number of training samples, which is difficult to guarantee good performance. Therefore, the SSE-ABL modifies the pseudo-labels and optimizes the internal parameters of the knowledge base at the same time, so that the consistency of them is maximized under the constraint of the knowledge base. Formally, the problem definition of the SSE-ABL can be summarized as an optimization problem of searching $\boldsymbol{Output}$ under a given $\boldsymbol{Input}$:

$$\min_{\boldsymbol{f}, \boldsymbol{W}, \boldsymbol{\theta}} \mathcal{L}oss_{label}(\boldsymbol{y}_{li}, \boldsymbol{f}_{li}) + \mathcal{L}oss_{unlabel}(\boldsymbol{\delta}(\boldsymbol{y}'_{uj}), \boldsymbol{f}_{uj}) \tag{1}$$

$$\text{s.t.} \operatorname*{argmax}_{\delta} constraint\left(\boldsymbol{\delta}\left(\boldsymbol{y}'_{uj}\right), \boldsymbol{f}_{li}, \boldsymbol{KB}_\theta\right) \tag{2}$$

where $\boldsymbol{y}'_{uj}$ is the pseudo-label corresponding to the jth unlabeled instance, which is generated by the perceptual module. $\boldsymbol{\delta}(\cdot)$ indicates a heuristic function obtained by optimization, which aims to revise pseudo-labels by logical abduction process. In addition to correcting inconsistent pseudo-labels, this goal also helps the knowledge base to learn accurate parameter $\boldsymbol{\theta}$. It can be seen from Eq. 1 and Eq. 2 that the major challenge is how to mine the effective information of massive unlabeled EEG data under the $\boldsymbol{KB}_\theta$ constraints and react to the iterative update of itself.

2.2 Architecture of the SSE-ABL Framework

The proposed method, as shown in Fig. 1, consists of four phases. A workflow outline of the proposed method in this paper is displayed in Algorithm 1.

1) Phase 1 (Sample Representation): For the purpose of promoting the ability to describe the local and global details of the brain activities, a time-frequency-space data representation method based on brain region division is proposed, which makes the receptive field of view cover the whole brain region in a fine-grained way.
2) Phase 2 (Multiscale Feature Fusion): Inspired by the neuroscience findings shown in Sect. 1, a multi-scale feature fusion model is proposed to adaptively integrate time-frequency-space information of EEG signals aiming at reducing the potential difference of feature distribution and selectively focusing on the MI-related materials.
3) Phase 3 (Motion Intention Estimation): Based on the common laws of EEG signals distribution, a motor intention evaluation model is constructed, in which the rules are used as the supervisory information to judge the authenticity of motor imagery recognition, so as to ensure that the overall process conforms to the actual criteria.
4) Phase 4 (Abductive Reasoning Optimization): When the classifier is insufficient-trained, the pseudo-labels could be wrong, the SSE-ABL method in this paper needs to correct the wrong pseudo-labels to achieve consistent abductions by using gradient free optimization method under the principle of minimal inconsistency [12, 13].

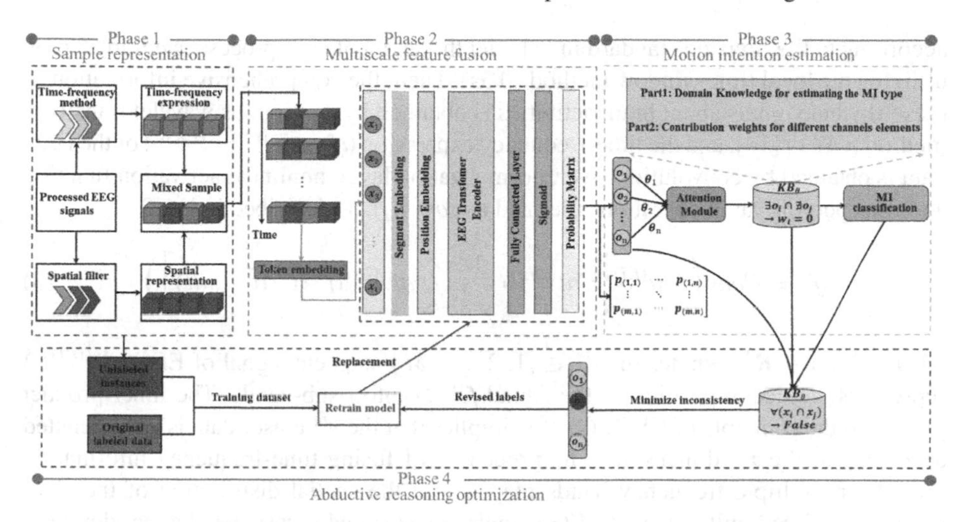

Fig. 1. Architecture of the proposed SSE-ABL method.

Algorithm 1 Small-Sample EEG Decoding Method Based on Abductive Learning

---

Input: Labeled EEG signals $\{(\boldsymbol{x}_{l1}, \boldsymbol{y}_{l1}), (\boldsymbol{x}_{l2}, \boldsymbol{y}_{l2}), \ldots, (\boldsymbol{x}_{li}, \boldsymbol{y}_{li})\}$; Unlabeled EEG signals $\{\boldsymbol{x}_{u1}, \boldsymbol{x}_{u2}, \ldots, \boldsymbol{x}_{uj}\}$; Knowledge base $\boldsymbol{KB}_{\theta}$

Parameters: Training parameters(batch size, learning rate, epoch ,etc.)

Output: Perception model $\boldsymbol{f}$; Parameters $\boldsymbol{W}$, $\boldsymbol{\theta}'$

1: **for** VarEpoch = 1→epoch **do** # Training epochs
2: $\boldsymbol{f}$ = TrainModel($\{(\boldsymbol{x}_{l1}, \boldsymbol{y}_{l1}), (\boldsymbol{x}_{l2}, \boldsymbol{y}_{l2}), \ldots, (\boldsymbol{x}_{li}, \boldsymbol{y}_{li})\}$) # Pre-train model
3: $\boldsymbol{\theta}$ = UpdateParameter($\{\boldsymbol{y}_{l1}, \boldsymbol{y}_{l2}, \ldots, \boldsymbol{y}_{li}\}$) # Pre-train learnable parameters
4: **end for**
5: **for** i = 1→E **do**
6: $\boldsymbol{Y}' = \boldsymbol{f}(\boldsymbol{X}_u)$ # Generate pseudo-labels $\boldsymbol{Y}'$
7: $\boldsymbol{\delta}(\boldsymbol{Y}')$ = Abduce($\boldsymbol{KB}_{\theta}$, $\boldsymbol{Y}'$) # Revise pseudo-labels
8: $\boldsymbol{\theta}'$ = UpdateParameter($\boldsymbol{Y}_l$, $\boldsymbol{\delta}(\boldsymbol{Y}')$) # Update parameters $\boldsymbol{\theta}'$
9: $\boldsymbol{f}$ = RetrainModel($\boldsymbol{f}$, $\{(\boldsymbol{x}_{l1}, \boldsymbol{y}_{l1}), (\boldsymbol{x}_{l2}, \boldsymbol{y}_{l2}), \ldots, (\boldsymbol{x}_{li}, \boldsymbol{y}_{li})\}$, $\boldsymbol{X}_u$, $\boldsymbol{Y}'$) # Reupdate $\boldsymbol{f}$
10: i = i + 1
11 : **end for**
12: **return Output**

2.3 Sample Representation

The preprocessed EEG signals are first divided into five bands containing delta (1–4 Hz), theta (4–8 Hz), alpha (8–14 Hz), beta (14–31 Hz) and gamma (31–50 Hz) wavebands by using digital band-pass filters whose corresponding cut-off frequency is performed according to the division standard in [1], and the segmentation process is completed by utilizing the local time window method (0.5s). Then, the comprehensive information of these rhythm signals about brain activities is obtained by continuous wavelet transform method (CWT) [14], and the time-frequency expression ($\boldsymbol{tf}_t \in \mathbb{R}^{N \times C \times H \times W}$) of the EEG data is obtained by convolution layer, normalization layer, nonlinear activation function and max pooling layer operations (defined as$Conv(\cdot)$), as described in Eq. 3.

$$\boldsymbol{tf}_t = Conv\left(Cat^{(t,i)}\left[Cat^{(ch,j)}\left(\frac{1}{\sqrt{s}}\int_{-\infty}^{+\infty} div^{(i,j)}(f^{(i)}(t))\overline{\psi}_{s,a}tdt\right)\right]\right) \tag{3}$$

where $f^{(i)}(t) \in \boldsymbol{R}^N$ denotes the $i^{th} \in \{1, 2, \ldots, M\}$ segment signal of EEG, $div^{(i,j)}(\cdot)$ represents the decomposition of the signal $f^{(i)}(t)$ into j sub-bands. The inner product process is the principle of CWT, $Cat^{(x,y)}$ implies that the yth tensor data is concatenated according to the xth dimension. The procedure of fusing time-frequency information directly in multiple frequency bands may ignore the spatial distribution of the electrode. Therefore, multi-channel EEG signals are executed across the channel direction through the channel-by-channel convolutional operations, which aims at capturing spatial dynamic correlation characteristics ($\boldsymbol{S}_t \in \mathbb{R}^{N \times C \times H \times W}$) among brain regions, as shown in Eq. 4.

$$\boldsymbol{S}_t = Cat\left(Append\left(g\left(\boldsymbol{W}f^{(i)}(t) + \boldsymbol{b}\right)\right)\right) \tag{4}$$

where $g(\cdot)$ implies 1-D convolutional operation, tensor $\boldsymbol{W}$ and $\boldsymbol{b}$ denotes convolution weights and bias, $Append(\cdot)$ means adding each element in turn. With the above configurations, the mixed sample ($\boldsymbol{MS}_t \in \mathbb{R}^{N \times 2C \times H \times W}$), which is composed of $\boldsymbol{tf}_t$ and$\boldsymbol{S}_t$, are represented as 4-D tensor, where (C, H, W) is the number of channels and the resolution of the feature map respectively.

2.4 Multiscale Feature Fusion

Algorithm 2 Multiscale Feature Fusion method

Input: the mixed sample $\boldsymbol{MS_t}$

Parameters: $\boldsymbol{W}$

Output: Multi-channel EEG sub-bands features ($\boldsymbol{SSF}$) about multiple brain regions

1: $\boldsymbol{Input} = Tok(\boldsymbol{MS_t}) + \mathrm{Seg}(\boldsymbol{MS_t}) + \mathrm{Pos}(\boldsymbol{MS_t})$ # Generate embedding
2: **for** n = 1→NUM **do**
3: $\mathbf{Q}_n = \boldsymbol{W}_{\boldsymbol{Q}_n} \times \mathrm{Norm}(\boldsymbol{MS}_{ti})$
4: $\mathbf{K}_n = \boldsymbol{W}_{\boldsymbol{K}_n} \times \mathrm{Norm}(\boldsymbol{MS}_{ti})$
5: $\mathbf{V}_n = \boldsymbol{W}_{\boldsymbol{V}_n} \times \mathrm{Norm}(\boldsymbol{MS}_{ti})$ # Calculate query (**Q**), key (**K**), value (**V**)
6: $\boldsymbol{A}_n = \boldsymbol{XSA}(\boldsymbol{Q}_n, \boldsymbol{K}_n, \boldsymbol{V}_n)$ #Forward temporal/spatial self-attention block with eeg data
7: **end for**
8: $\boldsymbol{SSF} = \boldsymbol{A}_{\mathrm{NUM}} + \mathrm{MLP}(\mathrm{Norm}(\boldsymbol{A}_{\mathrm{NUM}}))$ # Conduct residual operation
9: **return** $\boldsymbol{SSF}$

Given the mixed samples, the multiscale feature fusion model embeds them into feature vectors and extracts the time-frequency-space features of multi-channel EEG signals. We introduce the embedding layers in phase 2 to better describe the semantic and location information of EEG and improve the EEG transformer encoder to pay attention to seizing the global and local information of long-term EEG signals. Concretely, the normalized operation, the position mark of multiple $\boldsymbol{MS}_t$ and the order inside it are added to attain the input of the network by conducting token, segment and position embedding steps, respectively, as showed in Eq. 5.

$$\boldsymbol{Input} = Tok(\boldsymbol{MS}_t) + Seg(\boldsymbol{MS}_t) + Pos(\boldsymbol{MS}_t) \tag{5}$$

Note that the addition here is bitwise addition. Tok, Seg and Pos correspond to the above three operations. Then tensor-patches in $\boldsymbol{MS}_t$ are vectorized and carried to the network to calculate temporal and spatial self-attention block described in [15], as showed in Algorithm 2.

2.5 Motion Intention Estimation

We construct MI classification module based on attention mechanism and knowledge rules, which consists of two parts: domain knowledge expressed by first-order logic formula and learnable parameters weights, as shown in Fig. 2.

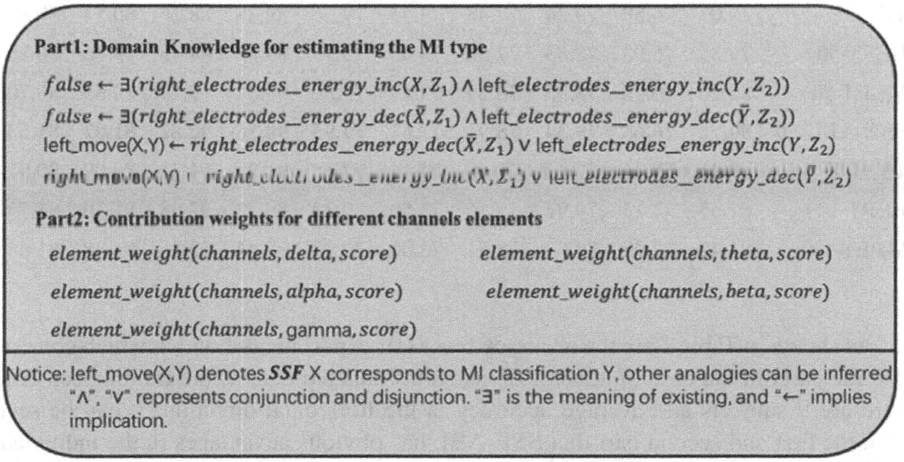

Part1: Domain Knowledge for estimating the MI type

$false \leftarrow \exists(right\_electrodes\_energy\_inc(X,Z_1) \wedge \text{left}\_electrodes\_energy\_inc(Y,Z_2))$

$false \leftarrow \exists(right\_electrodes\_energy\_dec(\bar{X},Z_1) \wedge \text{left}\_electrodes\_energy\_dec(\bar{Y},Z_2))$

$\text{left\_move(X,Y)} \leftarrow right\_electrodes\_energy\_dec(\bar{X},Z_1) \vee \text{left}\_electrodes\_energy\_inc(Y,Z_2)$

$\text{right\_move(X,Y)} \leftarrow right\_electrodes\_energy\_inc(X,Z_1) \vee \text{left}\_electrodes\_energy\_dec(\bar{Y},Z_2)$

Part2: Contribution weights for different channels elements

$element\_weight(channels, delta, score)$ $element\_weight(channels, theta, score)$

$element\_weight(channels, alpha, score)$ $element\_weight(channels, beta, score)$

$element\_weight(channels, \text{gamma}, score)$

Notice: left_move(X,Y) denotes ***SSF*** X corresponds to MI classification Y, other analogies can be inferred. "∧", "∨" represents conjunction and disjunction. "∃" is the meaning of existing, and "←" implies implication.

Fig. 2. MIR rules of the knowledge base *KB*. Part 1): the domain knowledge for MI type is to revise pseudo-labels and inference. 2) the contribution weights for different channels.

3 Results

We adopt the 2008 BCI competition IV-2a EEG dataset (https://www.bbci.de/competition/iv/) including 9 subjects with a 250 Hz sampling rate and band-pass filtered from 0.5 to 100 Hz, which consists of four MI tasks: imagine left hand (class 1), right hand (class

2), foot (class 3) and tongue (class 4) movements. And the preprocessing operations in this article refers to eliminating the wrong experiments and using digital band-pass filters of 1–50 Hz. In the experiments, all methods with self-teaching plan are set to the same number of modules using the default hyperparameters in the source code.

The following experimental tests of the proposed method (SSE-ABL) are designed to conduct: 1) Compared with the current mainstream algorithms containing transformer (TF) [16], BERT [17] and MEET [15] models to verify the performance of the SSE-ABL. 2) By changing proportion of training dataset size to verify the robustness of the SSE-ABL under the interference of different small-sample data, that is, the extrapolation ability. 3) The visualization of the underlying information flow of EEG decoding by the proposed method is displayed.

Table 1. Performance comparison of SSE-ABL method to state-of-the-art methods as the proportion change of training dataset in term of accuracy metrics (%) for BCI IV-2a dataset.

Method	Sub1	Sub2	Sub3	Sub4	Sub5	Sub6	Sub7	Sub8	Sub9	AVE
TF-10	74.10	73.27	73.41	74.69	70.63	76.07	89.27	81.61	75.55	76.51
BERT-10	72.30	69.72	76.01	71.27	65.12	72.91	82.97	82.42	75.31	74.23
MEET-10	78.63	66.80	73.62	74.70	67.91	75.92	86.73	82.51	73.29	75.57
SSE-ABL-10	**79.22**	**76.01**	**78.90**	**78.43**	**72.44**	**78.66**	**89.19**	**88.05**	**79.40**	**80.03**
TF-50	79.03	76.59	79.24	77.48	73.53	79.36	89.38	88.26	80.54	80.38
BERT-50	79.85	75.93	78.83	77.56	73.05	78.79	89.83	88.73	79.76	80.26
MEET-50	80.20	76.26	78.93	78.59	73.19	79.56	90.51	89.19	79.83	80.70
SSE-ABL-50	**80.78**	**76.68**	**79.34**	**78.91**	**73.62**	**79.95**	**90.56**	**92.82**	**81.07**	**81.53**
TF-100	80.91	75.36	79.12	78.68	74.13	77.38	91.21	91.32	80.09	80.91
BERT-100	81.65	77.13	79.92	79.16	73.59	80.34	89.72	92.68	81.39	81.73
MEET-100	81.42	77.98	79.53	78.51	73.06	79.98	90.44	92.46	81.38	81.64

As shown in Table 1 are the accuracy results of experiments 1 and 2, which consists of three parts in the case of 10%, 50% and100% labeled data as the training set. And there are 9 subjects and average accuracy in the horizontal direction. It can be seen from the first and second part that SSE-ABL has obvious advantages in the individual accuracy and overall average accuracy of different subjects, up to 89.19%, 80.03% and 92.8 2%, 81.53% respectively. By comparing the first two parts, can be found that when the labeling rate is low, the benefits of using both unlabeled data and MI knowledge are much higher during EEG decoding. Further comparison with the third part shows that the performance can even be competitive the methods with the 100% training set.

Then, the visualization results of experiment 3 is exhibited in Fig. 3,which is the analysis of FC_1, FC_3, CP_1, CP_3 and FC_2, FC_4, CP_2, CP_4 channels (corresponding to the left and right brain regions respectively) when imagining a left-handed task. And there are also three parts: the time domain EEG signal (P1), the time-frequency analysis (P2) and visualization analysis of the SSE-ABL (P3). From the time domain perspective shown in the P1, it seems that there is not much discrimination that can separate specific motor imagery tasks. However, it can be found in the P2 that the energy distribution on the left and right sides of the brain can be clearly found by the data representation method proposed in this paper, which can be distinguished by the brightness of the color. At the same time, there is also a phenomenon that the energy of the right brain region is lower than that of the left brain region, which is consistent with biological cognition. Moreover, it can be observed in the P3 that the high energy component of EEG signals is surrounded by a large amount of dark red, which is the information that the SS-ABL algorithm focuses on. And the contribution rates of the above channels to this recognition task are marked under the channels, respectively, which gives us an intuitive understanding of the EEG signal decoding process.

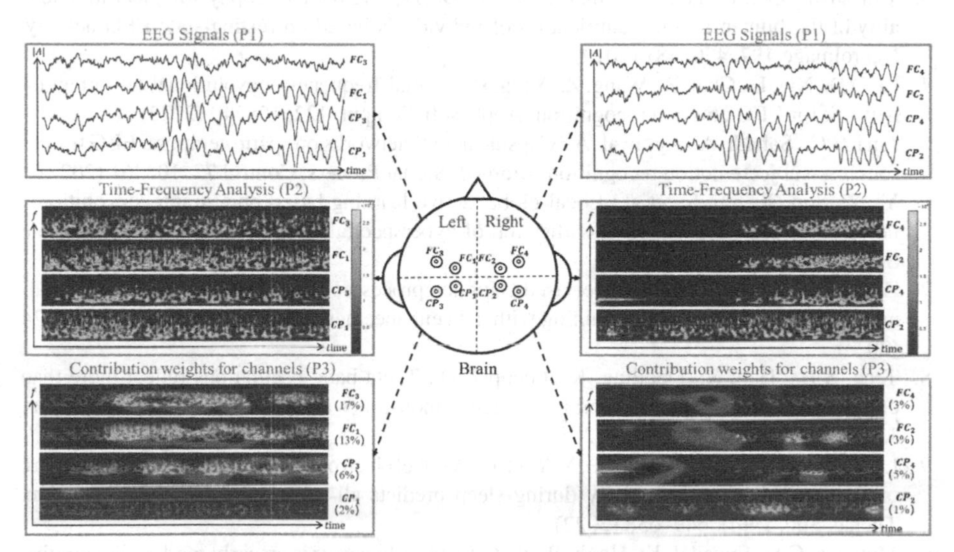

Fig. 3. Visual analysis of the typical channels (FC_{1-4} and CP_{1-4}) when imaging left hand movements through the SSE-ABL method.

4 Conclusion

A novel small-sample method with EEG signals based on abductive reasoning is studied for MI recognition. This method solves the problem of low precision and poor robustness ability of EEG decoding under a small amount of labeled data. The multiscale feature fusion module based on self-attention mechanism improves the ability of adaptive feature mining by capturing time-frequency-space information cover the whole brain region, which realizes the enhancement of abstract features. An effective knowledge base module of motor imagery is constructed and symbolized, which can upgrade the model space under this constraint by mining the potential facts of large-scale unlabeled EEG signals. Through the comparison experiments with other mainstream inversion methods, it can be founded that our method with 10% and 50% labeled EEG data can reach the accuracy of 80.03% and 81.53%, achieving the high precision standard of EEG decoding. In the future, we will consider a fine-grained visualization method of EEG signals in the case of partial data damage, such as signal coupling.

References

1. Li, D.L., Xu, J.C., Wang, J.H., Fang, X.K., Ji, Y.: A multi-scale fusion convolutional neural network based on attention mechanism for the visualization analysis of EEG signals decoding. IEEE Trans. Neur. Sys. Rehabil. Eng. **28**(12), 2615–2626 (2020)
2. Zhang, X., Yao, L.N, Wang, X.Z., et al.: A survey on deep learning-based non-invasive brain signals: recent advances and new frontiers. J. Neur. Eng. **18**(3), 031002 (2021)
3. Valizadeh, S.A., Riener, R., Elmer, S., et al.: Decrypting the electrophysiological individuality of the human brain: identification of individuals based on resting-state EEG activity. NeuroImage **197**, 470–481 (2019)
4. Li, D.D., Xie, L., Chai, B., Wang, Z., Yang, H.: Spatial-frequency convolutional self-attention network for EEG emotion recognition. Appl. Soft Comput. **122**, 108740 (2022)
5. Jana, G.C., Sabath, A., Agrawal, A.: Capsule neural networks on spatio-temporal EEG frames for cross-subjectemotion recognition. Biomed. Signal Process. Control **72**, 103361 (2022)
6. Ye, Z., Sun, T., Shi, S., et al.: Local-global active learning based on a graph convolutional network for semi-supervised classification of hyperspectral imagery. IEEE Geosci. Remote Sens. Lett. **20**, 1–5 (2023)
7. Zhou, C.H, Zou, L.Y.: Semi-supervised Gaussian processes active learning model for imbalanced small data based on tri-training with data enhancement. IEEE Access **11**, 17510–1724 (2023)
8. Teng, X.B., Tian, X., Doelling, K., Poeppel, D.: Theta band oscillations reflect more than entrainment: behavioral and neural evidence demonstrates an active chunking process. Eur. J. Neurosci. **48**(8), 2770–2782 (2018)
9. Lechat, B., Hansen, K.L., Melaku, Y.A., et al.: A novel electroencephalogram-derived measure of disrupted delta wave activity during sleep predicts all-cause mortality risk. Ann. Am. Thorac. Soc. **19**(4), 649–658 (2022)
10. Musaeus, C.S., Engedal, K., Høgh, P., et al.: EEG theta power is an early marker of cognitive decline in dementia due to Alzheimer's disease. J. Alzheimers Dis. **64**(4), 1359–1371 (2018)
11. Huang, Y.X., Dai, W.Z., Cai, L.W., et al.: Fast abductive learning by similarity-based consistency optimization. Adv. Neural. Inf. Process. Syst. **34**, 26574–26584 (2021)
12. Dai, W.Z., Xu, Q., Yu, Y., et al.: Bridging machine learning and logical reasoning by abductive learning. Adv. Neural Inf. Process. Syst. **32** (2019)
13. Zhou, Z.-H.: Abductive learning: towards bridging machine learning and logical reasoning. Sci. China Inf. Sci. **62**(7), 1–3 (2019). https://doi.org/10.1007/s11432-018-9801-4
14. Shi, X., Qin, P., Zhu, J., et al.: Feature extraction and classification of lower limb motion based on sEMG signal. IEEE access **8**, 132882–132892 (2020)
15. Shi, E.Z., Yu, S.G., Kang, Y.Q., et al.: MEET: Multi-band EEG Transformer. arXiv preprint (2023)
16. Song, Y., Jia, X., Yang, L., et al.: Transformer-based spatial-temporal feature learning for EEG decoding. arXiv preprint arXiv:2106.11170 (2021)
17. Devlin, J., Chang, M.W., Lee, K., et al.: Bert: Pre-training of deep bidirectional transformers for language understanding. arXiv preprint arXiv:1810.04805 (2018)

Multi-modal Variational Autoencoders for Normative Modelling Across Multiple Imaging Modalities

Ana Lawry Aguila(✉), James Chapman, and Andre Altmann

University College London, London WC1E 6BT, England
ana.aguila.18@ucl.ac.uk

Abstract. One of the challenges of studying common neurological disorders is disease heterogeneity including differences in causes, neuroimaging characteristics, comorbidities, or genetic variation. Normative modelling has become a popular method for studying such cohorts where the 'normal' behaviour of a physiological system is modelled and can be used at subject level to detect deviations relating to disease pathology. For many heterogeneous diseases, we expect to observe abnormalities across a range of neuroimaging and biological variables. However, thus far, normative models have largely been developed for studying a single imaging modality. We aim to develop a multi-modal normative modelling framework where abnormality is aggregated across variables of multiple modalities and is better able to detect deviations than uni-modal baselines. We propose two multi-modal VAE normative models to detect subject level deviations across T1 and DTI data. Our proposed models were better able to detect diseased individuals, capture disease severity, and correlate with patient cognition than baseline approaches. We also propose a multivariate latent deviation metric, measuring deviations from the joint latent space, which outperformed feature-based metrics.

Keywords: Unsupervised learning · Normative modelling · Multimodal modelling · multi-view VAEs

1 Introduction

Normative modelling is a popular method to study heterogeneous brain disorders. Normative models assume disease cohorts sit at the tails of a healthy population distribution and quantify individual deviations from healthy brain patterns. Typically, a normative analysis constructs a normative model per variable, e.g., using Gaussian Process Regression (GPR) [9]. Recently, to model complex non-linear interactions between features, deep-learning approaches using

Supplementary Information The online version contains supplementary material available at https://doi.org/10.1007/978-3-031-43907-0_41.

H. Greenspan et al. (Eds.): MICCAI 2023, LNCS 14220, pp. 425–434, 2023.
https://doi.org/10.1007/978-3-031-43907-0_41

adversarial (AAE) and variational autoencoder (VAE) models have been proposed [8,11]. These models have a uni-modal structure with a single encoder and decoder network. So far, almost all deep-learning normative models have modelled only one modality. However, many brain disorders show deviations from the norm in features of multiple imaging modalities to a varying degree. Often it is unknown which modality will be the most sensitive. Thus, it is advantageous to develop normative models suitable for multiple modalities.

Most previous deep-learning normative and anomaly detection models measure deviations in the feature space [4,7,11]. However, for multi-modal models built from modalities containing highly different, but complementary information (e.g., T1 and DTI features as used here), we may not expect to see significantly greater deviations in the feature space compared to uni-modal methods. Indeed previous work has shown that, when using VAEs, even for one modality, measuring deviation in the latent space outperforms metrics in the feature space [8] and provides a single measure of abnormality. As such, we develop a latent deviation metric suitable to measuring deviations in multi-modal data.

There are many approaches to extending VAEs to integrate information from multiple modalities and learn informative joint latent representations. Most multi-modal VAE frameworks learn separate encoder and decoder networks for each modality and aggregate the encoding distributions to learn a joint latent representation. Wu and Goodman [16] introduced a multi-modal VAE (mVAE) where each encoding distribution is treated as an 'expert' and the Product-of-Experts (PoE), which takes a product of the experts' densities, is used to approximate a joint encoding distribution. The PoE approach treats all experts as equally credible taking a uniform contribution from every modality. In practice, however, different levels of noise, complexity and information are present in different modalities. Furthermore, if we have an overconfident miscalibrated expert, i.e. a sharp, shifted probability distribution, the joint distribution will have low density in the region observed by the other experts and a biased mean prediction. This can result in a suboptimal latent space and data reconstruction. Shi et al. [13] address this problem by combining latent representations across modalities using a Mixture-of-Experts (MoE) approach. For MoE, the joint distribution is given by a mixture of the experts' densities so that the density is spread over all regions covered by the experts and overconfident experts do not monopolize the resulting prediction. However, MoE is less sensitive to consensus across modalities and will give lower probability to regions where experts are in agreement than PoE. Alternatively, we propose a mVAE modelling the joint encoding distribution as a generalised Product-of-Experts (gPoE) [2]. We optimise modality specific weightings to account for different information content between experts and enable the model to down-weight experts which cause erroneous predictions. Depending on the application, either MoE or gPoE will be most appropriate and so we consider both methods for normative modelling.

As far as we are aware, only one other multi-modal VAE normative modelling framework has been proposed in the literature which uses the PoE (PoE-normVAE) [7]. However, Kumar et al. [7] rely on measuring deviations in the feature space, which we argue does not leverage the benefits of multi-modal models. Here, we present an improved factorisation of the joint representation by modelling it as a weighted product or sum of each encoding distribution.

Our contributions are two-fold. Firstly, we present two novel multi-modal normative modelling frameworks, MoE-normVAE and gPoE-normVAE, which capture the joint distribution between different imaging modalities. Our proposed models outperform baseline methods on two neuroimaging datasets. Secondly, we present a deviation metric, based on the latent space, suitable for detecting deviations in multi-modal normative distributions. We show that our metric better leverages the benefits of multi-modal normative models compared to feature space-based metrics.

2 Methods

Multi-modal Variational Autoencoder (mVAE). Let $\mathbf{X}=\{\mathbf{x}_m\}_{m=1}^M$ be the observations of M modalities. We use a mVAE to learn a multi-modal generative model (Fig. 1c), where modalities are conditionally independent given a common latent variable, of the form $p_\theta(\mathbf{X}, \mathbf{z})=p(\mathbf{z}) \prod_{m=1}^M p_{\theta_m}(\mathbf{x}_m \mid \mathbf{z})$. The likelihood distributions $p_{\theta_m}(\mathbf{x}_m \mid \mathbf{z})$ are parameterised by decoder networks with parameters $\theta=\{\theta_1, \ldots, \theta_M\}$. The goal of VAE training is to maximise the marginal likelihood of the data. However, as this is intractable, we instead optimise an evidence lower bound (ELBO):

$$\mathcal{L}=\mathbb{E}_{q_\phi(\mathbf{z} \mid \mathbf{X})}\left[\sum_{m=1}^M \log p_\theta(\mathbf{x}_m \mid \mathbf{z})\right]-D_{KL}(q_\phi(\mathbf{z} \mid \mathbf{X}) | p(\mathbf{z})) \quad (1)$$

where the second term is the KL divergence between the approximate joint posterior $q_\phi(\mathbf{z} \mid \mathbf{X})$ and the prior $p(\mathbf{z})$. We model the posterior, likelihood, and prior distributions as isotropic gaussians.

Approximate Joint Posterior. To train the mVAE, we must specify the form of the joint approximate posterior $q_\phi(\mathbf{z} \mid \mathbf{X})$. Wu and Goodman [16] choose to factorise the joint posterior as a Product-of-Experts (PoE); $q_\phi(\mathbf{z} \mid \mathbf{X})=\frac{1}{K} \prod_{m=1}^M q_{\phi_m}(\mathbf{z} \mid \mathbf{x}_m)$, where the experts, i.e., individual posterior distributions $q_{\phi_m}(\mathbf{z} \mid \mathbf{x}_m)$, are parameterised by encoder networks with parameters $\phi=\{\phi_1, \ldots, \phi_M\}$. K is a normalisation term. Assuming each encoder network follows a Gaussian distribution $q(\mathbf{z} \mid \mathbf{x}_m)=\mathcal{N}(\boldsymbol{\mu}_m, \boldsymbol{\sigma}_m^2 \mathbf{I})$, the parameters of joint posterior distribution can be computed [5]; $\boldsymbol{\mu}=\frac{\sum_{m=1}^M \boldsymbol{\mu}_m / \boldsymbol{\sigma}_m^2}{\sum_{m=1}^M 1 / \boldsymbol{\sigma}_m^2}$ and $\boldsymbol{\sigma}^2=\frac{1}{\sum_{m=1}^M 1 / \boldsymbol{\sigma}_m^2}$ (see Supp. for proofs).

However, overconfident but miscalibrated experts may bias the joint posterior distribution (see Fig. 1b) which is undesirable for learning informative latent representations between modalities [13].

Shi et al. [13] instead factorise the approximate joint posterior as a Mixture-of-Experts (MoE); $q_\Phi(\mathbf{z} \mid \mathbf{X})=\frac{1}{K} \sum_{m=1}^M \frac{1}{M} q_{\phi_m}(\mathbf{z} \mid \mathbf{x}_m)$.

In the MoE setting, each uni-modal posterior $q_\phi(\mathbf{z} \mid \mathbf{x}_m)$ is evaluated with the generative model $p_\theta(\mathbf{X}, \mathbf{z})$ such that the ELBO becomes:

$$\mathcal{L} = \sum_{m=1}^{M} \left[\mathbb{E}_{q_\phi(\mathbf{z}|\mathbf{x}_m)} \left[\sum_{m=1}^{M} \log p_\theta(\mathbf{x}_m \mid \mathbf{z}) \right] - D_{KL}(q_\phi(\mathbf{z} \mid \mathbf{x}_m) | p(\mathbf{z})) \right]. \quad (2)$$

However, this approach only takes each uni-modal encoding distribution separately into account during training. Thus, there is no explicit aggregation of information from multiple modalities in the latent representation for reconstruction by the decoder networks. For modalities with a high degree of modality-specific variation, this enforces an undesirable upperbound on the ELBO potentially leading to a sub-optimal approximation of the joint distribution [3].

Generalised Product-of-Experts Joint Posterior. We propose an alternative approach to mitigate the problem of overconfident experts by factorising the joint posterior as a generalised Product-of-Experts (gPoE) [2]; $q_\phi(\mathbf{z} \mid \mathbf{X}) = \frac{1}{K} \prod_{m=1}^{M} q_{\phi_m}^{\alpha_m}(\mathbf{z} \mid \mathbf{x}_m)$ where α_m is a weighting for modality m such that $\sum_{m=1}^{M} \alpha_m = 1$ for each latent dimension and $0 < \alpha_m < 1$. We optimise α during training allowing the model to weight experts in such a way as to learn an approximate joint posterior $q_\phi(\mathbf{z} \mid \mathbf{X})$ where the likelihood distribution $p_\theta(\mathbf{X} \mid \mathbf{z})$ is maximised. This provides a means to down-weigh overconfident experts. Furthermore, as α is learnt per latent dimension, different modality weightings can be learnt for different vectors, thus explicitly incorporating modality specific variation in addition to shared information in different dimensions of the joint latent space. Similarly to the PoE approach, we can compute the parameters of the joint posterior distribution; $\boldsymbol{\mu} = \frac{\sum_{m=1}^{M} \boldsymbol{\mu}_m \alpha_m / \boldsymbol{\sigma}_m^2}{\sum_{m=1}^{M} \alpha_m / \boldsymbol{\sigma}_m^2}$ and $\boldsymbol{\sigma}^2 = \sum_{m=1}^{M} \frac{1}{\alpha_m / \boldsymbol{\sigma}_m^2}$.

Recently, a gPoE mVAE was proposed for learning joint representations of hand-poses and surgical videos [6]. However, we emphasize that our approach differs in application and offers a more lightweight implementation (Joshi et al. [6] require training of auxiliary networks to learn α per sample).

Multi-modal Normative Modelling. We propose two mVAE normative modelling frameworks shown in Fig. 1a. MoE-normVAE, which uses a MoE joint posterior distribution, and gPoE-normVAE, which uses a gPoE joint posterior distribution. For both models, the encoder ϕ and decoder θ parameters are trained to characterise a healthy population cohort. normVAE models assume abnormality due to disease effects can be quantified by measuring deviations in the latent space [8] or the feature space [11]. At test time, the clinical cohort is passed through the encoder and decoder networks. Deviations of test subjects from the multi-modal latent space of the healthy controls and data reconstruction errors are measured. We compare our methods to the previously proposed PoE-normVAE [7] and three uni-modal models; two single modality and one multi-modality with a concatenated input.

To compare our normVAE models to a classical normative approach, we trained one GPR (using the PCNToolkit) per feature on a sub-set of 2000 healthy

UK Biobank individuals and used extreme value statistics to calculate subject-level abnormality index [9]. We used a top 5% abnormality threshold (set using the healthy training cohort) to calculate a significance ratio (see Eq. 6).

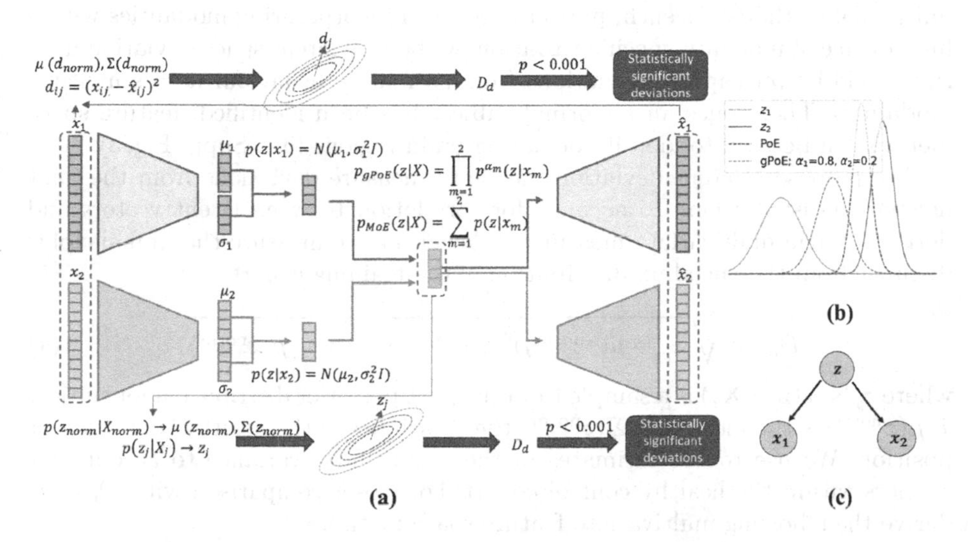

Fig. 1. (a) gPoE-normVAE and MoE-normVAE normative framework. All normVAE models were implemented using parameter settings; maximum epochs=2000, batch size=256, learning rate=10^{-4}, early stopping=50 epochs, encoder layers=[20, 40], decoder layers=[20, 40]. A ReLU activation function was applied between layers. Models were trained with a range of latent space sizes (L_{dim}) from 5 to 20. Models with L_{dim}=10 were fine-tuned (maximum 100 epochs) using the ADNI healthy cohort. Learnt α values are given in Supp. Table 1. (b) Example PoE and gPoE joint distributions. (c) Graphical model.

Multi-modal Latent Deviation Metric. Previous works using autoencoders as normative models mostly relied on feature-space based deviation methods [7,11]. That is, they compare the input value for subject j for the i-th brain region x_{ij} to the value reconstructed by the autoencoder $\widehat{x}_{ij}$: $d_{ij} = (x_{ij} - \widehat{x}_{ij})^2$. Kumar et al. [7] propose the following normalised z-score metric on the data reconstruction (a univariate feature space metric):

$$D_{\text{uf}} = \frac{d_{ij} - \mu_{\text{norm}}\left(d_{ij}^{\text{norm}}\right)}{\sigma_{\text{norm}}\left(d_{ij}^{\text{norm}}\right)} \tag{3}$$

where $\mu_{\text{norm}}\left(d_{ij}^{\text{norm}}\right)$ is the mean and $\sigma_{\text{norm}}\left(d_{ij}^{\text{norm}}\right)$ the standard deviation of the deviations d_{ij}^{norm} of a holdout healthy control cohort.

However, in the multi-modal setting, feature space-based deviation metrics may not highlight the benefits of multi-modal models over their uni-modal counterparts. The goal of the joint latent representation is to capture information from all modalities. Thus, decoders for each modality must extract the information from the joint latent representation, which now carries information from all other modalities as well. Therefore, data reconstructions capture only information relevant to a particular modality and may also be poorer compared to uni-modal methods. As such, particularly when incorporating modalities with a high degree of modality-specific variation, we believe latent space deviation metrics would better capture deviations from normative behaviour across multiple modalities. Then, once an abnormal subject has been identified, feature space metrics can be used to identify deviating brain regions (e.g. Supp. Fig. 3).

We propose a latent deviation metric to measure deviations from the joint normative distribution. To account for correlation between latent vectors and derive a single multivariate measure of deviation, we measure the Mahalanobis distance from the encoding distribution of the training cohort:

$$D_{\mathrm{ml}} = \sqrt{(z_j - \mu(z^{\mathrm{norm}}))^T \, \Sigma(z^{\mathrm{norm}})^{-1} \, (z_j - \mu(z^{\mathrm{norm}}))} \tag{4}$$

where $z_j \sim q\left(\mathbf{z}_j \mid \mathbf{X}_j\right)$ is a sample from the joint posterior distribution for subject j, $\mu(z^{\mathrm{norm}})$ is the mean and $\Sigma(z^{\mathrm{norm}})$ the covariance of the healthy cohort latent position. We use robust estimates of the mean and covariance to account for outliers within the healthy control cohort. For closer comparison with D_{ml}, we derive the following multivariate feature space metric:

$$D_{\mathrm{mf}} = \sqrt{(d_j - \mu(d^{\mathrm{norm}}))^T \, \Sigma(d^{\mathrm{norm}})^{-1} \, (d_j - \mu(d^{\mathrm{norm}}))} \tag{5}$$

where $d_j = \{d_{ij}, \ldots, d_{Ij}\}$ is the reconstruction error for subject j for brain regions $(i = 1, ..., I)$, $\mu(d^{\mathrm{norm}})$ is the mean and $\Sigma(d^{\mathrm{norm}})$ the covariance of the healthy cohort reconstruction error.

Assessing Deviation Metric Performance. For each model, we calculated D_{ml} and D_{mf} for a healthy holdout cohort and disease cohort. For each deviation metric, we identified individuals whose deviations were significantly different from the healthy training distribution ($p < 0.001$) [15]. Ideally, we want a model which correctly identifies disease individuals as outliers and healthy individuals as sitting within the normative distribution. As such, we use the following significance ratio (positive likelihood ratio) to assess model performance:

$$\text{significance ratio} = \frac{\text{True positive rate}}{\text{False positive rate}} = \frac{\text{TPR}}{\text{FPR}} = \frac{\frac{N_{\text{disease}}(\text{outliers})}{N_{\text{disease}}}}{\frac{N_{\text{holdout}}(\text{outliers})}{N_{\text{holdout}}}} \tag{6}$$

In order to calculate significance ratios, we calculated D_{uf} relative to the training cohort for the healthy holdout and disease cohorts (Bonferroni adjusted p=$0.05/N_{\text{features}}$) [7].

3 Experiments

Data Processing. To train the normVAE models, we used 10,276 healthy subjects from the UK Biobank [14] (application number: 70047). We used pre-processed (provided by the UK Biobank [1]) grey-matter volumes for 66 cortical (Desikan-Killiany atlas) and 16 subcortical brain regions, and Fractional Anisotropy (FA) and Mean Diffusivity (MD) measurements for 35 white matter tracts (John Hopkins University atlas). At test time, we used 2,568 healthy controls from a holdout cohort and 122 individuals with one of several neurodegenerative disorders; motor neuron disease, multiple sclerosis, Parkinson's disease, dementia/Alzheimer/cognitive-impairment and other demyelinating disease.

We also tested the models using an external dataset. We extracted 213 subjects from the Alzheimer's Disease Neuroimaging Initiative (ADNI)[1] [10] dataset with significant memory concern (SMC; N=27), early mild cognitive impairment (EMCI; N=63), late mild cognitive impairment (LMCI; N=34), Alzheimer's disease (AD; N=43) as well as healthy controls (HC; N=45). We used the healthy controls to fine-tune the models in a transfer learning approach. The same T1 and DTI features as for the UK Biobank were extracted for the ADNI dataset.

Rather than conditioning on covariates as done in some related work [7,8], we adjusted for confounding effects prior to analysis. Non-linear age and linear ICV affects where removed from the DTI and T1 MRI features of both datasets [12]. Each brain ROI was normalised by removing the mean and dividing by the standard deviation of the healthy control cohort brain regions.

UK Biobank Results. As expected, we see greater significance ratios for all models when using D_{ml} rather than D_{mf} (Table 1). When using D_{mf} or D_{uf},

Table 1. Significance ratio calculated from D_{ml}, D_{mf}, and D_{uf} for the UK Biobank. See Supp. for results in figure form. Using GPR, we observed a significance ratio of 6.01, poorer performance than our models (using D_{ml}).

	Significance ratio, D_{ml}				Significance ratio, D_{mf}				Significance ratio, D_{uf}			
Latent dimension	5	10	15	20	5	10	15	20	5	10	15	20
gPoE-normVAE (ours)	**7.89**	**9.24**	7.02	7.6	1.62	1.62	1.61	1.7	1.37	1.41	1.37	1.56
MoE-normVAE (ours)	7.25	8.77	**7.09**	**7.94**	1.67	1.7	1.59	1.68	1.44	1.46	1.46	1.44
PoE-normVAE	7.4	8.06	5.71	6.96	1.6	1.62	1.66	1.71	1.45	1.42	1.39	1.41
concatenated normVAE	7.63	6.21	5.43	3.55	1.61	1.59	1.57	1.58	1.48	1.4	1.39	1.44
T1 normVAE	5.26	4.43	2.63	2.44								
DTI normVAE	6.82	7.35	3.96	2.61								
Average T1&DTI normVAE					1.63	1.7	1.64	1.64	1.47	1.53	1.45	1.45

[1] Data used in preparation of this article were obtained from the Alzheimer's Disease Neuroimaging Initiative (ADNI) database (adni.loni.usc.edu). As such, the investigators within the ADNI contributed to the design and implementation of ADNI and/or provided data but did not participate in analysis or writing of this report. A complete listing of ADNI investigators can be found at: http://adni.loni.usc.edu/wp-content/uploads/how_to_apply/ADNI_Acknowledgement_List.pdf.

all models perform similiarly. Using D_{ml} over D_{mf} leads to a 4-fold increase in the signficance ratio. Further, our proposed models give the best overall performance across different L_{dim} with the highest significance ratio for gPoE-normVAE with L_{dim}=10. Generally, all multi-modal normVAE showed better performance than the uni-modal models suggesting that by modelling the joint distribution between modalities, we can learn better normative models.

ADNI Results. Previous work [8] explored the ability of a uni-modal T1 normVAE to detect deviations in the ADNI cohorts. Figure 2a shows the latent deviation D_{ml} for different diagnosis in the ADNI cohort for the T1 normVAE, DTI normVAE, PoE-normVAE and gPoE-normVAE models. All models reflect the increasing disease severity with increasing disease stage. The gPoE-normVAE model showed greater sensitivity to disease stage as suggested by the higher F statistic and p-values from an ANOVA analysis. We measured the Pearson correlation with composite measures of memory and executive function (Fig. 2b) and found that our proposed model exhibited greater correlation with both cognition scores than baseline approaches. Finally, we see that the sensitivity to disease severity for the gPoE-normVAE model extends to the feature space where we see a general increase in average D_{uf} from the LMCI to AD cohort (Supp. Figs. 3a and 3b respectively).

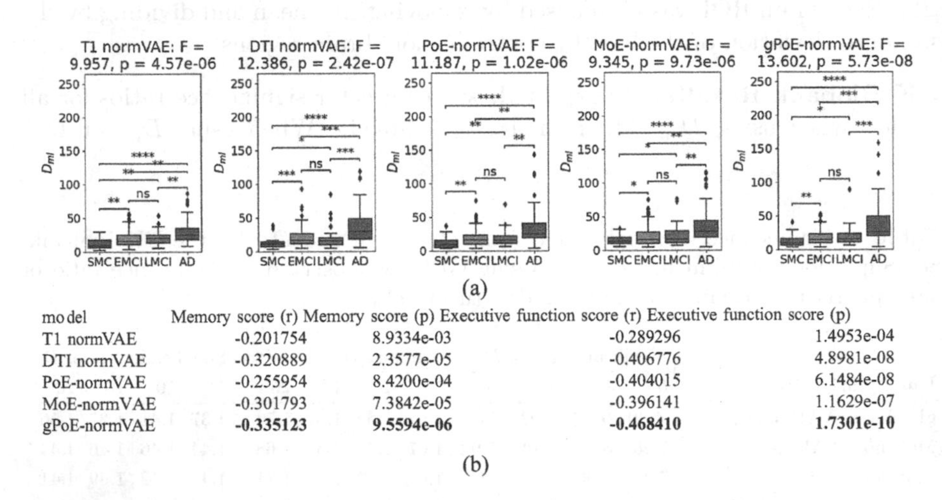

model	Memory score (r)	Memory score (p)	Executive function score (r)	Executive function score (p)
T1 normVAE	-0.201754	8.9334e-03	-0.289296	1.4953e-04
DTI normVAE	-0.320889	2.3577e-05	-0.406776	4.8981e-08
PoE-normVAE	-0.255954	8.4200e-04	-0.404015	6.1484e-08
MoE-normVAE	-0.301793	7.3842e-05	-0.396141	1.1629e-07
gPoE-normVAE	**-0.335123**	**9.5594e-06**	**-0.468410**	**1.7301e-10**

Fig. 2. (a) D_{ml} by disease label (L_{dim}=10). Statistical annotations were generated using Welch's t-tests between pairs of disease groups; ns : $0.05 < p <= 1, * : 0.01 < p <= 0.05, ** : 0.001 < p <= 0.01, *** : 0.0001 < p <= 0.001, **** : p <= 0.0001$. Robust estimates of the mean and covariance were not used to calculate D_{ml} due to the small healthy cohort size. (b) Pearson correlation between D_{ml} and patient cognition represented by age adjusted memory and executive function composite scores.

4 Discussion and Further Work

We have built on recent works [7,8,11] and introduced two novel mVAE normative models, which provide an alternative method of learning the joint normative distribution between modalities to address the limitations of current approaches. Our models provide a more informative joint representation compared to baseline methods as evidenced by the better significance ratio for the UK Biobank dataset and greater sensitivity to disease staging and correlation with cognitive measures in the ADNI dataset. We also proposed a latent deviation metric suitable for detecting deviations in the multivariate latent space of multi-modal normative models which gave an approximately 4-fold performance increase over metrics based on the feature space.

Further work will involve extending our models to more data modalities, such as genetic variants, to better characterise the behaviour of a physiological system. We note that, for fair comparison across models, we remove the effects of confounding variables prior to analysis. However, confounding effects could be removed during analysis via condition variables [8]. Another limitation of normVAE models introduced here is the use of ROI level data. Data processing software, such as FreeSurfer, may fail to accurately capture abnormality in images, particularly if large lesions are present. Further work involves creating normative models designed for voxel level data to better capture disease effects.

Normative models have been successfully applied to the study of a range of heterogeneous diseases. Diseases often present abnormalities across a range of neuroimaging, biological and physiological features which provide different information about the underlying disease process. Normative systems that incorporate features from different data modalities offer a holistic picture of the disease and will be capable of detecting abnormalities across a broad range of different diseases. Furthermore, multi-modal normative modelling captures the relationship between different modalities in healthy individuals, with disruption to this relationship potentially leading to a disease signal. Code is publicly available at https://github.com/alawryaguila/multimodal-normative-models.

Acknowledgements. This work is supported by the EPSRC-funded UCL Centre for Doctoral Training in Intelligent, Integrated Imaging in Healthcare (i4health) and the Department of Health's NIHR-funded Biomedical Research Centre at University College London Hospitals.

References

1. Alfaro-Almagro, F., et al.: Image processing and quality control for the first 10,000 brain imaging datasets from UK biobank. NeuroImage **166**, 400–424 (2018). https://doi.org/10.1016/j.neuroimage.2017.10.034, https://www.sciencedirect.com/science/article/pii/S1053811917308613
2. Cao, Y., Fleet, D.J.: Generalized product of experts for automatic and principled fusion of gaussian process predictions (2014). https://doi.org/10.48550/ARXIV.1410.7827, https://arxiv.org/abs/1410.7827

3. Daunhawer, I., Sutter, T.M., Chin-Cheong, K., Palumbo, E., Vogt, J.E.: On the limitations of multimodal vaes. CoRR abs/2110.04121 (2021). https://arxiv.org/abs/2110.04121
4. Hendrycks, D., Mazeika, M., Dietterich, T.G.: Deep anomaly detection with outlier exposure. CoRR abs/1812.04606 (2018). http://arxiv.org/abs/1812.04606
5. Hwang, H., Kim, G.H., Hong, S., Kim, K.E.: Multi-view representation learning via total correlation objective. In: Ranzato, M., Beygelzimer, A., Dauphin, Y., Liang, P., Vaughan, J.W. (eds.) Advances in Neural Information Processing Systems. vol. 34, pp. 12194–12207. Curran Associates, Inc. (2021), https://proceedings.neurips.cc/paper/2021/file/65a99bb7a3115fdede20da98b08a370f-Paper.pdf
6. Joshi, A., Gupta, N., Shah, J., Bhattarai, B., Modi, A., Stoyanov, D.: Generalized product-of-experts for learning multimodal representations in noisy environments. In: Proceedings of the 2022 International Conference on Multimodal Interaction, pp. 83–93. ICMI '22, Association for Computing Machinery, New York, NY, USA (2022). https://doi.org/10.1145/3536221.3556596, https://doi.org/10.1145/3536221.3556596
7. Kumar, S., Payne, P., Sotiras, A.: Normative modeling using multimodal variational autoencoders to identify abnormal brain structural patterns in alzheimer disease (2021). https://doi.org/10.48550/ARXIV.2110.04903, https://arxiv.org/abs/2110.04903
8. Lawry Aguila, A., Chapman, J., Janahi, M., Altmann, A.: Conditional vaes for confound removal and normative modelling of neurodegenerative diseases. In: Medical Image Computing and Computer Assisted Intervention - MICCAI 2022: 25th International Conference, Singapore, September 18–22, 2022, Proceedings, Part I. pp. 430–440. Springer-Verlag (2022)
9. Marquand, A., Rezek, I., Buitelaar, J., Beckmann, C.: Understanding heterogeneity in clinical cohorts using normative models: Beyond case-control studies. Biol. Psych. **80**(7), 552-561 (2016)
10. Petersen, R., et al.: Alzheimer's disease neuroimaging initiative (adni): clinical characterization. Neurology 74(3), 201–209 (2010). https://doi.org/10.1212/wnl.0b013e3181cb3e25, https://europepmc.org/articles/PMC2809036
11. Pinaya, W., et al.: Using normative modelling to detect disease progression in mild cognitive impairment and alzheimer's disease in a cross-sectional multi-cohort study. Sci. Reports **11**(1), 15746 (2021)
12. Pomponio, R., et al.: Harmonization of large mri datasets for the analysis of brain imaging patterns throughout the lifespan. NeuroImage **208**, 116450 (2020). https://doi.org/10.1016/j.neuroimage.2019.116450, https://www.sciencedirect.com/science/article/pii/S1053811919310419
13. Shi, Y., Siddharth, N., Paige, B., Torr, P.H.S.: Variational mixture-of-experts autoencoders for multi-modal deep generative models (2019). https://doi.org/10.48550/ARXIV.1911.03393, https://arxiv.org/abs/1911.03393
14. Sudlow, C., et al.: Uk biobank: an open access resource for identifying the causes of a wide range of complex diseases of middle and old age. PLoS Med. **12**, e1001779 (2015)
15. Tabachnick, B.G., Fidell, L.S.: Using multivariate statistics, 7th edn. Pearson, Upper Saddle River, NJ (2018)
16. Wu, M., Goodman, N.D.: Multimodal generative models for scalable weakly-supervised learning. CoRR abs/1802.05335 (2018). http://arxiv.org/abs/1802.05335

LOTUS: Learning to Optimize Task-Based US Representations

Yordanka Velikova[1(✉)], Mohammad Farid Azampour[1,2], Walter Simson[3], Vanessa Gonzalez Duque[1,4], and Nassir Navab[1,5]

[1] Computer Aided Medical Procedures, Technical University of Munich, Munich, Germany
dani.velikova@tum.de
[2] Department of Electrical Engineering, Sharif University of Technology, Tehran, Iran
[3] Department of Radiology, Stanford University School of Medicine, Stanford, USA
[4] LS2N Laboratory at Ecole Centrale Nantes, UMR CNRS 6004, Nantes, France
[5] Computer Aided Medical Procedures, John Hopkins University, Baltimore, USA

Abstract. Anatomical segmentation of organs in ultrasound images is essential to many clinical applications, particularly for diagnosis and monitoring. Existing deep neural networks require a large amount of labeled data for training in order to achieve clinically acceptable performance. Yet, in ultrasound, due to characteristic properties such as speckle and clutter, it is challenging to obtain accurate segmentation boundaries, and precise pixel-wise labeling of images is highly dependent on the expertise of physicians. In contrast, CT scans have higher resolution and improved contrast, easing organ identification. In this paper, we propose a novel approach for learning to optimize task-based ultrasound image representations. Given annotated CT segmentation maps as a simulation medium, we model acoustic propagation through tissue via ray-casting to generate ultrasound training data. Our ultrasound simulator is fully differentiable and learns to optimize the parameters for generating physics-based ultrasound images guided by the downstream segmentation task. In addition, we train an image adaptation network between real and simulated images to achieve simultaneous image synthesis and automatic segmentation on US images in an end-to-end training setting. The proposed method is evaluated on aorta and vessel segmentation tasks and shows promising quantitative results. Furthermore, we also conduct qualitative results of optimized image representations on other organs.

Keywords: Ultrasound · Unsupervised Domain Adaptation · Segmentation · Task Driven

---

Supplementary Information The online version contains supplementary material available at https://doi.org/10.1007/978-3-031-43907-0_42.

H. Greenspan et al. (Eds.): MICCAI 2023, LNCS 14220, pp. 435–445, 2023.
https://doi.org/10.1007/978-3-031-43907-0_42

1 Introduction

Ultrasound (US) imaging is a widely used modality in medical diagnosis for screening and follow-up examinations. Hence, precise segmentation of the target organs is crucial for diagnosing or tracking disease progression. Recently, the application of deep learning for ultrasound image segmentation has emerged as a powerful tool. However, accurate segmentation of US images remains a challenging task due to the complexity of the modality, as it has limited resolution and often contains clutter, shadowing and reverberation artefacts. This leads to a general lack of annotated data, and additionally, due to varying operator skills, there is high heterogeneity of ground truth data labels, which is the primary factor hampering solid segmentation performance [6].

On the other hand, large pixel-level labeled CT datasets are freely available online. Thus to overcome the lack of ground truth ultrasound data, researchers have utilized ultrasound simulators to generate large sets of ultrasound-like images from CT label maps and use them for training [10]. Simulated ultrasound data automatically provides a labeled pair of the tissue distribution and the resulting b-mode image and can be augmented with rotational, brightness, contrast, probe, and scanner variations.

Generally, ultrasound simulators can be categorized into two types based on their modeling techniques: finite difference models of the wave equation, modeling the mechanical propagation of sound waves through tissues, and simulating ray casting through tissue maps represented by ultrasound tissue properties [2,5,11]. Although the former can model higher-order non-linear effects, producing realistic images, generating a single image can take hours. The latter, on the other hand, is much faster and can be integrated into other systems [1,4]. While leveraging automatically generated ultrasound simulations with corresponding labels for training has benefits, models trained on simulations fail when applied directly to real, as they cannot perfectly simulate ultrasound images without distinguishable differences from real ones.

Thus, one main challenge when working with simulated data is reducing the domain shift between simulated and real data. In a supervised sense, many works have investigated the realistic parametrization of ultrasound simulators to reduce the domain shift between simulated and real data ultrasound data [12] and augmentation of ultrasound b-modes [13]. Recent domain adaptation models [8,19] employing generative adversarial networks have shown promise in improving image synthesis in an unsupervised manner. Moreover, recent works show their application in combination with segmentation or registration tasks between X-ray and CT or MRI scans [3,18]. Further works show their application in ultrasound by closing the real-simulation gap via translation from simulated images to "realistic" ones that match the target domain, thereby enabling the application of trained segmentation networks on real images [14,16].

However, those methods require separate training for each part of the architecture, limiting the models' flexibility. Notably, [15] proposes using an intermediate representation image with common properties between CT and US for the task of aorta segmentation. However, the intermediate image is not formulated differentiably but is statically calibrated, and the whole pipeline is not trained end-to-end.

Contributions. In this paper, we propose a novel approach for learning to optimize task-based ultrasound image representations. During training, we render an intermediate US image representation from segmented public CT scans and use it as input to a segmentation network. Our ultrasound renderer is fully differentiable and learns to optimize the parameters necessary for physics-based ultrasound simulation, guided by the downstream segmentation task. At the same time, we train an image style transfer network between real and simulated data to achieve simultaneous image synthesis as well as automatic segmentation on US images in an end-to-end training setting. In addition, no labels are required for the real ultrasound images, which are also unpaired with the simulated ultrasound images. We evaluate our method on aorta and vessel segmentation. Our quantitative and qualitative results demonstrate that our method learns the optimal image for the task of interest. The source code for our method is publicly available[1].

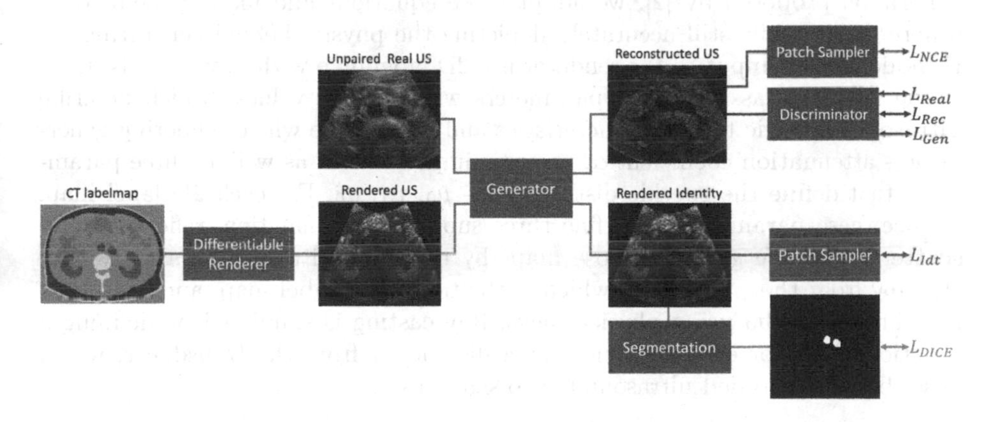

Fig. 1. Overview of the proposed framework. During training, we render online US simulation images from CT label maps and use them as input to a segmentation network. Our ultrasound renderer is fully differentiable and learns to optimize the parameters based on the downstream segmentation task. At the same time, we train an unpaired and unsupervised image style transfer network between real and rendered images to achieve simultaneous image synthesis as well as automatic segmentation on US images in an end-to-end training setting.

[1] https://github.com/danivelikova/lotus.

2 Methodology

2.1 Differentiable Ultrasound Renderer

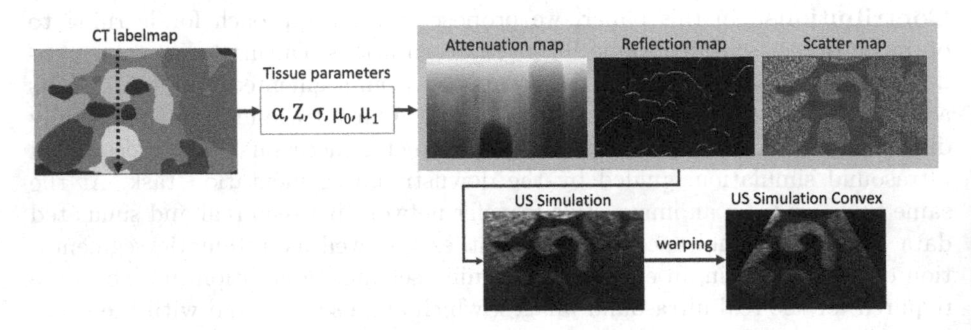

Fig. 2. Overview of the Differentiable Ultrasound Renderer.

Building on the mathematical foundations of ray tracing and ultrasound echo generation proposed by [2], we adopt those equations and modify them to be differentiable while still accurately depicting the physics behind generating US B-mode images. Input to the renderer is a 2D label map with tissue labels. Each tissue label has assigned five parameters with default values[2] which describe ultrasound-specific tissue characteristics and control the whole rendering generation - attenuation coefficient α, acoustic impedance Z, as well as three parameters that define the speckle distribution - μ_0, μ_1, σ_0. For each 2D label map, we use these parameters to define three sub-maps: attenuation, reflection, and scatter maps. We generate those maps by modeling ultrasound waves as rays starting from the transducer, which is the top of the label map, and propagating through media using physical laws. Ray casting is simulated by defining a function $E_i(d)$ for each scanline i at a distance d from the transducer, which describes the recorded ultrasound echo signal as:

$$E_i(d) = R_i(d) + B_i(d) \tag{1}$$

where $R_i(d)$ is the energy reflected from the interfaces between two tissues as the beam passes through them and $B_i(d)$ represents the energy backscattered from the scattering points along the scanline. The reflection of the ray is described as:

$$R_i(d) = |I_i(d) * Z_i(d)| * P(d) \otimes G(d) \tag{2}$$

where $I_i(d)$ is the remaining energy of the ray, which gets attenuated during tissue traversal. We model $I_i(d)$ by approximating the Beer-Lambert Law as: $I_i(d) = e^{-\alpha d}$, where α is the attenuation coefficient of the medium and d the distance travelled. To construct the final 2D attenuation map, we calculate, for

[2] https://github.com/Blito/burgercpp/blob/master/examples/ircad11/liver.scene.

each ray, the cumulative product of the attenuation as it traverses through various tissues, thereby modeling how the signal's strength diminishes. The reflection coefficient $Z = (Z_2 - Z_1)^2/(Z_2 + Z_1)^2$, is computed from the acoustic impedances of two adjacent tissues: Z_1 and Z_2. The $P(d)$ is the Point Spread Function (PSF) along the ray, and $G(d)$ is a boundary map, where 1 is assigned for points on the boundary of the surface and 0 otherwise. For simplicity, we model the PSF as a two-dimensional normalized Gaussian. The amount of the reflected signal, denoted by ϕ_r, equals the result of multiplying the reflection coefficient by the boundary condition. To build our final 2D reflection map, for each ray, we compute the cumulative product of the residual signal, defined as $1 - \phi_r$. The output represents the fraction of the signal that propagates forward.

In additionally to the reflection term, a backscattered energy term $B_i(d)$ in the returning echo is calculated:

$$B_i(d) = I_i(d) * P(d) \otimes \widetilde{T}(x, y) \tag{3}$$

the residual ultrasound wave energy $I_i(d)$ is multiplied with the PSF $P(d)$, which has been convolved with a texture $\widetilde{T}$ of random scatterers for each (x, y), where:

$$\widetilde{T}(x, y) = \begin{cases} S(x, y) & \text{if } T_1(x, y) \leq \mu_1(x, y) \\ 0 & \text{otherwise} \end{cases} \tag{4}$$

$$S(x, y) = T_0(x, y) * \sigma_0 + \mu_0 \tag{5}$$

This texture is constructed using two random textures $T_0(x, y)$ and $T_1(x, y)$ with Gaussian normalized distributions and the parameters μ_0, μ_1, and σ_0, which represent the brightness, density and standard deviation of scatterers respectively. To make the function fully differentiable, we replace the conditional operation $T_1 \leq \mu_1$ with a differentiable approximation:

$$\widetilde{T}(x, y) = \sigma(\beta \cdot (\mu_1(x, y) - T_1(x, y))) \cdot S(x, y) \tag{6}$$

where $\sigma(z) = \frac{1}{1+e^{-z}}$ is the sigmoid function and β is a scaling factor that adjusts its steepness. The resulting function is fully differentiable as the sigmoid function smoothly approximates the step function and all operations involved are differentiable. Additionally, we apply temporal gain compensation (TGC) to enhance tissues deeper in the image. The final rendered ultrasound image is constructed from the three sub-maps (see Fig. 2) and additionally warped to produce the desired fan shape. At the beginning of the training, we set the default tissue-specific values, which during the training, get changed, guided from the downstream task, and generate optimal US simulation.

2.2 End-to-End Learning

The proposed method's architecture is shown in Fig. 1. During training, our method follows two main paths: Real → Reconstructed US and CT label map → Segmentation. We explain the meaning of these paths in the order shown in the figure.

Real → Reconstructed US. Since there is an appearance gap between real and our rendered ultrasound images we incorporate an unpaired and unsupervised image-to-image translation network, CUT [8], which uses a contrastive learning scheme. Given a source image, the Generator learns a function $G : \mathcal{X} \mapsto \mathcal{Y}$ that translates the corresponding image into the target's appearance. We have two domains of unpaired instances: real US images as the source $\mathcal{X}$ and rendered US as the target $\mathcal{Y}$. The generator's encoder G_{enc} extracts relevant content characteristics, while the decoder G_{dec} learns to create the desired goal appearance. The Generator network employs an adversarial loss:

$$\mathcal{L}_{GAN} = \mathbb{E}_y \log D(y) + \mathbb{E}_x \log(1 - D(G(x))) \tag{7}$$

where the generated images $G(x)$ resemble images from domain $\mathcal{Y}$, and $D(.)$ differentiates between translated and real images y. However, the adversarial loss alone does not ensure that the translated image will preserve the structure of the anatomy. An additional contrastive loss must be imposed, which maximizes mutual information across corresponding image patches from the source and the output image. We use the Patch Sampler from CUT to extract image patches and calculate the contrastive NCE ($\mathcal{L}_{NCE}$) loss [8]. The final loss is defined as:

$$\mathcal{L}_{CUT}(X,Y) = \mathcal{L}_{GAN}(X,Y) + \mathcal{L}_{NCE}(X, G(X)) + \mathcal{L}_{NCE}(Y, G(Y)) \tag{8}$$

where, the $\mathcal{L}_{NCE}$ is calculated on two pairs, a sample from the source domain (x) paired with the generated output $G(x)$ and a sample from the target domain (y) paired with the $G(y)$ which we denote as the identity image. The loss over the second pair serves as an identity loss and prevents the generator from making unnecessary changes to the image.

CT Labelmap → Segmentation: The segmentation network forward pass has a nested structure. First, we obtain a 2D slice from the CT label map and pass it to the differentiable ultrasound renderer. The resulting rendered US is passed through the frozen Generator network, and the identity image output of the Generator is used as an input to the segmentation network to ensure the same distribution as the target domain. We update both the segmentation network and the Renderer using dice loss. The label for computing the dice loss comes directly from the input label map used for generating the rendered US.

Stopping Criterion: Once the segmentation network validation loss converges, we employ a small subset of 10 labeled images from the real US domain as a stopping indicator for the entire training pipeline.

3 Experimental Setup

CT Dataset: We use 12 CT volumes from a publicly available dataset Synapse[3] [7] for training. The data comes with labels for multiple organs. These

[3] https://www.synapse.org/#!Synapse:syn3193805/wiki/89480.

labels were additionally augmented with labels of bones, fat, skin, and lungs using TotalSegmentor [17] to complete the label maps.

In-vivo Images: We acquired abdominal ultrasound sweeps from eleven volunteers of age 26 ± 3 (m:7/f:4). For each person, one sweep was acquired with a convex probe[4]. Per sweep, 50 frames were randomly sampled and used for training the CUT network. To compare against a supervised approach, additional images were annotated (500 for the aorta, 400 for vessels) from all volunteers to train 5-fold cross-validation. From each set of annotated images, 100 images were randomly sampled as test sets for both segmentation tasks.

Training Details: We train the network with a learning rate of 10^{-5} for the segmentation network, 10^{-3} for the US Renderer, and 5^{-6} for the image adaptation network, with a batch size of 1, Adam optimizer and dice loss. We employ rotation, translation, and scaling augmentations on the CT label maps and split them randomly in an 80–20% ratio for training and validation, respectively. For the supervised approach, we trained the networks, for 120 epochs, with a learning rate of 10^{-3} and the Adam optimizer.

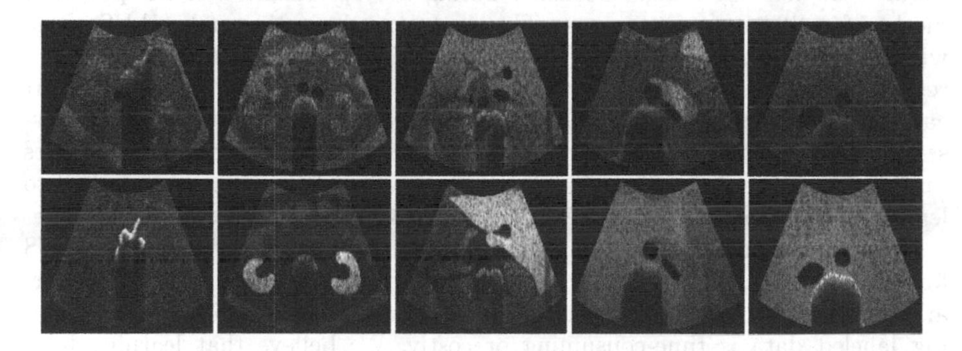

Fig. 3. Image representations of segmentation tasks for different target organs, learned during the optimization. Top row: rendered US with default parameters, bottom row: final optimized rendered US for each specific organ. From left to right: spine, kidney, liver, vessels, aorta only.

Experiments. We test the proposed framework quantitatively for two segmentation tasks: all vessels and the aorta only. We evaluate the accuracy of the proposed method by comparing it to a supervised network. For this, we train a 5-fold cross-validation U-Net [9], test on three hold-out subjects, and report the average DSC. We also compare to a fixed rendered image by freezing the US renderer instead of optimizing it. Additionally, we show qualitative results of the proposed method when the downstream tasks are changed Fig. 3.

[4] cQuest Cicada US scanner, Cephasonics, Santa Clara, CA, US.

4 Results and Discussion

In Table 1, we compare the performance of LOTUS against a fully supervised approach and against a frozen renderer's parameters and report the DSC and Hausdorff distance (HD) in mm for aorta segmentation and DSC only for all vessels segmentation. Our proposed method achieved the highest DSC score of 89.24 ± 0.13 and the lowest HD score of 2.52 ± 1.18 mm for aorta segmentation. For the task of vessels segmentation it also achieved the best DSC of 90.9 ± 0.06.

Table 1. Comparison of DSC and Hausdorff distance for the task of aorta and vessels segmentation of our proposed method with supervised network and with frozen renderer.

	Supervised	Frozen Renderer	LOTUS
DSC - Aorta	80.65 ± 2.35	84.67 ± 0.14	$\mathbf{89.24 \pm 0.13}$
Hausdorff (mm)	17.61 ± 1.32	11.08 ± 18.64	$\mathbf{2.52 \pm 1.18}$
DSC - Vessel	83.56 ± 4.16	89.05 ± 0.09	$\mathbf{90.9 \pm 0.06}$

Figure 3 depicts the images obtained during the optimization of the proposed method for different target organs. The upper row shows the rendered US image with default parameters, and the bottom row displays the optimized image representations for the corresponding target organ learned during optimization. It can be observed that the spine, kidney and liver appear brighter, while for vessel and aorta segmentation, the vessels darken and the background becomes uniformly homogeneous. This highlights the ability of the proposed method to learn optimal representations for each downstream task.

The results presented in this work demonstrate the effectiveness of LOTUS for segmenting organs in ultrasound images. Our physics-based simulator generates synthetic training data, which is especially useful in scenarios where obtaining labeled data is time-consuming or costly. We believe that learning from transferred labels from CT contributes to a more accurate model since CT data is more accessible and labels are more refined. Our quantitative results indicate that LOTUS can achieve accurate segmentation of aorta boundaries and other vessels. Furthermore, the end-to-end framework enables the differentiable US renderer and the unsupervised image translation to get optimized dynamically during the training. Thus, the intermediate representation image is not static but changes during the training. This illustrates the adaptivity of the proposed method to the downstream task, highlighting its prospective applicability across diverse applications and anatomies.

Moreover, rather than directly using the rendered US image as an input to the segmentation network, we use the identity image output from the Generator. This yielded significant improvement in the segmentation result as it learns from a distribution consistent with the reconstructed US while looking similar to the rendered US. As a result, during inference stage, the distribution of the translated real US is closer to the distribution the segmentation network was trained on, thereby improving the performance of the model.

One of the challenges when employing generative adversarial networks is that the loss is not an indicator of the best result. We determine the optimal model by utilizing a small subset of labeled images after the convergence of the segmentation network, to ensure robustness during inference. Further stopping criteria can be studied to achieve higher automation of the pipeline.

Currently, our model incorporates the basic physics of ultrasound imaging without considering artifacts explicitly. Thus, exploring the robustness of the method against artifacts could yield valuable future improvements.

5 Conclusion

This paper presents a novel approach to learning task-based ultrasound image representations. LOTUS leverages CT labelmaps to simulate ultrasound data via differentiable ray-casting. The proposed ultrasound simulator is fully differentiable and learns to optimize the parameters for generating physics-based ultrasound images guided by the downstream segmentation task. We also introduce an image adaptation network to achieve simultaneous image synthesis and automatic segmentation on US images in an end-to-end training setting without needing paired real and simulated images. Our method is evaluated on aorta and vessel segmentation tasks and shows promising quantitative results. Furthermore, we demonstrate the potential of our approach for other organs through qualitative results of optimized image representations. The ability to learn from unlabeled data and simulate the ultrasound modality has the potential for various clinical tasks beyond segmentation. We believe that our work has the potential to improve ultrasound imaging interpretation and learning.

Acknowledgements. We would like to thank Magdalena Wysocki for the insightful discussions and Dr. Magdalini Paschali for helping with refining and improving the manuscript. The authors were partially supported by the grant NPRP-11S-1219-170106 from the Qatar National Research Fund (a member of the Qatar Foundation). The findings herein are however solely the responsibility of the authors.

References

1. Brickson, L.L., Hyun, D., Jakovljevic, M., Dahl, J.J.: Reverberation noise suppression in ultrasound channel signals using a 3D fully convolutional neural network. IEEE Trans. Med. Imaging **40**(4), 1184–1195 (2021)
2. Burger, B., Bettinghausen, S., Radle, M., Hesser, J.: Real-time GPU-based ultrasound simulation using deformable mesh models. IEEE Trans. Med. Imaging **32**(3), 609–618 (2012)
3. Dou, Q., Ouyang, C., Chen, C., Chen, H., Heng, P.A.: Unsupervised cross-modality domain adaptation of convnets for biomedical image segmentations with adversarial loss. In: International Joint Conference on Artificial Intelligence (2018)

4. Hyun, D., Brickson, L.L., Looby, K.T., Dahl, J.J.: Beamforming and speckle reduction using neural networks. IEEE Trans. Ultrason. Ferroelectr. Freq. Control **66**(5), 898–910 (2019)
5. Jensen, J.A.: A new approach to calculating spatial impulse responses. In: 1997 IEEE Ultrasonics Symposium Proceedings. An International Symposium (Cat. No. 97CH36118), vol. 2, pp. 1755–1759. IEEE (1997)
6. Krönke, M., et al.: Tracked 3D ultrasound and deep neural network-based thyroid segmentation reduce interobserver variability in thyroid volumetry. PLoS ONE **17**(7), e0268550 (2022)
7. Landman, B., Xu, Z., Igelsias, J., Styner, M., Langerak, T., Klein, A.: MICCAI multi-atlas labeling beyond the cranial vault-workshop and challenge. In: Proceedings MICCAI Multi-Atlas Labeling Beyond Cranial Vault-Workshop Challenge. vol. 5, p. 12 (2015)
8. Park, T., Efros, A.A., Zhang, R., Zhu, J.Y.: Contrastive learning for unpaired image-to-image translation. In: European Conference on Computer Vision (2020)
9. Ronneberger, O., Fischer, P., Brox, T.: U-net: convolutional networks for biomedical image segmentation. In: Navab, N., Hornegger, J., Wells, W.M., Frangi, A.F. (eds.) Medical Image Computing and Computer-Assisted Intervention - MICCAI 2015, pp. 234–241. Springer International Publishing, Cham (2015)
10. Rubi, P., Vera, E.F., Larrabide, J., Calvo, M., D'Amato, J.P., Larrabide, I.: Comparison of real-time ultrasound simulation models using abdominal CT images. In: Romero, E., Lepore, N., Brieva, J., Brieva, J., and I.L. (eds.) 12th International Symposium on Medical Information Processing and Analysis. vol. 10160, p. 1016009. International Society for Optics and Photonics, SPIE (2017). https://doi.org/10.1117/12.2255741
11. Salehi, M., Ahmadi, S.-A., Prevost, R., Navab, N., Wein, W.: Patient-specific 3D ultrasound simulation based on convolutional ray-tracing and appearance optimization. In: Navab, N., Hornegger, J., Wells, W.M., Frangi, A.F. (eds.) MICCAI 2015. LNCS, vol. 9350, pp. 510–518. Springer, Cham (2015). https://doi.org/10.1007/978-3-319-24571-3_61
12. Simson, W.A., Paschali, M., Sideri-Lampretsa, V., Navab, N., Dahl, J.J.: Investigating pulse-echo sound speed estimation in breast ultrasound with deep learning. arXiv preprint arXiv:2302.03064 (2023)
13. Tirindelli, M., Eilers, C., Simson, W., Paschali, M., Azampour, M.F., Navab, N.: Rethinking ultrasound augmentation: a physics-inspired approach. In: de Bruijne, M., et al. (eds.) MICCAI 2021. LNCS, vol. 12908, pp. 690–700. Springer, Cham (2021). https://doi.org/10.1007/978-3-030-87237-3_66
14. Tomar, D., Zhang, L., Portenier, T., Goksel, O.: Content-preserving unpaired translation from simulated to realistic ultrasound images. In: de Bruijne, M., et al. (eds.) Medical Image Computing and Computer Assisted Intervention - MICCAI 2021, pp. 659–669. Springer International Publishing, Cham (2021)
15. Velikova, Y., Simson, W., Salehi, M., Azampour, M.F., Paprottka, P., Navab, N.: Cactuss: common anatomical CT-us space for us examinations. In: Wang, L., Dou, Q., Fletcher, P.T., Speidel, S., Li, S. (eds.) Medical Image Computing and Computer Assisted Intervention - MICCAI 2022, pp. 492–501. Springer Nature Switzerland, Cham (2022)
16. Vitale, S., Orlando, J.I., Iarussi, E., Larrabide, I.: Improving realism in patient-specific abdominal ultrasound simulation using cycleGANs. Int. J. Comput. Assist. Radiol. Surg. **15**, 183–192 (2019)

17. Wasserthal, J., et al.:Totalsegmentator: Robust segmentation of 104 anatomic structures in CT images. Radiology: Artificial Intell. 0(ja), e230024 (0). https://doi.org/10.1148/ryai.230024
18. Zhang, Y., Miao, S., Mansi, T., Liao, R.: Task driven generative modeling for unsupervised domain adaptation: application to X-ray image segmentation. In: Frangi, A.F., Schnabel, J.A., Davatzikos, C., Alberola-López, C., Fichtinger, G. (eds.) Medical Image Computing and Computer Assisted Intervention - MICCAI 2018, pp. 599–607. Springer International Publishing, Cham (2018)
19. Zhu, J.Y., Park, T., Isola, P., Efros, A.A.: Unpaired image-to-image translation using cycle-consistent adversarial networkss. In: Computer Vision (ICCV), 2017 IEEE International Conference on (2017)

Unsupervised 3D Out-of-Distribution Detection with Latent Diffusion Models

Mark S. Graham[1(✉)], Walter Hugo Lopez Pinaya[1], Paul Wright[1], Petru-Daniel Tudosiu[1], Yee H. Mah[1,2], James T. Teo[2,3], H. Rolf Jäger[4], David Werring[5], Parashkev Nachev[4], Sebastien Ourselin[1], and M. Jorge Cardoso[1]

[1] Department of Biomedical Engineering, School of Biomedical Engineering and Imaging Sciences, King's College London, London, UK
mark.graham@kcl.ac.uk
[2] King's College Hospital NHS Foundation Trust, Denmark Hill, London, UK
[3] Institute of Psychiatry, Psychology and Neuroscience, King's College London, London, UK
[4] Institute of Neurology, University College London, London, UK
[5] Stroke Research Centre, UCL Queen Square Institute of Neurology, London, UK

Abstract. Methods for out-of-distribution (OOD) detection that scale to 3D data are crucial components of any real-world clinical deep learning system. Classic denoising diffusion probabilistic models (DDPMs) have been recently proposed as a robust way to perform reconstruction-based OOD detection on 2D datasets, but do not trivially scale to 3D data. In this work, we propose to use Latent Diffusion Models (LDMs), which enable the scaling of DDPMs to high-resolution 3D medical data. We validate the proposed approach on near- and far-OOD datasets and compare it to a recently proposed, 3D-enabled approach using Latent Transformer Models (LTMs). Not only does the proposed LDM-based approach achieve statistically significant better performance, it also shows less sensitivity to the underlying latent representation, more favourable memory scaling, and produces better spatial anomaly maps. Code is available at https://github.com/marksgraham/ddpm-ood.

Keywords: Latent diffusion models · Out-of-distribution detection

1 Introduction

Methods for out-of-distribution (OOD) detection are a crucial component of any machine learning pipeline that is deployed in the real world. They are particularly necessary for pipelines that employ neural networks, which perform well on data drawn from the distribution they were trained on but can produce unexpected results when given OOD data. For medical applications, methods for

---

Supplementary Information The online version contains supplementary material available at https://doi.org/10.1007/978-3-031-43907-0_43.

H. Greenspan et al. (Eds.): MICCAI 2023, LNCS 14220, pp. 446–456, 2023.
https://doi.org/10.1007/978-3-031-43907-0_43

OOD detection must be able to detect both far-OOD data, such as images of a different organ or modality to the in-distribution data, and near-OOD data, such as in-distribution data corrupted by imaging artefacts. It is also necessary that these methods can operate on high-resolution 3D data. In this work, we focus on methods trained in a fully unsupervised way; without any labels or access to OOD data at train time.

Recently, Latent Transformer Models (LTMs) [9] have proven themselves to be effective for anomaly detection and synthesis in medical data [21,23,27]. These two-stage models first use a VQ-VAE [20] or VQ-GAN [9] to provide a compressed, discrete representation of the imaging data. An autoregressive Transformer [29] can then be trained on a flattened sequence of this representation. LTMs are particularly valuable in medical data, where the high input size makes training a Transformer on raw pixels infeasible. Recently, these models have been shown to be effective for 3D OOD detection by using the Transformer's likelihood of the compressed sequence to identify both far- and near-OOD samples [11]. These models can also provide spatial anomaly maps that highlight the regions of the image considered to be OOD, particularly valuable for highlighting localised artefacts in near-OOD data.

However, LTMs have some disadvantages. Firstly, likelihood models have well documented weaknesses when used for OOD detection [3,13,19], caused by focusing on low-level image features [12,26]. It can help to measure likelihood in a more abstract representation space, such as that provided by a VQ-VAE [8], but how to train models that provide optimal representations for assessing likelihood is still an open research problem. For example, [11] showed in an ablation study that LTMs fail at OOD when lower levels of VQ-VAE compression are used. Secondly, the memory requirements of Transformers mean that even with high compression rates, the technique cannot scale to very high-resolution medical data, such as a whole-body CT with an image dimension 512^3. Finally, the spatial anomalies maps produced by LTMs are low resolution, being in the space of the latent representation rather than that of the image itself.

A promising avenue for OOD detection is denoising diffusion probabilistic model (DDPM)-based OOD detection [10]. This approach works by taking the input images and noising them multiple times to different noise levels. The model is used to denoise each of these noised images, which are compared to the input; the key idea is that the model will only successfully denoise in-distribution (ID) data. The method has shown promising results on 2D data [10] but cannot be trivially extended to 3D; as even extending DDPMs to work on high-resolution 2D data is an area of active research. We propose to scale it to 3D volumetric data through the use of Latent Diffusion Models (LDMs). These models, analogous to LTMs, use a first-stage VQ-GAN to compress the input. The DDPM then learns to denoise these compressed representations, which are then decoded and their similarity to the input image is measured directly in the original image space.

The proposed LDM-based OOD detection offers the potential to address the three disadvantages of an LTM-based approach. Firstly, as the method is not likelihood based, it is not necessary that the VQ-GAN provides an ill-defined 'good representation'. Rather, the only requirement is that it reconstructs the inputs well, something easy to quantify using reconstruction quality metrics. Secondly, DDPMs have more favourable memory scaling behaviour than Transformers, allowing them to be trained on higher-dimensional representations. Finally, as the comparisons are performed at the native resolution, LDMs can produce high-resolution spatial anomaly maps. We evaluate both the LTM and the proposed LDM model on several far- and near-OOD detection tasks and show that LDMs overcome the three main failings of LTMs: that their performance is less reliant on the quality of the first stage model, that they can be trained on higher dimensional inputs, and that they produce higher resolution anomaly maps.

2 Methods

We begin with a brief overview of LDMs and relevant notation before describing how they are used for OOD detection and to estimate spatial anomaly maps.

2.1 Latent Diffusion Models

LDMs are trained in two stages. A first stage model, here a VQ-GAN, is trained to compress the input image into a latent representation. A DDPM [14] is trained to learn to sample from the distribution of these latent representations through iterative denoising.

VQ-GAN: The VQ-GAN operates on a 3D input of size $\mathbf{x} \in \mathbb{R}^{H \times W \times D}$ and consists of an encoder E that compresses to a latent space $\mathbf{z} \in \mathbb{R}^{h \times w \times d \times n}$, where n is the dimension of the latent embedding vector. This representation is quantised by looking up the nearest value of each representation in a codebook containing K elements and replacing the embedding vector of length d with the codebook index, k, producing $\mathbf{z_q} \in \mathbb{R}^{h \times w \times d}$. A decoder G operates on this quantised representation to produce a reconstruction, $\hat{\mathbf{x}} \in \mathbb{R}^{H \times W \times D}$.

In a VQ-VAE [20], E, G and the codebook are jointly learnt with a L_2 loss on the reconstructions and a codebook loss. The VG-GAN [9] aims to produce higher quality reconstructions by employing a discriminator D and training adversarially, and including a perceptual loss component [32] in addition to the L_2 reconstruction loss. Following [28], we also add a spectral loss component to the reconstruction losses [7].

The encoder and decoder are convolutional networks of l levels. There is a simple relationship between the spatial dimension of the latent space, the input, and number of levels: $h, w, d = \frac{H}{2^l}, \frac{W}{2^l}, \frac{D}{2^l}$, so the latent space is 2^{3l} times smaller spatially than the input image, with a 4×2^{3l} reduction in memory size when accounting for the conversion from a float to integer representation. In practice, most works use $l = 3$ ($512\times$ spatial compression) or $l = 4$ ($4096\times$ spatial compression); it is challenging to train a VQ-GAN at higher compression rates.

DDPM: A DDPM is then trained on the latent embedding $\mathbf{z}$ (the de-quantised latent). During training, noise is added to $\mathbf{z}$ according to a timestep t and a fixed Gaussian noise schedule defined by β_t to produce noised samples $\mathbf{z}_t$, such that

$$q(\mathbf{z}_t|\mathbf{z}_0) = \mathcal{N}\left(\mathbf{z}_t|\sqrt{\bar{\alpha}_t}\mathbf{z}_0, (1-\bar{\alpha})\mathbf{I}\right) \tag{1}$$

where we use $\mathbf{z}_0$ to refer to the noise-free latent $\mathbf{z}$, we have $0 \leq t \leq T$, and $\alpha_t := 1 - \beta_t$ and $\bar{\alpha}_t := \prod_{s=1}^{t} \alpha_s$. We design β_t to increase with t such that the latent $\mathbf{z}_T$ is close to an isotropic Gaussian. We seek to train a network that can perform the reverse or denoising process, which can also be written as a Gaussian transition:

$$p_\theta(\mathbf{z}_{t-1}|\mathbf{z}_t) = \mathcal{N}\left(\mathbf{z}_{t-1}|\boldsymbol{\mu}_\theta(\mathbf{z}_t, t), \boldsymbol{\Sigma}_\theta(z_t, t)\right) \tag{2}$$

In practice, following [14], we can train a network $\boldsymbol{\epsilon}_\theta(\mathbf{z}_t, t)$ to directly predict the noise used in the forward noising process, $\boldsymbol{\epsilon}$. We can train with a simplified loss $L_{\text{simple}}(\theta) = \mathbb{E}_{t,\mathbf{z}_0,\epsilon}\left[\|\boldsymbol{\epsilon} - \boldsymbol{\epsilon}_\theta\left(\mathbf{z}_t\right)\|^2\right]$, and denoise according to

$$\mathbf{z}_{t-1} = \frac{1}{\sqrt{\alpha_t}}\left(\mathbf{z}_t - \frac{\beta_t}{\sqrt{1-\bar{\alpha}_t}}\boldsymbol{\epsilon}_\theta\left(\mathbf{z}_t, t\right)\right) + \sigma_t\mathbf{n} \tag{3}$$

where $\mathbf{n} \sim \mathcal{N}(\mathbf{0}, \mathbf{I})$.

While in most applications an isotropic Gaussian is drawn and iteratively denoised to draw samples from the model, in this work, we take a latent input $\mathbf{z}_0$ and noise to $\mathbf{z}_t$ for a range of values of $t < T$ and obtain their reconstructions, $\hat{\mathbf{z}}_{0,t} = p_\theta(\mathbf{z}_0|\mathbf{z}_t)$.

2.2 OOD Detection with LDMs

In [10], an input image $\mathbf{x}$ that has been noised to a range of t-values spanning the range $0 < t < T$ is then denoised to obtain $\hat{\mathbf{x}}_{0,t}$, and we measure the similarity for each reconstruction, $\mathbf{S}(\hat{\mathbf{x}}_{0,t}, \mathbf{x})$. These multiple similarity measures are then combined to produce a single score per input, with a high similarity score suggesting the input is more in-distribution. Typically, reconstruction methods work by reconstruction through some information bottleneck - for an autoencoder, this might be the dimension of the latent space; for a denoising model, this is the amount of noise applied - with the principal that ID images will be successfully reconstructed through the bottleneck, yielding high similarity with the input, and OOD images will not. Prior works have shown the performance becomes dependent on the choice of the bottleneck - too small and even ID inputs are poorly reconstructed, too large and OOD inputs are well reconstructed [6,18,22,33]. Reconstructing from multiple t-values addresses this problem by considering reconstructions from multiple bottlenecks per image, outperforming prior reconstruction-based methods [10].

In order to scale to 3D data, we reconstruct an input $\mathbf{x}$ in the latent space of the VQ-GAN, $\mathbf{z} = E(\mathbf{x})$. Reconstructions are performed using the PLMS sampler [17], which allows for high-quality reconstructions with significantly fewer reconstruction steps. The similarity is measured in the original image space by decoding the reconstructed latents, $\mathbf{S}(G(\hat{\mathbf{z}}_{0,t}), \mathbf{x})$. As recommended by [10], we measure both the mean-squared error (MSE) and the perceptual similarity [32] for each reconstruction, yielding a total of $2N$ similarity measures for the N reconstructions performed. As the perceptual loss operates on 2D images, we measure it on all slices in the coronal, axial, and sagittal planes and average these values to produce a single value per 3D volume. Each similarity metric is converted into a z-score using mean and standard deviation parameters calculated on the validation set, and are then averaged to produce a single score.

2.3 Spatial Anomaly Maps

To highlight spatial anomalies, we aggregate a set of reconstruction error maps. We select reconstructions from t-values $= [100, 200, 300, 400]$, calculate the pixel-wise mean absolute error (MAE), z-score these MAE maps using the pixel-wise mean and standard deviation from the validation set, and then average to produce a single spatial map per input image.

3 Experiments

3.1 Data

We use three datasets to test the ability of our method to flag OOD values in both the near- and far-OOD cases. The **CROMIS** dataset [30,31] consists of 683 head CT scans and was used as the train and validation set for all models, with a 614/69 split. The **KCH** dataset consists of 47 head CTs acquired independently from CROMIS, and was used as the in-distribution test set. To produce near-OOD data, a number of corruptions were applied to this dataset, designed to represent a number of acquisition/ data preprocessing errors. These were: addition of Gaussian noise to the images at three levels ($\sigma = 0.01, 0.1, 0.2$), setting the background to values different to the 0 used during training (0.3, 0.6, 1), inverting the image through either of the three imaging planes, removing a chunk of adjacent slices from either the top or centre of the volume, skull-stripping (the models were trained on unstripped images), and setting all pixel values to either 1% or 10% of their true values (imitating an error in intensity scaling during preprocessing). Applying each corruption to each ID image yielded a total of 705 near-OOD images. The **Decathlon** dataset [1] comprises a range of 3D imaging volumes that are not head CTs and was used to represent far-OOD data. We selected 22 images from each of the ten classes. All CT head images were affinely registered to MNI space, resampled to 1 mm isotropic, and cropped to a $176 \times 208 \times 176$ grid. For the images in the Decathlon dataset, all were resampled to be 1mm isotropic and either cropped or zero-padded depending on size to produce a $176 \times 216 \times 176$ grid. All CT images had their intensities clamped between [−15,100] and then rescaled to lie in the range $[0, 1]$. All non-CT images were rescaled based on their minimum and maximum values to lie in the $[0, 1]$ range.

3.2 Implementation Details

All models were implemented in PyTorch v1.13.1 using the MONAI framework v1.1.0 [2]. Code is available at https://github.com/marksgraham/ddpm-ood. LTM model code can be found at https://github.com/marksgraham/transformer-ood.

LDMs: VQ-GANS were trained with levels $l = 2$, 3, or 4 levels with 1 convolutional layer and 3 residual blocks per level, each with 128 channels. Training with $l = 3/4$ represents standard practice, training with $l = 2$ ($64\times$ spatial compression) was done to simulate a situation with higher-resolution input data. All VQ-GANs had an embedding dim of 64, and the 2, 3, 4 level models have a codebook size of 64, 256, 1024, respectively. Models were trained with a perceptual loss weight of 0.001, an adversarial weight loss of 0.01, and all other losses unweighted. Models were trained with a batch size of 64 for 500 epochs on an A100, using the Adam optimizer [16] with a learning rate of 3×10^{-4} and early stopping if the validation loss did not decrease over 15 epochs. The LDM used a time-conditioned UNet architecture as in [25], with three levels with (128, 256, 256) channels, 1 residual block per level, and attention in the deepest level only. The noise schedule had $T = 1000$ steps with a scaled linear noise schedule with $\beta_0 = 0.0015$ and $\beta_T = 0.0195$. Models were trained with a batch size of 112 on an A100 with the Adam optimizer, learning rate 2.5×10^{-5} for 12,000 epochs, with early stopping. During reconstruction, the PLMS scheduler was used with 100 timesteps. Reconstructions were performed from 50 t values spaced evenly over the interval $[0, 1000]$.

LTM: The Latent Transformer Models were trained on the same VQ-GAN bases using the procedure described in [11], using a 22-layer Transformer with dimension 256 in the attention layers and 8 attention heads. The authors in [11] used the Performer architecture [4], which uses a linear approximation to the attention matrix to reduce memory costs and enable training on larger sequence lengths. Instead, we use the recently introduced memory efficient attention mechanism [24] to calculate exact attention with reduced memory costs. This enables us to train a full Transformer on a 3-level VQ-GAN embedding, with a sequence length of $22 \times 27 \times 22 = 13,068$. Neither the Performer nor the memory-efficient Transformer was able to train on the 2-level embedding, with a sequence length of $44 \times 52 \times 44 = 100,672$. Models were trained on an A100 with a batch size of 128 using Adam with a learning rate of 10^{-4}.

4 Results and Discussion

Results and associated statistical tests are shown in Table 1 as AUC scores, with tests for differences in AUC performed using Delong's method [5]. At 4-levels, the LDM and LTM both perform well, albeit with the proposed LDM performing better on certain OOD datasets. LTM performance degrades when trained on a 3-level model, but LDM performance remains high. The 3-level LTM result

Table 1. AUC scores for identifying OOD data, with the CT-2 dataset used as the in-distribution test set. Results shown split according to the number of levels in the VQ-GAN. Tests for difference in AUC compare each LTM and LDM models with the same VQ-GAN base, **bold values** are differences significant with $p < 0.001$ and underlined values significant with $p < 0.05$. Results are shown as N/A for the 2-level LTM as it was not possible to train a Transformer on such a long sequence.

	Dataset	Model					
		2-level		3-level		4-level	
		LTM	LDM	LTM	LDM	LTM	LDM
Far-OOD	Head MR	N/A	72	0	**100**	100	100
	Colon CT	N/A	100	100	100	100	100
	Hepatic CT	N/A	100	100	100	99.9	100
	Hippocampal MR	N/A	3.51	0	**100**	100	100
	Liver CT	N/A	100	100	100	99.8	100
	Lung CT	N/A	100	89	100	100	100
	Pancreas CT	N/A	100	100	100	99.3	100
	Prostate MR	N/A	99.9	0	**100**	100	100
	Spleen CT	N/A	100	100	100	99.6	100
	Cardiac MR	N/A	100	90	100	100	100
Near-OOD	Noise $\sigma = 0.01$	N/A	59.7	48.1	59.3	50.7	54.5
	Noise $\sigma = 0.1$	N/A	100	57.5	**100**	44.7	**100**
	Noise $\sigma = 0.2$	N/A	100	88.3	100	45.6	**100**
	BG value = 0.3	N/A	100	100	100	100	100
	BG value = 0.6	N/A	100	100	100	100	100
	BG value = 1.0	N/A	100	100	100	100	100
	Flip L-R	N/A	53.5	49.4	61.2	51.1	58.6
	Flip A-P	N/A	100	65.6	**100**	90.7	100
	Flip I-S	N/A	100	69.7	**100**	90.5	100
	Chunk top	N/A	46.1	28.6	**94.6**	97.6	99.8
	Chunk middle	N/A	94.4	22	**100**	96.2	100
	Skull stripped	N/A	98.1	0	**100**	100	100
	Scaling 1%	N/A	0.317	0	**100**	100	100
	Scaling 10%	N/A	100	0	**100**	100	100

is in agreement with the findings in [11]. This is likely caused by the previously discussed tendency for likelihood-based models, such as Transformers, to be sensitive to the quality of the underlying representation. For instance, [12] showed that likelihood-based models can fail unless forced to focus on high-level image features. We posit that at the high compression rates of a 4-level VQ-GAN the representation encodes higher-level features, but at 3-levels the representation can encode lower-level features, making it harder for likelihood-based models to perform well. By contrast, the LDM-based method only requires that the

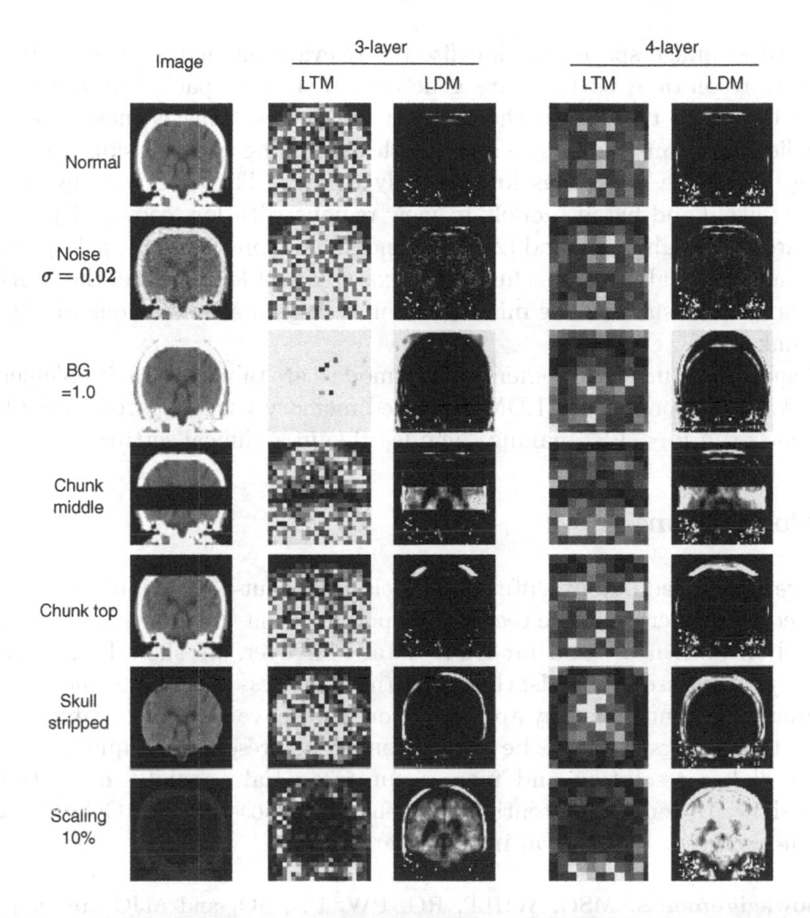

Fig. 1. Example anomaly maps for models based on 3- and 4-level VQ-GANs. Maps for each model are shown on the same colour scale, but the scales vary between each model to obtain the best display for each model. Brighter regions are more anomalous.

VG-GAN produces reasonable reconstructions. While memory constraints prevented training a 2-level LTM, the more modest requirements on the UNet-based LDM meant it was possible to train. This result has implications for the application of very high-resolution medical data: for instance, a whole-body CT with an image dimension 512^3 would have a latent dimension 32^3 even with 4-level compression, too large to train an LTM on but comfortably within the reach of a LDM. The 2-level LDM had reduced performance on two classes that have many pixels with an intensity close to 0 (Hippocampal MR, and Scaling 1%). Recent research shows that at higher resolutions, the effective SNR increases if the noise schedule is kept constant [15]. It seems this effect made it possible for the 2-level LDM to reconstruct these two OOD classes with low error for many values of t. In future work we will look into scaling the noise schedule with LDM input size.

Anomaly maps are shown in Fig. 1 for near-OOD cases with a spatially localised anomaly. The LDM-based maps are high-resolution, as they are generated in image space, and localise the relevant anomalies. The LTM maps are lower resolution, as they are generated in latent space, but more significantly often fail to localise the relevant anomalies. This is most obvious in anomalies that cause missing signal, such as missing chunks, skulls, or image scaling, which are flagged as low-anomaly regions. This is caused by the tendency of likelihood-based models to view regions with low complexity, such as blank areas, as high-likelihood [26]. The anomaly is sometimes picked up but not well localised, notable in the 'chunk top' example at 4-levels. Here, the transition between brain tissue and the missing chunk is flagged as anomalous rather than the chunk itself.

Memory and time requirements for all models are tabulated in Supplementary Table A. These confirm the LDM's reduced memory use compared to the LTM. All models run in $< 30s$, making them feasible in a clinical setting.

5 Conclusion

We have introduced Latent Diffusion Models for 3D out-of-distribution detection. Our method outperforms the recently proposed Latent Transformer Model when assessed on both near- and far-OOD data. Moreover, we show LDMs address three key weaknesses of LTMs: their performance is less sensitive to the quality of the latent representation they are trained on, they have more favourable memory scaling that allows them to be trained on higher resolution inputs, and they provide higher resolution and more accurate spatial anomaly maps. Overall, LDMs show tremendous potential as a general-purpose tool for OOD detection on high-resolution 3D medical imaging data.

Acknowledgements. MSG, WHLP, RG, PW, PN, SO, and MJC are supported by the Wellcome Trust (WT213038/Z/18/Z). MJC and SO are also supported by the Wellcome/EPSRC Centre for Medical Engineering (WT203148/Z/16/Z), and the InnovateUK-funded London AI centre for Value-based Healthcare. PTD is supported by the EPSRC (EP/R513064/1). YM is supported by an MRC Clinical Academic Research Partnership grant (MR/T005351/1). PN is also supported by the UCLH NIHR Biomedical Research Centre. Datasets CROMIS and KCH were used with ethics 20/ES/0005.

References

1. Antonelli, M., et al.: The medical segmentation decathlon. Nature Commu. **13**(1), 4128 (2022)
2. Cardoso, M.J., et al.: Monai: An open-source framework for deep learning in healthcare. arXiv preprint arXiv:2211.02701 (2022)
3. Choi, H., Jang, E., Alemi, A.A.: Waic, but why? generative ensembles for robust anomaly detection. arXiv preprint arXiv:1810.01392 (2018)
4. Choromanski, K., et al.: Rethinking attention with performers. arXiv preprint arXiv:2009.14794 (2020)

5. DeLong, E.R., DeLong, D.M., Clarke-Pearson, D.L.: Comparing the areas under two or more correlated receiver operating characteristic curves: a nonparametric approach. Biometrics, pp. 837–845 (1988)
6. Denouden, T., Salay, R., Czarnecki, K., Abdelzad, V., Phan, B., Vernekar, S.: Improving reconstruction autoencoder out-of-distribution detection with mahalanobis distance. arXiv preprint arXiv:1812.02765 (2018)
7. Dhariwal, P., Jun, H., Payne, C., Kim, J.W., Radford, A., Sutskever, I.: Jukebox: A generative model for music. arXiv preprint arXiv:2005.00341 (2020)
8. Dieleman, S.: Musings on typicality (2020). https://benanne.github.io/2020/09/01/typicality.html
9. Esser, P., Rombach, R., Ommer, B.: Taming transformers for high-resolution image synthesis. In: Proceedings of the IEEE/CVF Conference on Computer Vision and Pattern Recognition, pp. 12873–12883 (2021)
10. Graham, M.S., Pinaya, W.H., Tudosiu, P.D., Nachev, P., Ourselin, S., Cardoso, J.: Denoising diffusion models for out-of-distribution detection. In: Proceedings of the IEEE/CVF Conference on Computer Vision and Pattern Recognition, pp. 2947–2956 (2023)
11. Graham, M.S., et al.: Transformer-based out-of-distribution detection for clinically safe segmentation. In: International Conference on Medical Imaging with Deep Learning, pp. 457–476. PMLR (2022)
12. Havtorn, J.D., Frellsen, J., Hauberg, S., Maaløe, L.: Hierarchical vaes know what they don't know. In: International Conference on Machine Learning, pp. 4117–4128. PMLR (2021)
13. Hendrycks, D., Mazeika, M., Dietterich, T.: Deep anomaly detection with outlier exposure. In: International Conference on Learning Representations (2018)
14. Ho, J., Jain, A., Abbeel, P.: Denoising diffusion probabilistic models. Adv. Neural. Inf. Process. Syst. **33**, 6840–6851 (2020)
15. Hoogeboom, E., Heek, J., Salimans, T.: Simple diffusion: End-to-end diffusion for high resolution images. arXiv preprint arXiv:2301.11093 (2023)
16. Kingma, D.P., Ba, J.: Adam: A method for stochastic optimization. arXiv preprint arXiv:1412.6980 (2014)
17. Liu, L., Ren, Y., Lin, Z., Zhao, Z.: Pseudo numerical methods for diffusion models on manifolds. In: International Conference on Learning Representations (2021)
18. Lyudchik, O.: Outlier detection using autoencoders. Tech. rep. (2016)
19. Nalisnick, E., Matsukawa, A., Teh, Y.W., Gorur, D., Lakshminarayanan, B.: Do deep generative models know what they don't know? In: International Conference on Learning Representations (2018)
20. Oord, A.v.d., Vinyals, O., Kavukcuoglu, K.: Neural discrete representation learning. arXiv preprint arXiv:1711.00937 (2017)
21. Patel, A., et al.: Cross attention transformers for multi-modal unsupervised whole-body pet anomaly detection. In: MICCAI Workshop on Deep Generative Models, pp. 14–23. Springer (2022). https://doi.org/10.1007/978-3-031-18576-2_2
22. Pimentel, M.A., Clifton, D.A., Clifton, L., Tarassenko, L.: A review of novelty detection. Signal Process. **99**, 215–249 (2014)
23. Pinaya, W.H., et al.: Unsupervised brain imaging 3D anomaly detection and segmentation with transformers. Med. Image Anal. **79**, 102475 (2022)
24. Rabe, M.N., Staats, C.: Self-attention does not need $o(n^2)$ memory. arXiv preprint arXiv:2112.05682 (2021)
25. Rombach, R., Blattmann, A., Lorenz, D., Esser, P., Ommer, B.: High-resolution image synthesis with latent diffusion models. In: Proceedings of the IEEE/CVF Conference on Computer Vision and Pattern Recognition, pp. 10684–10695 (2022)

26. Serrà, J., Álvarez, D., Gómez, V., Slizovskaia, O., Núñez, J.F., Luque, J.: Input complexity and out-of-distribution detection with likelihood-based generative models. In: International Conference on Learning Representations (2019)
27. Tudosiu, P.D., et al.: Morphology-preserving autoregressive 3D generative modelling of the brain. In: International Workshop on Simulation and Synthesis in Medical Imaging, pp. 66–78. Springer (2022). https://doi.org/10.1007/978-3-031-16980-9_7
28. Tudosiu, P.D., et al.: Neuromorphologicaly-preserving volumetric data encoding using VQ-VAE. arXiv preprint arXiv:2002.05692 (2020)
29. Vaswani, A., et al.: Attention is all you need. Adv. Neural Inform. Process. Syst. **30**, 5998–6008 (2017)
30. Werring, D.: Clinical trial: Clinical relevance of microbleeds in stroke (cromis-2). Tech. Rep. NCT02513316, University College London (Nov 2017)
31. Wilson, D., et al.: Cerebral microbleeds and intracranial haemorrhage risk in patients anticoagulated for atrial fibrillation after acute ischaemic stroke or transient ischaemic attack (cromis-2): a multicentre observational cohort study. Lancet Neurol. **17**(6), 539–547 (2018)
32. Zhang, R., Isola, P., Efros, A.A., Shechtman, E., Wang, O.: The unreasonable effectiveness of deep features as a perceptual metric. In: Proceedings of the IEEE Conference on Computer Vision and Pattern Recognition, pp. 586–595 (2018)
33. Zong, B., et al.: Deep autoencoding gaussian mixture model for unsupervised anomaly detection. In: International Conference on Learning Representations (2018)

Improved Multi-shot Diffusion-Weighted MRI with Zero-Shot Self-supervised Learning Reconstruction

Jaejin Cho[1,2(✉)], Yohan Jun[1,2], Xiaoqing Wang[1,2], Caique Kobayashi[1,2,3,4], and Berkin Bilgic[1,2,5]

[1] Athinoula A. Martinos Center for Biomedical Imaging, Charlestown, MA 02129, USA

[2] Harvard Medical School, Boston, MA 02115, USA

jcho18@mgh.harvard.edu

[3] University of São Paulo, São Paulo, State of São Paulo 05508-060, Brazil

[4] Technische Universität München, 80333 München, Bavaria, Germany

[5] Harvard-MIT Health Sciences and Technology, Cambridge, MA 02139, USA

Abstract. Diffusion MRI is commonly performed using echo-planar imaging (EPI) due to its rapid acquisition time. However, the resolution of diffusion-weighted images is often limited by magnetic field inhomogeneity-related artifacts and blurring induced by T_2- and T_2^*-relaxation effects. To address these limitations, multi-shot EPI (msEPI) combined with parallel imaging techniques is frequently employed. Nevertheless, reconstructing msEPI can be challenging due to phase variation between multiple shots. In this study, we introduce a novel msEPI reconstruction approach called zero-MIRID (zero-shot self-supervised learning of Multi-shot Image Reconstruction for Improved Diffusion MRI). This method jointly reconstructs msEPI data by incorporating deep learning-based image regularization techniques. The network incorporates CNN denoisers in both k- and image-spaces, while leveraging virtual coils to enhance image reconstruction conditioning. By employing a self-supervised learning technique and dividing sampled data into three groups, the proposed approach achieves superior results compared to the state-of-the-art parallel imaging method, as demonstrated in an in-vivo experiment.

Keywords: Self-supervised learning · Multi-shot echo planar imaging · diffusion MRI

1 Introduction

Magnetic resonance imaging (MRI) is widely used for diagnosis and treatment monitoring as it provides structural and physiological information related to dis-

---

Supplementary Information The online version contains supplementary material available at https://doi.org/10.1007/978-3-031-43907-0_44.

H. Greenspan et al. (Eds.): MICCAI 2023, LNCS 14220, pp. 457–466, 2023.
https://doi.org/10.1007/978-3-031-43907-0_44

ease progression. Diffusion MRI (dMRI) measures molecular diffusion in biological tissues and provides microscopic details of tissue architecture, as molecules interact with many different obstacles while diffusing throughout tissues [16]. However, dMRI requires repeated acquisitions with different diffusion directions. Echo-planar imaging (EPI), which enables fast encoding per imaging slice, is commonly used for dMRI due to its fast acquisition time. However, single-shot (ss-) EPI is susceptible to severe susceptibility-induced geometric distortion and T_2- and T_2^*-induced voxel blurring. These artifacts worsen at higher in-plane resolutions as the time required to acquire each line of k-space increases approximately linearly.

Multi-shot (ms-) acquisition is an effective approach to mitigate EPI-related artifacts, which segments k-space into multiple portions covered across multiple repetition times (TRs) to reduce the effective echo spacing. However, potential shot-to-shot phase variations across multiple EPI shots can introduce additional artifacts. Recent algorithms, such as low-rank prior methods like low-rank modeling of local k-space neighborhoods (LORAKS) [7,8,14,15,17,18], and multi-shot sensitivity encoded diffusion data recovery using structured low-rank matrix completion (MUSSELS) [19], have successfully addressed this challenge by jointly reconstructing msEPI images through a low-rank constraint applied across the EPI shots.

In recent years, deep learning has emerged as a promising approach for image reconstruction, offering potential solutions to the challenges of existing techniques, including long reconstruction times, residual artifacts at high acceleration factors, and over-smoothing [6,9,10]. One notable development is model-based deep learning (MoDL), which leverages an unrolled convolutional neural network (CNN) and a parallel imaging (PI) forward model to denoise and unalias undersampled data [1]. MoDL has also been applied to multi-shot diffusion-weighted echo-planar imaging, known as MoDL-MUSSELS, effectively replacing MUSSELS and significantly reducing reconstruction times while achieving comparable results to state-of-the-art methods [2]. MoDL-MUSSELS includes CNN denoisers in both image- and k-spaces, as recent work has demonstrated that utilizing both domains has yielded improvement in performance based on metrics such as PSNR and SSIM [6]. However, it is worth noting that existing deep learning networks for dMRI have typically been trained in a supervised manner, which requires a significant amount of ground truth images that are not easily acquired in EPI acquisitions.

In contrast, self-supervised learning [3,24,25] does not rely on external training data and can be used in denoising, reconstruction, quantitative mapping, and other applications. Recent advancements in zero-shot self-supervised learning (ZS-SSL) have demonstrated successful scan-specific network training without any external database [24]. This approach has shown comparable or superior results to supervised networks. However, in dMRI, where the same volume is repeatedly acquired while changing diffusion directions, ZS-SSL typically requires training separate networks for different directions, which can be impractical.

The virtual coil (VC) approach is a highly effective technique for enhancing the performance of parallel MRI [5], particularly in the case of EPI that utilizes partial Fourier acquisition. VC generates virtual coils by incorporating conjugate symmetric k-space signals from actual coils. This integration provides supplementary information for missing data points in k-space, further being useful when combined with partial Fourier acquisition. Conceptually, the utilization of VC consistently ensures an image quality equivalent to or exceeds that of the image reconstructed without VC.

In this study, we propose a novel msEPI reconstruction method called zero-MIRID (zero-shot self-supervised learning of Multi-shot Image Reconstruction for Improved Diffusion MRI). Our method jointly reconstructs msEPI data by incorporating zero-shot self-supervised learning-based image reconstruction. Our key contributions are as follows:

- We jointly reconstruct multiple-shot images using self-supervised learning.
- We train one network for all diffusion directions, which accelerates training speed and improves performance.
- We used network denoisers in both k- and image-space and employed the VC [5] to improve the conditioning of the reconstruction.
- In the in-vivo experiment, the proposed method demonstrates more robust images and better diffusion metrics than the state-of-art PI technique for dMRI.
- To our best knowledge, this study proposes the first self-supervised learning reconstruction for dMRI.

Overall, our zero-MIRID method offers a promising approach to enhance msEPI reconstruction in dMRI, providing improved image quality and diffusion metrics through the integration of self-supervised learning techniques.

2 Method

2.1 PI Techniques for dMRI

For msEPI data, SENSE is commonly used for image reconstruction. SENSE individually reconstructs each shot's data using the spatial variation of the coil sensitivity profile. The mth shot image in the dth diffusion direction, $x_{d,m}$, can be reconstructed as follow.

$$x_{d,m} = \underset{x_{d,m}}{\text{argmin}} \|\mathcal{F}_m \mathbf{C} x_{d,m} - b_{d,m}\|_2^2 \tag{1}$$

where $\mathcal{F}_m$ is the undersampled Fourier transform for the mth shot, C is the coil sensitivity map, and $b_{d,m}$ is the acquired k-space data of dth direction and m^{th} shot.

On the other hand, MUSSELS and LORAKS jointly reconstruct multiple-shot images using the low-rank property among msEPI data. The images in the dth diffusion direction can be reconstructed using LORAKS as follows.

$$x_d = \underset{x_d}{\text{argmin}} \sum_{m=0}^{M} \|\mathcal{F}_m \mathbf{C} x_{d,m} - b_{d,m}\|_2^2 + \lambda \mathcal{J}(\mathcal{F} x_d) \tag{2}$$

where $\mathcal{J}$ is the LORAKS regularization. In this work, we utilized S-LORAKS, which employs phase information and k-space symmetry [14,15].

2.2 Network Design

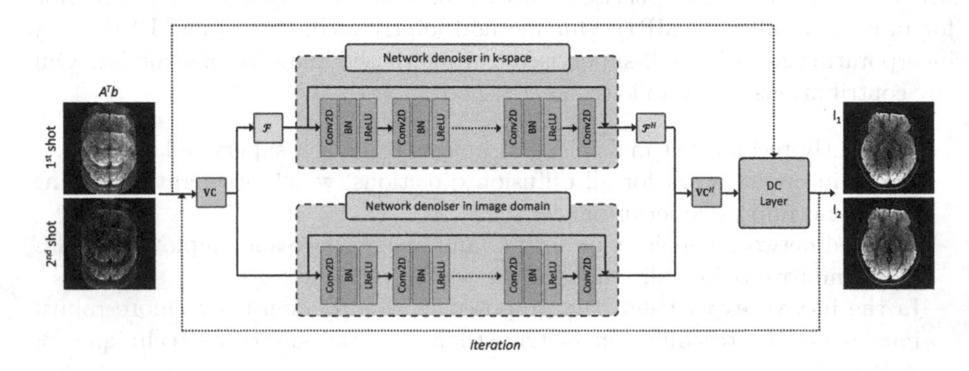

Fig. 1. The proposed image reconstruction diagram of zero-MIRID. The virtual coil (VC) layer was used to efficiently reconstruct the data accelerated by partial Fourier. The network denoisers in both the k-space and image domain were used. The DC layer enforces the consistency between the acquired data and the reconstructed images.

Figure 1 shows the proposed network diagram of zero-MIRID. The input of the network is $A_m^T b_d$, where $A_m = \mathcal{F}_m C$. The network consists of two CNNs in the k- and image-spaces. The VC was added and removed before and after the denoising CNNs, respectively. The images in the dth diffusion direction can be jointly reconstructed using zero-MIRID as follows.

$$\begin{aligned} x_d =& \underset{x_d}{\text{argmin}} \sum_{m=0}^{M} \|\mathcal{F}_m \mathbf{C} x_{d,m} - b_{d,m}\|_2^2 \\ &+ \lambda_1 \left\|\mathcal{V}_C^H N_i \mathcal{V}_C x_d\right\|_2^2 + \lambda_2 \left\|\mathcal{V}_C^H \mathcal{F}^H N_k \mathcal{F} \mathcal{V}_C x_d\right\|_2^2 \end{aligned} \tag{3}$$

where $\mathcal{V}_C$ is the VC operator, and N_i and N_k are denoising CNNs in the image- and k-space, respectively. We define $Nx = x - Dx$, where D is the CNN network, and modified the alternating minimization-based solution in [2] to get the solutions of equation (3), as follows.

$$\begin{aligned} x_{n+1} =& (A^H A + \lambda_1 I + \lambda_2 I)(A^H b + \lambda_1 \eta_n + \lambda_2 \zeta_n) \\ \zeta_{n+1} =& \mathcal{V}_C^H \mathcal{F}^H D_k \mathcal{F} \mathcal{V}_C x_{n+1} \\ \eta_{n+1} =& \mathcal{V}_C^H D_i \mathcal{V}_C x_{n+1} \end{aligned} \tag{4}$$

where n is the optimization step (iteration) number, η and ζ is the network denoising terms in k- and image-space, and $A = \mathcal{F}C$.

2.3 Zero-Shot Self-supervised Learning

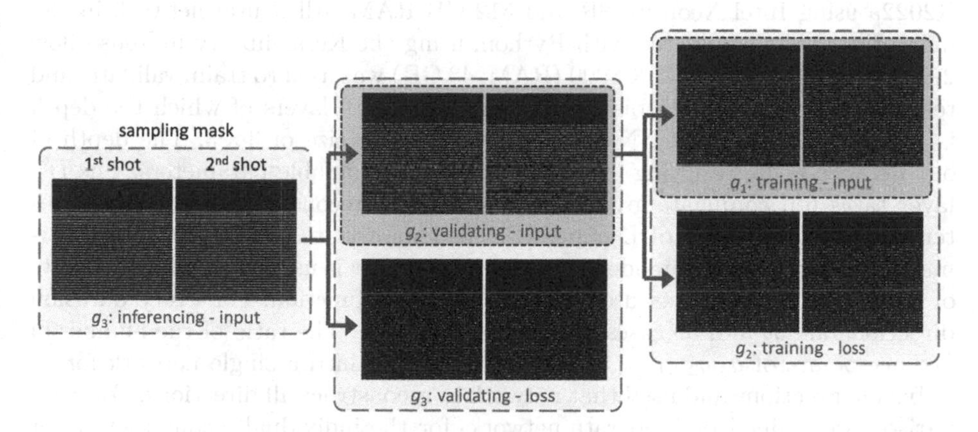

Fig. 2. The masks used for training, validation, and inference phases. The sampling mask was split into three different masks.

As proposed in the recent ZS-SSL study [24], we split the sampling mask into three different groups, as shown in Fig. 2, where g_3 is the entire sampling mask and $g_3 \supset g_2 \supset g_1$. In the training phase, g_1 was used for network input, while g_2 was used to calculate training losses. In the validating phase, g_2 was used for network input, while g_3 was used to calculate validating losses. In the inferencing phase, g_3 was used for network inputs. The loss in the dth direction in the training phase can be described as follows.

$$\mathcal{L}(g_2 \cdot b_d,\, g_2 \cdot Af(g_1 \cdot b_d; \theta)) \tag{5}$$

where $\mathcal{L}$ is the loss function, f is the zero-MIRID reconstruction, and θ is the trainable network parameters. Similarly, the loss in the dth direction in the validating phase can be described as follows.

$$\mathcal{L}(g_3 \cdot b_d,\, g_3 \cdot Af(g_2 \cdot b_d; \theta)) \tag{6}$$

In this study, we used the normalized root mean square error (NRMSE) and normalized mean absolute error (NMAE) as the loss functions.

2.4 Experiment Details

In-vivo experiments were conducted on a 3T Siemens Prisma system with a 32-channel head coil. For dMRI, we acquired the diffusion-weighted data in 32 different directions using 2-shot EPI, with each shot accelerated by 5-fold (R=5) and employing 75% Partial Fourier, resulting in 15% coverage of the k-space in each shot relative to a fully-sampled readout. Imaging parameters are; field of view (FOV)=224 $\times$ 224 $\times$ 128 mm^3, voxel size =1 $\times$ 1 $\times$ 4 mm^3, TR=3.5 s, and effective echo time (TE) =59 ms.

SENSE and S-LORAKS reconstructions were performed with MATLAB R2022a using Intel Xeon 6248R and 512 GB RAM. All neural network implementations were conducted with Python, using the Keras library in Tensorflow 2.4.1. NVIDIA Quadro RTX 8000 (RAM: 48 GB) was used to train, validate, and test the network. The denoising CNNs consist of 16 layers of which the depth is 46. For the 16 layer-CNN, we employed a filter size of 3×3. The depth of our network is 46, resulting in a total of 583,114 trainable parameters. The DC layer takes ten conjugate gradient steps, and the reconstruction block iterates ten times, where the MoDL paper [1] has demonstrated the saturated performance. For training the model, the Adam optimizer is used with a learning rate of 1e-3. Leaky ReLU was used as the activation function. For every diffusion direction, one g_2 and 50 cases of g_1 were generated. The ratio of the number of k-space points of g_3:g_2:g_1 = 1.00:0.80:0.48. We trained a single network for 32 diffusion directions and used that network to reconstruct all directions. For comparison, we trained two separate networks for the individual reconstruction for each shot (zero-SIRID, single-shot image reconstruction). We used the FSL toolbox for diffusion analysis [13,22,23]. To estimate multiple fiber orientations, we used the Bayesian Estimation of Diffusion Parameters Obtained using Sampling Techniques (BEDPOSTX) [4,11,12].

Example data and code can be found in the following link: https://github.com/jaejin-cho/miccai2023

3 Results

Figure 3 the reconstructed diffusion-weighted images at 5-fold acceleration per shot in the selected diffusion directions. The reference images were obtained from 5-shot EPI data that covers complementary k-space lines to each other with the S-LORAKS constraint. While SENSE shows severe noise amplification and remaining folding artifacts, zero-SIRID was able to partially mitigate the noise amplification. S-LORAKS jointly reconstructed two shots, considerably reduced noise, and improved the signal-to-noise ratio (SNR). Nonetheless, in the selected diffusion directions, folding artifacts were amplified, and the center of the image shows a dropped signal (pointed by yellow arrows). In contrast, zero-MIRID demonstrated robust image reconstruction even with a high reduction factor per shot. The NRMSE and NAE across the diffusion direction are provided in the supplementary material, demonstrating notable reductions in NRMSE and NMAE when the proposed method is compared to S-LORAKS.

Figure 4 presents the average diffusion-weighted image (DWI), fractional anisotropy (FA) map, and 2nd crossing fiber image calculated from the reconstructed images. S-LORAKS and zero-MIRID produced high-fidelity average

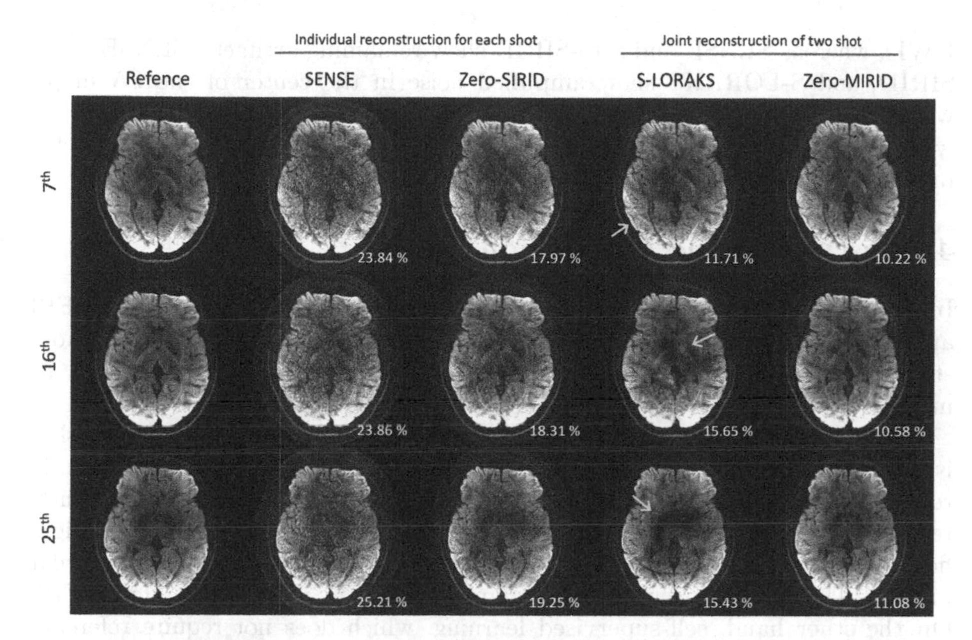

Fig. 3. The reconstructed diffusion-weighted images at R=5 per shot. Selected diffusion directions were shown. Reference images were obtained from 5-shot EPI data with S-LORAKS reconstruction. SENSE and zero-SIRID individually reconstruct each shot image, whereas S-LORAKS and zero-MIRID jointly reconstruct two shot images. NRMSE was shown at the bottom of each image.

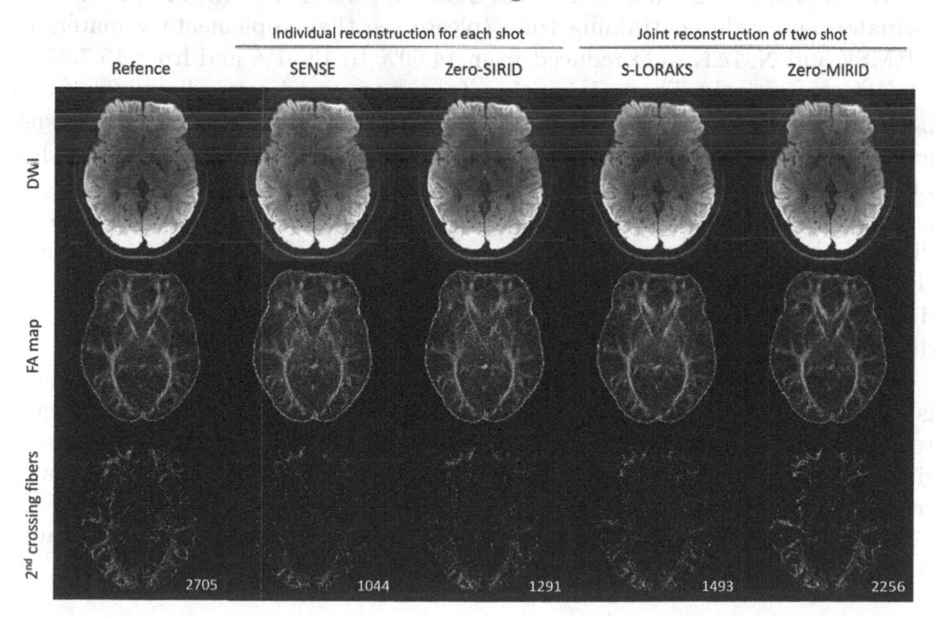

Fig. 4. Average DWI, FA map, and 2nd crossing fiber image from the reconstructed images in Fig. 3. The number of 2nd crossing fibers was shown at the bottom of each column.

DWIs, whereas SENSE and zero-SIRID show remaining artifacts. SENSE, zero-SIRID, and S-LORAKS show amplified noise in the center of the FA maps, whereas zero-MIRID effectively mitigated the noise. Furthermore, zero-MIRID well preserved the number of 2nd crossing fibers, often considered a crucial factor in evaluating successful dMRI acquisition [4,12].

4 Discussion and Conclusion

In this study, we proposed an improved image reconstruction method for msEPI and dMRI in a self-supervised deep learning manner. In-vivo experiment demonstrates the proposed method outperformed S-LORAKS, the state-of-art PI method for dMRI.

Acquiring reference images of msEPI can be challenging because each shot is typically highly accelerated and shot-to-shot phase variation prevents jointly reconstructing multiple shots efficiently. Advanced PI techniques that jointly reconstruct many EPI shots can improve the PI condition and provide high-fidelity images, but using a PI method may induce bias to that particular method. Therefore, supervised learning might not be an ideal solution for msEPI. On the other hand, self-supervised learning, which does not require reference images, could be a more suitable approach for msEPI. Due to the difficulty in obtaining reliable ground truth data, conventional quantitative metrics such as SSIM and NRMSE may be less reliable for evaluation. In dMRI, FA maps and 2nd crossing fibers could be used for obtaining more suitable metrics.

We trained a single network for all diffusion directions, which improved performance and reduced training time (please see the supplementary material). NRMSE and NMAE were reduced from 14.69% to 13.61% and from 15.73% to 14.41%, respectively. The training time for the proposed network was 22:30 min per diffusion direction/slice (on GPU). This is expected to be reduced by transfer learning. Inference took approximately 1 s per direction/slice, and 2-shot LORAKS took approximately 20 s per direction/slice (on CPU). Since the images are highly similar across diffusion directions, training on the entire diffusion direction has a similar effect to increasing the size of the training database, thereby enhancing network training. Moreover, using a single network for all directions reduces training time compared to training separate networks for each direction, from 40:01 min to 22:30 min per diffusion direction and slice.

As a future work, the simultaneous multi-slice (SMS) technique [21], which is often used for further acceleration, can be easily incorporated into the current network (please see the preliminary images in the supplementary material). At $R_{sms}=5 \times 2$-fold acceleration, NRMSE and NMAE were significantly reduced compared with SENSE, from 22.91% to 9.07% and from 26.09% to 11.12%, respectively. g-Slider could be a good application as well [20], because RF-encoded images also have highly similar image features.

Acknowledgment. This work was supported by research grants NIH R01 EB028797, R01 EB032378, R01 HD100009, R03 EB031175, U01 EB026996, U01 DA055353, P41 EB030006, and the NVidia Corporation for computing support.

References

1. Aggarwal, H.K., Mani, M.P., Jacob, M.: MoDL: model-based deep learning architecture for inverse problems. IEEE Trans. Med. Imaging **38**(2), 394–405 (2019)
2. Aggarwal, H.K., Mani, M.P., Jacob, M.: MoDL-MUSSELS: model-based deep learning for multishot Sensitivity-Encoded diffusion MRI. IEEE Trans. Med. Imaging **39**(4), 1268–1277 (2020)
3. Akçakaya, M., Moeller, S., Weingärtner, S., Uğurbil, K.: Scan-specific robust artificial-neural-networks for k-space interpolation (RAKI) reconstruction: database-free deep learning for fast imaging. Magn. Reson. Med. **81**(1), 439–453 (2019)
4. Behrens, T.E., Berg, H.J., Jbabdi, S., Rushworth, M.F., Woolrich, M.W.: Probabilistic diffusion tractography with multiple fibre orientations: What can we gain? neuroimage **34**(1), 144–155 (2007)
5. Blaimer, M., Gutberlet, M., Kellman, P., Breuer, F.A., Köstler, H., Griswold, M.A.: Virtual coil concept for improved parallel MRI employing conjugate symmetric signals. Mag. Resonance Med.: Off. J. Int. Society Mag. Resonance Med. **61**(1), 93–102 (2009)
6. Eo, T., Jun, Y., Kim, T., Jang, J., Lee, H.J., Hwang, D.: Kiki-net: cross-domain convolutional neural networks for reconstructing undersampled magnetic resonance images. Magn. Reson. Med. **80**(5), 2188–2201 (2018)
7. Haldar, J.P.: Low-rank modeling of local k-space neighborhoods (LORAKS) for constrained MRI. IEEE Trans. Med. Imaging **33**(3), 668–681 (2014)
8. Haldar, J.P., Zhuo, J.: P-LORAKS: Low-rank modeling of local k-space neighborhoods with parallel imaging data. Magn. Reson. Med. **75**(4), 1499–1514 (2016)
9. Hammernik, K., et al.: Learning a variational network for reconstruction of accelerated mri data. Magn. Reson. Med. **79**(6), 3055–3071 (2018)
10. Han, Y., Sunwoo, L., Ye, J.C.: *k*-space deep learning for accelerated MRI. IEEE Trans. Med. Imaging **39**(2), 377–386 (2019)
11. Hernández, M., et al.: Accelerating fibre orientation estimation from diffusion weighted magnetic resonance imaging using Gpus. PLoS ONE **8**(4), e61892 (2013)
12. Jbabdi, S., Sotiropoulos, S.N., Savio, A.M., Graña, M., Behrens, T.E.: Model-based analysis of multishell diffusion MR data for tractography: how to get over fitting problems. Magn. Reson. Med. **68**(6), 1846–1855 (2012)
13. Jenkinson, M., Beckmann, C.F., Behrens, T.E., Woolrich, M.W., Smith, S.M.: Fsl. Neuroimage **62**(2), 782–790 (2012)
14. Kim, T.H., Setsompop, K., Haldar, J.P.: Loraks makes better sense: phase-constrained partial fourier sense reconstruction without phase calibration. Magn. Reson. Med. **77**(3), 1021–1035 (2017)
15. Kim, T., Haldar, J.: Loraks software version 2.0: Faster implementation and enhanced capabilities. University of Southern California, Los Angeles, CA, Tech. Rep. USC-SIPI-443 (2018)
16. Le Bihan, D., Iima, M.: Diffusion magnetic resonance imaging: what water tells us about biological tissues. PLoS Biol. **13**(7), e1002203 (2015)
17. Lobos, R.A., et al.: Robust autocalibrated structured low-rank epi ghost correction. Magn. Reson. Med. **85**(6), 3403–3419 (2021)
18. Lobos, R.A., Kim, T.H., Hoge, W.S., Haldar, J.P.: Navigator-free epi ghost correction with structured low-rank matrix models: New theory and methods. IEEE Trans. Med. Imaging **37**(11), 2390–2402 (2018)

19. Mani, M., Jacob, M., Kelley, D., Magnotta, V.: Multi-shot sensitivity-encoded diffusion data recovery using structured low-rank matrix completion (MUSSELS). Magn. Reson. Med. **78**(2), 494–507 (2017)
20. Setsompop, K., et al.: High-resolution in vivo diffusion imaging of the human brain with generalized slice dithered enhanced resolution: Simultaneous multislice (g s lider-sms). Magn. Reson. Med. **79**(1), 141–151 (2018)
21. Setsompop, K., Gagoski, B.A., Polimeni, J.R., Witzel, T., Wedeen, V.J., Wald, L.L.: Blipped-controlled aliasing in parallel imaging for simultaneous multislice echo planar imaging with reduced g-factor penalty. Magn. Reson. Med. **67**(5), 1210–1224 (2012)
22. Smith, S.M., et al.: Advances in functional and structural MR image analysis and implementation as FSL. Neuroimage **23**, S208–S219 (2004)
23. Woolrich, M.W., et al.: Bayesian analysis of neuroimaging data in FSL. Neuroimage **45**(1), S173–S186 (2009)
24. Yaman, B., Hosseini, S.A.H., Akcakaya, M.: Zero-Shot Self-Supervised learning for MRI reconstruction. In: International Conference on Learning Representations (2022)
25. Yaman, B., Hosseini, S.A.H., Moeller, S., Ellermann, J., Uğurbil, K., Akçakaya, M.: Self-supervised learning of physics-guided reconstruction neural networks without fully sampled reference data. Magn. Reson. Med. **84**(6), 3172–3191 (2020)

Infusing Physically Inspired Known Operators in Deep Models of Ultrasound Elastography

Ali K. Z. Tehrani(✉) and Hassan Rivaz

Department of Electrical and Computer Engineering, Concordia University, Montreal, Canada
a_kafaei@encs.concordia.ca, hrivaz@ece.concordia.ca

Abstract. The displacement estimation step of Ultrasound Elastography (USE) can be done by optical flow Convolutional Neural Networks (CNN). Even though displacement estimation in USE and computer vision share some challenges, USE displacement estimation has two distinct characteristics that set it apart from the computer vision counterpart: high-frequency nature of RF data, and the physical rules that govern the motion pattern. The high-frequency nature of RF data has been well addressed in recent works by modifying the architecture of the available optical flow CNNs. However, insufficient attention has been placed on the integration of physical laws of deformation into the displacement estimation. In USE, lateral displacement estimation, which is highly required for elasticity and Poisson's ratio imaging, is a more challenging task compared to the axial one since the motion in the lateral direction is limited, and the sampling frequency is much lower than the axial one. Recently, Physically Inspired ConstrainT for Unsupervised Regularized Elastography (PICTURE) has been introduced which incorporates the physical laws of deformation by introducing a regularized loss function. PICTURE tries to limit the range of the lateral displacement by the feasible range of Poisson's ratio and the estimated high-quality axial displacement. Despite the improvement, the regularization was only applied during the training phase. Furthermore, only a feasible range for Poisson's ratio was enforced. We exploit the concept of known operators to incorporate iterative refinement optimization methods into the network architecture so that the network is forced to remain within the physically plausible displacement manifold. The refinement optimization methods are embedded into the different pyramid levels of the network architecture to improve the estimate. Our results on experimental phantom and *in vivo* data show that the proposed method substantially improves the estimated displacements.

Supplementary Information The online version contains supplementary material available at https://doi.org/10.1007/978-3-031-43907-0_45.

H. Greenspan et al. (Eds.): MICCAI 2023, LNCS 14220, pp. 467–476, 2023.
https://doi.org/10.1007/978-3-031-43907-0_45

Keywords: Ultrasound Elastography · Convolutional Neural Networks · Physically inspired constraint · Known operator · Poisson's ratio

1 Introduction

Ultrasound Elastography (USE) provides information related to the stiffness of the tissue. Ultrasound (US) data before and after the tissue deformation (which can be caused by an external or internal force) are collected and compared to calculate the displacement map, indicating each individual sample's relative motion. The strain is computed by taking the derivative of the displacement fields. In free-hand palpation, the force is external and applied by the operator by the probe [10].

Convolutional Neural Networks (CNN) have been successfully employed for USE displacement estimation [11,12,15]. Unsupervised and semi-supervised training methods have been proposed, which enable the networks to use real US images for training [1,14,17]. The proposed networks have achieved high-quality axial strains. In contrast to axial strain, lateral strain, which is highly required in Poisson's ratio imaging and elasticity reconstruction, has a poor quality due to the low sampling frequency, limited motion and lack of carrier signal in the lateral direction.

Recently, physically inspired constraint in unsupervised regularized elastography (PICTURE) has been proposed [5]. This method aims to improve lateral displacement by exploiting the high-quality axial displacement estimation and the relation between the lateral and axial strains defined by the physics of motion. Despite the substantial improvement, the regularization is only applied during the training phase. In addition, only a considerably large feasible range for Poisson's ratio was enforced, thereby providing further opportunities for the network to contravene the laws of physics.

Known operators, introduced by Maier *et al.* [7], have been widely utilized in deep neural networks. The core idea is that some known operations (for example inversion of a matrix) are embedded inside the networks to simplify the training and improving the generalization ability of the network. The known operator can be viewed as the prior knowledge related to the physics of the problem. Maier *et al.* investigated known operators in different applications such as computed tomography, magnetic resonance imaging, and vessel segmentation, and showed a substantial reduction in the maximum error bounds [7].

In this paper, we aim to embed two lateral displacement refinement algorithms in the CNNs to improve the lateral strains. The first algorithm limits the range of Effective Poisson's Ratio (EPR) inside the feasible range during the test time. It is important to note that in contrast to [5], the EPR range is enforced using the regularization during the training phase and the known operators framework during the test phase; therefore, it is enforced during both training and test phases. The second algorithm employs the refinement method proposed be Gou *et al.* [2] which exploits incompressibility constraint to refine the lateral displacement. The network weight and a demo code are publicly available online at http://code.sonography.ai.

2 Materials and Methods

In this section, we first provide a brief overview of PICTURE and underlie some differences to this work. We then introduce our method for incorporating known operators into our deep model and outline our unsupervised training technique. We then present the training and test datasets and finish the section by demonstrating the network architecture.

2.1 PICTURE

Let ε_x denote axial ($x = 1$), lateral ($x = 2$), and out-of-plane ($x = 3$) strains. Assuming linear elastic, isotropic, and homogeneous material that can move freely in the lateral direction, the lateral strain can be obtained from the axial strain and the Poisson's ratio by $\varepsilon_2 = -v \times \varepsilon$. Real tissues are inhomogeneous, and boundary conditions exist; therefore, the lateral strain cannot be directly obtained by the axial strain and the Poisson's ratio alone. In such conditions, EPR, which is defined as $v_e = \frac{-\varepsilon_{22}}{\varepsilon_{11}}$ can be employed [6]. EPR is spatially variant, and it is not equal to Poisson's ratio, particularly in the vicinity of inclusion boundaries or within inhomogeneous tissue. Its value tends to converge towards the Poisson's ratio in homogeneous regions, and it has a similar range of Poisson's ratio, i.e., between 0.2 and 0.5 [9]. In PICTURE, a regularization was defined to exploit this range and the out-of-range EPRs were penalized [5]. PICTURE loss can be obtained from the following procedure:

1- Detect out-of-range EPRs by:

$$M(i,j) = \begin{Bmatrix} 0 & v_{emin} < \widetilde{v_e}(i,j) < v_{emax} \\ 1 & otherwise \end{Bmatrix} \tag{1}$$

where $\widetilde{v_e}$ is the EPR obtained from the estimated displacements. v_{emin} and v_{emax} are two hyperparameters that specify the minimum and maximum accepted EPR values, which are assumed to be 0.1 and 0.6, respectively.

2- Penalize the out-of-range lateral strains using:

$$\begin{aligned} L_{vd} &= |(\varepsilon_{22} + < \widetilde{v_e} > \times \mathbf{S}(\varepsilon_{11}))|_2 \\ \bar{V_e} &= \frac{\sum_{i,j}(1 - M_{(i,j)})V_e(i,j)}{\sum_{i,j}(1 - M_{(i,j)})} \end{aligned} \tag{2}$$

where $< \widetilde{v_e} >$ is the average of EPR values within the feasible range. The operator $\mathbf{S}$ denotes stop gradient operation, which is employed to avoid the axial strain being affected by this regularization. It should be noted in contrast to [5] in which only out-of-range samples were contributing to the loss, in this work, all samples contribute to L_{vd} to reduce the estimation bias.

3- Smoothness of EPR is considered by:

$$L_{vs} = |\frac{\partial v_e}{\partial a}|_1 + \beta \times |\frac{\partial v_e}{\partial l}|_1 \quad (3)$$

4- PICTURE loss is defined as $L_V = L_{vd} + \lambda_{vs} \times L_{vs}$, where λ_{vs} is the weight of the smoothness loss. PICTURE loss is added to the data and smoothness losses of unsupervised training.

2.2 Known Operators

The known operators are added to the network in the inference mode only due to the high computational complexity of unsupervised training (outlined in the next section). We employ two known operators to impose physically known constraints on the lateral displacement.

The first known operator (we refer to it as Poisson's ratio clipper) limits the EPR to the feasible range of $v_{emin} - v_{emax}$. Although PICTURE tries to move all EPR values to the feasible range, in [5], it was shown that some samples in test time were still outside of the feasible range. Poisson's ratio clipper is an iterative algorithm since the lateral strains are altered by clipping the EPR values and affecting the neighbor samples' strain values.

The second algorithm employs the incompressibility of the tissue which can be formulated by:

$$\varepsilon_1 + \varepsilon_2 + \varepsilon_3 = 0 \quad (4)$$

In free-hand palpation, the force is approximately uniaxial ($\varepsilon_3 \simeq \varepsilon_2$); therefore Eq. 4 can be written as:

$$\varepsilon_1 + 2 \times \varepsilon_2 = 0 \quad (5)$$

Guo *et al.* enforced incompressibility in an iterative algorithm [2]. We made a few changes to increase the method's robustness by adding Gaussian filtering and using a hyper-parameter weight in each iteration. It should be noted that the algorithm can be employed for compressible tissues as well, and the incompressibility constraint is employed for the refinement of the obtained displacement. The proposed algorithms are outlined in Algorithm 1 and 2. The network architecture with the known operators is illustrated in Fig. 1. It is worth highlighting that known operators offer a compelling alternative to regularization. While the latter involves adjusting trained weights based on the training data and keeping them fixed during testing, the former relies on iterative refinement that is adaptable to the test data and does not require any learnable weights.

2.3 Unsupervised Training

We followed a similar unsupervised training approach presented in [5] for both PICTURE and kPICTURE methods. The loss function can be written as:

$$Loss = L_D + \lambda_S L_S + \lambda_V L_V \quad (6)$$

Algorithm 1: Poisson's ratio clipper

```
input : Lateral displacement w_l, axial displacement w_a, v_emin, v_emax, iteration
output: Refined lateral displacement w_ref
1 w_ref ← w_l
2 for q ← 1 to iteration do
3     e22 ← ∂w_l/∂l                                  // gradient in lateral direction.
4     e11 ← ∂w_a/∂a                                  // gradient in axial direction.
5     epr ← −e22/e11
6     epr(epr < v_emin) ← v_emin                     // Clip epr less than v_emin
7     epr(epr > v_emax) ← v_emax                     // Clip epr less than v_emax
8     w_ref(:, 2 to end) ← w_ref(:, 1 to end − 1) + epr × e11
      // use the displacement of previous line and the clipped epr to find the displacement of the next line
```

Algorithm 2: Guo *et al.* refinement [2] employed as known operator

```
input : Lateral displacement w_l, Axial displacement w_a of size w × h,
        iteration, λ1, λ2
output: Refined lateral displacement w_ref
1 w_ref ← w_l
2 for q ← 1 to iteration do
3     for i, j in w, h do
4         δ = W_l(i, j − 1) − 2W_l(i, j) + W_l(i, j + 1) + W_a(i + 1, j + 1) − W_a(i −
              1, j) − W_a(i, j − 1) + W_a(i − 1, j − 1) + λ1(W_ref^{q−1} − W_ref^{q−2})
5         w_ref^q = Gauss(w_ref^{q−1} + λ2 × δ)  // Gaussian filtering to reduce noise, λ2 controls the weight
              of updating w_ref^q
```

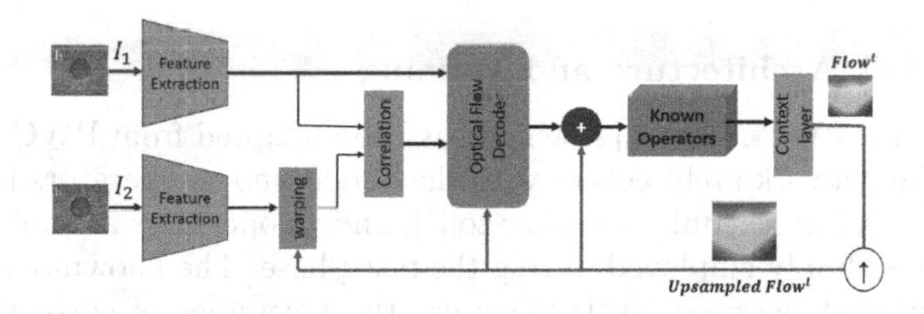

Fig. 1. MPWC-Net++ architecture with known operators. The network is iterative with 5 pyramid levels. The known operators are added after optical flow estimation, and refine the estimated lateral displacement in each pyramid level (added from level 3) to provide improved lateral displacement to the next pyramid level.

where L_D denotes photometric loss which is obtained by comparing the pre-compressed and warped compressed RF data, L_S is smoothness loss in both axial and lateral directions. λ_S and λ_V specify the weights of the smoothness loss and PICTURE loss, respectively.

2.4 Dataset and Quantitative Metrics

We use publicly available data collected from a breast phantom (Model 059, CIRS: Tissue Simulation & Phantom Technology, Norfolk, VA) using an Alpinion E-Cube R12 research US machine (Bothell, WA, USA). The center frequency was 8 MHz and the sampling frequency was 40 MHz. The Young's modulus of the experimental phantom was 20 kPa and contains several inclusions with Young's modulus of higher than 40 kPa. This data is available online at http://code.sonography.ai in [16].

In vivo data was collected at Johns Hopkins hospital from patients with liver cancer during open-surgical RF thermal ablation by a research Antares Siemens system using a VF 10-5 linear array with the sampling frequency of 40 MHz and the center frequency of 6.67 MHz. The institutional review board approved the study with the consent of the patients. We selected 600 RF frame pairs of this dataset for the training of the networks.

Two well-known metrics of Contrast to Noise Ratio (CNR) and Strain Ratio (SR) are utilized to evaluate the compared methods. Two Regions of Interest (ROI) are selected to compute these metrics and they can be defined as [10]:

$$CNR = \sqrt{\frac{2(\overline{s}_b - \overline{s}_t)^2}{{\sigma_b}^2 + {\sigma_t}^2}}, \qquad SR = \frac{\overline{s}_t}{\overline{s}_b}, \tag{7}$$

where the subscript t and b denote the target and background ROIs. The SR is only sensitive to the mean ($\overline{s}_X$), while CNR depends on both the mean and the standard deviation (σ_X) of ROIs. For stiff inclusions as the target, higher CNR correlates with better target visibility, and lower SR translates to a higher difference between the target and background strains.

2.5 Network Architecture and Training

We employed MPWC-Net++ [4] which has been adapted from PWC-Net-irr [3] for USE. The network architecture with the added known operators is shown in Fig. 1. The training schedule is similar to [5], known operators are not present in the training and only employed during the test phase. The known operators are added in different pyramid levels. This has the advantage of correcting lateral displacements in different pyramid levels. The known operators are added to the last 3 pyramid levels (there are 5 pyramid levels in this network) since the estimate in the first 2 pyramid levels are not accurate enough and adding the known operators would propagate the error. The hyper-parameters' values of unsupervised training and the known operators are given in Supplementary Materials.

3 Results and Discussions

3.1 Compared Methods

kPICTURE is compared to the following methods:

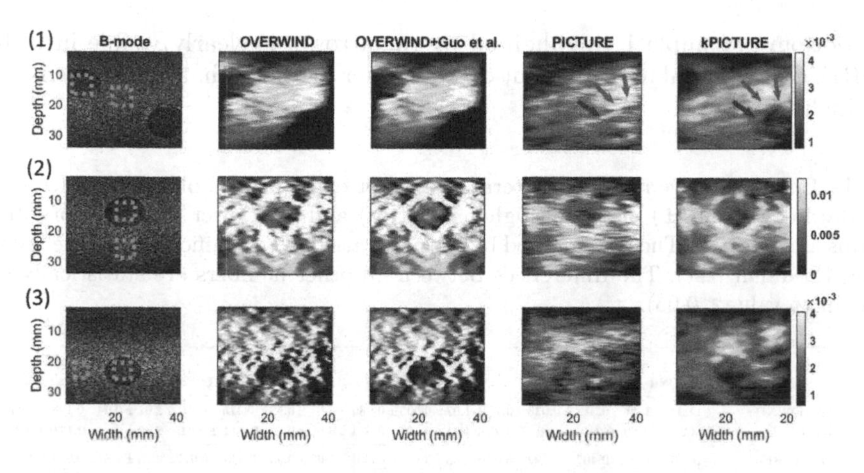

Fig. 2. Lateral strains in the experimental phantom obtained by different methods. The target and background windows for calculation of CNR and SR are marked in the B-mode images. The inclusion on the bottom of sample (1) is highlighted in PICTURE and kPICTURE strain images by purple and blue arrows. The samples 1, 2, and 3 are taken from different locations of the tissue-mimicking breast phantom. Axial strains are available in Supplementary Materials. (Color figure online)

- OVERWIND, an optimization-based USE method [8].
- The post-processing method of Guo *et al.* [2], which employs the output of OVERWIND as the initial displacement (OVERWIND+ Guo *et al.*).
- PICTURE, which penalize EPR values outside of feasible range [5].

We decided to compare with PICTURE instead of sPICTURE [13] (PICTURE with self-supervision) since self-supervision is not related to the physics of motion. To focus on the effectiveness of the known operators, we, therefore, provide a comparison to its corresponding method PICTURE. The proposed known operators can be applied to the network trained with sPICTURE method as well. We made the network's weight trained using both PICTURE and sPICTURE methods publicly available online at http://code.sonography.ai. We also employed a similar hyper-parameters and training schedule for experimental phantom and *in vivo* data.

3.2 Results and Discussions

The lateral strains of ultrasound RF data collected from three different locations of the tissue-mimicking breast phantom are depicted in Fig. 2, and the quantitative results are given in Table 1. Visual inspection of Fig. 2 denotes that the method proposed by Gou *et al.* [2] improves the displacement obtained by OVERWIND. For example, the inclusion borders in sample 2 are much more clearly visible. The strain images obtained by kPICTURE have a much higher quality than those of PICTURE. Furthermore, kPICTURE has the highest quality strain images among the compared methods. For example, the inclusion on the bottom in sample 1 (highlighted by the arrows) is clearly visible in kPICTURE, a substantial improvement over all other methods that do not even show the inclusion.

Table 1. Quantitative results of lateral strains for experimental phantoms. Mean and standard deviation (±) of CNR (higher is better) and SR (lower is better) of lateral strains are reported. The pair marked by † is not statistically significant (p-value > 0.05, using Friedman test). The differences between all other numbers are statistically significant (p-value < 0.05).

	sample (1)		sample (2)		sample (3)		*in vivo data*	
	CNR	SR	CNR	SR	CNR	SR	CNR	SR
OVERWIND	11.34 ± 1.32	0.318 ± 0.030	3.71 ± 1.07	0.505 ± 0.089	3.61 ± 0.58	0.415 ± 0.050	2.07 ± 0.94	0.196 ± 0.255
OVERWIND + Gou *et al.*	13.26 ± 1.89	0.313 ± 0.029	4.28 ± 1.31	0.503 ± 0.083	4.08 ± 0.62	**0.411** ± 0.049	2.39 ± 0.89	**0.170** ± 0.233
PICTURE	9.037 ± 0.88	0.407 ± 0.022	5.37 ± 1.33	0.449 ± 0.060†	1.63 ± 0.95	0.840 ± 0.077	4.36 ± 1.81	0.334 ± 0.149
kPICTURE	**24.40** ± 7.02	**0.290** ± 0.038	**7.81** ± 1.68	**0.446** ± 0.056†	**5.49** ± 2.20	0.598 ± 0.123	**5.54** ± 2.54	0.504 ± 0.141

The histograms of EPR values of OVERWIND+Gou *et al.*, PICTURE and kPICTURE are illustrated for the experimental phantom sample (1). To improve visualization, OVERWIND results are not included because the histogram was similar to that of OVERWIND+Gou *et al.*. Although PICTURE limits the range of EPR using a regularization (Eq. 2), some EPR values are outside the feasible range. kPICTURE further limits the EPR values; only a small number of samples are outside of the physically plausible range.

The lateral strain results of *in vivo* data are depicted in Fig. 3 (b), and axial strains are given in the Supplementary Materials (the quality of axial strains is high in all methods). While PICTURE may produce an adequate strain image, it still contains noisy regions. On the other hand, kPICTURE delivers exceptionally refined strain images and surpasses the other compared methods. The quantitative results given in Table 1 also confirm the visual inspection.

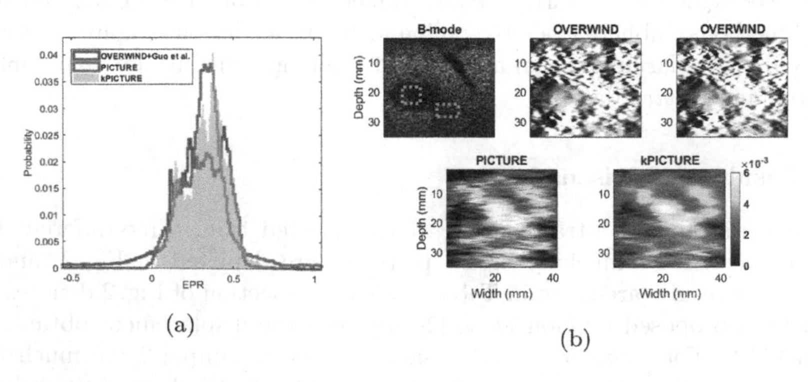

Fig. 3. The histogram of EPR values for experimental phantom sample 1 (a). The *in vivo* results of the compared methods (b).

The applied known operators and PICTURE assume that the material is isotropic. Their performance on anisotropic materials can be investigated by experiments on anisotropic tissues such as muscles. Furthermore, 3D imaging data can be collected from 2D arrays to have information in out-of-plane direction to be able to formulate known operators and PICTURE loss for anisotropic tissues.

It should be noted that after incorporating the known operators, the inference time of the network increased from an average of 195 ms to 240 ms (having 10 iterations for algorithm 1 and 100 iterations for algorithm 2).

4 Conclusions

In this paper, we proposed to incorporate two known operators inside a USE network. The network is trained by physically inspired constraints specifically designed to tackle the long-standing illusive problem of lateral strain imaging. The proposed operators provide a refinement in each pyramid level of the architecture and substantially improve the lateral strain image quality. Tissue mimicking phantom and *in vivo* results show that the method substantially outperforms previous displacement estimation method in the lateral direction.

Acknowledgments. This research was funded by Natural Sciences and Engineering Research Council of Canada (NSERC) Discovery Grant. The Alpinion ultrasound machine was purchased using funds from the Dr. Louis G. Johnson Foundation.

References

1. Delaunay, R., Hu, Y., Vercauteren, T.: An unsupervised approach to ultrasound elastography with end-to-end strain regularisation. In: Martel, A.L., et al. (eds.) Medical Image Computing and Computer Assisted Intervention – MICCAI 2020: 23rd International Conference, Lima, Peru, October 4–8, 2020, Proceedings, Part III, pp. 573–582. Springer, Cham (2020). https://doi.org/10.1007/978-3-030-59716-0_55
2. Guo, L., Xu, Y., Xu, Z., Jiang, J.: A PDE-based regularization algorithm toward reducing speckle tracking noise: A feasibility study for ultrasound breast elastography. Ultrason. Imaging **37**(4), 277–293 (2015)
3. Hur, J., Roth, S.: Iterative residual refinement for joint optical flow and occlusion estimation. In: Proceedings of the IEEE/CVF Conference on Computer Vision and Pattern Recognition, pp. 5754–5763 (2019)
4. Tehrani, A.K.Z., Rivaz, H.: MPWC-Net++: evolution of optical flow pyramidal convolutional neural network for ultrasound elastography. In: Medical Imaging 2021: Ultrasonic Imaging and Tomography, vol. 11602, p. 1160206. International Society for Optics and Photonics (2021)
5. Tehrani, A.K.Z., Rivaz, H.: Physically inspired constraint for unsupervised regularized ultrasound elastography. In: Wang, L., Dou, Q., Fletcher, P.T., Speidel, S., Li, S. (eds.) Medical Image Computing and Computer Assisted Intervention – MICCAI 2022: 25th International Conference, Singapore, September 18–22, 2022, Proceedings, Part IV, pp. 218–227. Springer, Cham (2022). https://doi.org/10.1007/978-3-031-16440-8_21

6. Ma, L., Korsunsky, A.M.: The principle of equivalent eigenstrain for inhomogeneous inclusion problems. Int. J. Solids Struct. **51**(25–26), 4477–4484 (2014)
7. Maier, A.K., et al.: Learning with known operators reduces maximum error bounds. Nat. Mach. Intell. **1**(8), 373–380 (2019)
8. Mirzaei, M., Asif, A., Rivaz, H.: Combining Total Variation Regularization with Window-Based Time Delay Estimation in Ultrasound Elastography. IEEE Trans. Med. Imaging **38**(12), 2744–2754 (2019). https://doi.org/10.1109/TMI.2019.2913194
9. Mott, P., Roland, C.: Limits to Poisson's ratio in isotropic materials-general result for arbitrary deformation. Phys. Scr. **87**(5), 055404 (2013)
10. Ophir, J., et al.: Elastography: ultrasonic estimation and imaging of the elastic properties of tissues. Proc. Inst. Mech. Eng. [H] **213**(3), 203–233 (1999)
11. Peng, B., Xian, Y., Zhang, Q., Jiang, J.: Neural-network-based motion tracking for breast ultrasound strain elastography: an initial assessment of performance and feasibility. Ultrason. Imaging **42**(2), 74–91 (2020)
12. Tehrani, A.K., Amiri, M., Rivaz, H.: Real-time and high quality ultrasound elastography using convolutional neural network by incorporating analytic signal. In: 2020 42nd Annual International Conference of the IEEE Engineering in Medicine & Biology Society (EMBC), pp. 2075–2078. IEEE (2020)
13. Tehrani, A.K.Z., Ashikuzzaman, Md., Rivaz, H.: Lateral strain imaging using self-supervised and physically inspired constraints in unsupervised regularized elastography. IEEE Trans. Med. Imaging **42**(5), 1462–1471 (2023). https://doi.org/10.1109/TMI.2022.3230635
14. K. Z. Tehrani, A., Mirzaei, M., Rivaz, H.: Semi-supervised training of optical flow convolutional neural networks in ultrasound elastography. In: Martel, A.L., et al. (eds.) MICCAI 2020. LNCS, vol. 12263, pp. 504–513. Springer, Cham (2020). https://doi.org/10.1007/978-3-030-59716-0_48
15. Tehrani, A.K., Rivaz, H.: Displacement estimation in ultrasound elastography using pyramidal convolutional neural network. IEEE Trans. Ultrason. Ferroelectr. Freq. Control **67**(12), 2629–2639 (2020)
16. Tehrani, A.K.Z., Sharifzadeh, M., Boctor, E., Rivaz, H.: Bi-directional semi-supervised training of convolutional neural networks for ultrasound elastography displacement estimation. IEEE Trans. Ultrason. Ferroelectr. Freq. Control **69**(4), 1181–1190 (2022). https://doi.org/10.1109/TUFFC.2022.3147097
17. Wei, X., et al.: Unsupervised convolutional neural network for motion estimation in ultrasound elastography. IEEE Trans. Ultrason. Ferroelectr. Freq. Control **69**(7), 2236–2247 (2022). https://doi.org/10.1109/TUFFC.2022.3171676

Weakly Supervised Lesion Localization of Nascent Geographic Atrophy in Age-Related Macular Degeneration

Heming Yao[1], Adam Pely[1], Zhichao Wu[2,3], Simon S. Gao[1], Robyn H. Guymer[2,3], Hao Chen[1], Mohsen Hejrati[1], and Miao Zhang[1(✉)]

[1] Genentech, South San Francisco, CA, USA
zhang.miao@gene.com

[2] Centre for Eye Research Australia, Royal Victorian Eye and Ear Hospital, East Melbourne, VIC, Australia

[3] Ophthalmology, Department of Surgery, The University of Melbourne, Melbourne, VIC, Australia

Abstract. The optical coherence tomography (OCT) signs of nascent geographic atrophy (nGA) are highly associated with GA onset. Automatically localizing nGA lesions can assist patient screening and endpoint evaluation in clinical trials. This task can be achieved with supervised object detection models, but they require laborious bounding box annotations. This study thus evaluated whether a weakly supervised method could localize nGA lesions based on the saliency map generated from a deep learning nGA classification model. This multi-instance deep learning model is based on 2D ResNet with late fusion and was trained to classify nGA on OCT volumes. The proposed method was cross-validated using a dataset consisting of 1884 volumes from 280 eyes of 140 subjects, which had volume-wise nGA labels and expert-graded slice-wise lesion bounding box annotations. The area under Precision-Recall curve (AUPRC) or correctly localized lesions was 0.72($\pm$0.08), compared to 0.77($\pm$0.07) from a fully supervised method with YOLO V3. No statistically significant difference is observed between the weakly supervised and fully supervised methods (Wilcoxon signed-rank test, $p = 1.0$).

Keywords: OCT · weakly supervised learning · object detection

1 Introduction

Nascent geographic atrophy (nGA), originally described by Wu et al. [1], describes features of photoreceptor degeneration seen on optical coherence tomography (OCT) imaging that are strongly associated with the development of geographic atrophy (GA), a late stage complication of age-related macular degeneration (AMD). A recent study reported that the development of nGA in individuals with intermediate AMD was associated with a 78-fold increased rate of GA development [2]. Thus, nGA could potentially

H. Yao and A. Pely—Contributed equally to this work.

H. Greenspan et al. (Eds.): MICCAI 2023, LNCS 14220, pp. 477–485, 2023.
https://doi.org/10.1007/978-3-031-43907-0_46

act as an earlier biomarker of AMD progression, or potentially as an earlier endpoint in intervention studies aiming to slow GA development [3]. Thus being able to easily identify eyes with nGA and localize nGA lesions is important in clinical trials and research.

However, identifying and grading the location of nGA lesions in OCT volume scans can be a laborious task, and would be an operationally expensive undertaking in clinical trials. Automation of this task would be invaluable when seeking to quantify the number of nGA lesions present, or when seeking to identify a smaller subset of B-scans for manual expert review (an "AI-assisted" approach). While the localization of nGA could be tackled by supervised object detection models – as demonstrated in other types of lesions [4–6] – it takes domain experts a large amount of time to provide sufficient number of lesion level annotations (e.g. with a bounding box). On the other hand, weakly supervised methods require only coarse annotations, and they have been popular in computer vision tasks where dense annotations are difficult to obtain [7].

In this work, we sought to develop a deep learning-based method to automate the localization of nGA lesions on OCT imaging, trained only on the information about the presence or absence of nGA at the volume level. A weakly supervised algorithm was developed that utilizes the saliency maps from Gradient Class Activation Maps (GradCAM) technique [8]. While existing literature has demonstrated the ability of the GradCAM in post-hoc model interpretation [9–12], it is unknown whether the saliency map can help further localize the class-related lesions or abnormalities that are often sparse anatomically. We thus explored the possibility of GradCAM in identifying the location of nGA-related abnormalities after training a model for classifying nGA in a 3D volume scan.

2 Methods

2.1 Dataset

This study included participants in the sham treatment arm of the Laser Intervention in the Early Stages of AMD study (LEAD, clinicaltrials.gov identifier, NCT01790802). The LEAD study was conducted according to the International Conference on Harmonization Guidelines for Good Clinical Practice and the tenets of the Declaration of Helsinki. Institutional review board approval was obtained at all sites and all participants provided written informed consent.

The participants of LEAD study were required to have bilateral large drusen and a best-corrected visual acuity of 20/40 or better in both eyes at baseline [13]. Participants were evaluated at the baseline and every 6-month follow up visits for up to 36-months. At each visit, OCT imaging was performed following pupillary dilation, by obtaining a 3D volume scan consisting of 49 B-scans (i.e. 2D slices along X-Z direction) covering a $20° \times 20° \times 1.9$ mm region of the macula, with $1024 \times 49 \times 496$ voxels anisotropically sampled along X, Y and Z d directions respectively.

Multimodal imaging was used to assess the development of late AMD as an endpoint in the LEAD study, which included nGA detected on OCT imaging. In order to evaluate the association between nGA and the subsequent development of GA as detected on color fundus photographs (CFP; the historical gold standard for atrophic AMD) in a previous study, OCT imaging and CFP were independently re-graded for the presence of nGA and GA respectively [14]. In this sub-study, we included individuals who did not have nGA at baseline based on the above independent re-grading of OCT imaging, and who had at least one follow-up visit.

A total of 1,884 OCT volumes from 280 eyes of 140 individuals were included in this analysis (1,910 volumes were collected, but 26 volumes were excluded from the study due to the development of neovascular AMD in the eye). In this study, the development of nGA was assessed by manual grading of all 49 B-scans of each OCT volume scans, and nGA was defined by the subsidence of the outer plexiform layer and inner nuclear layer, and/or the presence of a hyporeflective wedge-shaped band within Henle's fiber layer, as per the original definition [1]. All OCT volume scans were initially assessed by a senior grader, and all visits of any eye deemed to have questionable or definite nGA were then reviewed by two further experienced graders [2].

Overall, nGA was graded as being absent and present in 1,766 and 118 OCT volume scans respectively. In the context of this study, note that nGA also includes lesions that could also meet the criteria for having complete retinal pigment epithelium and outer retinal atrophy (cRORA), if the lesion also had choroidal signal hypertransmission and retinal pigment epithelium (RPE) attenuation or disruption of $\geq$250 μm [15]. Graders also concurrently graded the location of nGA lesions by identifying the B-scans with nGA lesions and by drawing bounding boxes on the B-scans horizontally covering the subsidence, vertically from the inner limiting membrane (ILM) to Bruch's membrane. For the weakly supervised model, the bounding boxes were used only in evaluating the weakly supervised lesion localization, not in model training. The bounding boxes were then used to train a fully supervised object detector to compare the results of the weakly supervised and fully supervised methods.

2.2 Deep Learning Architecture

A late-fusion model with a 2D ResNet backbone was developed to classify 3D OCT volumes, considering their anisotropic nature. As shown in Fig. 1a, B-scans from a 3D OCT volume were fed into a B-scan detector, and the outputs, which are vectors of classification logits for each B-scan, were averaged to generate prediction scores for the volume. Thinking of the B-scans as instances and the OCT volumes as bags, this framework can be categorized as simplified multi-instance learning [16] in which the network was trained on weakly labeled data, using labels on bags (OCT volumes) only. During the training process, given an OCT volume annotated as nGA, the network was forced to identify as many B-scans with nGA lesion to improve the final prediction of nGA, thus the trained model allows prediction of nGA labels on OCT volumes as well as on individual B-scans.

The details of the B-scan classifier are shown in Fig. 1b. An individual B-scan of size 1024 $\times$ 496 from the volume is downsampled to 512 $\times$ 496 and passed through the ResNet-18 backbone, which outputs activation maps of 512 $\times$ 16 $\times$ 16. A max-pooling layer and an average pooling layer are concatenated to generate a feature vector of 1024. Then a fully connected layer was applied to generate the classification logit for the B-scan.

The classification model was evaluated on its own both in terms of volume-wise and slice-wise performance in classifying nGA. After it was confirmed that the classification model worked well, the ability of the model to localize the lesions within individual OCT slices was evaluated.

Given an OCT volume and a trained model, saliency maps were generated with the GradCAM technique [8] to visualize regions making larger contributions to the final classification. In Fig. 2a, GradCAM output was overlaid as the yellow channel on the input images for easy visualization of the saliency as well as the original grayscale image. The saliency map from a legitimate model should highlight nGA lesions, thus GradCAM output can help localize nGA lesions. The objective was to localize nGA lesions in the 3D OCT volume, i.e. to identify which B-scans have nGA lesions and to generate a bounding box surrounding the lesions in those B-scans with confidence scores.

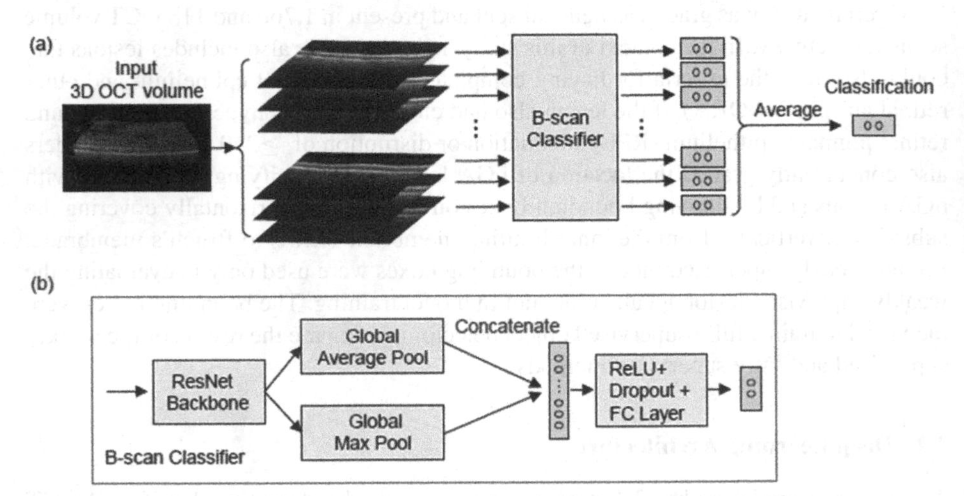

Fig. 1. Deep learning architecture of the nGA classification model. (a) Model network for 3D OCT volume. (b) The B-scan classifier. ReLU = rectified linear unit. FC = fully connected.

As illustrated in Fig. 2, the automated image processing pipeline was built upon adaptive thresholding and connected component analysis [17]. For each B-scan with positive logit, one or multiple bounding boxes covering potential lesions were detected. The confidence score for each bounding box was estimated from the individual classification logit of the B-scan classifier as Eq. (1).

$$S\left(\frac{l}{n}\frac{h}{\Sigma h}\right) \tag{1}$$

where S is the sigmoid function, l is the individual B-scan classification logit, and n is the number of B-scans in a volume, h is the mean saliency in the detected region and Σh is the total mean saliency of all detected regions within the B-scan. A higher confidence score implies a higher possibility that the detected region covers nGA lesions. Since only class labels of 3D OCT volume are required for training, the proposed lesion localization algorithm was weakly supervised.

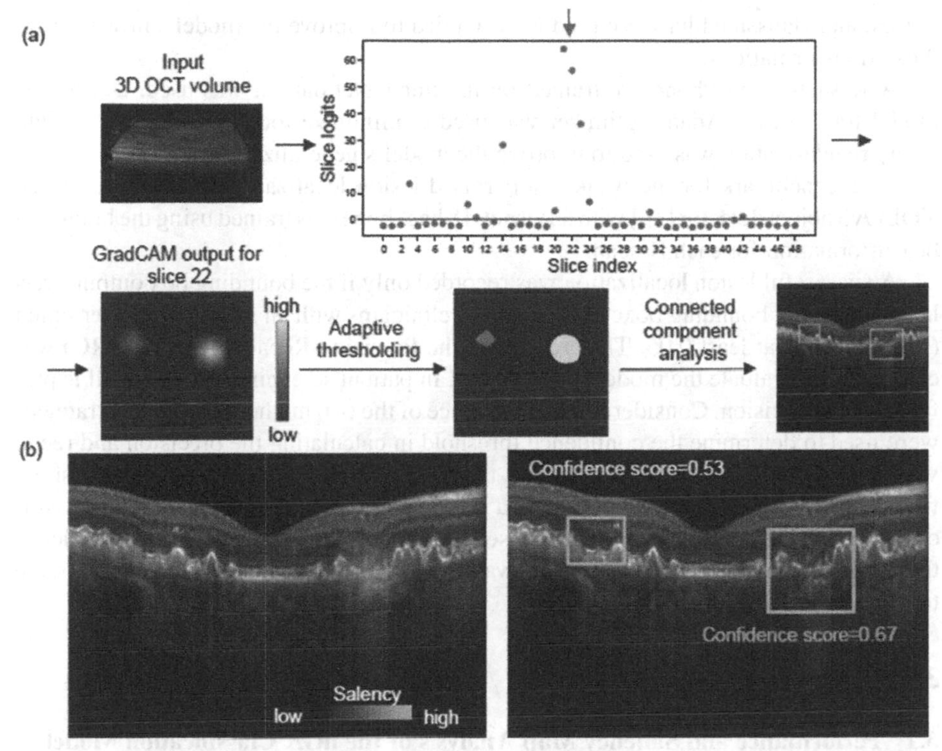

Fig. 2. (a) Illustration of locating nGA lesions with B-scan logit and GradCAM. B-scans with positive logits were selected, their GradCAM outputs (in the viridis colormap) were thresholded adaptively, then a bounding box was generated by the connected component analysis. A confidence score of the bounding box was estimated based on its average saliency and the corresponding B-scan logit. (b) Examples of a B-scan overlaid with GradCAM output (left) and bounding boxes with confidence scores (right), the yellow bounding box with confidence score below threshold was removed in following processing. (Color figure online)

2.3 Model Training, Tuning, and Validation Test

Considering the relatively small number of participants in the dataset, a five-fold cross-validation was applied to evaluate the proposed method's performance. We followed the nomenclature for data splitting as recommended previously [18]. For each fold, the validation test set of OCT volumes were obtained from roughly 20% of the participants stratified on whether the individual developed nGA. The OCT volumes from the remaining 80% individuals were further split into training (64%) and tuning sets (16%), with volumes from one individual only existing in one of the sets. With the proposed data split strategy, the corresponding validation test set was not used in the training or hyperparameter tuning process.

Pre-processing was performed on B-scans for standardization. The B-scans were first resized to 512 × 496, followed by rescaling the intensity range to [0, 1]. Data augmentation, including rotation of small angles, horizontal flips, add on of Gaussian noises, and Gaussian blur were randomly applied to improve the model's invariance to those transformations.

A Resnet-18 backbone pre-trained on the ImageNet dataset was used. During the model training, the Adam optimizer was used to minimize focal loss. The L2 weight decay regularization was used to improve the model's generalization.

As a benchmark for the weakly supervised lesion localization, a fully supervised YOLOv3 object detector [19] with a Resnet-18 backbone was trained using the bounding box information for each B-scan.

A successful lesion localization was recorded only if the bounding box output overlapped with the bounding boxes annotated by clinicians with an intersection over union (IoU) value of at least 0.05. The area under the Precision-Recall curve (AUPRC) was calculated to evaluate the model performance. In patient screening, a high recall is preferred over precision. Considering the difference of the two methods, different strategies were used to determine the confidence threshold in calculating the precision and recall values in the validation test dataset. For the weakly supervised method, the threshold for confidence score that would achieve a recall value of 0.98 for nGA volume classification in the training and tuning sets is used. For the supervised method, the confidence threshold which would achieve a recall value of 0.9 for bounding box detection in the turning set is used.

3 Results

3.1 Performance and Saliency Map Analysis of the nGA Classification Model on OCT Volumes

The deep learning based nGA classification model achieved an AUPRC of 0.83(±0.09) in classifying 3D OCT volumes. Based on the trained 3D OCT volume classification model and input OCT volumes, we generated the corresponding saliency map using GradCAM technique. Examples of GradCAM output are shown in Fig. 2. Values in the GradCAM output indicate the importance of the corresponding pixel in the input B-scan to the model's prediction. A higher (brighter) value means the corresponding pixel contributes more to the model's prediction that the input OCT volume is positive. A thresholding was applied to the pixel values to determine the region where the bounding box delineating nGA should be drawn.

3.2 Performance of the Weakly Supervised Localization of nGA Lesions

The weakly supervised algorithm achieved a similar level of performance for localizing nGA when compared to the YOLOv3 based fully supervised method, without utilizing bounding box annotations and these findings are illustrated in Fig. 3. The YOLOv3 based method achieved an AUPRC of 0.77(±0.07) compared to 0.72(±0.08) for the weakly supervised model; no statistically significant difference was observed between the two methods (Wilcoxon signed-rank test for AUPRC, $p = 1.0$).

In the patient screening setting described previously, the YOLOv3 based method achieved a precision and recall of 0.53(±0.16) and 0.88(±0.06), compared to 0.39(±0.13) and 0.88(±0.02) for the weakly supervised method.

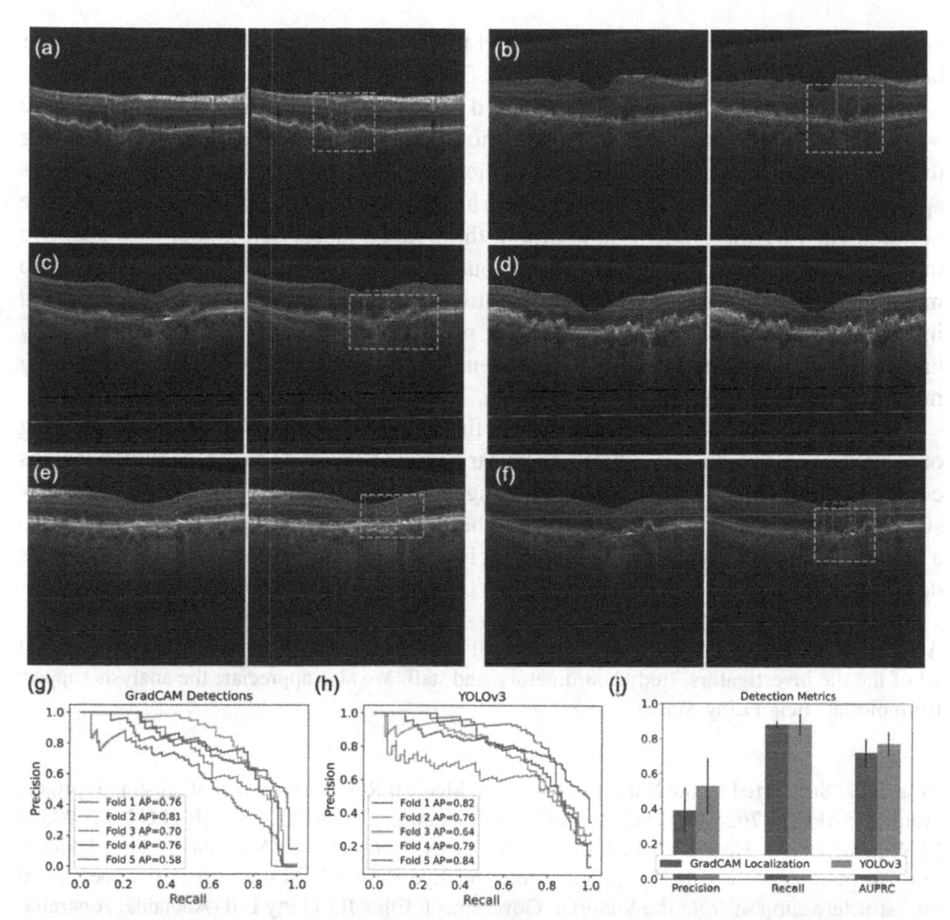

Fig. 3. (a)–(f) Example results from the weakly and fully supervised detection methods alongside the ground truth. The ground truth is in green, the YOLOv3 detector output in red, and the GradCAM based detector output is in yellow. (a–c) A true positive case for both methods. (d) A true positive case for the fully supervised method, but false negative for the weakly supervised method. (e) A true positive case for the weakly supervised method, but false negative for the fully supervised method. (f) A case where there are two ground truth lesions, but both methods detected only the left one. (g) The per fold precision-recall curve for the GradCAM based detector along with the AUPRC values. (h) The per fold precision-recall curve for YOLOv3 detector along with the AUPRC values. (i) A comparison between the GradCAM based detector and YOLOv3 on the metrics of AUPRC, as well as recall and precision in the patient screening setting described in Methods. (Color figure online)

4 Conclusion and Discussion

This study demonstrates that the performance for localizing nGA lesions by only using OCT volume-wise classification labels with the GradCAM technique was on par with a fully supervised approach using B-scan level annotations with the YOLOv3 detector. These findings therefore underscore the potential of a weakly supervised approach for enabling the development of a robust model for lesion localization without the need for laborious, lesion-level annotations on OCT B-scans.

One limitation of the GradCAM-based lesion localization is its relatively large bounding box size, often exceeding the annotated region. This is expected, considering the low spatial resolution of GradCAM saliency map, but also potentially because this approach identified contextual features that are distinguishing of nGA lesions that were not annotated by the graders. In addition, the weakly supervised model uses adaptive threshold of the saliency to determine the bounding box size, which was not optimized to match the ground truth grading. This limitation with the larger bounding box size could impact the quantification of the number of nGA lesions present, but it would unlikely have a substantial impact on the task of identifying a subset of OCT B-scans requiring manual review in an AI-assisted evaluation.

In conclusion, this study demonstrates that a weakly supervised method, requiring only volume-wise tags, can achieve a similar level of performance for localizing lesions compared to a fully supervised method using slice-wise bounding box labels. A weakly supervised approach could thus minimize the labeling burden when seeking to develop a lesion localization model, and could even leverage existing volume-wise labels for its development.

Acknowledgements. We would like to thank all of the study participants and their families, and all of the site investigators, study coordinators, and staff. We also appreciate the analysis support from biostatistician Ling Ma.

Funding. Supported by the National Health and Medical Research Council of Australia (project grant no.: APP1027624 [R.H.G.], and fellowship grant nos.: GNT1103013 [R.H.G.], #2008382 [Z.W.]; the BUPA Health Foundation (Australia) (R.H.G.) and the Macular Disease Foundation Australia (Z.W. and R.H.G.). The Centre for Eye Research Australia receives operational infrastructure support from the Victorian Government. Ellex R&D Pty Ltd (Adelaide, Australia) provided partial funding of the central coordinating center and the in-kind provision the Macular Integrity Assessment microperimeters for the duration of the LEAD study. The web-based Research Electronic Data Capture application and open- source platform OpenClinica allowed secure electronic data capture. The LEAD study was sponsored by the Centre for Eye Research Australia, East Melbourne, Australia, an independent medical research institute and a not-for-profit company. This sub-study was supported by Genentech, Inc.

Heming Yao, Adam Pely, Simon S. Gao, Hao Chen, Mohsen Hejrati, and Miao Zhang, are employees of Genentech, Inc. and shareholders in F. Hoffmann La Roche, Ltd.

References

1. Wu, Z., et al.: Optical coherence tomography-defined changes preceding the development of drusen-associated atrophy in age-related macular degeneration. Ophthalmology **121**(12), 2415–2422 (2014)
2. Wu, Z., et al.: Prospective longitudinal evaluation of nascent geographic atrophy in age-related macular degeneration. Ophthalmol. Retina **4**(6), 568–575 (2020)
3. Wu, Z., Guymer, R. H.: Can the onset of atrophic age-related macular degeneration be an acceptable endpoint for preventative trials?. ophthalmologica. J. Int. d'ophtalmologie. Int. J. Ophthalmol. Zeitschrift fur Augenheilkunde **243**(6), 399–403 (2020)

4. Derradji, Y., Mosinska, A., Apostolopoulos, S., Ciller, C., De Zanet, S., Mantel, I.: Fully-automated atrophy segmentation in dry age-related macular degeneration in optical coherence tomography. Sci. Rep. **11**(1), 21893 (2021)
5. Corradetti, G., et al.: Automated identification of incomplete and complete retinal epithelial pigment and outer retinal atrophy using machine learning. Investig. Ophthalmol. Vis. Sci. **63**(7), 3860 (2022)
6. Chiang, J.N., et al.: Automated identification of incomplete and complete retinal epithelial pigment and outer retinal atrophy using machine learning. Ophthalmol. Retina **7**(2), 118–126 (2023)
7. Yang, H.L., et al.: Weakly supervised lesion localization for age-related macular degeneration detection using optical coherence tomography images. PLoS ONE **14**(4), e0215076 (2019)
8. Selvaraju, R., Cogswell, M., Das, A., Vedantam, R., Parikh, D., Batra, D.: Grad-CAM: visual explanations from deep networks via gradient-based localization. In: Proceedings of the IEEE International Conference on Computer Vision, pp. 618–626 (2017)
9. Shi, X., et al.: Improving interpretability in machine diagnosis: detection of geographic atrophy in OCT scans. Ophthalmol. Sci. **1**(3), 100038 (2021)
10. Yoon, J., et al.: Optical coherence tomography-based deep-learning model for detecting central serous chorioretinopathy. Sci. Rep. **10**(1), 18852 (2020)
11. Wang, Y., Lucas, M., Furst, J., Fawzi, A.A., Raicu, D.: Explainable deep learning for biomarker classification of OCT images. In: 2020 IEEE 20th International Conference on Bioinformatics and Bioengineering (BIBE), Cincinnati, OH, pp. 204–210 (2020)
12. Li, Y., et al.: Development and validation of a deep learning system to screen vision-threatening conditions in high myopia using optical coherence tomography images. Br. J. Ophthalmol. **106**(5), 633–639 (2022)
13. Guymer, R.H., et al.: Subthreshold nanosecond laser intervention in age-related macular degeneration: the lead randomized controlled clinical trial. Ophthalmology **126**(6), 829–838 (2019)
14. Wu, Z., Bogunović, H., Asgari, R., Schmidt-Erfurth, U., Guymer, R.H.: Predicting progression of age-related macular degeneration using OCT and fundus photography. Ophthalmol. Retina **5**(2), 118–125 (2021)
15. Guymer, R.H., et al.: Incomplete retinal pigment epithelial and outer retinal atrophy in age-related macular degeneration: classification of atrophy meeting report 4. Ophthalmology **127**(3), 394–409 (2020)
16. Carbonneau, M.-A., Cheplygina, V., Granger, E., Gagnon, G.: Multiple instance learning: a survey of problem characteristics and applications. Pattern Recog. **77**, 329–353 (2018)
17. Otsu, N.: A threshold selection method from gray-level histograms. IEEE Trans. Syst. Man Cybern. **9**, 62–66 (1979)
18. Liu, X., et al.: A comparison of deep learning performance against health-care professionals in detecting diseases from medical imaging: a systematic review and meta-analysis. Lancet Digit. Health **1**(6), e271–e297 (2019)
19. Redmon, J., Farhadi, F.: YOLOv3: An Incremental Improvement. arXiv preprint arXiv:1804.02767 (2018)

Can Point Cloud Networks Learn Statistical Shape Models of Anatomies?

Jadie Adams[1,2](✉) and Shireen Y. Elhabian[1,2]

[1] Scientific Computing and Imaging Institute, University of Utah, Salt Lake, UT, USA
{jadie.adams,shireen}@utah.edu

[2] School of Computing, University of Utah, Salt Lake, UT, USA

Abstract. Statistical Shape Modeling (SSM) is a valuable tool for investigating and quantifying anatomical variations within populations of anatomies. However, traditional correspondence-based SSM generation methods have a prohibitive inference process and require complete geometric proxies (e.g., high-resolution binary volumes or surface meshes) as input shapes to construct the SSM. Unordered 3D point cloud representations of shapes are more easily acquired from various medical imaging practices (e.g., thresholded images and surface scanning). Point cloud deep networks have recently achieved remarkable success in learning permutation-invariant features for different point cloud tasks (e.g., completion, semantic segmentation, classification). However, their application to learning SSM from point clouds is to-date unexplored. In this work, we demonstrate that existing point cloud encoder-decoder-based completion networks can provide an untapped potential for SSM, capturing population-level statistical representations of shapes while reducing the inference burden and relaxing the input requirement. We discuss the limitations of these techniques to the SSM application and suggest future improvements. Our work paves the way for further exploration of point cloud deep learning for SSM, a promising avenue for advancing shape analysis literature and broadening SSM to diverse use cases.

Keywords: Statistical Shape Modeling · Point Cloud Deep Networks · Morphometrics

1 Introduction

Statistical Shape Modeling (SSM) enables population-based morphological analysis, which can reveal patterns and correlations between shape variations and clinical outcomes. SSM can help researchers understand the differences between healthy and pathological anatomy, assess the effectiveness of treatments, and

Supplementary Information The online version contains supplementary material available at https://doi.org/10.1007/978-3-031-43907-0_47.

H. Greenspan et al. (Eds.): MICCAI 2023, LNCS 14220, pp. 486–496, 2023.
https://doi.org/10.1007/978-3-031-43907-0_47

identify biomarkers for diseases (e.g., [5,8–10,16]). The traditional pipeline for constructing SSM entails the segmentation of 3D images to acquire either binary volumes or meshes and then aligning these shapes. SSM can then be constructed *explicitly* via finding surface-to-surface correspondences across the cohort or *implicitly* via deforming a predefined atlas to each shape sample. Correspondence-based shape models are widely used due to their intuitive interpretation [22]; they comprise of sets of landmarks or *correspondence points* that are defined consistently using invariant points across populations that vary in their form. Historically, correspondence points were established manually to capture biologically significant features. This cumbersome, subjective process has since been replaced via automated *optimization*, which defines dense sets of correspondence points, aka a Point Distribution Model (PDM). PDM optimization schemes have been defined using metrics such as entropy and minimum description length [12,13] and via parametric representations [17,24]. A significant drawback of these methods is that PDM optimization must be performed on the entire shape cohort of interest, which is time-consuming and hinders inference. To evaluate a new patient scan using PDM, the new scan must undergo segmentation and alignment, and the PDMs must be optimized again for the entire population of shapes. Moreover, current approaches require a complete, high-resolution mesh or binary volume representation of the shape that is free from noise and artifacts. Therefore, lightweight shape acquisition methods (such as thresholding clinical images, anatomical surface scanning, and shape acquired from stacked or orthogonal 2D slices) cannot be directly used for SSM [26,27]. Deep learning solutions have been proposed to mitigate these limitations by predicting PDMs directly from unsegmented 3D images using convolutional neural networks [1,2,7]. However, these frameworks require PDM supervision and, hence, need the traditional optimization-based workflow to acquire training data.

Effective SSM from point clouds would be widely applicable in clinical research, from artery disease progression from point clouds acquired via biplane angiographic and intravascular ultrasound image data fusion [26], to orthopedics implant design from point clouds acquired via 3D body scanning [27]. Applying existing methods for SSM generation from point clouds requires converting them to meshes or rasterizing them into segmentations, which is nontrivial given that point clouds are unordered and do not retain surface normals. SSM directly from point clouds would enable many clinical studies. Recently point cloud deep learning has gained attention, with significant efforts focused on effective point completion networks (PCNs) for generating complete point clouds from partial observations. Most point-completion methods use order-invariant feature encoding and two-stage coarse-to-fine decoding. An important outcome of such networks, which has been overlooked and not reported, is that the learned coarse point clouds are ordered and provide correspondence. In this work, we acknowledge this missed potential and explore the use of PCNs to predict PDM from 3D point clouds in an unsupervised manner. We investigate state-of-the-art PCNs as potential solutions for generating PDMs that (1) accurately represent shapes via uniformly distributed points constrained to the surface and (2) provide good correspondences that capture population-level statistics.[1] We discuss the benefits of this approach, its robustness to missingness and training size, current limitations, and possible improvements. This discussion will bring awareness to the community about the potential for learning SSM from point clouds and ultimately make SSM a more accessible, viable option in future clinical research.

2 Background

Point Distribution Models. The goal of SSM is to capture the inherent characteristics or underlying parameters of a shape class that remain when global geometrical information is removed. Given a PDM, correspondence points can be averaged across subjects to provide a mean shape, and principal component analysis (PCA) can be performed to compute the modes of shape variation, which can then be visualized and used in downstream medical tasks. Furthermore, if a PDM contains sub-populations, such as disease versus control, the differences in mean shapes can be quantified and visualized, providing group characterization.

Point Cloud Deep Learning. Deep learning from 3D point clouds is an emerging area of research with numerous applications in computer vision, robotics, and medicine (e.g., classification, object tracking, segmentation, registration, pose estimation) [3]. PointNet [20] pioneered a Multi-Layer Perceptron (MLP) and max-pooling-based approach for permutation invariant feature learning from raw point clouds. FoldingNet [31] proposed a point cloud auto-encoder with a folding-based decoder that utilizes 2D grid deformation for reconstruction. Several convolutional approaches have been proposed, including mapping point clouds to voxel grids to directly apply 3D CNNs [15] and graph-based methods [29].

The initial *point cloud completion* network, PCN [33], utilized PointNet [20] and FoldingNet [31] with a coarse-to-fine decoder. Since then, numerous point cloud completion approaches have been proposed, including point MLP and folding-based extensions, 3D convolution approaches, graph-based methods, generative modeling approaches (including generative adversarial network GAN-based and variational autoencoder VAE-based), and transformer-based methods. See [14] for a recent survey. Many approaches follow the general framework of first encoding the point cloud into a permutation-invariant feature representation, then decoding the encoded shape feature to acquire a coarse or sparse point cloud, and finally, refining the coarse point cloud to acquire the dense complete prediction. The general architecture of these methods is shown in Fig. 1.

3 Methods

3.1 Point Completion Networks for SSM

Our experiments demonstrate that when coarse-to-fine point completion networks are trained on anatomical shapes, the bottleneck captures a population-specific shape prior. Directly decoding the shape feature representation results in

[1] Source code is publicly available: https://github.com/jadie1/PointCompletionSSM.

a consistent ordering of the intermediate coarse point cloud across samples, providing a PDM. This phenomenon can be intuitively understood as an application of Occam's razor, where the model prefers to learn the simplest solution, resulting in consistent output ordering. Many point completion networks contain skip connections from the feature and/or input space to the refinement network. In the case where the unordered input point cloud is fed to the refinement network, the ordering of the output dense point cloud is understandably lost.

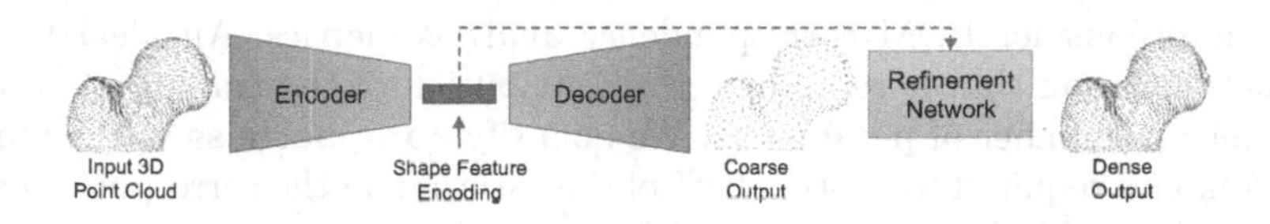

Fig. 1. General coarse-to-fine architecture of the considered PCNs.

To study whether point completion networks can learn anatomical SSM, we first extract point clouds from mesh vertices, then train point completion models, and finally evaluate the effectiveness of the predicted coarse point clouds as PDMs. Note this approach is not restricted to input point clouds obtained from meshes; point clouds from any acquisition process can be used. Global geometric information is factored out by aligning all shapes via iterative closest points [6] to a reference shape. We utilized the open-source toolkit ShapeWorks [11] for this step. In our experiments, the aligned, unordered mesh vertices serve as ground truth complete point clouds. The ground truth points are randomly downsampled to 2048 points and permuted to provide input point clouds. As is standard, the point clouds are uniformly scaled to be between -1 and 1 to assist network training. We consider a state-of-the-art model from the major point completion approach categories:

- **PCN** [33]: Point Completion Network (MLP-based)
- **ECG** [18]: Edge-aware Point Cloud Completion with Graph Convolution (convolution-based)
- **VRCnet** [19]: Variational Relational Point Completion Network (generative-modeling based)
- **SFN** [30]: SnowflakeNet (transformer-based)
- **PointAttN** [28]: Point Attention Network (attention-based)

Point completion network loss is based on the *L1* **Chamfer Distance (CD)** [14], which defines the minimum distance between two sets of points. The loss is typically defined as a combination of coarse and dense loss with weighting parameter α, which we consistently set across models. All model's hyperparameters are set to the original implementation values, and training is run until convergence (as assessed by training CD).

3.2 Evaluation Metrics

In addition to CD, **Fscore** [25] is typically used in point completion to quantify the percentage of points that are reconstructed correctly. In analyzing accurate surface sampling for SSM, we quantify the **point-to-face distance (P2F)** from each point in the predicted point cloud to the closest surface of the ground truth mesh. To analyze point **uniformity**, we quantify the variance in the distance of each point to its six nearest neighbors. A uniform PDM would result in a small variance in the point nearest neighbor distance.

We also consider PDM correspondence analysis metrics. An ideal PDM is compact, meaning that it represents the distribution of the training data using the minimum number of parameters. We quantify **compactness** as the number of PCA modes required to capture 99% of the variation in the correspondences. A good PDM should also generalize well from training examples to unseen examples. The **generalization** metric is defined as the reconstruction error ($L2$) between predicted correspondences of a held-out point cloud and the correspondences reconstructed via the training PDM. Finally, effective SSM is specific, generating only valid instances of the shape class in the training set. The average distance between correspondences sampled from the training PDM and the closest existing training correspondences provides the **specificity** metric.

4 Experiments

We use five datasets in experiments – one synthetic ellipsoid dataset for a proof-of-concept and four real anatomical shapes: proximal femurs, left atrium of the heart, spleen, and pancreas. These datasets vary greatly in size (see Table 1, Column 1) and shape variation. Details and visualization of these cohorts are provided in the supplementary materials. In all experiments, the input point cloud size is 2048, the coarse output size is 512, and the dense output size is 2048. The real datasets were split (90%/10%) into training and testing sets. Stratified sampling via clustering was used to define the test set to ensure it is representative, given the low sample size.

Proof-of-Concept: Ellipsoids. As a proof-of-concept, we generate 3D axis-aligned ellipsoid shapes with fixed z-radius and random x and y-radius. A testing set of 30 and a training set of just 50 were randomly defined to emulate the scarce data scenario. The results in Table 1 show that all of the model variants performed well with low S2F distance (<0.1mm), and all PDMs correctly captured just two modes of variation (x and y-radius). Results from the PCN [33] model are shown in Fig. 2.

Femur. The femur dataset is comprised of 56 femoral heads segmented from CT scans, nine of which have the cam-FAI pathology characterized by an abnormal bone growth lesion that causes hip osteoarthritis [5]. We utilize this pathology to analyze if PDM from point completion networks can correctly characterize

Table 1. Results: Evaluation metrics across all models and datasets. All metrics are quantified on coarse predictions, with the exception of the gray column, which reports the dense prediction CD. The CD values are scaled by 1000 for reporting and Fscore is calculated at 1% threshold. Best test values are shown in bold.

Data	Model	Point Accuracy Metrics (train/test)					SSM Evaluation Metrics		
		Dense CD ↓	CD ↓	Fscore ↑	P2F (mm) ↓	Uniformity ↓	Comp.↓	Gen. ↓	Spec.↓
Ellipsoids	PCN[33]	0.908/0.917	1.80/1.80	0.561/**0.582**	0.0142/**0.0168**	0.325/0.354	**2**	0.140	0.348
	ECG[18]	0.929/0.923	1.80/1.78	0.554/0.579	0.0171/0.0196	0.308/0.335	**2**	0.136	0.348
Train: 50	VRCnet[19]	0.923/0.919	1.78/**1.77**	0.563/**0.582**	0.0384/0.0367	0.295/0.327	**2**	0.143	0.348
Test: 30	SFN[30]	0.842/**0.838**	1.92/1.91	0.559/0.566	0.0960/0.0949	0.237/0.256	**2**	**0.103**	**0.332**
	PointAttN[28]	0.912/0.911	1.78/**1.77**	0.554/0.571	0.0255/0.0315	0.217/**0.243**	**2**	0.137	0.356
Femur	PCN[33]	1.27/1.80	2.30/**2.91**	0.501/**0.386**	0.235/**0.712**	1.77/1.63	18	0.28	1.30
	ECG[18]	1.59/2.29	2.37/3.02	0.485/0.377	0.297/0.750	1.71/1.63	15	0.255	1.23
Train: 51	VRCnet[19]	1.78/2.33	2.88/3.27	0.399/0.362	0.676/0.876	1.62/1.60	**5**	0.259	0.786
Test: 5	SFN[30]	1.04/**1.29**	2.36/3.13	0.464/0.365	0.341/0.774	1.01/**0.967**	17	0.297	1.37
	PointAttN[28]	1.27/1.50	3.25/3.94	0.352/0.295	0.759/1.03	2.32/2.29	9	**0.199**	**0.905**
Left Atrium	PCN[33]	0.405/0.773	0.707/1.09	0.941/0.865	0.245/1.02	1.87/2.01	82	0.932	5.85
	ECG[18]	0.571/0.868	0.814/1.16	0.919/0.848	0.440/1.07	1.85/2.04	57	0.929	5.22
Train: 987	VRCnet[19]	0.070/**0.085**	0.886/1.62	0.904/**0.768**	0.632/1.53	1.87/2.19	81	1.04	6.09
Test: 109	SFN[30]	0.310/0.311	0.822/**0.948**	0.927/0.902	0.511/**0.831**	1.30/**1.46**	**49**	**0.783**	**5.07**
	PointAttN[28]	0.332/0.360	0.875/1.20	0.909/0.840	0.523/1.08	1.65/1.79	82	0.807	5.46
Spleen	PCN[33]	3.59/7.88	4.51/9.77	0.326/0.155	1.67/3.73	15.3/12.0	7	1.47	4
	ECG[18]	1.84/6.46	3.0/9.69	0.456/0.174	1.07/3.94	10.2/7.98	14	1.62	5.57
Train: 36	VRCnet[19]	0.408/**0.583**	5.03/14.3	0.318/**0.117**	1.81/4.59	16.1/12.5	6	1.73	4.48
Test: 4	SFN[30]	0.986/1.28	4.57/**7.23**	0.265/0.189	1.64/**3.07**	17.8/14.7	**6**	**1.08**	**4.18**
	PointAttN[28]	1.13/3.70	3.49/11.5	0.380/0.158	1.33/4.22	7.67/**6.72**	15	1.86	6.52
Pancreas	PCN[33]	0.571/1.63	0.869/2.02	0.895/0.710	0.526/1.92	3.44/3.99	66	1.12	5.31
	ECG[18]	1.11/2.51	1.10/2.08	0.843/0.700	0.883/1.99	3.53/4.02	48	1.01	4.7
Train: 245	VRCnet[19]	0.100/**0.138**	2.40/3.79	0.614/**0.507**	2.13/3.17	4.75/5.25	**18**	**0.904**	**3.49**
Test: 28	SFN[30]	0.329/0.557	0.95/**1.85**	0.880/0.736	0.764/**1.83**	3.34/**3.61**	55	0.955	5.16
	PointAttN[28]	0.407/0.837	1.07/2.55	0.849/0.629	0.897/2.31	3.52/3.65	96	1.05	5.55

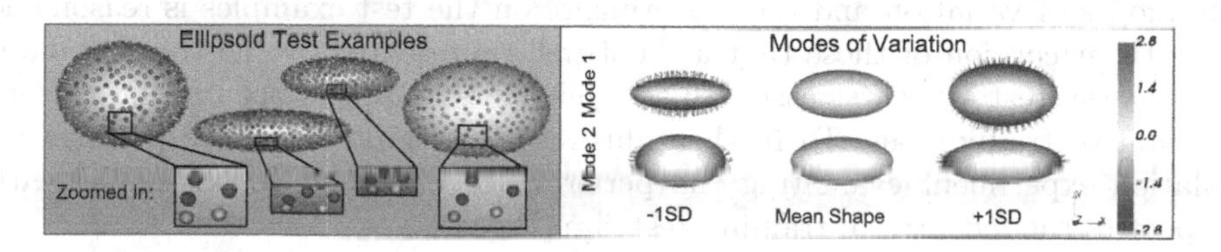

Fig. 2. Ellipsoid PDM from PCN [33] Left: Example test predictions on ground truth mesh with color-denoting point correspondence. Points are reasonably uniformly spread with good correspondence, shown via zoomed-in boxes. Right: The primary and secondary modes (training and testing combined) with ±1 standard deviation from the mean shape. Color and vectors denote the difference from the mean. The shape model correctly characterizes the x and y-radius as the only source of variation.

group differences. Table 1 shows the predicted coarse particles are close to the surfaces (P2F distance of 0.1mm), and the PCN [33] model performs best in this regard. The PCN [33] predictions were used to analyze the difference between the normal and CAM pathology mean shapes. Figure 3 shows the pathology is correctly characterized, and the *Linear Discrimination of Variation* (LDA) plot shows the difference in normal and CAM distributions captured by the PDM.

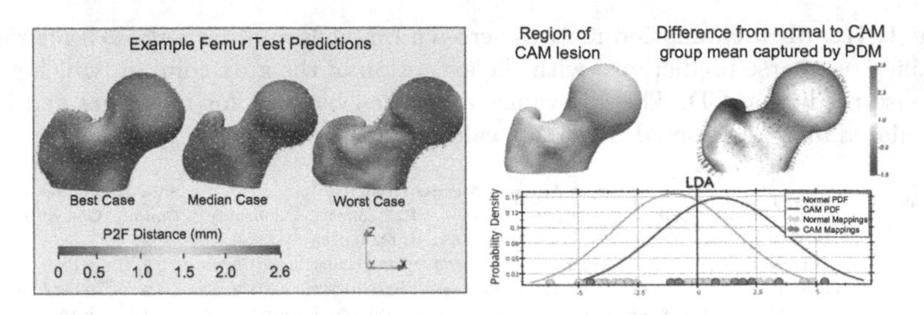

Fig. 3. Femur PDM from PCN [33] Left: Test examples with P2F distance displayed as a heatmap. Right: The expected region of CAM lesion correlates with the difference found from the normal to CAM group means.

Left Atrium. The left atrium dataset comprises of 1096 shapes segmented from cardiac LGE MRI images from unique atrial fibrillation patients. This cohort contains significant morphological variation in overall size, the size of the left atrium appendage, and the number and arrangement of the pulmonary veins. This variation is reflected in the large compactness values in Table 1. Despite this variation, the models achieve reasonable CD and P2F scores due to the large training size. Figure 4 shows the PDM predicted by SFN [30], which performed best. From the prediction examples, we can see the model represents the training data very well, even in the worst case, which has an extremely enlarged left atrium appendage. The size of the appendage is appropriately captured in the modes of variation and the performance on the test examples is reasonable with the exception of those that are not well represented by the training data, such as the test set worst case. This highlights the importance of a large, representative training set. To further illustrate this importance, we perform an ablation experiment evaluating the performance of the PCN [33] model with respect to the left atrium training test size (Fig. 4).

Pancreas. We utilize the pancreas dataset [23] to analyze the impact of incomplete input point clouds as the point completion networks were designed to address. Cases of incomplete observations frequently arise in clinical research. For example, in the analysis of bones where some are clipped due to scan size or in cases where 3D shape is interpreted from stacked or orthogonal 2D observations. Traditional methods of SSM generation are unable to handle such cases, but the point cloud learning approach has the potential to. Figure 5 shows how the test set error increase as the percentage of missing points increases. The SFN [30] model provides the best results given partial input.

Spleen. The spleen dataset [23] is included to provide an example of a small dataset with a large amount of variation to stress test the point completion models. Table 1 shows the models perform the worst on this dataset with regards to CD, Fscore, P2F, and Uniformity. This example illustrates the limitations of this approach to SSM generation.

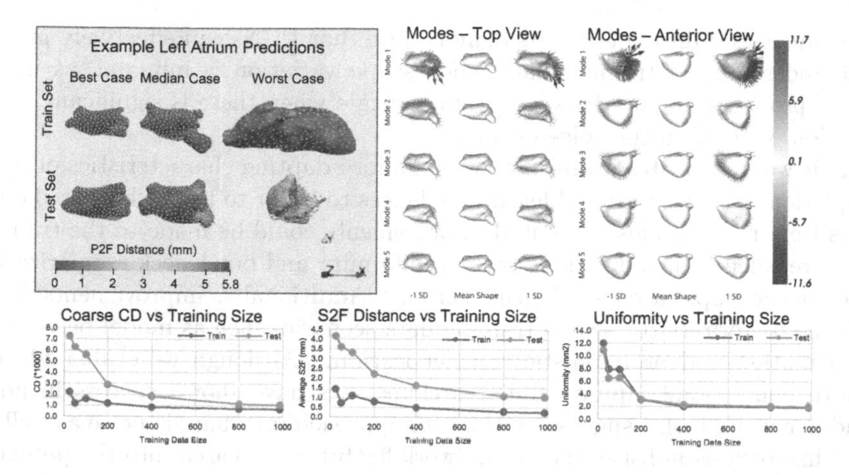

Fig. 4. Top: Left Atrium PDM from SFN [30] Left: Train and test examples shown from top view with P2F distance. Right: The first 5 modes of variation from top and anterior view. Bottom: Left Atrium PCN [33] results vs Training Size

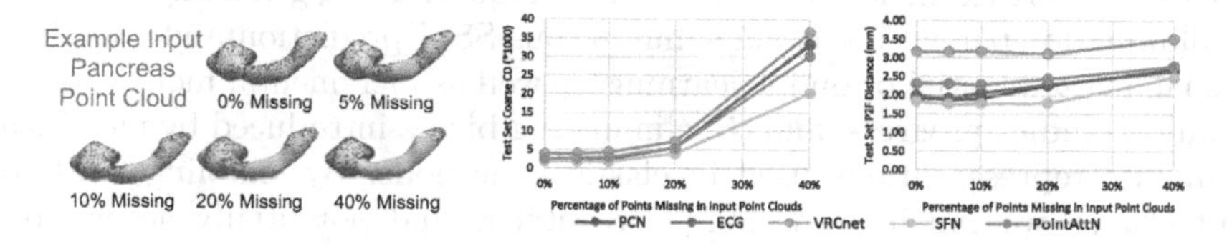

Fig. 5. Effect of Partial Input on Pancreas Prediction Accuracy

5 Discussion and Conclusion

Our experiments demonstrate that point cloud networks can learn accurate SSM of anatomy when provided with a sufficiently large and representative training dataset. The transformer-based SFN [30] architecture provided the best overall results among the models we explored. SFN [30] utilizes k-Nearest Neighbors (kNN) to capture geometric relations in the point cloud, while PointAttN [28] does not. PointAttN [28] has been shown to provide better point completion of complex man-made objects where kNN information could be misleading. However, in the case of anatomical SSM, it is likely that the kNN information assisted SFN [30] performance by providing accurate spatial information, given the more convex shape of organs and bones. Interestingly, the simplest model, PCN [33], achieved similar, and sometimes better, SSM accuracy than more current state-of-the-art methods, despite its inferior performance in point completion benchmarks [18,19,28,30]. This may be attributed to point completion benchmarks involving multiple object class point completion, which is a more challenging task. Another significant difference between our experiments and point completion benchmarks is the PCN datasets have tens of thousands of examples [19,32,33], while we worked with limited training data – the typical scenario in shape analysis. Our experiments demonstrate that PCNs can effectively predict SSM under limited training data when shape variation is minimal, as in the case of proximal femurs. However, they struggle when there is significant shape variation, such as in the spleen cohort.

This work indicates promising potential for adapting characteristics of point completion architectures and learning schemes to tailor to the task of predicting SSMs from point clouds. Potential improvements could be made to the training objective, such as penalization for non-uniformity and bottleneck regularization for compact population-statistical learning. Additionally, improvements could be made to address the scarce training data scenario, such as model-based data augmentation and probabilistic transfer learning. Although we evaluated only smooth point cloud inputs, similar architectures have shown success in point cloud denoising tasks, suggesting that our approach may handle noise as well [4, 21]. This work establishes the groundwork for future research into the potential of point cloud deep learning for SSM, offering significant benefits over traditional SSM generation, including: (1) reducing the input burden from complete, noise-free shape representations to point clouds, which significantly expands potential use cases, (2) providing fast inference and scalable training given any cohort size, (3) allowing for partial input via simultaneous SSM prediction and completion, (4) enabling sequential or online learning, as well as incremental model updating as clinical studies progress, and (5) eliminating biases introduced by metrics and parametric representations used in classical methods. By enabling SSM from point clouds, we can increase SSM accessibility and potentially accelerate its adoption as a widespread clinical tool.

Acknowledgements. This work was supported by the National Institutes of Health under grant numbers NIBIB-U24EB029011, NIAMS-R01AR076120, NHLBI-R01HL135568, and NIBIB-R01EB016701. The content is solely the responsibility of the authors and does not necessarily represent the official views of the National Institutes of Health. The authors would like to thank the University of Utah Division of Cardiovascular Medicine for providing left atrium MRI scans and segmentations from the Atrial Fibrillation projects as well as the Orthopaedic Research Laboratory (Andrew Anderson, PhD) at the University of Utah for providing femur CT scans and corresponding segmentations.

References

1. Adams, J., Bhalodia, R., Elhabian, S.: Uncertain-DeepSSM: from images to probabilistic shape models. In: Shape in Medical Imaging: International Workshop, ShapeMI 2020, Held in Conjunction with MICCAI 2020, Lima, Peru, October 4, 2020, Proceedings 12474, pp. 57–72 (2020)
2. Adams, J., Elhabian, S.: From images to probabilistic anatomical shapes: a deep variational bottleneck approach. In: Medical Image Computing and Computer Assisted Intervention-MICCAI 2022: 25th International Conference, Singapore, September 18–22, 2022, Proceedings, Part II. pp. 474–484. Springer (2022). https://doi.org/10.1007/978-3-031-16434-7_46

3. Akagic, A., Krivić, S., Dizdar, H., Velagić, J.: Computer vision with 3d point cloud data: methods, datasets and challenges. In: 2022 XXVIII International Conference on Information, Communication and Automation Technologies (ICAT), pp. 1–8. IEEE (2022)
4. Alliegro, A., Valsesia, D., Fracastoro, G., Magli, E., Tommasi, T.: Denoise and contrast for category agnostic shape completion. In: Proceedings of the IEEE/CVF Conference on Computer Vision and Pattern Recognition, pp. 4629–4638 (2021)
5. Atkins, P.R., et al.: Quantitative comparison of cortical bone thickness using correspondence-based shape modeling in patients with cam femoroacetabular impingement. J. Orthop. Res. **35**(8), 1743–1753 (2017)
6. Besl, P.J., McKay, N.D.: Method for registration of 3-D shapes. In: Sensor Fusion IV: Control Paradigms and Data Structures, vol. 1611, pp. 586–606. SPIE (1992)
7. Bhalodia, R., Elhabian, S.Y., Kavan, L., Whitaker, R.T.: DeepSSM: a deep learning framework for statistical shape modeling from raw images. In: Reuter, M., Wachinger, C., Lombaert, H., Paniagua, B., Lüthi, M., Egger, B. (eds.) ShapeMI 2018. LNCS, vol. 11167, pp. 244–257. Springer, Cham (2018). https://doi.org/10.1007/978-3-030-04747-4_23
8. Bischoff, J.E., Dai, Y., Goodlett, C., Davis, B., Bandi, M.: Incorporating population-level variability in orthopedic biomechanical analysis: a review. J. Biomech. Eng. **136**(2), 021004 (2014)
9. Bruse, J.L., et al.: A statistical shape modelling framework to extract 3D shape biomarkers from medical imaging data: assessing arch morphology of repaired coarctation of the aorta. BMC Med. Imaging **16**, 1–19 (2016)
10. Carriere, N., et al.: Apathy in parkinson's disease is associated with nucleus accumbens atrophy: a magnetic resonance imaging shape analysis. Mov. Disord. **29**(7), 897–903 (2014)
11. Cates, J., Elhabian, S., Whitaker, R.: ShapeWorks: particle-based shape correspondence and visualization software. In: Statistical Shape and Deformation Analysis, pp. 257–298. Elsevier (2017)
12. Cates, J., Fletcher, P.T., Styner, M., Shenton, M., Whitaker, R.: Shape modeling and analysis with entropy-based particle systems. In: Karssemeijer, N., Lelieveldt, B. (eds.) IPMI 2007. LNCS, vol. 4584, pp. 333–345. Springer, Heidelberg (2007). https://doi.org/10.1007/978-3-540-73273-0_28
13. Davies, R.H., Twining, C.J., Cootes, T.F., Waterton, J.C., Taylor, C.J.: A minimum description length approach to statistical shape modeling. IEEE Trans. Med. Imag. **21**(5), 525–537 (2002)
14. Fei, B., et al.: Comprehensive review of deep learning-based 3D point cloud completion processing and analysis. IEEE Trans. Intell. Transp. Syst. (2022)
15. Le, T., Duan, Y.: PointGrid: a deep network for 3D shape understanding. In: Proceedings of the IEEE Conference on Computer Vision and Pattern Recognition, pp. 9204–9214 (2018)
16. Merle, C., et al.: High variability of acetabular offset in primary hip osteoarthritis influences acetabular reaming-a computed tomography-based anatomic study. J. Arthroplasty **34**(8), 1808–1814 (2019)
17. Ovsjanikov, M., Ben-Chen, M., Solomon, J., Butscher, A., Guibas, L.: Functional maps: a flexible representation of maps between shapes. ACM Trans. Graph. (ToG) **31**(4), 1–11 (2012)
18. Pan, L.: ECG: edge-aware point cloud completion with graph convolution. IEEE Robot. Autom. Lett. **5**(3), 4392–4398 (2020)

19. Pan, L., et al.: Variational relational point completion network. In: Proceedings of the IEEE/CVF Conference on Computer Vision and Pattern Recognition, pp. 8524–8533 (2021)
20. Qi, C.R., Su, H., Mo, K., Guibas, L.J.: PointNet: deep learning on point sets for 3D classification and segmentation. In: Proceedings of the IEEE Conference on Computer Vision and Pattern Recognition, pp. 652–660 (2017)
21. Sahin, C.: CMD-Net: self-supervised category-level 3D shape denoising through canonicalization. Appl. Sci. **12**(20), 10474 (2022)
22. Sarkalkan, N., Weinans, H., Zadpoor, A.A.: Statistical shape and appearance models of bones. Bone **60**, 129–140 (2014)
23. Simpson, A.L., et al.: A large annotated medical image dataset for the development and evaluation of segmentation algorithms. arXiv preprint arXiv:1902.09063 (2019)
24. Styner, M., et al.: Framework for the statistical shape analysis of brain structures using SPHARM-PDM. Insight J., 242 (2006)
25. Tatarchenko, M., Richter, S.R., Ranftl, R., Li, Z., Koltun, V., Brox, T.: What do single-view 3D reconstruction networks learn? In: Proceedings of the IEEE/CVF Conference on Computer Vision and Pattern Recognition, pp. 3405–3414 (2019)
26. Timmins, L.H., Samady, H., Oshinski, J.N.: Effect of regional analysis methods on assessing the association between wall shear stress and coronary artery disease progression in the clinical setting. In: Biomechanics of Coronary Atherosclerotic Plaque, pp. 203–223. Elsevier (2021)
27. Treleaven, P., Wells, J.: 3D body scanning and healthcare applications. Computer **40**(7), 28–34 (2007). https://doi.org/10.1109/MC.2007.225
28. Wang, J., Cui, Y., Guo, D., Li, J., Liu, Q., Shen, C.: PointAttN: you only need attention for point cloud completion. arXiv preprint arXiv:2203.08485 (2022)
29. Wang, Y., Sun, Y., Liu, Z., Sarma, S.E., Bronstein, M.M., Solomon, J.M.: Dynamic graph CNN for learning on point clouds. ACM Trans. Graph. (tog) **38**(5), 1–12 (2019)
30. Xiang, P., et al.: SnowflakeNet: point cloud completion by snowflake point deconvolution with skip-transformer. In: Proceedings of the IEEE/CVF International Conference on Computer Vision, pp. 5499–5509 (2021)
31. Yang, Y., Feng, C., Shen, Y., Tian, D.: FoldingNet: point cloud auto-encoder via deep grid deformation. In: Proceedings of the IEEE Conference on Computer Vision and Pattern Recognition (CVPR) (2018)
32. Yu, X., Rao, Y., Wang, Z., Liu, Z., Lu, J., Zhou, J.: PoinTr: diverse point cloud completion with geometry-aware transformers. In: Proceedings of the IEEE/CVF International Conference on Computer Vision, pp. 12498–12507 (2021)
33. Yuan, W., Khot, T., Held, D., Mertz, C., Hebert, M.: PCN: point completion network. In: 2018 International Conference on 3D Vision (3DV), pp. 728–737. IEEE (2018)

CT-Guided, Unsupervised Super-Resolution Reconstruction of Single 3D Magnetic Resonance Image

Jiale Wang[1], Alexander F. Heimann[2], Moritz Tannast[2], and Guoyan Zheng[1](✉)

[1] Institute of Medical Robotics, School of Biomedical Engineering, Shanghai Jiao Tong University, No. 800, Dongchuan Road, Shanghai 200240, China
guoyan.zheng@sjtu.edu.cn

[2] Department of Orthopaedic Surgery, HFR Cantonal Hospital, University of Fribourg, Fribourg, Switzerland

Abstract. Deep learning-based algorithms for single MR image (MRI) super-resolution have shown great potential in enhancing the resolution of low-quality images. However, many of these methods rely on supervised training with paired low-resolution (LR) and high-resolution (HR) MR images, which can be difficult to obtain in clinical settings. This is because acquiring HR MR images in clinical settings requires a significant amount of time. In contrast, HR CT images are acquired in clinical routine. In this paper, we propose a CT-guided, unsupervised MRI super-resolution reconstruction method based on joint cross-modality image translation and super-resolution reconstruction, eliminating the requirement of high-resolution MRI for training. The proposed approach is validated on two datasets respectively acquired from two different clinical sites. Well-established metrics including Peak Signal-to-Noise Ratio (PSNR), Structural Similarity Index Metrics (SSIM), and Learned Perceptual Image Patch Similarity (LPIPS) are used to assess the performance of the proposed method. Our method achieved an average PSNR of 32.23, an average SSIM of 0.90 and an average LPIPS of 0.14 when evaluated on data of the first site. An average PSNR of 30.58, an average SSIM of 0.88, and an average LPIPS of 0.10 were achieved by our method when evaluated on data of the second site.

Keywords: Unsupervised image super-resolution · Cross-modality image translation · CT-Guided · Magnetic resonance imaging

1 Introduction

High-resolution magnetic resonance (MR) images (MRI) provide a wealth of structural details, which facilitate early and precise diagnosis [1]. However, images obtained in clinical practice are anisotropic due to the limitation of scan

This study was partially supported by the National Natural Science Foundation of China via project U20A20199.

H. Greenspan et al. (Eds.): MICCAI 2023, LNCS 14220, pp. 497–507, 2023.
https://doi.org/10.1007/978-3-031-43907-0_48

time and signal-noise ratio [2]. In order to speed up clinical scanning procedures, only a limited number of two-dimensional (2D) slices are acquired, despite the fact that the interested anatomical structures are in three-dimensional (3D). The acquired medical images have low inter-plane resolution, i.e., large spacing between slices. Such anisotropic images will lead to misdiagnosis and can greatly impact the performance of various clinical tasks, including computer-aided diagnosis and computer-assisted interventions. Therefore, we investigate the problem of reducing the slice spacing [3] via super-resolution (SR) reconstruction. Specifically, we refer to the image with large slice spacing as a low-resolution (LR) image and the image with small slice spacing as a high-resolution (HR) image. Our goal is to reconstruct the HR image from the LR input, which is an ill-posed inverse problem and presents significant challenges.

Deep learning-based algorithms for single MR image super-resolution show great potential in restoration of HR images from LR inputs [4]. Pham et al. [5] proposed the SRCNN method, which applied convolutional neural networks (CNN) to image super-resolution of MRI and achieved a better performance than the conventional methods, such as B-spline interpolation and low-rank total variation (LRTV) [6] method. Chaudhariet al. [7] proposed a 3D residual network, which learned the residual-based transformations between paired LR and HR images for the SR reconstruction of MRI. Chen et al. [8] proposed a densely connected super-resolution network (DCSRN), which reused the block features through the dense connection in the SR reconstruction of MRI. Chen et al. [9] extended this work by using generative adversarial network (GAN) [10] in SR reconstruction of MRI in order to improve the realism of the recovered images. Feng et al. [11,12] proposed a multi-contrast MRI SR method, which aimed to learn clearer anatomical structure and edge information with the help of auxiliary contrast MRI. Despite significant progress, however, there are still spaces for further improvement. Most networks require a large amount of paired LR and HR MR images for training, which are unrealistic in clinical practice. To address the challenge of organizing paired images, methods based on unpaired images have been proposed [13,14]. However, HR MR images are still difficult to obtain, as acquiring HR MR images in clinical settings requires a significant amount of time. In contrast, CT images are acquired in clinical routine. Therefore, it is of great significance to use HR CT images as a guidance to synthesize HR MR images from LR MR images.

To this end, we propose a CT-guided, unsupervised MRI super-resolution reconstruction method based on joint cross-modality image translation (CIT) and super-resolution reconstruction, eliminating the requirement of HR MR images for training. Specifically, our network design features a super-resolution Network (SRNet) and a cross-modality image translation network (CITNet) based on disentanged representation learning. After pretraining, the SRNet can generate pseudo HR MR images from LR MR images. The generated pseudo HR MR images are then taken together with the HR CT images as the input to the CITNet, which can generate quality-improved pseudo HR MR images by combining disentangled content code of the input CT data with the attribute

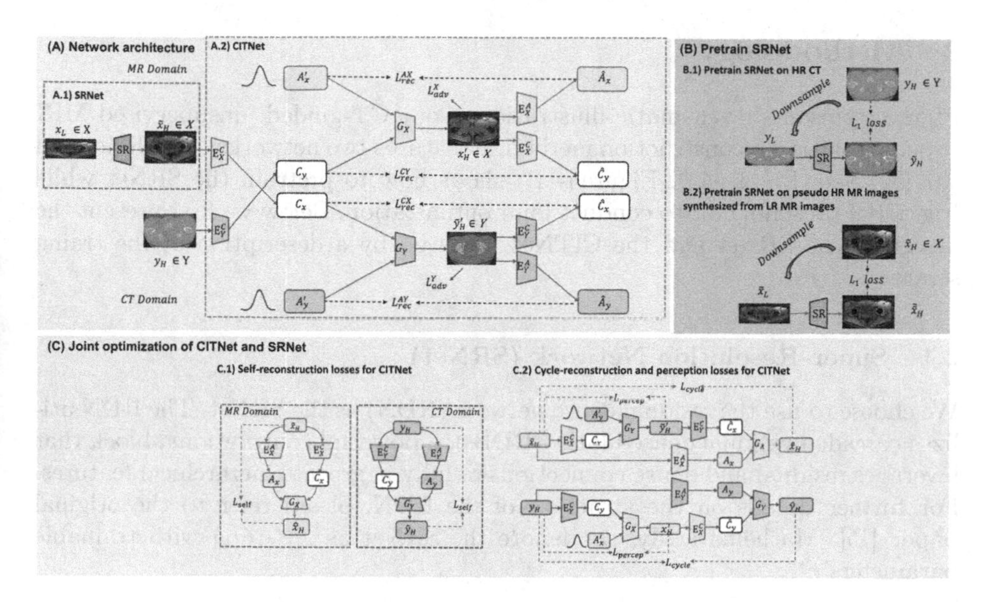

Fig. 1. A schematic illustration of our CT-guided, unsupervised MRI super-resolution reconstruction method. (A) Network architecture, including a SRNet and a CITNet; (B) Pretraining the SRNet; and (C) Joint optimization of the CITNet and the SRNet. Different colors represent different domains, i.e., orange represents the MR domain, green represents the CT domain, and white shows the shared content space. (Color figure online)

code of the input pseudo HR MR images. Joint optimization of the CITNet and the SRNet leads to better and better pseudo HR MR image generation. When converged, we can use the SRNet to generate high-quality pseudo HR MR images from given LR MR images. The contributions of our work can be summarized as follows:

- We propose a CT-guided, unsupervised MRI super-resolution reconstruction method based on joint cross-modality image translation and super-resolution reconstruction, eliminating the requirement of HR MRI for training. Our cross-modality image translation is based on disentangled representation leanring.
- Our network design features a SRNet and a CITNet. They work jointly to generate high-quality pseudo HR MR images from given LR MR images. Concretely, a better trained SRNet will help to generate a better input to the CITNet. On the other hand, the CITNet, taking the SRNet-generated pseudo HR MR images and the HR CT images as input, provides better supervision of the SRNet training. Joint optimization of the CITNet and the SRNet leads to the generation of high-quality pseudo HR MR image at the end.
- We validate the proposed method on two datasets collected from two different clinical centers.

2 Methodology

Figure 1 presents a schematic illustration of our CT-guided, unsupervised MRI super-resolution reconstruction method. It features two networks: the SRNet and the CITNet (Fig. 1-(A)). Figure 1-(B) shows how to pretrain the SRNet while Fig. 1-(C) presents how to conduct joint optimization. Below we first present the design of the SRNet and the CITNet, followed by a description of the traing strategy.

2.1 Super-Resolution Network (SRNet)

We choose to use the residual dense network (RDN) as the SRNet. The RDN utilizes cascaded residual dense blocks (RDBs), a powerful convolutional block that leverages residual and dense connections to fully aggregate hierarchical features. For further details on the structure of the RDN, please refer to the original paper [15]. Mathematically, we denote the SRNet as $\mathcal{F}_s(\cdot;\Theta_s)$ with trainable parameters Θ_s.

2.2 Cross-Modality Image Translation Network (CITNet)

The CITNet is inspired by MUNIT [16]. As depicted in Fig. 1-(A.2), it comprises two content encoders $\{E^{\mathcal{C}}_{\mathcal{X}}, E^{\mathcal{C}}_{\mathcal{Y}}\}$, two attribute encoders $\{E^{\mathcal{A}}_{\mathcal{X}}, E^{\mathcal{A}}_{\mathcal{Y}}\}$, and two generators $\{G_{\mathcal{X}}, G_{\mathcal{Y}}\}$. The encoder in each domain disentangles an input image separately into a domain-invariant content space $\mathcal{C}$ and a domain-specific attribute space $\mathcal{A}$. And the generator networks combine a content code with an attribute code to generate translated images in the target domain. For instance, when translating CT image $y_H \in \mathcal{Y}$ to MR image $x'_H \in \mathcal{X}$, we first randomly sample from the prior distribution $p(\mathcal{A}'_x) \sim \mathcal{N}(0, \mathbf{I})$ to obtain an MRI attribute code $\mathcal{A}'_x$, which is empirically set as a 8-bit vector. We then combine $\mathcal{A}'_x$ with the disentangled content code of the CT image $\mathcal{C}_y = E^{\mathcal{C}}_{\mathcal{Y}}(y_H)$ to generate the translated MRI image $x'_H \in \mathcal{X}$ through the generator $G_{\mathcal{X}}$. Similarly, we can get the the translated CT image $\tilde{y}'_H \in \mathcal{Y}$ through the generator $G_{\mathcal{Y}}(\mathcal{C}_x, \mathcal{A}'_y)$, where $\mathcal{C}_x = E^{\mathcal{C}}_{\mathcal{X}}(\mathcal{F}_s(x_L;\Theta_s))$ and $\mathcal{A}'_y$ is also sampled from the prior distribution $p(\mathcal{A}'_y) \sim \mathcal{N}(0, \mathbf{I})$.

Disentangled Representation Learning. Cross-modality image translation is based on disentangled representation learning, trained with self- and cross-cycle reconstruction losses. As shown in Fig. 1-(C.1, C.2), the self-reconstruction loss L_{self} is utilized to regularize the training when the content and attribute code originate from the same domain, whereas the cross-cycle consistency loss L_{cycle} is used when the content and attribute code come from different domains. The self-reconstruction and cross-cycle reconstruction losses are defined as follows:

$$L_{\text{self}} = \left\|G_{\mathcal{X}}\left(E^{\mathcal{C}}_{\mathcal{X}}(\tilde{x}_H), E^{\mathcal{A}}_{\mathcal{X}}(\tilde{x}_H)\right) - \tilde{x}_H\right\|_1 + \left\|G_{\mathcal{Y}}\left(E^{\mathcal{C}}_{\mathcal{Y}}(y_H), E^{\mathcal{A}}_{\mathcal{Y}}(y_H)\right) - y_H\right\|_1 \quad (1)$$

$$L_{\text{cycle}} = \|G_{\mathcal{X}}(E^{\mathcal{C}}_{\mathcal{Y}}(\tilde{y}'_H), E^{\mathcal{A}}_{\mathcal{X}}(\tilde{x}_H)) - \tilde{x}_H\|_1 + \|G_{\mathcal{Y}}(E^{\mathcal{C}}_{\mathcal{X}}(x'_H), E^{\mathcal{A}}_{\mathcal{Y}}(y_H)) - y_H\|_1 \quad (2)$$

where $\tilde{x}_H = \mathcal{F}_s(x_L; \Theta_s)$, $x'_H = G_{\mathcal{X}}(E^{\mathcal{C}}_{\mathcal{Y}}(y_H), \mathcal{A}'_x)$, $\tilde{y}'_H = G_{\mathcal{Y}}(E^{\mathcal{C}}_{\mathcal{X}}(\tilde{x}_H), \mathcal{A}'_y)$. Specially, in the cross-cycle translation processes, we employe a latent reconstruction loss to maintain the invertible mapping between the image and the latent space. In details, we have:

$$L_{\text{latent}} = \|\hat{\mathcal{C}}_x - \mathcal{C}_x\|_1 + \|\hat{\mathcal{C}}_y - \mathcal{C}_y\|_1 + \|\hat{\mathcal{A}}_x - \mathcal{A}'_x\|_1 + \|\hat{\mathcal{A}}_y - \mathcal{A}'_y\|_1 \quad (3)$$

We further use pretrained vgg16 network, denoted as $\phi(\cdot)$, to extract high-level features for computing the perceptual loss [17]:

$$L_{\text{percep}} = \frac{1}{CHW}\left\|\phi(\tilde{y}'_H) - \phi(\tilde{x}_H)\right\|_2^2 + \frac{1}{CHW}\left\|\phi(x'_H) - \phi(y_H)\right\|_2^2 \quad (4)$$

where C, H, W indicate the channel number and the image size, respectively.

Adversarial Learning. As shown in Fig. 1-(A.2), we use GAN [10] to learn the translation between MR and CT image domains better. A GAN typically contains a generation network and a discrimination network. We use the discriminator $D_{\mathcal{X}}$ to judge whether the image is from MR image domain, and the discriminator $D_{\mathcal{Y}}$ to judge whether the image is from CT image domain. The auto-encoders try to generate the image of the target domain to fool the discriminators so that the distribution of the translated images can match that of the target images. The minmax game is trained by:

$$L^{\mathcal{X}}_{adv} = \mathbb{E}_{\tilde{x}_H \sim P_{\mathcal{X}}(\tilde{x}_H)}\left[\log D_{\mathcal{X}}(\tilde{x}_H)\right] + \mathbb{E}_{y_H \sim P_{\mathcal{Y}}(y_H)}\left[\log(1 - D_{\mathcal{X}}(x'_H))\right] \quad (5)$$

$$L^{\mathcal{Y}}_{adv} = \mathbb{E}_{y_H \sim P_{\mathcal{Y}}(y_H)}\left[\log D_{\mathcal{Y}}(y_H)\right] + \mathbb{E}_{\tilde{x}_H \sim P_{\mathcal{X}}(\tilde{x}_H)}\left[\log(1 - D_{\mathcal{Y}}(\tilde{y}'_H))\right] \quad (6)$$

Joint Optimization. The SRNet and the CITNet are jointly optimized by minimizing following loss function:

$$L_{disentangle} = (L^{\mathcal{X}}_{adv} + L^{\mathcal{Y}}_{adv}) + \lambda_1(L_{self} + L_{cycle}) + \lambda_2 L_{latent} + \lambda_3 L_{percep} \quad (7)$$

where λ_1, λ_2, and λ_3 are parameters controlling the relative weights of different losses.

2.3 Training Strategy

Empirically, we found that training the network shown in Fig. 1-(A) end to end did not converge. We thus design the following three-stage training strategy.

Stage 1. Let's denote the downsampling function as $\mathcal{D}(\cdot)$. In this stage, we pretrain the SRNet using the HR CT images, as shown in Fig. 1-(B.1), for T iterations. At each iteration, we sample a batch of HR CT images. We then downsample the sampled HR CT images y_H to get the paired LR CT images $y_L = \mathcal{D}(y_H)$. The SRNet is trained with the paired LR-HR CT images by minimizing $L1$ loss $\|y_H - \mathcal{F}_s(\mathcal{D}(y_H); \Theta_s)\|_1$. In this stage, we are aiming to train the SRNet to learn the upsampling kernels.

Stage 2. As the SRNet is only pretrained with CT images in stage 1, we need to generalize the learned upsampling kernels to the MR image domain. We thus

Algorithm 1. Training procedure

(Stage1) Pretrain SRNet with CT based self-supervision:
  GET HR CT images y_H
  FOR $t = 1$ to T
    Train SRNet by minimizing $\|y_H - \mathcal{F}_s(\mathcal{D}(y_H); \Theta_s)\|_1$
  END FOR
(Stage2) Pretrain SRNet with pseudo MR based self-supervision:
  GET LR MR images x_L
  FOR $t = T$ to $2T$
    Train SRNet by minimizing $\|\mathcal{F}_s(x_L; \Theta_s) - \mathcal{F}_s(\mathcal{D}(\mathcal{F}_s(x_L; \Theta_s)); \Theta_s)\|_1$
  END FOR
(Stage3) Joint optimization of CITNet and SRNet:
  GET unpaired LR MR images x_L and HR CT images y_H
  FOR $t = 2T$ to $10T$
    Train $D_{\mathcal{X}}$, $D_{\mathcal{Y}}$ by maximizing $\left(L^{\mathcal{X}}_{adv} + L^{\mathcal{Y}}_{adv}\right)$
    Train $E^{\mathcal{C}}_{\mathcal{X}}$, $E^{\mathcal{C}}_{\mathcal{Y}}$, $E^{\mathcal{A}}_{\mathcal{X}}$, $E^{\mathcal{A}}_{\mathcal{Y}}$, $G_{\mathcal{X}}$, $G_{\mathcal{Y}}$ and SRNet by minimizing $L_{disentangle}$
  END FOR

further pretrain the SRNet with pseudo MR images, as shown in Fig. 1-(B.2), for another T iterations. At each iteration, we first sample a batch of LR MR images x_L and input them into the SRNet to get the pseudo HR MR images $\tilde{x}_H = \mathcal{F}_s(x_L; \Theta_s)$. We then downsample $\tilde{x}_H$ to get corresponding pseudo LR MR images $\tilde{x}_L = \mathcal{D}(\tilde{x}_H)$. The SRNet is trained with the paired pseudo LR-HR MR images by minimizing $L1$ loss $\|\mathcal{F}_s(x_L; \Theta_s) - \mathcal{F}_s(\mathcal{D}(\mathcal{F}_s(x_L; \Theta_s)); \Theta_s)\|_1$. The idea behind such a pretraining stategy is that since both CT and MR images share the common structural information, the model pretrained with CT images in stage 1 facilitates the super-resolution reconstruction of pseudo HR MR images in stage 2. On the other hand, the training done in stage 2 can help the SRNet to learn MRI-specific domain information.

Stage 3. The MR images generated by the model pretrained at the first two stages can be further improved. In stage 3, we conduct joint optimization of the SRNet and the CITNet as shown in Fig. 1-(C), for another $8 \times T$ iterations. At each iteration, we first train $D_{\mathcal{X}}$, $D_{\mathcal{Y}}$ by maximizing $\left(L^{\mathcal{X}}_{adv} + L^{\mathcal{Y}}_{adv}\right)$. We then train $E^{\mathcal{C}}_{\mathcal{X}}$, $E^{\mathcal{C}}_{\mathcal{Y}}$, $E^{\mathcal{A}}_{\mathcal{X}}$, $E^{\mathcal{A}}_{\mathcal{Y}}$, $G_{\mathcal{X}}$, $G_{\mathcal{Y}}$ and the SRNet by minimizing $L_{disentangle}$ as defined in Eq. (7).

The training procedure of our method is illustrated by Algorithm 1.

Implementation Details. To train the proposed network, each training sample is unpaired LR MRI and HR CT images. All images are normalized to the range between -1.0 and 1.0. Optimization is performed using Adam with a batch size of 1. The initial learning rate is set to 0.0001 and decreased by a factor of 5 every 2 epochs. We empirically set $\lambda_1 = 10$, $\lambda_2 = \lambda_3 = 1$ and $T = 100,000$.

Table 1. The mean and the standard deviation when the proposed method was compared with the state-of-the-art (SOTA) unsupervised [18–20] and supervised [15,21] methods on both datasets. Paired T-Tests of all evaluation metrics achieved by ours and other methods are all smaller than 0.0001.

Dataset	Site1			Site2		
Method	PSNR↑	SSIM↑	LPIPS↓	PSNR↑	SSIM↑	LPIPS↓
Bicubic	30.67±1.96	0.86±0.03	0.32±0.05	28.69±1.92	0.83±0.03	0.26±0.02
TSCN [18]	29.00±2.03	0.83±0.05	0.24±0.03	28.00±1.71	0.85±0.02	0.14±0.01
ZSSR [19]	30.95±2.12	0.88±0.03	0.16±0.02	28.94±1.69	0.83±0.03	0.16±0.01
SMORE [20]	31.78±1.98	0.89±0.03	0.21±0.03	29.93±2.00	0.86±0.03	0.14±0.02
Ours	**32.23±1.98**	**0.90±0.02**	**0.14±0.02**	**30.58±1.97**	**0.88±0.02**	**0.10±0.01**
Supervised [15]	32.99±2.07	0.91±0.02	0.10±0.02	31.66±1.72	0.90±0.02	0.08±0.01
ReconResNet [21]	32.93±3.12	0.88±0.05	0.09±0.02	29.97±1.50	0.84±0.03	0.07±0.01

Table 2. Results of ablation study on dataset from Site1.

Stage 1	Stage 2	Stage 3	PSNR↑	SSIM↑	LPIPS↓
√	–	–	31.01±2.18	0.87±0.03	0.20±0.03
√	√	–	31.62±2.08	0.89±0.03	0.16±0.02
√	√	√	**32.23±1.98**	**0.90±0.02**	**0.14±0.02**

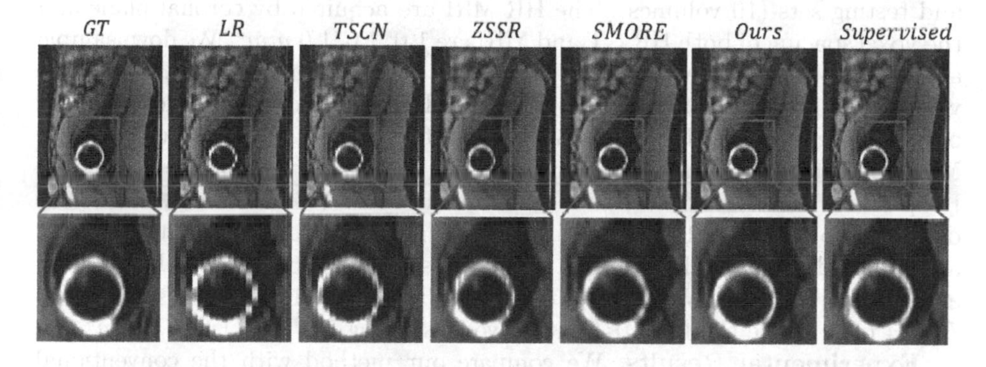

Fig. 2. Visual comparison of different methods when evaluated on dataset from Site1.

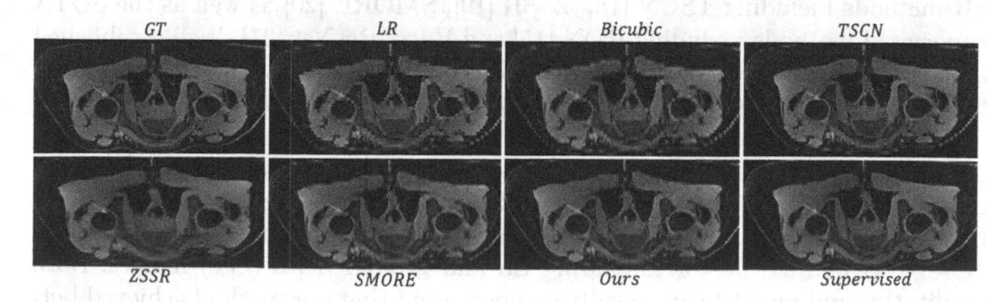

Fig. 3. Visual comparison of different methods when evaluated on dataset from Site2.

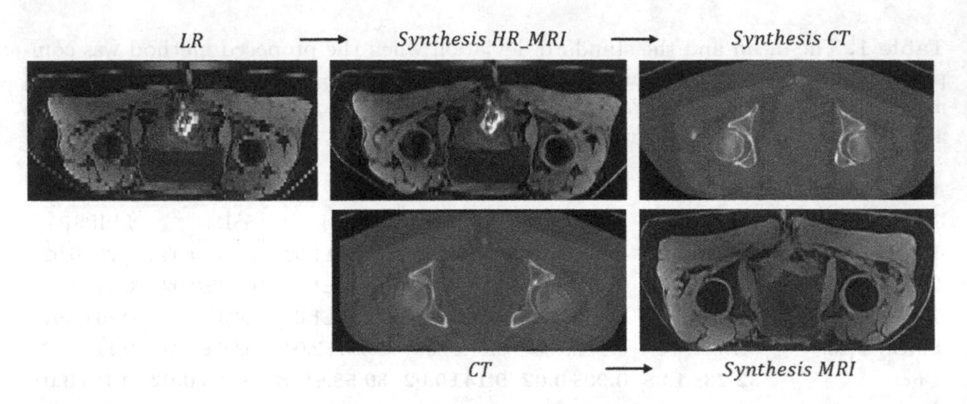

Fig. 4. Examples of cross-modality image translation between MRI and CT using data from Site2.

3 Experiments

Dataset. We conduct experiments to evaluate the proposed method on two datasets acquired from two different clinical centers. The dataset from HFR Cantonal Hospital, University of Fribourg (Site1) consists of 50 paired MR-CT volumes, which are divided into training (35 volumes), validation (5 volumes), and testing sets (10 volumes). The HR MRI are acquired by coronal plane and the voxel spacing of both HR CT and MRI are 1.0*1.0*1.0 mm^3. We downsample along the coronal axis with a scale factor $K = 4$ to generate the LR MRI with a voxel spacing 1.0*1.0*(1.0*K) mm^3. We shuffle the paired MR-CT volumes and only use the unpaired LR MRI and HR CT for training. Then we use the HR MRI to evaluate the reconstruction metrics. The dataset from the University Hospital of Bern (Site2) consists of 19 unpaired MR-CT volumes, which are divided into training (13 volumes) and testing sets (6 volumes). The HR MRI are acquired by coronal plane and the voxel spacing of both HR CT and MRI are 1.0*1.3*1.3 mm^3. We downsample along the coronal axis by a scale factor $K = 4$ to generate the LR MRI with a voxel spacing 1.0*1.3*(1.3*K) mm^3.

Experimental Results. We compare our method with the conventional algorithm bicubic interpolation, and the state-of-the-art (SOTA) unsupervised SR methods including TSCN [18], ZSSR [19], SMORE [20] as well as the SOTA supervised methods including RDN [15] and ReconResNet [21]. Well-established metrics including Peak Signal-to-Noise Ratio (PSNR) [22,23], Structural Similarity Index Metrics (SSIM) [24], and Learned Perceptual Image Patch Similarity (LPIPS) [25] are used to assess the performance of different methods.

Table 1 shows the mean and the standard deviation of the evaluation results of each method on both datasets. Figure 2 and Fig. 3 respectively show the super-resolution results on data from Site1 and Site2, when the scale factor is set as $K = 4$, as well as the corresponding LR and ground truth (GT) images. Both qualitative and quantitative results demonstrated that our method achieved better results than other SOTA unsupervised SR methods. It achieved comparable performance when compared with the supervised SR methods.

Our method is trained in two pretrain stages and one joint optimization stage. We thus conduct ablation study on dataset from Site1 to analyze the quality of the generated pseudo HR MR images at each stage. As shown in Table 2, quantitatively, the quality of the generated pseudo HR MR images is become better and better, demonstrating the effectiveness of the training strategy.

4 Conclusion

In this paper, we proposed a CT-guided, unsupervised MRI super-resolution reconstruction method based on joint cross-modality image translation and super-resolution reconstruction, eliminating the requirement of HR MRI for training. We conducted experiments on two datasets respectively acquired from two different clinical centers to validate the effectiveness of the proposed method. Quantitatively and qualitatively, the proposed method achieved superior performance over the SOTA unsupervised SR methods.

References

1. Jia, Y., Gholipour, A., He, Z., Warfield, S.K.: A new sparse representation framework for reconstruction of an isotropic high spatial resolution MR volume from orthogonal anisotropic resolution scans. IEEE Trans. Med. Imaging **36**(5), 1182–1193 (2017)
2. Plenge, E., et al.: Super-resolution methods in MRI: can they improve the trade-off between resolution, signal-to-noise ratio, and acquisition time? Magn. Reson. Med. **68**(6), 1983–1993 (2012)
3. Xuan, K., et al.: Reducing magnetic resonance image spacing by learning without ground-truth. Pattern Recogn. **120**, 108103 (2021)
4. Kim, J., Lee, J.K., Lee, K.M.: Accurate image super-resolution using very deep convolutional networks. In: Proceedings of the IEEE Conference on Computer Vision and Pattern Recognition, pp. 1646–1654 (2016)
5. Pham, C.H., Ducournau, A., Fablet, R., Rousseau, F.: Brain MRI super-resolution using deep 3D convolutional networks. In: 2017 IEEE 14th International Symposium on Biomedical Imaging (ISBI 2017), pp. 197–200. IEEE (2017)
6. Shi, F., Cheng, J., Wang, L., Yap, P.T., Shen, D.: LRTV: MR image super-resolution with low-rank and total variation regularizations. IEEE Trans. Med. Imaging **34**(12), 2459–2466 (2015)
7. Chaudhari, A.S., et al.: Super-resolution musculoskeletal MRI using deep learning. Magn. Reson. Med. **80**(5), 2139–2154 (2018)
8. Chen, Y., Xie, Y., Zhou, Z., Shi, F., Christodoulou, A.G., Li, D.: Brain MRI super resolution using 3d deep densely connected neural networks. In: 2018 IEEE 15th International Symposium on Biomedical Imaging (ISBI 2018), pp. 739–742. IEEE (2018)
9. Chen, Y., Shi, F., Christodoulou, A.G., Xie, Y., Zhou, Z., Li, D.: Efficient and accurate MRI super-resolution using a generative adversarial network and 3D multi-level densely connected network. In: Frangi, A.F., Schnabel, J.A., Davatzikos, C., Alberola-López, C., Fichtinger, G. (eds.) MICCAI 2018. LNCS, vol. 11070, pp. 91–99. Springer, Cham (2018). https://doi.org/10.1007/978-3-030-00928-1_11

10. Goodfellow, I., et al.: Generative adversarial nets. In: Advances in Neural Information Processing Systems 27 (2014)
11. Feng, C.M., Yan, Y., Yu, K., Xu, Y., Shao, L., Fu, H.: Exploring separable attention for multi-contrast MR image super-resolution. arXiv preprint arXiv:2109.01664 (2021)
12. Feng, C.-M., Fu, H., Yuan, S., Xu, Y.: Multi-contrast MRI super-resolution via a multi-stage integration network. In: de Bruijne, M., et al. (eds.) MICCAI 2021. LNCS, vol. 12906, pp. 140–149. Springer, Cham (2021). https://doi.org/10.1007/978-3-030-87231-1_14
13. You, C., et al.: CT super-resolution GAN constrained by the identical, residual, and cycle learning ensemble (GAN-circle). IEEE Trans. Med. Imaging **39**(1), 188–203 (2019)
14. Wang, J., Wang, R., Tao, R., Zheng, G.: UASSR: unsupervised arbitrary scale super-resolution reconstruction of single anisotropic 3D images via disentangled representation learning. In: Medical Image Computing and Computer Assisted Intervention-MICCAI 2022: 25th International Conference, Singapore, September 18–22, 2022, Proceedings, Part VI, pp. 453–462. Springer (2022). https://doi.org/10.1007/978-3-031-16446-0_43
15. Zhang, Y., Tian, Y., Kong, Y., Zhong, B., Fu, Y.: Residual dense network for image super-resolution. In: Proceedings of the IEEE Conference on Computer Vision and Pattern Recognition, pp. 2472–2481 (2018)
16. Huang, X., Liu, M.Y., Belongie, S., Kautz, J.: Multimodal unsupervised image-to-image translation. In: Proceedings of the European Conference on Computer Vision (ECCV), pp. 172–189 (2018)
17. Johnson, J., Alahi, A., Fei-Fei, L.: Perceptual losses for real-time style transfer and super-resolution. In: Leibe, B., Matas, J., Sebe, N., Welling, M. (eds.) ECCV 2016. LNCS, vol. 9906, pp. 694–711. Springer, Cham (2016). https://doi.org/10.1007/978-3-319-46475-6_43
18. Lu, Z., Li, Z., Wang, J., Shi, J., Shen, D.: Two-stage self-supervised cycle-consistency network for reconstruction of thin-slice MR images. In: de Bruijne, M., et al. (eds.) MICCAI 2021. LNCS, vol. 12906, pp. 3–12. Springer, Cham (2021). https://doi.org/10.1007/978-3-030-87231-1_1
19. Shocher, A., Cohen, N., Irani, M.: zero-shot super-resolution using deep internal learning. In: Proceedings of the IEEE Conference on Computer Vision and Pattern Recognition, pp. 3118–3126 (2018)
20. Zhao, C., Dewey, B.E., Pham, D.L., Calabresi, P.A., Reich, D.S., Prince, J.L.: Smore: a self-supervised anti-aliasing and super-resolution algorithm for MRI using deep learning. IEEE Trans. Med. Imaging **40**(3), 805–817 (2020)
21. Chatterjee, S., et al.: ReconResNet: regularised residual learning for MR image reconstruction of undersampled cartesian and radial data. Comput. Biol. Med. **143**, 105321 (2022)
22. Wu, Q., et al.: An arbitrary scale super-resolution approach for 3-dimensional magnetic resonance image using implicit neural representation. arXiv preprint arXiv:2110.14476 (2021)
23. Du, J., et al.: Super-resolution reconstruction of single anisotropic 3D MR images using residual convolutional neural network. Neurocomputing **392**, 209–220 (2020)
24. Wang, Z., Bovik, A.C., Sheikh, H.R., Simoncelli, E.P.: Image quality assessment: from error visibility to structural similarity. IEEE Trans. Image Process. **13**(4), 600–612 (2004)

25. Zhang, R., Isola, P., Efros, A.A., Shechtman, E., Wang, O.: The unreasonable effectiveness of deep features as a perceptual metric. In: Proceedings of the IEEE Conference on Computer Vision and Pattern Recognition, pp. 586–595 (2018)

Image2SSM: Reimagining Statistical Shape Models from Images with Radial Basis Functions

Hong Xu(✉) and Shireen Y. Elhabian

Scientific Computing and Imaging Institute, Kahlert School of Computing, University of Utah, Salt Lake, UT, USA
{hxu,shireen}@sci.utah.edu
http://www.sci.utah.edu , https://www.cs.utah.edu

Abstract. Statistical shape modeling (SSM) is an essential tool for analyzing variations in anatomical morphology. In a typical SSM pipeline, 3D anatomical images, gone through segmentation and rigid registration, are represented using lower-dimensional shape features, on which statistical analysis can be performed. Various methods for constructing compact shape representations have been proposed, but they involve laborious and costly steps. We propose Image2SSM, a novel deep-learning-based approach for SSM that leverages image-segmentation pairs to learn a radial-basis-function (RBF)-based representation of shapes directly from images. This RBF-based shape representation offers a rich self-supervised signal for the network to estimate a continuous, yet compact representation of the underlying surface that can adapt to complex geometries in a data-driven manner. Image2SSM can characterize populations of biological structures of interest by constructing statistical landmark-based shape models of ensembles of anatomical shapes while requiring minimal parameter tuning and no user assistance. Once trained, Image2SSM can be used to infer low-dimensional shape representations from new unsegmented images, paving the way toward scalable approaches for SSM, especially when dealing with large cohorts. Experiments on synthetic and real datasets show the efficacy of the proposed method compared to the state-of-art correspondence-based method for SSM.

Keywords: Statistical Shape Modeling · Deep Learning · Radial Basis Function Interpolation · Polyharmonic Splines

1 Introduction

Statistical Shape Modeling (SSM) or morphological analysis, is a widespread tool used to quantify anatomical shape variation given a population of segmented 3D anatomies. Quantifying such subtle shape differences has been crucial

---

Supplementary Information The online version contains supplementary material available at https://doi.org/10.1007/978-3-031-43907-0_49.

H. Greenspan et al. (Eds.): MICCAI 2023, LNCS 14220, pp. 508–517, 2023.
https://doi.org/10.1007/978-3-031-43907-0_49

in providing individualized treatments in medical procedures, detecting morphological pathologies, and advancing the understanding of different diseases [3,4,7,9,10,16,19–21,27–29].

The two principal shape representations for building SSMs and performing subsequent statistical analyses are *deformation fields* and *landmarks*. Deformation fields encode *implicit* transformations between cohort samples and a predefined (or learned) atlas. In contrast, landmarks are *explicit* points spread on shape surfaces that correspond across the population [22,23]. Landmark-based representations have been used extensively due to their simplicity, computational efficiency, and interpretability for statistical analyses [22,28]. Some applications use manually defined landmarks, however, this is labor-intensive, not reproducible, and requires domain expertise (e.g., radiologists). Computational methods (e.g., minimum description length – MDL [14], particle-based shape modeling – PSM [11,12], and frameworks based on Large Deformation Diffeomorphic Metric Mapping [15]) for automatically placing dense *correspondence points*, aka *point distribution models* (PDMs), have shifted the SSM field to data-driven characterization of population-level variabilities that is objective, reproducible, and scalable. However, this efficiency suffers when intricate shape surfaces require thousands of points representing localized, convoluted shape features that may live between landmarks. Furthermore, existing methods for landmark-based SSM must go through laborious and computationally expensive steps that require anatomical and technical expertise, starting from anatomy segmentation, shape data preprocessing, and correspondence optimization, to generate PDMs from 3D images. Existing methods (e.g., [1,2,5,6,24,25]) have been able to use deep learning to assuage the arduous process of building a PDM but still require the construction of PDMs (e.g., using a computational method such as PSM [11,12]) to supervise its learning task, making these deep learning based methods restricted and biased towards the shape statistics captured by the SSM method that is used to construct their training data.

To address the shortcomings of existing models, we propose Image2SSM, a novel deep-learning-based approach for SSM directly from images that, given pairs of images and segmentations, can produce a statistical shape model using an implicit, continuous surface representation. Once trained, Image2SSM can produce PDMs of new images without the need for anatomy segmentations. Unlike existing deep learning-based methods for SSM from images, Image2SSM only requires image-segmentation pairs and alleviates the need for constructing PDM to supervise learning shape statistics from images. Image2SSM leverages an implicit, radial basis function (RBF)-based, representation of shapes to construct a self-supervised training signal by tasking the network to estimate a sparse set of control points and their respective suface normals that best approximate the underlying surface in the RBF sense. This novel application of RBFs to build SSMs allows statistical analyses on representative points/landmarks, their surface normals, and the shape surfaces themselves due to its compact, informative, yet comprehensive nature. Combined with deep networks to directly learn such a representation from images, this method ushers a next step towards fully end-to-end SSM frameworks that can build better and less restrictive low-dimensional shape representations more conducive to SSM analysis. In summary, the proposed method for SSM has the following strengths.

- Using a continuous, but compact surface representation instead of only landmarks that allows performing analyses on points, normals, and surfaces alike.
- The RBF shape representation can adapt to the underlying surface geometry, spreading more landmarks over the more complex surface regions.
- A deep learning approach that bypasses any conventional correspondence optimization to construct training data for supervision, requiring virtually no hyperparameter tuning or preprocessing steps.
- This method uses accelerated computational resources to perform training and outperforms existing deep learning based methods that constructs PDMs from unsegmented images.

2 Methods

Image2SSM is a deep learning method that learns to build an SSM for an anatomical structure of interest directly from unsegmented images. It is trained on a population of $I-$3D images $\mathcal{I} = \{\mathbf{I}_i\}_{i=1}^{I}$ as input and is supervised by their respective segmentations $\mathcal{S} = \{\mathbf{S}_i\}_{i=1}^{I}$. Image2SSM learns an RBF-based shape representation, consisting of a set of J control points $\mathcal{P} = \{\mathbf{P}_i\}_{i=1}^{I}$, and their surface normals $\mathcal{N} = \{\mathbf{N}_i\}_{i=1}^{I}$ for each input shape, where the $i-$th shape point distribution model (PDM) is denoted by $\mathbf{P}_i = [\mathbf{p}_{i,1}, \mathbf{p}_{i,2}, \cdots, \mathbf{p}_{i,J}]$, the respective surface normals are $\mathbf{N}_i = [\mathbf{n}_{i,1}, \mathbf{n}_{i,2}, \cdots, \mathbf{n}_{i,J}]$, and $\mathbf{p}_{i,j}, \mathbf{n}_{i,j} \in \mathbb{R}^3$. The network is trained end-to-end to minimize a loss that (1) makes the learned control points and their surface normals adhere to the underlying surface, (2) approximates surface normals at each control point to encode a signed distance field to the surface, (3) promotes correspondence of these control points across shapes in the population, and (4) encourages a spread of control points on each surface that adapts to the underlying geometrical complexity. The learned control points define an anatomical mapping, or a *metric*, among the given shapes that enables quantifying subtle shape differences and performing shape statistics, for example, using principal component analysis (PCA) or other non-linear methods (e.g., [17]). More importantly, once trained, Image2SSM can generate PDMs for new unsegmented images, bypassing the conventional SSM workflow of the manual (or semi-automated) segmentation, data preprocessing, and correspondence optimization. Furthermore, the continuous, implicit nature of the RBF representation enables extracting a proxy geometry (e.g., surface mesh or signed distance transforms – SDFs) at an arbitrary resolution that can be rasterized trivially on graphics hardware [8,26].

In this section, we briefly elaborate on the RBF-shape representation, outline the network architecture, motivate the choices and design of the proposed losses, and detail the training protocol of Image2SSM.

2.1 Representing Shapes Using RBFs

Implicit surface representation based on radial basis functions, *RBF-shape* for short, has been proven effective at representing intricate shapes by leveraging both surface control points and normals to inform shape reconstructions [8,26]. It defines a set of control points at the zero-level set and a pair of off-surface points (aka *dipoles*) with a signed distance s and $-s$ along the surface normal of each control point. This is illustrated in Fig. 1. We refer to the set of control points and their dipoles as $\widetilde{\mathbf{P}}_i$ for shape i, where $\widetilde{\mathbf{P}}_i = [\mathbf{P}_i, \mathbf{P}_i^+, \mathbf{P}_i^-]$ with $\mathbf{p}_{i,j}^{\pm} = \mathbf{p}_{i,j} \pm s\mathbf{n}_{i,j}$. Using $\widetilde{\mathbf{P}}_i$, we define the shape's implicit function, a function that can query a distance to the surface given a point $\mathbf{x} \in \mathbb{R}^3$, as follows:

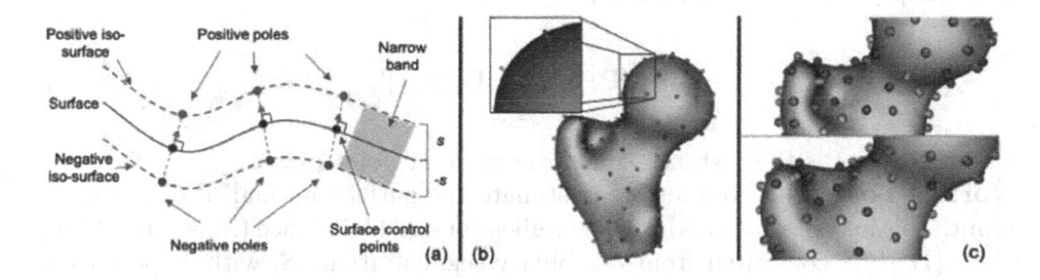

Fig. 1. (a) Concept of populating a surface using control points and the iso-surfaces using positive and negative pole points. (b) Same concept applied to an output three-dimensional reconstructed femur. (c) Normals can be used to describe very distinct features of the greater trochanter.

$$f_{\widetilde{\mathbf{P}}_i,\mathbf{w}_i}(\mathbf{x}) = \sum_{j \in \widetilde{\mathbf{P}}_i} w_{i,j}\phi(\mathbf{x}, \widetilde{\mathbf{p}}_{i,j}) + \mathbf{c}_i^T\mathbf{x} + c_i^0 \tag{1}$$

where ϕ is the chosen RBF basis function (e.g., the thin plate spline $\phi(\mathbf{x}, \mathbf{y}) = (|\mathbf{x} - \mathbf{y}|_2)^2 \log(|\mathbf{x} - \mathbf{y}|_2)$, the biharmonic $\phi(\mathbf{x}, \mathbf{y}) = |\mathbf{x} - \mathbf{y}|_2$ or the triharmonic $\phi(\mathbf{x}, \mathbf{y}) = (|\mathbf{x} - \mathbf{y}|_2)^3$) and $\mathbf{c}_i \in \mathbb{R}^3$ and $c_i^0 \in \mathbb{R}$ encodes the linear trend of the surface. We obtain $\mathbf{w}_i = [w_{i,1}, w_{i,2}, ..., w_{i,3J}, c_i^0, c_i^1, c_i^2, c_i^3] \in \mathbb{R}^{3J+4}$ by solving a system of equations formed by Eq. 1 over $\mathbf{x} \in \widetilde{\mathbf{P}}_i$, along with constraints to keep the linear part separate from the nonlinear deformations captured by the RBF term (first term in Eq. 1) to form a fully determined system. See [8,26] for more details. Ultimately, we can use this function f to query approximate distances to the surface to build a mesh or a signed distance transform for visualization and analysis.

This representation can better represent shapes with far fewer control points due to its built-in interpolation capabilities, even further enhanced by informing the system with the point normals. Furthermore, this continuous representation allows Image2SSM to adapt to the underlying surface geometry and correct for control point placement mistakes while training.

2.2 Loss Functions

Image2SSM uses four complementary loss functions to be trained on concurrently, illustrated in Fig. 2. These are (i) *surface loss*, which aims to promote control point and normal adherence to the shape surface, (ii) *normal loss*, which attempts to learn the correct normals at each control point, (iii) *correspondence loss*, which aims to enforce positional correspondence across shapes, and (iv) *sampling loss*, which promotes a spread of the control points that best describes the underlying surface.

Surface Loss: This loss guides control points to lie on the surface. We use l^1-norm to force control points to lie on the zero-level set of the distance transform $\mathbf{D}_i$ by minimizing the absolute distance-to-the-surface evaluated at it. For the $i-$th shape, this loss is defined as,

$$L_{\mathbf{D}_i}^{surf}(\mathbf{P}_i) = \sum_{j=1}^{J} |\mathbf{D}_i(\mathbf{p}_{i,j})|, \tag{2}$$

where $\mathbf{D}_i(\mathbf{p}_{i,j})$ is the distance transform value at point $\mathbf{p}_{i,j}$.

Normal Loss: This loss aims to estimate the surface normal of each control point. This loss is supervised by the gradient of signed distance transforms (SDF) $\mathcal{D} = \{\mathbf{D}_i\}_{i=1}^{I}$, computed from the binary segmentations $\mathcal{S}$, with respect to $\mathbf{x}$, $\partial\mathcal{D} = \{\partial\mathbf{D}_i\}_{i=1}^{I}$, which captures unnormalized surface normals. We use the cosine distance (in degrees) to penalize the deviation of the estimated normals from the normals computed from the distance transforms.

$$L_{\partial\mathbf{D}_i}^{norm}(\mathbf{P}_i, \mathbf{N}_i) = \frac{180}{\pi}\sum_{j=1}^{J} \cos^{-1}\left(1 - \frac{\mathbf{n}_{i,j}^T \partial\mathbf{D}_i(\mathbf{p}_{i,j})}{\|\mathbf{n}_{i,j}\|\|\partial\mathbf{D}_i(\mathbf{p}_{i,j})\|}\right). \tag{3}$$

Correspondence Loss: The notion of control points correspondence across the shape population can be quantified by the information content of the probability distribution induced by these control points in the shape space, the vector space defined by the shapes' PDMs [11,12]. The correspondence loss is triggered starting from the second epoch, where the mean shape $\boldsymbol{\mu} = \sum_{i=1}^{I} \mathbf{P}_i$ is allowed to lag behind the update of the control points. Given a minibatch of size K, the correspondence loss is formulated using the differential entropy H of the samples in the minibatch, assuming a Gaussian distribution.

$$L_{\boldsymbol{\mu}}^{corres}(\mathbf{P}_1, ..., \mathbf{P}_K) = H(\mathbf{P}) = \frac{1}{2}\log\left|\frac{1}{3JK}\sum_{k=1}^{K}(\mathbf{P}_k - \boldsymbol{\mu})(\mathbf{P}_k - \boldsymbol{\mu})^T\right|, \tag{4}$$

where $\mathbf{P}$ here indicates the random variable of the shape space.

Sampling Loss: This loss makes f encode the signed distance to the surface while encouraging the control points to be adapted to the underlying geometry. Here, we randomly sample $R-$ points $\mathbf{B}_i = [\mathbf{b}_{i,1}, ..., \mathbf{b}_{i,R}]$ that lie within a *narrow band* of thickness $2s$ around the surface (i.e., $\pm s$ from the zero-level set along the surface normal). The sampling loss minimizes distances between these narrow band points and the closest control point to each, scaled by the severity of the distance-to-surface approximation error. This objective guides control points to areas poorly described by f to progressively improve the signed distance-to-surface approximation and represent the shape more accurately.

Let $\mathbf{K}^i \in \mathbb{R}^{R \times M}$ define the pairwise distances between each narrow band point $\mathbf{b}_{i,r}$ and each control point $\mathbf{p}_{i,j}$ for the i−th shape, where its r, j−th element $k^i_{r,j} = \|\mathbf{b}_{i,r} - \mathbf{p}_{i,j}\|_2$. Let softmin($\mathbf{K}^i$) encode the normalized (over $\mathbf{P}_i$) spatial proximity of each narrow band point to each control point, where r, the jth element of softmin($\mathbf{K}^i$) is computed as $\exp(-k^i_{r,j}) / \sum_{j'=1}^{J} \exp(k^i_{r,j'})$. Let $\mathbf{e}_i \in \mathbb{R}^R_+$ captures the RBF approximation squared error at the narrow band points, where $e_{i,r} = [f_{\widetilde{\mathbf{P}_i},\mathbf{w}_i}(\mathbf{b}_{i,r}) - \mathbf{D}_i(\mathbf{b}_{i,r})]^2$. Let $\mathbf{E}_i = \mathbf{e}_i \mathbf{1}^T_M$, where $\mathbf{1}_M$ is a ones-vector of size M. The samples loss can then be written as,

$$L^{sampl}_{\mathbf{B}_i,\mathbf{D}_i,\mathbf{w}_i}(\mathbf{P}_i, \mathbf{N}_i) = \text{mean}\left(\text{softmin}(\mathbf{K}_i) \otimes \mathbf{K}_i \otimes \mathbf{E}_i\right), \tag{5}$$

where $\otimes$ indicates the Hadamard (elementwise) multiplication of matrices and mean computes the average over the matrix elements.

Image2SSM Loss: Given a minibatch of size K, the total loss of Image2SSM can be written as follows:

$$\begin{aligned} L_{\mathcal{I},\mathcal{D},\partial\mathcal{D}}(\mathcal{P}_K, \mathcal{N}_K) = \sum_{i=1}^{K} \Big(\alpha L^{surf}_{\mathbf{D}_i}(\mathbf{P}_i) + \beta L^{norm}_{\partial\mathbf{D}_i}(\mathbf{P}_i, \mathbf{N}_i) + \gamma L^{sampl}_{\mathbf{B}_i,\mathbf{D}_i,\mathbf{w}_i}(\mathbf{P}_i, \mathbf{N}_i) \Big) \\ + \zeta L^{corres}_{\mu}(\mathbf{P}_1, ..., \mathbf{P}_K) \end{aligned} \tag{6}$$

where α, β, γ, and $\zeta \in \mathbb{R}_+$ are weighting hyperparameters of the losses and $\mathcal{P}_K, \mathcal{N}_K$ are the control points and normals of the samples in the minibatch. Figure 2 gives a full overview of the network and its interaction with the loses. Image2SSM's network is trained end-to-end with $\mathbf{w}_i$s detached from the training so that the loss does not back-propagate through the volatile linear solver.

3 Results

We demonstrate Image2SSM's performance against the state-of-the-art correspondence optimization algorithm, namely the particle-based shape modeling (PSM), using its open-source implementation, ShapeWorks [11], and DeepSSM [5,6], a deep learning method that trains on an existing correspondence model (provided by the PSM in this case) to infer PDMs on new unsegmented images.

3.1 Datasets

We run tests on a dataset consisting of 50 proximal femur CT scans devoid of pathologies in the form of image-segmentation pairs. The femurs are reflected when appropriate and rigidly aligned to a common frame of reference. Due to

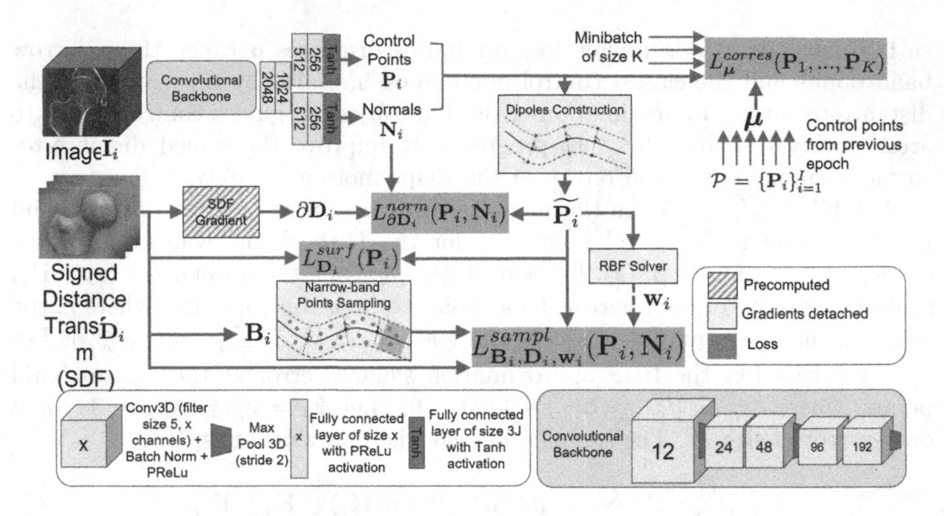

Fig. 2. The Image2SSM architecture. A 3D image is fed to the convolutional backbone, which produces a flattened output for the feature extractor to produce control points and their respective normals. These are then used to compute the losses of the network.

space limitations, we also show similar results for a large-scale left atrium MRI dataset in the supplementary materials. For ease of comparison, we build SSMs with 128 particles for all algorithms as is sufficient to cover important femur shape features (femoral head with its fovea and the lesser and greater trochanter).

Statistical Shape Model: We showcase Image2SSM in creating a statistical shape model on its training data and compare such a model with one optimized by PSM [11]. Figure 3 showcases the modes of variation, the surface-to-surface distances of Image2SSM against PSM, some representative reconstructions, and graphs for compactness (percentage of variance captured), specificity (ability to generate realistic shapes), and generalization (ability to represent unseen shape instance) [13]. We observe that the modes of variation and metrics match expectations in both approaches. We show the effectiveness of Image2SSM in adapting to surface details to achieve lower maximum surface-to-surface distance, and that, unlike PSM, we can achieve reasonable reconstructions using RBF-shape. More on adaptation to detail in Fig. 4.

We implement Image2SSM in PyTorch and leverage the Autograd functionality to perform gradient descent using the Adam optimizer [18]. We randomly sample 10,000 3D points within the narrow band of each surface at each iteration. We use the biharmonic kernel for the basis function. However, the performance of Image2SSM is not significantly influenced by the kernel choice. The hyperparameters we use for Image2SSM are $\alpha = 1e^2, \beta = 1e^2, \gamma = 1e^4$, and $\zeta = 1e^3$ for femurs and $\zeta = 1e^6$ for left atria, which were determined based on the validation set. In practice, the runtime of Image2SSM is comparable to PSM for the femurs and roughly 2X faster for the left atria.

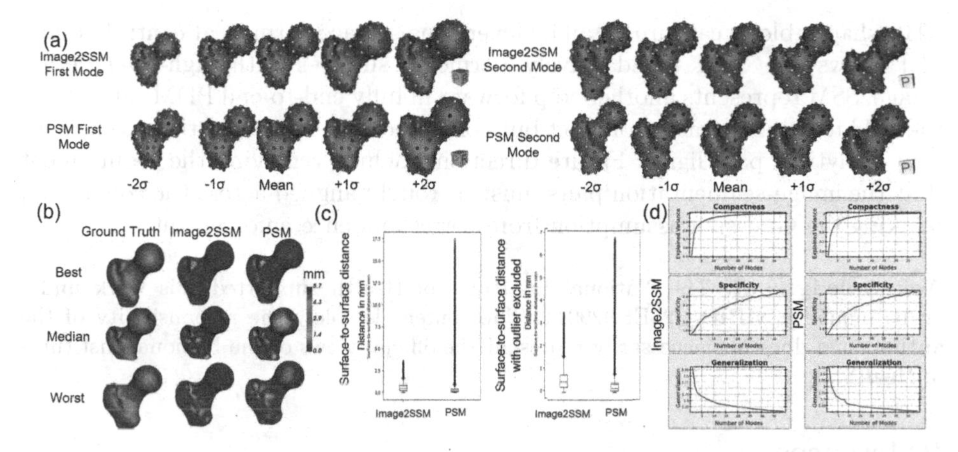

Fig. 3. (a) First and second modes of variation obtained from Image2SSM training data and PSM. (b) Surface-to-surface distance on a best, median, and worst training femur mesh. (c) The left image shows the surface-to-surface distance comparison on all the data used to train Image2SSM; the right shows it without outliers.

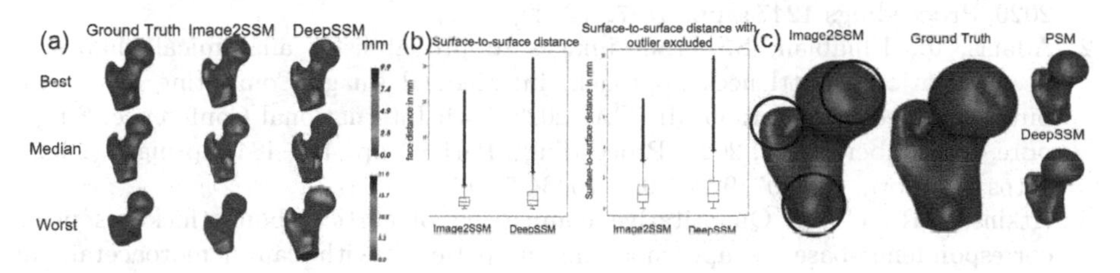

Fig. 4. (a) Surface-to-surface distance on a reconstructed femur mesh from particles of a few test samples. (b) Surface-to-surface distance plot between DeepSSM and Image2SSM, and the same plot without the outlier femur. (c) Illustrates Image2SSM's capacity to capture detail on an unseen test image. (d) Shows the compactness (higher is better), specificity (lower is better) and generalization (lower is better) graphs against the number of modes of variation.

Inference Results: We compare the inference capabilities of Image2SSM against DeepSSM on unseen test data. We train DeepSSM with a PDM generated by PSM as supervision. For a fair comparison, we use DeepSSM without its augmented data, since Image2SSM does not require augmentation to learn shape models. Nevertheless, it is possible to generate and train Image2SSM on augmented data with even more facility than with DeepSSM. Figure 4 shows that Image2SSM compares very favorably to DeepSSM qualitatively and in terms of surface-to-surface distance.

4 Conclusion

Image2SSM is a novel deep-learning framework that both builds PDMs from image-segmentation pairs and predicts PDMs from unseen images. It uses an

RBF-shape able to capture detail by leveraging surface normals at control points, and allows the SSM to adaptively permeate surfaces with high-level detail. Image2SSM represents another step forward in fully end-to-end PDMs and steers the field to utilizing more compact but comprehensive representations to achieve new analytical paradigms. Future directions include removing the requirement that the image-segmentation pairs must be rougly aligned across the cohort and relaxing the Gaussian assumption from correspondence enforcement.

Acknowledgment. The National Institutes of Health supported this work under grant numbers NIBIB-U24EB029011. The content is solely the responsibility of the authors and does not necessarily represent the official views of the National Institutes of Health.

References

1. Adams, J., Bhalodia, R., Elhabian, S.: Uncertain-DeepSSM: from images to probabilistic shape models. In: Shape in Medical Imaging: International Workshop, ShapeMI 2020, Held in Conjunction with MICCAI 2020, Lima, Peru, October 4, 2020, Proceedings 12474, pp. 57–72 (2020)
2. Adams, J., Elhabian, S.: From images to probabilistic anatomical shapes: a deep variational bottleneck approach. In: Medical Image Computing and Computer Assisted Intervention-MICCAI 2022: 25th International Conference, Singapore, September 18–22, 2022, Proceedings, Part II, pp. 474–484. Springer (2022). https://doi.org/10.1007/978-3-031-16434-7_46
3. Atkins, P.R., et al.: Quantitative comparison of cortical bone thickness using correspondence-based shape modeling in patients with cam femoroacetabular impingement. J. Orthop. Res. **35**(8), 1743–1753 (2017)
4. Bhalodia, R., Dvoracek, L.A., Ayyash, A.M., Kavan, L., Whitaker, R., Goldstein, J.A.: Quantifying the severity of metopic craniosynostosis: a pilot study application of machine learning in craniofacial surgery. J. Craniofacial Surg. (2020)
5. Bhalodia, R., Elhabian, S., Adams, J., Tao, W., Kavan, L., Whitaker, R.: DeepSSM: a blueprint for image-to-shape deep learning models (2021)
6. Bhalodia, R., Elhabian, S.Y., Kavan, L., Whitaker, R.T.: DeepSSM: a deep learning framework for statistical shape modeling from raw images. In: Reuter, M., Wachinger, C., Lombaert, H., Paniagua, B., Lüthi, M., Egger, B. (eds.) ShapeMI 2018. LNCS, vol. 11167, pp. 244–257. Springer, Cham (2018). https://doi.org/10.1007/978-3-030-04747-4_23
7. Bruse, J.L., et al.: A statistical shape modelling framework to extract 3D shape biomarkers from medical imaging data: assessing arch morphology of repaired coarctation of the aorta. BMC Med. Imaging **16**, 1–19 (2016)
8. Carr, J.C., et al.: Reconstruction and representation of 3D objects with radial basis functions. In: Proceedings of the 28th Annual Conference on Computer Graphics and Interactive Techniques, pp. 67–76. SIGGRAPH '01, Association for Computing Machinery, New York, NY, USA (2001). https://doi.org/10.1145/383259.383266
9. Carriere, N., et al.: Apathy in Parkinson's disease is associated with nucleus accumbens atrophy: a magnetic resonance imaging shape analysis. Mov. Disord. **29**(7), 897–903 (2014)
10. Cates, J., et al.: Computational shape models characterize shape change of the left atrium in atrial fibrillation. Clin. Med. Insights: Cardiol. **8**, CMC-S15710 (2014)

11. Cates, J., Elhabian, S., Whitaker, R.: ShapeWorks: particle-based shape correspondence and visualization software. In: Statistical Shape and Deformation Analysis, pp. 257–298. Elsevier (2017)
12. Cates, J., Fletcher, P.T., Styner, M., Shenton, M., Whitaker, R.: Shape modeling and analysis with entropy-based particle systems. In: Karssemeijer, N., Lelieveldt, B. (eds.) IPMI 2007. LNCS, vol. 4584, pp. 333–345. Springer, Heidelberg (2007). https://doi.org/10.1007/978-3-540-73273-0_28
13. Davies, R.H.: Learning shape: optimal models for analysing natural variability (2004)
14. Davies, R.H., Twining, C.J., Cootes, T.F., Waterton, J.C., Taylor, C.J.: A minimum description length approach to statistical shape modeling. IEEE Trans. Med. Imaging **21**(5), 525–537 (2002)
15. Durrleman, S., Prastawa, M., Charon, N., Korenberg, J.R., Joshi, S., Gerig, G., Trouvé, A.: Morphometry of anatomical shape complexes with dense deformations and sparse parameters. Neuroimage **101**, 35–49 (2014)
16. Harris, M.D., Datar, M., Whitaker, R.T., Jurrus, E.R., Peters, C.L., Anderson, A.E.: Statistical shape modeling of cam femoroacetabular impingement. J. Orthop. Res. **31**(10), 1620–1626 (2013). https://doi.org/10.1002/jor.22389
17. Kempfert, K.C., Wang, Y., Chen, C., Wong, S.W.: A comparison study on nonlinear dimension reduction methods with kernel variations: visualization, optimization and classification. Intell. Data Anal. **24**(2), 267–290 (2020)
18. Kingma, D., Ba, J.: Adam: a method for stochastic optimization. In: International Conference on Learning Representations (2014)
19. Lenz, A., et al.: Statistical shape modeling of the talocrural joint using a hybrid multi-articulation joint approach. Sci. Rep. **11**, 7314 (2021). https://doi.org/10.1038/s41598-021-86567-7
20. Merle, C., et al.: How many different types of femora are there in primary hip osteoarthritis? an active shape modeling study. J. Orthop. Res. **32**(3), 413–422 (2014)
21. Merle, C., et al.: High variability of acetabular offset in primary hip osteoarthritis influences acetabular reaming-a computed tomography-based anatomic study. J. Arthroplasty **34**(8), 1808–1814 (2019)
22. Sarkalkan, N., Weinans, H., Zadpoor, A.A.: Statistical shape and appearance models of bones. Bone **60**, 129–140 (2014)
23. Thompson, D.W., et al.: On growth and form (1942)
24. Tóthová, K., et al.: Probabilistic 3D surface reconstruction from sparse MRI information. In: Martel, A.L., et al. (eds.) MICCAI 2020. LNCS, vol. 12261, pp. 813–823. Springer, Cham (2020). https://doi.org/10.1007/978-3-030-59710-8_79
25. Tóthová, K., et al.: Uncertainty quantification in CNN-based surface prediction using shape priors. CoRR abs/1807.11272 (2018). http://arxiv.org/abs/1807.11272
26. Turk, G., O'Brien, J.F.: Variational implicit surfaces (1999)
27. van Buuren, M., et al.: Statistical shape modeling of the hip and the association with hip osteoarthritis: a systematic review. Osteoarthritis Cartilage **29**(5), 607–618 (2021). https://doi.org/10.1016/j.joca.2020.12.003, https://www.sciencedirect.com/science/article/pii/S106345842031219X
28. Zachow, S.: Computational planning in facial surgery. Facial Plast. Surg. **31**(05), 446–462 (2015)
29. Zadpoor, A.A., Weinans, H.: Patient-specific bone modeling and analysis: the role of integration and automation in clinical adoption. J. Biomech. **48**(5), 750–760 (2015)

MDA-SR: Multi-level Domain Adaptation Super-Resolution for Wireless Capsule Endoscopy Images

Tianbao Liu[1,2], Zefeiyun Chen[1,2], Qingyuan Li[3], Yusi Wang[3,4], Ke Zhou[4], Weijie Xie[1,2], Yuxin Fang[3], Kaiyi Zheng[1,2], Zhanpeng Zhao[4], Side Liu[3,4](✉), and Wei Yang[1,2](✉)

[1] School of Biomedical Engineering, Southern Medical University, Guangzhou, China
[2] Guangdong Provincial Key Laboratory of Medical Image Processing, Southern Medical University, Guangzhou, China
weiyanggm@gmail.com
[3] Department of Gastroenterology, Nanfang Hospital, Southern Medical University, Guangzhou, China
liuside2011@163.com
[4] Guangzhou SiDe MedTech Company Ltd, Guangzhou, China

Abstract. Super-resolution (SR) of wireless capsule endoscopy (WCE) images is challenging because paired high-resolution (HR) images are not available. An intuitive solution is to simulate paired low-resolution (LR) WCE images from HR electronic endoscopy images for supervised learning. However, the SR model obtained by this method cannot be well adapted to real WCE images due to the large domain gap between electronic endoscopy images and WCE images. To address this issue, we propose a Multi-level Domain Adaptation SR model (MDA-SR) in an unsupervised manner using arbitrary set of WCE images and HR electronic endoscopy images. Our approach implements domain adaptation at the image level and latent level during the degradation and SR processes, respectively. To the best of our knowledge, this is the first work to explore an unsupervised SR approach for WCE images. Furthermore, we design an Endoscopy Image Quality Evaluator (EIQE) based on the reference-free image evaluation metric NIQE, which is more suitable for evaluating WCE image quality. Extensive experiments demonstrate that our MDA-SR method outperforms state-of-the-art SR methods both quantitatively and qualitatively.

Keywords: Wireless capsule endoscopy · Super-resolution · Domain adaptation

1 Introduction

Wireless capsule endoscopy (WCE) is an emerging examination technique that offers several advantages over traditional electronic endoscopy, including non-invasiveness, safety, and non-cross-infection. It enables the examination of the

T. Liu and Z. Chen—contributed equally to this work.

H. Greenspan et al. (Eds.): MICCAI 2023, LNCS 14220, pp. 518–527, 2023.
https://doi.org/10.1007/978-3-031-43907-0_50

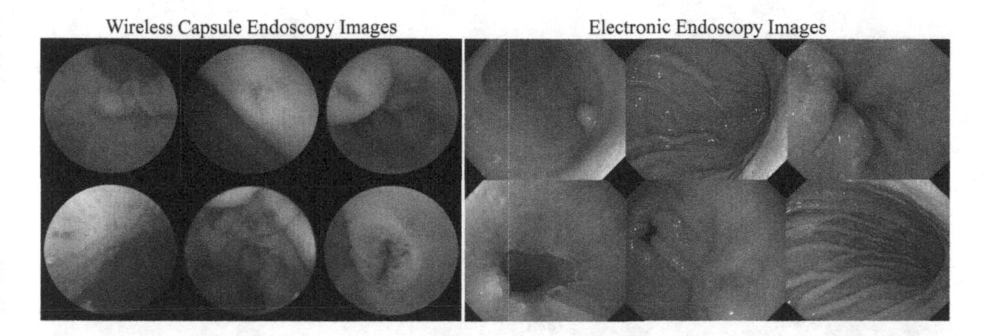

Fig. 1. The visual presentation of WCE images and electronic endoscopy images. The WCE images on the left are dim and blurred, while the electronic endoscopy images on the right are bright and clear.

entire human gastrointestinal tract and is widely used in clinical practice [15, 23]. Despite its successful clinical applications, the limited volume and data transmission bandwidth of WCE result in image quality drawbacks such as low resolution and poor quality [4]. Recent research has shown that high definition colonoscopy increases the identification of any polyps by 3.8% [19], and a 3-center prospective randomized trial has further proven the value of high resolution in invasive endoscopy [17]. Hence it is desirable to restore the image quality of WCE images via super-resolution techniques.

Most super-resolution methods based on deep learning are supervised by paired low-resolution (LR) and high-resolution (HR) images [8,21,26]. However, the corresponding HR WCE image is currently unavailable due to the hardware limitations in capsule size and transmission bandwidth. Alternatively, researchers have adopted predefined simple linear degradation assumptions (e.g., bicubic downsampling, gaussian downsampling) to feasibly synthesize corresponding LR samples from ground-truth HR images. Similarly, Almalioglu et al. [1] adopted this assumption to synthesis paired LR electronic endoscopy images from the HR ones for supervised super-resolution learning. However, this method is difficult to generalize to real WCE images due to the domain gap between WCE images and electronic endoscopy images.

What causes this domain gap? It might seem reasonable to adopt the simple linear degradation assumption by simply analogizing the domain gap between a mobile camera and a professional camera to a WCE and an electronic endoscope. However, what cannot be ignored is the different examination environment, where the WCE requires filling the stomach with water, while the electronic endoscopy inflates the stomach, which directly leads to the difference between the two image domains in terms of villi pose and speckle reflection, as shown in Fig. 1. This domain gap cannot be described by a simple linear degradation matrix.

Recently, many studies have utilized the CycleGAN [28] to combine explicit domain adaptation into SR. The basic idea is to generate LR versions of HR

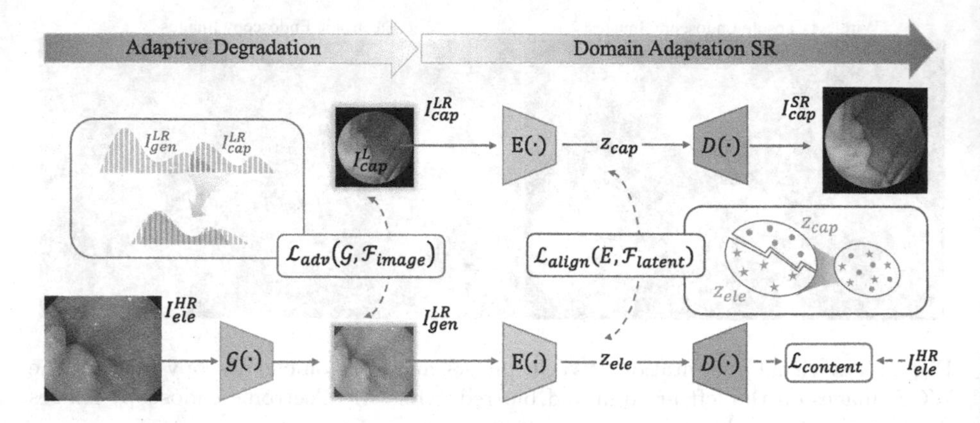

Fig. 2. Overview of the proposed MDA-SR, which consists of two parts: adaptive degradation and domian adaptation SR.

images with degenerate distributions similar to the real LR images, and then train the SR model on the generated LR-HR paired dataset [20,25,27]. The challenge is that the large domain gap between WCE and electronic endoscopic images makes the generator sensitive to learning shallow differences in content or style and unable to effectively bridge the domain gap.

In this work, we propose a Multi-level Domain Adaptation Super-Resolution (MDA-SR) for WCE images to bridge the domain gap between electronic endoscopy images and WCE images. MDA-SR leverages prior knowledge of HR electronic endoscopy images to guide the SR process of WCE images, as illustrated in Fig. 2. First, we train the adaptive degradation at the image level, employing a generative adversarial network to learn the complex and variable degradation distribution in WCE images, while incorporating an adaptive data loss [18] as the fidelity term of the image content. In contrast to previous methods that assume generated LR images are free from domain shift [3,9,25], we propose to minimize the domain gap during the SR process by aligning the latent feature distribution of electronic endoscopy images and WCE images. We further proposed EIQE for improving the reference-free image evaluation metric NIQE to be more suitable for endoscopy images. Through extensive experiments on real WCE images, we demonstrate the superiority of our method over other state-of-the-art SR methods, and its efficacy in reality.

2 Methods

2.1 Overview of the Proposed Method

Given a set of WCE images and HR electronic endoscopy images $\{I_{cap}^{LR}, I_{ele}^{HR}\}$, we aim to learn a SR function $\mathcal{R}(\cdot)$ that maps an observed I_{cap}^{LR} to its HR version according to the distribution defined by I_{ele}^{HR} in testing. As shown in Fig. 2, the proposed scheme consists of two major parts: an adaptive degradation that generates the LR version of I_{ele}^{HR}, denoted as I_{gen}^{LR}, and a domain adaptation SR that aligns the latent features of the WCE and electronic endoscopy datasets during the SR process.

2.2 Adaptive Degradation

The purpose of the adaptive degradation is to obtain I_{gen}^{LR} with a degradation distribution similar to I_{cap}^{LR}. To achieve this, we employ the architecture of GAN [5], where the degradation generator $\mathcal{G}$ maps the generated I_{gen}^{LR} to I_{cap}^{LR} and the image-level discriminator $\mathcal{F}_{image}$ learns to distinguish between generated samples I_{gen}^{LR} and realistic samples I_{cap}^{LR} with adversarial loss:

$$\mathcal{L}_{adv}\left(\mathcal{G}, \mathcal{F}_{image}\right) = \mathcal{F}_{image}\left(\mathcal{G}\left(I_{ele}^{HR}\right)\right)^2 + \left[1 - \mathcal{F}_{image}\left(I_{cap}^{LR}\right)\right]^2 \tag{1}$$

$\mathcal{G}$ is optimized by maximizing the loss in Eq. (1) against an adversarial $\mathcal{F}_{image}$ that tries to minimize the loss.

A critical requirement is that the LR image generated by $\mathcal{G}$ should be consistent with the low-frequency information of the HR image. To enforce this constraint, we incorporate data loss as an additional supervision information.

The process of degradation from HR images to LR images is unknown. We adopt an adaptive downsampling kernel $\bar{k}$ [18] to approximate the unknown degradation process, since a widely-used approach uses predefined downsampling assumptions, such as bicubic downsampling or $s \times s$ average pooling [3]. It has been observed in previous works [2,6,12] that an appropriate downsampling function consists of low-filtering and decimation, we linearize the degradation generator $\mathcal{G}$ to a corresponding 2D kernel $\bar{k}$:

$$\bar{k} = \arg\min_{k} \sum_{i=1}^{N} \left\| \left(I_{ele,i}^{HR} * k\right)_{\downarrow s} - \mathcal{G}\left(I_{ele,i}^{HR}\right) \right\|_2^2 \tag{2}$$

where $I_{ele,i}^{HR}$ denotes an i-th example to estimate the kernel, N is the total number of samples that have been used and $\downarrow s$ represents downsampling operation with scale factor s. Finally, the data fidelity term $\mathcal{L}_{data}$ is defined as follows:

$$\mathcal{L}_{data}\left(\mathcal{G}\right) = \left\| \left(I_{ele}^{HR} * \bar{k}\right)_{\downarrow s} - \mathcal{G}\left(I_{ele}^{HR}\right) \right\|_1 \tag{3}$$

Given the definitions of adversarial and data losses above, the training loss of our adaptive degradation is defined as:

$$\mathcal{L}_{LR}\left(\mathcal{G}, \mathcal{F}_{image}\right) = \mathcal{L}_{adv}\left(\mathcal{G}, \mathcal{F}_{image}\right) + \lambda \mathcal{L}_{data}\left(\mathcal{G}\right) \tag{4}$$

where λ is a hyperparameter.

2.3 Domain Adaptation SR

The SR function $\mathcal{R}(\cdot)$ can then be supervised by the aligned image pair set $\{I_{gen}^{LR}, I_{ele}^{HR}\}$ obtained from adaptive degradation. As shown in Fig. 2, we use the pixel-wise content loss on the SR results $\mathcal{R}\left(I_{gen}^{LR}\right)$, which ensures the accuracy of HR image composition:

$$\mathcal{L}_{content} = \left\|\mathcal{R}\left(I_{gen}^{LR}\right) - I_{ele}^{HR}\right\|_1 \tag{5}$$

To further improve the performance of the WCE SR, it is crucial to bridge the domain gap between WCE images and electronic endoscopy images, even though the adaptive degradation generator $\mathcal{G}$ already learns the degradation distribution of WCE images through domain adaptation at the image level. To achieve this, we improve the domain adaptation at the latent level during the SR process.

A straightforward way is to adopt a GAN [5] structure to reduce the distribution shift. As shown in Fig. 2, the SR function $\mathcal{R}(\cdot)$ consists of an encoder E and a decoder D. We use two encoders E with shared weights to generate the electronic endoscopy latent feature z_{ele} as well as the WCE latent feature z_{cap}. We introduce latent-level discriminator $\mathcal{F}_{latent}$ to distinguish the domain for each latent feature, while the encoder E is trained to deceive $\mathcal{F}_{latent}$. The optimization of E and $\mathcal{F}_{latent}$ is achieved via the adversarial way, we use LSGAN [11] here:

$$\mathcal{L}_{align}(E) = \left(\mathcal{F}_{latent}\left(E\left(I_{cap}^{LR}\right) - 0.5\right)\right)^2 + \left(\mathcal{F}_{latent}\left(E\left(I_{gen}^{LR}\right)\right) - 0.5\right)^2 \tag{6}$$

$$\mathcal{L}_{align}(\mathcal{F}_{latent}) = \left(\mathcal{F}_{latent}\left(E\left(I_{cap}^{LR}\right) - 1\right)\right)^2 + \left(\mathcal{F}_{latent}\left(E\left(I_{gen}^{LR}\right)\right) - 0\right)^2 \tag{7}$$

As a result, the discriminator $\mathcal{F}_{latent}$ is trained with its corresponding loss in Eq. (7). The SR function $\mathcal{R}(\cdot)$ is trained with the following loss function:

$$\mathcal{L}_{SR} = \mathcal{L}_{content} + \mu\mathcal{L}_{align}(E) \tag{8}$$

where μ is a hyperparameter.

3 Experiments

3.1 Experiment Settings

Datasets. Our proposed model is trained on WCE image dataset and electronic endoscopy image dataset, and tested on WCE images. The WCE image dataset contains 14090 images, and the electronic endoscopy image dataset contains 2033 images, including 1302 images from the public Kvasir dataset [16] and 731 images from the local hospital. We perform a strict quality selection and remove the problematic images such as blurry, low-resolution and poor quality images.

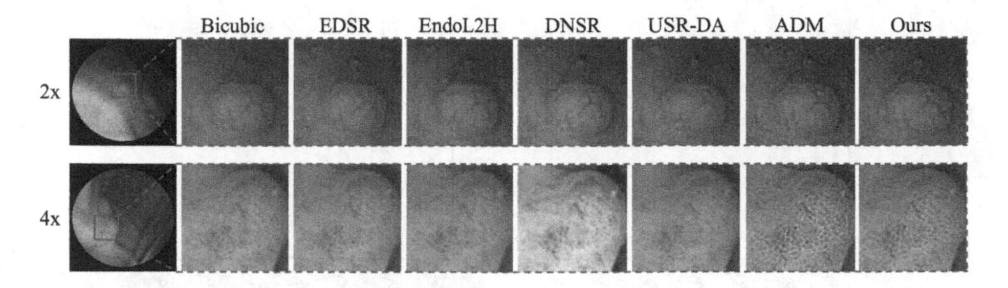

Fig. 3. The visual comparisons for 2x and 4x SR on the WCE images. Obviously, our result contains more natural details and textures suffering from less blur and artifacts.

Evaluation Metrics. In WCE SR problem, there is no corresponding ground-truth image used to calculate the evaluation metric. To address this issue, we design a no-reference Endoscopy Image Quality Evaluator (EIQE), which derives the quality-aware features from the Endoscopy Scene Statistic (ESS) model, inspired by the reference-free Natural Image Quality Evaluator (NIQE) [14]. The quality of a given test image is then expressed as the distance between a multivariate gaussian (MVG) fit of the ESS features extracted from the test image and a MVG model of the quality-aware features extracted from the corpus of HR electronic endoscopy images. Additionally, we use the no-reference metric BRISQUE [13] for evaluation purposes.

To better illustrate the subjective quality, we conduct a mean opinion score (MOS) test for comparison with other methods. We randomly select 100 different WCE images from the test set to subjectively evaluate the quality of the 2x and 4x WCE SR images. Four gastroenterology clinicians rate the visual perceptual qualities by assigning scores. Scores from 0 to 5 are used to indicate the qualities from low to high.

3.2 Training Details

Throughout the framework, the discriminators $\mathcal{F}_{image}$ and $\mathcal{F}_{latent}$ use the patch-based discriminator [7] with the instance normalization [22]. The generator $\mathcal{G}$, encoder E and decoder D in the model follow the residual block structure from the EDSR [8]. The parameters in Eq. (4) and Eq. (8) are set to be $\lambda = 1$ and $\mu = 0.001$, respectively. During training, we use the Adam optimizer and set the batch size and learning rate as 10 and 1×10^{-4}, respectively. To streamline the model training and reduce its complexity, we have divided the training process into two stages. In the first stage, we focus on training the adaptive degradation, which stabilizes the quality of generated LR images after 50 epochs. Following this, we incorporate the domain adaptation SR into the training process by training another 50 epochs. The experiments are implemented with Pytorch platform on NVIDIA GeForce RTX 2080 Ti.

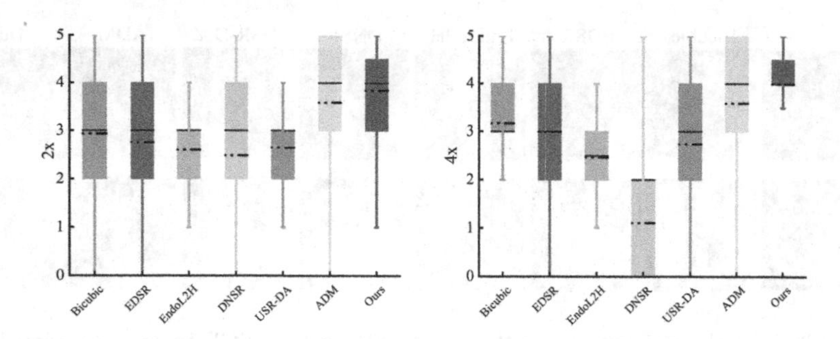

Fig. 4. Mean opinion of the subjective evaluation for different SR methods.

Table 1. Comparison results of the proposed MDA-SR model and other state-of-the-art methods on WCE image dataset, where DNSR, USR-DA and ADM incorporate domain adaptation but EDSR and EndoL2H do not. The best values are highlighted in bold font.

Method	2x		4x	
	EIQE↓	BRISQUE↓	EIQE↓	BRISQUE↓
Bicubic	5.65	58.03	7.36	76.75
EDSR [8]	6.06	59.06	6.81	75.89
EndoL2H [1]	6.63	57.42	6.66	69.92
DNSR [27]	5.50	58.16	6.71	74.72
USR-DA [24]	5.23	51.82	6.42	75.30
ADM [18]	5.19	52.62	6.21	**59.49**
MDA-SR(ours)	**5.14**	**50.70**	**6.16**	64.79

3.3 Results and Discussions

Comparison with Previous Methods. To validate the effectiveness of our proposed method, we compare it with existing state-of-the-art conventional SR methods without domain adaptation [1,8] and SR methods with domain adaptation [18,24,27]. Quantitative results are shown in Table 1, while visualization results are provided in Fig. 3. As EIQE is the most important metric in Table 1 and BRISQUE [13] is provided for reference since BRISQUE is based on natural scene statistic. As can be seen in Table 1, approaches that incorporated domain adaptation generally outperformed those that did not, thus highlighting the value of domain adaptation for the WCE SR problem. For the WCE image dataset, our MDA-SR method achieves the top EIQE performance.Note that in Fig. 3, the proposed method produces results containing clean and natural textures, while the result of ADM [18] are overly sharpened, producing unreal artifacts. The MOS results are shown in Fig. 4, indicating that our MDA-SR model produced the highest scores on average and with relatively smaller variance.

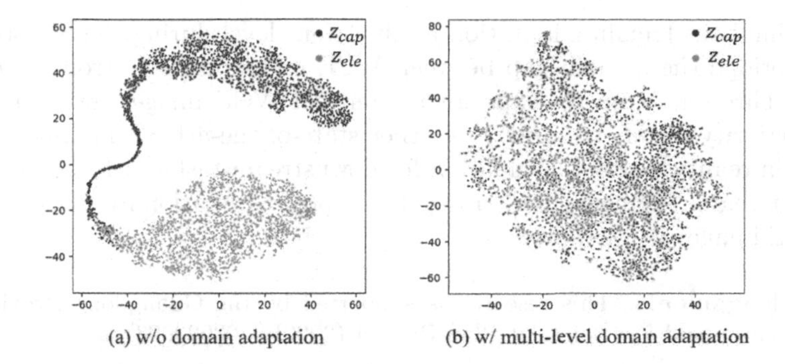

(a) w/o domain adaptation (b) w/ multi-level domain adaptation

Fig. 5. Visual representation of electronic endoscopy latent feature z_{ele} and WCE latent feature z_{cap} via t-SNE. z_{cap} in both (a) and (b) is obtained by encoding the I_{cap}^{LR}. z_{ele} in (a) is obtained by encoding the bicubic downsampled electronic endoscopy image I_{ele}^{LR}, while the z_{ele} in (b) is obtained by encoding the adaptively degraded electronic endoscopy image I_{gen}^{LR}

Table 2. Ablation study on our proposed method. We report four different levels of domain adaptation for the MDA-SR affect the SR results on WCE images.

Image Level	Latent Level	EIQE↓	
		2x	4x
		5.96	6.85
✓		5.76	6.35
	✓	5.48	6.26
✓	✓	**5.14**	**6.16**

Ablation Study. We use the t-SNE [10] for the visual representation of the image latent features distribution during SR process. It can be observed from Fig. 5(a) that there is a significant domain gap between WCE images and electronic endoscopy images, and from Fig. 5(b) show that our MDA-SR effectively bridges the domain gap. To better verify the effectiveness of our proposed model, we perform ablation experiments on WCE dataset. We obtain four different frameworks by removing different levels of domain adaptation structures. As shown in Table 2, domain adaptation at both the image and latent levels is effective. The worst framework is to train SR model on electronic endoscopy dataset and test it on WCE dataset. Our proposed MDA-SR model has the best SR effect on WCE images.

4 Conclusion

In this paper, we propose a multi-level domain adaptation SR for real WCE images. Our method first utilizes adaptive degradation to simulate the degradation distribution of WCE and generate LR electronic endoscopy images. We then employ implicit domain adaptation at the latent level during the SR process to further bridge the domain gap between WCE images and electronic endoscopy images. Through extensive experiments on real WCE images, we demonstrate the superiority of our method over other state-of-the-art SR methods, and its efficacy in reality. Further evaluation for downstream tasks such as disease classification, region segmentation, or depth and pose estimation from the generated SR WCE images is warranted.

Acknowledgements. This research is supported by the Guangdong Province Key Field Research and Development Plan Project (2022B0303020003).

References

1. Almalioglu, Y., et al.: EndoL2H: deep super-resolution for capsule endoscopy. IEEE Trans. Med. Imaging **39**(12), 4297–4309 (2020)
2. Bell-Kligler, S., Shocher, A., Irani, M.: Blind super-resolution kernel estimation using an internal-GAN. In: Advances in Neural Information Processing Systems 32 (2019)
3. Bulat, A., Yang, J., Tzimiropoulos, G.: To learn image super-resolution, use a GAN to learn how to do image degradation first. In: Ferrari, V., Hebert, M., Sminchisescu, C., Weiss, Y. (eds.) ECCV 2018. LNCS, vol. 11210, pp. 187–202. Springer, Cham (2018). https://doi.org/10.1007/978-3-030-01231-1_12
4. Fante, K.A., Abdurahman, F., Gemeda, M.T.: An ingenious application-specific quality assessment methods for compressed wireless capsule endoscopy images. Trans. Environ. Electr. Eng. **4**(1), 18–24 (2020)
5. Goodfellow, I., et al.: Generative adversarial networks. Commun. ACM **63**(11), 139–144 (2020)
6. Gu, J., Lu, H., Zuo, W., Dong, C.: Blind super-resolution with iterative kernel correction. In: Proceedings of the IEEE/CVF Conference on Computer Vision and Pattern Recognition, pp. 1604–1613 (2019)
7. Isola, P., Zhu, J.Y., Zhou, T., Efros, A.A.: Image-to-image translation with conditional adversarial networks. In: Proceedings of the IEEE Conference on Computer Vision and Pattern Recognition, pp. 1125–1134 (2017)
8. Lim, B., Son, S., Kim, H., Nah, S., Mu Lee, K.: Enhanced deep residual networks for single image super-resolution. In: Proceedings of the IEEE Conference on Computer Vision and Pattern Recognition Workshops, pp. 136–144 (2017)
9. Lugmayr, A., Danelljan, M., Timofte, R.: Unsupervised learning for real-world super-resolution. In: 2019 IEEE/CVF International Conference on Computer Vision Workshop (ICCVW), pp. 3408–3416. IEEE (2019)
10. Van der Maaten, L., Hinton, G.: Visualizing data using t-SNE. J. Mach. Learn. Res. **9**(11), 2579–2605 (2008)
11. Mao, X., Li, Q., Xie, H., Lau, R.Y., Wang, Z., Paul Smolley, S.: Least squares generative adversarial networks. In: Proceedings of the IEEE International Conference on Computer Vision, pp. 2794–2802 (2017)
12. Michaeli, T., Irani, M.: Nonparametric blind super-resolution. In: Proceedings of the IEEE International Conference on Computer Vision, pp. 945–952 (2013)
13. Mittal, A., Moorthy, A.K., Bovik, A.C.: Blind/referenceless image spatial quality evaluator. In: 2011 Conference Record of the Forty Fifth Asilomar Conference on SignAls, Systems And Computers (ASILOMAR), pp. 723–727. IEEE (2011)

14. Mittal, A., Soundararajan, R., Bovik, A.C.: Making a completely blind image quality analyzer. IEEE Signal Process. Lett. **20**(3), 209–212 (2012)
15. Muhammad, K., Khan, S., Kumar, N., Del Ser, J., Mirjalili, S.: Vision-based personalized wireless capsule endoscopy for smart healthcare: taxonomy, literature review, opportunities and challenges. Futur. Gener. Comput. Syst. **113**, 266–280 (2020)
16. Pogorelov, K., et al.: KVASIR: a multi-class image dataset for computer aided gastrointestinal disease detection. In: Proceedings of the 8th ACM on Multimedia Systems Conference, pp. 164–169 (2017)
17. Rex, D.K., et al.: High-definition colonoscopy versus Endocuff versus Endorings versus Full-spectrum Endoscopy for adenoma detection at colonoscopy: a multicenter randomized trial. Gastrointest. Endosc. **88**(2), 335–344 (2018)
18. Son, S., Kim, J., Lai, W.S., Yang, M.H., Lee, K.M.: Toward real-world super-resolution via adaptive downsampling models. IEEE Trans. Pattern Anal. Mach. Intell. **44**(11), 8657–8670 (2021)
19. Subramanian, V., Mannath, J., Hawkey, C., Ragunath, K.: High definition colonoscopy vs. standard video endoscopy for the detection of colonic polyps: a meta-analysis. Endoscopy **43**(06), 499–505 (2011)
20. Sun, W., Gong, D., Shi, Q., van den Hengel, A., Zhang, Y.: Learning to zoom-in via learning to zoom-out: real-world super-resolution by generating and adapting degradation. IEEE Trans. Image Process. **30**, 2947–2962 (2021)
21. Tong, T., Li, G., Liu, X., Gao, Q.: Image super-resolution using dense skip connections. In: Proceedings of the IEEE International Conference on Computer Vision, pp. 4799–4807 (2017)
22. Ulyanov, D., Vedaldi, A., Lempitsky, V.: Instance normalization: the missing ingredient for fast stylization. arXiv preprint arXiv:1607.08022 (2016)
23. Wang, A., et al.: Wireless capsule endoscopy. Gastrointest. Endosc. **78**(6), 805–815 (2013)
24. Wang, W., Zhang, H., Yuan, Z., Wang, C.: Unsupervised real-world super-resolution: a domain adaptation perspective. In: Proceedings of the IEEE/CVF International Conference on Computer Vision, pp. 4318–4327 (2021)
25. Yuan, Y., Liu, S., Zhang, J., Zhang, Y., Dong, C., Lin, L.: Unsupervised image super-resolution using cycle-in-cycle generative adversarial networks. In: Proceedings of the IEEE Conference on Computer Vision and Pattern Recognition Workshops, pp. 701–710 (2018)
26. Zhang, Y., Li, K., Li, K., Wang, L., Zhong, B., Fu, Y.: Image super-resolution using very deep residual channel attention networks. In: Proceedings of the European Conference on Computer Vision (ECCV), pp. 286–301 (2018)
27. Zhao, T., Ren, W., Zhang, C., Ren, D., Hu, Q.: Unsupervised degradation learning for single image super-resolution. arXiv preprint arXiv:1812.04240 (2018)
28. Zhu, J.Y., Park, T., Isola, P., Efros, A.A.: Unpaired image-to-image translation using cycle-consistent adversarial networks. In: Proceedings of the IEEE International Conference on Computer Vision, pp. 2223–2232 (2017)

PROnet: Point Refinement Using Shape-Guided Offset Map for Nuclei Instance Segmentation

Siwoo Nam, Jaehoon Jeong, Miguel Luna, Philip Chikontwe, and Sang Hyun Park(✉)

Department of Robotics and Mechatronics Engineering, Daegu Gyeongbuk Institute of Science and Technology (DGIST), Daegu, Korea
{siwoonam,shpark13135}@dgist.ac.kr

Abstract. Recently, weakly supervised nuclei segmentation methods using only points are gaining attention, as they can ease the tedious labeling process. However, most methods often fail to separate adjacent nuclei and are particularly sensitive to point annotations that deviate from the center of nuclei, resulting in lower accuracy. In this study, we propose a novel weakly supervised method to effectively distinguish adjacent nuclei and maintain robustness regardless of point label deviation. We detect and segment nuclei by combining a binary segmentation module, an offset regression module, and a center detection module to determine foreground pixels, delineate boundaries and identify instances. In training, we first generate pseudo binary masks using geodesic distance-based Voronoi diagrams and k-means clustering. Next, segmentation predictions are used to repeatedly generate pseudo offset maps that indicate the most likely nuclei center. Finally, an Expectation Maximization (EM) based process iteratively refines initial point labels based on the offset map predictions to fine-tune our framework. Experimental results show that our model consistently outperforms state-of-the-art methods on public datasets regardless of the point annotation accuracy.

Keywords: Weakly Supervised Nuclei Segmentation · Instance Segmentation · Point Refinement · Offset Map · Geodesic Distance

1 Introduction

Nuclei segmentation in histopathology images is an important task for cancer diagnosis and immune response prediction [1,13,18]. While several fully supervised deep learning approaches to segment nuclei exist [2,6,8,9,19,25], labeling

S. Nam and J. Jeong—Equal contribution.

Supplementary Information The online version contains supplementary material available at https://doi.org/10.1007/978-3-031-43907-0_51.

H. Greenspan et al. (Eds.): MICCAI 2023, LNCS 14220, pp. 528–538, 2023.
https://doi.org/10.1007/978-3-031-43907-0_51

thousands of instances are tedious and the ambiguous nature of nuclei boundaries requires high-level expert annotators. To address this, weakly-supervised nuclei segmentation methods [5,10,15,20,23,28] have emerged as an attractive alternative using cheap and inexact labels *e.g.*, center point annotations. As point labels alone do not provide sufficient foreground information, it is common to use Euclidean distance-based Voronoi diagrams and k-means clustering [7] to generate pseudo segmentation labels for training. However, since Euclidean distance-based schemes only use distance information while ignoring color, they often fail to capture nuclei shape information; resulting in inadequate boundary delineation between adjacent nuclei. Moreover, prior methods [17,21,22] typically assume that point labels are located precisely at the center of the nuclei. In real-world scenarios, point annotation locations may shift from nuclei centers as a result of the expert labeling process, leading to a lower performance after model training.

To overcome these challenges, we propose a novel weakly supervised instance segmentation method that effectively distinguishes adjacent nuclei and is robust to point shifts. The proposed model consists of three modules responsible for binary segmentation, boundary delineation, and instance separation. To train the binary segmentation module, we generate pseudo binary segmentation masks using geodesic distance-based Voronoi labels and cluster labels from point annotations. Geodesic distance provides more precise nuclei shape information than previous Euclidean distance-based schemes. To train the offset map module, we generate pseudo offset maps by computing the offset distance between binary segmentation pixel predictions and the point label. The offset information facilitates precise delineation of the boundaries between adjacent nuclei. To make the model robust to center point shifts, we introduce an Expectation Maximization (EM) [4] algorithm-based process to refine point labels. Note that previous approaches [17,21,22] optimize model parameters only using a fixed set of point labels, while we instead alternatively update model parameters and the center point locations. This refinement process ensures that the model maintains high performance even when the point annotation is not exactly located at the center of the nuclei.

The contributions of this paper are as follows: (1) We propose an end-to-end weakly supervised segmentation model that simultaneously predicts binary mask, offset map, and center map to accurately identify and segment nuclei. (2) By utilizing geodesic distance, we produce more detailed Voronoi and cluster labels that precisely delineate the boundary between adjacent nuclei. (3) We introduce an EM algorithm-based refinement process to encourage model robustness on center-shifted point labels. (4) Ablation and evaluation studies on two public datasets demonstrate our model's ability to outperform state-of-the-art techniques not only with ideal labels but also with shifted labels.

2 Methodology

We propose an end-to-end nuclei segmentation method that only uses point annotations P to predict nuclei instance segmentation masks $\hat{S}$. The proposed

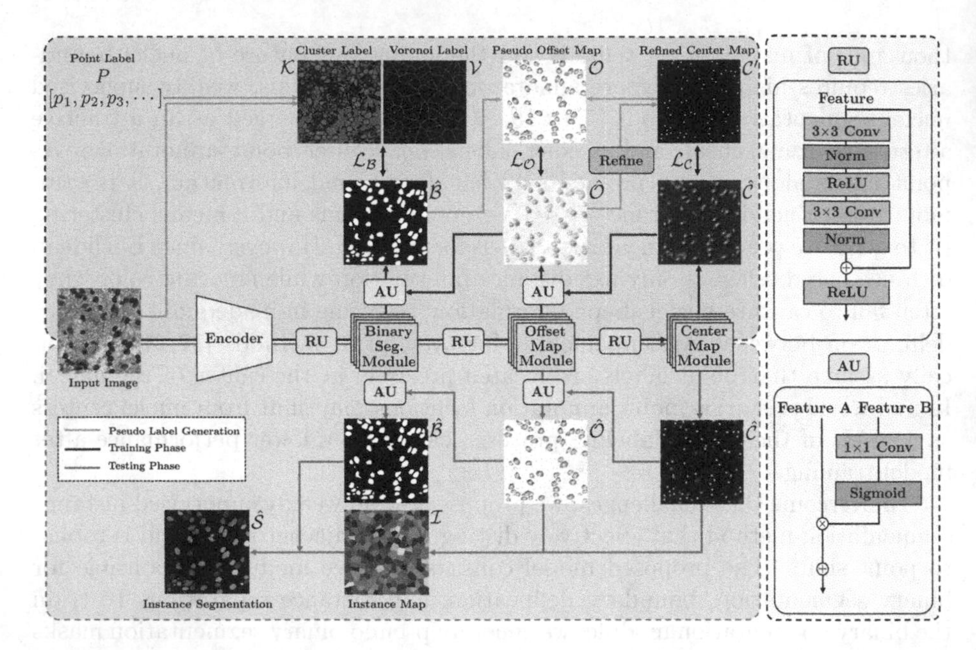

Fig. 1. Overview of the proposed method. It consists of an encoder and three modules for binary segmentation, offset map and center map prediction. To train offset map and center map modules(blue lines), pseudo labels are generated using point label and predicted binary segmentation mask(green lines). During inference, the instance map, obtained by predicted offset map and center map, is multiplied with predicted binary mask to produce instance segmentation prediction(orange lines). (Color figure online)

model consists of three modules: 1) binary segmentation module, 2) offset map module, and 3) center map module (Fig. 1). For a given input image, we extract feature maps with an ImageNet-pretrained VGG16 backbone encoder. The feature maps are further processed through a series of residual units (RUs) and attention units (AUs) to predict a binary segmentation mask $\hat{\mathcal{B}}$, an offset map $\hat{\mathcal{O}}$, and a center map $\hat{\mathcal{C}}$. The RUs are employed to maintain feature information so that subsequent modules can reuse the features from early-stage modules. In contrast, the AUs are used to refine the features of initial modules by using the predictions of later modules. In particular, the AUs use the point predictions to refine the features in the offset module, and the offset predictions to refine the features in the binary module.

In the training stage, we first generate a Voronoi label $\mathcal{V}$ and a cluster label $\mathcal{K}$ along the green lines in Fig. 1 to train the segmentation module. Then, we generate the pseudo offset map $\mathcal{O}$ by using $\hat{\mathcal{B}}$ and P. Next, following [29], we generate the center map $\mathcal{C}$ by expanding the point label P with Gaussian kernel within a radius r. Herein, our model is trained wih a segmentation loss $\mathcal{L}_{\mathcal{B}}(\mathcal{V},\mathcal{K},\hat{\mathcal{B}})$, an offset map loss $\mathcal{L}_{\mathcal{O}}(\mathcal{O},\hat{\mathcal{O}})$, and a center map loss $\mathcal{L}_{\mathcal{C}}(\mathcal{C},\hat{\mathcal{C}})$. Note that P can not sufficiently enable model robustness to imprecise point annotations. Thus, we employ an EM algorithm to search the optimal model parameters θ to obtain more reliable points P'.

In the inference stage, $\hat{\mathcal{B}}$, $\hat{\mathcal{O}}$, $\hat{\mathcal{C}}$ are predicted following the orange lines in Fig. 1. Then, we generate an instance map $\mathcal{I}$, which shares the same values among the same instances as follows:

$$\mathcal{I}(x, y) = \arg\min_{i} ||(x_{\hat{\mathcal{C}}_i}, y_{\hat{\mathcal{C}}_i}) - ((x, y) + \hat{\mathcal{O}}(x, y))||^2, \tag{1}$$

where (x, y) represents a coordinate and $(x_{\hat{\mathcal{C}}_i}, y_{\hat{\mathcal{C}}_i})$ means the location of ith point obtained from $\hat{\mathcal{C}}$. Finally, the instance segmentation output $\hat{\mathcal{S}}$ is obtained by $\hat{\mathcal{B}} \times \mathcal{I}$.

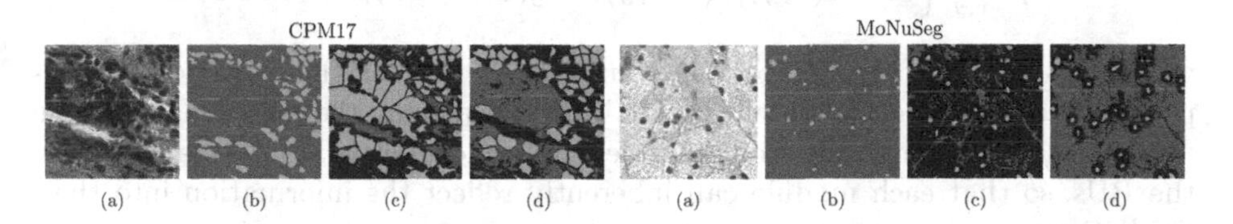

Fig. 2. Visualization of cluster label on CPM17(left) and MoNuSeg(right). (a) Input image; (b) ground truth; (c) the cluster labels generated by Euclidean distance, and (d) those by Geodesic distance. The green, red, and black colors are foreground, background, and ignored, respectively. (Color figure online)

2.1 Loss Functions Using Pseudo Labels

Segmentation Loss. We generate $\mathcal{V}$ and $\mathcal{K}$ to train the binary segmentation module. In [21], $\mathcal{V}$ was generated based on Euclidean distance between points without considering color information. As a result, the Voronoi boundaries are often created across nuclei instances, and the offset map's quality was limited. To mitigate this, we instead generate $\mathcal{V}$ using Geodesic distance [3,24] by computing distances d_i between all center points $p_i \in P$ and pixels. The boundaries of the diagram in $\mathcal{V}$ are defined as 0, while center points and the other regions are defined as 1 and 2, respectively.

For k-means clustering, we concatenate the RGB values and the geodesic distance value d_i truncated by d^* to generate the feature vectors $f_i = (d_i, r_i, g_i, b_i)$. We cluster f into three clusters (0 for background, 1 for foreground, and 2 for ignore) to generate $\mathcal{K}$ (Fig. 2d). To train the binary segmentation module using $\mathcal{V}$ and $\mathcal{K}$, we employ a Voronoi loss $\mathcal{L}_{\mathcal{V}}$ and a cluster loss $\mathcal{L}_{\mathcal{K}}$ based on the cross-entropy:

$$\mathcal{L}_{\mathcal{V}} = \frac{1}{N_{\mathbf{\Omega}_{\mathcal{V}}}} \sum_{x,y \in \mathbf{\Omega}_{\mathcal{V}}} \mathcal{V}(x, y)\log(\hat{\mathcal{B}}(x, y)) + (1 - \mathcal{V}(x, y))\log(1 - \hat{\mathcal{B}}(x, y)),$$

$$\mathcal{L}_{\mathcal{K}} = \frac{1}{N_{\mathbf{\Omega}_{\mathcal{K}}}} \sum_{x,y \in \mathbf{\Omega}_{\mathcal{K}}} \mathcal{K}(x, y)\log(\hat{\mathcal{B}}(x, y)) + (1 - \mathcal{K}(x, y))\log(1 - \hat{\mathcal{B}}(x, y)), \tag{2}$$

where $\mathbf{\Omega}_{\mathcal{V}}$ and $\mathbf{\Omega}_{\mathcal{K}}$ are the set of foreground and background pixels in $\mathcal{V}$ and $\mathcal{K}$, $N_{\mathbf{\Omega}_{\mathcal{V}}}$ and $N_{\mathbf{\Omega}_{\mathcal{K}}}$ denote the cardinality of $\mathbf{\Omega}_{\mathcal{V}}$ and $\mathbf{\Omega}_{\mathcal{K}}$. Following [17], we define the final segmentation loss as $\mathcal{L}_{\mathcal{B}} = \mathcal{L}_{\mathcal{V}} + \mathcal{L}_{\mathcal{K}}$.

Center Map Loss. To achieve instance-level predictions, we introduce a center map module. The module predicts a keypoint heatmap $\hat{\mathcal{C}} \in [0,1]^{W \times H}$ where $\hat{\mathcal{C}} = 1$ identifies nuclei centers and $\hat{\mathcal{C}} = 0$ for other pixels. W and H are the width and height of the input image. To train the module we employ a focal loss, commonly used in point detection problems. This loss can focus on a set of sparse hard examples while preventing easy negatives from dominating the model [16]:

$$\mathcal{L}_{\mathcal{C}} = \frac{-1}{N_P} \sum_{x,y} \begin{cases} (1-\hat{\mathcal{C}}(x,y))^{\alpha} \log(\hat{\mathcal{C}}(x,y)) & \text{if } \mathcal{C}(x,y) = 1 \\ (1-\mathcal{C}(x,y))^{\beta} (\hat{\mathcal{C}}(x,y))^{\alpha} \log(1-\hat{\mathcal{C}}(x,y)) & \text{otherwise,} \end{cases} \tag{3}$$

where N_P denotes the number of point labels. We set the focal loss hyper-parameters $\alpha = 2$ and $\beta = 4$ following [14,29]. By placing the center map module at the end of the model, the model is able to retain center point information along the RUs, so that each module can inherently reflect the information into their predictions.

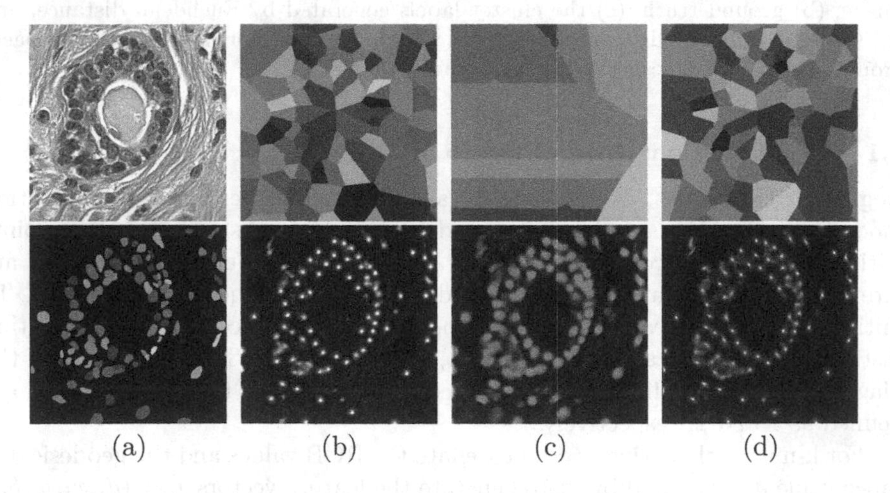

Fig. 3. (a) Input image (top) and ground truth (bottom), (b) Instance map (top) and center map (bottom) generated by the optimal nuclei center points, (c) those by shifted points (6–8), and (d) those by refined points.

Offset Map Loss. We employ an offset map module that considers the shape of each nucleus to improve boundary detection. Inspired by [2], we define an offset vector $\mathcal{O}(x,y)$ that indicates the displacement of a point (x,y) to the center of its corresponding nucleus. To train the offset module, we first compute $\mathcal{O}(x,y)$ of each nucleus segmented by $\hat{\mathcal{B}}$. Then, $\mathcal{L}_{\mathcal{O}}$ is defined as an L1 loss:

$$\mathcal{L}_{\mathcal{O}} = \frac{1}{W \times H} \sum_{x,y} |\mathcal{O}(x,y) - \hat{\mathcal{O}}(x,y)|. \tag{4}$$

It is worth noting that in the early stages of training, the pseudo offset map $\mathcal{O}$ generated by $\hat{B}$ and P is unreliable. Thus, we empirically use $\mathcal{L}_{\mathcal{O}}$ for back-propagation after 20 epochs. We optimize the entire model using the loss $\mathcal{L} = \lambda_{\mathcal{B}}\mathcal{L}_{\mathcal{B}} + \lambda_{\mathcal{O}}\mathcal{L}_{\mathcal{O}} + \lambda_{\mathcal{C}}\mathcal{L}_{\mathcal{C}}$, where $\lambda_{\mathcal{B}}$, $\lambda_{\mathcal{O}}$ and $\lambda_{\mathcal{C}}$ denote loss weights.

2.2 Refinement via Expectation Maximization Algorithm

Training with nuclei (center) shifted point labels can lead to blurry center map predictions (see Fig. 3c). This in turn limits model optimization and it's ability to distinguish objects, resulting in poor adjacent nuclei segmentation. To address this, we propose an EM based center point refinement process. Instead of the standard fixed-point label based model optimization, we alternatively optimize both model parameters and point labels.

In the E-step, we update the center of each nucleus according to $\hat{\mathcal{O}}$. We use $\hat{\mathcal{O}}$ to generate refined point labels P', since $\hat{\mathcal{O}}$ is reliable regardless of the point location *i.e.*, center of the nuclei or shifted.

$$p_i' = \arg\min_{x,y} \Big| \sum_{\bar{x},\bar{y} \in v_i} \hat{\mathcal{B}}(x,y) \times \hat{\mathcal{O}}(x+\bar{x}, y+\bar{y}) \Big|, \tag{5}$$

where v_i is i^{th} Voronoi region and p_i' is the refined center point. We repeat this for all Voronoi regions to obtain P', and replace P with P' if the distance between them is $< \delta$. In the M-step of iteration n, we generate $\mathcal{C}'$ by adapting the Gaussian mask to P', and then use it to train offset and center map modules. As maximizing a probability distribution is the same as minimizing the loss, the model parameter θ minimizing $\mathcal{L}$ is optimized as:

$$\theta^n := \arg\min_{\theta}(\mathcal{L}(\theta^n; \theta^{n-1}, X, \mathcal{V}, \mathcal{K}, \mathcal{O}, \mathcal{C}')). \tag{6}$$

Since reliable $\hat{\mathcal{O}}$ is necessary to refine nuclei centers, refinement starts after 30 epochs. E and M steps are alternately repeated to correct imprecise annotations bringing them closer to the real nuclei center points.

3 Experiments

Dataset. To validate the effectiveness of our model, we use two public nuclei segmentation datasets *i.e.*, CPM17 [26] & MoNuSeg [12]. CPM17 contains 64 H&E stained images with 7,570 annotated nuclei boundaries sized from 500×500 to 600×600. The set is split into 32/32 images for training and testing. Images were normalized and cropped to 300×300. MoNuSeg is a multi-organ nuclei segmentation dataset consisting of 30 H&E stained images (1000×1000) extracted from seven different organs. We used 16 images (4 images from the breast, liver, kidney, and prostate) as training and 14 images (2 images from each breast, liver, kidney, prostate, bladder, brain, and stomach) as testing. For a fair comparison, images were pre-processed before training/testing *i.e.*, normalized and cropped to 250×250 patches following the setting used in [17].

To make point labels, we use the center point of full mask annotations. For a realistic scenario, we generate shifted point label. The shift is performed in pixels and is randomly selected between the minimum and maximum values.

Implementation Details. For training, all evaluated models were run for 150 epochs with the Adam optimizer [11] using a learning rate of 1e-4, weight decay of 3e-2, and batch size of 4. The GeodisTK [27] library was used to compute geodesic distances. For clustering, we set the maximum distance d^* as 90 and 70 on CPM17 and MoNuSeg, respectively. The Gaussian kernel r was set as $r = 6$ and δ was set as 8 for refinement on CPM17. For MoNuSeg, $r = 8$ and $\delta = 8$, respectively. A threshold of 0.2 was applied to eliminate the noise and find important points in $\hat{\mathcal{C}}$. Finally, a variety of augmentations were employed *i.e.*, random resizing, cropping, and rotations etc., following [17], with loss weights $\lambda_{\mathcal{B}}, \lambda_{\mathcal{O}}$ and $\lambda_{\mathcal{C}}$ empirically set to 1. We used a NVIDIA RTX A5000 GPU and PyTorch version 1.7.1.

Table 1. Performance comparison of nuclei segmentation on two public datasets. Shift indicates the number of pixels point annotations deviate from the nuclei center.

	CPM17								MoNuSeg							
	shift0		shift2-4		shift4-6		shift6-8		shift0		shift2-4		shift4-6		shift6-8	
	Dice	AJI	Dice	AJI	Dice	AJI	Dice	AJI	Dice	AJI	Dice	AJI	Dice	AJI	Dice	AJI
MIDL [21]	75.0	55.5	75.3	56.9	74.4	53.7	72.2	49.9	70.1	44.9	69.9	45.0	66.3	39.9	61.0	31.5
Mixed Anno [22]	75.3	53.2	75.9	55.5	73.3	52.3	73.1	49.9	73.3	51.6	72.0	49.4	66.0	40.5	66.9	41.8
SPN+IEN [17]	74.3	54.3	72.9	52.1	70.1	47.9	69.4	46.8	74.0	53.4	72.3	50.4	69.1	46.5	65.6	39.4
PROnet	**78.7**	**62.7**	**78.2**	**61.8**	**77.4**	**60.7**	**77.0**	**60.2**	**75.0**	**55.5**	**74.8**	**54.8**	**73.3**	**53.2**	**72.5**	**50.9**

Main Results. Table 1 shows the performance of our method against state-of-the-art weakly supervised nuclei segmentation methods [17,21,22] based on Dice and Aggregated Jaccard Index (AJI) metrics. As opposed to the Dice score, AJI is key when evaluating adjacent nuclei separation in instance segmentation tasks. On CPM17, our method outperformed the prior approach by a large margin of +3.4% in Dice and +7.2% in AJI when the point label is located at the nuclei center. More importantly, our approach surpassed prior approaches by a substantial margin when the shift exists. We obtain statistically significant (p-value <0.05) for the AJI of all comparison methods on two datasets in all scenarios. Regarding refinement, we observed that our strategy is more beneficial when points exhibit significant shifts *i.e.*, on both CPM and MoNuSeg. Figure 3 showcases the effectiveness of the refinement process wherein the model generates precise instance and center maps. With the geodesic distance and the refinement process, our proposed method achieved state-of-the-art performance. This demonstrates that our method separates adjacent nuclei accurately, and maintains its robustness, achieving consistent performance even when the point annotations are not located at the center of the nuclei. Additionally, in Fig. 4, we qualitatively show the results to highlight how our method precisely separates adjacent nuclei.

Table 2. Evaluation on the effect of offset and center maps.

offset	geo	refine	CPM17								MoNuSeg							
			shift0		shift2-4		shift4-6		shift6-8		shift0		shift2-4		shift4-6		shift6-8	
			Dice	AJI	Dice	AJI	Dice	AJI	Dice	AJI	Dice	AJI	Dice	AJI	Dice	AJI	Dice	AJI
x	x	x	78.0	62.2	77.4	61.2	52.6	38.1	–	–	69.6	45.9	68.3	43.8	67.1	40.5	64.0	38.4
x	o	x	78.5	62.7	78.0	61.8	59.7	43.8	59.5	42.8	73.6	54.2	73.6	54.2	68.0	42.6	64.0	38.5
o	x	x	77.9	61.8	77.4	60.3	74.0	56.2	67.8	48.7	74.5	55.0	73.4	52.7	71.9	49.3	66.4	39.8
o	o	x	78.3	62.5	78.0	61.5	75.2	58.0	74.1	55.2	75.0	55.3	74.2	54.4	72.5	52.0	67.5	42.0
o	x	o	78.1	61.9	78.1	61.7	76.6	58.4	75.0	55.8	74.6	55.4	74.7	54.7	72.6	50.2	70.3	47.4
o	o	o	**78.7**	**62.7**	**78.2**	**61.8**	**77.4**	**60.7**	**77.0**	**60.2**	**75.0**	**55.5**	**74.8**	**54.8**	**73.3**	**53.2**	**72.5**	**50.9**

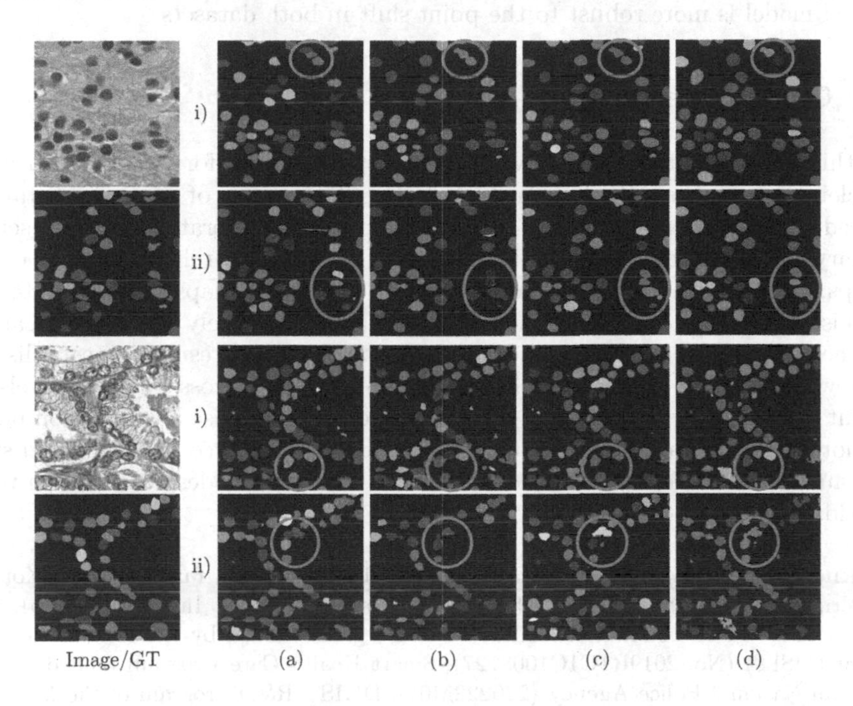

Fig. 4. Nuclei instance segmentation results on CPM17 (top 2 rows) and MoNuSeg (bottom 2 rows) images. The images and the ground truth (GT) are shown in the left column. The results using the precise point annotations are shown in i), while those using shifted (6–8) points are shown in ii). (a) PROnet (ours), (b) SPN+IEN [17], (c) Mixed Anno [22] and (d) MIDL [21]. The yellow circles indicate the major differences.

Ablation Studies. We conducted ablation studies to assess the impact of the offset regression module, geodesic distance, and point refinement process (Table 2). When the binary segmentation module is combined only with the center map module without the offset module, the model could separate nuclei only trained by the ideal label. On the other hand, since there was no refinement process due to the absence of the offset map, inaccurate points extracted from the center map are obtained in the real-world scenario. We also demonstrate that labels with Geodesic distance help improve overall performance. This is because it creates confident labels and more decent divides the boundaries between nuclei. Finally, using the full set of modules along with a complete instance map, the model was able to separate adjacent nuclei with precise boundaries, ultimately reporting higher scores. These findings validate the utility of the center map and offset map modules *i.e.*, they synergistically facilitate precise instance delineation and nuclei boundary prediction. The geodesic distance and refinement process also improved the accuracy by contributing to more accurate pseudo labels. Especially, most variants show a significant drop in performance when the annotations shift was over 4 pixels. Compared to other variants, our proposed model is more robust to the point shift in both datasets.

4 Conclusion

In this work, we proposed a novel and robust framework for weakly supervised nuclei segmentation. We demonstrated the effectiveness of geodesic distance-based Voronoi diagrams and k-means clustering to generate accurate pseudo binary segmentation labels. This allowed us to generate reliable pseudo offset maps, and then we iteratively improve the pseudo offset maps that facilitate the precise separation of adjacent nuclei as well as progressively refine the location of the center point labels. According to our experimental results, we established a new state-of-art on two publicly available datasets across different levels of point annotation imperfections. We believe being able to use low-precision point annotations while retaining good segmentation performance is an essential step for automatic nuclei segmentation models to become a widespread tool in real-world clinical practice.

Acknowledgment. This work was supported by IITP grant funded by the Korean government (MSIT) (No.2021-0-02068, Artificial Intelligence Innovation Hub), the National Research Foundation of Korea (NRF) grant funded by the Korean Government (MSIT) (No. 2019R1C1C1008727), Smart Health Care Program funded by the Korean National Police Agency (220222M01), DGIST R&D program of the Ministry of Science and ICT of KOREA (21-DPIC-08)

References

1. Alsubaie, N., Sirinukunwattana, K., Raza, S.E.A., Snead, D., Rajpoot, N.: A bottom-up approach for tumour differentiation in whole slide images of lung adenocarcinoma. In: Medical Imaging 2018: Digital Pathology, vol. 10581, pp. 104–113. SPIE (2018)
2. Cheng, B., et al.: Panoptic-DeepLab: a simple, strong, and fast baseline for bottom-up panoptic segmentation. In: Proceedings of the IEEE/CVF Conference on Computer Vision and Pattern Recognition, pp. 12475–12485 (2020)
3. Criminisi, A., Sharp, T., Blake, A.: GeoS: geodesic image segmentation. In: Forsyth, D., Torr, P., Zisserman, A. (eds.) ECCV 2008. LNCS, vol. 5302, pp. 99–112. Springer, Heidelberg (2008). https://doi.org/10.1007/978-3-540-88682-2_9

4. Dempster, A.P., Laird, N.M., Rubin, D.B.: Maximum likelihood from incomplete data via the EM algorithm. J. Royal Stat. Soc.: Series B (Methodological) **39**(1), 1–22 (1977)
5. Dong, M., et al.: Towards neuron segmentation from macaque brain images: a weakly supervised approach. In: Martel, A.L., et al. (eds.) MICCAI 2020. LNCS, vol. 12265, pp. 194–203. Springer, Cham (2020). https://doi.org/10.1007/978-3-030-59722-1_19
6. Graham, S., et al.: Hover-Net: simultaneous segmentation and classification of nuclei in multi-tissue histology images. Med. Image Anal. **58**, 101563 (2019)
7. Hartigan, J.A., Wong, M.A.: Algorithm as 136: a k-means clustering algorithm. J. Royal Stat. Soc. Ser. c (Applied Statistics) **28**(1), 100–108 (1979)
8. He, H., et al.: CDNet: centripetal direction network for nuclear instance segmentation. In: Proceedings of the IEEE/CVF International Conference on Computer Vision, pp. 4026–4035 (2021)
9. Kendall, A., Gal, Y., Cipolla, R.: Multi-task learning using uncertainty to weigh losses for scene geometry and semantics. In: Proceedings of the IEEE Conference on Computer Vision and Pattern Recognition, pp. 7482–7491 (2018)
10. Khoreva, A., Benenson, R., Hosang, J., Hein, M., Schiele, B.: Simple does it: weakly supervised instance and semantic segmentation. In: Proceedings of the IEEE Conference on Computer Vision and Pattern Recognition, pp. 876–885 (2017)
11. Kingma, D.P., Ba, J.: Adam: a method for stochastic optimization. arXiv preprint arXiv:1412.6980 (2014)
12. Kumar, N., et al.: A multi-organ nucleus segmentation challenge. IEEE Trans. Med. Imaging **39**(5), 1380–1391 (2019)
13. Kumar, N., Verma, R., Sharma, S., Bhargava, S., Vahadane, A., Sethi, A.: A dataset and a technique for generalized nuclear segmentation for computational pathology. IEEE Trans. Med. Imaging **36**(7), 1550–1560 (2017)
14. Law, H., Deng, J.: CornerNet: detecting objects as paired keypoints. In: Proceedings of the European Conference on Computer Vision (ECCV), pp. 734–750 (2018)
15. Lin, D., Dai, J., Jia, J., He, K., Sun, J.: ScribbleSup: scribble-supervised convolutional networks for semantic segmentation. In: Proceedings of the IEEE Conference on Computer Vision and Pattern Recognition, pp. 3159–3167 (2016)
16. Lin, T.Y., Goyal, P., Girshick, R., He, K., Dollár, P.: Focal loss for dense object detection. In: Proceedings of the IEEE International Conference on Computer Vision, pp. 2980–2988 (2017)
17. Liu, W., He, Q., He, X.: Weakly supervised nuclei segmentation via instance learning. In: 2022 IEEE 19th International Symposium on Biomedical Imaging (ISBI), pp. 1–5. IEEE (2022)
18. Lu, C., et al.: Nuclear shape and orientation features from h&e images predict survival in early-stage estrogen receptor-positive breast cancers. Lab. Investig. **98**(11), 1438–1448 (2018)
19. Neven, D., Brabandere, B.D., Proesmans, M., Gool, L.V.: Instance segmentation by jointly optimizing spatial embeddings and clustering bandwidth. In: Proceedings of the IEEE/CVF Conference on Computer Vision and Pattern Recognition, pp. 8837–8845 (2019)
20. Nishimura, K., Ker, D.F.E., Bise, R.: Weakly supervised cell instance segmentation by propagating from detection response. In: International Conference on Medical Image Computing and Computer-Assisted Intervention, pp. 649–657. Springer (2019)

21. Qu, H., et al.: Weakly supervised deep nuclei segmentation using points annotation in histopathology images. In: International Conference on Medical Imaging with Deep Learning, pp. 390–400. PMLR (2019)
22. Qu, H., Yi, J., Huang, Q., Wu, P., Metaxas, D.: Nuclei segmentation using mixed points and masks selected from uncertainty. In: 2020 IEEE 17th International Symposium on Biomedical Imaging (ISBI), pp. 973–976. IEEE (2020)
23. Tian, K., et al.: Weakly-supervised nucleus segmentation based on point annotations: a coarse-to-fine self-stimulated learning strategy. In: Martel, A.L., et al. (eds.) MICCAI 2020. LNCS, vol. 12265, pp. 299–308. Springer, Cham (2020). https://doi.org/10.1007/978-3-030-59722-1_29
24. Toivanen, P.J.: New geodosic distance transforms for gray-scale images. Pattern Recogn. Lett. **17**(5), 437–450 (1996)
25. Uhrig, J., Rehder, E., Fröhlich, B., Franke, U., Brox, T.: Box2Pix: single-shot instance segmentation by assigning pixels to object boxes. In: 2018 IEEE Intelligent Vehicles Symposium (IV), pp. 292–299. IEEE (2018)
26. Vu, Q.D., Graham, S., et al.: Methods for segmentation and classification of digital microscopy tissue images. Front. Bioeng. Biotech., p. 53 (2019)
27. Wang, G., et al.: DeepiGeoS: a deep interactive geodesic framework for medical image segmentation. IEEE Trans. Pattern Anal. Mach. Intell. **41**(7), 1559–1572 (2019). https://doi.org/10.1109/TPAMI.2018.2840695
28. Yoo, I., Yoo, D., Paeng, K.: PseudoEdgeNet: nuclei segmentation only with point annotations. In: Shen, D., et al. (eds.) MICCAI 2019. LNCS, vol. 11764, pp. 731–739. Springer, Cham (2019). https://doi.org/10.1007/978-3-030-32239-7_81
29. Zhou, X., Wang, D., Krähenbühl, P.: Objects as points. arXiv preprint arXiv:1904.07850 (2019)

Self-Supervised Domain Adaptive Segmentation of Breast Cancer via Test-Time Fine-Tuning

Kyungsu Lee[1], Haeyun Lee[2], Georges El Fakhri[3], Jonghye Woo[3], and Jae Youn Hwang[1(✉)]

[1] Department of Electrical Engineering and Computer Science, Daegu Gyeongbuk Institute of Science and Technology, Daegu 42988, South Korea
{ks_lee,jyhwang}@dgist.ac.kr

[2] Production Engineering Research Team, Samsung SDI, Yongin 17084, South Korea

[3] Gordon Center for Medical Imaging, Department of Radiology, Massachusetts General Hospital and Harvard Medical School, Boston, MA 02114, USA

Abstract. Unsupervised domain adaptation (UDA) has become increasingly popular in imaging-based diagnosis due to the challenge of labeling a large number of datasets in target domains. Without labeled data, well-trained deep learning models in a source domain may not perform well when applied to a target domain. UDA allows for the use of large-scale datasets from various domains for model deployment, but it can face difficulties in performing adaptive feature extraction when dealing with unlabeled data in an unseen target domain. To address this, we propose an advanced test-time fine-tuning UDA framework designed to better utilize the latent features of datasets in the unseen target domain by fine-tuning the model itself during diagnosis. Our proposed framework is based on an auto-encoder-based network architecture that fine-tunes the model itself. This allows our framework to learn knowledge specific to the unseen target domain during the fine-tuning phase. In order to further optimize our framework for the unseen target domain, we introduce a re-initialization module that injects randomness into network parameters. This helps the framework to converge to a local minimum that is better-suited for the target domain, allowing for improved performance in domain adaptation tasks. To evaluate our framework, we carried out experiments on UDA segmentation tasks using breast cancer datasets acquired from multiple domains. Our experimental results demonstrated that our framework achieved state-of-the-art performance, outperforming other competing UDA models, in segmenting breast cancer on ultrasound images from an unseen domain, which supports its clinical potential for improving breast cancer diagnosis.

Keywords: Unsupervised Domain Adaptation · Test-Time Tuning · Breast Cancer · Segmentation · Ultrasound Imaging

Supplementary Information The online version contains supplementary material available at https://doi.org/10.1007/978-3-031-43907-0_52.

H. Greenspan et al. (Eds.): MICCAI 2023, LNCS 14220, pp. 539–550, 2023.
https://doi.org/10.1007/978-3-031-43907-0_52

1 Introduction

In recent years, deep learning (DL) methods have demonstrated remarkable performance in detecting and localizing tumors on ultrasound images [2,27]. Compared with conventional image processing methods, DL methods provide an accurate feature extraction capability on ultrasound images, despite their low resolution and noise disturbance, leading to superior segmentation accuracy [2,5,14]. However, there are some limitations in developing a DL model in a source domain and deploying it in an unseen target domain. The primary limitation is that DL models require a large number of training samples to achieve accurate predictions [8,24]. Yet, acquiring large training datasets and their corresponding labels, especially from a cohort of patients, can be costly or even infeasible, which poses a significant challenge in developing a DL model with high performance [7]. Second, even when large-scale datasets are available through collaborative research from multiple sites, DL models trained on such datasets may yield sub-optimal solutions due to domain gaps caused by differences in images acquired from different sites [20]. Third, due to the small number of datasets from each domain, the images for each individual domain may not capture representative features, limiting the ability of DL models to generalize across domains [3].

Domain adaptation (DA) has been extensively studied to alleviate the aforementioned limitations, the goal of which is to reduce the domain gap caused by the diversity of datasets from different domains [12,20,26,29,33]. Example solutions include transfer learning- and style transfer-based methods. Nonetheless, unlike natural images, generating labels can be a challenging task, making it difficult to apply general DA methods; thus bridging domain gaps by DA methods remains limited [26,33]. This is due to sensitive privacy issues in patients' data, particularly in collaborative research, which restricts access to labels from different domains. As a result, conventional DA methods cannot be easily applied [10]. More recently, unsupervised domain adaptation (UDA) has been introduced to address this issue [16,33], aiming to generate semi-predictions (pseudo-labels) in target domains first, followed by producing accurate predictions using the pseudo-labels. One critical limitation of pseudo-label-based UDA is the possibility of error accumulation due to mispredicted pseudo-labels. This can lead to significant degradation of the performance of DL models, as errors can compound and become more pronounced over time [17,25].

To alleviate the problem of pseudo-label-based UDA, in this work, we propose an advanced UDA framework based on self-supervised DA with a test-time finetuning network. Test-time adaptation methods have been developed [4,11,13,23] to improve the learning of knowledge in target domains. The distinctive feature of our test-time self-supervised DA is that it enables the DL network (i) to learn knowledge about the features of target domains by fine-tuning the network itself during the test-time phase, rather than generating pseudo-labels and then (ii) to provide precise predictions on images in target domains, by using the fine-tuned network. Specifically, we adopt self-supervised learning and verify the model via thorough mathematical analysis. Our framework was tested on the task of breast cancer segmentation in ultrasound images, but it could also be applied to other lesion segmentation tasks.

To summarize, our contributions are three-fold:

- We design a self-supervised DA framework that includes a parameter search method and provide a mathematical justification for it. With our framework, we are able to identify the best-performing parameters that result in improved performance in DA tasks.
- Our framework is effective at preserving privacy, since it carries out DA using only pre-trained network parameters, without transferring any patient data.
- We applied our framework to the task of segmenting breast cancer from ultrasound imaging data, demonstrating its superior performance over competing UDA methods.

Our results indicate that our framework is effective in improving the accuracy of breast cancer segmentation from ultrasound images, which could have potential implications for improving the diagnosis and treatment of breast cancer.

2 Methodology

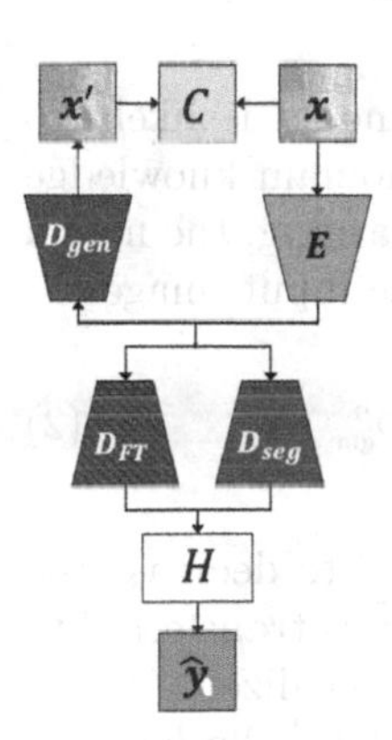

Algorithm 1: Test-Time Fine-Tuning Scheme

Input: E, H, C, and $D_{\text{gen}} = D_{\text{seg}}$
1: **def Training_on_Source:**
2: Sample batches of $(s, \bar{s}) \sim \mathcal{S}$
3: Update E and D_{seg} via $\mathcal{L}_{\text{BCE}}((H \circ D_{\text{seg}} \circ E)(s), \bar{s}))$
Update E and D_{gen} via $\mathcal{L}_{\text{GAN}}((D_{\text{gen}} \circ E)(s), s)$
4: **return** $E^{\mathcal{S}}$ and $D^{\mathcal{S}}_{\text{seg}} = D^{\mathcal{S}}_{\text{gen}}$
5: **End**
6: **def Fine_Tuning_on_Target:**
Sample batches of $(t, ?) \sim \mathcal{T}$
7: Update $D^{\mathcal{S}}_{\text{gen}}$ via $\mathcal{L}_{\text{GAN}}(D^{\mathcal{S}}_{\text{gen}}(E(t)), t)$, then $D^{\mathcal{S}\rightarrow\mathcal{T}}_{\text{gen}}$
8: Share parameters from $D^{\mathcal{S}\rightarrow\mathcal{T}}_{\text{gen}}$ to D_{FT}
9: **return** $D_{\text{FT}} = D^{\mathcal{S}\rightarrow\mathcal{T}}_{\text{seg}}$
10: **End**
11: **def Prediction_on_Target:**
12: Sample batches of $(t, ?) \sim \mathcal{T}$
13: $\hat{t} = \big(H \circ (D^{\mathcal{S}}_{\text{seg}} \oplus D_{\text{FT}}) \circ E\big)(t))$
14: **return** $\hat{y}$
15: **End**
Output: Predictions $(\hat{y})$ on $\mathcal{T}$

Fig. 1. Architecture of our TTFT network (**Left**) and its pipeline (**Right**).

2.1 Test-Time Fine-Tuning (TTFT) Network and Its Pipeline

Network Architecture. Our proposed TTFT network is based on self-supervised DA [31], which is a part of UDA and can be seen as multi-task learning, involving both the main and pretext tasks, as shown in Fig. 1. In the main task, an encoder (E), a decoder for segmentation (D_{seg}), and a segmentation header (H) are included. The main task is the segmentation task, $(H \circ D_{\text{seg}} \circ E)(x)$. In predicting segmentation labels in the target domain ($\mathcal{T}$), D_{FT} is also involved in the main task, and the final prediction after the fine-tuning is provided by $\big(H \circ (D_{\mathrm{seg}} \oplus D_{\mathrm{FT}}) \circ E\big)(x)$, where $\oplus$ is the concatenation operation. In the pretext task, E, a decoder for a generator, D_{gen}, and a discriminator, C, are involved. The pretext task aims to generate synthetic images, $(D_{\mathrm{gen}} \circ E)(t)$. Note that D_{gen} and D_{seg} share the same parameters to enable knowledge transfer. However, since the headers of image reconstruction and generating segmentation mask are different (different output), a new header incorporating D_{FT} and D_{seg} is devised and leverages the outputs of two decoders. Besides, $D_{gen} = D_{FT}$ is fine-tuned during the fine-tuning step, and the D_{FT} learns the knowledge of the input domain via image reconstruction. Two distinct knowledge (information) from D_{seg} and D_{FT} enable the network to utilize target domain knowledge and predict precise predictions.

Pre-training in Source Domain. The model M is first trained in $\mathcal{S}$ in a supervised manner with $(s, \bar{s}) \sim \mathcal{S}$ in both main and pretext tasks as below:

$$\Theta_{\mathcal{S}}^{m}, \Theta_{\mathcal{S}}^{p} = \underset{\Theta_{\mathcal{S}}^{m}, \Theta_{\mathcal{S}}^{p}}{\operatorname{argmin}} \sum_{s} \Big\{ \mathcal{L}_{\mathrm{BCE}}\big((H \circ D_{\mathrm{seg}} \circ E)(s), \bar{s}\big) + \mathcal{L}_{\mathrm{GAN}}\big((D_{\mathrm{gen}} \circ E)(s), s\big) \Big\}, \tag{1}$$

where $\mathcal{L}_{\mathrm{BCE}}$ and $\mathcal{L}_{\mathrm{GAN}}$ represent the loss functions for binary cross-entropy and generative adversarial network [6], respectively. $\Theta_{\mathcal{S}}^{m}$ includes $E^{\mathcal{S}}$, $D_{seg}^{\mathcal{S}}$, and $H^{\mathcal{S}}$, while $\Theta_{\mathcal{S}}^{p}$ includes $E^{\mathcal{S}}$, $D_{gen}^{\mathcal{S}}$, and $C^{\mathcal{S}}$. Additionally, $D_{\mathrm{seg}}^{\mathcal{S}} = D_{\mathrm{gen}}^{\mathcal{S}}$.

Fine-Tuning in Target Domain. Since the pre-trained model is likely to produce imprecise predictions in $\mathcal{T}$, the model should learn domain knowledge about $\mathcal{T}$. To this end, in the pretext task, for self-supervised learning, the model is fine-tuned in $\mathcal{T}$ to generate synthetic images identical to the input images as below:

$$\Theta_{\mathcal{T}}^{p} = \underset{\Theta_{\mathcal{T}}^{p}}{\operatorname{argmin}} \sum_{t} \mathcal{L}_{\mathrm{GAN}}\big((D_{\mathrm{gen}}^{\mathcal{S}} \circ E^{\mathcal{S}})(t), t\big) \;\Rightarrow\; \Theta_{\mathcal{T}}^{p} \supseteq E^{\mathcal{S}} \cup D_{\mathrm{gen}}^{\mathcal{S}\rightarrow\mathcal{T}}, \tag{2}$$

where only D_{gen} is fine-tuned to achieve memory efficiency and to decrease the fine-tuning time, and $D_{\mathrm{gen}}^{\mathcal{S}}$ is fine-tuned as $D_{\mathrm{gen}}^{\mathcal{S}\rightarrow\mathcal{T}}$. Then, $D_{\mathrm{gen}}^{\mathcal{S}\rightarrow\mathcal{T}}$ is transferred to D_{FT}, and knowledge distillation via self-supervised learning is realized. Hence, the precise predictions in $\mathcal{T}$ could be provided by $\big(H \circ (D_{\mathrm{seg}}^{\mathcal{S}} \oplus D_{\mathrm{FT}}^{\mathcal{T}}) \circ E\big)(x)$.

Benefits of Our Dual-Pipeline. Due to the symmetric property of mutual information in information entropy ($\mathbb{H}$), we have $I(X;Y) = H(X) + H(Y) - H(X,Y)$. As a result, the predictions made by the fine-tuned network in the target domain ($\mathcal{T}$) lead to reduced entropy, as shown below:

$$\mathbb{H}\big((H \circ (D_{\mathrm{seg}}^{\mathcal{S}} \oplus D_{\mathrm{FT}}^{\mathcal{T}}) \circ E)(t), \bar{t}\big) \leq \mathbb{H}\big((H \circ D_{\mathrm{seg}}^{\mathcal{S}} \circ E)(t), \bar{t}\big) + \mathbb{H}\big((H \circ D_{\mathrm{FT}}^{\mathcal{T}} \circ E)(t), \bar{t}\big). \tag{3}$$

Since $D_{\mathrm{seg}}^{\mathcal{S}}$ is fully optimized for $\mathcal{S}$ in a supervised manner, it guarantees a baseline segmentation performance. Furthermore, since $D_{\mathrm{FT}}^{\mathcal{T}}$ is fine-tuned in $\mathcal{T}$

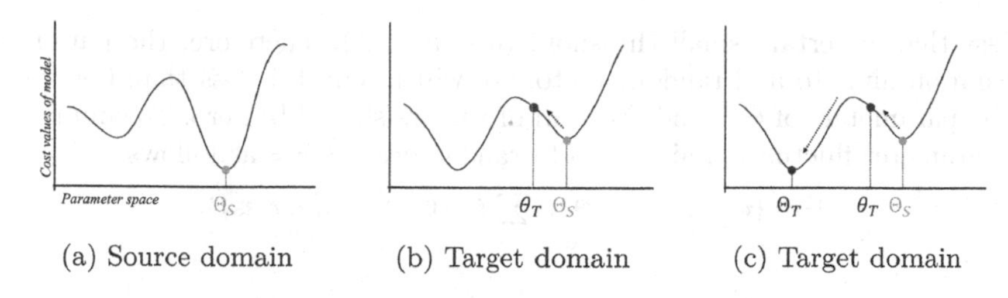

(a) Source domain (b) Target domain (c) Target domain

Fig. 2. Illustration of the local minimum of the source (a) and target (b) domains and parameter fluctuation (c)

using knowledge distillation, it can provide domain-specific information for $\mathcal{T}$. As a result, the predictions made by the fine-tuned model in $\mathcal{T}$ are jointly constrained by the expectations of $D^{\mathcal{S}}_{\text{seg}}$ and $D^{\mathcal{T}}_{\text{FT}}$. This enables the final model to provide precise predictions in $\mathcal{T}$ by taking into account both the source domain and target domain information.

2.2 Parameter Fluctuation: Parameter Randomization Method

Since the loss function and its values can vary based on the distribution of inputs, and different domains can have different distributions, the local minimum identified in the source domain ($\mathcal{S}$) cannot be considered as the same local minimum in $\mathcal{T}$, as illustrated in Fig. 2. The y-axis of Fig. 2 indicates $\frac{1}{|\mathcal{X}|}\sum_x \mathcal{L}(M(x;\theta), \bar{x})$, and the local minimum is different in $\mathcal{S}$ and $\mathcal{T}$ as $\Theta_{\mathcal{S}}$ in Fig. 2a and $\Theta_{\mathcal{T}}$ in Fig. 2c, respectively. A longer fine-tuning time is required to re-position $\Theta_{\mathcal{S}}$ to $\Theta_{\mathcal{T}}$ as in Fig. 2c than to re-position $\theta_{\mathcal{T}}$ to $\Theta_{\mathcal{T}}$. Therefore, efficient fine-tuning is necessary to re-position the local minimum in Fig. 2b and this process is known as parameter fluctuation. Note that the parameter fluctuation is followed by the fine-tuning step.

Suppose C_i be the ith convolution operator in D_{seg} with weight w_i, then $C_i(x) = w_i \cdot x$. Since $D^{\mathcal{S}}_{\text{seg}}$ provides the baseline segmentation performance, $D^{\mathcal{T}}_{FT}$ should provide similar feature maps to achieve the baseline performance. To this end, the mid-feature maps generated should be similar, i.e., $\forall_i C_i(F_i) \approx C'_i(F'_i)$, where C'_i represents the convolution in $D^{\mathcal{T}}_{FT}$, F_i represents i^{th} feature map, and $F_0 = E(x)$. Suppose $\forall_i |C_i(F_i) - C'_i(F'_i)| < \epsilon_i \ll 1$, such that $\forall_i F_i \approx F'_i$ by mathematical induction. Therefore, the sum of errors $(\sum |C_i(F_i) - C'_i(F'_i)|)$ is approximated by $\sum |w_i F_0 - w'_i F_0|$ *iff* $\forall_i F_i \approx F'_i$, which can be expressed as:

$$\sum |w_i F_0 - w'_i F_0| < \epsilon \ll 1 \Leftarrow \sum |w_i F_0 - w'_i F_0| \approx 0 \Leftrightarrow \sum |w_i - w'_i| = 0. \tag{4}$$

Here, we denote $w_i - w'_i = f_i$ as the fluctuation vector in the vector space, and the condition $\sum f_i = 0$ indicates that the sum of the fluctuation vectors should be zero under the condition of $|f_i| < r \ll 1$. Hence, we achieve the condition for the parameter fluctuation that the centers of parameters of $\Theta_{\mathcal{S}}$ and $\theta_{\mathcal{T}}$ should be the same in the vector space, and the length of the fluctuation vector should be less than a certain small threshold ($0 < r \ll 1$). Therefore, the parameter fluctuation aims to add random vectors of which length is less than $0 < r \ll 1$ on the parameters of Θ_S, and the sum of vectors should be zero. To summarize, the parameter fluctuation aims to add randomness on Θ_S as follows:

$$\theta_T = \{w_i + f_i |\ w_i \in \Theta_S,\ \sum f_i = 0,\ 0 < |f_i| < r \ll 1\}. \tag{5}$$

3 Experiments

3.1 Experimental Set-Ups

To evaluate the segmentation performance of our TTFT framework, we used three different ultrasound databases: BUS [32], BUSI [1], and BUV [18], which are considered to be different domains. All three databases contain ultrasound imaging data and segmentation masks for breast cancer, with the masks labeled as 0 (background) and 1 (lesion) using a one-hot encoding. The BUS database consists of 163 images along with corresponding labels. The BUSI database contains 780 images, with 133 images belonging to the NORMAL class and having labels containing only 0 values. The BUV database originally consists of ultrasound videos, providing a total of 21,702 frames. While the database also provides labels for the detection task, we processed these labels as segmentation masks using a region growing method [15].

We employed different deep-learning models for evaluation. Specifically, U-Net [22] and FusionNet [21] were employed as our baseline models, since U-Net is a widely used basic model for segmentation, and FusionNet contains advanced residual modules, compared with U-Net. *Ours I* and *Ours II* were based on U-Net and FusionNet as the baseline network, respectively. Additionally, MIB-Net [28], which is a state-of-the-art model for breast cancer segmentation using ultrasound images, was employed for comparison. Furthermore, CBST [33] and CT-Net [16] were employed as the comparison models for UDA methods. As the evaluation metrics, dice coefficient (D. Coef), PRAUC, which is an area under a precision-recall curve, and cohen kappa (κ) were employed [30]. Our experimental set-ups included: (i) individual databases were used to assess the baseline segmentation performance (Appendix); (ii) the domain adaptive segmentation performance was assessed using the three databases, where two databases were regarded as the source domain, and the remaining database was regarded as the target domain; and (iii) the ablation study was carried out to evaluate the proposed network architecture along with the randomized re-initialization method.

3.2 Comparison Analysis

Since all compared DL models show similar D. Coef, only UDA performance is comparable as a control in our experiments. In this experiment, two databases were used for training, and the remaining database was used for testing. For instance, *BUS* in Fig. 3 illustrates the *BUS* database was used for testing, and

Dataset	Model	κ	PRAUC	D. Ceof (95% CI)
BUS	U-Net	0.462	0.588	0.638 (0.628-0.648)
	FusionNet	0.504	0.620	0.664 (0.652-0.675)
	MIB-Net	0.554	0.653	0.695 (0.686-0.703)
	CBST	0.724	0.817	0.805 (0.791-0.820)
	CT-Net	0.734	0.819	0.812 (0.802-0.823)
	Ours I	0.790	0.864	0.850 (0.836-0.864)
	Ours II	0.813	0.876	0.866 (0.854-0.878)
BUSI	U-Net	0.250	0.190	0.338 (0.327-0.350)
	FusionNet	0.289	0.213	0.370 (0.356-0.385)
	MIB-Net	0.329	0.237	0.403 (0.392-0.415)
	CBST	0.565	0.424	0.606 (0.573-0.640)
	CT-Net	0.597	0.456	0.634 (0.608-0.661)
	Ours I	0.711	0.591	0.735 (0.692-0.778)
	Ours II	0.747	0.638	0.768 (0.728-0.808)
BUV	U-Net	0.225	0.189	0.329 (0.316-0.341)
	FusionNet	0.260	0.207	0.356 (0.341-0.372)
	MIB-Net	0.302	0.231	0.391 (0.379-0.402)
	CBST	0.536	0.411	0.585 (0.550-0.620)
	CT-Net	0.551	0.424	0.598 (0.576-0.620)
	Ours I	0.671	0.566	0.701 (0.655-0.748)
	Ours II	0.712	0.619	0.737 (0.699-0.776)

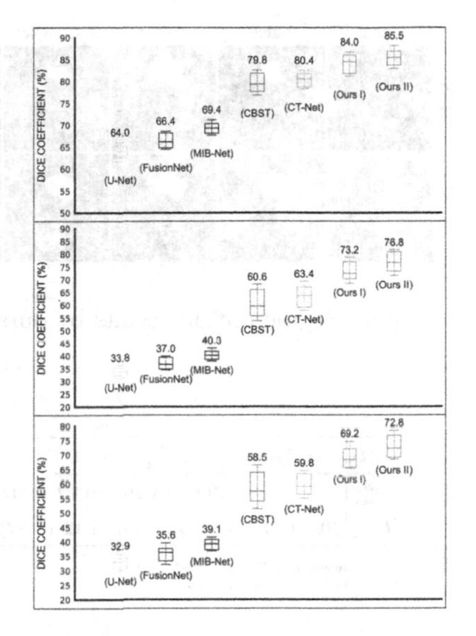

Fig. 3. Comparison analysis of our framework and comparison models: performance comparison table (Left) and Box-and-Whisker plot (Right).

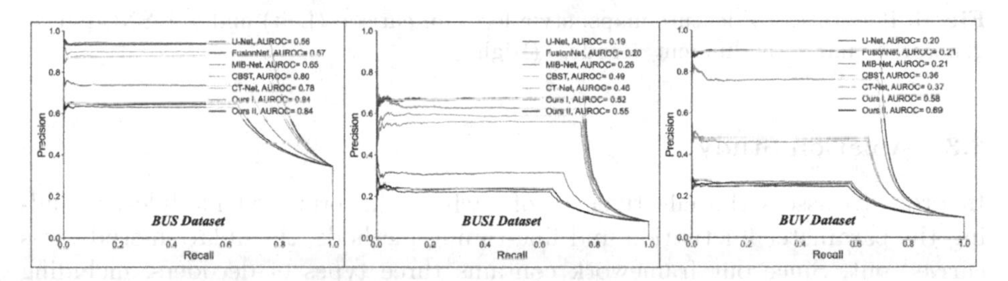

Fig. 4. Precision-Recall curves by ours and comparison models on each database. Area under the precision-recall curve (PR-AUC) values were reported.

the other two databases of *BUSI* and *BUV* were used for training. Figs. 3 and 4 show quantitative results, and Fig. 5 shows the sample segmentation results. Unlike the experiment using the individual database, U-Net, FusionNet, and MIB-Net showed significantly inferior scores due to domain gaps. In contrast, UDA methods of CBST and CT-Net showed superior scores, compared with others, and the scores were not strongly reduced, compared with the experiment with the single database. Note that, our TTFT framework achieved the best performance compared with other DL models. Additionally, *Ours II*, based on FusionNet, showed the best scores, potentially due to the advanced residual connection module. Furthermore, as illustrated in Fig 4, our framework provides superior precision scores in a long range of (0, 0.7), indicating that our frameworks estimated unnecessary mispredictions but precise predictions on cancer.

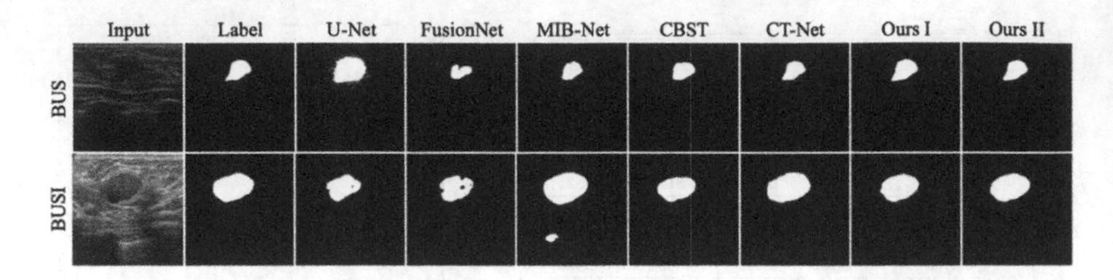

Fig. 5. Segmentation results by ours and comparison models on each database.

$E(t)$	$D^{\mathcal{S}}$seg	$D^{\text{fl}}_{\text{seg}}$	$D^{\mathcal{S}\rightarrow\mathcal{T}}_{\text{seg}}$
$D^{\mathcal{S}}_{\text{seg}}$	-	9.96 (3.26)	10.72 (3.02)
$D^{\mathcal{T}}_{\text{seg}}$	10.50 (3.87)	8.73 (2.95)	5.12 (0.87)

* Values are style loss of $D^{row}_{\text{seg}}(E(t))$ and $D^{col}_{\text{seg}}(E(t))$.
** The lower style loss, the more similar.

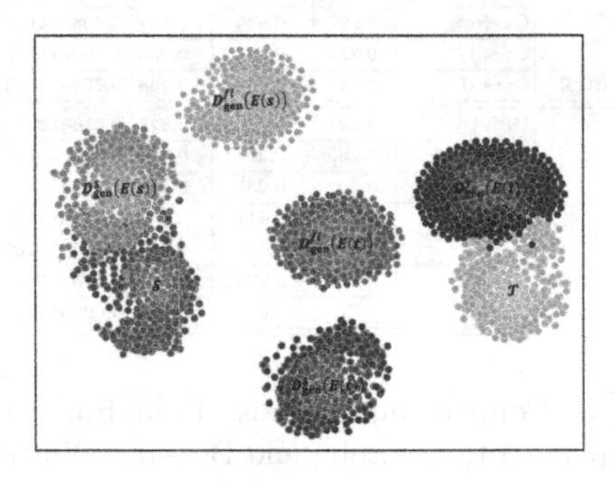

Fig. 6. Illustration of feature maps: style loss comparison (Left) and a T-SNE plot of generated images by different decoders (Right)

3.3 Ablation Study

In order to assess the effectiveness of each of the proposed modules, including the parameter fluctuation and fine-tuning methods, the ablation study was carried out. Since our framework contains three types of decoders, including $D^{\mathcal{S}}_{\text{seg}}$, D^{fl}_{seg}, and $D^{\mathcal{S}\rightarrow\mathcal{T}}_{\text{seg}}$ for the fine-tuning, we mainly targeted those decoders in our ablation study. Table 1 illustrates the quantitative results by different types of decoders. The higher D. coef value (+3.4%) of Pre-train + PF than that of Pre-train + Random Init and Pre-train + Offset confirms the effectiveness of the parameter fluctuation in the UDA performance. Additionally, the higher score (+11%) of Fine-tuning than Pre-train shows an outstanding UDA performance of the fine-tuning pipeline. Furthermore, the simultaneous utilization of the dual pipeline with $D^{\mathcal{S}}_{\text{seg}}$ and $D^{\mathcal{S}\rightarrow\mathcal{T}}_{\text{seg}}$ is justified by the scores of Pre-train + Fine-tuning. Using dual-pipeline and parameter fluctuation yielded the best performance. However, the utilization of ensemble pipelines of multiple fine-tuning modules was inefficient, since negligible performance improvements (+0.002) were observed, despite the heavy memory utilization.

Furthermore, Fig. 6 shows the effectiveness of the parameter fluctuation and fine-tuning methods. We first compared the similarity of feature-maps by decoders, including $D^{\mathcal{S}}_{\text{seg}}$, D^{fl}_{seg}, and $D^{\mathcal{S}\rightarrow\mathcal{T}}$, with $D^{\mathcal{S}}_{\text{seg}}$ and $D^{\mathcal{T}}_{\text{seg}}$, which was fully optimized decoder in $\mathcal{T}$. Here, a style loss [9] was employed to measure the similarity of feature maps. Our framework was fine-tuned as $D^{\mathcal{S}}_{\text{seg}} \rightarrow D^{fl}_{\text{seg}} \rightarrow D^{\mathcal{S}\rightarrow\mathcal{T}}_{\text{seg}}$

Table 1. Dice coefficients by different versions of our TTFT framework. Random Init is D_{FT} is randomly initialized, and Offset indicates D_{FT} is initialized with the value of D_{seg} added by the offset value.

D. Coef (95% CI)	BUS	BUSI	BUV
Pre-train	0.664 (0.653–0.675)	0.664 (0.653–0.675)	0.664 (0.653–0.675)
Fine-tuning	0.774 (0.763–0.785)	0.774 (0.763–0.785)	0.774 (0.763–0.785)
Pre-train + Random Init	0.663 (0.653–0.673)	0.663 (0.653–0.673)	0.663 (0.653–0.673)
Pre-train + Offset	0.676 (0.668–0.684)	0.676 (0.668–0.684)	0.676 (0.668–0.684)
Pre-train + PF	0.697 (0.686–0.707)	0.697 (0.686–0.707)	0.697 (0.686–0.707)
Pre-train + Fine-tuning	0.799 (0.789–0.809)	0.799 (0.789–0.809)	0.799 (0.789–0.809)
Pre-train + PF + Fine-tuning	0.855 (0.844–0.866)	0.855 (0.844–0.866)	0.855 (0.844–0.866)
Pre-train + PF + N Fine-tuning	0.857 (0.842–0.872)	0.857 (0.842–0.872)	0.857 (0.842–0.872)

along which the similarity with $D^{\mathcal{T}}_{\text{seg}}$ of those decoders were increasing, and the feature-maps by $D^{\mathcal{S}\to\mathcal{T}}_{\text{seg}}$ were similar to those of $D^{\mathcal{T}}_{\text{seg}}$, compared with $D^{\mathcal{S}}_{\text{seg}}$, indicating UDA was successfully performed. Additionally, the generated images by decoders, including $D^{\mathcal{S}}_{\text{seg}}$, $D^{\text{fl}}_{\text{seg}}$, and $D^{\mathcal{S}\to\mathcal{T}}_{\text{seg}}$ in $\mathcal{S}$ and $\mathcal{T}$ are plotted with T-SNE, where the short distance represents the similar features [19]. The generated images became similar to $\mathcal{T}$ in order of $D^{\mathcal{S}}_{\text{seg}}$, $D^{\text{fl}}_{\text{seg}}$, and $D^{\mathcal{S}\to\mathcal{T}}_{\text{seg}}$, which confirmed the effectiveness of the fine-tuning method in terms of knowledge distillation. Additionally, the parameters were successfully re-positioned from the local minimum in $\mathcal{S}$ by parameter fluctuation, which was confirmed by the distances from $\mathcal{S}$ to $D^{\mathcal{S}}_{\text{gen}}$ and D^{fl}_{gen}.

4 Discussion and Conclusion

In this work, we proposed a DL-based segmentation framework for multi-domain breast cancer segmentation on ultrasound images. Due to the low resolution of ultrasound images, manual segmentation of breast cancer is challenging even for expert clinicians, resulting in a sparse number of labeled data. To address this issue, we introduced a novel self-supervised DA network for breast cancer segmentation in ultrasound images. In particular, we proposed a test-time fine-tuning network to learn domain-specific knowledge via knowledge distillation by self-supervised learning. Since UDA is susceptible to error accumulation due to imprecise pseudo-labels, which can lead to degraded performance, we employed a self-supervised learning-based pretext task. Specifically, we utilized an auto-encoder-based network architecture to generate synthetic images that matched the input images. Moreover, we introduced a randomized re-initialization module that injects randomness into network parameters to reposition the network from the local minimum in the source domain to a local minimum that is better suited for the target domain. This approach enabled our framework to efficiently fine-tune the network in the target domain and achieve better segmentation performance. Experimental results, carried out with three ultrasound databases from different domains, demonstrated the superior segmentation performance of our framework over other competing methods. Additionally, our framework is well-suited to a scenario in which access to source domain data is limited, due to data privacy protocols. It is worth noting that we used vanilla U-Net [22] and FusionNet [21] as baseline models to evaluate the basic performance of our TTFT framework. However, the use of more advanced baseline models could lead to even better segmentation performance, which is a subject for our future work. Moreover, our proposed framework is not limited to breast cancer segmentation on ultrasound images acquired from different domains. It can also be applied to other disease groups or imaging modalities such as MRI or CT.

Acknowledgements. This work was partially supported by the Korea Medical Device Development Fund grant funded by the Korea government (the Ministry of Science and ICT, the Ministry of Trade, Industry and Energy, the Ministry of Health & Welfare, the Ministry of Food and Drug Safety) (Project Number: 1711174564, RS-2022-00141185). Also, this work was partially supported by the Technology Innovation Program(20014214) funded By the Ministry of Trade, Industry & Energy(MOTIE, Korea).

References

1. Al-Dhabyani, W., Gomaa, M., Khaled, H., Fahmy, A.: Dataset of breast ultrasound images. Data Brief **28**, 104863 (2020)
2. Badawy, S.M., Mohamed, A.E.N.A., Hefnawy, A.A., Zidan, H.E., GadAllah, M.T., El-Banby, G.M.: Automatic semantic segmentation of breast tumors in ultrasound images based on combining fuzzy logic and deep learning-a feasibility study. PLoS ONE **16**(5), e0251899 (2021)
3. Barbato, F., Toldo, M., Michieli, U., Zanuttigh, P.: Latent space regularization for unsupervised domain adaptation in semantic segmentation. In: Proceedings of the IEEE/CVF Conference on Computer Vision and Pattern Recognition, pp. 2835–2845 (2021)
4. Bateson, M., Kervadec, H., Dolz, J., Lombaert, H., Ayed, I.B.: Source-free domain adaptation for image segmentation. Med. Image Anal. **82**, 102617 (2022)
5. van Beers, F., Lindström, A., Okafor, E., Wiering, M.A.: Deep neural networks with intersection over union loss for binary image segmentation. In: ICPRAM, pp. 438–445 (2019)
6. Goodfellow, I., et al.: Generative adversarial nets. In: Advances in Neural Information Processing Systems, pp. 2672–2680 (2014)
7. Guan, H., Liu, M.: Domain adaptation for medical image analysis: a survey. IEEE Trans. Biomed. Eng. **69**(3), 1173–1185 (2021)
8. Ioffe, S., Szegedy, C.: Batch normalization: accelerating deep network training by reducing internal covariate shift. arXiv preprint arXiv:1502.03167 (2015)
9. Johnson, J., Alahi, A., Fei-Fei, L.: Perceptual losses for real-time style transfer and super-resolution. In: Leibe, B., Matas, J., Sebe, N., Welling, M. (eds.) ECCV 2016. LNCS, vol. 9906, pp. 694–711. Springer, Cham (2016). https://doi.org/10.1007/978-3-319-46475-6_43
10. Kaissis, G.A., Makowski, M.R., Rückert, D., Braren, R.F.: Secure, privacy-preserving and federated machine learning in medical imaging. Nat. Mach. Intell. **2**(6), 305–311 (2020)
11. Karani, N., Erdil, E., Chaitanya, K., Konukoglu, E.: Test-time adaptable neural networks for robust medical image segmentation. Med. Image Anal. **68**, 101907 (2021)
12. Kouw, W.M., Loog, M.: An introduction to domain adaptation and transfer learning. arXiv preprint arXiv:1812.11806 (2018)

13. Kundu, J.N., Kulkarni, A., Singh, A., Jampani, V., Babu, R.V.: Generalize then adapt: source-free domain adaptive semantic segmentation. In: Proceedings of the IEEE/CVF International Conference on Computer Vision, pp. 7046–7056 (2021)
14. Lee, H., Park, J., Hwang, J.Y.: Channel attention module with multi-scale grid average pooling for breast cancer segmentation in an ultrasound image. Ferroelectrics, and Frequency Control, IEEE Transactions on Ultrasonics (2020)
15. Lee, M.H., Kim, J.Y., Lee, K., Choi, C.H., Hwang, J.Y.: Wide-field 3D ultrasound imaging platform with a semi-automatic 3D segmentation algorithm for quantitative analysis of rotator cuff tears. IEEE Access **8**, 65472–65487 (2020)
16. Lee, S., Hyun, J., Seong, H., Kim, E.: Unsupervised domain adaptation for semantic segmentation by content transfer. In: Proceedings of the AAAI Conference on Artificial Intelligence, vol. 35, pp. 8306–8315 (2021)
17. Liang, J., He, R., Sun, Z., Tan, T.: Exploring uncertainty in pseudo-label guided unsupervised domain adaptation. Pattern Recogn. **96**, 106996 (2019)
18. Lin, Z., Lin, J., Zhu, L., Fu, H., Qin, J., Wang, L.: A new dataset and a baseline model for breast lesion detection in ultrasound videos. In: International Conference on Medical Image Computing and Computer-Assisted Intervention, pp. 614–623. Springer, Cham (2022). https://doi.org/10.1007/978-3-031-16437-8_59
19. Van der Maaten, L., Hinton, G.: Visualizing data using t-SNE. J. Mach. Learn. Res. **9**(11), 2579–2605 (2008)
20. Nam, H., Lee, H., Park, J., Yoon, W., Yoo, D.: Reducing domain gap by reducing style bias. In: Proceedings of the IEEE/CVF Conference on Computer Vision and Pattern Recognition, pp. 8690–8699 (2021)
21. Quan, T.M., Hildebrand, D.G., Jeong, W.K.: FusionNet: a deep fully residual convolutional neural network for image segmentation in connectomics. arXiv preprint arXiv:1612.05360 (2016)
22. Ronneberger, O., Fischer, P., Brox, T.: U-Net: convolutional networks for biomedical image segmentation. In: Navab, N., Hornegger, J., Wells, W.M., Frangi, A.F. (eds.) MICCAI 2015. LNCS, vol. 9351, pp. 234–241. Springer, Cham (2015). https://doi.org/10.1007/978-3-319-24574-4_28
23. Roy, S., Trapp, M., Pilzer, A., Kannala, J., Sebe, N., Ricci, E., Solin, A.: Uncertainty-guided source-free domain adaptation. In: Computer Vision-ECCV 2022: 17th European Conference, Tel Aviv, Israel, October 23–27, 2022, Proceedings, Part XXV. pp. 537–555. Springer, Cham (2022). https://doi.org/10.1007/978-3-031-19806-9_31
24. Ruder, S.: An overview of gradient descent optimization algorithms. arXiv preprint arXiv:1609.04747 (2016)
25. Sun, Y., Tzeng, E., Darrell, T., Efros, A.A.: Unsupervised domain adaptation through self-supervision. arXiv preprint arXiv:1909.11825 (2019)
26. Toldo, M., Maracani, A., Michieli, U., Zanuttigh, P.: Unsupervised domain adaptation in semantic segmentation: a review. Technologies **8**(2), 35 (2020)
27. Vakanski, A., Xian, M., Freer, P.E.: Attention-enriched deep learning model for breast tumor segmentation in ultrasound images. Ultrasound Med. Biol. **46**(10), 2819–2833 (2020)
28. Wang, J., et al.: Information bottleneck-based interpretable multitask network for breast cancer classification and segmentation. Med. Image Anal., 102687 (2022)
29. Wang, Q., Fink, O., Van Gool, L., Dai, D.: Continual test-time domain adaptation. In: Proceedings of the IEEE/CVF Conference on Computer Vision and Pattern Recognition, pp. 7201–7211 (2022)

30. Wang, Y., Yao, Y.: Breast lesion detection using an anchor-free network from ultrasound images with segmentation-based enhancement. Sci. Rep. **12**(1), 1–12 (2022)
31. Xu, J., Xiao, L., López, A.M.: Self-supervised domain adaptation for computer vision tasks. IEEE Access **7**, 156694–156706 (2019)
32. Yap, M.H., et al.: Automated breast ultrasound lesions detection using convolutional neural networks. IEEE J. Biomed. Health Inform. **22**(4), 1218–1226 (2017)
33. Zou, Y., Yu, Z., Kumar, B., Wang, J.: Unsupervised domain adaptation for semantic segmentation via class-balanced self-training. In: Proceedings of the European Conference on Computer Vision (ECCV), pp. 289–305 (2018)

Decoupled Consistency for Semi-supervised Medical Image Segmentation

Faquan Chen[1], Jingjing Fei[2], Yaqi Chen[1], and Chenxi Huang[1(✉)]

[1] Schoor of Informatics, Xiamen University, Xiamen, China
supermonkeyxi@xmu.edu.cn

[2] SenseTime Research, Shanghai, China

Abstract. By fully utilizing unlabeled data, the semi-supervised learning (SSL) technique has recently produced promising results in the segmentation of medical images. Pseudo labeling and consistency regularization are two effective strategies for using unlabeled data. Yet, the traditional pseudo labeling method will filter out low-confidence pixels. The advantages of both high- and low-confidence data are not fully exploited by consistency regularization. Therefore, neither of these two methods can make full use of unlabeled data. We proposed a novel decoupled consistency semi-supervised medical image segmentation framework. First, the dynamic threshold is utilized to decouple the prediction data into consistent and inconsistent parts. For the consistent part, we use the method of cross pseudo supervision to optimize it. For the inconsistent part, we further decouple it into unreliable data that is likely to occur close to the decision boundary and guidance data that is more likely to emerge near the high-density area. Unreliable data will be optimized in the direction of guidance data. We refer to this action as directional consistency. Furthermore, in order to fully utilize the data, we incorporate feature maps into the training process and calculate the loss of feature consistency. A significant number of experiments have demonstrated the superiority of our proposed method. The code is available at https://github.com/wxfaaaaa/DCNet.

Keywords: Semi-supervised Learning · Image Segementation · Consistency Regularization

1 Introduction

Deep learning technology can significantly assist clinicians in clinical diagnosis through accurate and robust segmentation of lesions or organs from medical

---

Supplementary Information The online version contains supplementary material available at https://doi.org/10.1007/978-3-031-43907-0_53.

H. Greenspan et al. (Eds.): MICCAI 2023, LNCS 14220, pp. 551–561, 2023.
https://doi.org/10.1007/978-3-031-43907-0_53

images [1]. In comparison to natural images, acquiring pixel-level labels of medical images involves input from clinical professionals, making such labels expensive to produce. In order to effectively alleviate the problem of data labeling, many attempts have been made towards semi-supervised medical image segmentation [2,3,5,6,11]. How to fully utilize unlabeled data in this situation becomes crucial.

Pseudo labeling [15,16] and consistency regularization [19–22] are two powerful techniques for using unlabeled data. The traditional pseudo labeling approach employs a high and fixed threshold to select the predicted pixels with high confidence as the pseudo label of unlabeled data [15]. Nevertheless, with this strategy, only a few samples can exceed the chosen threshold at the start of training. As a result, Zhang et al. [23]. present the Curriculum Pseudo Labeling (CPL) strategy, which adjusts the flexible threshold of each category dynamically during the training process. Despite the positive outcomes, low-confidence pixels will still be removed. Wang et al. [12]. demonstrated the importance of low-confidence pixels in model training. Given that consistency regularization appears to utilize data more, this research will focus more on this method.

Despite using all available prediction data, consistency regularization focuses more on how to get two similar predictions, such as data perturbation [9,10], model perturbation [25,28], feature perturbation [7,8], etc. Yet, when it comes to prediction optimization, all data is processed uniformly using a consistency loss function, such as L2 Loss. In light of this, we consider if we may decouple the prediction data into data with distinct functions, and optimize the prediction data by playing to each advantage's strengths. The benefit of this is that we can utilize all the data and, more significantly, fully utilize the advantages of various prediction data.

To sum up, this paper proposed a novel decoupled consistency regularization strategy. Specifically, inspired by CPL, we first designed a consistency threshold related to pixel confidence (Wang et al. [24] proved that the ideal dynamic threshold should be related to pixel confidence) to distinguish the consistent part from the inconsistent part, and the threshold increased as the training progressed to the maximum threshold. For the consistent part with high confidence, we use the method of cross pseudo supervision [18] to optimize. For the inconsistent part, we further decouple it into unreliable data that is likely to appear close to the decision boundary and guidance data that is more likely to be present near the high-density area. The guidance data plays the role of guiding the direction, and We don't propagate its gradient back. We refer to this action as directional consistency strategy. Additionally, we incorporate the feature map into the training process, suggest a feature consistency approach, and compute its loss in order to make better use of the data. We evaluated our method on the public PROMISE12 [14] and ACDC [13] datasets. Several experimental findings demonstrate that our strategy can significantly boost performance.

Overall, our contributions are four-fold: (1) we proposed a novel decoupled consistent semi-supervised medical image segmentation framework. The framework fully exploits prediction data, decoupled prediction data into data for various functions, and maximizes each function's advantages. (2) A dynamic threshold is proposed that can separate the prediction data into consistent and inconsistent parts. This threshold can effectively reflect the model's learning state and encourage more diversity of data at the start of training. (3) A novel direction consistency strategy is proposed to optimize the inconsistent part. This strategy focuses on optimizing the data around the decision boundary. The results of the experiment demonstrate that the direction consistency strategy is superior to the conventional consistent regularization. (4) The feature consistency approach is suggested and the feature map is incorporated into the training, allowing for better data utilization.

2 Method

Figure 1 shows the overall pipeline of our method. To describe this work precisely, we first define some mathematical terms. Let $D = D_l \cup D_u$ be the whole provided dataset. We denote an unlabeled image as $x_i \in D_u$ and a labeled image pair as $(x_i, y_i) \in D_l$, where y_i is ground-truth. θ_{dA} and θ_{dB} represent decoder A and decoder B, respectively. θ_{dA} employs bi-linear interpolation for up-sampling, and θ_{dB} employs original transpose convolution for up-sampling. o^A and o^B represent the outputs of θ_{dA} and θ_{dB}, respectively. $p^A(p^B)$ is the segmentation confidence map, which is the network output $o^A(o^B)$ after softmax normalization.

2.1 Dynamic Consistency Threshold

In recent years, the pseudo labeling method based on threshold has achieved great success. The sampling strategy of this method can be defined as follows:

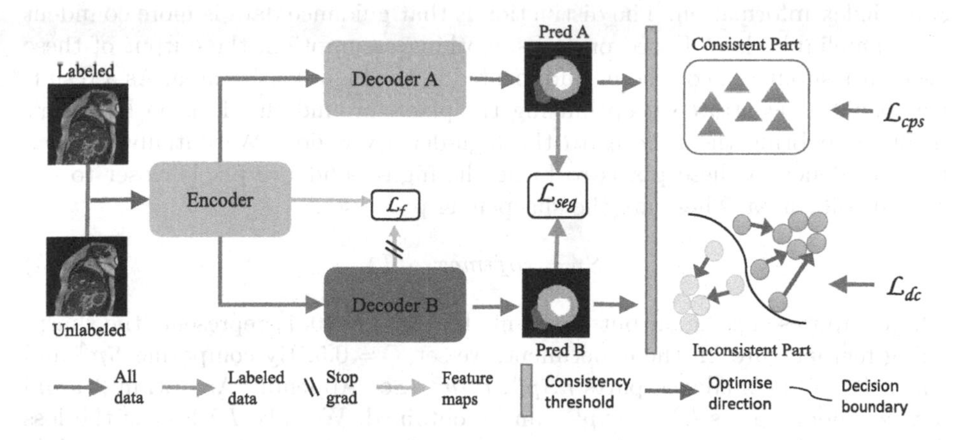

Fig. 1. Pipeline of our proposed DC-Net. DC-Net contains two decoders θ_{dA} and θ_{dB}, where θ_{dA} uses bi-linear interpolation for up-sampling and θ_{dB} uses transposed convolution for up-sampling. For the labeled data we calculate the loss $\mathcal{L}_{seg}$ between them and ground-truth, for the consistent part we calculate cross pseudo supervision loss $\mathcal{L}_{cps}$, for the inconsistent part we calculate directional consistency loss $\mathcal{L}_{dc}$, for feature maps, we calculate the feature consistency loss $\mathcal{L}_f$.

$$\tilde{y} = \mathbb{1}\left[p > \gamma\right] \tag{1}$$

where $\gamma \in (0,1)$ is a threshold used to select pseudo labels. p is the segmentation confidence map. FlexMatch [23] has proved that in the early stage of training, in order to improve the utilization of unlabeled data and promote the diversification of pseudo labels, γ should be relatively small. As the training progresses, γ should maintain a stable proportion of pseudo labels. Therefore, our consistency threshold is defined as follows:

$$\gamma_t = (1-\lambda)\gamma_{t-1} + \lambda \frac{1}{B} \sum_{b=1}^{B} max(p) \tag{2}$$

where B is the batch size, λ is the weight coefficient that increases with the training and we set $\lambda = \frac{t}{t_{max}}$. In order to sample more unlabeled data, we conduct threshold evaluation on p^A and p^B and select a smaller threshold as our consistency threshold. We initialize λ_t as $\frac{1}{C}$, where C indicates the number of classes. The consistency threshold γ_t is finally defined and adjusted as:

$$\gamma_t = \begin{cases} \frac{1}{C} & , if \quad t = 0 \\ min(\gamma_t^A, \gamma_t^B) & , otherwise \end{cases} \tag{3}$$

where γ_t^A and γ_t^B represent the threshold of p^A and p^B.

2.2 Decoupled Consistency

Inconsistent Part. We decouple the inconsistent part into unreliable data, which is probably going to appear at the decision boundary, and guidance data, which is more likely to occur near the high-density area. These two parts have the same index information. The distinction is that guidance data is more confident than unreliable data. Based on the smoothing assumption, the output of these two parts should be consistent and located in the high-density area. As a result, we should concentrate on optimizing the pixels around the decision boundary in order to bring them closer to the high-density region. We initially sharpen the confidence of these pixels to bring the high-confidence pixels closer to the high-density area. These are the sharpening processes:

$$Sp = softmax(o/T) \tag{4}$$

where o represents the output of the model and $T \in (0,1)$ represents the sharpening temperature. In the experiment, we set $T = 0.5$. By comparing Sp^A and Sp^B, the high-confidence parts hSp^A, hSp^B (See Appendix Algorithm 2.) and low-confidence parts lSp^A, lSp^B can be obtained. We take $L2$ loss as the loss function of this part. Note that we only optimize the low-confidence part and do not back-propagate the gradient of the high-confidence part. Therefore, our loss of directional consistency can be written as:

$$\mathcal{L}_{dc} = L2(lSp^B, detach(hSp^A)) + L2(lSp^A, detach(hSp^B)) \tag{5}$$

Consistent Part. Similar to CPS [18], the method of cross pseudo supervision is adopted for optimization. The details are as follows:

$$\mathcal{L}_{cps} = CE(o^B, PL^A) + CE(o^A, PL^B) \tag{6}$$

where PL^A and PL^B represent corresponding pseudo labels.

Feature Part. We incorporate the feature map into the training process to further utilize the data. In order to reduce the amount of computation, we have carried out an average mapping of the feature map to reduce its dimension($\mathbb{R}^{C_m \times H_m \times W_m} \longrightarrow \mathbb{R}^{H_m \times W_m}$). The mapping process is as follows:

$$\bar{f}_m = \frac{1}{C} \sum_i^C |f_{mi}|^p \tag{7}$$

where, $p > 1$, f_m represents the feature map of $m-$th layer, and f_{mi} denotes the $i-$th slice of f_m in the channel dimension, $\bar{f}_m$ represents the corresponding mapping result. In the experiment, we set $p = 2$. Our feature consistency loss is as follows:

$$\mathcal{L}_f = \sum_{m=1}^{n} \sum_{i=1}^{N} ||\bar{f}_{mi}^e - \bar{f}_{mi}^d||_2^2 \tag{8}$$

where N is the number of pixels of $\bar{f}_{mi}$, n is the number of network layers, $\bar{f}_{mi}^e$ and $\bar{f}_{mi}^d$ represent $i-$th pixels of $m-$th feature map of decoder and encoder respectively. In this paper, only feature maps of decoder B were used to calculate the loss.

Total Loss. The total loss is a weighted sum of the segmentation loss $\mathcal{L}_{seg}$ and the other three losses:

$$\mathcal{L}_{total} = \mathcal{L}_{seg} + \alpha\mathcal{L}_{cps} + \beta\mathcal{L}_{dc} + \lambda\mathcal{L}_f \tag{9}$$

where $\mathcal{L}_{seg}$ is a Dice loss, which is applied for the few labeled data. And hyperparameter β is set as an iteration-dependent warming-up function [29], $\beta = e^{(-5(1-\frac{t}{t_{max}})^2)}$, $\lambda = 1 - \beta$, $\alpha = 0.3$ in the experiment.

3 Experiment and Results

3.1 Dataset

We evaluated our methods on ACDC dataset [13] and PROMERE12 dataset [14]. For the ACDC dataset, we randomly selected 140 scans from 70 subjects, 20 scans from 10 subjects, and 40 scans from 20 subjects as training, validation, and test sets, ensuring that each set contains data from different subjects. For the PROMISE12 dataset, we randomly divided the data into 35, 5, and 10 cases as training, validation, and test sets. And according to the previous work on datasets [27], these two data sets are segmented in 2D (piece by piece). Each slice was resized to 256×256, and the intensities of pixels are normalized to the $[0, 1]$ range.

Table 1. Comparison with five recent methods on the ACDC dataset. All results are reproduced in the form of [18,25–28] in the same experimental environment for a fair comparison.

Method	#Scans used		Metrics			
	Labeled	Unlabeled	Dice(%)↑	Jaccard(%)↑	95HD(voxel)↓	ASD(voxel)↓
SupOnly	3 (5%)	67 (95%)	49.33	39.19	22.57	9.66
URPC [28]			55.87	44.64	13.60	3.74
CPS [18]			56.59	46.51	6.43	1.16
DTC [25]			56.90	45.67	23.36	7.39
MCNet [26]			62.85	52.29	7.62	2.33
SSNet [27]			65.82	55.38	6.67	2.28
Ours			**70.36**	**60.78**	**3.94**	**0.86**
SupOnly	7 (10%)	63 (90%)	79.37	67.78	10.73	3.12
URPC [28]			83.10	72.41	4.84	1.53
CPS [18]			84.78	74.69	7.56	2.39
DTC [25]			84.29	73.92	12.81	4.01
MCNet [26]			86.44	77.04	5.50	1.84
SSNet [27]			86.78	77.67	6.07	1.40
Ours			**89.42**	**81.37**	**1.28**	**0.38**

3.2 Implementation Details

All comparisons and ablation experiments are performed using the same experimental setting for a fair comparison. They are conducted on PyTorch using an Intel(R) Xeon(R) CPU and NVIDIA GeForce RTX 1080 Ti GPU. We adopt U-net [4] as our base network. And we use SGD as an optimizer, with a weight attenuation of 0.0005 and momentum of 0.9. The learning rate is 0.01. The batch size is set to 24, in which 12 images are labeled. All methods performed 30000 iterations during training. Moreover, a data augmentation strategy including random flipping and random rotation is exploited to alleviate overfitting.

3.3 Results

Comparison with Other Semi-supervised Methods. We use the metrics of Dice, Jaccard, 95% Hausdorff Distance (95HD), and Average Surface Distance (ASD) to evaluate the results. Table 1 gives the averaged performance of three-class segmentation including the myocardium, left and right ventricles on the ACDC dataset. Table 2 shows the segmentation results on the PROMISE12 dataset. It can be seen from the table that our method is superior to other methods. The visualized results in Fig. The visualization result in Fig. 2 shows that the segmentation result of our model is closer to the ground-truth and effectively eliminates most false-positive predictions on the ACDC (highlighted by yellow boxes at the bottom line).

Table 2. Comparison with five recent methods on the PROMISE12 dataset. All results are reproduced in the form of [18,25–28] in the same experimental environment for a fair comparison.

Method	#Scans used		Metrics			
	Labeled	Unlabeled	Dice(%)↑	Jaccard(%)↑	95HD(voxel)↓	ASD(voxel)↓
SupOnly	4 (10%)	31 (90%)	46.08	35.40	35.86	11.23
URPC [28]			52.96	39.93	37.53	11.43
CPS [18]			49.19	37.27	39.96	11.97
DTC [25]			57.87	43.80	81.54	25.38
MCNet [26]			56.91	43.82	23.44	5.91
SSNet [27]			61.10	47.07	23.73	7.44
Ours			**68.89**	**54.88**	**12.93**	**3.75**
SupOnly	7 (20%)	28 (80%)	62.28	50.42	16.55	3.56
URPC [28]			67.04	54.01	11.54	**2.11**
CPS [18]			64.50	50.71	12.07	3.09
DTC [25]			72.03	58.32	11.48	2.65
MCNet [26]			71.77	59.07	10.76	2.85
SSNet [27]			71.56	59.35	14.38	3.03
Ours			**78.68**	**65.44**	**10.65**	2.53

Ablation Study. In order to verify the effectiveness of different loss functions of the model, we conducted experiments on the ACDC dataset with 10% labeled data, and the experimental results are shown in Table 3. It can be seen from the table that the dice scores of $\mathcal{L}_{cps}$, $\mathcal{L}_f$ and $\mathcal{L}_{dc}$ are improved by 2.93%, 3.33%, 3.60%, respectively, compared with only using $\mathcal{L}_{seg}$. On the basis of $\mathcal{L}_{seg}$ and $\mathcal{L}_f$, $\mathcal{L}_{dc}$ is also better than $\mathcal{L}_{con}$, even close to our final results. To investigate the usefulness of directional consistency further, we do not employ a threshold and compare it to classical consistency regularization and cross pseudo supervision. The experimental results are shown in Table 4. It can be seen from the table that the $\mathcal{L}_{dc}$ is obviously superior to the other two methods, which fully proves the importance of the division of different pixel functions. Also, we discovered that utilizing threshold would improve the effect of $\mathcal{L}_{cps}$. This further demonstrates the significance of our dynamic threshold.

Table 3. Ablation studies on ACDC dataset with 10% labeled data.Where $\mathcal{L}_{cps}$ represents cross pseudo supervision, $\mathcal{L}_{dc}$ represents direction consistency, and $\mathcal{L}_f$ represents feature consistency.

$\mathcal{L}_{seg}$	$\mathcal{L}_{cps}$	$\mathcal{L}_{dc}$	$\mathcal{L}_f$	Dice(%)↑	Jaccard(%)↑	95HD(voxel)↓	ASD(voxel)↓
✓				84.98	75.00	12.03	3.31
✓	✓			87.91	79.13	5.09	1.47
✓		✓		88.58	80.10	5.35	1.49
✓	✓	✓		89.29	81.21	3.07	0.92
✓			✓	88.31	79.67	3.18	1.04
✓	✓		✓	89.12	80.82	2.63	0.96
✓		✓	✓	89.39	81.36	2.32	0.82
✓	✓	✓	✓	**89.42**	**81.37**	**1.28**	**0.38**

Table 4. Comparison of different optimization methods when there is no threshold. Here $\mathcal{L}_{mse}$ represents the classic consistent regularization using the L2 loss function, and $\mathcal{L}_{cps}$ represents the cross pseudo supervision.

$\mathcal{L}_{seg}$	$\mathcal{L}_{mse}$	$\mathcal{L}_{cps}$	$\mathcal{L}_{dc}$	Dice(%)↑	Jaccard(%)↑	95HD(voxel)↓	ASD(voxel)↓
✓				84.98	75.00	12.03	3.31
✓	✓			86.47	77.18	7.31	2.01
✓		✓		87.61	78.72	7.48	1.87
✓			✓	88.71	80.31	4.05	0.98
✓		✓	✓	**88.99**	**80.68**	**3.09**	**0.92**

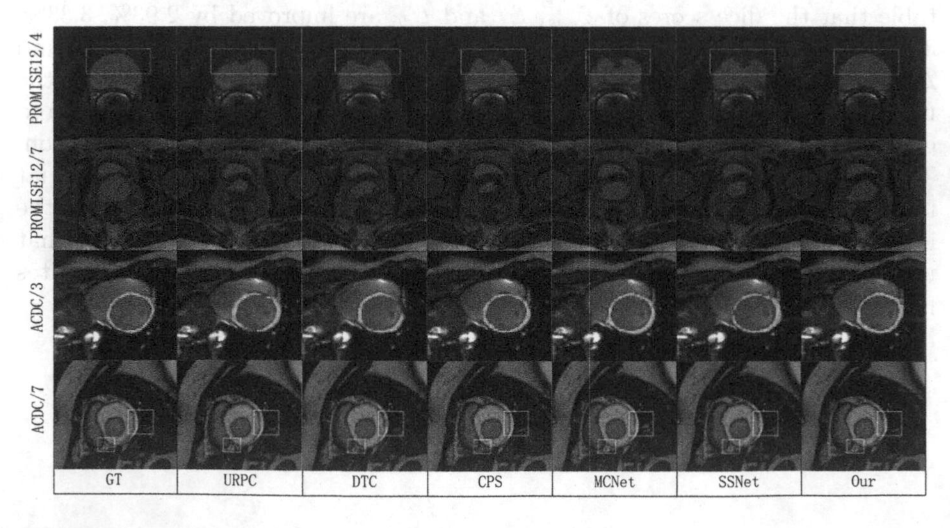

Fig. 2. Exemplar results of several semi-supervised segmentation methods and corresponding ground truth (GT) on PROMISE12 dataset (Top two rows) and ACDC dataset (Bottom two rows).

4 Conclusion

This paper proposed a framework DC-Net for semi-supervised medical image segmentation. In view of the current problem of insufficient utilization of unlabeled data, our fundamental concept is to fully exploit the benefits of data with various functionalities. Based on this, we decouple the prediction data into consistent and inconsistent parts through a dynamic threshold. Furthermore, the inconsistent part is further decoupled into guidance data and unreliable data, and optimized by a novel directional consistency strategy. Our method yielded excellent outcomes on both the ACDC and PROMISE12 datasets. In addition, directional consistency shows promising potential in the experiment, and future research will further explore the selection and treatment of directions.

Acknowledgement. This work was supported by the National Natural Science Foundation of China (Grant No. 62002304), and also by Anhui Province KevLaboratory of Translational Cancer Research (KFKT 202308), China.

References

1. Masood, S., Sharif, M., Masood, A., Yasmin, M., Raza, M.: A survey on medical image segmentation. Curr. Med. Imaging **11**(1), 3–14 (2015)
2. Li, X., Yu, L., Chen, H., Fu, C.W., Xing, L., Heng, P.A.: Transformation-consistent self-ensembling model for semisupervised medical image segmentation. IEEE Trans. Neural Netw. Learn. Syst. **32**(2), 523–534 (2020)
3. Qiao, S., Shen, W., Zhang, Z., Wang, B., Yuille, A.: Deep co-training for semi-supervised image recognition. In: Proceedings of the European Conference on Computer Vision (ECCV), pp. 135–152 (2018)
4. Ronneberger, O., Fischer, P., Brox, T.: U-Net: convolutional networks for biomedical image segmentation. In: Navab, N., Hornegger, J., Wells, W.M., Frangi, A.F. (eds.) MICCAI 2015. LNCS, vol. 9351, pp. 234–241. Springer, Cham (2015). https://doi.org/10.1007/978-3-319-24574-4_28
5. Zhou, Y., et al.: Semi-supervised 3d abdominal multi-organ segmentation via deep multi-planar co-training. In: 2019 IEEE Winter Conference on Applications of Computer Vision (WACV), pp. 121–140. IEEE (2019)
6. Xia, Y., et al.: 3D semi-supervised learning with uncertainty-aware multi-view co-training. In: Proceedings of the IEEE/CVF Winter Conference on Applications of Computer Vision, pp. 3646–3655 (2020)
7. Ouali, Y., Hudelot, C., Tami, M.: Semi-supervised semantic segmentation with cross-consistency training. In: Proceedings of the IEEE/CVF Conference on Computer Vision and Pattern Recognition, pp. 12674–12684 (2020)
8. Ke, Z., Qiu, D., Li, K., Yan, Q., Lau, R.W.H.: Guided collaborative training for pixel-wise semi-supervised learning. In: Vedaldi, A., Bischof, H., Brox, T., Frahm, J.-M. (eds.) ECCV 2020. LNCS, vol. 12358, pp. 429–445. Springer, Cham (2020). https://doi.org/10.1007/978-3-030-58601-0_26
9. Kim, J., Jang, J., Park, H.: Structured consistency loss for semi-supervised semantic segmentation. arXiv preprint arXiv:2001.04647 (2020)
10. French, G., Aila, T., Laine, S., Mackiewicz, M., Finlayson, G.: Semi-supervised semantic segmentation needs strong, high-dimensional perturbations (2019)

11. Masood, S., et al.: Automatic choroid layer segmentation from optical coherence tomography images using deep learning. Sci. Rep. **9**(1), 1–18 (2019)
12. Wang, Y., et al.: Semi-supervised semantic segmentation using unreliable pseudo-labels supplementary material (2022)
13. Bernard, O., et al.: Deep learning techniques for automatic MRI cardiac multi-structures segmentation and diagnosis: is the problem solved? IEEE Trans. Med. Imaging **37**(11), 2514–2525 (2018)
14. Litjens, G., et al.: Evaluation of prostate segmentation algorithms for MRI: the promise12 challenge. Med. Image Anal. **18**(2), 359–373 (2014)
15. Lee, D.H., et al.: Pseudo-label: the simple and efficient semi-supervised learning method for deep neural networks. In: Workshop on Challenges in Representation Learning, ICML, vol. 3, p. 896 (2013)
16. Arazo, E., Ortego, D., Albert, P., O'Connor, N.E., McGuinness, K.: Pseudo-labeling and confirmation bias in deep semi-supervised learning. In: 2020 International Joint Conference on Neural Networks (IJCNN), pp. 1–8. IEEE (2020)
17. Rizve, M.N., Duarte, K., Rawat, Y.S., Shah, M.: In defense of pseudo-labeling: an uncertainty-aware pseudo-label selection framework for semi-supervised learning. arXiv preprint arXiv:2101.06329 (2021)
18. Chen, X., Yuan, Y., Zeng, G., Wang, J.: Semi-supervised semantic segmentation with cross pseudo supervision. In: Proceedings of the IEEE/CVF Conference on Computer Vision and Pattern Recognition, pp. 2613–2622 (2021)
19. Bachman, P., Alsharif, O., Precup, D.: Learning with pseudo-ensembles. In: Advances in Neural Information Processing Systems 27 (2014)
20. Sajjadi, M., Javanmardi, M., Tasdizen, T.: Regularization with stochastic transformations and perturbations for deep semi-supervised learning. In: Advances in Neural Information Processing Systems 29 (2016)
21. Bortsova, G., Dubost, F., Hogeweg, L., Katramados, I., de Bruijne, M.: Semi-supervised medical image segmentation via learning consistency under transformations. In: Shen, D., et al. (eds.) MICCAI 2019. LNCS, vol. 11769, pp. 810–818. Springer, Cham (2019). https://doi.org/10.1007/978-3-030-32226-7_90
22. Xu, Y., et al.: Dash: Semi-supervised learning with dynamic thresholding. In: International Conference on Machine Learning, pp. 11525–11536. PMLR (2021)
23. Zhang, B., et al.: FlexMatch: boosting semi-supervised learning with curriculum pseudo labeling. Adv. Neural. Inf. Process. Syst. **34**, 18408–18419 (2021)
24. Wang, Y., et al.: FreeMatch: self-adaptive thresholding for semi-supervised learning. arXiv preprint arXiv:2205.07246 (2022)
25. Luo, X., Chen, J., Song, T., Wang, G.: Semi-supervised medical image segmentation through dual-task consistency. In: Proceedings of the AAAI Conference on Artificial Intelligence, vol. 35, pp. 8801–8809 (2021)
26. Wu, Y., Xu, M., Ge, Z., Cai, J., Zhang, L.: Semi-supervised left atrium segmentation with mutual consistency training. In: de Bruijne, M., et al. (eds.) MICCAI 2021. LNCS, vol. 12902, pp. 297–306. Springer, Cham (2021). https://doi.org/10.1007/978-3-030-87196-3_28
27. Wu, Y., Wu, Z., Wu, Q., Ge, Z., Cai, J.: Exploring smoothness and class-separation for semi-supervised medical image segmentation. arXiv preprint arXiv:2203.01324 (2022)

28. Luo, X., et al.: Efficient semi-supervised gross target volume of nasopharyngeal carcinoma segmentation via uncertainty rectified pyramid consistency. In: de Bruijne, M., et al. (eds.) MICCAI 2021. LNCS, vol. 12902, pp. 318–329. Springer, Cham (2021). https://doi.org/10.1007/978-3-030-87196-3_30
29. Laine, S., Aila, T.: Temporal ensembling for semi-supervised learning. arXiv preprint arXiv:1610.02242 (2016)

Combating Medical Label Noise via Robust Semi-supervised Contrastive Learning

Bingzhi Chen[1,3], Zhanhao Ye[1], Yishu Liu[2](✉), Zheng Zhang[2](✉), Jiahui Pan[1], Biqing Zeng[1], and Guangming Lu[2,3]

[1] School of Software, South China Normal University, Guangzhou, China
[2] School of Computer Science and Technology, Harbin Institute of Technology, Shenzhen, China
liuyishu@stu.hit.edu.cn, darrenzz219@gmail.com
[3] Guangdong Provincial Key Laboratory of Novel Security Intelligence Technologies, Shenzhen, China

Abstract. Deep learning-based AI diagnostic models rely heavily on high-quality exhaustive-annotated data for algorithm training but suffer from noisy label information. To enhance the model's robustness and prevent noisy label memorization, this paper proposes a robust Semi-supervised Contrastive Learning paradigm called SSCL, which can efficiently merge semi-supervised learning and contrastive learning for combating medical label noise. Specifically, the proposed SSCL framework consists of three well-designed components: the Mixup Feature Embedding (MFE) module, the Semi-supervised Learning (SSL) module, and the Similarity Contrastive Learning (SCL) module. By taking the hybrid augmented images as inputs, the MFE module with momentum update mechanism is designed to mine abstract distributed feature representations. Meanwhile, a flexible pseudo-labeling promotion strategy is introduced into the SSL module, which can refine the supervised information of the noisy data with pseudo-labels based on initial categorical predictions. Benefitting from the measure of similarity between classification distributions, the SCL module can effectively capture more reliable confident pairs, further reducing the effects of label noise on contrastive learning. Furthermore, a noise-robust loss function is also leveraged to ensure the samples with correct labels dominate the learning process. Extensive experiments on multiple benchmark datasets demonstrate the superiority of SSCL over state-of-the-art baselines. The code and pre-trained models are publicly available at https://github.com/Binz-Chen/MICCAI2023_SSCL.

Keywords: Medical Label Noise · Mixup · Semi-supervised Learning · Contrastive Learning

Supplementary Information The online version contains supplementary material available at https://doi.org/10.1007/978-3-031-43907-0_54.

H. Greenspan et al. (Eds.): MICCAI 2023, LNCS 14220, pp. 562–572, 2023.
https://doi.org/10.1007/978-3-031-43907-0_54

1 Introduction

The advancement of deep learning models heavily depends on the availability of large-scale datasets with high-quality annotated labels [4]. However, it is costly and time-consuming to obtain a sufficient number of accurate annotations from clinical systems, which inevitably introduces a certain level of noisy label [16]. This phenomenon seriously affects the stability and robustness of medical training and prediction procedures, leading to the production of corrupted representations and inaccurate classification boundaries [27]. Early approaches for combating the label noise mainly focused on the marginal improvements of model robustness, including designing robust loss functions [14,26], preprocessing the image [24,28], and estimating the noise transition matrix [8]. In recent years, some advanced methods [2,19] with semi-supervised learning [22] are designed to leverage the supervision signal provided by pseudo labels to re-correct the noise bias. With the development of contrastive learning technologies, some scholars [15,17] attempt to learn the contrastive representations between the clean and noisy labels by maximizing the similarity of positive pairs and minimizing the similarity of negative pairs. Despite the advances in conventional image processing, few studies have proposed to overcome the medical label noise.

To overcome this issue, this paper proposes a robust Semi-supervised Contrastive Learning (SSCL) paradigm, that simultaneously benefits from semi-supervised learning and contrastive learning for combating the medical noisy labels and promoting stability and robustness of the diagnostic model. Three important components, i.e., the Mixup Feature Embedding (MFE) module, the Semi-supervised Learning (SSL) module, and the Similarity Contrastive Learning (SCL) module, are proposed in the SSCL framework. The architecture of the SSCL framework is shown in Fig. 1. Specifically, the MFE module is built on a multi-branch architecture with the momentum update mechanism, which can effectively capture the abstract distributed feature representations from various levels of mixup augmented images. Based on the confidence scores provided by the initial classifier, a flexible pseudo-labeling promotion strategy is introduced into the SSL module to effectively select confident samples and generate the pseudo-labels, resulting in a more accurate supervision signal reconstruction. By calculating their similarity distribution of representation learning, a novel pair-wise selection strategy in the SCL module is designed to efficiently identify and select more reliable confident pairs out of noisy pairs for contrasting learning. In the training phase, we broaden the scope of penalization by incorporating loss functions, further reducing the impact of noise on statistical classification. The main contributions of our work are as follows:

- This paper presents a robust semi-supervised contrastive learning paradigm that effectively incorporates semi-supervised learning and contrastive learning to mitigate the effect of medical label noise. Our approach represents the first attempt to address this issue in the field of medical image analysis.
- The pseudo-labeling promotion strategy can re-correct the supervised information of noisy labels, while the pair-wise selection strategy can guide the confident pairs to dominate the contrasting learning process.

– The proposed SSCL framework is evaluated on multiple benchmark datasets, and extensive experiments demonstrate the generalization performance of our method in comparison with state-of-the-art baselines.

2 Related Work

2.1 Conventional Methods with Noisy Labels

To eliminate the memorization effect of noise labels in the training phase, recent works [14,26,29] were mainly devoted to exploring the effectiveness of robust loss function. Wang et al. [26] proposed a noise-robust loss function that combined with Cross-Entropy (CE) loss, to address the hard class learning problem and noisy label overfitting problem. Yi et al. [14] investigated the representational benefits of the contrastive regularization loss function to learn contrastive representations with noisy labels. With continuous in-depth research on image processing, data augmentation has been proven to have a significant role in combating noisy labels. Zhang et al. [28] presented a data-agnostic and straightforward data augmentation principle to mix up different images geometrically in the feature space, which has been widely used in the field of noise labeling. Moreover, researchers have attempted to leverage the label noise transfer matrix extracted from the data set to solve the noise label problem. Hendrycks et al. [8] directly used the matrix summarizing the probability of one class being flipped into another under noise. Ramaswamy et al. [18] proposed an efficient kernel mean embedding to overcome mixture proportion estimation.

2.2 Semi-supervised Learning

Compared with the existing learning methods, semi-supervised learning [3,12,23] has been recognized as an effective technique to solve the problem of noisy labels. As a semi-supervised learning technique, pseudo-labeling is frequently used in conjunction with confidence-based thresholding to retain unlabeled examples and increase the size of labeled training data. In recent years, some advanced works [2,19] demonstrate the capacity of pseudo-label based semi-supervised learning methods in combating medical label noise. Reed et al. [19] augmented the usual prediction objective with a notion of perceptual consistency, and the article referred to this approach as static hard guidance. Arazo et al. [2] proposed dynamic hard and soft bootstrapping losses by individual weight of each sample. Furthermore, the combination of semi-supervised learning and pseudo-labels can be used not only for single-label classification but also for negative learning and multi-label classification [20].

2.3 Contrastive Learning

With the advancement of deep learning, researchers have found the potential of contrastive-based similarity learning frameworks for representation learning [10,15,25]. Unsupervised contrastive learning [7,21] aims to maximize the similarity of positive pairs and minimize the similarity of negative pairs at the instance level. By incorporating clean label information, supervised contrastive learning [13] can obtain more supervised information and achieve better performance. Recently, many state-of-the-art works [15,17] have been proposed to make full use of contrastive learning for combating label noise. MOIT [17] adopted the method of interpolation contrastive learning, and used the supervised information obtained after semi-supervised learning for contrastive learning, so as to reduce the damage of noise labels on contrastive learning. Sel-CL [15] improved MOIT by introducing the concepts of confident use cases and confident pairs which improved the ability to filter noise labels with new detection strategies.

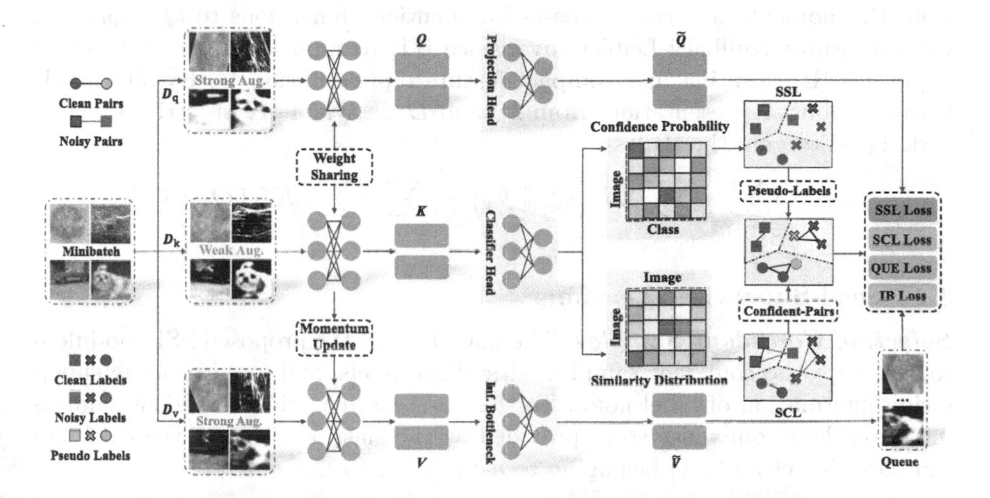

Fig. 1. Illustration of the proposed SSCL framework for combating medical label noise.

3 Semi-supervised Contrastive Learning

3.1 Mixup Feature Embedding

Mixup. Let $\boldsymbol{D} = \{(x_i, y_i)\}_i^N$ denotes the training minibatch of image-label pairs x_i and y_i, where N is the batch size. Initially, the operations of data augmentation are first conducted in the MFE module, to generate various levels of hybrid augmented images with Mixup [28]. As shown in Fig. 1, the obtained hybrid data set involves a group of weak augmentation images, i.e., $\boldsymbol{D}_k$, and two groups of strong augmentation images, i.e., $\boldsymbol{D}_q$ and $\boldsymbol{D}_v$. The principle of Mixup can be defined as:

$$x_i \leftarrow \lambda x_a + (1 - \lambda)\, x_b, \tag{1}$$

where $\lambda \in [0, 1] \sim Beta(\alpha_l, \alpha_r)$ is used to control the mixing strength of training samples, x_a and x_b are the training samples randomly drawn from each minibatch, and x_i is the enhanced image generated by the mixup preprocessing.

Feature Embedding. By taking the corresponding augmented images as inputs, the MFE module aims to capture abstract distributed feature representations. The MFE module consists of different branches, including a deep encoder with projection and classifier heads, and a momentum encoder with an information bottleneck (IB) [6]. Motivated from the physical perspective of optimization [7], the momentum update mechanism can be defined as:

$$\theta_v \leftarrow m\theta_v + (1-m)\theta_k, \tag{2}$$

where θ denotes the encoder parameters, and $m \in [0, 1)$ is a momentum coefficient. Specifically, the feature representations $\boldsymbol{Q}$ would be mapped to the same-dimensional representations $\tilde{\boldsymbol{Q}}$ by the projection head, while $\boldsymbol{K}$ is used to predict the categorical outputs with classifier. The feature representation $\boldsymbol{V}$ is derived from the momentum encoder, possessing identical dimensions to $\boldsymbol{Q}$. Moreover, we also utilize Kullback-Leibler divergence [11] to implement the criterion of exploiting IB, resulting in a compact feature representation $\tilde{\boldsymbol{V}}$. By alternately learning robust representations from $\boldsymbol{D}_q$ and $\boldsymbol{D}_v$, a symmetry objective function is designed to gain the IB loss,

$$\mathcal{L}_{IB} = \sum_{x_i \in \boldsymbol{D}_q} \sum_{x_j \in \boldsymbol{D}_v} KL\left(\tilde{\boldsymbol{Q}}_i \parallel \boldsymbol{V}_j\right) + \sum_{x_i \in \boldsymbol{D}_v} \sum_{x_j \in \boldsymbol{D}_q} KL\left(\tilde{\boldsymbol{Q}}_i \parallel \boldsymbol{V}_j\right). \tag{3}$$

3.2 Semi-Supervised Learning

Selecting Confident Samples. The main core of the proposed SSL module is to recognize the confident samples with clean labels, and re-correct the supervision information of label noise. To this end, we first select confident samples based on their confidence score provided by the classifier. Denote the confident samples with clean label belonging to n-th class as $\boldsymbol{D}_c^n$,

$$\boldsymbol{D}_c^n = \{(x_i, y_i) \mid y_i \cdot p_i > \gamma_n\}, \tag{4}$$

where $p_i \in [0, 1]$ is the classification probability of the enhanced image x_i, and γ_n is a dynamic confidence threshold for the n-th class to ensure a class-balanced set of identified confident examples.

Pseudo-Labeling. To generate accurate supervision signals, a flexible pseudo-labeling promotion strategy is introduced to replace noisy labels with pseudo-labels,

$$\tilde{y}_i = \begin{cases} y_i, & x_i \in \boldsymbol{D}_c, \\ p_i, & x_i \notin \boldsymbol{D}_c. \end{cases} \tag{5}$$

Benefiting from the unique semi-supervised learning structure, our SSL module can effectively reduce the impact of noise based on statistical classification. The objective function for semi-supervised learning is defined as:

$$\mathcal{L}_{SSL} = \sum_{x_i \in \boldsymbol{D}} \frac{1 - (\tilde{y}_i \cdot p_i)^{\omega}}{\omega}, \tag{6}$$

where $\omega \in (0, 1]$ is a tunable focusing parameter, which is utilized to exploit the benefits of both the noise-robustness and the implicit weighting scheme.

3.3 Similarity Contrastive Learning

Selecting Confident Pairs. To achieve a precise estimation of noisy pairs, a novel pair-wise selection strategy is proposed to identify the reliable confident pairs out of noisy pairs. By calculating their similarity distribution of representation learning, the SCL module can transform identified confident examples into a set of associated confident pairs $\mathcal{S}$ without knowing noise rates,

$$\mathcal{S} = \{(x_i, x_j) \mid p_i^\top p_j \geq \tau\}, \tag{7}$$

where τ is considered as a confidence threshold. Therefore, the objective loss on each sample pair for contrastive learning can be defined as:

$$\mathcal{L}(x_i, x_j) = \log\left(1 - \left\langle \tilde{\boldsymbol{Q}}_i, \boldsymbol{V}_j \right\rangle\right) \mathbb{1}\left[(x_i, x_j) \in \mathcal{S}\right]. \tag{8}$$

Consistent with Eq. 3, a symmetry loss function is applied for each minibatch,

$$\mathcal{L}_{SCL} = \sum_{x_q \in \boldsymbol{D}_q} \sum_{x_v \in \boldsymbol{D}_v} \mathcal{L}(x_q, x_v) + \sum_{x_v \in \boldsymbol{D}_v} \sum_{x_q \in \boldsymbol{D}_q} \mathcal{L}(x_v, x_q). \tag{9}$$

Queue. It is noted that blindly increasing the size of the minibatch will be limited by computing resources [15]. To overcome these issues, a queue with the length of L is also introduced into the SCL module, which can maintain a feature dictionary to store features and decouple the dictionary size from the mini-batch size [7,13]. As the new minibatch is added to the queue, the queue $\boldsymbol{M} = \{x_i, V_i\}_{i=1}^{L}$ is gradually replaced and the oldest features in the queue are removed. The objective loss of the queue is defined as:

$$\mathcal{L}_{QUE} = \sum_{x_q \in \boldsymbol{D}_q} \sum_{x_l \in \boldsymbol{M}} \mathcal{L}(x_q, x_l) + \sum_{x_v \in \boldsymbol{D}_v} \sum_{x_l \in \boldsymbol{M}} \mathcal{L}(x_v, x_l). \tag{10}$$

Objective Loss of SSCL. Based on the analysis of the above modules, the total objective loss for the proposed SSCL method can be obtained by,

$$\mathcal{L}_{SSCL} = \mathcal{L}_{SSL} + \alpha \cdot \mathcal{L}_{QUE} + \beta \cdot \mathcal{L}_{IB}, \tag{11}$$

where α and β are loss weight. In our experiments, both α and β are set to 0.25.

4 Experiments

4.1 Implementation Details and Settings

By randomly replacing labels for a percentage of the training data with all possible labels, the proposed SSCL method is extensively validated on four benchmarks with symmetric noise. ISIC-19 [30] dataset boasts a training set of 20,400

Table 1. Comparisons with the state-of-the-art baselines on benchmark datasets.

Methods		CE	Mixup	GCE	MOIT	Sel-CL	CTRR	SSCL	Improv.
Datasets	NR	–	–	–	CVPR'21	CVPR'22	CVPR'22	–	–
ISIC19	0.2	75.4	72.2	75.5	79.7	–	79.9	**81.1**	1.2 ↑
	0.3	73.1	70.2	72.6	75.9	–	75.8	**76.7**	0.9 ↑
	0.5	64.1	62.1	64.8	65.1	–	65.4	**66.8**	1.4 ↑
	Mean	70.9	68.2	71.0	73.6	–	73.7	**74.9**	1.2 ↑
BUSI	0.2	80.6	72.9	81.9	84.5	-	83.8	**86.9**	2.4 ↑
	0.3	76.7	67.8	77.4	77.2	–	78.3	**81.2**	2.9 ↑
	0.5	72.3	66.5	72.3	73.5	–	72.5	**74.9**	1.4 ↑
	Mean	76.5	69.1	77.2	78.4	–	78.2	**81.0**	2.6 ↑
CIFAR-10	0.2	82.7	92.3	86.6	94.1	**95.5**	93.9	95.0	0.5 ↓
	0.5	57.9	77.6	81.9	91.8	93.9	91.7	**94.2**	0.3 ↑
	0.8	26.1	46.7	54.6	81.1	89.2	88.1	**93.0**	3.8 ↑
	Mean	55.6	72.2	74.4	89.0	92.9	91.2	**94.1**	1.2 ↑
CIFAR-100	0.2	61.8	66.0	59.2	75.9	**76.5**	73.8	75.0	1.5 ↓
	0.5	37.3	46.6	47.8	70.6	72.4	72.2	**73.4**	1.0 ↑
	0.8	8.8	17.6	15.8	47.6	59.6	63.9	**66.5**	2.6 ↑
	Mean	36.0	43.4	40.9	64.7	69.5	70.0	**71.6**	1.6 ↑

dermoscopic images and a test set of 4,291 images, while BUSI [1] dataset encompasses 625 ultrasound images in the training set and 155 images in the test set. Both CIFAR-10 and CIFAR-100 contain 50,000 training images and 10,000 test images. The backbone of our SSCL framework is built on ResNet-18. In addition to Mixup, a series of augmentation techniques, such as random flipping, cropping, and Gaussian blur, are randomly applied to generate the hybrid augmented images during the training process. We compare the proposed SSCL method with several state-of-the-art baselines, including three conventional noise-robustness methods (i.e., CE [9], GCE [29], and Mixup [28]), three state-of-the-art baselines (i.e., MOIT [17], Sel-CL [15], and CTRR [14]). More implementation details are shown in the supplementary material.

4.2 Comparisons with the State-of-the-arts

In this part, we evaluate the performance of the proposed SSCL framework with classification accuracy under different label noise rates (NR). As shown in Table 1, the proposed SSCL significantly outperforms the state-of-the-art baselines on almost all evaluation metrics. For example, our SSCL achieves the highest classification accuracy on ISIC-19 and BUSI datasets, which can verify the effectiveness of our SSCL for medical image analysis with noisy supervision. Especially in the case of higher noise, SSCL has a more powerful capability to capture discriminative and robust features and minimize the effect of noisy

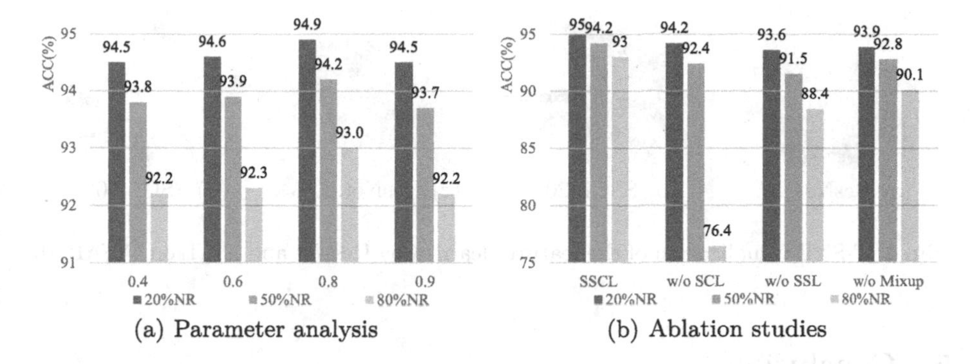

Fig. 2. Comparison of accuracy (%) in (a) parameter analysis and (b) ablation studies on CIFAR-10

labels. When the noise rate is 0.8, the proposed SSCL framework achieves a classification accuracy of 93.0% and 66.5% on CIFA-10 and CIFA-100, surpassing Sel-CL by 3.8% and 6.9%. The comparative results consistently demonstrate the superiority and generalizability of the proposed SSCL framework.

4.3 Parameter Analysis and Ablation Studies and Visualizations

As the key hyperparameter in Eq. 7, the threshold τ is designed to reduce the wrong sample pairs to achieve the best classification boundary construction. In this part, we empirically conduct the proposed SSCL framework with a range of different values τ. As shown in Fig. 2(a), our SSCL achieves the best performance when τ is increased to 0.8, which can avoid the adverse impact of noisy pairs. Moreover, we also conduct ablation studies by systematically removing each component within the SSCL. In Fig. 2(b), we can observe that all the modules are necessary for the function of the proposed SSCL framework. To further validate the discriminative power of the SSCL framework, we utilized t-SNE [5] to visualize the features extracted by vanilla ResNet-18 and SSCL. Compared with vanilla ResNet-18, the visualization results in Fig. 3 and Fig. 4 clearly demonstrate that SSCL has better characteristics for clustering and finer classification boundary, which can demonstrate the effectiveness of SSCL.

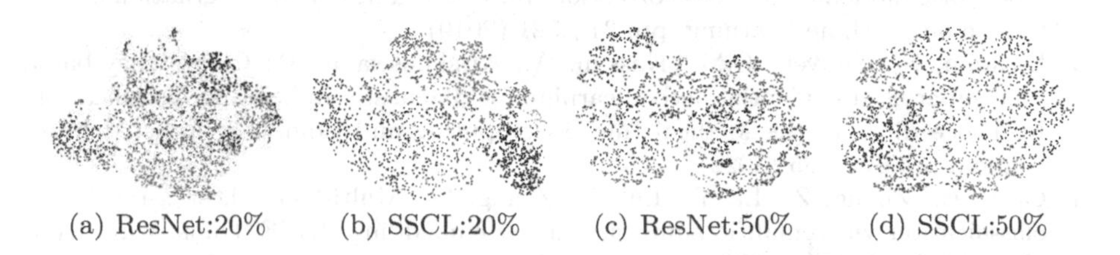

Fig. 3. T-SNE visualization of the features learned by ResNet and SSCL on ISIC19.

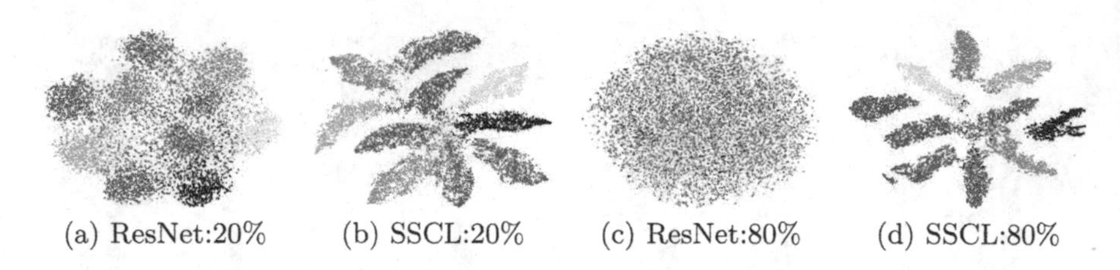

Fig. 4. T-SNE visualization of the features learned by ResNet and SSCL on CIFAR-10.

5 Conclusion

This paper presents a robust and reliable semi-supervised contrastive learning method that benefits greatly from the potential synergistic efficacy of semi-supervised learning and contrastive learning, which aims to tackle the challenge of learning with medical noisy labels. By explicitly selecting confident samples and pairs, our approach exhibits a powerful ability to learn discriminative feature representations, mitigating the impact of medical label noise. To demonstrate the effectiveness and versatility of our proposed approach in various practical scenarios, our future works would extend the proposed SSCL method to a broader range of real-world noisy datasets and tasks.

Acknowledgment. This work was supported in part by NSFC fund 62176077, in part by Guangdong Basic and Applied Basic Research Foundation under Grant 2023A1515010057, in part by Shenzhen Science and Technology Program (Grant No. RCYX20221008092852077), in part by the Shenzhen Key Technical Project under Grant 2022N001, in part by the Shenzhen Fundamental Research Fund under Grant JCYJ20210324132210025, and in part by the Guangdong Provincial Key Laboratory of Novel Security Intelligence Technologies (2022B1212010005).

References

1. Al-Dhabyani, W., Gomaa, M., Khaled, H., Fahmy, A.: Dataset of breast ultrasound images. Data Brief **28**, 104863 (2020)
2. Arazo, E., Ortego, D., Albert, P., O'Connor, N., McGuinness, K.: Unsupervised label noise modeling and loss correction. In: Proceedings of the International Conference on Machine Learning, pp. 312–321 (2019)
3. Balaram, S., Nguyen, C.M., Kassim, A., Krishnaswamy, P.: Consistency-based semi-supervised evidential active learning for diagnostic radiograph classification. In: Proceedings of the Medical Image Computing and Computer Assisted Intervention, pp. 675–685 (2022)
4. Chen, B., Zhang, Z., Li, Y., Lu, G., Zhang, D.: Multi-label chest x-ray image classification via semantic similarity graph embedding. IEEE Trans. Circ. Syst. Video Technol. **32**(4), 2455–2468 (2021)
5. Chen, B., Zhang, Z., Lu, Y., Chen, F., Lu, G., Zhang, D.: Semantic-interactive graph convolutional network for multilabel image recognition. IEEE Trans. Syst. Man Cybern. Syst. **52**(8), 4887–4899 (2021)

6. Harutyunyan, H., Reing, K., Ver Steeg, G., Galstyan, A.: Improving generalization by controlling label-noise information in neural network weights. In: Proceedings of the International Conference on Machine Learning, pp. 4071–4081 (2020)
7. He, K., Fan, H., Wu, Y., Xie, S., Girshick, R.: Momentum contrast for unsupervised visual representation learning. In: Proceedings of the IEEE Conference on Computer Vision and Pattern Recognition, pp. 9729–9738 (2020)
8. Hendrycks, D., Mazeika, M., Wilson, D., Gimpel, K.: Using trusted data to train deep networks on labels corrupted by severe noise. Adv. Neural Inf. Process. Syst. **31**, 1–10 (2018)
9. Hinton, G., et al.: Deep neural networks for acoustic modeling in speech recognition: the shared views of four research groups. IEEE Signal Process. Maga. **29**(6), 82–97 (2012)
10. Imran, A.A.Z., Wang, S., Pal, D., Dutta, S., Zucker, E., Wang, A.: Multimodal contrastive learning for prospective personalized estimation of CT organ dose. In: Proceedings of the Medical Image Computing and Computer Assisted Intervention, pp. 634–643 (2022)
11. Ji, S., Zhang, Z., Ying, S., Wang, L., Zhao, X., Gao, Y.: Kullback-leibler divergence metric learning. IEEE Trans. Cybern. **52**(4), 2047–2058 (2020)
12. Jiang, M., Yang, H., Li, X., Liu, Q., Heng, P.A., Dou, Q.: Dynamic bank learning for semi-supervised federated image diagnosis with class imbalance. In: Proceedings of the Medical Image Computing and Computer Assisted Intervention, pp. 196–206 (2022)
13. Khosla, P., et al.: Supervised contrastive learning. Adv. Neural Inf. Process. Syst. **33**, 18661–18673 (2020)
14. Lee, S., Lee, H., Yoon, S.: Adversarial vertex mixup: toward better adversarially robust generalization. In: Proceedings of the IEEE Conference on Computer Vision and Pattern Recognition, pp. 272–281 (2020)
15. Li, S., Xia, X., Ge, S., Liu, T.: Selective-supervised contrastive learning with noisy labels. In: Proceedings of the IEEE Conference on Computer Vision and Pattern Recognition, pp. 316–325 (2022)
16. Li, Y., Yang, J., Song, Y., Cao, L., Luo, J., Li, L.J.: Learning from noisy labels with distillation. In: Proceedings of the IEEE International Conference on Computer Vision, pp. 1910–1918 (2017)
17. Ortego, D., Arazo, E., Albert, P., O'Connor, N.E., McGuinness, K.: Multi-objective interpolation training for robustness to label noise. In: Proceedings of the IEEE Conference on Computer Vision and Pattern Recognition, pp. 6606–6615 (2021)
18. Ramaswamy, H., Scott, C., Tewari, A.: Mixture proportion estimation via kernel embeddings of distributions. In: Proceedings of the International Conference on Machine Learning, pp. 2052–2060 (2016)
19. Reed, S., Lee, H., Anguelov, D., Szegedy, C., Erhan, D., Rabinovich, A.: Training deep neural networks on noisy labels with bootstrapping. arXiv preprint arXiv:1412.6596 (2014)
20. Rizve, M.N., Duarte, K., Rawat, Y.S., Shah, M.: In defense of pseudo-labeling: an uncertainty-aware pseudo-label selection framework for semi-supervised learning. In: Proceedings of the International Conference on Learning Representations (2021)
21. Seyfioğlu, M.S., et al.: Brain-aware replacements for supervised contrastive learning in detection of alzheimer's disease. In: Proceedings of the Medical Image Computing and Computer Assisted Intervention, pp. 461–470 (2022)
22. Sohn, K., et al.: Fixmatch: simplifying semi-supervised learning with consistency and confidence. Adv. Neural Inf. Process. Syst. **33**, 596–608 (2020)

23. Tran, M., Wagner, S.J., Boxberg, M., Peng, T.: S5cl: unifying fully-supervised, self-supervised, and semi-supervised learning through hierarchical contrastive learning. In: Proceedings of the Medical Image Computing and Computer Assisted Intervention, pp. 99–108 (2022)
24. Venkataramanan, S., Kijak, E., Amsaleg, L., Avrithis, Y.: Alignmixup: improving representations by interpolating aligned features. In: Proceedings of the IEEE Conference on Computer Vision and Pattern Recognition, pp. 19174–19183 (2022)
25. Wang, X., Yao, L., Rekik, I., Zhang, Y.: Contrastive functional connectivity graph learning for population-based fmri classification. In: Proceedings of the Medical Image Computing and Computer Assisted Intervention, pp. 221–230 (2022)
26. Wang, Y., Ma, X., Chen, Z., Luo, Y., Yi, J., Bailey, J.: Symmetric cross entropy for robust learning with noisy labels. In: Proceedings of the IEEE Conference on Computer Vision and Pattern Recognition, pp. 322–330 (2019)
27. Xia, X., et al.: Robust early-learning: hindering the memorization of noisy labels. In: Proceedings of the International Conference on Learning Representations (2021)
28. Zhang, H., Cisse, M., Dauphin, Y.N., Lopez-Paz, D.: Mixup: beyond empirical risk minimization. arXiv preprint arXiv:1710.09412 (2017)
29. Zhang, Z., Sabuncu, M.: Generalized cross entropy loss for training deep neural networks with noisy labels, vol. 31 (2018)
30. Zhou, Y., et al.: Learning to bootstrap for combating label noise. arXiv preprint arXiv:2202.04291 (2022)

Multi-scale Self-Supervised Learning for Longitudinal Lesion Tracking with Optional Supervision

Anamaria Vizitiu[1(✉)], Antonia T. Mohaiu[1], Ioan M. Popdan[1], Abishek Balachandran[2], Florin C. Ghesu[3], and Dorin Comaniciu[3]

[1] Advanta, Siemens SRL, Brasov, Romania
anamaria.vizitiu@siemens.com
[2] Digital Technology and Innovation, Siemens Healthineers, Bangalore, India
[3] Digital Technology and Innovation, Siemens Healthineers, Princeton, NJ, USA

Abstract. Longitudinal lesion or tumor tracking is an essential task in different clinical workflows, including treatment monitoring with follow-up imaging or planning of re-treatments for radiation therapy. Accurately establishing correspondence between lesions at different timepoints, recognizing new lesions or lesions that have disappeared is a tedious task that only grows in complexity as the number of lesions or timepoints increase. To address this task, we propose a generic approach based on multi-scale self-supervised learning. The multi-scale approach allows the efficient and robust learning of a similarity map between multi-timepoint image acquisitions to derive correspondence, while the self-supervised learning formulation enables the generic application to different types of lesions and image modalities. In addition, we impose optional supervision during training by leveraging tens of anatomical landmarks that can be extracted automatically. We train our approach at large scale with more than 50,000 computed tomography (CT) scans and validate it on two different applications: 1) Tracking of generic lesions based on the DeepLesion dataset, including liver tumors, lung nodules, enlarged lymph-nodes, for which we report highest matching accuracy of 92%, with localization accuracy that is nearly 10% higher than the state-of-the-art; and 2) Tracking of lung nodules based on the NLST dataset for which we achieve similarly high performance. In addition, we include an error analysis based on expert radiologist feedback, and discuss next steps as we plan to scale our system across more applications.

Keywords: Self-supervised learning · Multi-scale · Longitudinal lesion tracking

Supplementary Information The online version contains supplementary material available at https://doi.org/10.1007/978-3-031-43907-0_55.

H. Greenspan et al. (Eds.): MICCAI 2023, LNCS 14220, pp. 573–582, 2023.
https://doi.org/10.1007/978-3-031-43907-0_55

1 Introduction

Longitudinal lesion or tumor tracking is a fundamental task in treatment monitoring workflows, and for planning of re-treatments in radiation therapy. Based on longitudinal imaging for a given patient it requires establishing which lesions are corresponding (i.e., same lesion, observed at different timepoints), which lesions have disappeared and which are new compared to prior scanning. This information can be leveraged to assess treatment response, e.g., by analyzing the evolution of size and morphology for a given tumor [1], but also for adaptation of (re-)treatment radiotherapy plans that take into account new tumors.

In practice, the development of automatic and reliable lesion tracking solutions is hindered by the complexity of the data (over different modalities), the absence of large, annotated datasets, and the difficulties associated with lesion identification (i.e., varying sizes, poses, shapes, and sparsely distributed locations). In this work, we present a multi-scale self-supervised learning solution for lesion tracking in longitudinal studies using the capabilities of contrastive learning [9]. Inspired by the pixel-wise contrastive learning strategy introduced in [5], we choose to learn pixel-wise feature representations that embed consistent anatomical information from unlabeled (i.e., without lesion-related annotations) and unpaired (i.e., without the use of longitudinal scans) data, overcoming barriers to data collection. To increase the system robustness and emulate the clinician's reading strategies, we propose to use multi-scale embeddings to enable the system to progressively refine the fine-grained location. In addition, as imaging offers contextual information about the human body that is naturally consistent, we design the model to benefit from biologically-meaningful points (i.e., anatomical landmarks). The reasoning behind this strategy is that simple data augmentation methods cannot faithfully model inter-subject variability or possible organ deformations. Hence, we ensure the spatial coherence of the tracked lesion location using well-defined anatomical landmarks.

Our proposed method brings two elements of novelty from a technical point of view: (1) the multi-scale approach for the anatomical embedding learning and (2) a positive sampling approach that incorporates anatomically significant landmarks across different subjects. With these two strategies, the goal is to ensure a high degree of robustness in the computation of the lesion matching across different lesion sizes and varying anatomies. Furthermore, a significant focus and contribution of our research is the experimental study at a very large scale: we (1) train a pixel-wise self-supervised system using a very large and diverse dataset of 52,487 CT volumes and (2) evaluate on two publicly available datasets. Notably, one of the datasets, NLST, presents challenging cases with 68% of lesions being very small (i.e., radius < 5 mm).

2 Background and Motivation

The problem of lesion tracking in longitudinal data is typically divided into two steps: (1) detection of lesions and (2) tracking the same lesion over multiple

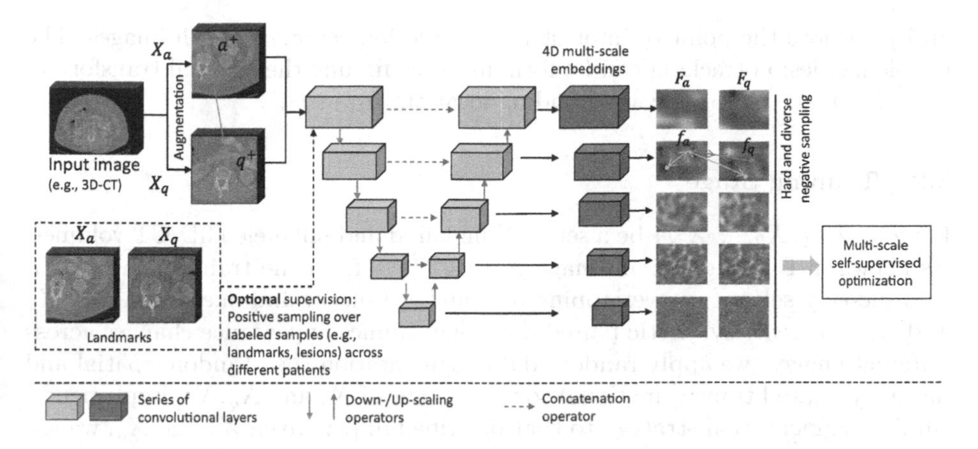

Fig. 1. Overview of the proposed multi-scale self-supervised learning system. During training, we randomly extract positive samples (optionally, include same anatomical landmarks from different volumes), hard-negative samples, and diverse negative samples of pixels from augmented 3D paired patches. During inference, the extracted embeddings are used to generate a cascade of cosine similarity maps that initially locate the corresponding location in a follow-up image within a larger area and subsequently improve the matching accuracy through gradual refinement.

time points. Classical methods to solve this problem rely on image registration, where tracking is performed via image alignment and rule-based correspondence matching [15,16,21]. These approaches are difficult to optimize, especially when scaling across different body regions and fields of view. Appearance-based trackers [19,20] adopt a different strategy by projecting lesions detected beforehand with dedicated detectors [17,18] onto a representation space and employing nearest neighbor analysis. One recent approach, Deep Lesion Tracker (DLT) [8], integrates both strategies to perform appearance-based recognition under anatomical constraints. As a more direct matching approach, Yan et al. [5] uses a self-supervised anatomical embedding model (SAM) to create semantic embeddings for each image pixel, avoiding the detection step. Training exclusively on augmented paired data prevents SAM from accurately representing anatomical changes and deformations that occur over time. This can influence the contextual information of a pixel, which in turn impacts the pixel-wise embeddings on which the similarity-based tracker depends. To overcome this, we propose to train a pixel-wise multi-scale embedding model that accounts for anatomical similarity among different subjects, making the embeddings more effective.

3 Method

3.1 Problem Definition

Let I_1 (i.e., template or baseline image) and I_2 (i.e., query or follow-up image) be two 3D-CT scans acquired at time t_1 and t_2, respectively, Additionally, let p_1

and p_2 denote the point of interest (i.e., the lesion center) in both images. The problem of lesion tracking can be formulated as finding the optimal transformation that maps p_1 to its corresponding location, p_2, in I_2.

3.2 Training Stage

Let $D = \{X_1, X_2, ..., X_N\}$ be a set of N unpaired and unlabeled 3D-CT volumes. As shown in Fig. 1, given an image $X \in \mathbb{R}^{d\times h\times w}$ from the training dataset D, we randomly select two overlapping 3D patches (anchor and query), namely X_a and X_q. To create synthetic paired data that mimics appearance changes across different images, we apply random data augmentation (i.e., random spatial and intensity-related transformations) to the content of X_a and X_q. We implement a similar augmentation strategy to that described in [5]. Given X_a and X_q, we use an embedding extraction model to construct a hierarchy of multi-scale semantic embeddings for each image pixel, labeled F_a and F_q respectively. The embedding at ith scale, $1 \le i \le s$, is denoted as F_a^i and F_q^i and is represented as a 4D feature map, with an embedding vector of length L associated with each pixel.

Given the nature of contrastive learning, the sampling strategy (extracting negative and positive pixel pairs from augmented 3D paired patches) is essential to achieving discriminative pixel-wise embeddings. We arbitrary sample n_{pos} positive pixel pairs from the overlapping area of X_a and X_q, denoted by $a^+ = \{a_1^+, ..., a_j^+ ..., a_{n_{pos}}^+\}$, $q^+ = \{q_1^+, ..., q_j^+, ..., q_{n_{pos}}^+\}$, $1 \le j \le n_{pos}$. To further enhance embeddings, 10% of the positive pixel pairs are derived from biologically-meaningful points across different volumes in the batch. We use data-driven models [14] to extract 37 anatomical landmarks, such as the top right lung, suprasternal notch, tracheal bifurcation, etc. Similar to [5], for each positive pixel pair (a_j^+, q_j^+), we select n_{neg} hard and diverse negative pixels, denoted by $h^- = \{h_1^-, ..., h_k^-, ..., h_{n_{neg}}^-\}$, $1 \le k \le n_{neg}$.

Next, at each scale, we extract the embedding vectors for positive and negative pixel pairs from F_a^i, F_q^i, guided by the corresponding locations, a^+, q^+, h^-, which are downsampled to match the scale. We denote the positive embeddings at ith scale at pixel location a_j^+, q_j^+ as $f_{aj}^{\,i}$, $f_{qj}^{\,i} \in \mathbb{R}^L$. Similarly, we denote the negative embeddings at pixel location h_k^- associated to a positive positive pixel pair (a_j^+, q_j^+) as $f_{jk}^i \in \mathbb{R}^L$. We use L2-norm to normalize the embedding vectors before the loss computation. We use pixel-wise InfoNCE loss [5,10] to enhance the similarity among similar pixels (i.e., positive pairs of pixels) and decrease the similarity among dissimilar pixels (i.e., negative pairs of pixels). Correspondingly, we set the contrastive loss at the ith scale:

$$L^i = -\sum_{j=1}^{n_{pos}} \log \frac{\exp(f_{aj}^{\,i} \cdot f_{qj}^{\,i}/\tau)}{\exp(f_{aj}^{\,i} \cdot f_{qj}^{\,i}/\tau) + \sum_{k=1}^{n_{neg}} \exp(f_{aj}^{\,i} \cdot f_{jk}^{i}/\tau)}, \tag{1}$$

where $\tau = 0.5$ is a temperature parameter. The final loss is then calculated as the average of all these individual losses.

3.3 Inference Stage

Let X_a be a 3D-CT volume template with an input point of interest $p_a \in X_a$, and X_q a corresponding query 3D-CT volume. The first step is to project the image X_a into a multi-scale feature space, creating a hierarchy of multi-scale semantic embeddings F_a for each pixel in the image (i.e., a 4D feature map). Next, we follow a similar process for the query image X_a and acquire the pixel-level embeddings F_q.

To measure the similarity between the embeddings of the input X_a at the point of interest p_a and the query embeddings F_q, we compute cosine similarity maps at each scale:

$$S^i = \frac{F_a^i(p_a) \cdot F_q^i}{\|F_a^i(p_a)\|_2 \cdot \|F_q^i\|_2} \tag{2}$$

Finally, we combine the multi-scale similarity maps through summation and select the voxel with the highest similarity as the matching point in the query volume.

4 Experiments

4.1 Datasets and Setup

Datasets: We train the universal and fine-grained anatomical point matching model using an in-house CT dataset (VariousCT). The training dataset contains 52,487 unlabeled 3D CT volumes capturing various anatomies, including chest, head, abdomen, pelvis, and more.

The evaluation is based on two datasets, the publicly released Deep Longitudinal Study (DLS) dataset [8] and the National Lung Screening Trial (NLST) dataset [12]. The DLS dataset is a subset of the DeepLesion [11] medical imaging dataset, containing 3891 pairs of lesions with information on their location and size. The dataset covers various types of lesions across different organs. We follow the official data split for DLS dataset and perform evaluation on the testing dataset which comprises 480 lesion pairs. For NLST, we randomly selected a subset of 1045 test images coming from 420 patients with up to 3 studies. A certified radiologist annotated the testing data by identifying the location and size of the pulmonary nodules, resulting in a total of 825 paired annotations. We evaluate lesion tracking in both directions, from baseline to follow-up and from follow-up to baseline [8]. This results in a total of 960 and 1650 testing lesion pairs in DLS and NLST test sets, respectively. The isotropic resolution of all CT volumes is adjusted to $2mm$ through bilinear interpolation.

System Training: Our learning model is implemented in PyTorch and uses the TorchIO library [13] for medical data manipulation and augmentation.

We employ a U-Net-based encoder-decoder architecture [2] that utilizes an inflated 3D ResNet-18 [3,4] as its encoder, which extends all 2D convolutions

Table 1. Comparison between the proposed solution and several state-of-the-art approaches (reference results are from [8]). The exact same test set was used to compute the performance of each approach listed in the table; however, we retrained only SAM.

Method	CPM@ 10 mm	CPM@ Radius	MEDx (mm)	MEDy (mm)	MEDz (mm)	MED (mm)
Affine [15]	48.33	65.21	4.1 ± 5.0	5.4 ± 5.6	7.1 ± 8.3	11.2 ± 9.9
VoxelMorph [16]	49.90	65.59	4.6 ± 6.7	5.2 ± 7.9	6.6 ± 6.2	10.9 ± 10.9
LENS-LesionG. [17,19]	63.85	80.42	2.6 ± 4.6	2.7 ± 4.5	6.0 ± 8.6	8.0 ± 10.1
VULD-LesionG. [18,19]	64.69	76.56	3.5 ± 5.2	4.1 ± 5.8	6.1 ± 8.8	9.3 ± 10.9
LENS-LesaNet [17,20]	70.00	84.58	2.7 ± 4.8	2.6 ± 4.7	5.7 ± 8.6	7.8 ± 10.3
DLT-SSL [8]	71.04	81.52	3.8 ± 5.3	3.7 ± 5.5	5.4 ± 8.4	8.8 ± 10.5
DEEDS [21]	71.88	85.52	2.8 ± 3.7	3.1 ± 4.1	5.0 ± 6.8	7.4 ± 8.1
DLT-Mix [8]	78.65	88.75	3.1 ± 4.4	3.1 ± 4.5	4.2 ± 7.6	7.1 ± 9.2
DLT [8]	78.85	86.88	3.5 ± 5.6	2.9 ± 4.9	4.0 ± 6.1	7.0 ± 8.9
SAM [5]	81.67	90.21	2.9 ± 8.0	2.5 ± 3.8	3.6 ± 5.2	6.5 ± 9.6
Ours	**83.13**†	**91.87**†	$\mathbf{2.9 \pm 6.0}$	$\mathbf{2.2 \pm 3.2}$	$\mathbf{3.1 \pm 3.9}$	$\mathbf{5.9 \pm 7.1}^{\dagger}$

† Improvement is statistically significant compared to SAM [5] (p-value $< 10^{-6}$).

in the standard ResNet to 3D convolutions and allows the use of pre-trained ImageNet weights. The multi-scale embedding model employs $s = 5$ scales, and the embedding length is fixed at $L = 128$ for each scale. Convolution with a stride of $(2, 2, 2)$ is used to reduce the feature map size at the first and fifth levels, while a stride of $(1, 2, 2)$ is employed for intermediary levels 2 to 4. The U-Net decoder uses a convolution layer with a $3 \times 3 \times 3$ kernel after every up-sampling layer to generate the final cascade of feature embeddings.

The model is trained with AdamW optimizer [6] for 64 epochs using an early stopping strategy with a patience of 5 epochs, a batch size of 8 augmented 3D paired patches of $32 \times 96 \times 96$, and a learning rate of 0.0001.

For data augmentation, we apply random cropping, scaling, rotation, and Gaussian noise injections. A windowing approach that covers the intensity ranges of lungs and soft tissues is used to scale CT intensity values to $[-1, 1]$. The sampling hyperparameters consist of 100 positive pixel pairs ($n_{pos} = 100$), 100 hard negative pixel pairs, and 200 diverse negative pixel pairs ($n_{neg} = 300$).

Evaluation Metrics: We use mean Euclidean distance (MED) to measure the distance between predicted lesion center and ground truth, and the center point matching accuracy (i.e., percentage of accurately matched lesions given the annotated lesion radius), denoted with CPM@Radius. For lesions of large sizes, we set a maximum distance limit of 10 mm as acceptance criteria [8], denoted with CPM@10 mm. The NLST testing dataset has a distinctive feature wherein nodules are relatively small, 68% of annotated lesions have a radius of less than 5 mm (compared to 6% in DLS dataset). To ensure that such small nodules are not missed during evaluation, we relax the minimum distance requirement and consider a distance of 6 mm as a permissible matching error.

4.2 Evaluation

For the lesion tracking task on DLS dataset, we quantitatively compare our system against existing trackers in Table 1. These include the Deep Lesion Tracker (DLT) and its variants [8], as well as registration-based trackers [15,16,21] and appearance-based trackers via detector learning [17–20]. Given the clear superiority of approach [5] compared to all reference solutions, we focus on achieving a direct comparison against SAM [5]. Hence, for performance comparison against self-supervised anatomical embedding tracker, we retrain SAM [5] with images from VariousCT dataset.

Table 2. Results on the NLST dataset related to the tracking of lung nodules.

Method	CPM@ 10 mm	CMP@ Radius	MEDx (mm)	MEDy (mm)	MEDz (mm)	MED(mm)
Ours	**90.05**	**92.12**	$\mathbf{2.0 \pm 2.6}$	$\mathbf{2.2 \pm 3.8}$	$\mathbf{2.7 \pm 4.8}$	$\mathbf{4.9 \pm 6.0}$

Our method achieves a matching accuracy of 91.87%, that is 1.84% higher than SAM and 5.74% higher than DLT. To confirm the significance of the improvement achieved by our method compared to SAM [5], we conduct a paired t-test for statistical analysis and show that the improvement is statistically significant (p-value $< 10^{-6}$). Compared to the self-supervised version of DLT, the difference in performance is significantly greater, the proposed systems outperforms DLT-SSL by more than 10%. When imposing a maximum distance limit of 10 mm between the ground truth and prediction, our method increases performance by 1.46%, showing the importance of the multi-scale approach in lesion

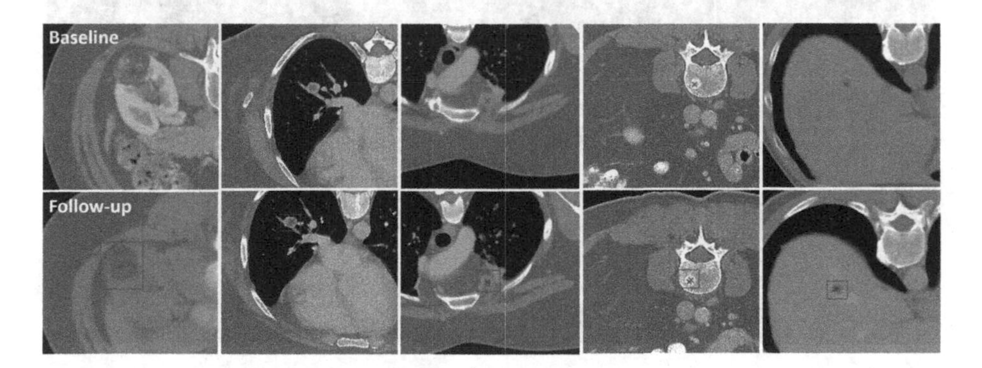

Fig. 2. Examples of lesion matching results on the DLS testing dataset. We denote the ground-truth points using green markers in both the baseline and follow-up images, whereas the predicted points are indicated by red markers. To illustrate the extent of the lesions, we also display the annotated bounding boxes on the follow-up images. For more clarity, we show only the axial view. (Color figure online)

location refinement. Consequently, superior accuracy is achieved compared to both registration and supervised appearance-based tracking methods. Qualitative examples are shown in Fig. 2.

On the NLST dataset, our proposed method obtains a center point matching accuracy of 92.12% (Table 2). In the case of longitudinal lung nodule tracking (Fig. 3), it is more frequent to observe significant changes in size and density. As our system relies on the concept of anatomical embedding matching, the most substantial errors in lesion matching for our system occur when there are significant pathological distortions that deviate greatly from one timepoint to another. Examples of such cases are depicted in Fig. 4, based on expert radiologist feedback.

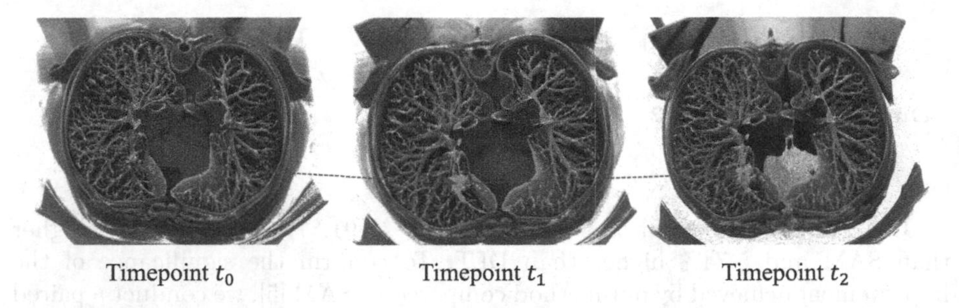

Fig. 3. Example case from the test set, highlighting the progression of a lung nodule over three timepoints: t_0, t_1 and t_3. Our system was robust to the axial rotations of the scans and the increasing size of the nodule and correctly established the correspondence.

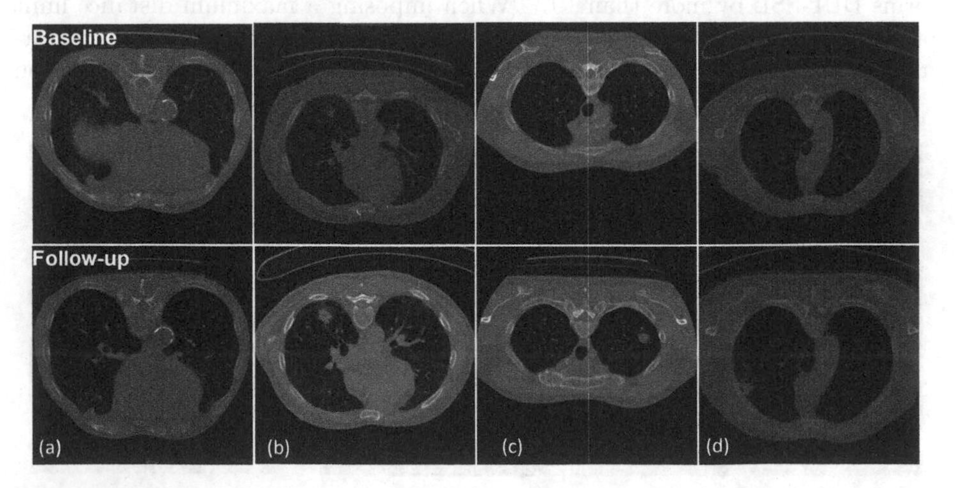

Fig. 4. Examples of lesion matching results on clinically challenging cases from NLST testing dataset: (a) bronchiectasis with mucus plugging adjacent to the nodule, (b) spiculated nodule in a setting of interstitial lung disease, (c), (d) small nodule progressed and increased significantly in size. The green and red markers denote the ground-truth and predicted lesion location. (Color figure online)

5 Conclusion

In conclusion, this paper presents an effective method for longitudinal lesion tracking based on multi-scale self-supervised learning. The method is generic, it does not require expert annotations or longitudinal data for training and can generalize to different types of tumors/organs/modalities. The multi-scale approach ensures a high degree of robustness and accuracy for small lesions. Through large-scale experiments and validation on two longitudinal datasets, we highlight the superiority of the proposed method in comparison to state-of-the-art. We found that adopting a multi-scale approach (instead of the global/local approach as proposed in [5]) can lead to embeddings that better capture the anatomical location and are able to handle lesions that vary in size or appearance at different scales. Moreover, the changes proposed in this work help to alleviate the confusion caused by left-right body symmetries (e.g., the apices of the lungs). This effect challenged the tracking of small nodules in the lungs using [5]. Our future work aims to enhance the matching accuracy by examining the implications of correlation magnitude, conducting robustness studies on slight variations in tracking initialization, and implementing a more advanced fusion strategy for the multi-scale similarity maps. In addition, we aim to expand to more applications, e.g., treatment monitoring for brain cancer using MRI.

Disclaimer: The concepts and information presented in this paper/presentation are based on research results that are not commercially available. Future commercial availability cannot be guaranteed.

Acknowledgements. The authors thank the National Cancer Institute for access to NCI's data collected by the National Lung Screening Trial (NLST). The statements contained herein are solely those of the authors and do not represent or imply concurrence or endorsement by NCI.

References

1. Eisenhauer, E.A., et al.: New response evaluation criteria in solid tumours: revised RECIST guideline (version 1.1). Eur. J. Cancer **45**(2), 228–247 (2009)
2. Ronneberger, O., Fischer, P., Brox, T.: U-Net: convolutional networks for biomedical image segmentation. In: Navab, N., Hornegger, J., Wells, W.M., Frangi, A.F. (eds.) MICCAI 2015. LNCS, vol. 9351, pp. 234–241. Springer, Cham (2015). https://doi.org/10.1007/978-3-319-24574-4_28
3. He, K., Zhang, X., Ren, S., Sun, J.: Deep residual learning for image recognition. In: CVPR, pp. 770–778 (2016)
4. Carreira, J., Zisserman, A.: Quo vadis, action recognition? a new model and the kinetics dataset. In: CVPR, pp. 4724–4733 (2017)
5. Yan, K., et al.: SAM: self-supervised learning of pixel-wise anatomical embeddings in radiological images. IEEE Trans. Med. Imaging **41**, 2658–2669 (2020)
6. Ilya, L., Hutter, F.: Decoupled weight decay regularization. In: International Conference on Learning Representations (2017)

7. Ilya, L., Hutter, F.: SGDR: stochastic gradient descent with warm restarts. ArXiv: Learning (2016)
8. Cai, J., et al.: Deep lesion tracker: monitoring lesions in 4D longitudinal imaging studies. In: CVPR, pp. 15154–15164 (2020)
9. Chen, T., Kornblith, S., Norouzi, M., Hinton, G.E.: A simple framework for contrastive learning of visual representations (2020)
10. Van Den Oord, A., Li, Y., Vinyals, O.: Representation learning with contrastive predictive coding (2018)
11. Yan, K., Wang, X., Lu, L., Summers, R.M.: DeepLesion: automated deep mining, categorization and detection of significant radiology image findings using large-scale clinical lesion annotations (2017)
12. National Lung Screening Trial Research Team: The National Lung Screening Trial: overview and study design. In: Radiology, pp. 243–253 (2011)
13. Perez-Garcia, F., Sparks, R., Ourselin, S.: TorchIO: a python library for efficient loading, preprocessing, augmentation and patch-based sampling of medical images in deep learning. Comput. Methods Prog. Biomed. **208**, 106236 (2020)
14. Ghesu, F.C., et al.: Multi-scale deep reinforcement learning for real-time 3D-landmark detection in CT scans. IEEE Trans. Pattern Anal. Mach. Intell. **41**, 176–189 (2019)
15. Marstal, K., Berendsen, F.F., Staring, M., Klein, S.: SimpleElastix: a user-friendly, multi-lingual library for medical image registration. In: IEEE Conference on Computer Vision and Pattern Recognition Workshops (CVPRW), pp. 574–582 (2016)
16. Balakrishnan, G., Zhao, A., Sabuncu, M.R., Guttag, J.V., Dalca, A.V.: An unsupervised learning model for deformable medical image registration. In: CVPR, pp. 9252–9260 (2018)
17. Yan, K., et al.: Learning from multiple datasets with heterogeneous and partial labels for universal lesion detection in CT. IEEE Trans. Med. Imaging **40**, 2759–2770 (2020)
18. Cai, J., et al.: Deep volumetric universal lesion detection using light-weight pseudo 3D convolution and surface point regression (2020)
19. Yan, K., et al.: Deep lesion graphs in the wild: relationship learning and organization of significant radiology image findings in a diverse large-scale lesion database. In: Conference on Computer Vision and Pattern Recognition, pp. 9261–9270 (2017)
20. Yan, K., Peng, Y., Sandfort, V., Bagheri, M., Lu, Z., Summers, R.M.: Holistic and comprehensive annotation of clinically significant findings on diverse CT images: learning from radiology reports and label ontology. In: CVPR, pp. 8515–8524 (2019)
21. Heinrich, M.P., Jenkinson, M., Brady, M., Schnabel, J.A.: MRF-based deformable registration and ventilation estimation of lung CT. IEEE Trans. Med. Imaging **32**, 1239–1248 (2013)

Tracking Adaptation to Improve SuperPoint for 3D Reconstruction in Endoscopy

O. León Barbed[1(✉)], José M. M. Montiel[1], Pascal Fua[2], and Ana C. Murillo[1]

[1] DIIS-i3A, University of Zaragoza, Zaragoza, Spain
leon@unizar.es

[2] CVLAB, École Polytechnique Fédérale de Lausanne, Lausanne, Switzerland

Abstract. Endoscopy is the gold standard procedure for early detection and treatment of numerous diseases. Obtaining 3D reconstructions from real endoscopic videos would facilitate the development of assistive tools for practitioners, but it is a challenging problem for current Structure From Motion (SfM) methods. Feature extraction and matching are key steps in SfM approaches, and these are particularly difficult in the endoscopy domain due to deformations, poor texture, and numerous artifacts in the images. This work presents a novel learned model for feature extraction in endoscopy, called SuperPoint-E, which improves upon existing work using recordings from real medical practice. SuperPoint-E is based on the SuperPoint architecture but it is trained with a novel supervision strategy. The supervisory signal used in our work comes from features extracted with existing detectors (SIFT and SuperPoint) that can be successfully tracked and triangulated in short endoscopy clips (building a 3D model using COLMAP). In our experiments, SuperPoint-E obtains more and better features than any of the baseline detectors used as supervision. We validate the effectiveness of our model for 3D reconstruction in real endoscopy data. Code and model: https://github.com/LeonBP/SuperPointTrackingAdaptation.

Keywords: deep learning · structure from motion · local features · endoscopy

1 Introduction

Endoscopy is an important medical procedure with many applications, from routine screening to detection of early signs of cancer and minimally invasive treatment. Automatic analysis and understanding of these videos raises many opportunities for novel assistive and automatization tasks on endoscopy procedures. Obtaining 3D models from the intracorporeal scenes captured in endoscopies is an essential step to enable these novel tasks and build applications,

Supplementary Information The online version contains supplementary material available at https://doi.org/10.1007/978-3-031-43907-0_56.

H. Greenspan et al. (Eds.): MICCAI 2023, LNCS 14220, pp. 583–593, 2023.
https://doi.org/10.1007/978-3-031-43907-0_56

for example, for improved monitoring of existing patients or augmented reality during training or real explorations.

3D reconstruction strategies have been studied for long, and one crucial step in these strategies is feature detection and matching which serves as input for Structure from Motion (SfM) pipelines. Endoscopic images are a challenging case for feature detection and matching, due to several well known challenges for these tasks, such as lack of texture, or the presence of frequent artifacts, like specular reflections. These problems are accentuated when all the elements in the scene are deformable, as it is the case in most endoscopy scenarios, and in particular in the real use case studied in our work, the lower gastrointestinal tract explored with colonoscopies. Existing 3D reconstruction pipelines are able to build small 3D models out of short clips from real and complete recordings [1]. One of the current bottle-necks to obtain better 3D models is the lack of more abundant and higher quality correspondences in real data.

This work introduces SuperPoint-E, a new model to extract interest points from endoscopic images. We build on the well known SuperPoint architecture [5], a seminal work that delivers state-of-the-art results when coupled with downstream tasks[1]. Our main contribution is a novel supervision strategy to train the model. We propose to automatically generate reliable training data from video sequences by tracking feature points from existing detection methods, which do not require training. We select *good features* with the COLMAP SfM pipeline [21], generating training examples with feature points that can be tracked across several images according to COLMAP result. When used to train SuperPoint, our approach yields a self-supervised method outperforming current ones.

2 Related Work

3D reconstruction is an open problem for laparoscopic and endoscopic settings [14] of high interest for the community. This idea is supported for example by recent efforts on collecting new public dataset, to further advance in this field, such as endoscopic recordings from EndoSLAM [16] and EndoMapper [1] datasets. Earlier works like Grasa *et al.* [7] have evaluated the performance of modern SLAM approaches on endoscopic sequences. Mahmoud *et al.* [13] improved the performance of such methods in laparoscopic sequences. More recent approaches attempt to tackle specific endoscopy challenges, such as the deformation [18] or the artifacts due to specular reflections in the feature extraction step [2].

Well known SfM and SLAM pipelines rely on accurate and robust feature extraction methods. COLMAP [21,22], a public SfM tool, uses SIFT [11] features while ORB-SLAM [15] extracts ORB [19] features because of their efficiency. Both these feature extraction methods count with classical, hand-crafted descriptors that allowed to build such complex applications. However, transferring that performance to endoscopy settings remains a difficult task due to

[1] https://www.cs.ubc.ca/research/image-matching-challenge/2021/leaderboard/.

several challenges. Artifacts or the lack of texture result in low amount of correspondences along real endoscopy videos, what motivates the need for improved strategies.

Deep learning methods for feature extraction and matching is a very active research field. The survey Ma *et al.* [12] shows the introduction of deep learning methods to feature detection and matching. Notable mentions are SuperPoint [5] for its self-supervised approach, R2D2 [17] for using reliability metrics as output of the network instead of the features themselves and D2-Net [6] that built a describe-and-detect strategy that aims to improve SfM applicability. Exporting this progress to the matching stage, DISK [24] proposes a formulation of the problem to optimize in an end-to-end manner. Other recent works have extended the networks to take advantage of the advances in attention for the matching task, as in SuperGlue [20] and LoFTR [23].

In this work we improve the performance of SuperPoint [5] on endoscopy images. We chose SuperPoint because it is a seminal work that has inspired many follow up works, and it is still among the top performers on current feature matching challenges [10]. Similar to DeTone *et al.* [4], we explore improvements on feature extraction that provide good properties for downstream tasks. They design an end-to-end method to optimize the visual odometry computed with their features. Differently, we propose to supervise our training with points that have been successfully used for 3D reconstruction using existing SfM pipelines. With this supervision, we train a model able to extract more features with good properties for SfM algorithms, e.g., being spread and out of large specularities.

3 *Tracking Adaptation* for Local Feature Learning

Superpoint supervision is referred to as *Homographic Adaptation* and assumes that the surfaces are locally plane, which is not the case in our data. Instead, we propose to use 3D reconstructions of points tracked along image sequences. This makes no assumptions about the local surface shapes and we will show in Sect. 4 that this yields a better trained network. We will refer to this as *Tracking Adaptation* and we will here describe how we obtain the tracks.

SfM as Supervision for Feature Extraction. We generate examples of *good features* by identifying features that were successfully reconstructed with existing methods for each sequence in our training set. Our training set contains short sequences (4–7 s) from the complete colonoscopy recordings in EndoMapper dataset where COLMAP software was able to obtain a 3D reconstruction. This is a very challenging domain, and existing SfM pipelines fail in longer videos.

3D Reconstruction of Training Set Videos. We generate 3D reconstructions for all our training sequences with out-of-the-box COLMAP. In particular, we use the following blocks: *feature_extractor*, *exhaustive_matcher* and *mapper*. Configuration parameters are detailed in the supplementary materials. We turn on the “guided_matching” option for the *exhaustive_matcher* module to find the best matches possible. We additionally compute the 3D reconstruction for the same sequences with a modified COLMAP pipeline that uses the official SuperPoint and SuperGlue[2] implementation with the *indoor* set of weights. All the parameters are left as default except for the keypoint_threshold= 0.015 and the nms_radius= 1. After providing the SuperGlue resulting matches to COLMAP, we execute only the *mapper* module with the same configuration as before.

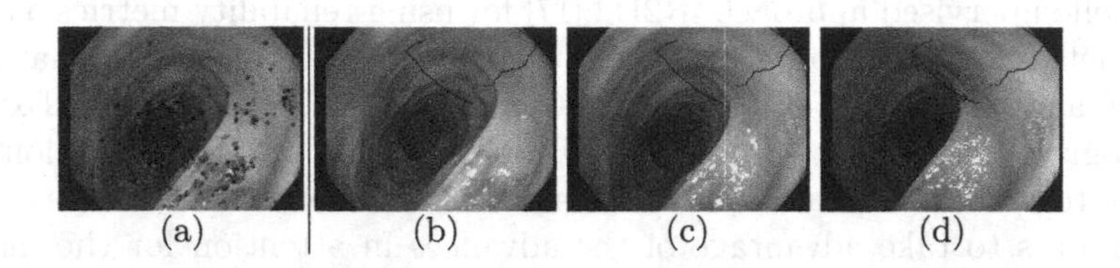

Fig. 1. Supervision points obtained from a COLMAP reconstruction. (a) All 3D points are reprojected into each video frame. We distinguish points that were originally detected in this frame (green) and points that were not (blue). (b–d) analyze a complete point track, i.e., all the positions of the same 3D point along the sequence. The *reliable track* for this point is the green segment. (b) The track starts when a point is first detected. (c) Movement of the point along the video. (d) When the feature is not detected anymore (e.g., because of occlusion), it is depicted in blue from then on. (Color figure online)

Re-project Good Features to the Training Set Frames. A successful 3D reconstruction includes the computed positions of the cameras that took the images and a point cloud with 3D coordinates of the triangulated points. We use the camera poses, the points' coordinates and the camera calibration parameters to reproject the 3D point cloud points into every image. Not all points were originally detected and triangulated at all frames, so we establish two types of reprojected points. If they were "originally" detected and matched in a particular image, we set them to green. Otherwise, we set them to blue (see Fig. 1).

For supervision, we only use reprojected points that fall within a *reliable track*. A *reliable track* is an interval bounded by green points. So, the reprojected points selected for training are either green or have preceding and subsequent green points along its track.

The different appearances of the same 3D point in different frames of the *track* are our **correspondences** for training our models. Figure 1 contains examples of reprojected points and an example of a *reliable track*.

Deep Feature Extraction for Endoscopy. SuperPoint uses a fully-convolutional network as backbone and learns to extract good features using *homographic adaptation*: extracting features that are robust to homographic deformations. It achieves this by using as supervision Y the average detections over several random homographic deformations of the same image. The feature extraction network then is run on an image I and a warped version I' of it with a new

[2] https://github.com/magicleap/SuperGluePretrainedNetwork.

homography. The network optimizes the loss function

$$\mathcal{L}_{SP}\left(\mathcal{X},\mathcal{X}',\mathcal{D},\mathcal{D}';Y,Y',S\right)=\mathcal{L}_p\left(\mathcal{X},Y\right)+\mathcal{L}_p\left(\mathcal{X}',Y'\right)+\lambda\mathcal{L}_d\left(\mathcal{D},\mathcal{D}',S\right), \quad (1)$$

where $\mathcal{X}$ and $\mathcal{D}$ are the detection and description heads' outputs, respectively. Y is the supervision for the detection. S is the correspondence between I and I' computed from the homography. $\mathcal{L}_p$ is the detection loss that measures the discrepancies between the supervision Y and the detection head's output $\mathcal{X}$. $\lambda = 1$ is a weighting parameter. $\mathcal{L}_d$ is the description loss that measures the discrepancies between both description head's outputs $\mathcal{D}$ and $\mathcal{D}'$ using S.

Using our new supervision from SfM in the form of *tracks* of points, we propose a new loss to train SuperPoint that is more aligned with our goal, called *tracking adaptation*. Instead of an image I and a warped version I', we use different images I_a and I_b from the same sequence. The supervision Y for the detection in this case is the set of points that have been reprojected on I_a and I_b from the 3D reconstruction. The detection loss $\mathcal{L}_p$ is calculated as in the original SuperPoint. We replace the description loss $\mathcal{L}_d$ for a new tracking loss

$$\mathcal{L}_t\left(\mathcal{D}_a,\mathcal{D}_b,\mathcal{T}\right)=\frac{1}{|\mathcal{T}|^2}\sum_{i=1}^{|\mathcal{T}|}\sum_{j=1}^{|\mathcal{T}|} l_t\left(\mathbf{d}_{a_i},\mathbf{d}_{b_j},i,j\right), \quad (2)$$

$$\text{with} \qquad l_t\left(\mathbf{d}_{a_i},\mathbf{d}_{b_j},i,j\right)=\begin{Bmatrix}\lambda_t\max(0,m_p-\mathbf{d}_{a_i}^T\mathbf{d}_{b_j}) & \text{if } i=j,\\ \max(0,\mathbf{d}_{a_i}^T\mathbf{d}_{b_j}-m_n) & \text{if } i\neq j\end{Bmatrix}, \quad (3)$$

where $\mathcal{D}_a$ and $\mathcal{D}_b$ are the description head's outputs for I_a and I_b, respectively. $\mathcal{T}$ is the set of all the *tracks* that appear in both images. l_t is a common triplet loss that measures the distance between positive pairs (weighting parameter $\lambda_t = 1$ and positive margin $m_p = 1$) and the distance between negative pairs (negative margin $m_n = 0.2$). Two descriptors from different images $\mathbf{d}_{a_i}$ and $\mathbf{d}_{b_j}$ are a positive pair if they belong to the same *track* ($i = j$), and negative pair otherwise ($i \neq j$).

4 Experiments

The following experiments demonstrate the proposed feature detection efficacy to obtain 3D models on real colonoscopy videos, comparing different variations of our approach and relevant baseline methods.

Dataset. We seek techniques that are applicable to real medical data, so we train and evaluate with subsequences from the EndoMapper dataset [1], which contains a hundred complete endoscopy recordings obtained during regular medical practice. We use COLMAP 3D reconstructions obtained from subsequences from this dataset (11260 frames from 65 reconstructions obtained along 14 different videos for training, and 838 frames from 7 reconstructions from 6 different videos for testing). The exact details are in the supplementary material.

Baselines and our Variations. We use COLMAP as our first baseline. It uses **SIFT** features and a standard guided matching algorithm to produce very accurate camera pose estimates. We also include as baseline the results of SuperPoint (**SP**) with SuperGlue matches and the COLMAP reconstruction module. The configuration for both baselines is the same as detailed in Sect. 3. We evaluate different variations of the original SuperPoint. All models were trained with a modification of a PyTorch implementation of SuperPoint [9]. Training parameters in supplementary material. The models differ in the supervision used and the loss applied in the training, as detailed in the first four columns of Table 1.

Table 1. Ablation study. Configuration of the training (left), and average reconstruction results, i.e., quality metrics (right). Best results highlighted in bold.

	Supervision & Train Config.			Reconstruction Test Results				
	point	*match*	*loss*	$\|3DIm\|$	$\|3DPts\|$	Err	Err-10K	len (Tr)
SP [5]	SP-O	H	SP (original)	93.9%	6421.3	**1.47**	1.47	6.86
SP-E v0	SF*	H	SP (original)	97.3%	12707.9	1.66	1.50	8.39
SP-E v1	SF*	TR	tr-2	98.6%	13255.1	1.69	1.51	8.95
SP-E v1	SF*+SP*	TR	tr-2	99.1%	28308.3	1.74	1.13	9.45
SP-E v2	SF*+SP*	TR	tr-N	99.1%	**34838.0**	1.75	**1.02**	9.53
SP-E v3	SF*+SP*	H + TR	SP + tr-N	**99.2%**	30777.6	1.74	1.09	**9.65**

point (Base point detector): SP-O: original superpoint detector; SF*/SP*: SIFT/SP points that were successfully reconstructed after the COLMAP optimization, reprojected in each video frame.
match (Matches Supervision): H: Homography based, i.e., *Homographic adaptation* from original SuperPoint work; TR: The proposed *Tracking adaptation*.
loss (Loss used for training): SP: original SuperPoint training loss; Tr-2 or Tr-N: *track*-based loss. Tr-2 means that the loss is computed for every pair of images in the track. Tr-N means we optimize simultaneously N views of the track (N=4 in our experiments).

Ablation Study. Table 1 (last five columns) summarizes the performance of our approach variations. We run all the models on the **Test set** subsequences to extract points. Matches between the points in two images are obtained with bi-directional nearest neighbor algorithm with L2 distance. Points and matches are given to COLMAP and the *mapper* module (configuration in supplementary material) attempts to generate a 3D reconstruction. The reconstruction quality statistics used to illustrate the performance of each detector are:

- $\|3DIm\|$: **Fraction of images** from the subsequence successfully introduced in the reconstruction. The closer to 100% the better.
- $\|3DPts\|$: **Number of points** that were successfully reconstructed. The more points the better, since it means a denser coverage of the scene.
- Err: Mean **reprojection error** of the 3D points after being reprojected onto the images of the subsequence.
- Err-10K: Mean **reprojection error of the best 10000** points of the reconstruction. Since all reconstructions have outliers that skew the average, this metric is more representative of the performance of the models.

- len(Tr): Mean **track length** represents the average number of images where a point is being consecutively matched, tracked.

SP-E v2 (SP-E moving forward) is our best variation, with the highest amount of reconstructed points and the lowest reprojection error for top 10000.

SfM Results Comparison. This experiment compares the performance of the considered baselines against the best configuration of our feature extraction model. Table 2 contains a summary of the results. In most metrics we observe a significant improvement using SP-E compared to the others. For example, the number of points at the final reconstruction is more than three times higher (see

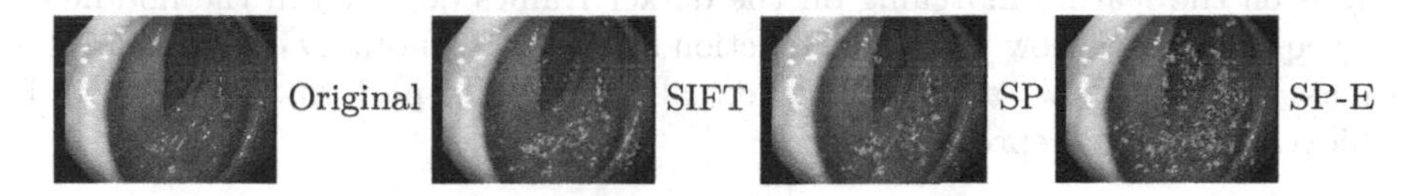

Fig. 2. Example of the points reconstructed by each method. Each point in each image has been reconstructed after the corresponding COLMAP reconstruction process.

Table 2. Reconstruction quality metrics for the comparison to the baselines.

Subsequence	001_1	002_1	014_1	016_1	017_1	095_1	095_2	Avg	(Std)
Reconstructed images (‖3*DIm*‖)									
Total^{+}	107	155	109	119	125	118	105	119.7	(15.9)
SIFT	98.1%	91.6%	71.6%	**100%**	52.0%	97.5%	99.0%	87.1%	(17.0)
SP	**100%**	**100%**	**93.6%**	**100%**	89.6%	99.2%	**100%**	97.5%	(3.9)
SP-E (Ours)	**100%**	**100%**	**93.6%**	**100%**	**100%**	**100%**	**100%**	**99.1%**	(2.3)
Reconstructed points (‖3*DPts*‖)									
SIFT	13470	6599	26225	5700	2505	7666	9608	10253.3	(7237.6)
SP	12941	9057	17451	6489	4093	8911	12535	10211.0	(4133.1)
SP-E (Ours)	**34851**	**45471**	**42727**	**33277**	**36403**	**19286**	**31851**	**34838.0**	(7846.5)
Mean reprojection error (Err)									
SIFT	**1.34**	**1.38**	**0.95**	**1.45**	**1.40**	**1.30**	**1.34**	**1.31**	(0.15)
SP	1.52	1.49	1.38	1.58	1.48	1.51	1.51	1.50	(0.06)
SP-E (Ours)	1.69	1.68	1.71	1.90	1.73	1.81	1.75	1.75	(0.07)
*Mean reprojection error of the best 10K points** (Err-10K)									
SIFT	1.08	1.38	**0.46**	1.45	1.40	**1.30**	1.34	1.20	(0.32)
SP	1.30	1.49	1.00	1.58	1.48	1.51	1.30	1.38	(0.19)
SP-E (Ours)	**0.92**	**0.73**	0.84	**1.30**	**0.91**	1.41	**1.06**	**1.02**	(0.23)
Mean track length (len(Tr))									
SIFT	6.57	5.73	**10.88**	12.48	7.74	**12.88**	7.56	9.12	(2.70)
SP	5.54	4.52	7.86	8.73	5.16	8.20	5.38	6.49	(1.59)
SP-E (Ours)	**7.05**	**6.78**	9.63	**14.73**	**8.42**	11.29	**8.78**	**9.53**	(2.55)

$^{+}$ Total number of images in the subsequence.
* If 10K points are not available, average is computed over all available reconstructed points.

the example in Fig. 2). The mean reprojection error of all the points is the lowest for SIFT, possibly due to it being more restrictive in all other aspects (number of images reconstructed, number of points, track length). However, the mean error for the top 10000 points is always lower for SP-E. The reprojection error plots in Figs. 3 and 4 provide more insight on this metric.

Figures 3 and 4 show a more detailed visualization of two representative reconstructions, including a summary of the sequence frames, the point cloud obtained by each method and a plot of the reprojection error for each point in the reconstruction, sorted in increasing error value. Note that even though SP-E obtains many more points, it is not at the cost of quality. Figure 3 shows a scenario where SIFT fails to reconstruct a large part of the subsequence, because it fails on the feature matching on the darker frames depicted in the middle of the sequence. Note how the reconstruction from SP-E is notably denser than the others. Figure 4 shows a scenario where all approaches perform well and SIFT achieves the lowest reprojection error.

Table 3. Analysis of the feature locations for each method.

	Spread of features ↑	% of features on specularities ↓
SIFT	43.9%	28.6%
SP	56.9%	19.6%
SP-E (Ours)	**67.5%**	**9.9%**

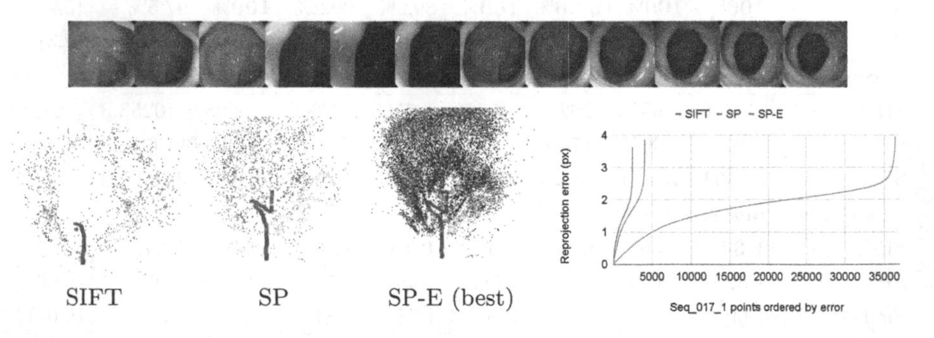

Fig. 3. Comparison of reconstructions obtained on Seq_017_1 by SIFT, SP, and our best model SP-E. The plot shows the reprojection error of each point reconstructed.

We analyze additional aspects of our detected features to showcase the higher quality with respect to other methods in Table 3. To measure the spread of the features over the images we defined a 16×16 grid over each image and computed the percentage of those cells that have at least one reconstructed point. We also measure how many extracted points fall on top of specularities (we consider a pixel as part of a specularity if its intensity is higher than 180). For both metrics, our detector achieves significantly better results, showcasing the better properties of our detector for 3D reconstruction.

To provide quantitative evaluation of the camera motion estimation, we use a simulated dataset [3] to have ground truth available for the camera trajectory. We took 5 sequences of 100-150 frames from this dataset, and we tested the baselines and our model. We align the ground truth trajectories with the reconstructed ones with Horn's method [8]. SP only reconstructed 3 out of the 5 sequences while SIFT and SP-E correctly reconstructed the 5 sequences, with an average RMSE of 4.61mm and 4.71 mm respectively. Simulated data lacks some of the biggest challenges of endoscopy images (e.g. specularities, deformations), but this experiment suggests that the camera motion estimation quality is similarly good for all methods when they manage to converge.

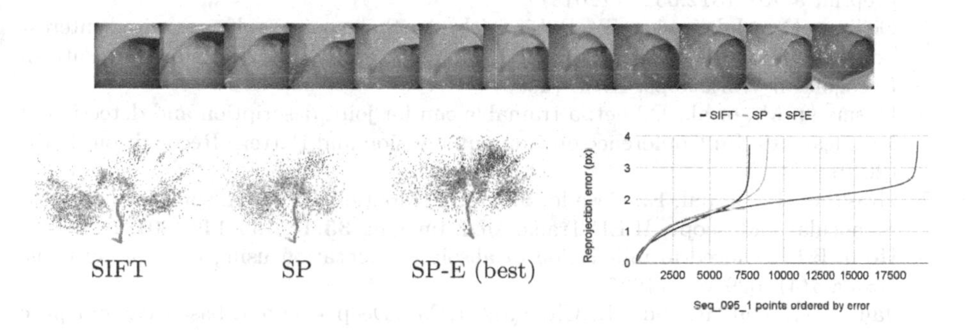

Fig. 4. Comparison of reconstructions obtained on Seq_095_1 by SIFT, SP, and our best model SP-E. The plot shows the reprojection error of each point reconstructed.

5 Conclusions

This work presents a novel training strategy for SuperPoint to improve its performance in SfM from endoscopy images. This strategy has two main benefits: we show how to use 3D reconstructions of endoscopy sequences as supervision to train feature extraction models; and we design a new tracking loss to perform *tracking adaptation* using this supervision. The benefits of our method are explored with an ablation study and against established baselines on SfM and feature extraction. Our proposed model is able to obtain more suitable features for 3D reconstruction, and to reconstruct larger sets of images with much denser point clouds.

Acknowledgements. This project has been funded by the European Union's Horizon 2020 research and innovation programme under grant agreement No 863146 and Aragón Government project T45_23R.

References

1. Azagra, P., et al.: Endomapper dataset of complete calibrated endoscopy procedures. arXiv preprint arXiv:2204.14240 (2022)
2. Barbed, O.L., Chadebecq, F., Morlana, J., Montiel, J.M., Murillo, A.C.: Superpoint features in endoscopy. In: Imaging Systems for GI Endoscopy, and Graphs in Biomedical Image Analysis: First MICCAI Workshop, ISGIE 2022, and Fourth MICCAI Workshop, GRAIL 2022, Held in Conjunction with MICCAI 2022, Singapore, 18 September 2022, Proceedings, pp. 45–55. Springer, Heidelberg (2022). https://doi.org/10.1007/978-3-031-21083-9_5
3. Bobrow, T.L., Golhar, M., Vijayan, R., Akshintala, V.S., Garcia, J.R., Durr, N.J.: Colonoscopy 3d video dataset with paired depth from 2d–3d registration. arXiv preprint arXiv:2206.08903 (2022)
4. DeTone, D., Malisiewicz, T., Rabinovich, A.: Self-improving visual odometry. arXiv preprint arXiv:1812.03245 (2018)
5. DeTone, D., Malisiewicz, T., Rabinovich, A.: Superpoint: self-supervised interest point detection and description. In: Conference on Computer Vision and Pattern Recognition Workshops. IEEE (2018)
6. Dusmanu, M., et al.: D2-net: a trainable cnn for joint description and detection of local features. In: Conference on Computer Vision and Pattern Recognition. IEEE (2019)
7. Grasa, O.G., Bernal, E., Casado, S., Gil, I., Montiel, J.: Visual slam for handheld monocular endoscope. IEEE Trans. Med. Imaging **33**(1), 135–146 (2013)
8. Horn, B.K.: Closed-form solution of absolute orientation using unit quaternions. Josa a **4**(4), 629–642 (1987)
9. Jau, Y.Y., Zhu, R., Su, H., Chandraker, M.: Deep keypoint-based camera pose estimation with geometric constraints. In: International Conference on Intelligent Robots and Systems. IEEE (2020). https://github.com/eric-yyjau/pytorch-superpoint
10. Jin, Y., et al.: Image matching across wide baselines: from paper to practice. Int. J. Comput. Vision **129**(2), 517–547 (2021)
11. Lowe, D.G.: Distinctive image features from scale-invariant keypoints. Int. J. Comput. Vision **60**(2), 91–110 (2004)
12. Ma, J., Jiang, X., Fan, A., Jiang, J., Yan, J.: Image matching from handcrafted to deep features: a survey. Int. J. Comput. Vision **129**, 1–57 (2020)
13. Mahmoud, N., Collins, T., Hostettler, A., Soler, L., Doignon, C., Montiel, J.M.M.: Live tracking and dense reconstruction for handheld monocular endoscopy. IEEE Trans. Med. Imaging **38**(1), 79–89 (2018)
14. Maier-Hein, L., et al.: Optical techniques for 3d surface reconstruction in computer-assisted laparoscopic surgery. Med. Image Anal. **17**(8), 974–996 (2013)
15. Mur-Artal, R., Montiel, J.M.M., Tardos, J.D.: Orb-slam: a versatile and accurate monocular slam system. IEEE Trans. Rob. **31**(5), 1147–1163 (2015)
16. Ozyoruk, K.B., et al.: Endoslam dataset and an unsupervised monocular visual odometry and depth estimation approach for endoscopic videos. Med. Image Anal. **71**, 102058 (2021)
17. Revaud, J., Weinzaepfel, P., de Souza, C.R., Humenberger, M.: R2D2: repeatable and reliable detector and descriptor. In: International Conference on Neural Information Processing Systems (2019)
18. Rodríguez, J.J.G., Tardós, J.D.: Tracking monocular camera pose and deformation for slam inside the human body. In: 2022 IEEE/RSJ International Conference on Intelligent Robots and Systems (IROS), pp. 5278–5285. IEEE (2022)

19. Rublee, E., Rabaud, V., Konolige, K., Bradski, G.: ORB: an efficient alternative to sift or surf. In: International Conference on Computer Vision. IEEE (2011)
20. Sarlin, P.E., DeTone, D., Malisiewicz, T., Rabinovich, A.: Superglue: learning feature matching with graph neural networks. In: Conference on Computer Vision and Pattern Recognition. IEEE (2020)
21. Schönberger, J.L., Frahm, J.M.: Structure-from-motion revisited. In: Conference on Computer Vision and Pattern Recognition (CVPR) (2016)
22. Schönberger, J.L., Zheng, E., Frahm, J.-M., Pollefeys, M.: Pixelwise view selection for unstructured multi-view stereo. In: Leibe, B., Matas, J., Sebe, N., Welling, M. (eds.) ECCV 2016. LNCS, vol. 9907, pp. 501–518. Springer, Cham (2016). https://doi.org/10.1007/978-3-319-46487-9_31
23. Sun, J., Shen, Z., Wang, Y., Bao, H., Zhou, X.: Loftr: detector-free local feature matching with transformers. In: CVPR. IEEE (2021)
24. Tyszkiewicz, M., Fua, P., Trulls, E.: Disk: learning local features with policy gradient. Adv. Neural Inf. Process. Syst. **33**, 14254–14265 (2020)

Structured State Space Models for Multiple Instance Learning in Digital Pathology

Leo Fillioux(✉), Joseph Boyd, Maria Vakalopoulou, Paul-henry Cournède, and Stergios Christodoulidis

MICS Laboratory, CentraleSupélec, Université Paris-Saclay, 91190 Gif-sur-Yvette, France
{leo.fillioux,joseph.boyd,maria.vakalopoulou, paul-henry.cournede,stergios.christodoulidis}@centralesupelec.fr

Abstract. Multiple instance learning is an ideal mode of analysis for histopathology data, where vast whole slide images are typically annotated with a single global label. In such cases, a whole slide image is modelled as a collection of tissue patches to be aggregated and classified. Common models for performing this classification include recurrent neural networks and transformers. Although powerful compression algorithms, such as deep pre-trained neural networks, are used to reduce the dimensionality of each patch, the sequences arising from whole slide images remain excessively long, routinely containing tens of thousands of patches. Structured state space models are an emerging alternative for sequence modelling, specifically designed for the efficient modelling of long sequences. These models invoke an optimal projection of an input sequence into memory units that compress the entire sequence. In this paper, we propose the use of state space models as a multiple instance learner to a variety of problems in digital pathology. Across experiments in metastasis detection, cancer subtyping, mutation classification, and multitask learning, we demonstrate the competitiveness of this new class of models with existing state of the art approaches. Our code is available at https://github.com/MICS-Lab/s4_digital_pathology.

Keywords: Multiple instance learning · Whole slide images · State space models

1 Introduction

Precision medicine efforts are shifting cancer care standards by providing novel personalised treatment plans with promising outcomes. Patient selection for such

L. Fillioux and J. Boyd—These authors contributed equally to this work.

Supplementary Information The online version contains supplementary material available at https://doi.org/10.1007/978-3-031-43907-0_57.

H. Greenspan et al. (Eds.): MICCAI 2023, LNCS 14220, pp. 594–604, 2023.
https://doi.org/10.1007/978-3-031-43907-0_57

treatment regimes is based principally on the assessment of tissue biopsies and the characterisation of the tumor microenvironment. This is typically performed by experienced pathologists, who closely inspect chemically stained histopathological whole slide images (WSIs). Increasingly, clinical centers are investing in the digitisation of such tissue slides to enable both automatic processing as well as research studies to elucidate the underlying biological processes of cancer. The resulting images are of gigapixel size, rendering their computational analysis challenging. To deal with this issue, multiple instance learning (MIL) schemes based on weakly supervised training are used for WSI classification tasks. In such schemes, the WSI is typically divided into a grid of patches, with general purpose features derived from pretrained ImageNet [18] networks extracted for each patch. These representations are subsequently pooled together using different aggregation functions and attention-based operators for a final slide-level prediction.

State space models are designed to efficiently model long sequences, such as the sequences of patches that arise in WSI MIL. In this paper, we present the first use of state space models for WSI MIL. Extensive experiments on three publicly available datasets show the potential of such models for the processing of gigapixel-sized images, under both weakly and multi-task schemes. Moreover, comparisons with other commonly used MIL schemes highlight their robust performance, while we demonstrate empirically the superiority of state space models in processing the longest of WSI sequences with respect to commonly used MIL methods.

2 Related Work

Using pretrained networks for patch wise feature extraction is a well established strategy for histopathology analysis [4,20]. An extension of this approach is with MIL, where the patch-wise features of an entire slide are digested simultaneously by an aggregator model, such as attention-based models CLAM [17] and TransMIL [19], the latter being a variant of self-attention transformers [21]. [3] proposes another transformer-based method in the form of a hierarchical ViT. Similar to our multitask experiments, [6] explores combining slide-level and tile-level annotations with a minimal point-based annotation strategy. One of the key components of MIL methods is the aggregation module that pools together the set of patch representations. Mean or max pooling operations are among the simplest and most effective for aggregating predictions over a whole slide [2]. In contrast, recurrent neural networks (RNN) with long short-term memory (LSTM) [14] model the patches more explicitly as a set of tokens in sequence. In particular, LSTM networks have been shown to work well in different MIL settings including both visual cognition [22] and computational pathology [1].

The state space model is a linear differential equation,

$$\begin{aligned} \dot{x}(t) &= \boldsymbol{A}x(t) + \boldsymbol{B}u(t) \\ y(t) &= \boldsymbol{C}x(t) + \boldsymbol{D}u(t) \end{aligned} \tag{1}$$

that is widely studied in control theory, and describes a continuous time process for input and output signals $u(t) \in \mathbb{R}^p$ and $y(t) \in \mathbb{R}^q$, and state signal $x(t) \in \mathbb{R}^n$, and where the process is governed by matrices $\boldsymbol{A} \in \mathbb{R}^{n \times n}$, $\boldsymbol{B} \in \mathbb{R}^{n \times p}$, $\boldsymbol{C} \in \mathbb{R}^{q \times n}$, $\boldsymbol{D} \in \mathbb{R}^{q \times p}$. In HiPPO [9] (high-order polynomial projection operator), continuous time memorisation is posed as a problem of function approximation in a Hilbert space defined by a probability measure μ. For a *scaled Legendre* probability measure, one obtains the HiPPO matrix $\boldsymbol{A}$, which enforces uniform weight in the memorisation of all previously observed inputs, in contrast to the exponentially decaying weighting of the constant error carousel of LSTMs [14]. The HiPPO mode of memorisation is shown empirically to be better suited to modeling long-range dependencies (LRD) than other neural memory layers, for which it serves as a drop-in replacement.

Whereas in HiPPO, the state matrix $\boldsymbol{A}$ is a fixed constant, the linear state space layer (LSSL) [12] incorporates $\boldsymbol{A}$ as a learnable parameter. However, this increased expressiveness introduces intractable powers of $\boldsymbol{A}$. In [10], the LSSL is instead reparameterised as the sum of diagonal and low-rank matrices, allowing for the efficient computation of the layer kernel in Fourier space. This updated formulation is known as the *structured* state space sequence layer (S4). Note that as a linear operator, the inverse discrete Fourier transform is amenable to backpropagation in the context of a neural network. Note also that under this formulation, the hidden state $x(t)$ is only computed implicitly. Finally, [11] presents a simplification of the S4 layer, known as diagonal S4 (S4D), in which $\boldsymbol{A}$ is approximated by a diagonal matrix.

3 Method

Given that the patch extraction of whole slide images at high magnifications results in long sequences of patches, we propose to incorporate a state space layer in a MIL aggregation network to better represent each patch sequence.

3.1 Neural State Space Models

In practice, neural state space models (SSM) simulate Eq. 1 in discrete time, invoking a recurrence relation on the discretised hidden state,

$$\begin{aligned} x_t &= \overline{\boldsymbol{A}} x_{t-1} + \overline{\boldsymbol{B}} u_t \\ y_t &= \overline{\boldsymbol{C}} x_t + \overline{\boldsymbol{D}} u_t \end{aligned} \tag{2}$$

where the sequences u_t, x_t, and y_t are the discretised $u(t)$, $x(t)$, and $y(t)$, and the modified model parameters arise from a bilinear discretisation [12]. As such, SSMs bear an inherent resemblance to RNNs, where the hidden representation x_t can be interpreted as a memory cell for the observed sequence over the interval $[0, t]$, and with $\overline{\boldsymbol{D}} u_t$ acting as a skip connection between the input and output at point t. Due to their lack of non-linearities, state space models can also be viewed as a convolution between two discrete sequences. Playing out the recurrence in Eq. 2, one obtains,

$$y = \overline{\boldsymbol{K}} * u + \overline{\boldsymbol{D}}u, \tag{3}$$

where $u \in \mathbb{R}^L$ and $y \in \mathbb{R}^L$ are the full input and output sequences, and the sequence $\overline{\boldsymbol{K}} \in \mathbb{R}^L$ is defined as,

$$\overline{\boldsymbol{K}} = (\overline{\boldsymbol{CB}}, \overline{\boldsymbol{CAB}}, \ldots, \overline{\boldsymbol{CA}}^{L-1}\overline{\boldsymbol{B}}), \tag{4}$$

which is computed efficiently by the S4D algorithm [11]. Note that although SSM layers are linear, they may be combined with other, non-linear layers in a neural network. Note also that although Eq. 3 is posed as modeling a one-dimensional signal, in practice multi-dimensional inputs are modelled simply by stacking SSM layers together, followed by an affine "mixing" layer.

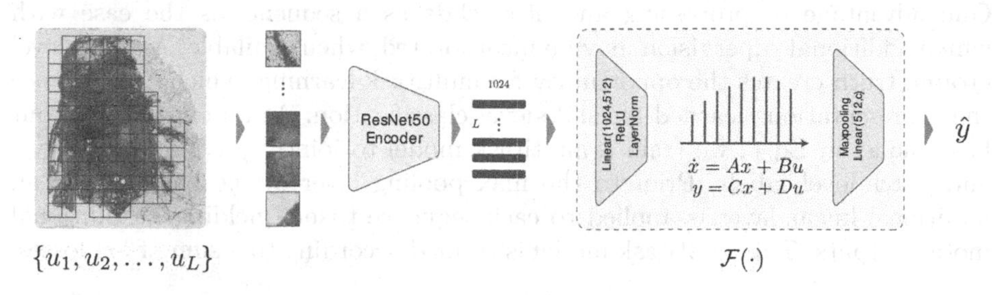

Fig. 1. Overview of the proposed pipeline. In the first step, patches are extracted from a regular grid on a WSI. These patches are embedded using a pre-trained ResNet50 and are aggregated by a sequence model based on a state space layer.

3.2 MIL Training

In our pipeline (Fig. 1) WSIs are first divided into a sequence of L patches $\{u_1, u_2, \ldots, u_L\}$, where L will vary by slide. A pretrained ResNet50 is then used to extract a 1024-dimensional feature vector from each patch $\{\mathbf{u}_1, \mathbf{u}_2, \ldots, \mathbf{u}_L\}$, which constitute the model inputs. We define a SSM-based neural network $\mathcal{F}$ to predict a WSI-level class probability given this input sequence,

$$\hat{y} = \mathcal{F}(\{\mathbf{u}_1, \mathbf{u}_2, \ldots, \mathbf{u}_L\}). \tag{5}$$

The architecture of $\mathcal{F}$ is composed of an initial linear projection layer, used to lower the dimensionality of each vector in the input sequence. A SSM layer is then applied feature-wise by applying the S4D algorithm. That is, Eq. 3, including the skip connection, transforms the sequence $\{u_{1,d}, u_{2,d}, \cdots, u_{L,d}\}$ for all features d, and the resulting sequences are concatenated. A linear "mixing" layer is applied token-wise, doubling the dimensionality of each token, followed by a gated linear unit [5] acting as an output gate, which restores the input dimensionality. For the SSM layer, we used the official implementation of S4D[1]. A max pooling layer merges the SSM layer outputs into a single vector, which is projected by a final linear layer and softmax to give the class probabilities $\hat{y}$. The model is trained according to,

$$\mathcal{L}_{MIL} = -\frac{1}{M}\sum_{m=1}^{M} \log \hat{y}_{c_m}, \tag{6}$$

where $\hat{y}_{c_m}$ denotes the probability corresponding to c_m, the slide-level label of the sequence corresponding to the mth of M whole slide images.

3.3 Multitask Training

One advantage of processing an entire slide as a sequence is the ease with which additional supervision may be incorporated, when available. A patch-level ground truth creates the opportunity for multitask learning, which can enhance the representations learned for slide-level classification. As an extension of our base model in Eq. 6, we train a multitask model to jointly predict a slide-level and patch-level labels. Prior to the max pooling layer of the base model, an additional linear layer is applied to each sequence token, yielding L additional model outputs. This multitask model is trained according to a sum of log losses,

$$\mathcal{L}_{MT} = -\frac{1}{M}\sum_{m=1}^{M}\left(\log \hat{y}_{c_m} + \frac{\lambda}{L}\cdot\sum_{l=1}^{L}\log \hat{y}_{c_{m,l}}\right), \tag{7}$$

where $c_{m,l}$ indexes the class of the lth patch in the mth training slide and λ is a tunable hyperparameter used to modulate the relative importance of each task.

3.4 Implementation Details

We extracted patches of size 256 × 256 from the tissue regions of WSIs at 20x magnification. Following CLAM [17], the third residual block of a pretrained ResNet50 [13] was used as a feature extractor, followed by a mean pooling operation, resulting in a 1024-dimensional representation for each patch. These features were used as inputs to all models. All model training was performed under a 10-fold cross-validation, and all reported results are averaged over the validation sets of the folds, aside from CAMELYON16, for which the predefined test set was utilized. Thus, for CAMELYON16, we report test set performances averaged over the validation.

Baseline models were chosen to be prior art CLAM [17] and TransMIL [19]. The official code of these two models was used to perform the comparison. In addition, we included a vanilla transformer, a LSTM RNN, and models based on mean and max pooling. Our vanilla transformer is composed of two stacked self-attention blocks, with four attention heads, a model dimension of 256, and

[1] https://github.com/HazyResearch/state-spaces.

a hidden dimension of 256. For the LSTM, we used an embedding size of 256 and a width of 256. The pooling models applied pooling feature-wise across each sequence, then used a random forest with 200 trees for classification. For the S4 models, the dimension of the state matrix $\boldsymbol{A}$ was tuned to 32 for CAMELYON16 and TCGA-RCC, and 128 for TCGA-LUAD. Our models were trained using the Adam [15] optimizer with the lookahead method [23], with a learning rate of $2 \cdot 10^{-4}$, and weight decay of 10^{-4} for TCGA-LUAD and TCGA-RCC and 10^{-3} for CAMELYON16. Early stopping with a patience of 10 was used for all our training. Our implementation is publicly available[2].

4 Experiments and Discussion

4.1 Data

CAMELYON16 [16] is a dataset that consists of resections of lymph nodes, where each WSI is annotated with a binary label indicating the presence of tumour tissue in the slide, and all slides containing tumors have a pixel-level annotation indicating the metastatic region. In multitask experiments, we use this annotation to give each patch a label indicating local tumour presence. There are 270 WSIs in the training/validation set, and 130 WSIs in the predefined test set. In our experiments, the average patch sequence length arising from CAMELYON16 is 6129 (ranging from 127 to 27444).

TCGA-LUAD is a TCGA lung adenocarcinoma dataset that contains 541 WSIs along with genetic information about each patient. We obtained genetic information for this cohort using Xena browser [7]. As a MIL task, we chose the task of predicting the patient mutation status of TP53, a tumor suppressor gene that is highly relevant in oncology studies. The average sequence length is 10557 (ranging from 85 to 34560).

TCGA-RCC is a TCGA dataset for three kidney cancer subtypes (denoted KICH, KIRC, and KIRP). It consists of 936 WSIs (121 KICH, 518 KIRC, and 297 KIRP). The average sequence length is 12234 (ranging from 319 to 62235).

4.2 Results

Multiple Instance Learning Results. We evaluate our method on each dataset by accuracy and area under receiver operating characteristic curve (AUROC). For multiclass classification, these were computed in a one-versus-rest manner.

Table 1 summarises the comparison between our proposed model and baselines. For the CAMELYON16 dataset, our method performs on par with TransMIL and the CLAM models, while it clearly outperforms the other methods. Similarly, in the TCGA-LUAD dataset the proposed model achieves comparable performance with both CLAM models, while outperforming TransMIL and the other methods. We note that TCGA-LUAD proves to be a more challenging

[2] https://github.com/MICS-Lab/s4_digital_pathology.

dataset for all models. Moreover, our method outperforms CLAM models on the TCGA-RCC dataset, while reporting very similar performance with respect to TransMIL. Overall, looking at the average metrics per model across all three datasets, our proposed method achieves the highest accuracy and the second highest AUROC, only behind CLAM-MB. A pairwise t-test between the proposed method, CLAM, and TransMIL shows that there is no statistical significance performance difference (see supplementary material).

We further compare our method with respect to model and time complexity. In Table 2 we report the number of trainable parameters, as well as the inference time for all models. The number of parameters is computed with all models configured to be binary classifiers, and the inference time is computed as the average time over 100 samples for processing a random sequence of 1024-dimensional vectors of length 30000. For our proposed method, we report both models with the different state dimensions (Ours (SSM_{32})) and (Ours (SSM_{128})). Compared

Table 1. Comparison of accuracy and AUROC on three datasets CAMELYON16, TCGA-LUAD, TCGA-RCC, and on average. All metrics in the table are the average of 10 runs. Best performing methods are indicated in **bold** and second best in *italics*. * indicates results from [19].

Dataset	CAMELYON16		TCGA-LUAD		TCGA-RCC		Average	
Metric	Acc.	AUROC	Acc.	AUROC	Acc.	AUROC	Acc.	AUROC
Mean-pooling	0.5969	0.5810	0.6261	0.6735	0.8608	0.9612	0.6946	0.7386
Max-pooling	0.7078	0.7205	0.6328	0.6686	0.8803	0.9659	0.7403	0.7850
Transformer [21]	0.5419	0.5202	0.5774	0.6214	0.7932	0.9147	0.6375	0.6854
LSTM [8]	0.5310	0.5053	0.5389	0.5208	0.6654	0.7853	0.5784	0.6038
CLAM SB [17]	0.8147	0.8382	0.6859	*0.7459*	0.8816*	0.9723*	0.7941	0.8532
CLAM MB [17]	*0.8264*	*0.8523*	**0.6901**	**0.7573**	0.8966*	0.9799*	*0.8044*	**0.8632**
TransMIL [19]	**0.8287**	**0.8628**	0.6348	0.7015	**0.9466***	*0.9882**	0.8034	0.8508
Ours	0.8217	0.8485	*0.6879*	0.7304	*0.9426*	**0.9885**	**0.8174**	*0.8558*

Table 2. Comparison of parameter count and inference time for all methods.

Model	Number of parameters	Inference time (ms)
Mean-pooling	1 025	5.60
Max-pooling	1 025	77.49
Transformer [21]	1 054 978	2.60
LSTM [8]	789 250	320.52
CLAM SB [17]	790 791	0.84
CLAM MB [17]	791 048	5.85
TransMIL [19]	2 672 146	8.58
Ours (SSM_{128})	1 184 258	2.01
Ours (SSM_{32})	1 085 954	1.97

with TransMIL, our method runs four times faster and has less than half the parameters. The CLAM models are more efficient in terms of number of trainable parameters, yet CLAM MB is slower.

Table 3 shows the effect of modifying parts of the architecture on the results for TCGA-RCC. Most modifications had very little impact on AUROC, but a more significant impact can be seen on the accuracy of the model. Models A and B show that stacking multiple SSM layers results in lower accuracy, which was observed over all three datasets, while models C and D show that modifying the state dimension of the SSM module can have an impact on the accuracy. The optimal state space dimension varies depending on the dataset.

Multitask Learning Results. We explored the ability of our model to combine slide- and patch-level information on the CAMEYLON16 dataset. We compared our model with the best performing model on CAMELYON16, TransMIL. Both models were trained according to Eq. 7 with $\lambda = 5$ tuned by hand. In Table 4 we give slide-level accuracy and AUROC for the two models. We observe that all accuracies and AUROC increase compared with those reported in Table 1. This indicates that the use of patch-level annotations complements the learning of the slide-level label. We furthermore observe that our model outperforms TransMIL when combining slide- and patch-level annotations. We map the sequence of output probabilities to their slide coordinates giving a heatmap localising metastasis (see supplementary material).

Table 3. Ablation study for the different SSM components on the TCGA-RCC dataset. Best results in **bold**.

Model	SSM layers	State dimension	Accuracy	AUROC
A	2	32	0.9236	0.9813
B	3	32	0.9179	0.9834
C	1	16	0.9352	0.9846
D	1	64	0.9352	0.9861
Ours	1	32	**0.9426**	**0.9885**

Table 4. Comparison of accuracy and AUROC for models trained as multitask classifiers on the CAMELYON16 dataset. Best results in **bold**.

Model	Accuracy	AUROC
TransMIL [19]	0.8403	0.8828
Ours	**0.8488**	**0.8998**

Performance on Longest Sequences. In order to highlight the inherent ability of SSM models to effectively model long sequences, we performed an experiment on only the largest WSIs of the TCGA-RCC dataset. Indeed, this dataset contains particularly long sequences (up to 62235 patches at 20x). We evaluated the trained models for each fold on a subset of the validation set, only containing sequences with a length in the 85th percentile. Table 5 shows the obtained average accuracy (weighted by the number of long sequences in each validation set) and AUROC on both CLAM models, TransMIL, and our proposed method. Both in terms of AUROC and accuracy, our method outperforms the other methods on long sequences, while the performances are comparable to Table 1, albeit slightly lower, illustrating the challenge of processing large WSIs.

Table 5. Results of CLAM SB, CLAM MB, TransMIL, and our proposed method on long sequences. Best results in **bold**.

Model	Accuracy	AUROC
CLAM SB [17]	0.9149	0.9635
CLAM MB [17]	0.8936	0.9654
TransMIL [19]	0.9007	0.9652
Ours	**0.9220**	**0.9737**

5 Conclusions

In this work we have explored the ability of state space models to act as multiple instance learners on sequences of patches extracted from histopathology images. These models have been developed for their ability to memorise long sequences, and they have proven competitive with state of the art MIL models across a range of pathology problems. Additionally, we demonstrated the ability of these models to perform multiclass classification, which furthermore allowed us to visualise the localisation of metastasic regions. Finally, we demonstrated that on the longest sequences in our datasets, state space models offer better performance than competing models, confirming their power in modeling long-range dependencies.

Acknowledgments. This work has benefited from state financial aid, managed by the Agence Nationale de Recherche under the investment program integrated into France 2030, project reference ANR-21-RHUS-0003. This work was partially supported by the ANR Hagnodice ANR-21-CE45-0007. Experiments have been conducted using HPC resources from the "Mésocentre" computing center of CentraleSupélec and École Normale Supérieure Paris-Saclay supported by CNRS and Région Île-de-France.

References

1. Agarwalla, A., Shaban, M., Rajpoot, N.M.: Representation-aggregation networks for segmentation of multi-gigapixel histology images. In: British Machine Vision Conference (BMVC) (2017)

2. Campanella, G., et al.: Clinical-grade computational pathology using weakly supervised deep learning on whole slide images. Nature Med. **25**(8), 1301–1309 (2019)
3. Chen, R.J., et al.: Scaling vision transformers to gigapixel images via hierarchical self-supervised learning. In: Proceedings of the IEEE/CVF Conference on Computer Vision and Pattern Recognition, pp. 16144–16155 (2022)
4. Coudray, N., et al.: Classification and mutation prediction from non-small cell lung cancer histopathology images using deep learning. Nature Med. **24**(10), 1559–1567 (2018)
5. Dauphin, Y.N., Fan, A., Auli, M., Grangier, D.: Language modeling with gated convolutional networks. In: International Conference on Machine Learning, pp. 933–941. PMLR (2017)
6. Gao, Z., et al.: A semi-supervised multi-task learning framework for cancer classification with weak annotation in whole-slide images. Med. Image Anal. **83**, 102652 (2023)
7. Goldman, M.J., et al.: Visualizing and interpreting cancer genomics data via the Xena platform. Nat. Biotechnol. **38**(6), 675–678 (2020)
8. Graves, A.: Long short-term memory. Supervised sequence labelling with recurrent neural networks, pp. 37–45 (2012)
9. Gu, A., Dao, T., Ermon, S., Rudra, A., Ré, C.: HiPPO: recurrent memory with optimal polynomial projections. Adv. Neural Inf. Process. Syst. **33**, 1474–1487 (2020)
10. Gu, A., Goel, K., Ré, C.: Efficiently modeling long sequences with structured state spaces. arXiv preprint arXiv:2111.00396 (2021)
11. Gu, A., Gupta, A., Goel, K., Ré, C.: On the parameterization and initialization of diagonal state space models. arXiv preprint arXiv:2206.11893 (2022)
12. Gu, A., et al.: Combining recurrent, convolutional, and continuous-time models with linear state space layers. Adv. Neural Inf. Process. Syst. **34**, 572–585 (2021)
13. He, K., Zhang, X., Ren, S., Sun, J.: Deep residual learning for image recognition. In: Proceedings of the IEEE Conference on Computer Vision and Pattern Recognition, pp. 770–778 (2016)
14. Hochreiter, S., Schmidhuber, J.: Long short-term memory. Neural Comput. **9**(8), 1735–1780 (1997)
15. Kingma, D.P., Ba, J.: Adam: A method for stochastic optimization. arXiv preprint arXiv:1412.6980 (2014)
16. Litjens, G., et al.: 1399 h&e-stained sentinel lymph node sections of breast cancer patients: the camelyon dataset. GigaScience **7**(6), giy065 (2018)
17. Lu, M.Y., Williamson, D.F., Chen, T.Y., Chen, R.J., Barbieri, M., Mahmood, F.: Data-efficient and weakly supervised computational pathology on whole-slide images. Nat. Biomed. Eng. **5**(6), 555–570 (2021)
18. Russakovsky, O., et al.: ImageNet large scale visual recognition challenge. Int. J. Comput. Vis. **115**(3), 211–252 (2015)
19. Shao, Z., Bian, H., Chen, Y., Wang, Y., Zhang, J., Ji, X., et al.: TransMIL: transformer based correlated multiple instance learning for whole slide image classification. Adv. Neural Inf. Process. Syst. **34**, 2136–2147 (2021)
20. Tellez, D., Litjens, G., van der Laak, J., Ciompi, F.: Neural image compression for gigapixel histopathology image analysis. IEEE Trans. Pattern Anal. Mach. Intell. **43**(2), 567–578 (2019)
21. Vaswani, A., et al.: Attention is all you need. In: Advances in Neural Information Processing Systems. vol. 30 (2017)

22. Wang, K., Oramas, J., Tuytelaars, T.: In defense of LSTMs for addressing multiple instance learning problems. In: Proceedings of the Asian Conference on Computer Vision (2020)
23. Zhang, M.R., Lucas, J., Hinton, G., Ba, J.: Lookahead optimizer: K steps forward, 1 step back. In: Proceedings of the 33rd International Conference on Neural Information Processing Systems. Curran Associates Inc., Red Hook, NY, USA (2019)

vox2vec: A Framework for Self-supervised Contrastive Learning of Voxel-Level Representations in Medical Images

Mikhail Goncharov[1](✉), Vera Soboleva[2], Anvar Kurmukov[3], Maxim Pisov[4], and Mikhail Belyaev[1,3]

[1] Skolkovo Institute of Science and Technology, Moscow, Russia
Mikhail.Goncharov2@skoltech.ru
[2] Artificial Intelligence Research Institute (AIRI), Moscow, Russia
[3] Institute for Information Transmission Problems, Moscow, Russia
[4] IRA-Labs, Moscow, Russia

Abstract. This paper introduces `vox2vec` — a contrastive method for self-supervised learning (SSL) of voxel-level representations. `vox2vec` representations are modeled by a Feature Pyramid Network (FPN): a voxel representation is a concatenation of the corresponding feature vectors from different pyramid levels. The FPN is pre-trained to produce similar representations for the same voxel in different augmented contexts and distinctive representations for different voxels. This results in unified multi-scale representations that capture both global semantics (e.g., body part) and local semantics (e.g., different small organs or healthy versus tumor tissue). We use `vox2vec` to pre-train a FPN on more than 6500 publicly available computed tomography images. We evaluate the pre-trained representations by attaching simple heads on top of them and training the resulting models for 22 segmentation tasks. We show that `vox2vec` outperforms existing medical imaging SSL techniques in three evaluation setups: linear and non-linear probing and end-to-end fine-tuning. Moreover, a non-linear head trained on top of the frozen `vox2vec` representations achieves competitive performance with the FPN trained from scratch while having 50 times fewer trainable parameters. The code is available at https://github.com/mishgon/vox2vec.

Keywords: Contrastive Self-Supervised Representation Learning · Medical Image Segmentation

1 Introduction

Medical image segmentation often relies on supervised model training [14], but this approach has limitations. Firstly, it requires costly manual annotations.

Supplementary Information The online version contains supplementary material available at https://doi.org/10.1007/978-3-031-43907-0_58.

H. Greenspan et al. (Eds.): MICCAI 2023, LNCS 14220, pp. 605–614, 2023.
https://doi.org/10.1007/978-3-031-43907-0_58

Secondly, the resulting models may not generalize well to unseen data domains. Even small changes in the task may result in a significant drop in performance, requiring re-training from scratch [18].

Self-supervised learning (SSL) is a promising solution to these limitations. SSL pre-trains a model backbone to extract informative representations from unlabeled data. Then, a simple linear or non-linear head on top of the frozen pre-trained backbone can be trained for various downstream tasks in a supervised manner (linear or non-linear probing). Alternatively, the backbone can be fine-tuned for a downstream task along with the head. Pre-training the backbone in a self-supervised manner enables scaling to larger datasets across multiple data and task domains. In medical imaging, this is particularly useful given the growing number of available datasets.

In this work, we focus on contrastive learning [8,12], one of the most effective approaches to SSL in computer vision. In contrastive learning, the model is trained to produce similar vector representations for augmented views of the same image and dissimilar representations for different images. Contrastive methods can also be used to learn dense, i.e., patch-level or even pixel- or voxel-level representations: pixels of augmented image views from the same region of the original image should have similar representations, while different pixels should have dissimilar ones [23].

Several works have implemented contrastive learning of dense representations in medical imaging [2,7,25,26,29]. Representations in [7,25] do not resolve nearby voxels due to the negative sampling strategy and the architectural reasons. This makes them unsuitable for full-resolution segmentation, especially in linear and non-linear probing regimes. In the current SotA dense SSL methods [2,26], authors employ restorative learning in addition to patch-level contrastive learning, in order to pre-train voxel-level representations in full-resolution. In [29], separate global and voxel-wise representations are learned in a contrastive manner to implement efficient dense image retrieval.

The common weakness of all the above works is that they do not evaluate their SSL models in linear or non-linear probing setups, even though these setups are de-facto standards for evaluation of SSL methods in natural images [8,13,23]. Moreover, fine-tuned models can deviate drastically from their pre-trained states due to catastrophical forgetting [11], while models trained in linear or non-linear probing regimes are more robust as they have several orders of magnitude fewer trainable parameters.

Our contributions are threefold. **First**, we propose `vox2vec`, a framework for contrastive learning of voxel-level representations. Our simple negative sampling strategy and the idea of storing voxel-level representations in a feature pyramid form result in high-dimensional, fine-grained, multi-scale representations suitable for the segmentation of different organs and tumors in full resolution. **Second**, we employ `vox2vec` to pre-train a FPN architecture on a diverse collection of six unannotated datasets, totaling over 6,500 CT images of the thorax and abdomen. We make the pre-trained model publicly available to simplify the reproduction of our results and to encourage practitioners to utilize this model as a starting point for the segmentation algorithms training. **Finally**, we compare the pre-trained model with the baselines on 22 segmentation tasks on seven CT datasets in three setups: linear probing, non-linear probing, and fine-tuning. We show that vox2vec performs slightly better than SotA models in the fine-tuning setup and outperforms them by a huge margin in the linear and non-linear probing setups. To the best of our knowledge, this is the first successful attempt to evaluate dense SSL methods in the medical imaging domain in linear and non-linear probing regimes.

2 Related Work

In recent years, self-supervised learning in computer vision has evolved from simple pretext tasks like Jigsaw Puzzles [22], Rotation Prediction [17], and Patch Position Prediction [10] to the current SotA methods such as restorative autoencoders [13] and contrastive [8] or non-contrastive [9] joint embedding methods.

Several methods produce dense or pixel-wise vector representations [6,23,28] to pre-train models for downstream tasks like segmentation or object detection. In [23], pixel-wise representations are learned by forcing local features to remain constant over different viewing conditions. This means that matching regions describing the same location of the scene on different views should be positive pairs, while non-matching regions should be negative pairs. In [28], authors define positive and negative pairs as spatially close and distant pixels, respectively. While in [6], authors minimize the mean square distance between matched pixel embeddings, simultaneously preserving the embedding variance along the batch and decorrelating different embedding vector components.

The methods initially proposed for natural images are often used to pre-train models on medical images. In [25], authors propose the 3D adaptation of Jigsaw Puzzle, Rotation Prediction, Patch Position Prediction, and image-level contrastive learning. Another common way for pre-training on medical images is to combine different approaches such as rotation prediction [26], restorative autoencoders [2,26], and image-level contrastive learning [2,26].

Several methods allows to obtain voxel-wise features. The model [29] maximizes the consistency of local features in the intersection between two differently augmented images. The algorithm [29] was mainly proposed for image retrieval and uses only feature representations in the largest and smallest scales in separate contrastive losses, while vox2vec produce voxels' representations via concatenation of feature vectors from a feature pyramid and pre-train them in a unified manner using a single contrastive loss. Finally, a number of works propose semi-supervised contrastive learning methods [20], however, they require additional task-specific manual labeling.

3 Method

In a nutshell, vox2vec pre-trains a neural network to produce similar representations for the same voxel placed in different contexts (positive pairs) and

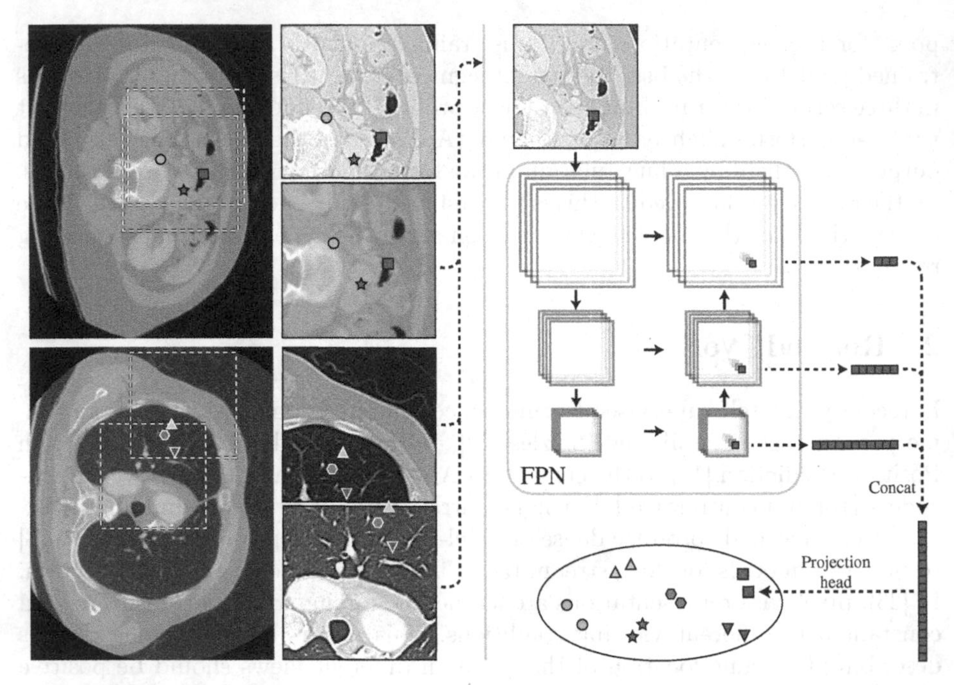

Fig. 1. Illustration of the vox2vec pre-training pipeline. Left: two overlapping augmented 3D patches are sampled from each volume in a batch. Markers of the same color and shape denote positive pairs of voxels. Right: voxel-level representations are obtained via the concatenation of corresponding feature vectors from different levels of the FPN. Finally, the representations are projected to the space where contrastive loss is computed.

predict distinctive representations for different voxels (negative pairs). In the following Sects. 3.1, 3.2, 3.3, we describe in detail the main components of our method: 1) definition and sampling of positive and negative pairs of voxels; 2) modeling voxel-level representations via a neural network; 3) computation of the contrastive loss. The whole pre-training pipeline is schematically illustrated in Fig. 1. We also describe the methodology of the evaluation of the pre-trained representations on downstream segmentation tasks in Sect. 3.4.

3.1 Sampling of Positive and Negative Pairs

We define a *positive pair* as any pair of voxels that correspond to the same location in a given volume. Conversely, we call a *negative pair* any pair of voxels that correspond to different locations in the same volume as well as voxels belonging to different volumes.

Figure 1 (left) illustrates our strategy for positive and negative pairs sampling. For a given volume, we sample two overlapping 3D patches of size (H, W, D). We apply color augmentations to them, including random gaussian blur, random gaussian sharpening, adding random gaussian noise, clipping the intensities to the random Hounsfield window, and rescaling them to the $(0, 1)$ interval. Next, we sample m different positions from the patches' overlapping region. Each position yields a pair of voxels — one from each patch, which results in a total of m positive pairs of voxels. At each pre-training iteration, we repeat this procedure for n different volumes, resulting in $2 \cdot n$ patches containing $N = n \cdot m$ positive pairs. Thus, each sampled voxel has one positive counterpart and forms negative pairs with all the remaining $2N - 2$ voxels.

In our experiments we set $(H, W, D) = (128, 128, 32)$, $n = 10$ and $m = 1000$.

We exclude the background voxels from the sampling and do not penalize their representations. We obtain the background voxels by using a simple two-step algorithm: 1) thresholding voxels with an intensity less than −500 HU; 2) keep voxels from the same connected component as the corner voxel of the CT volume, using a flood fill algorithm.

3.2 Architecture

A standard architecture for voxel-wise prediction is 3D UNet [24]. UNet's backbone returns a feature map of the same resolution as the input patch. However, our experiments show that this feature map alone is insufficient for modeling self-supervised voxel-level representations. The reason is that producing a feature map with more than 100 channels in full resolution is infeasible due to memory constraints. Meanwhile, to be suitable for many downstream tasks, representations should have a dimensionality of about 1000, as in [8].

To address this issue, we utilize a 3D FPN architecture instead of a standard 3D UNet. FPN returns voxel-level representations in the form of a feature pyramid. The pyramid's base is a feature map with 16 channels of the same resolution as the input patch. Each next pyramid level has twice as many channels and two times lower resolution than the previous one. Each voxel's representation is a concatenation of the corresponding feature vectors from all the pyramid levels. We use FPN with six pyramid levels, which results in 1008-dimensional representations. See Fig. 1 (right) for an illustration.

3.3 Loss Function

At each pre-training iteration, we fed $2 \cdot n$ patches to the FPN and obtain the representations for N positive pairs of voxels. We denote the representations in i-th positive pair as $h_i^{(1)}$ and $h_i^{(2)}$, $i = 1, \ldots, N$. Following [8], instead of penalizing the representations directly, we project them on 128-dimensional unit sphere via a trainable 3-layer perceptron $g(\cdot)$ followed by l2-normalization: $z_i^{(1)} = g(h_i^{(1)})/\|g(h_i^{(1)})\|$, $z_i^{(2)} = g(h_i^{(2)})/\|g(h_i^{(2)})\|$, $i = 1, \ldots, N$. Similar to [8] we use the InfoNCE loss as a contrastive objective: $\mathcal{L} = \sum_{i=1}^{N} \sum_{k \in \{1,2\}} \mathcal{L}_i^k$, where

$$\mathcal{L}_i^k = -\log \frac{\exp(\langle z_i^{(1)}, z_i^{(2)} \rangle / \tau)}{\exp(\langle z_i^{(1)}, z_i^{(2)} \rangle / \tau) + \sum_{j \in \{1, \ldots, N\} \setminus \{i\}} \sum_{l \in \{1,2\}} \exp(\langle z_i^{(k)}, z_j^{(l)} \rangle / \tau)}.$$

3.4 Evaluation Protocol

We evaluate the quality of self-supervised voxel-level representations on downstream segmentation tasks in three setups: 1) linear probing, 2) non-linear probing, and 3) end-to-end fine-tuning.

Linear or non-linear probing means training a voxel-wise linear or non-linear classifier on top of the frozen representations. If the representations are modeled by the UNet model, such classifier can be implemented as one or several 1×1 convolutional layers with a kernel size 1 on top of the output feature map. A linear voxel-wise head (linear FPN head) can be implemented as follows. Each pyramid level is separately fed to its own convolutional layer with kernel size 1. Then, as the number of channels on all pyramid levels has decreased, they can be upsampled to the full resolution and summed up. This operation is equivalent to applying a linear classifier to FPN voxel-wise representations described in Sect. 3.2. Linear FPN head has four orders of magnitude fewer parameters than FPN. The architecture of the non-linear voxel-wise head replicates the UNet's decoder but sets the kernel size of all convolutions to 1. It has 50 times fewer parameters than the entire FPN architecture.

In the end-to-end fine-tuning setup, we attach the voxel-wise non-linear head, but in contrast to the non-linear probing regime, we also train the backbone.

4 Experiments

4.1 Pre-training

We use `vox2vec` to pre-train both FPN and UNet models (further `vox2vec`-FPN and `vox2vec`-UNet) in order to ablate the effect of using a feature pyramid instead of single full-resolution feature map for modeling voxel-wise representations. For pre-training, we use 6 public CT datasets [1,3,5,15,21,27], totaling more than 6550 CTs, covering abdomen and thorax domains. We do not use the annotations for these datasets during the pre-training stage. Pre-processing includes the following steps: 1) cropping to the minimal volume containing all the voxels with the intensity greater than −500 HU; 2) interpolation to the voxel spacing of $1 \times 1 \times 2$ mm^3 (intensities are clipped and rescaled at the augmentation step, see Sect. 3.1). We pre-train both models for 100K batches using the Adam optimizer [16] with a learning rate of 0.0003. Both models are trained on a single A100-40Gb GPU for an average of 3 days. Further details about the pre-training setup can be found in Supplementary materials.

4.2 Evaluation

We evaluate our method on the Beyond the Cranial Vault Abdomen (BTCV) [19] and Medical Segmentation Decathlon (MSD) [4] datasets. The BTCV dataset consists of 30 CT scans along with 13 different organ annotations. We test our method on 6 CT MSD datasets, which include 9 different organ and tumor segmentation tasks. A 5 fold cross-validation is used for BTCV experiments, and a 3 fold cross-validation for MSD experiments. The segmentation performance of each model on BTCV and MSD datasets is evaluated by the Dice score.

For our method, the pre-processing steps are the same for all datasets, as at the pre-training stage, but in addition, intensities are clipped to $(-1350, 1000)$ HU window and rescaled to $(0, 1)$.

We compare our results with the current state-of-the-art self-supervised methods [2,26] in medical imaging. The pre-trained weights for the SwinUNETR encoder and TransVW UNet are taken from the official repositories of corresponding papers. In these experiments, we keep the crucial pipeline hyperparameters (e.g., spacing, clipping window, patch size) the same as in the original works. To evaluate the pre-trained SwinUNETR and TransVW in linear and non-linear probing setups, we use similar linear and non-linear head architectures as for vox2vec-FPN (see Sect. 3.4). SwinUNETR and TransVW cost 391 GFLOPs and 1.2 TFLOPS, correspondingly, compared to 115 GFLOPs of vox2vec-FPN.

We train all models for 45000 batches of size 7 (batch size for SwinUNETR is set to 3 due to memory constraints), using the Adam optimizer with a learning rate of 0.0003. In the fine-tuning setup, we freeze the backbone for the first 15000 batches and then exponentially increase the learning rate for the backbone parameters from 0.00003 up to 0.0003 during 1200 batches.

5 Results

The mean value and standard deviation of Dice score across 5 folds on the BTCV dataset for all models in all evaluation setups are presented in Table 1. vox2vec-FPN performs slightly better than other models in the fine-tuning setup. However, considering the standard deviation, all the fine-tuned models perform on par with their counterparts trained from scratch.

Nevertheless, vox2vec-FPN significantly outperforms other models in linear and non-linear regimes. On top of that, we observe that in non-linear probing regime, it performs (within the standard deviation) as well as the FPN trained from scratch while having x50 times fewer trainable parameters (see Fig. 2). We demonstrate an example of the excellent performance of vox2vec-FPN in both linear and non-linear probing regimes in Supplementary materials.

We reproduce the key results on MSD challenge CT datasets, which contain tumor and organ segmentation tasks. Table 2 shows that in the vox2vec representation space, organ voxels can be separated from tumor voxels with a quality comparable to the model trained from scratch. A t-SNE embedding of vox2vec representations on MSD is available in the Supplementary materials.

Table 1. Average cross validation Dice scores on BTCV multi-organ segmentation dataset.

model	Sp	Kid	Gb	Es	Li	St	Aor	IVC	PSV	Pa	AG	Avg
						from scratch						
TransVW UNet	79.2	82.7	43.9	65.9	83.7	62.1	86.6	76.9	61.3	56.7	51.4	68.0 ± 2.1
SwinUNETR	90.8	87.8	60.4	69.8	94.7	79.8	88.0	81.8	67.7	69.6	61.5	77.0 ± 2.5
UNet	91.1	88.5	58.8	72.3	96.0	**83.8**	89.0	83.2	68.3	70.4	63.2	78.2 ± 2.3
FPN	**92.4**	89.5	**60.9**	70.1	**96.3**	82.7	90.1	**83.9**	69.0	71.8	62.5	78.5 ± 2.2
						linear probing						
TransVW	34.4	25.7	8.9	34.4	56.8	12.1	47.2	19.0	18.8	8.2	20.6	25.6 ± 1.1
SwinUNETR	44.4	38.3	7.6	23.7	72.4	17.8	36.6	26.9	19.4	3.6	11.8	27.1 ± 2.4
random-FPN	68.0	61.2	30.0	38.0	81.6	45.3	65.0	52.4	27.7	22.9	26.0	46.6 ± 3.0
vox2vec-UNet	79.4	79.8	29.9	37.7	90.5	62.5	78.8	70.8	36.0	40.9	33.6	57.9 ± 2.0
vox2vec-FPN	83.7	84.0	43.7	58.0	93.1	67.5	85.6	77.5	56.6	58.8	53.3	69.2 ± 1.2
						non-linear probing						
TransVW	24.9	31.5	6.7	28.1	45.1	9.0	44.9	27.2	19.0	7.2	15.4	23.5 ± 2.7
random-FPN	76.7	67.0	34.1	47.1	83.7	52.8	70.2	57.5	30.2	28.6	31.5	52.1 ± 4.9
SwinUNETR	77.0	74.4	48.1	52.1	87.0	53.7	73.5	58.1	47.2	35.3	39.9	58.5 ± 2.6
vox2vec-UNet	80.3	81.4	34.1	42.7	91.1	64.0	79.6	71.6	42.7	43.3	37.6	60.6 ± 3.0
vox2vec-FPN	91.0	89.2	50.7	67.5	95.3	78.2	89.4	80.7	64.9	66.1	59.9	75.5 ± 1.7
						fine-tuning						
TransVW	77.8	80.7	42.9	66.5	83.6	59.3	86.2	77.3	63.7	54.4	54.0	67.8 ± 1.9
SwinUNETR	84.2	86.7	58.4	70.4	94.5	76.0	87.7	82.1	67.0	69.8	61.0	75.8 ± 3.3
vox2vec-UNet	91.4	90.1	52.3	72.5	95.8	83.0	89.9	82.6	66.5	71.1	61.8	77.6 ± 1.0
vox2vec-FPN	91.4	**90.7**	59.5	**72.7**	**96.3**	83.2	**91.3**	**83.9**	**69.2**	**73.9**	**65.2**	**79.5 ± 1.3**

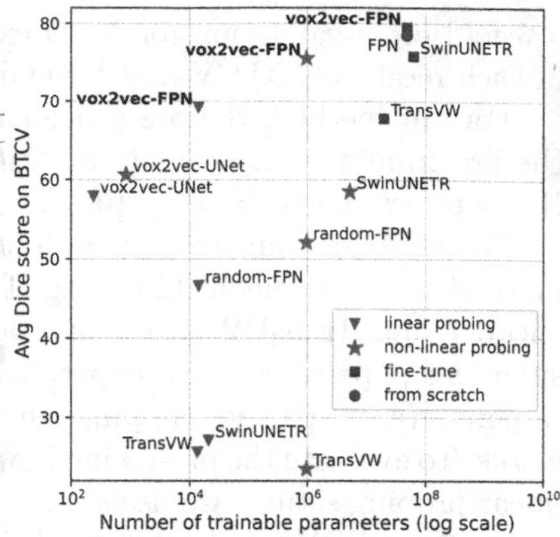

Fig. 2. Dice score on BTCV cross-validation averaged for all organs w.r.t. the number of trainable paramaters of different models in different evaluation setups.

Table 2. Cross validation Dice score on CT tasks of MSD challenge.

	Liver		Lung	Pancreas		Hepatic vessel		Spleen	Colon
model	organ	tumor	tumor	organ	tumor	organ	tumor	organ	cancer
				from scratch					
FPN	94.4	44.6	53.1	**77.1**	28.0	53.7	49.4	96.0	**32.2**
				non-linear probing					
vox2vec-FPN	94.7	43.9	49.5	71.4	28.5	58.1	54.8	95.1	24.8
				fine-tuning					
SwinUNETR	95.0	49.3	55.2	75.2	**35.9**	**60.9**	57.5	95.5	29.2
vox2vec-FPN	**95.6**	**51.0**	**56.6**	77.0	31.8	59.5	**62.4**	**96.1**	30.1

6 Conclusion

In this work, we present `vox2vec` — a self-supervised framework for voxel-wise representation learning in medical imaging. Our method expands the contrastive learning setup to the feature pyramid architecture allowing to pre-train effective representations in full resolution. By pre-training a FPN backbone to extract informative representations from unlabeled data, our method scales to large datasets across multiple task domains. We pre-train a FPN architecture on more than 6500 CT images and test it on various segmentation tasks, including different organs and tumors segmentation in three setups: linear probing, non-linear probing, and fine-tuning. Our model outperformed existing methods in all regimes. Moreover, `vox2vec` establishes a new state-of-the-art result on the linear and non-linear probing scenarios.

Still, this work has a few limitations to consider. We plan to investigate further how the performance of `vox2vec` scales with the increasing size of the pre-training dataset and the pre-trained architecture size. Another interesting research direction is exploring the effectiveness of vox2vec with regard to domain adaptation to address the challenges of domain shift between different medical imaging datasets obtained from different sources. A particular interest is a low-shot scenario when only a few examples from the target domain are available.

Acknowledgements. This work was supported by the Russian Science Foundation grant number 20-71-10134.

References

1. Data from the national lung screening trial (NLST) (2013). https://doi.org/10.7937/TCIA.HMQ8-J677, https://wiki.cancerimagingarchive.net/x/-oJY
2. Transferable visual words: exploiting the semantics of anatomical patterns for self-supervised learning. IEEE Trans. Med. Imaging **40**(10), 2857–2868 (2021). https://doi.org/10.1109/TMI.2021.3060634
3. Aerts, H., et al.: Data from NSCLC-radiomics. The cancer imaging archive (2015)
4. Antonelli, M., et al.: The medical segmentation decathlon. Nat. Commun. **13**(1), 4128 (2022)
5. Armato, S.G., III., et al.: The lung image database consortium (LIDC) and image database resource initiative (IDRI): a completed reference database of lung nodules on CT scans. Med. Phys. **38**(2), 915–931 (2011)
6. Bardes, A., Ponce, J., LeCun, Y.: VICRegL: Self-Supervised Learning of Local Visual Features. arXiv (2022). https://doi.org/10.48550/arXiv.2210.01571
7. Chaitanya, K., Erdil, E., Karani, N., Konukoglu, E.: Contrastive learning of global and local features for medical image segmentation with limited annotations. Adv. Neural Inf. Process. Syst. **33**, 12546–12558 (2020)
8. Chen, T., Kornblith, S., Norouzi, M., Hinton, G.: A simple framework for contrastive learning of visual representations. In: International Conference on Machine Learning, pp. 1597–1607. PMLR (2020)
9. Chen, X., He, K.: Exploring simple siamese representation learning. In: Proceedings of the IEEE/CVF Conference on Computer Vision and Pattern Recognition, pp. 15750–15758 (2021)
10. Doersch, C., Gupta, A., Efros, A.A.: Unsupervised visual representation learning by context prediction. In: Proceedings of the IEEE International Conference on Computer Vision, pp. 1422–1430 (2015)
11. French, R.M.: Catastrophic forgetting in connectionist networks. Trends Cogn. Sci. **3**(4), 128–135 (1999)
12. Hadsell, R., Chopra, S., LeCun, Y.: Dimensionality Reduction by Learning an Invariant Mapping. In: 2006 IEEE Computer Society Conference on Computer Vision and Pattern Recognition (CVPR 2006). vol. 2, pp. 1735–1742. IEEE (2006). https://doi.org/10.1109/CVPR.2006.100
13. He, K., Chen, X., Xie, S., Li, Y., Doll'ar, P., Girshick, R.B.: Masked autoencoders are scalable vision learners. 2022 IEEE/CVF Conference on Computer Vision and Pattern Recognition (CVPR), pp. 15979–15988 (2022)
14. Isensee, F., Jaeger, P.F., Kohl, S.A., Petersen, J., Maier-Hein, K.H.: nnU-Net: a self-configuring method for deep learning-based biomedical image segmentation. Nat. Methods **18**(2), 203–211 (2021)

15. Ji, Y., et al.: AMOS: a large-scale abdominal multi-organ benchmark for versatile medical image segmentation. arXiv preprint arXiv:2206.08023 (2022)
16. Kingma, D.P., Ba, J.: Adam: A Method for Stochastic Optimization. arXiv (2014). https://doi.org/10.48550/arXiv.1412.6980
17. Komodakis, N., Gidaris, S.: Unsupervised representation learning by predicting image rotations. In: International Conference on Learning Representations (ICLR) (2018)
18. Kondrateva, E., Druzhinina, P., Dalechina, A., Shirokikh, B., Belyaev, M., Kurmukov, A.: Neglectable effect of brain MRI data prepreprocessing for tumor segmentation. arXiv preprint arXiv:2204.05278 (2022)
19. Landman, B., Xu, Z., Igelsias, J., Styner, M., Langerak, T., Klein, A.: Miccai multi-atlas labeling beyond the cranial vault-workshop and challenge. In: Proceedings of MICCAI Multi-Atlas Labeling Beyond Cranial Vault-Workshop Challenge. vol. 5, p. 12 (2015)
20. Lee, C.E., Chung, M., Shin, Y.G.: Voxel-level siamese representation learning for abdominal multi-organ segmentation. Comput. Methods Programs Biomed. **213**, 106547 (2022). https://doi.org/10.1016/j.cmpb.2021.106547
21. Ma, J., et al.: Fast and low-GPU-memory abdomen CT organ segmentation: the flare challenge. Med. Image Anal. **82**, 102616 (2022)
22. Noroozi, M., Favaro, P.: Unsupervised learning of visual representations by solving jigsaw puzzles. In: Leibe, B., Matas, J., Sebe, N., Welling, M. (eds.) ECCV 2016. LNCS, vol. 9910, pp. 69–84. Springer, Cham (2016). https://doi.org/10.1007/978-3-319-46466-4_5
23. O Pinheiro, P.O., Almahairi, A., Benmalek, R., Golemo, F., Courville, A.C.: Unsupervised learning of dense visual representations. Adv. Neural Inf. Process. Syst. **33**, 4489–4500 (2020)
24. Ronneberger, O., Fischer, P., Brox, T.: U-Net: convolutional networks for biomedical image segmentation. In: Navab, N., Hornegger, J., Wells, W.M., Frangi, A.F. (eds.) MICCAI 2015. LNCS, vol. 9351, pp. 234–241. Springer, Cham (2015). https://doi.org/10.1007/978-3-319-24574-4_28
25. Taleb, A., et al.: 3D self-supervised methods for medical imaging. Adv. Neural Inf. Process. Syst. **33**, 18158–18172 (2020)
26. Tang, Y., et al.: Self-supervised pre-training of Swin transformers for 3D medical image analysis. In: Proceedings of the IEEE/CVF Conference on Computer Vision and Pattern Recognition, pp. 20730–20740 (2022)
27. Tsai, E., et al.: Medical imaging data resource center - RSNA international COVID radiology database release 1a - chest CT COVID+ (MIDRC-RICORD-1a) (2020). https://doi.org/10.7937/VTW4-X588, https://wiki.cancerimagingarchive.net/x/DoDTB
28. Xie, Z., Lin, Y., Zhang, Z., Cao, Y., Lin, S., Hu, H.: Propagate yourself: exploring pixel-level consistency for unsupervised visual representation learning. In: Proceedings of the IEEE/CVF Conference on Computer Vision and Pattern Recognition, pp. 16684–16693 (2021)
29. Yan, K., et al.: SAM: self-supervised learning of pixel-wise anatomical embeddings in radiological images. IEEE Trans. Med. Imaging **41**(10), 2658–2669 (2022)

Mesh2SSM: From Surface Meshes to Statistical Shape Models of Anatomy

Krithika Iyer[1,2(✉)] and Shireen Y. Elhabian[1,2]

[1] Scientific Computing and Imaging Institute, University of Utah, Salt Lake City, UT, USA
krithika.iyer@utah.edu, shireen@sci.utah.edu

[2] Kahlert School of Computing, University of Utah, Salt Lake City, UT, USA

Abstract. Statistical shape modeling is the computational process of discovering significant shape parameters from segmented anatomies captured by medical images (such as MRI and CT scans), which can fully describe subject-specific anatomy in the context of a population. The presence of substantial non-linear variability in human anatomy often makes the traditional shape modeling process challenging. Deep learning techniques can learn complex non-linear representations of shapes and generate statistical shape models that are more faithful to the underlying population-level variability. However, existing deep learning models still have limitations and require established/optimized shape models for training. We propose Mesh2SSM, a new approach that leverages unsupervised, permutation-invariant representation learning to estimate how to deform a template point cloud to subject-specific meshes, forming a correspondence-based shape model. Mesh2SSM can also learn a population-specific template, reducing any bias due to template selection. The proposed method operates directly on meshes and is computationally efficient, making it an attractive alternative to traditional and deep learning-based SSM approaches.

Keywords: Statistical Shape Modeling · Representation Learning · Point Distribution Models

1 Introduction

Statistical shape modeling (SSM) is a powerful tool in medical image analysis and computational anatomy to quantify and study the variability of anatomical structures within populations. SSM has shown great promise in medical research, particularly in diagnosis [12,23], pathology detection [19,25], and treatment planning [27]. SSM has enabled researchers to better understand the underlying biological processes, leading to the development of more accurate and personalized diagnostic and treatment plans [3,9,14,17].

Supplementary Information The online version contains supplementary material available at https://doi.org/10.1007/978-3-031-43907-0_59.

H. Greenspan et al. (Eds.): MICCAI 2023, LNCS 14220, pp. 615–625, 2023.
https://doi.org/10.1007/978-3-031-43907-0_59

Over the years, several SSM approaches have been developed that implicitly represent the shapes (deformation fields [8], level set methods [22]) or explicitly represent them as a ordered set of landmarks or *correspondence points* (aka point distribution models, PDMs). Here, we focus on the automated construction of PDMs because, compared to deformation fields, point correspondences are easier to interpret by clinicians, are computationally efficient for large datasets, and less sensitive to noise and outliers than deformation fields [5].

SSM performance depends on the underlying process used to generate shape correspondences and the quality of the input data. Various correspondence generation methods exist, including non-optimized landmark estimation and parametric and non-parametric correspondence optimization. Non-optimized methods manually label a reference shape and warp the annotated landmarks using registration techniques [10,16,18]. Parametric methods use fixed geometrical bases to establish correspondences [26], while group-wise non-parametric approaches find correspondences by considering the variability of the entire cohort during the optimization process. Examples of non-parametric methods include particle-based optimization [4] and Minimum Description Length (MDL) [7].

Traditional SSM methods assume that population variability follows a Gaussian distribution, which implies that a linear combination of training shapes can express unseen shapes. However, anatomical variability can be far more complex than this linear approximation, in which case nonlinear variations normally exist (e.g., bending fingers, soft tissue deformations, and vertebrae with different types). Furthermore, conventional SSM pipelines are computationally intensive, where inferring PDMs on new samples entail an optimization process. Deep learning-based approaches for SSM have emerged as a promising avenue to overcoming these limitations. Deep learning models can learn complex non-linear representations of the shapes, which can be used to generate shape models. Moreover, they can efficiently perform inference on new samples without computation overhead or re-optimization. Recent works such as FlowSSM [15], ShapeFlow [11], DeepSSM [2], and VIB-DeepSSM [1] have incorporated deep learning to generate shape models. FlowSSM [15] and ShapeFlow [11] operate on surface meshes and use neural networks to parameterize the deformations field between two shapes in a low dimensional latent space and rely on an encoder-free setup. Encoder-free methods randomly initialize the latent representations for each sample that are then optimized to produce the optimal deformations. One major caveat of an encoder-free setup is that inference on new meshes is no longer straightforward; the latent representation has to be re-optimized for every new sample. On the other hand, DeepSSM [2], TL-DeepSSM [2], and VIB-DeepSSM [1] learn the PDM directly from unsegmented CT/MRI images, and hence alleviate the need for PDM optimization given new samples and can bypass anatomy segmentation by operating directly on unsegmented images. However, these methods rely on supervised losses and require volumetric images, segmented images, and established/optimized PDMs for training. This reliance on supervised losses introduces linearity assumptions in generating ground truth PDMs. TL-DeepSSM [2], a variant of DeepSSM [2], differs from the others by not utilizing PCA scores as shape descriptors. Instead, it adopts an established correspondence model hence, similar to the vanilla DeepSSM [2] learns a linear model.

In this paper, we introduce Mesh2SSM[1], a deep learning method that addresses the limitations of traditional and deep learning-based SSM approaches. Mesh2SSM leverages *unsupervised, permutation-invariant representation learning* to learn the low dimensional nonlinear shape descriptor directly from mesh data and uses the learned features to generate a correspondence model of the population. Mesh2SSM also includes an analysis network that operates on the learned correspondences to obtain a data-driven template point cloud (i.e., template point cloud), which can replace the initial template, and hence reducing the bias that could arise from template selection. Furthermore, the learned representation of meshes can be used for predicting related quantities that rely on shape. Our main contributions are:

1. We introduce Mesh2SSM, a fully unsupervised correspondence generation deep learning framework that operates directly on meshes. Mesh2SSM uses an autoencoder to extract the shape descriptor of the mesh and uses this descriptor to transform a template point cloud using IM-Net [6].
2. The proposed method uses an autoencoder that combines geodesic distance features and EdgeConv [28] (dynamic graph convolution neural network) to extract meaningful feature representation of each mesh that is permutation-invariant.
3. Mesh2SSM also includes a variational autoencoder (VAE) [13,21] operating on the learned correspondence points and trained end-to-end with correspondence generation network. This VAE branch serves two purposes: (a) serves as a shape analysis module for the non-linear shape variations and (b) learns a data-specific template from the latent space of the correspondences that is fed back to the correspondence generation network.

To motivate the need for the mesh feature encoder and study the effect of the template selection, we considered the box-bump dataset, a synthetic dataset of 3D shapes of boxes with a moving bump. In Fig. 1, we compare Mesh2SSM (sans the VAE analysis branch) with FlowSMM [15] since this approach is the closest to Mesh2SSM. We performed experiments with three templates: medoid, sphere, and box without the bump. Although both methods show some sensitivity to the choice of template, FlowSSM is more sensitive toward the choice of the template than Mesh2SSM. Moreover, FlowSSM fails to identify the correct mode of variation, the horizontal movement of the bump as the primary variation, which can also be inferred by comparing the compactness curves in Fig. 1.c. Mesh2SSM performs best when the template is a medoid shape, which makes the case for learning a data-specific template. Since Mesh2SSM model uses an autoencoder, inference on unseen meshes only requires a single forward pass (1 s per sample); FlowSSM requires re-optimization, increasing the inference time drastically and require a convergence criteria to determine the best number of iterations per sample (0.15 s for one iterations per sample).

[1] Sourcecode: https://github.com/iyerkrithika21/mesh2SSM_2023.

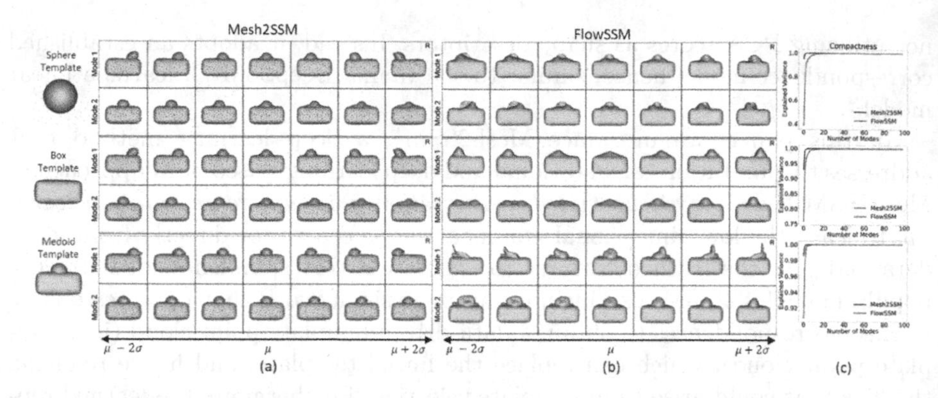

Fig. 1. Top two PCA modes of variations identified by (a) Mesh2SSM and (b) FlowSSM [15] with three templates: sphere, box without a bump, and medoid shape. FlowSSM fails to capture the horizontal movement as the primary mode of variation. (c) The compactness curves for both models with different templates.

2 Method

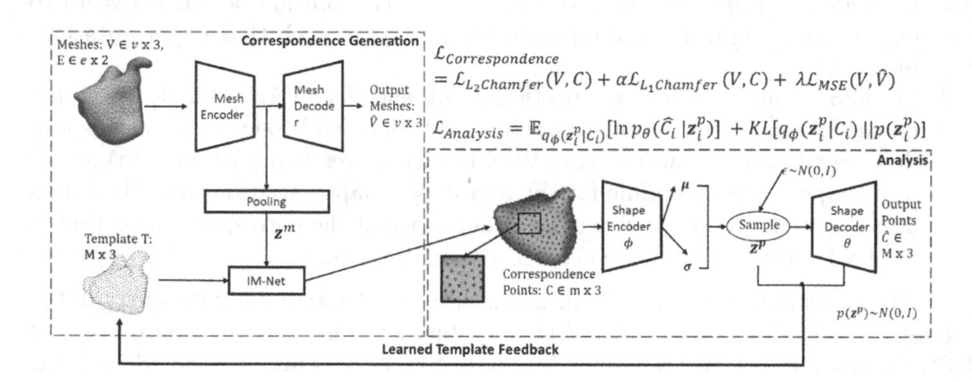

Fig. 2. Mesh2SSM: Architecture and loss of the proposed method.

The overview of the proposed pipeline is provided in Fig. 2. This section provides a brief description of each module.

2.1 Correspondence Generation

Given a set of N aligned surface meshes $\mathcal{X} = \{X_1, X_2, ...X_N\}$, each mesh $X_i = (V_i, E_i)$, where V_i and E_i represent the vertices and edge connectivity, respectively. The goal of the model is to predict a set C_i of M $3D$ correspondence points that fully describe each surface X_i and are anatomically consistent across all meshes. This goal is achieved by learning a low dimensional representation of the surface mesh $\mathbf{z}^m \in \mathbb{R}^L$ using the mesh autoencoder and then $\mathbf{z}^m$ is used to transform the template point cloud via the implicit field decoder (IM-Net) [6]. The network optimization is driven primarily by point-set to point-set two-way Chamfer distance between the learned correspondence point sets C_i and the vertex locations V_i of the original meshes. To ensure that the encoder learns useful features for the task, we regularize the optimization using the vertex reconstruction loss of the autoencoder between the input V_i and the predicted $\hat{V}_i$. The correspondence loss function is given by:

$$\mathcal{L}_C = \sum_{i=1}^{N} \left[\mathcal{L}_{L_2 Chamfer}(V_i, C_i) + \alpha \mathcal{L}_{L_1 Chamfer}(V_i, C_i) + \gamma \mathcal{L}_{MSE}(V_i, \hat{V}_i) \right] \quad (1)$$

where α, γ are the hyperparameters. We consider a combination of L_1 and L_2 two-way Chamfer distance for numerical stability as the magnitude of L_2 loss can be low over epochs and L_1 can compensate for it. The correspondence generation uses two networks:

Mesh Autoencoder (M-AE): We use EdgeConv [28] blocks, which are dynamic graph convolution neural network (DGCNN) blocks in the encoder and decoder to capture local geometric features of the mesh. The model takes vertices as input, computes an edge feature set of size k (using nearest neighbors) for each vertex at an EdgeConv layer, and aggregates features within each set to compute EdgeConv responses. The output features of the last EdgeConv layer are then globally aggregated to form a 1D global descriptor $\mathbf{z}_i^m$ of the mesh. The first EdgeConv block uses geodesic distance on the surface of the mesh to calculate the k features. The dynamic feature creation property of EdgeConv and the global pooling make this autoencoder permutation invariant.

Implicit Field Decoder (IM-NET): The IM-NET [6] architecture consists of fully connected layers with non-linearity and skip-layer connections. This network enforces the notion of correspondence across the samples. The network takes in two inputs, the latent representation of the mesh $\mathbf{z}^m$ and a template point cloud (a set of unordered points). IM-NET estimates the deformation of each point in the template required to deform the template to each sample, conditioned on $\mathbf{z}^m$. Based on the learned deformation, IM-NET directly produces the resultant displaced template point without the computational complexity of the deformation fields. Correspondence is established since the same template is deformed to all the samples.

2.2 Analysis

The Mesh2SSM model also consists of an analysis branch that acts as a shape analysis module to capture non-linear shape variations identified by the learned correspondences $\{C_i\}_{i=1}^N$ and also learns a data-informed template from the latent space of correspondences to be fed back into the correspondence generation network during training. This branch uses one network module:

Shape Variation Autoencoder (SP-VAE): The VAE [13,21] is a latent variable model parameterized by an encoder ϕ, decoder θ, and the prior $p(\mathbf{z}^p) \sim \mathcal{N}(0, \mathbf{I})$. The encoder maps the shape represented by the learned correspondence points C to the latent space and the decoder reconstructs the correspondences from the latent representation $\mathbf{z}^p$. By capturing the underlying structure of the PDM through a low-dimensional representation, SP-VAE allows for the estimation of the mean shape of the learned correspondences. The SP-VAE is trained using the loss function given by:

$$\mathcal{L}(\theta, \phi) = -\mathbb{E}_{q_\phi(\mathbf{z}_i^p|C_i)}[\log p_\theta(C_i|\mathbf{z}_i^p)] + KL(q_\phi(\mathbf{z}_i^p|C_i)||p(\mathbf{z}_i^p)) \quad (2)$$

The main difference between M-AE and a SP-VAE lies in the input and output representations they handle. SP-VAE operates directly on sets of landmarks or correspondences, aiding in the analysis of shape models. It takes a set of correspondences describing a shape as input and aims to learn a compressed latent representation of the shape. Importantly, the SP-VAE maintains the same ordering of correspondences at the input and output, so it does not use permutation-invariant layers or operations like pooling.

2.3 Training

We begin with a burn-in stage, where only the correspondence generation module is trained while the analysis module is frozen. After the burn-in stage, alternate optimization of the correspondence and analysis module begins. During the alternate optimization phase, we generate the data-informed template from the latent space of SP-VAE at regular intervals. The learned data-informed template is used in the correspondence generation module in the subsequent epochs. For the learned template, we sample 500 samples from the prior $p(\mathbf{z}^p) \sim \mathcal{N}(0, \mathbf{I})$ and pass it through the decoder of SP-VAE to get the reconstructed correspondence point set. The mean template is defined by taking the average of these generated samples. Inference with unseen meshes is straight forward; the meshes are passed through the mesh encoder and IM-NET of the correspondence generation module to get the predicted correspondences. All hyperparameters and network architecture details are mentioned in the supplementary material.

3 Experiments and Discussion

Dataset: We use the publicly available Decath-Pancreas dataset of 273 segmentations from patients who underwent pancreatic mass resection [24]. The shapes of the pancreas are highly variable and have thin structures, making it a good candidate for non-linear SSM analysis. The segmentations were isotropically resampled, smoothed, centered, and converted to meshes with roughly 2000 vertices. Although the DGCNN mesh autoencoder used in Mesh2SSM does not require the same number of vertices, uniformity across the dataset makes it computationally efficient; hence, we pad the smallest mesh by randomly repeating the vertices (akin to padding image for convolutions). The samples were randomly divided, with 218 used for training, 26 for validation, and 27 for testing.

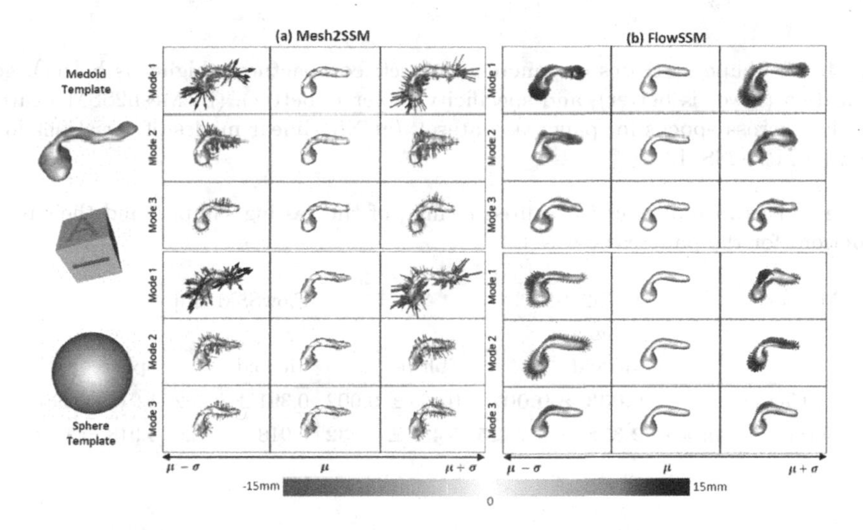

Fig. 3. Top three PCA modes of variations identified by (a) Mesh2SSM and (b) FlowSSM [15] with two templates: sphere, medoid. The color map and arrows show the signed distance and direction from the mean shape.

3.1 Results

We perform experiments with two templates: sphere and medoid. We compare the performance of FlowSSM [15] with Mesh2SSM with the template feedback loop. For Mesh2SSM template, we use 256 points uniformly spread across the surface of the sample. Mesh2SSM and FlowSSM do not have a equivalent latent space for comparison of the shape models, hence, we consider the deformed mesh vertices of FlowSSM as correspondences and perform PCA analysis. Figure 3 shows the top three PCA modes of variations identified by Mesh2SSM and FlowSSM. Similar to the observations made box-bump dataset, FlowSSM is affected by the choice of the template, and the modes of variation differ as the template changes. On the other hand, PDM predicted by Mesh2SSM identifies the same primary modes consistently. Pancreatic cancer mainly presents itself on the head of the structure [20] and for the Decath dataset, we can see the first mode identifies the change in the shape of the head. We evaluate the models based on compactness, generalization, and specificity. Compactness measures

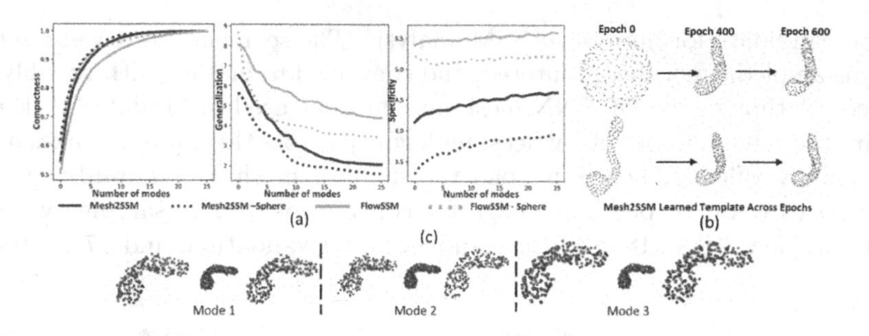

Fig. 4. (a) Shape statistics of pancreas dataset: compactness (higher is better), generalization (lower is better), and specificity (lower is better). (b) Mesh2SSM Learned template across epochs for pancreas dataset. (c) Non-linear modes of variations identified by Mesh2SSM.

Table 1. Distance metrics (measured in mm) of the testing samples and their reconstructions for the pancreas dataset

Metrics	Mesh2SSM		FlowSSM [15]	
	Template			
	Medoid	Sphere	Medoid	Sphere
L_1 Chamfer	**0.033 ± 0.002**	0.035 ± 0.002	0.391 ± 0.162	1.91 ± 0.687
Surface-to-Surface	**2.378 ± 0.7325**	5.436 ± 2.232	5.918 ± 2.026	4.918 ± 1.925

the ability of the model to reconstruct new shape instances with fewer parameters using PCA explained variance. Generalization measures the average surface distance between all test shapes and their reconstructions, and specificity measures the distance between randomly generated PCA samples. Figure 4.a shows the metrics for the pancreas dataset. Mesh2SSM outperforms FlowSSM in all three metrics, despite using only 256 correspondence points compared to FlowSSM's ∼2000 vertices. Mesh2SSM correspondence generation module efficiently parameterizes the surface of the pancreas with a minimum number of parameters. Mesh2SSM template, shown in Fig. 4.b, becomes more detailed as optimization continues, regardless of the starting template. The model can learn correct deformations in the correspondence generation module and identify the correct mean shape in the latent space of SP-VAE in the analysis module. Using the analysis module of Mesh2SSM, we visualized the top three modes of variation identified by sorting the latent dimensions of SP-VAE based on the standard deviations of the latent embeddings of the training dataset. Variations are generated by perturbing the latent representation of a sample in three directions, resulting in non-linear modes such as changes in the size and shape of the pancreas head and narrowing of the neck and body. This is shown in Fig. 4.c for MeshSSM model with medoid starting template. The distance metrics for the reconstructions of the testing samples were also computed. The results of the metrics are summarized in Table 1. The calculation involved the L_1 Chamfer loss between the predicted points (correspondences in the case of Mesh2SSM and the deformed mesh vertices in the case of FlowSSM) and the original mesh vertices. Additionally, the surface to surface distance of the mesh reconstructions (using the correspondences in Mesh2SSM and deformed meshes in FlowSSM) was included. For the pancreas dataset with the medoid as the initial template, Mesh2SSM with the template feedback produced more precise models.

3.2 Limitations and Future Scope

As SSM is included a part of diagnostic clinical support systems, it is crucial to address the drawbacks of the models. Like most deep learning models, performance of Mesh2SSM could be affected by small dataset size, and it can produce overconfident estimates. An augmentation scheme and a layer uncertainty calibration are could improve its usability in medical scenarios. Additionally, enforcing disentanglement in the latent space of SP-VAE can make the analysis module interpretable and allow for effective non-linear shape analysis by clinicians.

4 Conclusion

The paper presents a new systematic approach of generating non-linear statistical shape models using deep learning directly from meshes, which overcomes the limitations of traditional SSM and current deep learning approaches. The use of an autoencoder for meaningful feature extraction of meshes to learn the PDM provides a versatile and scalable framework for SSM. Incorporating template feedback loop via VAE [13,21] analysis module helps in mitigating bias and capturing non-linear characteristics of the data. The method is demonstrated to have superior performance in identifying shape variations using fewer parameters on synthetic and clinical datasets. To conclude, our method of generating highly accurate and detailed models of complex anatomical structures with reduced computational complexity has the potential to establish statistical shape modeling from non-invasive imaging as a powerful diagnostic tool.

Acknowledgements. This work was supported by the National Institutes of Health under grant numbers NIBIB-U24EB029011, NIAMS-R01AR076120, and NHLBI-R01HL135568. We thank the University of Utah Division of Cardiovascular Medicine for providing left atrium MRI scans and segmentations from the Atrial Fibrillation projects and the ShapeWorks team.

References

1. Adams, J., Elhabian, S.: From images to probabilistic anatomical shapes: a deep variational bottleneck approach. In: Wang, L., Dou, Q., Fletcher, P.T., Speidel, S., Li, S. (eds.) Medical Image Computing and Computer Assisted Intervention-MICCAI 2022: 25th International Conference, Singapore, 18–22 September 2022, Proceedings, Part II, pp. 474–484. Springer, Cham (2022). https://doi.org/10.1007/978-3-031-16434-7_46

2. Bhalodia, R., Elhabian, S.Y., Kavan, L., Whitaker, R.T.: DeepSSM: a deep learning framework for statistical shape modeling from raw images. In: Reuter, M., Wachinger, C., Lombaert, H., Paniagua, B., Lüthi, M., Egger, B. (eds.) Shape in Medical Imaging: International Workshop, ShapeMI 2018, Held in Conjunction with MICCAI 2018, Granada, Spain, 20 September 2018, Proceedings, vol. 11167, pp. 244–257. Springer, Cham (2018). https://doi.org/10.1007/978-3-030-04747-4_23
3. Bruse, J.L., et al.: A statistical shape modelling framework to extract 3D shape biomarkers from medical imaging data: assessing arch morphology of repaired coarctation of the aorta. BMC Med. Imaging **16**, 1–19 (2016)
4. Cates, J., Elhabian, S., Whitaker, R.: ShapeWorks: particle-based shape correspondence and visualization software. In: Statistical Shape and Deformation Analysis, pp. 257–298. Elsevier (2017)
5. Cerrolaza, J.J., et al.: Computational anatomy for multi-organ analysis in medical imaging: a review. Med. Image Anal. **56**, 44–67 (2019)
6. Chen, Z.: IM-NET: learning implicit fields for generative shape modeling (2019)
7. Davies, R.H.: Learning shape: optimal models for analysing natural variability. The University of Manchester (United Kingdom) (2002)
8. Durrleman, S., et al.: Morphometry of anatomical shape complexes with dense deformations and sparse parameters. Neuroimage **101**, 35–49 (2014)
9. Faghih Roohi, S., Aghaeizadeh Zoroofi, R.: 4D statistical shape modeling of the left ventricle in cardiac MR images. Int. J. Comput. Assist. Radiol. Surg. **8**, 335–351 (2013)
10. Heitz, G., Rohlfing, T., Maurer Jr., C.R.: Statistical shape model generation using nonrigid deformation of a template mesh. In: Medical Imaging 2005: Image Processing, vol. 5747, pp. 1411–1421. SPIE (2005)
11. Jiang, C., Huang, J., Tagliasacchi, A., Guibas, L.J.: ShapeFlow: learnable deformation flows among 3D shapes. Adv. Neural. Inf. Process. Syst. **33**, 9745–9757 (2020)
12. Khan, R.A., Luo, Y., Wu, F.X.: Machine learning based liver disease diagnosis: a systematic review. Neurocomputing **468**, 492–509 (2022)
13. Kingma, D.P., Welling, M.: Auto-encoding variational Bayes. arXiv preprint arXiv:1312.6114 (2013)
14. Lindberg, O., et al.: Hippocampal shape analysis in Alzheimer's disease and frontotemporal lobar degeneration subtypes. J. Alzheimers Dis. **30**(2), 355–365 (2012)
15. Lüdke, D., Amiranashvili, T., Ambellan, F., Ezhov, I., Menze, B.H., Zachow, S.: Landmark-free statistical shape modeling via neural flow deformations. In: Wang, L., Dou, Q., Fletcher, P.T., Speidel, S., Li, S. (eds.) Medical Image Computing and Computer Assisted Intervention-MICCAI 2022: 25th International Conference, Singapore, 18–22 September 2022, Proceedings, Part II, pp. 453–463. Springer, Cham (2022). https://doi.org/10.1007/978-3-031-16434-7_44
16. McInerney, T., Terzopoulos, D.: Deformable models in medical image analysis. In: Proceedings of the Workshop on Mathematical Methods in Biomedical Image Analysis, pp. 171–180. IEEE (1996)
17. Merle, C., et al.: High variability of acetabular offset in primary hip osteoarthritis influences acetabular reaming-a computed tomography-based anatomic study. J. Arthroplasty **34**(8), 1808–1814 (2019)
18. Paulsen, R., Larsen, R., Nielsen, C., Laugesen, S., Ersbøll, B.: Building and testing a statistical shape model of the human ear canal. In: Dohi, T., Kikinis, R. (eds.) MICCAI 2002. LNCS, vol. 2489, pp. 373–380. Springer, Heidelberg (2002). https://doi.org/10.1007/3-540-45787-9_47

19. Peiffer, M., et al.: Statistical shape model-based tibiofibular assessment of syndesmotic ankle lesions using weight-bearing CT. J. Orthop. Res.® **40**(12), 2873–2884 (2022)
20. Ralston, S.H., Penman, I.D., Strachan, M.W., Hobson, R.: Davidson's Principles and Practice of Medicine E-Book. Elsevier Health Sciences (2018)
21. Rezende, D.J., Mohamed, S., Wierstra, D.: Stochastic backpropagation and approximate inference in deep generative models. In: International Conference on Machine Learning, pp. 1278–1286. PMLR (2014)
22. Samson, C., Blanc-Féraud, L., Aubert, G., Zerubia, J.: A level set model for image classification. Int. J. Comput. Vision **40**(3), 187–197 (2000)
23. Schaufelberger, M., et al.: A radiation-free classification pipeline for craniosynostosis using statistical shape modeling. Diagnostics **12**(7), 1516 (2022)
24. Simpson, A.L., et al.: A large annotated medical image dataset for the development and evaluation of segmentation algorithms. arXiv preprint arXiv:1902.09063 (2019)
25. Sophocleous, F., et al.: Feasibility of a longitudinal statistical atlas model to study aortic growth in congenital heart disease. Comput. Biol. Med. **144**, 105326 (2022)
26. Styner, M., et al.: Framework for the statistical shape analysis of brain structures using SPHARM-PDM. Insight J. (1071), 242 (2006)
27. Vicory, J., et al.: Statistical shape analysis of the tricuspid valve in hypoplastic left heart syndrome. In: Puyol Antón, E., et al. (eds.) Statistical Atlases and Computational Models of the Heart. Multi-Disease, Multi-View, and Multi-Center Right Ventricular Segmentation in Cardiac MRI Challenge: 12th International Workshop, STACOM 2021, Held in Conjunction with MICCAI 2021, Strasbourg, France, 27 September 2021, Revised Selected Papers, pp. 132–140. Springer, Cham (2022). https://doi.org/10.1007/978-3-030-93722-5_15
28. Wang, Y., Sun, Y., Liu, Z., Sarma, S.E., Bronstein, M.M., Solomon, J.M.: Dynamic graph CNN for learning on point clouds. ACM Trans. Graph. (TOG) **38**(5), 1–12 (2019)

Graph Convolutional Network with Morphometric Similarity Networks for Schizophrenia Classification

Hye Won Park, Seo Yeong Kim, and Won Hee Lee(✉)

Kyung Hee University, Yongin, Republic of Korea
whlee@khu.ac.kr

Abstract. There is significant interest in using neuroimaging data for schizophrenia classification. Graph convolutional networks (GCNs) provide great potential to improve schizophrenia classification using brain graphs derived from neuroimaging data. However, accurate classification of schizophrenia is still challenging due to the heterogeneity of schizophrenia and their subtle differences in neuroimaging features. This paper presents a new graph convolutional framework for population-based schizophrenia classification that leverages graph-theoretical measures of morphometric similarity networks inferred from structural MRI scans and incorporates variational edges to reinforce the learning process. Specifically, we construct individual morphometric similarity networks based on inter-regional similarity of multiple morphometric features (cortical thickness, surface area, gray matter volume, mean curvature, and Gaussian curvature) extracted from T1-weighted MRI. We then formulate an adaptive population graph where each node is represented by the topological features of individual morphometric similarity networks and each edge models the similarity between the topological features of the subjects and incorporates the phenotypic information. An encode module is devised to estimate the associations between phenotypic data of the subjects and to adaptively optimize the edge weights. Our proposed method is evaluated on a large dataset collected from nine sites, resulting in a total sample of 366 patients with schizophrenia and 590 healthy individuals. Experimental results demonstrate that our proposed method improves the classification performance over traditional machine learning algorithms, with a mean classification accuracy of 81.8%. The most salient regions contributing to classification are primarily identified in the middle temporal gyrus and superior temporal gyrus.

Keywords: Graph Convolutional Networks · Morphometric Similarity Networks · Schizophrenia

Supplementary Information The online version contains supplementary material available at https://doi.org/10.1007/978-3-031-43907-0_60.

H. Greenspan et al. (Eds.): MICCAI 2023, LNCS 14220, pp. 626–636, 2023.
https://doi.org/10.1007/978-3-031-43907-0_60

1 Introduction

Schizophrenia is a severe and chronic psychiatric disorder characterized by psychotic episodes, cognitive impairment, and impaired functioning with high rates of disability [4]. Early diagnosis of schizophrenia is beneficial to patients as it leads to better prognosis and reduces symptoms more effectively [15]. However, early detection of schizophrenia is still challenging since the complex and heterogeneous symptoms are associated with schizophrenia, complicating the optimal treatment of patients and limiting positive outcomes. The current diagnosis of schizophrenia is mainly based on patient's clinical symptoms and clinical interviews. Therefore, there is an urgent need to establish an objective approach for an accurate diagnosis of schizophrenia.

There is an increasing interest in developing accurate and robust techniques to classify subjects into groups using neuroimaging data. Previous studies focused on handcrafted feature-based machine learning approach, which requires the process of feature extraction and selection prior to disease classification [28, 30]. Pre-selected features extracted from neuroimaging data might not be sufficient and generalizable across different disease datasets, thus yielding a substantial variation in classification performance.

With the rapid development of deep learning methods, recent studies have shown great potential for integrating neuroimaging data and deep learning models for an automatic diagnosis of brain diseases [12, 23]. In particular, graph convolutional network (GCN) has become a promising approach for medical image classification such as schizophrenia [16], Alzheimer's disease [10], and autism spectrum disorder [27]. GCN is capable of preserving graph topological properties while automatically learning features to perform various graph-related tasks such as graph classification and graph representation learning [2]. This is especially important for psychiatric disorders like schizophrenia which has complex and subtle differences in neuroimaging features compared to healthy individuals. Previous studies mainly focused on functional MRI data as the input of GCN for disease classification [18, 19]. Parisot *et al.* proposed a population GCN model combining subject-specific imaging and non-imaging information for autism spectrum disorder and Alzheimer's disease classification [18]. Qin *et al.* examined the effectiveness of GCN in distinguishing patients with major depressive disorder from healthy controls [19]. However, most studies adopted previously captured information based on functional brain networks which describe statistical dependence between functional MRI time series instead of modeling cortical networks based on similarity in regional cortical morphology estimated from structural MRI of the brain.

Recent studies have highlighted the potential of cortical networks, so-called morphometric similarity networks (MSNs), to predict individual differences in brain morphometry [13, 14], thereby allowing for their potential utility to capture individual cognitive variation [24]. Using the MSNs, it is also possible to unravel changes in the brain during normal cortical development [26] and psychiatric disorders [17] and to identify patterns of abnormal morphometric similarity that classify autism spectrum disorder [31] and Alzheimer's disease and mild cognitive impairment from healthy individuals [32]. However, much less is known about whether the MSNs are clinically useful phenotype and the ability of MSNs for schizophrenia classification.

To fill these gaps, we developed a generalizable graph convolutional framework for population-based schizophrenia classification using the MSNs inferred from structural MRI data. Our contributions are summarized as follows: 1) We propose a new population graph model for integrating MSN-driven features derived from structural MRI and phenotypic information; 2) A novel feature selection strategy is introduced to leverage graph-theoretical measures of the MSNs, which is new and generalizable for graph neural networks; 3) We validate the feasibility of our proposed method by conducting a comprehensive evaluation on a large schizophrenia dataset, which shows superior performance in classification over traditional machine learning approaches; 4) A complete sensitivity analysis for key parameters in our GCN-based classification framework is performed; 5) The most salient regions contributing to classification are primarily identified in the middle temporal gyrus and superior temporal gyrus.

2 Materials and Methods

Figure 1 shows the overview of our proposed classification framework. We construct individual MSNs based on inter-regional similarity of multiple morphometric features extracted from T1-weighted MRI. We formulate a population graph model where each node is represented by graph-theoretical measures of the MSNs and each edge models the similarity between the topological features of the subjects and incorporates the phenotypic information. We apply a threshold to the population graph to remove spurious connections. The edge weights are adaptively updated by using an MLP-based encoder based on non-imaging data of the subjects. Spectral graph convolutions are applied for learning, followed by the MLP-based classification for schizophrenia.

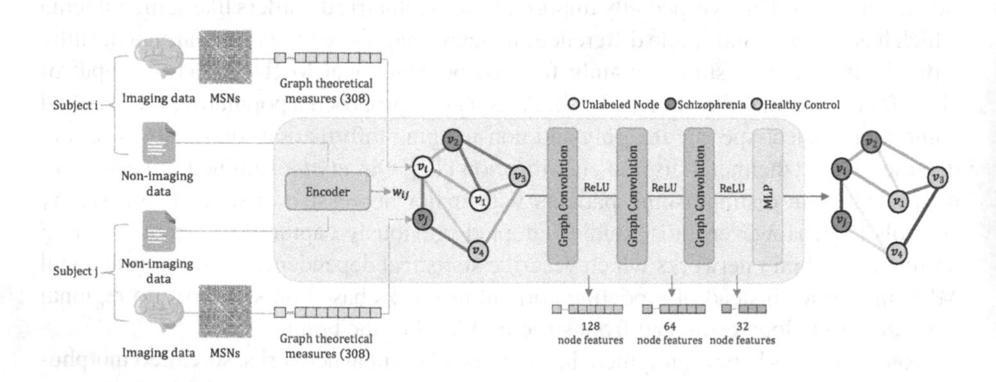

Fig. 1. Overview of our proposed classification framework.

2.1 Datasets

In this study, structural T1-weighted magnetic resonance imaging (MRI) scans were collected from six public databases, including DecNef [25], COBRE [1], CANDI [11],

MCICShare [7], UCLA CNP [3], and CCNMD [20]. A total of 956 subjects comprising 366 patients with schizophrenia and 590 healthy controls were included after quality control (for details, see Supplementary Materials). Ethical approvals and informed consents were obtained locally for each study, covering both participation and subsequent data sharing. Details about sample and demographic information are presented in Supplementary Table S1.

2.2 Morphometric Similarity Networks

The structural MRI scans from all subjects were preprocessed using the recon-all command from FreeSurfer (version 7.2.0) [6]. Cortical parcellation was based on an atlas with 308 cortical regions derived from the Desikan-Killiany atlas [21]. In each subject, this procedure generated 1540 morphometric features (308 regional measures of gray matter volume, surface area, cortical thickness, Gaussian curvature, and mean curvature). Morphometric features for each subject can be expressed as a set of 308 vectors of length 5. Each morphometric feature was standardized across 308 regions so that the data have a mean of zero and a standard deviation of one. A $N \times N$ correlation matrix, representing an individual MSN, was constructed by computing the Pearson's correlation coefficient between the z-normalized morphometric features of region i and region j for all (i, j) of regional properties. This procedure resulted in 308×308 MSN for each subject.

2.3 Graph Construction

We consider a population comprising N subjects; each subject being associated with a set of imaging and phenotypic information. We define the population graph as an undirected weighted graph $G = (V, E, W)$, where V is a set of $|V| = N$ nodes, E is a set of edges. $W \in \mathbb{R}^{N \times N}$ is a weighted adjacency matrix of population graph G. Each node v represents a subject, and the feature vectors for nodes are extracted from imaging data. Each edge is defined as the similarity between nodes. The two main decisions required to build the population model are the definition of the feature vector representing each node and the connectivity of the graph corresponding to its edges and weights.

Feature Vector. The MSN has a high dimensionality with 47,278 features considering only the upper triangle of the MSN. Using the Brain Connectivity Toolbox (BCT) [22], we computed the following graph-theoretical measures: strength, betweenness centrality, and clustering coefficient. These measures were considered as node features for subjects. The strength is the sum of weights of edges connected to the given node. The betweenness centrality is defined as the fraction of all shortest paths in the network that contain a given node. Nodes with high values of betweenness centrality participate in many shortest paths. The clustering coefficient is a measure of network segregation and is defined as the fraction of triangles around the given node.

We investigated two additional feature selection approaches. Firstly, we applied the ridge classifier to perform recursive feature elimination (RFE). RFE iteratively removes the irrelevant features to achieve the desired number of features while maximizing the ridge classification accuracy. Secondly, we used the k-best selection with ANOVA, a univariate feature selection algorithm that selects the k-best features with the highest F-values.

Graph Edge. The process of computing graph edges involved two steps: the calculation of initial edge weights and adaptive learning through encoders. The initial similarity between nodes were determined using imaging features and categorical information. Considering a set H categorical phenotypic measures $M = \{M_h\}$, such as sex and scan site, the initial edge weight W_I between node v and w are defined as follows:

$$W_I(v, w) = Sim(v, w) \sum_{h=1}^{H} \gamma(M_h(v), M_h(w)) \tag{1}$$

$$Sim(v, w) = exp\left(-\frac{[\rho(x(v), x(w)]^2}{2h^2}\right) \tag{2}$$

$$\gamma(M_h(v), M_h(w)) = \begin{cases} 0, if\ M_h(v) \neq M_h(w) \\ 1, if\ M_h(v) = M_h(w) \end{cases} \tag{3}$$

where $Sim(v, w)$ is a measure of similarity between node v and node w in the imaging features. γ is a measure of similarity between node v and node w in the categorical variables. ρ is the Pearson's correlation distance. $x(v)$ and $x(w)$ are the topological feature vectors of node v and w, respectively. h denotes the width of the kernel.

We set a threshold on $W_I(v, w)$ via quantile Q, , and adaptively calculated the edge weights only for the remaining edges. We investigated four different encoders (PAE, EA, L2, and Cosine + Tanh) that were used to determine the edge weights between node v and node w based on phenotypic information (e.g., sex, age, and scan site). The subject's normalized phenotypic inputs were projected into a latent space $\varphi \in \mathbb{R}^{64}$ using MLP. The pairwise association encoder (PAE) [9] scores the association between node v and w as follows: $W(v, w) = 0.5 * (S_C(\varphi_v, \varphi_w) + 1)$, where $S_C(\varphi_v, \varphi_w)$ is the cosine similarity between latent vector φ_v and φ_w. The Edge Adapter (EA) [8] scores the association between node v and w as follows: $W(v, w) = \sigma(\sum_{64} \alpha|\varphi_v - \varphi_w|)$, where $|\varphi_v - \varphi_w|$ is the absolute difference vector of two latent vectors. α is the 64-dimension trainable parameter used in fully connected layer and σ is the sigmoid function. We also designed the L2 encoder as follows: $W(v, w) = \sigma(1/||\varphi_v - \varphi_w||_2^2)$, where $||\varphi_v - \varphi_w||_2^2$ denotes L2 distance of two latent vectors. Finally, we designed the Cosine + Tanh encoder combining Tanh with cosine similarity as follows: $W(v, w) = Tanh(ReLU(S_C(\varphi_v, \varphi_w)))$. The encoders were optimized with graph convolution models using gradient descent algorithm.

2.4 Spectral Graph Convolutions

To extract spatial features from each graph node, we used the spectral graph convolution in the Fourier domain. A spectral convolution of signal x with a filter $g_\theta = diag(\theta)$ defined in the Fourier domain is defined as a multiplication in the Fourier domain: $g_\theta * x = U g_\theta U^T x$, where U is the matrix of the eigenvectors of the normalized Laplacian matrix. The normalized graph Laplacian $\mathcal{L}$ is defined as $\mathcal{L} = I - D^{-1/2} W D^{-1/2}$, where I and D are the identity matrix and the diagonal degree matrix, respectively. To reduce computational complexity of convolution, we used Chebyshev spectral convolutional network (ChebConv) [5] that approximates spectral graph convolution using Chebyshev polynomials. The Chebyshev polynomials are computed recursively as $T_k\left(\tilde{\mathcal{L}}\right) = 2\tilde{\mathcal{L}}T_{k-1}\left(\tilde{\mathcal{L}}\right) - T_{k-2}\left(\tilde{\mathcal{L}}\right)$, with $T_0\left(\tilde{\mathcal{L}}\right) = 1$ and $T_1\left(\tilde{\mathcal{L}}\right) = \tilde{\mathcal{L}}$. A K-order ChebConv is defined as $g_{\theta'} * x = \sum_{k=0}^{K} T_k(\tilde{\mathcal{L}})\theta'_k x$, where $\tilde{L}$ is the rescaled Laplacian and θ'_k are filter parameters. The output spatial features from l^{th} convolutional layer with input features $H^l \in \mathbb{R}^{N \times C^l}$ can be calculated as: $H^{l+1} = \sum_{k=0}^{K} T_k(\tilde{\mathcal{L}})H^l\Theta_k^l$, where $\Theta_k^l \in \mathbb{R}^{C^l \times C^{l+1}}$ are the trainable weights for the polynomial of order k in the lth layer.

The model consists of a GCN with L hidden layers. Each hidden layer is followed by a ReLU activation function to introduce non-linearity. The output layer is full connected and comprises two convolutional layers with 256 and 2 channels. The model was trained using the entire population graph. During training, the training nodes were labeled and the test nodes were masked. Cross-entropy loss computed on the training nodes was used to train the GCN and edge encoder. The performance of the model was evaluated using the test nodes.

2.5 Interpretability

We identified relevant node features that contribute to the classification using GNNExplainer [29]. Given a trained GCN model and its prediction on a test node, explainer maximizes the mutual information between a GCN's prediction and the distribution of subgraph structure. GNNExplainer identified a compact subgraph together with a small subset of node features that have a decisive role in the GCN's prediction. The node feature importance was measured by computing feature masks. Feature importance was determined as an average value across subjects to represent the contribution of each brain region. Finally, we reported the top 10 node features with the largest average value of feature importance.

3 Experiments and Results

Experimental Settings. We used pooled stratified cross-validation to split the samples into training and testing sets for the evaluation of the GCN. The pooled samples were randomly divided into 5-folds, of which 1-fold served as the testing set, and the remaining 4-folds were used as the training set. This strategy ensures that training and testing sets contain the equivalent proportions of each class. The model performance was evaluated on the testing set in terms of balanced accuracy (BAC), sensitivity (SEN), specificity (SPE), F1-score, and area under the receiver operating characteristic curve (AUC).

In our experiments, we set Chebyshev polynomial order $K = 3$, quantile $Q = 0.5$, and hidden layer $L = 3$. The number of channels was set to [128, 64, 32]. All models were trained using Adam optimizer with an initial learning rate 0.01, weight decay 5×10^{-5}, and dropout rate 0.2 for 300 epochs to avoid overfitting. Our model was implemented with the PyTorch framework and trained on NVIDIA RTX 3090 GPU.

Competing Methods. To validate the superiority of our proposed model, we compared the GCN model with traditional machine learning algorithms including support vector machine (SVM), random forest (RF), and K-nearest neighbor (KNN). The upper triangle of the MSN was used as input features. We tuned hyperparameters using 5-fold cross-validation to find the optimal parameters via grid search.

3.1 Results and Analysis

Results of Schizophrenia Classification. Table 1 presents the performance of our proposed model compared with those obtained by several machine learning models. The GCN model based on clustering coefficient achieved a superior performance compared with other machine learning models with an average accuracy of 80.2%.

Table 1. Classification comparisons between different methods.

Methods	BAC	SEN	SPE	F1-Score	AUC
RF	52.53 ± 1.17	56.37 ± 12.56	10.65 ± 1.56	17.77 ± 2.23	63.04 ± 2.19
SVM	54.03 ± 1.65	55.74 ± 8.62	17.22 ± 4.97	25.69 ± 5.92	61.04 ± 1.48
KNN	56.03 ± 2.81	50.35 ± 4.09	30.87 ± 7.48	37.81 ± 6.95	57.04 ± 2.58
Ours	**80.18 ± 2.85**	**70.32 ± 4.66**	**82.55 ± 8.44**	**75.50 ± 3.24**	**88.50 ± 1.83**

Identification of Discriminating Regions. Figure 2 shows the top 10 brain regions contributing to GCN classification. The most salient regions contributing to classification were primarily identified in the middle temporal gyrus, superior temporal gyrus, and inferior temporal gyrus. These findings suggest that these regions are crucial in distinguishing patients with schizophrenia from healthy controls. Detailed results are provided in Supplementary Table S2.

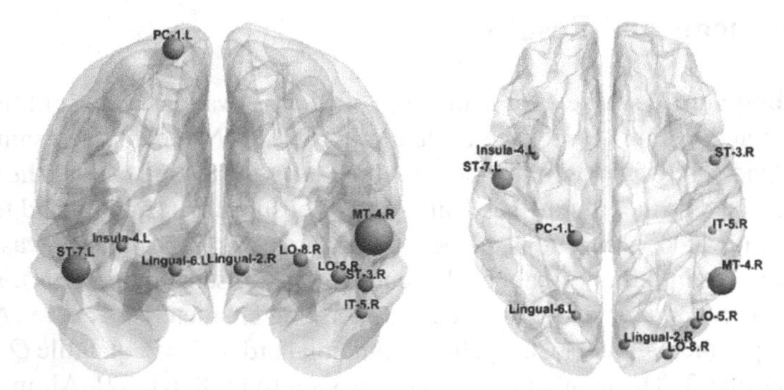

Fig. 2. The top 10 salient brain regions. Abbreviations: MT, middle temporal gyrus; ST, superior temporal gyrus; PC, postcentral gyrus; LO, lateral occipital cortex; IT, inferior temporal gyrus.

3.2 Ablation Studies

We provide detailed investigation of three key components (the density of MSNs, feature selection strategy, and edge encoder) and their influence on classification results.

Influence of the Connection Density of MSNs. We constructed a series of MSNs with different connection densities ranging from 10% to 100% in 10% increments. For each matrix, we computed strength, betweenness centrality, and clustering coefficient to examine their influence on classification performance. In Fig. 3(a), using GCN with strength provides decreased classification performance as the connection density increases. Betweenness centrality shows a lower performance as the density decreases. The use of clustering coefficient is most reliable and provides the highest performance (81.8% accuracy) at connection density of 30%.

Effect of the Feature Selection Strategy. We explored the influence of feature selection strategy on classification performance for the three methods: graph-theoretical measures, RFE, and k-best. The number of features for RFE and k-best considered was C = [308, 500, 1000]. Figure 3(b) shows that using graph-theoretical measures as node features yields improved classification performance. When using clustering coefficient, the GCN model achieved the best performance with 80.18% accuracy. The k-best method shows the lowest performance compared to the other methods.

Influence of the Encoder. We examined the influence of four different types of encoders (PAE, EA, L2, and Cosine + Tanh) on classification performance. Figure 3(c) presents that using PAE as an encoder provides the highest classification performance compared to the other encoders.

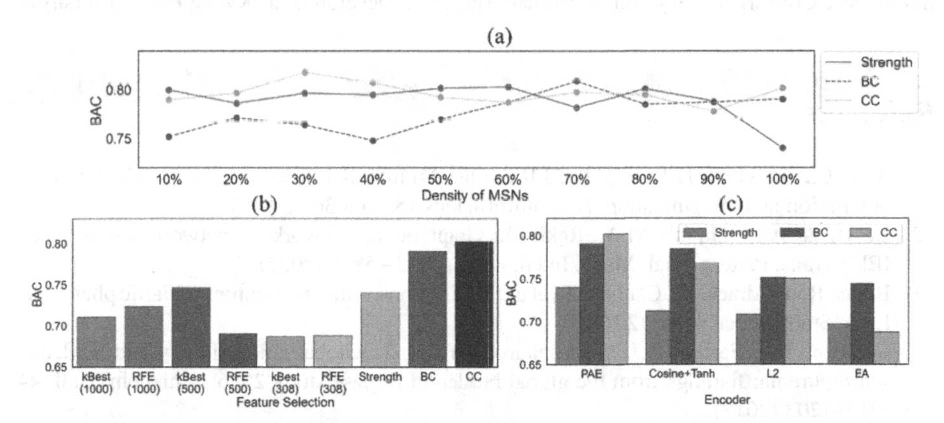

Fig. 3. Ablation studies on three different key components. Classification accuracy (a) for different density of MSNs, (b) for different features selection strategies, and (c) for different types of encoders. Abbreviations: BC, betweenness centrality; CC, clustering coefficient; EA, Edge Adapter.

4 Conclusion

In this study, we proposed an improved GCN structure, which combines graph-theoretical measures of MSNs derived from structural MRI data and non-imaging phenotypic measures for disease classification. This study explored the application value of our proposed GCN model based on cortical networks (MSNs) in differentiating patients with schizophrenia from healthy controls. Using a large multi-site dataset and various validation strategies, reliable and generalizable classification accuracy of 81.8% can be achieved. These results indicate that the MSNs serve as a useful and clinically-relevant phenotype and GCN modeling shows promise in detecting individual patients with schizophrenia. Further, we investigated different graph structures and their influence on classification performance. The GCN model making use of subject-specific clustering coefficient as imaging feature vectors and the PAE encoder performed best. By examining the saliency patterns contributing to GCN classification, we identified the most salient regions in the middle temporal gyrus and superior temporal gyrus. These findings suggest the potential utility of GCN for enhancing our understanding of the underlying neural mechanisms of schizophrenia by identifying clinically-relevant disruptions in brain network topology.

Acknowledgments. This work was supported by the National Research Foundation of Korea (NRF) grant funded by the Korea government (MSIT) (No. 2021R1C1C1009436), the Korea Health Technology R&D Project through the Korea Health Industry Development Institute (KHIDI) funded by the Ministry of Health and Welfare, Republic of Korea (grant number: HI22C0108), and the Institute for Information and Communications Technology Planning & Evaluation (IITP) grant funded by the Korea government (MSIT) (No. RS-2022–00155911, Artificial Intelligence Convergence Innovation Human Resources Development (Kyung Hee University)).

References

1. Aine, C., Bockholt, H.J., Bustillo, J.R., et al.: Multimodal neuroimaging in schizophrenia: description and dissemination. Neuroinformatics **15**, 343–364 (2017)
2. Bessadok, A., Mahjoub, M.A., Rekik, I.: Graph neural networks in network neuroscience. IEEE Trans. Pattern Anal. Mach. Intell. **45**(5), 5833−5848 (2022)
3. Bilder, R., Poldrack, R., Cannon, T., et al.: UCLA consortium for neuropsychiatric phenomics LA5c Study. OpenNeuro (2020)
4. Charlson, F.J., Ferrari, A.J., Santomauro, D.F., et al.: Global epidemiology and burden of schizophrenia: findings from the global burden of disease study 2016. Schizophr. Bull. **44**, 1195–1203 (2018)
5. Defferrard, M., Bresson, X., Vandergheynst, P.: Convolutional neural networks on graphs with fast localized spectral filtering. In: Advances in Neural Information Processing Systems, pp. 3844–3852. (2016)
6. Fischl, B.: FreeSurfer. Neuroimage **62**, 774–781 (2012)
7. Gollub, R.L., Shoemaker, J.M., King, M.D., et al.: The MCIC collection: a shared repository of multi-modal, multi-site brain image data from a clinical investigation of schizophrenia. Neuroinformatics **11**, 367–388 (2013)

8. Huang, Y., Albert, C.: Semi-supervised multimodality learning with graph convolutional neural networks for disease diagnosis. In: 2020 IEEE International Conference on Image Processing (ICIP), pp. 2451–2455 (2020)
9. Huang, Y., Chung, A.C.: Edge-variational graph convolutional networks for uncertainty-aware disease prediction. In: Medical Image Computing and Computer Assisted Intervention–MICCAI 2020: 23rd International Conference. Lima, Peru, October 4–8, 2020, Proceedings, Part VII 23, pp. 562–572. Springer, Cham (2020)
10. Jiang, H., Cao, P., Xu, M., Yang, J., Zaiane, O.: Hi-GCN: a hierarchical graph convolution network for graph embedding learning of brain network and brain disorders prediction. Comput. Biol. Med. **127**, 104096 (2020)
11. Kennedy, D.N., Haselgrove, C., Hodge, S.M., et al.: CANDIShare: a resource for pediatric neuroimaging data. Neuroinformatics **10**, 319–322 (2012)
12. Khodatars, M., Shoeibi, A., Sadeghi, D., et al.: Deep learning for neuroimaging-based diagnosis and rehabilitation of autism spectrum disorder: a review. Comput. Biol. Med. **139**, 104949 (2021)
13. Khundrakpam, B.S., Lewis, J.D., Jeon, S., et al.: Exploring individual brain variability during development based on patterns of maturational coupling of cortical thickness: a longitudinal MRI study. Cereb. Cortex **29**, 178–188 (2019)
14. Kong, R., Li, J., Orban, C., et al.: Spatial topography of individual-specific cortical networks predicts human cognition, personality, and emotion. Cereb. Cortex **29**, 2533–2551 (2019)
15. Larson, M.K., Walker, E.F., Compton, M.T.: Early signs, diagnosis and therapeutics of the prodromal phase of schizophrenia and related psychotic disorders. Expert Rev. Neurother. **10**, 1347–1359 (2010)
16. Lei, D., Qin, K., Pinaya, W.H., et al.: Graph convolutional networks reveal network-level functional dysconnectivity in schizophrenia. Schizophr. Bull. **48**, 881–892 (2022)
17. Morgan, S.E., Seidlitz, J., Whitaker, K.J., et al.: Cortical patterning of abnormal morphometric similarity in psychosis is associated with brain expression of schizophrenia-related genes. Proc. Natl. Acad. Sci. **116**, 9604–9609 (2019)
18. Parisot, S., Ktena, S.I., Ferrante, E., et al.: Disease prediction using graph convolutional networks: application to autism spectrum disorder and Alzheimer's disease. Med. Image Anal. **48**, 117–130 (2018)
19. Qin, K., Lei, D., Pinaya, W.H., et al.: Using graph convolutional network to characterize individuals with major depressive disorder across multiple imaging sites. EBioMedicine **78**, 103977 (2022)
20. Repovš, G., Barch, D.M.: Working memory related brain network connectivity in individuals with schizophrenia and their siblings. Front. Hum. Neurosci. **6**, 137 (2012)
21. Romero-Garcia, R., Atienza, M., Clemmensen, L.H., Cantero, J.L.: Effects of network resolution on topological properties of human neocortex. Neuroimage **59**, 3522–3532 (2012)
22. Rubinov, M., Sporns, O.: Complex network measures of brain connectivity: uses and interpretations. Neuroimage **52**, 1059–1069 (2010)
23. Sadeghi, D., Shoeibi, A., Ghassemi, N., et al.: An overview of artificial intelligence techniques for diagnosis of Schizophrenia based on magnetic resonance imaging modalities: methods, challenges, and future works. Comput. Biol. Med. 105554 (2022)
24. Seidlitz, J., Váša, F., Shinn, M., et al.: Morphometric similarity networks detect microscale cortical organization and predict inter-individual cognitive variation. Neuron **97**, 231–247 (2018)
25. Tanaka, S.C., Yamashita, A., Yahata, N., et al.: A multi-site, multi-disorder resting-state magnetic resonance image database. Scientific data **8**, 227 (2021)
26. Váša, F., Seidlitz, J., Romero-Garcia, R., et al.: Adolescent tuning of association cortex in human structural brain networks. Cereb. Cortex **28**, 281–294 (2018)

27. Wen, G., Cao, P., Bao, H., et al.: MVS-GCN: A prior brain structure learning-guided multi-view graph convolution network for autism spectrum disorder diagnosis. Comput. Biol. Med. **142**, 105239 (2022)
28. Winterburn, J.L., Voineskos, A.N., Devenyi, G.A., et al.: Can we accurately classify schizophrenia patients from healthy controls using magnetic resonance imaging and machine learning? a multi-method and multi-dataset study. Schizophr. Res. **214**, 3–10 (2019)
29. Ying, Z., Bourgeois, D., You, J., Zitnik, M., Leskovec, J.: Gnnexplainer: generating explanations for graph neural networks. Adv. Neural Inf. Process. Syst. 9244–9255 (2019)
30. Zhao, F., Zhang, H., Rekik, I., An, Z., Shen, D.: Diagnosis of autism spectrum disorders using multi-level high-order functional networks derived from resting-state functional MRI. Front. Hum. Neurosci. **12**, 184 (2018)
31. Zheng, W., Eilam-Stock, T., Wu, T., et al.: Multi-feature based network revealing the structural abnormalities in autism spectrum disorder. IEEE Trans. Affect. Comput. **12**, 732–742 (2019)
32. Zheng, W., Yao, Z., Xie, Y., Fan, J., Hu, B.: Identification of Alzheimer's disease and mild cognitive impairment using networks constructed based on multiple morphological brain features. Biol. Psychiatry: Cogn. Neurosci. Neuroimaging **3**, 887–897 (2018)

M-FLAG: Medical Vision-Language Pre-training with Frozen Language Models and Latent Space Geometry Optimization

Che Liu[1,2(✉)], Sibo Cheng[2,3], Chen Chen[3,5], Mengyun Qiao[2,4], Weitong Zhang[3], Anand Shah[6,7], Wenjia Bai[2,3,4], and Rossella Arcucci[1,2]

[1] Department of Earth Science and Engineering, Imperial College London, London, UK
che.liu21@imperial.ac.uk

[2] Data Science Institute, Imperial College London, London, UK

[3] Department of Computing, Imperial College London, London, UK

[4] Department of Brain Sciences, Imperial College London, London, UK

[5] Department of Engineering Science, University of Oxford, Oxford, UK

[6] Department of Infectious Disease Epidemiology, Imperial College London, London, UK

[7] Royal Brompton and Harefield Hospitals, London, UK

Abstract. Medical vision-language models enable co-learning and integrating features from medical imaging and clinical text. However, these models are not easy to train and the latent representation space can be complex. Here we propose a novel way for pre-training and regularising medical vision-language models. The proposed method, named **M**edical vision-language pre-training with **F**rozen language models and **L**atent sp**A**ce **G**eometry optimization (M-FLAG), leverages a frozen language model for training stability and efficiency and introduces a novel orthogonality loss to harmonize the latent space geometry. We demonstrate the potential of the pre-trained model on three downstream tasks: medical image classification, segmentation, and object detection. Extensive experiments across five public datasets demonstrate that M-FLAG significantly outperforms existing medical vision-language pre-training approaches and reduces the number of parameters by 78%. Notably, M-FLAG achieves outstanding performance on the segmentation task while using only 1% of the RSNA dataset, even outperforming ImageNet pre-trained models that have been fine-tuned using 100% of the data. The code can be found in https://github.com/cheliu-computation/M-FLAG-MICCAI2023.

Keywords: Vision-language model · Vision-language pre-training · Self-supervised learning

H. Greenspan et al. (Eds.): MICCAI 2023, LNCS 14220, pp. 637–647, 2023.
https://doi.org/10.1007/978-3-031-43907-0_61

1 Introduction

Deep learning has made significant progress in medical computer vision [2,7] but requires large annotated datasets, which are often difficult to obtain. Self-supervised learning (SSL) offers a solution by utilizing large unannotated medical image sets. It also enables vision-language pre-training (VLP), which learns representations for both imaging and text data and their relationships [5,21,26]. Several recent medical VLP approaches such as ConVIRT [32], GLoRIA [11], and MGCA [27] have shown the effectiveness of model pre-training with medical images and radiology reports together, which outperformed the conventionally pre-trained models using image only in downstream tasks [32]. However, training such models is not an easy task as they require extensive resources for training both vision and language models. In particular, most VLP approaches are based on pre-trained BERT [6,18], whose parameters are 5 times larger than a standard ResNet50 [10]. This indicates high computational cost, as well as training complexity and potential instability in joint training [13]. On the other hand, previous works [21,27,32] suggest a training strategy that forces image latent space to match language latent space, which can be sub-optimal with latent space collapse problem [14], reducing its performance for downstream tasks [34]. In this work, we would like to answer the following two questions: *(1) Is it necessary to tune pre-trained language models for medical VLP? (2) How to regularize the latent space in pre-training?*

We propose a novel VLP framework named **M**edical vision-language pre-training with **F**rozen language models and **L**atent sp**A**ce Geometry optimization method (M-FLAG). Different from most existing VLP approaches, M-FLAG is computationally efficient as it only requires training the vision model, while keeping the language model frozen. To harmonize the latent spaces in vision and language models, we relax the visual-language alignment objective with a orthogonality loss to alleviate the latent space collapse problem. The main contributions of this work include: **(1)** To the best of our knowledge, this is the first work to explore the collapsed latent space problem in medical VLP. **(2)** A novel and effective VLP framework is proposed to alleviate the collapsed latent space problem by explicitly optimizing the latent geometry towards orthogonal using our orthogonality loss in addition to the visual-language alignment loss, encouraging the in-dependency between latent variables and maximizing its informativeness for downstream tasks. **(3)** M-FLAG consistently outperforms existing medical VLP methods on three downstream tasks: medical image classification, segmentation, and object detection, while reducing 78% trainable parameters due to the frozen language model strategy.

Related Works: To connect vision and language modalities, the idea of VLP was proposed in CLIP [21], which involves learning mutual knowledge from two modalities by maximizing their feature similarity. CLIP [21] and later FLIP [19] focus on learning cross-representation in natural language and images. However, there is a significant lack of research in the medical domain due to the complexity of the medical text and the limited availability of large-scale paired medical image-text datasets. Recently, ConVIRT [32], GLoRIA [11], and MGCA [27] have made notable progress in aligning medical text and images. These methods require significant computational resources and are sometimes limited by the collapse issue of the latent space.

It has been suggested that optimal vision and language latent spaces should be of different geometry [8] and latent space uniformity is considered an essential indicator to evaluate the success of learning [28]. Yet, most existing VLP approaches rigorously align the vision latent space to the language space without considering the latent space geometry, which may lead to latent space collapse. As pointed out by [14], latent space collapse indicates significant information loss, which can crucially affect the robustness of the pre-trained model on downstream tasks when transferring the model to unseen domains [3]. To solve this problem, contrastive learning-based approaches can be used to spread visual features over the unit sphere with good uniformity [4,9]. However, it requires a large number of negative samples in each training batch, which inevitably increases computational costs. Differently, here we address this problem by employing a orthogonality loss, which directly aligns the geometry of the latent space towards a uniform hypersphere to tackle the collapse problem.

2 Methods

The proposed M-FLAG is a simple and light VLP framework that aims to learn visual and text representations by leveraging both medical images and radiology

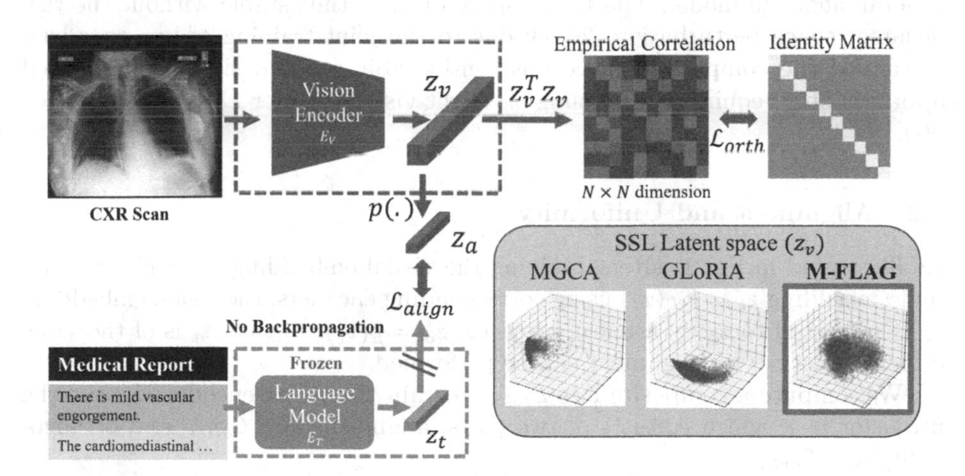

Fig. 1. M-FLAG overview. M-FLAG consists of a vision encoder E_V for learning vision latent z_v, a *frozen* language model E_T for extracting medical text latent z_t, and a projector $p(\cdot)$ to map z_v to z_a for alignment with z_t. M-FLAG employs an alignment loss $\mathcal{L}_{align}$ for vision-text latent space alignment between z_a and z_t and a orthogonality loss $\mathcal{L}_{orth}$ to encourage the orthogonality of z_v (Sect. 2.2). Visualization of the first 3 dominant dimensions of latent space z_v via PCA [30] shows the M-FLAG alleviates the dimensional collapse in the latent space, while MGCA [27] and GLoRIA [11] suffer the problem to different extents.

reports. We employ a *freeze* strategy for the text encoder E_T to mitigate ambiguity in vision-text latent space alignment. Additionally, we explicitly optimize the latent space geometry using a orthogonality loss. By doing so, we encourage the visual latent space to keep a stable geometry and reduce the risk of collapse. Figure 1 illustrates the workflow of M-FLAG and the learned latent space geometry compared to two recent medical VLP approaches.

Vision Encoder and Frozen Text Encoder: The paired medical image and text are denoted as x_v, x_t, respectively. As illustrated in Fig. 1, we obtain the image embedding z_v via the vision encoder E_V and the text embedding z_t via the frozen text encoder E_T.

Vision Embedding: E_V, the vision embedding $z_v \in \mathbb{R}^{B \times N}$ is extracted from the last pooling layer of E_V. N denotes the dimension of the latent space and B represents the batch size.

Text Embedding: A text encoder E_T extracts text embedding of word tokens from a medical report. Similar to BERT [6], a special token $[cls]$ is added, which aggregates the representations of all word tokens into one embedding. The embedding of $[cls]$ is used as the text report representation and denoted as z_t.

2.1 Frozen Language Model

In this work, we use a *frozen* text encoder E_T, which can be obtained from any general language model. The latent space of z_v is thus stable without the risk of latent space perturbation [20,33] due to the joint training of two encoders. Naturally, the computational cost is considerably reduced since the proposed approach only requires the training of a light vision encoder E_V and a projector $p(\cdot)$.

2.2 Alignment and Uniformity

As illustrated in Fig. 1, after obtaining the visual embedding $z_v = E_V(x_v)$ and text embedding $z_t = E_T(x_t)$ using corresponding encoders, the vision embedding z_v is projected to z_a by a linear projector $z_a = p(z_v)$, so that z_a is of the same dimension as z_t and alignment can be performed.

We compute a composite loss $\mathcal{L}_{total}$ to train the vision encoder E_V and the projector $p(\cdot)$, which consists of two parts, alignment loss $\mathcal{L}_{align}$ and orthogonality loss $\mathcal{L}_{orth}$:

$$\mathcal{L}_{total} = \mathcal{L}_{align} + \mathcal{L}_{orth} \tag{1}$$

$$\mathcal{L}_{align} = ||\bar{z_a} - \bar{z_t}||_2^2 = 2 - 2\bar{z_a}^T, \bar{z_t} \tag{2}$$

$$\mathcal{L}_{orth} = \sum_{i=1} \left(1 - (\bar{z_v}^T \cdot \bar{z_v})_{ii}\right)^2 + \sum_{i \neq j} (\bar{z}_v^T \cdot \bar{z}_v)_{ij}^2, \tag{3}$$

where $\{i, j\} \in \{1, ..., \dim(z_v)\}^2$. We implement ℓ_2-normalization on z_a, z_t, z_v to obtain $\bar{z_a}, \bar{z_t}, \bar{z_v}$. $\mathcal{L}_{align}$ minimizes the discrepancy between $\bar{z_a}$ and $\bar{z_t}$, while $\mathcal{L}_{orth}$ maximizes the independence among latent features in $\bar{z_v}$, forcing its empirical correlation matrix to be an identity matrix. In other words, we expect different latent feature dimensions to be independent. The objective of the first term on the right side in Eq. (3) aims to optimize the diagonal elements of the empirical correlation matrix to 1, while the second term on the right side aims to reduce all non-diagonal elements to 0. Here, $(\cdot)^T$ denotes the matrix transpose operation.

3 Experiments

3.1 Dataset for Pre-training

M-FLAG is pre-trained on the MIMIC-CXR (MIMIC) dataset [15,16], which contains 227,827 image-text pairs with chest X-ray (CXR) images and radiology reports. Following the preprocessing procedure of [11,27,32], 213,384 image-text pairs are used. We use ResNet50 [10] as E_V and frozen CXR-BERT [1] as E_T. Pre-training takes 100 epochs on 8 A100 GPUs, with a batch size of 128 for each GPU and a learning rate of 0.001 using the LARS [31] optimizer.

3.2 Datasets for Downstream Tasks

The pre-trained model is evaluated on 3 downstream tasks across 5 datasets:

Medical image classification is implemented on MIMIC, CheXpert (CXP), and NIH [12,16,29] datasets, each consisting of images from 14 disease categories.

To reduce sampling bias and maintain consistency, we follow the dataset split in CheXclusion [24] and evaluate the macro AUC scores.

Image segmentation is evaluated on two datasets, RSNA [25] (pneumonia segmentation) and SIIM [17] (pneumothorax segmentation). Following [11,27], we use U-Net [23] as the segmentation backbone. The pre-trained model is used as the frozen encoder of the U-net [23] and we only update the decoder of the U-net during fine-tuning. We evaluate segmentation performance using Dice scores.

Object detection is implemented on the RSNA [25] dataset for pneumonia detection, using the preprocessing techniques outlined in [27]. Following [27], we use YOLOv3 [22] as the detection framework. We employ the pre-trained vision encoder of M-FLAG as the backbone and only fine-tune the detection head. The detection task is evaluated using mean average precision (mAP) with the intersection of union (IoU) thresholds ranging from 0.4 to 0.75.

Table 1 reports the data split details. For all downstream tasks, we fine-tune using 1%, 10%, 100% of the train data on a single A100 GPU.

Table 1. Datasets are split following [11,24,27].

Task	Dataset	Split	Train	Valid	Test
Classification	MIMIC [16]	[24]	215,187	5,000	23,137
	CXP [12]	[24]	167,185	5,000	19,027
	NIH [29]	[24]	100,747	5,000	6,373
Segmentation	RSNA [25]	[11,27]	16,010	5,337	5,337
	SIIM [17]	[11,27]	8,433	1,807	1,807
Detection	RSNA [25]	[27]	16,010	5,337	5,337

3.3 Results

Medical Image Classification: The AUC scores on MIMIC, CXP, and CXR14 are reported in Table 2. It shows that M-FLAG consistently outperforms all baseline methods across almost all datasets and data fractions. Notably, M-FLAG achieves superior performance while using only 22% of the trainable parameters compared to other methods. While MGCA [27] slightly outperforms our method only when fine-tuning on 10% of the CXP dataset, it requires more than five times parameters than M-FLAG.

Segmentation and Object Detection: Table 3 shows that M-FLAG outperforms all SOTA methods across all datasets and data fractions in segmentation and detection tasks. In the segmentation task, M-FLAG achieves the highest Dice score across all fractions of both the SIIM and RSNA datasets. Interestingly, even when fine-tuned with only 1% of the data in RSNA, M-FLAG outperforms the ImageNet pre-trained model fine-tuned with 100% of the data. Similarly, in the object detection task, M-FLAG achieves the highest mean average precision (mAP) across all data fractions of the RSNA dataset. When fine-tuned with only 10% of the data, M-FLAG still outperforms the ImageNet pre-trained model with 100% fine-tuning.

These results indicate the advantages of using a frozen language model and introducing orthogonality loss during pre-training, which may yield more informative latent representations that are better suited for downstream tasks. Overall, the improvements achieved by M-FLAG across diverse downstream tasks demonstrate its effectiveness and versatility in medical image analysis.

3.4 Dimensional Collapse Analysis

Recent studies [14,28] have highlighted that latent space learned via self-supervised learning can suffer from issues such as complete collapse or dimensional collapse, which would lead to poor performance for downstream tasks. Figure 1 bottom right panel shows that both MGCA and GLoRIA [11,27] suffer from dimensional collapse. Figure 2 shows that if the last n layers of the language model in M-FLAG are not frozen, the latent space geometry would also exhibit varying degrees of collapse. This indicates the importance of using

Table 2. AUC scores (%) of image classification tasks on MIMIC, CXP, NIH datasets with 1%, 10%, 100% labeled data.

Method	Trainable parameters(M)	MIMIC			CXP			NIH		
		1%	10%	100%	1%	10%	100%	1%	10%	100%
Random	38.3	53.6	66.5	78.2	62.6	69.0	76.9	56.4	67.1	76.9
ImageNet	38.3	67.8	70.5	79.3	63.7	70.7	77.7	59.7	68.9	78.1
ConVIRT [32]	110.3	67.8	73.4	80.1	63.2	71.3	77.7	60.0	69.0	76.6
GLoRIA [11]	113.1	67.5	72.6	80.1	62.9	69.0	77.8	60.1	71.2	77.7
MGCA [27]	113.4	68.4	74.4	**80.2**	63.4	**72.1**	78.1	61.1	67.8	77.3
M-FLAG	**25.6**	**69.5**	**74.8**	**80.2**	**64.4**	71.4	**78.1**	**62.2**	**71.6**	**78.7**

Table 3. Dice (%) of segmentation tasks on SIIM, RSNA datasets. mAP (%) of detection task on RSNA dataset. All tasks are fine-tuned with 1%, 10%, 100% labeled data.

Method	Segmentation						Object Detection		
	SIIM (Dice%)			RSNA (Dice%)			RSNA (mAP%)		
	1%	10%	100%	1%	10%	100%	1%	10%	100%
Random	9.0	28.6	54.3	6.9	10.6	18.5	1.0	4.0	8.9
ImageNet	10.2	35.5	63.5	34.8	39.9	64.0	3.6	8.0	15.7
ConVIRT [32]	25.0	43.2	59.9	55.0	67.4	67.5	8.2	15.6	17.9
GLoRIA [11]	37.4	57.1	64.0	60.3	68.7	68.3	11.6	16.1	24.8
MGCA [27]	49.7	59.3	64.2	63.0	68.3	69.8	12.9	16.8	24.9
M-FLAG	**52.5**	**61.2**	**64.8**	**64.6**	**69.7**	**70.5**	**13.7**	**17.5**	**25.4**

a frozen language model. Quantitative results in Tables 2, 3, 4 and 5 and qualitative visualization in Figs. 1 and 2 further demonstrate that a collapsed latent space can impair the performance for various downstream tasks, especially for segmentation and detection. These findings highlight the usefulness of a frozen language model in preventing latent space collapse.

3.5 Ablation Study

Ablation Study: Table 4 presents the results of an ablation study to evaluate the impact of $\mathcal{L}_{orth}$ and $\mathcal{L}_{align}$ on model performance. Across all experiments, the proposed version of M-FLAG achieves the highest performance, with a clear advantage over implementations that only use $\mathcal{L}_{orth}$ or $\mathcal{L}_{align}$ in pre-training. The performance of the model pre-trained with only $\mathcal{L}_{align}$ drops dramatically in segmentation and detection tasks, although less severe in the classification tasks. On the other hand, the model pre-trained with only $\mathcal{L}_{orth}$ does not suffer severe performance drop across the three tasks, indicating that the uniform latent space could have a considerable contribution to the performance of M-FLAG. Overall, these results underscore the importance of both loss functions in M-FLAG and highlight their complementary contributions.

Table 4. Performance for ablation study of M-FLAG. "only $\mathcal{L}_{orth}/\mathcal{L}_{align}$" indicates that M-FLAG is pre-trained only with $\mathcal{L}_{orth}/\mathcal{L}_{align}$.

Method	MIMIC	CXP	NIH	SIIM	RSNA
	AUC (%)	AUC (%)	AUC (%)	Dice (%)	mAP (%)
	1%	1%	1%	1%	1%
only $\mathcal{L}_{align}$	69.3	62.6	61.4	45.7	12.1
only $\mathcal{L}_{orth}$	68.6	61.5	61.2	50.5	13.2
M-FLAG	**69.5**	**64.4**	**62.2**	**52.5**	**13.7**

Table 5. Performance of M-FLAG compared to its unfrozen variants. Unfreeze$_n$ indicates that the last n layers of the language model are unfrozen.

Method	Trainable Parameters(M)	MIMIC	CXP	NIH	SIIM	RSNA
		AUC (%)	AUC (%)	AUC (%)	Dice (%)	mAP (%)
		1%	1%	1%	1%	1%
Unfreeze$_1$	32.6	67.8	63.1	59.7	47.2	12.5
Unfreeze$_2$	39.7	68.7	63.3	60.6	48.9	12.3
Unfreeze$_3$	46.8	68.8	63.7	60.7	45.8	10.7
Unfreeze$_4$	53.9	68.7	62.6	60.1	50.3	11.4
Unfreeze$_5$	60.9	68.2	64.1	59.2	46.8	11.8
Unfreeze$_6$	68.1	68.2	63.7	59.9	50.1	11.5
M-FLAG	**25.6**	**69.5**	**64.4**	**62.2**	**52.5**	**13.7**

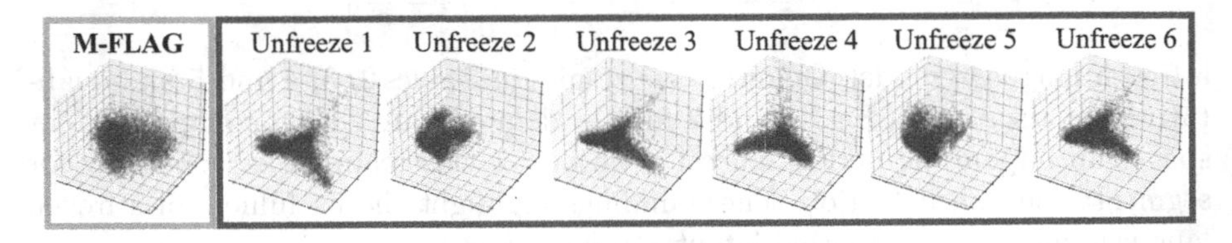

Fig. 2. Visualization of the first 3 dominant PCA dimensions of latent space on NIH dataset. M-FLAG (green) is compared to its variants (red) when the last n layers of the language model are not frozen. (Color figure online)

Comparing M-FLAG with Frozen vs. Unfrozen Language Models: We conducted further experiments to evaluate the performance of M-FLAG while unfreezing the last few layers of the language model. This not only increases the number of trainable parameters but also influences the model performance. Table 5 shows that when the language model is unfrozen, the performance slightly drops, compared to M-FLAG with the frozen language model (proposed). M-FLAG achieves a better performance with an average improvement of 2.18% than its Unfreeze$_{1-6}$ variants on the NIH dataset and an average improvement of 4.32% on the SIIM dataset.

4 Conclusion

Simple architecture means low computational cost and stable training. In this work, we propose a simple and efficient VLP framework that includes a frozen language model and a latent space orthogonality loss function. Extensive experiments show that M-FLAG outperforms SOTA medical VLP methods with 78% fewer parameters. M-FLAG also demonstrates its robustness by achieving the highest performance when transferred to unseen test sets and diverse downstream tasks for medical image classification, segmentation, and detection. This indicates the benefits of freezing the language model and regularizing the latent space. The results exhibit promising potential for improving the pre-training of vision-language models in the medical domain. In addition, the latent space geometry explored in this work provides useful insight for future work in VLP.

Acknowledgement. C. Liu and R. Arcucci were supported in part by EPSRC grant EP/T003189/1 Health assessment across biological length scales for personal pollution exposure and its mitigation (INHALE), EPSRC Programme Grant PREMIERE (EP/T000414/1). W. Bai and M. Qiao were supported by EPSRC Project Grant DeepGeM (EP/W01842X/1). A. Shah was supported by a MRC Clinical Academic Research Partnership award (MR/TOO5572/1) and by an MRC centre grant MRC (MR/R015600/1).

References

1. Boecking, B., Usuyama, N., Bannur, S., Castro, D.C., Schwaighofer, A., et al.: Making the most of text semantics to improve biomedical vision-language processing. In: Avidan, S., Brostow, G., Cissé, M., Farinella, G.M., Hassner, T. (eds.) Computer Vision – ECCV 2022. ECCV 2022. LNCS, vol. 13696, pp. 1–21. Springer, Cham (2022). https://doi.org/10.1007/978-3-031-20059-5_1
2. Chai, J., Zeng, H., Li, A., Ngai, E.W.: Deep learning in computer vision: a critical review of emerging techniques and application scenarios. Mach. Learn. Appl. **6**, 100134 (2021)
3. Chen, M., et al.: Perfectly balanced: improving transfer and robustness of supervised contrastive learning. In: International Conference on Machine Learning (2022)
4. Chen, T., Kornblith, S., Norouzi, M., Hinton, G.: A simple framework for contrastive learning of visual representations. In: International Conference on Machine Learning (2020)
5. Chen, Y., Liu, C., Huang, W., Cheng, S., Arcucci, R., Xiong, Z.: Generative text-guided 3D vision-language pretraining for unified medical image segmentation. arXiv preprint arXiv:2306.04811 (2023)
6. Devlin, J., Chang, M.W., Lee, K., Toutanova, K.: BERT: pre-training of deep bidirectional transformers for language understanding. In: NAACL-HLT (2019)
7. Esteva, A., Chou, K., Yeung, S., Naik, N., Madani, A., et al.: Deep learning-enabled medical computer vision. NPJ Digital Med. **4**(1), 1–9 (2021)
8. Fu, Y., Lapata, M.: Latent topology induction for understanding contextualized representations. arXiv preprint arXiv:2206.01512 (2022)

9. He, K., Fan, H., Wu, Y., Xie, S., Girshick, R.: Momentum contrast for unsupervised visual representation learning. In: IEEE/CVF Conference on Computer Vision and Pattern Recognition (2020)
10. He, K., Zhang, X., Ren, S., Sun, J.: Deep residual learning for image recognition. In: IEEE Conference on Computer Vision and Pattern Recognition (2016)
11. Huang, S.C., Shen, L., Lungren, M.P., Yeung, S.: GLoRIA: a multimodal global-local representation learning framework for label-efficient medical image recognition. In: IEEE/CVF International Conference on Computer Vision (2021)
12. Irvin, J., Rajpurkar, P., Ko, M., Yu, Y., Ciurea-Ilcus, S., et al.: CheXpert: a large chest radiograph dataset with uncertainty labels and expert comparison. In: AAAI Conference on Artificial Intelligence (2019)
13. Izsak, P., Berchansky, M., Levy, O.: How to train BERT with an academic budget. arXiv preprint arXiv:2104.07705 (2021)
14. Jing, L., Vincent, P., LeCun, Y., Tian, Y.: Understanding dimensional collapse in contrastive self-supervised learning. In: International Conference on Learning Representations (2021)
15. Johnson, A.E., et al.: MIMIC-CXR, a de-identified publicly available database of chest radiographs with free-text reports. Sci. Data **6**, 317 (2019)
16. Johnson, A.E., Pollard, T.J., Greenbaum, N.R., Lungren, M.P., Deng, C.Y., et al.: MIMIC-CXR-JPG, a large publicly available database of labeled chest radiographs. arXiv:1901.07042 (2019)
17. Langer, S.G., Shih, G.: SIIM-ACR Pneumothorax Segmentation (2019)
18. Li, J., Liu, C., Cheng, S., Arcucci, R., Hong, S.: Frozen language model helps ECG zero-shot learning. arXiv preprint arXiv:2303.12311 (2023)
19. Li, Y., Fan, H., Hu, R., Feichtenhofer, C., He, K.: Scaling language-image pre-training via masking. arXiv preprint arXiv:2212.00794 (2022)
20. Quan, D., et al.: Deep feature correlation learning for multi-modal remote sensing image registration. IEEE Trans. Geosci. Remote Sens. **60**, 1–16 (2022)
21. Radford, A., Kim, J.W., Hallacy, C., Ramesh, A., Goh, G., et al.: Learning transferable visual models from natural language supervision. In: International Conference on Machine Learning (2021)
22. Redmon, J., Farhadi, A.: YOLOv3: an incremental improvement. arXiv preprint arXiv:1804.02767 (2018)
23. Ronneberger, O., Fischer, P., Brox, T.: U-Net: convolutional networks for biomedical image segmentation. In: Navab, N., Hornegger, J., Wells, W.M., Frangi, A.F. (eds.) MICCAI 2015. LNCS, vol. 9351, pp. 234–241. Springer, Cham (2015). https://doi.org/10.1007/978-3-319-24574-4_28
24. Seyyed-Kalantari, L., Liu, G., McDermott, M., Chen, I.Y., Ghassemi, M.: CheXclusion: fairness gaps in deep chest X-ray classifiers. In: Biocomputing (2021)
25. Shih, G., Wu, C.C., Halabi, S.S., Kohli, M.D., Prevedello, L.M., et al.: Augmenting the national institutes of health chest radiograph dataset with expert annotations of possible Pneumonia. Radiol. Artif. Intell. **1**(1), e180041 (2019)
26. Wan, Z., et al.: Med-UniC: unifying cross-lingual medical vision-language pre-training by diminishing bias. arXiv preprint arXiv:2305.19894 (2023)
27. Wang, F., Zhou, Y., Wang, S., Vardhanabhuti, V., Yu, L.: Multi-granularity cross-modal alignment for generalized medical visual representation learning. Neural Inf. Process. Syst. **35**, 33536–33549 (2022)
28. Wang, T., Isola, P.: Understanding contrastive representation learning through alignment and uniformity on the hypersphere. In: International Conference on Machine Learning (2020)

29. Wang, X., Peng, Y., Lu, L., Lu, Z., Bagheri, M., Summers, R.M.: ChestX-ray8: hospital-scale chest X-ray database and benchmarks on weakly-supervised classification and localization of common thorax diseases. In: IEEE Conference on Computer Vision and Pattern Recognition (2017)
30. Wold, S., Esbensen, K., Geladi, P.: Principal component analysis. Chemom. Intell. Lab. Syst. **2**(1–3), 37–52 (1987)
31. You, Y., Li, J., Reddi, S., Hseu, J., Kumar, S., et al.: Large batch optimization for deep learning: training BERT in 76 minutes. In: International Conference on Learning Representations (2020)
32. Zhang, Y., Jiang, H., Miura, Y., Manning, C.D., Langlotz, C.P.: Contrastive learning of medical visual representations from paired images and text. arXiv preprint arXiv:2010.00747 (2020)
33. Zhou, T., Ruan, S., Canu, S.: A review: deep learning for medical image segmentation using multi-modality fusion. Array **3**, 100004 (2019)
34. Zhu, J.Y., et al.: Toward multimodal image-to-image translation. In: Advances in Neural Information Processing Systems (2017)

Machine Learning - Transfer Learning

Foundation Ark: Accruing and Reusing Knowledge for Superior and Robust Performance

DongAo Ma[1], Jiaxuan Pang[1], Michael B. Gotway[2], and Jianming Liang[1](✉)

[1] Arizona State University, Tempe, AZ 85281, USA
{dongaoma,jpang12,jianming.liang}@asu.edu
[2] Mayo Clinic, Scottsdale, AZ 85259, USA
Gotway.Michael@mayo.edu

Abstract. Deep learning nowadays offers expert-level and sometimes even super-expert-level performance, but achieving such performance demands massive annotated data for training (*e.g.,* Google's *proprietary* CXR Foundation Model (CXR-FM) was trained on 821,544 *labeled* and mostly *private* chest X-rays (CXRs)). *Numerous* datasets are *publicly* available in medical imaging but individually *small* and *heterogeneous* in expert labels. We envision a powerful and robust foundation model that can be trained by aggregating numerous small public datasets. To realize this vision, we have developed **Ark**, a framework that **a**ccrues and **r**euses **k**nowledge from *heterogeneous* expert annotations in various datasets. As a proof of concept, we have trained two Ark models on 335,484 and 704,363 CXRs, respectively, by merging several datasets including ChestX-ray14, CheXpert, MIMIC-II, and VinDr-CXR, evaluated them on a wide range of imaging tasks covering both classification and segmentation via fine-tuning, linear-probing, and gender-bias analysis, and demonstrated our Ark's superior and robust performance over the state-of-the-art (SOTA) fully/self-supervised baselines and Google's proprietary CXR-FM. This enhanced performance is attributed to our simple yet powerful observation that aggregating numerous public datasets diversifies patient populations and accrues knowledge from diverse experts, yielding unprecedented performance yet saving annotation cost. With all codes and pretrained models released at GitHub.com/JLiangLab/Ark, we hope that Ark exerts an important impact on open science, as accruing and reusing knowledge from expert annotations in public datasets can potentially surpass the performance of proprietary models trained on unusually large data, inspiring many more researchers worldwide to share codes and datasets to build open foundation models, accelerate open science, and democratize deep learning for medical imaging.

Supplementary Information The online version contains supplementary material available at https://doi.org/10.1007/978-3-031-43907-0_62.

H. Greenspan et al. (Eds.): MICCAI 2023, LNCS 14220, pp. 651–662, 2023.
https://doi.org/10.1007/978-3-031-43907-0_62

Keywords: Accruing and Reusing Knowledge · Large-scale Pretraining

1 Introduction

Deep learning nowadays offers expert-level and sometimes even super-expert-level performance, deepening and widening its applications in medical imaging and resulting in numerous public datasets for research, competitions, and challenges. These datasets are generally small as annotating medical images is challenging, but achieving superior performance by deep learning demands massive annotated data for training. For example, Google's *proprietary* CXR Foundation Model (CXR-FM) was trained on 821,544 *labeled* and mostly *private* CXRs [16]. We hypothesize that powerful and robust *open* foundation models can be trained by aggregating numerous small *public* datasets. To test this hypothesis, we have chosen CXRs because they are one of the most frequently used modalities, and our research community has accumulated copious CXRs (see Table 1). However, annotations associated with these public datasets are inconsistent in disease coverage. Even when addressing the same clinical issue, datasets created at different institutions tend to be annotated differently. For example, VinDr-CXR [13] is associated with global (image-level) and local (boxed-lesions) labels, while MIMIC-CXR [4] has no expert labels *per se* but comes with radiology reports. ChestX-ray14 [19] and CheXpert [4] both cover 14 conditions at the image level, and their 14 conditions have overlaps but are not exactly the same. Therefore, this paper seeks to address a critical need: *How to utilize a large number of publicly-available images from different sources and their readily-accessible but heterogeneous expert annotations to pretrain generic source (foundation) models that are more robust and transferable to application-specific target tasks.*

To address this need, we have developed a framework, called **Ark** for its ability of **a**ccruing and **r**eusing **k**nowledge embedded in heterogeneous expert annotations with numerous datasets, as illustrated in Fig. 1. We refer to the pretrained models with Ark as Foundation Ark or simply as Ark for short. To demonstrate Ark's capability, we have trained two models: Ark-5 on Datasets 1–5 and Ark-6 on Datasets 1–6 (Table 1), evaluated them on a wide range of 10 tasks via fine-tuning and on 6 tasks via linear probing, and demonstrated our Ark models outperform the SOTA fully/self-supervised baselines (Table 2) and Google CXR-FM[1] (Fig. 2). Ark also exhibits superior robustness over CXR-FM in mitigating underdiagnosis and reducing gender-related biases, with lower false-negative rates and greater robustness to imbalanced data (Fig. 3).

This performance enhancement is attributed to a simple yet powerful observation that aggregating numerous public datasets costs nearly nothing but enlarges data size, diversifies patient populations, and accrues expert knowledge from a large number of sources worldwide; thereby offering unprecedented performance yet reducing annotation cost. More important, Ark is fundamentally *different* from self-supervised learning (SSL) and federated learning (FL) in concept. SSL

[1] GitHub.com/Google-Health/imaging-research/tree/master/cxr-foundation.

can naturally handle images from different sources, but their associated expert annotations are left out of pretraining [10]. Clearly, every bit of expert annotation counts, conveying valuable knowledge. FL can utilize data with annotations from different sources, typically involving homogeneous labels, but it mainly concerns data privacy [12]. By contrast, Ark focuses on heterogeneous expert annotations with public data with no concern for data privacy and employs centralized training, which usually offers better performance with the same amount of data and annotation than distributed training as in FL.

Through this work, we have made the following contributions: (**1**) An idea that aggregates public datasets to enlarge and diversify training data; (**2**) A student-teacher model with multi-task heads via cyclic pretraining that accrues expert knowledge from existing heterogeneous annotations to achieve superior and robust performance yet reduce annotation cost; (**3**) Comprehensive experiments that evaluate our Ark via fine-tuning, linear-probing, and few-shot learning on a variety of target tasks, demonstrating Ark's better generalizability and transferability in comparison with SOTA methods and Google CXR-FM; and (**4**) Empirical analyses for a critical yet often overlooked aspect of medical imaging models—robustness to underdiagnosis and gender imbalance, highlighting Ark significantly enhances reliability and safety in clinical decision-making.

Table 1. Publicly available datasets are generally small and heterogeneously annotated. Our Ark (Fig. 1) aims to aggregate numerous datasets with heterogeneous annotations to diversify patient population, accrue knowledge from diverse experts, and meet the demand by deep learning for massive annotated training data, offering superior and robust performance (Table 2, Fig. 2 and Fig. 3) yet reducing annotation cost.

Abbrev.	Dataset	Task	Usage[a]	(Pre)train/val/test
1. CXPT	CheXpert [4]	classify 14 thoracic diagnoses	P|F|L|B	223414/-/234
2. NIHC	NIH ChestX-ray14 [19]	classify 14 thoracic diseases	P|F|L|B	75312/11212/25596
3. RSNA	RSNA Pneumonia [1]	classify lung opacity, abnormality	P|F|L	21295/2680/2709
4. VINC	VinDr-CXR [13]	classify 6 thoracic diagnoses	P|F|L	15000/-/3000
5. NIHS	NIH Shenzhen CXR [5]	classify tuberculosis	P|F|L	463/65/134
6. MMIC	MIMIC-II [6]	classify 14 thoracic diagnoses[b]	P	368879/2992/5159
7. NIHM	NIH Montgomery [5]	segment lungs	F	92/15/31
8. JSRT	JSRT [17]	segment lungs, heart, clavicles	F	173/25/49
9. VINR	VinDr-RibCXR [14]	segment 20 ribs	F	196/-/49
10. SIIM	SIIM-ACR PTX [2]	classify pneumothorax[c]	L	10675/-/1372

[a]The usage of each dataset in our experiments is denoted with P for pretraining, F for fine-tuning, L for linear probing, and B for bias study.
[b]The labels of CXRs in MIMIC-II are derived from their corresponding radiology reports using NegBio [15] and CheXpert [4].
[c]SIIM-ACR, originally for pneumothorax segmentation, is converted into a classification task for linear probing, as CXR-FM cannot be evaluated for segmentation using its only released API.

2 Accruing and Reusing Knowledge

Our Ark aims to learn superior and robust visual representations from large-scale *aggregated* medical images by accruing and reusing the expert knowledge embedded in all available *heterogeneous* labels. The following details our Ark.

Accruing Knowledge into the Student via Cyclic Pretraining. A significant challenge with training a single model using numerous datasets created for different tasks is label inconsistency (*i.e.,* heterogeneity) (see Table 3 in Appendix). Manually consolidating heterogeneous labels from different datasets would be a hassle. To circumvent this issue, for each task, we introduce a specific classifier, called task head, to learn from its annotation and encode the knowledge into the model. A task head can be easily plugged into Ark, making Ark scalable to additional tasks. With multi-task heads, Ark can learn from multiple tasks concurrently or cyclically. In concurrent pretraining, a mini-batch is formed by randomly sampling an equal number of images from each dataset, and the loss for each image is computed based on its associated dataset id and labels. This idea is intuitive, but the model hardly converges; we suspect that the loss summation over all task heads simultaneously weakens gradients for back-propagation, causing confusion in weight updating. We opt for cyclic pretraining by iterating through all datasets sequentially in each round to accrue expert knowledge from all available annotations, a strategy that, we have found, stabilizes Ark's pretraining and accelerates its convergence.

Accruing Knowledge into the Teacher via Epoch-Wise EMA. To further summarize the accrued knowledge and accumulate the learning experiences in the historical dimension, we introduce into Ark a teacher model that shares the same architecture with the student. The teacher is updated using exponential moving average (EMA) [18] based on the student's *one epoch of learning* at the end of each task. Eventually, the expert knowledge embedded in all labels and all historical learning experiences are accrued in the teacher model for further reuse in the cyclic pretraining and for future application-specific target tasks.

Reusing Accrued Knowledge from the Student to Bolster Cyclic Pretraining. If the model learns from multiple tasks sequentially, it may "forget" the previously learned knowledge, and its performance on an old task may degrade catastrophically [7]. This problem is addressed naturally in Ark by cyclic pretraining, where the model revisits all the tasks in each round and reuses all knowledge accrued from the previous rounds and tasks to strengthen its learning from the current and future tasks. That is, by regularly reviewing the accrued knowledge through task revisitation, Ark not only prevents forgetting but also enables more efficient and effective learning from multiple tasks iteratively.

Reusing Accrued Knowledge from the Teacher to Mitigate Forgetting. To leverage the accumulated knowledge of the teacher model as an additional self-supervisory signal, we incorporate a consistency loss between the student and the teacher, as shown in Fig. 1. To enhance this supervision, we introduce projectors in Ark that map the outputs of the student and teacher encoders to

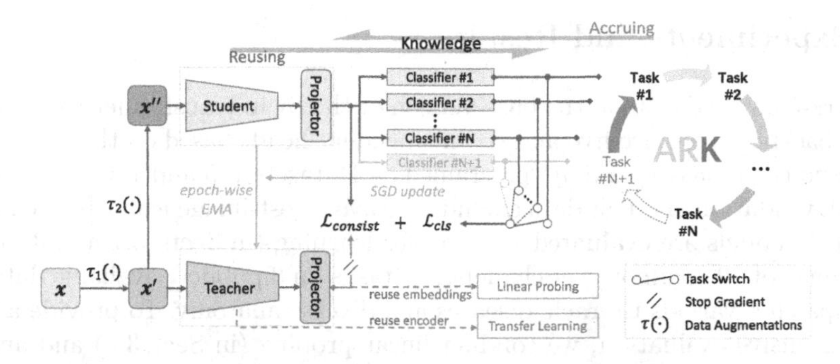

Fig. 1. Our Ark is built on a student-teacher model with multi-task heads and trained via cyclic pretraining, aiming to accrue and reuse the expert knowledge embedded in the *heterogeneous* labels with numerous public datasets (see Sect. 2 for details).

the same feature space. This further reinforces the feedback loop between the student and teacher models, facilitating the transfer of historical knowledge from the teacher to the student as a reminder to mitigate forgetting.

Ark has the following properties:

- **Knowledge-centric.** Annotating medical images by radiologists for deep learning is a process of transferring their in-depth knowledge and expertise in interpreting medical images and identifying abnormalities to a medium that is accessible for computers to learn. Ark's superior and robust performance is attributed to the accumulation of expert knowledge conveyed through medical imaging annotations from diverse expert sources worldwide. At the core of Ark is acquiring and sharing knowledge: "knowledge is power" (Mac Flecknoe) and "power comes not from knowledge kept but from knowledge shared" (Bill Gates).
- **Label-agnostic, task-scalable and annotation-heterogeneous.** Ark is label-agnostic as it does not require prior label "understanding" of public datasets, but instead uses their originally-provided labels. It is designed with pluggable multi-task heads and cyclic pretraining to offer flexibility and scalability for adding new tasks without manually consolidating heterogeneous labels or training task-specific controllers/adapters [22]. Therefore, Ark intrinsically handles the annotation heterogeneity across different datasets.
- **Application-versatile.** Ark trains versatile foundation models by utilizing a large number of publicly-available images from diverse sources and their readily-accessible diagnostic labels. As shown in Sect. 3, Ark models are more robust, generalizable, and transferable to a wide range of application-specific target tasks across diseases (*e.g.*, pneumothorax, tuberculosis, cardiomegaly) and anatomies (*e.g.*, lung, heart, rib), highlighting Ark's versatility.

3 Experiments and Results

Our Ark-5 and Ark-6 take the base version of the Swin transformer (Swin-B) [9] as the backbone, feature five and six independent heads based on the pretraining tasks and their classes, and are pretrained on Datasets 1–5 and 1–6, respectively, with all validation and test data excluded to avoid test-image leaks. In the following, both models are evaluated via transfer learning (in Sects. 3.1 and 3.2) on a wide range of 10 common, yet challenging, tasks on 8 publicly available datasets, encompassing various thoracic diseases and diverse anatomy. To provide a more comprehensive evaluation, we conduct linear probing (in Sect. 3.3) and analyze gender biases (in Sect. 3.4) on the Ark models in comparison with Google CXR-FM. Pretraining and evaluation protocols are detailed in Appendix E.

3.1 Ark Outperforms SOTA Fully/Self-supervised Methods on Various Tasks for Thoracic Disease Classification

Experimental Setup: To demonstrate the performance improvements achieved through Ark pretraining, we compare the Ark models with SOTA fully-supervised and self-supervised models [9,21] that were pretrained on ImageNet. We also include a comparison with a SOTA domain-adapted model [10] that was first pretrained on ImageNet and then on a large-scale domain-specific dataset comprising 926,028 CXRs from 13 different sources. All downstream models share the same Swin-B backbone, where the encoder is initialized using the pre-trained weights and a task-specific classification head is re-initialized based on the number of classes for the target task. We fine-tune all layers in the down-stream models under the same experimental setup. We also report the results of training the downstream models from scratch (random initialization) as the performance lower bound. Note that Google CXR-FM cannot be included for comparison as it is not publicly released for fine-tuning.

Results and Analysis: As shown in Table 2, our Ark models consistently outperform the SOTA fully/self-supervised ImageNet pretrained models on all target tasks. These results highlight the benefit of leveraging additional domain-relevant data in pretraining to reduce the domain gap and further improve the model's performance on target tasks. Furthermore, compared with the self-supervised domain-adapted model that utilizes 926K CXRs for pretraining, Ark models yield significantly superior performance on Dataset 1, 3–5 with only 335K CXRs, and on-par performance on 2.NIHC with 704K CXRs. These results demonstrate the superiority of Ark that accrues and reuses the knowledge retained in heterogeneous expert annotations from multiple datasets, emphasizing the importance of learning from expert labels. Moreover, we observe that Ark-6 consistently outperforms Ark-5, indicating the importance of incorporating more data and annotations from diverse datasets in pretraining.

Table 2. Our Ark-5 and Ark-6 outperform SOTA ImageNet pretrained models and the self-supervised domain-adapted model that utilizes even more training data, highlighting the importance of accruing and reusing knowledge in expert labels from diverse datasets for both classification and segmentation. With the best bolded and the second best underlined, a statistical analysis is conducted between the best vs. others, where green-highlighted boxes indicate no statistically significant difference at level $p = 0.05$.

Initialization	Pretraining	Classification task				
		1. CXPT	2. NIHC	3. RSNA	4. VINC	5. NIHS
Random	-	$83.39_{\pm 0.84}$	$77.04_{\pm 0.34}$	$70.02_{\pm 0.42}$	$78.49_{\pm 1.00}$	$92.52_{\pm 4.98}$
Supervised	IN	$87.80_{\pm 0.42}$	$81.73_{\pm 0.14}$	$73.44_{\pm 0.46}$	$90.35_{\pm 0.31}$	$93.35_{\pm 0.77}$
SimMIM	IN	$88.16_{\pm 0.31}$	$81.95_{\pm 0.15}$	$73.66_{\pm 0.34}$	$90.24_{\pm 0.35}$	$94.12_{\pm 0.96}$
SimMIM	IN→CXR(926K)	$88.37_{\pm 0.40}$	$\underline{83.04}_{\pm 0.15}$	$74.09_{\pm 0.39}$	$91.71_{\pm 1.04}$	$95.76_{\pm 1.79}$
Ark-5(ours)	IN→CXR(335K)	$\underline{88.73}_{\pm 0.20}$	$82.87_{\pm 0.13}$	$\underline{74.73}_{\pm 0.59}$	$\underline{94.67}_{\pm 0.33}$	$\underline{98.92}_{\pm 0.21}$
Ark-6(ours)	IN→CXR(704K)	$\mathbf{89.14}_{\pm 0.22}$	$\mathbf{83.05}_{\pm 0.09}$	$\mathbf{74.76}_{\pm 0.35}$	$\mathbf{95.07}_{\pm 0.16}$	$\mathbf{98.99}_{\pm 0.16}$
Initialization	Pretraining	Segmentation task				
		6. NIHM	7. JSRT$_{Lung}$	8. JSRT$_{Heart}$	9. JSRT$_{Clavicle}$	10. VINR
Random	-	$96.32_{\pm 0.18}$	$96.32_{\pm 0.10}$	$92.35_{\pm 0.20}$	$85.56_{\pm 0.71}$	$56.46_{\pm 0.62}$
Supervised	IN	$97.23_{\pm 0.09}$	$97.13_{\pm 0.07}$	$92.58_{\pm 0.29}$	$86.94_{\pm 0.69}$	$62.40_{\pm 0.80}$
SimMIM	IN	$97.12_{\pm 0.14}$	$96.90_{\pm 0.08}$	$93.53_{\pm 0.11}$	$87.18_{\pm 0.63}$	$61.64_{\pm 0.69}$
SimMIM	IN→CXR(926K)	$97.10_{\pm 0.40}$	$96.93_{\pm 0.12}$	$93.75_{\pm 0.36}$	$88.87_{\pm 1.06}$	$63.46_{\pm 0.89}$
Ark-5(ours)	IN→CXR(335K)	$\underline{97.65}_{\pm 0.17}$	$\underline{97.41}_{\pm 0.04}$	$\underline{94.16}_{\pm 0.66}$	$\underline{90.01}_{\pm 0.35}$	$\mathbf{63.96}_{\pm 0.30}$
Ark-6(ours)	IN→CXR(704K)	$\mathbf{97.68}_{\pm 0.03}$	$\mathbf{97.48}_{\pm 0.08}$	$\mathbf{94.62}_{\pm 0.16}$	$\mathbf{90.05}_{\pm 0.15}$	$\underline{63.70}_{\pm 0.23}$

3.2 Ark Provides Generalizable Representations for Segmentation Tasks

Experimental Setup: To evaluate the generalizability of Ark's representations, we transfer the Ark models to five segmentation tasks involving lungs, heart, clavicles, and ribs, and compare their performance with three SOTA fully/self-supervised models. We build the segmentation network upon UperNet [20], which consists of a backbone network, a feature pyramid network, and a decoder network. We implement the backbone network with Swin-B and initialize it with the pretrained weights from the Ark and those aforementioned SOTA models. The remaining networks are randomly initialized. We then fine-tune all layers in the segmentation models under the same experimental setup.

Results and Analysis: As seen in Table 2, Ark models achieve significantly better performance than the SOTA models, demonstrating that Ark learned generalizable representations for delineating organs and bones in CXR. This superior performance is achieved by pretraining using large-scale CXRs and various disease labels from diverse datasets. Clinically, certain thoracic abnormalities can be diagnosed by examining the edges of the lungs, heart, clavicles, or ribs in CXR. For instance, a pneumothorax can be detected by observing a visible "visceral pleural line" along part or all of the length of the lateral chest wall [11]. Cardiomegaly can be diagnosed when the heart appears enlarged, with maximum diameter of the heart exceeding a pre-defined cardiothoracic ratio [19]. Fractures can be identified when the edges of the clavicles or ribs appear abnormally displaced or the bone cortex appears offset [3]. Therefore, leveraging diagnostic information from disease labels during pretraining enables Ark models to better capture the nuanced and varied pathological patterns, strengthening the models' ability to represent anatomically specific features that reflect abnormal conditions in various oragns or bones. By contrast, the SimMIM (IN → CXR(926K)) model is pretrained with a self-supervised masked image modeling proxy task, which may use many clues to reconstruct the masked patches that are not necessarily related to pathological conditions, leading to lower performance despite training on more images.

3.3 Ark Offers Embeddings with Superior Quality over Google CXR-FM

Experimental Setup: To highlight the benefits of learning from more detailed diagnostic disease labels, we compare our Ark models with Google CXR-FM. CXR-FM was trained on a large dataset of 821,544 CXRs from three different sources, but with coarsened labels (normal or abnormal). By contrast, our Ark models are trained with less data, but aims to fully utilize all labels provided by experts in the original datasets. Furthermore, Ark models employ a much smaller backbone (88M parameters) compared with CXR-FM using EfficientNet-L2 (480M parameters). Since Google CXR-FM is not released and cannot be fine-tuned, we resorted to its released API to generate the embeddings (information-rich numerical vectors) for all images in the target tasks. For the sake of fairness, we also generated the embeddings from Ark's projector, whose dimension is the same as Google's. To evaluate the quality of the learned representations of these models, we conduct linear probing by training a simple linear classifier for each target task. The performance of both models is evaluated on six target tasks, including an unseen dataset, 10.SIIM, where the images have not been previously seen by the Ark models during pretraining. Additionally, we perform the same evaluation on 10.SIIM with partial training sets or even few-shot samples to further demonstrate the high quality of our Ark models' embeddings.

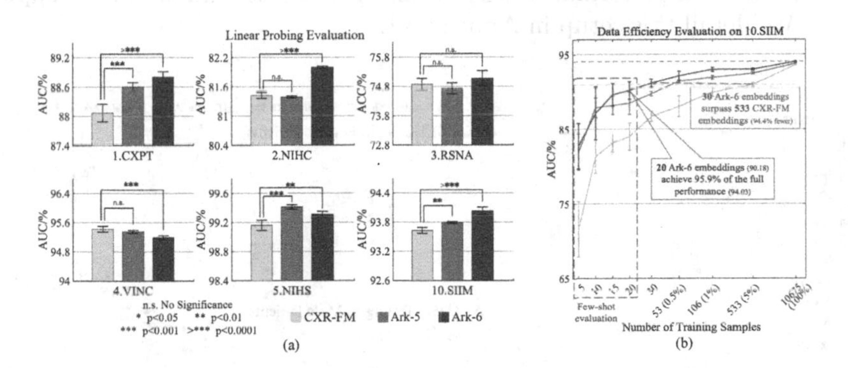

Fig. 2. Ark-5 and Ark-6 are compared with Google CXR-FM via linear probing (a) with complete training set on six target tasks, demonstrating Ark's superior or comparable performance and better embedding quality, and (b) with partial training sets or even few-shot samples, showcasing Ark's outstanding performance in terms of data efficiency.

Results and Analysis: Figure 2(a) shows that Ark-6 outperforms CXR-FM significantly on Dataset 1, 2, 5 and 10, and performs comparably to CXR-FM on 3.RSNA. Similarly, Ark-5 performs better than CXR-FM on Dataset 1, 5 and 10, while performing comparably on the remaining tasks. Moreover, Fig. 2(b) shows that both Ark-5 and Ark-6 consistently outperform CXR-FM in small data regimes, highlighting the superiority of Ark's embeddings, which carry richer information that can be utilized more efficiently. These results demonstrate that Ark models learn higher-quality representations with less pretraining data while employing a much smaller backbone than CXR-FM, highlighting that learning from more granular diagnostic labels, such as Ark, is superior to learning from coarsened normal/abnormal labels.

3.4 Ark Shows a Lower False-Negative Rate and Less Gender Bias

Experimental Setup: Underdiagnosis can lead to delayed treatment in healthcare settings and can have serious consequences. Hence, the false-negative rate (FNR) is a critical indicator of the robustness of a computer-aided diagnosis (CAD) system. Furthermore, population-imbalanced data can train biased models, adversely affecting diagnostic performance in minority populations. Therefore, a robust CAD system should provide a low false-negative rate and strong resilience to biased training data. To demonstrate the robustness of our Ark models in comparison with Google CXR-FM, we first compute the FNRs in terms of gender on 1.CXPT and 2.NIHC. We further investigate gender biases in Ark-6 and CXR-FM on 1.CXPT using gender-exclusive training sets. We follow the train/test splits in [8] to ensure a balanced number of cases per class in 40 male/female-only folds. We train linear classifiers on those folds using embeddings from Ark-6 and CXR-FM, and then evaluate these classifiers on the corresponding male/female-only test splits. The biased model will show significant differences in performance when training and test data are of the opposite gender. We detail this setup in Appendix E.

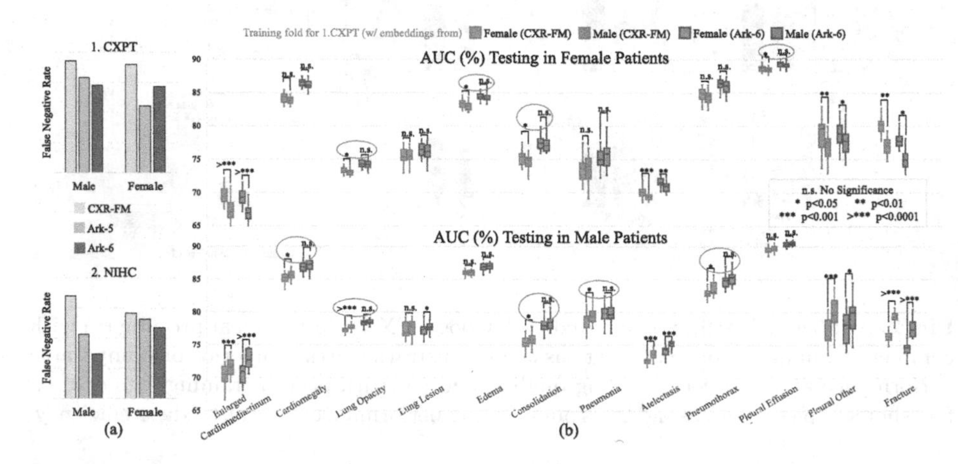

Fig. 3. Ark models are compared with Google CXR-FM as regards false-negative rate (FNR) and gender-related bias. (a) Ark models show lower FNRs, indicating superior underdiagnosis mitigation. (b) Ark-6 demonstrates greater resilience to gender-imbalanced data. Gender bias is characterized by a significant drop in performance when training and test data are of the opposite gender, compared to when they are of the same gender (*e.g.,* the orange whisker boxes are lower than the blue boxes in the lower-part (b)). Each green circle indicates a lung disease with gender bias by CXR-FM, as it performs differently between training on male and female data. But Ark exhibits a more robust performance, showing no significant difference on gender-segregated data. (Color figure online)

Results and Analysis: Figure 3(a) illustrates that Ark models have lower FNRs than CXR-FM for both genders on both tasks, demonstrating that Ark models are less likely to underdiagnose disease conditions than CXR-FM. In Fig. 3(b), the biases in the pretrained models are measured by performance differences between linear classifiers trained on male-only and female-only embeddings. The upper part of Fig. 3(b) depicts the results of *testing on female-only* sets, where the classifiers *trained on male-only* embeddings generally perform poorly compared with those trained on female embeddings, revealing gender biases due to data imbalance. Among the 12 diseases, the classifiers trained with Google's embeddings have unbiased performances for only 4 diseases, whereas those using Ark-6's embeddings perform in an unbiased fashion with no significant differences for the 8 diseases. The same situation occurs when testing is performed on male patients as shown in the lower part of Fig. 3(b). The gender bias analysis demonstrates that Ark has greater robustness to the extremely imbalanced data that contributes to gender bias in computer-aided diagnosis.

4 Conclusions and Future Work

We have developed Foundation Ark, the first open foundation model, that realizes our vision: accruing and reusing knowledge retained in heterogeneous expert annotations with numerous datasets offers superior and robust performance. Our experimental results are strong on CXRs, and we plan to extend Ark to other modalities. We hope Ark's performance encourages researchers worldwide to share codes and datasets big or small for creating open foundation models, accelerating open science, and democratizing deep learning for medical imaging.

References

1. RSNA pneumonia detection challenge (2018). https://www.kaggle.com/c/rsna-pneumonia-detection-challenge
2. SIIM-ACR pneumothorax segmentation (2019). https://kaggle.com/competitions/siim-acr-pneumothorax-segmentation
3. Collins, J.: Chest wall trauma. J. Thorac. Imaging **15**(2), 112–119 (2000)
4. Irvin, J., et al.: CheXpert: a large chest radiograph dataset with uncertainty labels and expert comparison. In: Proceedings of the AAAI Conference on Artificial Intelligence, vol. 33, pp. 590–597 (2019)
5. Jaeger, S., Candemir, S., Antani, S., Wáng, Y.X.J., Lu, P.X., Thoma, G.: Two public chest x-ray datasets for computer-aided screening of pulmonary diseases. Quant. Imaging Med. Surg. **4**(6), 475 (2014)
6. Johnson, A.E., et al.: MIMIC-CXR, a de-identified publicly available database of chest radiographs with free-text reports. Sci. Data **6**(1), 1–8 (2019)
7. Kemker, R., McClure, M., Abitino, A., Hayes, T., Kanan, C.: Measuring catastrophic forgetting in neural networks. In: Proceedings of the AAAI Conference on Artificial Intelligence, vol. 32 (2018)
8. Larrazabal, A.J., Nieto, N., Peterson, V., Milone, D.H., Ferrante, E.: Gender imbalance in medical imaging datasets produces biased classifiers for computer-aided diagnosis. Proc. Natl. Acad. Sci. **117**(23), 12592–12594 (2020)
9. Liu, Z., et al.: Swin transformer: hierarchical vision transformer using shifted windows. In: Proceedings of the IEEE/CVF International Conference on Computer Vision, pp. 10012–10022 (2021)
10. Ma, D., et al.: Benchmarking and boosting transformers for medical image classification. In: Kamnitsas, K., et al. (eds.) Domain Adaptation and Representation Transfer, DART 2022. Lecture Notes in Computer Science, vol. 13542, pp. 12–22. Springer, Cham (2022). https://doi.org/10.1007/978-3-031-16852-9_2

11. Mason, R.J., et al.: Murray and Nadel's Textbook of Respiratory Medicine E-Book: 2-Volume Set. Elsevier Health Sciences (2010)
12. McMahan, B., Moore, E., Ramage, D., Hampson, S., y Arcas, B.A.: Communication-efficient learning of deep networks from decentralized data. In: Artificial Intelligence and Statistics, pp. 1273–1282. PMLR (2017)
13. Nguyen, H.Q., et al.: VinDr-CXR: an open dataset of chest x-rays with radiologist's annotations. Sci. Data **9**(1), 429 (2022)
14. Nguyen, H.C., Le, T.T., Pham, H.H., Nguyen, H.Q.: VinDr-RibCXR: a benchmark dataset for automatic segmentation and labeling of individual ribs on chest x-rays. In: Medical Imaging with Deep Learning (2021)
15. Peng, Y., Wang, X., Lu, L., Bagheri, M., Summers, R., Lu, Z.: NegBio: a high-performance tool for negation and uncertainty detection in radiology reports. AMIA Summits Transl. Sci. Proc. **2018**, 188 (2018)
16. Sellergren, A.B., et al.: Simplified transfer learning for chest radiography models using less data. Radiology **305**(2), 454–465 (2022)
17. Shiraishi, J., et al.: Development of a digital image database for chest radiographs with and without a lung nodule: receiver operating characteristic analysis of radiologists' detection of pulmonary nodules. Am. J. Roentgenol. **174**(1), 71–74 (2000)
18. Tarvainen, A., Valpola, H.: Mean teachers are better role models: weight-averaged consistency targets improve semi-supervised deep learning results. In: Advances in Neural Information Processing Systems, vol. 30 (2017)
19. Wang, X., Peng, Y., Lu, L., Lu, Z., Bagheri, M., Summers, R.M.: ChestX-ray8: hospital-scale chest x-ray database and benchmarks on weakly-supervised classification and localization of common thorax diseases. In: Proceedings of the IEEE Conference on Computer Vision and Pattern Recognition, pp. 2097–2106 (2017)
20. Xiao, T., Liu, Y., Zhou, B., Jiang, Y., Sun, J.: Unified perceptual parsing for scene understanding. In: Ferrari, V., Hebert, M., Sminchisescu, C., Weiss, Y. (eds.) ECCV 2018. LNCS, vol. 11209, pp. 432–448. Springer, Cham (2018). https://doi.org/10.1007/978-3-030-01228-1_26
21. Xie, Z., et al.: SimMIM: a simple framework for masked image modeling. In: Proceedings of the IEEE/CVF Conference on Computer Vision and Pattern Recognition, pp. 9653–9663 (2022)
22. Zhu, Z., Kang, M., Yuille, A., Zhou, Z.: Assembling existing labels from public datasets to diagnose novel diseases: Covid-19 in late 2019. In: NeurIPS Workshop on Medical Imaging meets NeurIPS (2022)

Masked Frequency Consistency for Domain-Adaptive Semantic Segmentation of Laparoscopic Images

Xinkai Zhao[1(✉)], Yuichiro Hayashi[1], Masahiro Oda[1,2], Takayuki Kitasaka[3], and Kensaku Mori[1,4,5(✉)]

[1] Graduate School of Informatics, Nagoya University, Nagoya, Japan
xkzhao@mori.m.is.nagoya-u.ac.jp, kensaku@is.nagoya-u.ac.jp
[2] Information Strategy Office, Information and Communications, Nagoya University, Nagoya, Japan
[3] Department of Information Science, Aichi Institute of Technology, Toyota, Japan
[4] Information Technology Center, Nagoya University, Nagoya, Japan
[5] Research Center for Medical Bigdata, National Institute of Informatics, Tokyo, Japan

Abstract. Semantic segmentation of laparoscopic images is an important issue for intraoperative guidance in laparoscopic surgery. However, acquiring and annotating laparoscopic datasets is labor-intensive, which limits the research on this topic. In this paper, we tackle the Domain-Adaptive Semantic Segmentation (DASS) task, which aims to train a segmentation network using only computer-generated simulated images and unlabeled real images. To bridge the large domain gap between generated and real images, we propose a Masked Frequency Consistency (MFC) module that encourages the network to learn frequency-related information of the target domain as additional cues for robust recognition. Specifically, MFC randomly masks some high-frequency information of the image to improve the consistency of the network's predictions for low-frequency images and real images. We conduct extensive experiments on existing DASS frameworks with our MFC module and show performance improvements. Our approach achieves comparable results to fully supervised learning method on the CholecSeg8K dataset without using any manual annotation. The code is available at github.com/MoriLabNU/MFC.

Keywords: Unsupervised Domain Adaptation · Laparoscopic Image · Semantic Segmentation

1 Introduction

Laparoscopic surgery is a minimally invasive surgical technique in which a camera and surgical instruments are inserted through a series of small skin punctures.

Supplementary Information The online version contains supplementary material available at https://doi.org/10.1007/978-3-031-43907-0_63.

H. Greenspan et al. (Eds.): MICCAI 2023, LNCS 14220, pp. 663–673, 2023.
https://doi.org/10.1007/978-3-031-43907-0_63

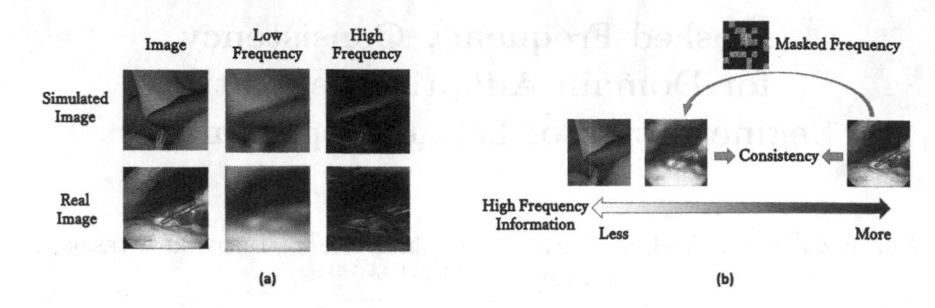

Fig. 1. Motivation for this work. (a) real images tend to contain more high frequency information than simulated images, and (b) our method aims to mitigate domain gaps by minimizing the discrepancy between high and low frequency information.

During this procedure, the surgeon relies heavily on a screen to visualize the surgical site, which can be a serious challenge. An inaccurate interpretation of abdominal anatomy can result in serious injury to the patient's bile ducts [25]. Therefore, deploying neural networks to accurately identify anatomical structures during laparoscopic surgery can markedly enhance both the quality and safety of the procedure [15]. Despite the marked achievements of deep neural networks in various medical computer vision tasks, the training of supervised models necessitates a substantial volume of precisely annotated images. Because the acquisition of large, high-quality datasets of laparoscopic images is labor-intensive and requires expert knowledge, the size and quality of publicly available datasets limit current research on semantic segmentation of laparoscopic images [21]. To overcome these limitations, several active learning [1,18] and domain generalization [14] methods have been developed to minimize the manual annotation required for network training. We take a step further and employ unsupervised domain adaptation (UDA) to eliminate the dependence on manual annotation.

The aim of UDA is to train a model on a labeled source and an unlabeled target domain for enhanced target domain performance. Various UDA methods exist, but we concentrate on two types that are relevant to our approach: self-training-based and Fourier transform-based. Self-training-based approaches [2,23,29] apply different data augmentations, multiple models or domain mixtures to the images and gauge the consistency regularization between them. On the other hand, Fourier transform-based approaches [12,27,28] exchange the low frequency components across domains to transform source domain images into target domain ones. While UDA has been extensively investigated in the medical field, the majority of existing research has focused only on domain migration between datasets from different sources [10,11] or segmentation of some distinct categories (e.g. instrument [11,19,20]). In contrast, we utilize computer-simulated images and unlabeled laparoscopic images to train a semantic segmentation network for laparoscopic images, which is more demanding and practical, as it deals with a more severe domain shift.

We propose a novel module for the UDA task in laparoscopic semantic segmentation, aiming to promote the network's exploration of consistency regularization between high-frequency and low-frequency images. Our approach is motivated by the observation in Fig. 1(a) that computer-generated images lack the high-frequency details present in real images. For example, the computer-generated abdominal wall appears smooth, whereas the real abdominal wall has rich textural information. Inspired by the effectiveness of masked image models [5,9], we randomly mask the high frequency information of the image in the frequency domain and train the network to predict the semantic segmentation result of the image lacking high frequency information. We use the pseudo-label generated by the exponential moving average (EMA) teacher network for supervision. During training, the masking of high frequency regions allows the network to discover a shared latent space for both high and low frequency images. Consequently, real laparoscopic images with high-frequency information and computer-generated images with low-frequency information can share the same feature space, facilitating the transfer of knowledge learned in the generated images to the real images, as illustrated in Fig. 1(b).

This paper's primary contributions are as follows: (1) We creatively address the domain-adaptive semantic segmentation task for laparoscopic images, which involves training the model with both unlabeled real images and computer-generated simulated images. (2) To bridge the severe domain shift between generated and real images, we propose a novel masking frequency consistency (MFC) module to reduce the domain gap. MFC encourages the network to learn shared features between high-frequency and low-frequency images. To our knowledge, MFC is the first UDA method that applies masking strategy on frequency domain. (3) We collect a vast number of image frames from public datasets and train them with computer-generated images. We evaluate our method, demonstrating that our MFC approach not only outperforms existing state-of-the-art UDA methods but also achieves performance comparable to fully supervised methods without the need for any manual annotation.

2 Method

This paper addresses the task of domain-adaptive semantic segmentation of laparoscopic images. Suppose we have a source domain of computer-generated simulated laparoscopic images, consisting of N images $\mathcal{X}^S = \{\boldsymbol{X}_i^S \mid i = 1, 2, \ldots, N\}$ with corresponding pixel-level annotations $\mathcal{Y}^S = \{\boldsymbol{Y}_i^S \mid i = 1, 2, \ldots, N\}$, and a target domain of M real laparoscopic images $\mathcal{X}^T = \{\boldsymbol{X}_j^T \mid j = 1, 2, \ldots, M\}$ without annotations. Our objective is to train a network f with robust semantic segmentation capability on the unlabeled target domain. To achieve this, we introduce a masked frequency consistency module for self-learning on the unlabeled target domain images $\boldsymbol{X}_i^T$, while supervised loss is used for training on the labeled source domain images $\boldsymbol{X}_i^S$. Our approach can integrate with different networks, effectively bridging the domain gap that occurs when applying networks to the target domain.

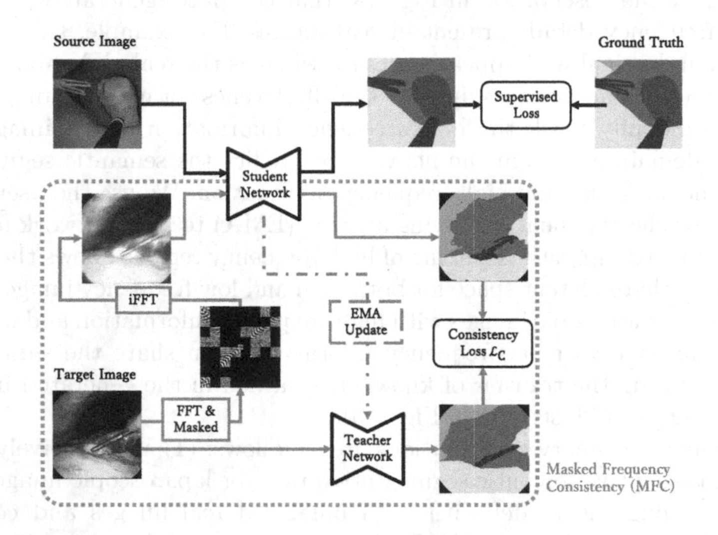

Fig. 2. The overview of our proposed Masked Frequency Consistency (MFC) module, which can be seamlessly integrated with different UDA methods and backbone networks. The MFC module works by augmenting the input image in the frequency domain using a mask, and then using a teacher-student structure to take both the original and the augmented image as inputs. A consistency loss is applied to facilitate the bridging of domains.

2.1 Image Frequency Representation

Considering an RGB image, $\boldsymbol{X} \in \mathbb{R}^{H \times W \times 3}$, we can generate its frequency representation map by applying the 2D Discrete Fourier Transform $\mathcal{F}$ for each channel $c \in \{0, 1, 2\}$, independently:

$$\mathcal{F}(\boldsymbol{X})_{(u,v,c)} = \sum_{h=0}^{H-1} \sum_{w=0}^{W-1} \boldsymbol{X}_{(h,w,c)} e^{-i2\pi\left(\frac{uh}{H} + \frac{vw}{W}\right)}, \text{with } i^2 = -1 \tag{1}$$

where (u, v) and (h, w) donate the coordinates in frequency map and image.

To facilitate subsequent operations, we rearrange the FFT data so that negative frequency terms precede positive ones, thereby centering the low frequency information. Furthermore, the inverse Fourier transform (iFFT) $\mathcal{F}^{-1}$ is utilized to transform the spectral signals back into the original image space:

$$\boldsymbol{X}_{(h,w,c)} = \frac{1}{HW} \sum_{u=0}^{H-1} \sum_{v=0}^{W-1} \mathcal{F}(\boldsymbol{X})_{(u,v,c)} e^{i2\pi\left(\frac{uh}{H} + \frac{vw}{W}\right)}, \text{with } i^2 = -1 \tag{2}$$

Computation of both the Fourier transform and its corresponding inverse is achieved through the Fast Fourier Transform (FFT) algorithm [16].

2.2 Masking Strategy

As illustrated in Fig. 2, MFC module perturbs the frequency information for target domain images. To do this, we define a mask $\mathcal{M} \in \{0,1\}^{H \times W}$ that randomly erases parts of the frequency map, thereby reducing the frequency data. Specifically, a patch mask $\mathcal{M}$ is randomly sampled as follows:

$$\mathcal{M}_{\substack{mb+1:(m+1)b, \\ nb+1:(n+1)b}} = [v > r], \text{with } v \sim \mathcal{U}(0,1) \tag{3}$$

where $[\cdot]$ is the Iverson bracket, $\mathcal{U}(0,1)$ the uniform distribution, b is the patch size, r represents the mask ratio, m and n are the patch indices. After this procedure, the patches in the mask are randomly masked.

However, using the random patch mask alone may result in the loss of all low frequency information, which would exacerbate the domain gap and lead to training instability. To avoid this, we set the central elements to 1, thus preserving the low frequency information from the images as:

$$\mathcal{M}_{\substack{H/2-h:H/2+h, \\ W/2-w:W/2+w}} = 1, \tag{4}$$

where h and w denote the size of the low-frequency information to be preserved.

We utilize the mask $\mathcal{M}$ to apply masking in the frequency domain and use the iFFT $\mathcal{F}^{-1}$ to transform the image back into the original spatial domain as the network input. The enhanced image can then be obtained as:

$$\boldsymbol{X}_m = \mathcal{F}^{-1}\left(\mathcal{F}\left(\boldsymbol{X}\right) \odot \mathcal{M}\right), \tag{5}$$

where $\odot$ is the Hadamard product between the matrices. Moreover, with conjugate symmetry's disruption inhibiting the imaginary component's cancellation, we employ complex number magnitudes as outputs.

2.3 Consistency Regularization

Consistency regularization is employed to extract common representations between high and low frequency images, thereby enhancing the generality of the network. Specifically, during training, the student segmentation network f_S takes the enhanced image $\boldsymbol{X}_m$ as input, whereas the original image $\boldsymbol{X}$ serves as the input for the teacher network f_T. The weight $\boldsymbol{\theta}_T$ of the teacher network undergoes updates using the exponential moving average (EMA) [22] of the weight $\boldsymbol{\theta}_S$ belonging to the student network:

$$\boldsymbol{\theta}_T = \alpha \boldsymbol{\theta}_S + (1 - \alpha)\boldsymbol{\theta}'_T\ , \tag{6}$$

where $\boldsymbol{\theta}'_T$ representing the weight from the previous training step. The EMA teacher generates a series of stable pseudo-labels over time, a tactic frequently utilized in both semi-supervised learning [22] and UDA [7–9,23].

To evaluate the prediction results, we employ the mean squared error (MSE) as as our loss function, which quantifies the divergence between the predictions:

$$\mathcal{L}_C = \boldsymbol{q}_T \odot MSE(f_T(\boldsymbol{X}), f_S(\boldsymbol{X}_m)), \tag{7}$$

where $\boldsymbol{q}_T$ denotes the quality weight. Due to potential inaccuracies in pseudo-labeling, like prior works [7,8,23], we only use confident pixels surpassing the maximum probability threshold τ, defined as:

$$\boldsymbol{q}_T = \frac{\sum_{h=0}^{H-1} \sum_{w=0}^{W-1} [\max_c f_T(\boldsymbol{X})_{(hwc)} > \tau]}{H \cdot W}. \tag{8}$$

3 Experiments

3.1 Datasets and Implementation

Datasets. Our experiments were conducted on three laparoscopic datasets, described as follows: (1) Simulation dataset [17] consists of 20,000 labeled images of 3D scenes assembled from CT data, with 6 categories. In addition, I2I [17] used generative adversarial networks (GAN) to translate these images into five different styles of realistic laparoscopic images, resulting in a total of 100,000 images. We used these two types of images (simulated and translated) as the source domain datasets with annotations. (2) CholecSeg8k [6] is a semantic segmentation dataset containing laparoscopic cholecystectomy images from 17 video clips of the Cholec80 dataset [24], labeled with 13 categories. We used these 17 video clips as a test set. (3) The Cholec80 dataset [24] consists of 80 videos of cholecystectomy procedures with annotations for phase and instrument presence, but no annotations related to segmentation. To train the UDA model, we selected 6819 images from 63 surgical videos, excluding the 17 videos used in CholecSeg8k [6]. Specifically, for the video of the preparation phase of surgery, we extracted one frame per second as an unlabeled target domain image. A more detailed description is available in supplementary material.

Dataset Partitioning. Our experiments were performed with two different settings: (1) simulated images to real images and (2) translated images to real images. Considering the 6 categories present in the simulated dataset and the 13 categories in the CholecSeg8k dataset, we performed semantic segmentation on the following 6 categories: Background (BG), Abdominal Wall (AW), Liver, Fat, Gallbladder (GB), and Instruments (INST).

Implementation. We used the mmsegmentation [4] codebase and trained each model on a single NVIDIA Tesla V100 GPU. We evaluated the Segformer [26] and DeepLabV2 [3] backbone networks, based on HRDA strategy [8], and initialized all backbone networks with pre-training on ImageNet. Training was performed using AdamW [13], with hyper-parameters taken from previous works [8,9,26]. In all experiments, we trained the models on a batch of randomly cropped $256px \times 256px$ images for 40k iterations, and the batch size is set to 4. As indicated in the ablation studies presented in the supplementary material, the optimal hyper-parameters for the MFC method vary according to the type of input datasets. Therefore, in all subsequent experiments, we set r to 0.7, b to 32, and h and w to 8, without further optimization of these hyper-parameters.

State-of-the-Art Methods. We benchmarked our method against contemporary leading approaches, which include UDA methods (DAFormer [7], HRDA [8], and MIC [9]), image translation method [17], and fully supervised network on the CholecSeg8k dataset using cross-validation. For equitable comparison, all methods employed the same segmentation network, initialization, and optimizer.

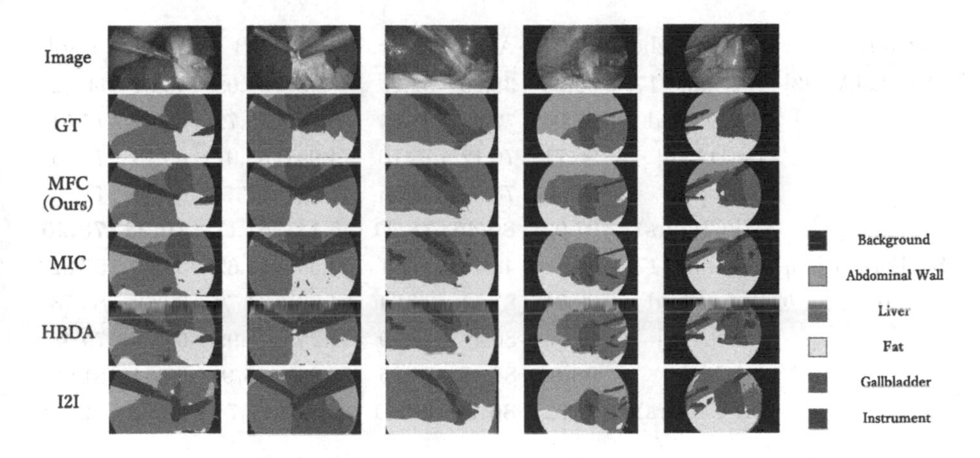

Fig. 3. Qualitative results of different methods applied to the CholecSeg8k [6] dataset.

3.2 Qualitative Evaluation

In Fig. 3, we present visualizations of the proposed MFC method and other compared methods on the CholecSeg8k dataset. All methods used only the simulated dataset for training, without any manual annotation. Our proposed method has two major advantages: (1) it effectively performs surgical instrument segmentation with minimal interference from reflections and shadows, and (2) it achieves a more accurate distinction between the boundaries of gallbladder and fat. However, we have found that our method exhibits imprecision in distinguishing the liver from the abdominal wall in certain cases.

Table 1. The results, alongside a comparison with other SOTA UDA methods utilizing simulated images as the source domain. The highest scores are emphasized in bold black text.

Network	UDA Method	BG	AW	Liver	Fat	GB	INST	mIoU
SegFormer [26]	Baseline	60.78	1.38	36.56	38.86	69.55	28.9	39.34
	Supervised	98.01	81.44	84.24	84.63	74.78	79.62	83.78
	I2I [17]	79.11	46.40	57.31	59.55	51.62	52.92	57.82
	DAFormer [7]	96.55	25.35	55.45	78.59	52.83	45.27	59.01
	HRDA [8]	97.80	44.04	63.23	78.90	55.39	55.28	65.77
	MIC [9]	98.19	**56.54**	**66.56**	85.09	54.49	59.26	70.02
	MFC(Ours)	**98.30**	43.48	60.40	**85.87**	**62.21**	**82.12**	**72.06**
	MFC w/o Eq.(3)	97.83	38.02	58.29	83.48	54.90	72.81	67.55
	MFC w/o Eq.(4)	98.11	34.35	56.53	85.81	61.10	83.37	69.88

Table 2. The results, alongside a comparison with other SOTA UDA methods utilizing translated images as the source domain. The highest scores are emphasized in bold black text.

Network	UDA Method	BG	AW	Liver	Fat	GB	INST	mIoU
DeepLabV2 [3]	Baseline [17]	62.25	38.20	52.29	57.48	51.08	43.61	50.82
	Supervised	97.04	72.45	85.64	68.79	58.72	82.42	77.51
	HRDA [8]	96.71	76.12	75.70	74.99	65.97	69.78	76.54
	MIC [9]	**97.63**	78.09	78.06	**75.36**	**67.91**	66.18	77.21
	MFC(Ours)	97.02	**80.46**	**78.51**	75.13	67.11	**70.98**	**78.20**
SegFormer [26]	Baseline [17]	79.11	46.40	57.31	59.55	51.62	52.92	57.82
	Supervised	98.01	81.44	84.24	84.63	74.78	79.62	83.78
	HRDA [8]	97.96	80.49	82.59	79.25	71.99	64.29	79.43
	MIC [9]	**98.66**	84.47	84.85	78.08	**75.94**	69.54	81.92
	MFC(Ours)	98.32	**86.77**	**87.21**	**80.10**	72.74	**70.71**	**82.64**

3.3 Quantitative Evaluation

Simulated Images → Real Images. For quantitative evaluation, we employed the intersection over union (IoU) and its overall mean of 6 categories. A performance comparison of our method with other SOTA method is presented in Table 1. MFC, trained on the simulated data as the source domain, outperform the existing SOTA methods in all categories except for the abdominal wall and liver. Notably, our method significantly improves surgical instrument segmentation. This improvement can be attributed to the fact that our method excludes the disturbing high frequency noise such as reflections and shadows, which are absent in the source domain dataset. Such results indicate that our approach effectively bridges the domain gap between the generated and real images by randomly masking high frequency information. Furthermore, DeepLabV2-based methods underperform SegFormer-based methods in this setting.

To verify the effectiveness of the patch mask outlined in Eq. (3) and the low frequency mask in Eq. (4), we also conduct ablation experiments. The results of variants of our method are presented in Table 1. It is observed that the adoption of either of the two masking strategies enhances the segmentation performance.

Translated Images → Real Images. Furthermore, we assessed the performance of various methods using translated images as the source domain, with results summarized in Table 2. These translated images reduced the domain disparity with real images, thereby boosting the performance of UDA methods on the target domain. To gauge the efficacy of our proposed module, we incorporated MFC into two different network backbones. The results show that our approach efficiently improves the mIoU performance of segmentation across different source domain datasets and network backbones.

4 Conclusion

This paper tackles the crucial issue of laparoscopic image segmentation, which is essential for surgical guidance and navigation. We propose a novel UDA module, called MFC, that leverages the consistency between high and low-frequency information in latent space. This consistency facilitates knowledge transfer from computer-simulated to real laparoscopic datasets for segmentation. Experimentally, MFC not only bolsters existing UDA models' performance but also outperforms leading methods, including fully supervised models that rely on annotated data. Our work unveils the potential of using computer-generated image data and UDA techniques for laparoscopic image segmentation. However, a limitation of our approach is that it does not account for the long-tail category distribution prevalent in real-world scenarios, such as venous vessels. Therefore, a future direction of our research is to extend our MFC module to handle rare category segmentation, thereby improving UDA models' generalization capabilities.

Acknowledgments. This work was supported in part by the JSPS KAKENHI Grant Numbers 17H00867, 21K19898, 26108006; in part by the JST CREST Grant Number JPMJCR20D5; and in part by the fellowship of the Nagoya University TMI WISE program from MEXT.

References

1. Aklilu, J., Yeung, S.: ALGES: active learning with gradient embeddings for semantic segmentation of laparoscopic surgical images. In: Proceedings of Machine Learning for Healthcare, pp. 892–911. PMLR (2022)
2. Araslanov, N., Roth, S.: Self-supervised augmentation consistency for adapting semantic segmentation. In: Proceedings of the IEEE/CVF Conference on Computer Vision and Pattern Recognition, pp. 15384–15394. IEEE (2021)

3. Chen, L.C., Papandreou, G., Kokkinos, I., Murphy, K., Yuille, A.L.: DeepLab: semantic image segmentation with deep convolutional nets, atrous convolution, and fully connected CRFs. IEEE Trans. Pattern Anal. Mach. Intell. **40**(04), 834–848 (2018)
4. Contributors, M.: MMSegmentation: openmmlab semantic segmentation toolbox and benchmark (2020). https://github.com/open-mmlab/mmsegmentation
5. He, K., Chen, X., Xie, S., Li, Y., Dollár, P., Girshick, R.: Masked autoencoders are scalable vision learners. In: Proceedings of the IEEE/CVF Conference on Computer Vision and Pattern Recognition, pp. 16000–16009. IEEE (2022)
6. Hong, W.Y., Kao, C.L., Kuo, Y.H., Wang, J.R., Chang, W.L., Shih, C.S.: CholecSeg8k: a semantic segmentation dataset for laparoscopic cholecystectomy based on Cholec80. arXiv preprint arXiv:2012.12453 (2020)
7. Hoyer, L., Dai, D., Van Gool, L.: DaFormer: improving network architectures and training strategies for domain-adaptive semantic segmentation. In: Proceedings of the IEEE/CVF Conference on Computer Vision and Pattern Recognition, pp. 9924–9935. IEEE (2022)
8. Hoyer, L., Dai, D., Van Gool, L.: HRDA: context-aware high-resolution domain-adaptive semantic segmentation. In: Avidan, S., Brostow, G., Cissé, M., Farinella, G.M., Hassner, T. (eds.) ECCV 2022. LNCS, vol. 13690, pp. 372–391. Springer, Cham (2022). https://doi.org/10.1007/978-3-031-20056-4_22
9. Hoyer, L., Dai, D., Wang, H., Van Gool, L.: MIC: masked image consistency for context-enhanced domain adaptation. In: Proceedings of the IEEE/CVF Conference on Computer Vision and Pattern Recognition, pp. 11721–11732. IEEE (2023)
10. Hu, S., Liao, Z., Xia, Y.: Domain specific convolution and high frequency reconstruction based unsupervised domain adaptation for medical image segmentation. In: Wang, L., Dou, Q., Fletcher, P.T., Speidel, S., Li, S. (eds.) MICCAI 2022. LNCS, vol. 13437, pp. 650–659. Springer, Cham (2022). https://doi.org/10.1007/978-3-031-16449-1_62
11. Liu, J., Guo, X., Yuan, Y.: Prototypical interaction graph for unsupervised domain adaptation in surgical instrument segmentation. In: de Bruijne, M., et al. (eds.) MICCAI 2021. LNCS, vol. 12903, pp. 272–281. Springer, Cham (2021). https://doi.org/10.1007/978-3-030-87199-4_26
12. Liu, Q., Chen, C., Qin, J., Dou, Q., Heng, P.A.: FedDG: federated domain generalization on medical image segmentation via episodic learning in continuous frequency space. In: Proceedings of the IEEE/CVF Conference on Computer Vision and Pattern Recognition, pp. 1013–1023. IEEE (2021)
13. Loshchilov, I., Hutter, F.: Decoupled weight decay regularization. arXiv preprint arXiv:1711.05101 (2017)
14. Lyu, J., Zhang, Y., Huang, Y., Lin, L., Cheng, P., Tang, X.: AADG: automatic augmentation for domain generalization on retinal image segmentation. IEEE Trans. Med. Imaging **41**(12), 3699–3711 (2022)
15. Madani, A., et al.: Artificial intelligence for intraoperative guidance: using semantic segmentation to identify surgical anatomy during laparoscopic cholecystectomy. Ann. Surg. **276**(2), 363–369 (2022)
16. Nussbaumer, H.J., Nussbaumer, H.J.: The fast fourier transform. Fast Fourier Transform and Convolution Algorithms, pp. 80–111 (1982)
17. Pfeiffer, M., et al.: Generating large labeled data sets for laparoscopic image processing tasks using unpaired image-to-image translation. In: Shen, D., et al. (eds.) MICCAI 2019. LNCS, vol. 11768, pp. 119–127. Springer, Cham (2019). https://doi.org/10.1007/978-3-030-32254-0_14

18. Qiu, J., Hayashi, Y., Oda, M., Kitasaka, T., Mori, K.: Class-wise confidence-aware active learning for laparoscopic images segmentation. Inter. J. Comput. Assisted Radiol. Surgery, 1–10 (2022)
19. Sahu, M., Mukhopadhyay, A., Zachow, S.: Simulation-to-real domain adaptation with teacher-student learning for endoscopic instrument segmentation. Int. J. Comput. Assist. Radiol. Surg. **16**(5), 849–859 (2021)
20. Sahu, M., Strömsdörfer, R., Mukhopadhyay, A., Zachow, S.: Endo-Sim2Real: consistency learning-based domain adaptation for instrument segmentation. In: Martel, A.L., et al. (eds.) MICCAI 2020. LNCS, vol. 12263, pp. 784–794. Springer, Cham (2020). https://doi.org/10.1007/978-3-030-59716-0_75
21. Silva, B., et al.: Analysis of current deep learning networks for semantic segmentation of anatomical structures in laparoscopic surgery. In: 2022 44th Annual International Conference of the IEEE Engineering in Medicine & Biology Society (EMBC), pp. 3502–3505. IEEE (2022)
22. Tarvainen, A., Valpola, H.: Mean teachers are better role models: weight-averaged consistency targets improve semi-supervised deep learning results. In: Advances in Neural Information Processing Systems 30 (2017)
23. Tranheden, W., Olsson, V., Pinto, J., Svensson, L.: DACS: domain adaptation via cross-domain mixed sampling. In: Proceedings of the IEEE/CVF Winter Conference on Applications of Computer Vision, pp. 1379–1389 (2021)
24. Twinanda, A.P., Shehata, S., Mutter, D., Marescaux, J., De Mathelin, M., Padoy, N.: Endonet: a deep architecture for recognition tasks on laparoscopic videos. IEEE Trans. Med. Imaging **36**(1), 86–97 (2016)
25. Way, L.W., et al.: Causes and prevention of laparoscopic bile duct injuries: analysis of 252 cases from a human factors and cognitive psychology perspective. Ann. Surg. **237**(4), 460–469 (2003)
26. Xie, E., Wang, W., Yu, Z., Anandkumar, A., Alvarez, J.M., Luo, P.: SegFormer: simple and efficient design for semantic segmentation with transformers. Adv. Neural. Inf. Process. Syst. **34**, 12077–12090 (2021)
27. Yang, Y., Soatto, S.: FDA: fourier domain adaptation for semantic segmentation. In: Proceedings of the IEEE/CVF Conference on Computer Vision and Pattern Recognition, pp. 4085–4095. IEEE (2020)
28. Zakazov, I., Shaposhnikov, V., Bespalov, I., Dylov, D.V.: Feather-light fourier domain adaptation in magnetic resonance imaging. In: Kamnitsas, K., et al. (eds.) DART 2022. LNCS, vol. 13542, pp. 88–97. Springer, Cham (2022). https://doi.org/10.1007/978-3-031-16852-9_9
29. Zhou, Q., et al.: Context-aware mixup for domain adaptive semantic segmentation. IEEE Trans. Circuits Syst. Video Technol. **33**, 804–817 (2021)

Pick the Best Pre-trained Model: Towards Transferability Estimation for Medical Image Segmentation

Yuncheng Yang[1], Meng Wei[3], Junjun He[3], Jie Yang[1(✉)], Jin Ye[3], and Yun Gu[1,2(✉)]

[1] Institute of Image Processing and Pattern Recognition, Shanghai Jiao Tong University, Shanghai, China
yungu@ieee.org
[2] Institute of Medical Robotics, Shanghai Jiao Tong University, Shanghai, China
[3] Shanghai AI Lab, Shanghai, China

Abstract. Transfer learning is a critical technique in training deep neural networks for the challenging medical image segmentation task that requires enormous resources. With the abundance of medical image data, many research institutions release models trained on various datasets that can form a huge pool of candidate source models to choose from. Hence, it's vital to estimate the source models' transferability (i.e., the ability to generalize across different downstream tasks) for proper and efficient model reuse. To make up for its deficiency when applying transfer learning to medical image segmentation, in this paper, we therefore propose a new **Transferability Estimation** (TE) method. We first analyze the drawbacks of using the existing TE algorithms for medical image segmentation and then design a source-free TE framework that considers both class consistency and feature variety for better estimation. Extensive experiments show that our method surpasses all current algorithms for transferability estimation in medical image segmentation. Code is available at here.

Keywords: Transferability Estimation · Model Selection · Medical Image Analysis · Deep Learning

1 Introduction

The development of deep neural networks has greatly promoted medical imaging-based computer-aided diagnosis. Due to the large amount of learnable parameters in neural networks, sufficient annotated training samples are required for training. However, the labeling process of medical images is tedious and time-consuming. To address this problem, the common paradigm of *transfer learning*,

Supplementary Information The online version contains supplementary material available at https://doi.org/10.1007/978-3-031-43907-0_64.

H. Greenspan et al. (Eds.): MICCAI 2023, LNCS 14220, pp. 674–683, 2023.
https://doi.org/10.1007/978-3-031-43907-0_64

which first pre-trains a model on upstream image datasets and then fine-tunes it on various target tasks, has been widely investigated in recent years [10,21,30]. Compared with the distributed training across multiple centers, there are no specific ethical issues or computational design of distributed/federated learning frameworks with the "pre-train-then-fine-tune" workflow.

Previous works mainly focused on the fine-tuning strategy to effectively adapt the knowledge from the pre-trained models to target tasks [4,12,19,26]. With the increasing number of pre-trained networks provided by the community, model repositories like Hugging Face [25] and PyTorch Hub [18] enable researchers to experiment across a large number of downstream datasets and tasks. These pre-trained models require less training time and have better performance and robustness compared with the learning-from-scratch models. However, it has been observed by recent works [23] that the pre-trained models cannot always benefit the downstream tasks. When the knowledge is transferred from a less relevant source, it may not improve the performance or even negatively affect the intended outcome [24]. A brute-force method is to fine-tune a set of pre-trained models with target datasets to find the optimal one. This process is time-consuming and laborious. Existing methods also measured the task-relatedness between source and target datasets [6,7,22,28]. However, most of these works require source information available while medical images have more privacy and ethical issues and fewer datasets are publicly available than natural images.

Considering the issues mentioned above, this work focused on source-free pre-trained model selection for segmentation tasks in the medical image. As shown in Fig. 1, models pre-trained by upstream data constitute the model bank. The main idea is to directly measure the transferability of the pre-trained models without fully training based on the downstream/target dataset. Among the recent works, LEEP [15] and its variant [1,13] were developed to utilize the log-likelihood between the target labels and the predictions from the source model. LogME [27] computed evidence based on the linear parameters assumption and efficiently leverages the compatibility between features and labels. GBC [17] applied the Gaussian distribution to each class, and estimate the separability between classes as the basis for transferability estimation. TransRatė [9] evaluated the transferability of models with the compactness and the completeness of embedding space. Cui et.al [5] contended that discriminability and transferability are crucial properties of representations and introduce the information bottleneck theory for transferability estimation. These methods have achieved promising performance on classification and regression tasks without fully considering the properties of medical image segmentation. First, unlike classification and regression problems that can use a single n-dimensional feature vector to represent each image, segmentation problems lack a global semantic representation, which poses difficulties for direct transferability estimation. In addition, most label-comparison-based methods [9,15,17,27] focus on the relationship between the embeddings and downstream labels without exploring the effectiveness of the features themselves. Third, medical images face severe class imbalance problems, with excessive differences between foreground and background. However, existing algorithms rarely give additional attention to the class imbalance problem.

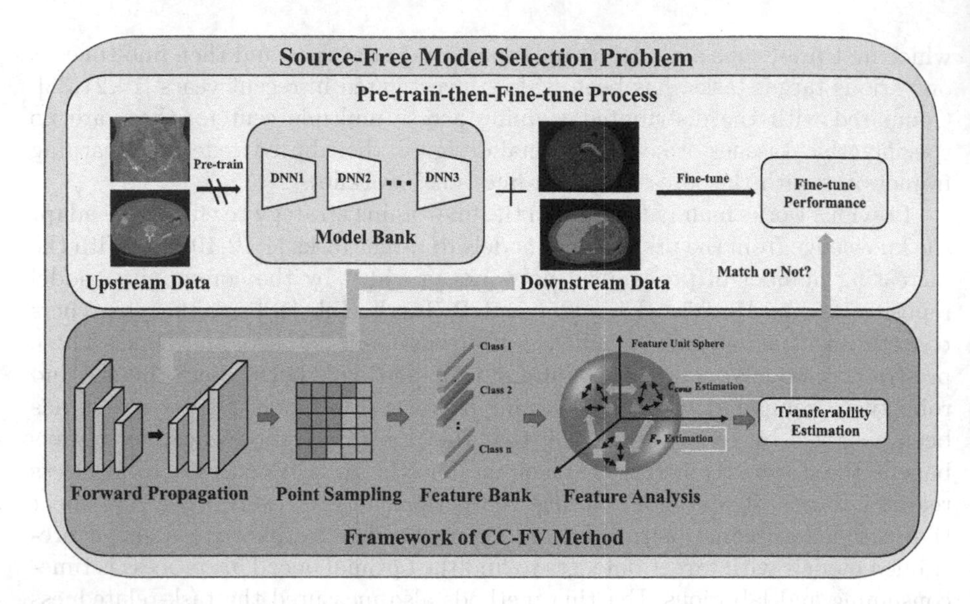

Fig. 1. Source-free model selection problem and the framework of our **C**lass **C**onsistency with **F**eature **V**ariety constraint(CC-FV) TE method. Our main goal is to predict the performance of models in the model bank after fine-tuning on downstream tasks without actually fine-tuning. Note that the upstream data are not available in our model selection process.

Besides, for semantic segmentation tasks, the feature pyramid is critical for the segmentation output of multi-scale objects while existing works neglect it. In our work, we propose a new method using class consistency and feature variety(CC-FV) with an efficient framework to estimate the transferability in medical image segmentation tasks. Class consistency employs the distribution of features extracted from foreground voxels of the same category in each sample to model and calculate their distance, the smaller the distance the better the result; feature diversity utilizes features sampled in the whole global feature map, and the uniformity of the feature distribution obtained by sampling is used to measure the effectiveness of the features themselves. Extensive experiments have proved the superiority of our method compared with baseline methods.

2 Methodology

2.1 Problem Formulation

In our work, a model bank $\mathbb{M}$ consisting of pre-trained models $\{M_i\}_{i=1}^{K}$ are available to be fine-tuned and evaluated with a target dataset $\mathbb{D} = \{X_j, Y_j\}_{j=1}^{N}$, where X_j is the image and Y_j is the ground truth of segmentation. After fine-tuning, the performance of M_i can be measured with the segmentation metric (e.g. Dice score), which is denoted by $\mathcal{P}_{s\to t}^{i}$ in this paper. Our work is to directly estimate the transferability score $\mathcal{T}^i_{s\to t}$ without fine-tuning the model on target datasets. A perfect transferability score should preserve the ordering, i.e. $\mathcal{T}^i_{s\to t} > \mathcal{T}^j_{s\to t}$ if and only if $\mathcal{P}^i_{s\to t} > \mathcal{P}^j_{s\to t}$.

2.2 Class Consistency with Feature Variety Constraint TE Method

The transferability of models from a weakly related source domain to a target domain can be compromised if the domains are not sufficiently comparable [24]. This intrigues us about the question of "what kind of models are transferable". The proposed method is intuitive and straightforward: features extracted by the pre-trained model should be consistent within the class of the target dataset while representative and various globally. Therefore, **C**lass **C**onsistency and **F**eature **V**ariety are considered to estimate the transferability between models and downstream data.

Class Consistency. The pre-trained models are trained with specific pretext tasks based on the upstream dataset. Therefore, features extracted by the pre-trained models cannot perfectly distinguish the foreground and background of target data. If the features are generalizable, foreground region features will likely follow a similar distribution even without fine-tuning.

Given a pair of target data X_j and $X_{j'}$, the distribution of the features is modeled with the n-dimensional Gaussian distribution. Since the size of the foreground class varies across the cases, we therefore randomly sample the pixels/voxels of X_j and $X_{j'}$ for each class and establish the feature distribution F^k_j, $F^k_{j'}$ based on the voxels of the kth class to approximate the case-wise distribution of different classes. The class consistency between the data pair is measured by the Wasserstein distance [16] as follows:

$$\mathcal{W}_2^2(F_j^k, F_{j'}^k) = \left\|\mu_{F_j^k} - \mu_{F_{j'}^k}\right\|^2 + \mathrm{Tr}\left(\Sigma_{F_j^k}\right) + \mathrm{Tr}\left(\Sigma_{F_{j'}^k}\right) - 2\,\mathrm{Tr}\left(\left(\Sigma_{F_j^k}\Sigma_{F_{j'}^k}\right)^{1/2}\right) \quad (1)$$

where $\mu_{F_j^k}$, $\mu_{F_{j'}^k}$ are the mean of Gaussian distribution F^k_j, $F^k_{j'}$ and $\Sigma_{F_j^k}$ and $\Sigma_{F_{j'}^k}$ are covariance matrices of F^k_j and $F^k_{j'}$. Compared to some commonly used metrics like KL-divergence or Bhattacharyya distance [17], Wasserstein distance is more stable during the computation of high-dimensional matrices because it is unnecessary to compute the determinant or inverse of a high-dimensional matrix, which can easily lead to an overflow in numerical computation. We calculate the Wasserstein distance of the distribution with voxels of the same class in a sample pair comprised of every two samples in the dataset, and obtained the following definition of class consistency C_{cons}

$$C_{cons} = \frac{1}{N(N-1)}\sum_{k-1}^{C}\sum_{i\neq j}\mathcal{W}_2(F_i^k, F_j^k) \quad (2)$$

Given that 3D medical images are computationally intensive, and prone to causing out-of-memory problems, in the sliding window inference process for each case, we do not concatenate the output of each patch into the final prediction result, but directly sample from the patched output and concatenate them into the final sampled feature matrix. In the calculation of class consistency, we only sample the foreground voxels with a pre-defined sampling number which is proportional to the voxel number of each class in the image because of the severe class imbalance problem.

Feature Variety. Class consistency is not the only criterion for transferability estimation. As a result of learning some trivial solutions, some overfitted models have limited generalization capacity and are difficult to apply to new tasks. We believe that the essential reason for this phenomenon is that class consistency is only concerned with local homogeneity of information while neglecting the integral feature quality assessment. Hence we propose the feature variety constraint, which measures the expressiveness of the features themselves and the uniformity of their probability distribution. Highly complex features are not easily overfitted in the downstream tasks and do not collapse to cause a trivial solution.

To calculate the variety of features we need to analytically measure the properties of the feature distribution over the full feature space. Besides, to prevent overfitting and trivial features, we expect the distribution of features in the feature space to be as uniform and dispersed as possible. Therefore we employ the following hyperspherical potential energy $\boldsymbol{E_s}$ as

$$\boldsymbol{E_s}(\boldsymbol{v}) = \sum_{i=1}^{L} \sum_{j=1, j \neq i}^{L} \boldsymbol{e_s}\left(\left\|\boldsymbol{v}_i - \boldsymbol{v}_j\right\|\right) = \begin{cases} \sum_{i \neq j} \left\|\boldsymbol{v}_i - \boldsymbol{v}_j\right\|^{-s}, & s > 0 \\ \sum_{i \neq j} \log\left(\left\|\boldsymbol{v}_i - \boldsymbol{v}_j\right\|^{-1}\right), & s = 0 \end{cases} \tag{3}$$

Here v is sampled feature of each image with point-wise embedding v_i and L is the length of the feature, which is also the number of sampled voxels. We randomly sample from the whole case so that the features can better express the overall representational power of the model. The feature vectors will be more widely dispersed in the unit sphere if the hyperspherical energy (HSE) is lower [3]. For the dataset with N cases, we choose $s = 1$ and the feature variety F_v is formulated as

$$\boldsymbol{F_v} = \frac{1}{N} \sum_{i=1}^{N} \boldsymbol{E_s}^{-1}(\boldsymbol{v}) \tag{4}$$

Overall Estimation. As for semantic segmentation problems, the feature pyramid structure is critical for segmentation results [14, 29]. Hence in our framework, different decoders' outputs are upsampled to the size of the output and can be used in the sliding window sampling process. Besides, we decrease the sampling ratio in the decoder layer close to the bottleneck to avoid feature redundancy. The final transferability of pre-trained model m to dataset t $\mathcal{T}_{m \to t}$ is

$$\mathcal{T}_{m \to t} = \frac{1}{D} \sum_{i=1}^{D} \log \frac{\boldsymbol{F_v^i}}{\boldsymbol{C_{cons}^i}} \tag{5}$$

where D is the number of decoder layers used in the estimation.

3 Experiment

3.1 Experiment on MSD Dataset

The Medical Segmentation Decathlon (MSD) [2] dataset is composed of ten different datasets with various challenging characteristics, which are widely used in the medical image analysis field. To evaluate the effectiveness of CC-FV, we conduct extensive experiments on 5 of the MSD dataset, including Task03 Liver(liver and tumor segmentation), Task06 Lung(lung nodule segmentation), Task07 Pancreas(pancreas and pancreas tumor segmentation), Task09 Spleen(spleen segmentation), and Task10 Colon(colon cancer segmentation). All of the datasets are 3D CT images. The public part of the MSD dataset is chosen for our experiments, and each dataset is divided into a training set and a test set at a scale of 80% and 20%. For each dataset, we use the other four datasets to pre-train the model and fine-tune the model on this dataset to evaluate the performance as well as the transferability using the correlation between two ranking sequences of upstream pre-trained models. We load all the pre-trained models' parameters except for the last convolutional layer and no parameters are frozen during the fine-tuning process. On top of that, we follow the nnUNet [11] with the self-configuring method to choose the pre-processing, training, and post-processing strategy. For fair comparisons, the baseline methods including TransRate [9], LogME [27], GBC [17] and LEEP [15] are also implemented. For these currently available methods, we employ the output of the layer before the final convolution as the feature map and sample it through the same sliding window as CC-FV to obtain different classes of features, which can be used for the calculation.

Figure 2 visualizes the average Dice score and the estimation value on Task 03 Liver. The TE results are obtained from the training set only. U-Net [20] and UNETR [8] are applied in the experiment and each model is pre-trained for 250k iterations and fine-tuned for 100k iterations with batch size 2 on a single NVIDIA A100 GPU. Besides, we use the model at the end of training for inference and calculate the final DSC performance on the test set. And we use weighted Kendall's τ [27] and Pearson correlation coefficient for the correlation between the TE results and fine-tuning performance. The Kendall's τ ranges from [-1, 1], and τ=1 means the rank of TE results and performance are perfectly correlated($\mathcal{T}^i_{s\to t} > \mathcal{T}^j_{s\to t}$ if and only if $\mathcal{P}^i_{s\to t} > \mathcal{P}^j_{s\to t}$). Since model selection generally picks the top models and ignores the poor performers, we assign a higher weight to the good models in the calculation, known as weighted Kendall's τ. The Pearson coefficient also ranges from [-1, 1], and measures how well the data can be described by a linear equation. The higher the Pearson coefficient, the higher the correlation between the variables. It is clear that the TE results of our method have a more positive correlation with respect to DSC performance.

Table 1 demonstrates that our method surpasses all the other methods. Most of the existing methods are inferior to ours because they are not designed for segmentation tasks with a serious class imbalance problem. Besides, these methods rely only on single-layer features and do not make good use of the hierarchical structure of the model.

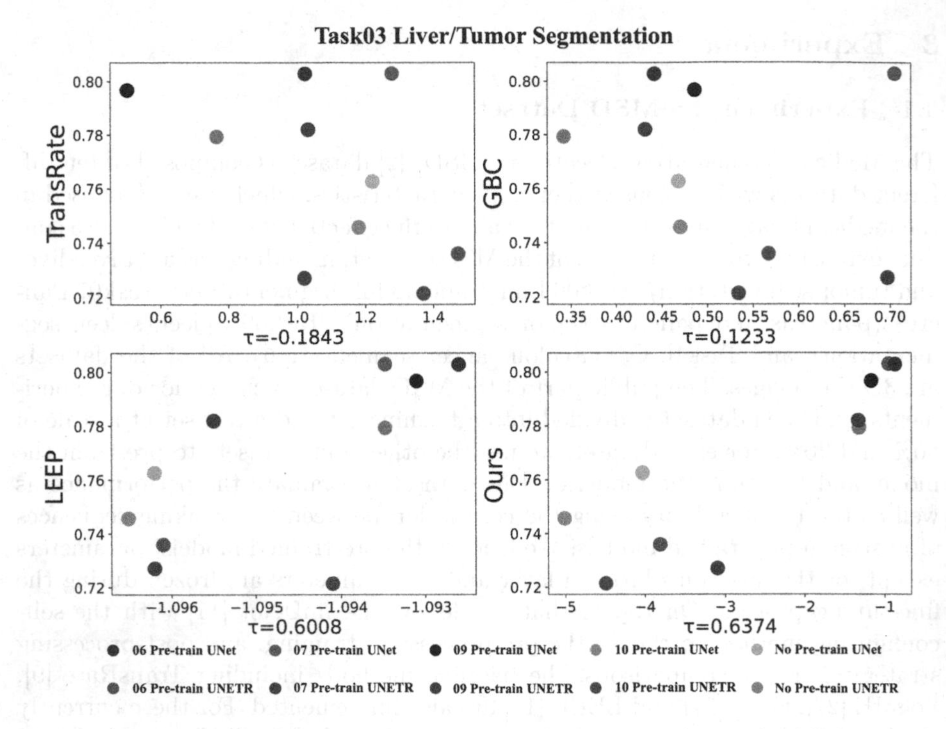

Fig. 2. Correlation between the fine-tuning performance and transferability metrics using Task03 as an example. The vertical axis represents the average Dice of the model, while the horizontal axis represents the transferability metric results. We have standardized the various metrics uniformly, aiming to observe a positive relationship between higher performance and higher transferability estimations.

Table 1. Pearson coefficient and weighted Kendall's τ for transferability estimation

Data/Method	Metrics	Task03	Task06	Task07	Task09	Task10	Avg
LogME	τ	−0.1628	−0.0988	0.3280	0.2778	−0.2348	0.0218
	pearson	0.0412	0.5713	0.3236	0.2725	−0.1674	0.2082
TransRate	τ	−0.1843	−0.1028	0.5923	0.4322	0.6069	0.2688
	pearson	−0.5178	−0.2804	0.7170	0.5573	0.7629	0.2478
LEEP	τ	0.6008	0.1658	0.2691	0.3516	0.5841	0.3943
	pearson	0.6765	−0.0073	0.7146	0.1633	0.4979	0.4090
GBC	τ	0.1233	−0.1569	0.6637	0.7611	0.6643	0.4111
	pearson	−0.2634	−0.3733	0.7948	0.7604	0.7404	0.3317
Ours CC-FV	τ	0.6374	0.0735	0.6569	0.5700	0.5550	**0.4986**
	pearson	0.8608	0.0903	0.9609	0.7491	0.8406	**0.7003**

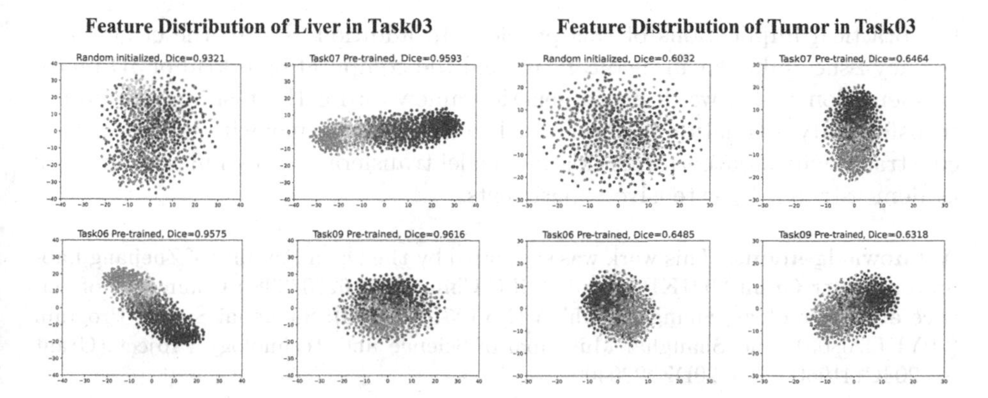

Fig. 3. Visualization of features with same labels using t-SNE. Points with different colors are from different samples. Pre-trained models tend to have a more consistent distribution within a class than the randomly initialized model and after fine-tuning they often have a better Dice performance than the randomly initialized models.

Table 2. Ablation on the effectiveness of different parts in our methods

Data/Method	Task03	Task06	Task07	Task09	Task10	Avg
Ours CC-FV	0.6374	0.0735	0.6569	0.5700	0.5550	**0.4986**
Ours w/o C_{cons}	0.1871	−0.2210	−0.2810	−0.0289	−0.2710	−0.1230
Ours w/o F_v	0.6165	0.3235	0.6054	0.2761	0.5269	0.4697
Single-scale	0.4394	0.0252	0.5336	0.5759	0.6007	0.4341
KL-divergence	−0.5658	−0.0564	0.2319	0.4628	−0.0323	0.0080
Bha-distance	0.1808	0.0723	0.2295	0.7866	0.4650	0.3468

3.2 Ablation Study

In Table 2 we analyze the different parts of our method and compare some other methods. First, we analyze the impact of class consistency C_{cons} and feature variety F_v. Though F_v can not contribute to the final Kendall's τ directly, C_{cons} with the constraint of F_v promotes the total estimation result. Then we compare the performance of our method at single and multiple scales to prove the effectiveness of our multi-scale strategy. Finally, we change the distance metrics in class consistency estimation. KL-divergence and Bha-distance are unstable in high dimension matrics calculation and the performance is also inferior to the Wasserstein distance. Figure 3 visualize the distribution of different classes using t-SNE methods. We can easily find that with models with a pre-training process have a more compact intra-class distance and a higher fine-tuning performance.

4 Conclusion

In our work, we raise the problem of model selection for upstream and downstream transfer processes in the medical image segmentation task and analyze the practical implications of this problem. In addition, due to the ethical and privacy issues inherent in medical care and the computational load of 3D image segmentation tasks, we design a generic framework for the task and propose a transferability estimation method based on class consistency with feature variety constraint, which outperforms existing model transferability estimation methods as demonstrated by extensive experiments.

Acknowledgement. This work was supported by the Open Funding of Zhejiang Laboratory under Grant 2021KH0AB03, NSFC China (No. 62003208); Committee of Science and Technology, Shanghai, China (No.19510711200); Shanghai Sailing Program (20YF1420800), and Shanghai Municipal of Science and Technology Project (Grant No.20JC1419500, No. 20DZ2220400).

References

1. Agostinelli, A., Uijlings, J., Mensink, T., Ferrari, V.: Transferability metrics for selecting source model ensembles. In: CVPR, pp. 7936–7946 (2022)
2. Antonelli, M., et al.: The medical segmentation decathlon. Nat. Commun. **13**(1), 4128 (2022)
3. Chen, W., et al.: Contrastive syn-to-real generalization. arXiv preprint arXiv:2104.02290 (2021)
4. Chen, X., Wang, S., Fu, B., Long, M., Wang, J.: Catastrophic forgetting meets negative transfer: batch spectral shrinkage for safe transfer learning. In: NIPS 32 (2019)
5. Cui, Q., et al.: Discriminability-transferability trade-off: an information-theoretic perspective. In: ECCV, pp. 20–37. Springer (2022). https://doi.org/10.1007/978-3-031-19809-0_2
6. Dwivedi, K., Huang, J., Cichy, R.M., Roig, G.: Duality diagram similarity: a generic framework for initialization selection in task transfer learning. In: Vedaldi, A., Bischof, H., Brox, T., Frahm, J.-M. (eds.) ECCV 2020. LNCS, vol. 12371, pp. 497–513. Springer, Cham (2020). https://doi.org/10.1007/978-3-030-58574-7_30
7. Dwivedi, K., Roig, G.: Representation similarity analysis for efficient task taxonomy & transfer learning. In: CVPR, pp. 12387–12396 (2019)
8. Hatamizadeh, A., et al.: Unetr: transformers for 3d medical image segmentation. In: CVPR, pp. 574–584 (2022)
9. Huang, L.K., Huang, J., Rong, Y., Yang, Q., Wei, Y.: Frustratingly easy transferability estimation. In: ICML, pp. 9201–9225. PMLR (2022)
10. Irvin, J., et al.: Chexpert: a large chest radiograph dataset with uncertainty labels and expert comparison. In: AAAI, vol. 33, pp. 590–597 (2019)
11. Isensee, F., Jaeger, P.F., Kohl, S.A., Petersen, J., Maier-Hein, K.H.: nnu-net: a self-configuring method for deep learning-based biomedical image segmentation. Nat. Methods **18**(2), 203–211 (2021)
12. Li, X., et al.: Delta: deep learning transfer using feature map with attention for convolutional networks. arXiv preprint arXiv:1901.09229 (2019)
13. Li, Y., et al.: Ranking neural checkpoints. In: CVPR, pp. 2663–2673 (2021)
14. Lin, T.Y., Dollár, P., Girshick, R., He, K., Hariharan, B., Belongie, S.: Feature pyramid networks for object detection. In: CVPR, pp. 2117–2125 (2017)

15. Nguyen, C., Hassner, T., Seeger, M., Archambeau, C.: Leep: a new measure to evaluate transferability of learned representations. In: ICML, pp. 7294–7305. PMLR (2020)
16. Panaretos, V.M., Zemel, Y.: Statistical aspects of wasserstein distances. Annual Rev. Stat. Appli. **6**, 405–431 (2019)
17. Pándy, M., Agostinelli, A., Uijlings, J., Ferrari, V., Mensink, T.: Transferability estimation using bhattacharyya class separability. In: CVPR, pp. 9172–9182 (2022)
18. Paszke, A., et al.: Pytorch: an imperative style, high-performance deep learning library. In: Advances in Neural Information Processing Systems 32 (2019)
19. Reiss, T., Cohen, N., Bergman, L., Hoshen, Y.: Panda: adapting pretrained features for anomaly detection and segmentation. In: CVPR, pp. 2806–2814 (2021)
20. Ronneberger, O., Fischer, P., Brox, T.: U-Net: convolutional networks for biomedical image segmentation. In: Navab, N., Hornegger, J., Wells, W.M., Frangi, A.F. (eds.) MICCAI 2015. LNCS, vol. 9351, pp. 234–241. Springer, Cham (2015). https://doi.org/10.1007/978-3-319-24574-4_28
21. Tajbakhsh, N., et al.: Convolutional neural networks for medical image analysis: full training or fine tuning? IEEE Trans. Med. Imaging **35**(5), 1299–1312 (2016)
22. Tong, X., Xu, X., Huang, S.L., Zheng, L.: A mathematical framework for quantifying transferability in multi-source transfer learning. NIPS **34**, 26103–26116 (2021)
23. Wang, T., Isola, P.: Understanding contrastive representation learning through alignment and uniformity on the hypersphere. In: ICML, pp. 9929–9939. PMLR (2020)
24. Wang, Z., Dai, Z., Póczos, B., Carbonell, J.: Characterizing and avoiding negative transfer. In: CVPR, pp. 11293–11302 (2019)
25. Wolf, T., et al.: Transformers: State-of-the-art natural language processing. In: EMNLP, pp. 38–45 (2020)
26. Xuhong, L., Grandvalet, Y., Davoine, F.: Explicit inductive bias for transfer learning with convolutional networks. In: ICML, pp. 2825–2834. PMLR (2018)
27. You, K., Liu, Y., Wang, J., Long, M.: Logme: practical assessment of pre-trained models for transfer learning. In: ICM, pp. 12133–12143. PMLR (2021)
28. Zamir, A.R., Sax, A., Shen, W., Guibas, L.J., Malik, J., Savarese, S.: Taskonomy: disentangling task transfer learning. In: CVPR, pp. 3712–3722 (2018)
29. Zhao, H., Shi, J., Qi, X., Wang, X., Jia, J.: Pyramid scene parsing network. In: CVPR, pp. 2881–2890 (2017)
30. Zhou, Z., Shin, J.Y., Gurudu, S.R., Gotway, M.B., Liang, J.: Active, continual fine tuning of convolutional neural networks for reducing annotation efforts. Med. Image Anal. **71**, 101997 (2021)

Source-Free Domain Adaptive Fundus Image Segmentation with Class-Balanced Mean Teacher

Longxiang Tang[1], Kai Li[2(✉)], Chunming He[1], Yulun Zhang[3], and Xiu Li[1(✉)]

[1] Tsinghua Shenzhen International Graduate School, Tsinghua University, Shenzhen, China
{lloong.x,chunminghe19990224}@gmail.com, li.xiu@sz.tsinghua.edu.cn

[2] NEC Laboratories America, Princeton, USA
li.gml.kai@gmail.com

[3] ETH Zurich, Zürich, Switzerland
yulun100@gmail.com

Abstract. This paper studies source-free domain adaptive fundus image segmentation which aims to adapt a pretrained fundus segmentation model to a target domain using unlabeled images. This is a challenging task because it is highly risky to adapt a model only using unlabeled data. Most existing methods tackle this task mainly by designing techniques to carefully generate pseudo labels from the model's predictions and use the pseudo labels to train the model. While often obtaining positive adaption effects, these methods suffer from two major issues. First, they tend to be fairly unstable - incorrect pseudo labels abruptly emerged may cause a catastrophic impact on the model. Second, they fail to consider the severe class imbalance of fundus images where the foreground (e.g., cup) region is usually very small. This paper aims to address these two issues by proposing the Class-Balanced Mean Teacher (CBMT) model. CBMT addresses the unstable issue by proposing a weak-strong augmented mean teacher learning scheme where only the teacher model generates pseudo labels from weakly augmented images to train a student model that takes strongly augmented images as input. The teacher is updated as the moving average of the instantly trained student, which could be noisy. This prevents the teacher model from being abruptly impacted by incorrect pseudo-labels. For the class imbalance issue, CBMT proposes a novel loss calibration approach to highlight foreground classes according to global statistics. Experiments show that CBMT well addresses these two issues and outperforms existing methods on multiple benchmarks.

Keywords: Source-free domain adaptation · Fundus image · Mean teacher

---

Supplementary Information The online version contains supplementary material available at https://doi.org/10.1007/978-3-031-43907-0_65.

H. Greenspan et al. (Eds.): MICCAI 2023, LNCS 14220, pp. 684–694, 2023.
https://doi.org/10.1007/978-3-031-43907-0_65

1 Introduction

Medical image segmentation plays an essential role in computer-aided diagnosis systems in different applications and has been tremendously advanced in the past few years [6,12,19,22]. While the segmentation model [3,10,11,21] always requires sufficient labeled data, unsupervised domain adaptation (UDA) approaches have been proposed, learning an adaptive model jointly with unlabeled target domain images and labeled source domain images [9], for example, the adversarial training paradigm [2,8,13,14,16].

Although impressive performance has been achieved, these UDA methods may be limited for some real-world medical image segmentation tasks where labeled source images are not available for adaptation. This is not a rare scenario because medical images are usually highly sensitive in privacy and copyright protection such that labeled source images may not be allowed to be distributed. This motivates the investigation of source-free domain adaptation (SFDA) where adapts a source segmentation model trained on labeled source data (in a private-protected way) to the target domain only using unlabeled data.

A few recent SFDA works have been proposed. OSUDA [17] utilizes the domain-specific low-order batch statistics and domain-shareable high-order batch statistics, trying to adapt the former and keep the consistency of the latter. SRDA [1] minimizes a label-free entropy loss guided with a domain-invariant class-ratio prior. DPL [4] introduces pixel-level and class-level pseudo-label denoising schemes to reduce noisy pseudo-labels and select reliable ones. U-D4R [27] applies an adaptive class-dependent threshold with the uncertainty-rectified correction to realize better denoising.

Although these methods have achieved some success in model adaptation, they still suffer from two major issues. First, they tend to be fairly unstable. Without any supervision signal from labeled data, the model heavily relies on the predictions generated by itself, which are always noisy and could easily make the training process unstable, causing catastrophic error accumulation after several training epochs as shown in Fig. 1(a). Some works avoid this problem by only training the model for very limited iterations (only 2 epochs in [4,27]) and selecting the best-performing model during the whole training process for testing. However, this does not fully utilize the data and it is non-trivial to select the best-performing model for this unsupervised learning task. Second, they failed to consider the severe foreground and background imbalance of fundus images where the foreground (e.g., cup) region is usually very small (as shown in Fig. 1(b)). This oversight could also lead to a model degradation due to the dominate background learning signal.

In this paper, we propose the Class-Balanced Mean Teacher (CBMT) method to address the limitations of existing methods. To mitigate the negative impacts of incorrect pseudo labels, we propose a weak-strong augmented mean teacher learning scheme which involves a teacher model and a student model that are both initialized from the source model. We use the teacher to generate pseudo label from a weakly augmented image, and train the student that takes strongly augmented version of the same image as input. We do not train the teacher

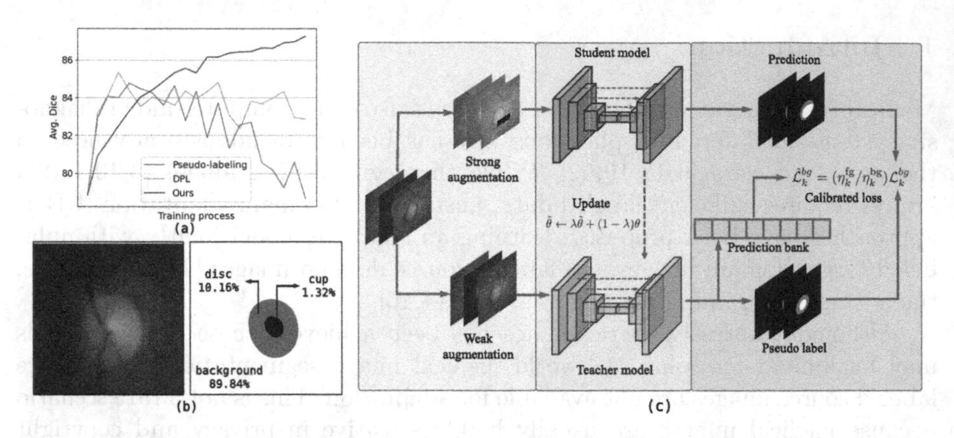

Fig. 1. (a) Training curve of vanilla pseudo-labeling, DPL [4] and our approach. (b) Fundus image and its label with class proportion from the *RIM-ONE-r3* dataset. (c) Illustrated framework of our proposed CBMT method.

model directly by back-propagation but update its weights as the moving average of the student model. This prevents the teacher model from being abruptly impacted by incorrect pseudo labels and meanwhile accumulates new knowledge learned by the student model. To address the imbalance between foreground and background, we propose to calibrate the segmentation loss and highlight the foreground class, based on the prediction statistics derived from the global information. We maintain a prediction bank to capture global information, which is considered more reliable than that inside one image.

Our contributions can be summarized as follows: (1) We propose the weak-strong augmented mean teacher learning scheme to address the stable issue of existing methods. (2) We propose the novel global knowledge-guided loss calibration technique to address the foreground and background imbalance problem. (3) Our proposed CBMT reaches state-of-the-art performance on two popular benchmarks for adaptive fundus image segmentation.

2 Method

Source-Free Domain Adaptive (SFDA) fundus image segmentation aims to adapt a source model h, trained with N_S labeled source images $\mathcal{S} = \{(X_i, Y_i)\}_{i=1}^{N_S}$, to the target domain using only N_T unlabeled target images $\mathcal{T} = \{X_i\}_{i=1}^{N_T}$. $Y_i \in \{0,1\}^{H \times W \times C}$ is the ground truth, and H, W, and C denote the image height, width, and class number, respectively. A vanilla pseudo-labeling-based method generates pseudo labels $\hat{y} \in \mathbb{R}^C$ from the sigmoided model prediction $p = h(x)$ for each pixel $x \in X_i$ with source model h:

$$\hat{y}_k = \mathbb{1}\left[p_k > \gamma\right], \tag{1}$$

where $\mathbb{1}$ is the indicator function and $\gamma \in [0,1]$ is the probability threshold for transferring soft probability to hard label. p_k and y_k is the k-th dimension of p

and y, respectively, denoting the prediction and pseudo label for class k. Then $(x, \hat{y})$ is utilized to train the source model h with binary cross entropy loss:

$$L_{bce} = \mathbb{E}_{x \sim X_i}[\hat{y} \log(p) + (1 - \hat{y}) \log(1 - p)] \tag{2}$$

Most existing SFDA works refine this vanilla method by proposing techniques to calibrate p and get better pseudo label $\hat{y}$, or measure the uncertainty of p and apply a weight when using $\hat{y}$ for computing the loss [4,27]. While achieving improved performance, these methods still suffer from the unstable issue because noisy $\hat{y}$ will directly impact h, and the error will accumulate since then the predictions of h will be used for pseudo labeling. Another problem with this method is that they neglect the imbalance of the foreground and background pixels in fungus images, where the foreground region is small. Consequently, the second term in Eq. (2) will dominate the loss, which is undesirable.

Our proposed CBMT model addresses the two problems by proposing the weak-strong augmented mean teacher learning scheme and the global knowledge-guided loss calibration technique. Figure 1(c) shows the framework of CBMT.

2.1 Weak-Strong Augmented Mean Teacher

To avoid error accumulation and achieve a robust training process, we introduce the weak-strong augmented mean teacher learning scheme where there is a teacher model h_t and a student model h_s both initialized from the source model h. We generate pseudo labels with h_t and use the pseudo labels to train h_s. To enhance generalization performance, we further introduce a weak-strong augmentation mechanism that feeds weakly and strongly augmented images to the teacher model and the student model, respectively.

Concretely, for each image X_i, we generate a weakly-augmented version X_i^w by using image flipping and resizing. Meanwhile, we generate a strongly-augmented version X_i^s. The strong augmentations we used include a random eraser, contrast adjustment, and impulse noises. For each pixel $x^w \in X_i^w$, we generate pseudo label $\hat{y}^w = h_t(x)$ by the teacher model h_t with Eq. (1). Then, we train the student model h_s with

$$\mathcal{L} = \mathbb{E}_{x^s \sim X_i^s, \hat{y}^w}[\tilde{\mathcal{L}}_{bce}], \tag{3}$$

where $\tilde{\mathcal{L}}_{bce}$ is the refined binary cross entropy loss which we will introduce later. It is based on Eq. (2) but addresses the fore- and back-ground imbalance problem.

The weakly-strong augmentation mechanism has two main benefits. First, since fundus image datasets are always on a small scale, the model could easily get overfitted due to the insufficient training data issue. To alleviate it, we enhance the diversity of the training set by introducing image augmentation techniques. Second, learning with different random augmentations performs as a consistency regularizer constraining images with similar semantics to the same class, which forms a more distinguishable feature representation.

We update the student model by back-propagating the loss defined in Eq. (3). But for the teacher model, we update it as the exponential moving average (EMA) of the student model as,

$$\tilde{\theta} \leftarrow \lambda\tilde{\theta} + (1 - \lambda)\theta, \tag{4}$$

where $\tilde{\theta}$, θ are the teacher and student model weights separately. Instead of updating the model with gradient directly, we define the teacher model as the exponential moving average of students, which makes the teacher model more consistent along the adaptation process. With this, we could train a model for a relatively long process and safely choose the final model without accuracy validation. From another perspective, the teacher model can be interpreted as a temporal ensemble of students in different time steps [18], which enhances the robustness of the teacher model.

2.2 Global Knowledge Guided Loss Calibration

For a fundas image, the foreground object (e.g., cup) is usually quite small and most pixel will the background. If we update the student model with Eq. (2), the background class will dominate the loss, which dilutes the supervision signals for the foreground class. The proposed global knowledge guided loss calibration technique aims to address this problem.

A naive way to address the foreground and background imbalance is to calculate the numbers of pixels falling into the two categories, respectively, within each individual image and devise a loss weighting function based on the numbers. This strategy may work well for the standard supervised learning tasks, where the labels are reliable. But with pseudo labels, it is too risky to conduct the statistical analysis based on a single image. To remedy this, we analyze the class imbalance across the whole dataset, and use this global knowledge to calibrate our loss for each individual image.

Specifically, we store the predictions of pixels from all images and maintain the mean loss for foreground and background as,

$$\eta_k^{\text{fg}} = \frac{\sum_i \mathcal{L}_{i,k} \cdot \mathbb{1}[\hat{y}_{i,k} = 1]}{\sum_i \mathbb{1}[\hat{y}_{i,k} = 1]}; \quad \eta_k^{\text{bg}} = \frac{\sum_i \mathcal{L}_{i,k} \cdot \mathbb{1}[\hat{y}_{i,k} = 0]}{\sum_i \mathbb{1}[\hat{y}_{i,k} = 0]} \tag{5}$$

where $\mathcal{L}$ is the segmentation loss mentioned above, and "fg" and "bg" represent foreground/background. The reason we use the mean of the loss, rather than the number of pixels, is that the loss of each pixel indicates the "hardness" of each pixel according to the pseudo ground truth. This gives more weight to those more informative pixels, thus more global knowledge is considered.

With each average loss, the corresponding learning scheme could be further calibrated. We utilize the ratio of η_k^{fg} to η_k^{bg} to weight background loss $\mathcal{L}_k^{bg}$:

$$\tilde{\mathcal{L}}_{bce} = \mathbb{E}_{x \sim X_i, k \sim C}[\hat{y}_k \log(p_k) + \eta_k^{\text{fg}}/\eta_k^{\text{bg}}(1 - \hat{y_k}) \log(1 - p_k)] \tag{6}$$

The calibrated loss ensures fair learning among different classes, therefore alleviating model degradation issues caused by class imbalance.

Since most predictions are usually highly confident (very close to 0 or 1), they are thus less informative. We need to only include pixels with relatively large loss scales to compute mean loss. We realize this by adopting constraint threshold α to select pixels: $\frac{|f(x_i)-\gamma|}{|\hat{y}_i-\gamma|} > \alpha$, where α is set by default to 0.2. α represents the lower bound threshold of normalized prediction, which can filter well-segmented uninformative pixels out.

3 Experiments

Implementation Details[1]. We apply the Deeplabv3+ [5] with MobileNetV2 [23] backbone as our segmentation model, following the previous works [4,26,27] for a fair comparison. For model optimization, we use Adam optimizer with 0.9 and 0.99 momentum coefficients. During the source model training stage, the initial learning rate is set to 1e-3 and decayed by 0.98 every epoch, and the training lasts 200 epochs. At the source-free domain adaptation stage, the teacher and student model are first initialized by the source model, and the EMA update scheme is applied between them for a total of 20 epochs with a learning rate of 5e-4. Loss calibration parameter η is computed every epoch and implemented on the class cup. The output probability threshold γ is set as 0.75 according to previous study [26] and model EMA update rate λ is 0.98 by default. We implement our method with PyTorch on one NVIDIA 3090 GPU and set batch size as 8 when adaptation.

Datasets and Metrics. We evaluate our method on widely-used fundus optic disc and cup segmentation datasets from different clinical centers. Following previous works, We choose the REFUGE challenge training set [20] as the source domain and adapt the model to two target domains: RIM-ONE-r3 [7] and Drishti-GS [24] datasets for evaluation. Quantitatively, the source domain consists of 320/80 fundus images for training/testing with pixel-wise optic disc and cup segmentation annotation, while the target domains have 99/60 and 50/51 images. Same as [26], the fundus images are cropped to 512×512 as ROI regions.

We compare our CBMT model with several state-of-the-art domain adaptation methods, including UDA methods BEAL [26] and AdvEnt [25] and SFDA methods: SRDA [1], DAE [15] and DPL [4]. More comparisons with U-D4R [27] under other adaptation settings could be found in supplementary materials. General metrics for segmentation tasks are used for model performance evaluation, including the Dice coefficient and Average Symmetric Surface Distance (ASSD). The dice coefficient (the higher the better) gives pixel-level overlap results, and ASSD (the lower the better) indicates prediction boundary accuracy.

3.1 Experimental Results

The quantitative evaluation results are shown in Table 1. We include the without adaptation results from [4] as a lower bound, and the supervised learning results

[1] The code can be found in https://github.com/lloongx/SFDA-CBMT.

Table 1. Quantitative results of comparison with different methods on two datasets, and the best score for each column is highlighted. - means the results are not reported by that method, ± refers to the standard deviation across samples in the dataset. S-F means source-free.

Methods	S-F	Optic Disc Segmentation		Optic Cup Segmentation	
		Dice[%] ↑	ASSD[pixel] ↓	Dice[%] ↑	ASSD[pixel] ↓
RIM-ONE-r3					
W/o DA [4]		83.18±6.46	24.15±15.58	74.51±16.40	14.44±11.27
Oracle [26]		96.80	–	85.60	–
BEAL [26]	×	89.80	–	81.00	–
AdvEnt [25]	×	89.73±3.66	9.84±3.86	77.99±21.08	**7.57**±4.24
SRDA [1]	✓	89.37±2.70	9.91±2.45	77.61±13.58	10.15±5.75
DAE [15]	✓	89.08±3.32	11.63±6.84	79.01±12.82	10.31±8.45
DPL [4]	✓	90.13±3.06	9.43±3.46	79.78±11.05	9.01±5.59
CBMT(Ours)	✓	**93.36±4.07**	**6.20±4.79**	**81.16±14.71**	8.37±6.99
Drishti-GS					
W/o DA [4]		93.84±2.91	9.05±7.50	83.36±11.95	11.39±6.30
Oracle [26]		97.40	–	90.10	–
BEAL [26]	×	96.10	–	**86.20**	–
AdvEnt [25]	×	96.16±1.65	4.36±1.83	82.75±11.08	11.36±7.22
SRDA [1]	✓	96.22±1.30	4.88±3.47	80.67±11.78	13.12±6.48
DAE [15]	✓	94.04±2.85	8.79±7.45	83.11±11.89	11.56±6.32
DPL [4]	✓	96.39±1.33	**4.08±1.49**	83.53±17.80	11.39±10.18
CBMT(Ours)	✓	**96.61±1.45**	**3.85±1.63**	84.33±11.70	**10.30±5.88**

Table 2. Ablation study results of our proposed modules on the *RIM-ONE-r3* dataset. P-L means vanilla pseudo-labeling method. * represents the accuracy is manually selected from the best epoch. The best results are highlighted.

P-L	EMA	Aug.	Calib.	Avg. Dice ↑	Avg. ASSD ↓
✓				64.19 (84.68*)	15.11 (9.67*)
✓	✓			83.63	8.51
✓	✓	✓		84.36	8.48
✓	✓		✓	86.04	8.26
✓	✓	✓	✓	**87.26**	**7.29**

from [26] as an upper bound, same as [4]. As shown in the table, both two quantitative metric results perform better than previous state-of-the-art SFDA methods and even show an improvement against traditional UDA methods on

Table 3. Loss calibration weight with different thresholds α on *RIM-ONE-r3* dataset. Our method is robust to the hyper-parameter setting.

α	0	0.1	0.2	0.3	0.4	0.5	0.6	0.7	0.8	0.9
$\eta_k^{\text{fg}}/\eta_k^{\text{bg}}$	2.99	0.24	0.24	0.24	0.24	0.24	0.23	0.23	0.22	0.22

some metrics. Especially in the RIM-ONE-r3 dataset, our CBMT gains a great performance increase than previous works (dice gains by 3.23 on disc), because the domain shift issue is severer here and has big potential for improvement.

Moreover, CBMT alleviates the need for precise tuning of hyper-parameters. Here we could set a relatively long training procedure (our epoch number is 10 times that of [4,27]), and safely select the last checkpoint as our final result without concerning about model degradation issue, which is crucial in real-world clinical source-free domain adaptation application.

3.2 Further Analyses

Ablation Study. In order to assess the contribution of each component to the final performance, we conducted an ablation study on the main modules of CBMT, as summarized in Table 2. Note that we reduced the learning rates by a factor of 20 for the experiments of the vanilla pseudo-labeling method to get comparable performance because models are prone to degradation without EMA updating. As observed in quantitative results, the EMA update strategy avoids the model from degradation, which the vanilla pseudo-labeling paradigm suffers from. Image augmentation and loss calibration also boost the model accuracy, and the highest performance is achieved with both. The loss calibration module achieves more improvement in its solution to class imbalance, while image augmentation is easy to implement and plug-and-play under various circumstances.

Hyper-parameter Sensitivity Analysis. We further investigate the impact of different hyper-parameter. Figure 2(a) presents the accuracy with different EMA update rate parameters λ. It demonstrates that both too low and too high update rates would cause a drop in performance, which is quite intuitive: a higher λ leads to inconsistency between the teacher and student, and thus teacher can hardly learn knowledge from the student; On the other hand, a lower λ will always keep teacher and student close, making it degenerated to vanilla pseudo-labeling. But within a reasonable range, the model is not sensitive to update rate λ.

To evaluate the variation of the loss calibration weight $\eta_k^{\text{fg}}/\eta_k^{\text{bg}}$ with different constraint thresholds α, we present the results in Table 3. As we discussed in Sect. 2.2, most pixels in an image are well-classified, and if we simply calculate with all pixels (i.e. $\alpha = 0$), as shown in the first column, the mean loss of background will be severely underestimated due to the large quantity of zero-loss pixel. Besides, as α changes, the calibration weight varies little, indicating the robustness of our calibration technique to threshold α.

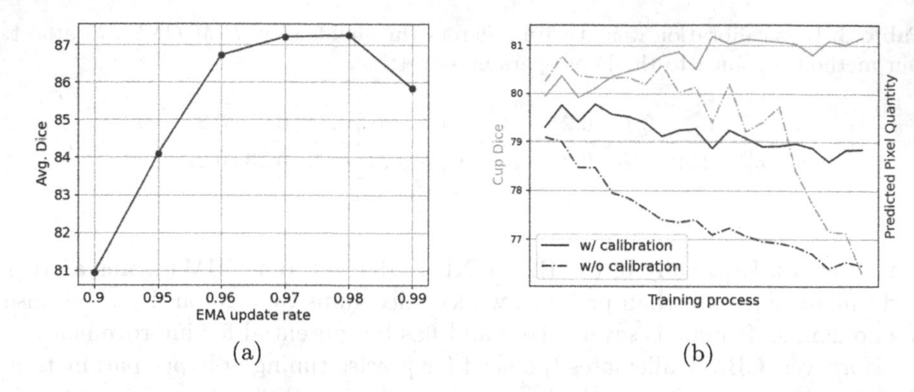

Fig. 2. (a) Model performance with different EMA update rate λ setting. (b) Training curves with and without our proposed loss calibration scheme.

The Effectiveness of Loss Calibration to Balance Class. The class imbalance problem can cause misalignment in the learning processes of different classes, leading to a gradual decrease of predicted foreground area. This can ultimately result in model degradation. As shown in Fig. 2(b), neglecting the issue of class imbalance can cause a significant drop in the predicted pixel quantity of the class "cup" during training, and finally leads to a performance drop. Loss calibration performs a theoretical investigation and proposes an effective technique to alleviate this issue by balancing loss with global context.

4 Conclusion

In this work, we propose a class-balanced mean teacher framework to realize robust SFDA learning for more realistic clinical application. Based on the observation that model suffers from degradation issues during adaptation training, we introduce a mean teacher strategy to update the model via an exponential moving average way, which alleviates error accumulation. Meanwhile, by investigating the foreground and background imbalance problem, we present a global knowledge guided loss calibration module. Experiments on two fundus image segmentation datasets show that CBMT outperforms previous SFDA methods.

Acknowledgement. This work was partly supported by Shenzhen Key Laboratory of next generation interactive media innovative technology (No: ZDSYS20210623092001004).

References

1. Bateson, M., Kervadec, H., Dolz, J., Lombaert, H., Ben Ayed, I.: Source-Relaxed domain adaptation for image segmentation. In: MICCAI 2020. LNCS, vol. 12261, pp. 490–499. Springer, Cham (2020). https://doi.org/10.1007/978-3-030-59710-8_48

2. Cai, J., Zhang, Z., Cui, L., Zheng, Y., Yang, L.: Towards cross-modal organ translation and segmentation: a cycle-and shape-consistent generative adversarial network. Med. Image Anal. **52**, 174–184 (2019)
3. Carion, N., Massa, F., Synnaeve, G., Usunier, N., Kirillov, A., Zagoruyko, S.: End-to-end object detection with transformers. In: Vedaldi, A., Bischof, H., Brox, T., Frahm, J.-M. (eds.) ECCV 2020. LNCS, vol. 12346, pp. 213–229. Springer, Cham (2020). https://doi.org/10.1007/978-3-030-58452-8_13
4. Chen, C., Liu, Q., Jin, Y., Dou, Q., Heng, P.-A.: Source-free domain adaptive fundus image segmentation with denoised pseudo-labeling. In: de Bruijne, M., et al. (eds.) MICCAI 2021. LNCS, vol. 12905, pp. 225–235. Springer, Cham (2021). https://doi.org/10.1007/978-3-030-87240-3_22
5. Chen, L.C., Zhu, Y., Papandreou, G., Schroff, F., Adam, H.: Encoder-decoder with atrous separable convolution for semantic image segmentation. In: Proceedings of the European Conference on Computer Vision (ECCV), pp. 801–818 (2018)
6. Drozdzal, M., Vorontsov, E., Chartrand, G., Kadoury, S., Pal, C.: The importance of skip connections in biomedical image segmentation. In: Carneiro, G., et al. (eds.) LABELS/DLMIA -2016. LNCS, vol. 10008, pp. 179–187. Springer, Cham (2016). https://doi.org/10.1007/978-3-319-46976-8_19
7. Fumero, F., Alayón, S., Sanchez, J.L., Sigut, J., Gonzalez-Hernandez, M.: Rim-one: an open retinal image database for optic nerve evaluation. In: 2011 24th International Symposium on Computer-based Medical Systems (CBMS), pp. 1–6. IEEE (2011)
8. Gadermayr, M., Gupta, L., Appel, V., Boor, P., Klinkhammer, B.M., Merhof, D.: Generative adversarial networks for facilitating stain-independent supervised and unsupervised segmentation: a study on kidney histology. IEEE Trans. Med. Imaging **38**(10), 2293–2302 (2019)
9. Ganin, Y., et al.: Domain-adversarial training of neural networks. J. Mach. Learn. Res. **17**(1), 2096–2030 (2016)
10. He, C., et al.: Camouflaged object detection with feature decomposition and edge reconstruction. In: Proceedings of the IEEE/CVF Conference on Computer Vision and Pattern Recognition, pp. 22046–22055 (2023)
11. He, C., et al.: Weakly-supervised concealed object segmentation with SAM-based pseudo labeling and multi-scale feature grouping. arXiv preprint: arXiv:2305.11003 (2023)
12. Ibtehaz, N., Rahman, M.S.: MultiResUNet: rethinking the u-net architecture for multimodal biomedical image segmentation. Neural Netw. **121**, 74–87 (2020)
13. Javanmardi, M., Tasdizen, T.: Domain adaptation for biomedical image segmentation using adversarial training. In: 2018 IEEE 15th International Symposium on Biomedical Imaging (ISBI 2018), pp. 554–558. IEEE (2018)
14. Kamnitsas, K., et al.: Unsupervised domain adaptation in brain lesion segmentation with adversarial networks. In: Niethammer, M., et al. (eds.) IPMI 2017. LNCS, vol. 10265, pp. 597–609. Springer, Cham (2017). https://doi.org/10.1007/978-3-319-59050-9_47
15. Karani, N., Erdil, E., Chaitanya, K., Konukoglu, E.: Test-time adaptable neural networks for robust medical image segmentation. Med. Image Anal. **68**, 101907 (2021)
16. Li, K., Zhang, Y., Li, K., Fu, Y.: Adversarial feature hallucination networks for few-shot learning. In: Proceedings of the IEEE/CVF Conference on Computer Vision and Pattern Recognition, pp. 13470–13479 (2020)

17. Liu, X., Xing, F., Yang, C., El Fakhri, G., Woo, J.: Adapting off-the-shelf source Segmenter for target medical image segmentation. In: de Bruijne, M., et al. (eds.) MICCAI 2021. LNCS, vol. 12902, pp. 549–559. Springer, Cham (2021). https://doi.org/10.1007/978-3-030-87196-3_51
18. Liu, Y.C., et al.: Unbiased teacher for semi-supervised object detection. arXiv preprint: arXiv:2102.09480 (2021)
19. Milletari, F., Navab, N., Ahmadi, S.A.: V-Net: fully convolutional neural networks for volumetric medical image segmentation. In: 2016 Fourth International Conference on 3D Vision (3DV), pp. 565–571. IEEE (2016)
20. Orlando, J.I., Fu, H., Breda, J.B., van Keer, K., Bathula, D.R., Diaz-Pinto, A., Fang, R., Heng, P.A., Kim, J., Lee, J., et al.: Refuge challenge: a unified framework for evaluating automated methods for glaucoma assessment from fundus photographs. Med. Image Anal. **59**, 101570 (2020)
21. Ren, S., He, K., Girshick, R., Sun, J.: Faster R-CNN: towards real-time object detection with region proposal networks. In: Advances in Neural Information Processing Systems, vol. 28 (2015)
22. Ronneberger, O., Fischer, P., Brox, T.: U-Net: convolutional networks for biomedical image segmentation. In: Navab, N., Hornegger, J., Wells, W.M., Frangi, A.F. (eds.) MICCAI 2015. LNCS, vol. 9351, pp. 234–241. Springer, Cham (2015). https://doi.org/10.1007/978-3-319-24574-4_28
23. Sandler, M., Howard, A., Zhu, M., Zhmoginov, A., Chen, L.C.: Mobilenetv 2: inverted residuals and linear bottlenecks. In: Proceedings of the IEEE Conference on Computer Vision and Pattern Recognition, pp. 4510–4520 (2018)
24. Sivaswamy, J., Krishnadas, S., Chakravarty, A., Joshi, G., Tabish, A.S., et al.: A comprehensive retinal image dataset for the assessment of glaucoma from the optic nerve head analysis. JSM Biomed. Imaging Data Pap. **2**(1), 1004 (2015)
25. Vu, T.H., Jain, H., Bucher, M., Cord, M., Pérez, P.: ADVENT: adversarial entropy minimization for domain adaptation in semantic segmentation. In: Proceedings of the IEEE/CVF Conference on Computer Vision and Pattern Recognition, pp. 2517–2526 (2019)
26. Wang, S., Yu, L., Li, K., Yang, X., Fu, C.-W., Heng, P.-A.: Boundary and entropy-driven adversarial learning for fundus image segmentation. In: Shen, D., et al. (eds.) MICCAI 2019. LNCS, vol. 11764, pp. 102–110. Springer, Cham (2019). https://doi.org/10.1007/978-3-030-32239-7_12
27. Xu, Z., et al.: Denoising for relaxing: unsupervised domain adaptive fundus image segmentation without source data. In: Wang, L., Dou, Q., Fletcher, P.T., Speidel, S., Li, S. (eds.) MICCAI 2022. Lecture Notes in Computer Science, vol. 13435, pp. 214–224. Springer, Cham (2022). https://doi.org/10.1007/978-3-031-16443-9_21

Unsupervised Domain Adaptation for Anatomical Landmark Detection

Haibo Jin[1], Haoxuan Che[1], and Hao Chen[1,2(✉)]

[1] Department of Computer Science and Engineering, The Hong Kong University of Science and Technology, Kowloon, Hong Kong
{hjinag,hche,jhc}@cse.ust.hk

[2] Department of Chemical and Biological Engineering, The Hong Kong University of Science and Technology, Kowloon, Hong Kong

Abstract. Recently, anatomical landmark detection has achieved great progresses on single-domain data, which usually assumes training and test sets are from the same domain. However, such an assumption is not always true in practice, which can cause significant performance drop due to domain shift. To tackle this problem, we propose a novel framework for anatomical landmark detection under the setting of unsupervised domain adaptation (UDA), which aims to transfer the knowledge from labeled source domain to unlabeled target domain. The framework leverages *self-training* and *domain adversarial learning* to address the domain gap during adaptation. Specifically, a self-training strategy is proposed to select reliable landmark-level pseudo-labels of target domain data with dynamic thresholds, which makes the adaptation more effective. Furthermore, a domain adversarial learning module is designed to handle the unaligned data distributions of two domains by learning domain-invariant features via adversarial training. Our experiments on cephalometric and lung landmark detection show the effectiveness of the method, which reduces the domain gap by a large margin and outperforms other UDA methods consistently.

1 Introduction

Anatomical landmark detection is a fundamental step in many clinical applications such as orthodontic diagnosis [11] and orthognathic treatment planning [6]. However, manually locating the landmarks can be tedious and time-consuming. And the results from manual labeling can cause errors due to the inconsistency in landmark identification [5]. Therefore, it is of great need to automate the task of landmark detection for efficiency and consistency.

In recent years, deep learning based methods have achieved great progresses in anatomical landmark detection. For supervised learning, earlier works [6,20,27] adopted heatmap regression with extra shape constraints. Later,

---

Supplementary Information The online version contains supplementary material available at https://doi.org/10.1007/978-3-031-43907-0_66.

H. Greenspan et al. (Eds.): MICCAI 2023, LNCS 14220, pp. 695–705, 2023.
https://doi.org/10.1007/978-3-031-43907-0_66

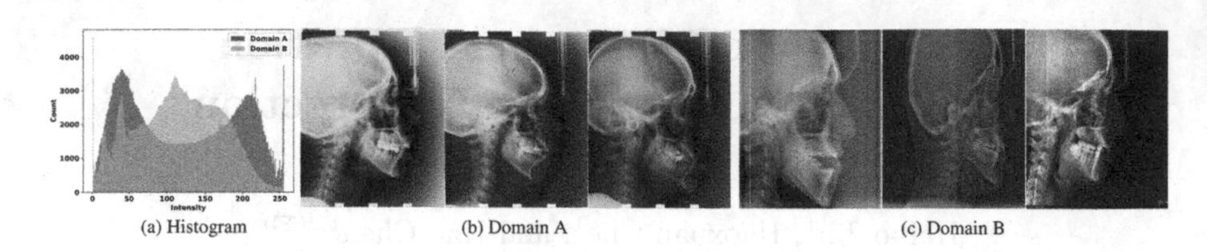

Fig. 1. Domain A vs. Domain B. (a) Image histogram. (b)–(c) Visual samples.

graph network [16] and self-attention [11] were introduced to model landmark dependencies in an end-to-end manner for better performance.

Despite the success of recent methods, they mostly focus on single-domain data, which assume the training and test sets follow the same distribution. However, such an assumption is not always true in practice, due to the differences in patient populations and imaging devices. Figure 1 shows that cephalogram images from two domains can be very different in both histogram and visual appearance. Therefore, a well trained model may encounter severe performance degradation in practice due to the domain shift of test data. A straightforward solution to this issue is to largely increase the size and diversity of training set, but the labeling is prohibitively expensive, especially for medical images. On the other hand, unsupervised domain adaptation (UDA) [10] aims to transfer the knowledge learned from the labeled source domain to the unlabeled target domain, which is a potential solution to the domain shift problem as unlabeled data is much easier to collect. The effectiveness of UDA has been proven in many vision tasks, such as image classification [10], object detection [7], and pose estimation [3,19,24]. However, its feasibility in anatomical landmark detection still remains unknown.

In this paper, we aim to investigate anatomical landmark detection under the setting of UDA. Our preliminary experiments show that a well-performed model will yield significant performance drop on cross-domain data, where the mean radial error (MRE) increases from 1.22 mm to 3.32 mm and the success detection rate (SDR) within 2 mm drops from 83.76% to 50.05%. To address the domain gap, we propose a unified framework, which contains a base landmark detection model, a self-training strategy, and a domain adversarial learning module. Specifically, self-training is adopted to effectively leverage the unlabeled data from the target domain via pseudo-labels. To handle confirmation bias [2], we propose landmark-aware self-training (LAST) to select pseudo-labels at the landmark-level with dynamic thresholds. Furthermore, to address the covariate shift [26] issue (i.e., unaligned data distribution) that may degrade the performance of self-training, a domain adversarial learning (DAL) module is designed to learn domain-invariant features via adversarial training. Our experiments on two anatomical datasets show the effectiveness of the proposed framework. For example, on cephalometric landmark detection, it reduces the domain gap in MRE by 47% (3.32 mm → 1.75 mm) and improves the SDR (2 mm) from 50.05% to 69.15%. We summarize our contributions as follows.

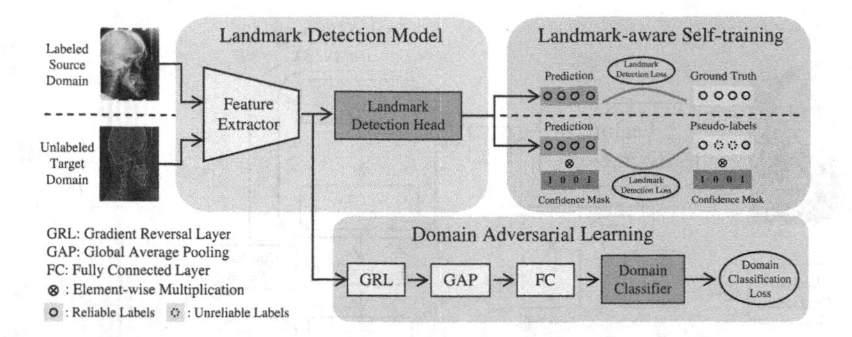

Fig. 2. The overall framework. Based on 1) the landmark detection model, it 2) utilizes LAST to leverage the unlabeled target domain data via pseudo-labels, and 3) simultaneously conducts DAL for domain-invariant features.

1. We investigated anatomical landmark detection under the UDA setting for the *first time*, and showed that domain shift indeed causes severe performance drop of a well-performed landmark detection model.
2. We proposed a novel framework for the UDA of anatomical landmark detection, which significantly improves the cross-domain performance and consistently outperforms other state-of-the-art UDA methods.

2 Method

Figure 2 shows the overall framework, which aims to yield satisfactory performance in target domain under the UDA setting. During training, it leverages both labeled source domain data $\mathcal{S} = \{x_i^S, y_i^S\}_{i=1}^N$ and unlabeled target domain data $\mathcal{T} = \{x_j^T\}_{j=1}^M$. For evaluation, it will be tested on a hold-out test set from target domain. The landmark detection model is able to predict landmarks with confidence, which is detailed in Sect. 2.1. To reduce domain gap, we further propose LAST and DAL, which are introduced in Sects. 2.2 and 2.3, respectively.

2.1 Landmark Detection Model

Recently, coordinate regression [11,16] has obtained better performance than heatmap regression [6,27]. However, coordinate based methods do not output confidence scores, which are necessary for pseudo-label selection in self-training [4,14]. To address this issue, we designed a model that is able to predict accurate landmarks while providing confidence scores. As shown in Fig. 3 (a), the model utilizes both coordinate and heatmap regression, where the former provides coarse but robust predictions via global localization, then projected to the local maps of the latter for prediction refinement and confidence measurement.

Global Localization. We adopt Transformer decoder [12,15] for coarse localization due to its superiority in global attentions. A convolutional neural network

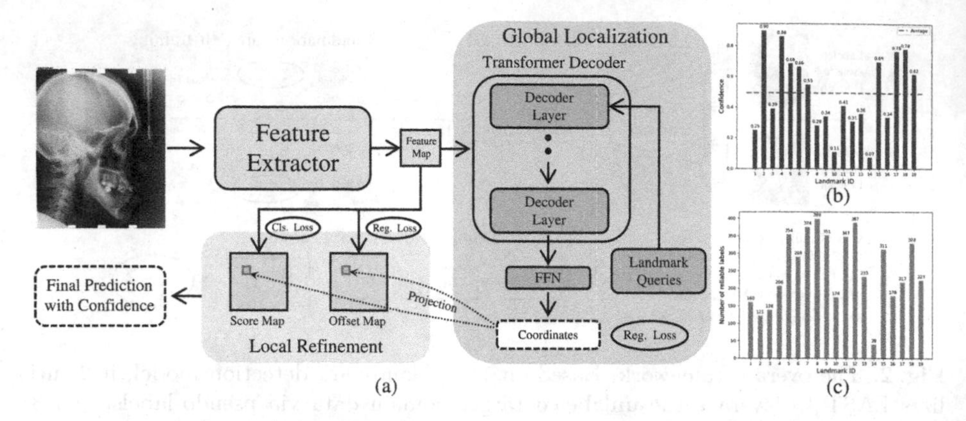

Fig. 3. (a) Our landmark detection model. (b) Confidence scores of different landmarks for a random target domain image. (c) Statistics of reliable landmark-level pseudo-labels with a fixed threshold $\tau = 0.4$ over 500 images.

(CNN) is used to extract feature $f \in \mathbb{R}^{C \times H \times W}$, where C, H, and W represents number of channels, map height and width, respectively. By using f as memory, the decoder takes landmark queries $q \in \mathbb{R}^{L \times C}$ as input, then iteratively updates them through multiple decoder layers, where L is the number of landmarks. Finally, a feed-forward network (FFN) converts the updated landmark queries to coordinates $\hat{y}_c \in \mathbb{R}^{L \times 2}$. The loss function L_{coord} is defined to be the L1 loss between the predicted coordinate $\hat{y}_c$ and the label y_c.

Local Refinement. This module outputs a score map $\hat{y}_s \in \mathbb{R}^{L \times H \times W}$ and an offset map $\hat{y}_o \in \mathbb{R}^{2L \times H \times W}$ via 1×1 convolutional layers by taking f as input. The score map indicates the likelihood of each grid to be the target landmark, while the offset map represents the relative offsets of the neighbouring grids to the target. The ground-truth (GT) landmark of the score map is smoothed by a Gaussian kernel [23], and L2 loss is used for loss function L_{score}. Since the offset is a regression problem, L1 is used for loss L_{offset}, and only applied to the area where its GT score is larger than zero. During inference, different from [6,18,23], the optimal local grid is not selected by the maximum score of $\hat{y}_s$, but instead the projection of the coordinates from global localization. Then the corresponding offset value is added to the optimal grid for refinement. Also, the confidence of each prediction can be easily obtained from the score map via projection.

The loss function of the landmark detection model can be summarized as

$$L_{\text{base}} = \sum_{\mathcal{S}} \lambda_{\text{s}} L_{\text{score}} + \lambda_{\text{o}} L_{\text{offset}} + L_{\text{coord}}, \tag{1}$$

where $\mathcal{S}$ is source domain data, λ_{s} and λ_{o} are balancing coefficients. Empirically, we set $\lambda_{\text{s}} = 100$ and $\lambda_{\text{o}} = 0.02$ in this paper.

2.2 Landmark-Aware Self-training

Self-training [14] is an effective semi-supervised learning (SSL) method, which iteratively estimates and selects reliable pseudo-labeled samples to expand the labeled set. Its effectiveness has also been verified on several vision tasks under the UDA setting, such as object detection [7]. However, very few works explored self-training for the UDA of landmark detection, but mostly restricted to the paradigm of SSL [9,21].

Since UDA is more challenging than SSL due to domain shift, reliable pseudo-labels should be carefully selected to avoid confirmation bias [2]. Existing works [9,19,21] follow the pipeline of image classification by evaluating reliability at the image-level, which we believe is not representative because the landmarks within an image may have different reliabilities (see Fig. 3 (b)). To avoid potential noisy labels caused by the image-level selection, we propose LAST, which selects reliable pseudo-labels at the landmark-level. To achieve this, we use a binary mask $m \in \{0,1\}^L$ to indicate the reliability of each landmark for a given image, where value 1 indicates the label is reliable and 0 the opposite. To decide the reliability of each landmark, a common practice is to use a threshold τ, where the l-th landmark is reliable if its confidence score $s^l > \tau$. During loss calculation, each loss term is multiplied by m to mask out the unreliable landmark-level labels. Thus, the loss for LAST is

$$L_{\text{LAST}} = \sum_{\mathcal{S} \cup \mathcal{T}'} M(L_{\text{base}}), \tag{2}$$

where M represents the mask operation, $\mathcal{T}' = \{x_j^{\mathcal{T}}, y'^{\mathcal{T}}_j\}_{j=1}^{M}$, and $y'^{\mathcal{T}}$ is the estimated pseudo-labels from the last self-training round. Note that the masks of the source domain data $\mathcal{S}$ always equal to one as they are ground truths.

However, the landmark-level selection leads to unbalanced pseudo-labels between landmarks, as shown in Fig. 3 (c). This is caused by the fixed threshold τ in self-training, which cannot handle different landmarks adaptively. To address this issue, we introduce percentile scores [4] to yield dynamic thresholds (DT) for different landmarks. Specifically, for the l-th landmark, when the pseudo-labels are sorted based on confidence (high to low), τ_r^l is used as the threshold, which is the confidence score of r-th percentile. In this way, the selection ratio of pseudo-labels can be controlled by r, and the unbalanced issue can be addressed by using the same r for all the landmarks. We set the curriculum to be $r = \Delta \cdot t$, where t is the t-th self-training round and Δ is a hyperparameter that controls the pace. We use $\Delta = 20\%$, which yields five training rounds in total.

2.3 Domain Adversarial Learning

Although self-training has been shown effective, it inevitably contains bias towards source domain because its initial model is trained with source domain data only. In other words, the data distribution of target domain is different from the source domain, which is known as covariate shift [26]. To mitigate it, we introduce DAL to align the distribution of the two by conducting an adversarial training between a domain classifier and the feature extractor. Specifically, the feature f further goes through a global average pooling (GAP) and a fully connected (FC) layer, then connects to a domain classifier D to discriminate the source of input x. The classifier can be trained with binary cross-entropy loss:

$$L_{\mathrm{D}} = -d \log D(F(x)) - (1-d)\log(1 - D(F(x))), \tag{3}$$

where d is domain label, with $d = 0$ and $d = 1$ indicating the images are from source and target domain, respectively. The domain classifier is trained to minimize L_{D}, while the feature extractor F is encouraged to maximize it such that the learned feature is indistinguishable to the domain classifier. Thus, the adversarial objective function can be written as $L_{\mathrm{DAL}} = \max_F \min_D L_{\mathrm{D}}$. To simplify the optimization, we adopt gradient reversal layer (GRL) [10] to mimic the adversarial training, which is placed right after the feature extractor. During backpropagation, GRL negates the gradients that pass back to the feature extractor F so that F is actually maximized. In this way, the adversarial training can be done via the minimization of L_{D}, i.e., $L_{\mathrm{DAL}} = L_{\mathrm{D}}$.

Finally, we have the overall loss function as follows:

$$L = \sum_{S \cup T'} L_{\mathrm{LAST}} + \lambda_D L_{\mathrm{DAL}}, \tag{4}$$

where λ_D is a balancing coefficient.

3 Experiments

3.1 Experimental Settings

In this section, we present experiments on cephalometric landmark detection. See lung landmark detection in Appendix A.

Source Domain. The ISBI 2015 Challenge provides a public dataset [22], which is widely used as a benchmark of cephalometric landmark detection. It contains 400 images in total, where 150 images are for training, 150 images are Test 1 data, and the remaining are Test 2. Each image is annotated with 19 landmarks by two experienced doctors, and the mean values of the two are used as GT. In this paper, we only use the training set as the labeled source domain data.

Target Domain. The ISBI 2023 Challenge provides a new dataset [13], which was collected from 7 different imaging devices. By now, only the training set is released, which contains 700 images. For UDA setting, we randomly selected 500 images as unlabeled target domain data, and the remaining 200 images are for evaluation. The dataset provides 29 landmarks, but we only use 19 of them, i.e., the same landmarks as the source domain [22]. Following previous works [11,16], all the images are resized to 640 × 800. For evaluation metric, we adopt MRE and SDR within four radius (2 mm, 2.5 mm, 3 mm, and 4 mm).

Table 1. Results on the target domain test set, under UDA setting.

Method	MRE↓	2 mm	2.5 mm	3 mm	4 mm
Base, Labeled Source	3.32	50.05	56.87	62.63	70.87
FDA [25]	2.16	61.28	69.73	76.34	84.57
UMT [8]	1.98	63.94	72.52	78.89	87.05
SAC [1]	1.94	65.68	73.76	79.63	87.81
AT [17]	1.87	66.82	74.81	80.73	88.47
Ours	**1.75**	**69.15**	**76.94**	**82.92**	**90.05**
Base, Labeled Target	1.22	83.76	89.71	92.79	96.08

Implementation Details. We use ImageNet pretrained ResNet-50 as the backbone, followed by three deconvolutional layers for upsampling to stride 4 [23]. For Transformer decoder, three decoder layers are used, and the embedding length $C = 256$. Our model has 41M parameters and 139 GFLOPs when input size is 640×800. The source domain images are oversampled to the same number of target domain so that the domain classifier is unbiased. Adam is used as the optimizer, and the model is trained for 720 epochs in each self-training round. The initial learning rate is 2×10^{-4}, and decayed by 10 at the 480th and 640th epoch. The batch size is set to 10 and λ_D is set to 0.01. For data augmentation, we use random scaling, translation, rotation, occlusion, and blurring. The code was implemented with PyTorch 1.13 and trained with one RTX 3090 GPU. The training took about 54 h.

3.2 Results

For the comparison under UDA setting, several state-of-the-art UDA methods were implemented, including FDA [25], UMT [8], SAC [1], and AT [17]. Additionally, the base model trained with source domain data only is included as the lower bound, and the model trained with equal amount of labeled target domain data is used as the upper bound. Table 1 shows the results. Firstly, we can see that the model trained on the target domain obtains much better performance than the one on source domain in both MRE (1.22 mm vs. 3.32 mm) and SDR (83.76% vs. 50.05%, within 2 mm), which indicates that the domain shift can cause severe performance degradation. By leveraging both labeled source domain and unlabeled target domain data, our model achieves 1.75 mm in MRE and 69.15% in SDR within 2 mm. It not only reduces the domain gap by a large margin (3.32 mm → 1.75 mm in MRE and 50.05% → 69.15% in 2 mm SDR), but also outperforms the other UDA methods consistently. However, there is still a gap between the UDA methods and the supervised model in target domain.

3.3 Model Analysis

We first do ablation study to show the effectiveness of each module, which can be seen in Table 2. The baseline model simply uses vanilla self-training [14] for

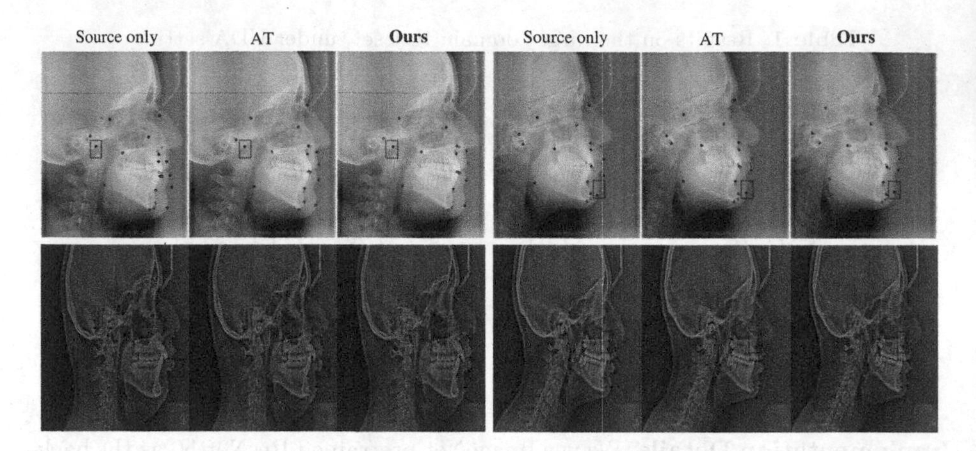

Fig. 4. Qualitative results of three models on target domain test data. Green dots are GTs, and red dots are predictions. Yellow rectangles indicate that our model performs better than the other two, while cyan rectangles indicate that all the three fail. (Color figure online)

Table 2. Ablation study of different modules.

Method	MRE↓	2 mm	2.5 mm	3 mm	4 mm
Self-training [14]	2.18	62.18	69.44	75.47	84.36
LAST w/o DT	1.98	65.34	72.53	78.03	86.11
LAST	1.91	66.21	74.39	80.23	88.42
DAL	1.96	65.92	74.18	79.73	87.60
LAST+DAL	**1.75**	**69.15**	**76.94**	**82.92**	**90.05**
LAST+DAL w/ HM	1.84	66.45	75.09	81.82	89.55

domain adaptation, which achieves 2.18 mm in MRE. By adding LAST but without dynamic thresholds (DT), the MRE improves to 1.98 mm. When the proposed LAST and DAL are applied separately, the MREs are 1.91 mm and 1.96 mm, respectively, which verifies the effectiveness of the two modules. By combining the two, the model obtains the best results in both MRE and SDR. To show the superiority of our base model, we replace it by standard heatmap regression [23] (HM), which obtains degraded results in both MRE and SDR. Furthermore, we conduct analysis on subdomain discrepancy, which shows the effectiveness of our method on each subdomain (see Appendix B).

3.4 Qualitative Results

Figure 4 shows the qualitative results of the source-only base model, AT [17], and our method on target domain test data. The green dots are GTs, and red dots are predictions. It can be seen from the figure that our model makes better predictions than the other two (see yellow rectangles). We also notice that some landmarks are quite challenging, where all the three fail to give accurate predictions (see cyan rectangles).

4 Conclusion

In this paper, we investigated anatomical landmark detection under the UDA setting. To mitigate the performance drop caused by domain shift, we proposed a unified UDA framework, which consists of a landmark detection model, a self-training strategy, and a DAL module. Based on the predictions and confidence scores from the landmark model, a self-training strategy is proposed for domain adaptation via landmark-level pseudo-labels with dynamic thresholds. Meanwhile, the model is encouraged to learn domain-invariant features via adversarial training so that the unaligned data distribution can be addressed. We constructed a UDA setting based on two anatomical datasets, where the experiments showed that our method not only reduces the domain gap by a large margin, but also outperforms other UDA methods consistently. However, a performance gap still exists between the current UDA methods and the supervised model in target domain, indicating more effective UDA methods are needed to close the gap.

Acknowledgments. This work was supported by the Shenzhen Science and Technology Innovation Committee Fund (Project No. SGDX20210823103201011) and Hong Kong Innovation and Technology Fund (Project No. ITS/028/21FP).

References

1. Araslanov, N., Roth, S.: Self-supervised augmentation consistency for adapting semantic segmentation. In: Proceedings of the IEEE/CVF Conference on Computer Vision and Pattern Recognition, pp. 15384–15394 (2021)
2. Arazo, E., Ortego, D., Albert, P., O'Connor, N.E., McGuinness, K.: Pseudo-labeling and confirmation bias in deep semi-supervised learning. In: 2020 International Joint Conference on Neural Networks (IJCNN), pp. 1–8. IEEE (2020)
3. Bigalke, A., Hansen, L., Diesel, J., Heinrich, M.P.: Domain adaptation through anatomical constraints for 3D human pose estimation under the cover. In: International Conference on Medical Imaging with Deep Learning, pp. 173–187. PMLR (2022)
4. Cascante-Bonilla, P., Tan, F., Qi, Y., Ordonez, V.: Curriculum labeling: revisiting pseudo-labeling for semi-supervised learning. In: Proceedings of the AAAI Conference on Artificial Intelligence, vol. 35, pp. 6912–6920 (2021)
5. Chen, M.H., et al.: Intraobserver reliability of landmark identification in cone-beam computed tomography-synthesized two-dimensional cephalograms versus conventional cephalometric radiography: a preliminary study. J. Dental Sci. **9**(1), 56–62 (2014)
6. Chen, R., Ma, Y., Chen, N., Lee, D., Wang, W.: Cephalometric landmark detection by attentive feature pyramid fusion and regression-voting. In: Shen, D., et al. (eds.) MICCAI 2019. LNCS, vol. 11766, pp. 873–881. Springer, Cham (2019). https://doi.org/10.1007/978-3-030-32248-9_97

7. Chen, Y., Li, W., Sakaridis, C., Dai, D., Van Gool, L.: Domain adaptive faster R-CNN for object detection in the wild. In: Proceedings of the IEEE Conference on Computer Vision and Pattern Recognition, pp. 3339–3348 (2018)
8. Deng, J., Li, W., Chen, Y., Duan, L.: Unbiased mean teacher for cross-domain object detection. In: Proceedings of the IEEE/CVF Conference on Computer Vision and Pattern Recognition. pp. 4091–4101 (2021)
9. Dong, X., Yang, Y.: Teacher supervises students how to learn from partially labeled images for facial landmark detection. In: Proceedings of the IEEE/CVF International Conference on Computer Vision, pp. 783–792 (2019)
10. Ganin, Y., Lempitsky, V.: Unsupervised domain adaptation by backpropagation. In: International Conference on Machine Learning, pp. 1180–1189. PMLR (2015)
11. Jiang, Y., Li, Y., Wang, X., Tao, Y., Lin, J., Lin, H.: CephalFormer: incorporating global structure constraint into visual features for general cephalometric landmark detection. In: Wang, L., Dou, Q., Fletcher, P.T., Speidel, S., Li, S. (eds.) MICCAI 2022. LNCS, vol. 13433, pp. 227–237. Springer, Cham (2022). https://doi.org/10.1007/978-3-031-16437-8_22
12. Jin, H., Li, J., Liao, S., Shao, L.: When liebig's barrel meets facial landmark detection: a practical model. arXiv preprint arXiv:2105.13150 (2021)
13. Khalid, M.A., et al.: Aariz: a benchmark dataset for automatic cephalometric landmark detection and CVM stage classification. arXiv:2302.07797 (2023)
14. Lee, D.H.: Pseudo-label: the simple and efficient semi-supervised learning method for deep neural networks. In: ICML Workshop on Challenges in Representation Learning (2013)
15. Li, J., Jin, H., Liao, S., Shao, L., Heng, P.A.: RepFormer: refinement pyramid transformer for robust facial landmark detection. In: IJCAI (2022)
16. Li, W., et al.: Structured landmark detection via topology-adapting deep graph learning. In: Vedaldi, A., Bischof, H., Brox, T., Frahm, J.-M. (eds.) ECCV 2020. LNCS, vol. 12354, pp. 266–283. Springer, Cham (2020). https://doi.org/10.1007/978-3-030-58545-7_16
17. Li, Y.J., et al.: Cross-domain adaptive teacher for object detection. In: Proceedings of the IEEE/CVF Conference on Computer Vision and Pattern Recognition, pp. 7581–7590 (2022)
18. Liu, W., Wang, Yu., Jiang, T., Chi, Y., Zhang, L., Hua, X.-S.: Landmarks detection with anatomical constraints for total hip arthroplasty preoperative measurements. In: Martel, A.L., et al. (eds.) MICCAI 2020. LNCS, vol. 12264, pp. 670–679. Springer, Cham (2020). https://doi.org/10.1007/978-3-030-59719-1_65
19. Mu, J., Qiu, W., Hager, G.D., Yuille, A.L.: Learning from synthetic animals. In: Proceedings of the IEEE/CVF Conference on Computer Vision and Pattern Recognition, pp. 12386–12395 (2020)
20. Payer, C., Štern, D., Bischof, H., Urschler, M.: Integrating spatial configuration into heatmap regression based CNNs for landmark localization. Med. Image Anal. (2019)
21. Wang, C., et al.: Pseudo-labeled auto-curriculum learning for semi-supervised keypoint localization. In: ICLR (2022)
22. Wang, C.W., et al.: A benchmark for comparison of dental radiography analysis algorithms. Med. Image Anal. (2016)
23. Xiao, B., Wu, H., Wei, Y.: Simple baselines for human pose estimation and tracking. In: Ferrari, V., Hebert, M., Sminchisescu, C., Weiss, Y. (eds.) ECCV 2018. LNCS, vol. 11210, pp. 472–487. Springer, Cham (2018). https://doi.org/10.1007/978-3-030-01231-1_29

24. Yang, W., Ouyang, W., Wang, X., Ren, J., Li, H., Wang, X.: 3D human pose estimation in the wild by adversarial learning. In: Proceedings of the IEEE Conference on Computer Vision and Pattern Recognition, pp. 5255–5264 (2018)
25. Yang, Y., Soatto, S.: FDA: Fourier domain adaptation for semantic segmentation. In: Proceedings of the IEEE/CVF Conference on Computer Vision and Pattern Recognition, pp. 4085–4095 (2020)
26. Zhao, S., et al.: A review of single-source deep unsupervised visual domain adaptation. IEEE Tran. Neural Netw. Learn. Syst. **33**(2), 473–493 (2020)
27. Zhong, Z., Li, J., Zhang, Z., Jiao, Z., Gao, X.: An attention-guided deep regression model for landmark detection in cephalograms. In: Shen, D., et al. (eds.) MICCAI 2019. LNCS, vol. 11769, pp. 540–548. Springer, Cham (2019). https://doi.org/10.1007/978-3-030-32226-7_60

MetaLR: Meta-tuning of Learning Rates for Transfer Learning in Medical Imaging

Yixiong Chen[1,2], Li Liu[3(✉)], Jingxian Li[4], Hua Jiang[1,2], Chris Ding[1], and Zongwei Zhou[5]

[1] The Chinese University of Hong Kong, Shenzhen, China
[2] Shenzhen Research Institute of Big Data, Shenzhen, China
[3] The Hong Kong University of Science and Technology (Guangzhou), Guangzhou, China
avrillliu@hkust-gz.edu.cn
[4] Software School, Fudan University, Shanghai, China
[5] Johns Hopkins University, Baltimore, USA

Abstract. In medical image analysis, transfer learning is a powerful method for deep neural networks (DNNs) to generalize on limited medical data. Prior efforts have focused on developing pre-training algorithms on domains such as lung ultrasound, chest X-ray, and liver CT to bridge domain gaps. However, we find that model fine-tuning also plays a crucial role in adapting medical knowledge to target tasks. The common fine-tuning method is manually picking transferable layers (*e.g.*, the last few layers) to update, which is labor-expensive. In this work, we propose a meta-learning-based learning rate (LR) tuner, named MetaLR, to make different layers automatically co-adapt to downstream tasks based on their transferabilities across domains. MetaLR learns LRs for different layers in an online manner, preventing highly transferable layers from forgetting their medical representation abilities and driving less transferable layers to adapt actively to new domains. Extensive experiments on various medical applications show that MetaLR outperforms previous state-of-the-art (SOTA) fine-tuning strategies. Codes are released.

Keywords: Medical image analysis · Meta-learning · Transfer learning

1 Introduction

Transfer learning has become a standard practice in medical image analysis as collecting and annotating data in clinical scenarios can be costly. The pre-trained parameters endow better generalization to DNNs than the models trained from scratch [8,23]. A popular approach to enhancing model transferability is by pre-training on domains similar to the targets [9,21,27–29]. However, utilizing specialized pre-training for all medical applications becomes impractical due to the

Supplementary Information The online version contains supplementary material available at https://doi.org/10.1007/978-3-031-43907-0_67.

H. Greenspan et al. (Eds.): MICCAI 2023, LNCS 14220, pp. 706–716, 2023.
https://doi.org/10.1007/978-3-031-43907-0_67

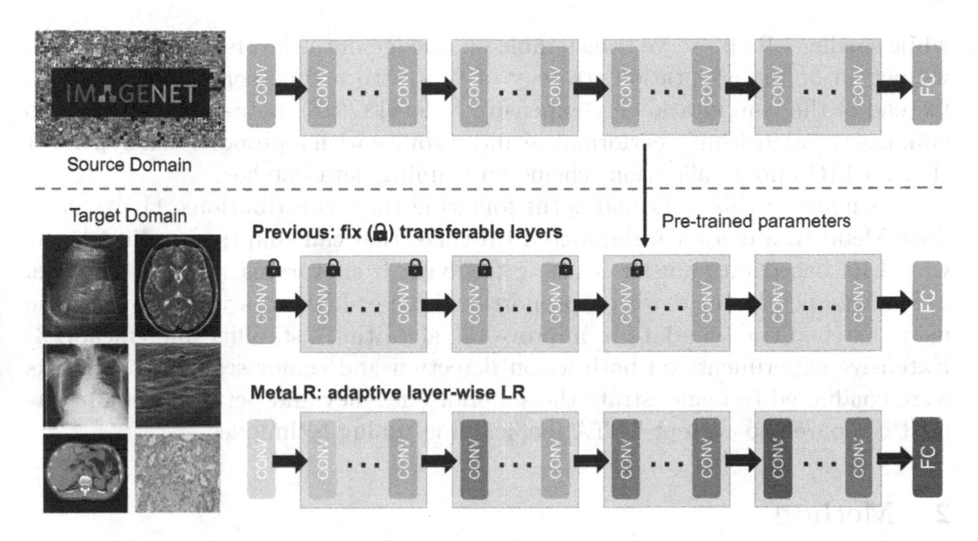

Fig. 1. The motivation of MetaLR. Previous works fix transferable layers in pre-trained models to prevent them from catastrophic forgetting. It is inflexible and labor-expensive for this method to find the optimal scheme. MetaLR uses meta-learning to automatically optimize layer-wise LR for fine-tuning.

diversity between domains and tasks and privacy concerns related to pre-training data. Consequently, recent work [2,6,14,22] has focused on improving the generalization capabilities of existing pre-trained DNN backbones through fine-tuning techniques.

Previous studies have shown that the transferability of lower layers is often higher than higher layers that are near the model output [26]. Layer-wise fine-tuning [23], was thus introduced to preserve the transferable low-level knowledge by fixing lower layers. But recent studies [7] revealed that the lower layers may also be sensitive to small domains like medical images. Given the two issues, transferability for medical tasks becomes more complicated [24,25]. It can even be irregular among layers for medical domains far from pre-training data [7]. *Given the diverse medical domains and model architectures, there is currently no universal guideline to follow to determine whether a particular layer should be retrained for a given target domain.*

To search for optimal layer combinations for fine-tuning, manually selecting transferable layers [2,23] can be a solution, but it requires a significant amount of human labor and computational cost. In order to address this issue and improve the flexibility of fine-tuning strategies, we propose controlling the fine-tuning process with layer-wise learning rates (LRs), rather than simply manually fixing or updating the layers (see Fig. 1). Our proposed algorithm, Meta Learning Rate (MetaLR), is based on meta-learning [13] and adaptively adjusts LRs for each layer according to transfer feedback. It treats the layer-wise LRs as meta-knowledge and optimizes them to improve the model generalization. Larger LRs indicate less transferability of corresponding layers and require more updating, while smaller LRs preserve transferable knowledge in the layers. Inspired by [20], we use an online adaptation strategy of LRs with a time complexity of $O(n)$, instead of the computationally-expensive bi-level $O(n^2)$ meta-learning. We also enhance the algorithm's performance and stability with a proportional hyper-LR (LR for LR) and a validation scheme on training data batches.

In summary, this work makes the following three contributions. 1) We introduce MetaLR, a meta-learning-based LR tuner that can adaptively adjust layer-wise LRs based on transfer learning feedback from various medical domains. 2) We enhance MetaLR with a proportional hyper-LR and a validation scheme using batched training data to improve the algorithm's stability and efficacy. 3) Extensive experiments on both lesion detection and tumor segmentation tasks were conducted to demonstrate the superior efficiency and performance of MetaLR compared to current SOTA medical fine-tuning techniques.

2 Method

This section provides a detailed description of the proposed MetaLR. It is a meta-learning-based [13,18] approach that determines the appropriate LR for each layer based on its transfer feedback. It is important to note that fixing transferable layers is a special case of this method, where fixed layers always have zero LRs. First, we present the theoretical formulation of MetaLR. Next, we discuss online adaptation for efficiently determining optimal LRs. Finally, we demonstrate the use of a proportional hyper-LR and a validation scheme with batched training data to enhance performance.

2.1 Formulation of Meta Learning Rate

Let (x, y) denote a sample-label pair, and $\{(x_i, y_i) \mid i = 1, ..., N\}$ be the training data. The validation dataset $\{(x_i^v, y_i^v) \mid i = 1, ..., M\}$ is assumed to be independent and identically distributed as the training dataset. Let $\hat{y} = \Phi(x, \theta)$ be the prediction for sample x from deep model Φ with parameters θ. In standard training of DNNs, the aim is to minimize the expected risk for the training set: $\frac{1}{N}\sum_{i=1}^{N} L(\hat{y}_i, y_i)$ with fixed training hyper-parameters, where $L(\hat{y}, y)$ is the loss function for the current task. The model generalization can be evaluated by the validation loss $\frac{1}{M}\sum_{i=1}^{M} L(\hat{y}_i^v, y_i^v)$. Based on the generalization, one can tune the hyper-parameters of the training process to improve the model. The key idea of MetaLR is considering the layer-wise LRs as self-adaptive hyper-parameters during the training and automatically adjusting them to achieve better model generalization. We denote the LR and model parameters for the layer j at the iteration t as α_j^t and θ_j^t. The LR scheduling scheme $\alpha = \{\alpha_j^t \mid j = 1, ..., d;\ t = 1, ..., T\}$ is what MetaLR wants to learn, affecting which local optimal $\theta^*(\alpha)$ the model parameters $\theta^t = \{\theta_j^t \mid j = 1, ..., d\}$ will converge to. The optimal parameters $\theta^*(\alpha)$ are given by optimization on the training data. At the same time, the best LR tuning scheme α^* can be optimized based on the feedback for $\theta^*(\alpha)$ from

Algorithm 1. Online Meta Learning Rate Algorithm

Input:
Training data $\mathcal{D}$, validation data $\mathcal{D}^v$, initial model parameter $\{\theta_1^0, ..., \theta_d^0\}$, LRs $\{\alpha_1^0, ..., \alpha_d^0\}$, batch size n, max iteration T;

Output:
Final model parameter $\theta^T = \{\theta_1^T, ..., \theta_d^T\}$;

1: **for** $t = 0 : T - 1$ **do**
2: $\{(x_i, y_i) \mid i = 1, ..., n\} \leftarrow$ TrainDataLoader($\mathcal{D}$, n) ;
3: $\{(x_i^v, y_i^v) \mid i = 1, ..., n\} \leftarrow$ ValidDataLoader($\mathcal{D}^v$, n) ;
4: Step forward for one step to get $\{\hat{\theta}_1^t(\alpha_1^t), ..., \hat{\theta}_d^t(\alpha_d^t)\}$ with Eq. (2);
5: Update $\{\alpha_1^t, ..., \alpha_d^t\}$ to become $\{\alpha_1^{t+1}, ..., \alpha_d^{t+1}\}$ with Eq. (3);
6: Update $\{\theta_1^t, ..., \theta_d^t\}$ to become $\{\theta_1^{t+1}, ..., \theta_d^{t+1}\}$ with Eq. (4);
7: **end for**

the validation loss. This problem can be formulated as the following bi-level optimization problem:

$$\min_{\alpha} \frac{1}{M} \sum_{i=1}^{M} L(\Phi(x_i^v, \theta^*(\alpha)), y_i^v),$$
$$s.t.\ \theta^*(\alpha) = \arg\min_{\theta} \frac{1}{N} \sum_{i=1}^{N} L(\Phi(x_i, \theta), y_i). \tag{1}$$

MetaLR aims to use the validation set to optimize α through an automatic process rather than a manual one. The optimal scheme α^* can be found by a nested optimization [13], but it is too computationally expensive in practice. A faster and more lightweight method is needed to make it practical.

2.2 Online Learning Rate Adaptation

Inspired by the online approximation [20], we propose efficiently adapting the LRs and model parameters online. The motivation of the online LR adaptation is updating the model parameters θ^t and LRs $\{\alpha_j^t \mid j = 1, 2, ..., d\}$ within the same loop. We first inspect the descent direction of parameters θ_j^t on the training loss landscape and adjust the α_j^t based on the transfer feedback. Positive feedback (lower validation loss) means the LRs are encouraged to increase.

We adopt Stochastic Gradient Descent (SGD) as the optimizer to conduct the meta-learning. The whole training process is summarized in Algorithm 1. At the iteration t of training, a training data batch $\{(x_i, y_i) \mid i = 1, ..., n\}$ and a validation data batch $\{(x_i^v, y_i^v) \mid i = 1, ..., n\}$ are sampled, where n is the size of the batches. First, the parameters of each layer are updated once with the current LR according to the descent direction on training batch.

$$\hat{\theta}_j^t(\alpha_j^t) = \theta_j^t - \alpha_j^t \nabla_{\theta_j} (\frac{1}{n} \sum_{i=1}^{n} L(\Phi(x_i, \theta_j^t), y_i)),\ j = 1, ..., d. \tag{2}$$

This step of updating aims to get feedback for LR of each layer. After taking derivative of the validation loss *w.r.t.* α_j^t, we can utilize the gradient to know how the LR for each layer should be adjusted. So the second step of MetaLR is to move the LRs along the meta-objective gradient on the validation data:

$$\alpha_j^{t+1} = \alpha_j^t - \eta \nabla_{\alpha_j} (\frac{1}{n} \sum_{i=1}^{n} L(\Phi(x_i^v, \hat{\theta}_j^t(\alpha_j^t)), y_i^v)), \tag{3}$$

where η is the hyper-LR. Finally, the updated LRs can be employed to optimize the model parameters through gradient descent truly.

$$\theta_j^{t+1} = \theta_j^t - \alpha_j^{t+1} \nabla_{\theta_j} (\frac{1}{n} \sum_{i=1}^{n} L(\Phi(x_i, \theta_j^t), y_i)). \tag{4}$$

For practical use, we constrain the LR for each layer to be $\alpha_j^t \in [10^{-6}, 10^{-2}]$. Online MetaLR optimizes the layer-wise LRs as well as the training objective on a single task, which differentiates it from traditional meta-learning algorithms [12, 19] that train models on multiple small tasks.

2.3 Proportional Hyper Learning Rate

In practice, LRs are often tuned in an exponential style (*e.g.*, 1e−3, 3e−3, 1e−2) and are always positive values. However, if a constant hyper-LR is used, it will linearly update its corresponding LR regardless of numerical constraints. This can lead to fluctuations in the LR or even the risk of the LR becoming smaller than 0 and being truncated. To address this issue, we propose using a proportional hyper-LR $\eta = \beta \times \alpha_j^t$, where β is a pre-defined hyper-parameter. This allows us to rewrite Eq. (3) as:

$$\alpha_j^{t+1} = \alpha_j^t (1 - \beta \nabla_{\alpha_j} (\frac{1}{n} \sum_{i=1}^{n} L(\Phi(x_i^v, \hat{\theta}_j^t(\alpha_j^t)), y_i^v))). \tag{5}$$

The exponential update of α_j^t guarantees its numerical stability.

2.4 Generalizability Validation on Training Data Batch

One limitation of MetaLR is that the LRs are updated using separate validation data, which reduces the amount of data available for the training process. This can be particularly problematic for medical transfer learning, where the amount of downstream data has already been limited. In Eq. 2 and Eq. 3, the update of model parameter θ_j^t and LR α_j^t is performed using different datasets to ensure that the updated θ_j^t can be evaluated for generalization without being influenced by the seen data. As an alternative, but weaker, approach, we explore using **another batch of training data** for Eq. 3 to evaluate generalization. Since this batch was not used in the update of Eq. 2, it may still perform well for validation in meta-learning. The effect of this approach is verified in Sect. 3.2, and the differences between the two methods are analyzed in Sect. 3.4.

3 Experiments and Analysis

3.1 Experimental Settings

We extensively evaluate MetaLR on four transfer learning tasks (as shown in Table 1). To ensure the reproducibility of the results, all pre-trained models (USCL [9], ImageNet [11], C2L [28], Models Genesis [29]) and target datasets (POCUS [5], BUSI [1], Chest X-ray [17], LiTS [4]) are publicly available. In our work, we consider models pre-trained on both natural and medical image datasets, with three target modalities and three target organs, which makes our experimental results more credible. For the lesion detection tasks, we used ResNet-18 [15] with the Adam optimizer. The initial learning rate (LR) and hyper-LR coefficient β are set to 10^{-3} and 0.1, respectively. In addition, we use 25% of the training set as the validation set for meta-learning. For the segmentation task, we use 3D U-Net [10] with the SGD optimizer. The initial LR and hyper-LR coefficient β are set to 10^{-2} and 3×10^{-3}, respectively. The validation set for the LiTS segmentation dataset comprises 23 samples from the training set of size 111. All experiments are implemented using PyTorch 1.10 on an Nvidia RTX A6000 GPU. We report the mean values and standard deviations for each experiment with five different random seeds. For more detailed information on the models and hyper-parameters, please refer to our supplementary material.

Table 1. Pre-training data, algorithms, and target tasks.

Source	Pre-train Method	Target	Organ	Modality	Task	Size
US-4 [9]	USCL [9]	POCUS [5]	Lung	US	COVID-19 detection	2116 images
ImageNet [11]	supervised	BUSI [1]	Breast	US	Tumor detection	780 images
MIMIC-CXR [16]	C2L [28]	Chest X-ray [17]	Lung	X-ray	Pneumonia detection	5856 images
LIDC-IDRI [3]	Models Genesis [29]	LiTS [4]	Liver	CT	Liver segmentation	131 volumes

3.2 Ablation Study

In order to evaluate the effectiveness of our proposed method, we conduct an ablation study *w.r.t.* the basic MetaLR algorithm, the proportional hyper-LR, and batched-training-data validation (as shown in Table 2). When applying only the basic MetaLR, we observe only marginal performance improvements for the four downstream tasks. We conjecture that this is due to two reasons: Firstly, the constant hyper-LR makes the training procedures less stable than direct training, which is evident from the larger standard deviation of performance. Secondly, part of the training data are split for validation, which can be detrimental to the performance. After applying the proportional hyper-LR, significant improvements are in both the performance and its stability. Moreover, although the generalization validation on the training data batch may introduce bias, providing sufficient training data ultimately benefits the performance.

Table 2. Ablation study for MetaLR, hyper-LR, and validation data. The baseline is the direct tuning of all layers with constant LRs. The default setting for MetaLR is a constant hyper-LR of 10^{-3} and a separate validation set.

MetaLR	Prop. hyper-LR	Val. on trainset	POCUS	BUSI	Chest X-ray	LiTS
			91.6 ± 0.8	84.4 ± 0.7	94.8 ± 0.3	93.1 ± 0.4
✓			91.9 ± 0.6	84.9 ± 1.3	95.0 ± 0.4	93.2 ± 0.8
✓	✓		93.6 ± 0.4	85.2 ± 0.8	95.3 ± 0.2	93.3 ± 0.6
✓		✓	93.0 ± 0.3	86.3 ± 0.7	95.5 ± 0.2	93.9 ± 0.5
✓	✓	✓	$\mathbf{93.9} \pm 0.4$	$\mathbf{86.7} \pm 0.7$	$\mathbf{95.8} \pm 0.3$	$\mathbf{94.2} \pm 0.5$

• Final MetaLR outperforms baseline with p-values of 0.0014, 0.0016, 0.0013, 0.0054.

Table 3. Comparative experiments on lesion detection. We report sensitivities (%) of the abnormalities, overall accuracy (%), and training time on each task.

Method	POCUS				BUSI				Chest X-ray		
	COVID	Pneu.	Acc	Time	Benign	Malignant	Acc	Time	Pneu.	Acc	Time
Last Layer	77.9 ± 2.1	84.0 ± 1.3	84.1 ± 0.2	15.8 m	83.5 ± 0.4	47.6 ± 4.4	66.8 ± 0.5	4.4 m	99.7 ± 1.3	87.8 ± 0.6	12.7 m
All Layers	85.8 ± 1.7	90.0 ± 1.9	91.6 ± 0.8	16.0 m	90.4 ± 1.5	77.8 ± 3.5	84.4 ± 0.7	4.3 m	98.8 ± 0.2	94.8 ± 0.3	12.9 m
Layer-wise	87.5 ± 1.0	92.3 ± 1.3	92.1 ± 0.3	2.4 h	90.8 ± 1.2	75.7 ± 2.6	85.6 ± 0.4	39.0 m	97.9 ± 0.3	95.2 ± 0.2	1.9 h
Bi-direc	90.1 ± 1.2	92.5 ± 1.5	93.6 ± 0.2	12.0 h	92.2 ± 1.0	77.1 ± 3.5	86.5 ± 0.5	3.2 h	98.4 ± 0.3	95.4 ± 0.1	9.7 h
AutoLR	89.8 ± 1.6	89.7 ± 1.5	90.4 ± 0.8	17.5 m	90.4 ± 1.8	76.2 ± 3.2	84.9 ± 0.8	4.9 m	95.4 ± 0.5	93.0 ± 0.8	13.3 m
MetaLR	94.8 ± 1.2	93.1 ± 1.5	$\mathbf{93.9} \pm 0.4$	24.8 m	92.2 ± 0.7	75.6 ± 3.6	$\mathbf{86.7} \pm 0.7$	6.0 m	97.4 ± 0.4	$\mathbf{95.8} \pm 0.3$	26.3 m

3.3 Comparative Experiments

In our study, we compare MetaLR with several other fine-tuning schemes, including tuning only the last layer / all layers with constant LRs, layer-wise fine-tuning [23], bi-directional fine-tuning [7], and AutoLR [22]. The U-Net fine-tuning scheme proposed by Amiri *et al.* [2] was also evaluated.

Results on Lesion Detection Tasks. MetaLR consistently shows the best performance on all downstream tasks (Table 3). It shows 1%–2.3% accuracy improvements compared to direct training (*i.e.*, tuning all layers) because it takes into account the different transferabilities of different layers. While manual picking methods, such as layer-wise and bi-directional fine-tuning, also achieve higher performance, they require much more training time (5×–50×) for searching the best tuning scheme. On the other hand, AutoLR is efficient, but its strong hypothesis harms its performance sometimes. In contrast, MetaLR makes no hypothesis about transferability and learns appropriate layer-wise LRs on different domains. Moreover, its performance improvements are gained with only 1.5×–2.5× training time compared with direct training.

Results on Segmentation Task. MetaLR achieves the best Dice performance on the LiTS segmentation task (Table 4). Unlike ResNet for lesion detection, the U-Net family has a more complex network topology. With skip connections, there are two interpretations [2] of depths for layers: 1) the left-most layers are the shallowest, and 2) the top layers of the "U" are the shallowest. This makes the handpicking methods even more computationally expensive. However, MetaLR

Table 4. Comparative experiments on LiTS liver segmentation task.

Method	PPV	Sensitivity	Dice	Time
Last Layer	26.1 ± 5.5	71.5 ± 4.2	33.5 ± 3.4	2.5 h
All Layers	94.0 ± 0.6	93.1 ± 0.7	93.1 ± 0.4	2.6 h
Layer-wise	92.1 ± 1.3	96.4 ± 0.4	93.7 ± 0.3	41.6 h
Bi-direc	92.4 ± 1.1	96.1 ± 0.2	93.8 ± 0.1	171.2 h
Mina *et al.*	92.7 ± 1.2	93.2 ± 0.5	92.4 ± 0.5	2.6 h
MetaLR	94.4 ± 0.9	93.6 ± 0.4	$\mathbf{94.2} \pm 0.5$	5.8 h

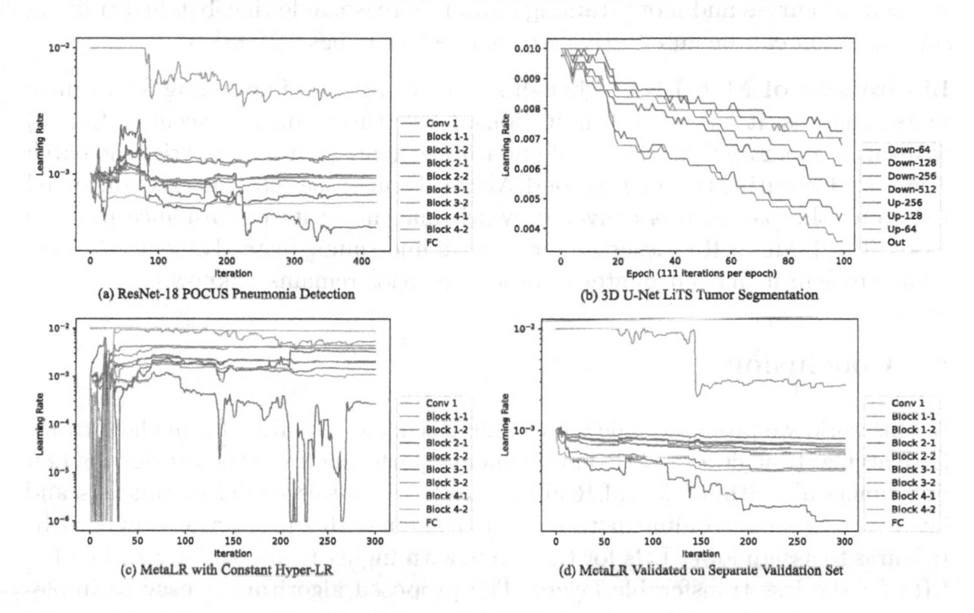

Fig. 2. The LR curves for MetaLR on POCUS detection (a), on LiTS segmentation (b), with constant hyper-LR (c), and with a separate validation set (d).

updates the LR for each layer according to their validation gradients, and its training efficiency is not affected by the complex model architecture.

3.4 Discussion and Findings

The LRs Learned with MetaLR. For ResNet-18 (Fig. 2 (a)), the layer-wise LRs fluctuate drastically during the first 100 iterations. However, after iteration 100, all layers except the first layer “Conv1” become stable at different levels. The first layer has a decreasing LR (from 2.8×10^{-3} to 3×10^{-4}) throughout the process, reflecting its higher transferability. For 3D U-Net (Fig. 2 (b)), the middle layers of the encoder “Down-128” and “Down-256” are the most transferable and have the lowest LRs, which is difficult for previous fine-tuning schemes to discover. As expected, the randomly initialized "FC" and "Out" layers have the largest LRs since they are not transferable.

The Effectiveness of Proportional Hyper-LR and Training Batches Validation. We illustrate the LR curves with a constant hyper-LR instead of a proportional one. The LR curves of "Block 3-1" and "Block 4-2" become much more fluctuated (Fig. 2 (c)). This instability may be the key reason for the instability of performance when using a constant hyper-LR. Furthermore, we surprisingly find that the learned LRs are similar to the curves learned when validated on the training set when using a separate validation set Fig. 2 (d)). With similar learned LR curves and more training data, it is reasonable that batched training set validation can be an effective alternative to the basic MetaLR.

Limitations of MetaLR. Although MetaLR improves fine-tuning for medical image analysis, it has several limitations. First, the gradient descent of Eq. (3) takes more memory than the usual fine-tuning strategy, it may restrict the batch size available during training. Second, MetaLR sometimes does not get converged LRs after the parameters converge, which may harm its performance in some cases. Third, MetaLR is designed for medical fine-tuning instead of general cases, what problem it may encounter in other scenarios remains unknown.

4 Conclusion

In this work, we proposed a new fine-tuning scheme, MetaLR, for medical transfer learning. It achieves significantly superior performance to the previous SOTA fine-tuning algorithms. MetaLR alternatively optimizes model parameters and layer-wise LRs in an online meta-learning fashion with a proportional hyper-LR. It learns to assign lower LRs for the layers with higher transferability and higher LRs for the less transferable layers. The proposed algorithm is easy to implement and shows the potential to replace manual layer-wise fine-tuning schemes. Future works include adapting MetaLR to a wider variety of clinical tasks.

Acknowledgement. This work was supported by the National Natural Science Foundation of China (No. 62101351) and the GuangDong Basic and Applied Basic Research Foundation (No.2020A1515110376).

References

1. Al-Dhabyani, W., Gomaa, M., Khaled, H., Fahmy, A.: Dataset of breast ultrasound images. Data Brief **28**, 104863 (2020)
2. Amiri, M., Brooks, R., Rivaz, H.: Fine-tuning u-net for ultrasound image segmentation: different layers, different outcomes. IEEE Trans. Ultrason. Ferroelectr. Freq. Control **67**(12), 2510–2518 (2020)
3. Armato, S.G., III., McLennan, G., Bidaut, L., et al.: The lung image database consortium (LIDC) and image database resource initiative (IDRI): a completed reference database of lung nodules on CT scans. Med. Phys. **38**(2), 915–931 (2011)

4. Bilic, P., et al.: The liver tumor segmentation benchmark (LITS). arXiv preprint arXiv:1901.04056 (2019)
5. Born, J., Wiedemann, N., Cossio, M., et al.: Accelerating detection of lung pathologies with explainable ultrasound image analysis. Appl. Sci. **11**(2), 672 (2021)
6. Chambon, P., Cook, T.S., Langlotz, C.P.: Improved fine-tuning of in-domain transformer model for inferring COVID-19 presence in multi-institutional radiology reports. J. Digit. Imaging, 1–14 (2022)
7. Chen, Y., Li, J., Ding, C., Liu, L.: Rethinking two consensuses of the transferability in deep learning. arXiv preprint arXiv:2212.00399 (2022)
8. Chen, Y., Zhang, C., Ding, C.H., Liu, L.: Generating and weighting semantically consistent sample pairs for ultrasound contrastive learning. IEEE TMI (2022)
9. Chen, Y., et al.: USCL: pretraining deep ultrasound image diagnosis model through video contrastive representation learning. In: de Bruijne, M., et al. (eds.) MICCAI 2021. LNCS, vol. 12908, pp. 627–637. Springer, Cham (2021). https://doi.org/10.1007/978-3-030-87237-3_60
10. Çiçek, Ö., Abdulkadir, A., Lienkamp, S.S., Brox, T., Ronneberger, O.: 3D U-net: learning dense volumetric segmentation from sparse annotation. In: Ourselin, S., Joskowicz, L., Sabuncu, M.R., Unal, G., Wells, W. (eds.) MICCAI 2016. LNCS, vol. 9901, pp. 424–432. Springer, Cham (2016). https://doi.org/10.1007/978-3-319-46723-8_49
11. Deng, J., Dong, W., Socher, R., Li, L.J., Li, K., Fei-Fei, L.: Imagenet: a large-scale hierarchical image database. In: CVPR, pp. 248–255 (2009)
12. Finn, C., Abbeel, P., Levine, S.: Model-agnostic meta-learning for fast adaptation of deep networks. In: ICML, pp. 1126–1135. PMLR (2017)
13. Franceschi, L., Frasconi, P., Salzo, S., Grazzi, R., Pontil, M.: Bilevel programming for hyperparameter optimization and meta-learning. In: ICML, pp. 1568–1577. PMLR (2018)
14. Guo, Y., Shi, H., Kumar, A., Grauman, K., Rosing, T., Feris, R.: Spottune: transfer learning through adaptive fine-tuning. In: CVPR, pp. 4805–4814 (2019)
15. He, K., Zhang, X., Ren, S., Sun, J.: Deep residual learning for image recognition. In: CVPR, pp. 770–778 (2016)
16. Johnson, A.E., et al.: Mimic-cxr-jpg, a large publicly available database of labeled chest radiographs. arXiv preprint arXiv:1901.07042 (2019)
17. Kermany, D., Zhang, K., Goldbaum, M.: Large dataset of labeled optical coherence tomography (OCT) and chest x-ray images. Mendeley Data **3**, 10–17632 (2018)
18. Li, Z., Zhou, F., Chen, F., Li, H.: Meta-SGD: learning to learn quickly for few-shot learning. arXiv preprint arXiv:1707.09835 (2017)
19. Nichol, A., Achiam, J., Schulman, J.: On first-order meta-learning algorithms. arXiv preprint arXiv:1803.02999 (2018)
20. Ren, M., Zeng, W., Yang, B., Urtasun, R.: Learning to reweight examples for robust deep learning. In: ICML, pp. 4334–4343. PMLR (2018)
21. Riasatian, A., et al.: Fine-tuning and training of densenet for histopathology image representation using TCGA diagnostic slides. Med. Image Anal. **70**, 102032 (2021)
22. Ro, Y., Choi, J.Y.: Autolr: layer-wise pruning and auto-tuning of learning rates in fine-tuning of deep networks. In: AAAI, vol. 35, pp. 2486–2494 (2021)
23. Tajbakhsh, N., et al.: Convolutional neural networks for medical image analysis: full training or fine tuning? IEEE TMI **35**(5), 1299–1312 (2016)
24. Vrbančič, G., Podgorelec, V.: Transfer learning with adaptive fine-tuning. IEEE Access **8**, 196197–196211 (2020)
25. Wang, G., et al.: Interactive medical image segmentation using deep learning with image-specific fine tuning. IEEE TMI **37**(7), 1562–1573 (2018)

26. Yosinski, J., Clune, J., Bengio, Y., Lipson, H.: How transferable are features in deep neural networks? In: NeurIPS, vol. 27 (2014)
27. Zhang, C., Chen, Y., Liu, L., Liu, Q., Zhou, X.: HiCo: hierarchical contrastive learning for ultrasound video model pretraining. In: Wang, L., Gall, J., Chin, T.J., Sato, I., Chellappa, R. (eds.) ACCV 2022. LNCS, vol. 13846, pp. 229–246. Springer, Cham (2022). https://doi.org/10.1007/978-3-031-26351-4_1
28. Zhou, H.-Y., Yu, S., Bian, C., Hu, Y., Ma, K., Zheng, Y.: Comparing to learn: surpassing imagenet pretraining on radiographs by comparing image representations. In: Martel, A.L., et al. (eds.) MICCAI 2020. LNCS, vol. 12261, pp. 398–407. Springer, Cham (2020). https://doi.org/10.1007/978-3-030-59710-8_39
29. Zhou, Z., et al.: Models genesis: generic autodidactic models for 3D medical image analysis. In: Shen, D., et al. (eds.) MICCAI 2019. LNCS, vol. 11767, pp. 384–393. Springer, Cham (2019). https://doi.org/10.1007/978-3-030-32251-9_42

Multi-Target Domain Adaptation with Prompt Learning for Medical Image Segmentation

Yili Lin[1], Dong Nie[3], Yuting Liu[2], Ming Yang[2], Daoqiang Zhang[1], and Xuyun Wen[1](✉)

[1] Department of Computer Science and Technology, Nanjing University of Aeronautics and Astronautics, Nanjing, China
wenzuyun@nuaa.edu.cn

[2] Department of Radiology, Children's Hospital of Nanjing Medical University, Nanjing, China

[3] Alibaba Inc, El Monte, USA

Abstract. Domain shift is a big challenge when deploying deep learning models in real-world applications due to various data distributions. The recent advances of domain adaptation mainly come from explicitly learning domain invariant features (e.g., by adversarial learning, metric learning and self-training). While they cannot be easily extended to multi-domains due to the diverse domain knowledge. In this paper, we present a novel multi-target domain adaptation (MTDA) algorithm, i.e., prompt-DA, through implicit feature adaptation for medical image segmentation. In particular, we build a feature transfer module by simply obtaining the domain-specific prompts and utilizing them to generate the domain-aware image features via a specially designed simple feature fusion module. Moreover, the proposed prompt-DA is compatible with the previous DA methods (e.g., adversarial learning based) and the performance can be continuously improved. The proposed method is evaluated on two challenging domain-shift datasets, i.e., the Iseg2019 (domain shift in infant MRI of different ages), and the BraTS2018 dataset (domain shift between high-grade and low-grade gliomas). Experimental results indicate our proposed method achieves state-of-the-art performance in both cases, and also demonstrates the effectiveness of the proposed prompt-DA. The experiments with adversarial learning DA show our proposed prompt-DA can go well with other DA methods. Our code is available at https://github.com/MurasakiLin/prompt-DA.

Keywords: Domain Adaptation · Prompt Learning · Segmentation

Supplementary Information The online version contains supplementary material available at https://doi.org/10.1007/978-3-031-43907-0_68.

H. Greenspan et al. (Eds.): MICCAI 2023, LNCS 14220, pp. 717–727, 2023.
https://doi.org/10.1007/978-3-031-43907-0_68

1 Introduction

Deep learning has brought medical image segmentation into the era of data-driven approaches, and has made significant progress in this field [1,2], i.e., the segmentation accuracy has improved considerably. In spite of the huge success, the deployment of trained segmentation models is often severely impacted by a distribution shift between the training (or labeled) and test (or unlabeled) data since the segmentation performance will deteriorate greatly in such situations. Domain shift is typically caused by various factors, including differences in acquisition protocols (e.g., parameters, imaging methods, modalities) and characteristics of data (e.g., age, gender, the severity of the disease and so on).

Domain adaptation (DA) has been proposed and investigated to combat distribution shift in medical image segmentation. Many researchers proposed using adversarial learning to tackle distribution shift problems [3–7]. These methods mainly use the game between the domain classifier and the feature extractor to learn domain-invariant features. However, they easily suffer from the balance between feature alignment and discrimination ability of the model. Some recent researchers begin to explore self-training based DA algorithms, which generate pseudo labels for the 'other' domain samples to fulfill self-training [8–11]. While it is very difficult to ensure the quality of pseudo labels in the 'other' domain and is also hard to build capable models with noise labels. However, most of these methods cannot well handle the situation when the domains are very diverse, since it is very challenging to learn domain-invariant features when each domain contains domain-specific knowledge. Also, the domain information itself is well utilized in the DA algorithms.

To tackle the aforementioned issues, we propose utilizing prompt learning to take full advantage of domain information. Prompt learning [12,13] is a recently emergent strategy to extend the same natural language processing (NLP) model to different tasks without re-training. Prompt learning models can autonomously tune themselves for different tasks by transferring domain knowledge introduced through prompts, and they can usually demonstrate better generalization ability across many downstream tasks. very few works have attempted to apply prompt learning to the computer vision field, and have achieved promising results. [14] introduced prompt tuning as an efficient and effective alternative to full fine-tuning for large-scale Transformer models. [15] exploited prompt learning to fulfill domain generalization in image classification tasks. The prompts in these models are generated and used in the very early stage of the models, which prevents the smooth combination with other domain adaptation methods.

In this paper, we introduce a domain prompt learning method (prompt-DA) to tackle distribution shift in multi-target domains. Different from the recent prompt learning methods, we generate domain-specific prompts in the encoding feature space instead of the image space. As a consequence, it can improve the quality of the domain prompts, more importantly, we can easily consolidate the prompt learning with the other DA methods, for instance, adversarial learning based DA. In addition, we propose a specially designed fusion module to reinforce the respective characteristics of the encoder features and domain-specific prompts, and thus generate domain-aware features. As a way to prove the prompt-DA is compatible with other DAs, a very simple adversarial learning module is jointly adopted in our method to further enhance the model's generalization ability (we denote this model as comb-DA). We evaluate our proposed method on two multi-domain datasets: 1). the infant brain MRI dataset for cross-age segmentation; 2). the BraTS2018 dataset for cross-grade tumor segmentation. Experiments show our proposed method outperforms state-of-the-art methods. Moreover, ablation study demonstrates the effectiveness of the proposed domain prompt learning and the feature fusion module. Our claim about the successful combination of prompt learning with adversarial learning is also well-supported by experiments.

2 Methodology

Our proposed prompt-DA network consists of three main components as depicted in Fig. 1(a): a typical encoder-decoder network (e.g., UNet) serving the segmentation baseline, a prompt learning network to learn domain-specific prompts, and a fusion module aggregating the image features and domain-specific prompts to build domain-aware feature representation, where the fused features are fed into the decoder. It is worth noting that our proposed method is compatible with other DA algorithms, and thus we can add an optional extra DA module to further optimize the domain generalization ability, in this paper, we choose an adversarial learning based DA as an example since it is the mostly used DA methods in medical image segmentation (as introduced in Sect. 1).

There are various encoder-decoder segmentation networks, many of which are well known. As a result, we donot introduce the details of the encoder-decoder and just choose two typical networks to work as the segmentation backbone, that is, 3D-UNet [16] (convolution based and 3D) and TransUNet [17] (transformer based and 2D). In the following subsections, our focus will be on domain-specific prompt generation and domain-aware feature learning with feature fusion.

2.1 Learning Domain-Specific Prompts

In our designed prompt learning based DA method, it is essential to learn domain-specific prompts. Moreover, the quality of generated prompts directly determines the domain-aware features. Therefore, we specially designed a prompt generation module to learn domain-specific prompts which mainly consists of two components, i.e., a classifier and a prompt generator.

Our approach incorporates domain-specific information into the prompts to guide the model in adapting to the target domain. To achieve this, we introduce a classifier $h(x)$ that distinguishes the domain (denoted as $\hat{d}$) of the input image, shown in Eq. 1.

$$\hat{d} = h(x) \tag{1}$$

where x is the image or abstracted features from the encoder.

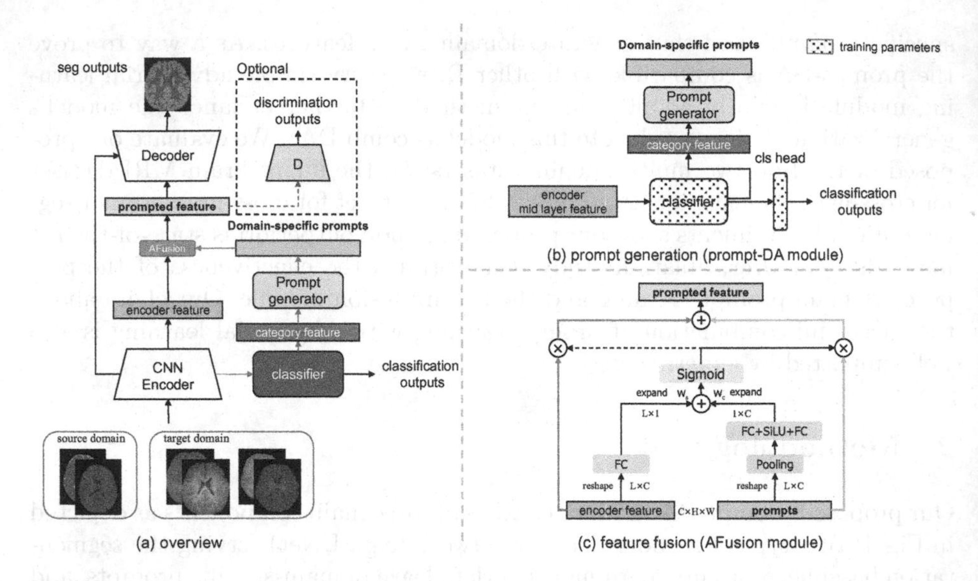

Fig. 1. a). Overview of the proposed prompt-DA network; b). The prompt generation module to learn domain-specific prompt; c). The feature fusion to learn domain-aware features.

To optimize the parameters, we adopt cross-entropy loss to train the classifier, as shown in Eq. 2.

$$L_{cls}(\hat{d}, d) = L_{ce}(\hat{d}, d) = -\sum_{i=1}^{C} d^{(i)} log\hat{d}^{(i)} \tag{2}$$

where $\hat{d}$ is the predicted domain information, and d is the ground truth domain information.

Prompt Generation: Instead of directly using $\hat{d}$ as the category information, we fed the second-to-last layer's features (i.e., z) of the classifier to a prompt generation, namely, $g(z)$. In particular, the $g(z)$ is a multi-layer-perception, as defined in Eq. 3.

$$g(z) = \phi_3(\phi_2(\phi_1(z))) \tag{3}$$

where ϕ can be a Conv+BN+ReLU sequence. Note this module does not change the size of the feature map, instead, it transforms the extracted category features into domain-specific prompts.

2.2 Learning Domain-Aware Representation by Fusion

The learned prompt captures clearly about a certain domain and the features from the encoder describe the semantics as well as spatial information for the images. We can combine them to adapt the image features to domain-aware representations.

Basically, suppose we have an image denoted as I, and the prompt encodings for the domain knowledge is $g(e(I))$ (where $e(I)$ is the features from a shallow layer), $E(I)$ is the encoder features for this image. Then the domain-aware features (i.e., F) are extracted by a fusion module as Eq. 4.

$$F = \psi(g(e(I)), E(I)) \tag{4}$$

As the learned prompt and encoder feature capture quite different aspects of the input data, we cannot achieve good effect by simply using addition, multiplication or concatenation to serve as the fusion function ψ. Specifically, while the encoder feature emphasizes spatial information for image segmentation, the prompt feature highlights inter-channel information for domain-related characteristics. To account for these differences, we propose a simple attention-based fusion (denoted as AFusion) module to smoothly aggregate the information. This module computes channel-wise and spatial-wise weights separately to enhance both the channel and spatial characteristics of the input. Figure 1(c) illustrates the structure of our proposed module.

Our module utilizes both channel and spatial branches to obtain weights for two input sources. The spatial branch compresses the encoder feature in the channel dimension using an FC layer to obtain spatial weights. Meanwhile, the channel branch uses global average pooling and two FC layers to compress the prompt and obtain channel weights. We utilize FC layers for compression and rescaling, denoted as f_{cp} and f_{re} respectively. The spatial and channel weights are computed according to Eq. 5.

$$\begin{aligned} \mathcal{W}_s &= f_{cp}(E(I)), \\ \mathcal{W}_c &= f_{re}(f_{cp}(avgpool(g(e(I))))) \end{aligned} \tag{5}$$

Afterward, we combine the weights from the spatial and channel dimensions to obtain a token that can learn both high level and low level features from both the encoder feature and the prompt, which guides the fusion of the two features. The process is illustrated as follows:

$$\begin{aligned} \mathcal{W} &= sigmoid(\mathcal{W}_c + \mathcal{W}_s), \\ \mathcal{F}_{out} &= g(e(I)) * \mathcal{W} + E(I) * (1 - \mathcal{W}) \end{aligned} \tag{6}$$

This module introduces only a few parameters, yet it can effectively improve the quality of the prompted domain-aware features after feature fusion. In the experimental section (i.e., Sect. 3.3), we conducted relevant experiments to verify that this module can indeed improve the performance of our prompt-DA method.

2.3 Adversarial Learning to Enhance the Generalization Ability

As aforementioned, our proposed prompt-DA is fully compatible with other DA algorithms. We thus use adversarial learning, which is widely adopted in medical image DA, to work as an optional component in our network to continuously enhance the domain adaptation ability.

Specially, inspired by the adversarial DA in [18], we adopt the classic GAN loss to train the discriminator and prompt generator (Note the adversarial loss, L_{adv}, for the generator will only be propagated to the prompt generator).

2.4 Total Loss

To optimize the segmentation backbone network, we use a combined loss function, L_{seg}, that incorporates both dice loss [19] and cross-entropy loss with a balance factor.

By summing the above-introduced losses, the total loss to train the segmentation network can be defined by Eq. 7.

$$L_{total} = L_{seg} + \lambda_{cls} L_{cls} + \lambda_{adv} L_{adv}, \tag{7}$$

where λ is the scaling factor to balance the losses.

2.5 Implementation Details

We use basic 3D-UNet [16] or TransUnet [17] as our segmentation network. We use a fully convolutional neural network consisting of four convolutional layers with 3×3 kernels and stride of 1 as the domain classifier, with each convolution layer followed by a ReLU parameterized by 0.2. We used three convolutional layers with ReLU activation function as the Prompt Generator and constructed a Discriminator with a similar structure to the Classifier.

We adopt Adam as the optimizer and set the learning rate to 0.0002 and 0.002 for the segmentation and domain classifier, respectively. The learning rate will be decayed by 0.1 every quarter of the training process.

3 Experiments and Results

3.1 Datasets

Our proposed method was evaluated using two medical image segmentation DA datasets. The first dataset, i.e., cross-age infant segmentation [20], was used for cross-age infant brain image segmentation, while the second dataset, i.e., Brats2018 [21], was used for HGG to LGG domain adaptation.

The first dataset is for infant brain segmentation (white matter, gray matter and cerebrospinal fluid). To build the cross-age dataset, we take advantage 10 brain MRIs of 6-month-old from iSeg2019 [20], and also build 3-month-old and 12-month-old in-house datasets. In this dataset, we collect 11 brain MRI for both the 3-month-old and 12-month-old infants. We take the 6-month-old data as the source domain, the 3-month-old and 12-month-old as the target domains.

The 2nd dataset is for brain tumor segmentation (enhancing tumor, peritumoral edema and necrotic and non-enhancing tumor core), which has 285 MRI samples (210 HGG and 75 LGG). We take HGG as the source domain and LGG as the target domain.

Table 1. Comparison with SOTA DA methods on the infant brain segmentation task. The evaluation metric shown is DICE.

Method	6 month				12 month				3 month			
	WM	GM	CSF	avg.	WM	GM	CSF	avg.	WM	GM	CSF	avg.
no-DA	82.47	88.57	93.84	88.29	68.58	72.37	76.45	72.47	69.29	61.92	62.84	64.68
nn-UNet	**83.44**	**88.97**	**94.76**	**89.06**	65.66	73.23	66.74	68.54	55.54	63.67	67.19	62.13
ADDA	80.88	87.36	92.96	87.07	69.98	75.12	75.78	73.62	70.02	68.13	62.94	67.03
CyCADA	81.12	87.89	93.06	87.36	70.12	75.24	77.13	74.16	70.12	70.54	62.91	67.86
SIFA	81.71	87.87	92.98	87.52	70.37	76.98	77.02	74.79	69.89	71.12	63.01	68.01
ADR	81.69	87.94	93.01	87.55	71.81	77.02	76.65	75.10	70.16	72.04	62.98	68.39
ours	81.77	88.01	93.04	87.61	**75.03**	**80.03**	**78.74**	**77.93**	**70.59**	**74.51**	**63.18**	**69.43**

3.2 Comparison with State-of-the-Art (SOTA) Method

We compared our method with four SOTA methods: ADDA [18], CyCADA [22], SIFA [23] and ADR [24]. We directly use the code from the corresponding papers. For fair comparison, we have replaced the backbone of these models with the same we used in our approach. The quantitative comparison results of cross-age infant brain segmentation is presented in Table 1, and due to space limitations, we put the experimental results of the brain tumor segmentation task in Table 1 of Supplementary Material, Sec.3.

As observed, our method demonstrates very good DA ability on the cross-age infant segmentation task, which improves about 5.46 DICE and 4.75 DICE on 12-month-old and 3-month-old datasets, respectively. When compared to the four selected SOTA DA methods, we also show superior transfer performance in all the target domains. Specially, we outperform other SOTA methods by at least 2.83 DICE and 1.04 DICE on the 12-month-old and 3-month-old tasks.

When transferring to a single target domain in the brain tumor segmentation task, our proposed DA solution improves about 3.09 DICE in the target LGG domain. Also, the proposed method shows considerable improvements over ADDA and CyCADA, but very subtle improvements to the SIFA and ADR methods (although ADR shows a small advantage on the Whole category).

We also visualize the segmentation results on a typical test sample of the infant brain dataset in Fig. 2, which once again demonstrates the advantage of our method in some detailed regions.

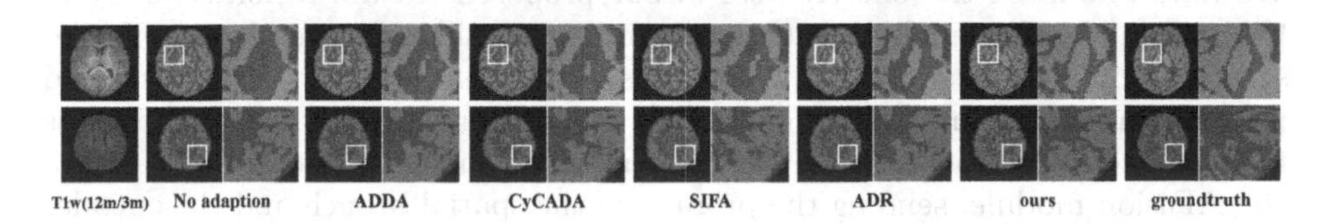

Fig. 2. Visualization of segmentation maps (details) for all the comparison methods.

Table 2. Ablation study about prompt-DA, adv-DA and comb-DA.

model	experiment	6 month				12 month				3 month			
		WM	GM	CSF	avg.	WM	GM	CSF	avg.	WM	GM	CSF	avg.
3D-Unet	no-DA	**82.47**	**88.57**	**93.84**	**88.29**	68.58	72.37	76.45	72.47	69.29	61.92	62.84	64.68
	adv-DA	80.88	87.36	92.96	87.07	69.98	75.12	75.78	73.62	70.02	68.13	62.94	67.03
	prompt-DA	81.57	87.90	93.06	87.51	71.3	77.82	77.16	75.06	70.69	69.51	62.83	67.68
	comb-DA	81.77	88.01	93.04	87.61	**75.03**	**80.03**	**78.74**	**77.93**	**70.59**	**74.51**	**63.18**	**69.43**
TransUnet	no-DA	**73.24**	**81.12**	**84.19**	**79.52**	66.04	70.12	54.94	63.72	39.70	59.49	59.25	52.81
	adv-DA	72.76	80.72	82.98	78.82	66.72	70.39	55.21	64.11	39.89	60.02	59.89	53.27
	prompt-DA	73.01	80.31	83.21	78.84	67.41	71.01	55.41	64.61	40.17	60.22	60.09	53.49
	comb-DA	72.98	80.59	83.61	79.06	**70.25**	**72.57**	**57.04**	**66.62**	**42.61**	**61.03**	**61.57**	**55.07**

3.3 Ablation Study

Prompt-DA vs. adv-DA: Since the performance reported in Table 1 is achieved with the method combining prompt-DA and adv-DA, we carry out more studies to investigate: 1). Does prompt-DA itself shows the transfer ability? 2). Is prompt-DA compatible with adv-DA?

The corresponding experiments are conducted on the infant brain dataset and experimental results are shown in Table 2. To make the table more readable, we denote: *no-DA* means only training the segmentation network without any DA strategies; *adv-DA* presents only using adversarial learning based DA; *prompt-DA* is the proposed prompt learning based DA and *comb-DA* is our final DA algorithm which combines both *adv-DA* and *prompt-DA*.

As observed in Table 2, both adv-DA and prompt-DA can improve the transfer performance on all the target domains. When looking into details, the proposed prompt-DA can improve more (1.44 DICE and 0.65 DICE respectively) compared to the adv-DA on both 12-month-old and 3-month-old with 3D-UNet segmentation backbone. When combined together (i.e., comb-DA), the performance can be further improved by a considerable margin, 2.87 DICE and 1.75 DICE on 12-month-old and 3-month-old respectively, compared to prompt-DA. With TransUNet segmentation backbone, we can find the similar phenomenon. To this end, we can draw conclusions that 1). Prompt-DA itself is beneficial to improve the transfer ability; 2). prompt-DA is quite compatible with adv-DA.

Fusion Strategy for Learning Domain-Aware Features: One of the key components of the prompt-DA is to learn domain-aware features through fusion. We have evaluated the effectiveness of our proposed feature fusion strategy in both 3D and 2D models. For comparison, we considered several other fusion strategies, including 'add/mul', which adds or multiplies the encoder feature and prompt directly, 'conv', which employs a single convolutional layer to process the concatenated features, and 'rAFusion', which utilizes a reverse version of the AFusion module, sending the prompt to the spatial branch and the encoder feature to the channel branch. The results of these experiments are presented in Table 3.

Table 3. Ablation study about fusion strategies to learn domain-aware features.

model	experiment	6 month				12 month				3 month			
		WM	GM	CSF	avg.	WM	GM	CSF	avg.	WM	GM	CSF	avg.
3D-Unet	no-DA	**82.47**	**88.57**	**93.84**	**88.29**	68.58	72.37	76.45	72.47	69.29	61.92	62.84	64.68
	add/mul	81.31	87.62	92.67	87.2	73.88	78.59	76.52	76.33	68.27	74.17	62.90	68.45
	conv	81.4	87.61	93.16	87.39	73.06	77.73	78.73	76.51	69.91	71.93	63.02	68.29
	rAFusion	81.72	88.17	93.33	87.74	74.82	79.75	77.84	77.47	69.42	74.54	62.98	68.98
	AFusion	81.77	88.01	93.04	87.61	**75.03**	**80.03**	**78.74**	**77.93**	**70.59**	**74.51**	**63.18**	**69.43**
TransUnet	no-DA	**73.24**	**81.12**	**84.19**	**79.52**	66.04	70.12	54.94	63.72	39.70	59.49	59.25	52.81
	add/mul	72.66	80.69	83.77	79.04	68.66	70.92	55.52	65.03	40.41	60.87	60.32	53.87
	conv	72.72	80.56	83.71	79.00	69.75	71.01	55.56	65.44	41.32	60.97	60.24	54.18
	rAFusion	72.96	80.72	83.74	79.14	69.80	72.44	56.18	66.14	42.12	61.01	60.50	54.54
	AFusion	72.98	80.59	83.61	79.06	**70.25**	**72.57**	**57.04**	**66.62**	**42.61**	**61.03**	**61.57**	**55.07**

Our experimental results demonstrate that the proposed AFusion module improves the model's performance significantly, and it is effective for both 3D and 2D models.

4 Conclusion

In this paper, we propose a new DA paradigm, namely, prompt learning based DA. The proposed prompt-DA uses a classifier and a prompt generator to produce domain-specific information and then employs a fusion module (for encoder features and prompts) to learn domain-aware representation. We show the effectiveness of our proposed prompt-DA in transfer ability, and also we prove that the prompt-DA is smoothly compatible with the other DA algorithms. Experiments on two DA datasets with two different segmentation backbones demonstrate that our proposed method works well on DA problems.

Acknowledgements. This work was supported by the National Natural Science Foundation of China (No. 62001222), the China Postdoctoral Science Foundation funded project (No. 2021TQ0150 and No. 2021M701699).

References

1. Ronneberger, O., Fischer, P., Brox, T.: U-net: convolutional networks for biomedical image segmentation. In: Navab, N., Hornegger, J., Wells, W.M., Frangi, A.F. (eds.) MICCAI 2015. LNCS, vol. 9351, pp. 234–241. Springer, Cham (2015). https://doi.org/10.1007/978-3-319-24574-4_28
2. Zhou, S., Nie, D., Adeli, E., Yin, J., Lian, J., Shen, D.: High-resolution encoder-decoder networks for low-contrast medical image segmentation. IEEE Trans. Image Process. **29**, 461–475 (2019)
3. Ouyang, C., Kamnitsas, K., Biffi, C., Duan, J., Rueckert, D.: Data efficient unsupervised domain adaptation for cross-modality image segmentation. In: Shen, D., et al. (eds.) MICCAI 2019. LNCS, vol. 11765, pp. 669–677. Springer, Cham (2019). https://doi.org/10.1007/978-3-030-32245-8_74

4. Xie, X., Chen, J., Li, Y., Shen, L., Ma, K., Zheng, Y.: MI^{2}GAN: generative adversarial network for medical image domain adaptation using mutual information constraint. In: Martel, A.L., et al. (eds.) MICCAI 2020. LNCS, vol. 12262, pp. 516–525. Springer, Cham (2020). https://doi.org/10.1007/978-3-030-59713-9_50
5. Dou, Q., et al.: Pnp-adanet: plug-and-play adversarial domain adaptation network at unpaired cross-modality cardiac segmentation. IEEE Access **7**, 99 065–99 076 (2019)
6. Chen, C., Dou, Q., Chen, H., Qin, J., Heng, P.-A.: Synergistic image and feature adaptation: Towards cross-modality domain adaptation for medical image segmentation. In: Proceedings of the AAAI Conference on Artificial Intelligence, vol. 33, no. 01, pp. 865–872 (2019)
7. Cui, H., Yuwen, C., Jiang, L., Xia, Y., Zhang, Y.: Bidirectional cross-modality unsupervised domain adaptation using generative adversarial networks for cardiac image segmentation. Comput. Biol. Med. **136**, 104726 (2021)
8. Kumar, A., Ma, T., Liang, P.: Understanding self-training for gradual domain adaptation. In: International Conference on Machine Learning, pp. 5468–5479. PMLR (2020)
9. Sheikh, R., Schultz, T.: Unsupervised domain adaptation for medical image segmentation via self-training of early features. In: International Conference on Medical Imaging with Deep Learning, pp. 1096–1107. PMLR (2022)
10. Xie, Q., et al.: Unsupervised domain adaptation for medical image segmentation by disentanglement learning and self-training. IEEE Trans. Med. Imaging (2022)
11. Yang, C., Guo, X., Chen, Z., Yuan, Y.: Source free domain adaptation for medical image segmentation with Fourier style mining. Med. Image Anal. **79**, 102457 (2022)
12. Liu, X., Ji, K., Fu, Y., Du, Z., Yang, Z., Tang, J.: P-tuning v2: prompt tuning can be comparable to fine-tuning universally across scales and tasks. arXiv preprint arXiv:2110.07602 (2021)
13. Zhou, K., Yang, J., Loy, C.C., Liu, Z.: Learning to prompt for vision-language models. Int. J. Comput. Vision **130**(9), 2337–2348 (2022)
14. Jia, M., et al.: Visual prompt tuning. In: Avidan, S., Brostow, G., Cissé, M., Farinella, G.M., Hassner, T. (eds.) Computer Vision - ECCV 2022. Lecture Notes in Computer Science, vol. 13693, pp. 709–727. Springer, Cham (2022)
15. Zheng, Z., Yue, X., Wang, K., You, Y.: Prompt vision transformer for domain generalization. arXiv preprint arXiv:2208.08914 (2022)
16. Çiçek, Ö., Abdulkadir, A., Lienkamp, S.S., Brox, T., Ronneberger, O.: 3D U-net: learning dense volumetric segmentation from sparse annotation. In: Ourselin, S., Joskowicz, L., Sabuncu, M.R., Unal, G., Wells, W. (eds.) MICCAI 2016. LNCS, vol. 9901, pp. 424–432. Springer, Cham (2016). https://doi.org/10.1007/978-3-319-46723-8_49
17. Chen, J., et al.: Transunet: transformers make strong encoders for medical image segmentation. arXiv preprint arXiv:2102.04306 (2021)
18. Tzeng, E., Hoffman, J., Saenko, K., Darrell, T.: Adversarial discriminative domain adaptation. In: Proceedings of the IEEE Conference on Computer Vision and Pattern Recognition, pp. 7167–7176 (2017)
19. Sudre, C.H., Li, W., Vercauteren, T., Ourselin, S., Jorge Cardoso, M.: Generalised dice overlap as a deep learning loss function for highly unbalanced segmentations. In: Cardoso, M.J., et al. (eds.) DLMIA/ML-CDS -2017. LNCS, vol. 10553, pp. 240–248. Springer, Cham (2017). https://doi.org/10.1007/978-3-319-67558-9_28
20. Sun, Y., et al.: Multi-site infant brain segmentation algorithms: the ISEG-2019 challenge. IEEE Trans. Med. Imaging **40**(5), 1363–1376 (2021)

21. Bakas, S., et al.: Identifying the best machine learning algorithms for brain tumor segmentation, progression assessment, and overall survival prediction in the brats challenge. arXiv preprint arXiv:1811.02629 (2018)
22. Hoffman, J., et al.: Cycada: cycle-consistent adversarial domain adaptation. In: International Conference on Machine Learning, pp. 1989–1998. PMLR (2018)
23. Chen, C., Dou, Q., Chen, H., Qin, J., Heng, P.A.: Unsupervised bidirectional cross-modality adaptation via deeply synergistic image and feature alignment for medical image segmentation. IEEE Trans. Med. Imaging **39**(7), 2494–2505 (2020)
24. Zeng, G., et al.: Semantic consistent unsupervised domain adaptation for cross-modality medical image segmentation. In: de Bruijne, M., et al. (eds.) MICCAI 2021. LNCS, vol. 12903, pp. 201–210. Springer, Cham (2021). https://doi.org/10.1007/978-3-030-87199-4_19

Spectral Adversarial MixUp for Few-Shot Unsupervised Domain Adaptation

Jiajin Zhang[1], Hanqing Chao[1], Amit Dhurandhar[2], Pin-Yu Chen[2], Ali Tajer[3], Yangyang Xu[4], and Pingkun Yan[1(✉)]

[1] Department of Biomedical Engineering and Center for Biotechnology and Interdisciplinary Studies, Rensselaer Polytechnic Institute, Troy, NY, USA
yanp2@rpi.edu
[2] IBM Thomas J. Watson Research Center, Yorktown Heights, NY, USA
[3] Department of Electrical, Computer, and Systems Engineering, Rensselaer Polytechnic Institute, Troy, NY, USA
[4] Department of Mathematical Sciences, Rensselaer Polytechnic Institute, Troy, NY, USA

Abstract. Domain shift is a common problem in clinical applications, where the training images (source domain) and the test images (target domain) are under different distributions. Unsupervised Domain Adaptation (UDA) techniques have been proposed to adapt models trained in the source domain to the target domain. However, those methods require a large number of images from the target domain for model training. In this paper, we propose a novel method for Few-Shot Unsupervised Domain Adaptation (FSUDA), where only a limited number of *unlabeled* target domain samples are available for training. To accomplish this challenging task, first, a spectral sensitivity map is introduced to characterize the generalization weaknesses of models in the frequency domain. We then developed a Sensitivity-guided Spectral Adversarial MixUp (SAMix) method to generate target-style images to effectively suppresses the model sensitivity, which leads to improved model generalizability in the target domain. We demonstrated the proposed method and rigorously evaluated its performance on multiple tasks using several public datasets. The source code is available at https://github.com/RPIDIAL/SAMix.

Keywords: Few-shot UDA · Data Augmentation · Spectral Sensitivity

1 Introduction

A common challenge for deploying deep learning to clinical problems is the discrepancy between data distributions across different clinical sites [6,15,20,28,29]. This discrepancy, which results from vendor or protocol differences, can cause a significant performance drop when models are deployed to a new site [2,21,23]. To solve this problem, many Unsupervised Domain Adaptation (UDA) methods [6] have been developed for adapting a model to a new site with only

H. Greenspan et al. (Eds.): MICCAI 2023, LNCS 14220, pp. 728–738, 2023.
https://doi.org/10.1007/978-3-031-43907-0_69

unlabeled data (target domain) by transferring the knowledge learned from the original dataset (source domain). However, most UDA methods require sufficient target samples, which are scarce in medical imaging due to the limited accessibility to patient data. This motivates a new problem of Few-Shot Unsupervised Domain Adaptation (FSUDA), where only a few *unlabeled* target samples are available for training.

Few approaches [11, 22] have been proposed to tackle the problem of FSUDA. Luo et. al [11] introduced Adversarial Style Mining (ASM), which uses a pre-trained style-transfer module to generate augmented images via an adversarial process. However, this module requires extra style images [9] for pre-training. Such images are scarce in clinical settings, and style differences across sites are subtle. This hampers the applicability of ASM to medical image analysis. SM-PPM [22] trains a style-mixing model for semantic segmentation by augmenting source domain features to a fictitious domain through random interpolation with target domain features. However, SM-PPM is specifically designed for segmentation tasks and cannot be easily adapted to other tasks. Also, with limited target domain samples in FSUDA, the random feature interpolation is ineffective in improving the model's generalizability.

In a different direction, numerous UDA methods have shown high performance in various tasks [4, 16–18]. However, their direct application to FSUDA can result in severe overfitting due to the limited target domain samples [22]. Previous studies [7, 10, 24, 25] have demonstrated that transferring the amplitude spectrum of target domain images to a source domain can effectively convey image style information and diversify training dataset. To tackle the overfitting issue of existing UDA methods, we propose a novel approach called Sensitivity-guided Spectral Adversarial MixUp (SAMix) to augment training samples. This approach uses an adversarial mixing scheme and a spectral sensitivity map that reveals model generalizability weaknesses to generate hard-to-learn images with limited target samples efficiently. SAMix focuses on two key aspects. **1)** *Model generalizability weaknesses*: Spectral sensitivity analysis methods have been applied in different works [26] to quantify the model's spectral weaknesses to image amplitude corruptions. Zhang et al. [27] demonstrated that using a spectral sensitivity map to weigh the amplitude perturbation is an effective data augmentation. However, existing sensitivity maps only use single-domain labeled data and cannot leverage target domain information. To this end, we introduce a Domain-Distance-modulated Spectral Sensitivity (DoDiSS) map to analyze the model's weaknesses in the target domain and guide our spectral augmentation. **2)** *Sample hardness*: Existing studies [11, 19] have shown that mining hard-to-learn samples in model training can enhance the efficiency of data augmentation and improve model generalization performances. Therefore, to maximize the use of the limited target domain data, we incorporate an adversarial approach into the spectral mixing process to generate the most challenging data augmentations.

This paper has three major contributions. **1)** We propose SAMix, a novel approach for augmenting target-style samples by using an adversarial spectral mixing scheme. SAMix enables high-performance UDA methods to adapt easily to FSUDA problems. **2)** We introduce DoDiSS to characterize a model's generalizability weaknesses in the target domain. **3)** We conduct thorough empirical analyses to demonstrate the effectiveness and efficiency of SAMix as a plug-in module for various UDA methods across different tasks.

2 Methods

We denote the labeled source domain as $\boldsymbol{X}_S = \{(\boldsymbol{x}_n^s, \boldsymbol{y}_n^s)\}_{n=1}^N$ and the unlabeled K-shot target domain as $\boldsymbol{X}_T = \{\boldsymbol{x}_k^t\}_{k=1}^K$, $\boldsymbol{x}_n^s$, $\boldsymbol{x}_k^t \in \mathbb{R}^{h \times w}$. Figure 1 depicts the framework of our method as a plug-in module for boosting a UDA method in the FSUDA scenario. It contains two components. First, a Domain-Distance-modulated Spectral Sensitivity (DoDiSS) map is calculated to characterize a source model's weaknesses in generalizing to the target domain. Then, this sensitivity map is used for Sensitivity-guided Spectral Adversarial MixUp (SAMix) to generate target-style images for UDA models. The details of the components are presented in the following sections.

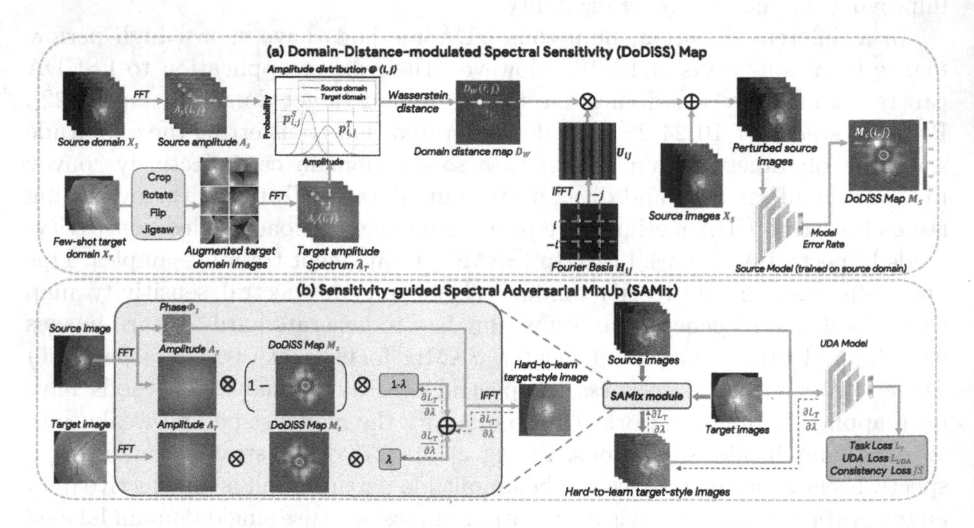

Fig. 1. Illustration of the proposed framework. **(a)** DoDiSS map characterizes a model's generalizability weaknesses. **(b)** SAMix enables UDA methods to solve FSUDA.

2.1 Domain-Distance-Modulated Spectral Sensitivity (DoDiSS)

The prior research [27] found that a spectral sensitivity map obtained using Fourier-based measurement of model sensitivity can effectively portray the generalizability of that model. However, the spectral sensitivity map is limited to single-domain scenarios and cannot integrate target domain information to assess model weaknesses under specific domain shifts. Thus, we introduce DoDiSS, extending the previous method by incorporating domain distance to tackle domain adaptation problems. Fig. 1 (**a**) depicts the DoDiSS pipeline. It begins by computing a domain distance map for identifying the amplitude distribution difference between the source and target domains in each frequency. Subsequently, this difference map is used for weighting amplitude perturbations when calculating the DoDiSS map.

Domain Distance Measurement. To overcome the limitations of lacking target domain images, we first augment the few-shot images from the target domain with random combinations of various geometric transformations, including random cropping, rotation, flipping, and JigSaw [13]. These transformations keep the image intensities unchanged, preserving the target domain style information. The Fast Fourier Transform (FFT) is then applied to all the source images and the augmented target domain images to obtain their amplitude spectrum, denoted as $\boldsymbol{A}_S$ and $\hat{\boldsymbol{A}}_T$, respectively. We calculate the probabilistic distributions $\boldsymbol{p}^S_{i,j}$ and $\boldsymbol{p}^T_{i,j}$ of $\boldsymbol{A}_S$ and $\hat{\boldsymbol{A}}_T$ at the $(i,j)_{th}$ frequency entry, respectively. The domain distance map at (i,j) is defined as $\boldsymbol{D}_W(i,j) = W_1(\boldsymbol{p}^S_{i,j}, \boldsymbol{p}^T_{i,j})$, where W_1 is the 1-Wasserstein distance.

DoDiSS Computation. With the measured domain difference, we can now compute the DoDiSS map of a model. As shown in Fig. 1 (**a**), a Fourier basis is defined as a Hermitian matrix $\boldsymbol{H}_{i,j} \in \mathbb{R}^{h \times w}$ with only two non-zero elements at (i,j) and $(-i,-j)$. A Fourier basis image $\boldsymbol{U}_{i,j}$ can be obtained by ℓ_2-normalized Inverse Fast Fourier Transform (IFFT) of $\boldsymbol{A}_{i,j}$, *i.e.*, $\boldsymbol{U}_{i,j} = \frac{\mathcal{IFFT}(\boldsymbol{A}_{i,j})}{\|\mathcal{IFFT}(\boldsymbol{A}_{i,j})\|_2}$. To analyze the model's generalization weakness with respect to the frequency (i,j), we generate perturbed source domain images by adding the Fourier basis noise $\boldsymbol{N}_{i,j} = r \cdot \boldsymbol{D}_W(i,j) \cdot \boldsymbol{U}_{i,j}$ to the original source domain image $\boldsymbol{x}^s$ as $\boldsymbol{x}^s + \boldsymbol{N}_{i,j}$. $\boldsymbol{D}_W(i,j)$ controls the ℓ_2-norm of $\boldsymbol{N}_{i,j}$ and r is randomly sampled to be either -1 or 1. The $\boldsymbol{N}_{i,j}$ only introduces perturbations at the frequency components (i,j) to the original images. The $\boldsymbol{D}_W(i,j)$ guarantees that images are perturbed across all frequency components following the real domain shift. For RGB images, we add $\boldsymbol{N}_{i,j}$ to each channel independently following [27]. The sensitivity at frequency (i,j) of a model F trained on the source domain is defined as the prediction error rate over the whole dataset $\boldsymbol{X}_S$ as in (1), where Acc denotes the prediction accuracy

$$\boldsymbol{M}_S(i,j) = 1 - \underset{(\boldsymbol{x}^s, \boldsymbol{y}^s) \in \boldsymbol{X}_S}{\text{Acc}} (F(\boldsymbol{x}^s + r \cdot \boldsymbol{D}_W(i,j) \cdot \boldsymbol{U}_{i,j}), \boldsymbol{y}^s). \tag{1}$$

2.2 Sensitivity-Guided Spectral Adversarial Mixup (SAMix)

Using the DoDiSS map $\boldsymbol{M}_S$ and an adversarially learned parameter λ^* as a weighting factor, SAMix mixes the amplitude spectrum of each source image with the spectrum of a target image. DoDiSS indicates the spectral regions where the model is sensitive to the domain difference. The parameter λ^* mines the heard-to-learn samples to efficiently enrich the target domain samples by maximizing the task loss. Further, by retaining the phase of the source image, SAMix preserves the semantic meaning of the original source image in the generated target-style sample. Specifically, as shown in Fig. 1 **(b)**, given a source image $\boldsymbol{x}^s$ and a target image $\boldsymbol{x}^t$, we compute their amplitude and phase spectrum, denoted as $(\boldsymbol{A}^s, \boldsymbol{\Phi}^s)$ and $(\boldsymbol{A}^t, \boldsymbol{\Phi}^t)$, respectively. SAMix mixes the amplitude spectrum by

$$\boldsymbol{A}^{st}_{\lambda^*} = \lambda^* \cdot \boldsymbol{M}_S \cdot \boldsymbol{A}^t + (1 - \lambda^*) \cdot (1 - \boldsymbol{M}_S) \cdot \boldsymbol{A}^s. \tag{2}$$

The target-style image is reconstructed by $\boldsymbol{x}^{st}_{\lambda^*} = \mathcal{IFFT}(\boldsymbol{A}^{st}_{\lambda^*}, \boldsymbol{\Phi}^s)$. The adversarially learned parameter λ^* is optimized by maximizing the task loss L_T using the projected gradient descent with T iterations and step size of δ:

$$\lambda^* = \arg\max_{\lambda} L_T(F(\boldsymbol{x}^{st}_{\lambda}; \theta), \boldsymbol{y}), \quad \text{s.t.} \quad \lambda \in [0, 1]. \tag{3}$$

In the training phase, as shown in Fig. 1 **(b)**, the SAMix module generates a batch of augmented images, which are combined with few-shot target domain images to train the UDA model. The overall training objective is to minimize

$$L_{tot}(\theta) = L_T(F(\boldsymbol{x}^s; \theta), \boldsymbol{y}) + \mu \cdot JS(F(\boldsymbol{x}^s; \theta), F(\boldsymbol{x}^{st}_{\lambda^*}; \theta)) + L_{UDA}, \tag{4}$$

where L_t is the supervised task loss in the source domain; JS is the Jensen-Shannon divergence [27], which regularizes the model predictions consistency between the source images $\boldsymbol{x}^s$ and their augmented versions $\boldsymbol{x}^{st}_{\lambda^*}$; L_{UDA} is the training loss in the original UDA method, and μ is a weighting parameter.

3 Experiments and Results

We evaluated SAMix on two medical image datasets. **Fundus** [5,14] is an optic disc and cup segmentation task. Following [21], we consider images collected from different scanners as distinct domains. The source domain contains 400 images of the REFUGE [14] training set. We took 400 images from the REFUGE validation set and 159 images of RIM-One [5] to form the target domain 1 & 2. We center crop and resize the disc region to 256×256 as network input. **Camelyon** [1] is a tumor tissue binary classification task across 5 hospitals. We use the training set of Camelyon as the source domain (302, 436 images from hospitals 1 – 3) and consider the validation set (34, 904 images from hospital 4) and test set (85, 054 images from the hospital 5) as the target domains 1 and 2, respectively. All the images are resized into 256×256 as network input. For all experiments, the source domain images are split into training and validation in the ratio of 4 : 1. We randomly selected K-shot target domain images for training, while the remaining target domain images were reserved for testing.

3.1 Implementation Details

SAMix is evaluated as a plug-in module for four UDA models: AdaptSeg [17] and Advent [18] for **Fundus**, and SRDC [16] and DALN [4] for **Camelyon**. For a fair comparison, we adopted the same network architecture for all the methods on each task. For **Fundus**, we use a DeepLabV2-Res101 [3] as the backbone

Table 1. 10-run average DSC (%) and ASD of models on REFUGE. The best performance is in **bold** and the second best is indicated with underline.

Method	Source Domain → Target Domain 1						Source Domain → Target Domain 2					
	DSC (↑)			ASD (↓)			DSC (↑)			ASD (↓)		
	cup	disc	avg	cup	disc	avg	cup	disc	avg	cup	disc	avg
Source Only	61.16*	66.54*	63.85*	14.37*	11.69*	13.03*	55.77*	58.62*	57.20*	20.95*	17.63*	19.30*
AdaptSeg	61.45*	66.61*	64.03*	13.79*	11.47*	12.64*	56.67*	60.50*	58.59*	20.44*	17.97*	19.21*
Advent	62.03*	66.82*	64.43*	12.82*	11.54*	12.18*	56.43*	60.56*	58.50*	20.31*	17.86*	19.09*
ASM	69.18*	71.91*	70.05*	8.92*	8.35*	8.64*	57.79*	61.86*	59.83*	19.26*	16.94*	18.10*
SM-PPM	74.55*	77.62*	76.09*	6.09*	5.66*	5.88*	59.62*	64.17*	61.90*	14.52*	12.22*	13.37*
AdaptSeg+SAMix	**76.56**	80.57	**78.57**	4.97	4.12	4.55	61.75	66.20	63.98	12.75	11.09	11.92
Advent+SAMix	76.32	**80.64**	78.48	**4.90**	**3.98**	**4.44**	**62.02**	**66.35**	**64.19**	**11.97**	**10.85**	**11.41**

* $p < 0.05$ in the one-tailed paired t-test with Advent+SAMix.

with SGD optimizer for 80 epochs. The task loss L_t is the Dice loss. The initial learning rate is 0.001, which decays by 0.1 for every 20 epochs. The batch size is 16. For **Camelyon**, we use a ResNet-50 [8] with SGD optimizer for 20 epochs. L_t is the binary cross-entropy loss. The initial learning rate is 0.0001, which decays by 0.1 every 5 epochs. The batch size is 128. We use the fixed weighting factor $\mu = 0.01$, iterations $T = 10$, and step size $\delta = 0.1$ in all the experiments.

3.2 Method Effectiveness

We demonstrate the effectiveness of SAMix by comparing it with two sets of baselines. *First*, we compare the performance of UDA models with and without SAMix. *Second*, we compare SAMix against other FSUDA methods [9,11].

Fundus. Table 1 shows the 10-run average Dice coefficient (DSC) and Average Surface Distance (ASD) of all the methods trained with the source domain and **1-shot** target domain image. The results are evaluated in the two target domains. Compared to the model trained solely on the source domain (Source only), the performance gain achieved by UDA methods (AdaptSeg and Advent) is limited. However, incorporating SAMix as a plug-in for UDA methods (AdaptSeg+SAMix and Advent+SAMix) enhances the original UDA performance significantly ($p < 0.05$). Moreover, SAMix+Advent surpasses the two FSUDA methods (ASM and SM-PPM) significantly. This improvement is primarily due to spectrally augmented target-style samples by SAMix.

To assess the functionality of the target-aware spectral sensitivity map in measuring the model's generalization performance on the target domain, we computed the DoDiSS maps of the four models (AdaptSeg, ASM, SM-PPM, and AdaptSeg+SAMix). The results are presented in Fig. 2(**a**). The DoDiSS map of AdaptSeg+SAMix demonstrates a clear suppression of sensitivity, leading to improved model performance. To better visualize the results, the model generalizability (average DSC) versus the averaged ℓ_1-norm of the DoDiSS map is presented in Fig. 2 (**b**). The figure shows a clear trend of improved model performance as the averaged DoDiSS decreases. To assess the effectiveness of

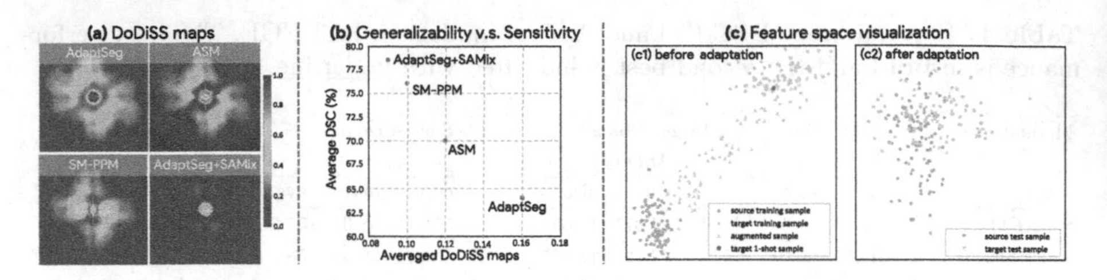

Fig. 2. Method effectiveness analysis. **(a)** The DoDiSS maps visualization; **(b)** Scattering plot of model generalizability v.s. sensitivity; **(c)** Feature space visualization.

SAMix-augmented target-style images in bridging the gap of domain shift, the feature distributions of Fundus images before and after adaptation are visualized in Fig. 2 **(c)** by t-SNE [12]. Figure 2**(c1)** shows the domain shift between the source and target domain features. The augmented samples from SAMix build the connection between the two domains with only a single example image from the target domain. Please note that, except the **1-shot** sample, all the other target domain samples are used here for visualization only but never seen during training/validation. Incorporating these augmented samples in AdaptSeg merges the source and target distributions as in Fig. 2 **(c2)**.

Table 2. 10-run average Acc (%) and AUC (%) of models on Camelyon. The best performance is in **bold** and the second best is indicated with underline.

Method	Source Domain → Target Domain 1		Source Domain → Target Domain 2	
	Acc (↑)	AUC (↑)	Acc (↑)	AUC (↑)
Source Only	75.42*	71.67*	65.55*	60.18*
DALN	78.63*	74.74*	62.57*	56.44*
ASM	83.66*	<u>80.43</u>	77.75*	73.47*
SRDC+SAMix	<u>84.28</u>	80.05	<u>78.64</u>	<u>74.62</u>
DALN+SAMix	**86.41**	**82.58**	**80.84**	**75.90**

* $p < 0.05$ in the one-tailed paired t-test with DALN+SAMix.

Camelyon. The evaluation results of the 10-run average accuracy (Acc) and Area Under the receiver operating Curve (AUC) of all methods trained with **1-shot** target domain image are presented in Table 2. The clustering-based SRDC is not included in the table, as the model crashed in this few-shot scenario. Also, the SM-PPM is not included because it is specifically designed for segmentation tasks. The results suggest that combining SAMix with UDA not only enhances the original UDA performance but also significantly outperforms other FSUDA methods.

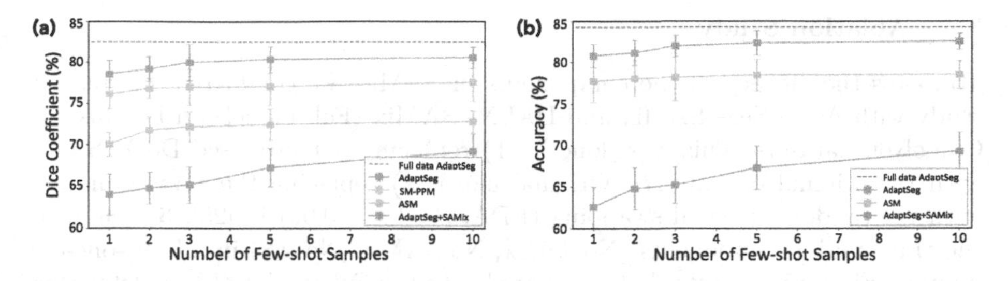

Fig. 3. Data efficiency of FSUDA methods on **(a) Fundus** and **(b) Camelyon**.

3.3 Data Efficiency

As the availability of target domain images is limited, data efficiency plays a crucial role in determining the data augmentation performance. Therefore, we evaluated the model's performance with varying numbers of target domain images in the training process. Figure 3 **(a)** and **(b)** illustrate the domain adaptation results on **Fundus** and **Camelyon** (both in target domain 1), respectively. Our method consistently outperforms other baselines with just a 1-shot target image for training. Furthermore, we qualitatively showcase the data efficiency of SAMix. Figure 4 **(a)** displays the generated image of SAMix given the target domain image. While maintaining the retinal structure of the source image, the augmented images exhibit a more similar style to the target image, indicating SAMix can effectively transfer the target domain style. Figure 4 **(b)** shows an example case of the segmented results. Compared with other baselines, the SAMix segmentation presents much less prediction error, especially in the cup region.

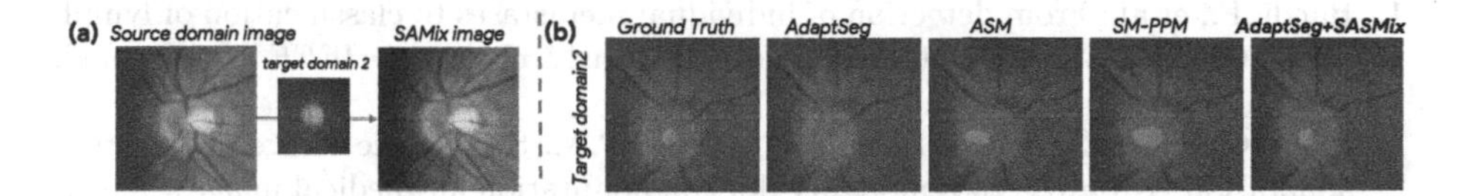

Fig. 4. **(a)** SAMix generated samples. **(b)** Case study of the Fundus segmentation.

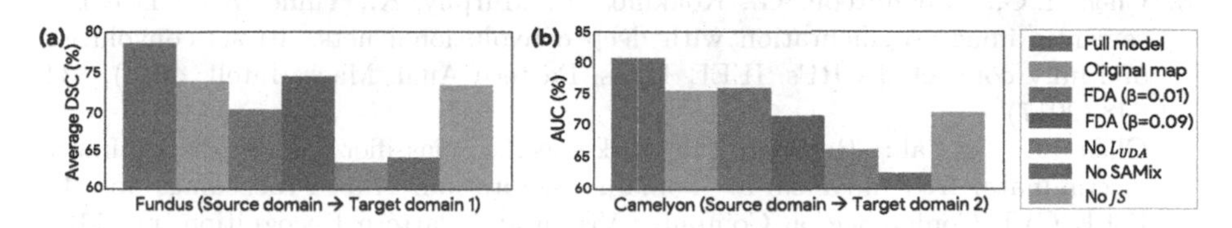

Fig. 5. Ablation study. **(a)** Average DSC on **Fundus**. **(b)** AUC on **Camelyon**.

3.4 Ablation Study

To assess the efficacy of the components in SAMix, we conducted an ablation study with AdaptSeg+SAMix and DALN+SAMix (Full model) on Fundus and Camelyon datasets. This was done by **1)** replacing our proposed DoDiSS map with the original one in [27] (Original map); **2)** replacing the SAMix module with the random spectral swapping (FDA, $\beta = 0.01$, 0.09) in [25]; **3)** removing the three major components (No L_{UDA}, No SAMix, No JS) in a leave-one-out manner. Figure 5 suggests that, compared with the Full model, the model performance degrades when the proposed components are either removed or replaced by previous methods, which indicates the efficacy of the SAMix components.

4 Discussion and Conclusion

This paper introduces a novel approach, Sensitivity-guided Spectral Adversarial MixUp (SAMix), which utilizes an adversarial mixing scheme and a spectral sensitivity map to generate target-style samples effectively. The proposed method facilitates the adaptation of existing UDA methods in the few-shot scenario. Thorough empirical analyses demonstrate the effectiveness and efficiency of SAMix as a plug-in module for various UDA methods across multiple tasks.

Acknowledgments. This research was partially supported by the National Science Foundation (NSF) under the CAREER award OAC 2046708, the National Institutes of Health (NIH) under award R21EB028001, and the Rensselaer-IBM AI Research Collaboration of the IBM AI Horizons Network.

References

1. Bandi, P., et al.: From detection of individual metastases to classification of lymph node status at the patient level: the camelyon17 challenge. IEEE Trans. Med. Imaging (2018)
2. Chen, C., Dou, Q., Chen, H., Qin, J., Heng, P.A.: Synergistic image and feature adaptation: Towards cross-modality domain adaptation for medical image segmentation. In: Proceedings of the AAAI Conference on Artificial Intelligence, vol. 33, pp. 865–872 (2019)
3. Chen, L.C., Papandreou, G., Kokkinos, I., Murphy, K., Yuille, A.L.: DeepLab: semantic image segmentation with deep convolutional nets, atrous convolution, and fully connected CRFs. IEEE Trans. Pattern Anal. Mach. Intell. **40**(4), 834–848 (2017)
4. Chen, L., et al.: Reusing the task-specific classifier as a discriminator: discriminator-free adversarial domain adaptation. In: Proceedings of the IEEE/CVF Conference on Computer Vision and Pattern Recognition, pp. 7181–7190 (2022)

5. Fumero, F., Alayón, S., Sanchez, J.L., Sigut, J., Gonzalez-Hernandez, M.: Rim-one: an open retinal image database for optic nerve evaluation. In: 2011 24th International Symposium on Computer-Based Medical Systems (CBMS), pp. 1–6. IEEE (2011)
6. Guan, H., Liu, M.: Domain adaptation for medical image analysis: a survey. IEEE Trans. Biomed. Eng. **69**(3), 1173–1185 (2021)
7. Guyader, N., Chauvin, A., Peyrin, C., Hérault, J., Marendaz, C.: Image phase or amplitude? Rapid scene categorization is an amplitude-based process. C.R. Biol. **327**(4), 313–318 (2004)
8. He, K., Zhang, X., Ren, S., Sun, J.: Deep residual learning for image recognition. In: Proceedings of the IEEE Conference on Computer Vision and Pattern Recognition, pp. 770–778 (2016)
9. Huang, X., Belongie, S.: Arbitrary style transfer in real-time with adaptive instance normalization. In: Proceedings of the IEEE International Conference on Computer Vision, pp. 1501–1510 (2017)
10. Liu, Q., Chen, C., Qin, J., Dou, Q., Heng, P.A.: Feddg: federated domain generalization on medical image segmentation via episodic learning in continuous frequency space. In: Proceedings of the IEEE/CVF Conference on Computer Vision and Pattern Recognition, pp. 1013–1023 (2021)
11. Luo, Y., Liu, P., Guan, T., Yu, J., Yang, Y.: Adversarial style mining for one-shot unsupervised domain adaptation. Adv. Neural. Inf. Process. Syst. **33**, 20612–20623 (2020)
12. Van der Maaten, L., Hinton, G.: Visualizing data using t-SNE. J. Mach. Learn. Res. **9**(11) (2008)
13. Noroozi, M., Favaro, P.: Unsupervised learning of visual representations by solving jigsaw puzzles. In: Leibe, B., Matas, J., Sebe, N., Welling, M. (eds.) ECCV 2016. LNCS, vol. 9910, pp. 69–84. Springer, Cham (2016). https://doi.org/10.1007/978-3-319-46466-4_5
14. Orlando, J.I., et al.: Refuge challenge: a unified framework for evaluating automated methods for glaucoma assessment from fundus photographs. Med. Image Anal. **59**, 101570 (2020)
15. Pan, S.J., Yang, Q.: A survey on transfer learning. IEEE Trans. Knowl. Data Eng. **22**(10), 1345–1359 (2009)
16. Tang, H., Chen, K., Jia, K.: Unsupervised domain adaptation via structurally regularized deep clustering. In: Proceedings of the IEEE/CVF Conference on Computer Vision and Pattern Recognition, pp. 8725–8735 (2020)
17. Tsai, Y.H., Hung, W.C., Schulter, S., Sohn, K., Yang, M.H., Chandraker, M.: Learning to adapt structured output space for semantic segmentation. In: Proceedings of the IEEE Conference on Computer Vision and Pattern Recognition, pp. 7472–7481 (2018)
18. Vu, T.H., Jain, H., Bucher, M., Cord, M., Pérez, P.: Advent: adversarial entropy minimization for domain adaptation in semantic segmentation. In: Proceedings of the IEEE/CVF Conference on Computer Vision and Pattern Recognition, pp. 2517–2526 (2019)
19. Wang, H., Xiao, C., Kossaifi, J., Yu, Z., Anandkumar, A., Wang, Z.: AugMax: adversarial composition of random augmentations for robust training. Adv. Neural. Inf. Process. Syst. **34**, 237–250 (2021)
20. Wang, J., et al.: Generalizing to unseen domains: a survey on domain generalization. IEEE Trans. Knowl. Data Eng. (2022)

21. Wang, S., Yu, L., Li, K., Yang, X., Fu, C.W., Heng, P.A.: Dofe: domain-oriented feature embedding for generalizable fundus image segmentation on unseen datasets. IEEE Trans. Med. Imaging (2020)
22. Wu, X., Wu, Z., Lu, Y., Ju, L., Wang, S.: Style mixing and patchwise prototypical matching for one-shot unsupervised domain adaptive semantic segmentation. In: Proceedings of the AAAI Conference on Artificial Intelligence, vol. 36, pp. 2740–2749 (2022)
23. Xie, Q., et al.: Unsupervised domain adaptation for medical image segmentation by disentanglement learning and self-training. IEEE Trans. Med. Imaging, 1 (2022). https://doi.org/10.1109/TMI.2022.3192303
24. Xu, Q., Zhang, R., Zhang, Y., Wang, Y., Tian, Q.: A Fourier-based framework for domain generalization. In: Proceedings of the IEEE/CVF Conference on Computer Vision and Pattern Recognition, pp. 14383–14392 (2021)
25. Yang, Y., Soatto, S.: FDA: Fourier domain adaptation for semantic segmentation. In: Proceedings of the IEEE/CVF Conference on Computer Vision and Pattern Recognition, pp. 4085–4095 (2020)
26. Yin, D., Gontijo Lopes, R., Shlens, J., Cubuk, E.D., Gilmer, J.: A Fourier perspective on model robustness in computer vision. In: Advances in Neural Information Processing Systems, vol. 32 (2019)
27. Zhang, J., et al.: When neural networks fail to generalize? a model sensitivity perspective. In: Proceedings of the AAAI Conference on Artificial Intelligence (2023)
28. Zhang, J., Chao, H., Xu, X., Niu, C., Wang, G., Yan, P.: Task-oriented low-dose CT image denoising. In: de Bruijne, M., et al. (eds.) MICCAI 2021. LNCS, vol. 12906, pp. 441–450. Springer, Cham (2021). https://doi.org/10.1007/978-3-030-87231-1_43
29. Zhang, J., Chao, H., Yan, P.: Toward adversarial robustness in unlabeled target domains. IEEE Trans. Image Process. **32**, 1272–1284 (2023). https://doi.org/10.1109/TIP.2023.3242141

Cross-Dataset Adaptation for Instrument Classification in Cataract Surgery Videos

Jay N. Paranjape[1(✉)], Shameema Sikder[2,3], Vishal M. Patel[1], and S. Swaroop Vedula[3]

[1] Department of Electrical and Computer Engineering, The Johns Hopkins University, Baltimore, USA
jparanj1@jhu.edu
[2] Wilmer Eye Institute, The Johns Hopkins University, Baltimore, USA
[3] Malone Center for Engineering in Healthcare, The Johns Hopkins University, Baltimore, USA

Abstract. Surgical tool presence detection is an important part of the intra-operative and post-operative analysis of a surgery. State-of-the-art models, which perform this task well on a particular dataset, however, perform poorly when tested on another dataset. This occurs due to a significant domain shift between the datasets resulting from the use of different tools, sensors, data resolution etc. In this paper, we highlight this domain shift in the commonly performed cataract surgery and propose a novel end-to-end Unsupervised Domain Adaptation (UDA) method called the Barlow Adaptor that addresses the problem of distribution shift without requiring any labels from another domain. In addition, we introduce a novel loss called the Barlow Feature Alignment Loss (BFAL) which aligns features across different domains while reducing redundancy and the need for higher batch sizes, thus improving cross-dataset performance. The use of BFAL is a novel approach to address the challenge of domain shift in cataract surgery data. Extensive experiments are conducted on two cataract surgery datasets and it is shown that the proposed method outperforms the state-of-the-art UDA methods by 6%. The code can be found at https://github.com/JayParanjape/Barlow-Adaptor.

Keywords: Surgical Tool Classification · Unsupervised Domain Adaptation · Cataract Surgery · Surgical Data Science

1 Introduction

Surgical instrument identification and classification are critical to deliver several priorities in surgical data science [21]. Various deep learning methods have been developed to classify instruments in surgical videos using data routinely

Supplementary Information The online version contains supplementary material available at https://doi.org/10.1007/978-3-031-43907-0_70.

H. Greenspan et al. (Eds.): MICCAI 2023, LNCS 14220, pp. 739–748, 2023.
https://doi.org/10.1007/978-3-031-43907-0_70

generated in institutions [2]. However, differences in image capture systems and protocols lead to nontrivial dataset shifts, causing a significant drop in performance of the deep learning methods when tested on new datasets [13]. Using cataract surgery as an example, Fig. 1 illustrates the drop in accuracy of existing methods to classify instruments when trained on one dataset and tested on another dataset [19,28]. Cataract surgery is one of the most common procedures [18], and methods to develop generalizable networks will enable clinically useful applications.

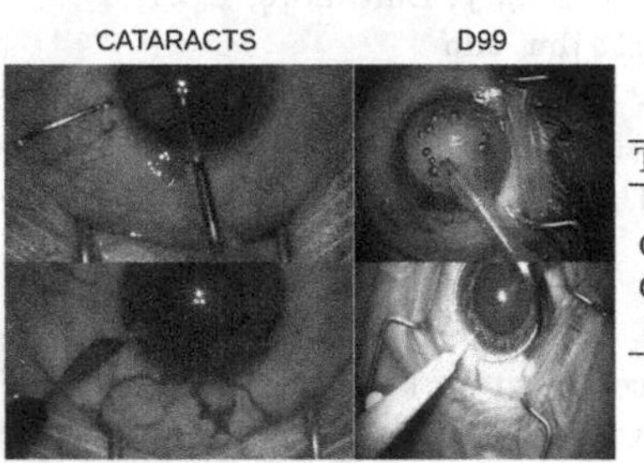

Train	Test	Macro Acc	Micro Acc
D99	D99	57.0	62.2
CAT	CAT	55.0	67.2
CAT	D99	14.25	16.9
D99	CAT	27.9	14.9

Fig. 1. Dataset shift between the CATARACTS dataset (CAT) [6] and D99 [7,9] dataset. Results for models trained on one dataset and tested on another show a significant drop in performance.

Domain adaptation methods aim to attempt to mitigate the drop in algorithm performance across domains [13]. Unsupervised Domain Adaptation (UDA) methods are particularly useful when the source dataset is labeled and the target dataset is unlabeled. In this paper, we describe a novel end-to-end UDA method, which we call the Barlow Adaptor, and its application for instrument classification in video images from cataract surgery. We define a novel loss function called the Barlow Feature Alignment Loss (BFAL) that aligns the features learnt by the model between the source and target domains, without requiring any labeled target data. It encourages the model to learn non-redundant features that are domain agnostic and thus tackles the problem of UDA. BFAL can be added as an add-on to existing methods with minimal code changes. The contributions of our paper are threefold:

1. We define a novel loss for feature alignment called BFAL that doesn't require large batch sizes and encourages learning non-redundant, domain agnostic features.
2. We use BFAL to generate an end-to-end system called the Barlow Adaptor that performs UDA. We evaluate the effectiveness of this method and compare it with existing UDA methods for instrument classification in cataract surgery images.
3. We motivate new research on methods for generalizable deep learning models for surgical instrument classification using cataract surgery as the test-bed. Our work proposes a solution to the problem of lack of generalizability of

deep learning models that was identified in previous literature on cataract surgery instrument classification.

2 Related Work

Instrument Identification in Cataract Surgery Video Images. The motivation for instrument identification is its utility in downstream tasks such as activity localization and skill assessment [3,8,22]. The current state-of-the-art instrument identification method called Deep-Phase [28] uses a ResNet architecture to identify instruments and then to identify steps in the procedure. However, a recent study has shown that while these methods work well on one dataset, there is a significant drop in performance when tested on a different dataset [16]. Our analyses reiterate similar findings on drop in performance (Fig. 1) and highlight the effect of domain shift between data from different institutions even for the same procedure.

Unsupervised Domain Adaptation. UDA is a special case of domain adaptation, where a model has access to annotated training data from a source domain and unannotated data from a target domain [13]. Various methods have been proposed in the literature to perform UDA. One line of research involves aligning the feature distributions between the source and target domains. Maximum Mean Discrepancy (MMD) is commonly used as a distance metric between the source and target distributions [15]. Other UDA methods use a convolutional neural network (CNN) to generate features and then use MMD as an additional loss to align distributions [1,11,12,20,25,27]. While MMD is a first-order statistic, Deep CORAL [17] penalizes the difference in the second-order covariance between the source and target distributions. Our method uses feature alignment by enforcing a stricter loss function during training.

Another line of research for UDA involves adversarial training. Domain Adaptive Neural Network (DANN) [5] involves a minimax game, in which one network minimizes the cross entropy loss for classification in the source domain, while the other maximizes the cross entropy loss for domain classification. Few recent methods generate pseudo labels on the target domain and then train the network on them. One such method is Source Hypothesis Transfer (SHOT) [10], which performs source-free domain adaptation by further performing information maximization on the target domain predictions. While CNN-based methods are widely popular for UDA, there are also methods which make use of the recently proposed Vision Transformer (ViT) [4], along with an ensemble of the above described UDA based losses. A recent approach called Cross Domain Transformer (CDTrans) uses cross-domain attention to produce pseudo labels for training that was evaluated in various datasets [24]. Our proposed loss function is effective for both CNN and ViT-based backbones.

3 Proposed Method

In the UDA task, we are given n_s observations from the source domain $\mathcal{D}_S$. Each of these observations is in the form of a tuple (x_s, y_s), where x_s denotes an image from the source training data and y_s denotes the corresponding label, which is the instrument index present in the image. In addition, we are given n_t observations from the target domain $\mathcal{D}_T$. Each of these can be represented by x_t, which represents the image from the target training data. However, there are no labels present for the target domain during training. The goal of UDA is to predict the labels y_t for the target domain data.

Barlow Feature Alignment Loss (BFAL). We introduce a novel loss, which encourages features between the source and target to be similar to each other while reducing the redundancy between the learnt features. BFAL works on pairs of feature projections of the source and target. More specifically, let $f_s \in \mathbb{R}^{BXD}$ and $f_t \in \mathbb{R}^{BXD}$ be the features corresponding to the source and target domain, respectively. Here B represents the batch size and D represents the feature dimension. Similar to [26], we project these features into a P dimensional space using a fully connected layer called the Projector, followed by a batch normalization to whiten the projections. Let the resultant projections be denoted by $p_s \in \mathbb{R}^{BXP}$ for the source and $p_t \in \mathbb{R}^{BXP}$ for the target domains. Next, we compute the correlation matrix $\mathbb{C}_1 \in \mathbb{R}^{PXP}$. Each element of $\mathbb{C}_1$ is computed as follows

$$\mathbb{C}_1^{ij} = \frac{\sum_{b=1}^{B} p_s^{bi} p_t^{bj}}{\sqrt{\sum_{b=1}^{B} (p_s^{bi})^2} \sqrt{\sum_{b=1}^{B} (p_t^{bj})^2}}. \tag{1}$$

Finally, the BFAL is computed using the L2 loss between the elements of $\mathbb{C}_1$ and the identity matrix $\mathbb{I}$ as follows

$$\mathbb{L}_{BFA} = \underbrace{\sum_{i=1}^{P} (1 - \mathbb{C}_1^{ii})^2}_{feature\ alignment} + \underbrace{\mu \sum_{i=1}^{P} \sum_{j \neq i} (\mathbb{C}_1^{ij})^2}_{redundancy\ reduction}, \tag{2}$$

where μ is a constant. Intuitively, the first term of the loss function can be thought of as a feature alignment term since we push the diagonal elements in the covariance matrix towards 1. In other words, we encourage the feature projections between the source and target to be perfectly correlated. On the other hand, by pushing the off-diagonal elements to 0, we decorrelate different components of the projections. Hence, this term can be considered a redundancy reduction term, since we are pushing each feature vector component to be independent of one another.

BFAL is inspired by a recent technique in self-supervised learning, called the Barlow Twins [26], where the authors show the effectiveness of such a formulation at lower batch sizes. In our experiments, we observe that even keeping a batch size of 16 gave good results over other existing methods. Furthermore, BFAL does not require large amounts of data to converge.

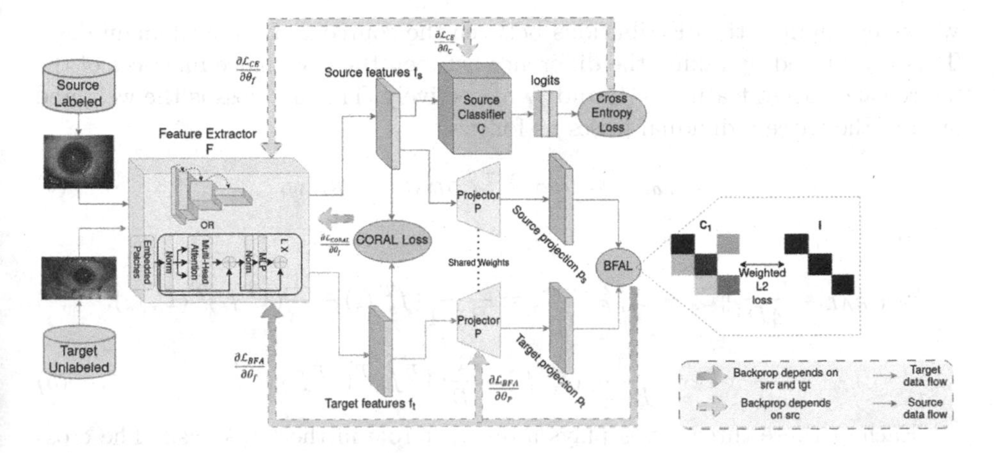

Fig. 2. Architecture corresponding to the Barlow Adaptor. Training occurs using pairs of images from the source and target domain. They are fed into the feature extractor, which generates features used for the CORAL loss. Further, a projector network P projects the features into a P dimensional space. These are then used to calculate the Barlow Feature Alignment Loss. One branch from the source features goes into the source classifier network that is used to compute the cross entropy loss with the labeled source data. [Backprop = backpropagation; src = source dataset, tgt = target dataset]

Barlow Adaptor. We propose an end-to-end method that utilizes data from the labeled source domain and the unlabeled target domain. The architecture corresponding to our method is shown in Fig. 2.

There are two main sub-parts of the architecture - the Feature Extractor F, and the Source Classifier C. First, we divide the training images randomly into batches of pairs $\{x_s, x_t\}$ and apply F on them, which gives us the features extracted from these sets of images. For the Feature Detector, we show the effectiveness of our novel loss using ViT and ResNet50 both of which have been pre-trained on ImageNet. The features obtained are denoted as f_s and f_t for the source and target domains, respectively. Next, we apply C on these features to get logits for the classification task. The source classifier is a feed forward neural network, which is initialized from scratch. These logits are used, along with the source labels y_s to compute the source cross entropy loss as $\mathbb{L}_{CE} = \frac{-1}{B}\sum_{b=1}^{B}\sum_{m=1}^{M} y_s^{bm}\log(p_s^{bm})$,
where M represents the number of classes, B represents the total mini-batches, while m and b represent their respective indices.

The features f_s and f_t are further used to compute the Correlation Alignment(CORAL) loss and the BFAL, which enforce the feature extractor to align its weights so as to learn features that are domain agnostic as well as non-redundant. The BFAL is calculated as mentioned in the previous subsection. The CORAL loss is computed as depicted in Eq. 4, following the UDA method Deep CORAL [17]. While the BFAL focuses on reducing redundancy, CORAL works by aligning the distributions between the source and target domain data. This is achieved by taking the difference between the covariance matrices of the source and target features - f_s and f_t respectively. The final loss is the weighted sum of the three individual losses as follows:

$$\mathbb{L}_{final} = \mathbb{L}_{CE} + \lambda(\mathbb{L}_{CORAL} + \mathbb{L}_{BFA}), \tag{3}$$

where

$$\mathbb{L}_{CORAL} = \frac{1}{4D^2}\|\mathbb{C}_s - \mathbb{C}_t\|_F^2, \quad \mathbb{C}_s = \frac{1}{B-1}(f_s^T f_s) - \frac{1}{B}(\mathbf{1}^T f_s)^T(\mathbf{1}^T f_s), \tag{4}$$

$$\mathbb{C}_t = \frac{1}{B-1}(f_t^T f_t) - \frac{1}{B}(\mathbf{1}^T f_t)^T(\mathbf{1}^T f_t). \tag{5}$$

Each of these three losses plays a different role in the UDA task. The cross entropy loss encourages the model to learn discriminative features between images with different instruments. The CORAL loss pushes the features between the source and target towards having a similar distribution. Finally, the BFAL tries to make the features between the source and the target non-redundant and same. BFAL is a stricter loss than CORAL as it forces features to not only have the same distribution but also be equal. Further, it also differs from CORAL in learning independent features as it explicitly penalizes non-zero non-diagonal entries in the correlation matrix. While using BFAL alone gives good results, using it in addition to CORAL gives slightly better results empirically. We note these observations in our ablation studies. Between the cross entropy loss and the BFAL, an adversarial game is played where the former makes the features more discriminative and the latter tries to make them equal. The optimal features thus learnt are different in aspects required to identify instruments but are equal for any domain-related aspect. This property of the Barlow Adaptor is especially useful for surgical domains where the background has similar characteristics for most of the images within a domain. For example, for cataract surgery images, the position of the pupil or the presence of blood during the usage of certain instruments might be used by the model for classification along with the instrument features. These features depend highly upon the surgical procedures and the skill of the surgeon, thus making them highly domain-specific and possibly unavailable in the target domain. Using BFAL during training attempts to prevent the model from learning such features.

4 Experiments and Results

We evaluate the proposed UDA method for the task of instrument classification using two cataract surgery image datasets. In our experiments, one dataset is used as the source domain and the other is used as the target domain. We use micro and macro accuracies as our evaluation metrics. Micro accuracy denotes the number of correctly classified observations divided by the total number of observations. In contrast, macro accuracy denotes the average of the classwise accuracies and is effective in evaluating classes with less number of samples.

Table 1. Mapping of surgical tools between CATARACTS(L) and D99(R)

CATARACTS	D99	CATARACTS	D99
Secondary Incision Knife	Paracentesis Blade	Bonn Forceps	0.12 Forceps
Charleux Cannula	Anterior Chamber Cannula	Irrigation	Irrigation
Capsulorhexis Forceps	Utrata Forceps	Cotton	Weckcell Sponge
Hydrodissection Cannula	Hydrodissection Cannula	Implant Injector	IOL Injector
Phacoemulsifier Handpiece	Phaco Handpiece	Suture Needle	Suture
Capsulorhexis Cystotome	Cystotome	Needle Holder	Needle Driver
Primary Incision Knife	Keratome	Micromanipulator	Chopper

Datasets. The first dataset we use is CATARACTS [6], which consists of 50 videos with framewise annotations available for 21 surgical instruments. The dataset is divided into 25 training videos and 25 testing videos. We separate 5 videos from the training set and use them as the validation set for our experiments. The second dataset is called D99 in this work [7,9], which consists of 105 videos of cataract surgery with annotations for 25 surgical instruments. Of the 105 videos, we use 65 videos for training, 10 for validation and 30 for testing. We observe a significant distribution shift between the two datasets as seen in Fig. 1. This is caused by several factors such as lighting, camera resolution, and differences in instruments used for the same steps. For our experiments in this work, we use 14 classes of instruments that are common to both datasets. Table 1 shows a mapping of instruments between the two datasets. For each dataset, we normalize the images using the means and standard deviations calculated from the respective training images. In addition, we resize all images to 224×224 size and apply random horizontal flipping with a probability of 0.5 before passing them to the model.

Experimental Setup. We train the Barlow Adaptor for multi-class classification with the above-mentioned 14 classes in Pytorch. For the Resnet50 backbone, we use weights pretrained on Imagenet [14] for initialization. For the ViT backbone, we use the base-224 class of weights from the TIMM library [23]. The Source Classifier C and the Projector P are randomly initialized. We use the validation sets to select the hyperparameters for the models. Based on these empirical results, we choose λ from Eq. 3 to be 0.001 and μ from Eq. 2 to be 0.0039. We use SGD as the optimizer with momentum of 0.9 and a batch size of 16. We start the training with a learning rate of 0.001 and reduce it by a factor of 0.33 every 20 epochs. The entire setup is trained with a single NVIDIA Quatro RTX 8000 GPU. We use the same set of hyperparameters for the CNN and ViT backbones in both datasets.

Results. Table 2 shows results comparing the performance of the Barlow Adaptor with recent UDA methods. We highlight the effect of domain shift by comparing the source-only models and the target-only models, where we observe a

Table 2. Macro and micro accuracies for cross domain tool classification. Here, source-only denotes models that have only been trained on one domain and tested on the other. Similarly, target-only denotes models that have been trained on the test domain and thus act as an upper bound. Deep CORAL [17] is similar to using CORAL with ResNet backbone, so we don't list the latter separately. Here, CAT represents the CATARACTS dataset.

	D99 → CAT		CAT → D99	
Method	Macro Acc	Micro Acc	Macro Acc	Micro Acc
Source Only (ResNet50 backbone)	27.9%	14.9%	14.25%	16.9%
MMD with ResNet50 backbone [15]	32.2%	15.9%	20.6%	24.3%
Source Only (ViT backbone)	30.43%	14.14%	13.99%	17.11%
MMD with ViT backbone [15]	31.32%	13.81%	16.42%	20%
CORAL with ViT backbone [17]	28.7%	16.5%	15.38%	18.5
DANN [5]	22.4%	11.6%	16.7%	19.5%
Deep CORAL [17]	32.8%	14%	18.6%	22
CDTrans [24]	29.1%	14.7%	20.9%	24.7%
Barlow Adaptor with ResNet50 (Ours)	**35.1%**	**17.1%**	**24.62%**	**28.13%**
Barlow Adaptor with ViT (Ours)	31.91%	12.81%	17.35%	20.8%
Target Only (ResNet50)	55%	67.2%	57%	62.2%
Target Only (ViT)	49.80%	66.33%	56.43%	60.46%

Table 3. Findings from ablation studies to evaluate the Barlow Adaptor. Here, Source Only is the case where neither CORAL nor BFAL is used. We use Macro Accuracy for comparison. Here, CAT represents the CATARACTS dataset.

	ViT Feature Extractor		ResNet50 Feature Extractor	
Method	D99 → CAT	CAT → D99	D99 → CAT	CAT → D99
Source Only($\mathbb{L}_{CE}$)	30.43%	16.7%	27.9%	14.9%
Only CORAL($\mathbb{L}_{CORAL}$)	28.7%	15.38%	32.8%	18.6%
Only BFAL($\mathbb{L}_{BFA}$)	29.8%	17.01%	32.3%	24.46%
Barlow Adaptor (Eq. 3)	**32.1%**	**17.35%**	**35.1%**	**24.62%**

significant drop of 27% and 43% in macro accuracy for the CATARACTS dataset and the D99 dataset, respectively. Using the Barlow Adaptor, we observe an increase in macro accuracy by 7.2% over the source only model. Similarly, we observe an increase in macro accuracy of 9% with the Barlow Adaptor when the source is CATARACTS and the target is the D99 dataset compared with the source only model. Furthermore, estimates of macro and micro accuracy are larger with the Barlow Adaptor than those with other existing methods. Finally, improved accuracy with the Barlow Adaptor is seen with both ResNet and ViT backbones.

Ablation Study. We tested the performance gain due to each part of the Barlow Adaptor. Specifically, the Barlow Adaptor has CORAL loss and BFAL as its two major feature alignment losses. We remove one component at a time and observe a decrease in performance with both ResNet and ViT backbones (Table 3). This shows that each loss has a part to play in domain adaptation. Further ablations are included in the supplementary material.

5 Conclusion

Domain shift between datasets of cataract surgery images limits generalizability of deep learning methods for surgical instrument classification. We address this limitation using an end-to-end UDA method called the Barlow Adaptor. As part of this method, we introduce a novel loss function for feature alignment called the BFAL. Our evaluation of the method shows larger improvements in classification performance compared with other state-of-the-art methods for UDA. BFAL is an independent module and can be readily integrated into other methods as well. BFAL can be easily extended to other network layers and architectures as it only takes pairs of features as inputs.

Acknowledgement. This research was supported by a grant from the National Institutes of Health, USA; R01EY033065. The content is solely the responsibility of the authors and does not necessarily represent the official views of the National Institutes of Health.

References

1. Baktashmotlagh, M., Harandi, M., Salzmann, M.: Distribution-matching embedding for visual domain adaptation. J. Mach. Learn. Res. **17**(1), 3760–3789 (2016)
2. Bouget, D., Allan, M., Stoyanov, D., Jannin, P.: Vision-based and marker-less surgical tool detection and tracking: a review of the literature. Med. Image Anal. **35**, 633–654 (2017)
3. demir, K., Schieber, H., Weise, T., Roth, D., Maier, A., Yang, S.: Deep learning in surgical workflow analysis: a review (2022)
4. Dosovitskiy, A., et al.: An image is worth 16×16 words: transformers for image recognition at scale. In: International Conference on Learning Representations (2021)
5. Ganin, Y., Lempitsky, V.: Unsupervised domain adaptation by backpropagation. In: Proceedings of the 32nd International Conference on International Conference on Machine Learning, ICML 2015, vol. 37, pp. 1180–1189. JMLR.org (2015)
6. Hajj, H., et al.: Cataracts: challenge on automatic tool annotation for cataract surgery. Med. Image Anal. **52**, 24–41 (2018)
7. Hira, S.: Video-based assessment of intraoperative surgical skill. Comput.-Assist. Radiol. Surg. **17**(10), 1801–1811 (2022)
8. Josef, L., James, W., Michael, S.: Evolution and applications of artificial intelligence to cataract surgery. Ophthalmol. Sci. **2**, 100164 (2022)
9. Kim, T., O'Brien, M., Zafar, S., Hager, G., Sikder, S., Vedula, S.: Objective assessment of intraoperative technical skill in capsulorhexis using videos of cataract surgery. Comput.-Assist. Radiol. Surg. **14**(6), 1097–1105 (2019)

10. Liang, J., Hu, D., Feng, J.: Do we really need to access the source data? source hypothesis transfer for unsupervised domain adaptation. In: III, H.D., Singh, A. (eds.) Proceedings of the 37th International Conference on Machine Learning. Proceedings of Machine Learning Research, vol. 119, pp. 6028–6039. PMLR (2020)
11. Long, M., Wang, J., Ding, G., Sun, J., Yu, P.S.: Transfer feature learning with joint distribution adaptation. In: 2013 IEEE International Conference on Computer Vision, pp. 2200–2207 (2013)
12. Pan, S.J., Tsang, I.W., Kwok, J.T., Yang, Q.: Domain adaptation via transfer component analysis. IEEE Trans. Neural Netw. **22**(2), 199–210 (2011)
13. Patel, V.M., Gopalan, R., Li, R., Chellappa, R.: Visual domain adaptation: a survey of recent advances. IEEE Signal Process. Maga. **32**(3), 53–69 (2015)
14. Russakovsky, O., et al.: ImageNet large scale visual recognition challenge. Int. J. Comput. Vision **115**(3), 211–252 (2015). https://doi.org/10.1007/s11263-015-0816-y
15. Schölkopf, B., Platt, J., Hofmann, T.: A kernel method for the two-sample-problem, pp. 513–520 (2007)
16. Sokolova, N., Schoeffmann, K., Taschwer, M., Putzgruber-Adamitsch, D., El-Shabrawi, Y.: Evaluating the generalization performance of instrument classification in cataract surgery videos. In: Ro, Y.M., et al. (eds.) MMM 2020. LNCS, vol. 11962, pp. 626–636. Springer, Cham (2020). https://doi.org/10.1007/978-3-030-37734-2_51
17. Sun, B., Saenko, K.: Deep CORAL: correlation alignment for deep domain adaptation, pp. 443–450 (2016)
18. Trikha, S., Turnbull, A., Morris, R., Anderson, D., Hossain, P.: The journey to femtosecond laser-assisted cataract surgery: new beginnings or a false dawn? Eye (London, England) **27** (2013)
19. Twinanda, A.P., Shehata, S., Mutter, D., Marescaux, J., de Mathelin, M., Padoy, N.: Endonet: a deep architecture for recognition tasks on laparoscopic videos. IEEE Trans. Med. Imaging **36**(1), 86–97 (2017)
20. Tzeng, E., Hoffman, J., Zhang, N., Saenko, K., Darrell, T.: Deep domain confusion: maximizing for domain invariance (2014)
21. Vedula, S.S., et al.: Artificial intelligence methods and artificial intelligence-enabled metrics for surgical education: a multidisciplinary consensus. J. Am. Coll. Surg. **234**(6), 1181–1192 (2022)
22. Ward, T.M., et al.: Computer vision in surgery. Surgery **169**(5), 1253–1256 (2021)
23. Wightman, R.: Pytorch image models. https://github.com/rwightman/pytorch-image-models (2019)
24. Xu, T., Chen, W., Wang, P., Wang, F., Li, H., Jin, R.: Cdtrans: cross-domain transformer for unsupervised domain adaptation (2021)
25. Yan, H., Ding, Y., Li, P., Wang, Q., Xu, Y., Zuo, W.: Mind the class weight bias: Weighted maximum mean discrepancy for unsupervised domain adaptation. In: 2017 IEEE Conference on Computer Vision and Pattern Recognition (CVPR), pp. 945–954 (2017)
26. Zbontar, J., Jing, L., Misra, I., LeCun, Y., Deny, S.: Barlow twins: self-supervised learning via redundancy reduction (2021)
27. Zhong, E., et al.: Cross domain distribution adaptation via kernel mapping. In: Proceedings of the 15th ACM SIGKDD International Conference on Knowledge Discovery and Data Mining, KDD 2009, pp. 1027–1036. Association for Computing Machinery, New York (2009)
28. Zisimopoulos, O., et al.: Deepphase: surgical phase recognition in cataracts videos (2018)

Black-box Domain Adaptative Cell Segmentation via Multi-source Distillation

Xingguang Wang[1], Zhongyu Li[1(✉)], Xiangde Luo[2], Jing Wan[1], Jianwei Zhu[1], Ziqi Yang[1], Meng Yang[3], and Cunbao Xu[4(✉)]

[1] School of Software Engineering, Xian Jiaotong University, Xi'an, China
zhongyuli@xjtu.edu.cn

[2] School of Mechanical and Electrical Engineering, University of Electronic Science and Technology of China, Chengdu, China

[3] Hunan Frontline Medical Technology Co., Ltd., Changsha, China

[4] Department of Pathology, Quanzhou First Hospital Affiliated to Fujian Medical University, Quanzhou, China
937340447@qq.com

Abstract. Cell segmentation plays a critical role in diagnosing various cancers. Although deep learning techniques have been widely investigated, the enormous types and diverse appearances of histopathological cells still pose significant challenges for clinical applications. Moreover, data protection policies in different clinical centers and hospitals limit the training of data-dependent deep models. In this paper, we present a novel framework for cross-tissue domain adaptative cell segmentation without access both source domain data and model parameters, namely Multi-source Black-box Domain Adaptation (MBDA). Given the target domain data, our framework can achieve the cell segmentation based on knowledge distillation, by only using the outputs of models trained on multiple source domain data. Considering the domain shift cross different pathological tissues, predictions from the source models may not be reliable, where the noise labels can limit the training of the target model. To address this issue, we propose two practical approaches for weighting knowledge from the multi-source model predictions and filtering out noisy predictions. First, we assign pixel-level weights to the outputs of source models to reduce uncertainty during knowledge distillation. Second, we design a pseudo-label cutout and selection strategy for these predictions to facilitate the knowledge distillation from local cells to global pathological images. Experimental results on four types of pathological tissues demonstrate that our proposed black-box domain adaptation approach can achieve comparable and even better performance in comparison with state-of-the-art white-box approaches. The code and dataset are released at: https://github.com/NeuronXJTU/MBDA-CellSeg.

Keywords: Multi-source domain adaptation · Black-box model · Cell segmentation · Knowledge distillation

H. Greenspan et al. (Eds.): MICCAI 2023, LNCS 14220, pp. 749–758, 2023.
https://doi.org/10.1007/978-3-031-43907-0_71

1 Introduction

Semantic segmentation plays a vital role in pathological image analysis. It can help people conduct cell counting, cell morphology analysis, and tissue analysis, which reduces human labor [19]. However, data acquisition for medical images poses unique challenges due to privacy concerns and the high cost of manual annotation. Moreover, pathological images from different tissues or cancer types often show significant domain shifts, which hamper the generalization of models trained on one dataset to others. Due to the abovementioned challenges, some researchers have proposed various white-box domain adaptation methods to address these issues.

Recently, [8,16] propose to use generative adversarial networks to align the distributions of source and target domains and generate source-domain lookalike outputs for target images. Source-free domain adaptation methods have been also widely investigated due to the privacy protection. [3,5,14] explore how to implicitly align target domain data with the model trained on the source domain without accessing the source domain data. There are also many studies on multi-source white-box domain adaptation. Ahmed et al. [1] propose a novel algorithm which automatically identifies the optimal blend of source models to generate the target model by optimizing a specifically designed unsupervised loss. Li et al. [13] extend the above work to semantic segmentation and proposed a method named model-invariant feature learning, which takes full advantage of the diverse characteristics of the source-domain models.

Nonetheless, several recent investigations have demonstrated that the domain adaptation methods for source-free white-box models still present a privacy risk due to the potential leakage of model parameters [4]. Such privacy breaches may detrimental to the privacy protection policies of hospitals. Moreover, the target domain uses the same neural network as the source domain, which is not desirable for low-resource target users like hospitals [15]. We thus present a more challenging task of relying solely on black-box models from vendors to avoid parameter leakage. In clinical applications, various vendors can offer output interfaces for different pathological images. While black-box models are proficient in specific domains, their performances greatly degrade when the target domain is updated with new pathology slices. Therefore, how to leverage the existing knowledge of black-box models to effectively train new models for the target domain without accessing the source domain data remains a critical challenge.

In this paper, we present a novel source-free domain adaptation framework for cross-tissue cell segmentation without accessing both source domain data and model parameters, which can seamlessly integrate heterogeneous models from different source domains into any cell segmentation network with high generality. To the best of our knowledge, this is the first study on the exploration of multi-source black-box domain adaptation for cross-tissue cell segmentation. In this setting, conventional multi-source ensemble methods are not applicable due to the unavailability of model parameters, and simply aggregating the black-box outputs would introduce a considerable amount of noise, which can be detrimental to the training of the target domain model. Therefore, we develop two

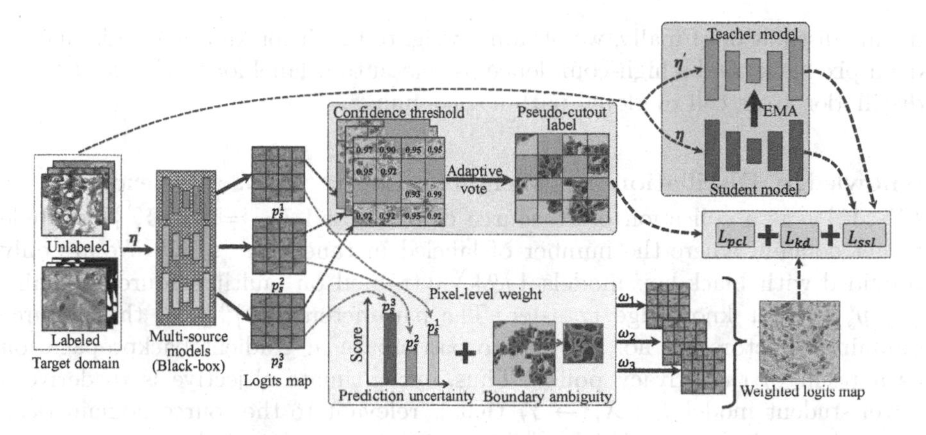

Fig. 1. Overview of our purposed framework, where logits maps denote the raw predictions from source models and ω denotes pixel-level weight for each prediction. The semi-supervised loss, denoted as L_{ssl}, encompasses the supervised loss, consistency loss, and maximize mutual information loss.

strategies within this new framework to address this issue. Firstly, we propose a pixel-level multi-source domain weighting method, which reduces source domain noise by knowledge weighting. This method effectively addresses two significant challenges encountered in the analysis of cellular images, namely, the uncertainty in source domain output and the ambiguity in cell boundary semantics. Secondly, we also take into account the structured information from cells to images, which may be overlooked during distillation, and design an adaptive knowledge voting strategy. This strategy enables us to ignore low-confidence regions, similar to Cutout [6], but with selective masking of pixels, which effectively balances the trade-off between exploiting similarities and preserving differences of different domains. As a result, we refer to the labels generated through the voting strategy as pseudo-cutout labels.

2 Method

Overview: Figure 1 shows a binary cell segmentation task with three source models trained on different tissues and a target model, i.e., the student model in Fig. 1. We only use the source models' predictions on the target data for knowledge transfer without accessing the source data and parameters. The η and η' indicate that different perturbations are added to the target images. Subsequently, we feed the perturbed images into the source domain predictor to generate the corresponding raw segmentation outputs. These outputs are then processed by two main components of our framework: a pixel-level weighting method that takes into account the prediction uncertainty and cell boundary ambiguity, and an adaptive knowledge voter that utilizes confidence gates and a dynamic ensemble strategy. These components we designed are to extract reliable knowledge from the predictions of source domain models and reduce noise during distillation. Finally, we obtain a weighted logit for knowledge distillation from pixel level and a high-confidence pseudo-cutout label for further structured distillation from cell to global pathological image.

Knowledge Distillation by Weighted Logits Map: We denote $\mathcal{D}_S^N = \{X_s, Y_s\}_N$ as a collection of N source domains and $\mathcal{D}_T = \{X_t^i, Y_t^j\}$ as single target domain, where the number of labeled instances $Y_t^j \ll X_t^i$. We are only provided with black-box models $\{f_s^n\}_{n=1}^N$ trained on multiple source domains $\{x_s^i, y_s^i\}_{n=1}^N$ for knowledge transfer. The parameters $\{\theta_s^n\}_{n=1}^N$ of these source domain predictors are not allowed to participate in gradient backpropagation as a result of the privacy policy. Thus, our ultimate objective is to derive a novel student model $f_t : X_t \to Y_t$ that is relevant to the source domain task. Accordingly, direct knowledge transfer using the output of the source domain predictor may lead to feature bias in the student model due to the unavoidable covariance [20] between the target and source domains. Inspired by [21], we incorporate prediction uncertainty and cell boundary impurity to establish pixel-level weights for multi-source outputs. We assume that *k-square-neighbors* of a pixel as a cell region, i.e., for a logits map with height H and width W, we define the region as follow:

$$\mathcal{N}_k\{(i,j) \mid (i,j) \in (H,W)\} = \{(u,v) \mid |u-i| \leq k, |v-j| \leq k\} \tag{1}$$

where (i, j) denotes centre of region, and k denotes the size of *k-square-neighbors.*

Firstly, we develop a pixel-level predictive uncertainty algorithm to aid in assessing the correlation between multiple source domains and the target domain. For a given target image $x_t \in X_t^i$, we initially feed it into the source predictors $\{f_s^n\}_{n=1}^N$ to obtain their respective prediction $\{p_s^n\}_{n=1}^N$. To leverage the rich semantic information from the source domain predictor predictions, we utilize predictive entropy of the softmax outputs to measure the prediction uncertainty scores. In the semantic segmentation scenario of C-classes classification, we define the pixel-level uncertainty score $\mathcal{U}_n^{(i,j)}$ as follow:

$$\mathcal{U}_n^{(i,j)} = -\sum_{c=1}^{C} O_s^{n(i,j,c)} \log O_s^{n(i,j,c)} \tag{2}$$

where O_s^n denotes softmax output,i.e.,$O_s^n = softmax(p_s^n)$ from nth source predictor.

Due to the unique characteristics of cell morphology, merely relying on uncertainty information is insufficient to produce high-quality ensemble logits map that accurately capture the relevance between the source and target domains. The target pseudo-label for the nth predictor f_s^n can be obtained by applying the softmax function to the output and selecting the category with the highest probability score, i.e., $\widehat{Y}_t = \arg\max_{c\in\{1,\ldots,C\}}(softmax(p_s^n))$. Then according to C-classes classification tasks, we divide the cell region into C subsets, $\mathcal{N}_k^c(i,j) = \{(u,v) \in \mathcal{N}_k(i,j) \mid \widehat{Y}_t = c\}$. After that, we determine the degree of impurity in an area of interest by analyzing the statistics of the boundary region, which represents the level of semantic information ambiguity. Specifically, the number of different objects within the area is considered a proxy for its impurity level, with higher counts indicating higher impurity.The boundary impurity $\mathcal{P}^{(i,j)}$ can be calculated as:

$$\mathcal{P}_n^{(i,j)} = -\sum_{c=1}^{C} \frac{|\mathcal{N}_k^c(i,j)|}{|\mathcal{N}_k(i,j)|} \log \frac{|\mathcal{N}_k^c(i,j)|}{|\mathcal{N}_k(i,j)|} \tag{3}$$

where $|\cdot|$ denotes the number of pixels in the area.

By assigning lower weights to the pixels with high uncertainty and boundary ambiguity, we can obtain pixel-level weight scores $\mathcal{W}^n$ for each p_s^n, i.e.,

$$\mathcal{W}^n = -\log\left(\frac{\exp\left(\mathcal{U}_n \odot \mathcal{P}_n\right)}{\sum_{n=1}^{N} \exp\left(\mathcal{U}_n \odot \mathcal{P}_n\right)}\right) \tag{4}$$

where $\odot$ denotes element-wise matrix multiplication. According to the pixel-level weight, we will obtain an ensemble logits map $\mathcal{M} = \sum_{n=1}^{N} \mathcal{W}^n \cdot p_s^n$. And the object of the knowledge distillation is a classical regularization term [9]:

$$\mathcal{L}_{kd}(f_t; X_t, \mathcal{M}) = \mathbb{E}_{x_t \in X_t} \mathcal{D}_{kl}(\mathcal{M} \,||\, f_t(x_t)) \tag{5}$$

where D_{kl} denotes the Kullback-Leibler (KL) divergence loss.

Adaptive Pseudo-Cutout Label: As previously mentioned, the outputs from the source domain black-box predictors have been adjusted by the pixel-level weight. However, they are still noisy and only pixel-level information is considered while ignoring structured information in the knowledge distillation process. Thus, we utilize the output of the black-box predictor on the target domain to produce an adaptive pseudo-cutout label, which will be employed to further regularize the knowledge distillation process. We have revised the method in [7] to generate high-quality pseudo labels that resemble the Cutout augmentation technique. For softmax outputs $\{O_s^n\}_{n=1}^N$ from N source predictors, we first set a threshold α to filter low-confidence pixels. To handle pixels with normalized probability values below the threshold, we employ a Cutout-like operation and discard these pixels. Subsequently, we apply an adaptive voting strategy to the N source domain outputs. Initially, during the training of the target model, if at least one source domain output exceeds the threshold, we consider the pixel as a positive or negative sample, which facilitates rapid knowledge acquisition by the model. As the training progresses, we gradually tighten the voting strategy and only retain regional pixels that have received adequate votes. The strategy can be summarised as follow:

$$\mathcal{V}_n^{((i,j) \mid (i,j) \in (H,W))} = \begin{cases} 1, & O_s^n(i,j) > \alpha, \\ 0, & \text{otherwise.} \end{cases} \tag{6}$$

where α is empirically set as 0.9.

Then we will aggregate the voting scores, i.e., $V^{(i,j)} = \sum_{n=1}^{N} \mathcal{V}_n^{(i,j)}$ and determine whether to retain each pixel using an adaptive vote gate $G \in \{1, 2, 3, etc.\}$. By filtering with a threshold and integrating the voting strategy, we generate high-confidence pseudo-labels that remain effective even when the source and target domains exhibit covariance. Finally, we define the ensemble result as a pseudo-cutout label $\hat{P}_s$ and employ consistency regularization as below:

$$\mathcal{L}_{pcl}(f_t; X_t, \hat{P}_s) = \mathbb{E}_{x_t \in X_t} l_{ce}(\hat{P}_s \mid\mid f_t(x_t)) \tag{7}$$

where l_{ce} denotes cross-entropy loss function.

Loss Functions: Finally, we incorporate global structural information about the predicted outcome of the target domain into both distillation and semi-supervised learning. To mitigate the noise effect of the source domain predictors, we introduce maximize mutual information targets to facilitate discrete representation learning by the network. We define $E(p) = -\sum_i p_i \log p_i$ as conditional entropy. The object can be described as follow:

$$\begin{aligned} \mathcal{L}_{mmi}(f_t; X_t) &= H(Y_t) - H(Y_t|X_t) \\ &= E\left(\mathbb{E}_{x_t \in X_t} f_t(x_t)\right) - \mathbb{E}_{x_t \in X_t} E\left(f_t(x_t)\right), \end{aligned} \tag{8}$$

where the increasing $H(Y_t)$ and the decreasing $H(Y_t|X_t)$ help to balances class separation and classifier complexity [15].

We adopt the classical and effective mean-teacher framework as a baseline for semi-supervised learning and update the teacher model parameters by exponential moving average. Also, we apply two different perturbations (η, η') to the target domain data and feed them into the student model and the mean-teacher model respectively. The consistency loss of unsupervised learning can be defined as below:

$$\mathcal{L}_{cons}(\theta_t, \theta_t') = \mathbb{E}_{x_t \in X_t} \left[||f_t(x_t, \theta_t', \eta') - f_t(x_t, \theta_t, \eta)||^2 \right] \tag{9}$$

Finally, we get the overall objective:

$$\mathcal{L}_{all} = \mathcal{L}_{kd} + \mathcal{L}_{pcl} + \mathcal{L}_{cons} - \mathcal{L}_{mmi} + \mathcal{L}_{sup} \tag{10}$$

where $\mathcal{L}_{sup}$ denotes the ordinary cross-entropy loss for supervised learning and we set the weight of each loss function to 1 in the training.

3 Experiments

Dataset and Setting: We collect four pathology image datasets to validate our proposed approach. Firstly, we acquire 50 images from a cohort of patients with Triple Negative Breast Cancer (TNBC), which is released by Naylor et al [18]. Hou et al. [10] publish a dataset of nucleus segmentation containing 5,060 segmented slides from 10 TCGA cancer types. In this work, we use 98 images from

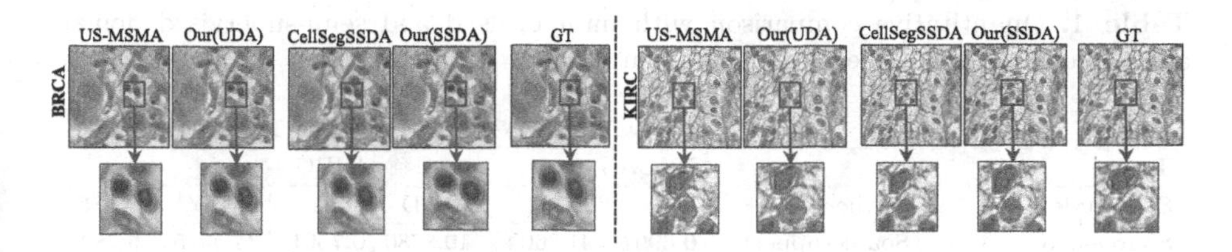

Fig. 2. Visualized segmentation on the BRCA and KIRC target domains respectively.

invasive carcinoma of the breast (BRCA). We have also included 463 images of Kidney Renal Clear cell carcinoma (KIRC) in our dataset, which are made publicly available by Irshad et al [11]. Awan et al. [2] publicly release a dataset containing tissue slide images and associated clinical data on colorectal cancer (CRC), from which we randomly select 200 patches for our study. In our experiments, we transfer knowledge from three black-box models trained on different source domains to a new target domain model (e.g.,from CRC, TNBC, KIRC to BRCA). The backbone network for the student model and source domain black-box predictors employ the widely adopted residual U-Net [12], which is commonly used for medical image segmentation. For each source domain network, we conduct full-supervision training on the corresponding source domain data and directly evaluate its performance on target domain data. The upper performance metrics (Source-only upper) are shown in the Table 1. To ensure the reliability of the results, we use the same data for training, validation, and testing, which account for 80%, 10%, and 10% of the original data respectively. For the target domain network, we use unsupervised and semi-supervised as our task settings respectively. In semi-supervised domain adaptation, we only use 10% of the target domain data as labeled data.

Experimental Results: To validate our method, we compare it with the following approaches: (1) CellSegSSDA [8], an adversarial learning based semi-supervised domain adaptation approach. (2) US-MSMA [13], a multi-source model domain aggregation network. (3) SFDA-DPL [5], a source-free unsupervised domain adaptation approach. (4) BBUDA [17], an unsupervised black-box model domain adaptation framework. A point worth noting is that most of the methods we compared with are white-box methods, which means they can obtain more information from the source domain than us. For single-source domain adaptation approach, CellsegSSDA and SFDA-DPL, we employ two strategies to ensure the fairness of the experiments: (1) single-source, i.e. performing adaptation on each single source, where we select the best results to display in the Table 1; (2) source-combined, i.e. all source domains are combined into a traditional single source. The Table 1 and Fig. 2 demonstrate that our proposed method exhibits superior performance, even when compared to these white-box methods, surpassing them in various evaluation metrics and visualization results. In addition, the experimental results also show that simply combining multiple

Table 1. Quantitative comparison with unsupervised and semi-supervised domain adaptation methods under 3 segmentation metrics.

Source		CRC&TNBC&KIRC			CRC&TNBC&BRCA		
Target		BRCA			KIRC		
Standards	Methods	Dice	HD95	ASSD	Dice	HD95	ASSD
Source-only	Source(upper)	0.6991	41.9604	10.8780	0.7001	34.5575	6.7822
Single-source(upper)	SFDA-DPL [5]	0.6327	43.8113	11.6313	0.6383	26.3252	4.6023
	BBUDA [17]	0.6620	39.6950	11.3911	0.6836	42.9875	7.4398
Source-Combined	SFDA-DPL [5]	0.6828	46.5393	12.1484	0.6446	25.4274	4.2998
	BBUDA [17]	0.6729	41.8879	11.5375	0.6895	46.7358	8.7463
Multi-source	US-MSMA [13]	0.7334	**37.1309**	8.7817	0.7161	**18.7093**	**3.0187**
Multi-source	Our(UDA)	**0.7351**	39.4103	**8.7014**	**0.7281**	30.9221	6.2080
Single-source(upper)	CellSegSSDA [8]	0.6852	45.2595	9.9133	0.6937	58.7221	12.5176
Source-Combined	CellSegSSDA [8]	0.7202	43.9251	**8.0944**	0.6699	55.1768	10.3623
Multi-source	Our(SSDA)	**0.7565**	**39.0552**	9.3346	**0.7443**	**31.7582**	**6.0873**
fully-supervised upper bounds		0.7721	35.1449	7.2848	0.7540	23.53767	4.1882

Table 2. Ablation study of three modules in our proposed method.

CRC&KIRC&BRCA to TNBC					
WL	PCL	MMI	Dice	HD95	ASSD
×	×	×	0.6708	56.9111	16.3837
✓	×	×	0.6822	**54.3386**	14.9817
✓	✓	×	0.6890	57.0889	12.9512
✓	✓	✓	**0.7075**	58.8798	**10.7247**

source data into a traditional single source will result in performance degradation in some cases, which also proves the importance of studying multi-source domain adaptation methods.

Ablation Study: To evaluate the impact of our proposed methods of weighted logits(WL), pseudo-cutout label(PCL) and maximize mutual information(MMI) on the model performance, we conduct an ablation study. We compare the baseline model with the models that added these three methods separately. We chose CRC, KIRC and BRCA as our source domains, and TNBC as our target domain. The results of these experiments, presented in the Table 2, show that our proposed modules are indeed useful.

4 Conclusion

Our proposed multi-source black-box domain adaptation method achieves competitive performance by solely relying on the source domain outputs, without the need for access to the source domain data or models, thus avoiding information leakage from the source domain. Additionally, the method does not assume the same architecture across domains, allowing us to learn lightweight target models from large source models, improving learning efficiency. We demonstrate the effectiveness of our method on multiple public datasets and believe it can be readily applied to other domains and adaptation scenarios. Moving forward, we plan to integrate our approach with active learning methods to enhance annotation efficiency in the semi-supervised setting. By leveraging multi-source domain knowledge, we aim to improve the reliability of the target model and enable more efficient annotation for better model performance.

Acknowledgements. This work is partially supported by the National Natural Science Foundation of China under grant No. 61902310 and the Natural Science Basic Research Program of Shaanxi, China under grant 2020JQ030.

References

1. Ahmed, S.M., Raychaudhuri, D.S., Paul, S., Oymak, S., Roy-Chowdhury, A.K.: Unsupervised multi-source domain adaptation without access to source data. In: Proceedings of the IEEE/CVF Conference on Computer Vision and Pattern Recognition, pp. 10103–10112 (2021)
2. Awan, R., et al.: Glandular morphometrics for objective grading of colorectal adenocarcinoma histology images. Sci. Rep. **7**(1), 16852 (2017)
3. Bateson, M., Kervadec, H., Dolz, J., Lombaert, H., Ben Ayed, I.: Source-relaxed domain adaptation for image segmentation. In: Martel, A.L., et al. (eds.) MICCAI 2020. LNCS, vol. 12261, pp. 490–499. Springer, Cham (2020). https://doi.org/10.1007/978-3-030-59710-8_48
4. Carlini, N., et al.: Extracting training data from large language models. In: USENIX Security Symposium, vol. 6 (2021)
5. Chen, C., Liu, Q., Jin, Y., Dou, Q., Heng, P.-A.: Source-free domain adaptive fundus image segmentation with denoised pseudo-labeling. In: de Bruijne, M., et al. (eds.) MICCAI 2021. LNCS, vol. 12905, pp. 225–235. Springer, Cham (2021). https://doi.org/10.1007/978-3-030-87240-3_22
6. DeVries, T., Taylor, G.W.: Improved regularization of convolutional neural networks with cutout. arXiv preprint arXiv:1708.04552 (2017)
7. Feng, H., et al.: Kd3a: unsupervised multi-source decentralized domain adaptation via knowledge distillation. In: ICML, pp. 3274–3283 (2021)
8. Haq, M.M., Huang, J.: Adversarial domain adaptation for cell segmentation. In: Medical Imaging with Deep Learning, pp. 277–287. PMLR (2020)
9. Hinton, G., Vinyals, O., Dean, J.: Distilling the knowledge in a neural network. arXiv preprint arXiv:1503.02531 (2015)
10. Hou, L., et al.: Dataset of segmented nuclei in hematoxylin and eosin stained histopathology images of ten cancer types. Sci. Data **7**(1), 185 (2020)
11. Irshad, H., et al.: Crowdsourcing image annotation for nucleus detection and segmentation in computational pathology: evaluating experts, automated methods, and the crowd. In: Pacific Symposium on Biocomputing Co-chairs, pp. 294–305. World Scientific (2014)
12. Kerfoot, E., Clough, J., Oksuz, I., Lee, J., King, A.P., Schnabel, J.A.: Left-ventricle quantification using residual U-Net. In: Pop, M., et al. (eds.) STACOM 2018. LNCS, vol. 11395, pp. 371–380. Springer, Cham (2019). https://doi.org/10.1007/978-3-030-12029-0_40

13. Li, Z., Togo, R., Ogawa, T., Haseyama, M.: Union-set multi-source model adaptation for semantic segmentation. In: Computer Vision-ECCV 2022: 17th European Conference, Tel Aviv, Israel, 23–27 October 2022, Proceedings, Part XXIX, pp. 579–595. Springer, Heidelberg (2022). https://doi.org/10.1007/978-3-031-19818-2_33
14. Liang, J., Hu, D., Feng, J.: Do we really need to access the source data? source hypothesis transfer for unsupervised domain adaptation. In: International Conference on Machine Learning, pp. 6028–6039. PMLR (2020)
15. Liang, J., Hu, D., Feng, J., He, R.: Dine: domain adaptation from single and multiple black-box predictors. In: Proceedings of the IEEE/CVF Conference on Computer Vision and Pattern Recognition, pp. 8003–8013 (2022)
16. Liu, X., et al.: Adversarial unsupervised domain adaptation with conditional and label shift: Infer, align and iterate. In: Proceedings of the IEEE/CVF International Conference on Computer Vision, pp. 10367–10376 (2021)
17. Liu, X., et al.: Unsupervised black-box model domain adaptation for brain tumor segmentation. Front. Neurosci. **16**, 837646 (2022)
18. Naylor, P., Laé, M., Reyal, F., Walter, T.: Segmentation of nuclei in histopathology images by deep regression of the distance map. IEEE Trans. Med. Imaging **38**(2), 448–459 (2018)
19. Scherr, T., Löffler, K., Böhland, M., Mikut, R.: Cell segmentation and tracking using cnn-based distance predictions and a graph-based matching strategy. Plos One **15**(12), e0243219 (2020)
20. Shimodaira, H.: Improving predictive inference under covariate shift by weighting the log-likelihood function. J. Stat. Plan. Infer. **90**(2), 227–244 (2000)
21. Xie, B., Yuan, L., Li, S., Liu, C.H., Cheng, X.: Towards fewer annotations: active learning via region impurity and prediction uncertainty for domain adaptive semantic segmentation. In: Proceedings of the IEEE/CVF Conference on Computer Vision and Pattern Recognition, pp. 8068–8078 (2022)

MedGen3D: A Deep Generative Framework for Paired 3D Image and Mask Generation

Kun Han[1(✉)], Yifeng Xiong[1], Chenyu You[2], Pooya Khosravi[1], Shanlin Sun[1], Xiangyi Yan[1], James S. Duncan[2], and Xiaohui Xie[1]

[1] University of California, Irvine, USA
khan7@uci.edu
[2] Yale University, New Haven, USA

Abstract. Acquiring and annotating sufficient labeled data is crucial in developing accurate and robust learning-based models, but obtaining such data can be challenging in many medical image segmentation tasks. One promising solution is to synthesize realistic data with ground-truth mask annotations. However, no prior studies have explored generating complete 3D volumetric images with masks. In this paper, we present MedGen3D, a deep generative framework that can generate paired 3D medical images and masks. First, we represent the 3D medical data as 2D sequences and propose the Multi-Condition Diffusion Probabilistic Model (MC-DPM) to generate multi-label mask sequences adhering to anatomical geometry. Then, we use an image sequence generator and semantic diffusion refiner conditioned on the generated mask sequences to produce realistic 3D medical images that align with the generated masks. Our proposed framework guarantees accurate alignment between synthetic images and segmentation maps. Experiments on 3D thoracic CT and brain MRI datasets show that our synthetic data is both diverse and faithful to the original data, and demonstrate the benefits for downstream segmentation tasks. We anticipate that MedGen3D's ability to synthesize paired 3D medical images and masks will prove valuable in training deep learning models for medical imaging tasks.

Keywords: Deep Generative Framework · 3D Volumetric Images with Masks · Fidelity and Diversity · Segmentation

1 Introduction

In medical image analysis, the availability of a substantial quantity of accurately annotated 3D data is a prerequisite for achieving high performance in tasks like segmentation and detection [7,15,23,26,28–36]. This, in turn, leads to more

Supplementary Information The online version contains supplementary material available at https://doi.org/10.1007/978-3-031-43907-0_72.

H. Greenspan et al. (Eds.): MICCAI 2023, LNCS 14220, pp. 759–769, 2023.
https://doi.org/10.1007/978-3-031-43907-0_72

precise diagnoses and treatment plans. However, obtaining and annotating such data presents many challenges, including the complexity of medical images, the requirement for specialized expertise, and privacy concerns.

Generating realistic synthetic data presents a promising solution to the above challenges as it eliminates the need for manual annotation and alleviates privacy risks. However, most prior studies [4,5,14,25] have focused on 2D image synthesis, with only a few generating corresponding segmentation masks. For instance, [13] uses dual generative adversarial networks (GAN) [12,34] to synthesize 2D labeled retina fundus images, while [10] combines a label generator [22] with an image generator [21] to generate 2D brain MRI data. More recently, [24] uses WGAN [3] to generate small 3D patches and corresponding vessel segmentations.

However, there has been no prior research on generating whole 3D volumetric images with the corresponding segmentation masks. Generating 3D volumetric images with corresponding segmentation masks faces two major obstacles. First, directly feeding entire 3D volumes to neural networks is impractical due to GPU memory constraints, and downsizing the resolution may compromise the quality of the synthetic data. Second, treating the entire 3D volume as a single data point during training is suboptimal because of the limited availability of annotated 3D data. Thus, innovative methods are required to overcome these challenges and generate high-quality synthetic 3D volumetric data with corresponding segmentation masks.

We propose MedGen3D, a novel diffusion-based deep generative framework that generates paired 3D volumetric medical images and multi-label masks. Our approach treats 3D medical data as sequences of slices and employs an autoregressive process to sequentially generate 3D masks and images. In the first stage, a Multi-Condition Diffusion Probabilistic Model (MC-DPM) generates mask sequences by combining conditional and unconditional generation processes. Specifically, the MC-DPM generates mask subsequences (i.e., several consecutive slices) at any position directly from random noise or by conditioning on existing slices to generate subsequences forward or backward. Given that medical images have similar anatomical structures, slice indices serve as additional conditions to aid the mask subsequence generation. In the second stage, we introduce a conditional image generator with a seq-to-seq model from [27] and a semantic diffusion refiner. By conditioning on the mask sequences generated in the first stage, our image generator synthesizes realistic medical images aligned with masks while preserving spatial consistency across adjacent slices.

The main contributions of our work are as follows: 1) Our proposed framework is the *first* to address the challenge of synthesizing complete 3D volumetric medical images with their corresponding masks; 2) we introduce a multi-condition diffusion probabilistic model for generating 3D anatomical masks with high fidelity and diversity; 3) we leverage the generated masks to condition an image sequence generator and a semantic diffusion refiner, which produces realistic medical images that align accurately with the generated masks; and 4) we present experimental results that demonstrate the fidelity and diversity of the generated 3D multi-label medical images, highlighting their potential benefits for downstream segmentation tasks.

2 Preliminary

2.1 Diffusion Probabilistic Model

A diffusion probabilistic model (DPM) [16] is a parameterized Markov chain of length T, which is designed to learn the data distribution $p(X)$. DPM builds the Forward Diffusion Process (FDP) to get the diffused data point X_t at any time step t by $q(X_t \mid X_{t-1}) = \mathcal{N}(X_t; \sqrt{1-\beta_t} X_{t-1}, \beta_t I)$, with $X_0 \sim q(X_0)$ and $p(X_T) = \mathcal{N}(X_T; 0, I)$. Let $\alpha_t = 1 - \beta_t$ and $\bar{\alpha}_t = \prod_{s=1}^{t}(1-\beta_s)$, Reverse Diffusion Process (RDP) is trained to predict the noise added in the FDP by minimizing:

$$Loss(\theta) = \mathbb{E}_{X_0 \sim q(X_0), \epsilon \sim \mathcal{N}(0,I), t}\left[\left\|\epsilon - \epsilon_\theta\left(\sqrt{\bar{\alpha}_t} X_0 + \sqrt{1-\bar{\alpha}_t}\epsilon, t\right)\right\|^2\right], \quad (1)$$

where ϵ_θ is predicted noise and θ is the model parameters.

2.2 Classifier-Free Guidance

Samples from conditional diffusion models can be improved with classifier-free guidance [17] by setting the condition c as $\emptyset$ with probability p. During sampling, the output of the model is extrapolated further in the direction of $\epsilon_\theta(X_t \mid c)$ and away from $\epsilon_\theta(X_t \mid \emptyset)$ as follows:

$$\hat{\epsilon}_\theta(X_t \mid c) = \epsilon_\theta(X_t \mid \emptyset) + s \cdot (\epsilon_\theta(X_t \mid c) - \epsilon_\theta(X_t \mid \emptyset)), \quad (2)$$

where $\emptyset$ represents a null condition and $s \geq 1$ is the guidance scale.

3 Methodology

We propose a sequential process to generate complex 3D volumetric images with masks, as illustrated in Fig. 1. The first stage generates multi-label segmentation, and the second stage performs conditional medical image generation. The details will be presented in the following sections.

3.1 3D Mask Generator

Due to the limited annotated real data and GPU memory constraints, directly feeding the entire 3D volume to the network is impractical. Instead, we treat 3D

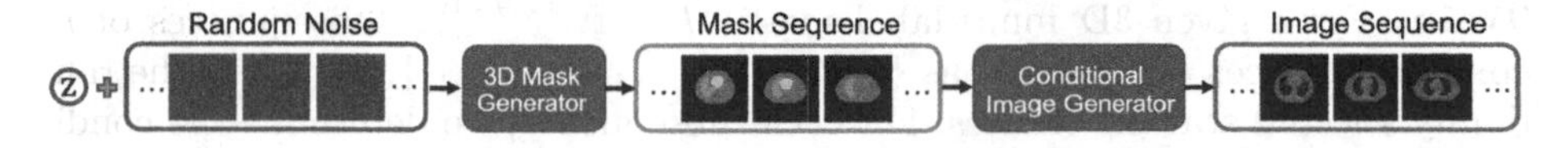

Fig. 1. Overview of the proposed **MedGen3D**, including a 3D mask generator to autoregressively generate the mask sequences starting from a random position z, and a conditional image generator to generate 3D images conditioned on generated masks.

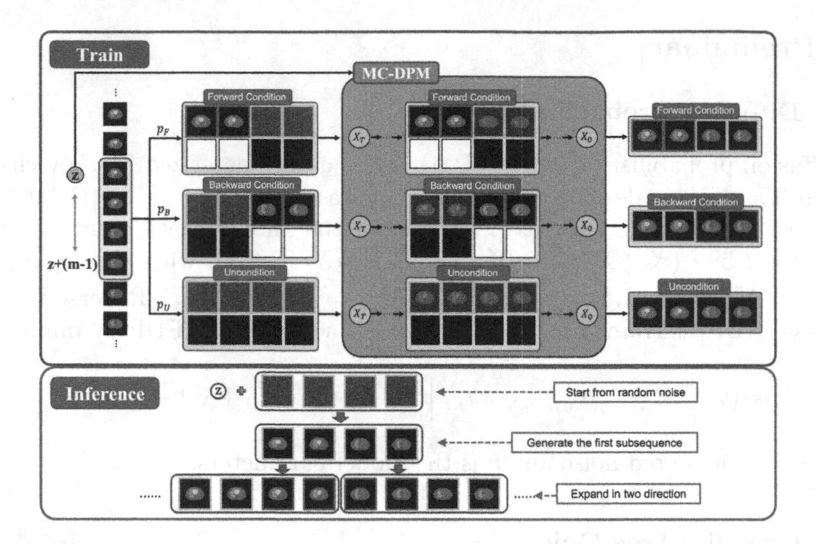

Fig. 2. Proposed 3D mask generator. Given target position z, MC-DPM is designed to generate mask subsequences (length of m) for specific region, unconditionally or conditioning on first or last n slices, according to the pre-defined probability $p^C \in \{p_F, p_B, p_U\}$. Binary indicators are assigned to slices to signify the conditional slices. We ignore the binary indicators in the inference process for clear visualization with red outline denoting the conditional slices and green outline denoting the generated slices.

medical data as a series of subsequences. To generate an entire mask sequence, an initial subsequence of m consecutive slices is **unconditionally** generated from random noise. Then the subsequence is expanded **forward** and **backward** in an autoregressive manner, conditioned on existing slices.

Inspired by classifier-free guidance in Sect. 2.2, we propose a general Multi-Condition Diffusion Probabilistic Model (MC-DPM) to unify all three conditional generations (unconditional, forward, and backward). As shown in Fig. 2, MC-DPM is able to generate mask sequences directly from random noise or conditioning on existing slices.

Furthermore, as 3D medical data typically have similar anatomical structures, slices with the same relative position roughly correspond to the same anatomical regions. Therefore, we can utilize the relative position of slices as conditions to guide the MC-DPM in generating subsequences of the target region and control the length of generated sequences.

Train: For a given 3D multi-label mask $M \in \mathbb{R}^{D \times H \times W}$, subsequneces of m consecutive slices are selected as $\{M_z, M_{z+1}, \ldots, M_{z+(m-1)}\}$, with z as the randomly selected starting indices. For each subsequence, we determine the conditional slices $X^C \in \{\mathbb{R}^{n \times H \times W}, \emptyset\}$ by selecting either the first or the last n slices, or no slice, based on a probability $p^C \in \{p_{Forward}, p_{Backward}, p_{Uncondition}\}$. The objective of the MC-DPM is to generate the remaining slices, denoted as $X^P \in \mathbb{R}^{(m-\text{len}(X^C)) \times H \times W}$.

To incorporate the position condition, we utilize the relative position of the subsequence $\tilde{z} = z/D$, where z is the index of the subsequence's starting slice. Then we embed the position condition and concatenate it with the time embedding to aid the generation process. We also utilize a binary indicator for each slice in the subsequence to signify the existence of conditional slices.

The joint distribution of reverse diffusion process (RDP) with the conditional slices X^C can be written as:

$$p_\theta(X_{0:T}^P | X^C, \tilde{z}) = p(X_T^P) \prod_{t=1}^{T} p_\theta(X_{t-1}^P \mid X_t^P, X^C, \tilde{z}). \tag{3}$$

where $p(X_T^P) = \mathcal{N}\left(X_T^P; 0, I\right)$, $\tilde{z} = z/D$ and p_θ is the distribution parameterized by the model.

Overall, the model will be trained by minimizing the following loss function, with $X_t^P = \sqrt{\bar{\alpha}_t} X_0^P + \sqrt{1-\bar{\alpha}_t}\epsilon$:

$$\text{Loss}(\theta) = \mathbb{E}_{X_0 \sim q(X_0), \epsilon \sim \mathcal{N}(0,I), p^C, z, t}\left[\left\|\epsilon - \epsilon_\theta\left(X_t^P, X^C, z, t\right)\right\|^2\right]. \tag{4}$$

Inference: During inference, MC-DPM first generates a subsequence of m slices from random noise given a random location z. The entire mask sequence can then be generated autoregressively by expanding in both directions, conditioned on the existing slices, as shown in Fig. 2. Please refer to the **Supplementary** for a detailed generation process and network structure.

3.2 Conditional Image Generator

In the second step, we employ a sequence-to-sequence method to generate medical images conditioned on masks, as shown in Fig. 3.

Image Sequence Generator: In the sequence-to-sequence generation task, new slice is the combination of the warped previous slice and newly generated texture, weighted by a continuous mask [27]. We utilize Vid2Vid [27] as our image sequence generator. We train Vid2Vid with its original loss, which includes GAN loss on multi-scale images and video discriminators, flow estimation loss, and feature matching loss.

Semantic Diffusion Refiner: Despite the high cross-slice consistency and spatial continuity achieved by vid2vid, issues such as blocking, blurriness and suboptimal texture generation persist. Given that diffusion models have been shown to generate superior images [9], we propose a semantic diffusion refiner utilizing a diffusion probabilistic model to refine the previously generated images.

For each of the 3 different views, we train a semantic diffusion model (SDM), which takes 2D masks and noisy images as inputs to generate images aligned with input masks. During inference, we only apply small noising steps (10 steps)

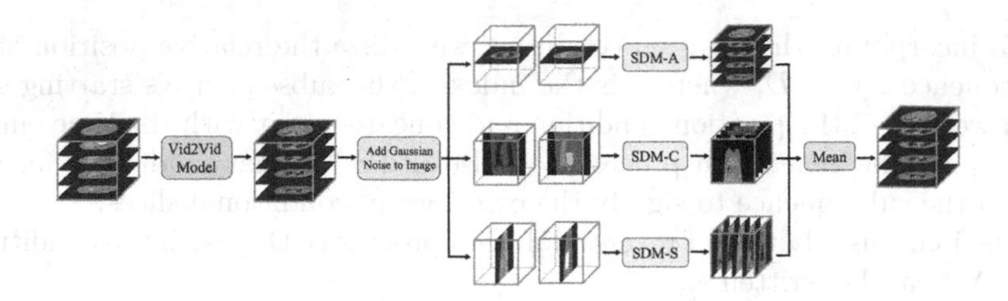

Fig. 3. Image Sequence Generator. Given the generated 3D mask, the initial image is generated by Vid2Vid model sequentially. To utilize the semantic diffusion model (SDM) to refine the initial result, we first apply small steps (10 steps) noise, and then use three SDMs to refine. The final result is the mean 3D images from 3 different views (Axial, Coronal, and Sagittal), yielding significant improvements over the initially generated image.

to the generated images so that the overall anatomical structure and spatial continuity are preserved. After that, we refine the images using the pre-trained semantic diffusion model. The final refined 3D images are the mean results from 3 views. Experimental results show an evident improvement in the quality of generated images with the help of semantic diffusion refiner.

4 Experiments and Results

4.1 Datasets and Setups

Datasets: We conducted experiments on the thoracic site using three thoracic CT datasets and the brain site with two brain MRI datasets. For both generative models and downstream segmentation tasks, we utilized the following datasets:

- SegTHOR [19]: 3D thorax CT scans (25 training, 5 validation, 10 testing);
- OASIS [20]: 3D brain MRI T1 scans (40 training, 10 validation, 10 testing);

For the downstream segmentation task only and the transfer learning, we utilized 10 fine-tuning, 5 validation, and 10 testing scans from each of the 3D thorax CT datasets of StructSeg-Thorax [2] and Public-Thor [7], as well as the 3D brain MRI T1 dataset from ADNI [1].

Implementation: For thoracic datasets, we crop and pad CT scans to ($96 \times 320 \times 320$). The annotations of six organs (left lung, right lung, spinal cord, esophagus, heart, and trachea) are examined by an experienced radiation oncologist. We also include a body mask to aid in the image generation of body regions. For brain MRI datasets, we use Freesurfer [11] to get segmentations of four regions (cortex, subcortical gray matter, white matter, and CSF), and then crop the volume to ($192 \times 160 \times 160$). We assign discrete values to masks of different regions or organs for both thoracic and brain datasets and then combine them into one 3D volume. When synthesizing mask sequences, we resize the width and height of the masks to 128×128 and set the length of the subsequence m to 6. We use official segmentation models provided by MONAI [6] along with standard data augmentations, including spatial and color transformations.

Setup: We compare the synthetic image quality with DDPM [16], 3D-α-WGAN [18] and Vid2Vid [27], and utilize four segmentation models with different training strategies to demonstrate the benefit for the downstream task.

4.2 Evaluate the Quality of Synthetic Image.

Synthetic Dataset: To address the limited availability of annotated 3D medical data, we used only 30 CT scans from SegTHOR (25 for training and 5 for validation) and 50 MRI scans from OASIS (40 for training and 10 for validation) to generate 110 3D thoracic CT scans and 110 3D brain MRI scans, respectively (Fig. 4).

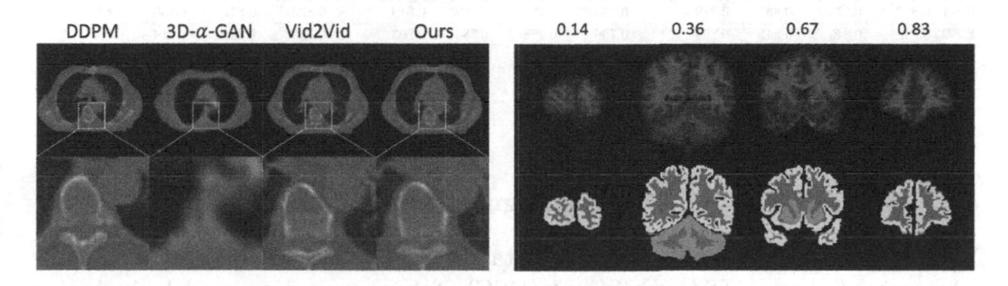

Fig. 4. Our proposed method produces more anatomically accurate images compared to 3D-α-WGAN and vid2vid, as demonstrated by the clearer organ boundaries and more realistic textures. Left: Qualitative comparison between different generative models. Right: Visualization of synthetic 3D brain MRI slices at different relative positions.

We compare the fidelity and diversity of our synthetic data with DDPM [16] (train 3 for different views), 3D-α-WGAN [18], and vid2vid [27] by calculating the mean Frèchet Inception Distance (FID) and Learned Perceptual Image Patch Similarity (LPIPS) from 3 different views.

According to Table 1, our proposed method has a slightly lower FID score but a similar LPIPS score compared to DDPM which directly generates 2D images from noise. We speculate that this is because DDPM is trained on 2D images without explicit anatomical constraints and only generates 2D images. On the other hand, 3D-α-WGAN [18], which uses much larger 3D training data (146 for thorax and 414 for brain), has significantly worse FID and LPIPS scores than our method. Moreover, our proposed method outperforms Vid2Vid, showing the effectiveness of our semantic diffusion refiner.

Table 1. Synthetic image quality comparison between baselines and ours.

	Thoracic CT		Brain MRI	
	FID ↓	LPIPS ↑	FID ↓	LPIPS ↑
DDPM [16]	**35.2**	**0.316**	**34.9**	0.298
3D-α-WGAN [18]	136.2	0.286	136.4	0.289
Vid2Vid [27]	47.3	0.300	48.2	0.324
Ours	39.6	0.305	40.3	**0.326**

Table 2. Experiment 2: DSC of different thoracic segmentation models. There are 5 training strategies, namely: **E2-1:** Training with real SegTHOR training data; **E2-2:** Training with synthetic data; **E2-3:** Training with both synthetic and real data; **E2-4:** Finetuning model from E2-2 using real training data; and **E2-5:** finetuning model from E2-3 using real training data. (* denotes the training data source.)

	SegTHOR*				StructSeg-Thorax				Public-Thor			
	Unet 2D	Unet 3D	UNETR	Swin UNETR	Unet 2D	Unet 3D	UNETR	Swin UNETR	Unet 2D	Unet 3D	UNETR	Swin UNETR
E2-1	0.817	0.873	0.867	0.878	0.722	0.793	0.789	0.810	0.822	0.837	0.836	0.847
E2-2	0.815	0.846	0.845	0.854	0.736	0.788	0.788	0.803	0.786	0.838	0.814	0.842
E2-3	0.845	0.881	0.886	0.886	0.772	0.827	0.824	0.827	0.812	0.856	0.853	0.856
E2-4	**0.855**	0.887	**0.894**	**0.899**	0.775	**0.833**	**0.825**	0.833	**0.824**	0.861	0.852	**0.867**
E2-5	0.847	**0.891**	0.890	0.897	**0.783**	**0.833**	0.823	**0.835**	0.818	**0.864**	**0.858**	**0.867**

4.3 Evaluate the Benefits for Segmentation Task

We explore the benefits of synthetic data for downstream segmentation tasks by comparing Sørensen-Dice coefficient (DSC) of 4 segmentation models, including Unet2D [23], UNet3D [8], UNETR [15], and Swin-UNETR [26]. In Table 2 and 3, we utilize real training data (from SegTHOR and OASIS) and synthetic data to train the segmentation models with 5 different strategies, and test on all 3 thoracic CT datasets and 2 brain MRI datasets. In Table 4, we aim to demonstrate whether the synthetic data can aid transfer learning with limited real finetuning data from each of the testing datasets (StructSeg-Thorax, Public-Thor and ADNI) with four training strategies.

According to Table 2 and Table 3, the significant DSC difference between 2D and 3D segmentation models underlines the crucial role of 3D annotated data. While purely synthetic data (**E2-2**) fails to achieve the same performance as real training data (**E2-1**), the combination of real and synthetic data (**E2-3**) improves model performance in most cases, except for Unet2D on the Public-Thor dataset. Furthermore, fine-tuning the pre-trained model with real data (**E2-4** and **E2-5**) consistently outperforms the model trained only with real data. Please refer to **Supplementary** for organ-level DSC comparisons of the Swin-UNETR model with more details.

According to Table 4, for transfer learning, utilizing the pre-trained model (**E3-2**) leads to better performance compared to training from scratch (**E3-1**).

Table 3. Experiment 2: DSC of brain segmentation models. Please refer to Table 2 for detailed training strategies. (* denotes the training data source.)

	OASIS*				ADNI			
	Unet 2D	Unet 3D	UNETR	Swin UNETR	Unet 2D	Unet 3D	UNETR	Swin UNETR
E2-1	0.930	0.951	0.952	0.954	0.815	0.826	0.880	0.894
E2-2	0.905	0.936	0.935	0.934	0.759	0.825	0.828	0.854
E2-3	0.938	0.953	0.953	0.955	0.818	0.888	0.898	**0.906**
E2-4	**0.940**	**0.955**	**0.954**	**0.956**	**0.819**	0.891	**0.903**	0.903
E2-5	**0.940**	0.954	**0.954**	**0.956**	**0.819**	**0.894**	0.902	**0.906**

Table 4. Experiment 3: DSC of Swin-UNETR finetuned with real dataset. There are 4 training strategies: **E3-1:** Training from scratch for each dataset using limited finetuning data; **E3-2** Finetuning the model E2-1 from experiment 2; **E3-3** Finetuning the model E2-4 from experiment 2; and **E3-4** Finetuning the model E2-5 from experiment 2. (* denotes the finetuning data source.)

	Thoracic CT		Brain MRI
	StructSeg-Thorax*	Public-Thor*	ADNI*
E3-1	0.845	0.897	0.946
E3-2	0.865	0.901	0.948
E3-3	0.878	0.913	**0.949**
E3-4	**0.882**	**0.914**	**0.949**

Additionally, pretraining the model with synthetic data (**E3-3** and **E3-4**) can facilitate transfer learning to a new dataset with limited annotated data.

We have included video demonstrations of the generated 3D volumetric images in the **supplementary material**, which offer a more comprehensive representation of the generated image's quality.

5 Conclusion

This paper introduces MedGen3D, a new framework for synthesizing 3D medical mask-image pairs. Our experiments demonstrate its potential in realistic data generation and downstream segmentation tasks with limited annotated data. Future work includes merging the image sequence generator and semantic diffusion refiner for end-to-end training and extending the framework to synthesize 3D medical images across modalities. Overall, we believe that our work opens up new possibilities for generating 3D high-quality medical images paired with masks, and look forward to future developments in this field.

References

1. https://adni.loni.usc.edu/
2. https://structseg2019.grand-challenge.org/dataset/
3. Arjovsky, M., Chintala, S., Bottou, L.: Wasserstein gan. arXiv preprint arXiv: Arxiv-1701.07875 (2017)
4. Baur, C., Albarqouni, S., Navab, N.: Melanogans: high resolution skin lesion synthesis with gans. arXiv preprint arXiv:1804.04338 (2018)
5. Bermudez, C., Plassard, A.J., Davis, L.T., Newton, A.T., Resnick, S.M., Landman, B.A.: Learning implicit brain mri manifolds with deep learning. In: Medical Imaging: Image Processing. SPIE (2018)
6. Cardoso, M.J., et al.: Monai: an open-source framework for deep learning in healthcare. arXiv preprint arXiv:2211.02701 (2022)
7. Chen, X., et al.: A deep learning-based auto-segmentation system for organs-at-risk on whole-body computed tomography images for radiation therapy. Radiother. Oncol. **160**, 175–184 (2021)
8. Çiçek, Ö., Abdulkadir, A., Lienkamp, S.S., Brox, T., Ronneberger, O.: 3D U-Net: learning dense volumetric segmentation from sparse annotation. In: Ourselin, S., Joskowicz, L., Sabuncu, M.R., Unal, G., Wells, W. (eds.) MICCAI 2016. LNCS, vol. 9901, pp. 424–432. Springer, Cham (2016). https://doi.org/10.1007/978-3-319-46723-8_49
9. Dhariwal, P., Nichol, A.: Diffusion models beat gans on image synthesis. In: NeurIPS (2021)
10. Fernandez, V., et al.: Can segmentation models be trained with fully synthetically generated data? In: Zhao, C., Svoboda, D., Wolterink, J.M., Escobar, M. (eds.) MICCAI Workshop. SASHIMI 2022, vol. 13570, pp. 79–90. Springer, Heidelberg (2022). https://doi.org/10.1007/978-3-031-16980-9_8
11. Fischl, B.: Freesurfer. In: Neuroimage (2012)
12. Goodfellow, I., et al.: Generative adversarial networks. Commun. ACM **63**, 139–144 (2020)
13. Guibas, J.T., Virdi, T.S., Li, P.S.: Synthetic medical images from dual generative adversarial networks. arXiv preprint arXiv:1709.01872 (2017)
14. Han, C., et al.: Gan-based synthetic brain MR image generation. In: ISBI. IEEE (2018)
15. Hatamizadeh, A., et al.: Unetr: transformers for 3d medical image segmentation. In: WACV (2022)
16. Ho, J., Jain, A., Abbeel, P.: Denoising diffusion probabilistic models. In: NeurIPS (2020)
17. Ho, J., Salimans, T.: Classifier-free diffusion guidance. arXiv preprint arXiv: Arxiv-2207.12598 (2022)
18. Kwon, G., Han, C., Kim, D.: Generation of 3D brain MRI using auto-encoding generative adversarial networks. In: Shen, D., et al. (eds.) MICCAI 2019. LNCS, vol. 11766, pp. 118–126. Springer, Cham (2019). https://doi.org/10.1007/978-3-030-32248-9_14
19. Lambert, Z., Petitjean, C., Dubray, B., Kuan, S.: Segthor: segmentation of thoracic organs at risk in ct images. In: IPTA. IEEE (2020)
20. Marcus, D.S., Wang, T.H., Parker, J., Csernansky, J.G., Morris, J.C., Buckner, R.L.: Open access series of imaging studies (oasis): cross-sectional mri data in young, middle aged, nondemented, and demented older adults. J. Cogn. Neurosci. **19**, 1498–1507 (2007)

21. Park, T., Liu, M.Y., Wang, T.C., Zhu, J.Y.: Semantic image synthesis with spatially-adaptive normalization. In: CVPR (2019)
22. Rombach, R., Blattmann, A., Lorenz, D., Esser, P., Ommer, B.: High-resolution image synthesis with latent diffusion models. In: CVPR (2022)
23. Ronneberger, O., Fischer, P., Brox, T.: U-Net: convolutional networks for biomedical image segmentation. In: Navab, N., Hornegger, J., Wells, W.M., Frangi, A.F. (eds.) MICCAI 2015. LNCS, vol. 9351, pp. 234–241. Springer, Cham (2015). https://doi.org/10.1007/978-3-319-24574-4_28
24. Subramaniam, P., Kossen, T., et al.: Generating 3d tof-mra volumes and segmentation labels using generative adversarial networks. Med. Image Anal. **78**, 102396 (2022)
25. Sun, L., Chen, J., Xu, Y., Gong, M., Yu, K., Batmanghelich, K.: Hierarchical amortized gan for 3d high resolution medical image synthesis. IEEE J. Biomed. Health Inf. **28**, 3966–3975 (2022)
26. Tang, Y., et al.: Self-supervised pre-training of swin transformers for 3d medical image analysis. In: CVPR (2022)
27. Wang, T.C., et al.: Video-to-video synthesis. arXiv preprint arXiv:1808.06601 (2018)
28. Yan, X., Tang, H., Sun, S., Ma, H., Kong, D., Xie, X.: After-unet: axial fusion transformer unet for medical image segmentation. In: WACV (2022)
29. You, C., et al.: Mine your own anatomy: revisiting medical image segmentation with extremely limited labels. arXiv preprint arXiv:2209.13476 (2022)
30. You, C., et al.: Rethinking semi-supervised medical image segmentation: a variance-reduction perspective. arXiv preprint arXiv:2302.01735 (2023)
31. You, C., Dai, W., Min, Y., Staib, L., Duncan, J.S.: Implicit anatomical rendering for medical image segmentation with stochastic experts. arXiv preprint arXiv:2304.03209 (2023)
32. You, C., Dai, W., Min, Y., Staib, L., Sekhon, J., Duncan, J.S.: Action++: improving semi-supervised medical image segmentation with adaptive anatomical contrast. arXiv preprint arXiv:2304.02689 (2023)
33. You, C., Dai, W., Staib, L., Duncan, J.S.: Bootstrapping semi-supervised medical image segmentation with anatomical-aware contrastive distillation. arXiv preprint arXiv:2206.02307 (2022)
34. You, C., et al.: Class-aware adversarial transformers for medical image segmentation. In: NeurIPS (2022)
35. You, C., Zhao, R., Staib, L.H., Duncan, J.S.: Momentum contrastive voxel-wise representation learning for semi-supervised volumetric medical image segmentation. In: Wang, L., Dou, Q., Fletcher, P.T., Speidel, S., Li, S. (eds.) Medical Image Computing and Computer Assisted Intervention – MICCAI, vol. 13434, pp. 639–652. Springer, Heidelberg (2022). https://doi.org/10.1007/978-3-031-16440-8_61
36. You, C., Zhou, Y., Zhao, R., Staib, L., Duncan, J.S.: Simcvd: simple contrastive voxel-wise representation distillation for semi-supervised medical image segmentation. IEEE Trans. Med.Imaging **41**, 2228–2237 (2022)

Unsupervised Domain Transfer with Conditional Invertible Neural Networks

Kris K. Dreher[1,2](✉), Leonardo Ayala[1,3], Melanie Schellenberg[1,4], Marco Hübner[1,4], Jan-Hinrich Nölke[1,4], Tim J. Adler[1], Silvia Seidlitz[1,4,5,6], Jan Sellner[1,4,5], Alexander Studier-Fischer[7], Janek Gröhl[8,9], Felix Nickel[7], Ullrich Köthe[4], Alexander Seitel[1,6], and Lena Maier-Hein[1,3,4,5,6]

[1] Intelligent Medical Systems, German Cancer Research Center (DKFZ), Heidelberg, Germany
{k.dreher,l.maier-hein}@dkfz-heidelberg.de
[2] Faculty of Physics and Astronomy, Heidelberg University, Heidelberg, Germany
[3] Medical Faculty, Heidelberg University, Heidelberg, Germany
[4] Faculty of Mathematics and Computer Science, Heidelberg University, Heidelberg, Germany
[5] Helmholtz Information and Data Science School for Health, Karlsruhe, Heidelberg, Germany
[6] National Center for Tumor Diseases (NCT) Heidelberg a Partnership Between DKFZ and Heidelberg University Hospital, Heidelberg, Germany
[7] Department of General, Visceral, and Transplantation Surgery, Heidelberg University Hospital, Heidelberg, Germany
[8] Cancer Research UK Cambridge Institute, University of Cambridge, Cambridge, UK
[9] Department of Physics, University of Cambridge, Cambridge, UK

Abstract. Synthetic medical image generation has evolved as a key technique for neural network training and validation. A core challenge, however, remains in the domain gap between simulations and real data. While deep learning-based domain transfer using Cycle Generative Adversarial Networks and similar architectures has led to substantial progress in the field, there are use cases in which state-of-the-art approaches still fail to generate training images that produce convincing results on relevant downstream tasks. Here, we address this issue with a domain transfer approach based on conditional invertible neural networks (cINNs). As a particular advantage, our method inherently guarantees cycle consistency through its invertible architecture, and network training can efficiently be conducted with maximum likelihood training. To showcase our method's generic applicability, we apply it to two spectral imaging modalities at different scales, namely hyperspectral imaging (pixel-level) and photoacoustic tomography (image-level). According to

Supplementary Information The online version contains supplementary material available at https://doi.org/10.1007/978-3-031-43907-0_73.

H. Greenspan et al. (Eds.): MICCAI 2023, LNCS 14220, pp. 770–780, 2023.
https://doi.org/10.1007/978-3-031-43907-0_73

comprehensive experiments, our method enables the generation of realistic spectral data and outperforms the state of the art on two downstream classification tasks (binary and multi-class). cINN-based domain transfer could thus evolve as an important method for realistic synthetic data generation in the field of spectral imaging and beyond. The code is available at https://github.com/IMSY-DKFZ/UDT-cINN.

Keywords: Domain transfer · invertible neural networks · medical imaging · photoacoustic tomography · hyperspectral imaging · deep learning

1 Introduction

The success of supervised learning methods in the medical domain led to countless breakthroughs that might be translated into clinical routine and have the potential to revolutionize healthcare [6,13]. For many applications, however, labeled reference data (ground truth) may not be available for training and validating a neural network in a supervised manner. One such application is spectral imaging which comprises various non-interventional, non-ionizing imaging techniques that can resolve functional tissue properties such as blood oxygenation in real time [1,3,23]. While simulations have the potential to overcome the lack of ground truth, synthetic data is not yet sufficiently realistic [9]. Cycle Generative Adversarial Networks (GAN)-based architectures are widely used for domain transfer [12,24] but may suffer from issues such as unstable training, hallucinations, or mode collapse [15]. Furthermore, they have predominantly been used for conventional RGB imaging and one-channel cross-modality domain adaptation, and may not be suitable for other imaging modalities with more channels. We address these challenges with the following contributions:

Domain Transfer Method: We present an entirely new sim-to-real transfer approach based on conditional invertible neural networks (cINNs) (cf. Fig. 1) specifically designed for data with many spectral channels. This approach inherently addresses weaknesses of the state of the art with respect to the preservation of spectral consistency and, importantly, does not require paired images.

Instantiation to Spectral Imaging: We show that our method can generically be applied to two complementary modalities: photoacoustic tomography (PAT; image-level) and hyperspectral imaging (HSI; pixel-level).

Comprehensive Validation: In comprehensive validation studies based on more than 2,000 PAT images (real: ~1,000) and more than 6 million spectra for HSI (real: ~6 million) we investigate and subsequently confirm our two main hypotheses: (H1) Our cINN-based models can close the domain gap between simulated and real spectral data better than current state-of-the-art methods

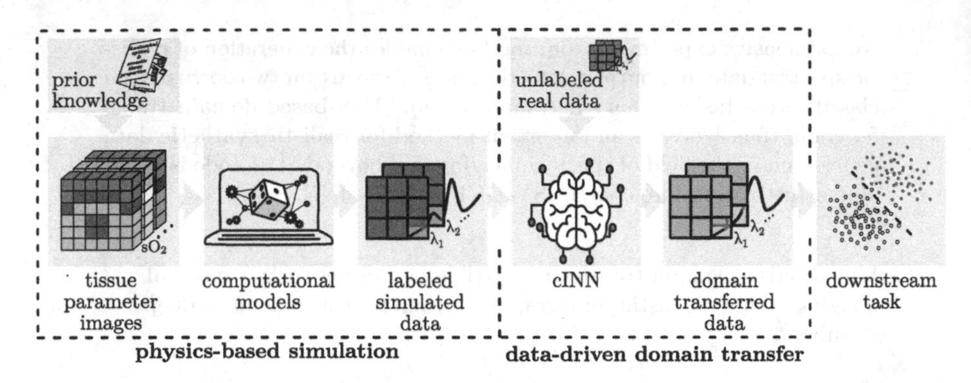

Fig. 1. Pipeline for data-driven spectral image analysis in the absence of labeled reference data. A physics-based simulation framework generates simulated spectral images with corresponding reference labels (e.g. tissue type or oxygenation (sO_2)). Our domain transfer method based on cINNs leverages unlabeled real data to increase their realism. The domain-transferred data can then be used for supervised training of a downstream task (e.g. classification).

regarding spectral plausibility. (H2) Training models on data transferred by our cINN-based approach can improve their performance on the corresponding (clinical) downstream task without them having seen labeled real data.

2 Materials and Methods

2.1 Domain Transfer with Conditional Invertible Neural Networks

Concept Overview. Our domain transfer approach (cf. Fig. 2) is based on the assumption that data samples from both domains carry domain-invariant information (e.g. on optical tissue properties) and domain-variant information (e.g. modality-specific artifacts). The invertible architecture, which inherently guarantees cycle consistency, transfers both simulated and real data into a shared latent space. While the domain-invariant features are captured in the latent space, the domain-variant features can either be filtered (during encoding) or added (during decoding) by utilizing a domain label D. To achieve spectral consistency, we leverage the fact that different tissue types feature characteristic spectral signatures and condition the model on the tissue label Y if available. For unlabeled (real) data, we use randomly generated proxy labels instead. To achieve high visual quality beyond spectral consistency, we include two discriminators Dis_{sim} and Dis_{real} for their respective domains. Finally, as a key theoretical advantage, we avoid mode collapse with maximum likelihood optimization. Implementation details are provided in the following.

cINN Model Design. The core of our architecture is a cINN [2] (cf. Fig. 2), comprising multiple (i) scales of N_i-chained affine conditional coupling (CC)

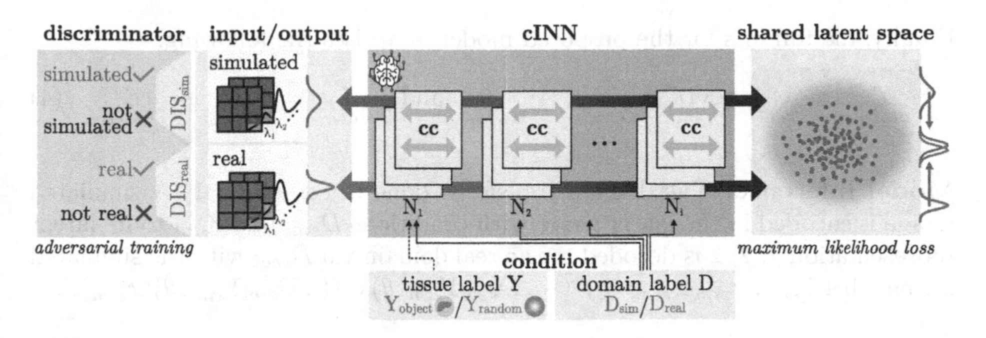

Fig. 2. Proposed architecture based on cINNs. The invertible architecture transfers both simulated and real data into a shared latent space (right). By conditioning on the domain D (bottom), a latent vector can be transferred to either the simulated or the real domain (left) for which the discriminator Dis_{sim} and Dis_{real} calculate the losses for adversarial training.

blocks [7]. These scales are necessary in order to increase the receptive field of the network and are achieved by Haar wavelet downsampling [11]. A CC block consists of subnetworks that can be freely chosen depending on the data dimensionality (e.g. fully connected or convolutional networks) as they are only evaluated in the forward direction. The CC blocks receive a condition consisting of two parts: domain label and tissue label, which are then concatenated to the input along the channel dimension. In the case of PAT, the tissue label is a full semantic and random segmentation map for the simulated and real data, respectively. In the case of HSI, the tissue label is a one-hot encoded vector for organ labels.

Model Training. In the following, the proposed cINN with its parameters θ will be referred to as $f(x, DY, \theta)$ and its inverse as f^{-1} for any input $x \sim p_D$ from domain $D \in \{D_{sim}, D_{real}\}$ with prior density p_D and its corresponding latent space variable z. The condition DY is the combination of domain label D as well as the tissue label $Y \in \{Y_{sim}, Y_{real}\}$. Then the maximum likelihood loss $\mathcal{ML}$ for a training sample x_i is described by

$$\underset{D}{\mathcal{ML}} = \mathbb{E}_i \left[\frac{||f(x_i, DY, \theta)||_2^2}{2} - log|J_i| \right] \text{ with } J_i = det\left(\left.\frac{\partial f}{\partial x}\right|_{x_i} \right). \quad (1)$$

For the adversarial training, we employ the least squares training scheme [18] for generator $Gen_D = f_D^{-1} \circ f_{D'}$ and discriminator Dis_D for each domain with $x_{D'}$ as input from the source domain and x_D as input from the target domain:

$$\underset{Gen_D}{\mathcal{L}} = \underset{x_{D'} \sim p_{D'}}{\mathbb{E}} \left[(Dis_D(Gen_D(x_{D'}) - 1))^2 \right] \quad (2)$$

$$\underset{Dis_D}{\mathcal{L}} = \underset{x_D \sim p_D}{\mathbb{E}} \left[(Dis_D(x_D) - 1)^2 \right] + \underset{x_{D'} \sim p_{D'}}{\mathbb{E}} \left[(Dis_D(Gen_D(x_{D'})))^2 \right]. \quad (3)$$

Finally, the full loss for the proposed model comprises the following:

$$\mathcal{L}_{Total_{Gen}} = \mathcal{ML}_{real} + \mathcal{ML}_{sim} + \mathcal{L}_{Gen_{real}} + \mathcal{L}_{Gen_{sim}} \text{ and } \mathcal{L}_{Total_{Dis}} = \mathcal{L}_{Dis_{real}} + \mathcal{L}_{Dis_{sim}}. \quad (4)$$

Model Inference. The domain transfer is done in two steps: 1) A simulated image is encoded in the latent space with conditions D_{sim} and Y_{sim} to its latent representation z, 2) z is decoded to the real domain via D_{real} with the simulated tissue label Y_{sim}: $x_{sim \to real} = f^{-1}(\cdot, D_{real} Y_{sim}, \theta) \circ f(\cdot, D_{sim} Y_{sim}, \theta)(x_{sim})$.

2.2 Spectral Imaging Data

Photoacoustic Tomography Data. PAT is a non-ionizing imaging modality that enables the imaging of functional tissue properties such as tissue oxygenation [22]. The **real PAT data** (cf. Fig. 3) used in this work are images of human forearms that were recorded from 30 healthy volunteers using the MSOT Acuity Echo (iThera Medical GmbH, Munich, Germany) (all regulations followed under study ID: S-451/2020, and the study is registered with the German Clinical Trials Register under reference number DRKS00023205). In this study, 16 wavelengths from 700 nm to 850 nm in steps of 10 nm were recorded for each image. The resulting 180 images were semantically segmented into the structures shown in Fig. 3 according to the annotation protocol provided in [20]. Additionally, a full sweep of each forearm was performed to generate more unlabeled images, thus

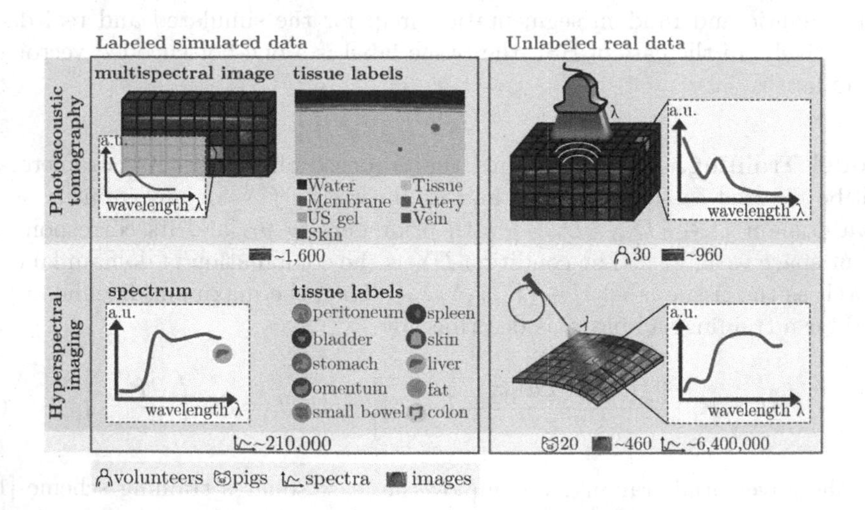

Fig. 3. Training data used for the validation experiments. For PAT, 960 real images from 30 volunteers were acquired. For HSI, more than six million spectra corresponding to 460 images and 20 individuals were used. The tissue labels PAT correspond to 2D semantic segmentations, whereas the tissue labels for HSI represent 10 different organs. For PAT, ~1600 images were simulated, whereas around 210,000 spectra were simulated for HSI.

amounting to a total of 955 real images. The **simulated PAT data** (cf. Fig. 3) used in this work comprises 1,572 simulated images of human forearms. They were generated with the toolkit for Simulation and Image Processing for Photonics and Acoustics (SIMPA) [8] based on a forearm literature model [21] and with a digital device twin of the MSOT Acuity Echo.

Hyperspectral Imaging Data. HSI is an emerging modality with high potential for surgery [4]. In this work, we performed pixel-wise analysis of HSI images. The **real HSI data** was acquired with the Tivita® Tissue (Diaspective Vision GmbH, Am Salzhaff, Germany) camera, featuring a spectral resolution of approximately 5 nm in the spectral range between 500 nm and 1000 nm nm. In total, 458 images, corresponding to 20 different pigs, were acquired (all regulations followed under study IDs: 35-9185.81/G-161/18 and 35-9185.81/G-262/19) and annotated with ten structures: bladder, colon, fat, liver, omentum, peritoneum, skin, small bowel, spleen, and stomach (cf. Fig. 3). This amounts to 6,410,983 real spectra in total. The **simulated HSI data** was generated with a Monte Carlo method (cf. algorithm provided in the supplementary material). This procedure resulted in 213,541 simulated spectra with annotated organ labels.

3 Experiments and Results

The purpose of the experiments was to investigate hypotheses H1 and H2 (cf. Sect. 1). As comparison methods, a CycleGAN [24] and an unsupervised image-to-image translation (UNIT) network [16] were implemented fully convolutionally for PAT and in an adapted version for the one-dimensional HSI data. To make the comparison fair, the tissue label conditions were concatenated with the input, and we put significant effort into optimizing the UNIT on our data.

Realism of Synthetic Data (H1) : According to qualitative analyses (Fig. 4) our domain transfer approach improves simulated PAT images with respect to key properties, including the realism of skin, background, and sharpness of vessels.

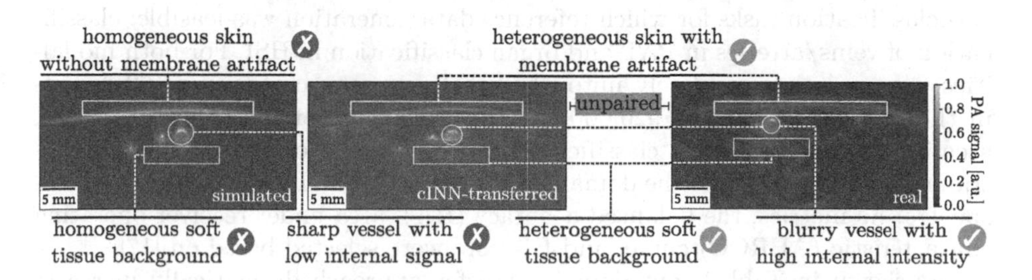

Fig. 4. Qualitative results. In comparison to simulated PAT images (left), images generated by the cINN (middle) resemble real PAT images (right) more closely. All images show a human forearm at 800 nm.

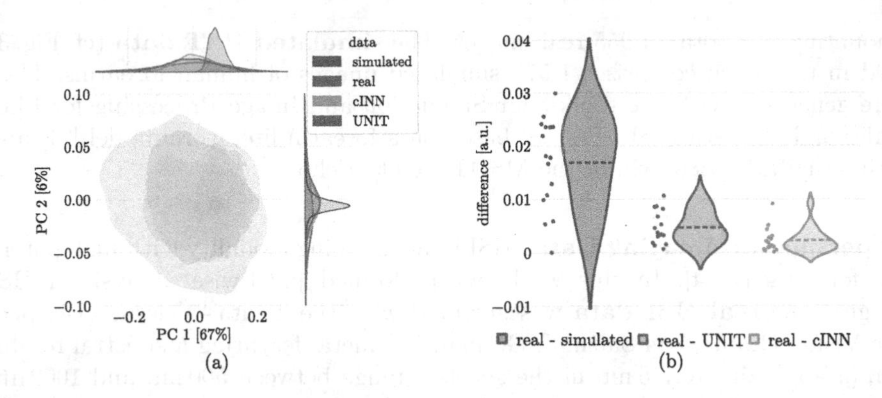

Fig. 5. Our domain transfer approach yields realistic spectra (here: of veins). The PCA plots in a) represent a kernel density estimation of the first and second components of a PCA embedding of the real data, which represent about 67% and 6% of the variance in the real data, respectively. The distributions on top and on the right of the PCA plot correspond to the marginal distributions of each dataset's first two components. b) Violin plots show that the cINN yields spectra that feature a smaller difference to the real data compared to the simulations and the UNIT-generated data. The dashed lines represent the mean difference value, and each dot represents the difference for one wavelength.

A principal component analysis (PCA) performed on all artery and vein spectra of the real and synthetic datasets demonstrates that the distribution of the synthetic data is much closer to the real data after applying our domain transfer approach (cf. Fig. 5a)). The same holds for the absolute difference, as shown in Fig. 5b). Slightly better performance was achieved with the cINN compared to the UNIT. Similarly, our approach improves the realism of HSI spectra, as illustrated in Fig. 6, for spectra of five exemplary organs (colon, stomach, omentum, spleen, and fat). The cINN-transferred spectra generally match the real data very closely. Failure cases where the real data has a high variance (translucent band) are also shown.

Benefit of Domain-Transferred Data for Downstream Tasks (H2): We examined two classification tasks for which reference data generation was feasible: classification of veins/arteries in PAT and organ classification in HSI. For both modalities, we used the completely untouched real test sets, comprising 162 images in the case of PAT and $\sim$ 920,000 spectra in the case of HSI. For both tasks, a calibrated random forest classifier (sklearn [19] with default parameters) was trained on the simulated, the domain-transferred (by UNIT and cINN), and real spectra. As metrics, the balanced accuracy (BA), area under receiver operating characteristic (AUROC) curve, and F1-score were selected based on [17].

As shown in Table 1, our domain transfer approach dramatically increases the classification performance for both downstream tasks. Compared to physics-based simulation, the cINN obtained a relative improvement of 37% (BA), 25% (AUROC), and 22% (F1 Score) for PAT whereas the UNIT only achieved a

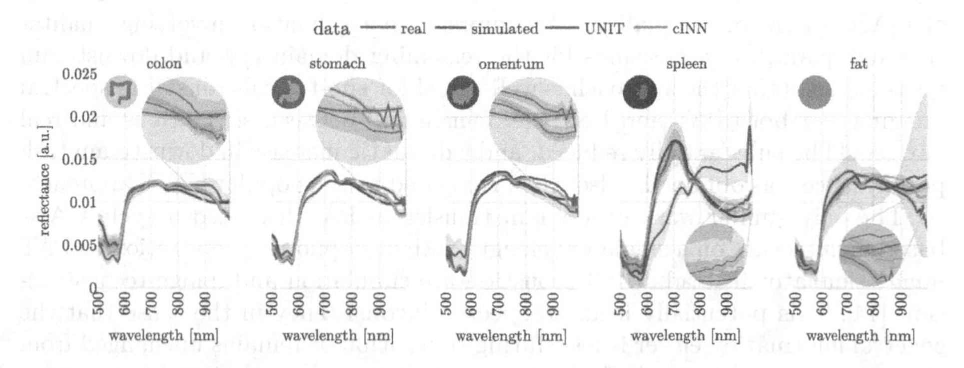

Fig. 6. The cINN-transferred spectra are in closer agreement with the real spectra than the simulations and the UNIT-transferred spectra. Spectra for five exemplary organs are shown from 500 nm to 1000 nm. For each subplot, a zoom-in for the near-infrared region (>900 nm) is shown. The translucent bands represent the standard deviation across spectra for each organ.

Table 1. Classification scores for different training data. The training data refers to real data, physics-based simulated data, data generated by a CycleGAN, by a UNIT without and with tissue labels (UNIT_Y), and by a cINN without (cINN_D) and with (proposed cINN_DY) tissue labels as condition. Additionally, $\text{cINN}_\text{DY without GAN}$ refers to a cINN_DY without the adversarial training. The best-performing methods, except if trained on real data, are printed in **bold**.

Classifier training data	PAT			HSI		
	BA	AUROC	F1-Score	BA	AUROC	F1-Score
Real	0.75	0.84	0.82	0.40	0.81	0.44
Simulated	0.52	0.64	0.64	0.24	0.75	0.18
CycleGAN	0.39	0.20	0.16	0.11	0.57	0.06
UNIT	0.50	0.44	0.65	0.20	0.72	0.20
UNIT_Y	0.64	**0.81**	0.77	0.24	0.74	0.25
cINN_D	0.66	0.73	0.72	0.25	0.72	0.20
$\text{cINN}_\text{DY without GAN}$	0.65	0.78	0.76	0.28	0.75	**0.26**
cINN_DY (proposed)	**0.71**	0.80	**0.78**	**0.29**	**0.76**	0.24

relative improvement in the range of 20%-27% (depending on the metric). For HSI, the cINN achieved a relative improvement of 21% (BA), 1% (AUROC), and 33% (F1 Score) and it scored better in all metrics except for the F1 Score than the UNIT. For all metrics, training on real data still yields better results.

4 Discussion

With this paper, we presented the first domain transfer approach that combines the benefits of cINNs (exact maximum likelihood estimation) with those of GANs (high image quality). A comprehensive validation involving qualitative and quantitative measures for the remaining domain gap and downstream tasks suggests that the approach is well-suited for sim-to-real transfer in spectral imaging. For both PAT and HSI, the domain gap between simulations and real data could be substantially reduced, and a dramatic increase in downstream task performance was obtained - also when compared to the popular UNIT approach.

The only similar work on domain transfer in PAT has used a cycle GAN-based architecture on a single wavelength with only photon propagation as PAT image simulator instead of full acoustic wave simulation and image reconstruction [14]. This potentially leads to spectral inconsistency in the sense that the spectral information either is lost during translation or remains unchanged from the source domain instead of adapting to the target domain. Outside the spectral/medical imaging community, Liu et al. [16] and Grover et al. [10] tasked variational autoencoders and invertible neural networks for each domain, respectively, to create the shared encoding. They both combined this approach with adversarial training to achieve high-quality image generation. Das et al. [5] built upon this approach by using labels from the source domain to condition the domain transfer task. In contrast to previous work, which used en-/decoders for each domain, we train a single network as shown in Fig. 2. with a two-fold condition consisting of a domain label (D) and a tissue label (Y) from the source domain, which has the advantage of explicitly aiding the spectral domain transfer.

The main limitation of our approach is the high dimensionality of the parameter space of the cINN as dimensionality reduction of data is not possible due to the information and volume-preserving property of INNs. This implies that the method is not suitable for arbitrarily high dimensions. Future work will comprise the rigorous validation of our method with tissue-mimicking phantoms for which reference data are available.

In conclusion, our proposed approach of cINN-based domain transfer enables the generation of realistic spectral data. As it is not limited to spectral data, it could develop into a powerful method for domain transfer in the absence of labeled real data for a wide range of image modalities in the medical domain and beyond.

Acknowledgements. This project was supported by the European Research Council (ERC) under the European Union's Horizon 2020 research and innovation programme (NEURAL SPICING, 101002198) and the Surgical Oncology Program of the National Center for Tumor Diseases (NCT) Heidelberg.

References

1. Adler, T.J., et al.: Uncertainty-aware performance assessment of optical imaging modalities with invertible neural networks. Int. J. Comput. Assist. Radiol. Surg. **14**(6), 997–1007 (2019). https://doi.org/10.1007/s11548-019-01939-9
2. Ardizzone, L., Lüth, C., Kruse, J., Rother, C., Köthe, U.: Conditional invertible neural networks for guided image generation (2020)

3. Ayala, L., et al.: Spectral imaging enables contrast agent-free real-time ischemia monitoring in laparoscopic surgery. Sci. Adv. (2023). https://doi.org/10.1126/sciadv.add6778
4. Clancy, N.T., Jones, G., Maier-Hein, L., Elson, D.S., Stoyanov, D.: Surgical spectral imaging. Med. Image Anal. **63**, 101699 (2020)
5. Das, H.P., Tran, R., Singh, J., Lin, Y.W., Spanos, C.J.: Cdcgen: cross-domain conditional generation via normalizing flows and adversarial training. arXiv preprint arXiv:2108.11368 (2021)
6. De Fauw, J., Ledsam, J.R., Romera-Paredes, B., Nikolov, S., Tomasev, N., Blackwell, S., et al.: Clinically applicable deep learning for diagnosis and referral in retinal disease. Nat. Med. **24**(9), 1342–1350 (2018)
7. Dinh, L., Sohl-Dickstein, J., Bengio, S.: Density estimation using real nvp. arXiv preprint arXiv:1605.08803 (2016)
8. Gröhl, J., et al.: Simpa: an open-source toolkit for simulation and image processing for photonics and acoustics. J. Biomed. Opt. **27**(8), 083010 (2022)
9. Gröhl, J., Schellenberg, M., Dreher, K., Maier-Hein, L.: Deep learning for biomedical photoacoustic imaging: a review. Photoacoustics **22**, 100241 (2021)
10. Grover, A., Chute, C., Shu, R., Cao, Z., Ermon, S.: Alignflow: cycle consistent learning from multiple domains via normalizing flows. In: Proceedings of the AAAI Conference on Artificial Intelligence, vol. 34, pp. 4028–4035 (2020)
11. Haar, A.: Zur theorie der orthogonalen funktionensysteme. Mathematische Annalen **71**(1), 38–53 (1911)
12. Hoffman, J., Tzet al.: Cycada: cycle-consistent adversarial domain adaptation. In: International Conference on Machine Learning, pp. 1989–1998 (2018)
13. Isensee, F., Jaeger, P.F., Kohl, S.A., Petersen, J., Maier-Hein, K.H.: nnu-net a self-configuring method for deep learning-based biomedical image segmentation. Nat. Methods **18**(2), 203–211 (2021)
14. Li, J., et al.: Deep learning-based quantitative optoacoustic tomography of deep tissues in the absence of labeled experimental data. Optica **9**(1), 32–41 (2022)
15. Li, K., Zhang, Y., Li, K., Fu, Y.: Adversarial feature hallucination networks for few-shot learning. In: Proceedings of the IEEE/CVF Conference on Computer Vision and Pattern Recognition, pp. 13470–13479 (2020)
16. Liu, M.Y., Breuel, T., Kautz, J.: Unsupervised image-to-image translation networks. Adv. Neural Inf. Process. Syst. **30** (2017)
17. Maier-Hein, L., Reinke, A., Godau, P., Tizabi, M.D., Büttner, F., Christodoulou, E., et al.: Metrics reloaded: pitfalls and recommendations for image analysis validation (2022). https://doi.org/10.48550/ARXIV.2206.01653
18. Mao, X., Li, Q., Xie, H., Lau, R.Y., Wang, Z., Paul Smolley, S.: Least squares generative adversarial networks. In: Proceedings of the IEEE International Conference on Computer Vision, pp. 2794–2802 (2017)
19. Pedregosa, F., et al.: Scikit-learn: machine learning in python. J. Mach. Learn. Res. **12**, 2825–2830 (2011)
20. Schellenberg, M., et al.: Semantic segmentation of multispectral photoacoustic images using deep learning. Photoacoustics **26**, 100341 (2022). https://doi.org/10.1016/j.pacs.2022.100341
21. Schellenberg, M., et al.: Photoacoustic image synthesis with generative adversarial networks. Photoacoustics **28**, 100402 (2022)
22. Wang, X., Xie, X., Ku, G., Wang, L.V., Stoica, G.: Noninvasive imaging of hemoglobin concentration and oxygenation in the rat brain using high-resolution photoacoustic tomography. J. Biomed. Opt. **11**(2), 024015 (2006)

23. Wirkert, S.J., et al.: Physiological parameter estimation from multispectral images unleashed. In: Descoteaux, M., Maier-Hein, L., Franz, A., Jannin, P., Collins, D.L., Duchesne, S. (eds.) MICCAI 2017. LNCS, vol. 10435, pp. 134–141. Springer, Cham (2017). https://doi.org/10.1007/978-3-319-66179-7_16
24. Zhu, J.Y., Park, T., Isola, P., Efros, A.A.: Unpaired image-to-image translation using cycle-consistent adversarial networks. In: Proceedings of the IEEE International Conference on Computer Vision, pp. 2223–2232 (2017)

Author Index

H. Greenspan et al. (Eds.): MICCAI 2023, LNCS 14220, pp. 781–785, 2023.
https://doi.org/10.1007/978-3-031-43907-0

Printed in the United States
by Baker & Taylor Publisher Services